Genetics and Analysis of Quantitative Traits

Genetics and Analysis of Quantitative Traits

Michael Lynch
University of Oregon

Bruce Walsh
University of Arizona

Sinauer Associates, Inc. *Publishers*
Sunderland, Massachusetts, 01375 U.S.A.

Cover Photo Credits (top to bottom):

- *Daphnia pulex.* Photo by Michael Lynch.
- Erin Lynch. Photo by Emília Martins.
- Long-term corn selection experiment at the University of Illinois. Photo by Michael Lynch.
- *Mimulus guttatus* (monkey flower). Photo by John Willis.
- Laboratory rat. © PhotoDisc.
- Coniferous forest. © PhotoDisc.
- *Geospiza magnirostris*, the large-beaked ground finch, from Isla Pinta, Galápagos. Photo by Dolph Schluter.
- *Drosophila melanogaster*. Photo by Edward B. Lewis.
- DNA double helix. Adapted from *Life: The Science of Biology*, Fourth Edition (Sinauer Associates, Inc. and W. H. Freeman and Company, 1995, page 246).

GENETICS AND ANALYSIS OF QUANTITATIVE TRAITS

For information address:
Sinauer Associates, Inc.,
23 Plumtree Road, Sunderland, MA 01375 U.S.A.

FAX 413-549-1118
Email: publish@sinauer.com

Library of Congress Cataloging-in-Publication Data

Lynch, Michael, 1951–
Genetics and analysis of quantitative traits / Michael Lynch, Bruce Walsh.
p. cm.
Includes bibliographical references and indexes.
ISBN 0-87893-481-2 (clothbound)
1. Quantitative genetics. I. Walsh, Bruce, 1957– . II. Title.
QH452.7.L96 1997
576.5′3--dc21 97-17666
CIP

Printed in Canada
5 4 3 2 1

Contents

PREFACE

With the emerging recognition that the expression of most characters is influenced by multiple genes and multiple environmental factors, quantitative genetics has become the central paradigm for the analysis of phenotypic variation and evolution. The historical development of the field is like that of a braided stream whose final destination has not been reached. Virtually all of quantitative genetics draws upon basic theoretical foundations laid down in the first third of this century, largely by Ronald Fisher and Sewall Wright. However, practical applications of this theory did not become common until the 1950s, and these were restricted almost entirely to agricultural settings. Plant and animal breeders subsequently diverged towards radically different modes of experimental design and analysis, perhaps because of the different population structures of crop plants and domesticated animals, but possibly also because of the historical segregation of the study of plants and animals in academia. Even today, at many major universities, separate courses in quantitative genetics are taught in departments of plant and animal science.

Only in the 1970s and 1980s did evolutionary biologists begin to fully embrace quantitative genetics as a major tool in both theoretical and empirical analysis. Many evolutionary quantitative geneticists are only vaguely aware of the extent to which the statistical machinery of the field traces to earlier work by animal and plant breeders as well as to work by the statistician Karl Pearson (an ardent non-Mendelian) at the turn of the century. Over the past few decades, human geneticists have also been progressively adopting quantitative-genetic approaches as the primary mode of analysis of genetic disorders. This work is largely unknown to (and generally uninformed by) those in the fields of breeding and evolutionary genetics.

In the mid 1980s, we realized that an integration of these disparate and semi-independent subdisciplines might be a useful contribution to the field of quantitative genetics at large. Our goal was to bring together the diverse array of theoretical and empirical applications of quantitative genetics under one cover, in a way that would be both comprehensive and accessible to anyone with a rudimentary understanding of statistics and genetics. As we ventured into a lot of unfamiliar territory, we gradually discovered that we had substantially underestimated the enormity of the task. So here we are, a decade later, about halfway to our final destination. What we originally envisioned as a single volume has now become two, with the focus of this first book being on the basic biology and methods of analysis of quantitative characters.

We have tried to write this book in a way that will encourage its use as a textbook in quantitative genetics. But the book also provides a thorough enough coverage of the literature so that it should be useful as a basic reference. Throughout, we have attempted to develop central theoretical concepts from first principles. To aid the less statistically sophisticated reader, we have included several chapters and appendices that review essentially all of the statistical tools employed in the book. Wherever possible, we have illustrated theoretical and analytical concepts with empirical examples from diverse settings. Both of our backgrounds are in evolutionary genetics, however, and a certain amount of bias may have crept in.

Today's quantitative genetics is not the science that it was 25 (or even 10) years ago. Three major developments are particularly noteworthy. First, largely motivated by the work of Russell Lande in the 1970s and early 1980s, there has been an explosive influx of quantitative-genetic thinking into evolutionary biology. It was, in fact, this dramatic refocusing of many evolutionary problems that first precipitated our interest in producing a book — most existing texts in quantitative genetics give little (and often no) attention to the great accomplishments that have been made in evolutionary biology. Thus, it is now ironic that much of our discussion of this work will be postponed to our second volume, *Evolution and Selection of Quantitative Traits.*

A second major development occurred in animal breeding. Here, enormous strides have been made in the development of new techniques for estimating breeding values (for the purposes of identifying elite individuals in selection programs) and for estimating variance components from samples of complex pedigrees. Although the foundation for many of these techniques was outlined in a remarkable series of papers by Charles Henderson in the 1960s and 1970s, their widespread application awaited the development of high-speed computers. Numerous technical treatises exist on these techniques, but their general absence from basic textbooks has endowed them with a certain mystique. In the last two chapters of this book, we have attempted to outline the basic principles of complex pedigree analysis, without getting greatly bogged down in technical details.

Third, in the past five years, as molecular markers have become widely available and economically feasible, there has been a rapid proliferation of new methods for detecting, locating, and characterizing quantitative-trait loci (QTLs). Currently one of the most active fields of quantitative-genetic research, QTL analysis was a mere dream when we embarked on this book. Thus, one benefit of our slow writing is the fact that we have been able to provide an up-to-date report on the exciting achievements of QTL analysis. Although a full integration of quantitative genetics and molecular genetics is still a long way off, with the recruitment of molecular biologists into the field we can anticipate great advances in the near future.

Over the past couple of years, we have heard a number of colleagues, some quite prominent, make statements like "quantitative genetics is dead," a rather hard thing to take when we have spent 10 years writing a treatise on the subject!

There are indeed some people who would dearly like to embrace this obituary as a rationale for ignoring a technically demanding field. However, the reality is that as a tool for the analysis of complex characters, quantitative genetics is as alive as it has ever been. What may be dead (or at least much less viable than we originally thought) is the simple caricature of traits being influenced by an effectively infinite number of loci with very small, additive effects. As we try to emphasize throughout this book, quantitative genetics is still fully capable of accommodating characters with small numbers of loci (even single loci), nonadditive genetic effects, non-Mendelian inheritance, and other genetic complexities. Indeed, the current machinery of quantitative genetics stands waiting (and its practitioners willing) to incorporate the fine genetic details of complex traits being elucidated by molecular and developmental biologists.

In spite of our best efforts, it is likely that a few errors have escaped scrutiny. Hopefully, they will be trivial and obvious, but Murphy's law suggests otherwise. Likewise, many of the methods we discuss can be computationally very demanding. To address both of these issues, we have set up a World Wide Web home page to post listings of detected errors and links to recent programs. The URL is **http://nitro.biosci.arizona.edu/zbook/book.html**; interested parties should contact BW at **jbwalsh@u.arizona.edu** for further information.

Acknowledgments

We have both profited considerably from interactions with numerous colleagues, most notably the clear-thinking, insights, and encouragement from Russ Lande. During the 1980s, while at the University of Illinois, ML was also highly fortunate to find company with an exceptional group of theoreticians in animal breeding (Rohan Fernando, Dan Gianola, Charles Henderson, and Mike Grossman), with whom joint participation in seminar courses was a defining experience. Collaborations with Steve Arnold, Reinhard Bürger, Wilfried Gabriel, Bill Hill, and Russ Lande also played critical roles in ML's movement into the field of quantitative genetics (from a background in limnology). BW's exposure to quantitative genetics arose from undergraduate, graduate, and postdoctoral experiences at the University of California at Davis, the University of Washington, and the University of Chicago, where he had the fortune to learn from the likes of Bob Allard, Steve Arnold, Joe Felsenstein, Russ Lande, Tom Nagylaki, Tim Prout, Monty Slatkin, Michael Turelli, and Mike Wade.

A very large number of individuals have aided in the development of this book. Most notable are Jim Crow and Bill Hill, both of whom critically evaluated nearly the entire manuscript, and in the process saved us from numerous embarrassing mistakes. The breadth of their understanding of population and quantitative genetics is a continual source of amazement to us. Large blocks of the book were reviewed by Bill Bradshaw, Doug Schemske, Ruth Shaw, and

Shizhong Xu, and comments on individual chapters were made by May Berenbaum, Hong-Wen Deng, Jim Fry, Sun-Wei Guo, Chris Haley, Jeff Hard, David Houle, Ritsert Jansen, Sara Knott, Trudy Mackay, Emília Martins, Patrick Phillips, Ahmed Rebaï, Ken Spitze, Tom Starmer, Sijne van der Beck, Sara Via, Joel Weller, John Willis, Mary Willson, and Zhao-Bang Zeng. To all of these individuals and to the many students who were victims of early drafts, we are extremely grateful for your insights.

We are especially grateful to our publisher, Andy Sinauer, who exhibited an extraordinary degree of patience (in the face of what, at times, must have been very serious doubt) as this project extended several years beyond its projected completion date. Marie Scavotto, Chris Small, and Carol Wigg were tremendously helpful during production.

As geneticists we certainly must acknowledge our parents, Bob and Bernice (ML), and Kevin and Phyllis (BW). We are especially grateful to them for providing us with an atmosphere conducive to learning and for putting up with our antics as young biologists, which can be a considerable strain to the faint-of-heart. Our sincere hope is that other budding biologists also find themselves in such supportive environments.

Finally, we thank our wives Emília Martins (ML) and Lee Fulmer (BW) for their continuous encouragement, patience, and friendship.

Mike Lynch, Eugene

J. Bruce Walsh, Tucson

October 1997

I

Foundations of Quantitative Genetics

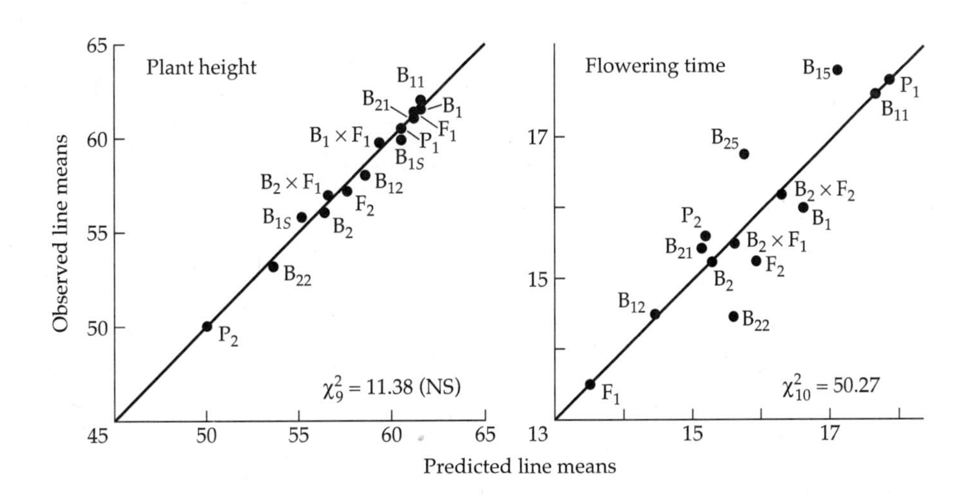

1

An Overview of Quantitative Genetics

Studies in most scientific disciplines can be classified into a "how" and "why" dichotomy. In biology, "how" questions are almost always concerned with proximate causes of observations. Since answers to such questions are generally sought at a lower level of organization, by its very nature the "how" approach to biology tends to be reductionistic. Questions of how muscles contract found their answers in the fields of cellular biology and neurobiology. The problem of how a plant develops its characteristic architecture is now being solved by molecular biologists. In principle, exact answers are available for "how" questions, the main limitation being the existence of the appropriate technology.

In contrast, "why" questions are concerned with ultimate causes. This naturally brings us to the study of adaptation, understanding of which requires an appeal to higher levels of organization. If, for example, one were interested in why a fish forages in a particular manner, it would be logical to examine the spatial and temporal distribution of food types and the rewards they offer, as well as to study the challenges from predators that confront the consumer. "Why" questions are the domain of evolutionary ecology, the general goal of which is to explain the diversity of morphology, physiology, behavior, and life histories both within and among populations and species. The study of proximate factors and ultimate causes are complementary approaches to biological investigation. Both are valid ways of doing science, and a knowledge of both is essential for a complete understanding of any biological system.

THE ADAPTATIONIST APPROACH TO PHENOTYPIC EVOLUTION

Most scientists test their ideas by phrasing them in the form of potentially falsifiable hypotheses, and in evolutionary biology, many such hypotheses are built around optimization principles. Having identified an interesting character or set of characters as well as one or more likely selective pressures, one asks the question, "What is the optimal phenotype?" An implicit assumption underlying the development of such hypotheses is that natural selection is the dominant, if not solitary, force molding the evolution of the phenotype.

Natural selection is the closest thing to a law that we have in biology. It is a universal force that applies, to varying degrees, at all levels of organization, from genes to individuals to populations to species. Thus, it is logical that we adopt it as a frame of reference in our attempt to understand nature. The relatively recent focus on selectionist thinking and hypothesis testing has revolutionized our understanding of essentially all areas of biology, ranging from molecular evolution to functional morphology. However, it is one thing to argue for the universal applicability of natural selection, and it is quite another to argue for its universal importance. To draw a physical analogy, while gravity is a universal force, it by no means provides a sufficient explanation for all motion. Points and counterpoints on the **adaptationist program** may be found in Williams (1966), Peters (1976), Brady (1979), Maynard Smith (1978, 1982), Popper (1978), Rosen (1978), Wassermann (1978), Gould and Lewontin (1979), Reed (1981), Willson (1981), Mayr (1983), Sober (1984), and Reeve and Sherman (1993).

While optimization theory can be highly successful at predicting what ought to evolve, it tells us very little if anything about how it should happen. The problem of how long it might take an optimal phenotype to evolve is usually ignored, as is the question of how the genetic variation necessary for adaptation arises. Generally, optimization theories are concerned entirely with the expected average phenotype, and thus give little consideration either to the expected phenotypic variance or to the problem of nonadaptive evolutionary change caused by forces such as random drift or mutation. In summary, although the adaptationist paradigm has forever transformed the way we think about biological phenomena, it has one very major limitation — it does not provide us with a mechanistic understanding of the evolutionary process. This is the primary goal of quantitative genetics.

QUANTITATIVE GENETICS AND PHENOTYPIC EVOLUTION

This book attempts to provide a "how" approach to "why" problems in evolutionary biology. The central issues to be discussed center on the fundamental nature of the processes that define the evolution of the sorts of morphological, physiological, life-history, and behavioral traits studied by ecologists and ethologists. These same types of traits are the focus of most plant and animal breeding programs (e.g., egg production, oil content, running speed), where man, rather than nature, defines the intensity and direction of selection.

Although cultural and/or maternal transmission of environmental effects play important roles in some cases, evolution is primarily a genetic process. Its study requires models that incorporate explicit genetic mechanisms. As we will soon see, most of the traits studied by whole-organism biologists are encoded by a large number of genetic loci, and for practical reasons, the individual loci are generally unobservable. The well-developed single-locus theory of population genetics (Crow and Kimura 1970, Hartl and Clark 1989) is of limited use in the study of such characters. In quantitative genetics, a statistical branch of genetics based upon fundamental Mendelian principles extended to polygenic (multilocus) characters, most final formulations are phrased in terms of phenotypic means and variances. Such measures are accessible to whole-organism biologists.

The science of quantitative genetics has been around for a long time, the main principles having been outlined independently by Ronald Fisher (1918) and Sewall Wright (1921a–d). It has served as the theoretical basis for most plant and animal breeding programs for well over a half century (Lush 1937, Hanson and Robinson 1963, Mayo 1980, Hallauer and Miranda 1981, Mather and Jinks 1982, Pirchner 1983, Henderson 1984a, Wricke and Weber 1986, Hill and Mackay 1989, Gianola and Hammond 1990, Falconer and Mackay 1996, Kearsey and Pooni 1996). It has also played an important role in our understanding of the inheritance of complex human genetic disorders such as heart disease, congenital malformations, and psychopathology. More recently (the past two decades), quantitative genetics has become well entrenched in the field of evolutionary biology (Bulmer 1980, Lande 1988, Barton and Turelli 1989, Boake 1994). The astonishing rate at which technical advances are now being made in molecular biology leaves little question that the field is now primed to enter a new frontier, in a sense an empirical return to its theoretical roots — the actual identification of the loci underlying quantitative variation. The practical implications of this ultimate synthesis for crop and livestock production and human medical practices are enormous.

The impact of early quantitative-genetic theory extends well beyond the biological sciences. It laid the foundations for modern theoretical and applied statistics. Out of a need for quantitative methods to describe the distributions of continuously distributed characters, Francis Galton provided the empirical motivation for Karl Pearson's formal development of the theory of regression and correlation (Provine 1971, Stigler 1986). Fisher's 1918 paper introduced the concept of variance-component partitioning, upon which the principles of analysis of variance (ANOVA) are based, and his subsequent contributions had a profound influence on the development of methods for experimental design and hypothesis testing (Fisher 1925, 1935, 1956). Wright's (1921a) method of path analysis has been broadly embraced by the social and ecological sciences.

The Mendelian principles upon which quantitative-genetic theory is built are universal, applying to domesticated species as well as to natural populations. Nevertheless, because the systems of mating and evolutionary forces found in natural populations are generally quite different than the controlled programs imposed on domesticated species, study of the inheritance of quantitative traits in natural populations presents a number of challenges. Most quantitative-genetic parameters are estimated through the comparison of phenotypes of individuals with known degrees of relatedness (Chapter 7). The basic idea here is that the resemblance between relatives is a function of the degree to which phenotypic expression is determined by shared genes as opposed to random environmental effects. Ideally, controlled genetic analyses should be performed with specific sets of relatives of specific ages in specific environmental backgrounds. While such situations are achievable with many crop plants, field biologists are often unable to perform controlled crosses and must let nature define the pattern of mating. This lack of experimental control can sometimes severely compromise the types of relatives that can be assayed and the degree of "balance" in the final data set. Breeders of large domesticated mammals (particularly cattle) have been confronted with similar problems for years, a situation that stimulated the development of new statistical procedures such as BLUP (best linear unbiased prediction, Chapter 26) and REML (restricted maximum likelihood, Chapter 27). Applications of these methods now extend well beyond the issues that motivated their development, although they still remain largely unexploited by evolutionary biologists.

Because Fisher and Wright are two of the most prominent evolutionary biologists of this century, it is puzzling that the principles of quantitative genetics were nearly immediately embraced by plant and animal breeders, while well over fifty years elapsed before they began playing a major role in evolutionary theory and analysis. Whatever the reason for the delay, the impact of quantitative-genetic thinking in evolutionary biology has been explosive, and this is where many of the recent advances in the field have been made. Although many challenges still exist, essentially all of the broad issues in evolution have now been explored in quantitative-genetic terms in one way or another.

The diverse roles played by selection present a unique challenge for studies of natural populations. For the breeder, the issue of evolutionary dynamics is relatively straightforward. By defining the population size, the mating system, and the intensity and direction of artificial selection, the breeder controls most of the evolutionary forces operating on a population. On the other hand, a fundamental goal of evolutionary studies is to identify those very forces, such as natural selection, that the breeder strives to eliminate. Thus, the development of methods for estimating the intensity and form of selection operating on quantitative traits has been largely restricted to the field of evolutionary biology. Because many of the predominant issues in evolutionary biology, such as mutation, stabilizing selection, random genetic drift, and sexual selection, have been of minimal relevance to short-term artificial selection programs, they too have only recently been incorporated into the theory by evolutionary biologists. Some of these theoretical advances are now being exploited by breeders, as attention turns to the long-term limits of artificial selection programs.

One of the goals of this book is to bring together the diverse results of quantitative genetics that have developed semi-independently in the fields of evolutionary biology, animal breeding, plant breeding, and human genetics. The primary emphasis is on the basic biological properties of quantitative characters and methods for their genetic analysis. An understanding of these issues is a prerequisite to studies of the evolutionary dynamics of quantitative traits under natural and/or artificial selection, random genetic drift, and mutation, all of which will be covered in our next book. As evolutionary biologists, our primary interest is in the natural world, and wherever possible, we have tried to emphasize the analysis of non-domesticated species (including humans). There are, however, many problems that can only be illustrated with data from laboratory or barnyard populations. Many extraordinary experiments performed in agronomy and animal breeding have been neglected unjustifiably by evolutionary purists.

On more than one occasion, we have heard criticism from molecular geneticists that quantitative genetics, due to its inattention to the structure and location of specific genes, is "phenomenological," "missing the boat," or "cheating." We sympathize with the desire for a more precise understanding of gene structure, function, and expression. However, molecular analysis is not necessary or even useful for explaining patterns of resemblance between relatives. It is not required to explain the shifts in the mean and variance of characters observed under inbreeding. As yet, molecular biology does not provide any means of describing or predicting the constraints on the joint expression of correlated characters, and it certainly has little bearing on problems in the area of natural selection. Quantitative genetics has dealt with all of these issues successfully. This is not meant to imply that molecular biology is irrelevant to problems in phenotypic evolution. Ultimately, observations at both levels of organization will have to answer to each other, and it is likely that they will soon do so (Chapters 14–16).

HISTORICAL BACKGROUND

Quantitative genetics developed early in this century in response to a major controversy that still has some remnants today. The rediscovery of Mendelian genetics in 1900 centered attention on the inheritance of discrete characters such as purple vs. white flower color, smooth vs. wrinkled seeds, and so on. This focus was in stark contrast to an independent branch of genetic analysis begun earlier by Francis Galton (1869, 1889), who concentrated on continuously varying characters (those not clearly separable into discrete classes). Galton's approaches marked the founding of the Biometrical school, from which most of modern statistics can be traced (Stigler 1986, Crow 1993a). A series of contentious debates ensued between the Mendelians (led by William Bateson) and the Biometricians (led by Karl Pearson and W. F. R. Weldon). The major issues were whether discrete characters have the same hereditary and evolutionary properties as continuously varying characters. The Mendelians held that variation in discrete characters drove evolution through the appearance of new macromutations (mutations with large effects), while the Biometricians viewed evolution as the result of natural selection acting upon continuously distributed characters. The Mendelian-Biometrician clash, often influenced more by personalities than facts, substantially delayed the fusion of genetics and Darwin's theory of evolution by natural selection (Provine 1971).

In order to explain the approximate intermediacy of progeny phenotypes

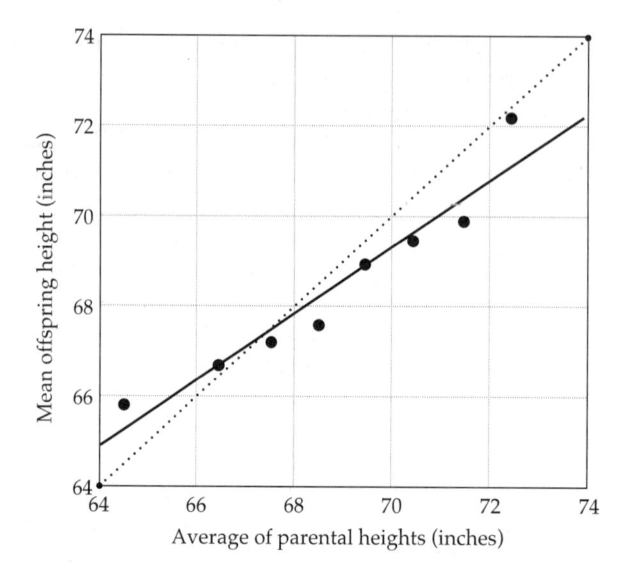

Figure 1.1 The relationship between height of adult children and the average height of their parents (in inches). Circles denote average offspring heights for different 1-inch classes of midparent heights. The best linear fit is given by the solid line, while the dotted line is the expected pattern if mean offspring height were equal to midparent height. (Data from Galton 1889.)

with respect to those of parents, Darwin (1859) had assumed that continuous characters exhibit blending inheritance. The problem with that view, as pointed out by the engineer Fleeming Jenkin (1867), is that half the existing variation is removed each generation under blending inheritance. Large amounts of new variation would have to be renewed each generation just to keep the level of existing variation constant, and without such variation there could be no further evolution.

Galton, a cousin of Darwin, pointed out another problem. When he plotted the mean heights of offspring (measured at adult age) against the average height of their parents (corrected for differences between the sexes), he obtained a linear relationship (Figure 1.1). However, the slope of the line indicated that offspring were, on average, less exceptional than their parents. Parents whose average height was below the mean tended to have offspring taller than themselves but still below the mean. Parents above the mean tended to have offspring shorter than themselves. Galton (1889) called this trend **regression toward mediocrity,** and argued that it would erode away any selective progress. He concluded that evolution must be based upon **sports** (mutations of large effects) rather than on selection acting upon continuous variation.

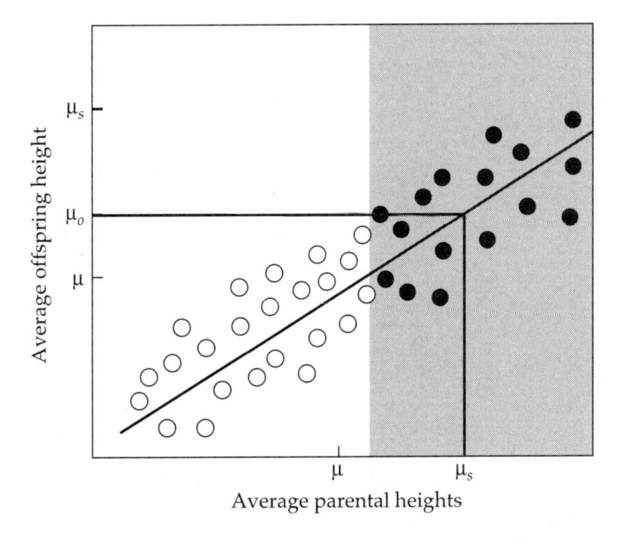

Figure 1.2 Response to selection. The solid line is Galton's regression. The closed circles represent selected parents. In the absence of selection, the mean height for both the parents and offspring is μ. However, if the average height of selected parents is μ_s, the expected height of offspring (μ_o), obtained by reading off the regression line, is greater than μ but less than μ_s.

Pearson (1903) pointed out the fault in Galton's logic. Imagine that selection acts in such a way that only the largest parents reproduce, so that the mean height of reproductive adults after selection (μ_s) is greater than the mean height before selection (μ) (Figure 1.2). If μ_o is the average adult height of the offspring of the selected parents, then the **response to selection** across generations is $\mu_o - \mu$. Since the slope of the parent-offspring relationship is less than one, the adult offspring are not expected to be as tall (on average) as their selected parents. However, the mean offspring height will be greater than the mean height in the parental base population. Furthermore, the new mean height is stable — if selection on height is stopped, the new height is not expected to decay back to the original value. Thus, the regression toward mediocrity does not pose a serious problem for the theory of evolution by natural selection. Galton apparently failed to realize that selection starts from a new mean each generation and that it is to this new mean that the next generation of selection regresses.

Although in disagreement with some of Galton's interpretations, Pearson was inspired greatly by Galton's quantitative approach to analysis and went on to develop many methods for the analysis of continuously distributed traits. Since Pearson was an ardent non-Mendelian, it is particularly noteworthy that many of the regression and correlation techniques that he invented serve today as a cornerstone in the analysis of selection on quantitative traits (Chapters 3 and 8).

A resolution to the problem of blending inheritance and the maintenance of variation was slower in coming. Mendel himself had actually suggested how this might be accomplished — the variation in continuous characters can be maintained by the independent segregation of multiple factors. The British mathematician G. Udny Yule gave formal proof for this idea in 1902. Unfortunately for Yule, the only thing that the Biometricians and the Mendelians could publicly agree on was the incompatibility of Mendelian genetics and the inheritance of continuous characters. The death of Weldon in 1906, followed by the publication of several key experiments from 1908 to 1916, resulted in the rapid emergence of the **multiple-factor hypothesis**.

George H. Shull, a major figure in American corn breeding, noted that self-fertilized corn strains are remarkably uniform in many continuous traits when compared to the outbred populations from which they are derived. His (Shull 1908) explanation (with modern terminology inserted and italicized) was

> The obvious conclusion to be reached is that an ordinary cornfield is a series of very complex hybrids (*genotypes*) produced by the combination of numerous elementary species (*alleles*). Self-fertilization soon eliminates the hybrid elements (*removes the heterozygosity*) and reduces the strain to its elementary components (*each locus becomes homozygous, resulting in each inbred strain being composed of a single genotype*).

The major implication of Shull's observation is that any variation for continuous characters that is lost upon inbreeding must have a Mendelian basis.

Additional support for the multiple-factor hypothesis came from H. Nilsson-Ehle (1909), a Swedish geneticist working with various cereal crops. Many of the characters that he examined yielded 3:1 ratios in the F_2 generation following the cross of two parental strains, consistent with expectations for a single segregating locus with one allele completely dominant over the other. However, there were some striking exceptions. For example, when red-seeded and white-seeded wheat strains were crossed, the F_1 progeny were identical in color, but in some of the F_2 crosses, a ratio of 63 red:1 white seeds was observed. Nilsson-Ehle interpreted this to be the result of the segregation of three independent factors, the initial parents being *AABBCC* and *aabbcc*, all members of the F_1 being *AaBbCc* and hence uniform in color, and the F_2 consisting of all possible genotypes, only one of which (*aabbcc*) gives rise to white seed. The probability of obtaining an *aabbcc* offspring from an *AaBbCc* $\times$ *AaBbCc* cross is $(1/2)^6 = 1/64$.

From these results, Nilsson-Ehle arrived at two general conclusions.First, *sexual reproduction can produce a huge diversity of genotypes.* For example, since a locus with two alleles *A* and *a* can produce three genotypes (*AA*, *Aa*, and *aa*), ten diallelic loci can produce $3^{10} \simeq 60,000$ genotypes. Second, given this huge potential diversity of genotypes, *apparently new types appearing within a population may be the result of rare segregants rather than new mutations.* Nilsson-Ehle, and independently East (1910), offered this as an explanation of **atavisms** — rare individuals that appear to be throwbacks to some previous population. Consider a hybrid population resulting from the cross of two pure lines differing at ten loci. The probability of obtaining a specific parental genotype in the F_2 or later generation is $(1/4)^{10}$. Hence, the population size has to be greater than $4^{10} \simeq 10^6$ for there to be much of a chance of observing a parental type.

Subsequent studies quickly confirmed the ideas of Shull and Nilsson-Ehle. East (1911, 1916) and Emerson (1910; Emerson and East 1913) examined quantitative variation in a large number of plants. Typically, strains differing widely in some character were crossed and the variance of the resulting F_1 and F_2 generations recorded. In most of these crosses, especially when the parental populations were formed by repeated self-fertilizations, an **outbreak of variation** was seen in the F_2 (Figure 1.3). Such outbreaks of variation, resulting from the segregation of multiple genotypes from the F_1 heterozygotes, are consistent with the Mendelian model, and they are completely inconsistent with any blending hypothesis. An extensive historical review of the experimental verification of the multiple-factor hypothesis is given in Chapter 15 of Wright (1968).

The Danish botanist Wilhelm Johannsen (1903, 1909) was among the first to demonstrate that some of the variation in continuous characters is due to environmental rather than genetic causes. Starting from a common stock of beans, he produced several inbred lines. For each line, parental seeds of different weights were planted and the mean seed weight in the offspring measured. Johannsen observed that the variation *within a pure line* was not heritable — the mean seed weights of the offspring were essentially independent of the weights of the parental seed (Figure 1.4). To clarify the distinction between genetic and environmental effects, he coined the terms **genotype** (to denote genetically identical members of a pure line) and **phenotype** (the actual observed value for an individual — a compounding of genetic and environmental effects). From his observations, Johannsen concluded that natural selection could never move a character value beyond the level of variation seen in the original population. Like Galton, he felt that macromutations were essential for evolution.

However, Payne (1918) soon demonstrated that selection on *Drosophila* bristle number can result in flies with more extreme phenotypes than those seen in the base population. Such a result has been observed in many selection programs involving economically important species of plants and animals — almost always, the range of observed variation underestimates the range of potential variation, often dramatically so. By increasing the frequency of favored genes, selection increases the probability of observing extreme genotypes. Recall from above that if the frequency of the favored allele at each of ten loci is 0.5, then the frequency of the most extreme genotype (a homozygote at all ten loci) is approximately one in a million. However, if selection advances the frequency of the favored allele at each locus to 0.9, the frequency of the most extreme genotype becomes $(0.9^2)^{10} \simeq 0.12$. In other words, about one in eight individuals would exhibit the most extreme genotype.

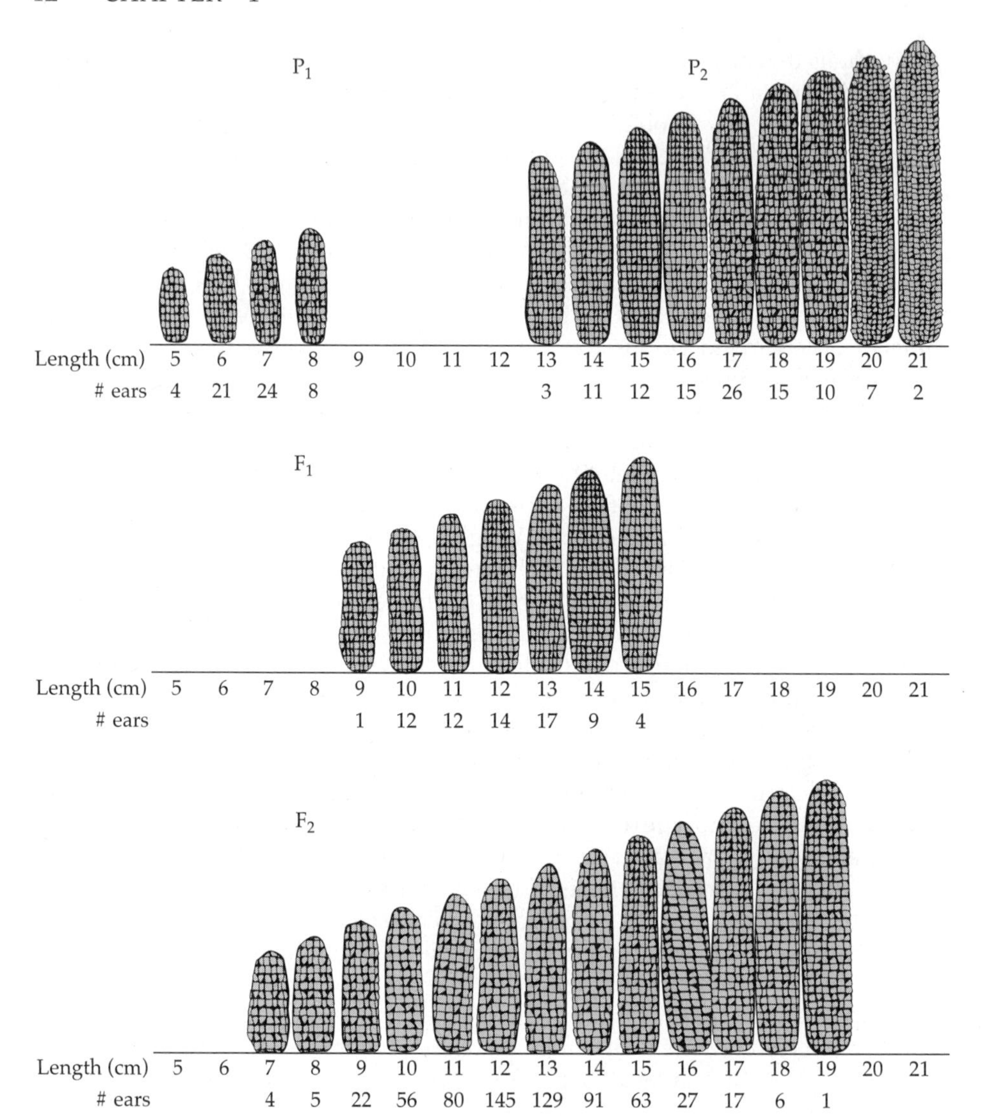

Figure 1.3 The distribution of ear size in the F_1 and F_2 generations formed by crossing two inbred lines of corn differing in ear length. The observed number of ears is given below each size class. The variation seen in the P_1, P_2 and F_1 populations is due entirely to environmental factors, as all individuals in each population have the same genotype. These three populations show roughly similar amounts of variation. In contrast, the F_2 generation shows considerably more variation, reflecting the diversity of genotypes in this population generated by segregation of genes in the F_1 parents. (Data from East 1911.)

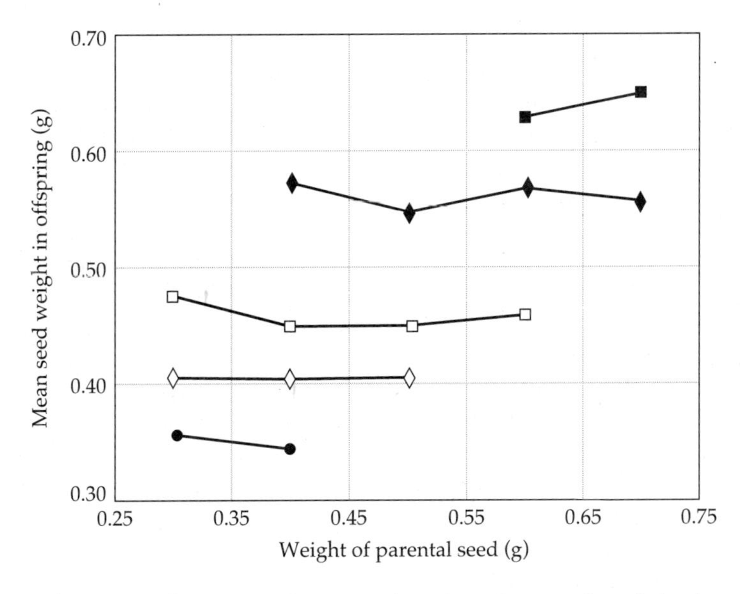

Figure 1.4 Mean offspring seed size as a function of parental seed size for some of Johannsen's pure lines. The data for the different lines are denoted by different symbols. If there is a heritable component to seed weight within a pure line, a line with positive slope is expected — larger parents should yield larger offspring. However, within each line, mean offspring size is essentially independent of the parental phenotype. (Data from Johannsen 1903.)

THE MAJOR GOALS OF QUANTITATIVE GENETICS

Before proceeding, we give a very brief summary of the main issues to be addressed in the remainder of the book:

The Nature of Quantitative-trait Variation

The preceding paragraphs have provided evidence that the expression of quantitative characters is typically influenced by both genetic and environmental factors, and that patterns of variation are qualitatively consistent with Mendelian expectations. However, many questions remain. An obvious one, of interest from many perspectives, is, How much of the standing variation in populations is due to genetic causes and how much to environmental ones? From the standpoint of evolution and applied breeding programs, genetic components of variance are of particular interest because they determine the rates at which characters respond to selection. Environmental variance reduces the efficiency of the response to selection by causing the phenotypes of selected individuals to deviate from their underlying genotypic values. From the perspective of human genetic disorders, the degree to which the expression of undesirable traits is determined by genetic vs. environmental causes has broad implications for the development of preventative procedures and genetic counseling strategies.

Many methods exist for partitioning phenotypic variance into its various components (Chapters 17–27), all of which are based on the principle that the phenotypic resemblance between relatives provides information on the degree of genetic differentiation among individuals (Chapter 7). However, virtually all of the methods have some undesirable theoretical and methodological features. For example, a simple dichotomy between genetic and environmental sources of variation is often overly simplistic, since genotype × environment interaction may exist (Chapter 22). More significantly, there are several components of both environmental and genetic variation (Chapters 4, 5, and 6). The different components influence the resemblance between relatives to different degrees, and as a consequence, have substantially different influences on the evolutionary process. The additive component of the genetic variance (also known as the variance of breeding values) is of particular interest because it is the primary determinant of the degree to which offspring resemble their parents, which governs the rate of response of a character to selection.

Ultimately, the pool of genetic variation in a population must be due to a quasi-balance between the forces of selection and random genetic drift, both of which tend to eliminate variation, and the replenishing force of mutation. Recent work has shown that the mutational rate of production of new variation for quantitative traits is remarkably high (Chapter 12). However, our understanding of the molecular basis of such variation is still very crude. There are only a handful of multilocus characters for which even one or two genetic loci are known, and in the vast majority of cases, we have no information on this issue. Observed patterns of variation in the progeny of line crosses can provide qualitative insight into whether the number of loci underlying a character is likely to be large or small (Chapters 9 and 13). With the development of genomic maps for many species on the horizon, molecular-marker-based approaches will rapidly refine our understanding of these issues, as statistical machinery for mapping and characterizing quantitative-trait loci (QTLs) is now in place (Chapters 14–16). Major issues that await resolution are the roles of nonadditive gene action in the expression of quantitative traits (Chapter 5), the extent to which alleles at different loci are distributed independently versus being associated statistically within populations (Chapter 5), and the mechanisms by which gene action maps developmentally into phenotypic expression (Chapter 11).

The Consequences of Inbreeding and Outcrossing

The mean phenotypes of progeny from consanguineous matings virtually always differ from those of progeny from random-bred parents. Such inbreeding effects are almost always deleterious, generally increasing linearly with the degree of relatedness between parents (Chapter 10). These observations are consistent with the presence of deleterious recessive alleles segregating at the loci underlying quantitative variation. The mechanisms and consequences of inbreeding have a number of practical implications, e.g., in the design of breeding programs for captive populations of endangered species, the maintenance of inbred lines for biomedical and agricultural research, and the genetic counseling of victims of incest. Patterns of fitness decline with inbreeding can also provide insight into the rate and average effects of deleterious mutation (Chapter 10).

Crosses between isolated lines or populations often exhibit "hybrid vigor" in the F_1 generation, only to be followed by substantial fitness decline in the next (F_2) generation. The pattern of change in mean phenotypes in line crosses can yield insight into the mode of gene action, particularly with regard to interaction between genes at different loci (Chapter 9). Such information is vital to attempts to understand the mechanisms of speciation (Chapter 14).

The Constraints on the Evolutionary Process

Fundamental questions arise in evolutionary biology and in selective breeding programs as to the factors that limit the rate of phenotypic evolution. As noted above, when selection operates on a single trait, the response to selection is roughly proportional to the additive genetic variance for the trait (Chapter 3). However, if the same genes influence the expression of different traits, then an evolutionary change in one trait will necessarily lead to changes in the correlated traits. This can impede the evolutionary process when there is a conflict in the fitness consequences of selection operating directly on a trait and that operating on correlated traits. Questions of evolutionary trade-offs have long been the focus of evolutionary ecology, but many of the ideas in that field (e.g., life-history theory) have developed out of simple energetic arguments or comparative surveys on the *phenotypes* of different species. An unambiguous understanding of the constraints on the evolution of systems of complex characters requires information on the magnitude and direction of *genetic* correlations between characters. Quantitative-genetic methodology provides a powerful, but data demanding, means of elucidating these issues (Chapter 21). Ultimately, we would like to know the extent to which patterns of genetic correlations within populations are reflected in multivariate patterns of differentiation among species.

The Estimation of Breeding Values

Plant and animal breeding programs are based on the principle that an individual's phenotype provides some insight into its underlying genotypic value. However, because the environment can contribute greatly to phenotypic expression, the amount of genetic information conveyed by a single phenotypic measurement is not necessarily high. On the other hand, if one considers jointly the phenotypes of all of the relatives of a focal individual, a more accurate assessment of that individual's breeding value can be acquired. The identification of elite individuals by BLUP (Chapter 26), in some cases supplemented by information from molecular markers, is now a major goal of applied quantitative genetics.

The Development of Predictive Models for Evolutionary Change

Much attention has been given to the predictive power of quantitative-genetic theory. For breeders facing a large economic investment in an artificial selection program, a fundamental issue is the magnitude of selection gain that is to be expected following a certain period of selection at a specified intensity. Models that provide reasonable predictions, at least for periods of a dozen generations or so, have long been available. These will be taken up in detail in our next book. For now, we simply emphasize that the application of all such models requires information on the components of variation and covariation of the selected traits, the primary topics of this volume.

MATHEMATICS IN BIOLOGY

Biologists often become alienated from mathematics at an early stage in their career under the misconception or self-deception that biology is not a mathematical science. The claim is also often made that mathematical theory is overly simplistic, ignoring the inherently complex nature of biological systems. This posture is not very productive, and ironically, those who make such claims frequently devote their entire careers to performing carefully controlled experiments in which only one or two factors are allowed to vary, effectively sweeping biological reality under the rug.

Most of the major advances in evolutionary biology over the last few decades have come from quantitative statements, many of which involve only elementary algebra or the most basic theorems of calculus. While it is understandable that most biologists do not want to spend their careers deriving theory, it is useful to maintain some degree of skill at evaluating the relevance of theory derived by others. It is equally important that theoreticians strive to describe their work in as simple and digestible a form as possible. There is no question that the contributions of Ronald Fisher to evolutionary biology and applied statistics are among the most brilliant ever, but his terse writing style, tendency to neglect mentioning the assumptions of his models, and frequent unstated use of mathematical approximations have not done anyone any favors.

The fact that quantitative genetics requires the use of some mathematics should come as no surprise. Some of the math is unavoidably tedious, but most of it is basic algebra. With the nonmathematician in mind, Chapters 2, 3, and 8 and the Appendices present the fundamental statistical concepts that will be used throughout the book. Although the technical reviews in those sections may be unnecessary for the mathematically sophisticated, numerous examples that provide direct contact with quantitative-genetic issues have also been incorporated. Throughout the remainder of the book, ample mathematical machinery is presented and supported by worked examples so that the diligent reader should emerge with some ability for developing practical methodology and quantitative theory. The rewards of mathematical theory can be enormous, particularly when what looks like mind-boggling complexity collapses into a relatively simple formulation. We will see, fortunately, that such simplifications often occur in quantitative genetics.

2

Properties of Distributions

Before delving into the genetics of quantitative variation, it is essential to have a basic understanding of statistics. The statistical concepts and techniques most frequently encountered in quantitative genetics are presented in this and the following chapter. For the reader with advanced training in statistical theory, much of what follows will probably be review, and some things may appear to be presented in a nonrigorous manner. Even so, it may still be profitable to skim the following pages to become familiar with the notation that will be used throughout the book. As an additional reward, a number of examples will provide some immediate contact with the field of quantitative genetics.

PARAMETERS OF THE UNIVARIATE DISTRIBUTIONS

Characters that are studied by biologists are of three types. Traits that are distributed into a range of discrete classes, such as scale counts in fish or leaf number in plants, are called **meristic characters**. Those that are measured on a continuous scale are known as **metric characters**. Length, weight, and growth rate attributes are examples of the latter. Attributes such as survival to a fixed age are known as **all-or-none** or **binary characters**. Of course, due to technical limitations, even measures of truly continuously distributed traits must always be artificially placed into discrete categories. Meter sticks, for example, are unable to distinguish between individuals that are 25.2 and 25.3 mm in length. Both would typically be placed in the 25–26 mm category, although the biological reality is that every conceivable length in the 25–26 mm range is possible.

Suppose that one performs a series of measurements on a collection of individuals. Compilation of the data provides some information on the relative incidence of different trait measures. A **univariate distribution** describes the relative frequencies of phenotypes for a single trait, whereas a **bivariate distribution** describes the mutual distribution of two characters. The joint distribution of more than two traits is referred to as a **multivariate distribution**. An example of a bivariate distribution for maternal weight and number of offspring is given for a population of rats in Table 2.1. The data are condensed into the univariate **marginal distributions** of the two traits in the last row and column.

Table 2.1 The bivariate distribution of mother's weight and number of offspring produced for a population of rats.

Maternal Weight (grams)	Number of Offspring *												Totals
	1	2	3	4	5	6	7	8	9	10	11	12	
50 -	—	—	—	1	3	1	—	—	—	—	—	—	5
60 -	—	—	—	1	6	2	—	—	—	—	—	—	9
70 -	—	—	2	10	17	12	4	—	1	—	—	—	46
80 -	1	1	11	8	18	10	9	3	2	—	—	—	63
90 -	2	5	7	18	30	28	12	5	1	—	—	—	108
100 -	3	5	10	25	37	35	21	7	2	1	—	—	146
110 -	1	4	12	19	38	37	29	6	2	—	—	—	148
120 -	2	6	9	21	36	26	30	14	6	—	1	—	151
130 -	4	4	9	12	35	29	17	17	6	1	1	1	136
140 -	1	4	6	9	12	27	15	6	2	1	—	—	83
150 -	—	3	—	2	13	11	6	6	2	—	—	—	43
160 -	—	2	—	1	11	11	9	3	4	—	—	—	41
170 -	1	—	1	1	2	4	2	2	1	—	1	—	15
180 -	—	—	1	1	—	2	2	2	—	—	—	—	8
190 -	—	—	—	—	—	—	—	—	—	1	—	—	1
Totals	15	34	68	129	258	235	156	71	29	4	3	1	1003

* Each number in the main body of the table refers to the number of observations in a particular bivariate class. For example, 38 animals weighed between 100 and 110 grams and produced 5 offspring. The final row and column are the marginal univariate distributions for the two traits. (From Pearson 1910.)

One of the goals of statistics is to fit fairly simple mathematical functions, known as **probability distributions,** to data. If a variable z takes on only discrete values (as with offspring number), the distribution of z is completely described by giving $P(z = z_i)$ for each possible outcome z_i, where P stands for probability. For example, for offspring number, letting $z_1 = 1$, then $P(z = z_1)$ is the proportion of mothers that produce a single offspring, which for the example in Table 2.1 is $15/1003$. Summing over all possible outcomes, $\sum_i P(z = z_i) = 1$, since the total probability of all possible events is one.

If, on the other hand, z is a continuously distributed variable (as with maternal weight), $P(z = z_i)$ makes no sense since the probability that z takes on any specific value is infinitesimally small. It is more meaningful to consider the probability that z lies within a specific range of values, say z_1 and z_2. This quantity is described by the **probability density function** $p(z)$, which satisfies the integral

$$P(z_1 \leq z \leq z_2) = \int_{z_1}^{z_2} p(z)\, dz \tag{2.1}$$

If z_{min} and z_{max} are the upper and lower bounds to z, then $p(z) = 0$ outside of this range, and over the entire range $\int_{z_{min}}^{z_{max}} p(z)\, dz = 1$. Both of these properties are in accord with common sense — a probability is never negative, and the total probability of all possible outcomes is one. A large number of functions fulfill these properties, and they have been studied in considerable detail (Johnson and Kotz 1970a,b, 1972; Kendall and Stuart 1977).

Example 1. Suppose that z is continuously distributed in the range of 0 to ∞ with probability density function

$$p(z) = \frac{1}{\lambda} e^{-z/\lambda}$$

This is the **negative exponential distribution** in which the density has a maximum at $z = 0$ and declines to zero as $z \rightarrow \infty$. Since the integral of $p(z)$ is $-e^{-z/\lambda}$,

$$\int_0^{\infty} p(z)\, dz = -e^{-z/\lambda}\Big|_0^{\infty} = 0 - (-1) = 1$$

showing that $p(z)$ fulfills the properties of a probability density.

What is the probability that a randomly drawn individual will have z in the range of $1/4$ to $1/2$?

$$P(1/4 \leq z \leq 1/2) = \int_{1/4}^{1/2} p(z)\, dz = -e^{-z/\lambda}\Big|_{1/4}^{1/2} = e^{-1/(4\lambda)} - e^{-1/(2\lambda)}$$

The numerical answer depends on the parameter λ. If, for example, $\lambda = 1/2$, then $P(1/4 \leq z \leq 1/2) = 0.239$.

Before moving on, we emphasize the importance of distinguishing between true **parameters** of distributions and **estimates** of those parameters obtained by sampling. True parameter values can only be obtained if every member of a population is measured with absolute accuracy. We must therefore almost always settle for approximations, the accuracy of which depends on the experimental setting, the measurement apparatus, and the sample size. Statisticians often denote parameters of a population with Greek symbols and to sample estimates with

Roman symbols. We will adhere to this protocol as much as possible, although there will be some instances where traditional quantitative-genetic notation prevents us from doing so.

The most useful probability density functions are defined completely by one or two parameters describing the central location and dispersion of the distribution. The most widely used measure of the location is the **arithmetic mean,** μ, also known as the **first moment about the origin**. If $p(z)$ is the probability density function of phenotype z, then weighting all values of z by their density leads to

$$\mu = \int_{-\infty}^{+\infty} z\, p(z)\, dz = E(z) \tag{2.2}$$

where $E(z)$ denotes the **expected value** or **expectation** of z. Here, we have arbitrarily put the limits $\pm\infty$ on the integral to ensure that the entire range of variation is covered. For discrete characters, $\mu = E(z) = \sum_i z_i P(z = z_i)$. For a character denoted by z, the sample estimate of the mean is generally denoted by $\bar{z}$, and estimated as the average of the n measures,

$$\bar{z} = \frac{1}{n} \sum_{i=1}^{n} z_i$$

Example 2. What is the mean of the distribution discussed in Example 1? Since the integral of $(z/\lambda)\, e^{-z/\lambda}$ is $-(z+\lambda)\, e^{-z/\lambda}$,

$$\mu = \int_0^{\infty} z\, p(z)\, dz = -(z+\lambda)\, e^{-z/\lambda} \Big|_0^{\infty} = \lambda$$

Thus, the parameter λ is the mean of the distribution defined by the density function $p(x) = (1/\lambda)\, e^{-z/\lambda}$.

Higher-order moments provide measures of the dispersion of a frequency distribution. The most familiar and useful such measure is the population **variance** (a term introduced in Fisher's 1918 paper). Also known as the **second moment about the mean**, the variance is the expected squared deviation of an observation from its mean,

$$\sigma^2 = \int_{-\infty}^{+\infty} (z-\mu)^2\, p(z)\, dz = E\left[\, (z-\mu)^2 \,\right] \tag{2.3}$$

Because $\mu = E(z)$, this quantity can be expressed more simply by expanding $(z - \mu)^2$ to obtain

$$\sigma^2 = E(z^2 - 2z\mu + \mu^2) = E(z^2) - 2\mu E(z) + \mu^2 = E(z^2) - \mu^2 \tag{2.4}$$

where we have used two useful properties of expectations,

$$E(x + y) = E(x) + E(y)$$
$$E(c\,x) = c\,E(x)$$

for a constant c. Several notations are used for the parametric variance of a distribution. When there is no ambiguity as to the variable being considered, σ^2 suffices. More generally, the variance of z is denoted by σ_z^2 or $\sigma^2(z)$.

A slight complication arises when one wishes to estimate the parameter σ^2 from a random sample of the population. As noted above, the true parameters μ and $E(z^2)$ cannot be known with certainty unless the entire population is sampled. Because the estimated mean $(\bar{z})$ is a function of the data, individual measures tend to be closer to the observed mean than to the true mean, and as a consequence, observed values of $\overline{z^2} - \bar{z}^2$ tend to be slightly less than the parametric value $[E(z^2) - \mu]$. Thus, the estimator $(\overline{z^2} - \bar{z}^2)$ is biased in the sense that it tends to underestimate the parameter $\sigma^2(z)$ to a degree that decreases with increasing sample size (n). A major goal of applied statistics is to obtain unbiased estimators that account for these kinds of small sample size limitations. In the case of the variance, the solution is simple (Example 2, Appendix 1), with

$$\text{Var}(z) = \frac{n\,(\overline{z^2} - \bar{z}^2)}{n - 1} \tag{2.5}$$

providing an unbiased estimate of $\sigma^2(z)$ (for the derivation of this expression, see Example 2, Appendix 1.) This equation should be used whenever the true population variance, $\sigma^2(z)$, is being estimated from actual sample data.

The variance is measured in units that are the square of those of the mean, but it is often desirable to describe the dispersion of a frequency distribution on the same scale as the mean. The square root of the variance of z is called the **standard deviation** of z. The parametric value is denoted by $\sigma(z)$, σ_z, or just σ, and the statistic by $\text{SD}(z) = \sqrt{\text{Var}(z)}$. The ratio of the standard deviation to the mean, the **coefficient of variation,** is frequently used as a relative measure of dispersion. It is known that the statistic $\text{CV}(z) = \text{SD}(z)/\bar{z}$ is a downwardly biased estimator of the parametric index (σ/μ), but the bias is expected to be negligible in most cases (Haldane 1955).

Quantitative geneticists generally rely on the variance as a measure of the dispersion of a distribution. However, additional moments can be informative. For example, the third moment about the mean (μ_3) is a useful measure of the asymmetry of a distribution. Also known as the **skewness**, μ_3 is the expected cubic deviation from the mean. As in the case of the variance, it can be expressed in terms of the moments about the origin,

$$\begin{aligned}\mu_3 &= \int_{-\infty}^{+\infty} (z-\mu)^3\, p(z)\, dz = E\left[(z-\mu)^3\right] \\ &= E(z^3) - 3\mu E(z^2) + 3\mu[E(z)]^2 - [E(z)]^3 \\ &= E(z^3) - 3\mu E(z^2) + 2\mu^3\end{aligned} \tag{2.6}$$

Thiele (1889) found that an unbiased sample estimator for μ_3 is

$$\text{Skw}(z) = \frac{n^2\,(\overline{z^3} - 3\,\overline{z^2}\,\bar{z} + 2\,\bar{z}^3\,)}{(n-1)(n-2)} \tag{2.7}$$

where $\overline{z^3}$ denotes the observed mean cubed value of z. The degree of asymmetry can also be described with a dimensionless index, the **coefficient of skewness,** which is estimated by the ratio

$$k_3 = \frac{\text{Skw}(z)}{\text{Var}(z)^{3/2}} \tag{2.8}$$

k_3 is positive when the longer tail of a distribution is to the right, negative when the tail is to the left, and zero for a perfectly symmetrical distribution.

From the above, it follows that

$$\mu_r = \int_{-\infty}^{+\infty} (z-\mu)^r p(z)\, dz \tag{2.9}$$

is a general expression for the rth moment about the mean. It also follows that μ_r can always be expressed in terms of moments about the origin $[\,E(z),\ E(z^2),\ \ldots,\ E(z^r)\,]$. As was shown for the variance and the skewness, these terms are obtainable from the binomial expansion of $(z-\mu)^r$.

Finally, we note that when moments are calculated from data that are grouped into classes, as in Table 2.1, a certain amount of bias is introduced because the true measures are assumed to be concentrated at the midpoints of the classes. Provided the total distribution is continuous and tails off smoothly at its extremities, this bias can often be eliminated by application of Sheppard's (1898) corrections. In the case of the variance, the corrected estimate is obtained by subtracting from Var(z) the quantity $\omega^2/12$, where ω is the width of the interval. No correction is required for the third moment about the mean. For details on higher-order moments, see Kendall and Stuart (1977, p. 77).

Example 3. Utilizing the data for maternal weight from Table 2.1, we now summarize the procedures for obtaining estimates of the first three moments.

grams *	z	$n(z)$	$z\,n(z)$	$z^2n(z)$	$z^3n(z)$
50 -	55	5	275	15,125	831,875
60 -	65	9	585	38,025	2,471,625
70 -	75	46	3,450	258,750	19,406,250
80 -	85	63	5,355	455,175	38,689,875
90 -	95	108	10,260	974,700	92,596,500
100 -	105	146	15,330	1,609,650	169,013,250
110 -	115	148	17,020	1,957,300	225,089,500
120 -	125	151	18,875	2,359,375	294,921,875
130 -	135	136	18,360	2,478,600	334,611,000
140 -	145	83	12,035	1,745,075	253,035,875
150 -	155	43	6,665	1,033,075	160,126,625
160 -	165	41	6,765	1,116,225	184,177,125
170 -	175	15	2,625	459,375	80,390,625
180 -	185	8	1,480	273,800	50,653,000
190 -	195	1	195	38,025	7,414,875
Totals		$n =$ 1,003	$\sum z\,n(z) =$ 119,255	$\sum z^2\,n(z) =$ 14,812,275	$\sum z^3\,n(z) =$ 1,913,429,875

* For each weight category, z is taken arbitrarily to be the midpoint of the measurement interval, so that for the interval 50-60, we take $z = 55$. The frequency of observations in each category, $f(z)$, is equal to $n(z)/n$, where $n(z)$ is the number of observations with phenotype z, and $n = \sum n(z)$ is the total sample size.

The moments about the origin are obtained by dividing the weighted sums in the table by n,

$$\bar{z} = \sum z f(z) = \sum z\,n(z)/n = \frac{119,255}{1,003} = 118.90$$

$$\overline{z^2} = \sum z^2 f(z) = \sum z^2 n(z)/n = \frac{14,812,275}{1,003} = 14,767.97$$

$$\overline{z^3} = \sum z^3 f(z) = \sum z^3 n(z)/n = \frac{1,913,429,875}{1,003} = 1,907,706.75$$

The variance estimated from the pooled data is

$$\text{Var}(z) = \frac{n\,(\,\overline{z^2} - \bar{z}^2\,)}{n-1} = 631.39$$

and application of Sheppard's correction, with $\omega = 10$, reduces this to

$$\text{Var}(z) = 631.39 - \frac{\omega^2}{12} = 623.06$$

The coefficient of variation is then

$$\text{CV}(z) = \frac{[\,\text{Var}(z)\,]^{1/2}}{\bar{z}} = 0.21$$

Finally, the skewness and coefficient of skewness are

$$\text{Skw}(z) = \frac{n^2\,(\,\overline{z^3} - 3\,\overline{z^2}\,\bar{z} + 2\,\bar{z}^3\,)}{(n-1)(n-2)} = 1,805.40$$

$$k_3 = \frac{\text{Skw}(z)}{[\,\text{Var}(z)\,]^{3/2}} = 0.12$$

THE NORMAL DISTRIBUTION

When large data sets of the type compiled in Table 2.1 are displayed in the form of frequency histograms (Figure 2.1), they often approximate a bell-shaped distribution. Three famous mathematicians, DeMoivre (1738), LaPlace (1778), and Gauss (1809), worked out the properties of a very useful description of this form — the **normal distribution**, also referred to as the **Gaussian distribution**. If z is a normally distributed variable, its density function is given by

$$p(z) = (2\pi\sigma^2)^{-1/2} \exp\left[-\frac{(z-\mu)^2}{2\sigma^2}\right] \tag{2.10}$$

where $\exp \simeq 2.7183$ is the base of natural logarithms, and $\pi \simeq 3.1416$. The normal distribution is a function of only two parameters, the population mean (μ) and variance (σ^2). The normal density attains a maximum when $z = \mu$ and declines continuously and symmetrically in both directions as z deviates from μ (Figure 2.1). A normally distributed variable with mean μ and variance σ^2 is often denoted by $z \sim \text{N}(\mu, \sigma^2)$, where $\sim$ means "is distributed as." In discussions in future chapters, we often use the notation $\varphi(z, \mu, \sigma^2)$ to denote the probability density of a normal, to remind the reader that it is also a function of the mean and variance.

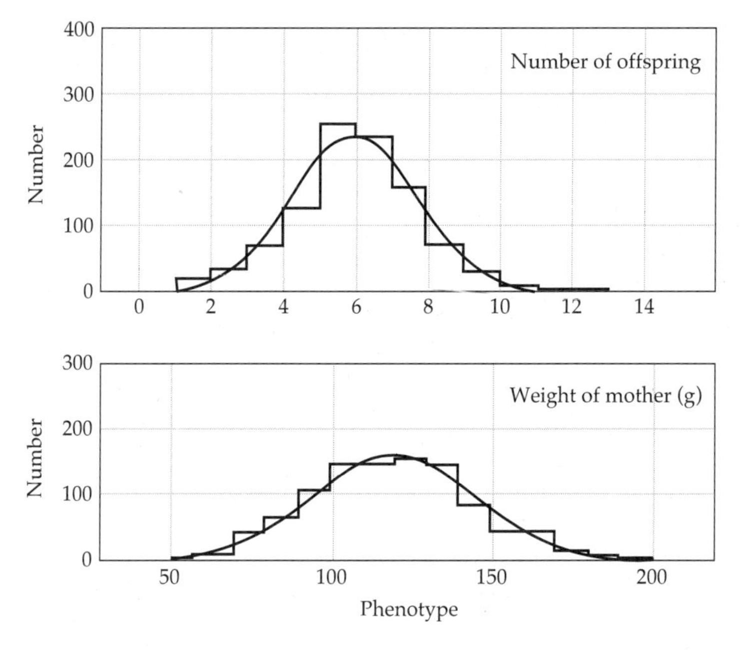

Figure 2.1 Frequency histograms for the two univariate distributions in Table 2.1 and their normal approximations based on the observed means and variances.

The normal distribution plays a central role in statistical theory for two reasons. First, the normal probability density function has many simple mathematical features that allow the derivation of practical statistical tests. Second, even when actual distributions of phenotypes are inconsistent with the normal density function, after an appropriate scale transformation (Chapter 11), many can be rendered approximately normal. A general reason why many traits are distributed normally or nearly so is provided by the **central limit theorem**, which states that the sum of a number of independent random variables approaches normality as the number of variables increases. This is expected to be the case, for example, for a metric character influenced by many environmental factors and a large number of unlinked genes, each with small additive effects. As a consequence, the normal distribution has been relied upon extensively in quantitative genetics. Whenever an assumption regarding the form of a phenotype distribution is necessary, the normal distribution is generally invoked as a first approximation. The normal density function is also often used to define a **Gaussian fitness function** in the theory of stabilizing selection, the "mean" serving as a measure of the optimum phenotype and the "variance" being inversely related to the intensity of selection (because the fitness function becomes flatter as the width increases).

There are, of course, limitations of the normal density function and of distribution functions in general. For instance, the normal distribution gives small positive values, rather than zero, for negative z, an unrealistic situation for traits such as body size or bone length, which cannot take on negative values. Nevertheless, if the mean of a distribution is sufficiently greater than zero, the theoretical incidence of negative values is minuscule and not problematical. It should also be emphasized that the normal distribution is a continuous function, giving positive values for noninteger values of z. It is, therefore, not strictly applicable to meristic traits such as egg number or spine count, although it provides a close approximation when the number of classes is large.

It is often convenient to work with a standardized form of Equation 2.10. A **standard normal deviate,** $z' = (z - \mu)/\sigma$, is the deviation of a measure from the population mean in units of standard deviations. Applying a useful property of distribution theory in the following example, we show that if z is normally distributed with mean μ and variance σ^2, then z' is normal with zero mean and unit variance, i.e.,

$$p(z') = (2\pi)^{-1/2} \exp\left[-\frac{(z')^2}{2}\right] \tag{2.11}$$

Example 4. It is known that if y is a function of z, denoted by $f(z)$, then its probability density function is

$$p(y) = \left|\frac{df(z)}{dz}\right|^{-1} p(z)$$

where $|\cdots|$ denotes absolute value. This transformation is valid provided that $df(z)/dz$ exists and is nonzero for all z values for which $p(z) > 0$. This criterion is met by the standard normal deviate.

Letting $z' = f(z) = (z - \mu)/\sigma$, then $df(z)/dz = \sigma^{-1}$. Substituting the normal probability density function for $p(z)$ recovers the **standard normal** or **(unit normal) distribution,**

$$p(z') = \left|\frac{1}{\sigma}\right|^{-1} \frac{1}{\sqrt{2\pi\sigma^2}} \exp\left[-\frac{(z-\mu)^2}{2\sigma^2}\right] = (2\pi)^{-1/2} \exp\left[-\frac{(z')^2}{2}\right]$$

Because the normal distribution is symmetrical, the third moment (μ_3) is equal to zero. The fourth moment has an expected value equal to $3\sigma^4$. Thus, if we let $\text{Kur}(z)$ be the sample estimate of μ_4, where Kur denotes **kurtosis**, the index

$$k_4 = \frac{\text{Kur}(z) - 3\,[\,\text{Var}(z)\,]^2}{[\,\text{Var}(z)\,]^2} \tag{2.12a}$$

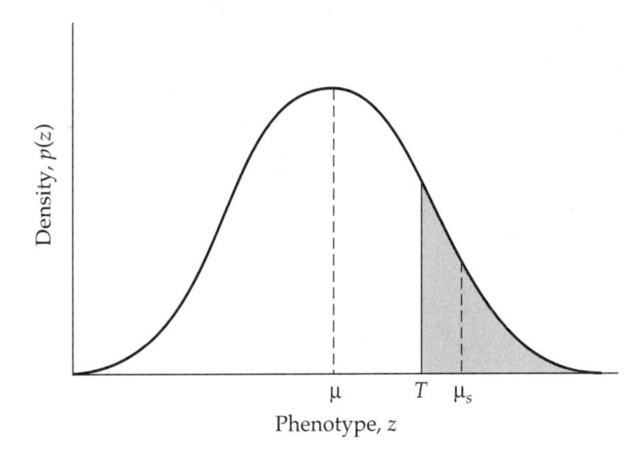

Figure 2.2 A truncated normal distribution (hatched area) with lower limit T and mean μ_s. The total area to the right of T is denoted by Φ_T.

where

$$\text{Kur}(z) = \frac{n^2(n+1)(\overline{z^4} - 4\,\overline{z^3}\,\overline{z} + 6\,\overline{z^2}\,\overline{z}^2 - 3\,\overline{z}^4)}{(n-1)(n-2)(n-3)} \tag{2.12b}$$

provides a measure of the peakedness of a distribution. For a truly normal distribution, $k_4 = 0$. A distribution with a high narrow peak relative to the normal $(k_4 > 0)$ is said to be **leptokurtic**. A broader peak than normal $(k_4 < 0)$ is referred to as **platykurtic.**

The Truncated Normal Distribution

We now consider the properties of a subset of the normal distribution, specifically a tail of the distribution (Figure 2.2). Such a consideration is important in the extreme form of directional selection that is used by most plant and animal breeders. Under **truncation selection**, all individuals below a certain phenotype are culled from the population and hence have zero fitness. The critical phenotype, T, is called the **truncation point**. For a normally distributed phenotype, the mean phenotype of the population above the threshold (i.e., after selection) can be written as

$$\mu_s = \frac{\int_T^\infty z p(z)\,dz}{\int_T^\infty p(z)\,dz} \tag{2.13}$$

In computing a mean, the phenotype frequencies must sum to one, and this is

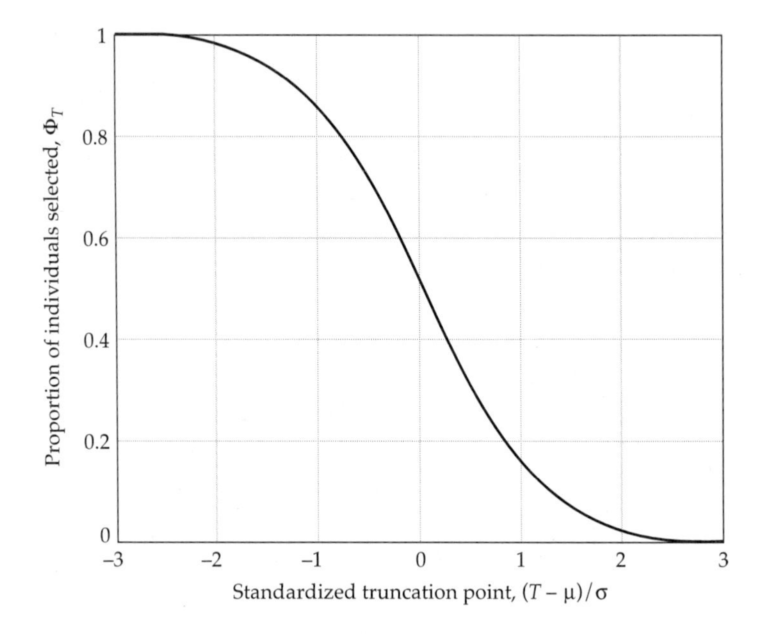

Figure 2.3 The proportion Φ_T of individuals selected from an underlying normal distribution with lower truncation point T, mean μ, and variance σ^2. As the truncation point moves to the right of the mean, i.e., as $(T-\mu)/\sigma$ becomes larger, the fraction of the population to the right of the truncation point approaches zero.

accomplished by letting the density of phenotype z after selection be

$$\frac{p(z)}{\int_T^\infty p(z)\,dz}$$

Since the denominator, $\int_T^\infty p(z)\,dz$, is the sum of frequencies for phenotypes greater than T (i.e., the fraction of individuals allowed to reproduce), it is a measure of the intensity of selection, and we hereafter denote it as Φ_T. After integration, the solution to Equation 2.13 is found to be

$$\mu_s = \mu + \frac{\sigma p_T}{\Phi_T} \tag{2.14}$$

where p_T is the height of the standard normal curve at the truncation point, obtained by setting $z' = (T - \mu)/\sigma$ in Equation 2.11,

$$p_T = (2\pi)^{-1/2} \exp\left[-\frac{(T-\mu)^2}{2\,\sigma^2}\right]$$

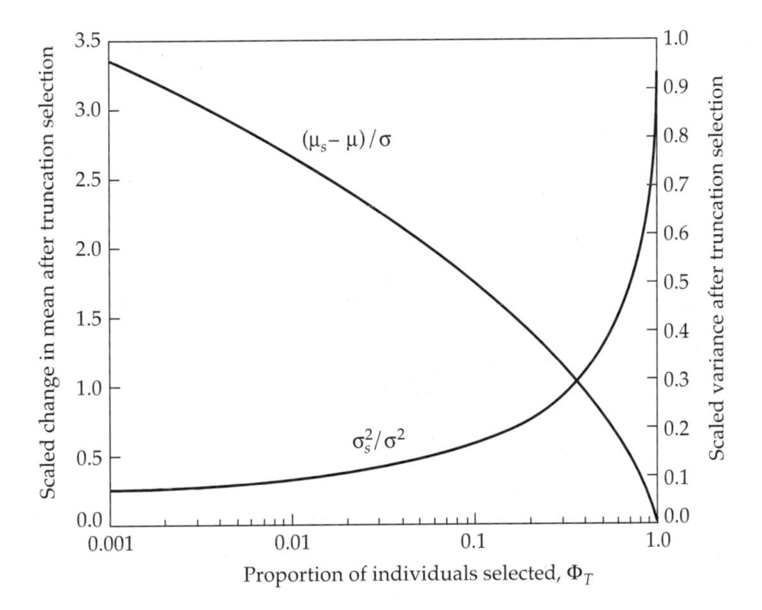

Figure 2.4 The change in the mean phenotype (in units of phenotypic standard deviations) and in the phenotypic variance (relative to the variance prior to selection) as a function of the selection intensity. As the proportion of the population above the threshold (Φ_T) increases, the selection intensity decreases.

Numerical values for Φ_T, which are functions of the standardized distance z' of T from μ, are obtainable from tables of the standard normal distribution (see also Figure 2.3).

Equation 2.14 states that if the upper proportion Φ_T of a normally distributed population is selected, the mean will advance $\mu_s - \mu = \sigma p_T/\Phi_T$ units. This change in the mean caused by selection is often denoted by S, the **directional selection differential** (Chapter 3). In units of phenotypic standard deviations, the standardized selection differential is simply $(\mu_s - \mu)/\sigma = p_T/\Phi_T$, which is plotted in Figure 2.4.

In a similar fashion, the variance of the selected population can be shown to be

$$\sigma_s^2 = \left[1 + \frac{p_T\, z'}{\Phi_T} - \left(\frac{p_T}{\Phi_T} \right)^2 \right] \sigma^2 \tag{2.15}$$

(Johnson and Kotz 1970a, p. 83). The quantity within brackets gives the fraction of phenotypic variance remaining after selection, σ_s^2/σ^2 (Figure 2.4).

Example 5. Equation 2.14 can be used to answer another interesting question. What is the average absolute deviation from the mean, $E(|z - \mu|)$, for an individual drawn randomly from a normally distributed population?

Since the normal distribution is symmetrical, we can set $\Phi_T = 0.5$ to obtain the average deviation to the right of the mean; the average deviation to the left of the mean will be identical in absolute value, but opposite in sign. Since the truncation point is the mean, $z' = 0$, which when applied to Equation 2.11 yields $p_T = (2\pi)^{-1/2}e^0 = 0.399$. Substituting into Equation 2.14, we obtain $(\mu_s - \mu) = 0.798\,\sigma$. Thus, the average absolute value of individual deviations from the mean is about 80% of the standard deviation. This quantity is known as the **most probable error.** Early in this century it was widely used by statisticians, but now the simple standard deviation is usually reported.

CONFIDENCE INTERVALS

Estimates such as $\bar{z}$ and $\text{Var}(z)$ vary from one sample to the next because of sampling error, so it is useful to know how far an observed statistic is likely to deviate from the true parameter that is being estimated. Although the true values are unknown, if something is known about the sampling error of the estimate, it is possible to evaluate the probability that the observed value lies within a specific range of the true value. Generally, we do not estimate the sampling error of statistics by sampling populations over and over again, but by using known algebraic expressions that themselves depend on sample statistics.

As an example, consider an estimate $\bar{z}$ of the mean of a distribution. An important issue here is the probability α that the parameter μ is within a certain range $\bar{z} \pm \Delta$. By symmetry, this is the same as the probability that $\bar{z}$ lies within the range $\mu \pm \Delta$. Transforming to standardized variables by letting $z' = (\bar{z} - \mu)/\sigma(\bar{z})$, where $\sigma(\bar{z})$ is the sampling variance of the mean, then the probability of interest is defined to be

$$\alpha = P[\,(\bar{z} - \Delta) \leq \mu \leq (\bar{z} + \Delta)\,] = \int_{-\Delta/\sigma(\bar{z})}^{+\Delta/\sigma(\bar{z})} p(z')dz' \tag{2.16}$$

The range $\bar{z} \pm \Delta$ defines the **confidence limits** or **interval** for the mean associated with the α probability level. In applications of Equation 2.16, it is generally assumed that the statistic is unbiased (so that the expected value of the statistic equals the true parameter value) and normally distributed. In the case of the mean, this implies that replicate estimates of the mean ($\bar{z}$) should be normally distributed about the parametric value (μ) with sampling variance $\sigma^2(\bar{z})$. Equation 2.16 is then simply an integration over the standardized normal density.

Although Equation 2.16 cannot be integrated directly, tables relating the standardized limits (Δ/σ) to α are provided in most statistics texts. The quantity Δ/σ, usually denoted as t, defines the distance (in standard errors) that the deviation between observed statistic and parametric value will lie with probability α. (Whereas the standard deviation is a measure of the dispersion of individual measures, the term **standard error** is usually reserved as a measure of the dispersion of statistics.) For any particular probability level, t decreases with increasing sample size, asymptotically approaching a constant. For sample sizes exceeding 50 or so, $t \simeq 1.96$ for $\alpha = 0.95$, and $t \simeq 2.58$ for $\alpha = 0.99$.

The remaining problem is to obtain an estimate of the sampling variance of the statistic (the square of the standard error). In the case of the mean, it is well known that an unbiased estimator of the sampling variance is $\text{Var}(z)/n$, where $\text{Var}(z)$ is the variance of individual measures, and n is the number of measures (Appendix 1). Thus, the 95% confidence interval for the mean is approximately $\bar{z} \pm 1.96\,[\,\text{Var}(z)/n\,]^{1/2}$.

Unfortunately, expressions for the sampling variances of other statistics (such as the variance, higher-order moments, coefficients of variation, etc.) are usually much more complicated than those for the mean. Appendix 1 outlines procedures that have been used extensively in quantitative genetics to obtain expressions for sampling variances for such statistics. These expressions are usually referred to as **large-sample variance** estimators because they are functions of observed statistics whose reliability increases with increasing sample size. A common procedure in statistics is to use twice the square root of the large-sample variance as a crude estimate of the 95% confidence limit. We emphasize that this assumes that the statistic has a sampling distribution that is close to normal, that the estimator is unbiased, and that the sample size is large enough that the large-sample variance (itself an estimate) is reasonably reliable.

3

Covariance, Regression, and Correlation

In the previous chapter, the variance was introduced as a measure of the dispersion of a univariate distribution. Additional statistics are required to describe the joint distribution of two or more variables. The **covariance** provides a natural measure of the association between two variables, and it appears in the analysis of many problems in quantitative genetics including the resemblance between relatives, the correlation between characters, and measures of selection.

As a prelude to the formal theory of covariance and regression, we first provide a brief review of the theory for the distribution of pairs of random variables. We then give a formal definition of the covariance and its properties. Next, we show how the covariance enters naturally into statistical methods for estimating the linear relationship between two variables (least-squares linear regression) and for estimating the goodness-of-fit of such linear trends (correlation). Finally, we apply the concept of covariance to several problems in quantitative-genetic theory. More advanced topics associated with multivariate distributions involving three or more variables are taken up in Chapter 8.

JOINTLY DISTRIBUTED RANDOM VARIABLES

The probability of joint occurrence of a pair of random variables (x, y) is specified by the **joint probability density function,** $p(x, y)$, where

$$P(\, y_1 \le y \le y_2,\, x_1 \le x \le x_2 \,) = \int_{y_1}^{y_2} \int_{x_1}^{x_2} p(x, y)\, dx\, dy \tag{3.1}$$

We often ask questions of the form: What is the distribution of y given that x equals some specified value? For example, we might want to know the probability that parents whose height is 68 inches have offspring with height exceeding 70 inches. To answer such questions, we use $p(y|x)$, the **conditional density** of y given x, where

$$P(\, y_1 \le y \le y_2 \,|\, x \,) = \int_{y_1}^{y_2} p(\, y \,|\, x \,)\, dy \tag{3.2}$$

Joint probability density functions, $p(x, y)$, and conditional density functions,

$p(y|x)$, are connected by

$$p(x,y) = p(\,y\,|\,x\,)\,p(x) \tag{3.3a}$$

where $p(x) = \int_{-\infty}^{+\infty} p(\,y\,|\,x\,)\,dy$ is the marginal (univariate) density of x.

Two random variables, x and y, are said to be **independent** if $p(x,y)$ can be factored into the product of a function of x only and a function of y only, i.e.,

$$p(x,y) = p(x)\,p(y) \tag{3.3b}$$

If x and y are independent, knowledge of x gives no information about the value of y. From Equations 3.3a and 3.3b, if $p(x,y) = p(x)\,p(y)$, then $p(\,y\,|\,x\,) = p(y)$.

Expectations of Jointly Distributed Variables

The expectation of a bivariate function, $f(x,y)$, is determined by the joint probability density

$$E[\,f(x,y)\,] = \int_{-\infty}^{+\infty}\int_{-\infty}^{+\infty} f(x,y)\,p(x,y)\,dx\,dy \tag{3.4}$$

Most of this chapter is focused on **conditional expectation,** i.e., the expectation of one variable, given information on another. For example, one may know the value of x (perhaps parental height), and wish to compute the expected value of y (offspring height) given x. In general, conditional expectations are computed by using the conditional density

$$E(\,y\,|\,x\,) = \int_{-\infty}^{+\infty} y\,p(\,y\,|\,x\,)\,dy \tag{3.5}$$

If x and y are independent, then $E(y|x) = E(y)$, the unconditional expectation. Otherwise, $E(\,y\,|\,x\,)$ is a function of the specified x value. For height in humans (Figure 1.1), Galton (1889) observed a linear relationship,

$$E(\,y\,|\,x\,) = \alpha + \beta x \tag{3.6}$$

where α and β are constants. Thus, the conditional expectation of height in offspring (y) is linearly related to the average height of the parents (x).

COVARIANCE

Consider a set of paired variables, (x,y). For each pair, subtract the population mean μ_x from the measure of x, and similarly subtract μ_y from y. Finally, for each pair of observations, multiply both of these new measures together to obtain $(x-\mu_x)(y-\mu_y)$. The **covariance** of x and y is defined to be the average of this quantity over all pairs of measures in the population,

$$\sigma(x,y) = E[\,(x-\mu_x)\,(y-\mu_y)\,] \tag{3.7}$$

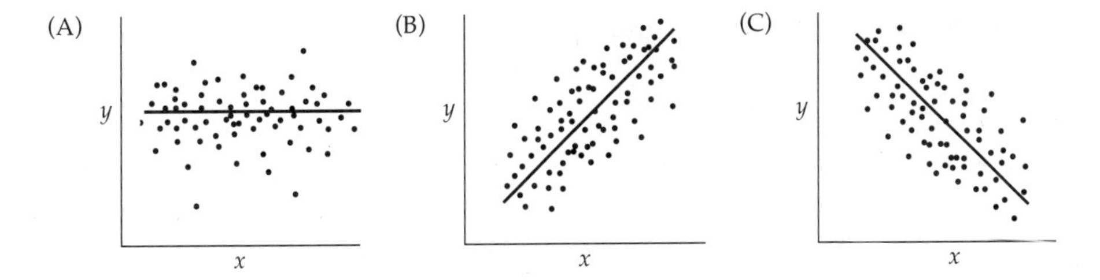

Figure 3.1 Scatterplots for the variables x and y. Each point in the x-y plane corresponds to a single pair of observations (x, y). The line drawn through the scatterplot gives the expected value of y given a specified value of x. (A) There is no linear tendency for large x values to be associated with large (or small) y values, so $\sigma(x, y) = 0$. (B) As x increases, the conditional expectation of y given x, $E(y|x)$, also increases, and $\sigma(x, y) > 0$. (C) As x increases, the conditional expectation of y given x decreases, and $\sigma(x, y) < 0$.

We often denote covariance by $\sigma_{x,y}$. Because $E(x) = \mu_x$ and $E(y) = \mu_y$, expansion of the product leads to further simplification,

$$\begin{aligned}\sigma(x, y) &= E[\,(x - \mu_x)\,(y - \mu_y)\,] \\ &= E\,(xy - \mu_y\, x - \mu_x\, y + \mu_x\, \mu_y\,) \\ &= E(x\, y) - \mu_y\, E(x) - \mu_x\, E(y) + \mu_x\, \mu_y \\ &= E(x\, y) - \mu_x\, \mu_y \end{aligned} \tag{3.8}$$

In words, the covariance is the mean of the pairwise cross-product $x\,y$ minus the cross-product of the means. The sampling estimator of $\sigma(x, y)$ is similar in form to that for a variance,

$$\text{Cov}(x, y) = \frac{n\,(\,\overline{xy} - \overline{x} \cdot \overline{y}\,)}{n - 1} \tag{3.9}$$

where n is the number of pairs of observations, and

$$\overline{xy} = \frac{1}{n} \sum_{i=1}^{n} x_i\, y_i$$

The covariance is a measure of association between x and y (Figure 3.1). It is positive if y increases with increasing x, negative if y decreases as x increases, and zero if there is no *linear* tendency for y to change with x. If x and y are independent, then $\sigma(x, y) = 0$, but the converse is not true — a covariance of zero does not necessarily imply independence. (We will return to this shortly; see Figure 3.3.)

Useful Identities for Variances and Covariances

Since $\sigma(x, y) = \sigma(y, x)$, covariances are symmetrical. Furthermore, from the definition of the variance and covariance,

$$\sigma(x, x) = \sigma^2(x) \tag{3.10a}$$

i.e., *the covariance of a variable with itself is the variance of that variable.* It also follows from Equation 3.8 that, for any constant a,

$$\sigma(a, x) = 0 \tag{3.10b}$$
$$\sigma(a\,x, y) = a\,\sigma(x, y) \tag{3.10c}$$

and if b is also a constant

$$\sigma(a\,x, b\,y) = a\,b\,\sigma(x, y) \tag{3.10d}$$

From Equations 3.10a and 3.10d,

$$\sigma^2(a\,x) = a^2\sigma^2(x) \tag{3.10e}$$

i.e., *the variance of the transformed variable ax is a^2 times the variance of x.* Likewise, for any constant a,

$$\sigma[\,(a + x), y\,] = \sigma(x, y) \tag{3.10f}$$

so that *simply adding a constant to a variable does not change its covariance with another variable.*

Finally, the covariance of two sums can be written as a sum of covariances,

$$\sigma[\,(x + y), (w + z)\,] = \sigma(x, w) + \sigma(y, w) + \sigma(x, z) + \sigma(y, z) \tag{3.10g}$$

Similarly, the variance of a sum can be expressed as the sum of all possible variances and covariances. From Equations 3.10a and 3.10g,

$$\sigma^2(x + y) = \sigma^2(x) + \sigma^2(y) + 2\sigma(x, y) \tag{3.11a}$$

More generally,

$$\sigma^2\left(\sum_i^n x_i\right) = \sum_i^n\sum_j^n \sigma(x_i, x_j) = \sum_i^n \sigma^2(x_i) + 2\sum_{i<j}^n \sigma(x_i, x_j) \tag{3.11b}$$

Thus, *the variance of a sum of uncorrelated variables is just the sum of the variances of each variable.*

We will make considerable use of the preceding relationships in the remainder of this chapter and in chapters to come. Methods for approximating variances and covariances of more complex functions are outlined in Appendix 1.

REGRESSION

Depending on the causal connections between two variables, x and y, their true relationship may be linear or nonlinear. However, regardless of the true pattern of association, a linear model can always serve as a first approximation. In this case, the analysis is particularly simple,

$$y = \alpha + \beta x + e \tag{3.12a}$$

where α is the y-intercept, β is the slope of the line (also known as the **regression coefficient**), and e is the **residual error**. Letting

$$\widehat{y} = \alpha + \beta x \tag{3.12b}$$

be the value of y predicted by the model, then the residual error is the deviation between the observed and predicted y value, i.e., $e = y - \widehat{y}$. When information on x is used to predict y, x is referred to as the **predictor** or **independent variable** and y as the **response** or **dependent variable**.

The objective of linear regression analysis is to estimate the model parameters, α and β, that give the "best fit" for the joint distribution of x and y. The true parameters α and β are only obtainable if the entire population is sampled. With an incomplete sample, α and β are approximated by sample estimators, denoted as a and b. Good approximations of α and β are sometimes obtainable by visual inspection of the data, particularly in the physical sciences, where deviations from a simple relationship are due to errors of measurement rather than biological variability. However, in biology many factors are often beyond the investigator's control. The data in Figure 3.2 provide a good example. While there appears to be a weak positive relationship between maternal weight and offspring number in rats, it is difficult to say anything more precise. An objective definition of "best fit" is required.

Derivation of the Least-Squares Linear Regression

The mathematical method of **least-squares linear regression** provides one such best-fit solution. Without making any assumptions about the true joint distribution of x and y, least-squares regression minimizes the average value of the squared (vertical) deviations of the observed y from the values predicted by the regression line. That is, the **least-squares** solution yields the values of a and b that minimize the mean squared residual, $\overline{e^2}$. Other criteria could be used to define "best fit." For example, one might minimize the mean absolute deviations (or cubed deviations) of observed values from predicted values. However, as we will now see, least-squares regression has the unique and very useful property of maximizing the amount of variance in y that can be explained by a linear model.

Consider a sample of n individuals, each of which has been measured for x and y. Recalling the definition of a residual

$$e = y - \widehat{y} = y - a - bx \tag{3.13a}$$

and then adding and subtracting the quantity $(\,\overline{y} + b\,\overline{x}\,)$ on the right side, we obtain

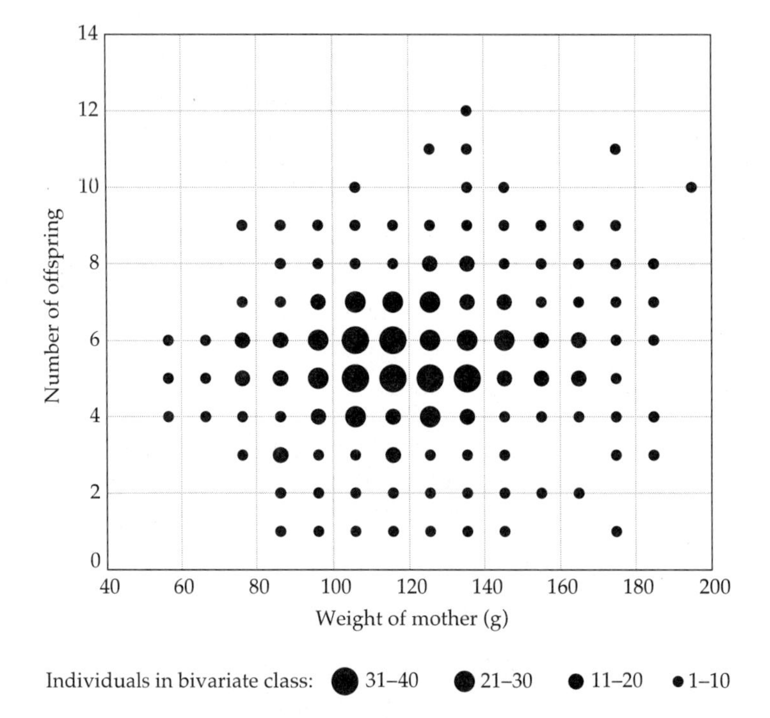

Figure 3.2 A bivariate plot of the relationship between maternal weight and number of offspring for the sample of rats summarized in Table 2.2. Different-sized circles refer to different numbers of individuals in the bivariate classes.

$$e = (y - \overline{y}) - b(x - \overline{x}) - (a + b\overline{x} - \overline{y}) \tag{3.13b}$$

Squaring both sides leads to

$$\begin{aligned} e^2 = &(y - \overline{y})^2 - 2b(y - \overline{y})(x - \overline{x}) + b^2(x - \overline{x})^2 + (a + b\overline{x} - \overline{y})^2 \\ &- 2(y - \overline{y})(a + b\overline{x} - \overline{y}) + 2b(x - \overline{x})(a + b\overline{x} - \overline{y}) \end{aligned} \tag{3.13c}$$

Finally, we consider the average value of e^2 in the sample. The final two terms in Equation 3.13b drop out here because, by definition, the mean values of $(x - \overline{x})$ and $(y - \overline{y})$ are zero. However, by definition, the mean values of the first three terms are directly related to the sample variances and covariance. Thus,

$$\overline{e^2} = \left(\frac{n-1}{n}\right)\left[\text{Var}(y) - 2b\,\text{Cov}(x,y) + b^2\,\text{Var}(x)\right] + (a + b\overline{x} - \overline{y})^2 \tag{3.13d}$$

The values of a and b that minimize $\overline{e^2}$ are obtained by taking partial derivatives of this function and setting them equal to zero:

$$\frac{\partial\,(\overline{e^2})}{\partial\,a} = 2\,(a + b\,\overline{x} - \overline{y}\,) = 0$$

$$\frac{\partial\,(\overline{e^2})}{\partial\,b} = 2\left[\left(\frac{n-1}{n}\right)[-\text{Cov}(x,y) + b\,\text{Var}(x)\,] + \overline{x}\,(\,a + b\,\overline{x} - \overline{y}\,)\right] = 0$$

The solutions to these two equations are

$$a = \overline{y} - b\,\overline{x} \tag{3.14a}$$

$$b = \frac{\text{Cov}(x,y)}{\text{Var}(x)} \tag{3.14b}$$

Thus, the least-squares estimators for the intercept and slope of a linear regression are simple functions of the observed means, variances, and covariances. From the standpoint of quantitative genetics, this property is exceedingly useful, since such statistics are readily obtainable from phenotypic data.

Properties of Least-squares Regressions

Here we summarize some fundamental features and useful properties of the least-squares approach to linear regression analysis:

1. *The regression line passes through the means of both x and y.* This relationship should be immediately apparent from Equation 3.14a, which implies $\overline{y} = a + b\,\overline{x}$.

2. *The average value of the residual is zero.* From Equation 3.13a, the mean residual is $\overline{e} = \overline{y} - a - b\,\overline{x}$, which is constrained to be zero by Equation 3.14a. Thus, the least-squares procedure results in a fit to the data such that the sum of (vertical) deviations above and below the regression line are exactly equal.

3. *For any set of paired data, the least-squares regression parameters, a and b, define the straight line that maximizes the amount of variation in y that can be explained by a linear regression on x.* Since $\overline{e} = 0$, it follows that the variance of residual errors about the regression is simply $\overline{e^2}$. As noted above, this variance is the quantity minimized by the least-squares procedure.

4. *The residual errors around the least-squares regression are uncorrelated with the predictor variable x.* This statement follows since

$$\begin{aligned}\text{Cov}(x,e) &= \text{Cov}[\,x,(y - a - b\,x)\,] = \text{Cov}(x,y) - \text{Cov}(x,a) - b\,\text{Cov}(x,x)\\ &= \text{Cov}(x,y) - 0 - b\,\text{Var}(x)\\ &= \text{Cov}(x,y) - \frac{\text{Cov}(x,y)}{\text{Var}(x)}\,\text{Var}(x) = 0\end{aligned}$$

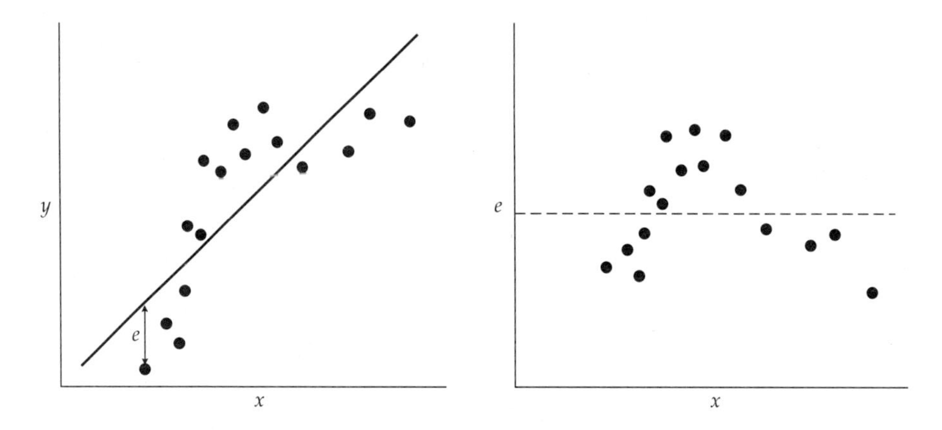

Figure 3.3 A linear least-squares fit to an inherently nonlinear data set. Although there is a systematic relationship between the residual error (e) and the predictor variable (x), the two are uncorrelated (show no net *linear* trend) when viewed over the entire range of x. The mean residual error ($\overline{e} = 0$) is denoted by the dashed line on the right graph.

Note, however, that $\text{Cov}(x, e) = 0$ does not guarantee that e and x are independent. In Figure 3.3, for example, because of a nonlinear relationship between y and x, the residual errors associated with extreme values of x tend to be negative while those for intermediate values are positive. Thus, if the true regression is nonlinear, then $E(\, e \,|\, x \,) \neq 0$ for some x values, and the predictive power of the linear model is compromised. Even if the true regression is linear, the variance of the residual errors may vary with x, in which case the regression is said to display **heteroscedasticity** (Figure 3.4). If the conditional variance of the residual errors given any specified x value, $\sigma^2(\, e \,|\, x \,)$, is a constant (i.e., independent of the value of x), then the regression is said to be **homoscedastic**.

5. There is an important situation in which *the true regression, the value of* $E(\, y \,|\, x \,)$, *is both linear and homoscedastic — when x are y are bivariate normally distributed*. The requirements for such a distribution are that the univariate distributions of both x and y are normal and that the conditional distributions of y given x, and x given y, are also normal (Chapter 8). Since statistical testing is simplified enormously, it is generally desirable to work with normally distributed data. For situations in which the raw data are not so distributed, a variety of transformations exist that can render the data close to normality (Chapter 11).

6. It is clear from Equations 3.14a,b that *the regression of y on x is different from the regression of x on y unless the means and variances of the two variables are equal*. This distinction is made by denoting the regression coefficient by $b(y, x)$ or $b_{y,x}$ when x is the predictor and y the response variable.

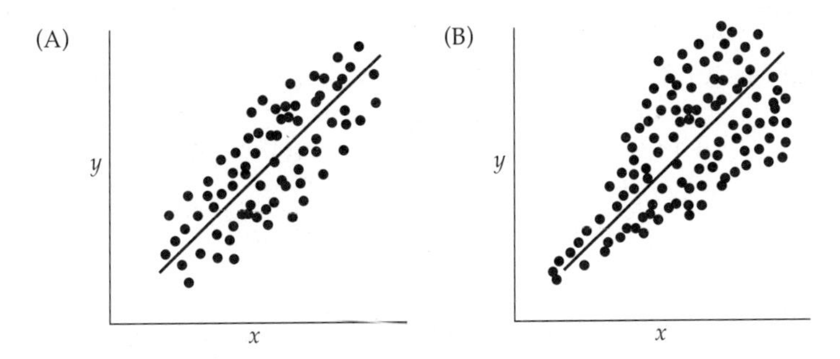

Figure 3.4 The dispersion of residual errors around a regression. (A) The regression is homoscedastic — the variance of residuals given x is a constant. (B) The regression is heteroscedastic — the variance of residuals increases with x. In this case, higher x values predict y with less certainty.

For practical reasons, we have expressed properties 1 – 6 in terms of the estimators a, b, Cov(x, y), and Var(x). They also hold when the estimators are replaced by the true parameters α, β, $\sigma(x, y)$, and $\sigma^2(x)$.

Example 1. Suppose Cov(x, y) = 10, Var(x) = 10, Var(y) = 15, and $\overline{x} = \overline{y} = 0$. Compute the least-squares regressions of y on x, and of x on y.

From Equation 3.14a, $a = 0$ for both regressions. However,

$$b(y, x) = \text{Cov}(x, y)/\text{Var}(x) = 10/10 = 1$$

while $b(x, y) = \text{Cov}(x, y)/\text{Var}(y) = 2/3$. Hence, $\widehat{y} = x$ is the least-squares regression of y on x, while $\widehat{x} = (2/3)y$ is the regression of x on y.

CORRELATION

For purposes of hypothesis testing, it is often desirable to use a dimensionless measure of association. The most frequently used measure in bivariate analysis is the **correlation coefficient,**

$$r(x, y) = \frac{\text{Cov}(x, y)}{\sqrt{\text{Var}(x)\,\text{Var}(y)}} \tag{3.15a}$$

Note that $r(x, y)$ is symmetrical, i.e., $r(x, y) = r(y, x)$. Thus, where there is no ambiguity as to the variables being considered, we abbreviate $r(x, y)$ as r. The parametric correlation coefficient is denoted by $\rho(x, y)$ (or ρ) and equals $\sigma(x, y)/\sigma(x)\sigma(y)$. The least-squares regression coefficient is related to the correlation coefficient by

$$b(y, x) = r\sqrt{\frac{\mathrm{Var}(y)}{\mathrm{Var}(x)}} \tag{3.15b}$$

An advantage of correlations over covariances is that the former are *scale independent*. This can be seen by noting that if w and c are constants,

$$r(w\,x, c\,y) = \frac{\mathrm{Cov}(w\,x, c\,y)}{\sqrt{\mathrm{Var}(w\,x)\,\mathrm{Var}(c\,y)}} = \frac{w\,c\,\mathrm{Cov}(x, y)}{\sqrt{w^2\,\mathrm{Var}(x)\,c^2\,\mathrm{Var}(y)}} = r(x, y) \tag{3.16a}$$

Thus scaling x and/or y by constants does not change the correlation coefficient, although the variances and covariances are affected. Since r is dimensionless with limits of ± 1, it gives a direct measure of the degree of association: if $|r|$ is close to one, x and y are very strongly associated in a linear fashion, while if $|r|$ is close to zero, they are not.

The correlation coefficient has other useful properties. First, *r is a standardized regression coefficient (the regression coefficient resulting from rescaling x and y such that each has unit variance)*. Letting $x' = x/\sqrt{\mathrm{Var}(x)}$ and $y' = y/\sqrt{\mathrm{Var}(y)}$ gives $\mathrm{Var}(x') = \mathrm{Var}(y') = 1$, implying

$$b(y', x') = b(x', y') = \mathrm{Cov}(x', y') = \frac{\mathrm{Cov}(x, y)}{\sqrt{\mathrm{Var}(x)\,\mathrm{Var}(y)}} = r \tag{3.16b}$$

Thus, when variables are standardized, the regression coefficient is equal to the correlation coefficient regardless of whether x' or y' is chosen as the predictor variable.

Second, *the squared correlation coefficient measures the proportion of the variance in y that is explained by assuming that $E(y|x)$ is linear.* The variance of the response variable y has two components: $r^2\,\mathrm{Var}(y)$, the amount of variance accounted for by the linear model (the **regression variance**), and $(1 - r^2)\,\mathrm{Var}(y)$, the remaining variance not accountable by the regression (the **residual variance**). To obtain this result, we derive the variance of the residual deviation defined in Equation 3.13a,

$$\begin{aligned}
\mathrm{Var}(e) &= \mathrm{Var}(\,y - a - bx\,) = \mathrm{Var}(\,y - bx\,) \\
&= \mathrm{Var}(y) - 2\,b\,\mathrm{Cov}(x, y) + b^2\,\mathrm{Var}(x) \\
&= \mathrm{Var}(y) - \frac{2\,[\,\mathrm{Cov}(x, y)\,]^2}{\mathrm{Var}(x)} + \frac{[\,\mathrm{Cov}(x, y)\,]^2\,\mathrm{Var}(x)}{[\,\mathrm{Var}(x)\,]^2} \\
&= \left(1 - \frac{[\,\mathrm{Cov}(x, y)\,]^2}{\mathrm{Var}(x)\,\mathrm{Var}(y)}\right)\mathrm{Var}(y) = (1 - r^2)\,\mathrm{Var}(y)
\end{aligned} \tag{3.17}$$

Example 2. Returning to Table 2.1, the preceding formulae can be used to characterize the relationship between maternal weight and offspring number in rats. Here we take offspring number as the response variable y and maternal weight as the predictor variable x. The mean and variance for maternal weight were found to be $\overline{x} = 118.90$ and $\text{Var}(x) = 623.06$ (Table 2.1). For offspring number, $\overline{y} = 5.49$ and $\text{Var}(y) = 2.94$. In order to obtain an estimate of the covariance, we first require an estimate of $E(x\,y)$. Taking the xy cross-product of all classes in Table 2.1 (using the midpoint of the interal for the value of x) and weighting them by their frequencies,

$$\overline{xy} = \frac{(1 \cdot 4 \cdot 55) + (3 \cdot 5 \cdot 55) + (1 \cdot 6 \cdot 55) + \cdots + (1 \cdot 10 \cdot 195)}{1003} = 660.14$$

The covariance estimate is then obtained using Equation 3.9,

$$\text{Cov}(x, y) = \frac{1003}{1002}\left[\,660.14 - (118.90 \times 5.49)\,\right] = 7.39$$

From Equation 3.14b, the slope of the regression is found to be

$$b(y, x) = \frac{7.39}{623.06} = 0.01$$

Thus, the expected increase in number of offspring per gram increase in maternal weight is about 0.01. How predictable is this change? From Equation 3.15a, the correlation coefficient is estimated to be

$$r = \frac{7.39}{\sqrt{623.06 \times 2.94}} = 0.17$$

Squaring this value, $r^2 = 0.03$. Therefore, only about 3 percent of the variance in offspring number can be accounted for with a model that assumes a linear relationship with maternal weight.

A TASTE OF QUANTITATIVE-GENETIC THEORY

Directional Selection Differentials and the Robertson-Price Identity

The evolutionary response of a character to selection is a function of the intensity of selection and the fraction of the phenotypic variance attributable to certain genetic effects. As noted at the end of last chapter, the **directional selection differential**, S, is defined to be the within-generation difference between the mean phenotype

μ_s after an episode of selection (but before reproduction) and the mean before selection μ,

$$S = \mu_s - \mu \tag{3.18}$$

The degree to which μ_s deviates from μ depends on the survivorship and reproductive rates of individuals with different phenotypes. If all individuals have equal fertility and viability, then $\mu_s = \mu$, $S = 0$, and the population mean phenotype is not expected to change between generations. Now, for simplicity, assume that individuals differ only in the probability of survival to maturity, so that fitness, $W(z)$, is the probability that individuals with phenotype z survive to reproduce. In what follows, no assumptions will be made about the general form of $W(z)$; it may be a continuous or discontinuous function of z, and it may take on values of 0 for some z. If $p(z)$ is the density of phenotype z before selection, then the density after selection is

$$p_s(z) = \frac{W(z)\,p(z)}{\int W(z)\,p(z)\,dz} \tag{3.19}$$

This expression is obtained by noting that $W(z)$ is a weighting factor for phenotype z. The denominator is the **mean individual fitness**, $\overline{W}$. Letting the **relative fitness** of phenotype z be $w(z) = W(z)/\overline{W}$, Equation 3.19 simplifies to $p_s(z) = w(z)\,p(z)$. It follows that the mean phenotype after selection is

$$\mu_s = \int z\,p_s(z)\,dz = \int z\,w(z)\,p(z)\,dz = E[\,z\,w(z)\,] \tag{3.20}$$

Note also that

$$\overline{w} = \int w(z)\,p(z)\,dz = \frac{1}{\overline{W}}\int W(z)\,p(z)\,dz = \overline{W}/\overline{W} = 1$$

i.e., the mean relative fitness in a population is always equal to one, and that since $\mu = E(z) \cdot E(w) = E(z) \cdot 1$, the directional selection differential may be rewritten as

$$S = \mu_s - \mu = E[\,z\,w(z)\,] - E(z)\,E(w) = \sigma[\,z, w(z)\,] \tag{3.21}$$

Thus, the *directional selection differential is equivalent to the covariance of phenotype and relative fitness*. This relationship, first noted by Robertson (1966), was greatly elaborated on by Price (1970, 1972). We refer to this very useful result as the **Robertson-Price identity**. It applies even when phenotypes vary in reproductive output, provided that the absolute fitnesses, $W(z)$, are weighted accordingly.

The importance of S can be seen by noting that if the regression of offspring phenotype on that of its average parent is linear with slope β (Figure 1.2), a change in the parental mean phenotype induces an expected change in the mean phenotype across generations equal to

$$\Delta\mu = \mu_o - \mu = \beta\,(\mu_s - \mu) = \beta\,S \tag{3.22}$$

where μ_o is the mean phenotype of the offspring of the selected parents. This fundamental relationship, known as the **breeders' equation,** combines information on the forces of selection (S) with that on inheritance (β) to yield a predictive equation for evolutionary change across generations. A genetic interpretation of the regression coefficient β will be provided in the final example of this chapter.

The Correlation between Genotypic and Phenotypic Values

Equation 3.22 shows that evolution by natural selection requires heritable variation, as no matter how large S is, the response to selection across generations is zero if $\beta = 0$. Quantification of the correspondence between phenotypic and genotypic values is related to one of the central goals of quantitative genetics — the partitioning of the phenotypic variance into genetic and nongenetic components. The standard approach is to consider the phenotypic value of an individual, z, to be the sum of the total effects of all loci on the trait, G (the **genotypic value**), and an environmental deviation E (analogous to the residual error above),

$$z = G + E$$

Using the properties of covariances noted above, the covariance between phenotypic and genotypic values may be written as

$$\sigma_{z,G} = \sigma[\,(G + E), G\,] = \sigma_G^2 + \sigma_{G,E} \tag{3.23}$$

The squared correlation coefficient is therefore

$$\rho^2(G, z) = \left(\frac{\sigma_{G,z}}{\sigma_G\,\sigma_z}\right)^2 = \frac{(\,\sigma_G^2 + \sigma_{G,E}\,)^2}{\sigma_G^2\,\sigma_z^2} \tag{3.24a}$$

which simplifies to

$$\rho^2(G, z) = \frac{\sigma_G^2}{\sigma_z^2} \tag{3.24b}$$

if there is no genotype-environment covariance, i.e., if $\sigma_{G,E} = 0$. In this special case, $\rho^2(G, z)$ is simply the proportion of the total phenotypic variance that is genetic. The quantity σ_G^2/σ_z^2 is generally referred to as **heritability in the broad sense** and abbreviated as H^2.

From Equation 3.23, it can be seen that covariance between genotypic values and environmental deviations causes the genotype-phenotype covariance to deviate from σ_G^2. Negative covariance between G and E causes a reduction in the correlation between phenotypic and genotypic values, and in extreme cases, can cause $\rho(G, z)$ to become negative. Further details on genotype-environment covariance are covered in Chapters 6 and 22.

Regression of Offspring Phenotype on Midparent Phenotype

Although the previous example has provided some insight into the genetic basis of phenotypic variation without getting bogged down in genetic complexities, in practice that approach is not very useful. Whereas phenotypic values are easily obtained (they are what we measure), the underlying genetic values are essentially unobservable without an extensive breeding program, and even then, they cannot be determined with complete accuracy (Chapter 26).

Fortunately, there are alternative ways to estimate levels of genetic variance of quantitative traits. All such methods are based on the simple fact that related individuals carry copies of many of the same alleles. Consider the resemblance between phenotypes of offspring (z_o) and their midparents (z_{mp}). A midparent value is simply the mean phenotype of a mother (z_m) and a father (z_f),

$$z_{mp} = \frac{z_m + z_f}{2}$$

We will confine our attention to a simple genetic situation — a single locus with purely additive gene effects, diploidy, random mating, and no selection. Let g_m and g_f, respectively, be the effects of the alleles that the offspring inherits from its mother and father, and g'_m and g'_f be the effects of the alleles that are not transmitted by the parents to this particular offspring. Further letting the environmental effects on the phenotypes of parents and offspring be E_m, E_f, and E_o, the three phenotypes may be expressed as

$$z_m = g_m + g'_m + E_m$$
$$z_f = g_f + g'_f + E_f$$
$$z_o = g_m + g_f + E_o$$

Because the equation for the offspring phenotype contains three terms and that for the midparent phenotype contains six, the complete algebraic expression for the midparent-offspring covariance, $\sigma(z_{mp}, z_o)$, is quite complex. It contains 18 terms. However, provided certain assumptions are met, most of these terms have expected values equal to zero. First, under the assumptions of random mating and no selection, there can be no covariance between the effects of alleles within individuals. Thus, the genes inherited by an offspring have zero covariance with the genes that are not inherited, and the genes in mothers are uncorrelated with those in fathers, i.e., $\sigma(g_m, g'_m) = \sigma(g_m, g_f) = \sigma(g_m, g'_f) = \sigma(g_f, g'_f) = \sigma(g_f, g_m) = \sigma(g_f, g'_m) = 0$. Second, provided there is no genotype-environment covariance, i.e., individuals are not assorted into environments on the basis of their genetic attributes, $\sigma(g_m, E_m) = \sigma(g_m, E_f) = \sigma(g_f, E_m) = \sigma(g_f, E_f) = \sigma(g_m, E_o) = \sigma(g'_m, E_o) = \sigma(g_f, E_o) = \sigma(g'_f, E_o) = 0$. Finally, provided the parents do not transmit their environmental effects to their progeny, i.e., there are no significant maternal or paternal environmental effects, then $\sigma(E_o, E_f) = \sigma(E_o, E_m) = 0$.

Most of these assumptions can be fulfilled in carefully designed experiments. Assuming this is the case, the only potential sources of covariance that exist between midparent and offspring phenotypes are those involving the inherited genes. Thus,

$$\begin{aligned}\sigma(z_{mp}, z_o) &= \sigma\left[\left(\frac{z_m + z_f}{2}\right), z_o\right] \\ &= \sigma\left[\left(\frac{g_m + g_f}{2}\right), (g_m + g_f)\right] \\ &= \frac{\sigma^2(g_m) + \sigma^2(g_f)}{2} \end{aligned} \tag{3.25a}$$

Recall that we assumed the genotypic value to be entirely defined by the additive effects of the two alleles. Thus, under random mating, the total genetic variance in the population is the sum of the variances of maternally and paternally derived genes, $\sigma^2(g_m) + \sigma^2(g_f)$. Since the gene effects are purely additive, this quantity may also be referred to as the **additive genetic variance,** σ_A^2. Thus, provided a number of assumptions are met, the phenotypic covariance between midparent and offspring is equivalent to half the additive genetic variance in the population,

$$\sigma(z_{mp}, z_o) = \frac{\sigma_A^2}{2} \tag{3.25b}$$

In Chapters 5 and 7, it will be shown that this equation holds for any number of loci provided they interact additively.

To obtain the expected least-squares regression of offspring on midparent phenotype, the covariance needs to be divided by the variance of midparent phenotypes. Using the properties of variances outlined above, and noting that the phenotypic covariance between parents, $\sigma(z_f, z_m)$, is zero under random mating,

$$\sigma^2(z_{mp}) = \sigma^2\left(\frac{z_m + z_f}{2}\right) = \frac{\sigma^2(z_m) + \sigma^2(z_f)}{4} \tag{3.26a}$$

Thus, provided the phenotypic variance in the two sexes is equal (or has been scaled to be so), the phenotypic variance of midparent values is half the phenotypic variance in the population,

$$\sigma^2(z_{mp}) = \frac{\sigma_z^2}{2} \tag{3.26b}$$

The slope of the least-squares linear regression of offspring phenotype on mid-parent phenotype is then

$$\beta_{o,mp} = \frac{\sigma_A^2}{\sigma_z^2} \tag{3.27}$$

Thus, for this special case, the slope of a midparent-offspring regression provides an estimate of the proportion of the phenotypic variance that is attributable to additive genetic factors. Obviously, we have made many assumptions in order to arrive at this expression, and the significance of these will be addressed in the remainder of the book. The salient issue here is that *inferences concerning the genetic basis of quantitative traits can be extracted from phenotypic measures of the resemblance between relatives.*

In closing, we note that there is an important distinction between the measures of genetic variance that appear in Equations 3.24b and 3.27. In Equation 3.24b, σ_G^2 refers to the total genetic variance, including that due to nonadditive interactions within and among loci. In Equation 3.27, σ_A^2 refers specifically to the additive component of genetic variation. The ratio σ_A^2/σ_z^2 is known as the **narrow-sense heritability** and is generally abbreviated as h^2. It is possible for the total genetic variance to be entirely additive, but often it is not. This distinction between broad- and narrow-sense heritability and the decomposition of genetic variance into additive and nonadditive components will be covered in detail in Chapters 4, 5, and 7.

The central importance of h^2 can be appreciated by recalling the breeders' equation, Equation 3.22, where β can now be seen to be h^2. Thus, if S is the change in mean phenotype caused by selection prior to reproduction, then the response to selection across generations is

$$\Delta\mu = h^2 S \tag{3.28}$$

The narrow-sense heritability can be thought of as the efficiency of the response to selection. If $h^2 = 0$, i.e., if there is no tendency for offspring to resemble their parents, there can be no evolutionary change, regardless of the strength of selection.

4

Properties of Single Loci

The fact that most principles of quantitative genetics can be expressed without reference to specific genes is precisely why quantitative-genetic analysis is so popular among those who study complex characters. Since this same feature can be cause for suspicion, a primary goal of the next few chapters is to clarify the ways in which quantitative genetics is grounded in fundamental Mendelian concepts. Prior to illustrating the connections between the properties of single genes and the expression and transmission of polygenic traits, we review some very basic and essential vocabulary.

It is well known that the genetic information encoding for characters resides on extremely long strands of deoxyribonucleic acid (DNA) called **chromosomes**. We still do not know the function of the vast majority of DNA in organisms, and many believe that a substantial portion of it has no function (Dover and Flavell 1982). DNA sequences that encode for particular products (proteins and RNAs) are referred to as **genes**, and their chromosomal locations are called **loci**. Most organisms have two copies of each of several chromosomes, in which case they are said to be **diploid**. Since DNA replication is an imperfect process, mutations arise, and as a consequence the two "copies" of each gene carried by diploid individuals need not be identical. The various forms of a gene are called **alleles**.

Gene loci that exhibit more than one allele are the subject of genetics. Such loci are said to be **polymorphic**, whereas loci at which all gene copies are identical are **monomorphic**. A substantial fraction of the gene pool in many species is polymorphic. The possible reasons for this are the subject of a long-standing debate in population genetics and molecular evolution (Kimura 1983, Gillespie 1991, Golding 1994). Many mutant alleles are extremely deleterious and are rapidly eliminated by natural selection, while others have only small or no effects at the phenotypic level and remain in the population until they are fixed or lost by chance. Still others are maintained at intermediate levels by a balance between opposing evolutionary forces.

Not all organisms are diploid. Prokaryotes have only a single copy of each gene and are referred to as **haploid**. Many of the lower plants (algae, mosses, and ferns) also have conspicuous haploid stages in their life cycles, as do the fungi and some animals (males of rotifers and haplo-diploid insects). Organisms with ploidy levels higher than diploid are known as **polyploids**. A **tetraploid** individual contains four sets of homologous chromosomes, whereas a **hexaploid** contains six. Polyploidy is extremely widespread among plants. It is relatively rare among sexual animals, but common among parthenogenetic species.

Even in diploids, some genes are effectively haploid. Such is the case for genes carried in organelles (mitochondria and chloroplasts). Although there may be hundreds of copies of organelle genes per cell, they are generally inherited uniparentally and are essentially all the same. Genes residing on the **sex chromosomes** of organisms with a genetic sex-determination mechanism also have a special ploidy status. In mammals, for example, males carry X and Y chromosomes, whereas females are XX, so that X-linked genes occur only in single "copies" in males. In some organisms, such as birds, moths, and butterflies, the heterogametic (WZ) sex is female. In order to distinguish sex chromosomes from the remaining pairs, the latter are referred to as **autosomes**. In this book, unless stated otherwise (see especially, Chapter 24), we will be dealing with autosomal loci in diploid populations.

The remainder of this chapter is concerned with the quantification of various properties of single loci. We start by reviewing the concepts of allele and genotype frequencies, showing how the two are connected in an ideal situation that is closely approximated in many natural settings. We next show how the phenotypic effects of different alleles can be described in terms of additive and dominance effects. The genotypic frequencies and effects are then incorporated into expressions for the additive and dominance components of genetic variance at the locus. Finally, we show how the additive effects of an individual's genes define its breeding value. These results provide a close mechanistic connection with the final example in the previous chapter. While several of the concepts covered in this chapter may seem rather abstract and far removed from the analysis of multilocus traits, their practical utility is becoming increasingly evident as molecular methods for locating and characterizing **quantitative-trait loci** (QTLs) become more refined (Chapters 13–16).

ALLELE AND GENOTYPE FREQUENCIES

When denoting the **genotype** at a single locus, we refer to the pair of alleles that a (diploid) individual carries at the locus. Individuals that have two identical alleles are called **homozygotes**, whereas those that have different alleles are **heterozygotes**. If, for example, we denote the alleles at a particular diallelic locus as B_1 and B_2, there are three possible genotypes: B_1B_1 and B_2B_2 homozygotes, and B_1B_2 heterozygotes. There may, of course, be more than two alleles, and hence more than three genotypes, present at a locus.

Allele frequencies are defined uniquely by genotype frequencies. Suppose that P_{11}, P_{12}, and P_{22} represent the proportions of the population that are B_1B_1,

B_1B_2, and B_2B_2. If these are the only possible genotypes at the locus, then by definition, $P_{11} + P_{12} + P_{22} = 1$. If there are N individuals in the population, then $P_{11}N$ individuals contain two B_1 alleles and $P_{12}N$ individuals contain a single B_1 allele. Since there are a total of $2N$ genes in the population for each autosomal locus, the frequency of the B_1 allele is simply

$$p_1 = \frac{2\,P_{11}\,N + P_{12}\,N}{2\,N} = P_{11} + \frac{1}{2}\,P_{12} \tag{4.1}$$

Thus, the general rule for a diploid, autosomal locus is that the frequency of an allele is estimated by the observed frequency of homozygotes plus one-half the observed frequency of all heterozygotes containing that allele.

For complex morphological and behavioral characters influenced by multiple genetic and environmental factors, it is usually impossible to be certain about the genotypic state of any particular locus. In some cases, however, the majority of the genetic variation for a character depends on a single locus with large effects, in which case the allele and genotype frequencies can be estimated directly. This was the fortuitous case in many of Mendel's classic experiments with peas, and some genetic disorders in humans appear to be products of mutant alleles at single loci. Data for a wing-color polymorphism in a British moth are examined in the following example.

Example 1. Fisher and Ford (1947) were able to distinguish three wing-color patterns in the moth *Panaxia dominula,* and through breeding experiments, the polymorphism was found to result from two alleles segregating at a single locus. The following table summarizes the distribution of genotype frequencies observed in a population in 1946.

Color Pattern	*dominula*	*medionigra*	*bimacula*	Total
Genotype	B_1B_1	B_1B_2	B_2B_2	
Sample Size (N_{ij})	905	78	3	$N = 986$
Frequency (P_{ij})	0.918	0.079	0.003	1.000

What are the estimated frequencies of the two alleles? Using Equation 4.1, the frequency of the B_1 allele is found to be

$$p_1 = 0.918 + \frac{0.079}{2} = 0.958$$

and since there are only two alleles, the frequency of B_2 is $p_2 = 1 - p_1 = 0.041$.

THE TRANSMISSION OF GENETIC INFORMATION

The Hardy-Weinberg Principle

From the standpoint of evolutionary analysis, it is crucial to understand how allele and genotype frequencies change from generation to generation. Such changes may result from natural selection, mutation, differential migration, inbreeding, or random drift due to gene sampling in finite populations. All of these forces will be considered in due course, but for now we will restrict our attention to a highly idealized situation — an autosomal locus uninfluenced by selection and mutation. By assuming the population to be effectively infinite in size and randomly mating, we also eliminate the possibility of inbreeding and random drift. We will further assume that generations are discrete and that the population is closed to immigrants.

Although such an idealized situation is never realized perfectly, in many cases it is close enough to the truth for practical purposes. Under the ideal model, simple and predictable relationships emerge between allele and genotype frequencies, within and between generations. It is therefore an essential point of departure, much like the ideal gas laws in physics.

In sexual populations, individuals do not necessarily produce offspring whose genotypes match their own. Prior to reproduction, sexual individuals produce haploid gametes by a special form of cell division called **meiosis** (Figure 4.1). Thus, with respect to a single locus, a B_1B_2 heterozygote produces two types of **gametes** — half B_1 and half B_2. The diploid state is restored when gametes from two parents fuse to form a **zygote**. Consequently, at a diallelic locus, a heterozygous parent can potentially produce three types of progeny (B_1B_1, B_1B_2, and B_2B_2), whereas homozygous parents can produce at most two.

Consider a population consisting of separate sexes (**dioecious**) with discrete, nonoverlapping generations. We denote the frequencies of B_1 and B_2 alleles in females in generation 0 by $p_{1f}(0)$ and $p_{2f}(0)$, and those in males by $p_{1m}(0)$ and $p_{2m}(0)$. Under random mating, the expected genotype frequencies in the next generation are obtained from the products of the respective gamete frequencies. For example, since the probability of drawing a B_1 female gamete is $p_{1f}(0)$ and that of drawing a B_1 male gamete is $p_{1m}(0)$, the expected frequency of B_1B_1 zygotes is $p_{1f}(0)\,p_{1m}(0)$. Similarly, the expected frequencies of B_1B_2 and B_2B_2 zygotes are $p_{1f}(0)\,p_{2m}(0)+p_{2f}(0)\,p_{1m}(0)$ and $p_{2f}(0)\,p_{2m}(0)$, respectively. Provided the locus is autosomal, the frequency of the B_1 allele will now be the same in both sexes, since the subpopulations of sons and daughters both acquire half their genes from mothers and half from fathers. Substituting into Equation 4.1, the B_1 allele frequency in generation 1 is

$$
\begin{aligned}
p_1 &= p_{1f}(0)\,p_{1m}(0) + \frac{p_{1f}(0)\,p_{2m}(0) + p_{1m}(0)\,p_{2f}(0)}{2} \\
&= \frac{p_{1f}(0)\,[\,p_{1m}(0) + p_{2m}(0)\,] + p_{1m}(0)\,[\,p_{1f}(0) + p_{2f}(0)\,]}{2}
\end{aligned}
$$

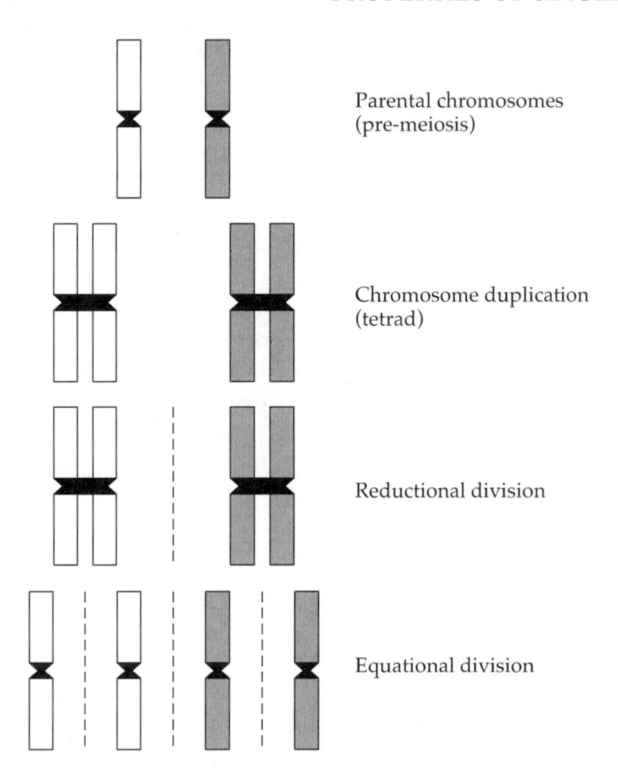

Figure 4.1 Idealized schematic of meiotic production of gametes. Only a single chromosome pair is shown. At the onset of meiosis, sister chromatids are formed by duplication and the homologous pairs come together to form a **tetrad**; although it is not shown, some exchange of material (**crossing-over**) between homologues may occur at this time. Two meiotic divisions (reductional and equational) then produce four haploid products. The maternal and paternal chromosomes migrate to opposite cells during the **reductional division**, and the sister chromatids are isolated into four potential haploid gametes after the **equational division**.

which is just

$$p_1 = \frac{p_{1f}(0) + p_{1m}(0)}{2}$$

The new frequency for the B_2 allele is $p_2 = 1 - p_1 = [\,p_{2f}(0) + p_{2m}(0)\,]/2$.

Under the conditions of our idealized population, in the next generation and in all subsequent generations, the B_1B_1, B_1B_2, and B_2B_2 genotypes will be found in frequencies p_1^2, $2\,p_1\,p_2$, and p_2^2. Such proportions are known as **Hardy-Weinberg frequencies**, after the two investigators who first pointed out the above relationship (Hardy 1908, Weinberg 1908). The Hardy-Weinberg frequencies can also be obtained directly by multiplying out the terms of the binomial expansion, $(\,p_1 + p_2)^2$. By this means, the Hardy-Weinberg law can be extended to any number of alleles. Suppose, for example, that four alleles (B_1, B_2, B_3, B_4) are present at the locus of interest. The Hardy-Weinberg frequencies for the various genotypes are obtained by squaring the quantity $(p_1 + p_2 + p_3 + p_4)$. The expected frequency of a genotype homozygous for the B_i allele is p_i^2, while that for a B_iB_j heterozygote is $2\,p_i\,p_j$.

Provided that all of the assumptions of the Hardy-Weinberg model are met, we can summarize as follows. First, it takes no more than a single generation to equilibrate and stabilize the gene frequencies in the two sexes. Second, only one additional generation is required for the stabilization of the genotype frequencies into the predictable Hardy-Weinberg proportions. These results have obvious implications for the analysis of natural populations. Even if genotype frequencies in a study population are vastly different from Hardy-Weinberg expectations, for example because of natural selection or population subdivision, they can be rendered close to the idealized proportions by imposing an artificial program of random mating for one or two generations.

Sex-Linked Loci

The preceding results do not extend to sex-linked loci. As noted above, when the male is the heterogametic sex, females are diploid for X linked loci, but males are haploid. Thus, for every mating pair, there are three X chromosomes, and the frequency of the B_1 allele in the population is $p_1 = [p_{1m}(0) + 2p_{1f}(0)]/3$. In the absence of any forces operating differentially on the alleles, this frequency will be maintained indefinitely. However, the gene frequency will not necessarily be p_1 in both of the sexes. Since males only receive an X chromosome from their mother, the male frequency of the B_1 allele in any generation (t) is necessarily equal to the frequency in females in the previous generation ($t-1$),

$$p_{1m}(t) = p_{1f}(t-1) \tag{4.2a}$$

On the other hand, fathers and mothers each contribute an X chromosome to their daughters, so the frequency of the B_1 allele in females is equal to the average of the gene frequency in the two sexes in the previous generation,

$$p_{1f}(t) = \frac{p_{1f}(t-1) + p_{1m}(t-1)}{2} \tag{4.2b}$$

The general solution to these equations is

$$p_{1f}(t) - p_1 = \left[-\frac{1}{2}\right]^t [p_{1f}(0) - p_1] \tag{4.2c}$$

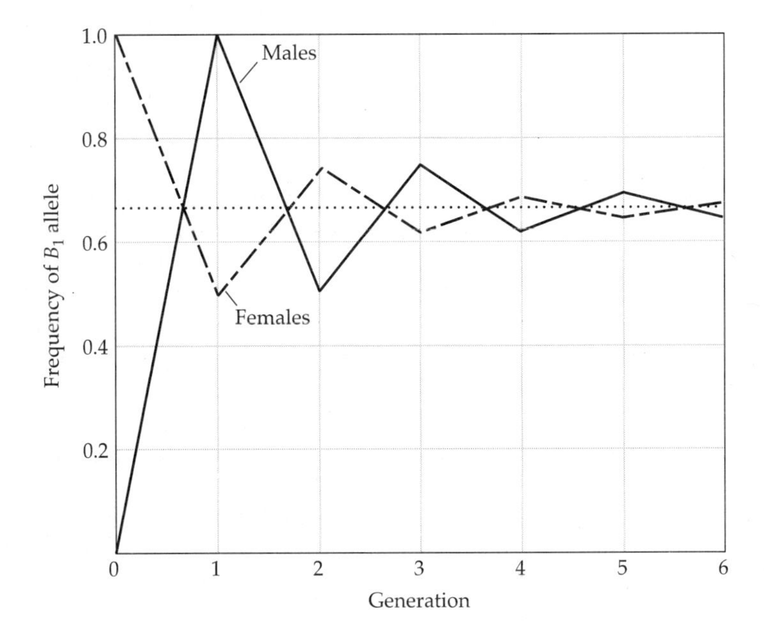

Figure 4.2 The dynamics of gene frequency change for an X-linked gene, B_1, under random mating . An extreme case is illustrated — initially, all females are homozygous for the B_1 gene, $p_{1f}(0) = 1$, while all of the males are haploid for the alternate allele, $p_{1m}(0) = 0$. Consequently, all males contain the B_1 allele in the following generation, while all females are heterozygous. The dotted line represents the population level gene frequency, $p_1 = [\,p_{1m}(0) + 2p_{1f}(0)\,]/3 = 0.67$, towards which both of the sexes converge over time.

Thus, the approach to the equilibrium allele frequency in the two sexes is gradual and oscillatory if the locus is X linked (Figure 4.2). The deviation of the allele frequency from p_1 is halved each generation for both males and females, but the sign changes from generation to generation.

Polyploidy

Another situation in which the Hardy-Weinberg principle is not met exactly arises in polyploid organisms. Because of the high frequency of polyploidy in plants, this case has been examined extensively by Fisher (1947) and Crow (1954) among others. It will only be considered briefly here for a tetraploid species, individuals of which propagate two genes per locus through gametes. The way in which sets of chromosomes assort during meiosis in polyploids depends on the degree of homology between ancestral chromosomes (Marsden et al. 1987). At one extreme are **allopolyploids** that originate by interspecific hybridization. In this case, provided the chromosomes of the parental species are sufficiently different, they will not pair. Meiosis is then identical to that for diploid organisms, except for the doubled number of chromosomes. At the other extreme, **autopolyploids** derive both chromosome sets from the same species.

For the remainder of our discussion of polyploidy, we will assume that the four sets of chromosomes are sufficiently similar that tetravalents (combinations of four homologues), rather than bivalents, are formed during meiosis. This condition raises the possibility that some gametes will contain two copies of one of the four genes carried by the parent (i.e., a parent with genotype $B_1B_2B_3B_4$ may produce a B_1B_1 gamete), a result that arises when a **crossover** (a reciprocal exchange of DNA) occurs between replicated arms of two of the four chromosomes during meiosis. The production of such a gamete is referred to as a **double reduction**, and we denote its probability by c. Of the $(1-c)$ gametes that are not doubly reduced, one-third will contain genes that came from the same parent, and the other two-thirds will contain one paternally derived and one maternally derived gene (Figure 4.3).

Here we assume the presence of only two alleles and random assortment of the four homologues. Letting, p_i be the frequency of the B_i allele and $p_{ij}(t)$ be the frequency of B_{ij} gametes in generation t, then the following dynamic equations hold:

$$p_{ii}(t) = c\,p_i + \frac{1-c}{3}\left[\,p_{ii}(t-1) + 2\,p_i^2\,\right] \tag{4.3a}$$

$$p_{ij}(t) = \frac{1-c}{3}\left[\,p_{ij}(t-1) + 2p_ip_j\,\right] \tag{4.3b}$$

(Crow and Kimura 1970, pp. 52–53). The equilibrium solution to these equations is obtained by setting $p_{ii}(t) = p_{ii}(t-1)$ and $p_{ij}(t) = p_{ij}(t-1)$,

$$p_{ii} = (1-f)\,p_i^2 + f\,p_i \tag{4.3c}$$

$$p_{ij} = (1-f)\,p_i\,p_j \tag{4.3d}$$

where $f = 3c/(2+c)$. This equilibrium is approached only gradually. The equilibrium genotype frequencies can be obtained as products of the appropriate equilibrium gametic frequencies.

In the absence of crossing-over between homologous pairs of chromosomes, $c = 0$, $f = 0$, and the equilibrium frequency of gamete types is simply equal to the product of the respective allele frequencies. However, if $c > 0$, the equilibrium genotype frequencies are not so simple. Consider the extreme case of free

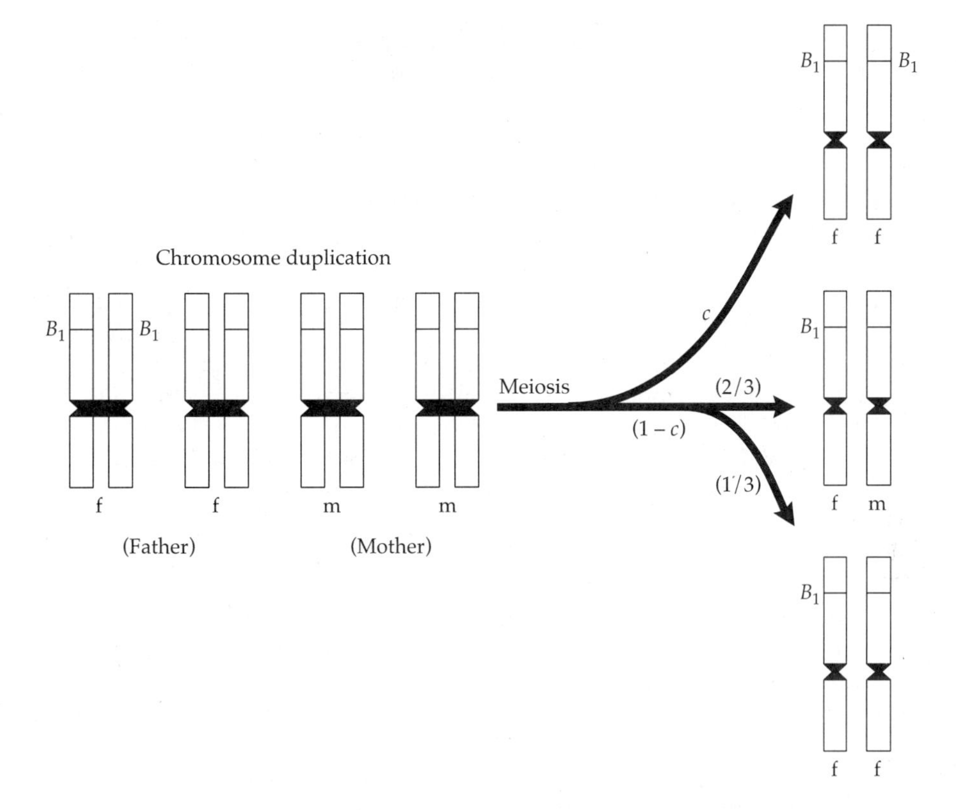

Figure 4.3 The production of three types of (diploid) gametes by a tetraploid individual. This example focuses upon a single paternally derived allele, B_1. The letters f and m refer to chromosomes derived from fathers and mothers. With four chromosomes (rather than the two of a diploid), the reductional division of meiosis isolates two chromosomes at random into each of the resulting two cells. Subsequent equational division generates the gamete types shown at the right. Here c is the probability that the allele of interest will become associated with itself during gametogenesis as a result of a double reduction. If this does not occur (probability $= 1 - c$), there is a 2/3 chance that chromosome B_1 will be associated in a gamete with a maternally derived chromosome and a 1/3 chance that it will be associated with the other paternally derived chromosome.

recombination. After chromosomal duplication during gametogenesis, eight chromosomes are assorted, two into each of four gametes. Conditional on one of these being transmitted to a gamete, then of the remaining seven possibilities, one will be identical by descent. Thus, for free recombination, $c = 1/7$, $f = 0.2$, and the equilibrium gamete frequencies are $p_{ii} = 0.2p_i(1 + 4p_i)$ and $p_{ij} = 0.8p_ip_j$. In essence, if there is any crossing-over, polyploidy results in a sort of "internal inbreeding," reducing the frequency of heterozygous gametes. Wricke and Weber (1986) provide a very useful coverage of the many complications that polyploidy introduces in quantitative-genetic formulations.

Age Structure

One final complication with respect to the idealized model is age structure. Up to now we have been assuming a population with discrete, nonoverlapping generations, such as an annual plant with no seed carry-over across years. In populations composed of several age classes (the majority of higher plants and animals), the generations overlap, and this causes the approach of genotype frequencies towards the Hardy-Weinberg expectations to be gradual, even in the case of an autosomal locus. This property arises because the genotypes of new recruits are a function of the allele frequencies specific to the reproductive age classes. Juvenile age classes only influence the change in genotype frequencies through mortality, but as they mature they begin to add copies of their genes to the population. The genotype frequencies become stable only after the allele frequencies become homogenized across age classes and sexes.

Of equal significance is the fact that the allele frequencies themselves can be unstable in an age-structured population even in the absence of genotypic differences in age-specific survival and reproduction. Further complexities are introduced by the scheme of mating between the various age classes. All of these subjects are taken up in detail by Charlesworth (1974, 1994) and Gregorius (1976). The important point to remember is that when newly founded populations have significant age structure, fluctuations in both gene and genotype frequencies may occur for a substantial period of time even in the absence of selection.

Testing for Hardy-Weinberg Proportions

When data are available on genotype frequencies in a population, it is standard practice to cross-check these with the Hardy-Weinberg expectations. Lack of concordance between the two implies that at least one of the assumptions of the Hardy-Weinberg model is violated and often instigates further investigation. Several different statistical techniques have been proposed (Weir 1996), the most popular by far being the χ^2 (Chi-square) test. However, the likelihood-ratio test is now becoming more common and appears to be at least as reliable as the former. Likelihood-based tests have a number of desirable statistical features (Appendix 4). Letting N_{ij} and $\widehat{N}_{ij}$ be the observed and expected numbers of genotype B_iB_j in a sample, then the likelihood-ratio test statistic

$$G = -2\sum_{i=1}^{n}\sum_{j\geq i}^{n} N_{ij}\ln\left(\frac{\widehat{N}_{ij}}{N_{ij}}\right) \tag{4.4}$$

has a sampling distribution very similar to the well-known χ^2 distribution. That is, if a population in Hardy-Weinberg equilibrium is sampled many different times and G calculated each time, the frequency distribution of the observed G values will be nearly χ^2 distributed. Thus, the test for Hardy-Weinberg proportions compares the observed statistic G with the cumulative χ^2 distribution. If G exceeds the level at which there is a 5% chance of obtaining a higher χ^2, then one can reject the null hypothesis of Hardy-Weinberg proportions with 95% confidence.

Regardless of which approach to testing for Hardy-Weinberg frequencies is taken, it should be kept in mind that some of the conditions underlying the Hardy-Weinberg theorem may be violated without causing detectable departures of observations from expectations. For example, if the product of the survivorships of the two homozygotes is equal to the square of the heterozygote survival, the zygotic frequencies after selection will still be in Hardy-Weinberg proportions (Lewontin and Cockerham 1959). Thus, a failure to reject the Hardy-Weinberg model should be interpreted with caution.

Example 2. As an example of the application of Equation 4.4, we return to the data in the table of Example 1.

The best estimates for the Hardy-Weinberg expectations are obtained from the observed allele frequencies: $\widehat{N}_{11} = p_1^2 N = 905$, $\widehat{N}_{12} = 2\,p_1 p_2 N = 79$, and $\widehat{N}_{22} = p_2^2 N = 2$. Applying these and the observed values (N_{11}, N_{12}, and N_{22}) from the table,

$$G = -2\,[\,905\,\ln(905/905) + 78\,\ln(79/78) + 3\,\ln(2/3)\,] = 0.446$$

Under the null hypothesis of Hardy-Weinberg frequencies, the sampling distribution of G is a function of the number of degrees of freedom, which in the case of the Hardy-Weinberg test is the number of genotypic classes minus the number of allele frequencies that must be estimated from the data minus one. Here, it was necessary to estimate one parameter (p_1) from the data, so there is $3 - 1 - 1 = 1$ degree of freedom. Referring to a χ^2 table in any statistics text, it can be found that, with one degree of freedom, G must exceed 3.841 to reject the null hypothesis at the 0.05 probability level. Therefore, the observed data are not significantly different from those expected under the Hardy-Weinberg expectations.

CHARACTERIZING THE INFLUENCE OF A LOCUS ON THE PHENOTYPE

In Chapter 3, we encountered the concept of partitioning the phenotype (z) of an individual into a genotypic value (G) and an environmental deviation (E),

$$z = G + E$$

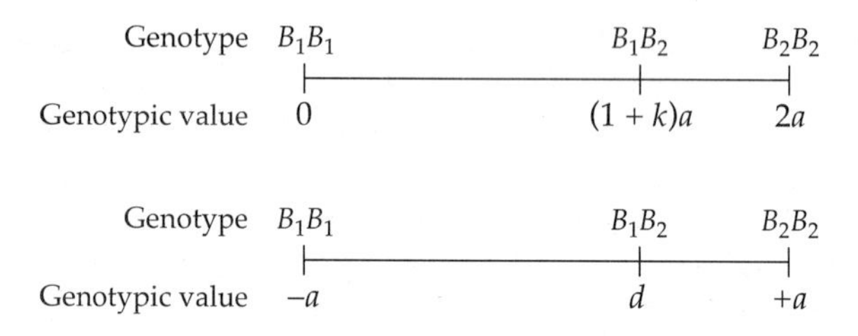

Figure 4.4 Two ways of representing genotypic values for a diallelic locus.

where G is the expected phenotype (for a given genotype) resulting from the joint expression of all of the genes underlying the trait. For a multilocus trait, G is a potentially complicated function. For now, however, we are concerned only with the direct contribution of a single autosomal locus, in which case things are quite tractable. We start with the special case in which there are only two alleles. The three genotypic values can then be represented by the scale at the top of Figure 4.4, with $2a$ representing the difference between the mean phenotypes of B_2B_2 and B_1B_1 homozygotes, and k providing a measure of dominance. Alleles B_1 and B_2 behave in a completely additive fashion when $k = 0$, whereas $k = +1$ implies complete dominance of the B_1 allele, and $k = -1$ implies complete dominance of the B_2 allele. If $k > 1$, the phenotypic expression of the heterozygote exceeds that of both homozygotes, and the locus is said to exhibit **overdominance**, whereas $k < -1$ implies **underdominance**.

The fact that we have set the genotypic value of the B_1B_1 homozygote equal to zero may seem troublesome, but it is desirable because it leads to some algebraic simplifications. Although phenotypic measures are often performed on scales where zeros are impossible, genotypic values can always be transformed to the above scale by simply subtracting the observed genotypic value of B_1B_1 from each measure. Such a transformation of a linear scale can be illustrated by considering an alternative scheme often used by quantitative geneticists (bottom of Figure 4.4). Although the genotypic values of the two homozygotes are now denoted by $-a$ and $+a$, the difference between them is still $2a$, as in the previous case. The previous scale can be completely recovered by adding a to all three measures on this new scale and letting $d = ak$. Generally, we will adhere to the first of these two scales.

Example 3. The scaling of genotypic values may be clarified by reference to a particular example — the Booroola (B) gene that influences fecundity in the Merino sheep of Australia (Piper and Bindon 1988).

Litter size in sheep has a polygenic basis, but in this particular breed, it is determined largely by a single polymorphic locus. The mean litter sizes for the *bb*,

Bb, and BB genotypes based on 685 total records are 1.48, 2.17, and 2.66, respectively. Taking these to be estimates of the genotypic values (G_{bb}, G_{Bb}, and G_{BB}), the homozygous effect of the B allele is estimated by $a = (2.66-1.48)/2 = 0.59$. The dominance coefficient is estimated by taking the difference between bb and Bb genotypes, $a(1+k) = 0.69$, substituting $a = 0.59$, and rearranging to obtain $k = 0.17$. This suggests slight dominance of the Booroola gene, but great confidence cannot be placed on this conclusion. Since the standard errors of the mean genotypic values are approximately 0.09, the midpoint between the two homozygotes, 2.07, is not significantly different from 2.17.

THE BASIS OF DOMINANCE

The presence of dominance complicates many formulations in quantitative genetics, but unfortunately it is a fact of life that cannot be ignored. Since the beginning of this century, there has been much debate on the genetic and physiological basis of dominance. In the early days, the only genes subject to detailed genetic analysis were those that had a major phenotypic effect. Loci involving such genes are usually characterized by striking levels of dominance. For example, the vast majority of genes with major, deleterious effects on fitness are recessive. Does this then indicate that new mutations are inherently recessive? Fisher (1928a,b, 1929, 1958) argued that since rare alleles are found almost entirely in the heterozygous state, selection should favor alleles at modifier loci that cause heterozygous carriers of deleterious alleles to resemble the normal homozygote. Implicit in this argument is the assumption that the heterozygote initially encodes for an intermediate phenotype. Using physiological arguments, Wright (1929a,b, 1934a,b) strongly disputed this idea. He also pointed out that although dominance relationships are subject to change, the intensity of selection operating on modifier loci is unlikely to ever be strong enough to be an important evolutionary force. The debate between Fisher and Wright was intense and at times bitter, and it scarred their relationship permanently.

Much later, Kacser and Burns (1981) developed a general explanation for dominance based on biochemical principles. Their model is in good accord with Wright's theory. Most gene products (enzymes) are involved in complex biochemical pathways such that the rate of production of a final end-product (phenotype) is regulated at many steps. Consequently, the relationship between enzyme activity (a function of allelic state) and end-product production is hyperbolic (Figure 4.5). Kacser and Burns showed that the "wild-type" activity normally lies on or near the plateau of this hyperbolic relationship. This leads to three predictions:

1. Mutations with large effects at the phenotypic level will be biased in a downward direction. Even if mutations that increase enzyme activity

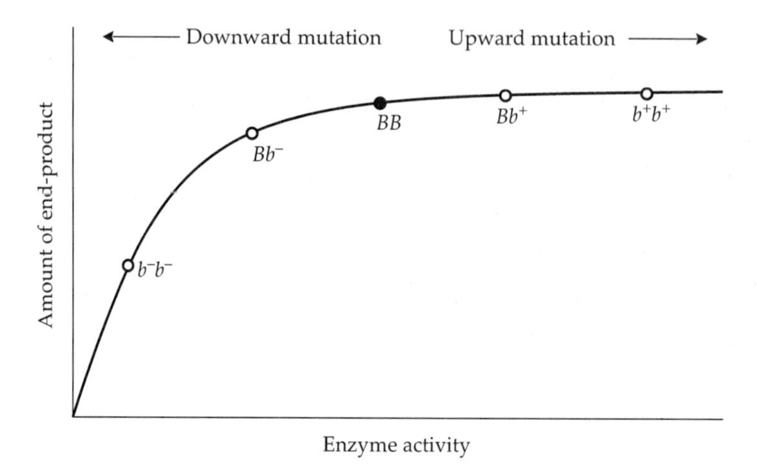

Figure 4.5 The relationship between the activity of a gene product and the flux or concentration of an end-product in an enzymatic pathway. BB represents the "wild-type" genotype. Upward and downward mutations with the same magnitude of change in enzyme activity are represented as b^+ and b^- alleles.

occur as frequently as those that decrease it, the former will usually cause imperceptible changes at the phenotypic level. Thus, if a high production rate or end-product concentration is beneficial, we can expect most individually *discernible* mutations to be detrimental.

2. The recessivity of downward mutations is an inevitable consequence of the hyperbolic enzyme-product relationship. If we take the heterozygote to be intermediate in enzyme activity, the allele producing the homozygote with greater activity will always exhibit dominance on the end-product scale, the degree of dominance diminishing out on the plateau.
3. The smaller the effect of a mutation, the less pronounced will be the level of dominance. Such a result is expected simply because the relationship between the BB, Bb, and bb genotypic values tends towards linearity as the deviations among their enzyme activities are reduced. In principle, dominance is much more likely to be a complicating factor for characters whose variation is influenced by one or two genes of large effect than for quantitative characters encoded by numerous loci whose individual effects are indiscernible.

Since the exact form of the relationship in Figure 4.5 can change with a shift in the genetic background, the Kacser-Burns model does not rule out the possibility of evolutionary changes in dominance relationships. It does, however, eliminate the necessity of ad hoc evolutionary explanations, such as modifier loci, to account for the existence of dominance. Careful empirical work in biochemical genetics will be required to test the model in its entirety, but two observations are already in good accord with the predictions. First, in a clever analysis of data on the haploid alga *Chlamydomonas reinhardtii,* Orr (1991) found that when mutations are observed in artificial diploid constructs, they are almost always recessive. Since the heterozygous state never exists in a haploid species, there can be no opportunity for the selection of dominance modifiers; the mutations must be "recessive" at first appearance. Second, in *Drosophila,* lethal alleles are almost nearly completely recessive, whereas mildly deleterious alleles, whose individual effects are indiscernible, interact in a nearly additive fashion (Chapter 12).

FISHER'S DECOMPOSITION OF THE GENOTYPIC VALUE

The number of copies of a particular allele (say B_2) in a genotype ($N_2 = 0$, 1, or 2 for diploids) is referred to as the **gene content**. As noted above, unless this allele interacts additively with all other alleles, there will be a nonlinear relationship between the gene content and the genotypic value. It is, nevertheless, useful to consider the best linear approximation to this relationship, as this leads to a partitioning of the genotypic values into their "expected" values based on additivity ($\widehat{G}$) and deviations from those expectations resulting from dominance (δ) (Figure 4.6).

The preceding points can be formalized by least-squares regression of genotypic values on the number of B_1 and B_2 alleles in the genotype (N_1 and N_2),

$$G_{ij} = \widehat{G}_{ij} + \delta_{ij} = \mu_G + \alpha_1\, N_1 + \alpha_2\, N_2 + \delta_{ij} \tag{4.5a}$$

The genotypic value of genotype B_iB_j is a function of μ_G, the mean genotypic value in the population, α_1 and α_2, the slopes of the regression, N_1 and N_2, the predictor variables, and δ_{ij}, the residual error. This partitioning of genotypic values into various components is one of several major advances developed in Fisher's 1918 paper. Many of the innovative ideas in this classic paper are presented in a characteristically cursory manner, but a useful interpretative guide has been produced by Moran and Smith (1966).

Unlike the univariate regression discussed in Chapter 3, Equation 4.5a is a **multiple regression**, the properties of which are discussed in Chapter 8. For the two-allele case, however, we can reduce the model to a standard univariate regression by noting that for any individual, $N_1 = 2 - N_2$, so that

$$\begin{aligned} G_{ij} &= \mu_G + \alpha_1\,(2 - N_2) + \alpha_2\, N_2 + \delta_{ij} \\ &= \iota + (\alpha_2 - \alpha_1)\, N_2 + \delta_{ij} \end{aligned} \tag{4.5b}$$

where $\iota = \mu_G + 2\alpha_1$ is the intercept. We denote the slope of this regression by

$$\alpha = \alpha_2 - \alpha_1 \tag{4.6}$$

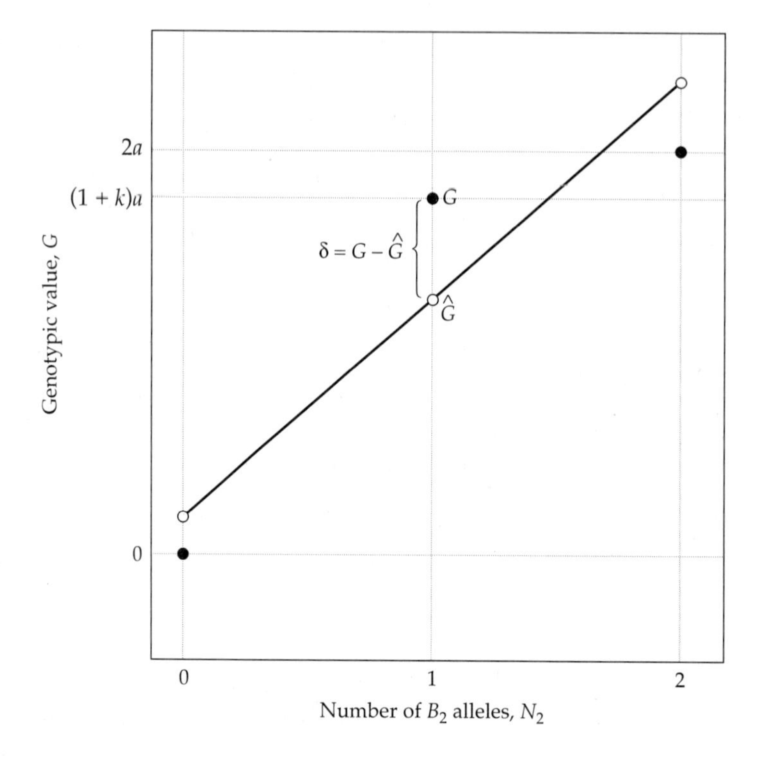

Figure 4.6 Linear least-squares regression of the genotypic value of a single locus G on the gene content (N_2). From left to right, the points represent the B_1B_1, B_1B_2, and B_2B_2 genotypes. Solid circles represent the true genotypic values, while open circles are the values expected on the basis of average effects ($\widehat{G}$). The deviation between G and $\widehat{G}$ for each genotype is δ, the dominance deviation.

and discuss its meaning shortly. The genotypic values predicted by the regression are

$$\widehat{G}_{ij} = \mu_G + \alpha_i + \alpha_j = \begin{cases} \mu_G + 2\,\alpha_1 & \text{for } G_{11} \\ \mu_G + \alpha_1 + \alpha_2 & \text{for } G_{21} \\ \mu_G + 2\,\alpha_2 & \text{for } G_{22} \end{cases} \tag{4.7}$$

We next show that the weighted mean of the coefficients α_1 and α_2 is equal to zero. To accomplish this, return to Equation 4.5a, and take expectations,

$$\mu_G = \mu_G + \alpha_1\,E(N_1) + \alpha_2\,E(N_2) + 0$$

The expected value of the residual δ_{ij} is equal to zero by the properties of least-squares regression, and $E(N_1)/2$ and $E(N_2)/2$ are equivalent, respectively, to p_1 and p_2, the frequencies of the B_1 and B_2 alleles. Thus, the previous expression simplifies to

$$p_1\,\alpha_1 + p_2\,\alpha_2 = 0 \tag{4.8}$$

showing that the mean value of α_i is indeed zero. Finally, from Equations 4.6 and 4.8 and the fact that $p_1 + p_2 = 1$, we obtain

$$\alpha_2 = p_1\,\alpha \qquad \text{and} \qquad \alpha_1 = -p_2\,\alpha \tag{4.9}$$

Now recall from Chapter 3 that the slope of a univariate regression is simply the covariance between response and predictor variable, divided by the variance of the predictor variable. Thus, the slope of the regression in Figure 4.6 is

$$\alpha = \frac{\sigma(G, N_2)}{\sigma^2(N_2)} \tag{4.10a}$$

The terms $\sigma(G, N_2)$ and $\sigma^2(N_2)$ are functions of the gene effects (a and k) and frequencies (p_1 and p_2). The steps leading up to their computation, under the assumption of random mating, are outlined in Table 4.1. Upon substitution,

Table 4.1 Properties of a single segregating diallelic locus under random mating.

Genotype	Gene Content (N)	Genotypic Value (G)	Freq.	$G \cdot N$	N^2	Regression Value ($\widehat{G}$)	Dominance Deviation ($\delta = G - \widehat{G}$)
B_1B_1	0	0	p_1^2	0	0	ι	$-\iota$
B_1B_2	1	$(1+k)a$	$2p_1p_2$	$(1+k)a$	1	$\iota + \alpha$	$(1+k)a - \iota - \alpha$
B_2B_2	2	$2a$	p_2^2	$4a$	4	$\iota + 2\alpha$	$2a - \iota - 2\alpha$

$$\mu_N = 2\,p_1p_2(1) + p_2^2(2) = 2p_2$$

$$E(N^2) = 2p_1p_2(1) + p_2^2(4) = 2p_2(1+p_2)$$

$$\mu_G = 2p_1p_2a(1+k) + 2p_2^2a = 2p_2a(1+p_1k)$$

$$E(GN) = 2p_1p_2a(1+k) + 4p_2^2a = 2p_2a[\,2p_2 + p_1(1+k)\,]$$

$$\sigma(G,N) = E(GN) - \mu_G\mu_N = 2p_1p_2a[\,1 + k(p_1 - p_2)\,]$$

$$\sigma^2(N) = E(N^2) - \mu_N^2 = 2p_1p_2$$

$$\mu_{\widehat{G}} = \iota + 2\,p_1p_2\alpha + 2p_2^2\,\alpha = \iota + 2p_2\alpha$$

$$\mu_\delta = -\iota + 2p_1p_2[(1+k)a - \alpha] + 2p_2^2(a-\alpha) = 0$$

$$E(\widehat{G}^2) = p_1^2\iota^2 + 2p_1p_2(\iota+\alpha)^2 + p_2^2(\iota+2\alpha)^2 = \iota^2 + 4p_2\alpha\iota + 2p_2\alpha^2(1+p_2)$$

$$E(\delta^2) = p_1^2\iota^2 + 2p_1p_2[\,(1+k)a - \iota - \alpha\,]^2 + p_2^2(2a - \iota - 2\alpha)^2 = (2p_1p_2ak)^2$$

$$\sigma_A^2 = E(\widehat{G}^2) - \mu_{\widehat{G}}^2$$

$$\sigma_D^2 = E(\delta^2) - \mu_\delta^2$$

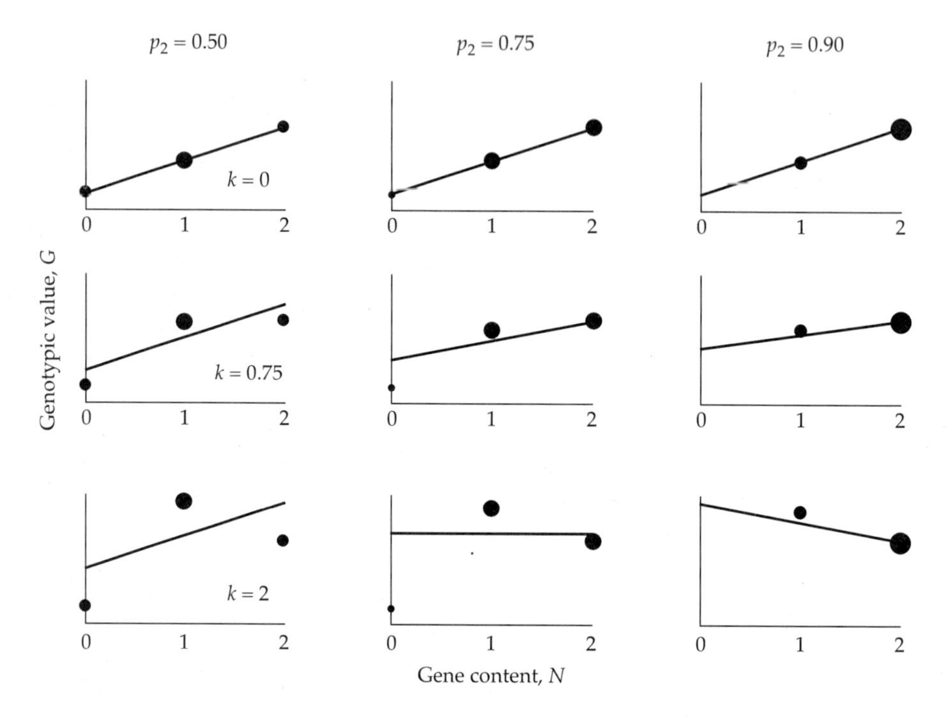

Figure 4.7 The slope α of the linear least-squares regression of genotypic value on gene content as a function of allele frequency, p_2, and degree of dominance, k. The lines denote the regressions, with each of the three points (representing genotypic values) being weighted by their frequency (denoted by the different-sized circles). The columns of graphs give results for different gene frequencies ($p_2 = 0.50, 0.75$, and 0.90), whereas the rows give results for different modes of gene action ($k = 0.00$, additivity; $k = 0.75$, partial dominance; and $k = 2.00$, overdominance). Note that, except for the case of complete additivity, the regressions differ with different allele frequencies. In the case of overdominance, the slope changes sign as the allele frequency changes; when $p_2 = 0.75$, the slope is zero, i.e., there is no additive genetic variance. For all cases when $p_2 = p_1 = 0.5$, the slope $\alpha = a$ regardless of the degree of dominance.

we obtain

$$\alpha = a\left[1 + k\left(p_1 - p_2\right)\right] \tag{4.10b}$$

Under the assumption of random mating, α is known as the **average effect of allelic substitution**. It represents the average change in genotypic value that results when a B_2 allele is randomly substituted for a B_1 allele. For the purely additive case ($k = 0$), α is simply equal to a. However, for all other cases, α is also a function of k and of the allele frequencies in the population (Figure 4.7). Such behavior results because, with dominance, the phenotypic effect of a gene substitution depends on the status of the unsubstituted allele. If B_2 is a dominant allele ($k > 0$), then α will be inflated relative to the case of additivity if B_2 is rare ($p_1 > p_2$), but diminished if B_2 is common ($p_1 < p_2$). Thus, except in the case of additivity, the average effect of allelic substitution is not simply a function of the inherent physiological properties of the allele. It can only be defined in the context of the population.

PARTITIONING THE GENETIC VARIANCE

Fisher (1918) showed that once the genotypic values have been partitioned in the above manner, it is a relatively simple step to partition the sources of genetic variation at a locus. Recalling the relationship $G = \widehat{G} + \delta$, the total genetic variance may be written as

$$\begin{aligned} \sigma_G^2 &= \sigma^2(\widehat{G} + \delta) \\ &= \sigma^2(\widehat{G}) + 2\sigma(\widehat{G}, \delta) + \sigma^2(\delta) \end{aligned}$$

From the property of least-squares regression (Chapter 3), the regression prediction (in this case, $\widehat{G}$) is uncorrelated with the residual error (in this case, δ). Thus, the total genetic variance attributable to a locus simplifies to the sum of additive and dominance components. Hereafter, we denote these components as σ_A^2 and σ_D^2,

$$\sigma_G^2 = \sigma_A^2 + \sigma_D^2 \tag{4.11}$$

Statistically speaking, σ_A^2 is the amount of the variance of G that is explained by the regression on N_2 (or equivalently, on N_1), whereas σ_D^2 is the residual variance for the regression. Biologically speaking, σ_A^2 is the genetic variance associated with the average additive effects of alleles (the **additive genetic variance**), and σ_D^2 is the genetic variance associated with dominance effects (the **dominance genetic variance**).

All of the information necessary to compute these two components of genetic variance for a diallelic locus is contained in Table 4.1, and leads to

$$\sigma_A^2 = 2p_1p_2\alpha^2 \tag{4.12a}$$

$$\sigma_D^2 = (2p_1p_2ak)^2 \tag{4.12b}$$

Both components of variance depend upon the gene frequencies, the dominance coefficient k, and the homozygous effect a (Figure 4.8). In the case of purely additive allelic effects ($k = 0$), the additive genetic variance reaches a maximum

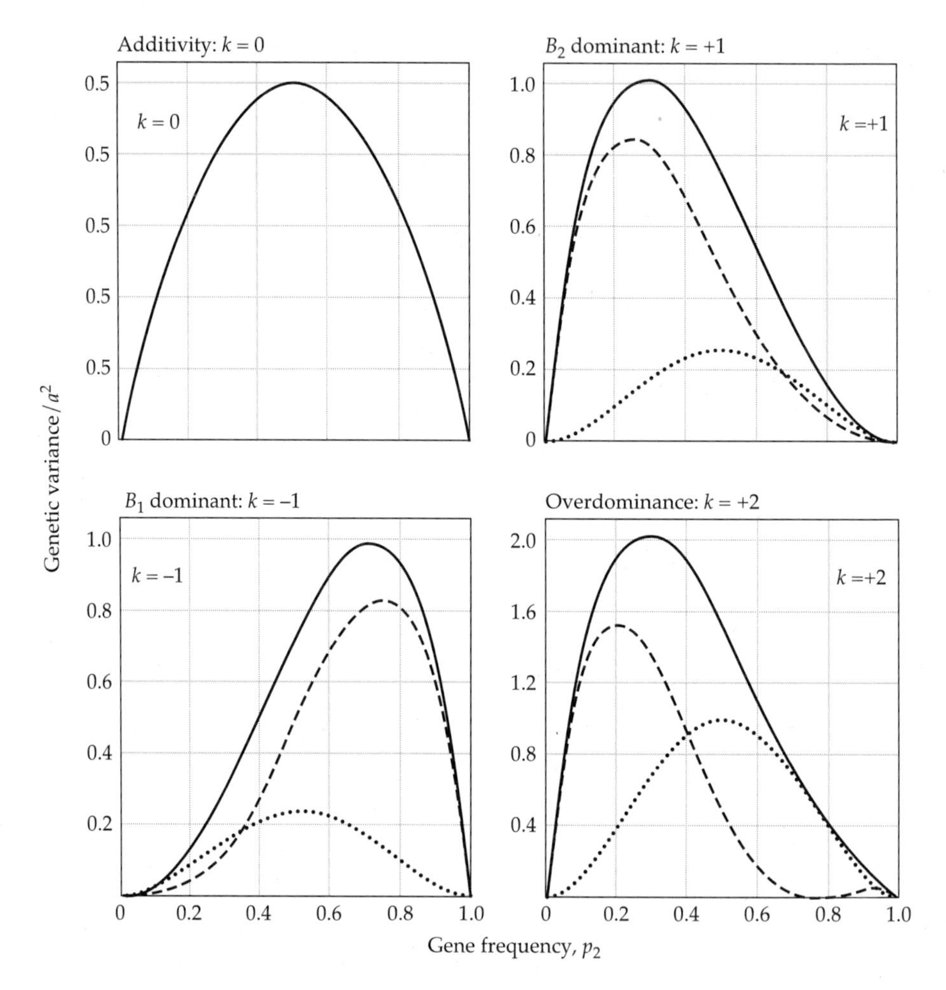

Figure 4.8 The dependence of the components of genetic variance at a locus on the frequency of the B_2 allele. The solid line denotes the total genetic variance, the dashed line (— — —) the additive genetic variance, and the dotted line ($\cdots$) the dominance genetic variance. Four cases are illustrated: $k = 0$ (additivity), $k = +1$ (dominance of the B_2 allele), $k = -1$ (dominance of the B_1 allele), and $k = +2$ (a case of overdominance). In the case of additivity, all of the genetic variance is of the additive type. The vertical axes are scaled such that, for any particular case, the actual variances are obtainable by multiplying by a^2, where a is half the difference between homozygous B_1B_1 and B_2B_2 genotypic values.

at $p_1 = p_2 = 0.5$, the gene frequency at which heterozygosity is most pronounced. With dominance, however, the additive genetic variance is maximized at a higher frequency of the recessive allele. This occurs because rare recessive alleles cause little genetic variance, due to their infrequent expression.

A common misconception is that the relative magnitudes of additive and dominance genetic variance provide information on the additivity of gene action. Equations 4.10b and 4.12a show that this generalization does not hold true. Through its influence on α, dominance contributes to the additive genetic variance, and for certain allele frequencies, can cause σ_A^2 to reach much higher levels than in the case of alleles with purely additive effects (Figure 4.8). Even in the case of complete dominance, σ_D^2 is unlikely to greatly exceed σ_A^2, and it is often substantially smaller. In the case of overdominance ($k > 1$), probably not a common situation (Chapter 10), there is always an intermediate gene frequency at which σ_A^2 is zero. This occurs when the least-squares regression of G on N_2 has a slope equal to zero (Figure 4.7).

ADDITIVE EFFECTS, AVERAGE EXCESSES, AND BREEDING VALUES

It may still be unclear why we have gone to the trouble of partitioning the genotypic value into additive and dominance components. Such a distinction is useful because, in randomly mating diploid species, a parent donates only one allele per locus to each of its offspring. The transmitted allele exhibits its additive effect when randomly combined with a gene from other parents. The dominance deviation of a parent, which is a function of the interaction between the two parental genes, is eliminated when gametes are produced. Thus, one can think of $\widehat{G}$ and δ as the heritable and nonheritable components of an individual's genotypic value. Before clarifying this concept further, however, we need some formal definitions. Two different measures of the effect of an allele were proposed by Fisher (1918, 1941): the average excess α_i^* and the additive effect α_i. As will be shown below, these two measures are equivalent in a randomly mating population, the first having a simple biological interpretation, the second being defined as a least-squares regression parameter.

The **average excess** α_2^* of allele B_2 is the difference between the mean genotypic value of individuals carrying at least one copy of B_2 and the mean genotypic value of a random individual from the entire population,

$$\alpha_2^* = \left(G_{12}\, P_{12\,|2} + G_{22}\, P_{22\,|2} \right) - \mu_G \tag{4.13a}$$

where $P_{ij|i}$ is the conditional probability of a B_iB_j genotype given that one allele is B_i. This is a completely general definition, but initially we will continue to focus on a diallelic locus under random mating, in which case $P_{ij|i} = p_j$, with p_j being the frequency of allele B_j. Under these conditions, Equation 4.13a becomes

$$\alpha_2^* = G_{12}\, p_1 + G_{22}\, p_2 - \mu_G \tag{4.13b}$$

This follows since of all individuals receiving a B_2 allele from one parent, a proportion p_2 (under random mating) receive another B_2 allele from the second parent, while a proportion p_1 receive a B_1 allele from the second parent. Since the genotypic values in these two cases are respectively $2a$ and $a(1+k)$, subtraction of the population mean μ_G (Table 4.1) from the conditional mean yields

$$\begin{aligned} \alpha_2^* &= \{\, p_1[\, a(1+k)\,] + p_2(2a)\,\} - 2\,a\,p_2\,(\,1 + p_1\,k\,) \\ &= p_1\,a\,[\,1 + k\,(\,p_1 - p_2\,)\,] = p_1\,\alpha \end{aligned} \tag{4.14a}$$

In the same manner, the average excess of allele B_1 is found to be

$$\alpha_1^* = -p_2\,a\,[\,1 + k\,(\,p_1 - p_2\,)\,] = -p_2\,\alpha \tag{4.14b}$$

Note that one of the average excesses is positive and the other negative because they are defined as deviations from the population mean genotypic value and hence have expected value zero.

The **additive effects**, α_i, on the other hand, are defined to be the least-squares regression coefficients of genotypic value on gene content. They are obtained by finding the α_1 and α_2 that miminize the mean-squared residual deviation

$$\begin{aligned} M &= E(\,\delta_{ij}^2\,) = E[\,(G_{ij} - \mu_G - \alpha_i - \alpha_j)^2\,] \\ &= (G_{11} - \widehat{G}_{11})^2\,P_{11} + (G_{12} - \widehat{G}_{12})^2\,P_{12} + (G_{22} - \widehat{G}_{22})^2\,P_{22} \end{aligned}$$

where P_{ij} is the frequency of the ijth genotype. Again, this is a general definition. For the special case of a randomly mating population (with $P_{11} = p_1^2$, $P_{12} = 2p_1p_2$, and $P_{22} = p_2^2$), setting the partial derivatives of M with respect to α_i equal to zero, and solving gives

$$\alpha_2 = p_1\,a\,[\,1 + k\,(p_1 - p_2)\,] = p_1\,\alpha \tag{4.15a}$$

$$\alpha_1 = -p_2\,a\,[\,1 + k\,(p_1 - p_2)\,] = -p_2\,\alpha \tag{4.15b}$$

Comparing these expressions with Equations 4.14a,b, we find that additive effects are identical to average excesses in randomly mating populations. The α_i are often referred to as **average effects**, but we use additive effects to discriminate them from average effects of higher-order gene actions (such as dominance).

An individual's **breeding value**, hereafter denoted by A, is the sum of the additive effects of its genes. In other words, the breeding value of a B_1B_1 homozygote is simply $2\alpha_1$, that of a heterozygote is $(\alpha_1 + \alpha_2)$, and that of a B_2B_2 individual is $2\alpha_2$. For random-mating populations, an extremely useful relationship emerges from these definitions for additive effects and breeding values. Consider the expected genotypic values of progeny produced by the parental genotypes. In the case of B_2B_2 parents, a proportion p_2 of the offspring will also be B_2B_2, in which case their genotypic value is $2a$, and a proportion p_1 will be B_1B_2 with genotypic value $a(1+k)$. The average genotypic value of offspring from a B_2B_2 parent is therefore $p_2(2a) + p_1\,a(1+k) = a\,[\,2p_2 + p_1\,(1+k)\,]$. When the population mean, μ_G, is subtracted, we obtain (after some simplification) α_2. Deviations of expected progeny phenotypes from the population mean are given for the other two parental genotypes in Table 4.2. The results in this table show that when mating is random *the breeding value of a genotype is equivalent to twice the expected deviation of its offspring mean phenotype from the population mean.* The deviation is multiplied by two because only one of the two parental genes is passed on to each offspring. Thus, we can estimate the breeding value of an individual by mating it to many randomly chosen individuals from the population and taking twice the deviation of its offspring mean from the population mean. Chapter 26 discusses the estimation of breeding values under very general settings.

Example 4. Consider the consequences of the Booroola gene (described in Example 3) in two hypothetical random-mating populations with gene frequencies of 0.5 and 0.1. We assume that the phenotypic means within genotypic classes are known without error, so that they are equivalent to the genotypic values. The additive and dominance genetic variances are, respectively, the mean-squared breeding values and the mean-squared dominance deviations because both types of effects have means equal to zero.

	$p_B = 0.5$			$p_B = 0.1$		
	bb	Bb	BB	bb	Bb	BB
Genotypic Value (G_{ij})	1.48	2.17	2.66	1.48	2.17	2.66
Genotype Frequency (P_{ij})	0.25	0.50	0.25	0.81	0.18	0.01
Mean Genotypic Value						
$\mu_G = P_{bb}G_{bb} + P_{Bb}G_{Bb} + P_{BB}G_{BB}$			2.120		1.616	
Additive Effects						
$\alpha_B = p_B G_{BB} + p_b G_{Bb} - \mu_G$			0.295		0.603	
$\alpha_b = p_b G_{bb} + p_B G_{Bb} - \mu_G$			−0.295		−0.067	
Breeding Values						
$A_{ij} = \alpha_i + \alpha_j$	−0.59	0.00	0.59	−0.134	0.536	1.206
$\overline{A} = P_{bb}A_{bb} + P_{Bb}A_{Bb} + P_{BB}A_{BB}$			0.00		0.00	
Dominance Deviations						
$\delta_{ij} = G_{ij} - (\mu_G + \alpha_i + \alpha_j)$	−0.05	0.05	−0.05	−0.002	0.018	−0.162
$\overline{\delta} = P_{bb}\,\delta_{bb} + P_{Bb}\,\delta_{Bb} + P_{BB}\,\delta_{BB}$			0.00		0.00	
Genetic Variance Components						
$\sigma^2_A = P_{bb}\,A^2_{bb} + P_{Bb}\,A^2_{Bb} + P_{BB}\,A^2_{BB}$			0.1740		0.0808	
$\sigma^2_D = P_{bb}\,\delta^2_{bb} + P_{Bb}\,\delta^2_{Bb} + P_{BB}\,\delta^2_{BB}$			0.0012		0.0003	
$\sigma^2_G = \sigma^2_A + \sigma^2_D$			0.1752		0.0811	

Although this example is somewhat artificial in that we employed arbitrary gene frequencies, the basic approach is now being widely exploited in the analysis of human genetic disorders. Biochemical studies are used to identify **candidate loci** that are potential contributors to the variation of the trait of interest, and the genotypes of random individuals are identified by use of molecular markers. The average phenotypic values within each genotypic class provide estimates of the genotypic values, which can then be used to estimate the fraction of the total phenotypic variance that is associated with the locus. Details on this **measured-genotype approach** are presented in Chapter 13.

Table 4.2 Conditional mean genotypic values of progeny under random mating, and their deviations from the mean genotypic value in the population, $\mu_G = 2ap_2(1 + p_1 k)$.

Parental Genotype	Breeding Value	Mean Genotypic Value of Progeny	Deviation of Expected Progeny Mean from μ_G
B_2B_2	$2\alpha_2$	$a[2p_2 + p_1(1+k)]$	α_2
B_1B_2	$\alpha_1 + \alpha_2$	$a[p_2 + (1+k)/2]$	$(\alpha_1 + \alpha_2)/2$
B_1B_1	$2\alpha_1$	$ap_2(1+k)$	α_1

EXTENSIONS FOR MULTIPLE ALLELES AND NONRANDOM MATING

Although the preceding results were obtained under the assumption of a diallelic locus, they are readily generalized to situations with an arbitrary number of alleles, as well as to nonrandomly mating populations. The algebra necessarily becomes more tedious, but some very useful principles emerge that will be relied upon heavily in subsequent chapters. In addition to presenting a more general treatment, the remainder of the chapter will serve as a review of the concepts introduced earlier in the chapter.

Average Excess

When n alleles are present, the average excess, α_i^*, for any allele B_i is given by

$$\alpha_i^* = \sum_{j=1}^{n} P_{ij|i}\, G_{ij} - \mu_G \tag{4.16a}$$

where $P_{ij|i}$ is the conditional probability of a B_iB_j genotype given that one allele is B_i. Under random mating, this reduces to

$$\alpha_i^* = \sum_{j=1}^{n} p_j \, G_{ij} - \mu_G \tag{4.16b}$$

where p_j is the frequency of the jth allele.

Example 5. Here we show how the average excess α_i^* of an allele i can be related to $\sigma(G, N_i)$, the covariance between genotypic value and the number of copies of that allele. This result will be useful in the following sections.

To compute $\sigma(G, N_i) = E(G \cdot N_i) - E(N_i) \cdot E(G)$, we start with the fact that $E(G) = \mu_G$, so we merely require expressions for $E(N_i)$ and $E(G \cdot N_i)$. The mean number of alleles of type i at the locus, $E(N_i)$, is straightforward. Since there are two genes at each locus, and the frequency of allele i is p_i, $E(N_i) = 2p_i$.

To obtain $E(G \cdot N_i)$, we use **ordered-genotype** notation, where $P_{ij(o)}$ is the probability of getting allele i from the mother and allele j from the father. We assume that $P_{ij(o)} = P_{ji(o)}$, so $P_{ij} = 2P_{ij(o)}$ when $i \neq j$. Because the variable N_i takes on only two nonzero values, two and one, the expected cross-product is

$$E(G \cdot N_i) = (G_{ii} \cdot 2) \cdot P_{ii(o)} + \sum_{j \neq i}^{n} (G_{ij} \cdot 1) \cdot 2P_{ij(o)} = 2p_i \sum_{j=1}^{n} P_{ij|i} G_{ij}$$

where the last step follows from the definition of a conditional genotype probability as $P_{ij\,|\,i} = P_{ij(o)}/p_i$. Putting the above results together, and recalling Equation 4.16a,

$$\sigma(G, N_i) = 2\,p_i \left[\sum_{j=1}^{n} P_{ij|i} G_{ij} - \mu_G \right] = 2\,p_i\,\alpha_i^* \tag{4.17a}$$

Under the assumption of random mating, average excesses are identical to additive effects, and

$$\sigma(G, N_i) = 2\,p_i\,\alpha_i \tag{4.17b}$$

Additive Effects

As in the diallelic case, with n alleles the additive effects are defined to be the set of α_i that minimizes $E(\delta_{ij}^2)$, obtained from the least-squares solution for the multiple regression

$$G = \mu_G + \sum_{i=1}^{n} \alpha_i \, N_i + \delta \tag{4.18}$$

This expression is the n-allele extension of Equation 4.5a, with N_i being the number of copies of allele i carried by an individual. For example, for the genotype G_{34}, $\sum \alpha_i N_i = \alpha_3 + \alpha_4$, and $\delta_{34} = G_{34} - \mu_G - \alpha_3 - \alpha_4$.

Multivariate regressions are covered in detail in Chapter 8, and here we simply cite the basic result — the regression coefficients (i.e., the α_i) are defined by the set of equations

$$\sigma(G, N_i) = \sum_{j=1}^{n} \alpha_j \, \sigma(N_i, N_j) \qquad \text{for } 1 \le i \le n \tag{4.19}$$

Expressed in this way, the definitions of the average effects are not immediately transparent, and the general solution to these equations is rather involved (Kempthorne 1957). However, under random mating, the solutions are simplified greatly and can be expressed in two ways. First, drawing from the previous example,

$$\alpha_i = \frac{\sigma(G, N_i)}{2p_i} \tag{4.20a}$$

Second, an equivalent and even more transparent solution follows from Equation 4.16b,

$$\alpha_i = \sum_{j=1}^{n} p_j \, G_{ij} - \mu_G \tag{4.20b}$$

i.e., under random mating, the average effects are equal to conditional mean deviations from μ_G.

If mating is nonrandom, but genotype frequencies are given by

$$P_{ii} = (1-f)p_i^2 + fp_i \tag{4.21a}$$

$$P_{ij} = 2(1-f)p_i p_j \tag{4.21b}$$

as occurs under regular inbreeding (Chapter 10), then

$$\alpha_i = \frac{\alpha_i^*}{1+f} \tag{4.22}$$

where f, the inbreeding coefficient, is the fractional reduction of heterozygote frequencies relative to those expected under random mating.

Additive Genetic Variance

To obtain the variance associated with the additive effects, we first need a result from regression theory. Consider the regression $y = \mu + \sum \beta_i \, x_i + e$. Since the total variance of a response variable y equals the variance accounted for by the regression plus the residual variance σ_e^2 (Chapter 8), it follows that the variance accounted for by the predictor variables is $\sum \beta_i \, \sigma(y, x_i)$. This can be immediately seen by noting

$$\sigma_y^2 = \sigma(y, y) = \sigma(y, \mu + \sum \beta_i \, x_i + e) = \sum_{i=1}^{N} \beta_i \, \sigma(y, x_i) + \sigma_e^2$$

Drawing the analogy with Equation 4.18, where the additive effects arise by considering the genotype G as a response variable and the gene contents N_i as predictor variables, the variance associated with the additive effects becomes

$$\sum_{i=1}^{n} \alpha_i \, \sigma(G, N_i)$$

Thus, recalling the result from Example 5 that $\sigma(G, N_i) = 2 \, p_i \, \alpha_i^*$, the additive genetic variance is

$$\sigma_A^2 = 2 \sum_{i=1}^{n} p_i \, \alpha_i \, \alpha_i^* \tag{4.23a}$$

as noted by Fisher (1941) and Kempthorne (1957). This general definition for the additive genetic variance holds for both randomly and nonrandomly mating populations. In the latter case, it reduces to

$$\sigma_A^2 = 2 \sum_{i=1}^{n} p_i \, \alpha_i^2 \tag{4.23b}$$

which with $n = 2$ (a diallelic locus) reduces further to Equation 4.12a. Thus, under random mating, σ_A^2 for a locus is simply equal to the mean-squared additive effect, multiplied by two to account for diploidy. More generally, since $E[\,\alpha\,] = 0$, $\sigma_A^2 = E[\,(\alpha_i + \alpha_j)^2\,]$.

From Equation 4.22, it follows that under regular inbreeding,

$$\sigma_A^2 = 2(1 + f) \sum_{i=1}^{n} p_i \, \alpha_i^2 \tag{4.23c}$$

In general, inbreeding inflates the additive genetic variance by causing correlations among the effects of alleles within the same individuals. However, because the additive effect itself is a function of f, inbreeding does not necessarily simply increase the additive genetic variance by the factor $(1 + f)$. From Kempthorne (1957),

$$\alpha_i = \left(\frac{1 - f}{1 + f} \right) \alpha_{ir} + \left(\frac{f}{1 + f} \right) (G_{ii} - \mu_G) \tag{4.24}$$

where α_{ir} and μ_G respectively denote the additive effect of allele i and the mean phenotype in the noninbred population. If gene action is additive, then $G_{ii} - \mu_G =$

$2a$, $\alpha_i = \alpha_{ir} = a$, and the additive genetic variance in an inbred population is, in fact, $(1+f)$ times greater than that under random mating. However, with any level of dominance, $\alpha_i \neq \alpha_{ir}$ under inbreeding, and the change in additive genetic variance with f is not likely to be linear.

Finally, we consider the general definition of the breeding value (A_{ij}) under random mating. Parents with genotype B_iB_j transmit alleles i and j with equal frequency, and the expected additive effect of the allele contributed by their mates is equal to zero. Thus, the expected deviation of the mean phenotype of offspring of genotype B_iB_j from the population mean is

$$\left(\mu_G + \frac{\alpha_i + \alpha_j}{2}\right) - \mu_G = \frac{A_{ij}}{2} \tag{4.25}$$

which is half the breeding value of the parental genotype. Returning to Equation 4.18,

$$\begin{aligned} G_{ij} &= \mu_G + \alpha_i + \alpha_j + \delta_{ij} \\ &= \mu_G + A_{ij} + \delta_{ij} \end{aligned} \tag{4.26}$$

Thus, the genotypic value at any locus can be decomposed into four quantities: the mean genotypic value for the population, the additive effects of the two genes (whose sum is the breeding value), and a dominance deviation due to the interaction between the genes. Since μ_G is a constant, and A and δ are (by the properties of least-squares regression) uncorrelated, it follows from Equation 4.26 that the genetic variance can be represented as

$$\sigma_G^2 = \sigma^2(\alpha_i + \alpha_j) + \sigma^2(\delta_{ij}) \tag{4.27a}$$

This is a completely general definition, applying even to the case of nonrandom mating (although as noted above, the definitions of the α_i and δ_{ij} change with the degree of inbreeding). For the special case of random mating, α_i and α_j are uncorrelated, and

$$\sigma_G^2 = \sigma_A^2 + \sigma_D^2 \tag{4.27b}$$

Comparing this with Equation 4.9, we find that σ_A^2 has a very specific and useful meaning. *Under random mating, the additive genetic variance is equivalent to the variance of breeding values of individuals in the population.*

Summing up, the additive effect of an allele, the breeding value of an individual, and the additive-genetic variance of a population are hierarchically related measures of genetic effects (Table 4.3). All of this notation can be quite confusing, particularly when the nonsubscripted α is used to denote the average effect of allelic substitution. We used the latter quantity in our introduction of the one-locus model for historical reasons and because it provides useful insight into the two-allele situation. However, we will not be using it much in the remainder of

Table 4.3 Summary of quantities used to measure genetic effects.

Quantity	Description
Homozygous effect, a Dominance coefficient, k	Intrinsic properties of allelic products. Not functions of allele frequencies, but may vary with genetic background.
Additive effect, α_i Average excess, α_i^*	Properties of *alleles* in a particular population. Functions of homozygous effects, dominance coefficients, and genotype frequencies.
Breeding value, A	Property of a particular *individual* in reference to a particular population. Sum of the additive effects of an individual's alleles.
Additive genetic variance, σ_A^2	Property of a particular *population*. Variance of the breeding values of individuals within the population.

the book, nor will we be utilizing the concept of average excess (the latter plays a significant role in considerations of selection response, which is covered in our next book). Unless otherwise noted, we will be dealing with randomly mating populations, so our reference to the additive effect of an allele will be consistent with the conditional mean deviation definition (Equation 4.20b), as well as formally equivalent to a least-squares regression coefficient. Further commentary on the relationship between average excesses and additive effects can be found in Falconer (1985) and Templeton (1987).

5

Sources of Genetic Variation for Multilocus Traits

As we proceed from single gene loci to quantitative characters, it becomes necessary to introduce several new concepts. First, we must consider whether the genotypic values defined for single loci can be combined additively to explain the phenotypic variation associated with multilocus genotypes or whether important nonlinear interactions exist. Second, we must consider whether the inheritance and distribution of genes at one locus are independent of those at other loci. Third, several sources of environmental variance influence the expression of polygenic traits, and this raises questions as to whether gene expression varies with the environmental context and whether specific genotypes are associated with particular environments.

The joint occurrence of all of these sources of variation usually hopelessly obscures our ability to identify the precise breeding value of any single individual. Nevertheless, it is possible to characterize populations with respect to the relative magnitudes of different sources of phenotypic variance. Such a partitioning of variance is useful because different sources of variation have predictable impacts on various evolutionary phenomena. As noted in Chapters 3 and 4, the response to selection is closely associated with the level of additive genetic variance. Environmental variance reduces the efficiency of the selection process by obscuring the relationship between genotypes and phenotypes. The response of a population to inbreeding depends primarily on the level of dominance genetic variance (Chapter 10).

In the previous chapter, we found that the genetic variance associated with a single locus can be partitioned into additive and dominance components. This approach will be generalized below to account for all of the loci contributing to the expression of a quantitative trait, as well as to allow for variance arising from gene interaction among loci. The sources of variation associated with the environment, alluded to above, are explored in Chapter 6, and the general strategy for the actual measurement of these sources of variance is covered in Chapter 7. Details of statistical methods are discussed in Chapters 17–27.

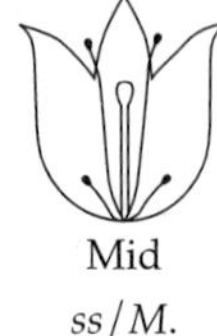

Figure 5.1 The three types of floral morphology in a tristylous plant. The central and lateral elongate structures denote stigmas and anthers. Floral morphology (stigma length) is a function of two-locus genotypes. There are no *SS* individuals because short morphs are reproductively incompatible with each other.

EPISTASIS

In the preceding chapter, the dominance effect at a locus was defined to be the deviation of the observed genotypic value from the expectation based on additive effects. Viewed in this manner, dominance is a measure of nonadditivity of allelic effects within loci. **Epistasis** describes the nonadditivity of effects between loci.

A striking example of epistasis is responsible for a polymorphism in flower morphology in tristylous plants, a group including species of *Eichhornia, Lythrum, Oxalis,* and *Pontederia* (Weller 1976, Ganders 1979). Long-style morphs have short- and medium-length stamens, mid-style morphs have short and long stamens, and short-style morphs have medium and long stamens (Figure 5.1). Each morph is incompatible with plants of its own type but can freely cross with members of the other types. The polymorphism is determined by two diallelic loci, *short* and *medium*. The presence of an *S* allele at the first locus results in short-style plants regardless of the genotype at the second locus, whereas *ss* individuals are mid-styles if they carry an *M* allele but long-styles if they are *mm*. Thus, *M* is dominant to *m*, but the epistatic interaction of an *S* allele overrules this relationship.

To provide a more quantitative description of epistasis, consider an individual with alleles A_i and A_j at one locus and B_k and B_l at another. Expanding from Equation 4.26, the genotypic value, G_{ijkl}, can be written as the sum of the effects within loci and a deviation ϵ due to interaction between loci,

$$G_{ijkl} = \mu_G + (\alpha_i + \alpha_j + \delta_{ij}) + (\alpha_k + \alpha_l + \delta_{kl}) + \epsilon_{ijkl} \tag{5.1}$$

When written in this manner, Equation 5.1 glosses over the considerable complexity of epistatic interactions. With only two loci, there are actually three ways in which interactions can arise between loci: additive $\times$ additive $(\alpha\alpha)$, additive $\times$ dominance $(\alpha\delta)$, and dominance $\times$ dominance $(\delta\delta)$. With three loci, there are four additional types of epistasis, $(\alpha\alpha\alpha)$, $(\alpha\alpha\delta)$, $(\alpha\delta\delta)$, and $(\delta\delta\delta)$, and the number steadily grows with additional loci.

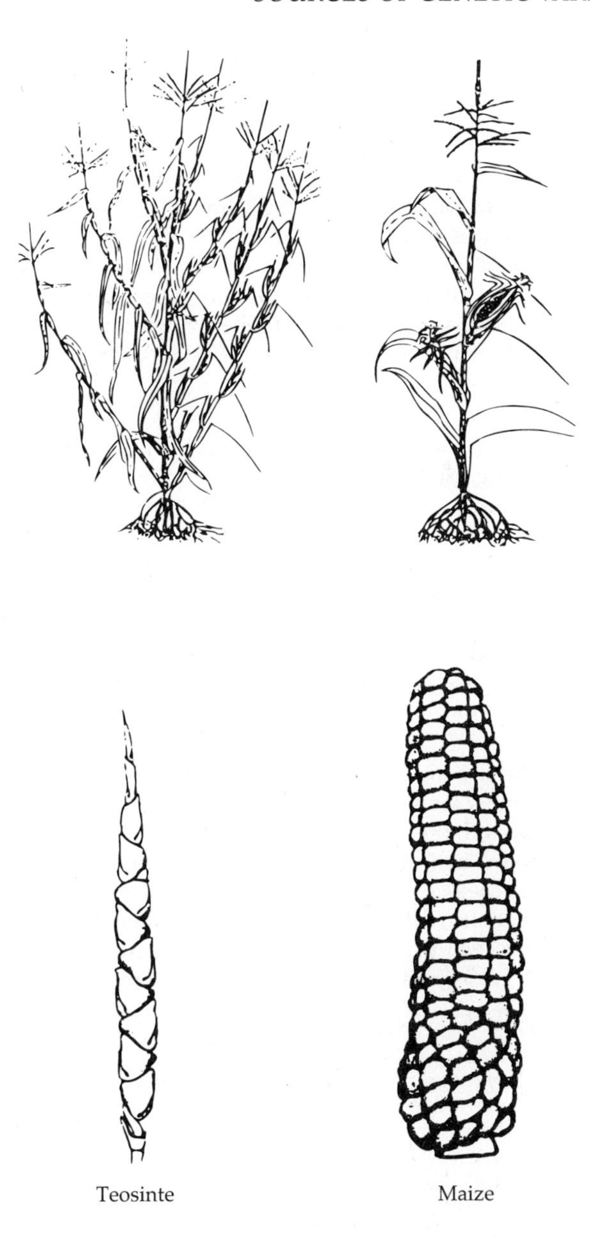

Figure 5.2 Phenotypic differences between domesticated maize *(Zea mays* sp. *mays)* and its hypothetical progenitor, teosinte *(Zea mays* sp. *parviglumis).* **Top:** Plant architecture. **Bottom:** Structure of the ear (the female inflorescence). (From Doebley et al. 1990.)

Example 1. As a straightforward numerical example of epistasis, we consider the genotypic values associated with two diallelic loci in an artificially constructed population of teosinte, the presumed wild progenitor of cultivated maize. The two species are fully interfertile, despite their dramatic differences in plant architecture and inflorescence structure (Figure 5.2). Beadle (1939) hypothesized that substitution of as few as five genes may have been responsible for the teosinte-maize transformation. Although this idea has not yet been fully confirmed, recent studies clearly implicate a small number of genes with major effects (Doebley and Stec 1993, Dorweiler et al. 1993, Doebley et al. 1995a,b). For further information, see Chapter 15.

A number of molecular markers (restriction fragment length polymorphisms) can be used to distinguish maize and teosinte. Through a progressive series of backcrosses and screening of genomic markers, Doebley et al. (1995a) introduced two maize markers, *UMC107* and *BV302*, into a teosinte genetic background. We will refer to the maize "alleles" for these two loci as U_M and B_M and the teosinte "alleles" as U_T and B_T. Each of these markers is associated with a tightly linked span of DNA from its parental origin (Chapters 14 and 15). Thus, the U_{TT}/B_{TT} genotype is essentially pure teosinte, whereas the U_{MM}/B_{MM} genotype is homozygous for two small stretches of maize DNA on an otherwise pure teosinte background.

For each of the nine two-locus marker genotypes, Doebley et al. assayed a large number of individuals for several morphological and reproductive traits. The following table gives the results for one such character, the average length of vegetative internodes in the lateral branch (in millimeters). Estimates of the homozygous-effect coefficients (a) and dominance coefficients (k) (defined in Figure 4.4), conditional on the genotypic states of the alternate locus, are given in the last two columns and rows. These are calculated by setting the three genotypes equal to u, $u + (1 + k)a$, and $u + 2a$, where u is a scaling factor.

		UMC107			
BV302	U_{MM}	U_{MT}	U_{TT}	a	k
B_{MM}	18.0	40.9	61.1	27.0	0.33
B_{MT}	54.6	47.6	66.5	6.0	−2.17
B_{TT}	47.8	83.6	101.7	21.6	0.06
a	14.9	21.4	20.3		
k	1.46	−0.69	−0.73		

Assuming that the observed mean phenotypes of the different marker classes are accurate estimates of the genotypic values, the fact that both a and k vary dramatically with the genetic background provides strong evidence for epistatic interaction between the genes associated with the markers. For example, when the genotype at the *UMC107* locus is U_{MT} or U_{TT}, B_M is quite dominant to B_T ($k \simeq -0.7$) and the homozygous effect of the B_T allele is $a \simeq 21$. However, on a U_{MM} background, the homozygous effect is reduced to $a \simeq 15$, and the *BV302* locus exhibits overdominance ($k > 1.0$). For this and other characters, the authors further demonstrated that the homozygous and dominance effects associated with these markers vary dramatically depending on whether they are expressed in teosinte, maize, or teosinte × maize genetic backgrounds.

A GENERAL LEAST-SQUARES MODEL FOR GENETIC EFFECTS

We now consider a statistical procedure for the definition of epistatic effects, which is a straightforward extension of the one-locus linear model introduced in Chapter 4. We will confine our attention to two loci, as the extension to higher numbers of loci should become obvious, and we assume random mating throughout.

In the previous chapter, the additive effect of an allele was shown to be equal to the deviation of members of the population with the allele from the population mean phenotype. This definition does not change with the addition of loci. Letting $G_{i...}$ represent the conditional mean phenotype of individuals with allele i at the first locus without regard to the other allele at the locus or to the genotype at the second locus, then

$$\alpha_i = G_{i...} - \mu_G \tag{5.2}$$

Analogous expressions follow for the three other additive effects (α_j, α_k, and α_l). Within each locus, the mean value of the average effects (weighted by the allele frequencies) is equal to zero.

The dominance effects are defined by considering $G_{ij..}$, the conditional mean phenotype of individuals with alleles i and j at the first locus without regard to genotypic state at the second locus. From Equation 4.26, it follows that

$$\delta_{ij} = G_{ij..} - \mu_G - \alpha_i - \alpha_j \tag{5.3a}$$
$$\delta_{kl} = G_{..kl} - \mu_G - \alpha_k - \alpha_l \tag{5.3b}$$

Here, we have subtracted the mean genotypic value and the additive effects, leaving the dominance effect as the only unexplained portion of the conditional mean at the locus. As in the case of additive effects, the mean dominance deviation at each locus is equal to zero.

The definition of epistatic effects proceeds in a similar fashion. Letting $G_{i.k.}$ be the mean phenotype of individuals with gene i at locus 1 and k at locus 2, without regard to the other two genes, the *ik*th additive × additive effect is

$$(\alpha\alpha)_{ik} = G_{i.k.} - \mu_G - \alpha_i - \alpha_k \tag{5.4}$$

In other words, $(\alpha\alpha)_{ik}$ is the deviation of the conditional mean $G_{i.k.}$ from the expectation based on the population mean μ_G and the additive effects α_i and α_k. An additive × dominance effect measures the interaction between an allele at one locus with a particular genotype of another locus. It is defined as the deviation of the conditional mean $G_{i.kl}$ from the expectation based on all lower-order effects, in this case the three additive effects, one dominance effect, and two additive × additive effects involving the constituent genes,

$$(\alpha\delta)_{ikl} = G_{i.kl} - \mu_G - \alpha_i - \alpha_k - \alpha_l - \delta_{kl} - (\alpha\alpha)_{ik} - (\alpha\alpha)_{il} \tag{5.5}$$

Finally, for a dominance × dominance effect,

$$\begin{aligned}(\delta\delta)_{ijkl} = G_{ijkl} &- \mu_G - \alpha_i - \alpha_j - \alpha_k - \alpha_l - \delta_{ij} - \delta_{kl} \\ &- (\alpha\alpha)_{ik} - (\alpha\alpha)_{il} - (\alpha\alpha)_{jk} - (\alpha\alpha)_{jl} \\ &- (\alpha\delta)_{ikl} - (\alpha\delta)_{jkl} - (\alpha\delta)_{ijk} - (\alpha\delta)_{ijl}\end{aligned} \tag{5.6}$$

This partitioning of the total genotypic value into a series of effects can be summarized as follows. We started with the lowest-order effects, the additive effects of alleles. These are defined in a least-squares sense (Chapter 3), in that they account for as much of the variance in genotypic values as possible. We then progressively defined higher-order effects, each time accounting for as much of the residual variation as possible. In the two-locus example, $(\delta\delta)_{ijkl}$ describes the final bit of variation not accounted for by additive, dominance, additive × additive, or additive × dominance effects. Summing up terms, we have a complete description of a genotypic value,

$$\begin{aligned}G_{ijkl....} &= \mu_G + [\alpha_i + \alpha_j + \alpha_k + \alpha_l] + [\delta_{ij} + \delta_{kl}] \\ &+ [(\alpha\alpha)_{ik} + (\alpha\alpha)_{il} + (\alpha\alpha)_{jk} + (\alpha\alpha)_{jl}] \\ &+ [(\alpha\delta)_{ikl} + (\alpha\delta)_{jkl} + (\alpha\delta)_{ijk} + (\alpha\delta)_{ijl}] + (\delta\delta)_{ijkl} + \cdots\end{aligned} \tag{5.7}$$

The open-ended equation merely implies that there are many more terms when more than two loci are involved in the expression of a trait. The parameters in this model depend very much upon the genotype frequencies in the population and can change as allele frequencies change (Example 4, Chapter 4). However, the mean value of each type of effect is always equal to zero.

At first glance, Equation 5.7 may seem to be of little practical use. After all, if the genotypic value of an individual cannot be ascertained without error because of environmental contributions to the phenotype, then what hope is there of identifying the individual components? Often there is none. However, a great conceptual advance is achieved when Equation 5.7 is extended to the population level. Provided that mating is random and segregation of loci is independent, there is no statistical relationship between the genes found within or among loci. Therefore, the total genetic variance is simply the sum of the variance of the individual effects. Letting $\sigma_A^2 = \sigma^2(\alpha_i)+\sigma^2(\alpha_j)+\sigma^2(\alpha_k)+\sigma^2(\alpha_l)$, $\sigma_D^2 = \sigma^2(\delta_{ij})+\sigma^2(\delta_{kl})$, $\sigma_{AA}^2 = \sigma^2[(\alpha\alpha)_{ik}] + \sigma^2[(\alpha\alpha)_{il}] + \sigma^2[(\alpha\alpha)_{jk}] + \sigma^2[(\alpha\alpha)_{jl}]$, and so on, the total genetic variance can be expressed as

$$\sigma_G^2 = \sigma_A^2 + \sigma_D^2 + \sigma_{AA}^2 + \sigma_{AD}^2 + \sigma_{DD}^2 + \cdots \tag{5.8}$$

This partitioning of the genetic variance into a series of components, the multilocus analog of Equation 4.11, was developed independently, but with very different approaches, by both Cockerham and Kempthorne in 1954.

Because of the hierarchical way in which genetic effects are defined, one might expect the magnitude of genetic variance components to become progressively smaller at higher stages in the hierarchy. Indeed, it is common for quantitative geneticists to use this argument as a rationalization for ignoring epistasis altogether. Unfortunately, such logic does not always hold up — as pointed out in the previous chapter, unless information on gene frequencies is available, variance components provide limited insight into the physiological mode of gene action. Recall that by modifying the additive effects of alleles, dominance (a higher-order effect than additivity) can inflate or deflate σ_A^2 relative to the situation expected under complete additivity (Figure 4.8). Qualitatively similar results apply to epistasis — depending on the frequencies of the genes involved, epistatic interactions (through their statistical influence on the additive and dominance effects of alleles) can greatly inflate the additive and/or dominance components of genetic variance. Thus, as will be seen in the next example and as cogently pointed out by Keightley (1989) and Cheverud and Routman (1995), relatively low magnitudes of epistatic components of variance are by no means incompatible with the existence of strong epistatic gene action.

A simple quantitative argument points further to the potential for significant epistasis in the expression of quantitative traits. With n loci underlying a trait, the number of additive effects is of order n, while the number of diallelic and triallelic epistatic effects are potentially of order n^2 and n^3. Thus, even if individual epistatic effects are relatively small, their summed effects may be large. This raises a serious issue for quantitative genetics that will become more apparent in later chapters. Although there are many possible types of epistasis, methods for evaluating their quantitative significance have substantial limitations from the standpoint of statistical power and experimental design.

In summary, simple statistical arguments suggest that epistatic interactions are likely to be important in the expression of many quantitative traits as well as in the determinance of levels of additive genetic variance. There is certainly no firm empirical evidence that epistasis is of negligible importance, and Wright (1968), for one, argued strongly that epistasis is the rule rather than the exception. We will be reaffirming that viewpoint throughout the book. For further reviews of the evidence, see Wright (1968), Barker (1979), and Moreno (1994).

Example 2. To acquire a deeper understanding of the connections between conditional genotypic values, genetic effects, and variance components, we return to the data in Example 1, making the assumption that the entries in the table are accurate estimates of the nine genotypic values associated with the marker loci. Suppose that this population randomly mates, with all alleles having frequency 0.5. What are the genetic effects associated with the different alleles (and their combinations), and what are the expected components of genetic variance associated with the two marker loci?

Under the assumption that the two loci are unlinked (in actuality, they are on different chromosomes), the frequencies of the two-locus genotypes are defined in the following table. They are simply equal to the products of the Hardy-Weinberg frequencies at the individual loci. Thus, each of the four double-homozygote classes has frequency $(1/4)^2$, while the classes with a single heterozygous locus have frequency $(1/2) \times (1/4)$, and the double-heterozygote has frequency $(1/2)^2$.

		UMC107	
BV302	U_{MM}	U_{MT}	U_{TT}
B_{MM}	1/16	1/8	1/16
B_{MT}	1/8	1/4	1/8
B_{TT}	1/16	1/8	1/16

In the following computations, we let G_{ijkl} denote the genotypic value of individuals with alleles i and j at locus *BV302* and alleles k and l at locus *UMC107*, where i, j, k, and l are either M or T. Letting P_{ijkl} be the frequency of the *ijkl*th genotype, the mean genotypic value in the population is

$$\begin{aligned}\mu_G &= \sum P_{ijkl}\, G_{ijkl} \\ &= [(1/16) \times (18.0 + 61.1 + 47.8 + 101.7)] \\ &\quad + [(1/8) \times (40.9 + 54.6 + 66.5 + 83.6)] + [(1/4) \times 47.6] \\ &= 56.8875\end{aligned}$$

Additive effects. Prior to computing the additive effects, we require the conditional genotypic means associated with each allele. These are obtained by weighting the phenotypic data in the table of Example 1 with conditional genotype frequencies. For example, the conditional genotypic mean for individuals containing a B_M allele is

$$G_{B_M\cdots} = \sum_{j,k,l} P(ijkl \,|\, i = B_M)\, G_{B_M jkl}$$

The conditional probability $P(ijkl \,|\, i = B_M)$ is the probability that an individual has alleles j, k, and l, given that it is has one B_M allele. This probability is simply the product of the appropriate (Hardy-Weinberg) genotype frequency at the *UMC107* locus and the appropriate allele frequency at locus *BV302*. For example, if $j = B_M$, $k = U_M$, and $j = U_M$, $P(ijkl \mid i = B_M) = 0.5 \times 0.25 = 0.125$. Thus, the genotypic mean conditional on having a B_M allele is

$$\begin{aligned} G_{B_M\cdots} &= (0.5 \times 0.25 \times 18.0) + (0.5 \times 0.5 \times 40.9) \\ &\quad + (0.5 \times 0.25 \times 61.1) + (0.5 \times 0.25 \times 54.6) \\ &\quad + (0.5 \times 0.5 \times 47.6) + (0.5 \times 0.25 \times 66.5) = 47.1500 \end{aligned}$$

The three other conditional means involving single alleles are obtained in a similar manner,

$$G_{B_T\cdots} = 66.6250 \qquad G_{\cdot\cdot U_M\cdot} = 49.3375 \qquad G_{\cdot\cdot U_T\cdot} = 64.4375$$

Recalling Equation 5.2, we obtain the additive effects by subtracting the population mean μ_G from the conditional means,

$$\begin{aligned} \alpha_{B_M} &= -9.7375 \qquad \alpha_{B_T} = 9.7375 \\ \alpha_{U_M} &= -7.5500 \qquad \alpha_{U_T} = 7.5500 \end{aligned}$$

Noting that each allele has frequency 0.5, the mean additive effect of the alleles at each locus is equal to zero.

Dominance effects. To obtain the dominance effects, we require the conditional genotypic means. In this case, for each single-locus genotype, we simply take the average of the three genotypic values associated with the second locus, weighted by their Hardy-Weinberg frequencies. For example, for the B_{MM} genotype,

$$\begin{aligned} G_{B_{MM}\cdot\cdot} &= \sum_{k,l} P(ijkl \mid i, j = B_M)\, G_{B_{MM}kl} \\ &= (0.25 \times 18.0) + (0.5 \times 40.9) + (0.25 \times 61.1) = 40.225 \end{aligned}$$

Similarly, the conditional means for the B_{MT} and B_{TT} genotypes are found to be

$$G_{B_{MT}\cdot\cdot} = 54.075 \qquad G_{B_{TT}\cdot\cdot} = 79.175$$

Substituting these values, the population mean, and the appropriate additive effects into Equation 5.3, we obtain the dominance effects,

$$\delta_{B_{MM}} = 2.8125 \qquad \delta_{B_{MT}} = -2.8125 \qquad \delta_{B_{TT}} = 2.8125$$

Weighting these by the Hardy-Weinberg frequencies associated with each genotype, it can be seen that the mean dominance effect at the *BV302* locus is equal to zero. We leave it to the reader to show that for the *UMC107* locus,

$$\delta_{U_{MM}} = 1.9625 \qquad \delta_{U_{MT}} = -1.9625 \qquad \delta_{U_{TT}} = 1.9625$$

Additive × additive effects. The same logic is used to obtain the additive × additive effects. Because there are two alleles at each locus, four conditional means (each involving one allele at each locus) must be considered. Each gametic type is found in four two-locus genotypes, in this case with equal frequency 0.25 because all alleles have frequency 0.5. Thus, for example, the conditional mean $G_{B_M \cdot U_M \cdot}$ is the average of the genotypic values for the genotypes $B_{MM}U_{MM}$, $B_{MT}U_{MM}$, $B_{MM}U_{MT}$, and $B_{MT}U_{MT}$. The four conditional means are

$$\begin{aligned} G_{B_M \cdot U_M \cdot} &= 40.275 & G_{B_M \cdot U_T \cdot} &= 54.025 \\ G_{B_T \cdot U_M \cdot} &= 58.400 & G_{B_T \cdot U_T \cdot} &= 74.850 \end{aligned}$$

The additive × additive effects are computed by substituting these values, the population mean, and the appropriate additive effects into Equation 5.4,

$$\begin{aligned} (\alpha\alpha)_{B_M \cdot U_M \cdot} &= 0.675 & (\alpha\alpha)_{B_M \cdot U_T \cdot} &= -0.675 \\ (\alpha\alpha)_{B_T \cdot U_T \cdot} &= 0.675 & (\alpha\alpha)_{B_T \cdot U_M \cdot} &= -0.675 \end{aligned}$$

Again, the mean value of these effects (weighted by their frequencies of occurrence) is zero.

Additive × dominance effects. At this point, we assume that the basic strategy for calculating conditional means is familiar, and we leave it to the reader to verify that the conditional means involving one allele at one locus and two at the other locus are

$$\begin{aligned} G_{B_M . U_{MM}} &= 36.30 & G_{B_M . U_{MT}} &= 44.25 & G_{B_M . U_{TT}} &= 63.80 \\ G_{B_T . U_{MM}} &= 51.20 & G_{B_T . U_{MT}} &= 65.60 & G_{B_T . U_{TT}} &= 84.10 \\ G_{B_{MM} U_M \cdot} &= 29.45 & G_{B_{MT} U_M \cdot} &= 51.10 & G_{B_{TT} U_M \cdot} &= 65.70 \\ G_{B_{MM} U_T \cdot} &= 51.00 & G_{B_{MT} U_T \cdot} &= 57.05 & G_{B_{TT} U_T \cdot} &= 92.65 \end{aligned}$$

The additive × dominance effects follow from substitution of quantities given above into Equation 5.5,

$$\begin{aligned} (\alpha\delta)_{B_M . U_{MM}} &= 0.9375 & (\alpha\delta)_{B_M . U_{MT}} &= -0.9375 & (\alpha\delta)_{B_M . U_{TT}} &= 0.9375 \\ (\alpha\delta)_{B_T . U_{MM}} &= -0.9375 & (\alpha\delta)_{B_T . U_{MT}} &= 0.9375 & (\alpha\delta)_{B_T . U_{TT}} &= -0.9375 \\ (\alpha\delta)_{B_{MM} U_M .} &= -4.5750 & (\alpha\delta)_{B_{MT} U_M .} &= 4.5750 & (\alpha\delta)_{B_{TT} U_M .} &= -4.5750 \\ (\alpha\delta)_{B_{MM} U_T .} &= 4.5750 & (\alpha\delta)_{B_{MT} U_T .} &= -4.5750 & (\alpha\delta)_{B_{TT} U_T .} &= 4.5750 \end{aligned}$$

Dominance × dominance effects. The nine dominance × dominance effects are computed from the observed genotypic values using Equation 5.6,

$$\begin{aligned} (\delta\delta)_{B_{MM} U_{MM}} &= -4.5125 & (\delta\delta)_{B_{MM} U_{MT}} &= 4.5125 & (\delta\delta)_{B_{MM} U_{TT}} &= -4.5125 \\ (\delta\delta)_{B_{MT} U_{MM}} &= 4.5125 & (\delta\delta)_{B_{MT} U_{MT}} &= -4.5125 & (\delta\delta)_{B_{MT} U_{TT}} &= 4.5125 \\ (\delta\delta)_{B_{TT} U_{MM}} &= -4.5125 & (\delta\delta)_{B_{TT} U_{MT}} &= 4.5125 & (\delta\delta)_{B_{TT} U_{TT}} &= -4.5125 \end{aligned}$$

Variance components. The final step in this example is to compute the components of genetic variance. Because the average effects are computed in such a way that their means are always equal to zero, the variances are simply equal to the mean squared effects (weighted by their frequency of occurrence) multiplied by the number of times the effect occurs. For example, for the additive genetic variance, the contribution associated with the *BV302* locus is $2 \times [(0.5 \times (-9.7375)^2) + (0.5 \times 9.7375^2)] = 189.6378$; the mean-squared effect is multiplied by two because there are two additive effects per diploid locus. In the same manner, the additive genetic variance associated with the *UMC107* locus is found to be 114.0050. The total additive genetic variance associated with these two loci is simply the sum of the locus-specific contributions,

$$\sigma_A^2 = 303.6428$$

(The same answer is obtained by defining the breeding values of the nine different genotypes as $A_{ijkl} = \alpha_i + \alpha_j + \alpha_k + \alpha_l$, and computing the mean squared breeding value; see Example 4, Chapter 4).

The dominance genetic variance associated with each locus is equal to the mean squared dominance effect, $2.8125^2 = 7.9101$ for the *BV302* locus and $1.9625^2 = 3.8514$ for the *UMC107* locus. Thus, the total dominance genetic variance is

$$\sigma_D^2 = 11.7615$$

The additive × additive variance is obtained by multiplying the mean squared additive × additive effect by four, because four such effects contribute to each genotypic value. In this example, all of the additive × additive effects have absolute value 0.675 (as in the case of the additive and dominance effects, the symmetry is a special consequence of all of the gene frequencies being equal to 0.5). Thus, the additive × additive variance is simply 4×0.675^2,

$$\sigma_{AA}^2 = 1.8225$$

For the additive × dominance variance, there are again four effects contributing to the genotypic value of each individual, two involving *BV302* additive effects and two involving *UMC107* additive effects. The additive × dominance variance is therefore $(2 \times 0.9375^2) + (2 \times 4.575^2)$,

$$\sigma_{AD}^2 = 43.6191$$

Finally, because there is only one dominance × dominance term per individual, the dominance × dominance variance is simply the mean squared effect, 4.5125^2,

$$\sigma_{DD}^2 = 20.3627$$

Summing the five components of genetic variance, we obtain the total genetic variance associated with the two markers,

$$\sigma_G^2 = 381.2086$$

As a check on the validity of this result, the total genetic variance can be computed directly from the data in the table in Example 1 as $\sum P_{ijkl} G_{ijkl}^2 - \mu_G^2$; the same answer is obtained.

These results show that despite the existence of strong epistatic interactions between the effects of genes associated with the two markers (Example 1), most of the interaction effects are associated statistically with the additive effects of alleles. The three epistatic components of variance account for only 17% of the total genetic variance, and the dominance genetic variance contributes only another 3%. Thus, as emphasized above, the relative magnitude of variance components provides little insight into the actual physiological mode of gene action.

In closing, we note that this example is a somewhat unconventional application of the linear model. Usually, the genes underlying quantitative traits are unknown, in which case it is impossible to estimate their individual effects. However, with the widespread application of molecular markers, this situation is changing rapidly (Chapters 13–16). In Chapter 7, we will begin to see that estimation of the components of genetic variance associated with the entire pool of constituent loci underlying a quantitative trait does not require any information about their identity.

Extension to Haploids and Polyploids

A large fraction of the "lower" organisms, most notably bacteria, fungi, algae, and mosses, have life cycles that are predominantly haploid. A very brief diploid stage occurs when gametes fuse and chromosomes reassort to form variable haploid progeny. Haploid species have long been used as model systems in genetic studies focused on single loci, but they also offer some interesting possibilities for studies on quantitative-genetic variation (Caten and Jinks 1976, Croft and Jinks 1977, Caten 1979). Because haploid organisms have only a single allele at each locus, dominance does not contribute to the genetic variation, so all variance due to nonadditive gene action involves epistasis between additive effects. Therefore, Equation 5.8 simplifies to

$$\sigma^2_G = \sigma^2_A + \sigma^2_{AA} + \sigma^2_{AAA} + \cdots \tag{5.9}$$

Almost all haploid organisms can be propagated vegetatively, and it is a simple matter to test whether genetic variation can be explained by an additive model. When measures are taken on a large number of replicates of the same genotype, the mean of these measures will be a good estimate of the genotypic value under the experimental setting. If there is no epistatic gene action, the mean phenotype of sexually produced progeny should not be significantly different from the average of the parents.

Figure 5.3 illustrates two analyses of growth rate resulting from crosses between strains of the bread mold, *Aspergillus*. In both cases, there is a wide dispersion of offspring phenotypes due to the reassortment of parental genes. The mean offspring phenotype is not significantly different from that of the midparent

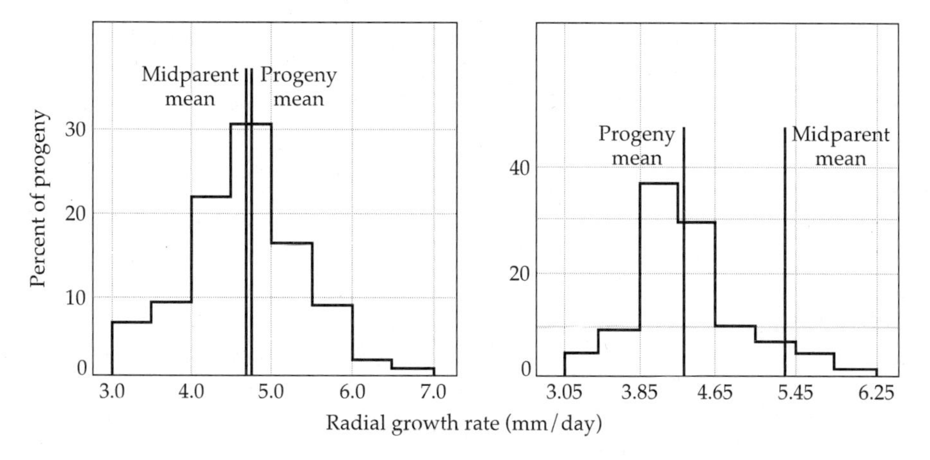

Figure 5.3 Frequency distributions of progeny phenotypes from single crosses in comparison to midparent means. **Left**: *Aspergillus amstelodami* cross 2 × 37. (From Caten and Jinks 1976.) **Right**: *Aspergillus nidulans* cross 7 × 43. (From Jinks et al. 1966.)

in one case, but in the second case there is a substantial depression in offspring growth rate. In this case, it appears that different combinations of genes with favorable epistatic effects on growth were sorted into the two parental lines. Sexual reproduction resulted in a breakup of these positive gene interactions. This type of result is observed commonly in crosses between fungal lines (Papa 1970, Merrick 1975), and we will consider it in a broader context in Chapter 9.

Partitioning the genetic variance becomes much more complicated in the case of polyploids because of additional gene effects that need to be taken into account. Consider the effects attributable to single loci. With haploids we need only consider the additive component of variance, while diploid analysis brings in a dominance component to account for diallelic interactions. In the case of a tetraploid with single-locus genotype B_{ijkl}, there are six potential diallelic interactions per locus ($i \times j,\ i \times k,\ i \times l,\ j \times k,\ j \times l,\ k \times l$), as well as four three-allele interactions ($i \times j \times k,\ i \times j \times l,\ i \times k \times l,\ j \times k \times l$), and one four-allele interaction ($i \times j \times k \times l$). Thus, in the absence of interactions between loci, the total genetic variance of a tetraploid can be written as

$$\sigma^2_G = \sigma^2_A + \sigma^2_B + \sigma^2_T + \sigma^2_Q \tag{5.10}$$

where σ^2_B, σ^2_T, and σ^2_Q denote the variances of effects involving bi-, tri-, and quadra-allelic interactions. With epistasis, this expression becomes much more complicated. Even if epistasis only involved pairs of loci, ten terms (σ^2_{AA}, σ^2_{AB}, σ^2_{AT}, σ^2_{AQ}, σ^2_{BB}, etc.) would need to be taken into account, compared with three in the case of diploidy. Further information on these matters can be found in

Kempthorne (1957), Killick (1971), and Mather and Jinks (1982). Wricke and Weber (1986) give a particularly lucid review and show how all of the concepts introduced in the previous chapter can be extended to tetraploids.

LINKAGE

Up to now we have been treating the transmission of genes at different loci as independent events. Such independence is generally true for genes that are on different chromosomes, but when loci are physically linked on the same chromosome, a statistical dependence can exist between the genes incorporated into gametes. Genes that lie on the same chromosome tend to be inherited as a group, a tendency that declines with increasing distance between the loci. **Crossing-over** is responsible for this decline.

During the meiotic production of gametes, each chromosome replicates to form two "sister" chromatids. Homologous nonsister chromatids then become intimately associated with each other, forming one or more physical connections known as **chiasmata** (Figure 5.4). Each chiasma represents a point of prior transfer of genetic information. Usually, the chromosomal material distal to a chiasma is exchanged completely, while that proximal to the centromere remains intact. Such an exchange is referred to as a cross-over.

Gamete frequencies are the fundamental census units in all considerations of linkage relationships. Consider two diallelic loci, *A* and *B*, and let the frequencies of the four gamete types be $P_{A_1B_1}$, $P_{A_1B_2}$, $P_{A_2B_1}$, and $P_{A_2B_2}$, and the frequencies of alleles be p_{A_1}, p_{A_2}, p_{B_1}, and p_{B_2}. If the allelic state at locus *A* is completely independent of that at locus *B*, then we expect $P_{A_1B_1} = p_{A_1}p_{B_1}$, $P_{A_1B_2} = p_{A_1}p_{B_2}$, and so on. However, many factors, including natural selection, founder effects, migration, and assortative mating, can cause gamete frequencies to depart from expectations based on allele frequencies. A natural measure of this deviation is the coefficient

$$D_{A_iB_j} = P_{A_iB_j} - p_{A_i}p_{B_j} \tag{5.11}$$

The index $D_{A_iB_j}$ is equivalent to the covariance of the frequencies of "nonalleles" A_i and B_j in the same gamete. To see this relationship, let $x = 1$ if A_i is present and $x = 0$ otherwise, and let $y = 1$ if B_j is present and $y = 0$ otherwise. Then, $xy = 1$ if both A_i and B_j are present, and $xy = 0$ otherwise, so $\sigma_{A_i,B_j} = E(xy) - E(x)E(y) = P_{A_iB_j} - p_{A_i}p_{B_j}$.

Note that $D_{A_iB_j}$ may be positive or negative depending on whether A_i and B_j are in **coupling** (A_iB_j gametes are overrepresented) or **repulsion** (A_iB_j gametes are underrepresented) disequilibrium. D is often referred to as the **coefficient of linkage disequilibrium,** but since disequilibrium can occur between unlinked genes, many authors prefer to call it the **coefficient of gametic phase disequilibrium**. Higher-order terms for three or more loci are defined in Bennett (1954), Geiringer (1944), and Slatkin (1972).

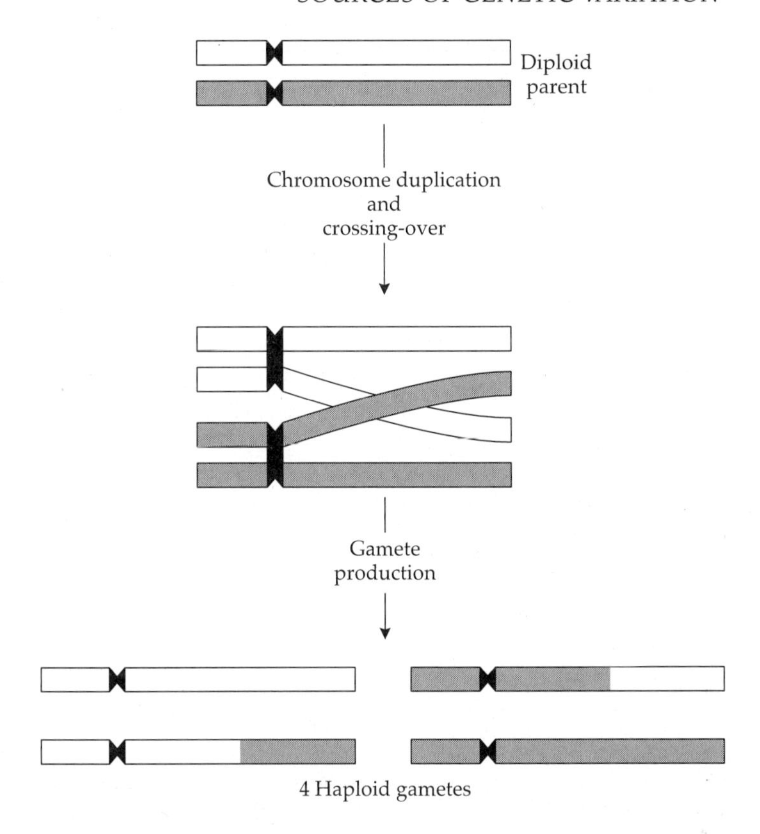

Figure 5.4 A recombination event caused by crossing-over during meiosis. Shaded and open bars represent maternally and paternally derived chromosomes; the darkened area represents the centromere. The two original homologous chromosomes give rise initially to two pairs of sister chromatids, which then recombine and sort independently into four haploid gametes.

There are many ways in which disequilibrium between loci can originate and many ways by which it can be maintained (Felsenstein 1965, 1974; Hill and Robertson 1966, 1968; Nei 1967; Franklin and Lewontin 1970; Barker 1979; Charlesworth and Charlesworth 1979; Feldman et al. 1980). For now, we simply note that, even in the absence of forces that tend to maintain gametic phase disequilibrium (selection, migration, mutation, and drift), once such disequilibrium is present, it may persist for many generations.

The expected dynamics of gametic phase disequilibrium depend on the recombination fraction between loci, c. This quantity takes on a minimum value of zero when two loci are so tightly linked that they effectively segregate as a single **supergene.** Free recombination between loci is denoted by $c = 0.5$. The upper limit is 0.5 rather than 1.0 because crossing-over normally involves one

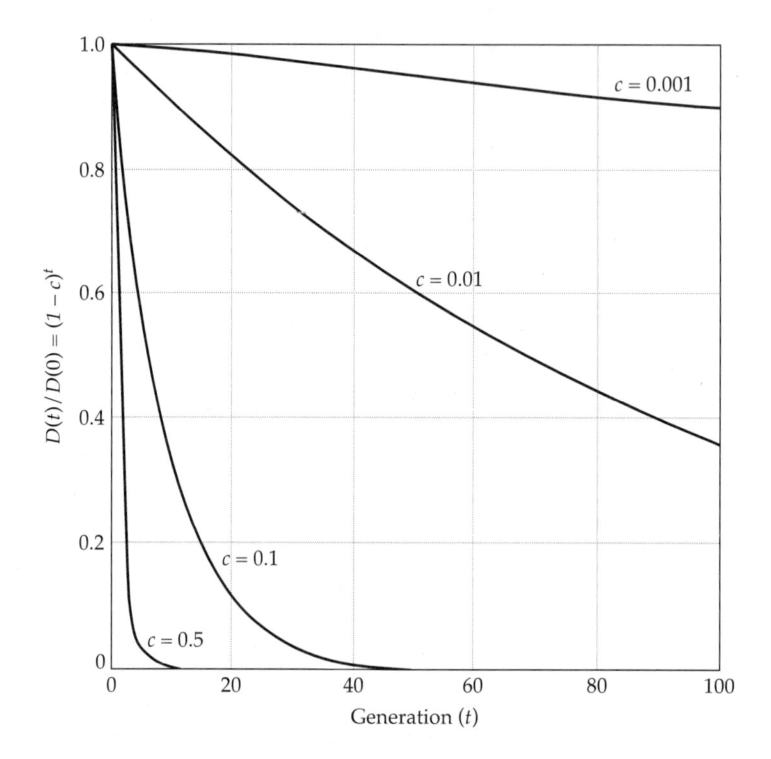

Figure 5.5 The decline, under random mating, of gametic phase disequilibrium relative to its initial value as a function of the recombination frequency, c.

rather than two pairs of nonsister chromatids. In principle, if cross-overs occurred between loci A and B on both pairs of chromatids, the recombination fraction could exceed 0.5, but this has been observed only rarely (Fisher and Mather 1936).

Consider the ideal case in which mating occurs randomly in a population that is effectively infinite in size and all systematic forces causing alleles at different loci to become statistically associated are absent. There are two ways to obtain gametes of a particular type: chromosomes of that type may be transmitted intact between generations, or different chromosomes may recombine to form new gene combinations of the appropriate type. If the frequency of gamete type A_iB_j in generation t is $P_{A_iB_j}(t)$, then $(1-c)P_{A_iB_j}(t)$ is the frequency that is passed on without recombination to the next generation. Among the proportion c of new gametes that are recombinants, $p_{A_i}p_{B_j}$ must contain both A_i and B_j genes, since under random mating, the maternally derived A gene is distributed independently of the paternally derived B gene. Summing terms,

$$P_{A_iB_j}(t+1) = (1-c)P_{A_iB_j}(t) + cp_{A_i}p_{B_j}$$

and after subtracting $p_{A_i}p_{B_j}$ from both sides,

$$D_{A_iB_j}(t+1) = (1-c)D_{A_iB_j}(t) \tag{5.12}$$

This equation generalizes to

$$D_{A_iB_j}(t) = (1-c)^t D_{A_iB_j}(0) \tag{5.13}$$

The main point of this illustration is that the disequilibrium between loci decays gradually. Even in the case of unlinked genes ($c = 0.5$), only 50% of the disequilibrium is removed each generation. With less frequent recombination, the time to attain gametic phase equilibrium ($D = 0$) can be quite long (Figure 5.5).

Estimation of Gametic Phase Disequilibrium

Although genes need not be linked to be in gametic phase disequilibrium, Equation 5.13 indicates that, all other things being equal, disequilibrium is more likely to persist for linked loci. Ecological factors can induce significant gametic phase disequilibrium between unlinked loci. Such an outcome is expected, for example, when a recent hybridization has occurred between two populations or when selection favors a particular multilocus genotype. Taken alone, estimates of gametic phase disequilibrium cannot discriminate among these and other hypotheses, but they do act as guides for further research.

There are two ways to estimate the degree of disequilibrium between loci. Since D is a function of gamete frequencies, an approach that allows a direct count of gamete types is preferable. There are only a few experimental systems, most notably *Drosophila*, in which this is possible. Male *Drosophila* do not exhibit recombination. Thus, when a homozygous laboratory stock is available, progeny from crosses to wild-caught males can yield the gametic states of the previous generation unambiguously. An example of this type of analysis is given below.

Example 3. Males from two wild populations of *Drosophila montana* were crossed with a laboratory stock, homozygous for unique marker alleles at four esterase loci (Baker and Kaeding 1981). The progeny were then scored electrophoretically for the presence of active or null alleles at the four loci. Only the results of esterase-3 and esterase-4, which exhibited the greatest amount of disequilibrium, are presented here. In the following table (A) and (O) denote, respectively, active and null alleles at both loci, so *AO*, for example, represents chromosomes with an active allele at the first locus (esterase-3) and a null at the second locus (esterase-4). The entries in the table represent the numbers of observations of each chromosome type, and N denotes the total sample size for each study.

The estimation procedure is from Hill and Robertson (1968) and Hill (1974). The formulae on the right in the table provide estimates for the active-allele frequencies at the esterase-3 and esterase-4 loci, and for the gametic phase disequilibrium between these loci. The estimates, obtained by substituting the observed incidences of the four chromosomal types, are given at the bottom of the table.

Chromosome	Utah	Colorado	
AA	4	71	
AO	50	737	$\widehat{p}_{A.} = (N_{AA} + N_{AO})/N$
OA	31	457	
OO	15	167	$\widehat{p}_{.A} = (N_{AA} + N_{OA})/N$
N	100	1432	
$\widehat{p}_{A.}$	0.540	0.564	
$\widehat{p}_{.A}$	0.350	0.368	$\widehat{D}_{AA} = P_{AA} - \widehat{p}_{A.}\widehat{p}_{.A}$
$\widehat{D}_{AA}$	−0.149	−0.158	

The sampling variance of $\widehat{D}$ is given by

$$\text{Var}(\widehat{D}) = \frac{\widehat{p}_{A.}(1-\widehat{p}_{A.})\widehat{p}_{.A}(1-\widehat{p}_{.A}) + (1-2\widehat{p}_{A.})(1-2\widehat{p}_{.A})\widehat{D} - \widehat{D}^2}{N}$$

(Weir 1996), and the standard error of $\widehat{D}$ is simply the square root of Var$(\widehat{D})$. For the two populations, SE$(\widehat{D}) = 0.028$ and 0.005. Thus, both populations exhibit significant excess frequencies of repulsion gametes (an active allele at one locus and a null at the other) and have essentially the same level of disequilibrium. Based on the estimated allele frequencies, the number of *AO* and *OA* chromosomes expected in the Utah sample under gametic phase equilibrium is 100 × [(0.54 × 0.65) + (0.46 × 0.35)] $\simeq$ 51, whereas 81 were observed. For the Colorado population, the equilibrium expectation (740) is also well below the observed value (1194). The existence of disequilibrium between these two loci is not too surprising, since these esterase loci are tightly linked in *Drosophila.*

For most natural populations, the luxury of an inbred marker line is not an option. Usually, the only available information is the frequency distribution of multilocus genotypes in the study population, which raises some conceptual difficulties. While the gametic composition of most zygotes can be resolved from the genotype (e.g., an *AaBB* individual must have arisen from *AB* and *aB* gametes), double heterozygotes, which can arise from the fusion of *AB* and *ab* or *Ab* and *aB* gametes, cannot be resolved definitively. If, however, one is justified in assuming random mating, it is not necessary to discriminate between coupling and repulsion heterozygotes. In this case, an unbiased estimator of D is given by

$$\widehat{D}_{AB} = \frac{N}{N-1}\left[\frac{4N_{AABB} + 2(N_{AABb} + N_{AaBB}) + N_{AaBb}}{2N} - 2\widehat{p}_A\widehat{p}_B\right] \quad (5.14)$$

where N is the total sample size, the terms in the numerator are observed numbers of the four genotypes, and $\widehat{p}_A$ and $\widehat{p}_B$ are gene frequency estimates. An expression for the sampling variance of $\widehat{D}$, derived by Weir and Cockerham (1979), is

$$\text{Var}(\widehat{D}) = \frac{\widehat{p}_A \widehat{p}_a \widehat{p}_B \widehat{p}_b}{N-1} + \frac{(2\widehat{p}_A - 1)(2\widehat{p}_B - 1)\widehat{D}}{2N} + \frac{\widehat{D}^2}{N(N-1)} \tag{5.15}$$

For loci in gametic phase equilibrium, this reduces to

$$\text{Var}(\widehat{D}) = \frac{\widehat{p}_A \widehat{p}_a \widehat{p}_B \widehat{p}_b}{N-1}$$

Example 4. Genotype frequency data for the M/N and S/s blood group loci for a sample of 1000 Caucasians in England are given in the following table (from Cleghorn 1960). These two loci are known to be very tightly linked; only a few well-established cases of recombination have appeared in thousands of analyses.

	MM	*MN*	*NN*
SS	57	39	3
Ss	140	224	54
ss	101	226	156

Using Equation 4.1, the estimated frequencies of the M and S alleles are

$$\widehat{p}_M = \frac{(57 + 140 + 101) + 0.5(39 + 224 + 226)}{1000} = 0.5425$$

$$\widehat{p}_S = \frac{(57 + 39 + 3) + 0.5(140 + 224 + 54)}{1000} = 0.3080$$

Substitution into Equations 5.14 and 5.15 gives the estimated disequilibrium and its sampling variance,

$$\begin{aligned}\widehat{D}_{MS} &= \frac{N}{N-1}\left[\frac{4N_{MMSS} + 2(N_{MMSs} + N_{MNSS}) + N_{MNSs}}{2N} - 2\widehat{p}_M\widehat{p}_S\right] \\ &= 0.071\end{aligned}$$

$$\text{Var}(\widehat{D}) = \frac{\widehat{p}_M \widehat{p}_N \widehat{p}_S \widehat{p}_s}{N-1} + \frac{(2\widehat{p}_M - 1)(2\widehat{p}_S - 1)\widehat{D}}{2N} + \frac{\widehat{D}^2}{N(N-1)} = 0.0000518$$

The square root of the sampling variance is the standard error of the estimate,

$$\text{SE}(\widehat{D}_{MS}) = 0.0072$$

Since the estimate $\widehat{D}_{MS}$ is approximately ten standard errors greater than zero, there appears to be significant disequilibrium between these two loci, with MS and Ns gametes being overrepresented relative to Ms and NS gametes.

Additional details regarding tests of hypotheses about gametic phase disequilibrium may be found in Brown (1975), Cockerham and Weir (1977a), Weir (1979), Weir and Cockerham (1979), Hedrick (1987a), Lewontin (1988), and Weir (1996). Brown (1975) has given considerable attention to the adequacy of sample sizes for detecting disequilibrium, providing useful tables. Unless gene frequencies are intermediate (0.4 to 0.6) at both loci and disequilibrium is strong ($|D| > 0.4$), several hundred to thousands of individuals should be assayed to achieve a reasonable level of statistical power. Estimation procedures for cases in which one or both loci exhibit complete dominance are developed in Hill (1974), and Hill (1975) considers the situation when more than two alleles are present at a locus.

EFFECT OF DISEQUILIBRIUM ON THE GENETIC VARIANCE

Studies using biochemical markers to detect gametic phase disequilibrium have been performed on numerous species, particularly those in the genus *Drosophila*. A predominance of statistically nonsignificant results from allozyme surveys has led a number of authors to conclude that gametic phase disequilibrium is the exception rather than the rule in large, random-mating populations (Langley 1977, Langley et al. 1978, Hedrick et al. 1978). However, the analyses of Brown (1975) suggest that the failure to detect disequilibria in most allozyme studies may be a simple consequence of the exceedingly low statistical power resulting from small sample sizes. This has been convincingly demonstrated in the case of *Drosophila* by Zapata and Alvarez (1992), who showed that, despite the lack of significance in most individual studies, when multiple studies are evaluated together, a significant negative relationship emerges between the degree of disequilibrium and the rate of recombination between pairs of loci. Such a pattern is consistent with the idea that events that cause gametic phase disequilibrium are more likely to have persistent effects with pairs of closely linked than weakly linked or unlinked loci. Fine resolution studies provided by RFLP (restriction fragment length polymorphism) and DNA-sequence analysis are now revealing extensive disequilibrium in many outcrossing species (Mitchell-Olds and Rutledge 1986, Zapata and Alvarez 1993).

Because we still have little knowledge of the loci underlying most quantitative characters, no further explicit statements regarding gametic phase disequilibrium between such loci can be made. However, theoretical arguments suggest

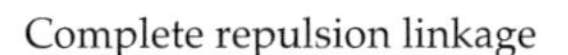

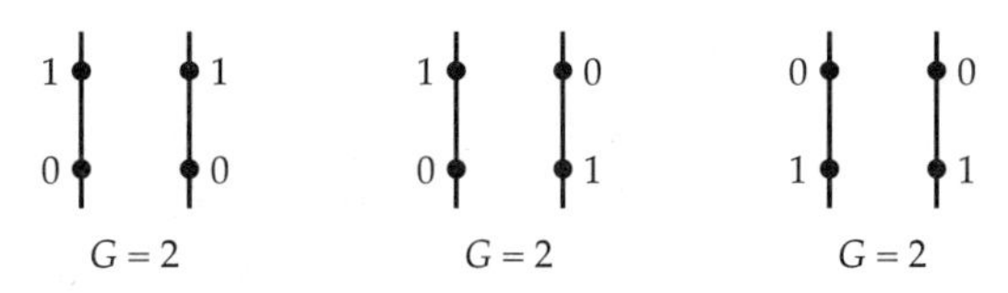

Complete coupling linkage

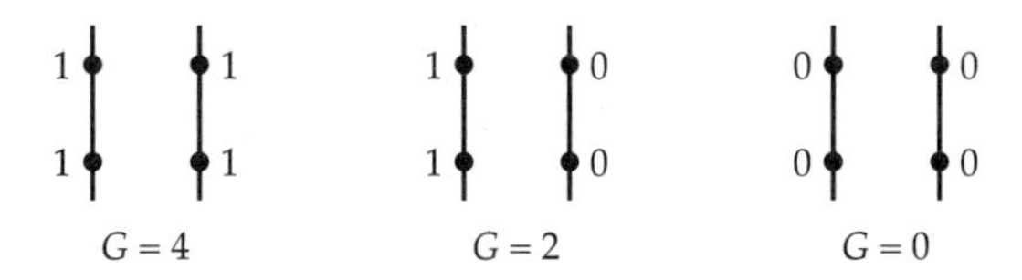

Figure 5.6 Possible two-locus genotypes for extreme cases of complete coupling and repulsion disequilibrium. The straight lines represent homologous chromosomes. The numbers are the additive effects of alleles. The sum of the four effects is the genotypic value.

that the aggregate effects of gametic phase disequilibrium might be extensive for quantitative traits whose expression is based on large numbers of loci, even if the average level of disequilibrium between pairs of loci is relatively small. If genes with a positive influence on a character tend to be associated on some chromosomes, and those with a negative influence on others (coupling disequilibrium), the observed genetic variation will be inflated relative to the expectation under random assortment. The opposite will occur if "plus" alleles at one locus tend to be associated with "minus" alleles at another (repulsion disequilibrium).

This point is made clear with the extreme example in Figure 5.6, where we assume purely additive allelic effects. In the first case, there are only two types of chromosomes — those with genes with additive effect 1 at locus 1 and effect 0 at locus 2, and vice versa. Although there are three types of two-locus genotypes, they all have the same genotypic value (in all cases, the sum of the additive effects is equal to two). Hence, there is no **expressed genetic variance**. The opposite situation is shown in the second example, where chromosomes have effects of 1 at both loci or effects of 0 at both loci. Assuming random mating and equal frequencies of chromosome types, genotypic values should be in the proportions: 25% 0, 50% 2, and 25% 4. The expressed genetic variance is therefore equal to 2. These extreme situations may be compared to the case in which the two loci are in complete gametic phase equilibrium, so that for each of its four genes, an individual has an equal probability of being a 1 or a 0, regardless of the other genic states. From Equation 4.12a, under random mating and with additive gene action, the variance at each locus is $2pqa^2$, so with $a = 1$ and $p = q = 0.5$ for both loci, the total genetic variance is equal to one in the absence of disequilibrium.

When dominance effects are present at both loci, gametic phase disequilibrium can have more complicated effects (Comstock and Robinson 1952; Robinson and Comstock 1955; Avery and Hill 1977, 1979; Weir et al. 1980). As in the purely additive case, the additive genetic variance can be either inflated or depressed. The same will be true for the dominance genetic variance, depending on the direction of dominance at pairs of loci. This point can be formalized by considering the multilocus analogs of Equations 4.12a,b. For n diallelic loci,

$$\sigma_A^2 = 2\sum_{i=1}^{n}\alpha(i)^2 p_i q_i + 2\sum_{i=1}^{n}\sum_{j\neq i}^{n}\alpha(i)\alpha(j)D_{ij} \tag{5.16a}$$

$$\sigma_D^2 = 4\sum_{i=1}^{n}(a_i k_i p_i q_i)^2 + 4\sum_{i=1}^{n}\sum_{j\neq i}^{n} a_i a_j k_i k_j D_{ij}^2 \tag{5.16b}$$

where $\alpha(i)$ is the average effect of allelic substitution at the ith locus (defined in Equation 4.10b) (Avery and Hill 1977). The equation for the additive genetic variance is obtained by recalling (Equation 3.11b) that the variance of a sum is equal to the sum of the variance of all the elements plus twice the sum of the covariances between all pairs of elements. In this case, there are $2n$ variance terms to consider, each with gene-frequency variance $p_i q_i$, and $2n(n-1)$ covariance terms, each with gene-frequency covariance D_{ij}. [There are $2n(n-1)$ rather than $2n(2n-1)$ covariance terms because $2n^2$ pairs of genes involve contributions from different gametes, whose effects must be uncorrelated under random mating.] Likewise, for σ_D^2, the terms in the double summation are covariances between dominance deviations at loci i and j. Such covariances require associations between both alleles at locus i and both alleles at locus j, and hence the D_{ij}^2 term appears.

A little more comfort with these formulations may be acquired by recalling Figure 5.6. In this additive case, $\alpha(1) = \alpha(2) = 1$, $p = q = 0.5$ for both loci, and each single-locus term in Equation 5.16a is equal to $2 \times 1 \times 0.5 \times 0.5 = 0.5$. In the case of repulsion linkage, there are no 1/1 gametes, so $D = 0-(0.5)(0.5) = -0.25$, and the two-locus summation term is $2\times 2\times 1\times 1\times(-0.25) = -1$. Summing up, we find that the observed additive genetic variance is equal to zero, in agreement with the verbal reasoning given above. The previous result for coupling linkage can also be recovered by noting that the contributions from single-locus terms are the same as in the previous example, 0.5 for each locus, whereas the disequilibrium is now $D = 0.5 - (0.5)(0.5) = 0.25$.

When epistatic interactions exist between loci in gametic phase disequilibrium, the situation becomes extremely complicated. Even when only pairs of loci are considered, the final expression for σ_G^2 requires nearly a printed page (Gallais 1974, Weir and Cockerham 1977), and there is little to be gained by reprinting it here. The main point is that, as in the case of the additive and dominance components of genetic variance, observed levels of epistatic genetic variance can be increased or decreased when gametic phase disequilibrium is important.

In summary, the components of expressed genetic variance for quantitative traits can be partitioned into expected values under gametic phase equilibrium (the first summations on the right side of Equations 5.16a,b) and deviations from these expectations caused by disequilibrium (the double summations). When the **disequilibrium covariance** is negative, we refer to it as **hidden genetic variance** because it is subject to conversion to expressed genetic variance via the breakdown of gametic phase disequilibria. Indeed, a conceptually simple way to test for the presence of disequilibrium covariance is to expand and randomly mate a population while minimizing selection over several consecutive generations. As gametic phase equilibrium is established, the phenotypic variance should converge on the expected equilibrium variance, with the deviation between initial and final levels of expressed genetic variance providing an estimate of the disequilibrium covariance in the base population. To our knowledge, such experiments have only been done with hybrid populations of maize (Gardner 1963).

The Evidence

Hidden genetic variation is expected to be a natural consequence of stabilizing selection, which favors linkage groups for their composite properties without regard to the alleles at individual loci. Returning to Figure 5.6, for example, the three genotypes under complete repulsion disequilibrium are selectively equivalent despite their different gene contents. A most extreme view of this idea was first invoked by Mather (1942, 1943) in his **polygenic balance** model. Mather suggested that the genes for characters under stabilizing selection would lie along chromosomes in such a way that the signs of their contribution to the phenotype would alter. He also suggested that stabilizing selection would lead to the evolution of pairs of chromosomes with complementary features, such that the predominant chromosomal types in a population would be of the forms $+ - + - + - \cdots$ and $- + - + - + \cdots$. Although this simple model is an extreme metaphor, subsequent theoretical work (Lewontin 1964; Bulmer 1971, 1974; Chevalet 1988; Gavrilets and Hastings 1993, 1994a, 1995) has upheld the idea that stabilizing selection encourages the development of substantial hidden genetic variance, potentially depressing the level of expressed genetic variance to 50% or less than its equilibrium expectation.

The data are still scant, but several lines of evidence suggest that levels of hidden genetic variance in natural populations may be substantial. First, in *Drosophila*, it is possible to isolate intact chromosomes from natural populations by mating wild males with females from a laboratory strain with a homologous chromosome carrying a dominant visible marker and cross-over suppressors (Figure 5.7). The wild chromosome can then be expanded into several full-sibs, each of which carries a marker chromosome. The progeny of crosses within full-sib families provide an assay of viability associated with the homozygous wild chromosome. Since the marker chromosome carries a lethal recessive, a 1:2 ratio of wild-type to marker progeny is expected if the homozygous wild-type is equivalent in viability

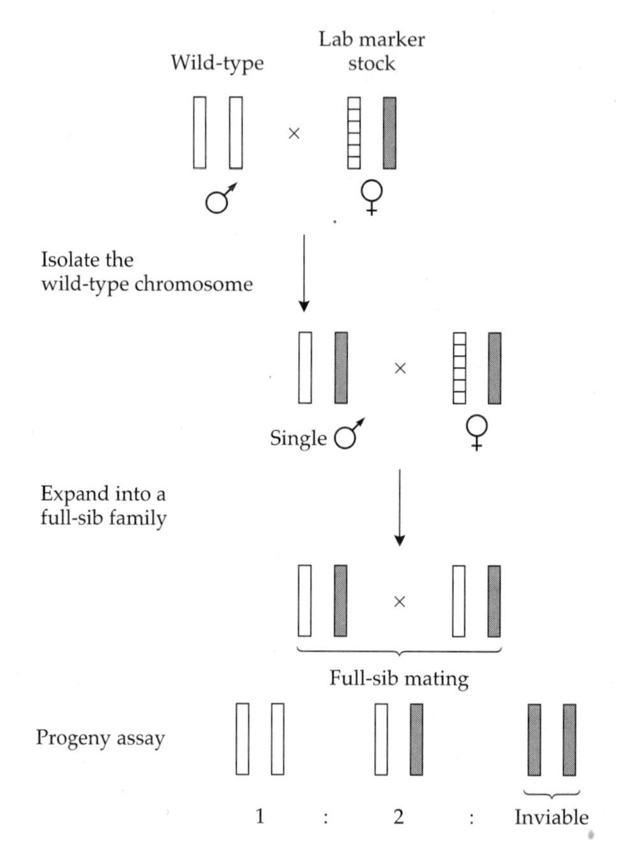

Figure 5.7 The procedure for isolating a drosophilid chromosome and assaying it for viability in the homozygous state. The chromosome is passed on intact between generations because *Drosophila* males do not exhibit recombination and because the marker chromosome suppresses crossing-over in females. If the chromosome has normal viability, a 1:2 ratio of wild to marker flies is expected in the progeny generation because the marker chromosome carries recessive lethals.

to the wild / marker heterozygote. Dobzhansky and his colleagues performed this sort of analysis with five species of *Drosophila*, both with wild intact chromosomes and with recombinant wild derivatives. (Recombinant chromosomes are obtained by extracting individuals from the wild and crossing them for one or more generations prior to subjecting males to the crossing scheme in Figure 5.7.) On average, the recombinant chromosomes were associated with lower viabilities than were the intact chromosomes (Table 5.1). This implies that the parental chromosomes extracted from the wild tend to carry (in disequilibrium) groups of genes with favorable epistatic effects on viability. Recombination tends to break up these favorable combinations.

Table 5.1 Viability properties of homozygous intact chromosomes relative to homozygous recombinant chromosomes for five species of *Drosophila*.

Species	Chromosome	% Viability Reduction	Reference
D. melanogaster	II	−10	Spiess & Allen 1961
	III	8	
D. persimilis			
South Fork	II	13	Spiess 1959
White Wolf	II	10	
D. prosaltans			
Pirassununga	II	8	Dobzhansky et al. 1959
Rio	II	15	
D. pseudoobscura			
Texas	II	13	Spassky et al. 1958
California	II	26	
D. willistoni	II	3	Krimbas 1961

Note: Only parental chromosomes with near normal viability scores were used.

In clonally reproducing organisms, the entire genome is inherited intact, acting as a single linkage group, and natural selection can result in the build-up of extremely high levels of hidden genetic variance (Lynch and Gabriel 1983). This has been demonstrated in populations of *Daphnia*, which periodically punctuate long phases of asexuality with a bout of sexual reproduction (Lynch 1984, Deng and Lynch 1996a). Such populations often exhibit annual cycles of expressed genetic variance for life-history traits (Figure 5.8). When repulsion disequilibrium is favored, clonal selection leads to the rapid erosion of expressed genetic variance and build-up of hidden genetic variance. Following the next phase of sexual reproduction, a fraction of the hidden genetic variance is then reconverted to expressed genetic variance. Qualitatively similar results have been observed when fungal isolates selected for the same phenotype have been crossed (Merrick 1975).

Selection need not always favor the development of hidden genetic variance. As noted above, if coupling disequilibria are favored, selection will cause the expressed genetic variance to exceed its equilibrium expectations. In this case, the disequilibrium covariance is positive, and recombination would be expected to result in a *reduction* of the expressed genetic variance. Such changes have been observed in a natural population of *Daphnia,* which exhibited an approximately 50% reduction in the levels of expressed genetic variance for life-history traits following a phase of sexual reproduction (Lynch and Deng 1994).

The previous arguments have considered how selection can cause the accumulation of gametic phase disequilibria for genes underlying a *single* quantitative trait. The same arguments apply to various aspects of multivariate phenotypes. For example, in populations of insects that exploit multiple host plants, one

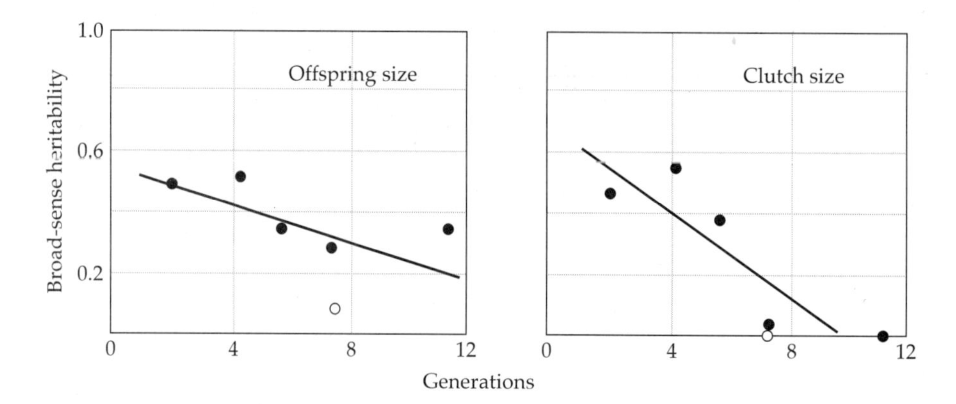

Figure 5.8 Temporal changes in the broad-sense heritability (the fraction of the total phenotypic variance that has a genetic basis) for two characters in a population of *Daphnia*. Open and solid circles refer to data obtained in consecutive years. Sexual reproduction occurs each year at generation 0, with subsequent generations being asexual. Notice that, for both traits, the expressed genetic variance is nearly zero at the end of the first year (open circle), recovers to high levels at the beginning of the next year (solid circles), and then declines again towards zero. (From Lynch 1984.)

might expect a genetic correlation to evolve such that individuals prefer to feed (or lay eggs) on the plant species upon which they perform best (Rausher 1983, Thompson 1988). Such correlations could result from gametic phase disequilibria between a set of genes influencing preference and another set influencing performance. (An alternative mechanism is that the genes that endow high performance on a particular host plant also encode for behavior that causes preference for that host, but because of the functional differences between these two types of traits, this mechanism seems unlikely). A number of studies have revealed positive genetic correlations between preference and performance, while others have not (see Chapters 6 and 21). In populations of garter snakes, a genetic correlation exists such that striped individuals tend to flee directly from predators while spotted individuals tend to use crypsis as a defense (Brodie 1989). Because this correlation involves behavioral and morphological traits, it again seems likely that the genetic correlation is a result of gametic phase disequilibrium — a consequence of natural selection favoring individuals that exhibit the right combinations of morphology and behavior.

6

Sources of Environmental Variation

Because all metabolic and developmental pathways are influenced to some degree by aspects of the environment, it stands to reason that the expression of most quantitative traits is not completely under genetic control. In some cases, phenotypic responses to changes in the environment are dramatic. For example, in some populations of the tiger salamander (*Ambystoma tigrinum nebulosum*), larvae develop into atypical cannibalistic morphs when raised at high densities (Collins and Cheek 1983). Clones of many planktonic cladocerans and rotifers can be induced to change their external morphology by means of appropriate physical or chemical stimuli (Hutchinson 1967, Havel 1987), and environmentally controlled flight polymorphisms are well documented in locusts (Kennedy 1956) and corixid bugs (Young 1965). However, more often than not, environmental effects are subtle, causing simple amplifications or reductions in sizes of parts, numbers of progeny, rates of growth, physiological performance, and so on.

As with the genetic variance, sources of environmental variance can be partitioned in various ways. It is useful to define two broad categories. **General environmental effects** refer to influential factors that are shared by groups of individuals, the size of the group depending on the context. The effects of an experimental treatment or of a patch of habitat are familiar examples. In addition, through maternal care, mothers will have general effects on their offspring (beyond the direct transmission of genes), usually referred to as maternal effects. **Special environmental effects** are residual deviations from the phenotype expected on the basis of genotype and general environmental effects. Such effects are unique to individuals — a consequence of microenvironmental variation and random developmental noise. There is one additional complication with respect to environmental effects. Ideally, the phenotype of an individual can be viewed simply as the sum of its genotypic value and the environmental effects with which it is associated, i.e., $z = G + E$. However, in some cases, different genotypes respond to environmental change in nonparallel ways, a phenomenon known as **genotype $\times$ environment interaction.**

The purposes of this chapter are twofold. First, we illustrate how various classes of environmental effects can be incorporated into the logical framework introduced in the previous chapter for genotypic values. This extension leads to a general linear model for the phenotype, which forms the foundation upon which much of the remainder of the book is built. Second, as background for the remaining chapters, we provide an overview of the biological basis for the various types of environmental effects. Further details will emerge in later chapters. For example, Chapter 22 is devoted entirely to issues related to genotype $\times$ environment interaction, and Chapter 23 focuses on maternal effects.

EXTENSION OF THE LINEAR MODEL TO PHENOTYPES

Here we let E and e denote the contributions of general and specific environmental effects to the phenotypic expression of a character, and let I denote the genotype $\times$ environment interaction effect. The phenotype of the kth individual of the ith genotype exposed to the jth general environmental effect can then be described as a linear function of four components,

$$z_{ijk} = G_i + I_{ij} + E_j + e_{ijk} \tag{6.1}$$

Each of these components may have subcomponents. For example, from Equation 5.7, the genotypic value, G_i, is a potentially complicated linear function of the population mean phenotype, the individual's breeding value, and various effects due to dominance and epistasis. As in the case of all of the genetic effects, the terms I_{ij}, E_j, and e_{ijk} are defined in a least-squares sense as deviations from lower-order expectations and, as a consequence, have mean values equal to zero. The population mean phenotype, $\mu_G = \overline{z}_{ijk}$, is the mean phenotype of all genotypes in the population, whereas G_i is the expected phenotype of the particular genotype i averaged over all possible environmental conditions within the experimental setting. The quantity $\mu_G + E_j$ is the mean phenotypic value expected if the entire population of genotypes were assayed in the jth macroenvironmental setting, whereas $G_i + I_{ij} + E_j$ is the expected phenotype of genotype i in that setting. Thus, I_{ij} is the residual deviation left after assuming that genotypic and environmental values act in an additive fashion. Finally, e_{ijk} is the deviation of an individual's phenotype from the expectation $G_i + I_{ij} + E_j$. As in any least-squares linear model (Chapter 3), the residual deviations e_{ijk} are defined to be uncorrelated with the explanatory variables G, I, and E.

Example 1. As an example of computing genotypic and environmental values, we consider a small data set from Strauss and Karban (1994) on three lines of thrips (*Apterothrix apteris*) grown on three clones of their host plant, the seaside daisy (*Erigeron glaucus*). The following table gives the mean performance of thrips, measured as population density after eight generations of growth, in the nine genotype-environment combinations (the thrips lines being denoted as I, II, and III, and the three environments (plant clones) as 1, 2, and 3). All nine treatments were replicated; the standard errors of the measures are small and are ignored in the following.

	Thrips Line			
Plant	I	II	III	E
1	77	34	47	$-19.22 = E_1$
2	61	159	51	$18.44 = E_2$
3	40	71	107	$0.78 = E_3$
G	59.33	88.00	68.33	

Assuming an equal weight for each cell, and averaging over all nine cells, the mean character value is found to be $\mu_G = 71.89$. The genotypic values of the thrips lines, obtained by averaging the elements within columns, are given in the bottom row of the table. The average environmental effects, each defined as the mean performance of all thrips lines in a specific environment (the average within rows) minus the grand mean (μ_G), are given in the final column. For example, $E_1 = [(77 + 34 + 47)/3] - 71.89$. Note that the three values of E average to zero.

To obtain the interaction effects, we rearrange Equation 6.1 to

$$I_{ij} = \bar{z}_{ij} - G_i - E_j$$

where i and j denote the thrips line and the daisy clone, and $\bar{z}_{ij}$ is the entry in the ith column and jth row of the table. The residual deviation drops out because we have assumed the data to have been obtained without error. Substituting into this equation,

$$\begin{array}{lll} I_{I,1} = 36.89 & I_{II,1} = -34.78 & I_{III,1} = -2.11 \\ I_{I,2} = -16.78 & I_{II,2} = 52.56 & I_{III,2} = -35.78 \\ I_{I,3} = -20.11 & I_{II,3} = -17.78 & I_{III,3} = 37.89 \end{array}$$

The interaction effects average to zero both within rows and within columns. Note also that the magnitudes of the interaction effects tend to be much greater than the magnitudes of the environmental effects, indicating strong genotype $\times$ environment interaction.

This study also provides strong evidence of genotype-environment covariance. In nature, thrips line I was found living on plant clone 1, line II on clone 2, and line III on clone 3. Thus, the individual clones were associated with the plants on which they best performed.

Since I and e are uncorrelated with the other variables (by construction), using the formula for the variance of a sum (Equation 3.11b), the total phenotypic variance of a population can be written as

$$\sigma_P^2 = \sigma_G^2 + \sigma_I^2 + 2\sigma_{G,E} + \sigma_E^2 + \sigma_e^2 \tag{6.2}$$

The expansion of the total genetic variance, σ_G^2, in terms of additive, dominance, and epistatic components was given in the previous chapter. σ_E^2 and σ_e^2 are the components of variance due to general and special environmental effects.

The greatest conceptual difficulty with Equation 6.2 is related to the interpretation of σ_I^2 and $\sigma_{G,E}$. The term $\sigma_{G,E}$ refers to **genotype-environment covariance**, which is quite distinct from genotype $\times$ environment interaction. Genotype $\times$ environment interaction is concerned with the variation in the phenotypic response of specific genotypes to specific environments, and σ_I^2 is a measure of that variation. Genotype-environment covariance is a measure of the physical association of particular genotypes with particular environmental general effects. If individuals are randomly distributed with respect to macroenvironments, then $\sigma_{G,E}$ is zero, but σ_I^2 will be nonzero if genotypic values and environmental effects are nonadditive. If, on the other hand, genotypes are nonrandomly distributed, $\sigma_{G,E} \neq 0$.

Even in the most carefully controlled situations, one cannot always rule out the presence of genotype-environment covariance. Nonrandom associations of genotypes and environment can result from limited seed or pollen dispersal in plants and from genetically based dominance hierarchies and other social interactions in animals. Maternal (or paternal) effects can also cause genotype-environment covariance if there is a correlation between parental genotype and ability to provision the young (Chapter 23). This latter source of genotype-environment covariance can occur in an environment that is otherwise completely homogeneous.

Whereas methods exist for the detection of genotype $\times$ environment interaction (Chapter 22), genotype-environment covariance is usually less tangible, contributing an unknown amount to estimates of genetic variance. Take, for example, the simple situation in which genotype $\times$ environment interaction is absent and an estimate of the environmental variance is obtained as the phenotypic variance within pairs of monozygotic twins (or among members of a clone), all of which must be genetically identical. Then, from Equation 6.2, it can be seen that the difference between the phenotypic variance and the environmental variance, the among-clone variance, is $\sigma_G^2 + 2\sigma_{G,E}$. Depending on whether $\sigma_{G,E}$ is positive or negative, the true genetic variance will be over- or underestimated.

Example 2. The results of an experiment with a clonal plant will help clarify the preceding concepts. The salt marsh cord grass (*Spartina patens*) occurs along much of the Atlantic coast on sand dunes, swale grasslands, and marshes. Silander

(1985) removed plants from these three environments and clonally propagated them via rhizomes in the greenhouse. After two years, sufficient material was available to perform a reciprocal transplant experiment in the field. Replicate progeny from each clone were grown at each of three sites — dune, swale, and marsh. A large number of vegetative and reproductive traits were measured, and the data were analyzed by two-way ANOVA (Chapter 20) with clone and site serving as the main factors.

Trait	Var(G)	Var(E)	Var(I)	Var(e)
Tillers / clone	0.2	34.2	19.9	45.8
Culm height	12.0	56.6	7.0	24.5
Leaves / culm	11.2	19.4	8.4	61.0
Culm diameter	20.6	4.7	0.0	74.7
Longest leaf length	26.7	29.5	6.1	37.7
Longest leaf width	27.5	7.8	3.6	61.3
Third leaf length	25.2	0.0	10.2	64.6
Third leaf width	23.4	3.4	3.6	69.7

The observed components of variance, given in the above table as percentages of the total phenotypic variance, can be interpreted as follows: the among-clone variance is an estimate of the total genetic variance σ_G^2; the variance among the three sites is an estimate of the general environmental effects variance σ_E^2; the variance within sites (more specifically, among members of the same clone within sites) is an estimate of σ_e^2; and the clone $\times$ site variance is an estimate of σ_I^2. Conceivably, some of the observed phenotypic variance may have been caused by maternal effects, but since all members of a clone had the same mother, any variance caused by such effects is compounded with the estimate of σ_G^2. In this experiment, $\sigma_{G,E}$ can be assumed to be zero because individual plants were distributed randomly within treatments.

When averages are taken over all of the traits in the study (including those not in the table), the vast majority (60%) of the phenotypic variance is found to be attributable to special environmental effects (σ_e^2). Genotype accounts for an additional 19% of the variance, whereas general environmental effects and genotype $\times$ environment interactions account for 6 and 5%, respectively. These results indicate that *Spartina* growth characters are relatively insensitive to what appear to be major changes in habitat.

SPECIAL ENVIRONMENTAL EFFECTS

As noted above, two sources contribute to the special environmental effects variance — internal developmental noise and external microenvironmental heterogeneity. Both sources of variation are generally unpredictable from the standpoint of the individual. Both are unique properties of the population under investigation and can be modified by changing the environmental setting.

Within-Individual Variation

In animals it is possible to gain some information about variance associated with special environmental effects by measuring the same attribute on the right and left sides of bilaterally symmetrical organisms. Differences may arise between such measures because of measurement error, but there is usually a small, but real, random component of asymmetry. This within-individual variation is the finest level at which phenotypic variance can be quantified, and it can sometimes constitute a substantial portion of the total variance of a trait. Leamy (1984, 1992), for example, found that differences between right and left measures account for up to 18% of the phenotypic variance for bone lengths in mice, and similarly high values have been noted for gill-raker and fin measures in rainbow trout (Leary et al. 1992), and for cranial measures in tamarins (Hutchison and Cheverud 1995).

Following the early suggestions of Mather (1953), Thoday (1953), and Van Valen (1962), numerous investigators have interpreted the variance of right-left measures to be a measure of developmental noise (or developmental instability). It is difficult to define the mechanistic basis of such variation, although we discuss some correlates below. Soulé (1982) suggested that it may be a simple consequence of random movement of molecules within developing individuals, and Emlen et al. (1993) have considered how variation at the cellular level might translate into differences at the levels of tissues or organs through cellular feedback mechanisms. However, it may be presumptuous to assume that variance in right-left measures is entirely due to internal factors. In no case has the involvement of subtle variation in the external environment from right to left sides been ruled out, although it is known that the extreme asymmetry that develops in lobster claws is a consequence of differential claw use (Govind and Pearce 1986). Thus, we prefer to simply denote the right-left variance as within-individual variance σ^2_{ew}, without invoking causality. The total variance resulting from special environmental effects can then be written as the sum of the among-individual and within-individual components,

$$\sigma^2_e = \sigma^2_{ew} + \sigma^2_{ea} \tag{6.3}$$

Van Valen (1962) pointed out the need to distinguish three types of asymmetry. **Directional asymmetry** refers to a consistent bias in one direction such as the tendency of the mammalian heart to be on the left side or of a particular coiling direction in the shells of snails. For a continuously distributed character, directional asymmetry is detectable as a mean difference between right and left measures that is significantly different from zero. **Antisymmetry** refers to situations in which asymmetry is the rule rather than the exception but is nondirectional. An example is provided by male fiddler crabs, which have equal probabilities of developing right or left signaling claws. **Fluctuating asymmetry** refers to the common situation in which the difference between the right and left measures is symmetrically (usually normally) distributed around a mean and mode of zero. Studies on developmental homeostasis focus upon the fluctuating form of asymmetry, and we devote the remainder of our attention to it.

Under the assumption that right and left measures are two of many possible random expressions of an individual's developmental program, an unbiased estimate of the within-individual variance for a trait is given by

$$\text{Var}(e_w) = \sum_{i=1}^{N} \frac{(r_i - l_i)^2}{2N} - \text{Var}(e_m) \tag{6.4}$$

where N is the number of individuals sampled, r_i and l_i are the right and left measures for the ith individual, and $\text{Var}(e_m)$ is the variance due to measurement error. The latter quantity is simply the variance among repeated measures of the same trait (on the same side) of the same individual; ideally, it should be estimated in a way that ensures that the investigator has no memory of previous measures. Palmer and Strobeck (1986) recommend making multiple measures on both sides of every individual, treating right and left measures as fixed effects, and performing a two-way analysis of variance. This approach allows one to test for directional asymmetry, as well as to extract $\text{Var}(e_w)$ and $\text{Var}(e_m)$ directly from the ANOVA mean squares (Chapter 20).

Provided an estimate of the total special environmental effects variance, $\text{Var}(e)$, is available, the among-individual component can be estimated as $\text{Var}(e_a) = \text{Var}(e) - \text{Var}(e_w)$. To our knowledge, this has not been done with any organism, but in principle it is straightforward for organisms that can be grown clonally, as the total variance within clones is simply $\text{Var}(e)$.

Interest in the within-individual component of variance derives from the idea that relatively asymmetrical individuals are victims of genetic and/or environmental circumstances that enhance the chances of random developmental errors on the two sides of the body. If this were true, $\text{Var}(e_w)$ would provide a useful measure of environmental/genetic stress. In an effort to test this idea, many investigators have searched for correlates of fluctuating asymmetry (hereafter FA), focusing particularly on the influence of the genetic background. Under the assumption that symmetry is adaptive, and therefore maximized in natural populations, it follows that perturbations to locally adapted gene pools should inflate the level of FA. Such perturbations can be induced artificially in two ways — the average homozygosity can be increased by mating individuals with their close relatives (inbreeding), or heterozygosity can be enhanced by hybridizing different populations or species (outcrossing).

A summary of some existing results indicates that enhanced FA in genetically perturbed individuals is by no means universal (Table 6.1). At least a third of

Table 6.1 Summary of studies on the relationship of fluctuating asymmetry to levels of inbreeding and crossbreeding.

Organism	Basis of Comparison	Character(s)	Reference
FA enhanced in inbreds			
Drosophila melanogaster	Inbred line crosses	Bristle number	Mather 1953 Reeve 1960
	Inbreeding variable base population	Wing length	Biémont 1983
Marine copepod	Inbreeding variable base population	Thoracic leg lengths	Clarke et al. 1986
Freshwater bivalves	Natural pops.	Plicae on palps	Kat 1982
Poeciliopsis monacha (a fish)	Natural pops.	Fin ray, scale, tooth number	Vrijenhoek & Lerman 1982
Trout (3 species)	Individuals varying in heterozygosity	Fin ray, gill raker, mandibular pores	Leary et al. 1983 1984, 1987
Rainbow trout	Inbreeding variable base population	Fin ray, gill raker, mandibular pores	Leary et al. 1985
Side-blotched lizard	Natural pops.	Scale counts	Soulé 1979
House mouse	Inbred line cross	Osteometric traits	Leamy 1984, 1992
Tamarins	Natural pops.	Cranial measures	Hutchison & Cheverud 1995
FA approximately equal in inbreds and outbreds, or variable results			
Drosophila melanogaster	Inbred line cross Inbreeding variable base population	Bristle number	Beardmore 1960 Fowler & Whitlock 1994
Honeybees	Inbreeding variable base population	Wing vein length	Brückner 1976 Clarke et al. 1986
Rainbow trout	Hatchery strain crosses	Fin ray, gill raker counts	Ferguson 1986
Bluegill sunfish	Subspecies crosses	Fin ray, scale counts	Felley 1980
Fence lizard	Interspecific hybrid zone	Scale counts	Jackson 1973
House mouse	Inbred line crosses natural pops.	Molar width	Bader 1965
Large cats	Species varying in heterozygosity	Craniometric traits Dental dimensions	Wayne et al. 1986 Modi et al. 1987 Kieser & Groeneveld 1991

Table 6.1 (Continued).

Organism	Basis of Comparison	Character(s)	Reference
FA enhanced in outbreds			
Sticklebacks	Hybrids between natural pops.	Scale counts	Zakharov 1981
Banded sunfish	Interspecific hybrid zone	Fin ray, scale, morph. traits	Graham & Felley 1985

the reported studies indicates no consistent change in asymmetry upon genetic perturbation. However, an objective evaluation of the results is difficult, since many types of crosses have been employed in these studies. Soulé (1982) argues that the reason that many inbreds exhibit enhanced FA is that inbreeding results in the production of extreme phenotypes that are more sensitive to developmental perturbations. If this were true, then one would expect the level of FA to be higher in individuals in the tails of phenotype distributions. However, attempts to find such a relationship have been unsuccessful (Soulé and Cuzin-Roudy 1982, Scheiner et al. 1991, Zakharov 1992, Livshits and Smouse 1993, Deng 1997).

There are many plausible explanations for the conflicts in the existing literature. For example, crosses between different strains need not be deleterious and can sometimes have the beneficial effects of masking the expression of deleterious recessives in the parental strains (Chapter 9). Second, many of the existing studies on FA may simply be statistically weak or flawed. The studies summarized in Table 6.1 have used a large number of indices other than Equation 6.4. Many investigators have simply employed the average absolute difference between the right and left measures, while others have attempted to correct for scaling that might occur with size changes by dividing the differences by the average measure. Very few studies have made any attempt to eliminate the bias caused by measurement error. Third, it is possible that the most adaptive level of FA is not the minimum level. Mather (1953) was able to increase and decrease FA for sternopleural bristle number in *Drosophila melanogaster* by artificial selection. After two generations of relaxed selection, the level of FA in both lines returned to the level in the original base population, suggesting that natural selection favors an intermediate level of FA. Presumably, the complete elimination of asymmetry can only be accomplished at the expense of important fitness characters such as development time.

The effects of environmental stress on FA are much more predictable — σ^2_{ew} tends to increase in extreme or novel environments (Hoffmann and Parsons 1991). Exposure of fish to DDT elevates levels of FA significantly (Valentine and Soulé 1973), and a variety of pesticides have been shown to have similar effects on several insect species (Clarke 1992, McKenzie and Yen 1995). When first exposed to diazinon, the sheep blowfly (*Lucilia cuprina*) initially exhibited an increase in FA, but as the population evolved insecticide resistance, FA returned to normal levels (McKenzie and Clarke 1988). Humans suffering from malnutrition exhibit increases in FA (Bailit et al. 1970), and various types of stress (cold, noise, behavioral modification) inflate FA for tooth morphology and limb bones in mice and rats (Siegel and Doyle 1975a,b,c, Siegel and Mooney 1987). In birds, parasite load increases the asymmetry in lengths of tail feathers (Møller 1992). Several studies have reported that individuals with developmental deformities tend to have elevated levels of FA for other traits (Bailit et al. 1970, Woolf and Gianas 1976, Barden 1980, Malina and Buschaung 1984, Leary et al. 1984).

In principle, the logic underlying the use of fluctuating asymmetry as a measure of σ^2_{ew} can be extended to certain aspects of organisms that are not bilaterally symmetrical. In plants, for example, there are many serially repeated organs, such as leaves, and flowers are often radially symmetrical. In cases such as these, more than two measures can be made within each individual, and the within-individual component of variance can be obtained by analysis of variance (Møller and Eriksson 1994, Freeman et al. 1993). As in the analysis of bilateral symmetry, a critical assumption underlying any such analysis is that the subjects of measurement are products of the same genes. An obvious violation of this assumption would involve a comparison of stem and basal leaves in a herbaceous plant, but the question remains open with more subtle comparisons such as the inner and outer leaves on a branch. For further details on the statistical analysis of FA, see Palmer and Strobeck (1986, 1992), Palmer et al. (1993), Swaddle et al. (1994), and Hutchison and Cheverud (1995).

Developmental Homeostasis and Homozygosity

In an influential book that reviewed much of the early literature, Lerner (1954) strongly endorsed the idea that the degree of developmental stability is positively correlated with the overall level of individual heterozygosity. The usual mechanistic explanation for this hypothesis is that heterozygosity acts as a buffer against environmental variation. This effect might occur, for example, if different allelic products have optimum activities under different environmental conditions.

Some of the logic behind Lerner's ideas were implicit in our discussion of FA, but here we focus on the total effects of heterozygosity on the within- and among-individual components of σ^2_e. Since the main prediction of the homeostasis-heterozygosity hypothesis is that σ^2_e will be higher for homozygous individuals than for heterozygotes, a conceptually simple test suggests itself. Consider two completely inbred lines and their hybrid (F_1) progeny. All members of the inbred lines will be 100% homozygous, while the F_1 individuals will be uniformly heterozygous at each locus for which the two inbred lines differed. Within each of the three groups, all individuals will be genetically identical, so the observed phenotypic variances within groups can be interpreted as genotype-

Table 6.2 Survey of studies of phenotypic variation in inbred parental lines and their F_1 hybrids.

Species	Characters	Reference
F_1 less variable		
Drosophila	Bristle number and wing length	Robertson & Reeve 1952
	Viability	Dobzhansky & Spassky 1954 Dobzhansky & Levene 1955
Mice	Behavioral traits	Hyde 1973
	Skeletal traits	Leamy & Thorpe 1984
	Time to maturity	Yoon 1955
	Weight	Chai 1957
Rats	Weight	Livesay 1930
Cotton	Fiber properties	Meredith et al. 1970
Corn	Vegetative and reproductive traits	Adams & Shank 1959 Shank & Adams 1960
Tobacco	Vegetative and reproductive traits	Jinks & Mather 1955
Rye	Grain production, and plant weight	Pfahler 1966
Sorghum	Grain production	Reich & Atkins 1970 Jowett 1972
Tomato	Growth rate	Lewis 1954
Equivocal results		
Guppies	Scale and fin ray counts	Angus & Schultz 1983
Arabidopsis	Plant weight	Pederson 1968
Tomato	Reproductive traits	Williams 1960
F_1 more variable		
Drosophila	Bristle number	Lewontin 1957
Corn	Grain production	Rowe & Anderson 1964

specific components of environmental variance. Thus, if all three genotypes are equally sensitive to environmental and developmental noise, they should exhibit the same level of phenotypic variance. The survey in Table 6.2 indicates that the phenotypic variance in F_1 hybrids is almost always lower than the average of the parental lines.

Although these kinds of results have been interpreted as strong support for the homeostasis-heterozygosity hypothesis, there are some unresolved issues. In most of the reported studies, the parental lines were not completely homozygous, so the observed phenotypic variances must contain a genetic component. This complicates matters greatly. If there is dominance at a locus, and the frequency of the dominant allele in the F_1 population is between 0.5 and 1.0, the average genetic variance in the parental lines will actually exceed that in the F_1 (Lerner 1954). This result is a simple consequence of the expression of the recessive allele being masked in the hybrid line. For a diallelic locus, the variance in the F_1 generation can be found by noting that the frequencies of individuals with genotypic values 0, $(1+k)a$, and $2a$ are q_1q_2, $p_1q_2 + p_2q_1$, and p_1p_2, where p_1 and p_2 are the frequencies of the dominant allele in the two parental populations. Following the procedures introduced in the previous chapter, the total genetic variance within an F_1 population is

$$\sigma_G^2 = a^2\Big\{ \left[(1+k)^2(p_1q_2 + p_2q_1) + 4p_1p_2\right] - \left[(1+k)(p_1q_2 + p_2q_1) + 2p_1p_2\right]^2 \Big\} \tag{6.5}$$

With certain allele frequencies, the mean genetic variance within the parental lines (obtained from Equations 4.12a,b) can be more than twice that in the F_1 (Figure 6.1). Thus, the observed patterns in Table 6.2 could be caused in part by residual heterozygosity within parental lines rather than by differences in environmental sensitivity between inbred and outbred individuals.

Several studies have attempted to test the homeostasis-heterozygosity hypothesis by comparing levels of phenotypic variation within groups of individuals with high vs. low heterozygosity for enzyme loci. Data from studies of oysters (Zouros et al. 1980), monarch butterflies (Eanes 1978), *Drosophila melanogaster* (Blanco and Sanchez-Prado 1986), killifish (Mitton 1978), ruffous-collared sparrows (Yezerinac et al. 1992), and humans (Livshits and Kobyliansky 1984) indicate a negative association between morphological variation and heterozygosity. However, studies on pitch pine (Ledig et al. 1983), the snail *Cerion* (Booth et al. 1990), plaice (McAndrew et al. 1982), herring (King 1984), and brook trout (Hutchings and Ferguson 1992) reveal no relationship, while there appears to be a weak positive relationship in *Daphnia magna* (Yampolsky and Scheiner 1994) and baboons (Bamshad et al. 1994).

Again, the discordancy between these results may be a statistical artifact. Not only does the problem outlined in the preceding paragraph apply to all of these studies, but there is another serious interpretative difficulty. In most of the analyses cited above, the homozygous groups contain all homozygous individuals regardless of the alleles they are carrying. Thus, for a single diallelic locus contributing to a quantitative trait, the homozygous group would contain *AA* and *aa*

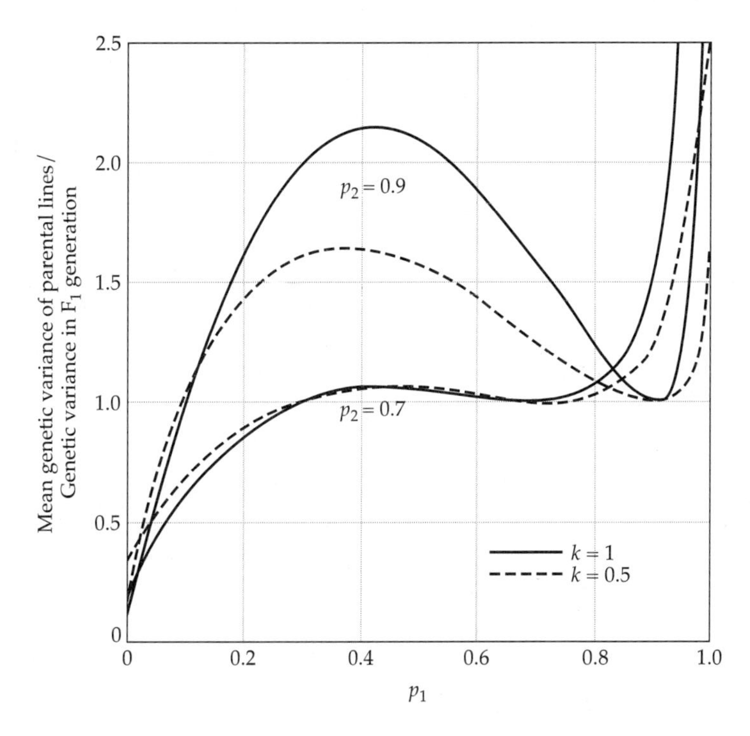

Figure 6.1 The mean single-locus genetic variance within two parental lines relative to that within the F_1 hybrid population. p_1 and p_2 are the frequencies of the dominant allele in the two populations. Results are given for complete $(k = 1)$ and partial $(k = 0.5)$ dominance and for two allele frequencies within parental population 2, $p_2 = 0.7$ and 0.9.

individuals with genotypic values 0 and $2a$, while the heterozygous group would only contain individuals with genotypic value $(1 + k)a$. When individuals are classified in this manner, for characters with an additive genetic basis, an inverse relationship is expected between multilocus heterozygosity and the genetic component of phenotypic variation, the most heterozygous group being the least variable (Chakraborty and Ryman 1983, Chakraborty 1987). No special appeal to homeostatic properties of heterozygosity are required to explain the results. On the other hand, when characters have a nonadditive genetic basis, the relationship between heterozygosity and the genetic component of phenotypic variance can be either positive or negative, depending on the allele frequencies and genetic effects (Bishop 1992, Gavrilets and Hastings 1994b; see also Chapter 4).

Interpopulation comparisons in humans (Kobyliansky and Livshits 1983), fox sparrows, and pocket gophers (Zink et al. 1985) have revealed an inverse relationship between population estimates of morphological variation and enzyme heterozygosity, consistent with the homeostasis-heterozygosity hypothesis, whereas the relationship appears to be positive in sculpins (Strauss 1991) and ruffous-collared sparrows (Yezerinac et al. 1992). With population level comparisons, the data are not biased by the artificial grouping of genotypic classes. However, there are still significant interpretative difficulties. The changes in morphological variability with heterozygosity may be due to shifts in the genetic as well as the environmental component of variance. If the molecular markers used to estimate heterozygosity levels are linked with the loci contributing to the genetic variance for the characters, then an increase in the latter associated with molecular heterozygosity may completely obscure changes in the environmental component of variance.

In summary, the fundamental problem with most descriptive studies of the relationship between heterozygosity and phenotypic variation is their inability to distinguish between the hypothetical stabilizing effects of heterozygosity on the environmental component of phenotypic variation and the very real effects of heterozygosity on the genetic component of variation. This issue can only be resolved by separating multilocus genotypes into homogeneous groups (based on allelic composition, rather than on total heterozygosity) and by separating the phenotypic component of variance into its genetic and environmental components. A simple way to resolve both of these problems exists for species that can be propagated clonally. All members within clonal groups are necessarily identical with respect to genotype, so the phenotypic variance within a clonal group is entirely due to environmental causes.

Deriving clones from natural populations of *Daphnia*, Deng (1997) used this approach to test the homeostasis-heterozygosity hypothesis. For each parental clone, a selfed offspring genotype was produced (by crowding, *Daphnia* can be induced to produce males, which then fertilize their genetically identical mother or sibs). In populations of *D. pulex* and *D. pulicaria*, the environmental components of variance for life-history characters within inbred clones averaged 75% and 88% higher than within parental clones. These results provide unambiguous support for Lerner's hypothesis.

Several studies with trees have addressed the relationship between annual variation in individual growth rate (a measure of environmental sensitivity) and enzyme heterozygosity. As in Deng's work, these studies are not complicated by the genetic artifacts mentioned above, since the units of comparison are the same individuals. Nevertheless, the results are equivocal. Heterozygosity and temporal variation in growth rate are positively correlated in quaking aspen (Mitton and Grant 1980), ponderosa pine (Knowles and Grant 1981), and knobcone pine (Strauss 1987), but negatively correlated in lodgepole pine (Knowles and Mitton 1980), and uncorrelated in pitch pine (Ledig et al. 1983). When all of the preceding results are considered, it appears that the acceptance of a general causal relationship between heterozygosity and developmental stability should be postponed until additional adequately designed experiments have been performed.

Repeatability

Because the variance among repeated measures on the same individual can only be due to environmental causes (or measurement errors), information on the within-individual component of variance can provide some insight into the possible magnitude of the environmental variance for a trait. For studies with a temporal component, there are two types of within-individual variation. The first, emphasized above, is the variance among measures of homologous characters expressed within individuals at the same time. The second is the variance in expression of a character across time, e.g., temporal variability in the weight of an individual. This second type of variation is not always logically distinct from the first. For example, for animals that are growing, the right and left measures of a bilaterally symmetrical trait may fluctuate on a daily basis. Moreover, one may question whether repeated measures made at wide intervals of time in the same individual should be treated as the measures of the same or different traits (Chapter 21). These issues aside, it follows that an upper-bound estimate of the genetic variance of a trait is provided by

$$\text{Var}(G)_{max} = \text{Var}(z) - \text{Var}(e_w) \tag{6.6}$$

where $\text{Var}(z)$ is an estimate of the total phenotypic variance for the trait, and $\text{Var}(e_w)$ is an estimate of the within-individual component of variance.

Measurement error, solely a function of the investigator, will always inflate estimates of the within-individual component of variance relative to its true value, but because it also contributes to the total phenotypic variance, it cancels out in Equation 6.6. Nevertheless, measurement error is still a problem, since we ordinarily would like to know the fraction of the true phenotypic variance that is accounted for by $\text{Var}(G)_{max}$. Thus, it is desirable to have an estimate of $\text{Var}(z)$ that is free of measurement error. As noted above, for morphological characters that have reached their final stage of development (and are not subject to wear) or for measures made on preserved samples, a simple correction for measurement error can be acquired from repeated measures of the same character. However, the problem is less tractable with behavioral and physiological traits. For such characters, repeated measures may differ because of temporal organismal changes as well as because of measurement error, rendering it essentially impossible to factor out the component of variance due to measurement error.

The expected value of $\text{Var}(G)_{max}$ is necessarily greater than the total genetic variance for the trait because it includes the among-individual component of the special environmental effects variance (σ^2_{ea}) as well as any variance due to general environmental effects (σ^2_E). Thus, letting $\text{Var}(e_m)$ denote the variance associated with measurement error, the **repeatability**

$$r = \frac{\text{Var}(z) - \text{Var}(e_w)}{\text{Var}(z) - \text{Var}(e_m)} \tag{6.7}$$

provides an upper-bound estimate of the broad-sense heritability of a trait (H^2), i.e., of the fraction of the total phenotypic variance that is genetic in basis. The degree to which r exceeds H^2 depends on the magnitude of $(\sigma^2_{ea} + \sigma^2_E)$ relative to σ^2_{ew}. In the unlikely event that all of the environmental variance is in the within-individual component, and measurement-error variance has been removed from the denominator, then Equation 6.7 provides an unbiased estimate of H^2. A large value of r offers the possibility that a considerable amount of the character variance is genetic, while a small value of r informs us that environmental variance dominates.

Repeatability is often computed as the correlation between two repeated measures (z_1 and z_2) on the same individuals (Falconer and Mackay 1996),

$$r_F = \frac{\text{Cov}(z_1, z_2)}{\text{SD}(z_1)\,\text{SD}(z_2)} \tag{6.8}$$

However, because the variance resulting from measurement error is contained in the denominator of this expression, r_F is downwardly biased, a problem since we are trying to obtain an upper bound on H^2. For a lucid account of the statistical procedures used to estimate repeatability and of common mistakes encountered in the literature, see Lessells and Boag (1987).

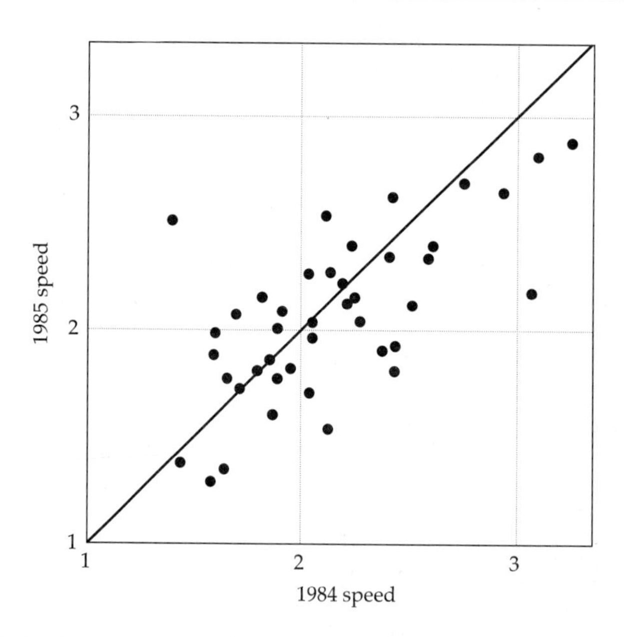

Example 3. In an attempt to determine the reliability of single measures of sprint speed as an assessment of performance, Huey and Dunham (1987) looked at

the correlation between measures (Falconer's method) on wild-caught lizards (*Sceloporus merriami*) in two consecutive years. The repeatability was quite high ($r_F \simeq 0.70$), despite injuries and changes in reproductive condition of the animals between years. Thus, since measurement error could deflate this estimate, 70% or more of the phenotypic variance in running speed in the study population could be a consequence of genetic differences among individuals.

On the other hand, in the sagebrush lizard (*Sceloporus graciosus*), repeatabilities of various aspects of push-up and head-bob displays average only 0.16 (Martins 1991). In this study, ten or more measures were made on each individual over a period of five weeks, and $\text{Var}(G)_{max}$ was estimated by analysis of variance, as the among-individual component of variance. Assuming that measurement error is of minor importance in this study, and noting that the within-individual variance is likely to be greater over longer time spans such that $\text{Var}(z)$ is underestimated, these behavioral traits cannot have very high broad-sense heritabilities.

GENERAL ENVIRONMENTAL EFFECTS OF MATERNAL ORIGIN

It is difficult to conceive of an organism for which some form of maternal effect is not a potential source of variation. The quality of postnatal care in many vertebrates has obvious effects on many aspects of the offspring phenotype. Other types of maternal effect, such as egg quality or endosperm quantity, are more subtle but nevertheless important (Schaal 1984, Roach and Wulff 1987).

Because of maternal effects, the phenotypic composition of a population can depend greatly on the ecological setting in the previous generation. Consider the results shown in Figure 6.2 for an experiment in which mothers of a single clone of *Daphnia* were grown on high and low nutritional levels and their progeny were raised at each of the two conditions, to yield four treatments. While the growth trajectory of offspring raised on high food depended only slightly on the maternal environment, low-food progeny whose mothers were well fed grew to substantially larger sizes than those whose mothers were food-limited. The message of this example is that, unless a study population is raised in the environment of interest for at least a generation prior to analysis, one runs the risk that the observed phenotypes are more a product of the past than the current environment.

Although it is useful to think of maternal effects as environmental sources of variation from the standpoint of the recipients (the progeny), it does not follow that the variance of such effects always has an environmental basis. It is well known, for example, that considerable genetic variance exists for milk production in mammals. In plants, the seed coat, which influences dispersal and dormancy in many species, is under the genetic control of the mother but not the father. In the snail, *Lymnaea peregra*, the direction of shell coiling is determined entirely by the maternal genotype, with dextrality being completely dominant (Freeman and Lundelius 1982).

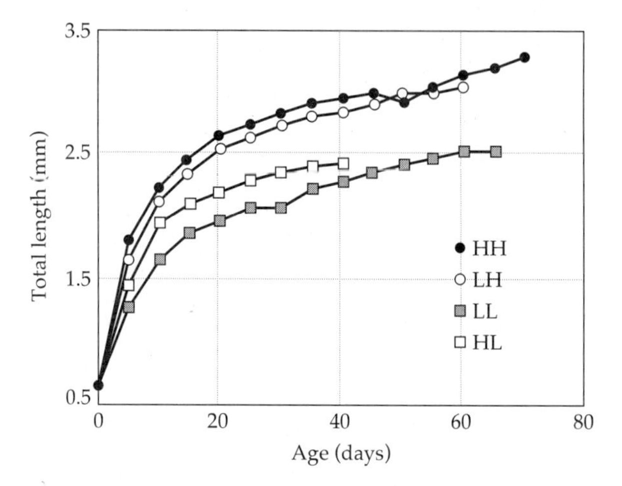

Figure 6.2 Growth trajectories for members of a clone of *Daphnia pulex* raised under various nutritional conditions. The first letter of the key indicates the diet of the mother (H = high, and L = low food conditions); the second letter refers to the diet of the measured progeny. (From Lynch and Ennis 1983.)

A rather striking example of a genetically based maternal effect is provided by Reznick's (1981, 1982) work on the mosquito fish, *Gambusia affinis*. Since the eggs in this species are fully provisioned prior to fertilization, size at birth depends primarily on the maternal phenotype and hardly at all on the genes inherited through the father. The results of various crosses involving small-egged Illinois fish and large-egged North Carolina fish are shown in Figure 6.3. Note that the mean size at birth of hybrid progeny is purely a function of the maternal parent. This also holds when F_1 females are back-crossed to the parental lines. Here, offspring size is intermediate, as expected, since the F_1 females contain genes from both parental populations.

The relative importance of maternal effects can vary with the age of the mother. Working with highly inbred lines of guinea pigs, Wright (1926) found that the incidence of spotting of the fur increased by approximately 20% with maternal age, while the incidence of polydactyly (extra digits) exhibited a nearly fivefold decline. A dramatic change in the area of daughter fronds accompanies maternal age in duckweed (Ashby and Wangermann 1954). The deviations in fingerprint ridge counts between monozygotic twins decline with age of the mother in humans (Lints and Parisi 1981). Since the units of comparison in all of these studies are genetically identical, the results can only be explained as a change in environmental sensitivity with maternal age.

Age-specific maternal effects can also influence the genetic properties of populations. For example, in *Drosophila melanogaster*, the genetic variance for bristle number increases with maternal age (Beardmore et al. 1975), and the same is true

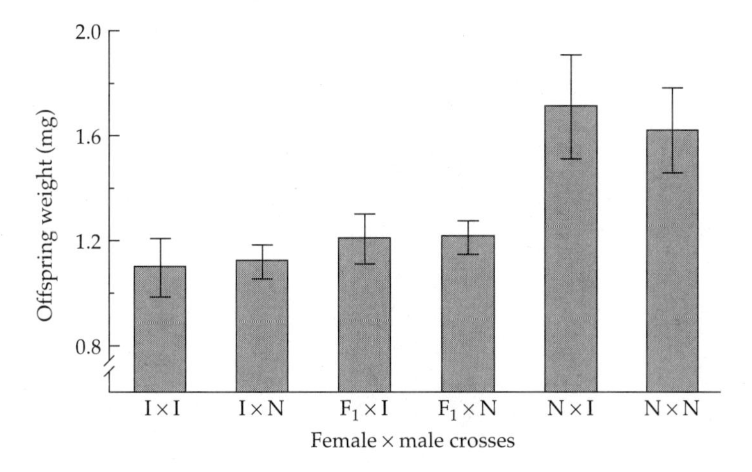

Figure 6.3 Mean offspring sizes for crosses involving Illinois (I) and North Carolina (N) populations of mosquito fish (*Gambusia affinis*). F_1 refers to a hybrid mother. Vertical lines denote ± 2SE. (From Reznick 1981.)

for caudal fin ray numbers in guppies (Figure 6.4). However, an investigation in pines revealed little influence of maternal age on the components of variance in the progeny (Lints and Baeten 1981).

In insects, a number of cases are known in which sexually transmitted microorganisms have a pronounced influence on progeny phenotypes and parental reproductive performance (Hoffmann and Turelli 1988, Stouthamer et al. 1993, Moran and Baumann 1994, Wade and Chang 1995). For example, nearly half of

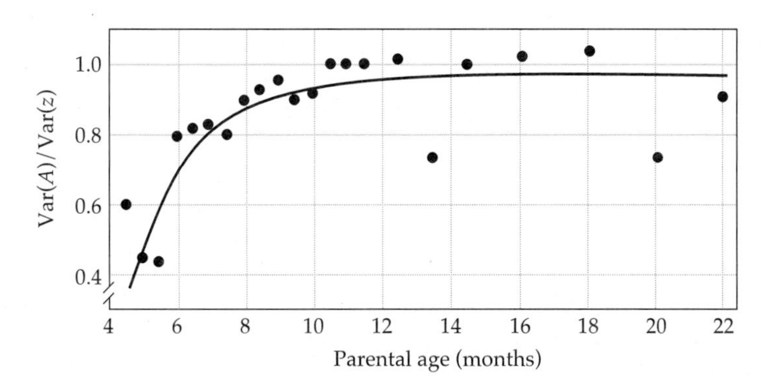

Figure 6.4 Estimated fractions of the phenotypic variance attributable to additive gene action for caudal fin ray number in a laboratory population of guppies. Cohorts of progeny were obtained from groups of parents of various ages. (From Beardmore and Shami 1976.)

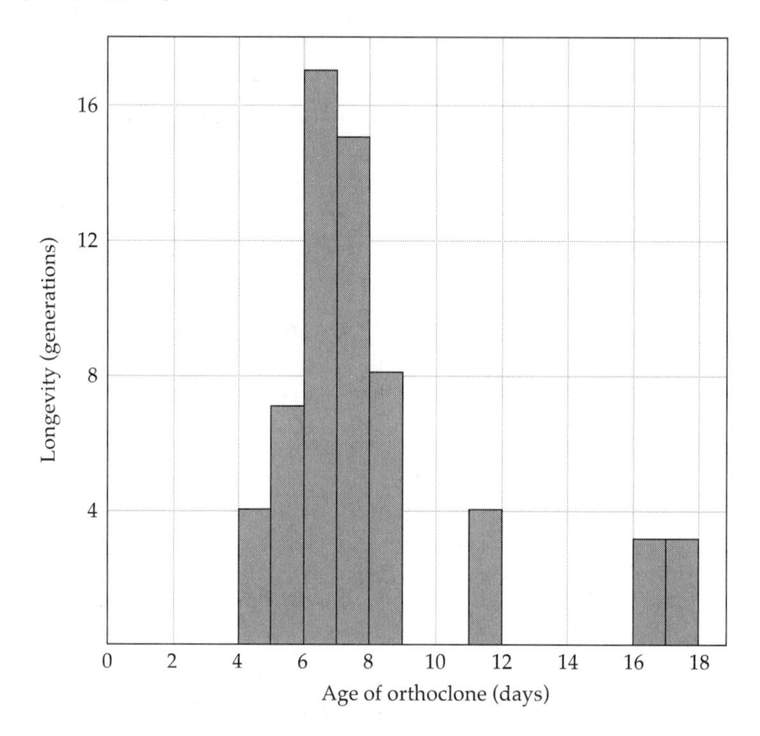

Figure 6.5 The number of generations until extinction as a function of the age at reproduction of orthoclones of the rotifer *Philodina citrina* (Lansing 1947, 1948). The nth orthoclone is a line in which each generation is started from progeny produced by mothers of age n. (After Lints 1978.)

the wild-caught individuals of the parasitic wasp *Nasonia vitripennis* carry one of three sex-ratio distorter organisms (Werren et al. 1981, 1986). One of these, *psr* (paternal sex ratio), is paternally inherited and causes all-male families, while *msr* (maternal sex ratio) is maternally inherited and results in nearly pure female families. A third, *sk* (son-killer), is maternally and contagiously transmitted and causes death of male eggs.

Once one accepts the significance of maternal effects, the question arises as to whether such effects are transmissible over more than a single generation. A fair amount of work on this subject appears in the gerontological literature. To demonstrate the cumulative effects of maternal age on longevity, Lansing (1947, 1948) worked with the clonal rotifer, *Philodina citrina*. He produced a series of **orthoclones** by propagating successive generations with progeny from mothers of constant age. All of the lines died out eventually, but young and old orthoclones went extinct most rapidly (Figure 6.5). Lansing found that extinction could be avoided in the late-age orthoclones by allowing them to reproduce at younger ages. Thus, the mechanism of the cumulative age effect was most likely cytoplasmic in basis. Cumulative, reversible maternal age effects, now known as **Lansing effects,** have been studied in a number of other organisms, with several characters other than longevity, and with mixed results (Lints 1978, Finch 1990). Aside from these types of studies, however, there is a glaring absence of data on multigenerational transmission of environmental effects. Intuition may suggest that such effects are unlikely to be of significance, but empirical observation would provide a more convincing argument.

GENOTYPE × ENVIRONMENT INTERACTION

The existence of genotype × environment interaction in a population indicates that different genotypes respond to environmental change in different ways. In extreme cases, the ranking of genotypes may be altered by a shift in the environment. These problems are of great concern to breeders of economically important plants and animals since substantial genotype × environment interaction necessitates the development of locally adapted breeds. They are also at the heart of many studies of species adaptations, although few such studies have ever been couched formally in terms of genotype × environment interaction.

A field experiment with the leaf-mining insect, *Liriomyza sativae*, a serious pest of vegetable crops, further illustrates the principal aspects of genotype × environment interaction. Via (1984) obtained animals from adjacent cowpea and tomato fields, produced half-sib families in the lab, and then monitored their performance on both plants. Matings were restricted to pairs taken from the same field. Figure 6.6 (left) illustrates the response of the half-sib family means to the larval food plant. In general, the larvae develop more rapidly on tomato, and the variance between families is approximately equal on both hosts. However, the nonparallel responses indicate the existence of genotype × environment interaction. Some families even develop more rapidly on cowpea than on tomato. Figure 6.6 (right) illustrates that the mean responses of four *Liriomyza* populations to treatment are not greatly different. Thus, while the within-population analysis indicates that host specialization could evolve in this species, it has not occurred, possibly because of the close proximity of the field sites and the lack of migration barriers between them.

A second example of genotype × environment interaction involves a selection experiment on longevity in *Drosophila melanogaster* (Clare and Luckinbill 1985). Selection for increased and decreased life span was imposed for about 30 generations, and then the two divergent lines and their F_1 hybrids were assayed for longevity under two conditions — high and low larval density. Under high densities, the same conditions under which selection was practiced, the genes for longevity appear to behave in a perfectly additive manner, the mean phenotype of the hybrids being intermediate to that of the parents (Figure 6.7). However, at

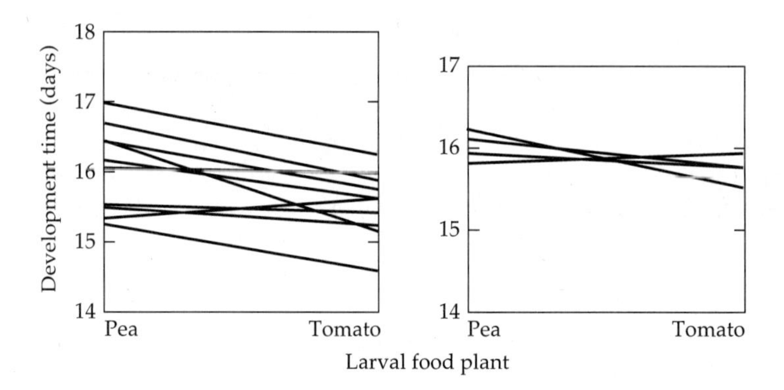

Figure 6.6 The response of development time in the leaf-mining insect *Liriomyza sativae* to a change in larval food plant. **Left:** Paternal half-sib families of one population. **Right:** Four population means. Differences in the slopes, which result from genotype × environment interaction, are significant for half-sib families, but not for population means. (From Via 1984.)

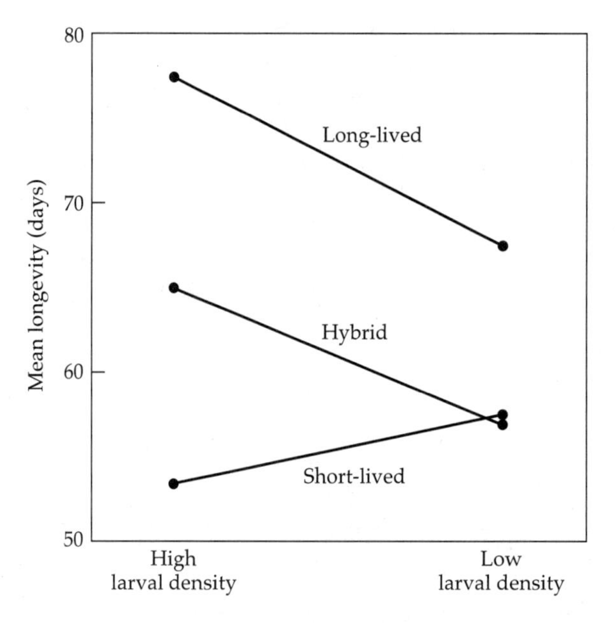

Figure 6.7 The response of adult longevity to density treatment for lines of *Drosophila melanogaster* selected for short and long life spans and for their F_1 hybrids. Genotype × environment interaction is indicated by the nonparallel response of the short-lived line. (From Clare and Luckinbill 1985.)

low densities, the genes for short life span appear to be completely dominant to those for high life span.

In both of these examples, it was possible to make some inference as to the existence of genotype × environment interaction because members of the same genetic groups were evaluated under well-defined treatments. In the case of field studies, however, individuals cannot usually be assigned to discrete environmental categories. This does not mean that genotype × environment interaction is not important, only that it is unmeasurable. Unless discrete treatments are employed in an experiment, any genotype × environment interaction will be confounded with the environmental source of variance.

7

Resemblance Between Relatives

In the previous two chapters, we found that the phenotypic variance of a trait can theoretically be partitioned into a number of genetic and environmental components. However, the significant practical issue of how these components can be estimated remains. The key to this matter, first pointed out by Fisher (1918) and Wright (1921b), is the fact that various genetic and environmental sources of variance contribute differentially to the resemblance between different types of relatives.

Assuming for the time being an absence of genotype × environment interaction, and recalling Equation 6.1, let $z_x = G_x + E_x + e_x$ and $z_y = G_y + E_y + e_y$ be the phenotypic values of two members of a particular relationship, such as parent and offspring. As in previous chapters, G, E, and e denote genotypic values, general environmental effects, and special environmental effects, respectively. The phenotypic covariance between relatives x and y thus becomes

$$\begin{aligned}\sigma_z(x,y) &= \sigma[(G_x + E_x + e_x), (G_y + E_y + e_y)] \\ &= \sigma_G(x,y) + \sigma_{G,E}(x,y) + \sigma_{G,E}(y,x) + \sigma_E(x,y)\end{aligned} \tag{7.1a}$$

Because the special environmental effects are random residual deviations, they are uncorrelated among individuals and do not contribute to the resemblance between relatives, i.e., $\sigma_e(x,y) = 0$. The middle two terms, $\sigma_{G,E}(x,y) + \sigma_{G,E}(y,x)$, refer to the covariance of the genotypic value of one member of the pair and the general environmental effect of the other, while the final term, $\sigma_E(x,y)$, is the covariance between general environmental effects. Experiments can often be designed so that all three terms involving E have expected values equal to zero. For now, we ignore the issue of genotype-environment covariance, i.e., we assume $\sigma_{G,E}(x,y) = \sigma_{G,E}(y,x) = 0$. This assumption reduces Equation 7.1a to

$$\sigma_z(x,y) = \sigma_G(x,y) + \sigma_E(x,y) \tag{7.1b}$$

The genetic covariance between relatives, $\sigma_G(x,y)$, merits special attention. Such covariance is a natural consequence of relatives inheriting copies of the same genes. As in the case of the genetic variance, the genetic covariance between relatives can be partitioned into components attributable to additive, dominance, and various epistatic effects. Each term consists of one of the familiar components of genetic variance (Chapter 5) weighted by a coefficient that describes the joint distribution of genetic effects in pairs of relatives. These coefficients are the first focus of our attention. Once they are understood, it is a relatively simple step to use the results in Chapter 5 to derive a general expression for the genetic covariance between relatives.

We will first consider the ideal situation in which mating is random and loci are unlinked and in gametic phase equilibrium. The complications that are introduced with linkage, gametic phase disequilibrium, and assortative mating will then be evaluated. Some of these complications make for difficult reading, but they are realities that should not be ignored. Next, the environmental causes of the resemblance between relatives will be discussed. This does not really complete the picture, as still other complications such as sex-linkage, maternal genetic effects, and inbreeding may be of substantial importance in particular cases, but we leave most of our discussion of these to later chapters. The chapter ends with a broad overview of the concept of heritability, a central parameter in many quantitative-genetic formulations.

MEASURES OF RELATEDNESS

Many relatedness measures have found their way into the population-genetic and sociobiological literature (Wright 1922, Cotterman 1940, Malécot 1948, Denniston 1974, Jacquard 1974, Orlove and Wood 1978, Michod and Hamilton 1980, Grafen 1985). Not all of these play a central role in quantitative-genetic formulations, but they all share two essential features.

First, relatedness can only be defined with respect to a *specified frame of reference*. Technically speaking, all members of a species or population are related to each other to some degree for the simple reason that they contain copies of genes that were present in some remote ancestor in the phylogeny. We avoid this problem by letting the reference population be the base of an observed pedigree. If, for example, no individuals further back than the parental generation have been observed, the usual procedure is to treat that generation as the base and to assume that its members are unrelated.

Second, all measures of relatedness are based upon the concept of **identity by descent**. Genes that are identical by descent are *direct* descendents of a specific gene carried in some ancestral individual. The distinction between identity by descent and **identity in state** is critical. Two genes that have identical nucleotide sequences but have descended from different copies in the reference population are identical in state but not by descent. On the other hand, genes that are identical by descent are necessarily identical in state, barring mutation. The distinction between these two types of identity is clarified in Figure 7.1, where the parent generation is treated as the base population. Although the first offspring contains two A_1 genes, only one of them is identical by descent with the A_1 gene in the second offspring.

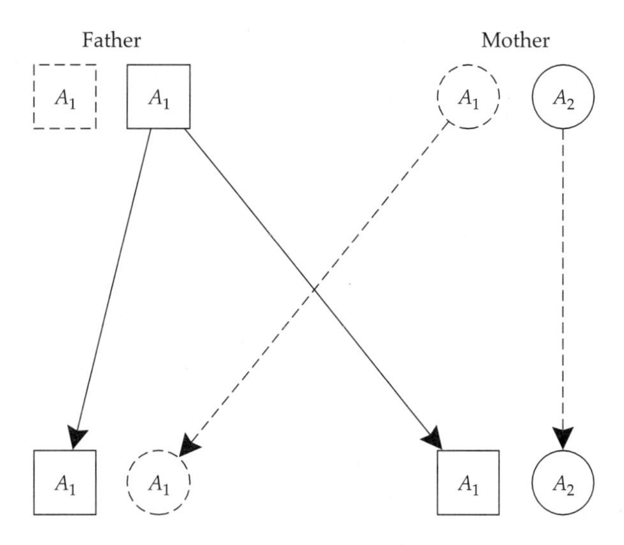

Figure 7.1 The transmittance of genes of two parents to two offspring. All A_1 alleles are identical in *state*. However, of the two A_1 genes carried by the offspring on the left, only the one in the square is identical by *descent* with the A_1 allele carried by the offspring on the right.

Coefficients of Identity

Consider a single locus in two diploid individuals. For the four genes involved, there are 15 possible configurations of identity by descent due to the fact that identity may exist *within* as well as *between* individuals (Gillois 1964; Figure 7.2). Individuals that contain pairs of alleles that are identical by descent are said to be **inbred**. If we ignore the distinction between maternally and paternally derived genes, the 15 possible configurations reduce to nine **identity states**. These range from state 1 in which the two individuals are inbred and share a gene that is identical by descent (so that all four genes are identical by descent) to state 9 in which none of the four genes are identical by descent. Many situations exist in which further simplification can be justified. For example, in large random-mating populations, the probability of the first six identity states is essentially zero.

Associated with each of the nine identity states are probabilities, Δ_1 to Δ_9, which Jacquard (1974) called **condensed coefficients of identity** (Figure 7.2). These coefficients provide a complete description of the probability distribution of identity by descent between single loci of two individuals. The values that they take on depend on the relationship. For example, suppose that x is a parent and y its offspring, and that neither is inbred. Then, since an individual inherits one and only one gene from its parent, $\Delta_8 = 1$ and all other $\Delta_i = 0$. For noninbred full sibs, there is a 0.5 probability of inheriting the same paternal

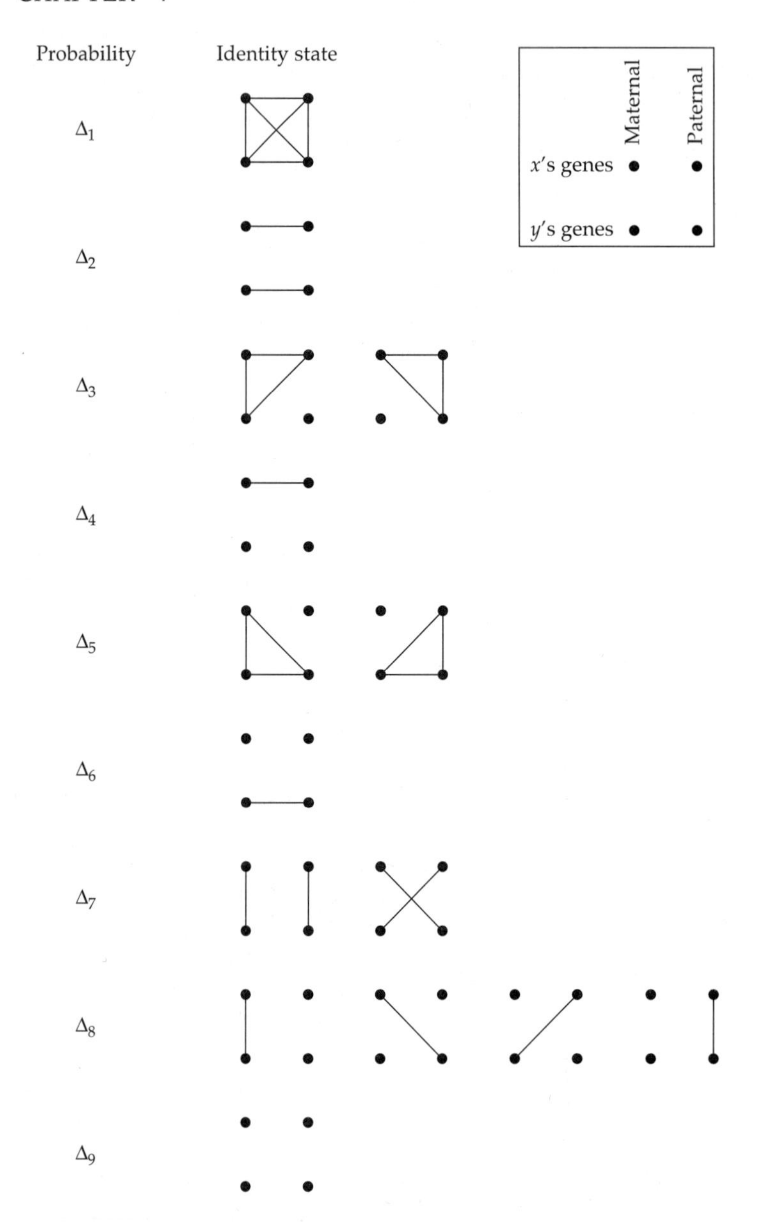

Figure 7.2 The 15 possible states of identity by descent for a locus in individuals x and y, condensed into nine classes. Genes that are identical by descent are connected by lines.

gene and, independently, a 0.5 probability of inheriting the same maternal gene. Thus, $\Delta_7 = 0.25$ (both pairs of genes are identical by descent), $\Delta_8 = 0.50$ (one pair of genes is identical by descent), and $\Delta_9 = 0.25$ (there is no identity by descent at the locus), and all other $\Delta_i = 0$.

Coefficients of Coancestry and Inbreeding

Suppose now that single genes are drawn randomly from individuals x and y. The probability that these two genes are identical by descent, Θ_{xy}, is called the **coefficient of coancestry.** (In other publications, Θ_{xy} is sometimes referred to as the **coefficient of consanguinity, coefficient of kinship,** or **coefficient de parente**). In terms of the condensed coefficients of identity,

$$\Theta_{xy} = \Delta_1 + \frac{1}{2}(\Delta_3 + \Delta_5 + \Delta_7) + \frac{1}{4}\Delta_8 \tag{7.2}$$

Here, we have simply weighted each condensed coefficient of identity by the conditional probability that a randomly drawn gene from x is identical by descent with a randomly drawn gene from y (at the same locus). Another way to look at the problem is to consider a hypothetical offspring (z) of x and y. By the above definition, Θ_{xy} is the probability that the two genes at a locus in individual z are identical by descent. The latter quantity is Wright's (1922) **inbreeding coefficient**, f_z. Thus, an individual's inbreeding coefficient is equivalent to its parents' coefficient of coancestry, $f_z = \Theta_{xy}$.

We now proceed by example to demonstrate how estimates of Θ_{xy} are derived. The first problem to be tackled is the coefficient of coancestry of an individual with itself, Θ_{xx}. This may seem like a nonsensical task. However, we will soon see that Θ_{xx} is an essential element of all coancestry estimates. Denote the two genes carried by individual x as A_1 and A_2, and then randomly draw a gene from the locus, replace it, and randomly draw another. Θ_{xx} is the probability that the two genes drawn are identical by descent. There are four ways, each with probability $1/4$, in which the genes can be drawn: A_1 both times, A_1 first and A_2 second, A_2 first and A_1 second, and A_2 both times. If two A_1 genes are drawn, they must be identical by descent since they are copies of the same gene. The same applies to a draw of two A_2 genes. Thus, provided that genes A_1 and A_2 are not identical to each other by descent, then Θ_{xx} is simply $(1/4)(1)+(1/4)(1) = 1/2$. We should, however, recognize the possibility that individual x is inbred, in which case the probability that the gene A_1 is identical by descent with the gene A_2 is f_x. Thus, a general expression for the coefficient of coancestry of an individual with itself is

$$\Theta_{xx} = \frac{1}{4}(1 + f_x + f_x + 1) = \frac{1}{2}(1 + f_x) \tag{7.3}$$

A slightly more complicated situation arises in calculating the coefficient of coancestry between a parent and its offspring. In order to simplify the discussion,

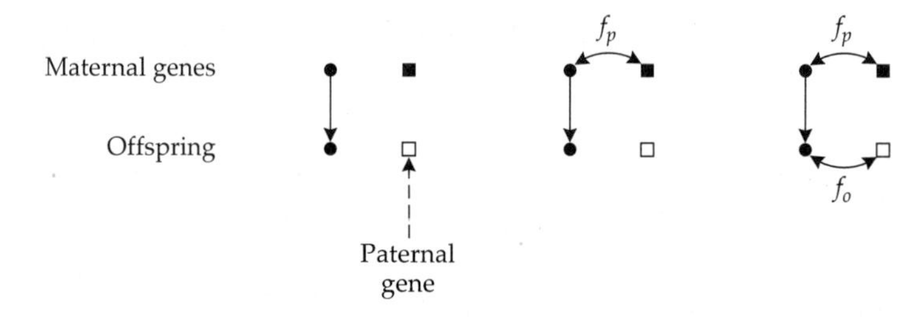

Figure 7.3 The identity of genes by descent for a parent and offspring. Circles and squares represent, respectively, maternally and paternally derived genes. **Left:** The mother is not inbred and her mate is not a relative (so the offspring is not inbred). **Center:** The mother is inbred but unrelated to her mate. **Right:** In addition to the mother being inbred, she is related to her mate, so that her offspring is also inbred.

we will call the parent (p) of interest the mother, but the same results apply to fathers provided the locus is autosomal. We first consider the situation in which neither the mother nor her offspring (o) are inbred, i.e., the mother's parents are unrelated, and she is unrelated to her mate. In that case, of the four ways in which single genes can be drawn from the mother and the child, only one involves a pair that is identical by descent (Figure 7.3, left). Therefore, $\Theta_{po} = 1/4$. Suppose, however, that the mother is inbred (Figure 7.3, center), so that the probability that both of her alleles are identical by descent is f_p. This is the same as the probability that the maternal gene inherited by the offspring is identical by descent with the maternal gene not inherited. The probability of drawing such a gene combination is $1/4$. Therefore, inbreeding in the parent inflates Θ_{po} to $(1 + f_p)/4$. With complete inbreeding ($f_p = 1$), both parental alleles are identical by descent, increasing Θ_{po} to $1/2$. Finally, we allow for the possibility that the parents of o are related, so that the offspring is inbred with coefficient f_o (Figure 7.3, right). It is now necessary to consider the implications of drawing a paternally derived gene from the offspring, the probability of which is $1/2$. Since f_o is equivalent to the probability that maternally and paternally derived genes are identical by descent, the additional parent-offspring identity induced by inbreeding is $f_o/2$. In summary, the most general expression for the coefficient of coancestry for a parent and offspring is

$$\Theta_{po} = \frac{1}{4}(1 + f_p + 2f_o) \tag{7.4}$$

Often in the literature, Θ_{po} is simply considered to be $1/4$. It should now be clear that this implicitly assumes the absence of matings between relatives.

We now move on to the coefficient of coancestry of two individuals that share the same father and mother (full sibs). We assume a species with separate sexes so that the mother and father are different individuals, and we again start with

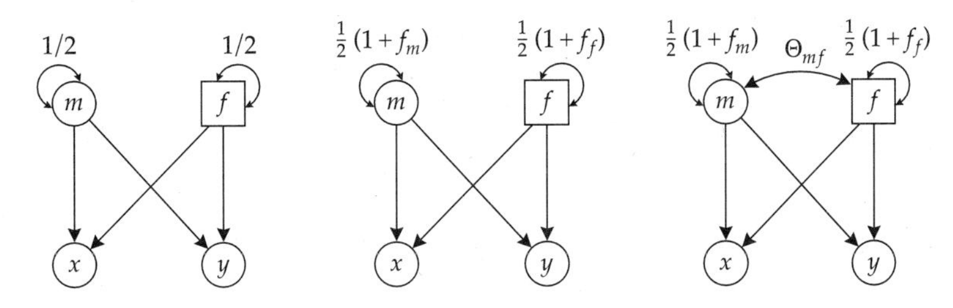

Figure 7.4 Path diagrams for analyzing the probability that random genes from two full sibs are identical by descent. The path coefficients along single-headed arrows are always equal to $1/2$. **Left:** The parents, m and f, are neither related nor inbred. **Center:** The parents are unrelated, but inbred. **Right:** In addition to being inbred, the parents are related with coefficient of coancestry Θ_{mf}.

the simplest situation, progressively allowing the parents to be inbred and/or related (Figure 7.4). For the analysis of full sibs as well as more complicated degrees of relationship, the method of path analysis (Appendix 2) provides a useful tool. The elements in Figure 7.4 no longer represent gametes (as in Figure 7.3) but individuals.

Let m represent the mother, f the father, and x and y their two offspring. When the parents are neither inbred nor related, there are two paths by which the same gene can be passed to both x and y: $x \leftarrow m \rightarrow y$ and $x \leftarrow f \rightarrow y$. Since both paths have identical consequences, we will simply consider the first of them. First, we note that the probability that both x and y receive the same maternal gene is $1/2$. This is the coefficient of coancestry of the (noninbred) mother with herself, Θ_{mm}, and is represented by the double-headed arrow in the figure. Second, we note that the probability of randomly drawing a maternal gene from individual x is $1/2$, and that the same is true for individual y. Thus, the probability of drawing two maternal genes, identical by descent, one from x and the other from y, is $\Theta_{mm}/4 = 1/8$. Adding the same contribution from the paternal path, $x \leftarrow f \rightarrow y$, we obtain the coefficient of coancestry $\Theta_{xy} = 1/4$. Path analysis (Appendix 2) provides a simple way to obtain this result. First, set the path coefficients on all of the single-headed arrows in Figure 7.4 equal to $1/2$. Then, note that the contribution of a path to a correlation between two variables is equal to the product of the path coefficients and the correlation coefficient associated with the common factor (in this case, Θ_{mm} or $\Theta_{ff} = 1/2$).

We now allow for the possibility that the parents are inbred with inbreeding coefficients f_m and f_f, a condition that inflates the coefficient of coancestry of an individual with itself. This is the only necessary change for the path diagram in Figure 7.4 (center). There are still only two paths that lead to genes identical by descent in x and y, and their sum is

$$\Theta_{xy} = \frac{1}{4}(\Theta_{mm} + \Theta_{ff}) = \frac{1}{4}\left(\frac{1+f_m}{2} + \frac{1+f_f}{2}\right) = \frac{1}{8}(2 + f_m + f_f) \tag{7.5a}$$

Finally, we allow for the possibility that m and f are related, such that the probability of drawing two genes (one from each of them) that are identical by descent is Θ_{mf}. It is then necessary to consider two additional paths between x and y: $x \leftarrow m \leftrightarrow f \rightarrow y$ and $x \leftarrow f \leftrightarrow m \rightarrow y$ (Figure 7.4, right). Again taking the coefficients on the single-headed arrows to be $1/2$, it can be seen that each of these two new paths makes a contribution $\Theta_{mf}/4$ to Θ_{xy}. Adding these to our previous result, we obtain a general expression for the coefficient of coancestry of full sibs,

$$\Theta_{xy} = \frac{1}{8}(2 + f_m + f_f + 4\Theta_{mf}) \tag{7.5b}$$

which reduces to $\Theta_{xy} = 1/4$ under random mating.

The preceding techniques are extended readily to more distant relationships and more complicated schemes of relatedness. The coefficient of coancestry is always the sum of a series of two types of paths between x and y. The first type of path leads from a single common ancestor to the two individuals of interest, while the second type passes through two remote ancestors that are related to each other. Neither type of path is allowed to pass through the same ancestor more than once. This procedure is summarized by the following equation

$$\Theta_{xy} = \sum_i \Theta_{ii}\left(\frac{1}{2}\right)^{n_i-1} + \sum_j \sum_{j\neq k} \Theta_{jk}\left(\frac{1}{2}\right)^{n_{jk}-2} \tag{7.6}$$

where n_i is the number of individuals (including x and y) in the path leading from common ancestor i, and n_{jk} is the number of individuals (including x and y) on the path leading from two different but related ancestors, j and k. Formal proof of this equation can be found in Boucher (1988).

Up to now, we have assumed an autosomal locus. The rules change when the locus of interest is sex-linked. Assuming the female is the heterogametic sex, a male cannot receive an X-linked gene from his father, whereas fathers pass on their X-linked genes to daughters with probability one. Females have two X chromosomes and pass each one on to sons or daughters with the usual probability of one-half. Thus, the protocol for obtaining the coefficient of coancestry for an X-linked locus is similar to that used for autosomal loci except that path coefficients leading from fathers to daughters are replaced by a 1 and those leading from fathers to sons are replaced by a 0. For all paths containing two consecutive males, the probability that X-linked genes are identical by descent is zero.

Example 1. One of the first pedigrees to which Wright (1922) applied his theory of inbreeding is that of Roan Gauntlet, an English bull. In the following figure, rectangles and ovals refer to bulls and cows, respectively. We wish to compute the coefficient of coancestry of the Royal Duke of Gloster and Princess Royal. This is the same as the inbreeding coefficient of their son, Roan Gauntlet. The four possible paths by which alleles identical by descent can be inherited by the Royal Duke and Princess Royal are indicated by the coded lines adjacent to the arrows in the pedigree. Two of these paths contain four individuals and two contain seven. Thus, assuming that the remote ancestors, Lord Raglan and Champion of England, are not inbred (so that for both, $\Theta_{ii} = 1/2$), the coefficient of coancestry of the Royal Duke and Princess Royal is $[2(1/2)^4 + 2(1/2)^7] = 0.141$. This is a slightly closer relationship than that for half sibs (for which $\Theta = 0.125$). Relative to the base population, the alleles at 14% of the autosomal loci in the offspring, Roan Gauntlet, are expected to be identical by descent.

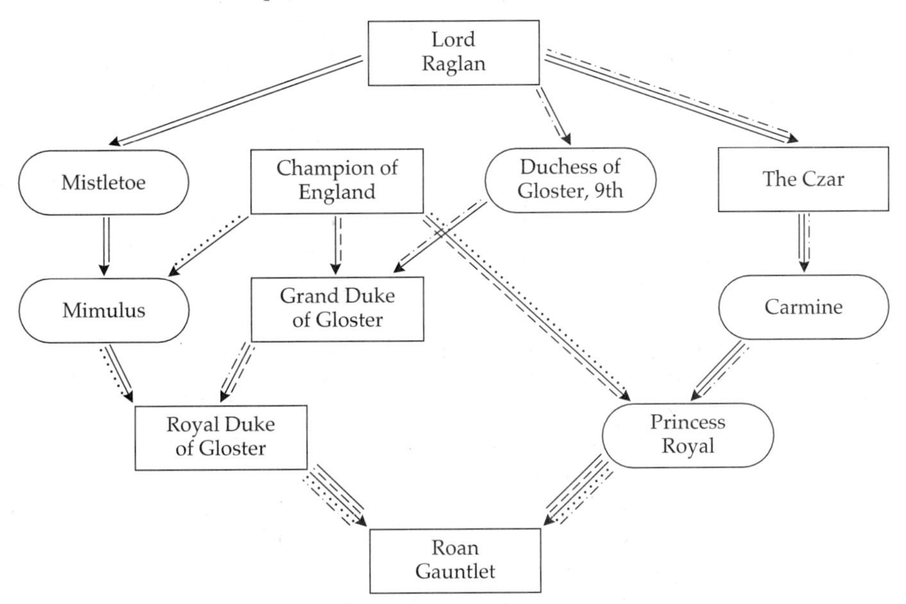

What is the coefficient of coancestry of the Royal Duke of Gloster and Princess Royal with respect to X-linked loci? Of the four paths between these two individuals, only the one traced by the dotted line does not contain consecutive males. Champion of England passes on his X chromosome to each of his daughters, Mimulus and Princess Royal, with probability 1. Mimulus passes that chromosome on to the Royal Duke of Gloster with probability 1/2. The probability of drawing a specific X chromosome from the Royal Duke is 1 (since he is a male) and from Princess Royal is 1/2. The coefficient of coancestry for X-linked loci is therefore $1 \cdot 1/2 \cdot 1 \cdot 1/2 = 1/4$. This is substantially greater than the coefficient for autosomal loci over the same path, which is $(1/2)^4 = 1/16$.

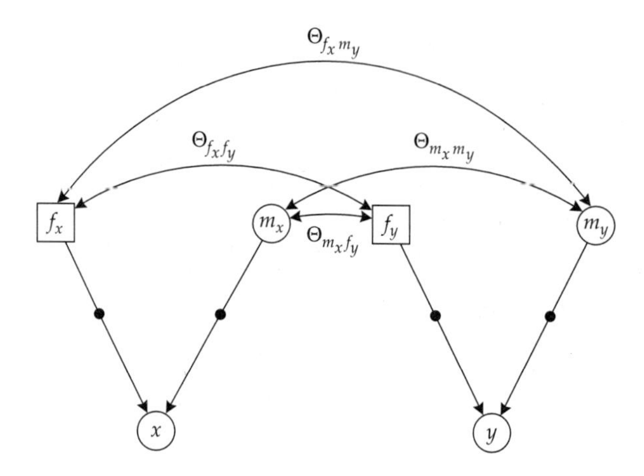

Figure 7.5 The analysis of the identity by descent of genotypes of individuals x and y. f_x and f_y represent fathers (which may be the same individual) of x and y, respectively, whereas m_x and m_y represent their mothers. Double-headed arrows between two parents represent coefficients of coancestry.

The Coefficient of Fraternity

Up to now we have been considering the identity of single genes by descent. Another useful measure is the probability that single-locus genotypes (both genes) of two individuals are identical by descent. The formulation of such a measure, which we denote as Δ_{xy}, is attributable to Cotterman (1954) and was called the **coefficient of fraternity** by Trustrum (1961). The problem is set out in Figure 7.5. Here we denote the mothers of individuals x and y as m_x and m_y, and the fathers as f_x and f_y. The coefficients of coancestry $\Theta_{m_x m_y}$, $\Theta_{m_x f_y}$, $\Theta_{f_x m_y}$, and $\Theta_{f_x f_y}$ provide measures of the probability of drawing genes identical by descent from all four combinations of parents.

There are two ways by which the genotype of x can be identical by descent with that of y: (1) the gene descending from m_x may be identical by descent with that descending from m_y, and that from f_x identical by descent with that from f_y, or (2) the gene from m_x may be identical by descent with that from f_y, and that from f_x identical by descent with that from m_y. Thus, the coefficient of fraternity is defined as

$$\Delta_{xy} = \Theta_{m_x m_y}\Theta_{f_x f_y} + \Theta_{m_x f_y}\Theta_{f_x m_y} \tag{7.7}$$

In terms of the condensed coefficients of identity, $\Delta_{xy} = \Delta_1 + \Delta_7$, which reduces to Δ_7 in the absence of inbreeding.

Two examples will suffice to illustrate the use of this equation. First, consider the situation when x and y are full sibs, in which case the mothers are the same individual ($m_x = m_y = m$), as are the fathers $f_x = f_y = f$. Equation 7.7 then reduces to

$$\Delta_{xy} = \Theta_{mm}\Theta_{ff} + \Theta_{mf}^2 \tag{7.8}$$

If the parents are unrelated, then $\Theta_{mf} = 0$; and if the parents are not inbred, then $\Theta_{mm} = \Theta_{ff} = 1/2$. Substituting these values into the above expression, we obtain $\Delta_{xy} = 1/4$.

Now consider the case of paternal half sibs, in which case the fathers are the same individual, but the mothers are different. Now,

$$\Delta_{xy} = \Theta_{m_x m_y}\Theta_{ff} + \Theta_{m_x f}\Theta_{f m_y} \tag{7.9}$$

Provided that the parents are unrelated, then $\Theta_{ff} = 1/2$ and $\Theta_{m_x m_y} = \Theta_{m_x f} = \Theta_{f m_y} = 0$, which yields $\Delta_{xy} = 0$. The genotypes of two individuals cannot be identical by descent if their maternally (or paternally) derived genes come from unrelated individuals.

Example 2. Returning to the figure in Example 1, what is Δ_{xy} for x = Royal Duke of Gloster and y = Princess Royal?

Designate the parents as f_x = Grand Duke of Gloster, m_x = Mimulus, f_y = Champion of England, and m_y = Carmine. Noting that Champion of England is the father of Grand Duke of Gloster and Mimulus, $\Theta_{f_x f_y} = \Theta_{m_x f_y} = (1/4)$. Counting the number of individuals in the paths of descent between the remaining two pairs of parents, $\Theta_{m_x m_y} = \Theta_{f_x m_y} = (1/2)^5$. Substituting into Equation 7.7, the probability that x and y have identical genotypes by descent at an arbitrary autosomal locus is

$$\Delta_{xy} = (1/4)(1/2)^5 + (1/2)^5(1/4) = (1/2)^6$$

We now have a complete system for describing the identity by descent at an arbitrary locus for any two individuals. For complex pedigrees, this can be a rather tedious process, but relatively simple algorithms exist for the computation of Θ_{xy} from simple information on parentage (Chapter 26). Karigl (1981) gives a general recursive procedure for obtaining all nine condensed coefficients of identity for arbitrary pedigrees. The identity coefficients for several common relationships are summarized in Table 7.1.

THE GENETIC COVARIANCE BETWEEN RELATIVES

The bulk of the credit goes to Fisher (1918) and Wright (1921b) for elucidating the connection between the phenotypic resemblance of relatives and the types of genetic variance in populations. However, despite the tremendous advances in these

Table 7.1 Identity coefficients for common relationships under the assumption of no inbreeding, in which case Δ_1 to $\Delta_6 = 0$.

Relationship	Δ_7	Δ_8	Δ_9	Θ_{xy}	Δ_{xy}
Parent–offspring	0	1	0	$\frac{1}{4}$	0
Grandparent–grandchild	0	$\frac{1}{2}$	$\frac{1}{2}$	$\frac{1}{8}$	0
Great grandparent–great grandchild	0	$\frac{1}{4}$	$\frac{3}{4}$	$\frac{1}{16}$	0
Half sibs	0	$\frac{1}{2}$	$\frac{1}{2}$	$\frac{1}{8}$	0
Full sibs, dizygotic twins	$\frac{1}{4}$	$\frac{1}{2}$	$\frac{1}{4}$	$\frac{1}{4}$	$\frac{1}{4}$
Uncle(aunt)–nephew(neice)	0	$\frac{1}{2}$	$\frac{1}{2}$	$\frac{1}{8}$	0
First cousins	0	$\frac{1}{4}$	$\frac{3}{4}$	$\frac{1}{16}$	0
Double first cousins	$\frac{1}{16}$	$\frac{6}{16}$	$\frac{9}{16}$	$\frac{1}{8}$	$\frac{1}{16}$
Second cousins	0	$\frac{1}{16}$	$\frac{15}{16}$	$\frac{1}{64}$	0
Monozygotic twins (clonemates)	1	0	0	$\frac{1}{2}$	1

papers, a number of problems remained. Wright did not pursue nonadditive gene action beyond dominance and even then ran into difficulties. Fisher incorporated epistasis but only of the additive $\times$ additive type. In 1954, Cockerham and Kempthorne independently published papers outlining how the earlier results could be generalized to include any type of gene action. Both authors arrived at the same result by very different routes, but Kempthorne's approach is simpler by far and is the one that we will pursue. Cockerham's name will appear many times throughout the book in other contexts.

At the outset, it must be emphasized that the simple results that will emerge below are not obtained without making a lot of assumptions: (1) all of the genetic variation is attributable to diploid, autosomal loci; (2) mating is random; (3) all loci are unlinked and in gametic phase equilibrium; (4) there is no genetic variation for maternal effects; (5) genotype-environment covariance and interaction are unimportant; (6) there is no sexual dimorphism; and (7) selection is not operating on the population. In due course, we will relax all of these conditions.

Our first task is to decompose the total genetic covariance between relatives into fundamental components that describe the various types of gene action. We accomplish this by following the logic used in Chapter 5 to partition the genetic variance. Consider a collection of pairs of individuals all of the same type of relationship, and let x and y represent the members of a random pair. From Equation 5.7, taking things out only to the two-locus effects, the genotypic values of the two individuals may be written as

$$\begin{aligned} G_{ijkl..}(x) &= \mu_G + [\alpha_i^x + \alpha_j^x + \alpha_k^x + \alpha_l^x + \cdots] + [\delta_{ij}^x + \delta_{kl}^x + \cdots] \\ &\quad + [(\alpha\alpha)_{ik}^x + (\alpha\alpha)_{il}^x + (\alpha\alpha)_{jk}^x + (\alpha\alpha)_{jl}^x + \cdots] \\ &\quad + [(\alpha\delta)_{ikl}^x + (\alpha\delta)_{jkl}^x + (\alpha\delta)_{kij}^x + (\alpha\delta)_{lij}^x + \cdots] + (\delta\delta)_{ijkl}^x + \cdots \\ G_{ijkl..}(y) &= \mu_G + [\alpha_i^y + \alpha_j^y + \alpha_k^y + \alpha_l^y + \cdots] + [\delta_{ij}^y + \delta_{kl}^y + \cdots] \\ &\quad + [(\alpha\alpha)_{ik}^y + (\alpha\alpha)_{il}^y + (\alpha\alpha)_{jk}^y + (\alpha\alpha)_{jl}^y + \cdots] \\ &\quad + [(\alpha\delta)_{ikl}^y + (\alpha\delta)_{jkl}^y + (\alpha\delta)_{kij}^y + (\alpha\delta)_{lij}^y + \cdots] + (\delta\delta)_{ijkl}^y + \cdots \end{aligned} \tag{7.10}$$

where i, j and k, l represent genes at the first and second loci. Fisher (1918) showed that just as the different types of effects are uncorrelated within individuals, they are also uncorrelated between individuals, provided the preceding assumptions are met. Consequently, the genetic covariance between relatives can be expanded into a series of terms, each describing the covariance between the same kinds of effects in two individuals:

$$\sigma_G(x,y) = \sigma_A(x,y) + \sigma_D(x,y) + \sigma_{AA}(x,y) + \sigma_{AD}(x,y) + \sigma_{DD}(x,y) + \cdots \tag{7.11}$$

Note that if $x = y$, Equation 7.11 reduces to Equation 5.8, the usual expression for the genetic variance.

The remaining task is to express the terms in Equation 7.11 in terms of variance components and coefficients of relationship. We will do this only for the first three terms and then give the general result. First, we evaluate the additive genetic covariance at locus 1. Since the mean value of the effects is zero by definition (Chapter 5), the covariance between x and y caused by the additive effects is equal to the expectation of the cross-product, $E[(\alpha_i^x + \alpha_j^x)(\alpha_i^y + \alpha_j^y)]$. Consider one of the four terms in the expansion, $E(\alpha_i^x \alpha_i^y)$. The two genes of interest may be identical by descent, with probability Θ_{xy}, in which case $E(\alpha_i^x \alpha_i^y) = E(\alpha_i^2)$, which is half the additive genetic variance attributable to locus 1. If the two genes are not identical by descent, then they must be distributed independently so that $E(\alpha_i^x \alpha_i^y) = [E(\alpha_i)]^2 = 0$. These same arguments can be applied to the remaining three terms in $E[(\alpha_i^x + \alpha_j^x)(\alpha_i^y + \alpha_j^y)]$. Thus, the additive genetic covariance is $4\Theta_{xy}E(\alpha_i^2)$, which is twice the additive genetic variance at the locus times the probability that randomly drawn genes from x and y are identical by descent. Noting that this result applies to all loci and that the distributions of effects at different loci are independent under the assumptions of the model, the additive genetic covariance, obtained by summing over loci, reduces to $\sigma_A(x,y) = 2\Theta_{xy}\sigma_A^2$.

We now move on to the dominance genetic covariance, which for locus 1 is $E(\delta_{ij}^x \delta_{ij}^y)$. If x and y are identical by descent for both genes at this locus, then $E(\delta_{ij}^x \delta_{ij}^y) = E(\delta_{ij}^2)$, which is the dominance genetic variance attributable to locus 1.

The probability of such identity is Δ_{xy}. On the other hand, if x and y do not have identical genotypes by descent, the dominance effects must be distributed independently, and hence $E(\delta_{ij}^x \delta_{ij}^y) = E^2(\delta_{ij}) = 0$. Again, since these arguments apply to every locus, the dominance genetic covariance, obtained by summing over all loci, is $\sigma_D(x, y) = \Delta_{xy}\sigma_D^2$.

Finally, we consider the epistatic genetic covariance caused by additive $\times$ additive effects. For any pair of loci, this involves the 16 cross-product terms in the expectation $E\{[(\alpha\alpha)_{ik}^x + (\alpha\alpha)_{il}^x + (\alpha\alpha)_{jk}^x + (\alpha\alpha)_{jl}^x][(\alpha\alpha)_{ik}^y + (\alpha\alpha)_{il}^y + (\alpha\alpha)_{jk}^y + (\alpha\alpha)_{jl}^y]\}$. Since the results are the same for all terms, it is sufficient to evaluate only one of them. The term $E[(\alpha\alpha)_{ik}^x (\alpha\alpha)_{ik}^y]$ is equivalent to $E[(\alpha\alpha)^2]$ provided that two conditions hold — a random gene drawn from the first locus in x must be identical by descent with one drawn from y, and the same condition must hold at the second locus. If identity by descent does not arise simultaneously for gene pairs drawn from both loci, $E[(\alpha\alpha)_{ik}^x (\alpha\alpha)_{ik}^y] = [E(\alpha\alpha)]^2 = 0$. Now, under the assumption of gametic phase equilibrium, the probability of identity by descent at locus 1 is independent of that at locus 2. Both probabilities are Θ_{xy}. Thus, the probability of joint identity by descent is Θ_{xy}^2, and the covariance caused by additive $\times$ additive epistasis between loci 1 and 2 is $16\,\Theta_{xy}^2 E[(\alpha\alpha)^2]$. Noting that $E[(\alpha\alpha)^2]$ is one-fourth the additive $\times$ additive genetic variance for a single pair of loci, and summing over all pairs, $\sigma_{AA}(x, y) = (2\Theta_{xy})^2\sigma_{AA}^2$.

All three of the terms evaluated reduce to simple functions of identity coefficients and a component of genetic variance, and the same is true for all higher-order terms. Due to the independent distributions of genes at different loci under the assumptions of the model, any covariance due to higher-order epistatic effects is equal to the product of the identity for each component additive effect, the probability of identity for each component dominance effect, and the corresponding variance component. Thus, the covariance attributable to additive $\times$ dominance epistasis is $2\Theta_{xy}\Delta_{xy}\sigma_{AD}^2$ because there is one additive and one dominance effect involved, while that caused by dominance $\times$ dominance epistasis is $\Delta_{xy}^2\sigma_{DD}^2$ because there are two dominance but no additive effects involved. Letting n be the number of additive effects and m be the number of dominance effects in a type of gene action, the expression for the covariance between relatives becomes

$$\begin{aligned}\sigma_G(x, y) &= \sum (2\Theta_{xy})^n \Delta_{xy}^m \sigma_{A^n D^m}^2 \\ &= 2\Theta_{xy}\sigma_A^2 + \Delta_{xy}\sigma_D^2 + (2\Theta_{xy})^2\sigma_{AA}^2 \\ &\quad + 2\Theta_{xy}\Delta_{xy}\sigma_{AD}^2 + \Delta_{xy}^2\sigma_{DD}^2 + (2\Theta_{xy})^3\sigma_{AAA}^2 + \cdots \end{aligned} \tag{7.12}$$

Drawing from the coefficients Θ_{xy} and Δ_{xy} given in Table 7.1, explicit expressions for the genetic covariances of common types of relatives are given in Table 7.2. Although these expressions are only expanded to include two-locus epistasis, several things are immediately apparent. First, gene action involving dominance only rarely contributes to the covariance between relatives. It requires that each parent of x be related to a different parent of y. Such relationships (full sibs,

Table 7.2 Coefficients for the components of genetic covariance between different types of relatives under the assumptions of random mating, free recombination, and gametic phase equilibrium.

Relationship	σ_A^2	σ_D^2	σ_{AA}^2	σ_{AD}^2	σ_{DD}^2
Parent–offspring	$\frac{1}{2}$		$\frac{1}{4}$		
Grandparent–grandchild	$\frac{1}{4}$		$\frac{1}{16}$		
Great grandparent–great grandchild	$\frac{1}{8}$		$\frac{1}{64}$		
Half sibs	$\frac{1}{4}$		$\frac{1}{16}$		
Full sibs, dizygotic twins	$\frac{1}{2}$	$\frac{1}{4}$	$\frac{1}{4}$	$\frac{1}{8}$	$\frac{1}{16}$
Uncle (aunt)–nephew (neice)	$\frac{1}{4}$		$\frac{1}{16}$		
First cousins	$\frac{1}{8}$		$\frac{1}{64}$		
Double first cousins	$\frac{1}{4}$	$\frac{1}{16}$	$\frac{1}{16}$	$\frac{1}{64}$	$\frac{1}{256}$
Second cousins	$\frac{1}{32}$		$\frac{1}{1024}$		
Monozygotic twins (clonemates)	1	1	1	1	1

Note: To obtain the covariance expression for a particular type of relationship, multiply each variance component by its coefficient and sum. For example, the genetic covariance between half sibs is $(\sigma_A^2/4) + (\sigma_{AA}^2/16)$. Blanks indicate values of zero.

double first cousins, and monozygotic twins) are said to be **collateral**. Second, the coefficient for σ_{AA}^2 declines more rapidly with the distance of the relationship than does that for σ_A^2. As noted in Chapters 4 and 5, because the additive genetic variance is a function of all higher-order types of gene action, these results should not be misconstrued to mean that the resemblance between relatives is influenced only slightly by dominance and epistatic gene action.

The most useful feature of the expressions in Table 7.2 involves their different coefficients, which permit the estimation of the different variance components from linear combinations of different observed genetic covariances between relatives. For example, ignoring higher-order epistasis and environmental sources of covariance, 8 × [parent-offspring covariance − (2 × half-sib covariance)] has an expected value of $8[(\sigma_A^2/2 + \sigma_{AA}^2/4) - 2(\sigma_A^2/4 + \sigma_{AA}^2/16)] = \sigma_{AA}^2$. Similarly, 2 × [(4 × half-sib covariance) − (parent-offspring covariance)] has an expected value of σ_A^2. Subsequent examples in this chapter will illustrate the utility of these kinds of manipulations.

THE EFFECTS OF LINKAGE AND GAMETIC PHASE DISEQUILIBRIUM

In deriving the Kempthorne-Cockerham equation (7.12), we assumed that the constituent loci are freely recombining and in gametic phase equilibrium. We now consider the extent to which the interpretation of observed covariances between relatives needs to be modified in the face of violations of these assumptions. In most practical situations, we have little if any information on either linkage or gametic phase disequilibrium, so the following theoretical results provide our only guidance as to the potential seriousness of the matter.

The most complete analyses of this problem were developed by Gallais (1974) and Weir and Cockerham (1977), who allowed for inbreeding as well as linkage and gametic phase disequilibrium. Neither study went beyond two-locus relationships, as even then the algebraic and notational complexities are enormous. We will take a simpler approach than these authors, first considering the consequences of linkage under the assumption of gametic phase equilibrium and then evaluating some of the consequences of disequilibrium. In both cases, we will assume that mating is random and that all of the remaining assumptions of the Kempthorne-Cockerham model are fulfilled.

Linkage

Under the assumption of gametic phase equilibrium, linkage influences only the epistatic components of genetic covariance, which depend upon the multilocus gene combinations inherited through gametes. Recall that in Equation 7.12, each component of epistatic variance is weighted by a coefficient of the form $(2\Theta_{xy})^n\Delta_{xy}^m$. The term Θ_{xy}^n is the probability that randomly drawn pairs of genes (one from x and one from y) will be simultaneously identical by descent at n loci, the 2^n coming in because of diploidy. Similarly, Δ_{xy}^m is the probability that the genotypes of x and y are simultaneously identical by descent at m loci. These definitions of the joint probability of events at multiple loci as the products of probabilities at individual loci assume that identity by descent is distributed independently among loci. However, for linked loci, identity by descent is expected to be positively correlated since the genes at such loci tend to be inherited together. Thus, for linked loci, we can anticipate that the multilocus coefficients $(2\Theta_{xy})^n\Delta_{xy}^m$ must be too low, to a degree depending on the recombination frequency.

A general solution to this problem requires the use of **digenic descent coefficients,** which define the probability that two nonalleles (genes from different loci) are copies of genes that were originally contained in the same gamete. Such a condition is known as **equivalence by descent.** The formal theory of digenic descent, developed in great detail by Cockerham and Weir (1968, 1973, 1977a) and Weir and Cockerham (1968, 1969, 1973, 1974, 1977), is algebraically and notationally complex. We will rely upon an approach that is less general but more transparent (Schnell 1961, 1963, Van Aarde 1975). There is no easy entry into this field, but for the adventuresome we suggest the reviews of Weir and Cockerham

(1977, 1989) as a starting point.

For our purposes, it will suffice to examine an arbitrary pair of loci, A and B. Consider an individual whose two-locus genotype resulted from the fusion of $A_m B_m$ and $A_f B_f$ gametes (m for mother, f for father). If the recombination fraction for the two loci is c, then this individual will produce gametes in frequencies: $p(A_m B_m) = p(A_f B_f) = (1-c)/2$ and $p(A_m B_f) = p(A_f B_m) = c/2$. Two gametes randomly drawn from this individual can have four possible identity-by-descent relationships. Identity exists at both loci with probability $p(A,B) = p^2(A_m B_m) + p^2(A_f B_f) + p^2(A_m B_f) + p^2(A_f B_m)$, at neither locus with probability $p(0,0) = 2p(A_m B_m)p(A_f B_f) + 2p(A_m B_f)p(A_f B_m)$, at only the A locus with probability $p(A,0) = 2p(A_m B_m)p(A_m B_f) + 2p(A_f B_m)p(A_f B_f)$, and at only the B locus with probability $p(0,B) = 2p(A_f B_f)p(A_m B_f) + 2p(A_f B_m)p(A_m B_m)$. Letting $\lambda = (1-2c)^2$, these gametic-identity probabilities simplify to

$$p(A,B) = p(0,0) = \frac{1+\lambda}{4} \tag{7.13a}$$

$$p(A,0) = p(0,B) = \frac{1-\lambda}{4} \tag{7.13b}$$

all of which reduce to $1/4$ with free recombination ($c = 0.5$).

As a specific application of gametic identity probabilities, we will consider the case of full sibs, first evaluating the covariance caused by additive $\times$ additive epistasis. As noted previously, for each pair of loci, the genotypic value of each sib contains four $(\alpha\alpha)$ terms, so there are 16 combinations of these terms between individuals. For each of the 16 combinations, the quantity of interest is the joint probability that randomly drawn A genes (one from each sib) are identical by descent and that randomly drawn B genes (one from each sib) are identical by descent. We first draw a random A gene from the two sibs. As noted earlier, these are identical by descent with probability $1/4$. Given that identity by descent was obtained at the A locus, we now draw the B genes (again, one from each sib). These genes may be identical by descent through two routes — they may both derive from the parent from which the A genes were drawn (with probability $1/4$) or both from the opposite parent (with probability $1/4$). If they come from the same parent as the A gene, the B genes will be identical by descent with conditional probability $p(A,B)/[p(A,B)+p(A,0)] = (1+\lambda)/2$. If they come from the opposite parent, they are identical by descent with probability $1/2$. Summing up, the probability of digenic identity by descent for two full sibs is $1/4 \cdot 1/4 \cdot [\,(1+\lambda)/2 + 1/2\,] = (2+\lambda)/32$. Given such identity, the contribution to the genetic covariance is $E[(\alpha\alpha)^2]$, which is equivalent to one-fourth of the additive $\times$ additive genetic variance. Thus, after taking account of all 16 pairs of additive $\times$ additive effects, the covariance between full sibs resulting from additive $\times$ additive epistasis is $16 \cdot [(2+\lambda)/32] \cdot [\sigma^2_{AA}/4] = (2+\lambda)\sigma^2_{AA}/8$. With free recombination ($\lambda = 0$), this reduces to $\sigma^2_{AA}/4$, our previous result (Table 7.2). Linkage ($\lambda > 0$) causes the additive $\times$ additive genetic variance to be inflated.

Table 7.3 Coefficients for the components of genetic covariance between relatives modified to account for linkage (first five columns) and gametic phase disequilibrium (last two columns).

Relationship	σ_A^2	σ_D^2	σ_{AA}^2	σ_{AD}^2	σ_{DD}^2	$\sigma_{A,A}(0)$	$\sigma_{D,D}(0)$
Parent–offspring	$\frac{1}{2}$		$\frac{1}{4}$			$\frac{(1-c)^t}{2}$	
GP – GC	$\frac{1}{4}$		$\frac{1+\lambda^{1/2}}{16}$			$\frac{(1-c)^t}{4}$	
Great GP – great GC	$\frac{1}{8}$		$\frac{(1+\lambda^{1/2})^2}{64}$			$\frac{(1-c)^t}{8}$	
Half sibs	$\frac{1}{4}$		$\frac{1+\lambda}{16}$			$\frac{(1-c)^t}{4}$	
Full sibs, dizygotic twins	$\frac{1}{2}$	$\frac{1}{4}$	$\frac{2+\lambda}{8}$	$\frac{1+\lambda}{8}$	$\frac{(1+\lambda)^2}{16}$	$\frac{(1-c)^t}{2}$	$\frac{(1-c)^{2t}}{4}$
Uncle(aunt)–nephew(niece)	$\frac{1}{4}$		$\frac{1+\lambda(1+\lambda^{1/2})/2}{16}$			$\frac{(1-c)^t}{4}$	
First cousins	$\frac{1}{8}$		$\frac{1+(1+\lambda)^3}{128}$			$\frac{(1-c)^t}{8}$	
Double first cousins	$\frac{1}{4}$	$\frac{1}{16}$	$\frac{3+(1+\lambda)^3}{64}$	$\frac{1+(1+\lambda)^3}{128}$	$\frac{(1+\lambda)^4}{256}$	$\frac{(1-c)^t}{4}$	$\frac{(1-c)^{2t}}{16}$
Second cousins	$\frac{1}{32}$		$\frac{1+(1+\lambda)^5}{1024}$			$\frac{(1-c)^t}{32}$	
Monozygotic twins	1	1	1	1	1	$(1-c)^t$	$(1-c)^{2t}$

Note: GP represents grandparent, GC represents grandchild, $\lambda = (1 - 2c)^2$, and t is the generation number for the common ancestors. Blanks denote values of zero.

We next consider the additive × dominance covariance between full sibs. Again, from Equation 7.10, for any pair of loci, there are 16 combinations of $(\alpha\delta)$ terms in the two sibs. If the two individuals possess an additive × dominance effect that is identical by descent, the contribution to the genetic covariance will be $\sigma_{AD}^2/4$. To evaluate the probability of such an event, we need to determine the joint probability of drawing single genes that are identical by descent at one locus and genotypes that are identical by descent at the second locus. Such identity is only possible in the eight comparisons for which the additive effects involve the same locus in x and y. As usual with full sibs, the probability of drawing two genes (one from each sib) that are identical by descent is $1/4$. Also as noted above, given that the genes drawn at one locus are identical by descent, the genes for the second locus that descended from gametes from the same parent are identical by descent with probability $(1+\lambda)/2$. The other pair of genes at the second locus are identical by descent with probability $1/2$. Summing up, the additive × dominance covariance becomes $8 \cdot (1/4) \cdot [(1+\lambda)/2] \cdot (1/2) \cdot (\sigma_{AD}^2/4) = (1+\lambda)\sigma_{AD}^2/8$, which reduces to $\sigma_{AD}^2/8$ with free recombination.

Finally, we note that covariance between the single $(\delta\delta)$ term for a pair of loci requires that both sibs inherit maternal gametes that are identical by descent at

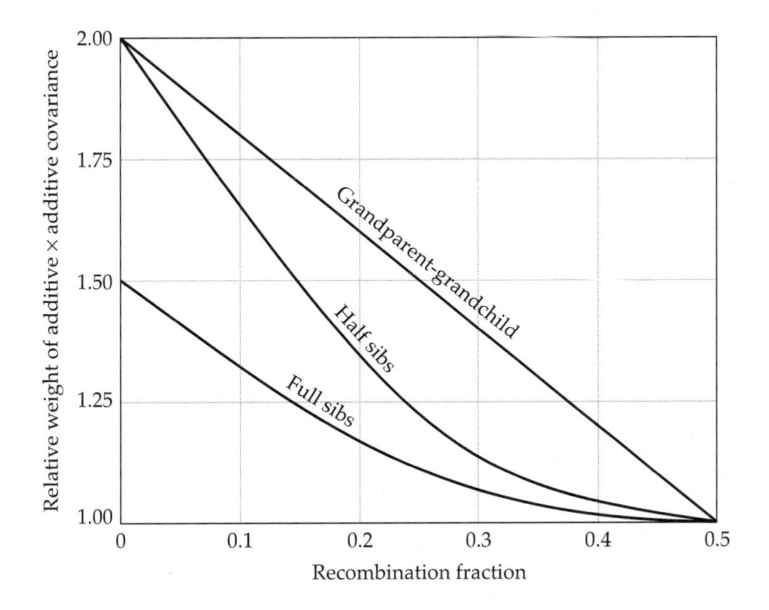

Figure 7.6 Inflation of the additive × additive covariance between relatives caused by linkage, obtained by use of the coefficients in Table 7.3.

both loci, and similarly for the paternal gametes. Each of these events occurs with probability $p(A, B)$. Therefore, the dominance × dominance covariance between full sibs is $p^2(A, B)\sigma^2_{DD} = (1 + \lambda)^2\sigma^2_{DD}/16$.

The coefficients for full sibs and for other common relationships are summarized in Table 7.3. The main conclusion to be drawn from these results is that *except for parent-offspring and monozygotic twin relationships, linkage inflates the covariance between relatives, unless there are no epistatic sources of genetic variance.* Linkage has no influence on the parent-offspring covariance because the two individuals always have exactly one gene identical by descent at each locus, and therefore share exactly one of the four additive × additive effects. For grandparent-grandchild and half-sib relations, relative to the situation with free recombination, the additive × additive covariance can be inflated as much as twofold with completely linked loci ($\lambda = 1$), whereas with full sibs the inflation can be no greater than 50% (Figure 7.6). It is of interest to note that whereas the expressions for the covariance of half sibs and for grandparent-grandchild are identical under free recombination, the latter is inflated to a greater degree by linkage unless linkage is complete ($\lambda = 1$). Thus, at least in principle, a comparison of these two types of covariance may shed some light on the presence of linkage for quantitative-trait loci.

Gametic Phase Disequilibrium

Further complications arise when loci are in gametic phase disequilibrium since this can cause a covariance between the effects of genes carried in the same gamete. The problem has been addressed by Weir et al. (1980), who ignored epistasis and assumed random mating, as we do below. In Chapter 5 (Equations 5.16a,b), we saw that in the presence of gametic phase disequilibrium, the total genetic variance can be expressed as

$$\sigma_G^2 = \sigma_A^2 + \sigma_{A,A} + \sigma_D^2 + \sigma_{D,D} \tag{7.14}$$

where $\sigma_{A,A}$ is the contribution due to covariance of additive effects of nonalleles within gametes (the **additive disequilibrium covariance**), and $\sigma_{D,D}$ is the contribution of covariance due to dominance effects of different loci within individuals (the **dominance disequilibrium covariance**).

We now consider the genetic covariance between relatives derived from a base population displaying gametic phase disequilibria. The covariance between relatives attributable to equilibrium additive and dominance genetic variance (σ_A^2 and σ_D^2) is the same as given above, so it is only necessary to consider the additional contributions resulting from $\sigma_{A,A}$ and $\sigma_{D,D}$. For situations in which the study population is being maintained in a steady state of disequilibrium from generation to generation (by processes such as natural selection, migration, and/or nonrandom mating), the modifications to the theory are simple — all of the preceding formulations still apply, except that $(\sigma_A^2 + \sigma_{A,A})$ replaces σ_A^2, and $(\sigma_D^2 + \sigma_{D,D})$ replaces σ_D^2.

The more interesting situation arises when a study population, initially in gametic phase disequilibrium, is allowed to mate randomly in an environment in which the forces maintaining disequilibria are relaxed. In this case, the covariances between relatives change through time as recombination causes the disequilibrium components of covariance to decay towards zero. In the following, we will consider only the simple (and most common) situation in which the ancestors leading to a relationship are all members of the same generation; more general results can be found in Weir et al. (1980).

We start by considering the additive component of disequilibrium covariance. Returning to Equation 7.10, we make the distinction that allele i at the first locus and allele k at the second locus are inherited in one gamete, whereas genes j and l are inherited in the other gamete. By using the expression for the variance of a sum, Equation 3.11b, the additive equilibrium genetic variance is defined to be

$$\sigma_A^2 = E(\alpha_i^2) + E(\alpha_j^2) + E(\alpha_k^2) + E(\alpha_l^2) \tag{7.15a}$$

On the other hand, the additive disequilibrium covariance in the base population is

$$\sigma_{A,A}(0) = 2E(\alpha_i\alpha_k) + 2E(\alpha_j\alpha_l) \tag{7.15b}$$

Note that terms involving cross-gamete expectations, e.g., $E(\alpha_i\alpha_j)$ and $E(\alpha_i\alpha_l)$, do not appear in this expression because their expected values are zero under the assumption of random mating (as there is no correlation between gametes). In Chapter 5, we found that in the absence of restoring forces, gametic phase disequilibrium is reduced by a fraction c after each generation of random mating. Thus, the additive disequilibrium covariance remaining after t generations is

$$\sigma_{A,A}(t) = (1-c)^t \sigma_{A,A}(0) \tag{7.15c}$$

To obtain the covariance between relatives associated with the additive equilibrium component of variance, we take expectations of the cross-products of additive effects of genes in two individuals. A similar procedure is followed in obtaining the covariance due to $\sigma_{A,A}$, except that we now focus on pairs of genes at different loci, one in each member of the relationship,

$$\begin{aligned}\sigma_{A,A}(x,y,t) = \big[&E(\alpha_i^x\alpha_k^y) + E(\alpha_i^x\alpha_l^y) + E(\alpha_j^x\alpha_k^y) + E(\alpha_j^x\alpha_l^y)\\ &+ E(\alpha_k^x\alpha_i^y) + E(\alpha_k^x\alpha_j^y) + E(\alpha_l^x\alpha_i^y) + E(\alpha_l^x\alpha_j^y)\big]\end{aligned} \tag{7.15d}$$

For any of the terms in this equation to be nonzero, the genes in them must be equivalent by descent. The key to solving Equation 7.15d is the fact that *the probability of equivalence by descent for genes at different loci in different gametes is the same as the probability of identity by descent for alleles at the same locus.* The reason for this equivalence is that when parents produce a gamete pool, although the two loci (A and B) may be linked, under random mating the gene at the A locus in one parental gamete is independent of the gene at the B locus in a second gamete. Thus, the probability of any term in Equation 7.15d being nonzero is Θ_{xy}.

From Equation 7.15b, we see that nonzero terms of the form $E(\alpha_i\alpha_k)$ initially have expected values equal to $\sigma_{A,A}(0)/4$. At time t, however, each of the nonzero terms in Equation 7.15d has expected value $\sigma_{A,A}(t)/4$. Because there are eight terms, the covariance between relatives due to additive disequilibrium covariance is $8\,\Theta_{xy}\,\sigma_{A,A}(t)/4$, or from Equation 7.15c,

$$\sigma_{A,A}(x,y,t) = 2\Theta_{xy}(1-c)^t\sigma_{A,A}(0) \tag{7.16}$$

where t denotes the number of generations that the common ancestors are removed from the base population. Thus, for example, the parent-offspring covariance (with $\Theta_{xy} = 1/4$) resulting from additive disequilibrium covariance is $\sigma_{A,A}(0)/2$ if the parents are members of the base population, $(1-c)\sigma_{A,A}(0)/2$ if the parents are second-generation individuals, and $(1-c)^2\sigma_{A,A}(0)/2$ if they are third-generation individuals.

Derivation of the covariance between relatives resulting from dominance disequilibrium covariance follows the same logic just presented. However, in this case, the dominance disequilibrium covariance declines each generation to $(1-c)^2$ of its previous value. This quadratic decline occurs because dominance disequilibria are only maintained if neither of the gametes fusing to form a zygote have undergone recombination. Thus, the dominance disequilibrium covariance is

$$\sigma_{D,D}(t) = (1-c)^{2t}\sigma_{D,D}(0) \tag{7.17a}$$

and the covariance between relatives due to this disequilibrium is

$$\sigma_{D,D}(x,y,t) = \Delta_{xy}(1-c)^{2t}\sigma_{D,D}(0) \tag{7.17b}$$

For example, the covariance between full sibs resulting from dominance disequilibrium covariance is $\sigma_{D,D}(0)/4$, $(1-c)^2\sigma_{D,D}(0)/4$, and $(1-c)^4\sigma_{D,D}(0)/4$, respectively, when the parents are members of the base population, second, and third generations.

The general coefficients for the disequilibrium covariances for common types of relationships, assuming the common ancestor to be a member of generation t (where $t = 0$ denotes the base population), are given in Table 7.3. Unlike all of the equilibrium components of genetic variance, which always cause positive phenotypic covariance between relatives, the components resulting from disequilibrium covariance may be positive or negative depending upon whether loci are in coupling or repulsion disequilibrium.

Strictly speaking, the modified coefficients in Table 7.3 apply to a single pair of loci with recombination frequency c. These expressions could be refined further to account for the influence of linkage and/or gametic phase disequilibrium of all loci on the covariance between relatives by summing terms over all pairs of loci, weighting each locus by its specific set of variances and covariances. However, without detailed information on the map structure of the constituent loci for a quantitative trait, such refinements would be of little practical value. An alternative approximation can be obtained by assuming that the loci underlying the trait are distributed randomly throughout the genome and using the average recombination frequency $\bar{c}$ in place of c. Estimates of $\bar{c}$ are given for a number of species in Table 9.2.

It may be argued that linkage is of little consequence for the resemblance between relatives in organisms with high chromosome numbers since most pairs of loci will lie on different chromosomes. In this case, $\lambda \simeq 0$, and there is no inflation of the epistatic components of covariance, regardless of the magnitude of gametic phase disequilibrium. However, regardless of the degree of linkage, gametic phase disequilibrium is of special concern, as theoretical arguments and some empirical data suggest that it may reach significant levels for characters under selection (Chapter 5). In this case, even with free recombination ($c = 0.5$), the contribution of disequilibrium covariance to the resemblance between relatives is nonzero. In the following section, we consider a common situation in which the build-up of positive $\sigma_{A,A}$ is inevitable, even with unlinked loci.

Example 3. There is one case where the probability of equivalence by descent is unequal to the coefficient of coancestry, which renders Equations 7.17a,b inappropriate. In monozygotic twins, both members of the pair are products of the same gametes, so the genes inherited in one twin are not independent of those inherited in the other twin through the same parent. Because monozygotic twins have genetic effects at all loci identical by descent, the covariance between monozygotic twins is equivalent to the expressed genetic variance in the population in the present generation. For twins whose parents are members of the base population, the probability that each ancestral gamete contributing to the twin progeny has not experienced a recombination event (between two loci of interest) is $(1-c)$. Therefore, the covariance between monozygotic twins resulting from additive and dominance effects is

$$\sigma_A(MZ) = \sigma_A^2 + (1-c)\sigma_{A,A}(0)$$

$$\sigma_D(MZ) = \sigma_D^2 + (1-c)^2\sigma_{D,D}(0)$$

where the disequilibrium covariances refer to the levels in the parental generation. The sum of these two quantities is the total expressed genetic variance in the population in the twin's generation (ignorning epistasis).

ASSORTATIVE MATING

Although we have assumed a randomly mating population though this chapter, it is not unusual for mate choice to be based on aspects of the phenotype. Often, individuals will choose mates whose phenotype resembles their own (positive assortative mating). In humans, for example, there is significant correlation between mates with respect to height, skin color, IQ, social status, religion, and some conditions such as deafness (Spuhler 1968, Vandenberg 1972). In plants, assortative mating based on flowering time is common since individuals often produce viable pollen for only short periods of time. Negative or disassortative mating is less common.

Many assortative mating systems are selective, such that some phenotypes of one or both sexes have a greater ability to attract mates than others (discussed in our next volume). For the time being, however, we will confine our attention to nonselective assortative mating, i.e., we assume that all individuals have an equal opportunity to reproduce, the phenotypic distribution of their available mates being the only limitation. This is the type of assortative mating that Fisher (1918) and Wright (1921d) had in mind when they first attacked the problem from a quantitative-genetic perspective. Almost all of the basic results of the theory were produced by these two men. Fisher's (1918) elaborate treatment of the subject is notoriously difficult, and some uncertainty still exists as to exactly what he meant to say (Moran and Smith 1966, Wilson 1973, Vetta and Smith 1974).

In some ways, positive assortative mating is like inbreeding, but in other ways it is very different. A simple way to discriminate between the two is to note that inbreeding is choice of mates based on similar genotypes, while assortative mating is based on phenotypes. If there is genetic variance for the characters that are the targets of mate choice, then positive assortative mating must cause identical alleles to come together more often than in the case of random mating. This will cause an increase in the homozygosity of the population, just as inbreeding does. However, while prolonged inbreeding can result in a completely homozygous population, positive assortative mating will not generally induce such an extreme genetic structure. It definitely will not if there are many loci underlying the character upon which mate choice depends, if there is any variation for the trait due to things other than additive genetic variance, or if the correlation between mates is less than perfect.

It was shown in Chapter 4 (Equation 4.23c) that inbreeding can inflate the genetic variance of a population up to twofold in the absence of epistasis. Strong positive assortative mating has the potential to cause an even greater inflation, while disassortative mating induces a reduction of the genetic variance. The change in variance brought about by assortative mating is primarily a consequence of a directional build-up of gametic phase disequilibria. Positive assortative mating increases the coupling of genes with similar effects, i.e., induces a positive covariance between allelic effects at different loci, while disassortative mating leads to the proliferation of repulsion gametes, in which positive effects at one locus are balanced by negative effects at another. Inbreeding also causes a build-up of gametic phase disequilibrium, but there is no tendency for extreme gamete types to be formed at the expense of balanced ones, or vice versa.

We now take a more quantitative look at these matters, starting with the consequences of assortative mating for the additive genetic variance. We will restrict our attention to the situation in which interactions between loci are additive, since epistasis has not yet been incorporated into the theory. We define ρ_z and ρ_g to be the phenotypic and genetic correlations between mates and assume that the regression of phenotypes of mates is linear. We also suppose that there are n loci, each contributing equally to the genetic variance of the trait, and define the parameter $\gamma = 1 - [1/(2n)]$.

Starting from a random-mating base population in gametic phase equilibrium with additive genetic variance σ_A^2, a single generation of assortative mating will shift the additive genetic variance to

$$\sigma_A^2(1) = \left(1 + \frac{\rho_z h^2}{2}\right)\sigma_A^2 \tag{7.18}$$

where $h^2 = \sigma_A^2/\sigma_z^2$ (Crow and Kimura 1970). With continued assortative mating,

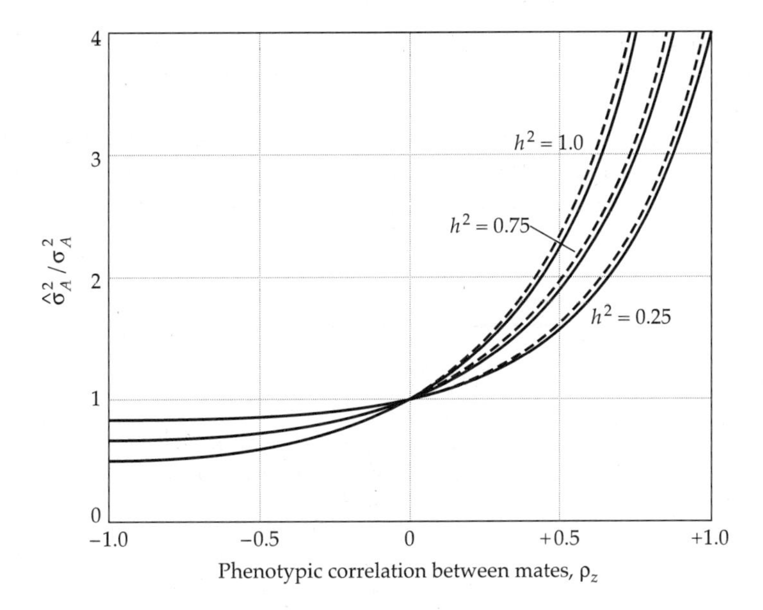

Figure 7.7 Inflation of the additive genetic variance at equilibrium under assortative mating relative to that in an otherwise identical random-mating population. Solid and dashed lines refer to 10 and 20 effective loci, respectively.

the variance asymptotically approaches the equilibrium

$$\widehat{\sigma}_A^2 = \frac{\sigma_A^2}{1 - \gamma\widehat{\rho}_g} \tag{7.19a}$$

where $\widehat{\rho}_g = \rho_z \widehat{\sigma}_A^2 / \widehat{\sigma}_z^2$ is the equilibrium genetic correlation between mates. This solution was first obtained by Wright (1921d) for unlinked loci with equivalent effects. Later, Crow and Felsenstein (1968), Bulmer (1980), and Nagylaki (1982) showed that the same result holds with linked loci with arbitrary effects, as long as n is replaced by n_e, the effective number of loci. Thus, the general form of Equation 7.19a holds regardless of the map structure of loci. Gimelfarb (1984) proved that the equilibrium is stable. Equation 7.19a can be rewritten as

$$\frac{\widehat{\sigma}_A^2}{\sigma_A^2} = \frac{2 + \left[\sqrt{1 - 4\gamma\rho_z h^2(1 - h^2)} - 1\right]/h^2}{2(1 - \gamma\rho_z)} \tag{7.19b}$$

which gives the inflation of the additive equilibrium genetic variance relative to that in the random-mating base population (Figure 7.7). The difference $\widehat{\sigma}_A^2 - \sigma_A^2$ is the additive disequilibrium covariance, $\sigma_{A,A}$, maintained by assortative mating.

Three points can be gleaned from Figure 7.7. First, there is an asymmetry in the response to positive and negative assortative mating. Very strong positive assortative mating can inflate the additive genetic variance to nearly $2n_e\sigma_A^2$ if h^2 is also very high. However, negative assortative mating can depress the variance to no less than $\sigma_A^2/2$. The little empirical work that has been done in this area is qualitatively consistent with this expectation (Breese 1956, McBride and Robertson 1963; see also the following example). Second, assortative mating must be fairly strong ($\rho_z^2 \geq 0.2$) and combined with high h^2 to induce much change in the variance. Third, the effective number of loci has a negligible effect unless it is very small.

Example 4. Gimelfarb (1984) performed an experiment in which lines of *Drosophila melanogaster* were artificially maintained under an absolute regimen of positive and negative assortative mating ($\rho_z \simeq \pm 1.0$). Each generation, 300 male and 300 female flies from each line were scored for abdominal bristle number, and the line was then propagated by performing 300 rank-ordered matings. The change in the phenotypic variance of abdominal bristle number over an eight generation period is shown in the following figure.

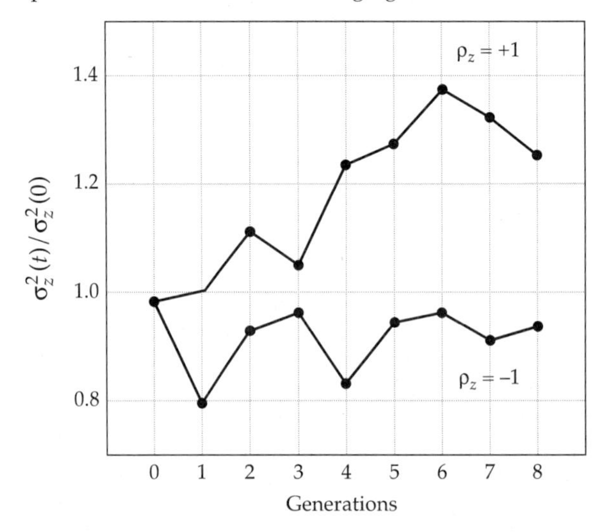

To what extent can these results be reconciled with the theory? Abdominal bristle number in *D. melanogaster* has been the subject of many quantitative-genetic studies. It appears to be nearly completely lacking in nonadditive genetic variance, and random-mating laboratory populations generally exhibit heritabilities of approximately $h^2 = \sigma_A^2/(\sigma_A^2 + \sigma_e^2) = 0.5$. Assuming that the effective number of loci underlying the trait is at least five or so, then $\gamma \simeq 1$, and Equation 7.19b predicts that complete negative assortative mating should ultimately reduce the additive genetic variance to approximately 70% of the level in the base population. Thus, scaling the original phenotypic variance to be one so that $\sigma_A^2 = \sigma_e^2 = 0.50$, we expect complete negative assortative mating to depress the additive genetic variance to approximately $\widehat{\sigma}_A^2 = 0.70 \times 0.50 = 0.35$. From the figure, it can be seen that the *phenotypic* variance declined to approximately 90% of the level in the base population. This is close to the theoretical expectation that the phenotypic variance should be 85% of the base-population level, $\widehat{\sigma}_z^2/\sigma_z^2 = (\widehat{\sigma}_A^2 + \sigma_e^2)/(\sigma_A^2 + \sigma_e^2) = (0.35 + 0.50)/(0.50 + 0.50) = 0.85$.

Such an analysis cannot be performed for the line under positive assortative mating because it is less clear that an equilibrium phenotypic variance had been obtained. Nevertheless, a simple expectation can be pointed out for an experiment of this type. With absolute positive assortative mating, $\rho_z = 1.0$, and a heritability of 0.5, the expected inflation of the additive genetic variance predicted by Equation 7.19b is simply $(2n_e)^{1/2}$. Thus, if this experiment had been carried out to the point at which $\widehat{\sigma}_z^2$ had stabilized, an estimate of the effective number of factors for abdominal bristle number would have been possible.

To this point, we have said nothing about the dominance component of variance. When only two loci contribute to the trait, σ_D^2 can change with assortative mating (Reeve 1961). However, Fisher (1918) argued that if the number of loci is even moderate, the effect will be negligible, and this was later confirmed by Vetta (1976). Thus, for all practical purposes, we can assume that the dominance genetic variance is unaltered by assortative mating.

In addition to creating gametic phase disequilibrium, assortative mating causes the genotype frequencies for the selected trait to deviate from Hardy-Weinberg expectations. Positive assortative mating leads to a heterozygote deficit, while negative assorative mating leads to a heterozygote excess. Wright (1921d) and Nagylaki (1982) showed that if the effective number of loci is large, Hardy-Weinberg deviations will be minor unless both the phenotypic correlation between mates and h^2 are very large. This implies that the vast majority of the change in genetic variance induced by assortative mating is a consequence of gametic phase disequilibrium rather than Hardy-Weinberg disequilibrium, i.e., of allelic associations within rather than between gametes.

The covariance between relatives in a population undergoing assortative mating has been considered by Fisher (1918), Crow and Felsenstein (1968), and Gimelfarb (1981). Nagylaki (1978) uses path analysis to yield a particularly lucid overview for the case of additive gene action. The coefficients of the additive and dominance components of covariance are given for common types of relationships in Table 7.4. Positive assortative mating inflates the additive genetic covariance between all types of relatives, while disassortative mating depresses it. Moreover, for some types of relationships (uncle-nephew, first and second cousins), assortative mating induces dominance genetic covariance where there otherwise would be none. We emphasize that the additive variance appearing in the resemblance

Table 7.4 Coefficients for the additive and dominance components of the covariance between relatives for an equilibrium population undergoing assortative mating.

Relationship	$\widehat{\sigma}^2_A$	σ^2_D
Parent–offspring	$\frac{1}{2}(1+\rho_z)$	
Grandparent–grandchild	$\frac{1}{4}(1+\rho_z)(1+\rho_z\widehat{h}^2)$	
Great grandparent–great grandchild	$\frac{1}{8}(1+\rho_z)(1+\rho_z\widehat{h}^2)^2$	
Half sibs	$\frac{1}{4}(1+2\rho_z\widehat{h}^2+\rho_z^2\widehat{h}^2)$	
Full sibs, dizygotic twins	$\frac{1}{2}(1+\rho_z\widehat{h}^2)$	$\frac{1}{4}$
Uncle(aunt)–nephew(niece)	$\frac{1}{4}(1+\rho_z\widehat{h}^2)^2$	$\frac{1}{8}\rho_z\widehat{h}^2$
First cousins	$\frac{1}{8}(1+\rho_z\widehat{h}^2)^3$	$\frac{1}{16}(\rho_z\widehat{h}^2)^2$
Double first cousins	$\frac{1}{4}(1+3\rho_z\widehat{h}^2)$	$\frac{1}{16}$
Second cousins	$\frac{1}{32}(1+\rho_z\widehat{h}^2)^5$	$\frac{1}{64}(\rho_z\widehat{h}^2)^4$
Monozygotic twins	1	1

Note: The equilibrium heritability is $\widehat{h}^2 = \widehat{\sigma}^2_A/\widehat{\sigma}^2_z$. In the absence of assortative mating, $\rho_z = 0$, and all coefficients reduce to those in Table 7.2.

between relatives is $\widehat{\sigma}^2_A$, the variance in the current assortatively mating population, not σ^2_A, the variance that would exist in a random-mating population. The latter can always be obtained by rearranging Equation 7.19a. Assuming large n_e,

$$\sigma^2_A = \left(1 - \frac{\rho_z\widehat{\sigma}^2_A}{\widehat{\sigma}^2_z}\right)\widehat{\sigma}^2_A \tag{7.20}$$

Example 5. As an example of the application of the preceding theory, we consider an early data set on height in British families (Pearson and Lee 1903). Pearson recruited college students to obtain data from approximately 1300 families, recording whenever possible the stature of father, mother, and eldest son and daughter (ignoring offspring less than 18 years of age). This was a very large data set for the precomputer era, and it took two years to calculate the statistics by hand. The data are remarkable for their essentially normal distribution and for the linearity of the regressions between relatives.

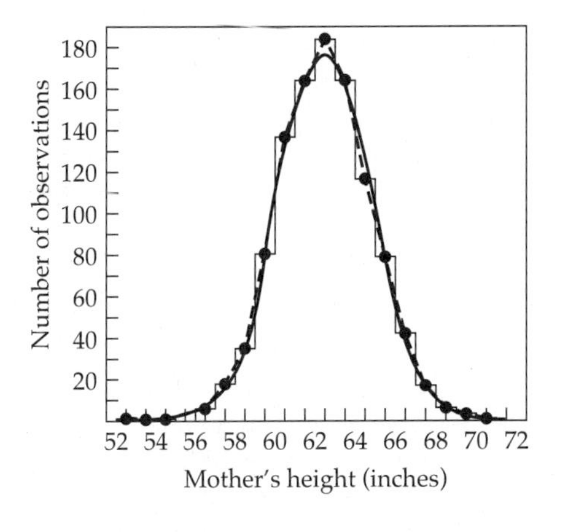

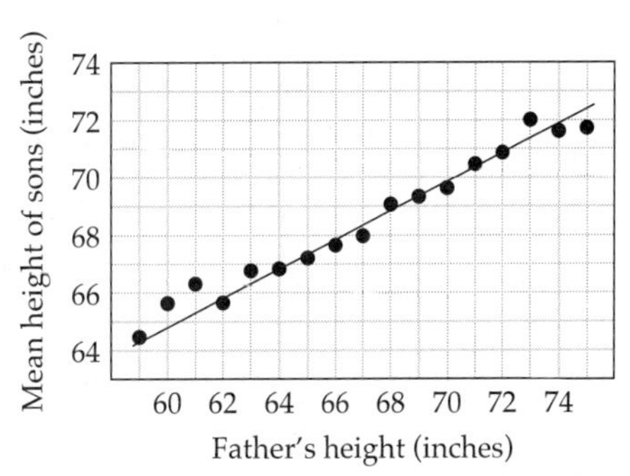

In the figure on the left, the histogram gives the observed data for maternal height, while the smooth curve is the fitted normal distribution. On the right, the straight line is the least-squares regression of sons' height on paternal height, the data having been pooled into one-inch size classes for paternal height.

Pearson and Lee do not report the covariances between relatives, but rather the correlations, but this turns out to be quite convenient for the following analysis. The variance of height tends to be larger for males than for females, and this may be expected to influence the regressions depending upon which sexes are involved. However, since a correlation coefficient is equivalent to the covariance between two standardized variables (each with unit variance), we can ignore this problem. The data in the following table indicate that the phenotypic correlations for all four sex-specific parent–offspring combinations are quite consistent, yielding a pooled value of 0.506 ± 0.011. The three full-sib correlations are also consistent with each other and give the pooled value of 0.534 ± 0.019. Finally, we note that there is highly significant assortative mating for height, the estimated value of ρ_z being 0.280 ± 0.028.

Relationship	r	SE
Parent–offspring:		
Father–son	0.514	0.022
Father–daughter	0.510	0.019
Mother–son	0.494	0.024
Mother–daughter	0.507	0.021
Full sibs:		
Brother–brother	0.511	0.042
Brother–sister	0.553	0.019
Sister–sister	0.537	0.033
Husband–wife	0.280	0.028

We first consider the additive genetic variance in the population. Recall that $\widehat{\sigma}_z^2 = 1$ on the scale of analysis. From Table 7.4, we see that the expected covariance between parent and offspring is $(1 + \rho_z)\widehat{\sigma}_A^2/2$. Setting this quantity equal to 0.506, substituting 0.280 for ρ_z, and rearranging, we obtain an estimate for $\widehat{\sigma}_A^2$ of 0.791 ± 0.024 (the standard error being obtained by the Taylor expansion method, Appendix 1). Thus, unless epistasis is very strong (we have assumed it to be equal to zero), it appears that approximately 80% of the variance in human height is attributable to additive gene action.

Suppose assortative mating were to be eliminated. What is the expected equilibrium value of the additive genetic variance? Assuming a large effective number of loci ($\gamma \simeq 1$), and substituting the estimates of $\widehat{\sigma}_A^2$ and ρ_z into Equation 7.20, we obtain $\sigma_A^2 \simeq 0.616$. The additive disequilibrium covariance for height is estimated as $0.791 - 0.616 = 0.175$, showing that through the creation of gametic phase disequilibrium, assortative mating has induced an approximately 28% increase in the additive genetic variance.

We now consider whether the full-sib data are consistent with an additive genetic model. Again from Table 7.4, we see that in the absence of dominance, the covariance between full sibs has an expected value of $\widehat{\sigma}_A^2(1 + \rho_z\widehat{h}^2)/2$. Since we are operating on a scale for which $\widehat{\sigma}_z^2 = 1$, this value is also the expected correlation coefficient. Substituting 0.280 for ρ_z and 0.791 for $\widehat{\sigma}_A^2$, we obtain the expectation 0.483 ± 0.024. The difference between the observed and expected value $(0.534 - 0.483 = 0.051)$ has a standard error equal to $[(0.024)^2 + (0.019)^2]^{1/2} = 0.031$ and cannot be considered to be significant. Thus, on the basis of the existing data, there are no grounds for rejecting the purely additive model.

In a more recent study, Roberts et al. (1978) performed a census of adult heights in families of a West African population, where assortative mating by height is relatively weak, $\rho_z \simeq 0.10 \pm 0.10$. The correlations between parents and offspring, (0.434 ± 0.015), and between full sibs, (0.378 ± 0.048), are somewhat lower than those observed in the study of Pearson and Lee. Using the procedures outlined above, these data yield estimates of $\widehat{\sigma}_A^2 \simeq 0.789$ and $\sigma_A^2 = 0.727$. Thus, the expressed additive genetic variances in the two populations are very similar, although a smaller fraction is associated with disequilibrium covariance in the West African population. Roberts et al. (1978) also report a half-sib correlation equal to 0.198 ± 0.059, which can be used to provide a further check on the theory. Substituting the estimate of ρ_z and $\widehat{h}^2 = 0.789$ into the half-sib expression in Table 7.4, the predicted correlation is 0.229, which is not significantly different from the observed value.

There is an interesting historical note regarding the data of Pearson and Lee. Fisher's (1918) demonstration that the data were quite consistent with a Mendelian hypothesis flew in the face of Pearson's notorious non-Mendelian philosophy. Fisher also extended the analyses to other types of relatives, illustrating their consistency with resemblances between first cousins and between grandparents and their grandchildren. However, the literature that Fisher cites for the latter correlations actually involves characters other than height (in one case, eye color)! Thus, at the time, Fisher apparently believed that there was a universal correlation for all characters within a species. This is an interesting twist since Pearson had similar feelings, as amply documented in Pearson and Lee (1903, p. 379): "Thus for most practical purposes we may assume parental heredity for all species and all characters to be approximately represented by a correlation of 0.5." We now know that this is far from the truth.

POLYPLOIDY

Because of the high incidence of polyploidy in plants, some attention needs to be given to its effects on the covariance between relatives. We will confine our comments to single-locus effects in a randomly mating tetraploid population and assume that the four alleles at each locus assort independently during meiosis. The results for this special case were first worked out by Kempthorne (1953, 1957).

As in the case of diploids, the genotypic values of tetraploids can be partitioned into several independent effects defined in a least-squares sense. However, because four alleles are present at each locus, in addition to additive (single-gene) and dominance (gene-pair) effects, triallelic and quadra-allelic effects must be considered (Chapter 5). Thus, the single-locus genotypic value of individual x is written

$$G_{ijkl}(x) = \mu_G + [\alpha_i^x + \alpha_j^x + \alpha_k^x + \alpha_l^x] + [\delta_{ij}^x + \delta_{ik}^x + \delta_{il}^x + \delta_{jk}^x + \delta_{jl}^x + \delta_{kl}^x] + [\gamma_{ijk}^x + \gamma_{ijl}^x + \gamma_{ikl}^x + \gamma_{jkl}^x] + \tau_{ijkl}^x \tag{7.21}$$

As in the case of diploids, the expected value of each of the effects is zero. Thus, the additive and dominance genetic variances for a locus are defined to be $\sigma_A^2 = 4E(\alpha^2)$ and $\sigma_D^2 = 6E(\delta^2)$, while the trigenic and quadragenic variances are $\sigma_T^2 = 4E(\gamma^2)$ and $\sigma_Q^2 = E(\tau^2)$.

The covariances between tetraploid relatives are obtained in the same manner as in the case of diploids. For example, the additive genetic covariance involves 16 cross-products of additive effects (four in individual $x \times$ four in individual y). Each of these cross-products has expectation $E(\alpha^2)$ if the two genes are identical by descent, the probability of which is Θ_{xy}, and expectation zero otherwise. Thus, the expected additive genetic covariance between relatives is $16\Theta_{xy}E(\alpha^2) = 4\Theta_{xy}\sigma_A^2$. In the case of digenic interaction, there are 36 cross-products to consider, each of which has expectation $E(\delta^2)$ if both members of the pair of alleles in x are identical by descent with those in the pair from y (probability Δ_{xy}). Therefore, the digenic covariance is $36\Delta_{xy}E(\delta^2) = 6\Delta_{xy}\sigma_D^2$. Letting φ_{xy} be the probability that all members of random three-gene sets are identical by descent, and ϕ_{xy} be

Table 7.5 Coefficients for Equation 7.22 for describing the covariance between tetraploid relatives.

Relationship	Θ_{xy}	Δ_{xy}	φ_{xy}	ϕ_{xy}
Parent–offspring	$\frac{1}{8}$	$\frac{1}{36}$	0	0
Grandparent–grandchild	$\frac{1}{16}$	$\frac{1}{216}$	0	0
Full sibs, dizygotic twins	$\frac{1}{8}$	$\frac{1}{27}$	$\frac{1}{48}$	$\frac{1}{36}$
Half sibs	$\frac{1}{16}$	$\frac{1}{216}$	0	0
Uncle(aunt)–nephew(neice)	$\frac{1}{16}$	$\frac{1}{648}$	0	0
Monozygotic twins	$\frac{1}{4}$	$\frac{1}{6}$	$\frac{1}{4}$	1

Source: Kempthorne 1957.

the probabilty that the complete single-locus tetraploid genotypes of x and y are identical by descent, the trigenic and quadragenic covariances between relatives are $4\varphi_{xy}\sigma_T^2$ and $\phi_{xy}\sigma_Q^2$. Thus, assuming random mating, and no epistasis, linkage, or gametic phase disequilibrium, the covariance between tetraploid relatives can be summarized as

$$\sigma_G(x,y) = 4\Theta_{xy}\sigma_A^2 + 6\Delta_{xy}\sigma_D^2 + 4\varphi_{xy}\sigma_T^2 + \phi_{xy}\sigma_Q^2 \tag{7.22}$$

The coefficients for a number of common relationships are given in Table 7.5.

An interesting consequence of polyploidy is that the covariance between noncollateral relatives (individuals that do not share both parents) can be influenced by dominance. This effect occurs because parents pass two alleles on to their progeny. Thus, the covariance between parent and offspring is $\sigma_A^2/2 + \sigma_D^2/6$, as compared to $\sigma_A^2/2$ in the case of diploidy. Only the resemblance between collateral relatives is influenced by trigenic and quadragenic variance.

ENVIRONMENTAL SOURCES OF COVARIANCE BETWEEN RELATIVES

Up to now, we have focused entirely on the genetic causes of resemblance between relatives. There are, however, many circumstances in which the environmental effects on the phenotypes of relatives are correlated. The most obvious situation arises when full sibs are raised in a common familial environment, but the resemblance between other types of relatives can also be modified by shared aspects of the environment. For example, in a spatially heterogeneous environment, the resemblance between parents and offspring may be exaggerated by environmental effects if the latter do not randomly disperse following birth. In humans, and probably in other vertebrates, cultural transmission can create a continuity in the behavioral phenotypes of parents and offspring as well as among more distant relatives. In some circumstances, the environment may actually deflate the resemblance between relatives. Consider, for example, trees with limited seed-dispersal abilities. Parent trees that have grown large due to fortuitous circumstances, such as germination in a light gap, may tend to create relatively poor microhabitats for their offspring as a consequence of shading, attraction of herbivores, and so on.

In an effort to deal with the complications of cultural transmission, human quantitative geneticists have gone to great extremes to incorporate various types of environmental covariance into expressions for the resemblance between relatives (Rao et al. 1974, Eaves 1976, Cloninger et al. 1979a,b, Eaves et al. 1988). The result is a family of models that contain large numbers of parameters. Since the number of observable relationships must equal or exceed the number of parameters for there to be any hope of estimating the latter, these models are of little practical utility for most organisms. However, in humans the difficulties are not insurmountable. Due to the frequency of divorce, adoption, and fostering by relatives, and to the availability of records on parentage, many kinds of familial relationships are observable.

Here we consider some of the possibilities, under the assumption of additive gene action, and negligible linkage, gametic phase disequilibrium, and inbreeding. To illustrate the major points, we will rely on path analysis (Appendix 2), drawing specifically from examples given in Rao et al. (1974). The basic feature of path analysis that we will exploit is its ability to partition the correlation between any two variables into a series of pathways connecting the two variables through causal components. As noted above with coefficients of coancestry, the proportional contribution of each pathway to the correlation between relatives is simply equal to the product of various path coefficients and correlation coefficients along the pathway. We start by providing some general definitions, and then show, for several types of relationships, how path analysis can yield expressions for the expected phenotypic correlations in terms of causal genetic and environmental components. We then complete our discussion of environmental effects by providing a worked example involving a large data set on human birth weight.

Before proceeding, we consider how the phenotypic variance can be partitioned in terms of path coefficients. As usual, we consider an individual's phenotype (z) to be the sum of its genotypic value (G), general (shared) environmental effects (E), and special environmental effects (e). However, we now partition G into three quantities: the population mean (μ_G), the mean genotypic (breeding) value of the parents ($\bar{A}$), and the deviation of G from $\bar{A}$ caused by segregation of parental genes (S). Thus, an individual's phenotypic value is expressed as

$$z = \mu_G + \bar{A} + S + E + e \tag{7.23}$$

The path coefficients from $\bar{A}$, S, E, and e to the phenotype are denoted by g, s, c, and d, respectively (for expected genotypic value, deviation due to segregation, common environmental effect, and residual deviation). Since S is a random genetic deviation, it is uncorrelated with the other components of Equation 7.23. However, genotype-environment correlation may exist between $\bar{A}$ and E when offspring are raised by their own parents, and we denote it by ρ_{GE}.

The phenotypic variance among individuals raised by their biological parents is

$$\sigma_z^2 = \sigma_{\bar{A}}^2 + \sigma_S^2 + \sigma_E^2 + \sigma_e^2 + 2\sigma_{\bar{A},E} \tag{7.24a}$$

Dividing all terms by σ_z^2 provides an expression for the partitioning of the phenotypic variance of offspring raised by their own parents,

$$1 = g^2 + s^2 + c^2 + d^2 + 2g\rho_{GE}c \tag{7.24b}$$

In this expression, g^2, s^2, c^2, and d^2, are the fractions of the phenotypic variance attributable to midparent breeding values, segregational, general and special environmental effects, while $2g\rho_{GE}c = 2\sigma(G,E)/\sigma_z^2$ is the fraction resulting from genotype-environment covariance.

This expression requires modification in the case of progeny raised in adoptive homes. Under the assumption that there is no correlation between offspring genotype and the environment provided by an adoptive home, then $\rho_{GE} = 0$, and the phenotypic variance of adopted children can differ from that of progeny raised by their own parents. For example, if $\rho_{GE} > 0$ in intact families, the variance among adopted children will be reduced. To account for this, Equation 7.24b needs to be modified to ensure that the sum of the various paths is equal to one. For adopted children, the modification is accomplished by letting

$$1 = \theta^2(g^2 + s^2 + c^2 + d^2) \tag{7.24c}$$

where $\theta^2 = 1/(1 - 2g\rho_{GE}c)$ is the ratio of the phenotypic variances for the two types of offspring. Multiplication by θ^2 has the effect of dividing each of the causal path contributions by the phenotypic variance of progeny raised in adoptive homes rather than by the phenotypic variance of progeny raised by their biological parents.

Finally, we note that it is useful to define

$$h^2 = g^2 + s^2 \tag{7.25}$$

as the heritability of the trait, i.e., the fraction of the total phenotypic variance attributable to additive genetic differences among individuals. With assortative mating,

$$g = h\sqrt{(1 + \rho_g)/2} \tag{7.26a}$$

$$s = h\sqrt{(1 - \rho_g)/2} \tag{7.26b}$$

Monozygotic twins

Raised by own parents

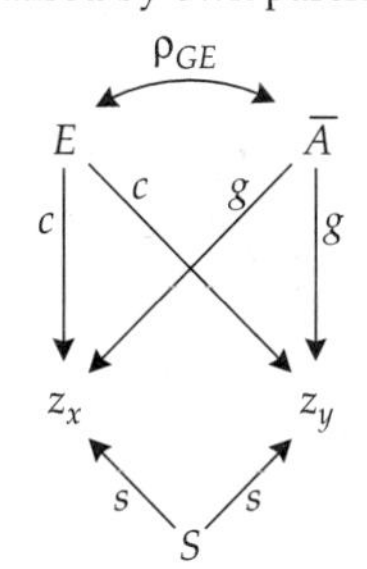

Raised by different parents

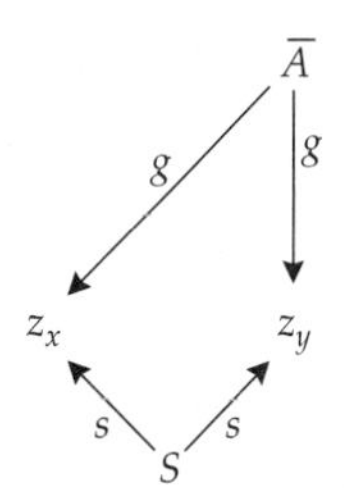

Full sibs

Raised by own parents

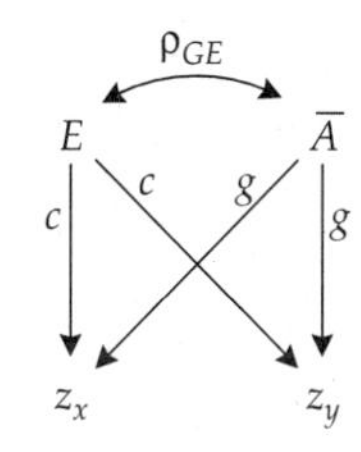

X raised by parents,
Y by foster-parents

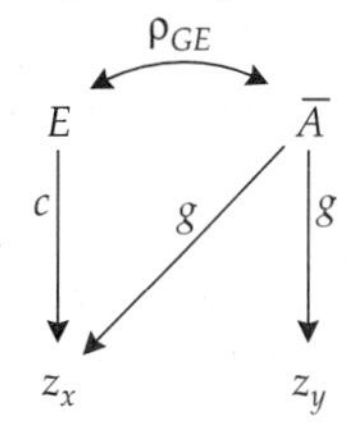

Half sibs raised
by common parent

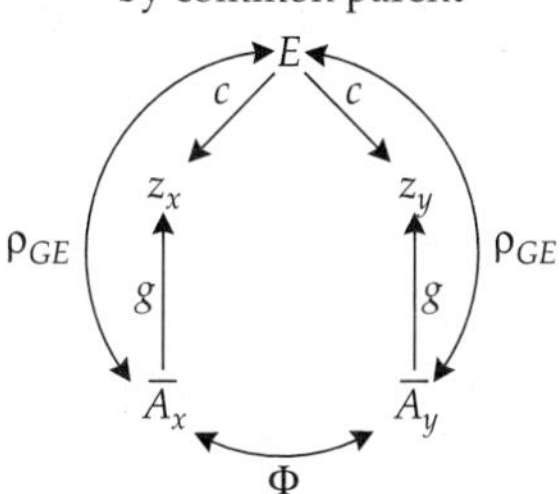

Unrelated foster sibs,
one raised by own parent

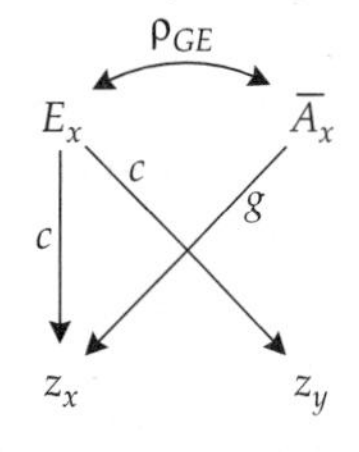

Figure 7.8 Path diagrams for the phenotypic correlation between individuals x and y. All terms are defined in the text.

where, as before, ρ_g is the genetic correlation between mates. Under random mating, $g = s = h/\sqrt{2}$.

To illustrate how these general relationships can be extended to the description of the expected phenotypic correlation between individuals, we now focus on the path diagrams for four specific relationships (Figure 7.8). For simplicity, the diagrams only include those factors that contribute jointly to the phenotypes of both members of a pair of individuals. Thus, the residual environmental deviation (e) never appears, while S (as a random genetic deviation) is only relevant in the case of monozygotic twins.

Because they are genetically identical, monozygotic twins (or clonemates) raised by their natural parents share the same general environmental effects, midparent value, and segregation deviation. Allowing for genotype-environment covariance, and summing over all pathways between z_x and z_y, we obtain the phenotypic correlation,

$$\rho(MZ) = s^2 + g^2 + c^2 + 2g\rho_{GE}c = h^2 + c^2 + 2g\rho_{GE}c \tag{7.27}$$

Thus, the correlation between monozygotic twins raised by their biological parents is not particularly informative, since it is a function of additive genetic variance (h^2), variance due to shared environment (c^2), and genotype-environment correlation $2g\rho_{GE}c$.

Now consider the situation in which monozygotic twins are separated at birth, with each being raised in a different adoptive home. The removal of the common-environment effect eliminates the path $z_x \leftarrow E \rightarrow z_y$ (Figure 7.8), and as noted above, the absence of genotype-environment correlation changes the phenotypic variance in the subpopulation of such twins. Thus, the expected phenotypic correlation between twins raised by different foster parents is $h^2\theta^2$. An estimate of the heritability can be acquired after factoring out θ^2 (the ratio of phenotypic variances, defined above). Expressions for twins living in other combinations of home environments are given in Table 7.6. (Note that when one member of a relationship is living with its parents and the other is living in an adoptive home, the correlation is multiplied by θ, rather than θ^2, because only one member of the pair is from a subpopulation with modified variance.) Chapter 19 treats the issue of twin analysis in considerable detail.

We next consider the correlation between full sibs. The path diagram in this case is identical to that for monozygotic twins, except that the sibs, being products of different gametes, do not share the segregational deviation (S). Thus, the expected correlation between full sibs raised by their biological parents is

$$\rho(FS) = g^2 + c^2 + 2g\rho_{GE}c \tag{7.28}$$

As in the case of monozygotic twins, this expression is simplified in situations where one or both sibs are raised in adoptive environments. For example, suppose one sib is raised by its biological parents, while the other is raised by unrelated foster parents (Figure 7.8). This eliminates the paths $z_x \leftarrow E \rightarrow z_y$ and $z_x \leftarrow \bar{A} \leftrightarrow E \rightarrow z_y$. Moreover, since the second sib comes from a segment of the population without genotype-environment correlation, all of the path coefficients leading to it must be multiplied by θ. Summing over the two paths between x and y, the phenotypic correlation becomes $\theta(g^2 + g\rho_{GE}c)$. The issue of sib analysis will be covered in detail in Chapter 18.

Table 7.6 Expected phenotypic correlations between sibs in terms of path coefficients.

Relationship	Phenotypic correlation
Monozygotic twins:	
Reared by own parents	$h^2 + c^2 + 2g\rho_{GE}c$
One reared by own parents, one by foster parents	$(h^2 + g\rho_{GE}c)\theta$
Raised by different foster parents	$h^2\theta^2$
Reared together by foster parents	$(h^2 + c^2)\theta^2$
Full sibs:	
Reared by own parents	$g^2 + c^2 + 2g\rho_{GE}c$
One reared by own parents, one by foster parents	$(g^2 + g\rho_{GE}c)\theta$
Reared by different foster parents	$g^2\theta^2$
Half sibs:	
Raised by common parent	$\phi g^2 + c^2 + 2g\rho_{GE}c$
One reared by own parents, one by foster parents	$(\phi g^2 + g\rho_{GE}c)\theta$
Reared by different foster parents	$\phi g^2\theta^2$
Reared apart by own parents	$\phi g^2 + c^2 b + g\rho_{GE}c$
Unrelated foster sibs:	
Reared together by same foster parents	$c^2\theta^2$
Reared together by parents of one of them	$(c^2 + g\rho_{GE}c)\theta$

Source: Rao et al. 1974.
Note: All coefficients are defined in the text except b, which is the correlation of environments provided by parents of half sibs. Completely additive gene action is assumed.

A slight complication arises in the case of half sibs. The correlation between midparent values ($\bar{A}_x$ and $\bar{A}_y$) is no longer one, since there is only one common parent. Rao et al. (1974) showed it to be $\phi = (1 + 3\rho_g)/[2(1 + \rho_g)]$, which reduces to 1/2 under random mating. Summing over all paths between z_x and z_y (Figure 7.8), the correlation between half sibs raised by the parent creating the common environmental effect is

$$\rho(HS) = \phi g^2 + c^2 + 2g\rho_{GE}c \tag{7.29}$$

Again, various simplifications arise when one or both members of the sib pair are raised by foster parents (Table 7.6). For example, when each sib is raised in a different adoptive environment, the only path between sib phenotypes is $z_x \leftarrow \bar{A}_x \leftrightarrow \bar{A}_y \rightarrow z_y$, so the correlation is simply $\phi g^2\theta^2$.

Finally, we note that unrelated individuals fostered by the same set of parents can resemble each other as a consequence of the common environment in the adoptive home. The expected phenotypic correlation between such individuals depends on whether the adoptive parents are the biological parents of one of the foster sibs (Figure 7.8, Table 7.6).

Even more complicated scenarios, for additional types of relatives, have been considered by the authors cited above. However, we assume that the basic principles are clear at this point, and will not pursue these any further. One very notable aspect of the models outlined above is their ability to provide estimates of genotype-environment correlation when data are available on sibs raised in various types of home environments. For example, when covariances are observed for monozygotic twins living in the four types of environmental settings outlined in Table 7.6, joint estimates of h^2, c^2, and ρ_{GE} can be obtained by setting the observed correlations equal to their expected values and solving. Modifications of all the expressions in Table 7.6 are necessary, however, in the presence of significant sources of nonadditive genetic variance. We close this section with an example of a character whose expression is strongly influenced by shared environmental effects.

Example 6. Several large and independent studies have been performed on human birth weight. As can be seen in the following table, the estimated correlations between relatives are quite consistent among studies. For example, the five available full-sib correlations have a narrow range of 0.47 to 0.52. Mi et al. (1986) performed large analyses on several ethnic groups in Hawaii and found only minor differences among them for the correlations between relatives. We will therefore pool the independent estimates where they exist. In keeping with the linear model just outlined, we will assume that dominance and epistatic sources of variance are of negligible importance. In the absence of conflicting data, we will also assume that assortative mating and genotype-environment correlation are negligible, and that general environmental effects are only transmitted through mothers. Under these assumptions, $\rho_G = 0$, $g^2 = h^2/2$, $\rho_{GE} = 0$, and $\theta = 1$.

Relationship	Estimated Correlations	Prediction
Full sibs[2-6]	0.50, 0.52, 0.47, 0.48, 0.48	$\frac{h^2}{2} + c^2 = 0.50$
Maternal half sibs[3]	0.58	$\frac{h^2}{4} + c^2 = 0.42$
Paternal half sibs[3]	0.10	$\frac{h^2}{4} = 0.08$
Maternal first cousins[2,6]	0.14, 0.13	$\frac{h^2}{8} + \frac{c_G^2}{2} = 0.15$
Paternal first cousins[2,6]	0.02, 0.06	$\frac{h^2}{8} = 0.04$

Relationship	Estimated Correlations	Prediction
Monozygotic twins[1]	0.67	$h^2 + c^2 = 0.65$
Dizygotic twins[2,3]	0.59, 0.66	$\frac{h^2}{2} + c^2 = 0.50$
Half sibs via monozygotic twin parents:		
Maternal[5]	0.31	$\frac{h^2}{4} + c_G^2 = 0.30$
Paternal[5,7]	−0.03, 0.12	$\frac{h^2}{4} = 0.08$

References: 1. Penrose (1954a); 2. Robson (1955); 3. Morton (1955a); 4. Billewicz (1972); 5. Nance et al. (1983); 6. Mi et al. (1986); 7. Magnus (1984).

We first consider the additive genetic variance. Inferences about it must be derived from relationships for which shared environmental effects do not influence the covariance. Paternal half sibs and paternal first cousins satisfy these conditions. The expected correlations for these types of relatives are $h^2/4$ and $h^2/8$, respectively, where $h^2 = \sigma_A^2/\sigma_z^2$. Since the observed correlations are 0.10 and 0.04, we obtain independent estimates of h^2 of $4 \times 0.10 = 0.40$ and $8 \times 0.04 = 0.32$. Also available is an average correlation of 0.05 for offspring of monozygotic twin brothers. Such individuals are genetically equivalent to paternal half sibs (the fathers are different individuals, but identical genetically), so this result yields an additional estimate of $h^2 = 4 \times 0.05 = 0.20$. Averaging over all three types of relationship, $h^2 \simeq 0.30$, i. e., additive genetic variance appears to account for approximately 30% of the phenotypic variance.

The data make it very clear that aspects of the maternal environment have a substantial influence on birth weight. For example, the correlation between maternal half sibs is several times greater than that between paternal half sibs, and the same pattern is seen for maternal vs. paternal first cousins. The total variation caused by the maternal environment can be obtained from the maternal half-sib correlation, 0.58, whose expectation is $(h^2/4) + c^2$. Subtracting out the additive genetic contribution, $c^2 = 0.58 - (0.30/4) = 0.50$. Pooling this with an independent estimate of $c^2 = 0.20$ obtained by Magnus (1984), we estimate $c^2 \simeq 0.35$. Thus, aspects of the mother (in excess of the genes that she contributes to her offspring) account for approximately 35% of the variance in birth weight.

There are two ways to partition the maternal effects variance into genetic and environmental components, $c^2 = c_G^2 + c_E^2$. First, monozygotic twin sisters provide the same genetic environment but different home settings for their progeny, which are genetically equivalent to maternal half sibs. The expected correlation between half sibs via this route is therefore $(h^2/4) + c_G^2$. Subtracting $h^2/4$ from the observed correlation, we obtain an estimate of the variance caused by genetic maternal effects, $c_G^2 = 0.31 - (0.30/4) = 0.23$. Second, the covariance between maternal first cousins is unaffected by common maternal environment, but is influenced by half the genetic maternal variance since the mothers are full sibs; their expected correlation is therefore $(h^2/8) + (c_G^2/2)$. Again equating observed and expected correlations, we obtain a second estimate $c_G^2 = 2[0.14 - (0.30/8)] = 0.20$. Thus, of the maternal effects variance, approximately two-thirds $(0.215/0.35)$ appears to be caused by the effects of the maternal genotype on the uterine environment.

The causes of approximately 35% of the variance remain to be identified. Offspring sex, birth order, and gestation age account for 2, 3, and 4% of the variance, respectively (Penrose 1954a, Morton 1955a, Billewicz 1972, Magnus 1984). Some of these sources of variance presumably fall in the environmental maternal-effects category. Relatively low weights for first-born offspring account for another 5% of the variation. Morton (1955a) has argued for the existence of dominance genetic variance, but the following argument suggests that this source of variance is negligible. In principle, the correlation between full sibs is $(\sigma_A^2/2+\sigma_D^2/4+\sigma_E^2)/\sigma_z^2$. However, from the above $(\sigma_A^2/2 + \sigma_E^2)/\sigma_z^2 \simeq (0.30/2) + 0.35 = 0.50$, which accounts for the mean observed correlation of 0.49.

The approach that we have taken to analyze these data is not very rigorous from a statistical standpoint, our main objective having simply been to provide a heuristic guide to understanding how correlations derived from several types of relatives can be used to estimate components of variance. Nevertheless, when the estimates of h^2, c^2, and c_G^2 are substituted into the expressions for the expected correlations between relatives, the overall fit to the data is quite good (last column in the preceding table). Thus, variation in human birth weight appears to be largely a function of additive gene action, maternal effects, and special environmental effects, each of which accounts for about a third of the total variance.

THE HERITABILITY CONCEPT

We have now seen, in theory and by example, that the analysis of a series of relationships provides the basis for partitioning the phenotypic variance into its elementary components. In practice, however, we are often confronted with difficulties, aside from the problem of finite resources, that prevent us from ever obtaining exact estimates of variance components. Some of the variance, such as that caused by higher-order epistatic interactions, is essentially beyond reach in a statistical sense. Nevertheless, with appropriate experimental designs, most of the fundamental sources of variance (additive and dominance genetic variance, and environmental variance due to common familial environments) can be approximated to a good degree, and levels of confidence attached to them. Most practical applications of quantitative genetics have been concerned with only the additive genetic component of the phenotypic variance, with the remaining components being treated as noise. The ratio σ_A^2/σ_z^2 has come to be known as the **heritability** of a trait (more precisely, the **narrow-sense heritability**).

This brings us to an important conceptual issue that has plagued the field of quantitative genetics almost since its inception (Feldman and Lewontin 1975, Bell 1977, Jacquard 1983). The preoccupation with the additive component of genetic variance stems from the desire for a parameter that describes the genetic resemblance between parents and offspring. At the close of Chapter 3 it was shown that the slope of a regression of offspring phenotype on average parental phenotype has expected value σ_A^2/σ_z^2, provided that gene action is purely additive and all of the assumptions underlying the Kempthorne-Cockerham model are met. Moreover, we showed that if selection changes the mean phenotype in the parental generation by S units, the expected evolutionary advance in the offspring generation (relative to that of the parents before selection) is $S\sigma_A^2/\sigma_z^2$. Based on this reasoning, many studies have accepted uncritically the slope of a midparent-offspring regression ($b_{o\bar{p}}$) (or equivalently, twice the slope of mother-offspring or father-offspring regression, $2b_{op}$) as an estimate of σ_A^2/σ_z^2. However, in the last several pages, we have found that the validity of this interpretation requires, among other things, random mating, gametic phase equilibrium, absence of additive $\times$ additive epistatic genetic variance, and absence of common environmental effects. Certainly, we cannot expect all of these conditions to be fulfilled in many natural populations.

Jacquard (1983) provides a useful discussion of the problems of interpretation of $b_{o\bar{p}}$ and $2b_{op}$ and suggests that these statistics simply be labeled **biometric heritability** without prejudice regarding the mechanisms causing similarity. However, because of the fundamental importance of the ratio σ_A^2/σ_z^2, particularly in selection theory, we will continue to call the latter quantity the heritability, denoting it as h^2 in keeping with Wright's original usage of h as the path coefficient σ_A/σ_z (Appendix 2). Providing an explicit definition eliminates the ambiguity of the usage of h^2 in theoretical contexts, but highlights the practical problems of estimation.

It should now be clear that heritabilities can often be approximated by reference to sets of relatives other than parents and offspring. The general logic behind this approach is that the first term in any genetic covariance expression is $2\Theta_{xy}\sigma_A^2$. Thus, under the assumption that the additive genetic variance is the dominant source of phenotypic covariance,

$$h^2 \simeq \frac{\text{Cov}(z_x, z_y)}{2\Theta_{xy}\text{Var}(z)} \tag{7.30}$$

should provide a good approximation to the heritability. Violations of the assumptions of the ideal additive model will usually cause $\text{Cov}(z_x, z_y)/2\Theta_{xy}$ to be an upwardly biased estimator of σ_A^2.

A simple means of evaluating the likelihood of bias in heritability estimates arises when estimates of the phenotypic covariance are available for more than one type of relative. Such a test was performed by Clayton et al. (1957) on abdominal

Table 7.7 Independent estimates of h^2, obtained with Equation 7.30, for temperature tolerance in the marine copepod *Eurytemora affinis*.

Relationship	Females	Males
Parent–offspring	0.14 ± 0.44	0.72 ± 0.26
Full sibs	0.20 ± 0.05	0.82 ± 0.04
Paternal half sibs	0.40 ± 0.18	0.84 ± 0.35
Maternal half sibs	0	0.73 ± 0.32

Source: Bradley 1978.

bristle number in a laboratory population of *Drosophila melanogaster*. The estimates of h^2 derived from four types of relatives were consistent with each other: mother–daughter (0.54 ± 0.11), mother–son (0.48 ± 0.11), half sibs (0.48 ± 0.11), and full sibs (0.53 ± 0.07). For this population, the evidence is strong that approximately 50% of the total variance for abdominal bristle number is attributable to additive genetic variance and that the remainder is a function of special environmental effects.

A second example in which heritability estimates are consistent across relationships involves a study of the susceptibility of the marine copepod *Eurytemora affinis* to high temperature shock (Table 7.7). Within each sex, four different relationships give fairly consistent results, but there is a clear sexual dimorphism — approximately four times as much variance in males is accounted for by additive genetic variance as in females. The fact that estimates from full sibs and maternal half sibs are not inflated implies that dominance genetic variance and maternal-effects variance are of minor significance. Thus, the data are consistent with the hypothesis that $h^2 \simeq 0.2$ in females and 0.8 in males with the residual variance being attributable to special environmental effects.

Such results in which additive genetic variance is the only source of resemblance between relatives are by no means universal in quantitative-genetic analyses. We have already encountered a striking exception with human birth weight, and more will appear in the following chapters. While there is a general tendency for heritability estimates based on parent-offspring and full-sib analyses to be consistent with each other (Figure 7.9), in any particular study, it is incumbent upon the investigator to evaluate whether the inconsistencies between different estimates of h^2 are significant. High levels of dominance genetic variance often exist for fitness-related characters (Crnokrak and Roff 1995).

For practical reasons, the components of variance of natural populations are frequently estimated by assaying a segment of the population in a laboratory setting. Although the goal of such studies is generally to infer the genetic properties of the wild population, laboratory settings often impose a rather substantial

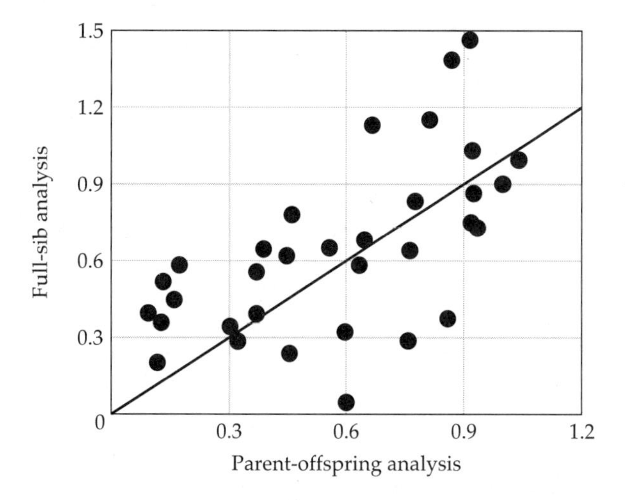

Figure 7.9 The relationship between heritability estimated as 2 Cov(PO)/Var(z) from parent-offspring analysis and as 2 Cov(FS)/Var(z) from full-sib analysis, for studies in which both estimates are available. Individual data points are for physiological and morphological characters for various natural populations of animals. The straight line gives the expected pattern under a perfect correspondence of the two estimates. (From Mousseau and Roff 1987.)

change in the environment. It is tempting to speculate that heritability estimates derived from controlled laboratory experiments will be inflated relative to those expressed in the natural environment, where the environmental component of variance might be expected to be magnified by spatial and temporal heterogeneity. However, other outcomes are possible. For example, homeostatic mechanisms, such as habitat selection, which are operable in the field may be rendered inoperable in the laboratory. It is also conceivable that a shift in the environment may induce a change in the additive genetic variance by altering gene expression.

The few attempts that have been made to evaluate the sensitivity of heritability estimates to environmental change have yielded a diversity of results. Contrary to expectations, Mackay (1981) found that parent-offspring regressions for sternopleural bristle number and body weight in *Drosophila melanogaster* were increased by varying the environment temporally and spatially in the laboratory. For the same species, Coyne and Beecham (1987) found that the parent-offspring regression for abdominal bristle number was not affected by raising the parents in the laboratory (as opposed to the field), while that for wing length increased by 150%. The change was a consequence of a reduction in the environmental component of variance in the lab-reared parents. Simons and Roff (1994) also observed a general increase in the heritabilities of life-history traits when crickets were raised in the lab as opposed to the field. In their study, however, the increase was due

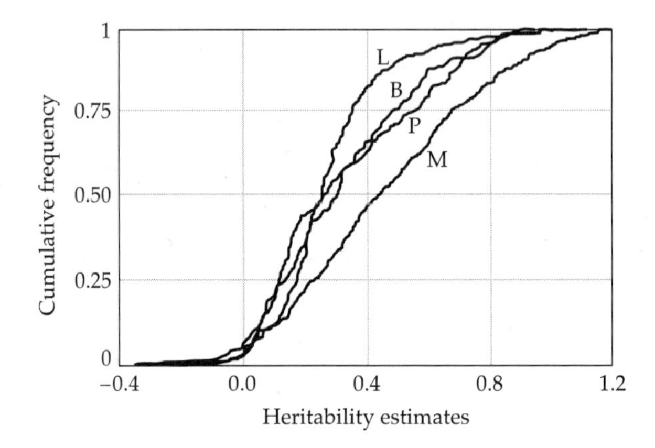

Figure 7.10 Cumulative frequency distributions for heritability estimates derived from numerous wild animal populations. L = life history, B = behavior, P = physiology, M = morphology. These data, from Mousseau and Roff (1987), do not include *Drosophila* studies, which yield a similar pattern (Roff and Mousseau 1987).

to a reduction in the environmental component of variance as well as an increase in the genetic component. In a broad survey of the existing data on a diversity of organisms, Weigensberg and Roff (1996) found that there are no *systematic* differences in heritability estimates obtained in the laboratory and in the field; if anything, the latter tend to be slightly higher on average.

In another review, Hoffmann and Parsons (1991) found that heritabilities tend to increase in stressful environments. This observation may be of relevance to the interpretation of some laboratory analyses, in that the laboratory may constitute a form of stress in some cases. However, there are many exceptions to the pattern suggested by Hoffmann and Parsons. In natural populations of birds, for example, heritabilities of bone lengths and body size tend to decline, sometimes to undetectable levels, under poor growth conditions (Gebhardt-Henrich and van Noordwijk 1991, Larsson 1993). Thus, no strong generalizations emerge from existing studies as to how heritabilities are likely to change in laboratory vs. field situations, benign vs. harsh environments, novel vs. usual conditions, and so forth. The best that can be said is that heritabilities do respond to environmental change, and that substantial care should be taken in extrapolating results beyond the environment in which they are obtained.

If one's sole interest in performing a quantitative-genetic analysis is to demonstrate that the character of interest is heritable, there is probably little point in expending the effort. The outcome is virtually certain. Almost every character in almost every species that has been studied intensely exhibits nonzero heritability. This should come as no surprise, since mutation brings in a small amount of new variation each generation (Chapter 12). The interesting questions remaining are, How does the magnitude of h^2 different among characters and species, and why?

One weak generalization that has emerged is that morphological characters tend to have higher heritabilities than life-history traits, with behavioral and physiological characters falling at intermediate levels (Figure 7.10). Although there are plenty of exceptions, these results are consistent with the intuitive concept that natural selection will most efficiently reduce the genetic variation for characters closely related to fitness by rapidly driving beneficial genes to fixation and eliminating deleterious ones (Robertson 1955, Fisher 1958, Falconer 1989). However, there are other explanations. As emphasized by Price and Schluter (1991), the relatively low heritabilities of life-history traits may be as much a consequence of relatively high levels of environmental variance as of unusually low levels of genetic variance for such traits. One possible reason for this is that the environmental variance of life-history traits is a function of the variance of all of the other morphological, physiological, and behavioral characters that influence their expression. Alternatively, characters closely related to fitness (life-history traits) may be relatively canalized genetically (Stearns and Kawecki 1994; Chapter 11), such that their expression is relatively insensitive to new mutations. This would result in low levels of genetic variation maintained under selection-mutation balance.

Evolvability

In comparing evolvabilities of different traits / species, it is clearly desirable to use a dimensionless parameter, and one such measure is the heritability. Recall the traditional expression for the rate of evolution of a trait,

$$\Delta\mu = h^2 S \tag{7.31a}$$

This equation neatly separates the forces of selection from the properties of inheritance, such that h^2 can be viewed as the efficiency of the response to selection. The change in the mean relative to the phenotypic standard deviation provides another useful descriptor of evolvability, and in this case,

$$\frac{\Delta\mu}{\sigma_z} = h^2 i \tag{7.31b}$$

where $i = S/\sigma_z$ is the standardized selection differential, i.e., the change in the mean caused by selection in units of phenotypic standard deviations. Again, the heritability provides a measure of the efficiency of response to selection.

Houle (1992) has suggested that heritability may not be the best measure of the **evolvability** of a trait, arguing for the use of the coefficient of additive genetic variation, i.e., σ_A/μ, where σ_A is the square root of the additive genetic variance and μ is the mean of the trait. Here we show that the utility of this metric is limited to a very special situation. Dividing both sides of Equation 7.31a by the mean phenotype yields an expression for the proportional change of the trait,

$$\frac{\Delta\mu}{\mu} = \left(\frac{\sigma_A^2}{\mu\sigma_z}\right) i \tag{7.31c}$$

This suggests that the dimensionless parameter $\sigma_A^2/(\mu\sigma_z)$, the ratio of the additive genetic variance to the product of the phenotypic mean and standard deviation, is an appropriate parameter for comparing evolvability when proportional change is the measure of interest. Unlike heritability, this ratio does not have a simple biological interpretation.

Now consider the special case in which the character is fitness (W). Recall from Chapter 3 that the selection differential is equivalent to the phenotypic covariance between the character and relative fitness ($w = W/\overline{W}$), i.e., $S = \sigma(z, W)/\overline{W}$, where $\overline{W}$ is mean fitness on the absolute scale. If z is fitness, then $S = \sigma_z^2(W)/\overline{W}$, where $\sigma_z^2(W)$ is the phenotypic variance of fitness. Equation 7.31c then reduces to

$$\frac{\Delta\overline{W}}{\overline{W}} = \frac{\sigma_A^2(W)}{\overline{W}^2} = \sigma_A^2(w) \tag{7.31d}$$

where $\sigma_A^2(W)$ and $\sigma_A^2(w)$ are the additive genetic variances of absolute and relative fitness. Thus, the proportional rate of evolution in mean fitness is equal to the squared coefficient of additive genetic variation of absolute fitness, or equivalently, to the additive genetic variance of relative fitness. This is **Fisher's** (1958) **fundamental theorem of natural selection.**

These alternative formulations merely serve to illustrate that there are several ways to measure evolvability, each of which has its own merits in particular contexts. All of the measures are interchangeable provided that information is available on the phenotypic variance, additive genetic variance, and mean phenotype. As emphasized by Houle (1992), however, many quantitative-genetic studies simply report the heritability of a trait, with no mention of the mean or variance components. This greatly limits the scope of investigation that can be performed with published data.

8

Introduction to Matrix Algebra and Linear Models

We have already encountered several examples of models in which response variables are linear functions of two or more explanatory (or predictor) variables. For example, we have been routinely expressing an individual's phenotypic value as the sum of genotypic and environmental values. A more complicated example is the use of linear regression to decompose an individual's genotypic value into average effects of individual alleles and residual contributions due to interactions between alleles (Chapters 4 and 5). Such **linear models** form the backbone of parameter estimation in quantitative genetics (Chapters 17–27).

This chapter provides a more formal introduction to the general features of linear models, which will be used extensively throughout the rest of this volume, most notably in Chapters 9, 26, and 27. We start by introducing multiple regression, wherein two or more variables are used to make predictions about a response variable. A review of elementary matrix algebra then follows, starting with matrix notation and building up to matrix multiplication and solutions of simultaneous equations using matrix inversion. We next use these results to develop tools for statistical analysis, considering the expectations and covariance matrices of transformed random vectors. After introducing the multivariate normal distribution, which is by far the most important distribution in quantitative-genetics theory, we discuss parameter estimation via both ordinary and generalized least squares. Those with strong statistical backgrounds will find little new in this chapter, other than perhaps some immediate contact with quantitative genetics in the examples. Additional material on matrix algebra and linear models is given in Appendix 3.

MULTIPLE REGRESSION

As a point of departure, consider the multiple regression

$$y = \alpha + \beta_1 z_1 + \beta_2 z_2 + \cdots + \beta_n z_n + e \tag{8.1}$$

where y is the **response variable**, and the z_i are the **predictor** (or **explanatory**) **variables** used to predict the value of the response variable. This multivariate equation is similar to the expression for a simple linear regression, Equation 3.12a, except that y is now a function of n predictor variables, rather than of one. The variables $y, z_1, \ldots, z_n$ represent observed measures, whereas α and $\beta_1, \ldots, \beta_n$ are constants to be estimated. As in the case of simple linear regression, e (the **residual error**) is the deviation between the observed and fitted value of y. Recall that the use of a linear model involves no assumptions regarding the true form of relationship between y and $z_1, \ldots, z_n$. It simply gives the best linear approximation. Many statistical techniques, including path analysis (Appendix 2) and analysis of variance (Chapter 17), are based on versions of Equation 8.1.

The terms $\beta_1, \ldots, \beta_n$ are known as **partial regression coefficients**. Notice that when all but the ith predictor variable are held constant in Equation 8.1, the formula reduces to a univariate model similar to Equation 3.12a but with slope β_i. This partial regression coefficient often differs from the simple regression coefficient, β_i. Suppose, for example, that a simple regression of y on z_1 has a slope of zero. This might lead to the suggestion that there is no relationship between z_1 and y. However, it is conceivable that z_1 actually has a strong positive effect on y that is obscured by positive correlations of z_1 with other variables that have negative influences on y. A multiple regression that included the appropriate variables would clarify this situation by yielding a positive β_1.

Since it is usually impossible for biologists to evaluate partial regression coefficients by empirically imposing constancy on all extraneous variables, we require a more indirect approach to the problem. From Chapter 3, the covariance of y and a predictor variable is

$$\begin{aligned}\sigma(y, z_i) &= \sigma\left[(\alpha + \beta_1 z_1 + \beta_2 z_2 + \cdots + \beta_n z_n + e), z_i\right] \\ &= \beta_1\sigma(z_1, z_i) + \beta_2\sigma(z_2, z_i) + \cdots + \beta_n\sigma(z_n, z_i) + \sigma(e, z_i)\end{aligned} \tag{8.2}$$

The term $\sigma(\alpha, z_i)$ has dropped out because the covariance of z_i with a constant (α) is zero. By applying Equation 8.2 to each predictor variable, we obtain a set of n equations in n unknowns $(\beta_1, \ldots, \beta_n)$,

$$\begin{aligned}\sigma(y, z_1) &= \beta_1\sigma^2(z_1) + \beta_2\sigma(z_1, z_2) + \cdots + \beta_n\sigma(z_1, z_n) + \sigma(z_1, e) \\ \sigma(y, z_2) &= \beta_1\sigma(z_1, z_2) + \beta_2\sigma^2(z_2) + \cdots + \beta_n\sigma(z_2, z_n) + \sigma(z_2, e) \\ &\vdots \\ \sigma(y, z_n) &= \beta_1\sigma(z_1, z_n) + \beta_2\sigma(z_2, z_n) + \cdots + \beta_n\sigma^2(z_n) + \sigma(z_n, e)\end{aligned} \tag{8.3}$$

As in univariate regression, our task is to find the set of constants (α and the partial regression coefficients β_i) that gives the best linear fit of the conditional expectation of y given $z_1, \cdots, z_n$. Again, the criterion we choose for "best" relies on the **least-squares** approach, which minimizes the squared differences between observed and expected values. Thus, our task is to find that set of $\alpha, \beta_1, \cdots, \beta_n$ giving $\widehat{y} = \alpha + \sum \beta_i z_i$ such that $E[(y - \widehat{y})^2 | z_1, \cdots, z_n]$ is minimized. Taking derivatives of this expectation with respect to α and the β_i and setting each equal to zero, it can be shown that the set of equations given by Equation 8.3 is, in fact, the least-squares solution to Equation 8.1. If the appropriate variances and covariances are known, the β_i can be obtained exactly. If these are unknown, as is usually the case, the least-squares estimates b_i are obtained from Equation 8.3 by substituting the observed (estimated) variances and covariances for their (unknown) population values.

The properties of least-squares multiple regression are analogous to those for simple regression. First, the procedure yields a solution such that the average deviation of y from $\widehat{y}$, $E(e)$, is zero. Hence $E(\,y\,) = E(\,\widehat{y}\,)$, implying

$$\bar{y} = a + b_1\bar{z}_1 + \cdots + b_n\bar{z}_n$$

Thus, once the fitted values $b_1, \ldots, b_n$ are obtained from Equation 8.3, the intercept is defined by $a = \bar{y} - b_1\bar{z}_1 - \cdots - b_n\bar{z}_n$. Second, least-squares analysis gives a solution in which the residual errors are uncorrelated with the predictor variables. Thus, the terms $\sigma(e, z_i)$ can be dropped from Equation 8.3. Third, the partial regression coefficients are entirely defined by variances and covariances. However, unlike simple regression coefficients, which depend on only a single variance and covariance, each partial regression coefficient is a function of the variances and covariances of all measured variables. Notice that if $n = 1$, then $\sigma(y, z_1) = \beta_1\sigma^2(z_1)$, and we recover the univariate solution $\beta_1 = \sigma(y, z_1)/\sigma^2(z_1)$.

A simple pattern exists in each of the n equations in 8.3. The *i*th equation defines the covariance of y and z_i as the sum of two types of quantities: a single term, which is the product of the *i*th partial regression coefficient and the variance of z_i, and a set of $(n-1)$ terms, each of which is the product of a partial regression coefficient and the covariance of z_i with the corresponding predictor variable. This general pattern suggests an alternative way of writing Equation 8.3,

$$\begin{pmatrix} \sigma^2(z_1) & \sigma(z_1, z_2) & \ldots & \sigma(z_1, z_n) \\ \sigma(z_1, z_2) & \sigma^2(z_2) & \ldots & \sigma(z_2, z_n) \\ \vdots & \vdots & \ddots & \vdots \\ \sigma(z_1, z_n) & \sigma(z_2, z_n) & \ldots & \sigma^2(z_n) \end{pmatrix} \begin{pmatrix} \beta_1 \\ \beta_2 \\ \vdots \\ \beta_n \end{pmatrix} = \begin{pmatrix} \sigma(y, z_1) \\ \sigma(y, z_2) \\ \vdots \\ \sigma(y, z_n) \end{pmatrix} \tag{8.4}$$

The table of variances and covariances on the left is referred to as a **matrix**, while the columns of partial regression coefficients and of covariances involving y are called **vectors**. If these matrices and vectors are abbreviated as $\mathbf{V}$, $\boldsymbol{\beta}$, and $\mathbf{c}$, Equation 8.4 can be written even more compactly as

$$\mathbf{V}\boldsymbol{\beta} = \mathbf{c} \tag{8.5}$$

The standard procedure of denoting matrices as bold capital letters and vectors as bold lowercase letters is adhered to in this book. Notice that $\mathbf{V}$, which is generally called a **covariance matrix**, is symmetrical about the main diagonal. As we shall see shortly, the *i*th equation in 8.3 can be recovered from Equation 8.4 by multiplying the elements in β by the corresponding elements in the ith horizontal row of the matrix $\mathbf{V}$. Although a great deal of notational simplicity has been gained by condensing the system of Equations 8.3 to matrix form, this does not alter the fact that the solution of a large system of simultaneous equations is a tedious task if performed by hand. Today, such solutions are rapidly accomplished on computers. Before considering matrix methods in more detail, we present an application of Equation 8.1 to quantitative genetics.

An Application to Multivariate Selection

Karl Pearson developed the technique of multiple regression in 1896, although some of the fundamentals can be traced to his predecessors (Pearson 1920, Stigler 1986). Pearson is perhaps best known as one of the founders of statistical methodology, but his intense interest in evolution may have been the primary motivating force underlying many of his theoretical endeavors. Almost all of his major papers, including the one of 1896, contain rigorous analyses of data gathered by his contemporaries on matters such as resemblance between relatives, natural selection, correlation between characters, and assortative mating (recall the assortative mating example in Chapter 7). The foresight of these studies is remarkable considering that they were performed prior to the existence of a genetic interpretation for the expression and inheritance of polygenic traits.

Pearson's (1896, 1903) invention of multiple regression developed out of the need for a technique to resolve the observed directional selection on a character into its direct and various indirect components. In Chapter 3 we defined the selection differential S (the within-generation change in the mean phenotype due to selection) as a measure of the total directional selection on a character. However, S cannot be considered to be a measure of the direct forces of selection on a character unless that character is uncorrelated with all other selected traits. An unselected character can appear to be under selection if other characters with which it is correlated are under directional selection. Alternatively, a character under strong directional selection may exhibit a negligible selection differential if the indirect effects of selection on correlated traits are sufficiently compensatory.

Because he did not employ matrix notation, some of the mathematics in Pearson's papers can be rather difficult to follow. Lande and Arnold (1983) did a great service by extending this work and rephrasing it in matrix notation. Suppose that a large number of individuals in a population have been measured for n characters and for fitness. Individual fitness can then be approximated by the linear model

$$w = \alpha + \beta_1 z_1 + \cdots + \beta_n z_n + e$$

where w is relative fitness (observed fitness divided by the mean fitness in the population), and $z_1, \ldots, z_n$ are the phenotypic measures of the n characters. Recall from Chapter 3 that the selection differential for the ith trait is defined as the covariance between phenotype and relative fitness, $S_i = \sigma(z_i, w)$. Thus, we have

$$\begin{aligned} S_i = \sigma(z_i, w) &= \sigma(z_i, \alpha + \beta_1 z_1 + \cdots + \beta_n z_n + e) \\ &= \beta_1 \sigma(z_i, z_1) + \cdots + \beta_n \sigma(z_i, z_n) + \sigma(z_i, e) \end{aligned}$$

Note that this expression is of the same form as Equation 8.3, so that by taking the β_i to be the partial regression coefficients we have $\sigma(z_i, e) = 0$. Note also that the selection differential of any trait may be partitioned into a component estimating the **direct selection** on the character and the sum of components from **indirect selection** on all correlated characters,

$$S_i = \beta_i \sigma^2(z_i) + \sum_{j \neq i}^{n} \beta_j \sigma(z_i, z_j)$$

It is important to realize that the labels "direct" and "indirect" apply strictly to the specific set of characters included in the analysis; the partial regression coefficients are subject to change if a new analysis includes additional correlated characters that are under selection.

Example 1. A morphological analysis of a pentatomid bug (*Euschistus variolarius*) population performed by Lande and Arnold (1983) provides a good example of the insight that can be gained from a multivariate approach. The bugs were collected along the shore of Lake Michigan after a storm. Of the 94 individuals that were recovered, 39 were alive. All individuals were measured for four characters: head and thorax width, and scutellum and forewing length. The data were then logarithmically transformed to more closely approximate normality (Chapter 11). All surviving bugs were assumed to have equal fitness ($W = 1$), and all dead bugs to have zero fitness ($W = 0$). Hence, mean fitness is the fraction p of individuals that survived, giving **relative fitnesses**, $w = W/\overline{W}$, as

$$w = \begin{cases} 1/p & \text{if the individual survived} \\ 0 & \text{if the individual did not survive} \end{cases}$$

The selection differential for each of the characters is simply the difference between the mean phenotype of the 39 survivors and the mean of the entire sample. These are reported in units of phenotypic standard deviations in the following table, along with the partial regression coefficients of relative fitness on the four morphological characters. Here * and ** indicate significance at the 5% and 1% levels. All of the phenotypic correlations are highly significant.

Character	Selection Differential	Partial Regression Coef. of Fitness	Phenotypic Correlations			
z_i	S_i	b_i	H	T	S	F
Head (H)	–0.11	–0.7	1.00	0.72	0.50	0.60
Thorax (T)	–0.06	11.6**		1.00	0.59	0.71
Scutellum (S)	–0.28*	–2.8			1.00	0.62
Forewing (F)	–0.43**	–16.6**				1.00

The estimates of the partial regression coefficients nicely illustrate two points discussed earlier. First, despite the strong directional selection operating directly on thorax size, the selection differential for thorax size is negligible. This lack of apparent selection results because the positive correlation between thorax width and wing length is coupled with negative forces of selection on the latter character. Second, there is a significant negative selection differential on scutellum length even though there is no significant direct selection on the character. The negative selection differential is largely an indirect consequence of the strong selection for smaller wing length.

ELEMENTARY MATRIX ALGEBRA

The solutions of systems of linear equations generally involve the use of matrices and vectors of variables. For those with little familiarity with such constructs and their manipulations, the next few pages provide an overview of the basic tools of matrix algebra.

Basic Notation

A matrix is simply a rectangular array of numbers. Some examples are:

$$\mathbf{a} = \begin{pmatrix} 12 \\ 13 \\ 47 \end{pmatrix} \qquad \mathbf{b} = \begin{pmatrix} 2 & 0 & 5 & 21 \end{pmatrix} \qquad \mathbf{C} = \begin{pmatrix} 3 & 1 & 2 \\ 2 & 5 & 4 \\ 1 & 1 & 2 \end{pmatrix} \qquad \mathbf{D} = \begin{pmatrix} 0 & 1 \\ 3 & 4 \\ 2 & 9 \end{pmatrix}$$

A matrix with r rows and c columns is said to have **dimensionality** $r \times c$ (a useful mnemonic for remembering this is *r*ailroad *c*ar). In the examples above, **D** has three rows and two columns, and is thus a 3×2 matrix. An $r \times 1$ matrix, such as **a**, is a **column vector**, while a $1 \times c$ matrix, such as **b**, is a **row vector**. A matrix in which the number of rows equals the number of columns, such as **C**, is called a **square matrix**. Numbers are also matrices (of dimensionality 1×1) and are often referred to as **scalars**.

A matrix is completely specified by the **elements** that comprise it, with M_{ij} denoting the element in the ith row and jth column of matrix **M**. Using the sample matrices above, $C_{23} = 4$ is the element in the second row and third column of **C**. Likewise, $C_{32} = 1$ is the element in the third row and second column. Two matrices are equal if and only if all of their corresponding elements are equal.

Partitioned Matrices

It is often useful to work with **partitioned matrices** wherein each element in a matrix is itself a matrix. There are several ways to partition a matrix. For example, we could write the matrix **C** above as

$$\mathbf{C} = \begin{pmatrix} 3 & 1 & 2 \\ 2 & 5 & 4 \\ 1 & 1 & 2 \end{pmatrix} = \begin{pmatrix} 3 & \vdots & 1 & 2 \\ \cdots & \cdots & \cdots & \cdots \\ 2 & \vdots & 5 & 4 \\ 1 & \vdots & 1 & 2 \end{pmatrix} = \begin{pmatrix} \mathbf{a} & \mathbf{b} \\ \mathbf{d} & \mathbf{B} \end{pmatrix}$$

where

$$\mathbf{a} = (\,3\,), \quad \mathbf{b} = (\,1 \quad 2\,), \quad \mathbf{d} = \begin{pmatrix} 2 \\ 1 \end{pmatrix}, \quad \mathbf{B} = \begin{pmatrix} 5 & 4 \\ 1 & 2 \end{pmatrix}$$

Alternatively, we could partition **C** into a single row vector whose elements are themselves column vectors,

$$\mathbf{C} = (\,\mathbf{c}_1 \quad \mathbf{c}_2 \quad \mathbf{c}_3\,) \quad \text{where} \quad \mathbf{c}_1 = \begin{pmatrix} 3 \\ 2 \\ 1 \end{pmatrix}, \quad \mathbf{c}_2 = \begin{pmatrix} 1 \\ 5 \\ 1 \end{pmatrix}, \quad \mathbf{c}_3 = \begin{pmatrix} 2 \\ 4 \\ 2 \end{pmatrix}$$

or **C** could be written as a column vector whose elements are row vectors,

$$\mathbf{C} = \begin{pmatrix} \mathbf{b}_1 \\ \mathbf{b}_2 \\ \mathbf{b}_3 \end{pmatrix} \quad \text{where} \quad \mathbf{b}_1 = (\,3 \quad 1 \quad 2\,), \quad \mathbf{b}_2 = (\,2 \quad 5 \quad 4\,), \quad \mathbf{b}_3 = (\,1 \quad 1 \quad 2\,)$$

Addition and Subtraction

Addition and subtraction of matrices is straightforward. To form a new matrix $\mathbf{A} + \mathbf{B} = \mathbf{C}$, **A** and **B** must have the same dimensions. One then simply adds the corresponding elements, $C_{ij} = A_{ij} + B_{ij}$. Subtraction is defined similarly. For example, if

$$\mathbf{A} = \begin{pmatrix} 3 & 0 \\ 1 & 2 \end{pmatrix} \quad \text{and} \quad \mathbf{B} = \begin{pmatrix} 1 & 2 \\ 2 & 1 \end{pmatrix}$$

then

$$\mathbf{C} = \mathbf{A} + \mathbf{B} = \begin{pmatrix} 4 & 2 \\ 3 & 3 \end{pmatrix} \quad \text{and} \quad \mathbf{D} = \mathbf{A} - \mathbf{B} = \begin{pmatrix} 2 & -2 \\ -1 & 1 \end{pmatrix}$$

Multiplication

Multiplying a matrix by a scalar is also straightforward. If $\mathbf{M} = a\mathbf{N}$, where a is a scalar, then $M_{ij} = aN_{ij}$. Each element of **N** is simply multiplied by the scalar. For example,

$$(-2)\begin{pmatrix} 1 & 0 \\ 3 & 1 \end{pmatrix} = \begin{pmatrix} -2 & 0 \\ -6 & -2 \end{pmatrix}$$

Matrix multiplication is a little more involved. We start by considering the **dot product** of two vectors, as this forms the basic operation of matrix multiplication. Letting **a** and **b** be two n-dimensional vectors (the first a column vector, the second a row vector), their dot product $\mathbf{a} \cdot \mathbf{b}$ is a scalar given by

$$\mathbf{a} \cdot \mathbf{b} = \sum_{i=1}^{n} a_i b_i$$

For example, for the two vectors

$$\mathbf{a} = \begin{pmatrix} 1 \\ 2 \\ 3 \\ 4 \end{pmatrix} \quad \text{and} \quad \mathbf{b} = \begin{pmatrix} 4 & 5 & 7 & 9 \end{pmatrix}$$

the dot product is $\mathbf{a} \cdot \mathbf{b} = (1 \times 4) + (2 \times 5) + (3 \times 7) + (4 \times 9) = 71$. Note that the dot product is not defined if the vectors have different lengths.

Now consider the matrix $\mathbf{L} = \mathbf{MN}$ produced by multiplying the $r \times c$ matrix **M** by the $c \times b$ matrix **N**. Partitioning **M** as a column vector of r row vectors,

$$\mathbf{M} = \begin{pmatrix} \mathbf{m_1} \\ \mathbf{m_2} \\ \vdots \\ \mathbf{m_r} \end{pmatrix} \quad \text{where} \quad \mathbf{m_i} = \begin{pmatrix} M_{i1} & M_{i2} & \cdots & M_{ic} \end{pmatrix}$$

and **N** as a row vector of b column vectors,

$$\mathbf{N} = \begin{pmatrix} \mathbf{n_1} & \mathbf{n_2} & \cdots & \mathbf{n_b} \end{pmatrix} \quad \text{where} \quad \mathbf{n_j} = \begin{pmatrix} N_{1j} \\ N_{2j} \\ \vdots \\ N_{cj} \end{pmatrix}$$

the ijth element of **L** is given by the dot product

$$L_{ij} = \mathbf{m_i} \cdot \mathbf{n_j} = \sum_{k=1}^{c} M_{ik} N_{kj} \tag{8.6a}$$

Hence the resulting matrix $\mathbf{L}$ is of dimension $r \times b$ with

$$\mathbf{L} = \begin{pmatrix} \mathbf{m_1} \cdot \mathbf{n_1} & \mathbf{m_1} \cdot \mathbf{n_2} & \cdots & \mathbf{m_1} \cdot \mathbf{n_b} \\ \mathbf{m_2} \cdot \mathbf{n_1} & \mathbf{m_2} \cdot \mathbf{n_2} & \cdots & \mathbf{m_2} \cdot \mathbf{n_b} \\ \vdots & \vdots & \ddots & \vdots \\ \mathbf{m_r} \cdot \mathbf{n_1} & \mathbf{m_r} \cdot \mathbf{n_2} & \cdots & \mathbf{m_r} \cdot \mathbf{n_b} \end{pmatrix} \tag{8.6b}$$

Note that using this definition, the matrix product given by Equation 8.4 recovers the set of equations given by Equation 8.3.

Example 2. Compute the product $\mathbf{L} = \mathbf{MN}$ where

$$\mathbf{M} = \begin{pmatrix} 3 & 1 & 2 \\ 2 & 5 & 4 \\ 1 & 1 & 2 \end{pmatrix} \quad \text{and} \quad \mathbf{N} = \begin{pmatrix} 4 & 1 & 0 \\ 1 & 1 & 3 \\ 3 & 2 & 2 \end{pmatrix}$$

Writing $\mathbf{M} = \begin{pmatrix} \mathbf{m_1} \\ \mathbf{m_2} \\ \mathbf{m_3} \end{pmatrix}$ and $\mathbf{N} = (\,\mathbf{n_1} \quad \mathbf{n_2} \quad \mathbf{n_3}\,)$, we have

$$\mathbf{m_1} = (\,3 \quad 1 \quad 2\,), \quad \mathbf{m_2} = (\,2 \quad 5 \quad 4\,), \quad \mathbf{m_3} = (\,1 \quad 1 \quad 2\,)$$

and

$$\mathbf{n_1} = \begin{pmatrix} 4 \\ 1 \\ 3 \end{pmatrix}, \quad \mathbf{n_2} = \begin{pmatrix} 1 \\ 1 \\ 2 \end{pmatrix}, \quad \mathbf{n_3} = \begin{pmatrix} 0 \\ 3 \\ 2 \end{pmatrix}$$

The resulting matrix $\mathbf{L}$ is 3×3. Applying Equation 8.6b, the element in the first row and first column of $\mathbf{L}$ is the dot product of the first row vector of $\mathbf{M}$ with the first column vector of $\mathbf{N}$,

$$L_{11} = \mathbf{m_1} \cdot \mathbf{n_1} = (\,3 \quad 1 \quad 2\,) \begin{pmatrix} 4 \\ 1 \\ 3 \end{pmatrix} = \sum_{k=1}^{3} M_{1k} N_{k1}$$

$$= M_{11}N_{11} + M_{12}N_{21} + M_{13}N_{31} = (3 \times 4) + (1 \times 1) + (2 \times 3) = 19$$

Computing the other elements gives

$$\mathbf{L} = \begin{pmatrix} \mathbf{m_1} \cdot \mathbf{n_1} & \mathbf{m_1} \cdot \mathbf{n_2} & \mathbf{m_1} \cdot \mathbf{n_3} \\ \mathbf{m_2} \cdot \mathbf{n_1} & \mathbf{m_2} \cdot \mathbf{n_2} & \mathbf{m_2} \cdot \mathbf{n_3} \\ \mathbf{m_3} \cdot \mathbf{n_1} & \mathbf{m_3} \cdot \mathbf{n_2} & \mathbf{m_3} \cdot \mathbf{n_3} \end{pmatrix} = \begin{pmatrix} 19 & 8 & 7 \\ 25 & 15 & 23 \\ 11 & 6 & 7 \end{pmatrix}$$

Certain dimensional properties must be satisfied when two matrices are to be multiplied. Specifically, since the dot product is defined only for vectors of the same length, for the matrix product **MN** to be defined, the number of columns in **M** must equal the number of rows in **N**. Thus, while

$$\begin{pmatrix} 3 & 0 \\ 1 & 2 \end{pmatrix} \begin{pmatrix} 4 \\ 3 \end{pmatrix} = \begin{pmatrix} 12 \\ 10 \end{pmatrix}, \qquad \begin{pmatrix} 4 \\ 3 \end{pmatrix} \begin{pmatrix} 3 & 0 \\ 1 & 2 \end{pmatrix} \quad \text{is undefined.}$$

Writing $\mathbf{M}_{r \times c}\mathbf{N}_{c \times b} = \mathbf{L}_{r \times b}$ shows that the inner indices must match, while the outer indices (r and b) give the number of rows and columns of the resulting matrix. The order in which matrices are multiplied is critical. In general, **AB** is not equal to **BA**. For example, when the order of the matrices in Example 2 is reversed,

$$\mathbf{NM} = \begin{pmatrix} 4 & 1 & 0 \\ 1 & 1 & 3 \\ 3 & 2 & 2 \end{pmatrix} \begin{pmatrix} 3 & 1 & 2 \\ 2 & 5 & 4 \\ 1 & 1 & 2 \end{pmatrix} = \begin{pmatrix} 14 & 9 & 12 \\ 8 & 9 & 12 \\ 15 & 15 & 18 \end{pmatrix}$$

Since order is important in matrix multiplication, it has specific terminology. For the product **AB**, we say that matrix **B** is **premultiplied** by the matrix **A**, or that matrix **A** is **postmultiplied** by the matrix **B**.

Transposition

Another useful matrix operation is **transposition**. The transpose of a matrix **A** is written $\mathbf{A}^T$ (while not used in this book, the notation $\mathbf{A}'$ is also widely used), and is obtained simply by switching rows and columns of the original matrix. For example,

$$\begin{pmatrix} 3 & 1 & 2 \\ 2 & 5 & 4 \\ 1 & 1 & 2 \end{pmatrix}^T = \begin{pmatrix} 3 & 2 & 1 \\ 1 & 5 & 1 \\ 2 & 4 & 2 \end{pmatrix}$$

$$\begin{pmatrix} 7 & 4 & 5 \end{pmatrix}^T = \begin{pmatrix} 7 \\ 4 \\ 5 \end{pmatrix}$$

A useful identity for transposition is that

$$(\mathbf{AB})^T = \mathbf{B}^T\mathbf{A}^T \tag{8.7a}$$

which holds for any number of matrices, e.g.,

$$(\mathbf{ABC})^T = \mathbf{C}^T\mathbf{B}^T\mathbf{A}^T \tag{8.7b}$$

Vectors of statistics are generally written as column vectors and we follow this convention by using lowercase bold letters, e.g., **a**, for a column vector and $\mathbf{a}^T$

for the corresponding row vector. With this convention, we distinguish between two vector products, the **inner product** (the dot product) which yields a scalar and the **outer product** which yields a matrix. For the two n-dimensional column vectors $\mathbf{a}$ and $\mathbf{b}$,

$$\mathbf{a} = \begin{pmatrix} a_1 \\ \vdots \\ a_n \end{pmatrix} \qquad \mathbf{b} = \begin{pmatrix} b_1 \\ \vdots \\ b_n \end{pmatrix}$$

the inner product is given by

$$\begin{pmatrix} a_1 & \cdots & a_n \end{pmatrix} \begin{pmatrix} b_1 \\ \vdots \\ b_n \end{pmatrix} = \mathbf{a}^T\mathbf{b} = \sum_{i=1}^{n} a_i b_i \tag{8.8a}$$

while the outer product yields the $n \times n$ matrix

$$\begin{pmatrix} a_1 \\ \vdots \\ a_n \end{pmatrix} \begin{pmatrix} b_1 & \cdots & b_n \end{pmatrix} = \mathbf{a}\mathbf{b}^T = \begin{pmatrix} a_1b_1 & a_1b_2 & \cdots & a_1b_n \\ a_2b_1 & a_2b_2 & \cdots & a_2b_n \\ \vdots & \vdots & \ddots & \vdots \\ a_nb_1 & a_nb_2 & \cdots & a_nb_n \end{pmatrix} \tag{8.8b}$$

Inverses and Solutions to Systems of Equations

While matrix multiplication provides a compact way of writing systems of equations, we also need a compact notation for expressing the solutions of such systems. Such solutions utilize the **inverse** of a matrix, an operation analogous to scalar division. The essential utility of matrix inversion can be noted by first considering the solution of the simple scalar equation $ax = b$ for x. Multiplying both sides by a^{-1}, we have $(a^{-1}a)x = 1 \cdot x = x = a^{-1}b$. Now consider a square matrix $\mathbf{A}$. The **inverse of** $\mathbf{A}$, denoted $\mathbf{A}^{-1}$, satisfies $\mathbf{A}^{-1}\mathbf{A} = \mathbf{I} = \mathbf{A}\mathbf{A}^{-1}$, where $\mathbf{I}$, the **identity matrix**, is a square matrix with diagonal elements equal to one and all other elements equal to zero. The identity matrix serves the role that 1 plays in scalar multiplication. Just as $1 \times a = a \times 1 = a$ in scalar multiplication, for any matrix $\mathbf{A} = \mathbf{IA} = \mathbf{AI}$. A matrix is called **nonsingular** if its inverse exists. Conditions under which this occurs are discussed in the next section. A useful property of inverses is that if the matrix product $\mathbf{AB}$ is a square matrix (where $\mathbf{A}$ and $\mathbf{B}$ are square), then

$$(\mathbf{AB})^{-1} = \mathbf{B}^{-1}\mathbf{A}^{-1} \tag{8.9}$$

The fundamental relationship between the inverse of a matrix and the solution of systems of linear equations can be seen as follows. For a square nonsingular matrix $\mathbf{A}$, the unique solution for $\mathbf{x}$ in the matrix equation $\mathbf{Ax} = \mathbf{c}$ is obtained by premultiplying by $\mathbf{A}^{-1}$,

$$\mathbf{x} = \mathbf{A}^{-1}\mathbf{Ax} = \mathbf{A}^{-1}\mathbf{c} \tag{8.10a}$$

When $\mathbf{A}$ is either singular or nonsquare, solutions for $\mathbf{x}$ can still be obtained using **generalized inverses** in place of $\mathbf{A}^{-1}$ (Appendix 3), but such solutions are not unique, applying instead to certain linear combinations of the elements of $\mathbf{x}$. (See Appendix 3 for details.) Recalling Equation 8.5, the solution of the multiple regression equation can be expressed as

$$\boldsymbol{\beta} = \mathbf{V}^{-1}\mathbf{c} \tag{8.10b}$$

Likewise, for the Pearson-Lande-Arnold regression giving the best linear predictor of fitness,

$$\boldsymbol{\beta} = \mathbf{P}^{-1}\mathbf{s} \tag{8.10c}$$

where $\mathbf{P}$ is the covariance matrix for phenotypic measures $z_1, \ldots, z_n$, and $\mathbf{s}$ is the vector of selection differentials for the n characters.

Before developing the formal method for inverting a matrix, we consider two extreme (but very useful) cases that lead to simple expressions for the inverse. First, if the matrix is **diagonal** (all off-diagonal elements are zero), then the matrix inverse is also diagonal, with $\mathbf{A}_{ii}^{-1} = 1/A_{ii}$. For example,

$$\text{for} \quad \mathbf{A} = \begin{pmatrix} a & 0 & 0 \\ 0 & b & 0 \\ 0 & 0 & c \end{pmatrix} \qquad \text{then} \qquad \mathbf{A}^{-1} = \begin{pmatrix} a^{-1} & 0 & 0 \\ 0 & b^{-1} & 0 \\ 0 & 0 & c^{-1} \end{pmatrix}$$

Note that if any of the diagonal elements of $\mathbf{A}$ are zero, $\mathbf{A}^{-1}$ is not defined, as $1/0$ is undefined. Second, for any 2×2 matrix $\mathbf{A}$,

$$\mathbf{A} = \begin{pmatrix} a & b \\ c & d \end{pmatrix} \qquad \text{then} \qquad \mathbf{A}^{-1} = \frac{1}{ad - bc} \begin{pmatrix} d & -b \\ -c & a \end{pmatrix} \tag{8.11}$$

To check this result, note that

$$\begin{aligned} \mathbf{A}\mathbf{A}^{-1} &= \frac{1}{ad - bc} \begin{pmatrix} a & b \\ c & d \end{pmatrix} \begin{pmatrix} d & -b \\ -c & a \end{pmatrix} \\ &= \frac{1}{ad - bc} \begin{pmatrix} ad - bc & 0 \\ 0 & ad - bc \end{pmatrix} = \mathbf{I} \end{aligned}$$

If $ad = bc$, the inverse does not exist, as division by zero is undefined.

Example 3. Consider the multiple regression of y on two predictor variables, z_1 and z_2, so that $y = \alpha + \beta_1 z_1 + \beta_2 z_2 + e$. In the notation of Equation 8.5, we have

$$\mathbf{c} = \begin{pmatrix} \sigma(y, z_1) \\ \\ \sigma(y, z_2) \end{pmatrix} \qquad \mathbf{V} = \begin{pmatrix} \sigma^2(z_1) & \sigma(z_1, z_2) \\ \\ \sigma(z_1, z_2) & \sigma^2(z_2) \end{pmatrix}$$

Recalling that $\sigma(z_1, z_2) = \rho_{12}\,\sigma(z_1)\sigma(z_2)$, Equation 8.11 gives

$$\mathbf{V}^{-1} = \frac{1}{\sigma^2(z_1)\sigma^2(z_2)\,(1-\rho_{12}^2)} \begin{pmatrix} \sigma^2(z_2) & -\sigma(z_1, z_2) \\ -\sigma(z_1, z_2) & \sigma^2(z_1) \end{pmatrix}$$

The inverse exists provided both characters have nonzero variance and are not completely correlated ($|\rho_{12}| \neq 1$). Recalling Equation 8.10b, the partial regression coefficients are given by $\boldsymbol{\beta} = \mathbf{V}^{-1}\mathbf{c}$, or

$$\begin{pmatrix} \beta_1 \\ \beta_2 \end{pmatrix} = \frac{1}{\sigma^2(z_1)\sigma^2(z_2)\,(1-\rho_{12}^2)} \begin{pmatrix} \sigma^2(z_2) & -\sigma(z_1, z_2) \\ -\sigma(z_1, z_2) & \sigma^2(z_1) \end{pmatrix} \begin{pmatrix} \sigma(y, z_1) \\ \sigma(y, z_2) \end{pmatrix}$$

Again using $\sigma(z_1, z_2) = \rho_{12}\,\sigma(z_1)\sigma(z_2)$, this equation reduces to

$$\beta_1 = \frac{1}{1-\rho_{12}^2} \left[\frac{\sigma(y, z_1)}{\sigma^2(z_1)} - \rho_{12}\,\frac{\sigma(y, z_2)}{\sigma(z_1)\sigma(z_2)} \right]$$

and

$$\beta_2 = \frac{1}{1-\rho_{12}^2} \left[\frac{\sigma(y, z_2)}{\sigma^2(z_2)} - \rho_{12}\,\frac{\sigma(y, z_1)}{\sigma(z_1)\sigma(z_2)} \right]$$

Note that only when the predictor variables are uncorrelated ($\rho_{12} = 0$), do the partial regression coefficients β_1 and β_2 reduce to the univariate regression slopes,

$$\beta_1 = \frac{\sigma(y, z_1)}{\sigma^2(z_1)} \quad \text{and} \quad \beta_2 = \frac{\sigma(y, z_2)}{\sigma^2(z_2)}$$

Determinants and Minors

For a 2×2 matrix, the quantity

$$|\mathbf{A}| = A_{11}A_{22} - A_{12}A_{21} \tag{8.12a}$$

is called the **determinant**, which more generally is denoted by $\det(\mathbf{A})$ or $|\mathbf{A}|$. As with the 2- dimensional case, $\mathbf{A}^{-1}$ exists for a square matrix $\mathbf{A}$ (of any dimensionality) if and only if $\det(\mathbf{A}) \neq 0$. For square matrices with dimensionality greater than two, the determinant is obtained recursively from the general expression

$$|\mathbf{A}| = \sum_{j=1}^{n} A_{ij}(-1)^{i+j}|\mathbf{A}_{ij}| \tag{8.12b}$$

where i is any fixed row of the matrix $\mathbf{A}$ and $\mathbf{A}_{ij}$ is a submatrix obtained by deleting the ith row and jth column from $\mathbf{A}$. Such a submatrix is known as a **minor.** In words, each of the n quantities in this equation is the product of three components: the element in the row around which one is working, -1 to the $(i+j)$th power, and the determinant of the ijth minor. In applying Equation 8.12b, one starts with the original $n \times n$ matrix and works down until the minors are reduced to 2×2 matrices whose determinants are scalars of the form $A_{11}A_{22} - A_{12}A_{21}$. A useful result is that the determinant of a diagonal matrix is the product of the diagonal elements of that matrix, so that if

$$A_{ij} = \begin{cases} a_i & i = j \\ 0 & i \neq j \end{cases} \qquad \text{then} \qquad |\mathbf{A}| = \prod_{i=1}^{n} a_i$$

The next section shows how determinants are used in the computation of a matrix inverse.

Example 4. Compute the determinant of

$$\mathbf{A} = \begin{pmatrix} 1 & 1 & 1 \\ 1 & 3 & 2 \\ 1 & 2 & 1 \end{pmatrix}$$

Letting $i = 1$, i.e., using the elements in the first row of $\mathbf{A}$,

$$|\mathbf{A}| = 1 \cdot (-1)^{1+1} \begin{vmatrix} 3 & 2 \\ 2 & 1 \end{vmatrix} + 1 \cdot (-1)^{1+2} \begin{vmatrix} 1 & 2 \\ 1 & 1 \end{vmatrix} + 1 \cdot (-1)^{1+3} \begin{vmatrix} 1 & 3 \\ 1 & 2 \end{vmatrix}$$

Using Equation 8.12a to obtain the determinants of the 2×2 matrices, this simplifies to

$$|\mathbf{A}| = [1 \times (3 - 4)] - [1 \times (1 - 2)] + [1 \times (2 - 3)] = -1$$

The same answer is obtained regardless of which row is used, and expanding around a column, instead of a row, produces the same result. Thus, in order to reduce the number of computations required to obtain a determinant, it is useful to expand using the row or column that contains the most zeros.

Computing Inverses

The general solution of a matrix inverse is

$$A_{ij}^{-1} = \left[\frac{(-1)^{i+j}|\mathbf{A}_{ij}|}{|\mathbf{A}|} \right]^T \tag{8.13}$$

where A_{ij}^{-1} denotes the ijth element of $\mathbf{A}^{-1}$, and $\mathbf{A}_{ij}$ denotes the ijth minor of $\mathbf{A}$. It can be seen from Equation 8.13 that a matrix can only be inverted if it has a nonzero determinant. Thus, a matrix is singular if its determinant is zero. This occurs whenever a matrix contains a row (or column) that can be written as a weighted sum of any other rows (or columns). In the context of our linear model, Equation 8.4, this happens if one of the n equations can be written as a combination of the others, a situation that is equivalent to there being n unknowns but less than n independent equations.

Example 5. Compute the inverse of

$$\mathbf{A} = \begin{pmatrix} 3 & 1 & 2 \\ 2 & 5 & 4 \\ 1 & 1 & 2 \end{pmatrix}$$

First, find the determinants of the minors,

$$|\mathbf{A}_{11}| = \begin{vmatrix} 5 & 4 \\ 1 & 2 \end{vmatrix} = 6 \qquad |\mathbf{A}_{23}| = \begin{vmatrix} 3 & 1 \\ 1 & 1 \end{vmatrix} = 2$$

$$|\mathbf{A}_{12}| = \begin{vmatrix} 2 & 4 \\ 1 & 2 \end{vmatrix} = 0 \qquad |\mathbf{A}_{31}| = \begin{vmatrix} 1 & 2 \\ 5 & 4 \end{vmatrix} = -6$$

$$|\mathbf{A}_{13}| = \begin{vmatrix} 2 & 5 \\ 1 & 1 \end{vmatrix} = -3 \qquad |\mathbf{A}_{32}| = \begin{vmatrix} 3 & 2 \\ 2 & 4 \end{vmatrix} = 8$$

$$|\mathbf{A}_{21}| = \begin{vmatrix} 1 & 2 \\ 1 & 2 \end{vmatrix} = 0 \qquad |\mathbf{A}_{33}| = \begin{vmatrix} 3 & 1 \\ 2 & 5 \end{vmatrix} = 13$$

$$|\mathbf{A}_{22}| = \begin{vmatrix} 3 & 2 \\ 1 & 2 \end{vmatrix} = 4$$

Using Equation 8.12b and expanding using the first row of $\mathbf{A}$ gives

$$|\mathbf{A}| = 3|\mathbf{A}_{11}| - |\mathbf{A}_{12}| + 2|\mathbf{A}_{13}| = 12$$

Returning to the matrix in brackets in Equation 8.13, we obtain

$$\frac{1}{12}\begin{pmatrix} 1 \times 6 & -1 \times 0 & 1 \times -3 \\ -1 \times 0 & 1 \times 4 & -1 \times 2 \\ 1 \times -6 & -1 \times 8 & 1 \times 13 \end{pmatrix} = \frac{1}{12}\begin{pmatrix} 6 & 0 & -3 \\ 0 & 4 & -2 \\ -6 & -8 & 13 \end{pmatrix}$$

and then taking the transpose,

$$\mathbf{A}^{-1} = \frac{1}{12}\begin{pmatrix} 6 & 0 & -6 \\ 0 & 4 & -8 \\ -3 & -2 & 13 \end{pmatrix}$$

To verify that this is indeed the inverse of $\mathbf{A}$, multiply $\mathbf{A}^{-1}$ by $\mathbf{A}$,

$$\frac{1}{12}\begin{pmatrix} 6 & 0 & -6 \\ 0 & 4 & -8 \\ -3 & -2 & 13 \end{pmatrix}\begin{pmatrix} 3 & 1 & 2 \\ 2 & 5 & 4 \\ 1 & 1 & 2 \end{pmatrix} = \frac{1}{12}\begin{pmatrix} 12 & 0 & 0 \\ 0 & 12 & 0 \\ 0 & 0 & 12 \end{pmatrix} = \begin{pmatrix} 1 & 0 & 0 \\ 0 & 1 & 0 \\ 0 & 0 & 1 \end{pmatrix}$$

EXPECTATIONS OF RANDOM VECTORS AND MATRICES

Matrix algebra provides a powerful approach for analyzing linear combinations of random variables. Let $\mathbf{x}$ be a column vector containing n random variables, $\mathbf{x} = (x_1, x_2, \cdots, x_n)^T$. We may wish to construct a new univariate (scalar) random variable y by taking some linear combination of the elements of $\mathbf{x}$,

$$y = \sum_{i=1}^{n} a_i x_i = \mathbf{a}^T \mathbf{x}$$

where $\mathbf{a} = (a_1, a_2, \cdots, a_n)^T$ is a column vector of constants. Likewise, we can construct a new k-dimensional vector $\mathbf{y}$ by premultiplying $\mathbf{x}$ by a $k \times n$ matrix $\mathbf{A}$ of constants, $\mathbf{y} = \mathbf{A}\mathbf{x}$. More generally, an $(n \times k)$ matrix $\mathbf{X}$ of random variables can be transformed into a new $m \times \ell$ dimensional matrix $\mathbf{Y}$ of elements consisting of linear combinations of the elements of $\mathbf{X}$ by

$$\mathbf{Y}_{m\times\ell} = \mathbf{A}_{m\times n}\mathbf{X}_{n\times k}\mathbf{B}_{k\times\ell} \tag{8.14}$$

where the matrices $\mathbf{A}$ and $\mathbf{B}$ are constants with dimensions as subscripted.

If $\mathbf{X}$ is a matrix whose elements are random variables, then the expected value of $\mathbf{X}$ is a matrix $E(\mathbf{X})$ containing the expected value of each element of $\mathbf{X}$. If $\mathbf{X}$ and $\mathbf{Z}$ are matrices of the same dimension, then

$$E(\mathbf{X} + \mathbf{Z}) = E(\mathbf{X}) + E(\mathbf{Z}) \tag{8.15}$$

This easily follows since the ijth element of $E(\mathbf{X} + \mathbf{Z})$ is $E(x_{ij} + z_{ij}) = E(x_{ij}) + E(z_{ij})$. Similarly, the expectation of $\mathbf{Y}$ as defined in Equation 8.14 is

$$E(\mathbf{Y}) = E(\mathbf{AXB}) = \mathbf{A}E(\mathbf{X})\mathbf{B} \tag{8.16a}$$

For example, for $\mathbf{y} = \mathbf{Xb}$ where $\mathbf{b}$ is an $n \times 1$ column vector,

$$E(\mathbf{y}) = E(\mathbf{Xb}) = E(\mathbf{X})\mathbf{b} \tag{8.16b}$$

Likewise, for $y = \mathbf{a}^T\mathbf{x} = \sum_i^n a_i x_i$,

$$E(y) = E(\mathbf{a}^T\mathbf{x}) = \mathbf{a}^T E(\mathbf{x}) \tag{8.16c}$$

COVARIANCE MATRICES OF TRANSFORMED VECTORS

To develop expressions for variances and covariances of linear combinations of random variables, we must first introduce the concept of quadratic forms. Consider an $n \times n$ square matrix $\mathbf{A}$ and an $n \times 1$ column vector $\mathbf{x}$. From the rules of matrix multiplication,

$$\mathbf{x}^T\mathbf{A}\mathbf{x} = \sum_{i=1}^{n}\sum_{j=1}^{n} a_{ij} x_i x_j \tag{8.17}$$

Expressions of this form are called **quadratic forms** (or **quadratic products**) and yield a scalar. A generalization of a quadratic form is the **bilinear form**, $\mathbf{b}^T\mathbf{A}\mathbf{a}$, where $\mathbf{b}$ and $\mathbf{a}$ are, respectively, $n \times 1$ and $m \times 1$ column vectors and $\mathbf{A}$ is an $n \times m$ matrix. Indexing the matrices and vectors in this expression by their dimensions, $\mathbf{b}^T_{1\times n}\mathbf{A}_{n\times m}\mathbf{a}_{m\times 1}$, shows that the resulting matrix product is a 1×1 matrix — in other words, a scalar. As scalars, bilinear forms equal their transposes, giving the useful identity

$$\mathbf{b}^T\mathbf{A}\mathbf{a} = \left(\mathbf{b}^T\mathbf{A}\mathbf{a}\right)^T = \mathbf{a}^T\mathbf{A}^T\mathbf{b} \tag{8.18}$$

Again let $\mathbf{x}$ be a column vector of n random variables. A compact way to express the n variances and $n(n-1)/2$ covariances associated with the elements of $\mathbf{x}$ is the matrix $\mathbf{V}$, where $V_{ij} = \sigma(x_i, x_j)$ is the covariance between the random variables x_i and x_j. We will generally refer to $\mathbf{V}$ as a **covariance matrix**, noting that the diagonal elements represent the variances and off-diagonal elements the covariances. The $\mathbf{V}$ matrix is symmetric, as

$$V_{ij} = \sigma(x_i, x_j) = \sigma(x_j, x_i) = V_{ji}$$

Now consider a univariate random variable $y = \sum c_k x_k$ generated from a linear combination of the elements of $\mathbf{x}$. In matrix notation, $y = \mathbf{c}^T\mathbf{x}$, where $\mathbf{c}$ is a column vector of constants. The variance of y can be expressed as a quadratic form involving the covariance matrix $\mathbf{V}$ for the elements of $\mathbf{x}$,

$$\begin{aligned}
\sigma^2\left(\mathbf{c}^T\mathbf{x}\right) &= \sigma^2\left(\sum_{i=1}^{n} c_i x_i\right) = \sigma\left(\sum_{i=1}^{n} c_i\, x_i\, , \sum_{j=1}^{n} c_j\, x_j\right) \\
&= \sum_{i=1}^{n}\sum_{j=1}^{n} \sigma\left(c_i\, x_i, c_j\, x_j\right) = \sum_{i=1}^{n}\sum_{j=1}^{n} c_i\, c_j\, \sigma\left(x_i, x_j\right) \\
&= \mathbf{c}^T\mathbf{V}\,\mathbf{c}
\end{aligned} \tag{8.19}$$

Likewise, the covariance between two univariate random variables created from different linear combinations of $\mathbf{x}$ is given by the bilinear form

$$\sigma(\mathbf{a}^T\mathbf{x}, \mathbf{b}^T\mathbf{x}) = \mathbf{a}^T\mathbf{V}\,\mathbf{b} \tag{8.20}$$

If we transform $\mathbf{x}$ to two new vectors $\mathbf{y}_{\ell\times 1} = \mathbf{A}_{\ell\times n}\mathbf{x}_{n\times 1}$ and $\mathbf{z}_{m\times 1} = \mathbf{B}_{m\times n}\mathbf{x}_{n\times 1}$, then instead of a single covariance we have an $\ell \times m$ dimensional covariance matrix, denoted $\boldsymbol{\sigma}(\mathbf{y},\mathbf{z})$. Letting $\boldsymbol{\mu}_\mathbf{y} = \mathbf{A}\boldsymbol{\mu}$ and $\boldsymbol{\mu}_\mathbf{z} = \mathbf{B}\boldsymbol{\mu}$, with $E(\mathbf{x}) = \boldsymbol{\mu}$, then $\boldsymbol{\sigma}(\mathbf{y},\mathbf{z})$ can be expressed in terms of $\mathbf{V}$, the covariance matrix of $\mathbf{x}$,

$$\begin{aligned}\boldsymbol{\sigma}(\mathbf{y},\mathbf{z}) &= \boldsymbol{\sigma}(\mathbf{A}\mathbf{x},\mathbf{B}\mathbf{x}) \\ &= E\left[(\mathbf{y}-\boldsymbol{\mu}_\mathbf{y})(\mathbf{z}-\boldsymbol{\mu}_\mathbf{z})^T\right] \\ &= E\left[\mathbf{A}(\mathbf{x}-\boldsymbol{\mu})(\mathbf{x}-\boldsymbol{\mu})^T\mathbf{B}^T\right] \\ &= \mathbf{A}\mathbf{V}\,\mathbf{B}^T \end{aligned} \tag{8.21a}$$

In particular, the covariance matrix for $\mathbf{y} = \mathbf{A}\mathbf{x}$ is

$$\boldsymbol{\sigma}(\mathbf{y},\mathbf{y}) = \mathbf{A}\mathbf{V}\,\mathbf{A}^T \tag{8.21b}$$

so that the covariance between y_i and y_j is given by the ijth element of the matrix product $\mathbf{AVA}^T$.

Finally, note that if $\mathbf{x}$ is a vector of random variables with expected value $\boldsymbol{\mu}$, then the expected value of the scalar quadratic product $\mathbf{x}^T\mathbf{A}\mathbf{x}$ is

$$E(\mathbf{x}^T\mathbf{A}\mathbf{x}) = \text{tr}(\mathbf{A}\mathbf{V}) + \boldsymbol{\mu}^T\mathbf{A}\boldsymbol{\mu} \tag{8.22}$$

where $\mathbf{V}$ is the covariance matrix for the elements of $\mathbf{x}$, and the **trace** of a square matrix, $\text{tr}(\mathbf{M}) = \sum M_{ii}$, is the sum of its diagonal values (Searle 1971).

THE MULTIVARIATE NORMAL DISTRIBUTION

As we have seen above, matrix notation provides a compact way to express vectors of random variables. We now consider the most commonly assumed distribution for such vectors, the multivariate analog of the normal distribution discussed in Chapter 2. Much of the theory for the evolution of quantitative traits is based on this distribution, which we hereafter denote as the MVN.

Consider the probability density function for n independent normal random variables, where x_i is normally distributed with mean μ_i and variance σ_i^2. In this case, because the variables are independent, the joint probability density function is simply the product of each univariate density,

$$\begin{aligned} p(\mathbf{x}) &= \prod_{i=1}^{n}(2\pi)^{-1/2}\sigma_i^{-1}\exp\left(-\frac{(x_i-\mu_i)^2}{2\,\sigma_i^2}\right) \\ &= (2\pi)^{-n/2}\left(\prod_{i=1}^{n}\sigma_i\right)^{-1}\exp\left(-\sum_{i=1}^{n}\frac{(x_i-\mu_i)^2}{2\,\sigma_i^2}\right) \end{aligned} \tag{8.23}$$

We can express this equation more compactly in matrix form by defining the matrices

$$\mathbf{V} = \begin{pmatrix} \sigma_1^2 & 0 & \cdots & 0 \\ 0 & \sigma_2^2 & \cdots & 0 \\ \vdots & \vdots & \ddots & \vdots \\ 0 & \cdots & \cdots & \sigma_n^2 \end{pmatrix} \quad \text{and} \quad \boldsymbol{\mu} = \begin{pmatrix} \mu_1 \\ \mu_2 \\ \vdots \\ \mu_n \end{pmatrix}$$

Since $\mathbf{V}$ is diagonal, its determinant is simply the product of the diagonal elements

$$|\mathbf{V}| = \prod_{i=1}^{n} \sigma_i^2$$

Likewise, using quadratic products, note that

$$\sum_{i=1}^{n} \frac{(x_i - \mu_i)^2}{\sigma_i^2} = (\mathbf{x} - \boldsymbol{\mu})^T \mathbf{V}^{-1} (\mathbf{x} - \boldsymbol{\mu})$$

Putting these together, Equation 8.23 can be rewritten as

$$p(\mathbf{x}) = (2\pi)^{-n/2} |\mathbf{V}|^{-1/2} \exp\left[-\frac{1}{2} (\mathbf{x} - \boldsymbol{\mu})^T \mathbf{V}^{-1} (\mathbf{x} - \boldsymbol{\mu}) \right] \tag{8.24}$$

We will also write this density as $p(\mathbf{x}, \boldsymbol{\mu}, \mathbf{V})$ when we wish to stress that it is a function of the mean vector $\boldsymbol{\mu}$ and the covariance matrix $\mathbf{V}$.

More generally, when the elements of $\mathbf{x}$ are correlated, Equation 8.24 gives the probability density function for a vector of multivariate normally distributed random variables, with mean vector $\boldsymbol{\mu}$ and covariance matrix $\mathbf{V}$. We denote this by

$$\mathbf{x} \sim \text{MVN}_n(\boldsymbol{\mu}, \mathbf{V})$$

where the subscript indicating the dimensionality of $\mathbf{x}$ is usually omitted. The multivariate normal distribution is also referred to as the **Gaussian distribution**.

Properties of the MVN

As in the case of its univariate counterpart, the MVN is expected to arise naturally when the quantities of interest result from a large number of underlying variables. Since this condition seems (at least at first glance) to describe many biological systems, the MVN is a natural starting point in biometrical analysis. Further details on the wide variety of applications of the MVN to multivariate statistics can be found in the introductory texts by Morrison (1976) and Johnson and Wichern (1988) and in the more advanced treatment by Anderson (1984). The MVN has a number of useful properties, which we summarize below.

1. *If* $\mathbf{x} \sim$ MVN, *then the distribution of any subset of the variables in* $\mathbf{x}$ *is also MVN*. For example, each x_i is normally distributed and each pair (x_i, x_j) is bivariate normally distributed.

2. *If* $\mathbf{x} \sim$ MVN, *then any linear combination of the elements of* $\mathbf{x}$ *is also MVN.* Specifically, if $\mathbf{x} \sim \text{MVN}_n(\boldsymbol{\mu}, \mathbf{V})$, $\mathbf{a}$ is a vector of constants, and $\mathbf{A}$ is a matrix of constants, then

$$\text{for} \quad \mathbf{y} = \mathbf{x} + \mathbf{a}, \qquad \mathbf{y} \sim \text{MVN}_n(\boldsymbol{\mu} + \mathbf{a}, \mathbf{V}) \tag{8.25a}$$

$$\text{for} \quad y = \mathbf{a}^T\mathbf{x} = \sum_{k=1}^{n} a_i x_i, \qquad y \sim \text{N}(\mathbf{a}^T\boldsymbol{\mu}, \mathbf{a}^T\mathbf{V}\,\mathbf{a}) \tag{8.25b}$$

$$\text{for} \quad \mathbf{y} = \mathbf{A}\mathbf{x}, \qquad \mathbf{y} \sim \text{MVN}_m\left(\mathbf{A}\boldsymbol{\mu}, \mathbf{A}^T\mathbf{V}\mathbf{A}\right) \tag{8.25c}$$

3. *Conditional distributions associated with the MVN are also multivariate normal.* Consider the partitioning of $\mathbf{x}$ into two components, an $(m \times 1)$ column vector $\mathbf{x}_1$ and an $[(n-m)\times 1]$ column vector $\mathbf{x}_2$ of the remaining variables, e.g.,

$$\mathbf{x} = \begin{pmatrix} \mathbf{x}_1 \\ \mathbf{x}_2 \end{pmatrix}$$

The mean vector and covariance matrix can be partitioned similarly as

$$\boldsymbol{\mu} = \begin{pmatrix} \boldsymbol{\mu}_1 \\ \boldsymbol{\mu}_2 \end{pmatrix} \quad \text{and} \quad \mathbf{V} = \begin{pmatrix} \mathbf{V}_{\mathbf{x}_1\mathbf{x}_1} & \mathbf{V}_{\mathbf{x}_1\mathbf{x}_2} \\ \mathbf{V}^T_{\mathbf{x}_1\mathbf{x}_2} & \mathbf{V}_{\mathbf{x}_2\mathbf{x}_2} \end{pmatrix} \tag{8.26}$$

where the $m \times m$ and $(n-m) \times (n-m)$ matrices $\mathbf{V}_{\mathbf{x}_1\mathbf{x}_1}$ and $\mathbf{V}_{\mathbf{x}_2\mathbf{x}_2}$ are, respectively, the covariance matrices for $\mathbf{x}_1$ and $\mathbf{x}_2$, while the $m \times (n-m)$ matrix $\mathbf{V}_{\mathbf{x}_1\mathbf{x}_2}$ is the matrix of covariances between the elements of $\mathbf{x}_1$ and $\mathbf{x}_2$. If we condition on $\mathbf{x}_2$, the resulting conditional random variable $\mathbf{x}_1|\mathbf{x}_2$ is MVN with $(m \times 1)$ mean vector

$$\boldsymbol{\mu}_{\mathbf{x}_1|\mathbf{x}_2} = \boldsymbol{\mu}_1 + \mathbf{V}_{\mathbf{x}_1\mathbf{x}_2}\mathbf{V}^{-1}_{\mathbf{x}_2\mathbf{x}_2}(\mathbf{x}_2 - \boldsymbol{\mu}_2) \tag{8.27}$$

and $(m \times m)$ covariance matrix

$$\mathbf{V}_{\mathbf{x}_1|\mathbf{x}_2} = \mathbf{V}_{\mathbf{x}_1\mathbf{x}_1} - \mathbf{V}_{\mathbf{x}_1\mathbf{x}_2}\mathbf{V}^{-1}_{\mathbf{x}_2\mathbf{x}_2}\mathbf{V}^T_{\mathbf{x}_1\mathbf{x}_2} \tag{8.28}$$

A proof can be found in most multivariate statistics texts, e.g., Morrison (1976).

4. *If* $\mathbf{x} \sim$ MVN, *the regression of any subset of* $\mathbf{x}$ *on another subset is linear and homoscedastic.* Rewriting Equation 8.27 in terms of a regression of the predicted value of the vector $\mathbf{x}_1$ given an observed value of the vector $\mathbf{x}_2$, we have

$$\mathbf{x}_1 = \boldsymbol{\mu}_1 + \mathbf{V}_{\mathbf{x}_1\mathbf{x}_2}\mathbf{V}^{-1}_{\mathbf{x}_2\mathbf{x}_2}(\mathbf{x}_2 - \boldsymbol{\mu}_2) + \mathbf{e} \tag{8.29a}$$

where

$$\mathbf{e} \sim \text{MVN}_m\left(\mathbf{0}, \mathbf{V}_{\mathbf{x}_1|\mathbf{x}_2}\right) \tag{8.29b}$$

Example 6. Consider the regression of the phenotypic value of an offspring (z_o) on that of its parents (z_s and z_d for sire and dam, respectively). Assume that the joint distribution of z_o, z_s, and z_d is multivariate normal. For the simplest case of noninbred and unrelated parents, no epistasis or genotype-environment correlation, the covariance matrix can be obtained from the theory of correlation between relatives (Chapter 7), giving the joint distribution as

$$\begin{pmatrix} z_o \\ z_s \\ z_d \end{pmatrix} \sim \text{MVN} \left[\begin{pmatrix} \mu_o \\ \mu_s \\ \mu_d \end{pmatrix}, \sigma_z^2 \begin{pmatrix} 1 & h^2/2 & h^2/2 \\ h^2/2 & 1 & 0 \\ h^2/2 & 0 & 1 \end{pmatrix} \right]$$

Let

$$\mathbf{x_1} = (z_o), \quad \mathbf{x_2} = \begin{pmatrix} z_s \\ z_d \end{pmatrix}$$

giving

$$\mathbf{V_{x_1,x_1}} = \sigma_z^2, \quad \mathbf{V_{x_1,x_2}} = \frac{h^2\sigma_z^2}{2} (1 \ \ 1), \quad \mathbf{V_{x_2,x_2}} = \sigma_z^2 \begin{pmatrix} 1 & 0 \\ 0 & 1 \end{pmatrix}$$

From Equation 8.29a, the regression of offspring value on parental values is linear and homoscedastic with

$$\begin{aligned} z_o &= \mu_o + \frac{h^2\sigma_z^2}{2} (1 \ \ 1) \sigma_z^{-2} \begin{pmatrix} 1 & 0 \\ 0 & 1 \end{pmatrix} \begin{pmatrix} z_s - \mu_s \\ z_d - \mu_d \end{pmatrix} + e \\ &= \mu_o + \frac{h^2}{2} (z_s - \mu_s) + \frac{h^2}{2} (z_d - \mu_d) + e \end{aligned} \tag{8.30a}$$

where, from Equations 8.28 and 8.29b, the residual error is normally distributed with mean zero and variance

$$\begin{aligned} \sigma_e^2 &= \sigma_z^2 - \frac{h^2\sigma_z^2}{2} (1 \ \ 1) \sigma_z^{-2} \begin{pmatrix} 1 & 0 \\ 0 & 1 \end{pmatrix} \frac{h^2\sigma_z^2}{2} \begin{pmatrix} 1 \\ 1 \end{pmatrix} \\ &= \sigma_z^2 \left(1 - \frac{h^4}{2} \right) \end{aligned} \tag{8.30b}$$

Example 7. The previous example dealt with the prediction of the phenotypic value of an offspring given parental phenotypic values. The same approach can be used to predict an offspring's additive genetic value A_o given knowledge of the parental values (A_s, A_d). Again assuming that the joint distribution is multivariate normal and that the parents are unrelated and noninbred, the joint distribution can be written as

$$\begin{pmatrix} A_o \\ A_s \\ A_d \end{pmatrix} \sim \text{MVN}\left[\begin{pmatrix} \mu_o \\ \mu_s \\ \mu_d \end{pmatrix}, \sigma_A^2 \begin{pmatrix} 1 & 1/2 & 1/2 \\ 1/2 & 1 & 0 \\ 1/2 & 0 & 1 \end{pmatrix} \right]$$

Proceeding in the same fashion as in Example 6, the conditional distribution of offspring additive genetic values, given the parental values, is normal, so that the regression of offspring additive genetic value on parental value is linear and homoscedastic with

$$A_o = \mu_o + \frac{A_s - \mu_s}{2} + \frac{A_d - \mu_d}{2} + e \tag{8.31a}$$

and

$$e \sim \text{N}(0, \sigma_A^2/2) \tag{8.31b}$$

OVERVIEW OF LINEAR MODELS

Linear models form the backbone of most estimation procedures in quantitative genetics and will be extensively used throughout the rest of this book. They are generally structured such that a vector of observations of one variable (y) is modeled as a linear combination of other variables observed along with y. The remainder of this chapter introduces some of the basic tools and key concepts underlying the use of linear models. Advanced topics are examined in detail in Chapters 26 and 27, and further comments are given in Appendix 3.

In multiple regression, the commonest type of linear model, the predictor variables $x_1, \cdots, x_n$ represent observed values for n traits of interest. More generally, some or all of the predictor variables could be **indicator variables**, with values of 0 or 1 indicating whether an observation belongs in a particular category or grouping of interest. As an example, consider the half-sib design wherein each of p unrelated sires is mated at random to a number of unrelated dams and a single offspring is measured from each cross. The simplest model for this design is

$$y_{ij} = \mu + s_i + e_{ij}$$

where y_{ij} is the phenotype of the jth offspring from sire i, μ is the population mean, s_i is the **sire effect**, and e_{ij} is the residual error (the "noise" remaining in the data after the sire effect is removed). Although this is clearly a linear model, it differs significantly from the regression model described above in that while there are parameters to estimate (the sire effects s_i), the only measured values are the y_{ij}. Nevertheless, we can express this model in a form that is essentially identical to the standard regression model by using p indicator (i.e., zero or one) variables to classify the sires of the offspring. The resulting linear model becomes

$$y_{ij} = \mu + \sum_{k=1}^{p} s_k \, x_{ik} + e_{ij}$$

where

$$x_{ik} = \begin{cases} 1 & \text{if sire } k = i \\ 0 & \text{otherwise} \end{cases}$$

By the judicious use of indicator variables, an extremely wide class of problems can be handled by linear models. Models containing only indicator variables are usually termed ANOVA (**analysis of variance**) models, while regression usually refers to models in which predictor variables can take on a continuous range of values. Both procedures are special cases of the **general linear model** (GLM), wherein each observation (y) is assumed to be a linear function of p observed and/or indicator variables plus a residual error (e),

$$y_i = \sum_{k=1}^{p} \beta_k \, x_{ik} + e_i \tag{8.32a}$$

where $x_{i1}, \cdots, x_{ip}$ are the values of the p predictor variables for the ith individual. For a vector of n observations, the GLM can be written in matrix form as

$$\mathbf{y} = \mathbf{X}\boldsymbol{\beta} + \mathbf{e} \tag{8.32b}$$

where the **design** or **incidence matrix X** is $n \times p$, and **e** is the vector of residual errors. It is important to note that **y** and **X** contain the observed values, while $\boldsymbol{\beta}$ is a vector of parameters (usually called **factors** or **effects**) to be estimated.

Example 8. Suppose that three different sires used in the above half-sib design have two, one, and three offspring, respectively. This can be expressed in GLM form, $\mathbf{y} = \mathbf{X}\boldsymbol{\beta} + \mathbf{e}$ with

$$\mathbf{y} = \begin{pmatrix} y_{11} \\ y_{12} \\ y_{21} \\ y_{31} \\ y_{32} \\ y_{33} \end{pmatrix}, \quad \mathbf{X} = \begin{pmatrix} 1 & 1 & 0 & 0 \\ 1 & 1 & 0 & 0 \\ 1 & 0 & 1 & 0 \\ 1 & 0 & 0 & 1 \\ 1 & 0 & 0 & 1 \\ 1 & 0 & 0 & 1 \end{pmatrix}, \quad \boldsymbol{\beta} = \begin{pmatrix} \mu \\ s_1 \\ s_2 \\ s_3 \end{pmatrix}, \quad \text{and} \quad \mathbf{e} = \begin{pmatrix} e_{11} \\ e_{12} \\ e_{21} \\ e_{31} \\ e_{32} \\ e_{33} \end{pmatrix}$$

Likewise, the multiple regression

$$y_i = \alpha + \sum_{j=1}^{p} \beta_j \, x_{ij} + e_i$$

can be written in GLM form with

$$\mathbf{y} = \begin{pmatrix} y_1 \\ \vdots \\ y_n \end{pmatrix}, \quad \mathbf{X} = \begin{pmatrix} 1 & x_{11} & \cdots & x_{1p} \\ 1 & x_{21} & \cdots & x_{2p} \\ \vdots & \vdots & \ddots & \vdots \\ 1 & x_{n1} & \cdots & x_{np} \end{pmatrix}, \quad \boldsymbol{\beta} = \begin{pmatrix} \alpha \\ \beta_1 \\ \vdots \\ \beta_p \end{pmatrix}, \quad \text{and} \quad \mathbf{e} = \begin{pmatrix} e_1 \\ \vdots \\ e_n \end{pmatrix}$$

Ordinary Least Squares

Estimates of the vector $\boldsymbol{\beta}$ for the general linear model are usually obtained by the method of least-squares, which uses the observations $\mathbf{y}$ and $\mathbf{X}$ and makes special assumptions about the covariance structure of the vector of residual errors $\mathbf{e}$. The method of **ordinary least squares** assumes that the residual errors are homoscedastic and uncorrelated, i.e., $\sigma^2(e_i) = \sigma_e^2$ for all i, and $\sigma(e_i,\, e_j) = 0$ for $i \neq j$.

Let $\mathbf{b}$ be an estimate of $\boldsymbol{\beta}$, and denote the vector of y values predicted from this estimate by $\widehat{\mathbf{y}} = \mathbf{Xb}$, so that the resulting vector of residual errors is

$$\widehat{\mathbf{e}} = \mathbf{y} - \widehat{\mathbf{y}} = \mathbf{y} - \mathbf{Xb}$$

The ordinary least-squares (OLS) estimate of $\boldsymbol{\beta}$ is the $\mathbf{b}$ vector that minimizes the residual sum of squares,

$$\sum_{i=1}^{n} \widehat{e}_i^{\,2} = \widehat{\mathbf{e}}^T \widehat{\mathbf{e}} = (\mathbf{y} - \mathbf{Xb})^T (\mathbf{y} - \mathbf{Xb})$$

Taking derivatives, it can be shown that our desired estimate satisfies

$$\mathbf{b} = (\mathbf{X}^T \mathbf{X})^{-1} \mathbf{X}^T \mathbf{y} \tag{8.33a}$$

Under the assumption that the residual errors are uncorrelated and homoscedastic (i.e., the covariance matrix of the residuals is $\sigma_e^2 \cdot \mathbf{I}$), the covariance matrix of the elements of $\mathbf{b}$ is

$$\mathbf{V_b} = (\mathbf{X}^T \mathbf{X})^{-1} \sigma_e^2 \tag{8.33b}$$

Hence, the OLS estimator of β_i is the ith element of the column vector $\mathbf{b}$, while the variance of this estimator is the ith diagonal element of the matrix $\mathbf{V_b}$. Likewise, the covariance of this estimator with the OLS estimator for β_j is the ijth element of $\mathbf{V_b}$.

If the residuals follow a multivariate normal distribution with $\mathbf{e} \sim \text{MVN}(\mathbf{0}, \sigma_e^2 \cdot \mathbf{I})$, the OLS estimate is also the maximum-likelihood estimate. If $\mathbf{X}^T\mathbf{X}$ is singular, Equations 8.33a,b still hold when a generalized inverse is used, although only certain linear combinations of fixed factors can be estimated (see Appendix 3 for details).

Example 9. Consider a univariate regression where the predictor and response variable both have expected mean zero, so that the regression passes through the origin. The appropriate model becomes

$$y_i = \beta\, x_i + e_i$$

With observations on n individuals, this relationship can be written in GLM form with $\boldsymbol{\beta} = \beta$ and design matrix $\mathbf{X} = (x_1,\, x_2,\, \cdots x_n)^T$, implying

$$\mathbf{X}^T\mathbf{X} = \sum_{i=1}^n x_i^2 \qquad \text{and} \qquad \mathbf{X}^T\mathbf{y} = \sum_{i=1}^n x_i\, y_i$$

Applying Equations 8.33a,b gives the OLS estimate of β and its sample variance (assuming the covariance matrix of $\mathbf{e}$ is $\mathbf{I} \cdot \sigma_e^2$) as

$$b = \left(\mathbf{X}^T\mathbf{X}\right)^{-1} \mathbf{X}^T\mathbf{y} = \frac{\sum x_i\, y_i}{\sum x_i^2}, \qquad \sigma^2(b) = \left(\mathbf{X}^T\mathbf{X}\right)^{-1} \sigma_e^2 = \frac{\sigma_e^2}{\sum x_i^2}$$

This estimate of β differs from the standard univariate regression slope (Equation 3.14b) where the intercept value is not assumed to be equal to zero.

Example 10. Recall from Equation 8.10b that the vector of partial regression coefficients for a multivariate regression is defined to be $\mathbf{b} = \mathbf{V}^{-1}\,\mathbf{c}$ (where $\mathbf{V}$ is the estimated covariance matrix, and $\mathbf{c}$ is the vector of estimated covariances between $\mathbf{y}$ and $\mathbf{z}$). Here we show that this expression is equivalent to the OLS estimator $\mathbf{b} = (\mathbf{X}^T\mathbf{X})^{-1}\mathbf{X}^T\mathbf{y}$. Using the notation from Example 8, for the ith individual we observe y_i and the values of p predictor variables, $z_{i1}, \cdots, z_{ip}$. Since the regression satisfies $\bar{y} = \alpha + \beta_1\bar{z}_1 + \cdots + \beta_p\bar{z}_p$, subtracting the mean from each observation removes the intercept, with

$$y_i^* = (y_i - \bar{y}) = \beta_1(z_{i1} - \bar{z}_1) + \cdots + \beta_p(z_{ip} - \bar{z}_p) + e_i$$

For n observations, the resulting linear model $\mathbf{y}^* = \mathbf{X}\boldsymbol{\beta} + \mathbf{e}$ has

$$\mathbf{y}^* = \begin{pmatrix} y_1 - \overline{y} \\ \vdots \\ y_n - \overline{y} \end{pmatrix}, \quad \boldsymbol{\beta} = \begin{pmatrix} \beta_1 \\ \vdots \\ \beta_p \end{pmatrix}, \quad \mathbf{X} = \begin{pmatrix} (z_{11} - \overline{z}_1) & \cdots & (z_{1p} - \overline{z}_p) \\ \vdots & \ddots & \vdots \\ (z_{n1} - \overline{z}_1) & \cdots & (z_{np} - \overline{z}_p) \end{pmatrix}$$

where z_{ij} is the value of character j in the ith individual. Partitioning the design matrix $\mathbf{X}$ into p column vectors corresponding to the n observations on each of the p predictor variables gives

$$\mathbf{X} = (\,\mathbf{x}_1, \quad \cdots, \quad \mathbf{x}_p\,) \quad \text{where} \quad \mathbf{x}_j = \begin{pmatrix} z_{1j} - \overline{z}_j \\ z_{2j} - \overline{z}_j \\ \vdots \\ z_{nj} - \overline{z}_j \end{pmatrix}$$

giving the jth element of the vector $\mathbf{X}^T\mathbf{y}^*$ as

$$\left(\mathbf{X}^T\mathbf{y}^*\right)_j = \mathbf{x}_j^T\mathbf{y}^* = \sum_{i=1}^{n}(y_i - \overline{y})(z_{ij} - \overline{z}_j) = (n-1)\text{Cov}(y, z_j)$$

and implying $\mathbf{X}^T\mathbf{y}^* = (n-1)\,\mathbf{c}$. Likewise, the jkth element of $\mathbf{X}^T\mathbf{X}$ is

$$\mathbf{x}_j^T\mathbf{x}_k = \sum_{i=1}^{n}(z_{ij} - \overline{z}_j)(z_{ik} - \overline{z}_k) = (n-1)\text{Cov}(z_j, z_k)$$

implying $\mathbf{X}^T\mathbf{X} = (n-1)\mathbf{V}$. Putting these results together gives

$$(\mathbf{X}^T\mathbf{X})^{-1}\mathbf{X}^T\mathbf{y}^* = \mathbf{V}^{-1}\mathbf{c}$$

showing that Equation 8.10b does indeed give the OLS estimates of the partial regression coefficients.

Generalized Least Squares

Under OLS, the unweighted sum of squared residuals is minimized. However, if some residuals are inherently more variable than others (have a higher variance), less weight should be assigned to the more variable data. Correlations between residuals can also influence the weight that should be assigned to each individual, as the data are not independent. Thus, if the residual errors are heteroscedastic and/or correlated, ordinary least-squares estimates of regression parameters and standard errors of these estimates are potentially biased.

A more general approach to regression analysis expresses the covariance matrix of the vector of residuals as $\sigma_e^2\,\mathbf{R}$, with $\sigma(e_i, e_j) = R_{ij}\sigma_e^2$. Lack of independence between residuals is indicated by the presence of nonzero off-diagonal elements in **R**, while heteroscedasticity is indicated by differences in the diagonal elements of **R**. **Generalized** (or **weighted) least squares** (GLS) takes these complications into account. As shown in Appendix 3, if the linear model is

$$\mathbf{y} = \mathbf{X}\boldsymbol{\beta} + \mathbf{e} \qquad \text{with } \mathbf{e} \sim (0, \mathbf{R}\,\sigma_e^2)$$

the GLS estimate of $\boldsymbol{\beta}$ is

$$\mathbf{b} = \left(\mathbf{X}^T\mathbf{R}^{-1}\mathbf{X}\right)^{-1}\mathbf{X}^T\mathbf{R}^{-1}\mathbf{y} \tag{8.34}$$

(Aitken 1935). The covariance matrix for the GLS estimates is

$$\mathbf{V_b} = \left(\mathbf{X}^T\mathbf{R}^{-1}\mathbf{X}\right)^{-1}\sigma_e^2 \tag{8.35}$$

If residuals are independent and homoscedastic, $\mathbf{R} = \mathbf{I}$, and GLS estimates are the same as OLS estimates. If $\mathbf{e} \sim \text{MVN}(\mathbf{0},\, \mathbf{R}\,\sigma_e^2)$, the GLS estimate of $\boldsymbol{\beta}$ is also the maximum-likelihood estimate.

Example 11. A common situation requiring weighted least-squares analysis occurs when residuals are independent but heteroscedastic with $\sigma^2(e_i) = \sigma_e^2/w_i$, where w_i are known positive constants. For example, if each observation y_i is the mean of n_i independent observations (each with uncorrelated residuals with variance σ_e^2), then $\sigma^2(e_i) = \sigma_e^2/n_i$, and hence $w_i = n_i$. Here

$$\mathbf{R} = \text{Diag}(w_1^{-1}, w_2^{-1}, \ldots, w_n^{-1})$$

where Diag denotes a diagonal matrix, giving

$$\mathbf{R}^{-1} = \text{Diag}(w_1, w_2, \ldots, w_n)$$

With this residual variance structure, consider the weighted least-squares estimate for the simple univariate regression model $y = \alpha + \beta\,x + e$. In GLM form,

$$\mathbf{y} = \begin{pmatrix} y_1 \\ \vdots \\ y_n \end{pmatrix}, \qquad \mathbf{X} = \begin{pmatrix} 1 & x_1 \\ \vdots & \vdots \\ 1 & x_n \end{pmatrix}, \qquad \text{and} \qquad \boldsymbol{\beta} = \begin{pmatrix} \alpha \\ \beta \end{pmatrix}$$

Define the following weighted means and cross products,

$$w = \sum_{i=1}^{n} w_i, \quad \overline{x}_w = \sum_{i=1}^{n} \frac{w_i x_i}{w}, \quad \overline{x^2}_w = \sum_{i=1}^{n} \frac{w_i x_i^2}{w}$$

$$\overline{y}_w = \sum_{i=1}^{n} \frac{w_i y_i}{w}, \quad \overline{xy}_w = \sum_{i=1}^{n} \frac{w_i x_i\, y_i}{w}$$

With these definitions, matrix multiplication and a little simplification give

$$\mathbf{X}^T\mathbf{R}^{-1}\mathbf{y} = w \begin{pmatrix} \overline{y}_w \\ \overline{xy}_w \end{pmatrix} \qquad \text{and} \qquad \mathbf{X}^T\mathbf{R}^{-1}\mathbf{X} = w \begin{pmatrix} 1 & \overline{x}_w \\ \overline{x}_w & \overline{x^2}_w \end{pmatrix}$$

Applying Equation 8.34, the GLS estimates of α and β are

$$a = \overline{y}_w - b\overline{x}_w \tag{8.36a}$$

$$b = \frac{\overline{xy}_w - \overline{x}_w\, \overline{y}_w}{\overline{x^2}_w - \overline{x}_w^2} \tag{8.36b}$$

If all weights are equal ($w_i = c$), these expressions reduce to the standard (OLS) least-squares estimators given by Equation 3.14. Applying Equation 8.35, the sampling variances and covariance for these estimates are

$$\sigma^2(a) = \frac{\sigma_e^2 \cdot \overline{x^2}_w}{w\,(\,\overline{x^2}_w - \overline{x}_w^2\,)} \tag{8.37a}$$

$$\sigma^2(b) = \frac{\sigma_e^2}{w\,(\,\overline{x^2}_w - \overline{x}_w^2\,)} \tag{8.37b}$$

$$\sigma(a,b) = \frac{-\sigma_e^2\,\overline{x}_w}{w\,(\,\overline{x^2}_w - \overline{x}_w^2\,)} \tag{8.37c}$$

9

Analysis of Line Crosses

Distinct populations, such as the isolated demes that comprise some natural populations, often exhibit remarkable phenotypic divergence. Such differences are sometimes a simple consequence of environmental influences on phenotypic expression, but genetic differences may arise as a result of local adaptation and/or random genetic drift. The genetic basis of interdemic differentiation is of interest for several reasons. With nonadditive gene action, the mean phenotypes of progeny of interdemic crosses (F_1 hybrids) will not be intermediate to those of their parents. Depending upon the relation between phenotype and fitness, hybridization can lead to outbreeding depression (reduced fitness) or outbreeding enhancement (increased fitness) in the F_1 or later generations. For populations that do not normally have an opportunity to interbreed in nature, such properties may evolve passively as an indirect consequence of local adaptation. However, when opportunities for interdemic exchange are common, natural selection may favor specific mating system properties, including dispersal strategies or reproductive isolation, which enhance or discourage outcrossing. Thus, understanding the genetic basis of interdemic differentiation is a key to deciphering the mechanisms of speciation.

The mechanisms of genetic differentiation also have important practical implications. For example, in artificial selection programs, the mean phenotypes of selected lines often evolve well beyond the range of variation seen in the base population prior to selection (Chapter 1). The extent to which such changes are caused by a large number of genes of relatively small effects, as opposed to a few major segregating factors, is an important determinant of whether a search for informative molecular markers is likely to be successful (Chapters 13–16). In addition, the degree to which selection advances made in different lines can be successfully integrated into a single crossbred line is a function of the ways in which genes from the isolated lines interact.

In this chapter, we show how the judicious choice of line crosses can be used to reveal the relative contributions of additive, dominance, and epistatic effects to population differentiation. A statistical test of the adequacy of alternative genetic models will be presented, and its application to a variety of data sets will be used to show that nonadditive gene action is commonly associated with population differentiation. Several methods for estimating the minimum number of loci responsible for population differentiation will then be discussed. Their application firmly supports the conviction that most characters of interest to evolutionary biologists and breeders are influenced by multiple loci. This, however, does not rule out the possibility that a small number of loci are responsible for the majority of the differentiation between species and/or lines within species.

EXPECTATIONS FOR LINE-CROSS MEANS

We start with two parental populations (P_1 and P_2), each with loci assumed to be in Hardy-Weinberg and gametic phase equilibrium. For the time being, we also assume that the loci differentiating the two populations are unlinked. An F_1 population is obtained by crossing the P_1 and P_2 lines, and subsequent random mating of F_1 individuals results in an F_2 generation. Since the F_2 population will be in Hardy-Weinberg and gametic phase equilibrium for unlinked loci, both for genes derived within and between populations, it is logical to treat it as a point of reference for the definition of genetic effects.

In previous chapters, we have presented definitions of additive, dominance, and epistatic effects for specific genes and gene combinations within randomly mating populations. The statistical machinery underlying line-cross analysis has parallels with this approach. The questions here, however, are whether there is a net difference between the additive effects of the genes in the P_1 and P_2 populations, whether the genes in P_1 tend on average to be dominant over those in P_2, and whether there are net directional epistatic interactions between P_1 and P_2 genes. Before explicitly defining these **composite effects**, it will be useful to have some indices to describe the gene content and the degree of hybridity of line-cross derivatives.

For any pair of lines and their derivatives, the genotypes at a locus can be partitioned into three classes: (1) both alleles are from a random sample of P_1 genes, (2) both alleles are from a random sample of P_2 genes, and (3) one allele is from a random sample of P_1 genes, while the other is from the P_2 pool of genes. Let S be the fraction of P_1 genes in a line, and H be the probability that a member of the line has one P_1 and one P_2 gene at a locus. These two indices uniquely specify the expected frequencies of the three classes of genotypes at any autosomal locus:

$$S - \frac{H}{2} = \text{frequency of individuals containing only } P_1 \text{ alleles}$$

$$H = \text{frequency of individuals containing one } P_1 \text{ and one } P_2 \text{ allele}$$

$$1 - S - \frac{H}{2} = \text{frequency of individuals containing only } P_2 \text{ alleles}$$

Composite effects are formally defined in a least-squares framework, similar to that used for effects at single loci (Chapter 4). Consider first the composite additive effects, defined as the difference of additive effects of P_1 vs. P_2 alleles summed over all loci. A simple expression for this can be obtained by recalling Equation 4.9—for a diallelic locus, the additive effects of the B_1 and B_2 alleles can be written as $-p_2\alpha$ and $p_1\alpha$, where α is the average effect of allelic substitution, and p_1 and p_2 are the frequencies of the B_1 and B_2 alleles. These expressions apply to a randomly mating population, precisely the situation in the F_2 generation. Noting that the F_2 consists of 50% P_1 and 50% P_2 genes, the frequencies of all contrasting alleles (of P_1 vs. P_2 origin) are 0.5. Thus, the composite additive effects of P_1 and P_2 genes in the F_2 reference population are equal in absolute value but opposite in sign, and we denote them respectively as $+\alpha^c/2$ and $-\alpha^c/2$, where the superscript c denotes composite (as opposed to single-gene) effects. The total composite additive effect in the F_2 generation, $[(\alpha^c - \alpha^c)/2]$, is then equal to zero, which is what we desire for a reference population. The total composite additive effect in the F_1 generation is also equal to zero, since every locus contains one P_1 and one P_2 allele. In the P_1 population, however, there are only P_1 alleles, each of which contributes $\alpha^c/2$, giving the total composite additive effect as $(\alpha^c + \alpha^c)/2 = \alpha^c$. In contrast, as each P_2 allele contributes $-\alpha^c/2$, the total composite effect in the P_2 population is $-\alpha^c$. More generally, the contribution of composite additive effects to the mean genotypic value of any line-cross derivative is

$$\left(S - \frac{H}{2}\right)(+\alpha^c) + (H)(0) + \left(1 - S - \frac{H}{2}\right)(-\alpha^c) = (2S - 1)\alpha^c = \theta_S \alpha^c$$

where $\theta_S = 2S - 1$ denotes the **source index.** The three terms on the left represent, respectively, the contributions from P_1 homozygotes, P_1P_2 heterozygotes, and P_2 homozygotes. The index θ_S contrasts the expected number of P_1 alleles at a locus in a particular line $(2S)$ with that in the F_2 reference population (1).

The composite dominance effect is obtained in a similar manner, again treating the F_2 population as a reference. Returning to Table 4.2, it can be shown that when the frequencies of the two alleles are equal to 0.5, the dominance deviations from the regression on gene content are $-ak/2$ for both homozygotes and $ak/2$ for the heterozygote. An analogous situation exists for the F_2 of a line cross, which consists of 25% P_1P_1, 50% P_1P_2, and 25% P_2P_2 individuals at each locus. Thus, we denote the composite dominance effects associated with crossbred (P_1P_2) and purebred $(P_1P_1,\ P_2P_2)$ loci as $+\delta^c$ and $-\delta^c$, respectively. More generally, the contribution of composite dominance effects to the genotypic mean of a particular line is

$$\left(S - \frac{H}{2}\right)(-\delta^c) + (H)(+\delta^c) + \left(1 - S - \frac{H}{2}\right)(-\delta^c) = (2H - 1)\delta^c = \theta_H \delta^c$$

where $\theta_H = 2H - 1$ is the **hybridity index.** Thus, $\theta_H \delta^c$ yields a value of δ^c for the F_1 population, which consists entirely of hybrids, $-\delta^c$ for the parental lines, and zero for the F_2, maintaining the property that the average composite effects are defined to be zero in the reference population.

There are three important points to note about this approach to interpreting line-cross means. First, the composite effects are denoted as such because they summarize the total effects over all loci. Since some of the effects of individual genes in population P_1 may be positive and others negative, there is a possibility of considerable cancellation of locus-specific effects. A comparison of the variances within different line-cross derivatives can shed some light on this problem (Mather and Jinks 1982), but we will not take this up here.

Second, provided there is no mating with close relatives, the definition of composite effects does not require that the parental populations be pure (completely homozygous) lines. For any line-cross derivative, the subset of individuals with two P_1 genes at a particular locus will have the same expected Hardy-Weinberg genotype distribution of P_1 genotypes as the P_1 generation. The same argument applies to loci with two P_2 genes. Thus, in the absence of inbreeding (which alters the genotypic frequencies within classes, see Chapter 10), differences between the means of various line-cross derivatives cannot be due to a shift in the genotype frequencies within the groups P_1P_1, P_1P_2, and P_2P_2. It can only be caused by a shift in the relative abundances of these three groups.

Third, the source and hybridity indices are all that are needed to define the contributions of composite epistatic effects to line means. To obtain the general expression, we let $(\alpha_n^c \delta_m^c)$ denote the composite effect involving the interaction of n additive and m dominance effects. (Note that this notation does not imply that $(\alpha_n^c \delta_m^c) = \alpha_n^c \cdot \delta_m^c$.) The general expression for the mean genotypic value of a line is then

$$\mu = \mu_0 + \theta_S \alpha_1^c + \theta_H \delta_1^c + \theta_S^2 \alpha_2^c + \theta_S \theta_H (\alpha_1^c \delta_1^c) + \theta_H^2 \delta_2^c + \cdots \tag{9.1}$$

Note that the coefficients for the composite effects are of the form $\theta_S^n \theta_H^m$. Hill (1982a) provides a formal derivation of Equation 9.1, and alternative modes of presentation appear in Cockerham (1980), Lynch (1991), and Schnell and Cockerham (1992). Example 1 (below) provides a derivation of the result for the additive $\times$ additive component of epistasis. The compositions of the expected line means for common types of crossbreds are given in Table 9.1.

The preceding model applies in the presence of linkage as long as there is no epistasis. However, linked genes tend to be inherited as a unit, although they are gradually rendered independent by crossing-over. This process has the effect of altering the likelihood of specific epistatic interactions through progressive rounds of recombination (see the following example). Provided the parental lines are in gametic phase equilibrium, the expressions for the P_1, P_2, and F_1 lines still hold, since there is no opportunity for recombination between chromosomes of different parental lines, but those for all subsequent line-cross derivatives will be biased to an extent depending on the map structure of the constituent

Table 9.1 Expected mean phenotypes for various line-cross derivatives in terms of composite additive, dominance, and two-locus epistatic effects, taking the F_2 as a point of reference.

Line	θ_S	θ_H	Expected Mean Phenotype
P_1	1	-1	$\mu_0 + \alpha_1^c - \delta_1^c + \alpha_2^c - \alpha_1^c\delta_1^c + \delta_2^c + \cdots$
P_2	-1	-1	$\mu_0 - \alpha_1^c - \delta_1^c + \alpha_2^c + \alpha_1^c\delta_1^c + \delta_2^c + \cdots$
F_1	0	1	$\mu_0 + \delta_1^c + \delta_2^c + \cdots$
F_2	0	0	μ_0
$B_1 = (P_1 \times F_1)$	$\frac{1}{2}$	0	$\mu_0 + \frac{1}{2}\alpha_1^c + \frac{1}{4}\alpha_2^c + \cdots$
$B_2 = (P_2 \times F_1)$	$-\frac{1}{2}$	0	$\mu_0 - \frac{1}{2}\alpha_1^c + \frac{1}{4}\alpha_2^c + \cdots$
$F_2 \times P_1$	$\frac{1}{2}$	0	$\mu_0 + \frac{1}{2}\alpha_1^c + \frac{1}{4}\alpha_2^c + \cdots$
$F_2 \times P_2$	$-\frac{1}{2}$	0	$\mu_0 - \frac{1}{2}\alpha_1^c + \frac{1}{4}\alpha_2^c + \cdots$
$F_2 \times F_1$	0	0	μ_0
$B_1 \times F_1$	$\frac{1}{4}$	0	$\mu_0 + \frac{1}{4}\alpha_1^c + \frac{1}{16}\alpha_2^c + \cdots$
$B_2 \times F_1$	$-\frac{1}{4}$	0	$\mu_0 - \frac{1}{4}\alpha_1^c + \frac{1}{16}\alpha_2^c + \cdots$
Second backcrosses			
$B_1 \times P_1$	$\frac{3}{4}$	$-\frac{1}{2}$	$\mu_0 + \frac{3}{4}\alpha_1^c - \frac{1}{2}\delta_1^c + \frac{9}{16}\alpha_2^c - \frac{3}{8}\alpha_1^c\delta_1^c + \frac{1}{4}\delta_2^c + \cdots$
$B_1 \times P_2$	$-\frac{1}{4}$	$\frac{1}{2}$	$\mu_0 - \frac{1}{4}\alpha_1^c + \frac{1}{2}\delta_1^c + \frac{1}{16}\alpha_2^c - \frac{1}{8}\alpha_1^c\delta_1^c + \frac{1}{4}\delta_2^c + \cdots$
$B_2 \times P_1$	$\frac{1}{4}$	$\frac{1}{2}$	$\mu_0 + \frac{1}{4}\alpha_1^c + \frac{1}{2}\delta_1^c + \frac{1}{16}\alpha_2^c + \frac{1}{8}\alpha_1^c\delta_1^c + \frac{1}{4}\delta_2^c + \cdots$
$B_2 \times P_2$	$-\frac{3}{4}$	$-\frac{1}{2}$	$\mu_0 - \frac{3}{4}\alpha_1^c - \frac{1}{2}\delta_1^c + \frac{9}{16}\alpha_2^c + \frac{3}{8}\alpha_1^c\delta_1^c + \frac{1}{4}\delta_2^c + \cdots$
Selfed backcrosses			
B_{1s}	$\frac{1}{2}$	$-\frac{1}{4}$	$\mu_0 + \frac{1}{2}\alpha_1^c - \frac{1}{4}\delta_1^c + \frac{1}{4}\alpha_2^c - \frac{1}{8}\alpha_1^c\delta_1^c + \frac{1}{16}\delta_2^c + \cdots$
B_{2s}	$-\frac{1}{2}$	$-\frac{1}{4}$	$\mu_0 - \frac{1}{2}\alpha_1^c - \frac{1}{4}\delta_1^c + \frac{1}{4}\alpha_2^c + \frac{1}{8}\alpha_1^c\delta_1^c + \frac{1}{16}\delta_2^c + \cdots$
Continued selfing from the F_2			
F_3	0	$-\frac{1}{2}$	$\mu_0 - \frac{1}{2}\delta_1^c + \frac{1}{4}\delta_2^c + \cdots$
F_4	0	$-\frac{3}{4}$	$\mu_0 - \frac{3}{4}\delta_1^c + \frac{9}{16}\delta_2^c + \cdots$

Note: These expressions assume freely recombining loci. For situations involving self-fertilization, it is further assumed that the parental lines are completely homozygous.

loci. For a pair of loci with recombination fraction c, the following modifications need to be applied to the expressions for the F_2 and backcross means,

$$\mu(\mathrm{F}_2) = \mu_0 + \left(\frac{1-2c}{2}\right)\alpha_2^c + (1-2c)^2\delta_2^c + \cdots \tag{9.2a}$$

$$\mu(\mathrm{B}_1) = \mu_0 + \frac{\alpha_1^c}{2} + \left(\frac{1-c}{2}\right)\alpha_2^c + \left(\frac{2c-1}{2}\right)\alpha_1^c\delta_1^c + (1-2c)\delta_2^c + \cdots \tag{9.2b}$$

$$\mu(\mathrm{B}_2) = \mu_0 - \frac{\alpha_1^c}{2} + \left(\frac{1-c}{2}\right)\alpha_2^c - \left(\frac{2c-1}{2}\right)\alpha_1^c\delta_1^c + (1-2c)\delta_2^c + \cdots \tag{9.2c}$$

(see the following example for a derivation). The expressions for more advanced crosses (e.g., $B_1 \times F_1$) are complicated because additional generations of recombination must be accounted for.

Assuming that the epistatic effects between loci are independent of c, these expressions also apply to the total composite effects of all loci when $\bar{c}$, the mean recombination frequency between all pairs of loci, is substituted for c. Since we generally do not know c for any pair of loci, let alone for all of the loci underlying a quantitative trait, the best we can provide is a heuristic guide to the potential significance of linkage. Assuming that the genes are uniformly distributed across all chromosomes, then

$$\bar{c} = 0.5 - \frac{2L - N + \sum_{i=1}^{N} e^{-2L_i}}{4L^2} \tag{9.3}$$

where N is the number of chromosomes in a haploid set, L_i is the genetic map length of the ith chromosome (in Morgans), and $L = \sum L_i$ is the total map length (Zeng et al. 1990). This expression is based on Haldane's mapping function, $c_{ij} = 0.5(1 - e^{-2L_{ij}})$, which relates the recombination frequency to the genetic map length between two linked loci, i and j (Chapter 14). Estimates of $\bar{c}$ are given in Table 9.2 for several species for which the genetic maps are reasonably well resolved. For species with a very small number of chromosomes and/or restricted recombination in males, as in *Drosophila* and the mosquito *Aedes*, $\bar{c}$ can be somewhat less than 0.40, but when N exceeds six or so, it tends to be greater than 0.45. For most mammals, N is typically on the order of 15 or more, so most pairs of genes are on different chromosomes, and $\bar{c}$ is very close to 0.5. With $\bar{c} > 0.45$, Equations 9.2a-c are quite close to the expressions in Table 9.1. Thus, unless the genes underlying quantitative traits tend to be aggregated on chromosomes, linkage is unlikely to cause much bias in the interpretation of line-cross means, except perhaps in the case of species such as *Drosophila.*

Table 9.2 Genetic map structure and the average recombination frequency ($\bar{c}$) between random pairs of loci throughout the genome.

Species	N	L	Lengths of Individual Chromosomes, L_i	$\bar{c}$
Drosophila melanogaster	3	2.77	0.66, 1.03, 1.08	0.365
Drosophila pseudoobscura	4	4.46	0.68, 0.69, 1.01, 2.08	0.386
Aedes aegypti	3	2.28	0.62, 0.80, 0.86	0.380
Caenorhabditis elegans	5	1.61	0.27, 0.31, 0.33, 0.34, 0.36	0.418
Arabidopsis thaliana	5	5.24	0.63, 0.83, 0.98, 1.36, 1.44	0.443
Hordeum vulgare (barley)	7	9.49	0.70, 1.12, 1.15, 1.26, 1.27, 1.59, 2.40	0.465
Neurospora crassa	7	10.02	1.07, 1.13, 1.18, 1.33, 1.49, 1.52, 2.30	0.466
Zea mays (maize)	10	12.10	0.42, 0.78, 0.95, 1.07, 1.12, 1.37, 1.41, 1.55, 1.55, 1.67, 1.76	0.474
Phaseolus vulgaris (bean)	11	8.92	1.05, 1.04, 0.95, 0.92, 0.86, 0.78, 0.74, 0.71, 0.71, 0.60, 0.56	0.471
Pinus pinaster (maritime pine)	12	18.56	1.90, 1.75, 1.69, 1.66, 1.63, 1.61, 1.57, 1.54, 1.54, 1.41, 1.36, 0.90	0.481
Lycopersicon esculentum (tomato)	12	14.91	0.90, 0.92, 0.98, 1.01, 1.04, 1.04, 1.23, 1.34, 1.42, 1.63, 2.11	0.479
Mus musculus (mouse)	20	14.25	0.36, 0.36, 0.49, 0.56, 0.57, 0.57, 0.68, 0.70, 0.71, 0.73, 0.74, 0.78, 0.78, 0.80, 0.81, 0.84, 0.87, 0.89, 1.00, 1.01	0.483
Homo sapiens	23	40.00	0.69, 0.76, 1.12, 1.12, 1.22, 1.24, 1.25, 1.26, 1.38, 1.39, 1.40, 1.47, 1.62, 1.67, 1.67, 1.67, 1.74, 1.75, 1.75, 1.77, 1.92, 2.21, 2.49	0.490
Danio rerio (zebrafish)	25	28.06	0.59, 0.80, 0.84, 0.85, 0.86, 0.87, 0.95, 1.00, 1.02, 1.05, 1.05, 1.09, 1.13, 1.13, 1.14, 1.16, 1.22, 1.22, 1.33, 1.34, 1.39, 1.39, 1.45, 1.53, 1.66	0.489

Source: All data are from O'Brien (1990), except that for *Phaseolus* (Vallejos et al. 1992), *Pinus* (Plomion et al. 1995), and *Danio* (supplied by J. Postlethwait).
Note: N is the haploid number of chromosomes per genome, and L is the total map length (in Morgans) in females. For *Drosophila,* we ignore a tiny dot chromosome for which little mapping data are available. In the mosquito *Aedes* and in humans, respectively, recombination in males is approximately 50% and 65% as frequent as that in females; and there is no recombination within chromosomes in male *Drosophila.* For these three taxa, the reported values of $\bar{c}$ are averages for the two sexes. All results are approximations, as genomic maps are continuously being refined with the addition of molecular markers (Chapter 14).

Example 1. All of the composite effects described above are defined in a least-squares sense, and the nice symmetry whereby all effects have the same absolute value but differ in sign is a consequence of all contrasting pairs of alleles having frequency 0.5 in the F_2 generation. We now provide a formal derivation of the additive $\times$ additive composite effects in the context of a reference population that is both in Hardy-Weinberg and gametic phase equilibrium. We denote the four gamete types as C_{11}, C_{12}, C_{21}, and C_{22}, where the subscripts refer to the parental sources of alleles at the first and second locus. All four gamete types have frequencies equal to 0.25. Let the additive $\times$ additive effects associated with these gametes be α_{11}, α_{12}, α_{21}, and α_{22}. The effect α_{ij} is defined to be the average residual effect associated with a gamete containing a P_i -derived allele at the first locus and a P_j -derived allele at the second locus, after the additive effects of the two genes have been accounted for (see Equation 5.4).

Under a least-squares framework, the mean residual error is defined to be zero (Chapter 3), which implies

$$\alpha_{11} + \alpha_{12} + \alpha_{21} + \alpha_{22} = 0 \tag{1a}$$

Furthermore, the mean squared error is minimized. Noting that the previous expression implies that $\alpha_{22} = -\alpha_{11} - \alpha_{12} - \alpha_{21}$, the function to be minimized is

$$M = \alpha_{11}^2 + \alpha_{12}^2 + \alpha_{21}^2 + (-\alpha_{11} - \alpha_{12} - \alpha_{21})^2 \tag{1b}$$

Taking the partial derivative with respect to α_{11} and setting it equal to zero, we obtain

$$2\alpha_{11} + \alpha_{12} + \alpha_{21} = 0$$

Subtracting Equation 1a from this expression, we find that the epistatic effects associated with each of the parental chromosome types are equal, i.e., $\alpha_{11} = \alpha_{22}$. By similar means, it can be shown that $\alpha_{12} = \alpha_{21}$, which when applied to Equation 1a implies that

$$\alpha_{11} = \alpha_{22} = -\alpha_{12} = -\alpha_{21}$$

Thus, the additive $\times$ additive effects associated with both recombinant chromosome types are equal and opposite in sign to those of the parental chromosomes. Since, in a diploid, there are four combinations of genes at two loci (two within and two between the uniting gametes), we define the effects as

$$\alpha_{11} = \alpha_{22} = +\alpha_2^c/4$$
$$\alpha_{12} = \alpha_{21} = -\alpha_2^c/4$$

In both the P_1 and P_2 populations, all additive $\times$ additive interactions within and between chromosomes are of parental type (α_{ii}), and the composite effect of such interactions is $4(\alpha_2^c/4) = \alpha_2^c$. In the F_1 generation, the two interactions within chromosomes are of parental type, but the pairs of nonalleles between chromosomes are of different parental type, so the composite effect is $2(\alpha_2^c/4) + 2(-\alpha_2^c/4) = 0$.

Now consider the situation in the F_2 generation. Under free recombination, all four gametes are equally frequent, and with random mating, the average additive $\times$ additive epistatic effect within and between uniting gametes is equal to zero, as noted above. Suppose, however, that the two loci are linked, so that a fraction c of gametes are recombinant, and a fraction $(1-c)$ are nonrecombinant. The gamete frequencies are then $p(C_{12}) = p(C_{21}) = c/2$, and $p(C_{11}) = p(C_{22}) = (1-c)/2$. Assuming random mating, the paternal source of an individual's allele at one locus is independent of the maternal source of an allele at a second locus. Thus, the average composite effects associated with additive $\times$ additive interactions between uniting gametes is equal to zero. However, because the parental sources of genes *within* gametes will not have been completely randomized, the composite additive $\times$ additive effect within gametes is not zero, but

$$2[c(-\alpha_2^c/4) + (1-c)(\alpha_2^c/4)] = (1-2c)\alpha_2^c/2$$

giving the term in Equation 9.2a.

These results can be extended to the backcross generations. Consider, for example, the situation when F_1 individuals are crossed to members of parental line P_1, creating the B_1 backcross generation. The F_1 parent can contribute each of the four possible gametes, while the P_1 parent contributes only C_{11} gametes. With probability $(1-c)/2$, the offspring genotype is C_{11}/C_{11}, and each of the four possible two-locus interactions involves two P_1 alleles; each such interaction contributes $\alpha_2^c/4$, giving $(1-c)\alpha_2^c/2$. Likewise, with probability $c/2$ the genotype is C_{12}/C_{11}. Here, there are two P_1/P_1 interactions $(2\alpha_2^c/4)$ and two P_1/P_2 interactions $(-2\alpha_2^c/4)$, yielding a total contribution for this genotype of $(c/2)(2\alpha_2^c/4 - 2\alpha_2^c/4) = 0$. In a similar fashion, expectations for the other two genotypes are found to be equal to zero, giving the total contribution in the B_1 backcross as $(1-c)\alpha_2^c/2$.

ESTIMATION OF COMPOSITE EFFECTS

With the preceding model, the expected line means are linear functions of the composite effects. Thus, straightforward procedures can be used to estimate the latter from the observed line means. Generally, when the estimates of k parameters are desired, k types of lines can be identified that allow a solution using simultaneous equations. For example, if the epistatic effects are ignored, the expectations for the P_1, P_2, and F_1 means are simply

$$\mu(P_1) = \mu_0 + \alpha_1^c - \delta_1^c \tag{9.4a}$$
$$\mu(P_2) = \mu_0 - \alpha_1^c - \delta_1^c \tag{9.4b}$$
$$\mu(F_1) = \mu_0 + \delta_1^c \tag{9.4c}$$

Table 9.3 Coefficients for observed line means in expressions for estimated composite effects using the six-parameter model.

Parameter	P_1	P_2	F_1	F_2	B_1	B_2
μ_0	0	0	0	1	0	0
α_1^c	0	0	0	0	1	–1
δ_1^c	$\frac{1}{4}$	$\frac{1}{4}$	$-\frac{1}{2}$	2	–1	–1
α_2^c	0	0	0	–4	2	2
$\alpha_1^c\delta_1^c$	$\frac{1}{2}$	$-\frac{1}{2}$	0	0	–1	1
δ_2^c	$\frac{1}{4}$	$\frac{1}{4}$	$\frac{1}{2}$	1	–1	–1

Rearranging and substituting observed for expected means, we obtain the three estimators,

$$\widehat{\mu}_0 = \frac{\bar{z}(\mathrm{P}_1) + \bar{z}(\mathrm{P}_2) + 2\bar{z}(\mathrm{F}_1)}{4} \tag{9.5a}$$

$$\widehat{\alpha}_1^c = \frac{\bar{z}(\mathrm{P}_1) - \bar{z}(\mathrm{P}_2)}{2} \tag{9.5b}$$

$$\widehat{\delta}_1^c = \frac{2\bar{z}(\mathrm{F}_1) - \bar{z}(\mathrm{P}_1) - \bar{z}(\mathrm{P}_2)}{4} \tag{9.5c}$$

With a model that includes all three forms of two-locus epistasis, there are six unknowns, so the mean phenotypes of at least six types of lines need to be evaluated. This is most easily accomplished by assaying both parental lines, their F_1 and F_2 derivatives, and the two backcrosses of F_1 individuals to the parental lines (B_1 and B_2). As in the previous example, the estimated composite effects can then be written as simple linear functions of the observed means (Table 9.3). For example, the composite additive $\times$ additive effect is estimated by

$$\widehat{\alpha}_2^c = -4\bar{z}(\mathrm{F}_2) + 2\bar{z}(\mathrm{B}_1) + 2\bar{z}(\mathrm{B}_2)$$

Although we do not consider it in any detail here, it is worth noting that reciprocal crosses between parental lines can be used to estimate maternal effects and/or the effect of sex chromosomes. For example, when males of the P_1 line are crossed to females of the P_2 line, and vice versa, assuming that males are the heterogametic sex, the difference in mean phenotypes of daughters from the two lines provides an estimate of the difference between maternal effects associated with each parental line. In F_1 males, the effect of the X chromosome is superimposed on the maternal-effect difference. However, by making four possible types of crosses between F_1 individuals (two maternal sources of cytoplasm in females $\times$ two maternal sources of the X chromosome in males), a clean partitioning of these two sources of composite effects can be obtained (Carson and Lande 1984, Hard et al. 1992).

Since the estimates of the composite effects are linear functions of the line-cross means, the sampling variances of the estimates can be obtained from the expression for the variance of a sum (Equation 3.11b). Because the mean phenotype of each line is estimated independently of the others, there is no covariance between the different mean estimates. Thus, the sampling variance of a composite effect is simply the sum of the sampling variances of the line means used in its estimation, each weighted by the squared coefficient in the estimating equation. For example, for the preceding estimator of $\widehat{\alpha}_2^c$,

$$\mathrm{Var}(\widehat{\alpha}_2^c) = 16\mathrm{Var}[\bar{z}(\mathrm{F}_2)] + 4\mathrm{Var}[\bar{z}(\mathrm{B}_1)] + 4\mathrm{Var}[\bar{z}(\mathrm{B}_2)]$$

where $\mathrm{Var}[\bar{z}(\cdots)]$ is the squared standard error of a line mean.

Hypothesis Testing

When the number of observations equals the number of parameters to be estimated, the solution of simultaneous equations cannot be used to evaluate the adequacy of the genetic model, as the parameter estimates are constrained to yield the observed line means exactly. This problem is eliminated when the number of observed line means exceeds the number of parameters to be estimated. For example, if data are available for the P_1, P_2, F_1, and F_2 lines, the statistic

$$\Delta = \bar{z}(\mathrm{F}_2) - \left(\frac{\bar{z}(\mathrm{P}_1) + \bar{z}(\mathrm{P}_2)}{4} + \frac{\bar{z}(\mathrm{F}_1)}{2}\right) \tag{9.6a}$$

provides a simple test for epistasis. In the absence of epistasis, the expected value of Δ is zero because at every locus, by the Hardy-Weinberg law, the F_2 is 25% $\mathrm{P}_1\mathrm{P}_1$, 50% $\mathrm{P}_1\mathrm{P}_2$, and 25% $\mathrm{P}_2\mathrm{P}_2$. The sampling variance of this test statistic is

$$\mathrm{Var}(\Delta) = \mathrm{Var}[\bar{z}(\mathrm{F}_2)] + \frac{\mathrm{Var}[\bar{z}(\mathrm{F}_1)]}{4} + \frac{\mathrm{Var}[\bar{z}(\mathrm{P}_1)] + \mathrm{Var}[\bar{z}(\mathrm{P}_2)]}{16} \tag{9.6b}$$

Under the reasonable assumption that the sampling distribution of Δ is approximately normal, the ratio $|\Delta|/\sqrt{\mathrm{Var}(\Delta)}$ provides a simple t test for epistasis. For large samples, if this ratio is greater than 1.96, then the null hypothesis of no epistasis can be rejected with 95% confidence.

A more general approach to hypothesis testing uses least-squares regression to estimate the model parameters and then compares the observed means with the model predictions. With this approach, one can start with a very simple model, evaluate its significance, and gradually add higher-order composite effects to the model, until no further significant improvement in the model fit occurs. This procedure, first suggested by Cavalli (1952) and Hayman (1960a), is known as the **joint-scaling test**.

Consider the simple additive model,

$$\bar{z}_i = \mu_0 + \theta_{Si}\alpha_i^c + e_i$$

where the ith line mean $(\bar{z}_i)$ has coefficient θ_{Si}, and e_i denotes the deviation of the observed mean from the prediction of the model. In matrix form, letting $\bar{\mathbf{z}}$ be the vector of observed line means, $\mathbf{a}$ be the (2×1) vector of effects μ_0 and α_1^c, and $\mathbf{M}$ be the matrix of coefficients, the linear model becomes

$$\bar{\mathbf{z}} = \mathbf{Ma} + \mathbf{e} \tag{9.7}$$

where $\mathbf{e}$ is the column vector of residual errors, i.e., the vector of deviations between observed and predicted line means. Note that all of the elements in the first column of $\mathbf{M}$ are equal to one, as they are all multipliers for μ_0, whereas the second column contains the coefficients θ_{Si} for the various lines.

Since the line means may vary with respect to accuracy (reflecting, for example, different sample sizes), they should not be weighted equally in the computation of $\mathbf{a}$. From Equation 8.34, the weighted least-squares solution is

$$\widehat{\mathbf{a}} = (\mathbf{M}^T\mathbf{V}^{-1}\mathbf{M})^{-1}\mathbf{M}^T\mathbf{V}^{-1}\bar{\mathbf{z}} \tag{9.8a}$$

where the covariance matrix $\mathbf{V}$ for the residuals is diagonal with diagonal elements equal to the squared standard errors of the means. The treatment of $\mathbf{V}$ as a diagonal matrix assumes that the measured individuals from the different lines are unrelated, i.e., that there is no sampling covariance between the observed means.

From Equation 8.35, the sampling variances and covariances of the two parameter estimates, $\widehat{\mu}_0$ and $\widehat{\alpha}_1^c$, are given by the elements of the (2×2) covariance matrix

$$\text{Var}(\widehat{\mathbf{a}}) = \mathbf{C} = (\mathbf{M}^T\mathbf{V}^{-1}\mathbf{M})^{-1} \tag{9.8b}$$

The two diagonal elements are $\text{Var}(\widehat{\mu}_0)$ and $\text{Var}(\widehat{\alpha}^c)$, whereas both off-diagonal elements are equal to $\text{Cov}(\widehat{\mu}_0, \widehat{\alpha}^c)$. Sampling covariance arises between estimates of the parameters because they are jointly estimated from a common data set. Letting $\widehat{z}_i = \widehat{\mu}_0 + \theta_{Si}\widehat{\alpha}^c$ be the fitted (predicted) mean phenotype for the ith line, the sampling variance of $\widehat{z}_i$ is simply the variance of a sum,

$$\text{Var}(\widehat{z}_i) = \text{Var}(\widehat{\mu}_0) + 2\theta_{Si}\text{Cov}(\widehat{\mu}_0, \widehat{\alpha}^c) + \theta_{Si}^2\text{Var}(\widehat{\alpha}^c) \tag{9.9a}$$

Using Equations 8.21b and 9.8b, the covariance matrix for the predicted means is

$$\text{Var}(\widehat{\mathbf{z}}) = \text{Var}(\mathbf{M}\widehat{\mathbf{a}}) = \mathbf{MCM}^T = \mathbf{M}(\mathbf{M}^T\mathbf{V}^{-1}\mathbf{M})^{-1}\mathbf{M}^T \tag{9.9b}$$

Although the least-squares solution is completely general, significance testing requires the assumption of normality. If that assumption is met, the weighted error sum of squares

$$\chi^2 = \sum_{i=1}^{k} \frac{(\bar{z}_i - \widehat{z}_i)^2}{\text{Var}(\bar{z}_i)} \tag{9.10}$$

where k is the number of observed lines, provides a test statistic for the adequacy of the model (Appendix 3). Under the null hypothesis of purely additive gene action, this test statistic will be χ^2 distributed with degrees of freedom equal to the number of lines minus the number of estimated parameters, giving $(k-2)$ df.

If the test statistic is large enough to reject the additive model, the logical next step is to evaluate the additive-dominance model. In this case, the vector **a** contains a third element, δ_1^c, and the matrix **M** contains a third row consisting of the elements θ_{Hi}. The solution again follows from Equation 9.8a, and the new fit is evaluated by use of Equation 9.10, where the χ^2 statistic is now distributed with $k-3$ degrees of freedom, as three parameters are fitted.

Letting χ_A^2 and χ_{AD}^2 denote the test statistics associated with the additive and additive-dominance models, then the difference

$$\Lambda = \chi_A^2 - \chi_{AD}^2 \tag{9.11}$$

is equivalent to a likelihood-ratio test statistic (see Example 5 from Appendix 4). Such statistics are asymptotically χ^2 distributed (with large sample sizes), with degrees of freedom equal to the difference in the number of parameters included in the two models (in this case, one). While the inclusion of dominance in the model will definitely improve the fit, Equation 9.11 provides a test for whether the improvement is significant.

If the additive-dominance model is rejected on the basis of an overly large value for χ_{AD}^2, the next step is to proceed with the analysis of models containing epistatic effects, assuming that enough line means are available for such analysis. At this point, δ_1^c may or may not be dropped from the model depending on its degree of significance in the previous analysis. The significance of the improvement of fit with models containing epistasis can again be evaluated by use of the appropriate likelihood-ratio test, i.e., by the difference in χ^2 test statistics between the modified model and the previous restricted model.

Example 2. We now use the joint-scaling test to study the genetic basis of human skin color. The sample consists primarily of residents of Liverpool, England (Harrison and Owen 1964). Pigmentation was measured as the reflectance of the medial aspect of the right upper arm at 545 mμ. The P_1 consists of individuals of West African origin and the P_2 of individuals of European descent.

	P_1	P_2	F_1	F_2	B_1	B_2
$\bar{z}_i$	14.4	41.0	28.4	30.3	24.2	34.7
$SE(\bar{z}_i)$	0.611	0.453	0.581	1.483	1.334	1.122

There is a threefold range of variation in the standard errors of the mean phenotypes, and this translates into a tenfold range in the sampling variances. Clearly, weighted least-squares regression is desirable in this situation.

We start by considering the simplest genetic model, assuming that all gene action is additive within and between loci. The coefficients for the effects μ_0 and α^c are given by

$$\mathbf{M} = \begin{pmatrix} 1 & 1 \\ 1 & -1 \\ 1 & 0 \\ 1 & 0 \\ 1 & 0.5 \\ 1 & -0.5 \end{pmatrix}$$

The sampling covariance matrix for the line means is diagonal with $V_{ii} = [\text{SE}(\bar{z}_i)]^2$,

$$\mathbf{V} = \begin{pmatrix} 0.373 & 0.000 & 0.000 & 0.000 & 0.000 & 0.000 \\ 0.000 & 0.205 & 0.000 & 0.000 & 0.000 & 0.000 \\ 0.000 & 0.000 & 0.338 & 0.000 & 0.000 & 0.000 \\ 0.000 & 0.000 & 0.000 & 2.199 & 0.000 & 0.000 \\ 0.000 & 0.000 & 0.000 & 0.000 & 1.780 & 0.000 \\ 0.000 & 0.000 & 0.000 & 0.000 & 0.000 & 1.259 \end{pmatrix}$$

These lead to

$$\mathbf{M}^T\mathbf{V}^{-1}\mathbf{M} = \begin{pmatrix} 12.325 & -2.311 \\ -2.311 & 7.891 \end{pmatrix}$$

which yields the sampling covariance matrix for the parameter estimates (Equation 9.8b),

$$\mathbf{C} = (\mathbf{M}^T\mathbf{V}^{-1}\mathbf{M})^{-1} = \begin{pmatrix} 0.086 & 0.025 \\ 0.025 & 0.134 \end{pmatrix}$$

and from Equation 9.8a, the parameter estimates

$$\widehat{\mathbf{a}} = \begin{pmatrix} \widehat{\mu}_0 \\ \widehat{\alpha}^c \end{pmatrix} = \begin{pmatrix} 28.17 \\ -13.07 \end{pmatrix}$$

The standard errors of the parameter estimates are equal to the square roots of the diagonal elements of $\mathbf{C}$,

$$\text{SE}(\widehat{\mu}_0) = (0.086)^{1/2} = 0.29$$
$$\text{SE}(\widehat{\alpha}^c) = (0.134)^{1/2} = 0.37$$

The line means predicted by the model are obtained as $\widehat{\mathbf{z}} = \mathbf{M}\widehat{\mathbf{a}}$, and the sampling variances and covariances of predicted values by Equation 9.9b,

$$\text{Var}(\widehat{\mathbf{z}}) = \mathbf{M}(\mathbf{M}^T\mathbf{V}^{-1}\mathbf{M})^{-1}\mathbf{M}^T$$

The square roots of the diagonal elements of this latter matrix are the estimated standard errors of the predicted means. In all cases, the predicted values are very close to the observed means:

	P_1	P_2	F_1	F_2	B_1	B_2
$\widehat{z}$	15.2	41.2	28.2	28.2	21.6	34.7
$\bar{z} - \widehat{z}$	−0.8	−0.2	0.2	2.1	2.6	0.0
$\text{SE}(\widehat{z})$	0.52	0.41	0.29	0.29	0.38	0.31

The test statistic, $\chi^2_A = 7.510$, with four degrees of freedom, is not significant as $\Pr(\chi^2_4 \geq 7.510) = 0.11$. Thus, the fitted model with $\widehat{\mu}_0 = 28.17$ and $\widehat{\alpha}^c = -13.07$ appears to adequately explain the data. Reevaluation of the data with the additive-dominance model confirms this conclusion. In this case, the analysis proceeds with

$$\mathbf{M} = \begin{pmatrix} 1 & 1 & -1 \\ 1 & -1 & -1 \\ 1 & 0 & 1 \\ 1 & 0 & 0 \\ 1 & 0.5 & 0 \\ 1 & -0.5 & 0 \end{pmatrix}$$

and

$$\mathbf{a} = \begin{pmatrix} \mu_0 \\ \alpha^c \\ \delta^c \end{pmatrix}$$

yielding the parameter estimates (and associated standard errors):

$$\widehat{\mu}_0 = 28.32\,(0.32), \quad \widehat{\alpha}^c = -13.14\,(0.37), \quad \widehat{\delta}^c = 0.44\,(0.34)$$

Notice that the estimates $\widehat{\mu}_0$ and $\widehat{\alpha}^c$ are very close to those obtained under the purely additive model, and that $\widehat{\delta}^c$ is only slightly greater than its standard error. The test statistic for this analysis is $\chi^2_{AD} = 5.879$. The likelihood-ratio test statistic, $\Lambda = \chi^2_A - \chi^2_{AD} = 1.631$, provides a test of the hypothesis that dominance accounts for a significant proportion of the variance among line means. With one degree of freedom, Λ is not significant, as $\Pr(\chi^2_1 \geq 1.631) = 0.20$.

Line Crosses in *Nicotiana rustica*

Few attempts have been made to estimate more than two-locus epistatic effects using line-cross analysis. However, Mather, Jinks, and their associates have done

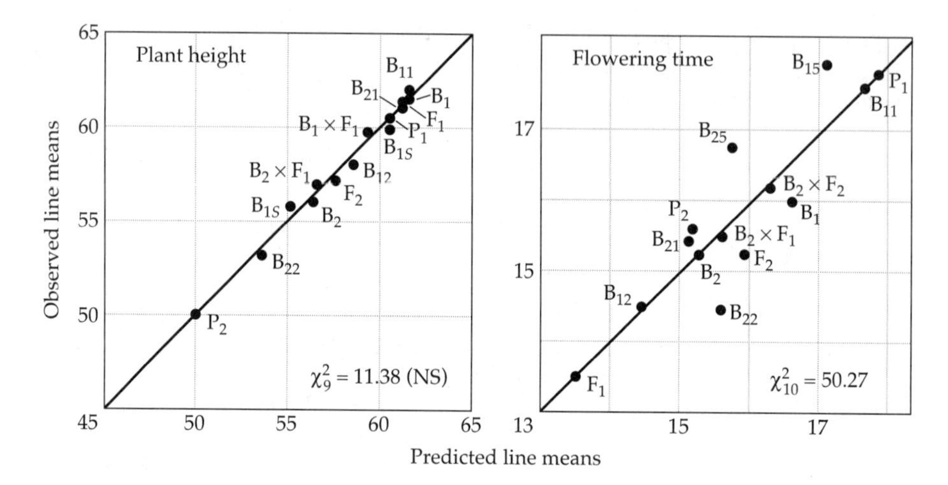

Figure 9.1 Observed vs. fitted mean phenotypes for 14 line-cross derivatives from two pure parental lines of tobacco. The diagonal line gives the expected pattern if observed and predicted line means were identical. (From Jinks and Perkins 1969.)

an enormous amount of work with highly inbred lines of tobacco to evaluate the relative importance of various types of higher-order gene action (Mather and Jinks 1982). An experiment by Jinks and Perkins (1969) is particularly noteworthy. In addition to the six fundamental line crosses, they created four second-generation backcrosses ($B_1 \times P_1$, $B_1 \times P_2$, $B_2 \times P_1$, $B_2 \times P_2$), the crosses $B_1 \times F_1$ and $B_2 \times F_1$, and the selfed backcrosses (B_{1s} and B_{2s}), and they assayed all of them simultaneously. With such a large number of lines, there are enough degrees of freedom to test for the significance of three-locus epistatic effects.

In the case of final plant height, a model with fitted parameters (and associated standard errors) $\widehat{\mu}_0 = 57.64 \pm 0.23$, $\widehat{\alpha}_1^c = 5.32 \pm 0.20$, $\widehat{\delta}_1^c = 3.55 \pm 0.38$, $\widehat{\alpha}_2^c = 4.85 \pm 1.28$, and $\widehat{\alpha}_2^c\widehat{\delta}_1^c = 3.73 \pm 1.03$ provides an excellent fit to the data (Figure 9.1). Thus, for this trait, there are significant additive $\times$ additive and additive $\times$ additive $\times$ dominance interactions involving genes from different lines, but no significant additive $\times$ dominance, dominance $\times$ dominance, or higher-order interactions. The most parsimonious three-locus fit for the data on flowering time is obtained with $\widehat{\mu}_0 = 15.91 \pm 0.09$, $\widehat{\alpha}_1^c = 1.36 \pm 0.15$, $\widehat{\delta}_1^c = -2.39 \pm 0.22$, and $\widehat{\alpha}_2^c\widehat{\delta}_1^c = 1.81 \pm 0.37$. There are significant differences between observed and predicted line means in this case (Figure 9.1), however, suggesting the existence of even higher-order epistatic interactions. Analyses with other parental lines (Smith 1937, Hill 1966) led to similar results — final height was usually described adequately by a model incorporating dominance and at least one form of two-locus

Table 9.4 Composite effects estimated from phenotypic means of two line crosses involving inbred parental stocks of tobacco, $V_1 \times V_5$ (upper lines of data for each trait) and $V_2 \times V_{12}$ (lower lines).

Character	$\widehat{\mu}_0$	$\widehat{\alpha}_1^c$	$\widehat{\delta}_1^c$	$\widehat{\alpha}_2^c$	$\widehat{\alpha}_1^c\widehat{\delta}_1^c$	$\widehat{\delta}_2^c$
Height — 2 wk	5.50	1.19	1.10	1.27		
	4.34	1.97	0.48	0.55		
— 4 wk	15.74	3.53	3.32	3.97		
	15.22	9.96	2.58	3.48		
— 6 wk	49.84	5.02	−2.48	9.95		
	49.14	30.22	11.14	9.21	4.12	
— flowering	73.75	7.99		−5.02		
	83.40	5.84	3.12	−6.21		
— final	126.80	12.57	3.72	−7.58		
	143.44	8.79	21.91			−3.33
Leaf length	19.10	0.35		−1.89		

Source: Pooni et al. 1985.
Note: Only significant effects are given, and the resulting models provide an adequate fit to the data in all cases. Leaf length was not analyzed in the $V_2 \times V_{12}$ experiment.

epistasis, while flowering time was influenced by epistatic interactions between pairs, triplets, and higher numbers of loci. The results from two additional line-cross experiments, involving only the six fundamental generations, indicate that epistasis contributes to line differences in other characters in *Nicotiana* (Table 9.4). In these additional cases, however, effects involving more than pairs of loci need not be invoked to explain the data, since in no case are the observed means significantly different from the final model predictions.

Additional Data

A summary of results from some other line-cross studies is given in Table 9.5. In most cases, the parental lines are conspecific isolates known at the outset to differ (often substantially) in mean phenotypes. The main message is consistent with the *Nicotiana* results — differentiation of divergent lines almost always involves epistatic effects. We will see in Chapter 11 that epistatic interactions can sometimes be removed by a suitable scale transformation, but it is unlikely that this would be successful for all the tabulated studies.

Table 9.5 A survey of the composite effects estimated in line-cross analyses.

Character	$\widehat{\mu}_0$	$\widehat{\alpha}^c_1$	$\widehat{\delta}^c_1$	$\widehat{\alpha}^c_2$	$\widehat{\alpha}^c_1\widehat{\delta}^c_1$	$\widehat{\delta}^c_2$	Reference
Corn							
time to silking	65.19	0.20	−6.16	−4.40	7.33	5.92	Mohamed 1959
time to shed pollen	62.88	−1.57	−4.36	−1.94	4.74	3.69	
Lima beans							
seed size	0.57	−0.25	−0.16	0.04	0.30	0.19	Ryder 1958
Tomatos							
$\log_{10}$(fruit weight)	0.69	−0.87	0.02	0.12	−0.03		Powers 1951
Pitcher-plant mosquito							
$\log_{10}$(critical photoperiod)	−1.84	−4.45	0.40		1.79	4.18	Hard et al. 1992
Drosophila melanogaster							
ln(longevity)	1.79	−0.06	0.01	0.03			Luckinbill et al. 1988
Drosophila tripunctata							
ovipos. site pref.	0.31	−0.12	0.05	0.18			Jaenike 1987
D. heteroneura × *D. sylvestris*							
ln(head length)	3.00	0.09					Templeton 1977
ln(head width)	3.96	−0.01					
Astyanax (cave fish)							
eye diameter	4.72	−2.43	1.62	1.10	−1.57	−1.24	Wilkens 1971
Mice							
$\log_{10}$(body weight)	1.39	0.18	−0.01	−0.07	−0.01	0.01	Chai 1956
Chickens							
weight	3.49	−0.12	−0.06	−0.08	0.08	0.05	Waters 1931

Note: The results reported for each analysis describe the most parsimonious genetic model, and all recorded effects are statistically significant. The motivation for the use of logarithmic transformations in some cases is discussed in Chapter 11.

THE GENETIC INTERPRETATION OF HETEROSIS AND OUTBREEDING DEPRESSION

Agronomists and animal breeders have long known that crossbreeding of two lines often has positive fitness-related effects in the F_1 progeny (Darwin 1876, Lerner 1954, Sheridan 1981, Turton 1981, Sprague 1983). An F_1 performance that exceeds the average parental performance is generally referred to as **hybrid vigor** or **heterosis** (Shull 1914). However, outcrossing does not always increase fitness. For example, crosses between different species or distantly related populations frequently lead to substantial or complete loss of viability and/or fecundity (Barton and Hewitt 1981; Templeton 1981; Shields 1982; Coyne and Orr 1989a, 1997; Wu and Palopoli 1994). Complicating matters further is the frequent observation that when heterosis arises in an F_1 population, much of it is lost in the subsequent F_2 generation. In some cases, there is more than just a loss of heterosis — rather, the F_2 progeny are significantly less fit than the members of the original parental lines, a phenomenon known as **outbreeding depression.**

The altered response of progeny phenotypes with increasing genetic distance between parents suggests that there is a fundamental change in predominant gene interactions as mates become more and more distantly related, and has led to the suggestion that there must be an optimal degree of outbreeding (Shields 1982, Bateson 1983, Waser and Price 1983, Waser 1993b). Dominance is generally believed to be the primary agent of inbreeding depression within populations (Chapter 10). However, the decline in fitness under outcrossing is usually attributed to a breakup of coadapted gene complexes (favorable epistatic interactions) in the parental lines (Dobzhansky 1948, Templeton 1986, Lynch 1991). Thus, in proceeding from issues of outbreeding enhancement to those of outbreeding depression, at least among theoreticians, there is often a shift in emphasis from interactions within loci (dominance) to those among loci (epistasis).

In the following chapter, we will consider the substantial body of theory and data that bears on the genetic mechanisms of inbreeding depression. An understanding of the genetic basis of outbreeding depression has been more elusive, although the issues bear significantly on our understanding of the mechanisms of speciation and local adaptation. Just as the external environment molds the evolution of local adaptations by natural selection, the internal genetic environment of populations is expected to lead to the evolution of local complexes of genes that interact in a mutually favorable manner. The particular gene combinations that evolve in any local population may be largely fortuitous, depending in the long run on the chance variants that mutation creates. By breaking up coadapted gene complexes, hybridization can lead to the production of individuals that have lower fitness than either parental type, even in the case of populations that are adapted to identical extrinsic environments.

The line-cross theory presented above provides a framework for investigating the genetic basis of heterosis and outbreeding depression, and shows that both dominance and epistasis can have effects on the expression of mean phenotypes in the F_1 and F_2 generations. Returning to the expressions developed above, we find that heterosis as expressed in the F_1 and F_2 generations (as a deviation from the mean of the parental lines) is defined to be

$$\mathrm{F}_1 - \overline{P} = 2\delta_1^c - \alpha_2^c \tag{9.12a}$$

$$\mathrm{F}_2 - \overline{P} = \frac{\mathrm{F}_1 - \overline{P}}{2} - \bar{c}\,[\,\alpha_2^c + 4(1-\bar{c})\delta_2^c\,] \tag{9.12b}$$

Several general conclusions can be drawn from these expressions. First, the deviation between the mean phenotype in the F_1 and parental lines is a function of both dominance (δ_1^c) and additive $\times$ additive epistasis (α_2^c). Ignoring epistasis, a simple explanation for heterosis is the presence of complementary sets of deleterious recessive genes in both parental lines and the masking of their effects in the F_1 heterozygotes. However, the loss of favorable additive $\times$ additive effects (α_2^c) that exist within populations must also be considered. Although the two interactions *within* gametes are preserved in the F_1, those *between* gametes are not. An F_1 line will only exhibit heterosis if the gain in favorable between-population dominance effects exceeds the loss in favorable additive $\times$ additive interactions within populations.

Second, of the heterosis (or outbreeding depression) in the F_1 generation, half is lost by segregation when gametes leading to the F_2 generation are produced, and an additional fraction is lost due to the recombination between parental line genes. This latter quantity, termed **recombination loss** by Dickerson (1969), is entirely a function of epistatic effects, since recombination does not influence the transmittance of effects associated with single loci. Depending on the signs of α_2^c and δ_2^c, it may be positive or negative.

Third, under free recombination, $F_2 - \overline{P} = \delta_1^c - \alpha_2^c - \delta_2^c$, and this is also true in later generations with restricted recombination, provided enough time has passed to eliminate gametic phase disequilibria. Thus, a hybrid population will ultimately experience outbreeding depression (in the absence of selection) if the net gains due to dominance are less than the net losses due to the breakup of favorable additive $\times$ additive and dominance $\times$ dominance interactions.

Our current understanding of the time scales (and geographic scales) over which outbreeding depression evolves is extremely crude. However, a number of recent empirical studies have uncovered striking examples of F_1 and/or F_2 breakdown in crosses among populations of the same species. For example, Burton (1987, 1990a,b) and Brown (1991) have obtained extensive evidence for the breakdown of physiological competence in crosses between populations of the marine copepod *Tigriopus californicus* inhabiting rock pools separated by only tens of kilometers. Other dramatic evidence of outbreeding depression comes from observations of reduced fitness in crosses of inbred lines of *Drosophila* (Templeton et al. 1976) and plants (Parker 1992) adapted to identical environments. Crosses between outbreeding plants separated by a few to several tens of meters can exhibit substantial reductions in fitness (Waser 1993b, Waser and Price 1989, 1994), as can crosses between fish derived from different sites in the same drainage basin (Leberg 1993) and crosses between clones of *Daphnia* from the same pond (Lynch and Deng 1994, Deng and Lynch 1996a).

In most of these examples, a decline in performance was observed in the F_1 generation, and data on the F_2 progeny were not obtained. Such results strongly implicate a breakup of favorable additive $\times$ additive epistatic effects as a factor contributing to outbreeding depression. On the other hand, studies that report heterosis in the F_1 generation with no evaluation of the F_2 can be very misleading. A particularly difficult issue underlying assessments of the potential for outbreeding depression concerns the time scale over which outbreeding depression is revealed. There will always be some loss of heterosis in the F_2 generation, and a breakup of favorable epistatic gene complexes is always implicated when the F_2 performance is not intermediate to that of the F_1 and the mean of the parental lines. However, with low rates of recombination between pairs of epistatically interacting genes, it may take several generations for the negative consequences of mixing coadapted gene complexes to emerge fully. Line-cross analysis provides a powerful statistical framework for investigating these issues.

A fundamental issue that arises in the interpretation of reduced hybrid performance is whether it is a consequence of the intrinsic effects of interactions between genes from different parental sources, the loss of local adaptation resulting from a 50% dilution of the local gene pool, or both. Local adaptation has been repeatedly documented through reciprocal transplant experiments with plants, which often exhibit adaptive divergence on spatial scales as small as a few meters (Schemske 1984, Waser and Price 1985, McGraw 1987, Schmitt and Gamble 1990, Galen et al. 1991). Similar results have been obtained with herbivorous insects residing on adjacent, long-lived hosts (Edmunds and Alstad 1978, Karban 1989). In most of these studies the environmental differences perceived by the organism were not apparent to the investigators, so an absence of obvious ecological differentiation does not provide a compelling argument for ruling out local adaptation.

Isolating the contributions of local adaptation and coadapted gene complexes to line-cross performance is very difficult, if not impossible in many cases. At the very least, both parental and hybrid lines need to be evaluated in both parental environments. If the responses of all lines to environmental change are parallel (or can be made to be parallel by an appropriate scale transformation), then genotype $\times$ environment interaction (local adaptation) can be ruled out, and the relative performances of the different lines must be due to intrinsic genetic differences.

Example 3. Moll et al. (1965) produced F_1 and F_2 generations from crosses between several lines of maize with varying degrees of genetic divergence "based on ancestral relationships and differences in adaptation." As illustrated in the following figure, when assayed in a common environment, all crosses exhibited heterosis for grain yield in the F_1 and F_2 generations, but this was most pronounced in crosses involving lines with intermediate degrees of divergence. Moreover, at the highest levels of divergence, the performance of the F_1 and F_2 lines converged (in the figure, the data from all crosses have been standardized so that $\bar{z}(F_2) = 1$).

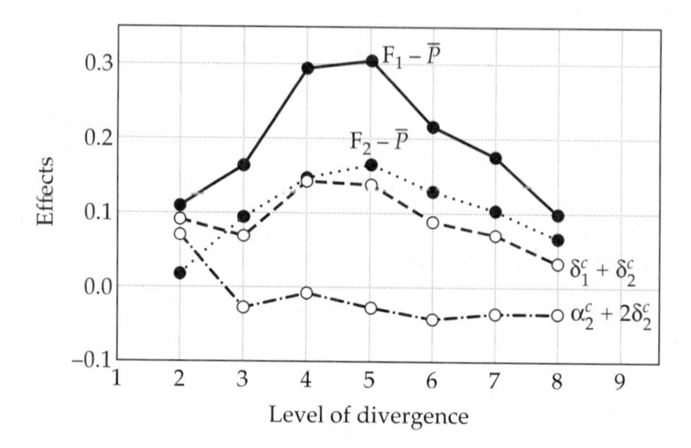

Can any insight into the mode of gene action be inferred from these results? Assuming that $\bar{c} \simeq 0.5$, as suggested by the results in Table 9.2, then $\bar{z}(\mathrm{F}_1) - \bar{z}(\mathrm{F}_2) \simeq \delta_1^c + \delta_2^c$, the sum of the composite dominance and dominance × dominance effects. In addition, $\bar{z}(\mathrm{F}_1) - 2\bar{z}(\mathrm{F}_2) + \bar{z}(\bar{P}) \simeq \alpha_2^c + 2\delta_2^c$, which is the net loss of performance due to segregation and recombination of parental line gene combinations. As shown in the figure, application of these two formulae suggests that the net effects of dominance between parental lines have a positive influence on grain yield at all levels of divergence, but that the magnitude of this effect is maximized at an intermediate genetic distance. On the other hand, except at the lowest levels of divergence, the estimates of $\alpha_2^c + 2\delta_2^c$ are negative and roughly constant, suggesting favorable epistatic effects between genes from *different* sources, contrary to the expectation if individual lines were harboring coadapted gene complexes.

VARIANCE OF LINE-CROSS DERIVATIVES

In principle, the joint-scaling test can be used to interpret the variances as well as the means obtained from line crosses. One of the most important applications of such an analysis is in the interpretation of the outbreaks of variation often seen in an F_2 generation (Figure 1.3). To keep things simple, let us assume that gene action is additive and that the environmental variance (here denoted as σ_E^2) is independent of the genetic background. The expected phenotypic variances for the parental lines and their F_1 offspring are then

$$\sigma^2(\mathrm{P}_1) = \sigma_E^2 + \sigma_{A_1}^2 \tag{9.13a}$$

$$\sigma^2(\mathrm{P}_2) = \sigma_E^2 + \sigma_{A_2}^2 \tag{9.13b}$$

$$\sigma^2(\mathrm{F}_1) = \sigma_E^2 + \frac{1}{2}\sigma_{A_1}^2 + \frac{1}{2}\sigma_{A_2}^2 \tag{9.13c}$$

where $\sigma_{A_1}^2$ and $\sigma_{A_2}^2$ are the additive genetic variances in the P_1 and P_2 lines. The genotypic variance in the F_1 generation follows from the fact that, for each locus,

F_1 individuals contain exactly one P_1 allele and one P_2 allele, and that the two haploid sets of alleles contribute variance $\sigma^2_{A_1}/2$ and $\sigma^2_{A_2}/2$.

An additional source of genetic variance will appear in any line-cross derivative for which there is variation among individuals in the proportion of P_1 and P_2 genes. For example, in the F_2 generation there is a 50% probability of being P_1P_2 and 25% probabilities of being P_1P_1 or P_2P_2 at any locus. This variation is in contrast to that in the F_1 generation where all individuals are P_1P_2. Letting $-\alpha_i^c/2$ and $+\alpha_i^c/2$ be the mean additive effects of alleles at the ith locus in the P_1 and P_2 lines, then the variance among F_2 individuals attributable to differences between parental lines at the locus is

$$\sigma^2_S(i) = 0.25(-\alpha_i^c)^2 + 0.5\left(\frac{\alpha_i^c}{2} - \frac{\alpha_i^c}{2}\right)^2 + 0.25(\alpha_i^c)^2 = \frac{(\alpha_i^c)^2}{2} \tag{9.14}$$

Summing over all loci, we obtain the **segregational variance,**

$$\sigma^2_S = \frac{1}{2}\sum_{i=1}^{n}(\alpha_i^c)^2 \tag{9.15}$$

which describes the excess variance that appears in the F_2 generation as a consequence of the segregation of parental-line genes. The expected variance in the F_2 generation is then

$$\sigma^2(\mathrm{F}_2) = \sigma^2_E + \frac{1}{2}\sigma^2_{A_1} + \frac{1}{2}\sigma^2_{A_2} + \sigma^2_S \tag{9.16}$$

To obtain a general expression for the variances within backcross and more advanced generations, we take the following approach, again assuming additive gene action. Recall that if a line has associated parameters H and S, then the proportions of P_1-purebred, P_2-purebred, and crossbred genotypes at any locus are $S - (H/2)$, $1 - S - (H/2)$, and H, respectively. For the ith locus, these classes have genotypic means and variances equal to $(\alpha_i^c,\ \sigma^2_{A_1i})$, $(-\alpha_i^c,\ \sigma^2_{A_2i})$, and $(0,\ [\sigma^2_{A_1i}+\sigma^2_{A_2i}]/2)$, respectively. For any line, the genetic variance associated with the locus can be expressed as

$$\sigma^2_A(i) = E(A_i^2) - \mu^2_{A_i}$$

where A_i denotes the breeding value of an individual at the locus, and from arguments given above, $\mu_{A_i} = \theta_S\alpha_i^c = (1-2S)\alpha_i^c$. An expression for $E(A_i^2)$ is obtained by averaging over the three possible classes of genotypes,

$$\begin{aligned} E(A_i^2) &= \left(S - \frac{H}{2}\right)[\sigma^2_{A_1i} + (\alpha_i^c)^2] + H\left(\frac{\sigma^2_{A_1i} + \sigma^2_{A_2i}}{2}\right) \\ &\qquad + \left(1 - S - \frac{H}{2}\right)[\sigma^2_{A_2i} + (-\alpha_i^c)^2] \\ &= S\sigma^2_{A_1i} + (1-S)\sigma^2_{A_2i} + (1-H)(\alpha_i^c)^2 \end{aligned}$$

Putting these results together, for any derivative line with indices S and H, the variance associated with the ith locus is

$$\sigma_A^2(i) = S\sigma_{A_1 i}^2 + (1-S)\sigma_{A_2 i}^2 + [4S(1-S) - H](\alpha_i^c)^2 \tag{9.17}$$

Adding the environmental variance, summing over all loci, and recalling Equation 9.15, the expected phenotypic variance for a line with properties S and H is

$$\sigma^2 = \sigma_E^2 + S\sigma_{A_1}^2 + (1-S)\sigma_{A_2}^2 + 2[4S(1-S) - H]\sigma_S^2 \tag{9.18}$$

For example, for the B_1 backcross ($H = 1/2$ and $S = 3/4$) and the B_2 backcross ($H = 1/2$ and $S = 1/4$),

$$\sigma^2(B_1) = \sigma_E^2 + \frac{3}{4}\sigma_{A_1}^2 + \frac{1}{4}\sigma_{A_2}^2 + \frac{1}{2}\sigma_S^2 \tag{9.19a}$$

$$\sigma^2(B_2) = \sigma_E^2 + \frac{1}{4}\sigma_{A_1}^2 + \frac{3}{4}\sigma_{A_2}^2 + \frac{1}{2}\sigma_S^2 \tag{9.19b}$$

We are now equipped with all of the statistical machinery necessary to develop a predictive model for line variances. Analogous to the model developed above for line means, we let $\mathbf{v}$ be the vector of observed phenotypic variances for the various lines, $\mathbf{a}$ be the vector of variance components (σ_E^2, σ_{A1}^2, σ_{A2}^2, and σ_S^2), and $\mathbf{M}$ be the matrix of coefficients for the lines, with all elements in the first column being equal to one, and values of S, $(1-S)$, and $2[4S(1-S) - H]$ being entered in the remaining three columns. The linear model for line variances can then be summarized as

$$\mathbf{v} = \mathbf{Ma} + \mathbf{e} \tag{9.20}$$

where $\mathbf{e}$ is the vector of residual errors. The weighted least-squares estimates for the model parameters are given by

$$\widehat{\mathbf{a}} = (\mathbf{M}^T\mathbf{V}^{-1}\mathbf{M})^{-1}\mathbf{M}^T\mathbf{V}^{-1}\mathbf{v} \tag{9.21}$$

where $\mathbf{V}$ is the sampling covariance matrix for the line variances. Provided the individuals of the various lines are unrelated, then all of the elements of $\mathbf{V}$ are zero except those on the diagonal. These are set equal to $2v_j^2/(n_j+2)$, the unbiased estimator of the sampling variance of a variance under the assumption of normality (Equation A1.10c), where n_j is the sample size, and v_j is the observed phenotypic variance of the jth line.

There is one small remaining problem. Unless one is willing to assume that the parental lines are completely homozygous, i.e. $\sigma_{A1}^2 = \sigma_{A2}^2 = 0$, Equation 9.21 cannot be solved directly. The problem is that $\mathbf{M}$ is singular, since for any line, the coefficient for σ_E^2 is equal to the sum of the coefficients for σ_{A1}^2 and σ_{A2}^2. This difficulty can be circumvented by deleting the first column from $\mathbf{M}$ and reducing the variance component vector to $\mathbf{a}^T = [\sigma^2(P_1), \sigma^2(P_2), \sigma_S^2]$, where

$\sigma^2(\mathrm{P}_1) = \sigma_E^2 + \sigma_{A1}^2$ and $\sigma^2(\mathrm{P}_2) = \sigma_E^2 + \sigma_{A2}^2$ are the phenotypic variances for the two parental lines.

Hayman (1960b) took this analysis a step further in producing a maximum-likelihood procedure. The diagonal elements of the matrix $\mathbf{V}$ have expectations equal to $2\sigma_j^4/(n_j + 2)$, where σ_j^2 represents the expectation of the appropriate entry in $\mathbf{v}$. Under the assumption that the additive model is correct, the projected least-squares values $\widehat{\mathbf{v}} = \mathbf{M}\widehat{\mathbf{a}}$ should actually be better estimates of the within-line variances than the original elements of $\mathbf{v}$. This implies that the elements of $\widehat{\mathbf{a}}$ should be computed a second time by use of Equation 9.21 after substituting the elements of $\widehat{\mathbf{v}}$ into $\mathbf{V}$. This procedure is then iterated until the estimates of $\widehat{\mathbf{a}}$ stabilize. (Note that during the iterative process, it is only the elements of the covariance matrix $\mathbf{V}$, and not those of the vector of observed variances $\mathbf{v}$, that are modified recursively.)

Since $\mathbf{V}$ is diagonal with $V_{ii} = 2\,\sigma_j^4/(n_j + 2)$, Equation A3.11a gives the χ^2 statistic for goodness of fit of the observed variances to the predictions of the additive model as

$$\chi^2 = \sum_{j=1}^{k} \frac{(v_j - \widehat{v}_j)^2}{2\,\widehat{v}_j^{\,2}/(n_j + 2)} \tag{9.22}$$

where the $\widehat{v}_j$ are the final estimates of the σ_j^2, and the degrees of freedom associated with the test statistic equal the number of lines (k) minus three (the number of variance components estimated).

Example 4. We now apply the joint-scaling test for variances to Harrison and Owen's (1964) data on human skin color. Recall from Example 2 that the analysis of means supports the idea that this character has an additive genetic basis. On the other hand, the phenotypic variances, recorded in the following table, appear to be rather inconsistent with the additive model. For example, the F_1 variance is much higher than the average of the P_1 and P_2 variances, and even exceeds that of the F_2. However, since the sampling variance of a variance is quite large (as can be seen in the third row of the following table), there is some question as to the significance of these differences.

	P_1	P_2	F_1	F_2	B_1	B_2
v_j	14.918	21.098	31.748	26.382	37.366	37.766
n_j	40	103	94	12	21	30
$2v_j^2/(n_j+2)$	10.597	8.479	20.999	99.430	121.410	89.142

The coefficients for the variance components in the model, $\sigma^2(\mathrm{P}_1)$, $\sigma^2(\mathrm{P}_2)$ and σ_S^2, are obtained from Equation 9.18,

$$\mathbf{M} = \begin{pmatrix} 1 & 0 & 0 \\ 0 & 1 & 0 \\ 0.5 & 0.5 & 0 \\ 0.5 & 0.5 & 1 \\ 0.75 & 0.25 & 0.5 \\ 0.25 & 0.75 & 0.5 \end{pmatrix}$$

Inserting the values from the table above, the initial sampling covariance matrix is

$$\mathbf{V} = \begin{pmatrix} 10.597 & 0.000 & 0.000 & 0.000 & 0.000 & 0.000 \\ 0.000 & 8.479 & 0.000 & 0.000 & 0.000 & 0.000 \\ 0.000 & 0.000 & 20.999 & 0.000 & 0.000 & 0.000 \\ 0.000 & 0.000 & 0.000 & 99.430 & 0.000 & 0.000 \\ 0.000 & 0.000 & 0.000 & 0.000 & 121.410 & 0.000 \\ 0.000 & 0.000 & 0.000 & 0.000 & 0.000 & 89.142 \end{pmatrix}$$

Substituting into Equation 9.21, we obtain the initial set of least-squares parameter estimates,

$$\widehat{\mathbf{a}} = (\mathbf{M}^T\mathbf{V}^{-1}\mathbf{M})^{-1}\mathbf{M}^T\mathbf{V}^{-1}\mathbf{v} = \begin{pmatrix} 18.120 \\ 23.681 \\ 14.441 \end{pmatrix}$$

The following table shows how the parameter estimates change over the next several rounds of iterations,

	Iteration						
Estimates	2	3	5	10	15	20	Final SE
Var(P_1)	22.199	22.930	23.446	23.583	23.586	23.586	4.207
Var(P_2)	25.994	25.505	25.185	25.103	25.102	25.102	3.166
Var(S)	17.163	17.038	16.876	16.830	16.829	16.829	10.411
χ^2	14.135	10.432	10.175	10.102	10.101	10.101	

The standard errors are the square roots of the diagonal elements of the final estimate of $(\mathbf{M}^T\mathbf{V}^{-1}\mathbf{M})^{-1}$. Using the final set of parameter estimates, the predicted line variances ($\widehat{v}_j$), obtained from Equation 9.18, and their standard errors, obtained as the square roots of the diagonal elements of $\mathbf{M}(\mathbf{M}^T\mathbf{V}^{-1}\mathbf{M})^{-1}\mathbf{M}^T$, are

	P_1	P_2	F_1	F_2	B_1	B_2
$\widehat{v}_j$	23.586	25.102	24.344	41.172	32.379	33.137
SE	4.207	3.166	2.306	9.845	5.042	5.187

The final χ^2 value (10.101) is rather large, but because of possible nonnormality of the data, its statistical interpretation is somewhat questionable. Since the difference between observed and expected line variances is less than two standard errors $(2[\text{SE}(v_j)^2 + \text{SE}(\widehat{v}_j)^2]^{1/2})$ for all lines, there seems to be no strong justification for rejecting the additive model.

A graphical comparison of the observed and predicted means and variances of human skin color (Figure 9.2) serves to illustrate two important points. First, due to the large standard errors of variance estimates, scaling tests based on variances are much less powerful than those based on means. Thus, while the preceding methodology can be generalized to compute the dominance and epistatic components of the segregational variance (Mather and Jinks 1982), the statistical reliability of such analyses is very low, and we will not pursue it.

Second, the additive model leads to some very simple geometric relationships (Figure 9.2). The expected line means are linear functions of the proportion of genes derived from each parental line. The expected line variances fall on a triangle, the vertices of which represent the expected P_1, P_2, and F_2 variances. The expected F_1 variance lies on the midpoint of the line leading from P_1 to P_2, while the expected B_1 and B_2 variances lie on the midpoints of the lines from P_1 to F_2 and P_2 to F_2, respectively.

BIOMETRICAL APPROACHES TO THE ESTIMATION OF GENE NUMBER

We now turn to a second application of line-cross analysis — estimation of the number of segregating loci responsible for quantitative variation. The subject is of importance for several reasons. First, since the beginning of this century (Chapter 1), there has been considerable debate as to whether most large evolutionary changes are due to a small number of macromutations or to gradual substitution of minor allelic variants at a large number of loci (Gould 1980, Charlesworth et al. 1982, Gottlieb 1984, Coyne and Lande 1985, Orr and Coyne 1992, Wu and Palopoli 1994). Second, from a more statistical point of view, the nice properties of normal theory that facilitate quantitative-genetic analysis become violated to a greater degree as the number of segregating loci becomes small. The ideal setting for much of quantitative-genetic theory is a very large number of loci, all with small effects, the so-called **infinitesimal model.** Third, for situations in which most of the genetic variation is a function of one or two genes with major effects, a fine-scale Mendelian analysis (through the direct observation of segregation ratios) will often be possible, provided the environmental component of variance is not of overwhelming importance (Chapter 13).

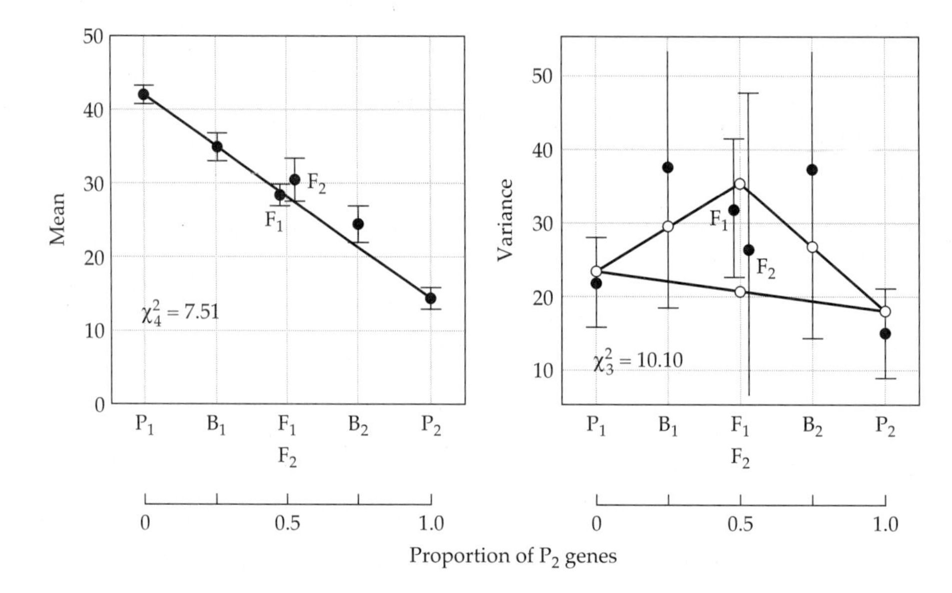

Figure 9.2 Observed means and variances (± twice the standard errors) of human skin color in relation to the predictions of the additive model. The maximum-likelihood predictions are given by the line on the left graph and the open circles on the triangle on the right graph. (From Harrison and Owens 1964.)

There are two approaches to estimating gene number. The biometrical approach, which is the subject of this chapter, is based on statistical properties (means and variances) of phenotype distributions, and uses these properties to infer indirectly the number of segregating factors that are likely to be responsible for them. The second approach involves the search for associations between segregating molecular markers and quantitative-genetic variation. The molecular-marker approach is now rapidly supplanting the first in species of economic and biomedical importance and will be considered in detail in Chapters 14–16.

Although we are ultimately interested in the total number of loci (n) contributing to the variance in trait expression, most estimates of n are actually measures of the **effective number of factors** (n_e) by which the characters in two lines differ. This quantity is equivalent to the number of freely segregating loci with equal effects that would yield the observed pattern of line means and variances. It is important to realize that estimates of n_e do not include the potentially large fraction of loci that do not vary between lines, yet could lead to phenotypic differences given the right kinds of mutations. In addition, n_e cannot exceed the number of independently segregating chromosomal segments, i.e., the number of chromosomes plus the mean number of recombination events per gamete (the **segregation index**). In eukaryotes, there are usually one to two recombination events per chromosome. Thus, the maximum possible value of n_e is usually two to three times the haploid chromosome number, although each segregating unit can contain many loci.

In the following section, we will first describe an estimator $\widehat{n}_e$ for the effective number of factors. We will then describe a more refined estimator $\widehat{n}$, which provides estimates that are closer to the actual number of loci. Procedures for estimating n_e involve a number of assumptions, the most important of which is additivity of gene action. Thus, prior to analysis, a serious attempt should be made to find a scale on which the observed line means and variances are consistent with the additive model. The joint-scaling tests described on the previous pages provide an essential means of testing the utility of various scale transformations (see also Chapter 11).

The Castle-Wright Estimator

The most widely used method for estimating n_e utilizes information on the phenotypic means and variances of two parental lines and their line-cross derivatives (F_1, F_2, B_1, B_2, etc.) As first developed by Castle (1921) with his graduate student Sewall Wright (1968), the method was intended for use with inbred parental lines. Lande (1981) generalized it for use with genetically variable base populations, and the theory developed below is based on his modifications, and those of Cockerham (1986). In addition to additive gene action, the Castle-Wright technique assumes unlinked loci and equality of allelic effects, although we relax the latter assumption in the following derivation. It also assumes that all genes with a positive influence on the trait are sorted into one line and all those with negative influences into the other.

Letting α_i^c be the composite additive effect for the ith locus, the mean phenotype of the P_1 line can be written as $\mu(P_1) = \mu_0 + \sum_{i=1}^{n} \alpha_i^c = \mu_0 + n\,\overline{\alpha_i^c}$, where $\overline{\alpha_i^c}$ is the average composite additive effect of a locus. That for the P_2 line is $\mu(P_2) = \mu_0 - n\,\overline{\alpha_i^c}$. Thus, the expected difference in mean phenotypes is $\mu(P_1) - \mu(P_2) = 2n\,\overline{\alpha_i^c}$. The segregational variance, defined in Equation 9.15, can also be written as

$$\sigma_S^2 = n[\sigma^2(\alpha_i^c) + (\overline{\alpha_i^c})^2]/2$$

where $\sigma^2(\alpha_i^c)$ is the variance of composite effects among loci. The trick to deriving an estimator of n_e is to note that the expected squared difference between parental line means is

$$\begin{aligned}[\mu(P_1) - \mu(P_2)]^2 &= E[\,(2\sum_{i=1}^{n} \alpha_i^c)^2\,] \\ &= 4n[\,\overline{(\alpha_i^c)^2} + (n-1)(\,\overline{\alpha_i^c}\,)^2\,] \\ &= 4n[\,n(\,\overline{\alpha_i^c}\,)^2 + \sigma^2(\alpha_i^c)\,]\end{aligned}$$

Taking the ratio of these two results

$$\frac{[\mu(P_1) - \mu(P_2)]^2}{\sigma_S^2} = \frac{8[\,n(\overline{\alpha}_i^c)^2 + \sigma^2(\alpha_i^c)\,]}{(\overline{\alpha}_i^c)^2 + \sigma^2(\alpha_i^c)}$$

which upon rearrangement yields

$$n_e = \frac{[\mu(\text{P}_1) - \mu(\text{P}_2)]^2(1 + C_\alpha)}{8\sigma_S^2} - C_\alpha \tag{9.23}$$

where $C_\alpha = \sigma^2(\alpha_i^c)/(\overline{\alpha_i^c})^2$ is the squared coefficient of variation of the locus-specific additive effects. In the very unlikely event that all of the assumptions of the model hold, then Equation 9.23 would define the actual number of loci. In recognition of the fact that one or more assumptions are likely to be violated, Equation 9.23 is denoted as a measure of the effective number of factors, n_e. In effect, Equation 9.23 states that the line means and segregational variance are distributed in the same way that would occur if the two populations were differentiated at n_e freely recombining loci with equal and additive effects.

In general, C_α is an unobservable quantity, but it must be positive. Ignoring this term for the time being, and substituting observed quantities for expectations, we obtain a biased estimator for the effective number of factors,

$$\widehat{n}_e = \frac{[\bar{z}(\text{P}_1) - \bar{z}(\text{P}_2)]^2 - \text{Var}[\bar{z}(\text{P}_1)] - \text{Var}[\bar{z}(\text{P}_2)]}{8\text{Var}(S)} \tag{9.24}$$

hereafter referred to as the **Castle-Wright estimator,** where the $\bar{z}(\text{P}_i)$ and $\text{Var}[\bar{z}(\text{P}_i)]$ are the observed means and sampling variances of the means for the ith parental line. Estimates of n_e in the literature often ignore the two variance terms in the numerator, but these are required to correct for the sampling error of the estimates of the line means (Cockerham 1986).

When data are available for the backcross generations, the segregational variance estimate, $\text{Var}(S)$, can be obtained by the weighted least-squares procedure described above. In the absence of backcrosses, it can be computed as a linear function of the observed phenotypic variances within lines, either as $\text{Var}(\text{F}_2) - \text{Var}(\text{F}_1)$ or $\text{Var}(\text{F}_2) - \{[2\text{Var}(\text{F}_1) + \text{Var}(\text{P}_1) + \text{Var}(\text{P}_2)]/4\}$. The importance of using least-squares estimates whenever possible is seen from Example 4, where $\text{Var}(\text{F}_2) - \text{Var}(\text{F}_1)$ is negative, but the least-squares estimate of σ_S^2 is 17.

The large-sample variance of n_e, obtained from the equation for the variance of a ratio under the assumption of normality (Appendix 1), is approximately

$$\text{Var}(\widehat{n}_e) = \widehat{n}_e^2 \left[\frac{4\{\text{Var}[\bar{z}(\text{P}_1)] + \text{Var}[\bar{z}(\text{P}_2)]\}}{[\bar{z}(\text{P}_1) - \bar{z}(\text{P}_2)]^2} + \frac{\text{Var}[\text{Var}(S)]}{[\text{Var}(S)]^2} \right] \tag{9.25}$$

If $\text{Var}(S)$ is estimated by least-squares analysis, its sampling variance, $\text{Var}[\text{Var}(S)]$, is obtained directly from the matrix $(\mathbf{M}^T\mathbf{V}^{-1}\mathbf{M})^{-1}$, as described above. Otherwise, it is estimated by the sum of the variances of the variances used to compute $\text{Var}(S)$, each weighted by the square of the appropriate coefficient. For example, if $\text{Var}(S)$ is estimated by $\text{Var}(\text{F}_2) - \text{Var}(\text{F}_1)$, then

$$\text{Var}[\text{Var}(S)] = \frac{2[\text{Var}(\text{F}_2)]^2}{n_{\text{F}_2} + 2} + \frac{2[\text{Var}(\text{F}_1)]^2}{n_{\text{F}_1} + 2}$$

where n_{F_2} and n_{F_1} are the sample sizes. This follows since $\text{Var}(F_1)$ and $\text{Var}(F_2)$ are independent estimates, and since under the assumption of normality, the large-sample variance of a variance is $2\text{Var}^2/(n+2)$ (Appendix 1).

We noted above that failure to account for the variation in composite effects among segregating loci, i.e., ignoring C_α, will tend to depress $\widehat{n}_e$ below the true number of segregating loci. Additional factors will usually result in a further downward bias. For example, if the genes with positive effects are distributed among both parental lines, the difference between the parental line means will be less than the maximum value because the positive effects of genes at some loci will be canceled by negative effects at others. Consider the case of one line being fixed for $+1$ genes at locus 1 and -1 genes at locus 2 and another line being fixed for -1 genes at locus 1 and $+1$ genes at locus 2. The number of segregating loci in the F_2 generation is two, but since both parental lines have mean phenotypes equal to zero, the expected value of n_e yielded by Equation 9.24 would equal zero. Such problems become immediately apparent when F_2 individuals exhibit phenotypes outside of the range of variation in both parental lines, a phenomenon known as **transgressive segregation** (Chapter 15), but the absence of such individuals does not rule out the possibility of transgression. In order to minimize such interpretative difficulties, most investigators utilize parental lines with the maximum range of variation between mean phenotypes. This is often accomplished by artificially selecting lines in the upward and downward direction for several generations prior to crossing.

By inflating the estimated segregational variance, linkage will also cause $\widehat{n}_e$ to be downwardly biased. Letting c_{ij} be the recombination fraction between loci i and j, a more general formula for the segregational variance in the F_2 generation is

$$\sigma_S^2 = \frac{1}{2}\left[\sum_{i=1}^{n}(\alpha_i^c)^2 + \sum_{i=1}^{n}\sum_{j\neq i}^{n}\alpha_i^c\alpha_j^c(1-2c_{ij})\right] \tag{9.26a}$$

where the term on the right is the disequilibrium covariance. Assuming that the effects of pairs of alleles are uncorrelated with their map distances, then Equation 9.26a simplifies to

$$\sigma_S^2 = \frac{n}{2}\left[\sigma^2(\alpha_i^c) + (\,\overline{\alpha_i^c}\,)^2 + (n-1)(1-2\bar{c})(\,\overline{\alpha_i^c}\,)^2\right] \tag{9.26b}$$

In principle, the disequilibrium contribution to σ_S^2 can be removed by taking the F_2 generation through several additional generations of random mating, since this reduces the disequilibrium covariance by the factor $(1-c)$ each generation (Chapters 5 and 16). A modified estimate of the segregational variance can then be obtained as the difference between the variance in the advanced generation and that in the F_1 line.

An alternative way to deal with the problem of linkage is to use this more general expression for σ_S^2, Equation 9.26b, combined with our previous expression for the expected squared difference between line means to define the relationship between the effective number of factors and the actual number of loci. Substituting Equation 9.26b into 9.23 and rearranging leads to the expression

$$n = \frac{2\bar{c}n_e + C_\alpha(n_e - 1)}{1 - n_e(1 - 2\bar{c})} \tag{9.27}$$

which reduces to Equation 9.23 for the special case in which $\bar{c} = 0.5$. Zeng (1992) suggested that by substituting the estimate $\widehat{n}_e$, obtained by use of Equation 9.24, into this expression, nearly unbiased estimates of the *actual* number of loci (n) are achievable.

In order to take advantage of Zeng's suggestion, we require, at the very least, estimates of $\bar{c}$ and C_α. We have already provided a number of estimates of $\bar{c}$ in Table 9.2, showing how these can be obtained from genetic maps of chromosomes under the assumption of randomly distributed loci. For many species, such detailed information is not available. However, provided the haploid chromosome number M is known, then a downwardly biased estimate of $\bar{c}$ is given by

$$\bar{c} = \frac{M - 1}{2M} \tag{9.28}$$

which assumes that recombination only occurs between pairs of genes on different chromosomes (by independent assortment), and that all chromosomes contain equal numbers of genes. Fortunately, estimates of $\bar{c}$ using this approximation are not greatly different than the more refined estimates obtained by use of Equation 9.3. For example, for humans $(M = 23)$, maize $(M = 10)$, and *Arabidopsis* $(M = 5)$, Equation 9.28 yields estimates of $\bar{c} = 0.478$, 0.450, and 0.400, respectively, which contrast with the more exact computations, 0.490, 0.474, and 0.443. Thus, provided the minimum amount of information exists on the cytology of the organism, fairly reasonable estimates of $\bar{c}$ are achievable.

The squared coefficient of variation of effects, C_α, is much more elusive. While estimates of the distribution of allelic effects are expected to be generated in the future as QTL mapping continues (Chapters 14–16), the only available estimates of this parameter derive from Keightley's (1994) analysis of data from mutation-accumulation experiments performed on lines of *Drosophila melanogaster* (Chapter 12). For abdominal bristle number, sternopleural bristle number, and viability, C_α is on the order of 6, 24, and 17, respectively. Unfortunately, aside from the fact that these estimates have very large sampling variances, it is unclear how similar the spectrum of effects of spontaneous mutations is to that of the effects of alleles normally segregating in natural populations. To the extent that they are representative, such high values of C_α suggest a very leptokurtic (L-shaped) distribution of allelic effects, with a very high density of small effects and a long tail to the right. For comparison, with a half-normal distribution (truncated at the mean), $C_\alpha = 0.57$, and with a negative exponential distribution, $C_\alpha = 1.0$.

The modifications suggested above are not trivial, as the magnitude of bias that variation in allelic effects and/or linkage causes with the Castle-Wright estimator can be quite large. Consider, for example, the situation in which $C_\alpha = 15$, the average of the results reported above, and suppose that the Castle-Wright estimator yields $\widehat{n}_e = 4$. Substituting into Equation 9.27, for humans, for which $\bar{c} = 0.49$, such an estimate would be compatible with the presence of 53 actual loci, and for *C. elegans*, for which $\bar{c} = 0.42$, it would imply the presence of 134 loci. Assuming constant allelic effects ($C_\alpha = 0$), the estimate for humans would be essentially unbiased, as $\widehat{n}$ still equals four, but for *C. elegans*, $\widehat{n} = 9$.

By inflating the segregational variance in the F_2, nonadditive gene action is still another factor that has a downward influence on estimates derived by the Castle-Wright estimator. However, provided dominance is the primary source of nonadditivity, then the use of $2\text{Var}(F_2) - \text{Var}(B_1) - \text{Var}(B_2)$ as the estimate of the segregational variance in Equation 9.24 can eliminate most of the problem (Wright 1968, Ollivier and Janss 1993). Since the expectation of this quantity is identical to σ_S^2 defined in Equation 9.26b, Equation 9.27 applies as well.

In summary, we find that violations of the various assumptions of the Castle-Wright model usually conspire to ensure that $\widehat{n}_e$ is an underestimate of the actual number of loci contributing to the divergence of lines. However, although the bias can be very substantial (Zeng et al. 1990), most of it can be eliminated by making the modifications suggested above, i.e., by first computing $\widehat{n}_e$ by use of Equation 9.24, then substituting this estimate and estimates of C_α and $\bar{c}$ into Equation 9.27, and solving. An approximate expression for the sampling variance of the improved estimate of the actual number of loci is given by

$$\text{Var}(\widehat{n}) = \frac{4\bar{c}^2(1 + C_\alpha)^2 \text{Var}(\widehat{n}_e)}{[1 - \widehat{n}_e(1 - 2\bar{c})]^4} \tag{9.29}$$

(Zeng 1992), where $\text{Var}(\widehat{n}_e)$ is defined in Equation 9.25. Simulations by Zeng (1992) suggest that $\widehat{n}$ provides much more reasonable estimates of the number of loci (n), than does $\widehat{n}_e$. However, the sampling variance of $\widehat{n}$ can be quite large. Even negative estimates are possible, when by chance the estimate of σ_S^2 is negative. (Negativity can occur with the Castle-Wright estimator as well.) Thus, any attempt to estimate n by either approach should be based on large sample sizes (ideally, with hundreds of individuals measured in each line).

Example 5. In previous examples involving human skin color, we found that $\bar{z}(P_1) = 14.4$ and $\bar{z}(P_2) = 41.0$. Squaring the standard errors of the means, $\text{Var}[\bar{z}(P_1)] = 0.205$ and $\text{Var}[\bar{z}(P_2)] = 0.373$. The estimated segregational variance (Example 3) is $\text{Var}(S) = 17.264$, and its sampling variance is obtained by squaring its standard error, $\text{Var}[\text{Var}(S)] = 11.033^2 = 121.724$. Substituting into Equation 9.24, we obtain $\widehat{n}_e = 5.1$. Substituting into Equation 9.25,

$\text{Var}(\widehat{n}_e) = 10.703$, giving the standard error of $\widehat{n}_e$ as $10.703^{1/2} \simeq 3.3$. Thus, the data suggest the hypothesis that the majority of the genetic difference in skin color between the major races of man is a consequence of a very small number of segregating factors. It should be kept in mind, however, that because of the low degree of accuracy of the estimated segregational variance, $\widehat{n}_e$ is a highly uncertain measure of the effective number of factors.

Supposing, for heuristic purposes, that the estimate $\widehat{n}_e$ is accurate, what might the actual number of loci (n) contributing to the character be? From Table 9.2, we know that the mean recombination fraction for randomly distributed genes is extremely high in humans ($\bar{c} = 0.49$). Substituting this and $n_e = 5$ into Equation 9.27, we obtain

$$n = \frac{4.9 + 4C_\alpha}{0.9}$$

Assuming that all loci have equal effects ($C_\alpha = 0$), which seems unlikely, then $n = 10$. For $C_\alpha = 1$, 10, and 100, $n = 10$, 50, and 450. Thus, if the squared coefficient of variation of effects is much greater than one, the actual number of loci may greatly exceed the effective number of factors.

A survey of estimates of $\widehat{n}_e$ is given in Table 9.6. Here it should be emphasized that each estimate only applies to the specific pair of parental lines and that substantial differences would be likely if other parental stocks were used. Furthermore, the data are adequately described by an additive model in only a few cases, so most of the estimates are definitely biased in the downward direction by nonadditive gene action. Despite these limitations, while several of the analyses imply that a dozen or more loci are responsible for the differentiation of characters between parental lines, a number of cases suggest the possibility that a single major factor is involved. The latter conclusion may, of course, be substantially in error due to the approximate and biased nature of the biometrical approach. Nevertheless, the Castle-Wright model serves as a flag for situations in which a leading-factor (major gene) hypothesis (Chapter 13) warrants consideration.

Effect of the Leading Factor

If the assumptions of additive gene action and unlinked loci hold, then n_e provides some information on the effect of the **leading factor** (the locus accounting for the largest amount of the difference between parental means). Let $\phi_{max} = 2\alpha^c_{max}/[\mu(\text{P}_1) - \mu(\text{P}_2)]$ be the proportion of the difference between parental means attributable to the largest factor, and denote its effect α^c_{max}. Equation 9.26a yields the inequality $\sigma^2_S \geq (\alpha^c_{max})/2$, which upon substitution into Equation 9.23 gives

$$\phi_{max} \leq \sqrt{\frac{1 + C_\alpha}{n_e + C_\alpha}}$$

Table 9.6 A sample of estimates ($\pm$ their standard errors) of the effective number of segregating factors differentiating parental lines, obtained by use of Equation 9.24.

Species	Character	$\widehat{n}_e$	Additive Model	Reference
Corn	log(% oil + 1.87)	18 ± 2	+	Sprague and Brimhall 1949
	time to silking	1 ± 1	−	Mohamed 1959
	time to shed pollen	1 ± 1	−	
	ln (height)	4 ± 1	−	Emerson and East 1913
	ln (nodes)	5 ± 1	+	
	ln (internode length)	1 ± 1	−	
	ln (ear length)	13 ± 3	−	
	ln (seed weight)	13 ± 3	−	
Lima beans	seed size	17 ± 2	−	Ryder 1958
Red pepper	fruit shape	3 ± 1	−	Khambononda 1950
	fruit weight	13 ± 1	−	
Rice	plant height	1 ± 1	−	Mohamed and Hanna 1964
Goldenrod (*Solidago*)	date of anthesis	6 ± 2	−	Goodwin 1944
Nicotiana Langsdorffii × *N. Sanderae*	corolla length	13 ± 1	−	Smith 1937
Mimulus nasutus × *M. guttatus*	ln(flowering time)	1 ± 1	+	Fenster and Ritland 1994
	corolla width	2 ± 1	+	
	stamen level	3 ± 1	+	
Mimulus guttatus × *M. cupriphilus*	flower width	5 ± 2	+	Macnair and Cumbes 1989
	flower height	4 ± 1	+	
	pistil length	18 ± 18	+	
	corolla length	6 ± 3	+	
Tomato	$\log_{10}$ (fruit weight)	12 ± 1	−	Powers 1942
Pearl millet (*Pennisetum*)	height	4 ± 1	−	Burton 1951
		4 ± 1	−	
	internode length	2 ± 1	−	
	leaves/stem	5 ± 1	−	
		7 ± 1		
Drosophila melanogaster	ln (longevity)	1 ± 1	−	Luckinbill et al. 1988
Drosophila tripunctata	ovipos. site pref.	1 ± 1	−	Jaenike 1987
Drosophila heteroneura × *D. silvestris*	head length	7 ± 4	+	Templeton 1977
	head width	1 ± 1	+	
Cave fish (*Astyanax*)	eye diameter	6 ± 1	−	Wilkens 1971
Chickens	weight	5 ± 1	−	Waters 1931
Mice	$\log_{10}$ (weight)	12 ± 1	−	Chai 1956

Note: Whenever possible, the segregational variance was obtained by least-squares analysis. + and − denote agreement and incompatibility with an additive model.

as an estimate of the upper bound on the effect of the leading segregating factor. Since each segregating factor contains one or more loci, ϕ_{max} is also an upper bound on the effect of the leading locus. Lander and Botstein (1989) proposed a simple idea that yields a lower bound estimate for the effect of the leading factor. Under the assumption that one strain contains all "positive" genes and the other all "negative" genes (which might be approximated if the two strains were obtained by intense selection in opposite directions from a common stock), there must be at least one segregating factor with an effect at least as great as $[\mu(\mathrm{P}_2) - \mu(\mathrm{P}_1)]/n_e$. Expressed in terms of the proportion of the total difference, this effect is simply $\phi_{min} = 1/n_e$. Alternatively, from the standpoint of individual loci, an estimate of the minimum effect of the leading locus is $1/n$. Finally, since (from Equation 9.27) there are at least $2\bar{c}n_e$ loci, the maximum value of the *average* allelic effect is $\bar{\phi}_{max} = 1/(2\bar{c}n_e)$.

Estimates of the statistical bounds on the effects of leading factors are of practical importance, since they can provide insight into the likely utility of molecular marker-based searches for loci with major effects (Chapters 14–16). Unfortunately, none of the above-mentioned statistics is particularly informative in this regard. Even when they are reliable, large estimates of the upper bound of the leading factor do not necessarily imply that any locus actually has a large effect, and although a large ϕ_{min} implies that at least one locus has a major effect, a small ϕ_{min} does not rule out the possibility of several loci with major effects.

Zeng (1992) suggested an alternative approach to predicting the effects of leading factors. Given an estimate of C_α, one first derives the estimate of the number of loci, $\widehat{n}$. This provides an estimate of the mean allelic effect as $\widehat{\alpha} = [\bar{z}(\mathrm{P}_1) - \bar{z}(\mathrm{P}_2)]/\widehat{n}$. Together, $\widehat{\alpha}$ and C_α then provide the first two moments of the distribution of allelic effects. If the form of the distribution is specified and uniquely defined by the mean and variance, one can then draw $\widehat{n}$ random effects from the distribution, order them, and evaluate the relative contributions of the various factors to the parental line divergence. As we have noted above, different values of C_α can lead to very different estimates of n. However, Zeng (1992) found that the number of significant loci (for example, the number that account for 90% of the total divergence) is extremely insensitive to changes in C_α, changing by only five or so over a range in which n changes by hundreds. Thus, for large surveys in which highly reliable estimates of phenotypic means and variances can be acquired, Zeng's approach has promise as a means of estimating the number of major factors.

Example 6. Here we present an alternative analytical approach for estimating the number of leading factors and their effects. This approach assumes that something is known about the form of the distribution of allelic effects. Let $p(\alpha)$ be the probability density function of the effects, α_i, and let $F(\alpha)$ be the cumulative frequency distribution, the probability that the effect of a randomly drawn gene is less than α. By definition, $dF(\alpha)/d\alpha = p(\alpha)$. Suppose now that n genes are randomly drawn from the distribution $p(\alpha)$, and rank ordered in terms of increasing effect, such that α_1 is the smallest effect and α_n is the largest effect (the leading factor). From the perspective of genetic analysis, one would like to know the expected effects of α_n, α_{n-1}, α_{n-2}, and so on. The theory of **order statistics** provides a potential solution.

Consider the rth smallest value in the set of random draws of n genes. The probability that at least r draws in a sample do not exceed the value α is

$$F_r(\alpha) = \sum_{i=r}^{n} \binom{n}{i} [F(\alpha)]^i [1 - F(\alpha)]^{n-i}$$

which leads to the probability density function for the rth order statistic,

$$p_r(\alpha) = \frac{dF_r(\alpha)}{d\alpha} = \frac{n!}{(r-1)!(n-r)!}[F(\alpha)]^{r-1}[1 - F(\alpha)]^{n-r} p(\alpha)$$

Thus, the expected value of the rth smallest factor is given by

$$E(\alpha_r) = \int_{-\infty}^{+\infty} \alpha p_r(\alpha) d\alpha$$

which, for the leading factor, reduces to

$$E(\alpha_n) = n \int_{-\infty}^{+\infty} \alpha [F(\alpha)]^{n-1} p(\alpha) d\alpha$$

An estimate of the proportional contribution of the leading factor to the line differentiation is $2E(\alpha_n)/[\bar{z}(\mathrm{P}_1) - \bar{z}(\mathrm{P}_2)]$.

These expressions only outline a general approach. Their actual implementation requires that one define the form of the probability density function $p(\alpha)$ (for example, a normal or a gamma distribution), and then characterize the function in terms of its parameters (usually the mean and variance of effects). Once an estimate of the number of loci (n) has been acquired (for example, by the use of Equation 9.27), these parameters are specified. The mean effect is estimated by $\bar{\alpha} = [\bar{z}(\mathrm{P}_1) - \bar{z}(\mathrm{P}_2)]/2n$, and the variance is defined by $\bar{\alpha}^2 C_\alpha$.

Extension to Haploids

The Castle-Wright model can be extended to the estimation of gene number in haploid organisms without great difficulty (Chovnick and Fox 1953). Since most

Table 9.7 Expected means and variances for line crosses derived from two haploid parental lines, under the assumption of zero linkage and additive gene action.

Line	Mean	Variance
P_1	$\mu_0 - \alpha_1^c$	σ_E^2
P_2	$\mu_0 + \alpha_1^c$	σ_E^2
F_1	μ_0	$\sigma_E^2 + \sigma_S^2$
F_2	μ_0	$\sigma_E^2 + \sigma_S^2$
B_1	$\mu_0 - 0.5\alpha_1^c$	$\sigma_E^2 + \frac{3}{4}\sigma_S^2$
B_2	$\mu_0 + 0.5\alpha_1^c$	$\sigma_E^2 + \frac{3}{4}\sigma_S^2$

Note: Composite additive effects are defined as haploid effects, so that parental line divergence is still $2\alpha_1^c$ as in the diploid model.

haploids can be maintained clonally, we assume that the environment is the sole source of variation within the parental lines. The expected means and variances of the derived generations are laid out in Table 9.7. Note that the F_1 generation exhibits the same segregational variance as the F_2 due to the complete segregation of parental genes following fertilization and the production of haploid progeny. This segregational variance is

$$\sigma_S^2 = \sum_{i=1}^{n} 0.5[(\alpha_i^c - 0)^2 + (-\alpha_i^c - 0)^2] = n[\sigma^2(\alpha_i^c) + (\overline{\alpha_i^c})^2]$$

which is twice the expectation in the case of diploidy. Thus, the estimation equation for n_e with haploid organisms is the same as in the case of diploidy except that $4\text{Var}(S)$, rather than $8\text{Var}(S)$, appears in the denominator of Equations 9.23 and 9.24. The equation for the sampling variance for $\widehat{n}_e$ needs to be multiplied by four, but the estimators for gene number, Equations 9.27 and 9.29, still apply provided the segregational variance is estimated in the F_1 generation.

A very similar strategy can be employed with **doubled-haploid lines**. Such lines, which can be produced by a variety of cytological techniques (Kermicle 1969, Nitzsche and Wenzel 1977, Choo 1981), are homozygous at all loci. If two such parental lines are crossed to produce an F_1 generation, and a random sample of F_1 gametes is used to produce a new series of doubled haploids, the effective number of factors differentiating any two lines can be obtained by computing the segregational variance as half the difference between the F_1 doubled-haploid variance and the average variance in the P_1 and P_2, and employing Equation 9.24 (Choo and Reinbergs 1982). (The segregational variance is inflated twofold by the enforcement of homozygosity at each locus in doubled haploids.) The most reasonable estimate of n_e obtained with this approach utilizes parental lines with the highest and lowest mean phenotypes in a random sample from the population. However, with even a moderate number of segregating loci, the probability of obtaining the two most extreme lines possible is low unless the number of lines assayed is very large. Choo and Reinbergs (1982) used this technique to show that at least eight segregating factors contribute to the variation in grain yield, heading date, and plant height in barley. For additional biometrical approaches to gene number estimation in doubled haploids, see Snape et al. (1984).

Example 7. Croft and Simchen (1965) isolated dikaryotic mycelia from wild populations of the fungus *Collybia velutipes* and from these extracted asexually and sexually derived monokaryotic spores. (A dikaryotic mycelium is a filament comprised of fused cells of two different parental origins, each containing a haploid nucleus). The growth rates of germinating spores were then assayed on a laboratory medium. Barring mutations, the growth rate of each asexual propagule is expected to be representative of one of the parental lines, since these propagules contain a single, nonrecombinant nucleus. On the other hand, the sexually derived progeny will exhibit segregational variance. Frequency distributions are given below for both types of offspring for one particular isolate. The mean growth rates of the two parental types differ by $\bar{z}(\mathrm{P}_1) - \bar{z}(\mathrm{P}_2) = 39.45$ (mm/10 days), and the sampling variances of the two means are $\mathrm{Var}[\bar{z}(\mathrm{P}_1)] = 2.43$ and $\mathrm{Var}[\bar{z}(\mathrm{P}_2)] = 0.53$. The variance of growth rate among haploid replicates is 26.80, while the excess variance among sexual propagules is 277.80. Taking the latter quantity to be an estimate of the segregational variance, $\mathrm{Var}(S)$, substitution into Equation 9.24 (multiplied by 2) yields $\widehat{n}_e = 1.4$. The standard error is approximately 0.5. These results are reasonably consistent with those obtained from four other isolates: $1 \pm 0.2, 5 \pm 0.8, 3 \pm 0.3,$ and 1 ± 0.1. Thus, it seems likely that most of the growth rate differences among parental strains may be attributable to one to three loci.

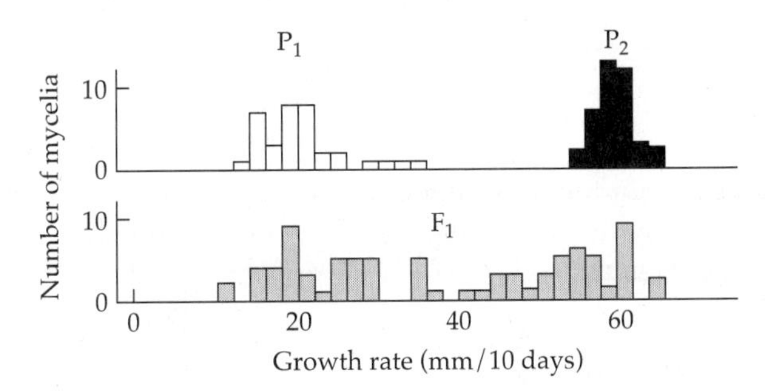

OTHER BIOMETRICAL APPROACHES TO GENE NUMBER ESTIMATION

In outlining the Castle-Wright model, we emphasized several assumptions, violations of which tend to result in underestimation of the actual number of segregating loci. Although two potential problems, gametic phase disequilibrium and dominance, appear to be reconcilable, two others are less tractable — transgressive segregation and variation among loci for allelic effects. We now consider three approaches that have been developed to circumvent these problems, all involving the use of species that can be self-fertilized.

The Inbred-Backcross Technique

Wehrhahn and Allard (1965) developed a useful technique that yields estimates of both the minimum number of genetic factors responsible for the differentiation of two lines and the magnitude of the locus-specific effects. Two pure lines are crossed to form F_1 individuals, each of which is then backcrossed to one of the parental populations (say the P_1) for k generations (i.e., an $F_1 \times P_1$ cross, followed by a cross of their progeny to the P_1, etc.) The backcross descendants are then inbred for several generations to fix any segregating factors. The rationale for this breeding scheme is that as k becomes large, the probability that any inbred backcross line will retain more than one allele from the donor parent (the P_2) becomes small. The effects of individual genes can then be ascertained by comparing the phenotypic means of the recurrent parent (P_1) and the derived lines. A unique advantage of the inbred-backcross technique is that, after single-gene deviant lines have been isolated, lines with pairs of various genes can be constructed to evaluate epistatic effects between specific isolated factors. Likewise, genotype $\times$ environment interaction involving individual genes can be examined by growing lines in different environments.

A more formal statement of these arguments follows. The probability that a specific gene from the P_2 is incorporated into a specific line after k generations of backcrossing is $p_k = (1/2)^{k+1}$. If n freely segregating factors are responsible for the character difference between the P_1 and P_2, then the probability that a derived line retains just one of the P_2 alleles is

$$p(1) = np_k(1 - p_k)^{n-1} \tag{9.30a}$$

This follows directly from the binomial distribution, since each gene is retained or lost independently with probability p_k. The probability that a derived line contains at least one P_2 gene is

$$p(r \geq 1) = 1 - (1 - p_k)^n \tag{9.30b}$$

The ratio $p(1)/p(r \geq 1)$ is the conditional probability that any deviant derivative line contains only a single P_2 gene. Table 9.8 shows that unless n is very large, this probability is very high after only three or four generations of backcrossing.

Table 9.8 The proportion of deviant derived lines that are expected to contain a single gene from the donor parent (last column).

n	k	p_k	$p(1)$	$p(r \geq 1)$	$p(1)/p(r \geq 1)$
4	2	$\frac{1}{8}$	0.335	0.414	0.809
	3	$\frac{1}{16}$	0.206	0.227	0.905
	4	$\frac{1}{32}$	0.114	0.119	0.953
10	2	$\frac{1}{8}$	0.376	0.624	0.510
	3	$\frac{1}{16}$	0.350	0.650	0.735
	4	$\frac{1}{32}$	0.235	0.765	0.863

Source: Wehrhahn and Allard 1965.

If, by statistical comparison of phenotypic means, a fraction $\widehat{p}(r \geq 1)$ of the inbred-backcross lines is found to differ significantly from the P_1, then a minimum estimate of the number of effective factors can be found by rearrangement of Equation 9.30b,

$$\widehat{n}_{WA} = \frac{\ln[1 - \widehat{p}(r \geq 1)]}{\ln(1 - 0.5^{k+1})} \tag{9.31}$$

(Mulitze and Baker 1985a,b). Unlike the Castle-Wright estimator, Equation 9.31 yields estimates of n_e that are essentially independent of the degree of transgression of gene effects in the P_1 and P_2, i.e., a gene with a low effect in an otherwise high-performing line can be detected when substituted into a low-performing line.

Example 8. Wehrhahn and Allard (1965) crossed two pure lines of wheat (Ramona and Baart 46) with very different heading dates (flowering times). Two successive backcrosses were made to Ramona, followed by three generations of selfing, to produce 69 inbred backcross lines. How many lines are expected to contain any specific Baart 46 gene?

Since $k = 2$, we have $p_k = (1/2)^3 = 1/8$, and since there are 69 total lines, $69/8 = 8.6$ of these are expected to carry a Baart 46 gene at a specified locus. From the properties of the binomial distribution, the standard error of the estimate is $[69p_k(1 - p_k)]^{1/2} = 2.8$. The distribution of heading date in the derived lines shows three groups of deviants from the Ramona distribution: (1) a group of eight very late lines that appear to contain a factor with major effect, (2) a group of 14 lines with slightly late heading dates, and (3) a group of three lines with earlier heading dates. (The means and 95% confidence limits for the parental lines are given by the vertical and horizontal lines, and the three groups of deviants from the Ramona (recurrent) line are differentially shaded.)

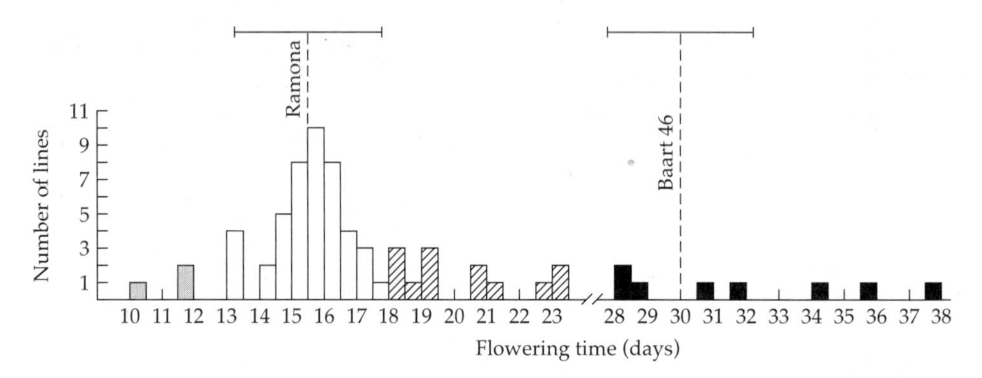

By other means, the authors showed rather convincingly that the group of 14 was heterogeneous for two factors. Thus, the difference in heading date between the two parental lines is caused by at least four effective factors, one of which operates in a direction opposite to the others. These four factors accounted for 96% of the line differentiation (80% was due to the leading factor), so if additional loci are involved, their effects must be very small. To see that the observations in the figure are consistent with Equation 9.31, let $\widehat{p}(r \geq 1) = (8 + 14 + 3)/69 = 0.362$. We then obtain $\widehat{n}_{WA} = 3.4$, which rounds up to 4.

Genotype Assay

Jinks and Towey (1976, Towey and Jinks 1977) developed a method for estimating n_e that is similar in philosophy to the inbred-backcross technique. In this case, however, the F_1 progeny of a cross between two pure lines are self-fertilized (instead of backcrossed). Their descendants are also self-fertilized until generation F_k, where k is usually 2 to 5. Two random (selfed) progeny are then raised from each F_k individual and selfed, and their offspring are assayed in a randomized design (Figure 9.3). A comparison of means (t test) and variances (F test) between the two families is used to detect whether one or more loci were segregating in the F_k grandparent. Not all heterozygotes are detected by this approach, since the probability that two randomly chosen offspring of a heterozygous parent will differ in genotype is only 5/8 (the probability that one is BB and the other is Bb is $2 \times (1/4) \times (1/2) = 1/4$, that one is BB and the other is bb is $2 \times (1/4) \times (1/4) = 1/8$, and that one is Bb and the other is bb is $2 \times (1/2) \times (1/4) = 1/4$). However, the observed fraction of segregating grandparents can still be used to make inferences about n.

The logic behind this idea is as follows. Under continuous self-fertilization, heterozygosity is reduced by 50% each generation, so that the probability that an F_k individual is a heterozygote at a particular locus is $(1/2)^{k-1}$. The probability that two assayed progeny from a random F_k individual differ at this locus is then

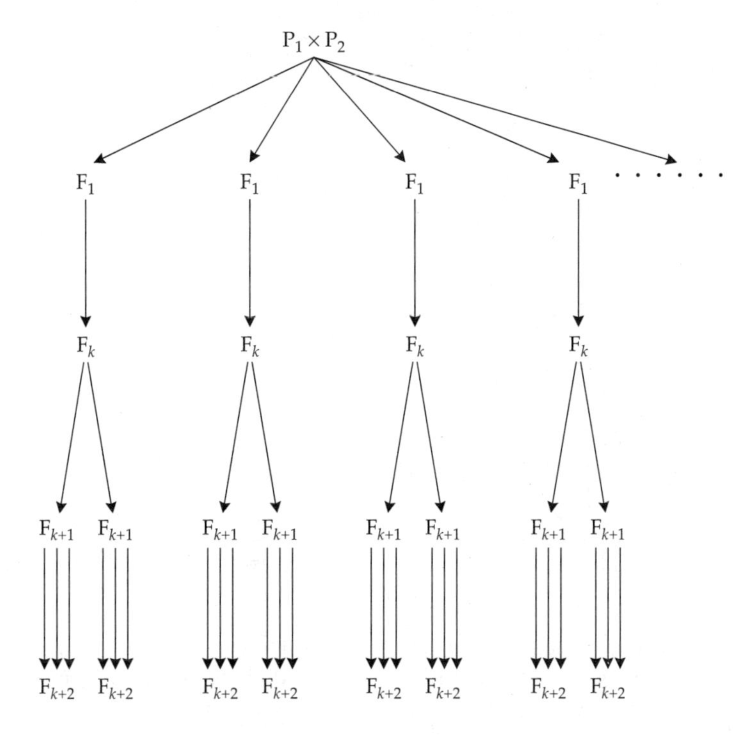

Figure 9.3 The crossing scheme involved in the genotype assay technique of Jinks and Towey (1976). Progeny from all generations beyond the F_1 are obtained by selfing. In generation $k + 2$, multiple progeny are assayed from each of two sublines for each of the isolated selfed lineages.

$(5/8) \cdot (1/2)^{k-1} = 5/2^{k+2}$. It follows that the probability that the two descendant sublines differ at least at one segregating locus is

$$P = 1 - \left(1 - \frac{5}{2^{k+2}}\right)^n \tag{9.32}$$

Rearranging and substituting observations for expectations, we obtain another estimator for the effective number of segregating factors,

$$\widehat{n}_{JT} = \frac{\ln(1 - \widehat{P})}{\ln\left(1 - \dfrac{5}{2^{k+2}}\right)} \tag{9.33}$$

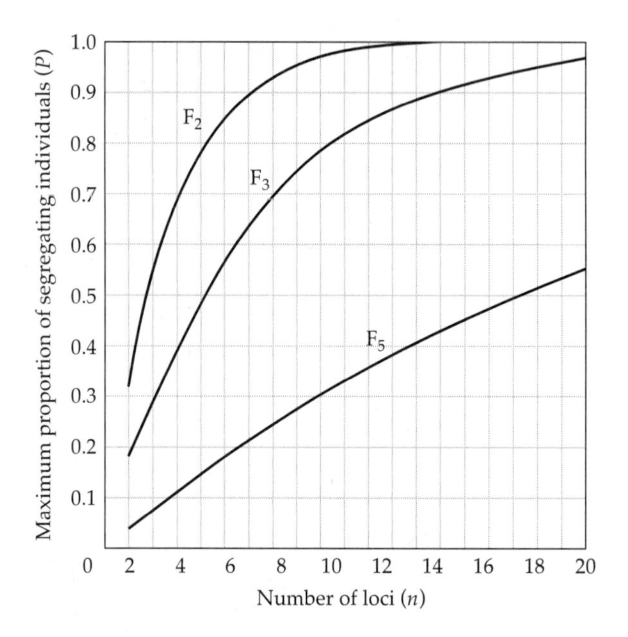

Figure 9.4 The expected fraction of individuals in the F_2, F_3, and F_5 generations detected as segregating by a genotype assay on their grandchildren, under the assumptions leading to Equation 9.32. (From Jinks and Towey 1976.)

There are several potential causes of bias in this estimator, most of which will lead to the usual underestimation of the actual number of loci. First, as in the inbred-backcross technique, the observed fraction of intrapair differences ($\widehat{P}$) is a matter of statistical power, increasing with the size of the assayed families, but decreasing with more stringent criteria for statistical significance. Second, the number of segregating factors will be depressed below the actual number of loci by linkage, the magnitude of this bias decreasing with increasing k (more opportunity for recombination). Hill and Avery (1978) consider this issue in some detail. Third, dominance, epistasis, and gametic phase disequilibrium can cause the expected phenotypes associated with different genotypes to be the same. Taking this masking problem into consideration, Jinks and Towey (1976) and Mulitze and Baker (1985a) have derived an alternative to Equation 9.33 that yields an upper (rather than lower) bound to n_e.

The relationship between P and n_{JT} varies rather substantially with the number of generations (k) prior to the genotype assay (Figure 9.4). If there are only a few effective factors, the greatest sensitivity is achieved (i.e., there is a strong response of P to n) when $k = 2$, provided linkage is unimportant. However, if there are more than five segregating loci, an F_2 assay is of little use, whereas an F_5 assay is quite sensitive.

Table 9.9 Minimum number of effective factors responsible for the differentiation of two lines of tobacco (*Nicotiana rustica*) as determined by genotype assay in progressive generations (F_k).

Grandparent Generation:	F_2	F_3	F_4	F_5	F_6
Flowering time					
$\widehat{P}$	0.200	0.450	0.314	0.233	0.306
$\widehat{n}_{JT}$	1	4	5	7	19
Final height					
$\widehat{P}$	0.400	0.300	0.257	0.200	0.306
$\widehat{n}_{JT}$	2	3	4	6	19

Source: Towey and Jinks 1977.
Note: n_{JT} is computed with Equation 9.33 and rounded up to the nearest unit.

Towey and Jinks (1977) applied the genotype assay to five generations of a cross between two lines of *Nicotiana rustica*. Even after six generations of selfing, there was no obvious decline in the fraction of descendant pairs exhibiting variation, for either flowering time or plant height (Table 9.9). Consequently, the estimates of n_{JT} for both characters increased approximately tenfold throughout the study.

The authors interpreted this increase to be a consequence of the gradual elimination of gametic phase disequilibrium through progressive rounds of recombination. If nothing else, this interpretation justifies our earlier discussion about the bias caused by linkage.

10

Inbreeding Depression

The previous chapter reviewed how the mean phenotypes of progeny from crosses *between* populations often exceed the average of the parents, a phenomenon known as heterosis. A related phenomenon arises *within populations* — inbred individuals are almost always less fit than progeny of nonrelatives. The decline in the mean phenotype with increasing homozygosity within populations, known as **inbreeding depression**, is often interpreted as heterosis-in-reverse. However, as will be seen below, there are some important distinctions between the genetic mechanisms contributing to inbreeding depression within populations and heterosis between populations.

The near universal existence of inbreeding depression bears importantly on many basic issues in evolutionary biology as well as on a number of practical issues in agriculture and conservation biology. For example, the deleterious consequences of self-fertilization are likely to be the leading selective forces responsible for the evolution of various aspects of mating systems in plants (Darwin 1876, Lande and Schemske 1985, Schemske and Lande 1985, Charlesworth and Charlesworth 1987, Uyenoyama 1993, Waller 1993), and of behavioral mechanisms for avoiding mating with close relatives in animals (Shields 1982, Thornhill 1993). The observations of maize breeders that crosses between inbred lines yield substantially more grain than the inbreds themselves (East 1908, Shull 1908, Jones 1918, Sprague 1983) has given rise to a situation in which corn farmers are now almost entirely reliant on seed-producing companies for hybrid seed (Figure 10.1). Finally, the loss of fitness due to the development of inbreeding depression in small populations is a major concern in endangered species management (Templeton and Read 1984, Lacy et al. 1993, Hedrick 1994, Lynch et al. 1995a,b, Lynch 1996).

It is widely appreciated that inbreeding depression is an inevitable consequence of dominance. When gene action is purely additive, the average phenotypic effects associated with alleles are independent of the genetic background. Hence, inbreeding depression cannot occur for characters with a purely additive genetic basis. With dominance, however, the average phenotypic effect of an allele changes with a change in genotype frequencies, even in the absence of allele frequency change, because allelic expression is a function of the genetic background.

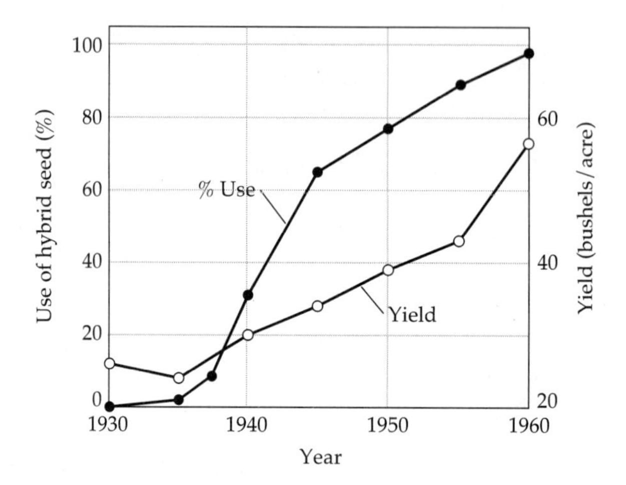

Figure 10.1 Historical change in the reliance on hybrid corn by United States farmers and the increase in mean annual harvest. Because usage of fertilizers, pesticides, and herbicides has shifted over this period, the substantial gain in yield is not solely attributable to heterosis. (From Sprague 1983.)

Because there are many types of dominance, this simple explanation for inbreeding depression by no means provides a complete understanding of the process. We start by showing how the two competing hypotheses on the genetic mechanism of inbreeding depression, partially recessive deleterious alleles vs. overdominance, lead to some very similar predictions that are in accordance with empirical observation, but also to some major differences that are less easy to resolve empirically. Second, we provide a brief outline of the basic statistical issues that arise in the analysis of inbreeding depression. Third, we review the large body of existing data, showing that inbreeding depression exists, at least to some degree, for essentially all characters in all populations of diploid organisms. We close by reviewing how molecular-marker analysis is starting to refine our understanding of the issues.

THE GENETIC BASIS OF INBREEDING DEPRESSION

In the absence of selection, inbreeding shifts the genotype frequencies in a population in a very simple way. Let f denote the inbreeding coefficient for the population, i.e., the probability that an individual carries two alleles that are identical by descent at a locus (Chapter 7). At any locus, a partially inbred population has a fraction $(1 - f)$ of noninbred individuals, whose genotype frequencies are in the Hardy-Weinberg proportions. The remaining proportion of the population

Table 10.1 Genotypic frequencies and fitnesses under the two dominance hypotheses for inbreeding depression.

Genotype	Frequency	Fitness: Partial Dominance	Fitness: Overdominance	Phenotype for Arbitrary Character
BB	$p^2(1-f)+pf$	1	$1-t$	$2a$
Bb	$2pq(1-f)$	$1-hs$	1	$(1+k)a$
bb	$q^2(1-f)+qf$	$1-s$	$1-s$	0

Note: Two alleles (*B*, *b*) are assumed to be present, with respective frequencies p and q.

that is inbred (f) consists entirely of homozygous classes, each of which has a frequency equal to the respective allele frequency (Table 10.1). With this information in hand, it is straightforward to derive quantitative expressions for the two mechanistic hypotheses for inbreeding depression.

The **dominance hypothesis** (Davenport 1908, Bruce 1910, Keeble and Pellew 1910, Jones 1917) argues that inbreeding depression is caused by the expression of deleterious recessive genes in homozygous individuals. (We retain the use of the term dominance to describe this hypothesis only for historical reasons. It is a misnomer in that the hypothesis focuses explicitly on partially to completely recessive genes). Consider a diallelic locus, where the frequency of the deleterious allele is q, and the fitnesses of the three genotypes are denoted as 1, $1-hs$, and $1-s$ (Table 10.1). Here, s measures the selection against homozygotes for the deleterious allele, and h is a measure of dominance, with $h = 0.5$ implying additivity and $0 < h < 0.5$ implying that the deleterious allele is partially recessive. The mean fitness in a population inbred to level f is

$$\overline{W}_f = \overline{W}_0 - fpqs(1-2h) \tag{10.1a}$$

where

$$\overline{W}_0 = 1 - 2pqsh - q^2 s \tag{10.1b}$$

is the mean fitness in the random-mating base population. Note that provided $h < 0.5$, $(1-2h)$ is necessarily positive. Thus, with partially recessive deleterious alleles, mean fitness is expected to decline linearly with increasing inbreeding coefficient f.

Unless the mutation rate is very high, deleterious alleles are expected to be maintained at low frequency by selection, so it can be assumed in Equation 10.1a that $p = 1 - q \simeq 1$, showing that the expected decline in fitness due to complete inbreeding at a locus is approximately $qs(1-2h)$. For a randomly mating population in selection-mutation balance, if u is the mutation rate from the beneficial to the deleterious allele and $u < h^2 s$, then $q \simeq u/(hs)$ (Haldane 1927). Thus, for large randomly mating populations, the decline in fitness resulting from complete inbreeding at a locus is approximately $u(1-2h)/h$. This result is independent of the intensity of selection at the locus (s) because of the inverse relationship between the equilibirum frequency of a deleterious allele and its selection coefficient.

A second potential explanation for inbreeding depression is referred to as the **overdominance hypothesis** (East 1908, Shull 1908, Hull 1946). The idea here is that something special about the heterozygous state causes increased vigor relative to both homozygotes. Letting s and t denote the proportional reduction in fitness of the two homozygotes relative to that of the heterozygote (Table 10.1),

$$\overline{W}_f = \overline{W}_0 - fpq(s+t) \tag{10.2a}$$

where

$$\overline{W}_0 = 1 - p^2t - q^2s \tag{10.2b}$$

As in the case of partial recessives, the overdominance hypothesis leads to the prediction that mean fitness will decline linearly with increasing f. However, contrary to the situation with partial recessives, the loss of fitness increases with the strength of selection maintaining the polymorphism in the random-mating population. In a large randomly mating population, heterozygote superiority leads to a balanced polymorphism with $p = s/(s+t)$ and $q = t/(s+t)$ (Haldane 1927). Thus, the term on the right of Equation 10.2a is $fst/(s+t)$. If, for example, $s = t$, the loss of fitness per locus under complete inbreeding is $s/2$.

These alternative hypotheses for inbreeding depression have extremely different evolutionary implications. Under the dominance hypothesis, inbreeding depression is an inevitable consequence of recurrent mutation at the genomic level, implying that much of the genetic variation within populations must be associated with the constant influx of deleterious alleles. Although selection removes some of these alleles each generation, mutation replaces them. Under the overdominance hypothesis, variation is maintained by selection favoring the heterozygous state at multiple loci. Here, variation is maintained even in the absence of mutation pressure.

Considerable uncertainty exists as to whether overdominance with respect to fitness is a common phenomenon. Only rarely has it been suggested by studies with molecular markers, and most of those studies are open to alternative interpretations (discussed below). Nevertheless, as cogently pointed out by Crow (1948, 1952), even if overdominance is quite rare, it warrants serious consideration as a contributing factor in inbreeding depression. To see why, consider the expected reduction in fitness under both hypotheses when gene frequencies are in equilibrium. Under the dominance hypothesis, the maximum inbreeding depression per locus, arising with very small h, is approximately u/h. Since u is likely to be on the order of 10^{-5} or smaller for most loci, and the evidence suggests that h is usually greater than 0.1 or so (discussed below), the per-locus inbreeding depression resulting from partial dominance is expected to be quite small. On

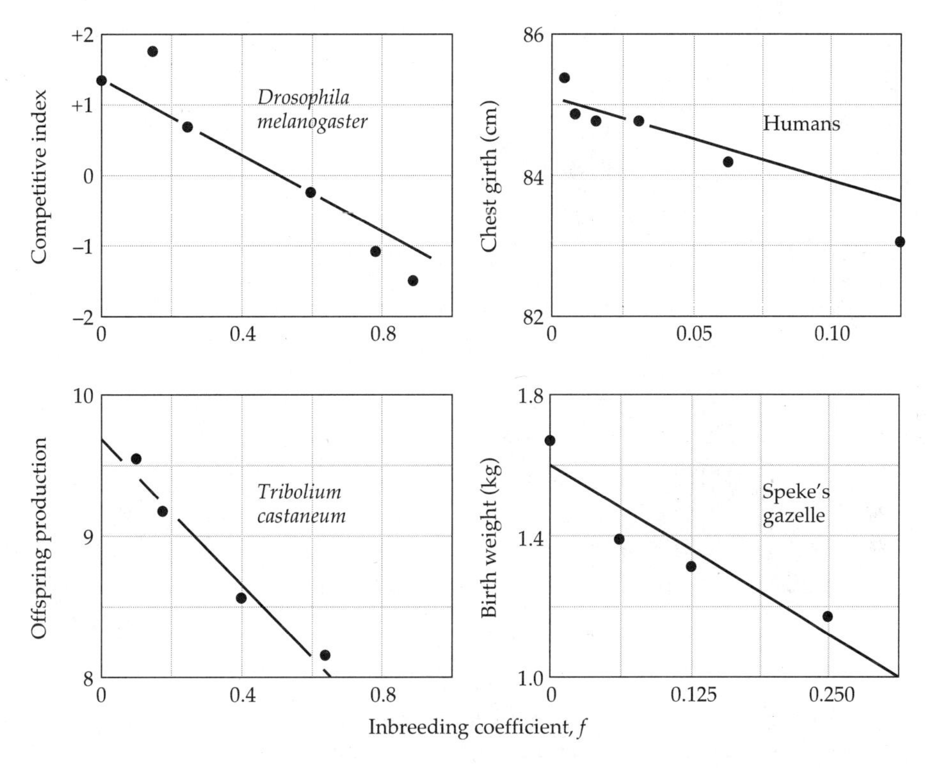

Figure 10.2 Change in mean phenotypes as a function of inbreeding. References: *Drosophila* (Latter and Robertson 1962); humans (Barrai et al. 1964); *Tribolium* (Rich et al. 1984); Speke's gazelle (Templeton and Read 1983).

the other hand, with overdominance, complete inbreeding leads to the loss of the fittest genotype, so the reduction in fitness is potentially quite large. Thus, even if overdominance is a rare phenomenon, only a few such loci need to exist for its contribution to inbreeding depression to rival that caused by a much larger number of loci displaying partial dominance.

The linear decline in the means of fitness-related characters with an increase in the inbreeding coefficient, observed in many sets of data (Figure 10.2), is consistent with both the partial dominance and overdominance hypotheses. There is, however, a major distinction between the two hypotheses with respect to the expected distribution of mean phenotypes among inbred lines. If overdominance is the major cause of inbreeding depression, all inbred lines must eventually perform below the mean of the randomly mating base population, because a pure line of the best-performing genotype (a heterozygote) cannot be attained. If, on the other hand, partial dominance is the major factor, it should be possible to produce

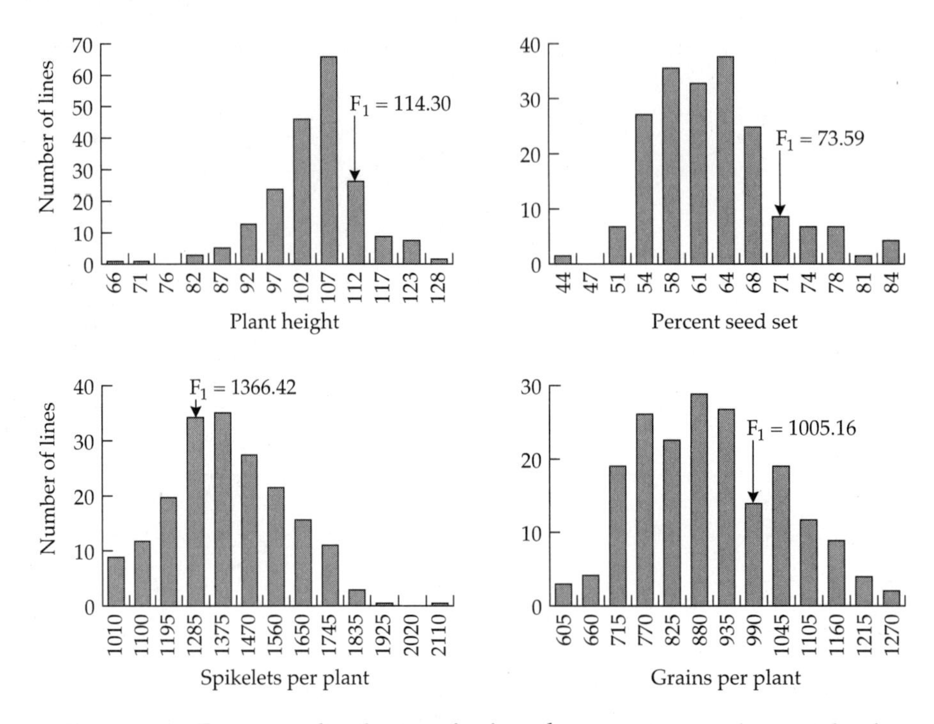

Figure 10.3 Frequency distributions for four characters in recombinant inbred lines of rice, compared to the mean of the F_1 progeny obtained from a cross between two homozygous lines. In all four cases, the F_1 performance exceeds that of both parental lines, and some individual inbred lines exceed the performance of the parents. The recombinant inbred lines were obtained by randomly sampling 194 individuals from the F_2 population, and taking each of them through six rounds of selfing and single-seed descent. (From Xiao et al. 1995.)

a pure inbred line that performs at least as well as the most outstanding member of the base population. If large numbers of loci contribute to the trait of interest, the probability of producing such a line may be quite low. Nevertheless, such lines have been obtained in several studies (Smith 1952, Williams 1959, Wienhues 1968, Busch et al. 1971, Pooni et al. 1994, Uddin et al. 1994) (Figure 10.3). These results raise serious questions about the necessity of relying upon commercial sources of hybrid seed in agricultural programs.

A More General Model

The preceding paragraphs have focused on the consequences of inbreeding for fitness. A more general account of the change in the mean of an arbitrary character under inbreeding will now be given. Recalling the genotypic frequencies for an inbred population (Table 10.1) and multiplying them by their respective genotypic values (scaled as in Chapter 4), a general expression for the mean genotypic value for a single diallelic locus is

$$\mu_f = (1-f)[\,p^2(2a) + 2pqa(1+k) + q^2(0)\,] + f[\,p(2a) + q(0)\,]$$
$$= \mu_0 - (2pqak)f \tag{10.3}$$

where $\mu_0 = 2ap(1+qk)$ is the mean genotypic value in the randomly mating base population. Summing over all loci, the total inbreeding depression is $2f\sum p_i q_i a_i k_i$. Recalling that the dominance genetic variance in a randomly mating population in gametic phase equilibrium is $\sum(2p_i q_i a_i k_i)^2$, it is clear that dominance variance is necessary for inbreeding depression to occur. However, since the sign of $a_i k_i$ may vary from locus to locus, it is possible for considerable canceling to occur among the effects at different loci, leading to negligible inbreeding depression in spite of substantial dominance genetic variance. In other words, significant inbreeding depression requires **directional dominance.**

Example 1. A large empirical study with the flour beetle *Tribolium castaneum* provides some perspective on this principle. López-Fanjul and Jódar (1977) derived 105 lines from a large base population and maintained them by single brother-sister matings for 8 generations (to $f = 0.785$). Despite the large sample sizes, the authors could find no evidence that inbreeding causes a shift in the mean rate of egg laying by virgin females at 33 or 28°C. Independent estimates of the heritabilities for these traits, obtained by full-sib correlation and daughter-mother regression, were 0.34 ± 0.02 and 0.33 ± 0.01 at 33°C, and 0.33 ± 0.02 and 0.26 ± 0.02 at 28°C. Recalling from Chapter 7 that heritabilities estimated from full-sib analysis are inflated by dominance genetic variance relative to those obtained by parent-offspring analysis, only for the second temperature is there any evidence of dominance genetic variance, and this is slight. Thus, the absence of inbreeding depression for rate of egg laying by virgins is not surprising. However, the study population was not immune to the effects of inbreeding, since two other traits, the rate of egg laying by fertilized females and egg viability, exhibited substantial declines with inbreeding.

Since inbreeding depression is a consequence of nonlinear interactions between gene effects, it stands to reason that epistasis may complicate matters. However, provided the base population is in gametic phase equilibrium, only epistasis involving dominance contributes to inbreeding depression within populations (Anderson and Kempthorne 1954, Bulmer 1980, Hill 1982a, Lynch 1991). This result arises because although inbreeding causes a change in genotypic frequencies within loci, in the absence of selection and gametic phase disequilibrium, it does not alter the gametic frequencies in the population. With this in mind, a general expression for inbreeding depression can be acquired as follows.

Letting $-\delta_1$ be the expected change in the mean caused by single-locus dominance effects (summed over all loci) under complete inbreeding, then $-f\delta_1$ is the expected change at inbreeding level f. The composite additive $\times$ dominance effect may also be altered under inbreeding. It depends only on the inbreeding at single loci and may be represented as $-f(\alpha\delta)$. The composite dominance $\times$ dominance effect depends on whether one or two loci are inbred. Assuming unlinked loci, the probabilities of these two situations are, respectively, $2f(1-f)$ and f^2. Thus, the shift in the mean through the alteration of dominance $\times$ dominance effects can be represented by $-2f(1-f)\delta_2^1$ in the first case and $-f^2\delta_2^2$ in the second. Summing up terms,

$$\mu_f = \mu_0 - f[\delta_1 + 2\delta_2^1 + (\alpha\delta)] - f^2(\delta_2^2 - 2\delta_2^1), \tag{10.4a}$$

or more succinctly,

$$\mu_f = \mu_0 - f\Delta_1 - f^2\Delta_2 + \cdots \tag{10.4b}$$

Thus, the expected mean phenotype under inbreeding can be written as a polynomial function of f, with the coefficients Δ_1 and Δ_2 being functions of multiple types of nonadditive gene action.

This relationship indicates that a net contribution of epistasis to inbreeding depression may sometimes be detected as a nonlinear relationship between the mean phenotype and level of inbreeding. Since the composite inbreeding effects can be positive or negative, a variety of forms of this relationship is possible. Nevertheless, the possibility that the epistatic effects involving different loci may cancel each other means that a lack of nonlinearity cannot be taken as definitive evidence for the absence of important dominance epistatic interactions between loci. Even in the absence of any canceling effect, large departures from linearity are unlikely unless epistasis is very pronounced for the simple reason that f^2 is small relative to f, especially with small f. Moreover, when a nonlinear response is observed, care must be taken to ensure that it is not simply due to the selective elimination of lines.

Finally, we note that although it is often stated that heterosis (the tendency for F_1 phenotypes to exceed the mean phenotypes of two parental lines) and inbreeding depression are the same phenomenon, this equivalency is not strictly correct. In Chapter 9, it was shown that heterosis is genetically equivalent to $(2\delta_1^c - \alpha_2^c)$, where δ_1^c and α_2^c are the composite dominance and composite additive $\times$ additive effects of genes in the two parental lines. On the other hand, under complete inbreeding within populations, the decline in the mean phenotype is defined by the sum $[\delta_1 + (\alpha\delta) + \delta_2^2]$, assuming the base population is in gametic phase equilibrium. Thus, dominance is a factor in both heterosis and inbreeding depression, but it is not a necessary condition for heterosis, which can arise entirely as function of additive $\times$ additive epistasis. In addition, inbreeding depression, but not F_1 heterosis, is a function of additive $\times$ dominance and dominance $\times$ dominance interactions. These differences in the genetic underpinnings of heterosis and inbreeding depression are a consequence of the extreme degree of gametic phase disequilibrium that exists in the first generation of a line cross. Further complexities arise when the F_1 progeny of a line cross are subsequently selfed (Lynch 1991).

METHODOLOGICAL CONSIDERATIONS

A number of difficulties arise in attempts to test for inbreeding depression. Some of these are associated with the selective consequences of the inbreeding depression itself. In humans, for example, there is evidence that consanguineous couples, whose early offspring die from the expression of lethal recessives, compensate by reproducing until viable replacements have been born (Schull and Neel 1972). Some plants may behave in a similar manner by selective abortion of embryos (Willson and Burley 1983). Lack of knowledge of such compensation can lead to underestimates of the deleterious consequences of inbreeding. Keeping these difficulties in mind, we will now consider the statistical aspects of two common approaches to quantifying inbreeding depression. These matters are taken up in more detail in Lynch (1988a).

Before proceeding, a brief introduction to the temporal dynamics of the inbreeding coefficient f under regular systems of mating is necessary. The general theory is covered elsewhere (Crow and Kimura 1970, Hartl and Clark 1989), and since the vast majority of studies on inbreeding depression involve either self-fertilization or full-sib mating, we simply give the results for these special cases. Starting from a random-mating base population at time 0, the average inbreeding coefficient at a locus after t generations of self-fertilization in the absence of selection is

$$f(t) = 1 - \left(\frac{1}{2}\right)^t \tag{10.5a}$$

The quantity $[1-f(t)] = (1/2)^t$ is equivalent to the fraction of the heterozygosity in the base population that is still present after t generations of selfing. Thus, a single generation of selfing reduces the number of heterozygous loci within individuals by 50%. Thereafter, the heterozygosity declines exponentially towards zero, such that only 1.6% of the original heterozygosity remains after six generations of selfing. With full-sib mated lines, the inbreeding coefficient must be computed with the recurrence equation,

$$f(t) = \frac{f(t-1)}{2} + \frac{f(t-2)}{4} + \frac{1}{4} \tag{10.5b}$$

letting $f(-1) = f(0) = 0$. Thus, under full-sib mating, the first generation of inbred progeny has $f(1) = 0.25$, i.e, the heterozygosity within individuals is reduced by 25% relative to that in the random-mating base population. The inbreeding coefficient then progressively approaches one, although more slowly than in the case of selfing.

Single-generation Analysis

A common short-term test of inbreeding depression involves the comparison of the mean phenotypes of offspring from random matings with those from a specific class of consanguineous mating. In any such analysis, both types of individuals should be raised simultaneously in a random design to eliminate the possibility that the differences in means are a product of the environment. Ideally, the offspring of both types of matings should be derived from several mothers to minimize the importance of maternal effects, and all mothers certainly should be derived from the same base population.

With such an experimental design, an approximate t test can be constructed for the null hypothesis of no inbreeding depression. Here we take the null model to be one of purely additive gene action. Consider the situation in which n progeny are assayed from each of L independent families, both in the control and in the inbred population. Under the null hypothesis of no inbreeding depression, the difference between the observed mean phenotypes of noninbred and inbred offspring ($\Delta\bar{z} = \bar{z}_O - \bar{z}_I$) has expectation zero, and the observed difference must be evaluated against its sampling variance, the sum of the variances of $\bar{z}_O$ and $\bar{z}_I$. Several factors contribute to this variance, as can be seen by referring to the definition of the sample mean

$$\bar{z} = \frac{1}{Ln} \sum_{i=1}^{L} \sum_{j=1}^{n} (A_{i\cdot} + a_{ij} + E_{i\cdot} + e_{ij})$$

where $A_{i\cdot}$ is the mean genotypic value associated with the ith family, a_{ij} is the deviation from that value for the jth member of the family, $E_{i\cdot}$ is the maternal effect associated with the ith family, and e_{ij} is the residual environmental effect on the ijth individual.

We start by considering the control. First, the variance in the control mean caused by environmental effects specific to individuals is $\sigma_e^2/(Ln)$ because all such effects are distributed independently among the Ln individuals. Second, the variance caused by maternal (or general environmental) effects is σ_E^2/L; it is only divided by L because L mothers contribute to the control mean. Third, because the segregational variance within families in the random-mating base population is $\sigma^2(a_{ij}) = \sigma_A^2/2$, and the effects of such residual variation are random with respect to individuals, the contribution of the within-family genetic variance to the variance of the mean is $\sigma_A^2/(2Ln)$. (To ease the passage through this difficult area, we give this and a few other results without proof.) Finally, the among-family variance, $\sigma^2(A_{i\cdot})$, is also $\sigma_A^2/2$, and it contributes $\sigma_A^2/(2L)$ to the sampling variance of the control mean. Summing up terms, the variance of the control mean phenotype is

$$\sigma^2(\bar{z}_O) = \frac{1}{L}\left[\frac{1}{2}\left(1+\frac{1}{n}\right)\sigma_A^2 + \sigma_E^2 + \frac{\sigma_e^2}{n}\right] \tag{10.6a}$$

Now consider the situation for a sample of progeny derived by selfing L mothers with n progeny sampled per mother. The sampling variance of the mean resulting from general and specific environmental effects is exactly the same as in the control. In the first generation of selfing, the within-family segregational variance is also identical to that within the control, $\sigma_A^2/2$. However, because of inbreeding, the variance among families is σ_A^2, twice that in the control families, where σ_A^2 is still defined as the genetic variance in the base (control) population. Thus, the expected variance of the mean of the sample of selfed progeny is

$$\sigma^2(\bar{z}_S) = \frac{1}{L}\left[\left(1 + \frac{1}{2n}\right)\sigma_A^2 + \sigma_E^2 + \frac{\sigma_e^2}{n}\right] \tag{10.6b}$$

The situation for full-sib mating is a little more complicated, but assuming that all progeny within lines are derived from a single brother-sister mating,

$$\sigma^2(\bar{z}_{FS}) = \frac{1}{L}\left[\left(\frac{7}{8} + \frac{3}{8n}\right)\sigma_A^2 + \sigma_E^2 + \frac{\sigma_e^2}{n}\right] \tag{10.6c}$$

The above expressions are at slight variance with those in Lynch (1988a) and appear to be more accurate.

Although these formulae give exact expectations of the variances of control and inbred line means under the additive model, they are difficult to implement unless one has prior information on the components of variance in the base population. However, the basic structures of the formulae yield a very useful result. Note that for the cases of selfing and sib-mating, $\sigma^2(\Delta\bar{z}_S) = \sigma^2(\bar{z}_o) + \sigma^2(\bar{z}_S)$ and $\sigma^2(\Delta\bar{z}_{FS}) = \sigma^2(\bar{z}_o) + \sigma^2(\bar{z}_{FS})$, respectively. In both of these cases, provided the sample size within families (n) is at least two, then $\sigma^2(\Delta\bar{z}) \leq 2\sigma^2(z_O)/L$, where $\sigma^2(z_O)$ is the phenotypic variance within the control line. Thus, a conservative test for inbreeding depression based on a single generation of consanguineous mating employs the test statistic

$$t = \frac{|\Delta\bar{z}|}{\text{SD}(z_O)\sqrt{2/L}} \tag{10.7}$$

where SD(z_O) is the observed phenotypic standard deviation in the random-mating population. Sampling distributions of means are usually approximately normally distributed, so t may be treated as t-distributed with $L - 1$ degrees of freedom.

In the case of self-compatible plants that produce multiple flowers, there is a simple way to further increase the power of a test of inbreeding depression. For any pair of parent plants (A and B), both reciprocal outcrosses (A $\times$ B and B $\times$ A) and two inbreds (A $\times$ A and B $\times$ B) can be produced. Because the two parents contribute equal numbers of genes to both inbred and outbred progeny, variance from general (maternal) environmental effects and parent sampling do not contribute to $\sigma^2(\Delta\bar{z})$ in this case, and for any pair of parent individuals, the test statistic

$$\Delta\bar{z}_{A,B} = \frac{(\bar{z}_{AA} + \bar{z}_{BB}) - (\bar{z}_{AB} + \bar{z}_{BA})}{2} \tag{10.8a}$$

has an expected value equal to zero under the null hypothesis of no inbreeding depression. If n replicates are assayed within each of the four groups of progeny, the expected sampling variance of $\Delta\bar{z}_{A,B}$ is

$$\sigma^2(\Delta\bar{z}_{A,B}) = \frac{(\sigma_A^2/2) + \sigma_c^2}{n} \tag{10.8b}$$

Because the numerator of this expression is less than $\sigma^2(z_O)$, a conservative test for inbreeding depression associated with any pair of parents is provided by

$$t = \frac{|\Delta\bar{z}_{A,B}|}{\text{SD}(z_O)/\sqrt{n}} \tag{10.9}$$

where $\text{SD}(z_O)$ is again the phenotypic standard deviation of outcrossed individuals.

Multigenerational Analysis

A common long-term approach to quantifying inbreeding depression is to regress the mean phenotype on the inbreeding coefficient (Figure 10.2). In studies of this sort, the data usually represent progressively inbred generations derived from the same population, a protocol that introduces a series of statistical problems. First, when the different classes of inbreeding are assayed in different generations (the usual case in animals), the possibility arises that any trend in the mean may be caused by a shift in the environment. Second, the usual assumptions underlying hypothesis testing in regression theory are violated in at least two ways: (1) since the means are based upon individuals that are descendants of each other, the data are not independent, and (2) the sampling variance of the means varies with f because of the loss of genetic variance with inbreeding. The problem of nonindependence of data is a particularly serious one, since it diminishes the effective degrees of freedom in an analysis. For example, once a population of selfers is almost completely inbred, the subsequent generations are no longer free to vary genetically except by mutation. Nonindependence of data can also cause spurious nonlinearities in the apparent response to inbreeding—if one data point lies above the regression line, the preceding and subsequent ones are also likely to.

A partial resolution of the nonindependence problem is given below. First, however, some attention needs to be given to the correction of data for environmental shifts between generations. Plant breeders have been able to avoid the statistical complexities of this issue by storing seed from progressive generations of inbreeding and then growing representatives of all generations simultaneously in a randomized design (Russell et al. 1963, Hallauer and Sears 1973, Cornelius and Dudley 1974). Even here, it is assumed implicitly that seed storage time does not influence performance and that general environmental effects experienced by

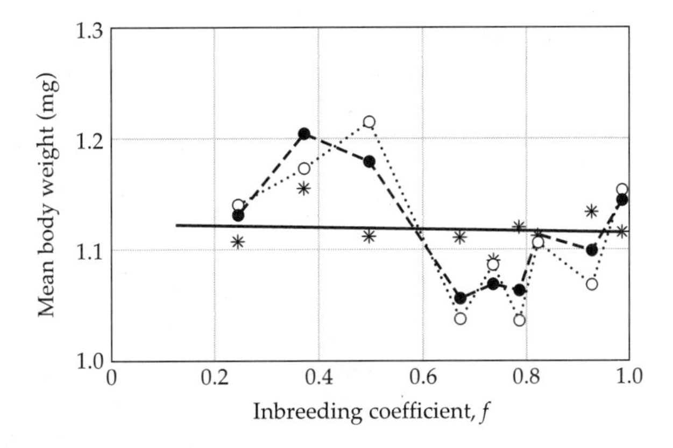

Figure 10.4 Observed phenotypic means in inbred (•) and control (○) lines of *Drosophila melanogaster*, and corrected values for the inbred lines (∗) obtained by use of Equation 10.12. The partial regression is represented by the solid line. (From Kidwell and Kidwell 1966.)

the parents are not transmitted to the progeny. With most animals, embryo storage is either not currently reliable or not economically feasible, so there is need for a statistical means of correcting the data. The question of interest is whether a trend in the inbred-line means is influenced by a temporal shift in the environment. The issue is not trivial as can be seen from the striking parallel directional trend in control and full-sib mated lines of *Drosophila melanogaster* shown in Figure 10.4.

A solution to the environmental trend problem was promoted by Muir (1986a,b), who suggested the use of a parallel control as a means of assaying changes in the environment. There are two important considerations in the choice of a control. First, it is essential that the temporal phenotypic changes in the control are entirely attributable to environmental causes. This condition will essentially hold if clones or highly inbred lines are relied upon. A random-bred base population may also serve as an adequate control, provided the character of interest is not modified by selection during the course of the experiment and provided the population is large enough that significant genetic drift is unlikely to occur. Second, given a choice of control lines, the one that provides the strongest signal of the environment, i.e., explains a maximum amount of the variance in the inbred line means, is most desirable.

We start by considering the mean phenotypes of the control (C) and inbred (I) lines at generation t to be functions of general environmental effects common to both of them (E), special environmental effects unique to each of them (e_C and e_I), and genetic change confined to the inbred population, $\Delta\mu_G(t)$,

$$\bar{z}_I(t) = \mu_I(0) + E(t) + e_I(t) + \Delta\mu_G(t) \tag{10.10a}$$

$$\bar{z}_C(t) = \mu_C(0) + E(t) + e_C(t) \tag{10.10b}$$

Since the general environmental effects, $E(t)$, are the only common components of the inbred and control line means, a partial regression of the observed inbred line means, $\bar{z}_I(t)$, on the observed control line means, $\bar{z}_C(t)$, and the inbreeding coefficient, $f(t)$, provides a way of factoring out any general trend of the environment,

$$\bar{z}_I(t) = a + b\bar{z}_C(t) + If(t) + e(t) \tag{10.11}$$

where I is the estimated inbreeding depression (i.e., the expected difference in mean phenotypes of noninbred and completely inbred individuals), and $e(t)$ is the deviation of the tth generation mean from the multiple regression. Applying Equation 10.11, after removing any environmental trend, the corrected means for the inbred lines become

$$\bar{z}_I^*(t) = \bar{z}_I(t) - b[\bar{z}_C(t) - \bar{z}_C] \tag{10.12}$$

where $\bar{z}_C$ is the mean phenotype of the control lines over all generations. The estimated inbreeding depression (I) is equivalent to the regression of the $\bar{z}_I^*(t)$ on $f(t)$. Figure 10.4 shows a rather striking example of how the application of Muir's approach can overcome a trend obscured by environmental factors.

We finally return to the problem of hypothesis testing, assuming now that the means have been corrected adequately for general environmental trends prior to analysis. Because it ignores the nonindependence of data, ordinary least-squares regression of $\bar{z}_I^*(t)$ on $f(t)$ leads to downwardly biased estimates of the standard error of I, often by a factor of three or four (Lynch 1988a). An expression for the sampling variance of I, which fully accounts for the correlational structure of the data, under the null hypothesis of a neutral character with an additive genetic basis is worked out in Lynch (1988a). The solution, portrayed graphically in Figure 10.5, assumes that the regression is performed on a progressive series of inbred lines (e.g., self-fertilization, full-sib mating, or first-cousin mating), starting with $f = 0$ and proceeding for k generations to a final level of inbreeding $f(k)$. The plotted values are minimum estimates of the sampling variance of I, since it is assumed that the variance in the environment makes no contribution to the sampling variance of the means. The sampling variance of I depends primarily on the additive genetic variance in the base population, the number of inbred families, and the level of inbreeding in the final generation. Since the sampling variance of I declines with increasing L and $f(k)$, it is clear that for a fixed amount of resources, the smallest unit of inbreeding (selfing or full-sib mating) should be employed while maximizing the number of lines.

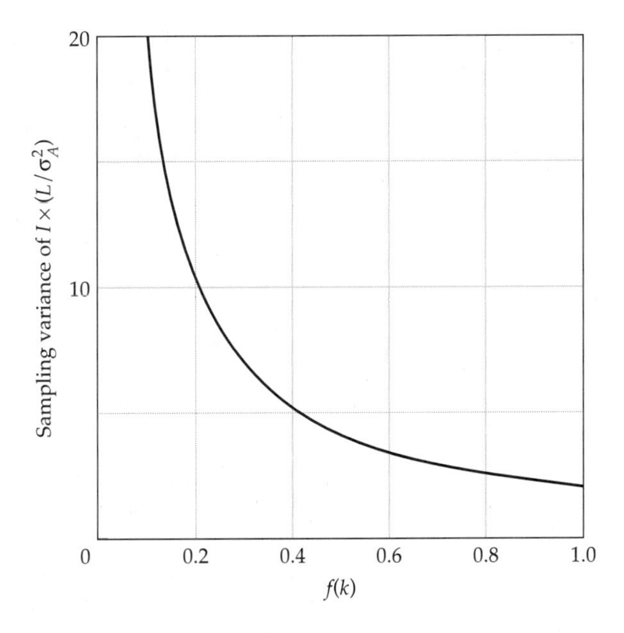

Figure 10.5 The minimum sampling variance of the regression coefficient of consecutive line means on their respective inbreeding coefficients, under the assumption of purely additive gene action and ignoring environmental effects. $f(k)$ is the inbreeding coefficient in the final generation. To obtain the actual sampling variance of I, the points on the ordinate must be multiplied by σ_A^2/L, the ratio of the additive genetic variance in the base population to the number of inbred lines. (From Lynch 1988a.)

Many of the preceding statistical problems can be avoided when data are available for contemporaneous groups of individuals inbred to various degrees. Such is typically the case in the analysis of human populations where pedigrees are known, and the same can be accomplished in experiments that simultaneously mate various classes of relatives and assay their progeny in a common environment. In both cases, provided the individuals with different levels of f are unrelated, the problem of nonindependent data is eliminated, and provided all individuals are assayed contemporaneously, the need for a temporal control is removed. Ordinary least-squares regression of the group mean phenotypes on f then provides a simple approximation of I and its standard error.

Example 2. Consider an experimental design in which the means of $L = 10$ full-sib mated lines are assayed from generation 0 with $f = 0$ to generation

9 with $f(k) = 0.859$ (obtained using Equation 10.5b). Reading off Figure 10.5 at $f(k) = 0.859$, we find the point on the ordinate to be 2.5. The expected sampling variance of the slope I under the null model of no inbreeding depression is obtained by multiplying 2.5 by σ_A^2/L, which gives $\sigma_A^2/4$. If an estimate of σ_A^2 is available, then in this case, two standard errors of the slope is estimated by $\sqrt{\text{Var}(A)}$. Since our treatment ignores environmental sources of variance, it is clear that with this design any regression coefficient whose absolute value is less than the square root of the additive genetic variance in the base population must be considered consistent with the null model of no inbreeding depression.

Ritland's Method

Most empirical attempts to measure inbreeding depression involve assays of individuals in controlled environments. Because lab conditions often deviate substantially from the situation in nature, one is then left wondering how generalizable the results are to the field situation. To eliminate this problem, Ritland (1990a,b) proposed a technique for partially selfing populations of plants that involves essentially no disturbance of individuals in nature and requires no direct estimation of individual fitness. Applying neutral molecular markers to progeny arrays, it is possible to estimate the fraction of seed that adults produce by self-fertilization (ϕ), as well as to survey the change in genotype frequencies in a population within and between generations. Genotype frequencies change across generations are a function of the degree of selfing in the parents, while the within-generation changes are a function of genotype-specific fitnesses.

Ritland (1990a,b) suggested several ways in which marker information can be exploited to infer indirectly the fitness consequences of inbreeding. Here, we simply point out the simplest situation, which arises when a population has attained an equilibrium state of inbreeding, i.e., a balance between the production of excess homozygosity by selfing and its loss by selection. From genotypic assays of neutral molecular markers in adult individuals, the inbreeding coefficient (f) of surviving individuals can be computed as $f = 1 - [(\text{observed heterozygosity})/(2pq)]$ (obtained using the principles outlined in Table 10.1). The ratio of fitnesses of selfed to outcrossed individuals is then estimated by

$$w = \frac{2(1-\phi)f}{\phi(1-f)} \tag{10.13}$$

Note that when $\phi > 0$ and $f = 0$ (i.e., the adult population is in Hardy-Weinberg equilibrium), $w = 0$, implying that selfed progeny have zero fitness. More general estimators that allow for generational changes in ϕ and f, which appear to be common (Dole and Ritland 1993), are provided in Ritland (1990a,b).

Applications of Equation 10.13 to partially selfing plant populations have generally yielded estimates of w that are slightly lower than those obtained by direct observations of the performance of selfed and outcrossed progeny in experimental populations (Eckert and Barrett 1994, Kohn and Biardi 1995, Schultz and Ganders 1996). Although violations in the assumptions of Ritland's model (such as an absence of biparental inbreeding and an absence of linkage between marker and fitness loci) can lead to biased estimates of w, the bias does not generally appear to be large. Thus, the empirical results tentatively suggest that the inbreeding depression observed in manipulated populations may generally be lower than that expressed in natural settings. This difference may occur because manipulative studies often fail to fully account for all components of fitness (such as seedling survival) or because the effects of deleterious genes are ameliorated in more benign environments (discussed below).

Epistasis and Inbreeding Depression

An unresolved issue is the extent to which epistasis is involved in inbreeding depression. As noted above, a nonlinear relationship between the mean phenotype and the inbreeding coefficient is an indicator of the presence of epistasis involving dominance effects. Least-squares quadratic regressions are frequently performed to test for such nonlinearities, but these are saddled with all of the statistical problems discussed above, most notably the extreme nonindependence of data at high levels of f, precisely where nonlinearities are most likely to show up.

A simple way to test for nonlinearity, which avoids the pitfalls of regression, is to compare the change in mean phenotype (per increment in f) between two low levels of f and that between two high levels of f. Provided the two ranges of f are nonoverlapping, the two observed *changes* are statistically independent, even if the individuals at all four points in time are related. Letting the four observed mean phenotypes, in order of increasing f, be $\bar{z}_1$, $\bar{z}_2$, $\bar{z}_3$, and $\bar{z}_4$, a measure of nonlinearity is then given by

$$\Delta I = \frac{\bar{z}_2 - \bar{z}_1}{\Delta f_L} - \frac{\bar{z}_4 - \bar{z}_3}{\Delta f_H} \tag{10.14a}$$

where $\Delta f_L = f_2 - f_1$, and $\Delta f_H = f_4 - f_3$. A conservative estimate of the sampling variance of ΔI is given by

$$\text{Var}(\Delta I) = \frac{[\text{SE}(\bar{z}_2)]^2 + [\text{SE}(\bar{z}_1)]^2}{(\Delta f_L)^2} + \frac{[\text{SE}(\bar{z}_4)]^2 + [\text{SE}(\bar{z}_3)]^2}{(\Delta f_H)^2} \tag{10.14b}$$

A test statistic for nonlinearity is then

$$t = \frac{|\Delta I|}{\sqrt{\text{Var}(\Delta I)}} \tag{10.14c}$$

which under the null hypothesis of linearity should be t-distributed with degrees of freedom equal to the number of inbred lines in the analysis. To ensure that ΔI

is not a function of the differential extinction of lines, only the lines surviving to contribute to $\bar{z}_4$ should be used in such an analysis.

Although it can only detect epistatic effects involving dominance, this test is one of the only ways that we currently have to quantify directional epistasis within populations, short of employing molecular markers. Willis (1993) used a very similar approach to test for epistasis for life-history characters in the monkey flower (*Mimulus guttatus*). Although he did not correct for line loss, he found very little evidence for epistasis.

Variance in Inbreeding Depression

Evolutionary biologists interested in the origins of diverse mating systems, particularly in plants, have reason to be concerned with the potential for variance in inbreeding depression among members of the same population (Holsinger 1988, Johnston and Schoen 1994, Uyenoyama et al. 1994, Schultz and Willis 1995). Such variation would seem to be necessary to foster the evolution of alternative forms of mating. To obtain information on this matter, plant population biologists often use ratios of fitness of selfed progeny to outcrossed progeny as a measure of inbreeding depression. This practice raises some statistical problems in that the ratio of two estimates is biased. Some of the issues are discussed by Johnston and Schoen (1994), but some of the formulae in their paper are incorrect. Using Equation A1.19a, an unbiased estimate of the performance of selfed relative to outcrossed progeny is given by

$$w_i = \frac{\overline{W}_{Si}}{\overline{W}_{Oi}} \left[\frac{1}{1 + [\sigma^2(W_{Oi})/(n_i \overline{W}_{Oi}^2)]} \right] \tag{10.15}$$

where $\overline{W}_{Si}$ and $\overline{W}_{Oi}$ are the observed mean fitnesses of selfed and outcrossed progeny derived from individual i, $\sigma^2(W_{Oi})$ is the variance in fitness of outcrossed progeny, and n_i is the number of outcrossed progeny assayed, all for the ith individual.

Using measures such as Equation 10.15 to quantify variance in inbreeding depression among individuals raises a number of difficult and unresolved issues. A central problem is that inbreeding depression is not just a property of the individual, but of the individual's prospective mates as well. It is straightforward enough to estimate an individual's fitness through selfing, but what about the situation with species with separate sexes? An individual's sibs will generally differ with respect to fitness, so the fitness of progeny from full-sib matings will depend on which sibs are employed as mates. The situation is even more extreme when one considers the fitness of individuals produced through outcrossing. Ideally, one would like an estimate of the fitness of outcrossed progeny averaged over all potential mates, but with most species (other than plants), only a small number of matings per individual are possible. Presumably, variance in inbreeding depression can be estimated using ANOVA approaches, treating differences between replicate pairs of inbred and outcrossed matings within lineages as the units of observation, but the procedures remain to be worked out.

THE EVIDENCE

Although few of the existing studies of inbreeding depression have fully accounted for all of the difficulties pointed out above, the aggregate of evidence for inbreeding depression is overwhelming. While substantial variation of inbreeding depression exists among species (and among characters within species), almost all organisms exhibit it to some degree. Here we only summarize some of the better-documented cases. An extensive survey of the early literature is available in Wright (1978), and recent reviews may be found in Shields (1982), Charlesworth and Charlesworth (1987), Thornhill (1993), and Husband and Schemske (1996).

More and better data on the phenotypic consequences of inbreeding are available for maize than for any other organism. Hallauer and Miranda (1981) review the evidence, which was recognized as early as 1876 by Darwin. There have been some very well conceived experiments involving prolonged selfing and full-sib mating in lines derived from a genetically diverse base population (Sing et al. 1967, Hallauer and Sears 1973, Cornelius and Dudley 1974, Good and Hallauer 1977, Lamkey and Smith 1987, Benson and Hallauer 1994). The experiments are very large (involving up to 250 independent lines), and the potential influence of temporal changes in the environment has been minimized by the simultaneous analysis of stored seed. Almost without exception, vegetative, reproductive, and physiological characters exhibit significant shifts in the mean phenotype with inbreeding. Cases have arisen in which the regressions of $\bar{z}$ on f appear to be nonlinear (Hallauer and Sears 1973, Good and Hallauer 1977), but in all cases the departure from linearity is small. Two characters that give no evidence of nonlinearity are total grain yield and plant height (Figure 10.6). Starting from a genetically diverse base population, complete inbreeding results in an approximately 65% decline in yield and an approximately 25% decline in plant height.

Several independent investigations of inbreeding depression have been performed with laboratory stocks of *Drosophila* (Table 10.2). Although substantial variation exists among the results from different studies, a pattern emerges. Primary fitness characters such as viability, fertility, and egg production tend to exhibit very high levels of inbreeding depression (averaging approximately 50%), while morphological characters (bristle numbers, body weight and length), which are perhaps more remotely related to fitness, change by only a few percent, if at all. The latter traits are known to exhibit substantial levels of additive genetic variance, while the fitness characters tend to have lower heritabilities (Mousseau and Roff 1987). Thus, in *Drosophila* there appears to be a major difference in the

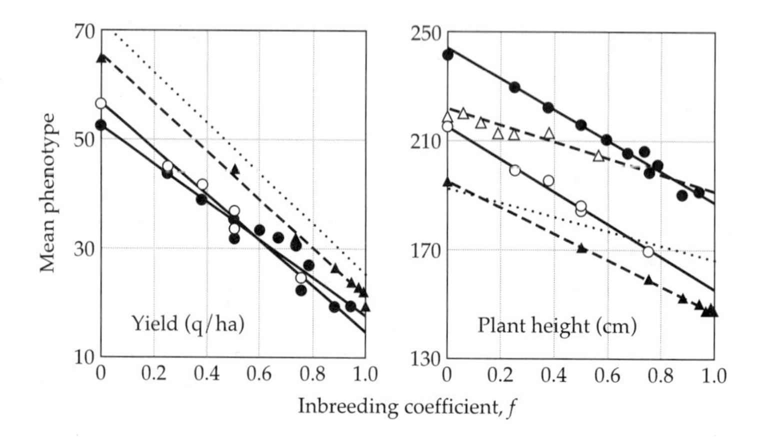

Figure 10.6 The response of mean grain yield and plant height to inbreeding in maize. Data are from: (•, ○) Cornelius and Dudley (1974); (closed triangles) Hallauer and Sears (1973); (open triangles) Sing et al. (1967); (· · ·) Good and Hallauer (1977) — only the regression line is available. The two studies of Cornelius and Dudley are for the same lines grown in different years. The variation in intercepts is presumably due to differences among base populations as well as among environments in which the experiments were performed.

way the genetic variance for morphological and fitness characters is partitioned — mostly additive for the former, mostly dominant for the latter.

Selection theory helps explain why the additive genetic variance for fitness should be low and why dominance should be directional for fitness-related characters. Alleles with favorable effects on fitness should move rapidly towards fixation, regardless of their degree of dominance, and dominant alleles with deleterious effects will be eliminated rapidly. However, deleterious recessive alleles will be maintained at low frequencies by mutation pressure. For characters only weakly related to fitness or under stabilizing selection for an intermediate optimum, directional dominance may be less pronounced since mutations that cause a shift in the mean in either direction will be selectively equivalent.

Several large surveys provide firm empirical justification for the incest taboos that exist in humans. Rarely is it possible to obtain data for more extreme situations than first- and second-cousin marriages, but by linear extrapolation the data are sufficient to demonstrate that more extreme inbreeding would lead to substantial depression in body size and IQ (Table 10.3). The consequences of inbreeding for juvenile mortality and the incidence of congenital effects in man are well known and are examined in the next section from a somewhat different perspective. One precautionary note is in order here. Analyses based on inbreeding depression that rely on natural mating assemblages (as is always true in human

Table 10.2 A survey of the inbreeding depression observed in laboratory populations of *Drosophila*.

Character	I. D.	Reference
Competitive ability	0.84	Latter et al. 1995
	0.97	Latter and Sved 1994
Egg-to-adult viability	0.57	Garcia et al. 1994
	0.44	Mackay 1985a
	0.66*	Malogolowkin-Cohen et al. 1964
	0.48*	Dobzhansky et al. 1963
	0.06	Tantaway and Reeve 1956
Female fertility	0.81	Mackay 1985a
	0.18	Tantaway and Reeve 1956
	0.35	Hollingsworth and Maynard Smith 1955
Female rate of reproduction	0.32	Latter et al. 1995
	0.56	Mackay 1985a
	0.96	Hollingsworth and Maynard Smith 1955
	0.57	Marinkovic 1967
Male mating ability	0.52*	Hughes 1995
	0.92	Partridge et al. 1985
	0.76	Sharp 1984
Male longevity	0.18*	Hughes 1995
Male fertility	0.00*	Hughes 1995
	0.22*	Dobzhansky and Spassky 1963
Male weight	0.07*	Hughes 1995
	0.10	Mackay 1985a
Female weight	−0.10	Kidwell and Kidwell 1966
Abdominal bristle number	0.05	Mackay 1985a
	0.06	Kidwell and Kidwell 1966
	0.00	Rasmuson 1952
Sternopleural bristle number	−0.01	Mackay 1985a
	0.00	Rasmuson 1952
Wing length	0.03	Tantaway 1957
	0.01	Tantaway and Reeve 1956
Thorax length	0.02	Tantaway 1957

Source: All data are for *D. melanogaster*, except for *D. subobscura* (Hollingsworth and Maynard Smith 1955), *D. pseudoobscura* (Dobzhansky and Spassky 1963, Dobzhansky et al. 1963, and Marinkovic 1967), and *D. willistoni* (Malogolowkin-Cohen et al. 1964).

Note: Records are given as I.D. $= 1 - (\bar{z}_I/\bar{z}_O)$, where $\bar{z}_O$ and $\bar{z}_I$ are the means of the random mating base and the completely inbred population (obtained by linear extrapolation). Results marked with an asterisk were obtained from studies involving only one or two chromosomes; in these cases, extrapolation to the entire genome was done by assuming that each major chromosome arm constitutes 20% of the genome, and that the effects are multiplicative across chromosomes. Negative values imply an increase in character value with inbreeding.

Table 10.3 Decrease in the mean expected upon complete inbreeding in humans, equal to the mean for noninbred individuals minus the expectation at $f = 1$ (obtained by linear extrapolation).

Trait	Site	I	Reference
Birth weight (kg)	Japan	5.4*	Morton 1958
	United States	1.7	Slatis and Hoene 1961
Adult height (cm)	Hutterites, U. S.	56	Barrai et al. 1964
	Italy	3	Mange 1964
	Japan	20	Schull 1962
		21	Neel et al. 1970
IQ	Japan	42	Neel et al. 1970
		43	Kudo et al. 1972
		73	Schull and Neel 1965
	United States	42	Slatis and Hoene 1961
Prereproductive survival (%)	Global	70	Bittles and Neel 1994

*This value is obviously too high since it gives a birth weight less than zero.

studies) run the potential risk that progeny with different levels of f are products of genotypically different groups of parents. If parents that tend to inbreed also tend to be genetically low on the fitness scale, the apparent level of inbreeding depression in the progeny may be exaggerated substantially.

Extensive reviews exist on the deleterious consequences of inbreeding in domesticated mammals: beef cattle (Dinkel et al. 1968), dairy cattle (Turton 1981), dogs (Scott and Fuller 1965), horses (Cothran et al. 1986), sheep (Lamberson and Thomas 1984, Wiener et al. 1992a–c), and swine (Dickerson et al. 1954, Bereskin et al. 1968). A nonlinear response of the mean phenotype to the inbreeding coefficient has been seen in many of these studies. However, since few studies incorporate appropriate controls or account for the nonindependence of data in inbred lines, it is difficult to say whether the apparent nonlinearities in the data are simply statistical artifacts, as opposed to real reflections of directional epistasis. In organisms with extensive parental care, still another explanation exists. If maternal performance (the ability to raise young) is adversely affected by inbreeding, then an individual's phenotype will be influenced not only by its own level of inbreeding but also by that of its mother.

An elegant experiment performed with laboratory mice (White 1972) illustrates this point. When progeny with several different levels of inbreeding were crossfostered with mothers inbred to different degrees, maternal inbreeding was found to have approximately half the impact on juvenile weight as individual inbreeding (Figure 10.7). Other experiments with mice have verified the effects of maternal inbreeding on progeny performance (Bowman and Falconer 1960,

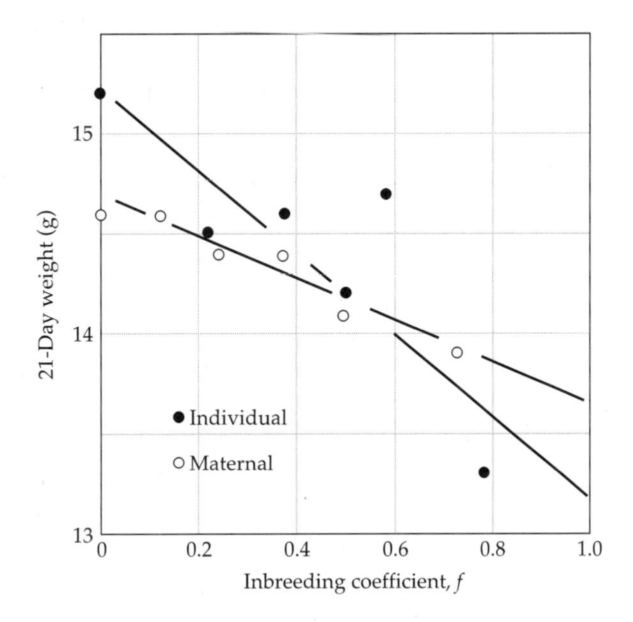

Figure 10.7 The decline in offspring size in laboratory mice as a function of individual and maternal inbreeding. The results of two experiments have been combined after adjusting for mean differences in the control lines. (From White 1972.)

Falconer and Roberts 1960, McCarthy 1967, Nagai et al. 1971), and convincing but less extensive data exist for humans (Schull et al. 1970), sheep (Wiener et al. 1992a–c), and birds (Sittmann et al. 1966, van Noordwijk and Scharloo 1981).

A second point that has received substantial attention in the animal breeding literature is heterosis × environment interaction (Orozco 1976, Barlow 1981, Sheridan 1981, Cunningham 1982). Although the general opinion is that heterosis is more pronounced in suboptimal environments, there are many exceptions to this pattern, and few of the data bearing on the subject come from well-designed experiments. As noted above, heterosis between breeds need not have the same genetic basis as inbreeding depression within populations. However, experiments with *Drosophila* support the contention that inbreeding depression is more severe in stressful environments (Hoffmann and Parsons 1991, Miller 1994), as do those with the flour beetle *Tribolium* (Pray et al. 1994) and with mice (Jiménez et al. 1994). On the other hand, in a very large study comparing 38 human populations, Bittles and Neel (1994) found that the effects of inbreeding on survival to age 10 were independent of the mortality rate of noninbred progeny, which ranged from 3 to 40%. Likewise, while numerous studies with plants have documented increased inbreeding depression under extreme conditions (Antonovics 1968, Schemske

1983, Dudash 1990, Schmitt and Ehrhardt 1990, Wolfe 1993), there are also many cases in which the influence of the environment is negligible (Johnston 1992, Heywood 1993, Charlesworth et al. 1994, Ouborg and Van Treuren 1994, Nason and Ellstrand 1995, Norman et al. 1995). We know of no cases in which inbreeding depression is more intense in benign laboratory or greenhouse environments than in natural settings, so it seems reasonably safe to assume that the former estimates tend to err on the conservative side with respect to the situation in nature.

PURGING INBREEDING DEPRESSION

In recent years, concern has arisen that inbreeding in small captive populations of endangered species has led to a loss in fitness, possibly an irretrievable one (Ralls et al. 1979; Ralls and Ballou 1982a,b; Lacy et al. 1993). Almost every species for which breeding records have been kept shows an elevated mortality rate in inbred versus outbred progeny. For this reason, special efforts are now being made to restrict matings of endangered species to nonrelatives. However, the options are limited in many cases because essentially all remaining members of the species are descendants of the same pedigree. Similar situations must eventually arise in closed populations of elite breeds of domesticated species.

As a potential strategy for dealing with such extreme situations, Templeton and Read (1983, 1984) advocated intentional inbreeding, combined with rapid population expansion, as a means for exposing deleterious recessive genes, and purging them from a remnant population. Implicit in such a management program is the assumption that rare deleterious recessives, not overdominant alleles, are the primary agents of inbreeding depression.

Evidence for purging of deleterious genes has been recorded in several studies with flies (Ehiobu et al. 1989, Bryant et al. 1990, Garcia et al. 1994, Latter et al. 1995) and mice (Bowman and Falconer 1960, Lynch 1977, Connor and Bellucci 1979), a fact that provides further support for the dominance hypothesis. However, care needs to be taken in extrapolating the idea of purging as a conservation tool too far (Hedrick 1994, Lynch et al. 1995a, Lynch 1996). When inbreeding is associated with small population size, it is accompanied by an enhanced probability of fixation of mildly deleterious genes by random genetic drift, which can permanently reduce mean population fitness. In full-sib mated lines of mice, for example, purging is only accomplished at the expense of extreme selection (extinction) among replicate lines (Figure 10.8). In addition, because the rate of production of new deleterious alleles by spontaneous mutation is quite high (Chapter 12), any purging of the mutation load from a population is at best a transient phenomenon.

Related to Templeton and Read's (1983, 1984) notion of purging is the idea that obligately self-fertilizing plants should be largely free of inbreeding depression (Wright 1969, Lande and Schemske 1985, Husband and Schemske 1996). Here

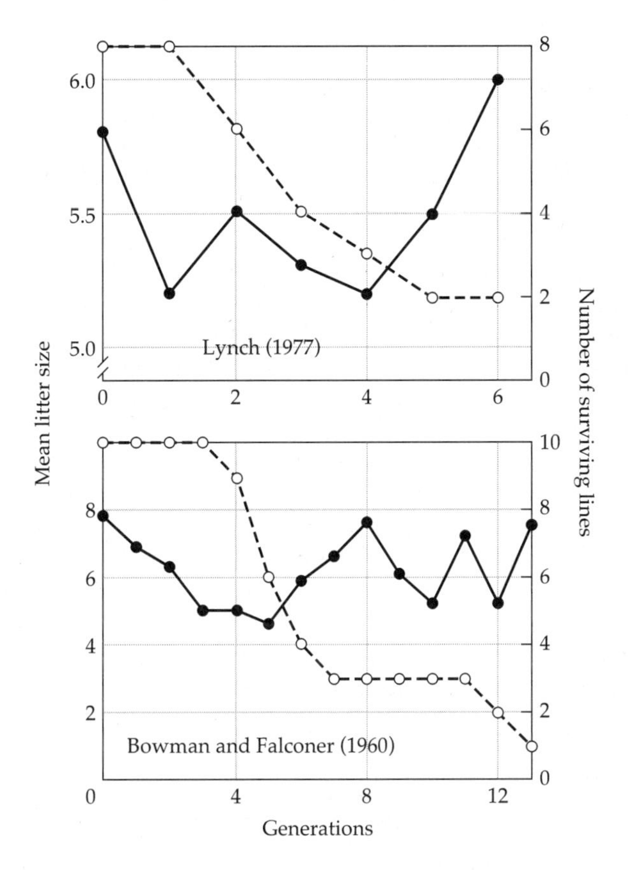

Figure 10.8 Purging of deleterious alleles through inbreeding — the response of mean litter size (•) to inbreeding in the mouse. In both studies, replicate lines were established from the same random-mating base population, and subsequently maintained by single brother-sister matings. Differential extinction of the lines (○) is caused by the failure to produce at least one reproductively competent male and female offspring. In both studies, there is initially a sharp drop in the litter size averaged over all lines, but as the less prolific lines go extinct, the grand mean returns to its initial level. Such results are not expected under the overdominance hypothesis, since with population sizes of only two individuals, most loci are expected to lose all but one allele by random genetic drift within a few generations.

too, some precautionary notes are in order. The effects of close inbreeding on the evolution of inbreeding depression needs to be considered from two frames of reference. By eliminating the genetic variation within an isolated population (or within a lineage of obligate selfers), long-term inbreeding can lead to a situation in which any further inbreeding has essentially no genetic consequences, simply because the progeny genotypes are essentially identical to those of their mothers. Once this situation has been reached, from the standpoint of the current population, there is no inbreeding depression. However, as noted above, this need not be the case from the standpoint of the ancestral population if, during the inbreeding process, different deleterious recessive genes have become fixed within different selfing lineages.

The latter point is nicely illustrated by a study of self-fertilized lines of the normally outcrossing aquatic plant *Eichhornia paniculata* (Barrett and Charlesworth 1991). Inbreeding caused an immediate depression in fitness, but after only two generations of selfing, there was no further decline, suggesting that the vast majority of loci affecting fitness had become homozygous within lines. Nevertheless, despite the absence of inbreeding depression *within* the derived lines, crosses *between* lines exhibited a substantial increase in mean fitness, as expected if the different lines had become fixed for different deleterious recessives. Over time, as new deleterious mutations become sequestered in different selfing lineages, the beneficial effects of outcrossing are expected to increase further (Lynch et al. 1995b). Enhanced fitness in outcrossed progeny of normally self-fertilizing plants has been demonstrated in numerous species (Levin 1989, Charlesworth et al. 1990, Holtsford and Ellstrand 1990, Ågren and Schemske 1993, Van Treuren et al. 1993, Latta and Ritland 1994, Johnston and Schoen 1995). Clearly, the absence of local inbreeding depression provides no information about the mutation load harbored by a population.

THE NUMBER OF LETHAL EQUIVALENTS

Morton et al. (1956) developed a simple regression technique for summarizing the deleterious consequences of inbreeding for attributes classified by incidence, such as survival, expression of mental retardation, and so on. They defined a **lethal equivalent** (**detrimental equivalent** for traits other than survival) as any group of genes "that if dispersed in different individuals . . . would cause on average one death." Thus, a lethal equivalent can consist of a single lethal gene or of a large number of mildly deleterious genes.

Assuming that the environment and all loci act independently in determining survivorship, the probability of survival at inbreeding level f can be written as

$$S_f = (1 - P_e) \prod_{i=1}^{n} [1 - P_i(f)] \tag{10.16}$$

where P_e is the genotype-independent probability of dying from environmental causes, and $P_i(f)$ is the probability of dying as a result of deleterious genes at the ith locus when the inbreeding coefficient is equal to f. From Table 10.1,

$$P_i(f) = f q_i s_i + (1 - f)[\, q_i^2 s_i + 2q_i(1 - q_i)(sh)_i \,] \tag{10.17}$$

where q_i is the frequency of the deleterious allele, and s_i and $(sh)_i$ are the probabilities of mortality for homozygotes and heterozygotes. If the probability of dying from any single cause is small, then the approximation $(1-x) \simeq e^{-x}$ gives

$$S_f \simeq \exp\left[-P_e - \sum_{i=1}^{n} P_i(f)\right] = \exp[-(A + Bf)] \tag{10.18a}$$

where

$$A = P_e + \sum_{i=1}^{n} q_i[q_i s_i + 2(1-q_i)(sh)_i] \tag{10.18b}$$

is the sum of probabilities of mortality in the random-mating population, and

$$B = \sum_{i=1}^{n} q_i[s_i - q_i s_i - 2(1-q_i)(sh)_i] \tag{10.18c}$$

is the excess sum of probabilities of mortality that would exist in a completely inbred population. Logarithmic transformation of Equation 10.18a leads to

$$\ln S_f \simeq -A - Bf \tag{10.19}$$

Thus, the composite measures A and B can be estimated by regressing the natural logarithm of survivorship on the inbreeding coefficient.

A slight problem with this approach arises in cases in which the observed survivorship for certain inbreeding classes is zero, since the logarithm of zero is undefined. Templeton and Read (1984) suggest the small sample-size correction

$$S'_f = \frac{1 + N'_f}{2 + N_f} \tag{10.20}$$

where N_f and N'_f are the numbers of total and surviving individuals at inbreeding level f. With no observed survivors, this quantity rapidly approaches zero as the total sample size increases. Morton et al. (1956) also recommend the use of weighted least-squares analysis, weighting the data by $N_f S_f/[1-S_f]$, the inverse of the sampling variance of $\ln S_f$, and iterating the regression by substituting expected for observed S_f in the weights. (Recall that a similar procedure is used in the estimation of line-cross variances; Chapter 9.)

Unfortunately, the mean number of lethal equivalents per gamete, $\sum q_i s_i$, cannot be separated cleanly from other terms in the definitions of A and B. However, since $A+B = P_e + \sum q_i s_i$, the number of lethal equivalents per gamete must be between B and $A+B$, assuming $(sh)_i \geq 0$. As will be seen below, estimates of $A+B$ are usually not much greater than B, so the use of B as an approximation of the effective number of lethals is not greatly troubling.

Practical situations often arise in which one only has data for noninbred individuals and a single class of inbred individuals. A regression is not possible in this case, but the parameters can still be estimated by

$$A = -\ln S_0 \tag{10.21a}$$

$$B = -\frac{\ln(S_f/S_0)}{f} \tag{10.21b}$$

where, for example, f is $1/2$ for self-fertilization and $1/4$ for full-sib mating. Using the methods of Appendix 1, the large-sample variance for B in this case is found to be

$$\text{Var}(B) \simeq \frac{1}{f^2}\left(\frac{1-S_f}{S_f N_f} + \frac{1-S_0}{S_0 N_0}\right) \tag{10.22}$$

Results from Vertebrates

The regression method of Morton et al. (1956) has been extensively applied to humans, with several independent studies indicating that the average number of lethal equivalents per gamete is on the order of one to two (Table 10.4). Results from other vertebrate species are, for the most part, very similar, suggesting on the order of 0.5 to 3 lethal equivalents per gamete (Table 10.4). This translates into one to six lethal equivalents per zygote, enough to kill the average individual a few times over if fully expressed in the homozyogous state.

Of all of the existing data, those for the European bison and Holstein cattle, which exhibit no significant lethal load, are the most anomalous. There is no obvious explanation for the Holstein data; they are quite inconsistent with those from other breeds. However, it is known that earlier in this century the European bison was reduced to only a dozen individuals. Thus, it is possible that the heterozygosity in this species was largely eliminated by extensive inbreeding during the population bottleneck. We also note that a study on congenital birth defects, birth weight, and gestational age for people of the Indian state of Tamil Nada revealed no evidence of inbreeding depression (Rao and Inbaraj 1980). These results, which are quite unusual for humans, may also be related to the decline in the lethal load caused by previous inbreeding. Approximately 40% of the marriages in this population were between second cousins or closer relatives.

A critical assumption underlying lethal-equivalent analysis is that the effects of different loci on survivorship are independent. Directional epistatic interactions between pairs of loci will give rise to nonlinearities in plots of log survival vs. inbreeding coefficient, but as noted above, these can be detected only if survivorship estimates are available for several levels of inbreeding. Other than the data in Figure 10.9, which yield no evidence of nonlinearity, few data sets are extensive enough to evaluate this matter.

Table 10.4 Estimates of the effective number of lethals per gamete for vertebrates (bounded by B to $A + B$).

Species	Trait	A	B	Reference
Humans	Survival to maturity			
	France, 1919–1925	0.16	2.87	Morton et al. 1956
	Chicago, 1936–1956	0.18	1.55	Slatis et al. 1958
	Fukuoka, Japan	0.07	0.67	Yamaguchi et al. 1970
	Nagasaki, Hiroshima, and Hirado, 1948–65	0.10	0.67	Schull and Neel 1972
	Survival to age 10	0.20	0.70	Bittles and Neel 1994*
	Conspicuous abnormalities	0.10	1.16	Slatis et al. 1958
	Mental retardation	0.01	0.80	Morton 1978
	Congenital heart disease	0.01	0.32	Gev et al. 1986
Speke's gazelle	1-year viability	0.42	3.75	Templeton and Read 1984
European bison	2-year viability	0.26	0.13	Slatis 1960
Sheep	Survival, 1.5–5 years	0.09	0.39	Wiener et al. 1992c*
Swine	Embryo survival	0.27	1.01	Pisani and Kerr 1961
Cattle	Survival through calving			
Holstein		0.16	0.02	Pisani and Kerr 1961
Jersey		0.18	1.15	
Hereford		0.19	0.64	MacNeil et al. 1989*
Great tit	Survival to fledging	0.36	0.84	van Noordwijk and Scharloo 1981
Japanese quail	16-week survival	0.60	1.91	Sittmann et al. 1966
Chicken	18-month survival	0.82	2.10	Pisani and Kerr 1961

Note: All estimates were obtained by regression, except those marked by an asterisk, which were obtained with Equations 10.21a,b.

Results from *Drosophila*

Through the use of balancer-chromosome techniques, unfortunately still only available for *Drosophila*, it is possible to get a finer picture of the types of deleterious genes that contribute to inbreeding depression. Recall (Figure 5.7) that this procedure enables one to isolate intact chromosomes from natural populations and to assay their homozygous performance with respect to a control chromosome (the balancer). By crossing two lines, each one carrying a different chromosome, it is also possible to assay the relative performance of chromosomal heterozygotes.

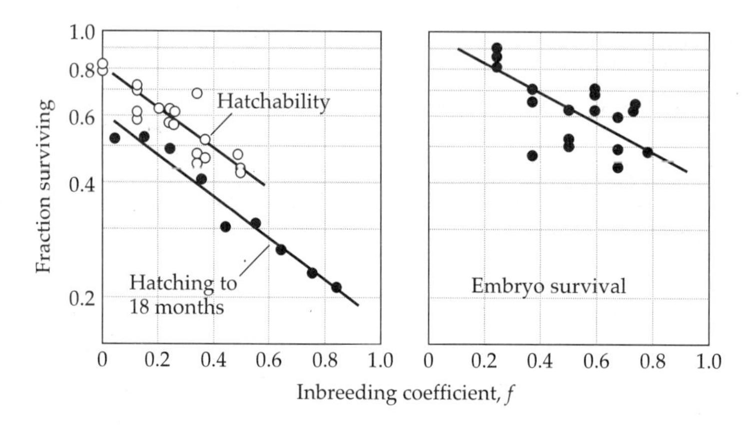

Figure 10.9 Survivorship of white-leghorn chickens (left) and Poland-China pigs (right) as a function of level of inbreeding. (From Pisani and Kerr 1961.)

The ratio of the two relative performances provides a measure of the fitness of chromosomal homozygotes ($f = 1$) relative to that of heterozygotes ($f = 0$).

Greenberg and Crow (1960) reasoned that this approach might be exploited to partition the deleterious load in populations into components due to alleles with various magnitudes of effects. The partitioning of fitness classes is arbitrary, but the technique is nevertheless general. The usual procedure has been to classify as **lethals** those chromosomes that, when homozygous, yield less than 10% of the viability observed in random heterozygotes. Chromosomes with relative viabilities greater than 10% but less than one are referred to as **detrimentals.** Nearly all chromosomes extracted from natural *Drosophila* populations have relative viabilities less than one when in the homozygous state, so this categorization encompasses essentially all chromosomes. Denoting the mean viabilities of chromosomal heterozygotes, detrimental chromosomal homozygotes, and all homozygotes as S_0, S_D, and S_T, then from Equation 10.21b,

$$B_T = \ln S_0 - \ln S_T \qquad (10.23a)$$

$$B_D = \ln S_0 - \ln S_D \qquad (10.23b)$$

$$B_L = \ln S_D - \ln S_T \qquad (10.23c)$$

The different components of the deleterious load are additive, since they are the summations of effects of individual viability mutations, i.e., $B_T = B_D + B_L$. In other words, B_D (the deleterious load) estimates the total number of lethal equivalents resulting from the cumulative effects of all deleterious genes that are individually nonlethal, whereas B_L (the lethal load) estimates the additional number of lethal equivalents resulting from recessive lethals being present on a subset of the chromosomes.

Table 10.5 Partitioning of the total number of lethal equivalents (B_T) into the subcomponents resulting from detrimental (B_D) and lethal (B_L) factors.

Species	Chromosome	N	B_T	B_D	B_L
Drosophila melanogaster	II	16	0.483	0.236	0.247
	III	3	0.691	0.284	0.407
Drosophila pseudoobscura	II	3	0.450	0.246	0.204
	III	1	0.578	0.352	0.226
Drosophila willistoni	II	1	0.766	0.380	0.386
	III	1	0.690	0.506	0.184

Note: N is the number of studies, over which the data, summarized from Simmons and Crow (1977), are averaged.

Simmons and Crow (1977) have summarized a large number of studies employing this approach with the second and third chromosomes in *Drosophila.* The data are remarkably consistent (Table 10.5). The total number of lethal equivalents associated with each chromosome, in each species, ranges from 0.5 to 0.8. Noting that chromosomes II and III each comprise approximately 40% of the *Drosophila* genome, these observations suggest that the average drosophilid carries approximately three lethal equivalents, similar to the situation in vertebrates. Averaging over all of the data, the detrimental and lethal loads per chromosome are 0.33 and 0.28. Thus, about half of the total lethal equivalents are associated with lethal recessives, and about one in three chromosomes carries such a gene.

The idea that deleterious alleles do not interact epistatically can be checked with the balancer-chromosome technique. In the absence of average interchromosomal epistasis, the number of lethal equivalents expressed in individuals homozygous for both chromosomes II and III should not be significantly different from the sum of the loads obtained for individuals homozygous for just chromosome II and for just chromosome III. Only a few studies of this nature have been undertaken, and the results are somewhat mixed. The overall picture, albeit a weak one, is that if epistasis exists among the genes on the two chromosomes, it is weak and synergistic (positively reinforcing) (Simmons and Crow 1977).

Results from Plants

Equation 10.21b has been used extensively in estimating the number of lethal equivalents in coniferous trees of economic importance. The usual approach has been to compare self-pollinations to outcrosses using a mixture of pollen from several distant trees. Most of the emphasis has been on embryonic mortality, which is easily assayed by counting unfilled seeds. The number of lethal equiv-

Table 10.6 Estimates of lethal equivalents per gamete affecting early embryonic survival in conifers and herbaceous angiosperms.

Species	B	Reference
Conifers		
Nobèl fir *(Abies procera)*	1.7	Sorensen et al. 1976
Tamarack *(Larix laricina)*	5.4	Park and Fowler 1982
Norway spruce *(Picea abies)*	4.8	Koski 1971
White spruce *(Picea glauca)*	5.0	Fowler and Park 1983
	4.4	Coles and Fowler 1976
Black spruce *(Picea mariana)*	2.4	Park and Fowler 1984
Ponderosa pine *(Pinus ponderosa)*	2.0	Sorensen 1970
Scots pine *(Pinus sylvestris)*	4.4	Koski 1971
	3.6	Savolainen et al. 1992
Loblolly pine *(Pinus taeda)*	4.2	Franklin 1972
	4.8	Bishir and Namkoong 1987
Virginia pine *(Pinus virginiana)*	5.0	Bishir and Namkoong 1987
Douglas fir *(Pseudotsuga menziesii)*	5.0	Sorensen 1969
Short-lived angiosperms		
Begonia hirsuta	0.04	Ågren and Schemske 1993
Begonia semiovata	0.11	Ågren and Schemske 1993
Clarkia tembloriensis	0.07	Holtsford and Ellstrand 1990
Lychnis flos-cuculi	0.39	Hauser and Loeschcke 1994
Mimulus guttatus	0.16	Latta and Ritland 1994
Raphanus sativus	0.01	Nason and Ellstrand 1995
Salvia pratensis	0.67	Ouborg and Van Treuren 1994
Schiedea lydgatei	0.91	Norman et al. 1995

Note: Estimates for conifers were taken directly from the literature, while those for angiosperms were computed from data on percent germination for outcrossed and selfed seed.

alents expressed at this stage is exceptionally high, ranging from one to five per gamete (Table 10.6). Most studies show a high variance in the number of lethal equivalents per individual, with few individuals completely free of them and some carrying as many as 30 (Figure 10.10). Longer-term studies (Park and Fowler 1982, 1984, Fowler and Park 1983) indicate that most of the lethal equivalents affecting survival are expressed at the embryonic stage, with approximately one to two additional lethal equivalents per gamete influencing subsequent survivorship. It is conceivable that the extraordinarily high mutation load in conifers is a consequence of their long generation time, which may magnify the mutation rate on a per generation basis. Fundamentally different results arise with short-lived herbaceous plants, where B for probability of germination is consistently less than one

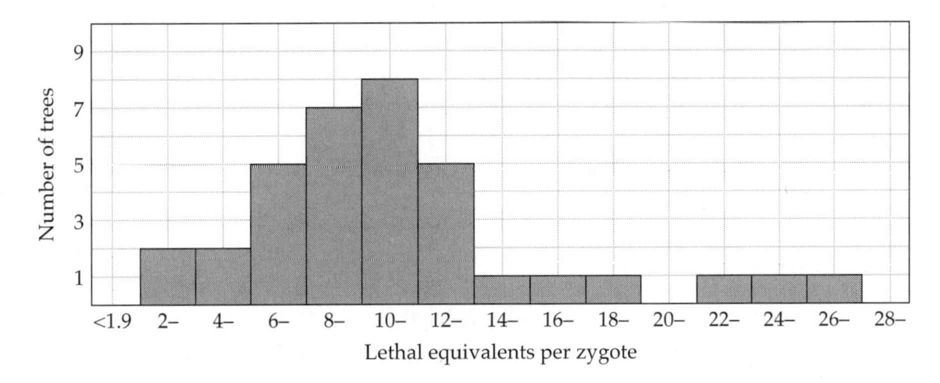

Figure 10.10 Frequency distribution of the mean number of lethal equivalents per zygote for 35 trees in a population of Douglas fir. (From Sorensen 1969.)

(Table 10.6). Such plants do, however, express additional lethal equivalents in the form of survival to maturity and reproductive performance (see Charlesworth and Charlesworth 1987 for a summary).

Hedrick (1987b) has reviewed the extensive literature on genetic load in ferns. The majority of studies have been performed by selfing gametophytes and counting the proportion of spores that germinate, ignoring the load expressed subsequent to germination. Again, the mean number of lethal equivalents per gamete, which ranges from 0 to 1.3, appears to be substantially lower than that found in conifers. Although the fern data are still limited, they suggest that the species-specific loads are inversely proportional to the frequency of self-fertilization in nature, as expected when inbreeding purges lethals from a population.

PARTIAL RECESSIVES vs. OVERDOMINANCE

The observation of inbred lines that equal or exceed the average performance of individuals in outcrossed populations is a serious challenge to the contention that overdominance is the primary mechanism of inbreeding depression. Nevertheless, there is still a substantial amount of controversy on the subject, fostered to a large degree by a number of puzzling observations with allozyme loci. Here, we summarize results from biometrical analyses that bear on the question of mode of dominance. None of these results supports the idea that overdominance is a common mode of gene action. We then close by scrutinizing the results of molecular analysis.

The $(A+B)/A$ Ratio

In the previous section, we defined the model for lethal equivalents in terms of partially recessive deleterious alleles. The same logic can be used to redefine the model in terms of overdominant gene action. This approach again gives rise to Equation 10.19, but with the definitions of A and B altered to

$$A = P_e + \sum_{i=1}^{n} [\, q_i^2 s_i + p_i^2 t_i \,] \tag{10.24a}$$

$$B = \sum_{i=1}^{n} q_i p_i (s_i + t_i) \tag{10.24b}$$

Morton et al. (1956) noticed a useful feature of this model. Recall that for a balanced polymorphism maintained by overdominance, the equilibrium allele frequencies are $q_i = t_i/(s_i + t_i)$ and $p_i = s_i/(s_i + t_i)$. Substituting these into the previous expressions,

$$B = \sum_{i=1}^{n} \frac{s_i t_i}{s_i + t_i} \tag{10.25a}$$

$$A = P_e + B \tag{10.25b}$$

Thus, if inbreeding depression is primarily a consequence of overdominance, then the ratio $(A + B)/A$ is constrained to be less than or equal to two in populations that are in a state of balanced polymorphism. Strictly speaking, this result applies to a diallelic locus. With k alleles per locus, the constraint is $(A + B)/A \leq k$ (Crow 1958, Lewontin 1974). Returning to Table 10.4, we see that this ratio is usually on the order of 10 or more. Thus, unless a very large number of alleles are maintained in a delicately balanced polymorphism at each locus contributing to inbreeding depression, which seems quite unlikely, the results from lethal-equivalent analyses seem generally inconsistent with the overdominance model.

Estimating the Average Degree of Dominance

Accepting that the linear decline in log fitness in a lethal-equivalent analysis is, in fact, a consequence of multiple partially recessive alleles, then some further inference about the mode of gene action can be made. Using Equations 10.18b,c, it can be shown that

$$\frac{B}{A+B} \leq 1 - \frac{\sum 2 q_i s_i h_i}{\sum q_i s_i} \tag{10.26}$$

Recalling that under selection-mutation balance, $q_i = u_i/(h_i s_i)$, and rearranging, Equation 10.26 implies that for a population in equilibrium,

$$\tilde{h}_1 = \frac{\sum u_i}{\sum (u_i/h_i)} \leq \frac{A}{2(A+B)} \tag{10.27}$$

The expression on the left is equal to the harmonic mean of the dominance coefficients among *newly arising mutations.* An upwardly biased estimate of this quantity is provided by the ratio $A/[2(A + B)]$, the bias approaching zero as the environmental contribution to mortality becomes negligible. Application of Equation 10.27 to the data in Table 10.4 shows that, in the vast majority of cases, $\tilde{h}_1$ is in the range of 0.02 to 0.15. It should be kept in mind that $\tilde{h}_1$ will tend to exceed the average dominance coefficient of *segregating* deleterious alleles because mutant alleles with higher degrees of expression are more easily removed by selection. However, because a harmonic mean is always less than the arithmetic mean, these two sources of bias may approximately cancel, leaving $A/[2(A + B)]$ as a reasonable estimator of the arithmetic mean h of segregating alleles. In any event, the data clearly suggest that the majority of deleterious alleles influencing early survival are quite recessive.

When highly inbred lines are available, less biased methods for estimating the average degree of dominance exist, as pointed out by Mukai et al. (1974). Consider a single diallelic locus with the relative fitnesses of the BB, Bb, and bb genotypes being 1, $(1 - hs)$, and $(1 - s)$. If a randomly mating base population is inbred (through a series of lines) to complete homozygosity, then (from Table 10.1) a fraction q of the lines will have genotype bb and fitness $(1 - s)$ at the locus, while the remaining fraction p will have genotype BB and fitness 1. Now suppose that the inbred lines are randomly paired and mated. $BB \times BB$ matings will then occur with frequency p^2, giving rise to BB progeny with fitness 1. Similarly, $bb \times bb$ matings will occur with frequency q^2, giving rise to bb progeny with fitness $(1 - s)$, and $BB \times bb$ matings will occur with frequency $2pq$, giving rise to Bb progeny with fitness $(1 - hs)$. Summing over all loci, the genetic variances for log fitness among inbred lines, among midparent values, and among F_1 progeny are, respectively,

$$\sigma^2(G_p) = \sum p_i q_i s_i^2 \tag{10.28a}$$

$$\sigma^2(G_{mp}) = \sum p_i q_i s_i^2/2 \tag{10.28b}$$

$$\begin{aligned} \sigma^2(G_o) &= \sum 2p_i q_i s_i^2 [\,(1 - 2p_i q_i)h_i^2 - 2q_i^2 h_i + q_i(1 + q_i)/2\,] \\ &\simeq \sum 2p_i q_i s_i^2 h_i^2 \end{aligned} \tag{10.28c}$$

the approximation in Equation 10.28c following from the reasonable assumption that the frequencies of deleterious alleles are kept low by selection, i.e., $q_i << 1$. In addition, the covariance among offspring and midparent genotypic values is

$$\begin{aligned} \sigma(G_o, G_{mp}) &- \sum p_i q_i s_i^2 [\,h_i(1 - 2q_i) + q_i] \\ &\simeq \sum p_i q_i s_i^2 h_i \end{aligned} \tag{10.28d}$$

Thus, half the genetic regression of offspring on midparent values has the expected value

$$\frac{b(o, mp)}{2} = \frac{\sigma(G_o, G_{mp})}{2\sigma^2(G_{mp})} = \frac{\sum p_i q_i s_i^2 h_i}{\sum p_i q_i s_i^2} \tag{10.29a}$$

Recalling that under selection-mutation balance, $q_i = u_i/(h_i s_i)$ and $p_i \simeq 1$, this expression reduces to

$$\frac{b(o,mp)}{2} = \tilde{h}_2 \simeq \frac{\sum u_i s_i}{\sum u_i s_i / h_i} \tag{10.29b}$$

Like Equation 10.27, this expression is a harmonic mean estimate of the average degree of dominance. In this case, however, each allele is weighted by $(u_i s_i)$, the product of the mutation pressure to the allele and the homozygous mutational effect. If s_i and h_i are uncorrelated, then Equation 10.29b, like 10.27, provides an estimate of the harmonic mean dominance coefficient of new mutations (Watanabe et al. 1976). However, data on newly arisen mutations suggest that such independence is unlikely (Chapter 12).

Keeping in mind these interpretative limitations, we now consider the situation for viability in *Drosophila.* Averaging over four studies for which the appropriate data are available (Tantaway 1957, Dobzhansky et al. 1963, Malogolowkin-Cohen et al. 1964, Garcia et al. 1994), we estimate $\tilde{h}_1 \leq 0.14 \pm 0.05$. Thus, consistent with our broader interpretation of the data in Table 10.4, most deleterious segregating genes that influence viability appear to be quite recessive. On the other hand, three studies with *D. melanogaster* (Mukai et al. 1972, Mukai and Yamaguchi 1974, Watanabe et al. 1976) yield estimates of $\tilde{h}_2$ that average 0.30 ± 0.05. Working with the same species, Hughes (1995) obtained estimates of $\tilde{h}_2$ for additional characters in males: 0.08 for body size, 0.17 for mating ability, 0.14 for mortality rate, and 0.30 for longevity. Finally, data from Wills (1966) and Strickberger (1972) yield estimates of $\tilde{h}_2 = 0.27$ and 0.18 for viability in *D. pseudoobscura.* None of these estimates of $\tilde{h}_2$ are strictly comparable to those for $\tilde{h}_1$, not only because different weights are employed in the two definitions, but also because all of the studies employing Mukai's regression method have excluded chromosomal lines with highly deleterious effects. Thus, to the extent that lethal or semilethal alleles have lower dominance coefficients than do mildly deleterious alleles, as the data clearly suggest (Chapter 12), $\tilde{h}_2$ estimates are expected to be higher than those for $\tilde{h}_1$. Summing up the extensive data for *Drosophila,* most segregating deleterious mutations appear to be recessive, with the average h for all deleterious alleles being on the order of perhaps 0.1, and that for mildly deleterious alleles being more on the order of 0.15 to 0.3.

There is an unfortunate void on information on the average degree of dominance in other organisms. However, the regression technique of Mukai et al. (1974) is easily extended to certain species, most notably those that reproduce in nature by obligate self-fertilization. Individuals within such populations are as close to being completely homozygous as one can get, and provided that forced outcrosses can be implemented, it is possible to use Equation 10.29b to obtain an estimate of the dominance coefficient averaged over the entire genome. Under obligate selfing, most lethal recessives should be purged, leaving most of the deleterious mutation load in the detrimental class. Johnston and Schoen (1995) used the regression method to obtain estimates of the average degree of dominance in four populations of the annual plant *Amsinckia*. Their average estimates of $\tilde{h}_2$ are: 0.30 for flower number, 0.36 for survivorship from germination to flowering, 0.10 for seed production, and 0.21 for total fitness. These results are close to those for detrimentals in *Drosophila*. Additional methods for estimating the average degree of dominance are covered in Chapter 20.

Inferences from Molecular Markers

In Chapter 6, we reviewed the evidence relating molecular heterozygosity to phenotypic *variation,* illustrating the substantial degree of inconsistency in results that exists among different studies. The situation is not much different in studies that have attempted to relate an individual's overall heterozygosity to *mean* phenotypes.

In search for a mechanism that might explain the maintenance of molecular variation within natural populations, numerous studies have attempted to correlate biochemical heterozygosity, usually as revealed by allozyme polymorphisms, with fitness-related characters. Under the overdominance hypothesis, one would expect individual fitness to increase with the number of heterozygous allozyme loci per individual, either because the loci themselves are heterotic or because they mark heterotic regions of the genome. Since heterozygosity is assumed to be unconditionally advantageous, this correlation should hold in populations with any level of inbreeding. The pattern is expected to be more pronounced in highly structured populations where, due to individual variation in inbreeding, some individuals are highly homozygous and others highly heterozygous. **Identity disequilibrium** refers to situations in which there is a correlation across loci for the probability of alleles identical by descent as a consequence of variance in inbreeding.

In contrast, under the dominance hypothesis, a correlation between multilocus heterozygosity (MLH) and fitness should arise *only* if there is a correlation between MLH and the level of individual inbreeding f. In large, randomly mating populations, the association between MLH and f is expected to be negligible because essentially all individuals trace through pedigrees with similar (and very low) levels of inbreeding, rendering the variance in f among individuals insignificant. Thus, if the dominance hypothesis is correct, such populations should not exhibit a correlation between individual measures of MLH and fitness. On the other hand, populations with significant levels of gametic phase and/or identity disequilibrium can exhibit positive correlations between MLH and fitness for reasons that are totally unrelated to overdominance.

Consider the situation for a molecular-marker locus with two alleles, M_1 and M_2, neither of which has any direct influence on fitness, and assume that each marker allele is tightly linked (and in complete gametic phase disequilibrium) with a deleterious allele at a different locus. Letting lowercase letters denote deleterious alleles, then the gametic states associated with the marker locus are AM_1b

and aM_2B. This is an extreme case of repulsion disequilibrium (Chapter 5), as there are only three genotypes in the population associated with the marker: AM_1b/AM_1b, AM_1b/aM_2B, and aM_2B/aM_2B. Now suppose that each deleterious allele reduces fitness by the fraction s in the homozygous state, by hs in the heterozygous state, and that the effects of the two loci are independent. Then, the fitnesses associated with the marker locus are $(1 - s)$ for the two homozygous classes and $(1 - hs)^2$ for the heterozygous class. Under this scenario, the heterozygous marker class will exhibit greater fitness than the homozygous classes provided $h < (1 + s)/2$.

This apparent heterozygote superiority, solely an artifact of linked loci being in repulsion disequilibrium for deleterious alleles, is known as **associative overdominance** (Frydenberg 1963). Notice that when fitness is analyzed on the direct scale of measurement, associative overdominance can arise even with additivity ($h = 0.5$) or with slight dominance of the deleterious allele. If $h > (1 + s)/2$, **associative underdominance** occurs — the heterozygotes exhibit reduced fitness. These latter peculiarities disappear if fitness is measured on a logarithmic scale. Associative overdominance or underdominance can still arise in this case, but assuming small s, it depends more simply upon whether h is less than or greater than 0.5, i.e., on whether the deleterious alleles are partially recessive or partially dominant. We point this scaling property out because many molecular-marker analyses are performed on non-log transformed data. The important point is that even for marker loci in Hardy-Weinberg equilibrium (i.e., with no evidence of inbreeding), with no direct effects on fitness, and with no functional overdominance elsewhere in the genome, associative overdominance arises if loci carrying partially recessive deleterious alleles are linked to the marker and in repulsion disequilibrium.

Now consider the opposite situation — coupling disequilibrium, such that the two marked stretches of DNA are aM_1b and AM_2B. The fitnesses associated with the M_1M_1, M_1M_2, and M_2M_2 genotypes are then $(1 - s)^2$, $(1 - hs)^2$, and 1. Most molecular-marker studies simply consider whether the fitness of heterozygotes exceeds the average of the homozygous classes (without respect to homozygous genotype). Thus, the relevant observation is that associative overdominance with respect to the mean logarithm of fitness for the two homozygous classes will arise if $h < 0.5$, the same conclusion that we arrived at with repulsion disequilibrium. For non-log transformed fitness, the requirement is $h < (2 + s)/4$.

Thus, the general conclusion is that linked loci carrying partially recessive deleterious alleles in disequilibrium, whether in repulsion or in coupling, will always lead to the appearance of overdominance. The fact that situations fully in accord with the dominance hypothesis can lead to observations fully compatible with the predictions of the overdominance hypothesis is an obvious problem. Because essentially all populations have some degree of structure (either variance in inbreeding and/or gametic phase disequilibria, both of which can be difficult to quantify), it is extremely difficult to draw rigorous conclusions about the mechanism of inbreeding depression (or about the advantages of heterozygosity at individual loci) from descriptive surveys of the relationship between MLH and fitness-related characters. Not surprisingly, studies of this nature have raised more questions than they have resolved.

Perhaps the most that can be gained from molecular surveys comes from observations on historically large, random-mating populations, for which gametic phase and identity disequilibria are likely to be minimized. Studies of this nature with adequate statistical power raise serious questions about the generality of the overdominance hypothesis when they do not yield a positive correlation between MLH and fitness. What do the data tell us? For most organisms that have been studied, there is a positive relationship between allozyme heterozygosity and fitness-related characters (Mitton and Grant 1984, Zouros and Foltz 1987). There are, however, numerous exceptions, and even when large numbers of loci are assayed, the biochemical data account for no more than 5-10% of the variance in fitness. A more detailed account for some intensely studied organisms follows.

A substantial body of data for marine bivalves suggests that most populations exhibit a positive correlation between growth rate and individual heterozygosity (MLH) (Singh and Zouros 1978; Zouros et al. 1980, 1988; Koehn et al. 1988; Gaffney 1990; Gaffney et al. 1990; David et al. 1995). Surprisingly, however, samples from these same populations almost always exhibit a deficiency of heterozygotes, at least in early-age cohorts. In addition, there appears to be substantial multilocus disequilibrium — an excess of highly homozygous and highly heterozygous individuals. Many studies have documented the elimination of the heterozygote deficiency as cohorts age, which suggests that MLH is positively correlated with survival as well as with growth rate. The presence of disequilibria in new recruits clearly indicates the potential for these results to be a simple consequence of associative overdominance, rather than an intrinsic advantage to allozyme heterozygosity, a conclusion that is bolstered by the fact that increases in heterozygosity with cohort aging only appear at loci that initially have heterozygote deficiencies. Pogson and Zouros (1994) have argued that the failure of random DNA-based markers to show the correlations seen in allozymes is a point in favor of functional overdominance of allozymes. However, the markers used in their study exhibited extremely high heterozygosities (which reduces the power of an MLH survey) and showed only small heterozygote deficiencies, bringing them in line with the fraction of allozyme loci that are also in Hardy-Weinberg equilibrium.

Given that most marine bivalves are broadcast spawners, it is difficult to envision how a high degree of genetic structure can arise within populations via restricted mating. An alternative explanation for both the heterozygote deficiencies and the multilocus disequilibria at some loci is the presence of either null alleles or of aneuploidy (missing chromosomes). If either situation is common, as appears to be the case (see Gaffney et al. 1990 for a summary), then a fraction of the individuals that are scored electrophoretically as homozygotes will actually be either active/null heterozygotes or chromosomal haploids. If such individuals have reduced activity for important metabolic and / or developmental functions, as seems likely, their undetected presence will cause a downward bias in the estimated fitnesses of homozygous classes (Foltz 1986). Thus, in addition to yielding apparent heterozygote deficiencies, null alleles and / or aneuploidy will promote the appearance of heterozygote advantage.

Apparent heterozygote advantages have also been recorded for growth rate in some species of trees (Mitton et al. 1981, Ledig et al. 1983, Strauss 1986, Strauss and Libby 1987). However, other extensive surveys have failed to find any such relationship (Bush and Smouse 1991, Savolainen and Hedrick 1995). In *Pinus radiata,* stands with greater heterozygote deficiencies exhibit higher correlations between individual heterozygosity and growth rate (Strauss and Libby 1987), results that are reminiscent of those obtained in marine bivalves. If not an artifact of null alleles or aneuploidy, such results may be a simple consequence of multilocus homozygosity acting as a marker for the variation in degree of inbreeding within individual stands of trees.

Attempts have been made to find an association between MLH and fitness in many other organisms, some with success, others not. Most studies are based on small numbers of individuals and loci, and those that have found positive correlations are subject to the challenge that the results are an artifact of population structure, rather than a consequence of true functional overdominance (Houle 1989a). Three of the largest studies that have been performed, *D. melanogaster* by Houle (1989a), fungus beetles by Whitlock (1993), and brook trout by Hutchings and Ferguson (1992), all failed to find an MLH-fitness association.

A number of attempts have been made to develop statistical approaches that could definitively resolve the associative overdominance issue. For example, Smouse (1986) reasoned that a deeper understanding of the mechanisms of inbreeding depression would be obtained by looking at the fitnesses of the alternative homozygous classes within loci. Assuming two alleles per locus, under the overdominance hypothesis, the rarer of the two homozygous classes should have the lowest fitness. Smouse's adaptive distance model transiently attracted some followers, until Houle (1994) showed that the model fit cases of associative overdominance as well as cases of functional overdominance.

Fu and Ritland (1994) have recently suggested an approach that may have more utility. Their idea is to identify a group of heterozygous individuals at a particular locus, self-fertilize them, and then assay the frequencies of the three genotypes after selection has acted. (In principle, the method can also be applied to a group of identically heterozygous individuals allowed to outcross.) The expected frequencies of the three marker genotypes in the offspring generation will be a function of their own direct effects on fitness, the fitness properties of linked polymorphic loci, and the recombination frequency between marker and associated loci. In a study of the monkey flower (*Mimulus guttatus*), Fu and Ritland applied a model that assumes a neutral marker linked to one selected locus and surprisingly found that most progeny arrays were consistent with partial or complete dominance of deleterious alleles or with underdominance. Only a few results were consistent with partially recessive deleterious alleles or with overdominance. In other words, most of the data were inconsistent with both of the traditional explanations for inbreeding depression. It may be premature to make too much of this result. An analysis of the robustness of the model's predictions when at least two loci are in disequilibrium with the marker seems essential.

A promising approach to more clearly defining the mode of gene action within and between loci is now developing with the resolution of dense molecular maps for economically important species (Chapters 14–16). Consider the cross between two highly inbred lines subsequently expanded to produce a diverse F_2 population. If such a population is kept large with minimal selection and randomly mated for several generations, then all alleles would be randomly distributed over many genetic backgrounds, and all individuals would be equally heterozygous at the genome level. Thus, any increase in the performance of heterozygotes at a marker locus would necessarily be due to overdominance associated with the locus itself or with loci linked tightly enough to maintain significant disequilibrium over the period of random mating.

Two recent studies have taken a related approach. Xiao et al. (1995) crossed two elite lines of rice, extracted 200 lines from the F_2 generation, and subsequently purified each of them by five generations of self-fertilization. They then backcrossed each inbred line to each of the parents, and looked for phenotypic differences between individuals identified as heterozygotes or homozygous genotypes at a large number of molecular-marker loci. Thirty-seven out of 141 markers were informative, and in every case, the heterozygote was inferred to have a phenotype intermediate to that of the parental lines. Thus, there was no evidence of overdominant gene action, and this was confirmed more generally by a lack of correlation between phenotypes of the backcross progeny and MLH. Nor was there any evidence of epistasis. Although these results are contrary to the conclusions reached in a similarly designed study with maize, where almost every informative marker appeared to exhibit overdominance (Stuber et al. 1992), a rigorous reanalysis of the data in the latter study supports the dominance hypothesis (Cockerham and Zeng 1996).

In summary, the vast majority, perhaps all, of the results that have been cited in support of the overdominance hypothesis appear to be compatible with associative overdominance. On the other hand, some compelling results that support the dominance hypothesis are not easily accommodated by the overdominance hypothesis. Ultimately, the controversy can be resolved by cloning of alleles of individual genes and using the tools of molecular biology to place the various genotypes onto a constant genetic background. However, fine-scale analysis with molecular markers may soon tell us whether studies of that sort are even warranted.

11

Matters of Scale

Generally, the scales on which we take direct measurements are selected more for their convenience than for their biological relevance or for their amenability to statistical analysis. Common artifacts of scale that complicate the analysis and interpretation of results are the dependence of the variance on the mean, departures from normality, and nonadditive interactions. Such complications can often be eliminated by transformation of the raw data to a new scale. A change in scale does not alter the information content of the original data. It simply changes the relationship of character values to one another.

Often, transformations can only be found that satisfy one or two of the desired properties of normality, additivity, and variance independent of the mean. For polygenic traits, it can sometimes be rather difficult to completely eliminate interaction effects since a scale transformation that is successful in eliminating dominance from one locus may create it at another locus, or may create epistasis, and so on. Similarly, a transformation that successfully yields a normal distribution for one population may cause another to deviate from normality substantially. The utility of a particular transformation can also change dramatically with a shift in the environmental background. When these kinds of conflicts arise, the investigator must decide which criteria can be sacrificed in light of the objectives of the analysis.

Scale can have biological as well as statistical consequences. For example, a change in body size can result in disproportionate (**allometric**) changes in other correlated characters. Features of development such as **canalization** and **genetic assimilation** can also be direct consequences of scale. Our discussion of scale in this chapter thus considers both statistical and biological issues.

TRANSFORMATIONS TO ACHIEVE NORMALITY

Since many statistical methods, such as hypothesis testing using regression analysis and analysis of variance, are predicated on the assumption that the data are normally distributed, a standard procedure in most quantitative-genetic investigations is to transform the data to resemble normality as closely as possible prior to analysis. We first consider the logarithmic transform, as it is one of the most common, and successful, normalizing transformations.

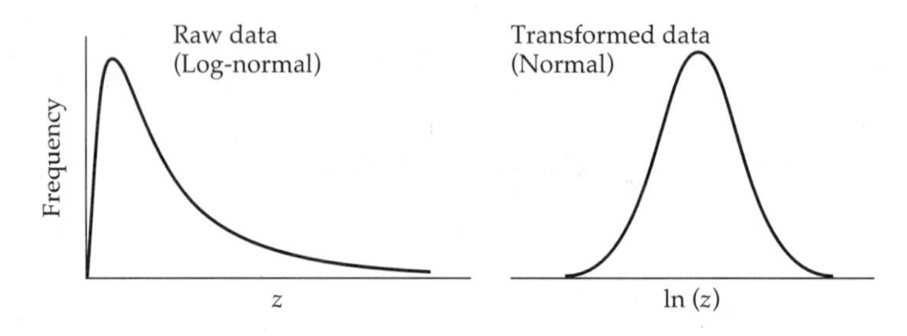

Figure 11.1 Conversion of a log-normal distribution on the original scale to a normal distribution on the log-transformed scale.

Log-normal Distributions and the Log Transform

A common departure from normality is positive skewness, such that the tail to the right of the mean is longer than that to the left. Such a pattern is often indicative of a **log-normal distribution**, in which case a simple logarithmic transformation gives a normal distribution (Figure 11.1).

The log-normal distribution adequately describes many biological attributes (Wright 1968) and is worth dwelling upon briefly. Suppose that a variable z is log-normally distributed with mean μ and variance σ^2 on the original scale of measurement. It follows that $y = \ln(z)$ is normally distributed, and that the moments of y are related directly to the moments of z by

$$\mu_y = \ln \mu - \frac{1}{2} \ln\left(1 + \frac{\sigma^2}{\mu^2}\right) \tag{11.1a}$$

$$\sigma_y^2 = \ln\left(1 + \frac{\sigma^2}{\mu^2}\right) \tag{11.1b}$$

(Aitchison and Brown 1966). Since $\ln(1 + x) \simeq x$ for $|x| << 1$, when the **coefficient of variation** (σ/μ) is small (less than ~ 0.3), these expressions are closely approximated by

$$\mu_y = \ln \mu - \frac{\sigma^2}{2\mu^2} \tag{11.2a}$$

$$\sigma_y^2 = \frac{\sigma^2}{\mu^2} \tag{11.2b}$$

In estimating the moments of a distribution on a log-transformed scale, it is desirable to work directly with the raw data (rather than to simply apply Equations 11.1a,b) because the original distribution may, in fact, not be log-normal. Occasionally, however, one only has recourse to the sample mean, $\overline{z}$, and sample standard deviation, $\text{SD} = \sqrt{\text{Var}(z)}$, on the original scale. In this case, using a Taylor-series approximation (Appendix 1), we can still approximate the mean and variance of the log-transformed variables by

$$\overline{y} = \overline{\ln z} \simeq \ln \overline{z} - \frac{1}{2} \ln\left(1 + \text{CV}^2\right) \simeq \ln \overline{z} - \frac{\text{CV}^2}{2} \tag{11.3a}$$

$$\text{Var}(y) = \text{Var}(\ln z) \simeq \ln\left(1 + \text{CV}^2\right) \simeq \text{CV}^2 \tag{11.3b}$$

where CV $=$ SD$/\,\overline{z}$ is the estimated coefficient of variation on the original scale. Note that these first-order approximations (identical in form to Equations 11.2a,b) apply to *any* distribution on the original scale, not just the log-normal. Thus, logarithmic transformation will successfully **stabilize the variance** (removing its dependence on the mean) under a broad range of distributions on the original scale, *provided* the coefficients of variation are roughly constant. On the other hand, logarithmic transformation will only **normalize** the data if the untransformed data are actually log-normally distributed. Methods for testing for normality are discussed below.

Galton (1879) first pointed out a possible explanation for the commonness of the log-normal distribution. Suppose the phenotype can be represented as a product of a large number of independent factors $z = x_1 \cdot x_2 \cdot x_3 \cdots x_n$. Upon logarithmic transformation, $y = \ln z = \ln x_1 + \ln x_2 + \ln x_3 + \ldots + \ln x_n$, and under the central limit theorem, $\ln z$ will tend to normality as n becomes large. Thus, a simple explanation for a log-normal distribution is the existence of multiplicative interaction between a large number of factors.

A number of other scale transformations, such as power functions and trigonometric functions, are available if the original data depart from both normality and log-normality (Wright 1968). If, for example, length measures in a population are known to be normally distributed, then it is likely that a cube root transformation would be required to normalize the distribution of weights, since weight is usually proportional to the cube of the length. The **Box-Cox transformation,**

$$y = \frac{z^\lambda - 1}{\lambda} \tag{11.4}$$

provides a fairly general class of transformations for achieving normality, and Box and Cox (1964) developed an approach to find the λ that gives the best fit to normality. This type of transformation usually deals quite adequately with skewness on the original scale ($\lambda = 0$, for example, is equivalent to the logarithmic transformation), but is less effective with kurtosis. John and Draper (1980) present alternative functions to deal with long tails on symmetrical distributions, and Atkinson (1982) discusses all of these transformations in some detail.

Tests for Normality

There are a variety of methods for determining whether a character is normally distributed — see Chapter 6 of Sokal and Rohlf (1995) for an introduction, and

Chapter 8 of Wetherill (1986) for more advanced methods. Graphical tests for departures from normality are especially informative. For example, a simple frequency histogram may immediately reveal departures from normality, such as multiple peaks or strong asymmetries. If a character is determined by one (or a few) genes of major effect, a bimodal (or multimodal) distribution can sometimes result (Chapter 13). Such distributions can also result when a character is strongly influenced by a few distinct environmental effects.

With only sample moments in hand, deviations from normality are indicated by significant skewness and/or kurtosis. Bowman and Shenton (1975) proposed a joint test of this, employing the statistic

$$S = \frac{n \cdot k_3^2}{6} + \frac{n \cdot k_4^2}{24} \tag{11.5}$$

where n is the sample size, and from Equations 2.8 and 2.12a, $k_3 = \text{Skw}(z)/\text{Var}(z)^{3/2}$ and $k_4 = [\text{Kur}(z)/\text{Var}(z)^2] - 3$ are, respectively, the standardized sample skewness and kurtosis, both of which have expected value zero under the assumption of normality. For large sample sizes, S is distributed as a χ^2 with two degrees of freedom. Thus, the hypothesis that a distribution is normal is rejected at the 5% level if $S > 5.99$ and at the 1% level if $S > 9.21$.

A more powerful approach for evaluating normality involves **normal probability plots**, which can be generated either by plotting cumulative frequency on special normal probability graph paper or by transforming cumulative frequency to a **normal probability** (or **probit**) **scale**. In Chapter 2, we defined Φ_T as the fraction of measures in a distribution that are greater than value T. Here we let $q = (1 - \Phi_T)$ be the cumulative frequency to point T. A given cumulative frequency is rescaled to a normal probability scale by using the **probit transform** prb(q), which is defined as the solution of

$$\Pr[U < \text{prb}(q)] = q \tag{11.6}$$

where U is a standardized (or **unit**) normal random variable with zero mean and unit variance. Given q, prb(q) is obtained from tables of the unit normal (e.g., Beyer 1968) or from common statistical packages. For example, since $\Pr(U < -1) = 0.1587$, the normal probability scale value associated with a cumulative frequency of $q = 0.1587$ is $\text{prb}(0.1587) = -1$. Since U is symmetrically distributed about zero, it follows that

$$-\text{prb}(0.5 - \delta) = \text{prb}(0.5 + \delta) \qquad \text{for} \quad 0 \leq \delta \leq 0.5 \tag{11.7}$$

Table 11.1 gives the normal probability scale measures for different values of δ. For example, suppose that only 1.2% of the population is below a given

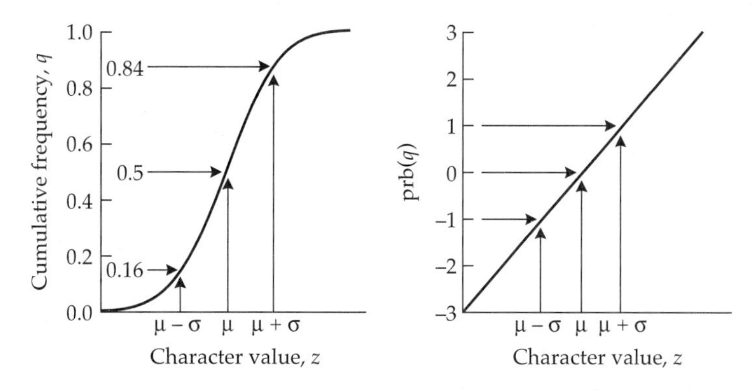

Figure 11.2 The normal probability scale, applied to a normally distributed random variable, z, with mean μ and variance σ^2. **Left**: A plot of cumulative frequency, q, as a function of z gives a sigmoidal curve. **Right**: Rescaling cumulative frequency using the transformation prb(q) (Equation 11.6) gives a straight line. The scale given by prb(q) is called a *normal probability scale*. As this figure shows, a character measure with associated $q = 0.50$ (which for a normal distribution corresponds to the mean) returns a normal probability scale value of zero. A character with $q = 0.84$ (which for a normal distribution corresponds to one standard deviation above the mean) returns a value of $+1$, and a character with $q = 0.16$ (which for a normal distribution corresponds to one standard deviation below the mean) returns a value of -1.

character value, so that $q = 0.012$. Expressing this in the form used by Table 11.1 gives $\text{prb}(0.012) = \text{prb}(0.5 - 0.488) = -\text{prb}(0.5 + 0.488) = -2.25$. Zero on the probit scale corresponds to the population median ($q = 0.50$), which is also the mean for a normal. For a normally distributed variable, a unit change on the probit scale corresponds to one standard deviation (Figure 11.2).

As shown in Figure 11.2, with a normal distribution, cumulative frequency as a function of z is sigmoidal, but it is linear when transformed to a normal probability scale. Nonnormal distributions deviate from linearity on such plots,

Table 11.1 Normal probability scale (probit) values, $\text{prb}(0.5 + \delta)$, as a function of δ.

δ	0.000	0.099	0.191	0.273	0.341	0.394	0.433	0.460	0.477
$\text{prb}(0.5 + \delta)$	0.000	0.250	0.500	0.750	1.000	1.250	1.500	1.750	2.000

δ	0.488	0.494	0.497	0.4987	0.4994	0.4998	0.4999	0.5000
$\text{prb}(0.5 + \delta)$	2.250	2.500	2.750	3.0000	3.2500	3.5000	3.7500	∞

Note: Values of $\delta < 0$ follow from the identity $\text{prb}(0.5 - \delta) = -\text{prb}(0.5 + \delta)$.

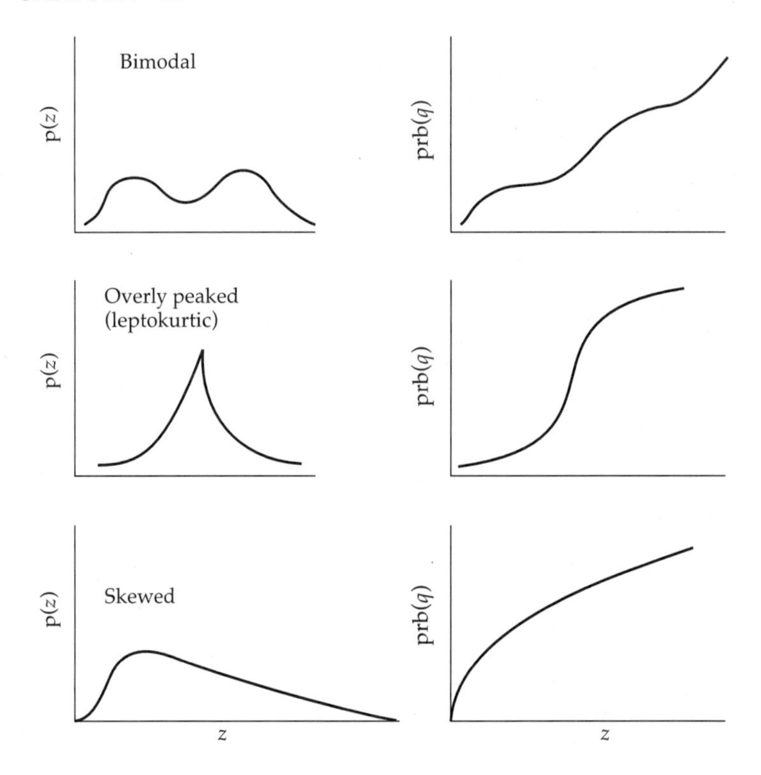

Figure 11.3 Diagnosis of three types of departure from normality using normal probability plots. On the left are probability density functions. On the right are cumulative frequencies plotted on the normal probability scale, prb(q). **Top**: Bimodal distributions show flat regions when plotted on a normal probability scale. **Middle**: Leptokurtic distributions (overly peaked relative to a normal scale) give a sigmoidal curve on a normal probability scale. **Bottom**: Skewed distributions depart from linearity on a normal probability scale by curving downward if the distribution is skewed to the right (as in the figure), or by curving upward if the distribution is skewed to the left.

with multimodal, overly peaked, and skewed distributions all exhibiting characteristic departures (Figure 11.3). The normal probability scale forms the basis for the most powerful tests for departures from normality, such as the small-sample W test of Shapiro and Wilk (1965) and the large-sample D test of D'Agostino (1971).

Example 1. Fisher (1958) gives the following data set (based on unpublished work of Ford and Bull) on the number of vertebrae in herrings. Ignoring the discrete nature of the data, does the normal distribution give a reasonable fit?

	Number of vertebrae					
	53	54	55	56	57	58
Population frequency (%)	0.08	1.06	28.36	61.30	8.91	0.29
Cumulative frequency, q (%)	0.08	1.14	29.50	90.80	99.71	100.00
prb(q)	–3.14	–1.22	–0.54	1.33	2.76	∞

The cumulative frequency associated with a particular character value is the sum of all frequencies up to that point (e.g., the cumulative frequency associated with 55 vertebrae is 0.08 + 1.06 + 28.36 = 29.50%). Plotting cumulative frequency as a function of vertebrae number gives a sigmoidal plot, as shown in the accompanying figure, suggesting a normal distribution. Using unit normal tables, the cumulative frequencies can be transformed to a normal probability scale. For example, the cumulative frequency associated with 55 vertebrae is $q = 0.2950$. Interpolating from the normal distribution table, we find that for a unit normal U, $\Pr(U < -0.54) = 0.2950$, so the prb(q) value associated with 55 vertebrae is -0.54 (0.54 standard deviations below the mean assuming a normal distribution). Plotting prb(q) versus vertebrae number gives a good linear relationship (see the following figure), suggesting that the normal distribution provides a reasonable description of the data.

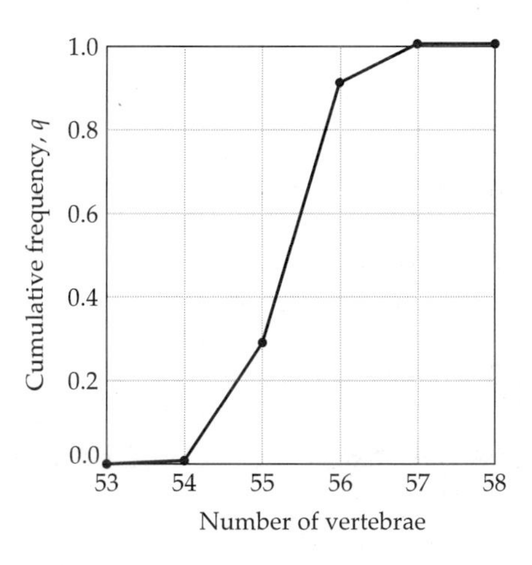

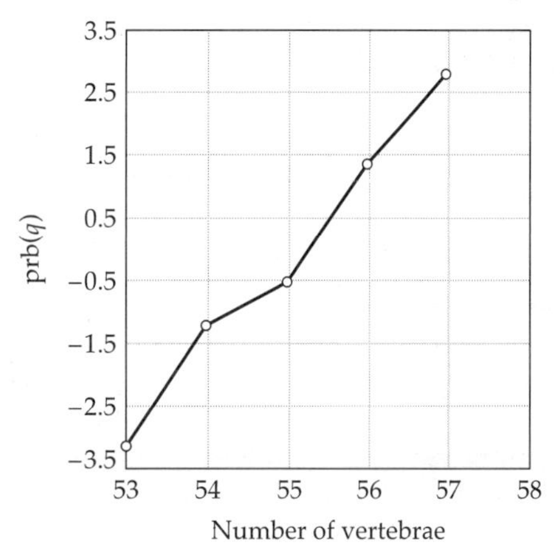

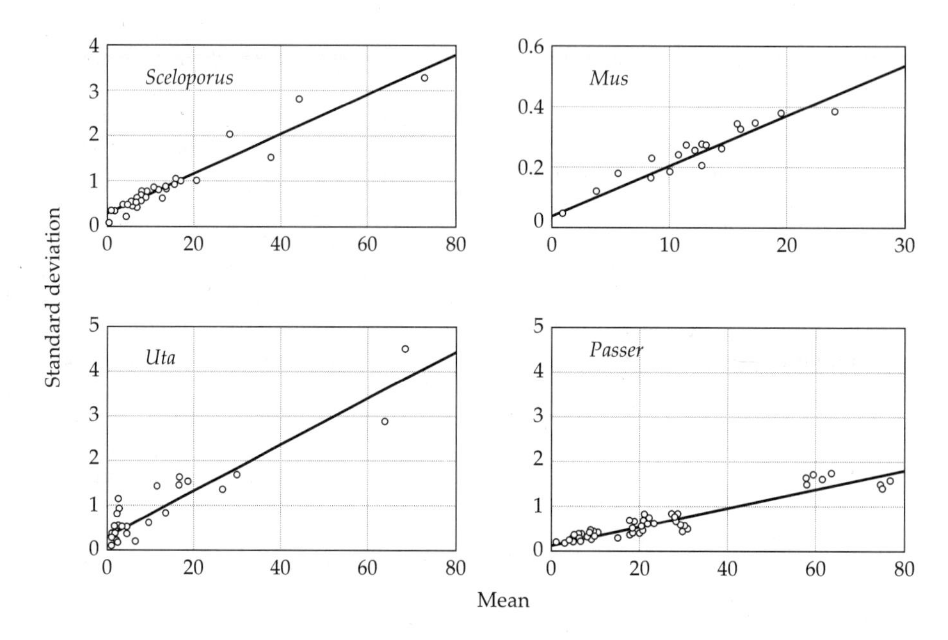

Figure 11.4 Relationship of the standard deviation and the mean for four sets of morphological characters in vertebrates. *Sceloporus* (rough-scaled lizard): meristic traits, several species (Kluge and Kerfoot 1973); *Uta* (side-blotched lizard): scale counts and ratios of morphometric characters (Kluge and Kerfoot 1973); *Mus* (mouse): skeletal and weight characters (several sources given in Soulé 1982); *Passer* (house sparrow): skeletal and feather characters (described in Kluge and Kerfoot 1973).

STABILIZING THE VARIANCE

For studies concerned with relative levels of variation in different populations, it is useful to know whether the observed differences are simply a consequence of scale. For example, it is not uncommon for the variance to increase as the population mean increases (Figure 11.4). By operating on a scale for which there is no discernible relation between the mean and variance, one can be secure that any significant differences in observed levels of variance between samples must be attributable to something other than differences in the mean. Here, we consider the use of **variance-stabilizing transformations** to render the variance on the transformed scale independent of the mean. Wright (1968, Chapters 10 and 11) gives an excellent discussion of the application of such transformations to distributions of quantitative characters.

Kleckowski's Transformation

We have already encountered the idea that a log transformation can render the variance of different samples independent of the mean when the CV is approximately constant on the underlying scale. For this to be strictly valid, the SD must be directly proportional to the mean, $\text{SD}(z) = b\bar{z}$. However, a fairly common situation is for the regression of standard deviations on means to have an intercept significantly greater than zero (Figure 11.4). This causes the coefficient of variation to increase dramatically as the mean approaches zero, which in turn implies an inflation of the variance on the logarithmic scale.

Kleckowski (1949) suggested a simple way to eliminate this problem. If the standard deviation on the original scale can be described adequately by the linear regression $\text{SD}(z) = a + b\bar{z}$, then it follows by rearrangement that $\text{SD}(z)/(\bar{z} + a/b) = b$. Because a/b is a constant, $\text{SD}(z) = \text{SD}[z + (a/b)]$, so the previous ratio is equivalent to the coefficient of variation of $y = z + a/b$. Thus, the use of the transformed variable $y = z + a/b$ in place of z yields an expected coefficient of variation that is equal to b and independent of the mean (also see Example 3). This results in the independence of the mean and variance for the log-transformed variable $\ln(y) = \ln(z + a/b)$.

Example 2. Consider the *Uta* data in Figure 11.4. The intercept and slope of the least-squares regression, $\text{SD}(z) = 0.33 + 0.052\,\bar{z}$, are highly significant (W. C. Kerfoot, pers. comm.). Since the relationship between the standard deviation and the mean closely approximates linearity, the transformation

$$y = \ln(z + 0.33/0.052) = \ln(z + 6.35)$$

renders the variance of different populations independent of the mean.

General Variance-stabilizing Transformations

For more complicated relationships between means and standard deviations, a general formula exists for ascertaining the correct variance-stabilizing transformation. Given the relationship $\sigma_z = f(\mu_z)$, the appropriate transform is given by a result due to Fisher,

$$y = C \int \frac{dz}{f(z)} \tag{11.8}$$

where C is an arbitrary nonnegative constant, generally chosen to set the rescaled variance equal to one. The rescaled character y has a variance that is independent of the mean in any particular population. Thus, the procedure for obtaining a variance-stabilizing transform for any standard deviation-mean relationship, $f(\bar{z})$, is simple. A fit for the function $f(\bar{z})$ is obtained by a least-squares polynomial (or some other nonlinear) regression, and Equation 11.8 is solved using the estimated regression coefficients.

Example 3. Suppose $\text{SD}(z) = a + b\,\bar{z}$. Applying Equation 11.8, the variance-stabilizing transform is given by

$$y = C \int \frac{dz}{a + b \cdot z} = \frac{C}{b} \ln\left(z + \frac{a}{b}\right)$$

Note that there is no unique solution for y, as we can multiply y by any constant and still have a variance-stabilizing transform. Letting $C = b$ recovers Kleckowski's correction, $y = \ln(z + a/b)$.

The Roginskii-Yablokov Effect

Inverse relationships between the coefficients of variation and means of functionally related traits, and their biological implications, have been discussed by many investigators (Pearson and Davin 1924, Roginskii 1959, Yablokov 1974, Rohlf et al. 1983, Kerfoot 1988). We refer to such a relationship as a **Roginskii-Yablokov effect**. Whether such a scaling has important biological underpinnings, as opposed to being a simple statistical artifact, merits some scrutiny.

The most obvious problem in attempting to relate coefficients of variation and means is that the former are inverse functions of the latter, since by definition, $\text{CV}(z) = \text{SD}(z)/\,\bar{z}$. Thus, for the extreme case in which the standard deviation is a constant independent of mean, the expected relationship between the CV and the mean is an inverse hyperbola. On a log-log plot, this is revealed as a linear relationship with a slope of minus one. The data for a number of characters closely approximate this pattern (Figure 11.5), suggesting that in some cases the Roginskii-Yablokov effect is nothing more than a mathematical consequence of the regression of a ratio on its denominator.

A second statistical artifact that can lead to a Roginskii-Yablokov effect is a standard deviation-mean relationship, such as that pointed out in the previous section. In reality, for characters such as length measurements, which take on only positive values, the SD-mean relationship is constrained to pass through the origin since the standard deviation of measurements must be zero if the mean is zero. However, a linear approximation to the data generally yields positive y-intercepts, suggesting that the SD versus mean relationship is slightly bowed upward near the origin. Such behavior can be a simple consequence of measurement error. Suppose that the SD versus mean relationship is adequately described by the function $\text{SD}(z) = \sqrt{k^2 \cdot \bar{z}^{\,2} + V}$, where k is the CV if we measure without error, and V is the additional variance caused by measurement error (assumed to be independent of the mean). With this relationship, the coefficient of variation approaches k as $\bar{z}$ becomes large, but rapidly increases as $\bar{z}$ becomes smaller than $\sqrt{V}/k$. This explanation is probably more relevant to metric characters than

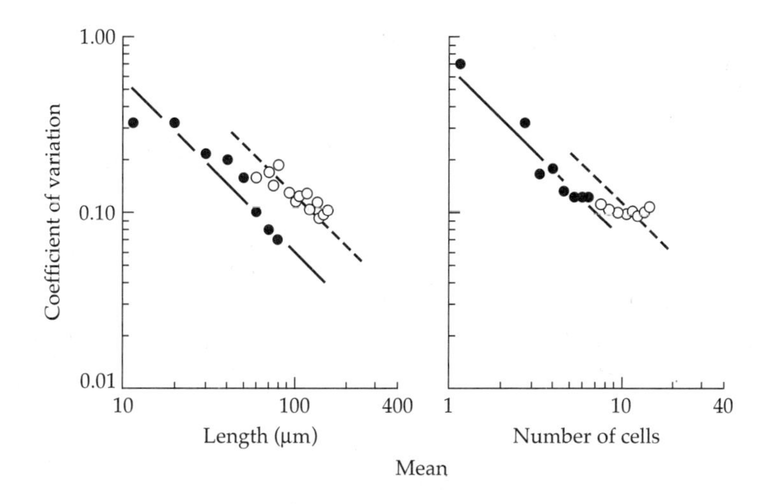

Figure 11.5 Coefficients of variation versus mean phenotypes for lengths and cell counts of tail spines (solid points) and antennules (open points) for samples of the cladoceran *Bosmina longirostris.* As mentioned in the text, a straight line with a slope of minus one is the expected relationship if the standard deviation is independent of the mean. Only the antennule cell counts deviate greatly from this expectation. (From Kerfoot 1988.)

meristic ones. Presumably, as the mean number of counts per individual declines, so does the sampling variance.

Rohlf et al. (1983) pointed out a third statistical artifact that can lead to a Roginskii-Yablokov effect when the data are log-normally distributed. The discretization of log-normally distributed data into classes (delimited by counts or measurement constraints) leads to a reduction of the mean and an inflation of the variance relative to that expected under a continuous log-normal distribution. This in turn causes an inflation of the CV to a degree that increases with decreasing means. In principle, the bias caused by discretization can be corrected by applications analogous to Sheppard's correction (Chapter 2) with normally distributed data (Thompson 1951), but this has not been done with most published data.

Despite the numerous statistical artifacts that may be largely (and in some cases, entirely) responsible for observed negative relationships between means and CVs, there are biological reasons for suspecting that Roginskii-Yablokov effects are sometimes real. A simple explanation was offered by Pearson and Davin (1924). Consider a character that can be represented as the sum of n developmental units,

$$Z = z_1 + z_2 + \ldots + z_n$$

For example, Z could be composed of a number of smaller bones or a number of specific cells. The variance of Z is then

$$\sigma_Z^2 = \sigma_{z_1}^2 + \sigma_{z_2}^2 + \cdots + \sigma_{z_n}^2 + 2\sigma_{z_1,z_2} + \cdots + 2\sigma_{z_{n-1},z_n}$$

Letting $\overline{\sigma}_{z_i,z_j}$ be the mean of the covariance terms, the squared coefficient of variation is

$$\mathrm{CV}_Z^2 = \frac{\sigma_Z^2}{\mu_Z^2} = \frac{\sum_{i=1}^{n} \sigma_{z_i}^2 + n(n-1) \cdot \overline{\sigma}_{z_i,z_j}}{\left(\sum_{i=1}^{n} \mu_i\right)^2} \tag{11.9}$$

Now, for simplicity, let each developmental unit share the same mean (μ_z) and variance (σ_z^2). Then

$$\mathrm{CV}_Z^2 = \mathrm{CV}_z^2 \cdot \left[\frac{1 + (n-1)\overline{\rho}}{n}\right] \tag{11.10}$$

where $\mathrm{CV}_z = \sigma_z/\mu_z$ is the coefficient of variation for each developmental unit, and $\overline{\rho}$ is the mean correlation between units. Except in the unlikely event that all of the component parts are perfectly correlated, the fraction on the right is less than one. Thus, the CV of Z is less than the average CV of the component parts — variation in individual components averages out as the sum increases. Provided that developmental units are similar in size and variance, the Pearson-Davin argument leads to the prediction that the CV of size of a structure is expected to decline as the number of component parts increases.

The Pearson-Davin argument is supported in a number of cases. Analyzing data of Bader and Hall (1960) on osteometric characters in bats, Lande (1977) found that characters consisting of several bones have CVs of 0.02–0.03, while the individual bones themselves have CVs of around 0.03–0.06. Similarly, Kerfoot (1988) found that while the CV for tail spine length in the cladoceran *Bosmina* is 0.10, the CV for length of the component cells ranges from 0.28 to 0.36.

A second, and related, biological explanation for the Roginskii-Yablokov effect applies to analyses that compare the same character in different populations. During development the variance in size of morphological characters often increases initially and then declines as compensatory growth focuses most individuals into a narrow range of phenotypes. This pattern of development, known as **targeted growth** (Riska et al. 1984), would generate a Roginskii-Yablokov effect if the individuals from different samples varied in average age. Those samples containing the oldest, and presumably largest, individuals would be expected to exhibit the lowest CVs.

While much work remains to be done before the underlying determinants of the Roginskii-Yablokov effect can be deciphered, if indeed any generalities are possible, its existence has serious implications for the use of scale transformations. If the CV is not independent of the mean, the routine procedure of log-transforming data prior to analysis will not be successful in stabilizing the variance. Rather, it will cause a negative correlation between the mean and variance on the logarithmic scale. Failure to appreciate the importance of such a scaling effect can lead to interpretative difficulties in long-term studies of natural selection, leading, for example, to the conclusion that directional selection for larger size is accompanied by stabilizing selection about the optimum (Halbach and Jacobs 1971).

The Kluge-Kerfoot Phenomenon

Drawing from morphological data on a number of vertebrates, Kluge and Kerfoot (1973) suggested that traits with high phenotypic variance within populations tend to exhibit high levels of divergence among populations. They argued that this pattern is due, at least in part, to the variance of a trait within a population being inversely proportional to the intensity of stabilizing selection. Their conclusion appears to require the assumption that characters that are under strong stabilizing selection within populations have similar optimal phenotypes in different populations. To support their case, Kluge and Kerfoot plotted for various characters the range of means from different populations against the within-population standard deviation, after first dividing both statistics by the overall population mean.

As in the case of the Roginskii-Yablokov effect, the Kluge-Kerfoot phenomenon may be largely a statistical artifact (Sokal 1976, Rohlf et al. 1983). The statistics employed by Kluge and Kerfoot are expected to be intrinsically correlated due to the fact that both contain the same variable (the mean) in the denominator. This problem is exacerbated by the fact that the range of population means is positively related to the within-population variance that causes sampling error of the means. In an attempt to eliminate the latter problem, Sokal (1976) employed coefficients of within- and among-population variation extracted by analysis of variance. However, since both CVs still share the same denominator, this procedure has the same problem as Kluge and Kerfoot's analysis. Moreover, as Rohlf et al. (1983) have pointed out, there appears to be a Roginskii-Yablokov effect for the among-population CV just as there is for the within-population CV. Thus, the correlation of within- and among-population variance in many existing studies may be an indirect effect of both measures being correlated with a third (the mean). In a study of morphometric differentiation among house sparrow populations, Baker (1980) showed that the Kluge-Kerfoot phenomenon disappeared after the scaling with the mean was eliminated. Until these numerous statistical problems have been accounted for properly with several independent data sets, the biological and evolutionary significance of the Kluge-Kerfoot phenomenon must remain in doubt.

ALLOMETRY: THE SCALING IMPLICATIONS OF BODY SIZE

Many aspects of shape, life histories, behavior, and physiology scale proportionately with body size, both within individuals through development and among different individuals (Huxley 1932; Thompson 1917, 1943; Gould 1966; Vogel 1981, 1989; Wainwright et al. 1982; Peters 1983; Calder 1984; Schmidt-Nielsen 1984; Fleagle 1985). A simple consequence of such scaling is that the phenotypic variation seen for many characters may be little more than an indirect consequence of variation in body size. Huxley (1932) noted that the relationship of most characters with body size can be summarized with a particularly simple mathematical expression — a log-log plot of character y versus some measure of body size x (such as weight or length) generally yields a straight line,

$$\ln y = \ln a + b \ln x \tag{11.11a}$$

This implies a power relationship on the original scale of measurement,

$$y = a\,x^b \tag{11.11b}$$

Characters that scale with size according to Equation 11.11a are said to display **allometry,** with the **allometric coefficient** b providing a measure of the scaling of character value with body size. If $b = 1$, **isometric** growth occurs and the ratio of character value to body size is constant. If $b > 1$, **positive allometry** occurs, with the character becoming proportionately larger as size increases. Characters with $b < 1$ display **negative allometry**, becoming proportionately smaller as size increases. Thus, unless $b = 1$, the shape of an organism changes with size, potentially generating very different morphologies. Figure 11.6 demonstrates allometric scaling for Huxley's classic analysis of the different morphologies of ant castes. The proportionately larger heads associated with soldier ants appear to be a simple scale effect of body size.

Three very different types of data can display allometry, and unfortunately these are often confused and/or treated interchangeably. In **ontogenetic** or **growth allometry**, a character is followed through the development of an individual — each data point corresponds to the character at a different developmental stage, ideally in the same individual. **Static** or **intraspecific allometry** compares the character in different-sized adults (or individuals of the same age) from the same species or population — each data point represents a different adult from a common taxon. **Evolutionary** or **interspecific allometry** examines the character in different species or divergent populations — each data point represents the mean for a different taxon. These distinctions are critical, as a character displaying one form of allometry need not display another, e.g., ontogenetic allometry does not necessarily imply evolutionary allometry, and so forth. For further discussions of these differences, see Lande (1979), Cheverud (1982), and Fleagle (1985).

Allometric equations are routinely used to correct for the effects of size by replacing the character value by the deviation from the fitted allometric equation, i.e., by $y - \widehat{y}$, where $\widehat{y}$ is the predicted character value based on Equation 11.11b. Such residuals can then be used to study variation in character values that is

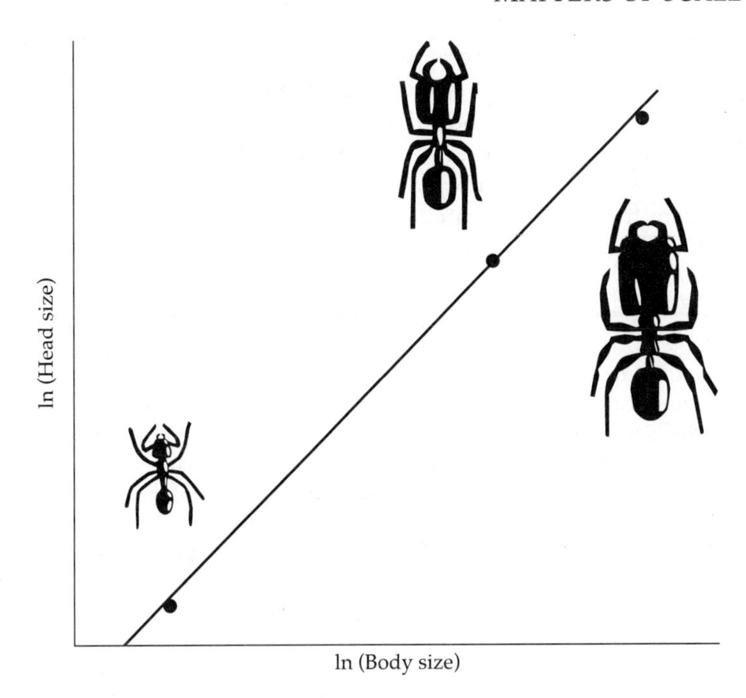

Figure 11.6 Castes of the ant *Pheidole instabilis* display positive allometry for head size. The proportionately larger heads of soldiers can be accounted for simply by changes in size, with head size growing at a faster rate than the rest of the body. (Based on Huxley 1932.)

independent of the allometric consequences of size variation. Packard and Boardman (1987) give a highly readable account of this approach with numerous examples applied to physiological data.

An alternate approach to size correction, especially common in systematics, involves ratios of the character over some measure of body size, e.g., head length/total body length. Unfortunately, there are serious problems with this approach. Unless the character displays isometric growth, taking ratios does not remove aspects of body size, as $y/x = ax^{b-1}$, which is a function of x unless $b = 1$. Further, as noted above, ratios introduce spurious correlations — the variable y/x is usually negatively correlated with x, even if x and y are themselves independent (Pearson 1897, Tanner 1949, Atchley et al. 1976, Albrecht 1978, Atchley and Anderson 1978, Strauss 1985). Thus, character/size ratios often confound, rather than remove, size effects and should be avoided.

REMOVAL OF INTERACTION EFFECTS

Because nonadditive interactions (e.g., genotype $\times$ environment interaction, dominance, and epistasis) can only complicate statistical analysis, additivity is a simplifying assumption underlying much of quantitative-genetic theory. Therefore, in practical applications, it is very desirable to find a scale on which genetic and environmental effects are additive. Since there is no compelling reason to expect that a scale removing one aspect of nonadditivity will also eliminate others, nor to expect that the optimal scale for one population will be suitable for all others, it is rare that a wholly satisfactory solution can be found. Nevertheless, experience has shown that considerable simplification is often attainable.

Example 4. Suppose the genotypes *BB*, *Bb*, and *bb* have genotypic values of 1, 4, and 9 on the original scale of measurement. Applying the square-root transformation, the genotypic values on this new scale of measurement become 1, 2, and 3. Thus, while *B* is slightly dominant to *b* on the original scale, the gene action is perfectly additive on the transformed scale. The square-root transformation condensed the scale between *Bb* and *bb* relative to that between *BB* and *Bb*, resulting in additivity on the new scale. Provided the genotypic value of *Bb* lies within the range of *BB* and *bb*, a scale will always exist for which *Bb* can be made exactly intermediate to *BB* and *bb*.

In the genetic analysis of populations, nonadditive interactions between alleles (dominance) do not usually cause any insurmountable difficulties. However, because epistasis comes in many forms, all of which are quite difficult to quantify with precision, it is highly desirable to work on a scale for which a model involving no more than additive and dominance effects can be shown to be adequate. In some cases, the appropriateness of a particular transformation in removing nonadditive effects can be examined using joint-scaling tests (Chapter 9).

Genotype $\times$ environment interaction can be very difficult to detect in natural populations, since its measurement requires that several distinct genetic groups be grown in a discrete set of environments (Chapter 22). Such an analysis can often be performed in the laboratory or in a common garden experiment, and the results used as a guide to the choice of scale for the field population. However, care should be taken in extending the knowledge gained from controlled experiments too far, since the mode of action of microenvironmental variation in the field may be fundamentally different from that involving the macroenvironmental treatment in the laboratory.

Table 11.2 gives an example of removing a genotype $\times$ environment interaction using a transformed scale. The character considered is the mean weight of offspring for two inbred lines of guinea pigs as a function of litter size, the latter being regarded as a component of the offspring's environment. For both strains, there is a substantial decline in offspring weight with increasing litter size,

Table 11.2 Mean weight (in grams) of offspring (age 33 days) from litter sizes of 1 to 4 for two inbred lines of guinea pigs, for untransformed and transformed [$\log_{10}(x - 80)$] data.

	Litter Size			
	1	2	3	4
Untransformed data				
Strain 13	299.5	264.0	226.8	203.6
Strain 2	213.4	195.0	171.8	155.4
Difference	86.1	69.0	55.0	48.2
Transformed data				
Strain 13	2.341	2.265	2.167	2.092
Strain 2	2.125	2.061	1.963	1.877
Difference	0.216	0.204	0.204	0.215

Source: From Wright 1968.
Note: The untransformed data show a G × E interaction with the difference in strain means being a function of litter size. This dependency is largely eliminated on the transformed scale.

presumably due to competition for maternal care. On the scale of raw measurements, genotype × environment interaction is indicated by the steady decline in the difference between line means with increasing litter size. In the absence of a genotype × environment interaction, this difference should be constant across environments (across different litter sizes). Such behavior is seen upon logarithmic transformation. On the other hand, Wright found that the variance in weight among litter mates was independent of litter size on the original scale but positively correlated with it on the transformed scale. Thus, while logarithmic transformation was able to eliminate genotype × environment interaction of the means, it destabilized the variance.

DEVELOPMENTAL MAPS, CANALIZATION, AND GENETIC ASSIMILATION

We hope that by now the reader has a sense that scale is more than a statistical issue in that by choosing an appropriate scale of analysis, we can often simplify the biological interpretation of characters. For example, allometry shows us that complex changes in character shape can be simple consequences of changes in size. Another potentially important aspect of scale concerns the interpretation of character states as the outcome of the mapping (or rescaling) of some underlying variable into a complex phenotypic space. In concluding our discussion of scale, we first examine **developmental maps** for discrete (meristic) characters and then explore some of the potential implications for continuously distributed characters.

Estimating Developmental Maps

To motivate the idea behind developmental maps, consider hypertension (high blood pressure) in humans. When blood pressure exceeds some critical (or **threshold**) value, either for genetic or environmental reasons, the individual displays hypertension. Thus, classification of individuals as either normal or hypertensive transforms a continuous underlying measure (blood pressure) into a dichotomous character (normal vs. hypertensive). Many other diseases appear to result when some underlying physiological variable exceeds a threshold value (Chapter 25).

Letting y be the value of the underlying variable (the **liability**), the observed phenotypic value z can be regarded as a function of the liability, $z = \phi(y)$, where ϕ is the **developmental map** that rescales the underlying liability value y into the phenotypic value z. For a simple threshold character, letting y_c be the critical value, and denoting the alternate character states by 0 and 1,

$$\phi(y) = \begin{cases} 0 & \text{for} \;\; y < y_c \\ 1 & \text{for} \;\; y \geq y_c \end{cases}$$

This type of scaling can easily be extended to characters displaying multiple states. For example, a character with three discrete states can be represented as the mapping of some underlying continuous value y by

$$\phi(y) = \begin{cases} \text{character state 1} & \text{for} \;\; y < y_{c_1} \\ \text{character state 2} & \text{for} \;\; y_{c_1} \geq y < y_{c_2} \\ \text{character state 3} & \text{for} \;\; y \geq y_{c_2} \end{cases}$$

Following Waddington (1949), a character (or character state) is said to be **canalized** if the resulting phenotype is fairly insensitive to changes in the underlying environmental and/or genetic factors. For example, a continuous character can show a **zone of canalization** — a region of ϕ that is rather flat and within which changes in the underlying genetic and environmental factors result in relatively little (or no) change in the phenotype (Figure 11.7). Assuming that the distribution of the underlying liability values is normal with variance σ^2, the width of the developmental map associated with a particular character state can be estimated in units of σ (Figure 11.8). This procedure is often referred to as **probit analysis** because cumulative frequencies are measured on the probit (normal probability) scale.

Suppose the character of interest has k discrete states, which we order from 1 to k. Let $q(m)$ be the cumulative frequency to (and including) class m, and define $x_m = \text{prb}[q(m)]$ to be the probit-scale value associated with $q(m)$, i.e., x_m satisfies $\Pr(U \leq x_m) = q(m)$ where U is a unit normal. The width of liability values that map to class m is estimated by

$$D_m = x_m - x_{m-1} \qquad \text{for} \;\; 2 \leq m \leq k-1 \tag{11.12a}$$

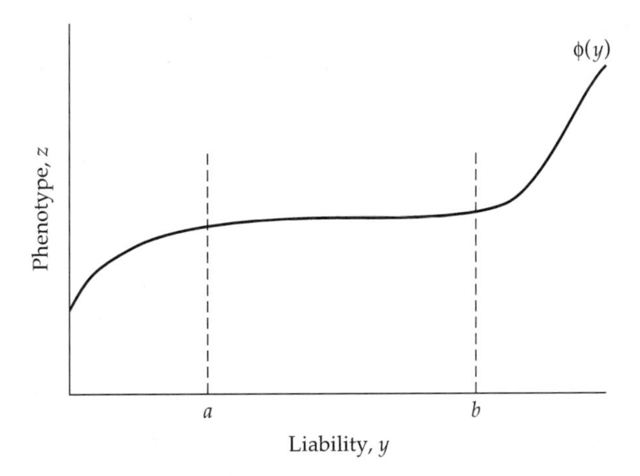

Figure 11.7 A nonlinear developmental map. This map shows a zone of strong canalization for $a \leq y \leq b$, as liability values in this range give essentially the same phenotype.

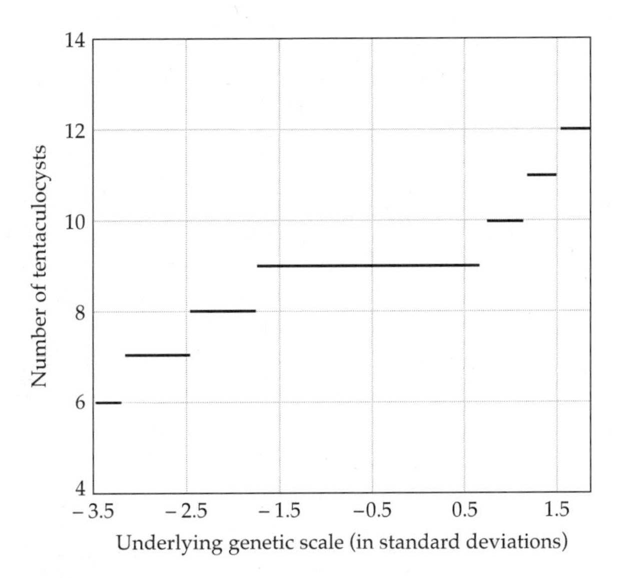

Figure 11.8 Reconstruction of the development map for tentaculocyst number in the jellyfish *Ephyra* using probit analysis under the assumption that liability is normally distributed on some appropriate underlying scale. Strong canalization is seen, with a larger range of liability mapping to the nine-tentaculocyst character state relative to the range mapping to other character states. (Original data of Browne, from Wright 1968.)

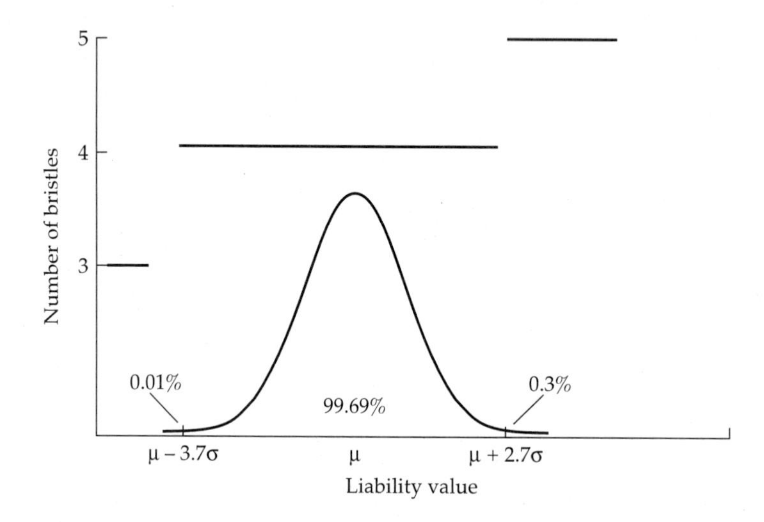

Figure 11.9 Reconstructing a developmental map for *Drosophila* scutellar bristle number. The bulk of the population (99.69%) had four bristles, while 0.01% had three, and 0.3% had five. Assuming a normal distribution of liability on an appropriate underlying scale, the developmental map translating liability into character value (here bristle number) can be estimated via probit analysis. Individuals with a liability value less than 3.7 standard deviations below the mean have three bristles, while individuals whose liability exceeds 2.7 standard deviations above the mean have five bristles. Details are given in Example 5.

where the units of D_m are in terms of number of standard deviations of liability values (Figure 11.9). Since $q(k) = 1$ and $q(0) = 0$, the widths for the first and last classes are not defined, as $\text{prb}[1] = -\text{prb}[0] = \infty$. Using standard results (Kendall and Stuart 1977, p. 254), the sampling variance for the estimator D_m is given by

$$\sigma^2(D_m) = \frac{1}{n}\left\{ \frac{q(m-1)[1-q(m-1)]}{[p(x_m)]^2} + \frac{q(m)[1-q(m)]}{[p(x_{m-1})]^2} - 2\,\frac{q(m-1)[1-q(m)]}{p(x_m)\,p(x_{m-1})} \right\} \tag{11.12b}$$

where $p(x) = \exp(-x^2/2)/\sqrt{2\pi}$ is the unit normal density function.

Example 5. Sheldon et al. (1964) examined scutellar bristle number in *Drosophila melanogaster* in a number of isogenic lines derived from a common ancestral population. Consider the following data for females from their isogenic line 5, which have 3, 4, or 5 bristles with the following frequencies:

Class	Observed	Frequency	Cumulative frequency	prb[$q(m)$]
3	1	0.00012	0.00012	−3.70
4	8124	0.99693	0.99705	2.74
5	24	0.00295	1.00000	—

Here $n = 8149$, $x_3 = \text{prb}[0.00012] = -3.70$, and $x_4 = \text{prb}[0.99705] = +2.74$. Applying Equation 11.12a, the estimated width of the developmental map for class 4 is

$$D_4 = x_4 - x_3 = 2.74 - (-3.70) = 6.44$$

Noting that $q(3) = 0.00012$, $p(-3.70) = 0.0004$, $q(4) = 0.99705$, and $p(2.74) = 0.0093$, Equation 11.12b gives the sample variance for D_4 as

$$\begin{aligned}\text{Var}(D_4) &= \frac{1}{8149}\left[\frac{0.00012 \times 0.99988}{0.0004^2} + \frac{0.99705 \times 0.00295}{0.0093^2}\right.\\ &\quad \left. - 2\,\frac{0.00012 \times 0.00295}{0.0004 \times 0.0093}\right]\\ &= 0.096\end{aligned}$$

yielding a standard error of $\sqrt{0.096} = 0.31$ and an estimated width of liability values mapping to the four-bristle class as $6.44\sigma \pm 0.31\sigma$. Sheldon et al. found considerable variation in the strength of canalization across isogenic lines. The mean D_4 of all lines was 5.39, but ranged from 6.44 ± 0.31 down to 3.34 ± 0.10.

One reason for using probit analysis to define developmental maps is that untransformed frequencies can give a misleading picture of the strength of canalization. Consider two populations with identical developmental maps, and with normal distributions of liability values with the same variance but different means. For the first population, suppose that individuals with liability values within one standard deviation of the mean map to the four-bristle class, resulting in 68% of the population having four bristles. In the second population, individuals with liability values between one and three standard deviations above the mean map to the four-bristle class, resulting in only 15.7% of the population having four bristles. In both cases, the width of the four-bristle class is 2σ, but on an untransformed scale, the first population appears to be more canalized for the four-bristle class.

While probit analysis allows for comparisons of the strength of canalization, interpreting these results is not always straightforward. An increase in the width of a canalized class can occur through the evolution of genotypes with wider developmental maps. However, increased class width can also occur by a reduction

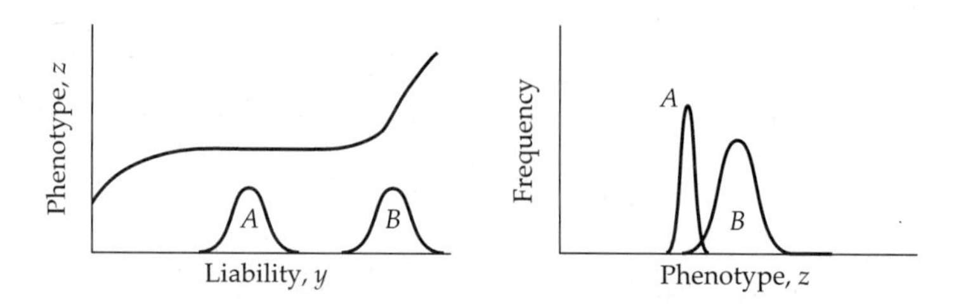

Figure 11.10 When the developmental map is nonlinear, a simple change in the mean of the liability distribution (**left**) can result in complex changes in the character distribution (**right**). Here, populations *A* and *B* have different means, but otherwise identical liability distributions. Besides changing the character value, this simple increase in mean liability also results in an increase in character variance as liability is selected outside the zone of canalization.

of the variance of the distribution of liability values without any change in the developmental map on the absolute scale. Because class widths are estimated in terms of liability standard deviations, as the variance of liability decreases, the class width increases.

Selection and Canalization

If the developmental map is highly nonlinear, a simple change in the mean of the underlying liability distribution can result in complex changes in the character distribution. For example, if a population has liability values distributed around a canalized region, as the population is selected away from this region, the character becomes much more variable (Figure 11.10). As shown in the figure, this increase in character variance can occur without any change in the variance of the underlying liability distribution.

Given genetic variation in canalization (due to genetic differences in developmental maps), the strength of canalization is itself a selectable character. For example, Waddington (1959) selected for increased canalization of facet number for bar-eyed mutants of *Drosophila melanogaster*. Families were split, with one group being reared at 18°C, the other at 25°C. Families showing the smallest difference in facet number between temperatures were selected to form the next generation. After only five generations of selection, Waddington observed a significant increase in the amount of canalization (Table 11.4).

An interesting example of natural selection for increased canalization is given by Clarke and McKenzie (1987) for Australian sheep blowflies (*Lucilia cuprina*). These flies are a major pest, and in 1955, the insecticide dieldrin was introduced in an attempt to control them. By 1957, most flies in Australia were resistant to dieldrin, and it was replaced by diazinon. Although resistance to diazinon

Table 11.4 Mean number of eye facets in families ($\pm$ one standard error) for the control and three lines selected for increased canalization.

Line	Reared at 18°C	Reared at 25°C	Difference
Unselected	156.3 ± 4.8	55.5 ± 4.5	100.8 ± 4.5
Selected A	106.0 ± 1.2	96.5 ± 1.7	14.5 ± 2.3
Selected B	100.7 ± 1.4	95.3 ± 1.7	5.5 ± 2.0
Selected C	111.4 ± 1.7	99.5 ± 2.7	12.9 ± 3.1

Source: Waddington 1959.

Table 11.5 Level of fluctuating asymmetry, measured as the absolute value of the differences in three bristle characters on the left and right side for different strains of *Lucilia.*

Strain	Mean Asymmetry
SWT	1.83 ± 0.08
M15	1.81 ± 0.07
Rop-1 / Rop-1	1.92 ± 0.07
Rdl / Rdl	3.23 ± 0.08

Source: Clarke and McKenzie 1987.
Note: The SWT and M15 lines are susceptible to the insecticides diazinon and dieldrin; the other two lines are fixed for one of the resistance alleles (*Rop-1* for diazinon, *Rdl* for dieldrin) and wild-type (susceptible) for the other.

developed in 1976, it is still used. A single locus is responsible for resistance to each insecticide — *Rop-1* for diazinon, *Rdl* for dieldrin. Clarke and McKenzie measured developmental stability for different lines using fluctuating asymmetry (Chapter 6) by comparing the absolute difference in the numbers of three bristle characters on the left versus right sides of flies. *Rdl* alleles cause a significant disruption of developmental canalization, as indicated by a much higher value of fluctuating asymmetry relative to the other strains, while *Rop-1* alleles apparently do not (Table 11.5). However, after 12 generations of continuously backcrossing the *Rop-1 / Rop-1* strain to the susceptible line M15, while selecting for the retention of *Rop-1,* the amount of fluctuating asymmetry associated with *Rop-1 / Rop-1* genotypes increased to 5.09 ± 0.08. Clarke and McKenzie interpret these results as implying that when the *Rop-1* allele arose, it had deleterious pleiotropic effects on development. Strong selection retained the *Rop-1* allele in the population, allowing selection for modifier loci that reduce *Rop-1*'s pleiotropic effects on developmental stability. These modifiers are lost during the continual backcrossing to a strain lacking them, hence the increase in mean asymmetry. The lack of comparable modifiers for *Rdl* alleles may be due to the fact that dieldrin was used only for a limited period, insufficient time for the selection of modifiers.

This observation of a major gene disrupting the apparent canalization of a character is fairly common, as revealed in several artificial-selection experiments involving characters that are normally very highly canalized. For example, scutellar bristle number in *Drosophila* (which is highly canalized at four bristles in almost all *Drosophila* species) could be down-selected to two bristles using the character variance exposed when the mutant *scute* was introduced (Rendel 1965, 1977, 1979; Rendel and Sheldon 1960; Rendel et al. 1966; Fraser 1963, 1967, 1970). Likewise, vibrissae number (whiskers on the nose and fore limbs) in the house mouse, which is normally highly canalized at 19 was considerably down-selected using the variance exposed by the sex-linked gene *Tabby* (Dun and Fraser 1958, 1959; Fraser and Kindred 1960). Still another example is ocellar bristle number in *Drosophila subobscura*, normally highly canalized at 8, which responded to selection for decreased number when the mutation *ocelliless* was introduced (Sondhi 1961). Again, this change in mean was accompanied by an increase in the variance. In all of these systems, the increase in variance presumably results from the population being moved outside of a zone of canalization (e.g., Figure 11.10). This idea is supported by the fact that, for all three systems, back-selection to return the character values to the original mean value reduced the character variance. Thus, a return of the mean back to that seen in normal populations also returns the population to the zone of canalization.

Finally, we note that zones of canalization are themselves subject to selection. Changes in the developmental map can occur through changes in the variance of the underlying liability. More subtly, the form of the map itself can be modified by selection if the population displays genetic variation in developmental maps. Through patient directional selection, Scharloo and his associates (reviewed in Scharloo 1988) obtained populations of *D. melanogaster* with a mean of eight scutellar bristles per individual, which is double the normally canalized value of four. As in previous examples, this shift in the mean was accompanied by a substantial increase in genetic variance for the trait as the population moved out of the original zone of canalization. Artificial stabilizing selection around the new mean eventually doubled the width of the eight-bristle class on the underlying scale.

Genetic Assimilation

Another potential consequence of moving a population outside of a zone of canalization is **genetic assimilation**, wherein a character state that originally appears to be environmentally determined apparently becomes genetically determined following several generations of selection. This term was coined by Waddington (1953) to account for the behavior of crossveinless, a defect in the wing venation pattern of *Drosophila melanogaster*. In the lines studied by Waddington (1952, 1953), no crossveinless flies were found in base populations raised under normal temperatures. However, when temperature-shocked as pupae, some fraction of

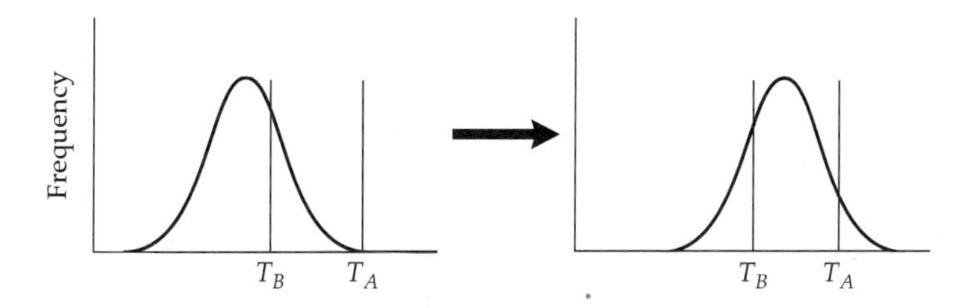

Figure 11.11 A model for the genetic assimilation of the crossveinless character state. Suppose flies display the crossveinless phenotype when the value on some underlying liability scale exceeds some threshold value T_A. **Left**: The effect of the temperature shock is to reduce the threshold to $T_B < T_A$, exposing variation in liabilities that are otherwise masked. **Right**: Selection under temperature shock is effective in increasing liability values, eventually shifting the distribution far enough to the right such that crossveinless flies are seen at normal temperatures.

flies displayed the crossveinless (*cvl*) phenotype. These environmentally induced *cvl* flies were used to form the subsequent generation. Not only did the frequency of *cvl* flies induced by temperature shock increase, but *cvl* flies appeared in the untreated populations, reaching frequencies in excess of 95% after 18 generations of selection. Waddington had genetically "assimilated" the crossveinless character state from an apparently environmentally determined character in the base population.

Waddington (1949, 1953, 1957) suggested that changes in canalization could account for genetic assimilation, but was rather vague in terms of an explicit model. As Figure 11.11 shows, assimilation can follow as a simple consequence of moving a population outside of a zone of canalization. One environment exposes more of the underlying genetic variation in liability than another, making selection in that environment more effective.

II

Quantitative Trait Loci

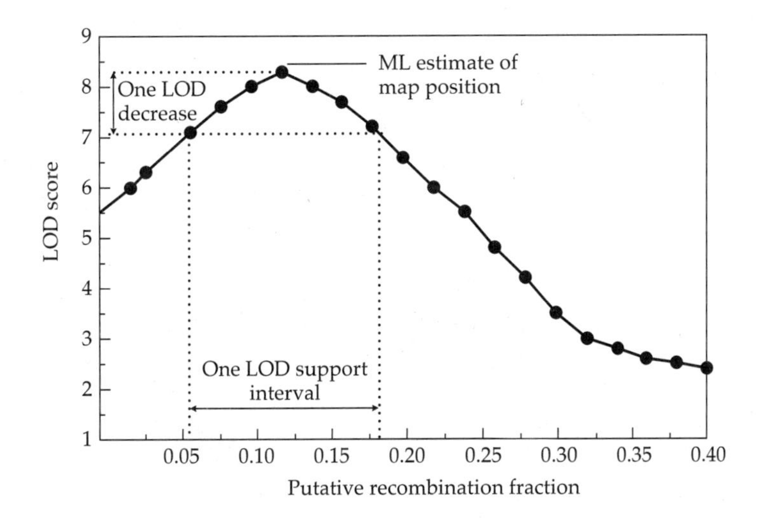

12

Polygenes and Polygenic Mutation

As discussed in Chapters 9 and 10, line-cross analysis and inbreeding experiments can provide a crude characterization of the **genetic architecture** of a character, yielding information on the nature of genetic interactions (dominance and epistasis) as well as some insight into the number of loci accounting for the difference in mean character value between lines. For a variety of reasons, more precise analysis of the individual loci underlying a character is often desirable. For example, from a population-genetics standpoint, we are interested in the distribution of allelic effects within and among loci. If genetic variation in a character is mainly due to a few genes with large allelic effects, the response to selection can be considerably different from that predicted from standard models of quantitative genetics that assume a large number of loci with small effects. From a developmental genetics standpoint, we are interested in how variation at particular loci translates into variation in complex characters.

An ultimate understanding of the mechanisms responsible for expressed variation in quantitative characters requires information at the molecular level. Are the genes (and alleles) underlying quantitative-trait variation the same as those with large and obvious qualitative effects that molecular biologists are attracted to? If so, does most of the expressed variation result from DNA sequence variation in coding regions or in nontranscribed regulatory regions? How much quantitative-trait variation is a consequence of variation in timing vs. magnitude of gene expression? To what extent are the genes underlying the expression of a quantitative trait present in redundant sets that can complement and/or replace each other?

These questions, still largely unanswered, represent a large and formidable gap in our understanding of the genetics of quantitative traits. However, with the advent of fine-scale molecular mapping techniques (Chapters 14–16), we are now poised for rapid advances in these areas. This chapter summarizes the current state of our knowledge of the nature of the loci underlying quantitative traits. We devote considerable attention to the rate at which mutation generates variation for quantitative traits, as this must ultimately dictate the potential rate of phenotypic evolution. Although the mutational mechanisms responsible for quantitative-trait variation remain largely unexplored, phenotypic data strongly support the idea that mutation is a powerful force at the level of polygenic traits.

THE GENETIC BASIS OF QUANTITATIVE-GENETIC VARIATION

Problems with the actual molecular characterization of loci aside, precise analysis of the phenotypic effects of individual loci underlying quantitative traits has remained elusive for a purely statistical reason. As noted in Chapter 9, limitations introduced by linkage restrict analysis to effective factors (linked regions of the genome that may contain several loci influencing the trait), which can lead to biased estimates of the effects of individual loci. For example, if a factor contains linked loci at which alleles influencing character expression are in coupling phase ($++$ and $--$), the effects of individual loci will be overestimated, while if the alleles are in repulsion phase ($+-$ and $-+$), the effects will be underestimated. The larger the average size of an unrecombined region, the greater the opportunity for bias from lumping the effects of several loci.

Because of this empirical restriction of dealing with factors rather than individual loci, we need a reasonable term for a locus at which segregation contributes to the variance of a quantitative character. Mather (1941) suggested the term **polygene** for a gene of small effect underlying a character. However, for reasons to be discussed below, the distinction between major genes (sometimes referred to as **oligogenes**) and polygenes is by no means straightforward. A less ambiguous approach is to refer to a locus underlying a quantitative character as a **QTL**, for **quantitative-trait locus**, as suggested by Geldermann (1975).

Major Genes and Isoalleles

Is there a real distinction between genes of major and minor effects or are they just extremes on a continuous distribution? The presence of **isoalleles** at major loci provides some insight. Stern and Schaeffer (1943) coined this term to describe alleles whose effects are essentially indistinguishable except under certain environmental settings. They found that three allelic variants of the *cubitus interruptus* locus, which has a major influence on the wing venation pattern of *Drosophila melanogaster*, displayed the same wild-type phenotypes at 26° C, but exhibited different wing venation patterns at lower temperatures. Similarly, Muller (1935) and Green (1959) found a large number of isoalleles affecting eye color at the *white* locus of *D. melanogaster*. Since both of these loci were originally detected from alleles of major effect (**major alleles**), the subsequent discovery of isoalleles suggests the possibility of a continuum of allelic effects. Isoalleles have been found at a variety of other loci in a number of organisms, with alleles having a large effect on one character often showing a wide range of **pleiotropic effects** on the expression of other traits. Thus, alternate alleles at a locus can display a spectrum of effects, which depends on the genetic and environmental background (Coen et al. 1986, Romeo and McKusick 1994). The alleles present at a locus, not the locus itself, determine if that locus has a major effect on character variation.

Modifier loci provide a mechanism for "fine tuning" a major allele by altering its main and/or pleiotropic effects. As noted in Chapter 11, modifiers can sometimes magnify the expression of a major allele, thereby inflating the response to selection. They can also mask the effect of a major allele, producing a zone of canalization. In natural populations of *Drosophila,* extensive genetic variation has been found for loci that modify the effects of the alleles at the *crossveinless* (Milkman 1970), *Beadex-3, notchoid,* and *scalloped* loci (Thompson and Spivey 1984), all of which influence wing venation. Modifiers also appear to be involved in the evolution of mimicry in butterflies. Initially, a major allele causes wing patterning to crudely resemble a distasteful model species, and the pattern is subsequently refined by a number of modifier loci (Turner 1977, 1981). Modifiers can also alter the dominance relationships of major alleles, as demonstrated for the melanic form of the famous black-peppered moth *Biston betularia* (Ford 1975; other examples are discussed in Thompson and Thoday 1972).

Modifiers can have either general or specific effects. An example of the former is given by Thompson (1973, 1974), who found that modifiers selected to alter a specific wing vein in *Drosophila* had similar effects on the expression of major alleles at three different loci: *cubitus interruptus, short vein,* and *veinlet.* On the other hand, Fraser (1968) found that the expression of *Drosophila* bristle loci *scute* and *extravert* is influenced by different sets of modifiers. In this case, the modifiers appear to be more gene-specific than pathway-specific. Through artificial selection experiments, Weber (1992) was able to alter the relative sizes of two very small regions (less than 100 cells across) at the base of the *Drosophila* wing. If this result is general, the control of developmental detail may be modulated at extremely localized levels.

Whether loci exist at which *all* allelic variation results in only small phenotypic changes is still an open question. The very nature of this question and the bias towards isolating factors of fairly large effect make this question extremely difficult to answer.

The Molecular Nature of QTL Variation

While variation at "standard" loci (those that make a transcript) may account for most genetic variation, there are clear situations in which "nonstandard" genetic regions make significant contributions to the variance of a character. For example, Mather (1944) presented evidence that heterochromatic regions (containing highly condensed DNA, typically devoid of transcribable genes) can contribute to phenotypic variance. An extreme view, argued cogently by Cavalier-Smith (1978, 1985), is that genome size, independent of actual sequence composition, can influence certain aspects of the cellular phenotype (such as cell size and cell-division rate), which have cascading effects on the emergent properties of the phenotype. In plants, for example, there is compelling evidence that genome size variation within populations influences growth rate and morphology (Bachmann et al. 1985, Cullis 1985, Schneeberger and Cullis 1991). In this sense, variance in genome size translates into a continuum of allelic effects at a single, albeit ill-defined, super-locus, which may or may not be in disequilibrium with the effects of underlying "standard" loci. Diffuse effects like these are essentially impossible to map genetically.

Phenotypes are the end result of a multiplicity of genomic interactions. Structural variation refers to differences in actual gene products that result from differences in amino acid sequences of a protein or in nucleotide sequences of a functional RNA. Regulatory changes refer to allelic variation that alters the timing and/or expression of gene products. This distinction is not sharp, as changes in the structure of one gene product can influence the regulation of another. One immediate question is whether the majority of quantitative variation is due to structural or regulatory variation. Although this question is unresolved, it is generally felt (e.g., Wright 1968, Wilson 1976) that regulatory variation resulting in differences in the amounts and distributions of gene products contributes more than variation in the actual structure of gene products. This view is based, in part, on differences in the mutational "target size" for structural versus regulatory regions — the region that actually codes for the structural part of a gene is generally small relative to regions in which mutations can alter the regulation of that gene. As will be discussed shortly, mutational data in *Drosophila* are consistent with this view.

In fairly general terms, the current picture is that a large component of gene regulation involves control of transcription by a small number of short sequences surrounding the transcribed region. Control regions act in a modular fashion to evoke a specific regulatory program (e.g., Struhl 1987), and single nucleotide changes within them often result in radical (qualitative) changes in gene expression, sometimes creating major alleles. For example, in *Drosophila*, loss-of-function mutations in the *ultrabithorax* gene result in a duplication of the thorax, implicating regulatory properties of this gene in the development of segmental identity. Waddington (1957) found that bithorax **phenocopies** can be obtained in otherwise normal flies by exposing early embryos to ether, and that the tendency to express such a phenotype can be advanced by artificial selection. A dramatic example of genetic assimilation (Chapter 11), such evolutionary change appears to result from allele frequency change at the *ultrabithorax* locus itself (Gibson and Hogness 1996). This example provides clear evidence for the potential of segregating polymorphisms in a gene encoding for transcription factors to contribute to morphological variation.

Molecular biology has shattered our view of a rather quiescent genome, replacing it with a picture of a genome undergoing a large number of seemingly bizarre processes, including the insertion and excision of mobile gene sequences, changes in gene copy number by unequal crossing over, and gene conversion of large tracts of DNA (Figure 12.1). This emerging view of a dynamic genome blurs the distinction between mutation and recombination, supporting the idea that the mutational origin of quantitative-genetic variation may only rarely involve simple point mutations or short insertions/deletions (Mackay 1988, 1989; Frankham 1988).

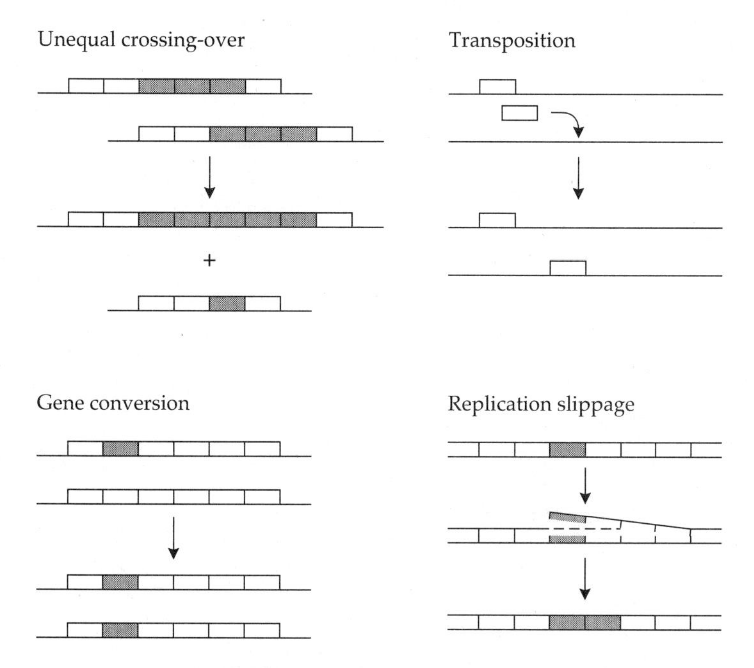

Figure 12.1 Some of the molecular processes that influence genome structure. In **unequal crossing-over**, copy-number changes result from recombination between tandemly repeated sequences (usually on sister chromosomes) that are aligned out of register. **Gene conversion**, the nonreciprocal exchange of DNA between two sequences, can change the number of copies of a particular allele that an individual carries. **Transposition** occurs when a mobile genetic element inserts into a new site, usually (but not exclusively) while still retaining a copy in the original position. **Replication slippage** is a rough analog of unequal crossing-over, with the single stands of an individual DNA mispairing during replication.

In addition to producing loss-of-function alleles by direct insertion into a coding region, insertion of mobile genetic elements (**transposons**) near a gene can alter gene expression, since the elements themselves often contain regulatory sequences. In addition, when transposable elements excise, they frequently leave altered sequences in their wake. Considerable regulatory variation (resulting in a wide range of phenotypic effects) has been found in mutant alleles arising from such excision events, e.g., at the *bronze-1* locus in maize (Schiefelbein et al. 1988), the *pallida* locus in snapdragons (Coen et al. 1986), the *An1*/*An2* anthocyanin synthesis loci in petunias (Gerats et al. 1984), and the *rudimentary* and *rosy* loci in *Drosophila* (Daniels et al. 1985, Tsubota and Schedl 1986). More than 50% of all known mutations with phenotypic effects in *Drosophila* are associated with transposable element insertions (Finnegan 1992).

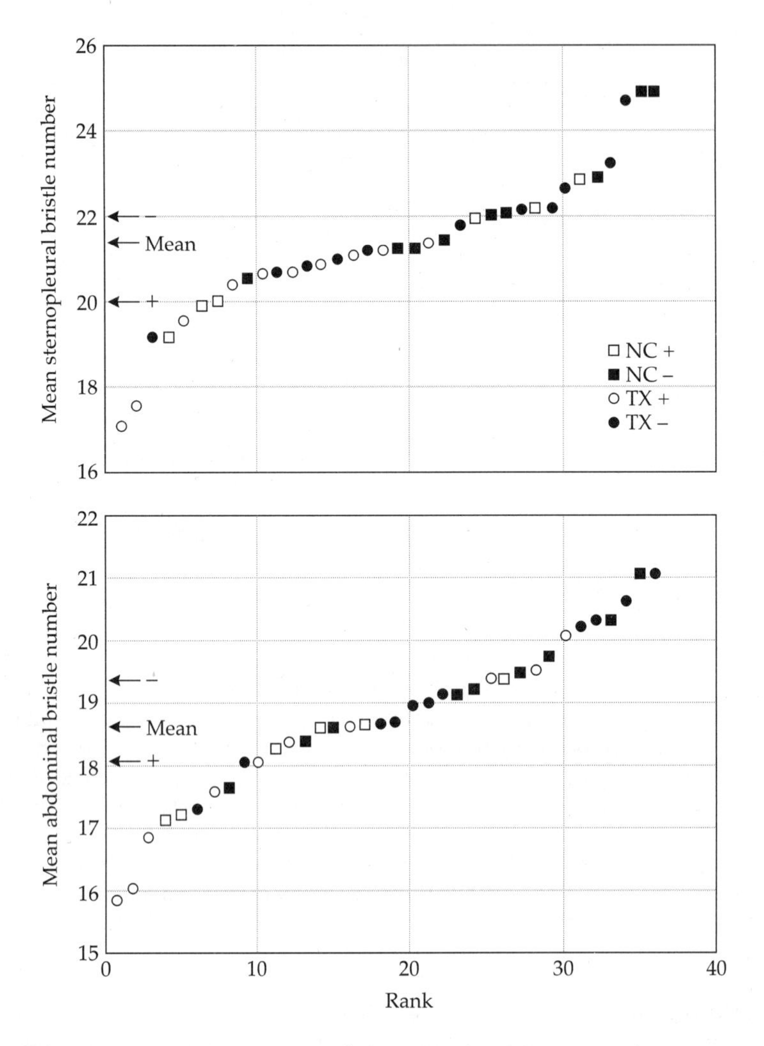

Figure 12.2 Association between chromosomes containing at least one insertional variant in the *achaete-scute* region and sternopleural (**top**) and abdominal (**bottom**) bristle numbers. Squares denote samples from a natural population at Raleigh, NC, and circles denote samples from Texas. Open symbols denote extracted chromosomes containing at least one insertional variant (+), while filled symbols denote chromosomes without insertions (−). The arrows denote the mean bristle number associated with + and − chromosomes and the mean for the entire sample. (From Mackay and Langley 1990.)

In an investigation of the importance of mobile-element insertions as contributors to quantitative variation in natural populations, Mackay and Langley (1990) asked whether abdominal and sternopleural bristle numbers in *Drosophila* are correlated with insertional variation in the *achaete-scute* region. Because this locus is known to be capable of mutating to major alleles for the presence / absence of sensory hairs, it is a reasonable **candidate locus** for variation in bristle numbers. The authors simply asked whether chromosomes with at least one insertion in the *achaete-scute* region were associated with altered bristle numbers. A strong association was found, with chromosomes bearing inserts having lower numbers of both types of bristles (Figure 12.2). Further analysis suggested that these insertions account for about 5% of the total genetic variation observed in natural populations for both bristle number characters. Variation involving insertions, deletions, and nucleotide substitutions within a 45 kb region surrounding a second neurogenic locus, *scabrous,* accounts for another 10% or so of the genetic variance (Lai et al. 1994). Thus, we again see that loci known to generate mutant alleles with major qualitative effects are also responsible for quantitative variation.

Natural populations of *Drosophila mercatorum* are known to be polymorphic for a complex syndrome known as abnormal abdomen, characterized by reduced age at maturity, reduced longevity, increased early-age fecundity, and retention of juvenile abdominal cuticle in the adult stage (Templeton and Rankin 1978). The syndrome is associated with the insertion of the R1 transposable element into the coding regions of the 28S ribosomal genes, which exist in a tandem array (multiple copies) on the X chromosome (DeSalle et al. 1986). The expression of abnormal abdomen requires that at least a third of the ribosomal gene copies contain the insert, and the inserted genes must not be underreplicated in the larval fat body (DeSalle and Templeton 1986, Templeton et al. 1989). A population survey indicated a broad distribution of insertion number variants, ranging from chromosomes with almost no insertions to those in which all genes had insertions (Figure 12.3). Moreover, there was an association between numbers of rDNA inserts and the allelic status at a locus determining underreplication, such that both of the components necessary for abnormal abdomen expression tended to cosegregate on the same chromosomes (Hollocher et al. 1992). Thus, the two tightly linked loci act as a multiallelic supergene to determine the expression of the abnormal abdomen syndrome. As in the preceding examples for bristle numbers, these results support the contention that a substantial fraction of the standing variation for morphological traits may be associated with a relatively small number of loci. This particular example may also be fairly generalizable — R1 elements interrupt the 28S rDNA coding region in a wide variety of insects (Jakubczak et al. 1991).

Since multigene families, such as the ribosomal gene families, are common features of all eukaryotic genomes, variation in copy number (in addition to copy type) may also contribute to quantitative variation. The clearest indication of this again comes from evidence on ribosomal genes, which typically exhibit copy number variation due to unequal crossing-over (Figure 12.1). Frankham and colleagues (reviewed in Frankham 1988) demonstrated that selection for reduced bristle number in *Drosophila* often leads to elevated frequencies of chromosomes

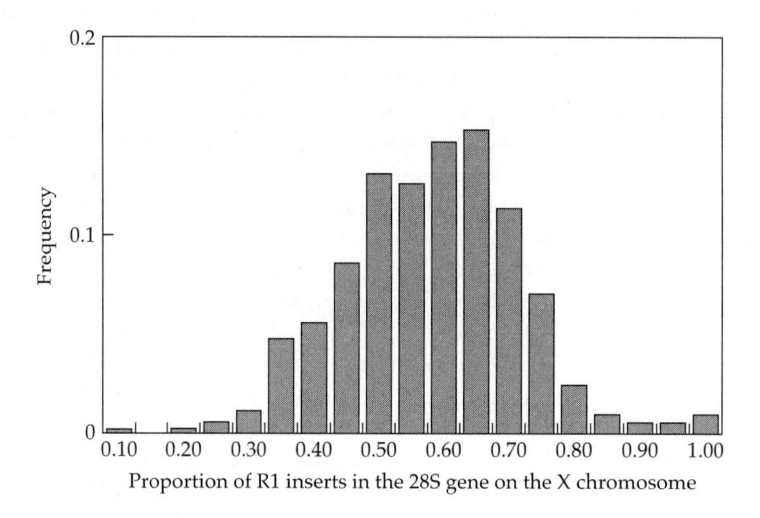

Figure 12.3 The frequency distribution for the fraction of 28S rDNA repeats on the X chromosome that contain an insertion of the mobile element R1 in a population of *Drosophila mercatorum* near Kamuela, Hawaii. 1036 X chromosomes were assayed. (From Hollocher et al. 1992.)

with reduced rDNA copy number. In addition, several studies with plant populations selected for grain yield have reported significant associations between rDNA copy number and morphological and life-history features (Saghai-Maroof et al. 1984, Rocheford et al. 1990, Powell et al. 1992). Remarkably, in flax (*Linum grandiflorum*), copy number variation in the rDNA, as well as in other regions of the genome, can be induced by a change in growing conditions. Such changes have substantial and heritable effects on plant morphology (Cullis 1981). These results, quite contrary to the general wisdom that genome expression, but not the genome itself, responds to environmental change, are unlikely to be unique to flax.

THE MUTATIONAL RATE OF PRODUCTION OF QUANTITATIVE VARIATION

Estimates of the rate of production of new quantitative variation by mutation are normally computed on a per character, rather than a per locus, basis. The measure generally sought is the amount of additive genetic variation produced by mutation per generation. This measure compounds three issues: the genic mutation rate, u; the number of loci underlying the character, n; and the mean squared effect of a mutant allele in the heterozygous state. Here, as in Chapter 4, we scale the genotypic values at a locus such that the ancestral homozygous state is 0, the heterozygous state is $a(1+k)$, and the homozygous mutant state is $2a$. Noting that the frequency of newly arisen mutations at a locus in any generation $(2u_i)$ is very small so that the change in the mean effect at the locus is negligible, then after summing over all loci, the mutational variance becomes

$$\begin{aligned}\sigma_m^2 &= 2\sum_{i=1}^{n} u_i\, E[a_i^2(1+k_i)^2] \\ &= \sigma_{m0}^2 + 2\sigma_{m1}^2 + \sigma_{m2}^2 \end{aligned} \tag{12.1a}$$

where $\sigma_{m0}^2 = UE(a^2)$, $\sigma_{m1}^2 = UE(a^2k)$, and $\sigma_{m2}^2 = UE(a^2k^2)$, with $U = 2\sum_{i=1}^{n} u_i$ being the zygotic mutation rate for the character (i.e., the expected number of mutations influencing the trait that arise in a newborn each generation), and $E(x)$ denoting the average value of x. For mutations with purely additive effects $(k=0)$, the mutational variance simplifies to

$$\sigma_m^2 = \sigma_{m0}^2 = UE(a^2) \tag{12.1b}$$

A cautionary note in the interpretation of Equation 12.1a is necessary. Strictly speaking, the mutational variance defined in this expression is **nonsegregational mutational variance** in that it describes the variance produced by a cohort of new mutations that exist entirely as heterozygotes, rather than at Hardy-Weinberg frequencies where the mutations appear in both heterozygotes and homozygotes. In the case of mutations with additive effects, this definition is not a problem because the variance produced by an allelic effect is independent of the genetic background and uninfluenced by segregation. With completely recessive $(k = -1)$ mutations, however, Equation 12.1a defines the mutational variance to be zero, despite the fact that variation will be exposed once the mutant alleles have segregated into their homozygous states. For autosomal loci, Hardy-Weinberg frequencies are reached in one generation under monoecy and in two generations under dioecy.

A reasonable way to define the relevant measure of mutational variance for a segregating population is to consider the situation in which the only determinants of the dynamics of genetic variance are mutation and random genetic drift. Under this **neutral model,** considered in detail in Lynch and Hill (1986), populations with constant size eventually reach an equilibrium state at which point the input of new variation by mutation is balanced by the loss via random genetic drift. For populations of size one, maintained by self-fertilization and single-seed descent, the equilibrium genetic variance among the progeny is

$$\sigma_W^2 = \frac{3}{2}\sigma_{m0}^2 + \left(\frac{8\sigma_{m1}^2 + 5\sigma_{m2}^2}{4}\right) \tag{12.2a}$$

With full-sib mating, the equilibrium variance is

$$\sigma_W^2 = 4\sigma_{m0}^2 + 2\left(\frac{6\sigma_{m1}^2 + 5\sigma_{m2}^2}{3}\right) \tag{12.2b}$$

and with all larger population sizes (with effective size denoted by N_e), it is very close to

$$\sigma_W^2 = 2N_e \left(\sigma_{m0}^2 + \frac{2}{3}(\sigma_{m1}^2 + \sigma_{m2}^2) \right) \tag{12.2c}$$

In this latter case, all of the genetic variance is additive, except for $2N_e\sigma_{m2}^2/3$, which defines the dominance component of variance (Lynch and Hill 1986). Note that for recessive mutations ($k < 0$), which are likely to be the rule (Chapters 4 and 10), σ_{m1}^2 is negative while σ_{m2}^2 is positive, so the contributions from these terms partially cancel (completely canceling in the case of fully recessive mutations). Moreover, the absolute values of both terms are necessarily less than that of σ_{m0}^2. Thus, although not exact in all cases, we may expect the equilibrium level of genetic variance in the absence of selection to be close to $2N_e\sigma_{m0}^2$ in most cases. This result justifies the reliance on σ_{m0}^2 as a measure of mutational variance, a useful relationship since, as we will soon see, σ_{m0}^2 is much easier to estimate than either σ_{m1}^2 or σ_{m2}^2.

Estimates of the mutational variance are often scaled by the environmental variance to give the **mutational heritability**,

$$h_m^2 = \frac{\sigma_{m0}^2}{\sigma_E^2} \tag{12.3}$$

a dimensionless measure suitable for comparing different characters. When mutations have additive effects, h_m^2 approximates the heritability of an initially homozygous population following one generation of mutation. (The fact that the denominator in h_m^2 is the environmental, rather than the total, variance is not problematical because, as will be shown below, $\sigma_{m0}^2 << \sigma_E^2$).

Estimation from Divergence Experiments

Estimation of the mutational variance is straightforward, but labor intensive. The simplest approach is a divergence experiment. Replicate lines, extracted from the same ancestral population, are maintained at a small enough size so that the fate of essentially all newly arisen mutations will be determined by random genetic drift. In the cases of selfed lines maintained by single-seed descent and of full-sib mated lines, all but extremely deleterious alleles will be impervious to selection, so that any census of mutational effects will yield nearly unbiased results. Over time, the mean phenotypes of the replicate lines are expected to drift apart as they randomly accumulate unique mutations. If the mutations have directional effects, a shift in the grand mean phenotype (averaged over all lines) is also expected to occur.

Provided that, prior to expansion, the base population is kept at the same size as the derived lines for a long enough period to be in mutation-drift equilibrium, the expected increment in the among-line component of variance per generation is equal to $2\sigma_{m0}^2$ (Bailey 1959, Lande 1976, Chakraborty and Nei 1982, Lynch and

Hill 1986). The simplest way to obtain this result is to note from the neutral theory (Kimura 1983) that the expected number of mutational substitutions per locus differentiating two populations separated for t generations is $2u_i t$. (Letting N denote the population size, then each generation in each population, an expected $2Nu_i$ new mutations arise. Each of these has an ultimate probability of fixation equal to the initial frequency, $1/(2N)$, so that at equilibrium, the expected number of fixations in a pair of populations each generation is $2 \times 2Nu_i \times 1/(2N) = 2u_i$ at the ith locus.) Upon fixation, each mutation causes an expected increment in the among-population variance of $E[(2a - 0)^2]/2 = 2E(a^2)$. Thus, summing over all loci, the expected among-line variance after t generations of divergence is

$$\sigma_B^2(t) = 2\sum_{i=1}^{n}(2u_i t) \cdot E(a^2) = 2\sigma_{m0}^2 t \tag{12.4a}$$

This is a fairly robust result in that it does not depend on the population size, on the degree of dominance of new mutations, or on the linkage relationships among loci (Lynch and Hill 1986). Dominance has no influence on the rate of divergence because it has a transient effect; once a mutant allele goes to fixation, a permanent shift in the mean phenotype to $2a_i$ occurs at the locus, regardless of k_i. In principle, the asymptotic rate of divergence might accelerate (or decelerate) when mutations have general epistatic effects (Tachida and Cockerham 1990), but the time scale over which such complications would arise is likely to be quite long (dozens to hundreds of generations), and the details of such a model have not yet been worked out.

To summarize, a conceptually simple way to estimate the mutational variance is to perform a long-term divergence experiment with replicate lines kept at small population sizes, extracting the among-line component of variance (σ_B^2) by analysis of variance. Division of the latter by twice the number of generations of divergence ($2t$) provides an estimate of σ_{m0}^2. These results apply strictly to diploid sexual populations, but only slight modifications are required for other forms of reproduction. For example, for haploid populations (sexual or asexual), the preceding expression becomes

$$\sigma_B^2(t) = \sigma_{m0}^2 t/2 \tag{12.4b}$$

whereas for diploid asexuals,

$$\sigma_B^2(t) = \sigma_m^2 t \tag{12.4c}$$

where σ_m^2 is defined as in Equation 12.1a (Lynch 1994). Terms involving dominance arise in the case of diploid asexuals because, in the absence of segregation, mutations are kept permanently in the heterozygous state with effects $a(1 + k)$.

There are several potential sources of error in the estimation of the mutational variance with a divergence experiment. A common misconception is that estimates of the mutational variance will be upwardly biased by the segregation of genetic variance in the base population if the latter is not 100% homozygous. On the contrary, the expected increase in among-line variance of $2\sigma^2_{m0}$ per generation strictly applies to situations in which the base population has been kept at the same size as the descendent lines for about $6N_e$ or so generations prior to the initiation of a divergence experiment (Lynch and Hill 1986). If the base population is totally homozygous, the rate of divergence will be less than $2\sigma^2_{m0}$ during the early generations of the experiment due to the lack of segregating polymorphisms. If the base population contains excess heterozygosity relative to the equilibrium expectations within the descendent lines, the divergence rate will exceed $2\sigma^2_{m0}$ in the early generations of the experiment. Eventually, under the neutral model, all lines will asymptotically diverge at the rate $2\sigma^2_{m0}$ regardless of the status of the base population.

A second potentially serious issue in the analysis of a divergence experiment concerns the possibility that environmental effects artificially inflate the among-line component of variance. To guard against such inflation, a set of control lines should be assayed in parallel with the mutation-accumulation lines whenever possible. The control lines should be genetically identical (and maintained in an evolutionarily inert state by, for example, freezing or seed storage) so that any variance among them can only be attributable to environmental causes. Any variance among control lines can then be subtracted from that observed among the mutation-accumulation lines. Maternal effects can be a significant source of inflation of the among-line variance when all of the progeny assayed within a subline are derived from a single mother. Such problems can be eliminated by deriving all assay individuals from different mothers, as this ensures that any variance associated with maternal effects will contribute to the within-line component of variance (Lynch 1985).

Third, contamination events can be a major problem in mutation-accumulation experiments, either deflating estimates of σ^2_{m0} when gene flow accidentally occurs among lines, or inflating it if foreign genotypes immigrate into the experiment. Periodic screening of lines for the genomic distribution of mobile elements or for other informative molecular markers, such as highly mutable microsatellite loci, can provide an excellent control against contamination (Mackay et al. 1992a, Keightley and Hill 1992).

Finally, we note that an alternative approach to estimating the mutational variance involves the application of artificial selection to an initially homozygous base population. Recall from Chapter 3 that the single-generation response to selection is equal to $R = h^2 S$, where S is the selection differential imposed by the investigator. The idea behind the selection approach to estimating mutational variance is that the standing heritability in a population can be estimated from the ratio of selection response to selection differential, $\widehat{h}^2 = R/S$. As in the case of the neutral model, expressions can be derived for the expected equilibrium heritability of a quantitative trait under truncation selection as a function of the population size and the mutational variance (Hill 1982b,c). Thus, by equating observed and expected heritabilities, an estimate of σ^2_{m0} can be obtained.

Bristle Numbers in *Drosophila*

The vast majority of our knowledge of the properties of polygenic mutation derives from studies on one set of characters in one species — abdominal and sternopleural bristle numbers in *Drosophila melanogaster* (Figure 14.1). Averaging over the results from numerous studies (reviewed in Lynch 1988b, Keightley et al. 1993, Houle et al. 1996), the mutational heritabilities for these traits appear to be on the order of 0.0035 and 0.0043, respectively. The effects of individual bristle-number mutations tend to be nonadditive, with mutations with large effects being nearly recessive, and those with small effects being nearly additive on average but with variable levels of dominance (Mackay et al. 1992a, Caballero and Keightley 1994, Fry et al. 1995). These results are particularly notable since quantitative-genetic analyses of *Drosophila* bristle numbers have frequently led to the suggestion that these traits have a purely additive genetic basis (Chapter 7). The discrepancy can be reconciled if most mutations for bristle number are deleterious and hence kept at low frequency in natural populations by selection. Because the dominance component of genetic variance is a function of the squared heterozygosity, the vast majority of the genetic variance is additive when recessive alleles are kept at low frequency (Equation 4.12b).

Genetic analyses do indeed imply that most bristle-number mutations with large effects have substantial negative pleiotropic effects on fitness, some being recessive lethals (Mackay et al. 1992a, 1994; Fry et al. 1995). Evidence for the general negative fitness consequences of such mutations comes from long-term divergence experiments involving initially homozygous lines kept subsequently at average population sizes of approximately 10 individuals (Mackay et al. 1992b). In the absence of any artificial selection, for the first 100 generations or so of isolation, the lines diverged via random genetic drift, but then the rate of divergence declined dramatically (Mackay et al. 1995), presumably because of the deleterious fitness consequences of mutant bristle-number alleles with extreme effects.

Long-term selection experiments, again starting with homozygous lines, provide further evidence for the negative pleiotropic effects of bristle-number mutations on fitness. Such experiments often culminate in dramatic divergence between high- and low-selected lines (Clayton and Robertson 1955, 1964; Caballero et al. 1991; Santiago et al. 1992; López and López-Fanjul 1993). For example, after only 125 generations of selection, differences of 12 abdominal bristles and 8 sternopleural bristles (equivalent to 7 to 8 phenotypic standard deviations) have been observed (Figure 12.4). As in the case of the drift

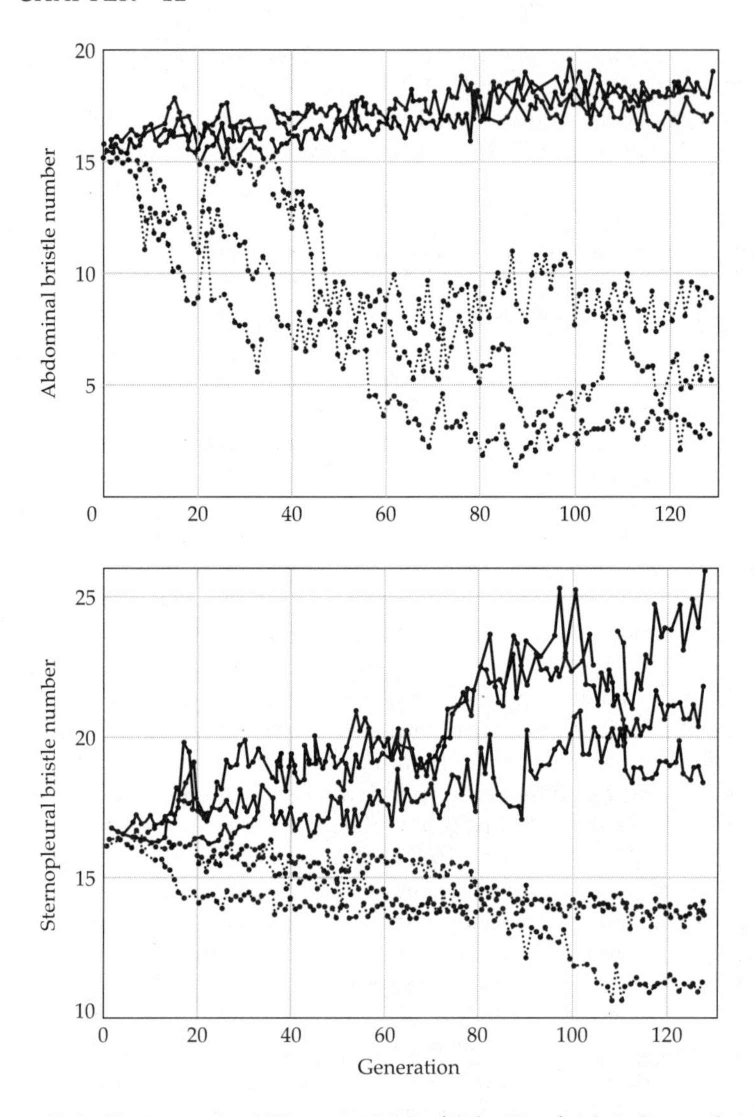

Figure 12.4 Responses to 125 generations of selection for bristle numbers in *Drosophila melanogaster.* For each character, results are given for three upwardly selected and three downwardly selected lines. Up to generation 63, the 10 extreme males and 10 extreme females were selected from a total of 40 measured individuals of each sex; thereafter, 10 pairs were selected from 20 measured individuals of each sex. (From Mackay et al. 1994.)

experiments, substantial reductions in fitness are associated with such selective changes (Mackay et al. 1994, Nuzhdin et al. 1995). Line-cross analysis (Chapter

9) reveals that the selection response typically involves multiple mutations on all chromosomes (Fry et al. 1995). Responses to artificial selection on bristle number, which must reflect the distributions of mutational effects (as well as their pleiotropic effects on fitness) to a large degree, are almost always asymmetrical — a downward skew in the case of abdominal bristle number, and an upward skew in the case of sternopleural bristle number (Figure 12.4).

An attractive feature of *D. melanogaster* as a model system for studying aspects of polygenic mutation is its amenability to transposon tagging. Through the use of appropriate stocks (H. Robertson et al. 1988), it is possible to mobilize the transposable element *P* and then to subsequently stabilize it. Hybridization of radiolabeled *P*-element DNA to chromosomal squashes can then be used to ascertain the approximate locations of insertional events. Lyman et al. (1996) used this approach to study the effects of single *P*-element insertions on the expression of bristle-number characters. By assaying multiple singly mutated chromosomes on an otherwise constant genetic background, they found that the distributions of mutational effects were symmetrical but highly leptokurtic, with a few inserts having very large effects and the average effect being approximately equal to zero (Figure 12.5). On average, the mutational effects were partially recessive — $\bar{k} = -0.37$ for abdominal bristles and $\bar{k} = -0.65$ for sternopleural bristles. There was a weak correlation between viability and the absolute magnitude of effect of a bristle-number mutation, with the viability effects being highly recessive ($\bar{k} = -1.00$).

The results of Lyman et al. (1996) imply that the mean-squared homozygous effects of individual *P*-element insertion events relative to the environmental variance, $E[(2a)^2]/\sigma_E^2$, are approximately 0.28 for abdominal bristles and 0.82 for sternopleural bristles. Recalling the values for the mutational heritability, $h_m^2 = UE(a^2)/\sigma_E^2$, noted above, these single-mutation estimates imply genomic mutation rates of approximately 0.05 and 0.02 per zygote per generation, respectively. This extrapolation assumes that *P*- element induced mutations have effects that are comparable to the average effects of mutations from all sources. Since many of the mutant chromosomes examined by Lyman et al. (1996) were likely to have had inserts at loci that have little, if any, influence on bristles, their estimates of $E[(2a)^2]/\sigma_E^2$ are almost certainly too low. This suggests that zygotic mutation rates for bristle numbers may be as large as 0.1 or so.

Additional Data

Table 12.1 summarizes estimates of the mutational heritabilities for a number of organisms and a variety of characters. In invertebrates, nearly all of the estimates are in the range of 10^{-3} to 10^{-2}, whereas estimates for the mouse are somewhat higher, ranging up to 0.025. On average, the estimates for plants are similar to those for invertebrates, with a slightly broader range perhaps due to sampling error. In highly mutagenic situations, as when transposable elements are mobilized

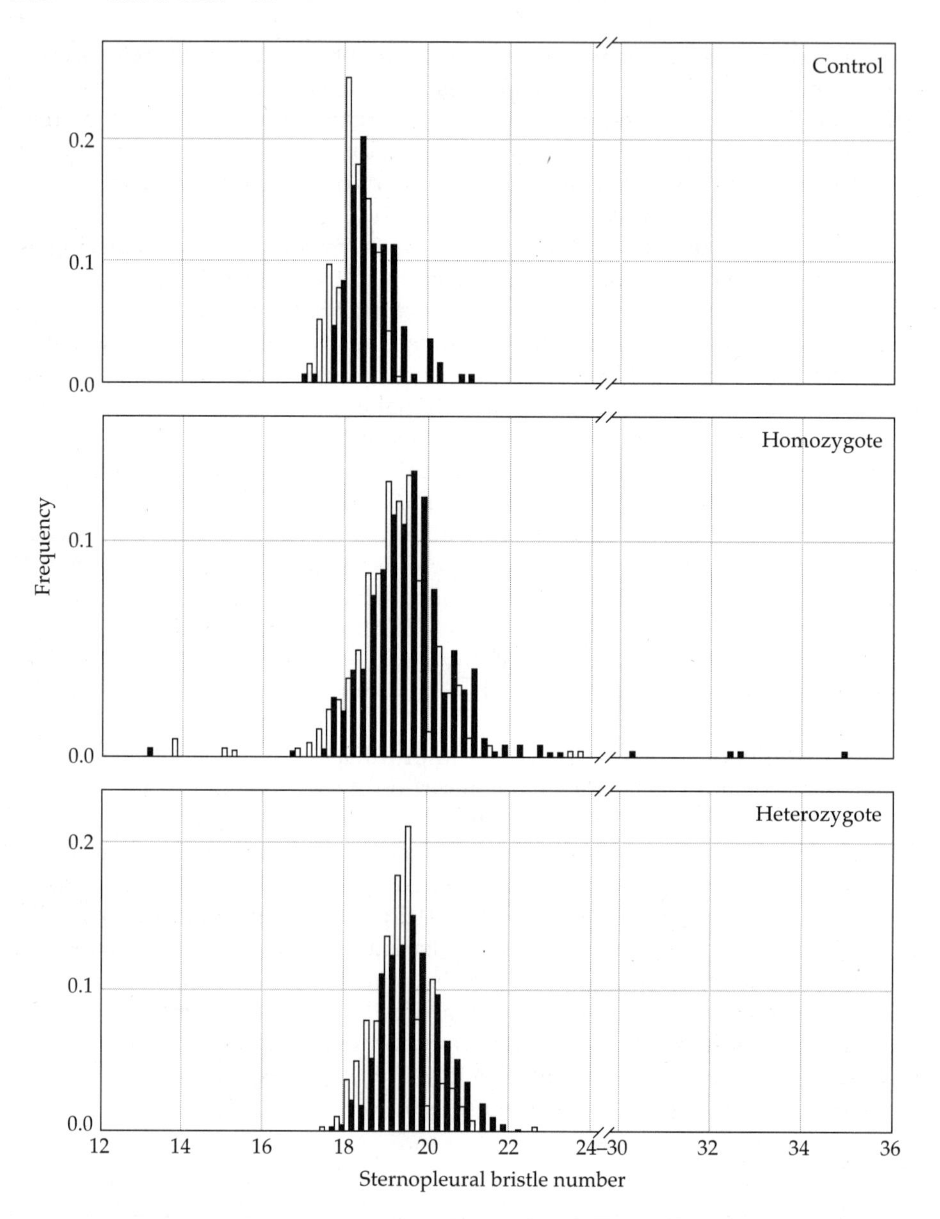

Figure 12.5 The distributions of sternopleural bristle numbers for control, homozygous *P*-element insert lines, and heterozygous *P*-element insert lines, in *Drosophila melanogaster.* Each insert line carries a single *P*-element insertion, as revealed by cytogenetic analysis. The open bars denote second chromosome line means, while the closed bars denote the results for chromosome three. All chromosomes were evaluated on a constant genetic background. (From Lyman et al. 1996.)

following hybrid dysgenesis in *Drosophila*, mutational heritabilities can be as high as 0.1 (Mackay 1985b, 1987, 1988). Within *Drosophila*, all estimates of h_m^2 are within the range of 0.001 to 0.005, except that for viability, which is certainly downwardly biased because chromosomes with highly deleterious mutations were excluded from the analyses.

Using the *P*-element tagging approach, Clark et al. (1995a) obtained scaled single mutational effects, $E[(2a)^2]/\sigma_E^2$, equal to 0.064 ± 0.005 for body weight and 0.062 ± 0.018 for the activities of 15 enzymes. Comparing these values with the mutational heritabilities in Table 12.1, we obtain minimum zygotic mutation rates of 0.29 for body weight and 0.14 for activities of individual enzymes, several times higher than those noted above for bristle numbers. Such high mutation rates cannot be attributable to mutations at the actual enzyme loci, which are expected to occur at the rate of 10^{-6} to 10^{-5} per gene per generation (Mukai and Cockerham 1977). Thus, most of the mutations must arise in regions of the genome involved in the regulation of enzyme activity, implying that such regions must be rather numerous. As in the case of bristle numbers, the distributions of mutational effects in the study of Clark et al. were quite symmetrical, although highly leptokurtic, with mutations being equally likely to cause an increase or a decrease in the character. Most of the enzyme-activity mutations had pleiotropic effects on multiple traits.

It should be kept in mind that all of the above estimates of the mutational heritability may substantially overestimate the input of genetic variation available to natural selection. If most spontaneous mutations have negative pleiotropic effects on fitness, as suggested by the *Drosophila* bristle-number data, much of the recurrent mutational input of variation may simply represent a price that organisms pay for the privilege of producing rare adaptive mutations.

While the estimation of composite measures of mutation such as σ_{m0}^2 is nontrivial, direct estimation of the actual mutation rate for quantitative characters ($U = 2\sum \mu_i$) is even more difficult. The indirect extrapolations that we presented above for *P*-element tagging experiments suggest that U for individual traits may often be on the order of 0.1 or higher. The few direct estimates suggest the same. For example, Grewal (1962) examined 27 skeletal traits in inbred strains of lab mice and found that 21 detectable mutations had accumulated in the roughly 72 total generations since the strains were separated (summing the number of generations over all branches). This gives a rate of approximately $2 \times [(21/27)/72] = 0.022$ per character per generation per zygote. In a more extensive study using the same traits, Hoi-Sen (1972) obtained essentially the same answer. Studies examining a number of reproductive characters in maize using doubled-haploids (Sprague et al. 1960) and inbred lines (Russell et al. 1963) yield estimates of 0.090 and 0.056 mutations per character per zygote. These estimates of per character mutation rates are certainly too low, perhaps substantially so, since mutations of insufficiently large effect were statistically undetectable and hence not counted.

Table 12.1 Estimates of the mutational heritability for a variety of organisms and characters.

Species	Character	h_m^2	Reference
Drosophila melanogaster	Abdominal bristle number	0.0035	See text
	Sternopleural bristle number	0.0043	See text
	Enzyme activities	0.0022	Clark et al. 1995b Harada 1995
	Ethanol resistance	0.0009	Weber and Diggins 1990
	Body weight	0.0047	Clark et al. 1995b
	Wing dimensions	0.0020	Santiago et al. 1992
	Viability	0.0003	Mukai 1964 Mukai et al. 1972 Cardellino and Mukai 1975 Ohnishi 1977
Tribolium castaneum	Pupal weight	0.0091	Goodwill and Enfield 1971
Daphnia pulex	Life-history traits	0.0017	Lynch 1985
Mouse	Lengths of limb bones	0.0234	Bailey 1959
	Mandible measures	0.0231	Festing 1973
	Skull measures	0.0052	Carpenter et al. 1957 Deol et al. 1957 Hoi-Sen 1972
	6-week weight	0.0034	Caballero et al. 1995
Arabidopsis thaliana	Life-history traits	0.0039	Schultz et al. (in prep.)
Maize	Plant size	0.0112	Russell et al. 1963
	Reproductive traits	0.0073	Russell et al. 1963
Rice	Plant size	0.0030	Oka et al. 1958
	Reproductive traits	0.0028	Sakai and Suzuki 1964
Barley	Life-history traits	0.0002	Cox et al. 1987

Note: Where possible the results represent averages over multiple studies, ranging from analyses of drift among initially homozygous lines to artificial selection experiments. Detailed analyses of experiments performed prior to 1985 can be found in Lynch (1988b), and a more recent survey is contained in Houle et al. (1996).

If estimates of U in the neighborhood of 0.1 are indeed correct, then typical quantitative traits must be influenced by a large number of loci and / or have very high per locus mutation rates. If the typical reported estimates of genic mutation rates, usually on the order of 10^{-5} to 10^{-6} (Griffiths et al. 1996), are correct, then 5,000 to 50,000 loci are required to account for a per character mutation rate of 0.1. While one might imagine that thousands of loci can affect a character like viability or fitness, such a high number seems unlikely for most other characters, although there is little evidence to support or refute this conjecture. This apparent discrepancy has been discussed by a number of workers (Grewal 1962, Russell et al. 1963, Hoi-Sen 1972, Grüneberg 1970, Beardmore 1970, Turelli 1984). In principle, the high reported estimates of U may be inflated by experimental artifacts, such as the presence of residual heterozygosity in the base population, but evidence on this issue is also lacking.

An alternative explanation for the unexpectedly high estimates of U is that the mutation rate to alleles of small effects is much higher than the rate to alleles of large effect (on which most of the data on genic mutation rates are based). We have already reviewed evidence that support this idea for *Drosophila* bristle number and enzyme activities, and it is bolstered by additional data. In a descriptive analysis of highly inbred peanut populations, Gregory (1965a,b) found that deviations of small effect are much more frequent than those of larger effect, although his approach was not statistically rigorous. Mukai (1964) found that the mutation rate over the entire second chromosome of *D. melanogaster* is 20 times higher for mutations with minor viability effects (0.15) than for recessive lethals (0.006). Kibota (1996) found that the vast majority of spontaneous mutations for fitness in *E. coli* have very small effects, on the order of a 1% decline in fitness per generation, although mutations with larger effects do occur.

Finally, we note that the per character mutation rates cited above are not out of line with what we know about the rates of various types of genic mutation if most characters are functions of gene expression at large numbers of loci (Crow 1992, 1993b; Kondrashov 1995). In *Drosophila*, for example, a survey of 18 transposable element families suggests that the total number of transpositions per zygote is on the order of 0.8 per generation (Nuzhdin and Mackay 1995, Lyman et al. 1996). In addition, the rate of nucleotide substitutional mutation in *Drosophila* has been estimated to be about 1.6×10^{-8} per site per year (Sharp and Li 1989). Assuming flies go through approximately five generations each year, and noting that the *Drosophila* genome contains approximately 4×10^8 bases, these results imply that the average number of nucleotide substitutional mutations per genome is at least 1.3 per generation. Summing these two types of mutation rates and noting that we have ignored other types of mutations (e.g., single nucleotide insertions/deletions, unequal crossing-over, and so on), it is clear that the total genomic mutation rate in *Drosophila* must be at least 2.0 per generation.

Although quantitative estimates of the transposition rate in humans are lacking, the rate of nucleotide substitutional mutation appears to be close to 2.5×10^{-9} per site per year (Li and Graur 1991, Easteal and Collet 1994, Ohta 1995). Assuming a diploid genome size of 6×10^9 nucleotides and a generation time of 20 years, this implies a nucleotide substitution rate on the order of 300 per individual per generation. Similarly, using the estimated nucleotide substitution rate of 6×10^{-9} for plants (Wolfe et al. 1987) and the known genome sizes for *Arabidopsis* and maize, it can be shown that the genomic mutation rate for single-site substitutions alone is on the order of 1 to 40 per generation. Thus, there is little question that mutagenic activities operating at the genome level are sufficient to generate high rates of mutation for multilocus traits.

THE DELETERIOUS EFFECTS OF NEW MUTATIONS

Evolutionary biologists are especially interested in the rate of mutation for fitness and its underlying components such as viability. As first noted by Haldane (1937), large outcrossing populations of diploids will evolve towards a situation where the recurrent input of deleterious mutations is balanced by their selective elimination. In large populations, at mutation-selection balance, the frequencies of deleterious alleles equilibrate to levels such that the reduction in fitness is solely a function of the mutation rate and independent of the effects of the mutations (Chapter 10). The cumulative fitness consequences of a mutation are independent of the mutational effect because the equilibrium frequency of a deleterious allele is inversely proportional to its effect in heterozygotes. Thus, for an understanding of the population-level consequences of deleterious mutation, the genomic mutation rate is of central importance.

Characterization of the properties of deleterious mutations is also important for other issues in evolutionary biology and conservation genetics. In small populations, mildly deleterious mutations do not simply attain equilibrium levels of heterozygosity. Rather, mildly deleterious mutant alleles have an appreciable chance of randomly drifting to fixation. The accumulation of such mutations can eventually imperil a population, driving it to extinction when the mutation load exceeds the point at which individuals can replace themselves (Lynch et al. 1993, 1995a,b; Lande 1994). It is also well known that pressure from recurrent deleterious mutations can foster the evolution of alternative modes of reproduction (Pamilo et al. 1987; Kondrashov 1988; D. Charlesworth et al. 1992, 1993), with outcrossing populations being much more efficient at eradicating deleterious genes than asexual or obligately selfing populations. The risk of extinction via deleterious mutation accumulation depends not just on the genomic deleterious mutation rate, but also on the distribution of mutational effects. Mutations with selection coefficients on the order of $1/(2N)$, where N is the population size, are of particular concern because such mutations are highly vulnerable to the vagaries of random genetic drift while still having significant effects on fitness (Lynch and Gabriel 1990, Gabriel et al. 1993, Lande 1994).

To obtain a more than qualitative understanding of these issues, estimates of the genomic deleterious mutation rate, the selective coefficient s, and the dominance coefficient h of new mutations are required. Two approaches to the problem have been suggested: (1) the analysis of laboratory mutation-accumulation lines, expanding on the procedures outlined above for the estimation of mutational heritabilities, and (2) the analysis of the fitness properties of natural populations under the assumption that they are in mutation-selection balance.

The Bateman-Mukai Technique

Using estimates of the per generation changes in the grand mean and the among-line variance for a trait caused by mutation, it is possible to place bounds on the mutation rate and the average effect of a new mutation (Bateman 1959, Mukai 1964, Mukai et al. 1972, Crow and Simmons 1983). Although Bateman and Mukai were primarily interested in the deleterious consequences of mutation, the statistical approach that they advocated applies to any character for which mutation causes a directional change in the mean. We will first derive the general results following Lynch (1994), and then apply them specifically to the analysis of fitness characters.

We start with the assumption that the mutation-accumulation lines are kept at a small enough size that essentially all new mutations are effectively neutral, i.e., the vagaries of random genetic drift overwhelm the forces of selection. Roughly speaking, this requires that the selection coefficients of mutant alleles be less than $1/(2N)$, e.g., $s < 0.5$ for lines maintained by self-fertilization and single-seed descent, or $s < 0.25$ for lines maintained by single brother-sister mating. Letting U denote the genomic mutation rate for the trait, then NU new mutations enter a line each generation, and each of these has a fixation probability equal to its initial frequency, $1/(2N)$. The expected number of fixations per generation is simply the product of these two terms, $U/2$ (the gametic mutation rate). Following the notation used earlier in this chapter, conditional on fixation, each mutation causes an expected change in the mean genotypic value equal to $E(2a)$. Thus, assuming there are no substantial epistatic interactions among new mutations, the expected dynamics of change in the mean genotypic value in a mutation-accumulation experiment are given by

$$\bar{g}(t) = \bar{g}(0) + U\overline{a}t \tag{12.5}$$

where $\bar{g}(t)$ is the mean genotypic value in generation t. In the following, we let ΔM denote an estimate of the rate of change in the mean, $U\overline{a}$. It may be obtained from a regression of line means on time, or as the total change in the mean $\bar{g}(t) - \bar{g}(0)$ divided by t.

As noted above, the expected rate of increase of the among-line variance in a mutation-accumulation experiment is equal to $2\sigma^2_{m0} = 2UE(a^2)$. Thus, twice the squared rate of change in the mean divided by the rate of change in the variance is equal to $U\overline{a}^2/\overline{a^2}$. This quantity reduces to U in the unlikely event that mutations have constant effects. More generally, the ratio is equal to $U/(1 + C_m)$, where $C_m = \sigma^2_a/\overline{a}^2$ is the squared coefficient of variation of mutational effects. Letting ΔV be an estimate of the rate of line divergence, $2\sigma^2_{m0}$, and accounting for the bias produced by sampling error in the estimators ΔM and ΔV, a lower-bound estimate for the genomic mutation rate is

$$\widehat{U}_{min} = \frac{2(\Delta M)^2}{\Delta V(1 + C_{\Delta M})(1 + C_{\Delta V})} \tag{12.6}$$

where $C_{\Delta M}$ and $C_{\Delta V}$ are squared coefficients of sampling variance (ratios of sampling variance to squared estimates) of ΔM and ΔV, e.g.,

$$C_{\Delta M} = [\mathrm{SE}(\Delta M)/\Delta M]^2$$

By similar means, it can be shown that an upper-bound estimate of the average mutational effect is given by

$$\widehat{a}_{max} = \frac{\Delta V}{2(\Delta M)(1 + C_{\Delta M})} \tag{12.7}$$

The expected value of $\widehat{a}_{max}$ is actually $(1 + C_m)\overline{a}$, so the factor by which it overestimates $\overline{a}$ is identical to the factor by which the mutation rate estimator underestimates U.

By crossing lines in which mutations have accumulated independently, some insight can be gained into the dominance of mutational effects. For example, starting from a base population that is in mutation-drift equilibrium, the expected difference between mean phenotypes of hybrid progeny and that of their contemporary (mutation-accumulation line) parents after t generations of divergence is $UE(ak)t$ (Lynch 1994). Letting ΔD denote an estimate of $UE(ak)$, then

$$\widehat{k} = \frac{\Delta D}{\Delta M(1 + C_{\Delta M})} \tag{12.8}$$

provides an estimate of $\overline{k} \cdot [1 + (\sigma_{a,k}/\overline{a}\overline{k})]$, where $\sigma_{a,k}$ is the covariance of effects a and k in new mutations. Since the signs of all of the terms in this expression can be positive or negative, $\widehat{k}$ can be either an upwardly or downwardly biased estimator of $\overline{k}$, the bias being negligible if $(\sigma_{a,k}/\overline{a}\overline{k})$ is small. Other approaches to estimating the dominance properties of mutations and their influence on the mutational variance are discussed in Mukai (1979) and Lynch (1994).

The preceding expressions can be interpreted as yielding estimates of the genomic mutation rate and mutational effects that would explain the pattern of change in the mean and variance of a trait in a mutation-accumulation experiment if all mutations had constant effects. The formulae are general, in that they can be applied to any quantitative trait, so long as $\Delta M \neq 0$. In principle, the bias inherent in these estimators can be removed if information is available on the variance of mutational effects, but such information is not ordinarily available. An additional precaution that should be taken in the interpretation of the Bateman-Mukai estimators (Equations 12.6 and 12.7) is that, like all estimators, they are subject to sampling error. Thus, if the estimates ΔM and/or ΔV are not very accurate, then the estimate of U_{min} can easily be substantially above or below its actual value.

The derivations presented above apply to diploid sexual lines, but they are readily modified for alternative reproductive systems. For example, for haploids (either sexual or asexual), the expected rate of change of the mean is $U\overline{a}$, where U is the genomic mutation rate of the haploid individual and $\overline{a}$ is the mean effect of a mutational change. As noted above, the rate of divergence of haploid line means is $\sigma_{m0}^2/2$. Again letting ΔM and ΔV denote estimates of the rates of change of mean and variance, estimators for the genomic mutation rate and the average mutational effect are

$$\widehat{U}_{min} = \frac{(\Delta M)^2}{\Delta V(1 + C_{\Delta M})(1 + C_{\Delta V})} \tag{12.9a}$$

$$\widehat{a}_{max} = \frac{\Delta V}{\Delta M(1 + C_{\Delta M})} \tag{12.9b}$$

(These are the estimators used in the analysis of *Drosophila* mutation-accumulation experiments involving haploid chromosomes, although the terms involving sampling error have routinely been ignored.) These same estimators apply to diploid asexual populations, except that Equation 12.9b estimates the upper bound on the average heterozygous effect, $E[a(1 + k)]$.

Finally, we note that the among-line genetic covariances involving pairs of traits, like the among-line genetic variances, increase proportionally with the genomic mutation rate and generation number. Thus, the pleiotropic effects of mutations on pairs of traits can be quantified by the genetic correlation,

$$r_m(x, y) = \frac{\Delta C_{x,y}}{\sqrt{\Delta V_x \, \Delta V_y}} \tag{12.10}$$

where $\Delta C_{x,y}$ is the rate of change in the among-line covariance between traits x and y (Lynch and Hill 1986, Lynch 1994). For diploid sexual lines and for haploid lines, $r_m(x, y)$ provides an estimate of the correlation between homozygous effects (a_x and a_y), while for diploid asexuals, it estimates the correlation between heterozygous effects ($[a(1 + k)]_x$ and $[a(1 + k)]_y$). In the absence of selection and epistasis, the genetic correlation is expected to remain constant through time, although later estimates are expected to be much more accurate as more mutations are sampled.

Results from Flies, Plants, and Bacteria

The preceding approaches have been applied extensively to viability in *Drosophila.* As can be seen in Figure 12.6, when second chromosomes are passed clonally through males (using the techniques outlined in Figure 5.7), where they do not experience recombination, replicate lines exhibit a gradual decline in mean fitness and a gradual increase in among-line variance. Letting hs and s denote the reductions in viability in mutant heterozygotes and homozygotes, and extrapolating results to the whole genome, the results of Mukai and associates (Mukai 1964, 1969, 1979; Mukai and Yamazaki 1968; Mukai et al. 1965, 1972) and of Ohnishi (1977) lead to average estimates of $\widehat{U}_{min} = 0.6$, $s_{max} = 0.06$, and $\overline{h} = 0.36$

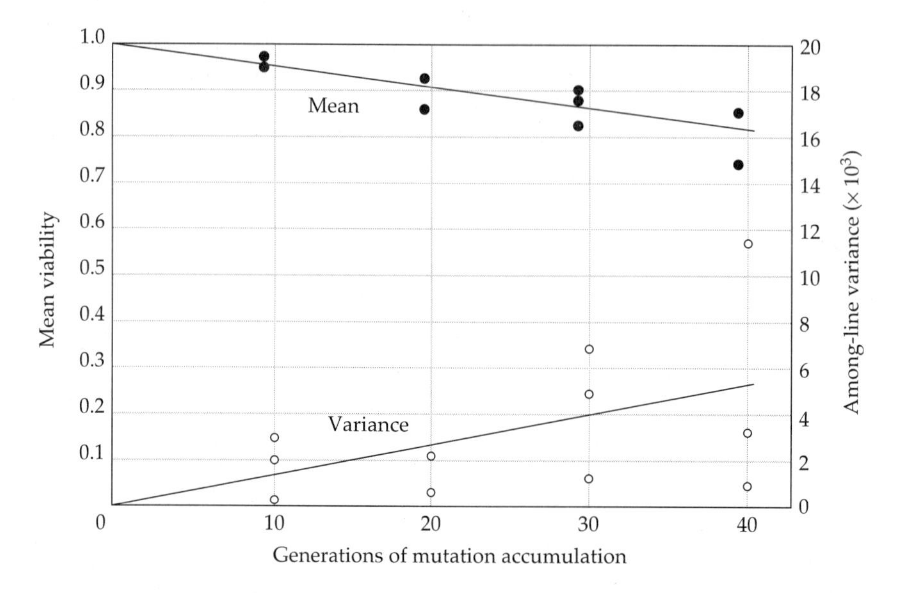

Figure 12.6 Decrease in mean viability and increase in genotypic variance in viability starting from a uniform base population. Second chromosomes of *Drosophila melanogaster* were kept sheltered over a balancer (see Figure 5.7) and passed through males (which do not exhibit recombination in *Drosophila*). By this means, mutations were accumulated on essentially clonally propagated chromosomes for a number of generations before being extracted and made homozygous. The mean viability of these sheltered chromosomes decreased with increasing number of generations of mutation accumulation, while the variance among lines increased. ΔM is estimated from the slope of the regression of mean on numbers of generations of mutation accumulation, while ΔV is estimated from the slope of the variance regression. (From Mukai et al. 1972.)

(summarized in Simmons and Crow 1977, Crow and Simmons 1983, Lynch et al. 1995b). That is, newly arisen mutations for viability arise at a rate of nearly one per individual per generation, and on average these mutations are slightly recessive, individually causing a reduction of viability of approximately 6% in homozygotes and approximately 2% ($0.06 \times 0.36 = 0.023$) in heterozygotes.

These results indicate that the per generation input of spontaneous mutations is such that in the absence of effective selection, a *Drosophila* population is expected to experience an approximately 2% reduction in individual viability per generation. There are two reasons why this value is almost certainly an underestimate. First, the assays reported above were performed in relatively benign environments. When mutation-accumulation lines of *Drosophila* were assayed in a variety of environments, their fitness (relative to controls) declined much

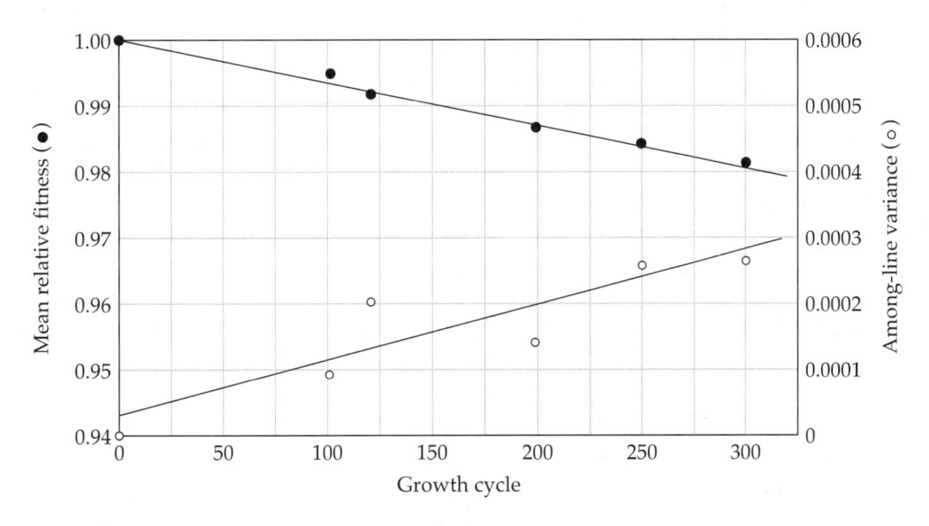

Figure 12.7 Similar plots to Figure 12.6 for 50 lines of the bacterium *E. coli* taken through daily single-cell bottlenecks for 300 days (with approximately 22 cell divisions per day). Here fitness is measured as the exponential rate of clonal expansion in liquid medium. (From Kibota and Lynch 1996.)

more dramatically in the more extreme environments (involving dilution of the medium and crowding), in some cases approaching zero (Kondrashov and Houle 1994). Future experiments are needed to evaluate whether this genotype × environment interaction is a consequence of an enhancement of the selective effects (s) of individual mutations in harsh environments, of the expression of a larger number of loci (U) in such environments, or both. Second, all of the preceding analyses excluded chromosomes carrying lethal recessive mutations. Such mutations appear to arise at a rate of about 0.02 per genome per generation (Crow and Simmons 1983), and results from a number of studies suggest that they cause an average viability reduction of 1% to 5% in heterozygotes, which is very similar to the heterozygous effects of mild detrimentals (Simmons and Crow 1977).

Unfortunately, all of the *Drosophila* experiments focus entirely on one aspect of fitness — egg-to-adult viability, and we still do not have an estimate of the mutation parameters for total fitness. Houle et al. (1992) had generated such estimates, but their control lines were later found to be contaminated (Houle et al. 1994a). Results from Mukai and Yamazaki (1971), Yoshimaru and Mukai (1985), and Houle et al. (1994b) suggest the possibility that most deleterious mutations have pleiotropic effects on all components of fitness. In surveys of fitness components such as developmental rate, fecundity, longevity, and mating ability, in no case is the genetic correlation caused by mutation significantly different from one, implying that the total impact of deleterious mutations on fitness in *Drosophila* can be no less than that for viability, and possibly substantially more.

Two other mutation-accumulation experiments provide insight into the magnitude of the deleterious mutation process at the level of total fitness. Starting from a highly inbred line of the self-fertilizing plant *Arabidopsis thaliana,* Schultz et al. (in prep.) maintained 1000 replicate lines by single-seed descent for 10 generations. Compared to the initial (control) population, stored as seed, total fitness (seed set assayed in a fairly benign environment) exhibited a decline of about 1% per generation. This total impact of mutations on fitness was a function of cumulative, but smaller, effects on individual fitness components — 0.06%, 0.05%, and 0.4% per generation, respectively, for germination probability, numbers of fruits set, and numbers of seeds per fruit. Application of the Bateman-Mukai technique to the observations on total fitness yields estimates of $U_{min} = 0.1$ and $\overline{s}_{max} = 0.1$.

If real, the lower genomic mutation rate in *Arabidopsis,* relative to *Drosophila,* may be due to a number of factors, including differences in genome size and differences in rates of spontaneous mutation per locus. The genome size of *Arabidopsis* is only about a third the size of that of *Drosophila* (Table 14.3). On a per year basis, the mutational rate of substitution (estimated from rates of synonymous substitutions) in plants is approximately 2.7 times higher than in *Drosophila* (Wolfe et al. 1987, Sharp and Li 1989). However, if one assumes that most mutations occur during meiosis, and that *Drosophila* has two to three times more generations per year than *Arabidopsis,* then the substitutional mutation rate in the two species may be quite similar on a per generation basis. As noted above, transposable-element activity is a significant source of mutation in *Drosophila,* but transposable elements are unusually rare and sedentary in *Arabidopsis.*

Kibota and Lynch (1996) performed a mutation-accumulation experiment on the bacterium *Escherichia coli,* taking 50 lines through daily single-cell bottlenecks for 300 days, and periodically assaying the lines for performance (relative to the initial control line, stored at $-80°$ C to prevent its evolution between assays). Qualitatively, the results were very similar to those for Mukai's *Drosophila* lines — there was a gradual decline in mean fitness (measured as the rate of clonal expansion) and a gradual increase in the variance among lines (Figure 12.7). The estimate of $\overline{s}_{max} = 0.012$ (in this case the haploid effect of a mutation) is not greatly different from that observed for viability in *Drosophila.* The genomic deleterious mutation rate estimate, $U_{min} = 1.7 \times 10^{-4}$, is about 10% of various estimates of the total genomic mutation rate for this and related species (Drake 1991, Andersson and Hughes 1996). Although this estimate of U_{min} for *E. coli* is orders of magnitude lower than the haploid rate for *Drosophila,* 0.3, the two can be brought more in line by the following considerations. There are about 25 germline cell divisions per generation in *Drosophila* (Lindsley and Tokuyasy 1980), so the genomic mutation rates per cell division are closer — 0.00017 vs. 0.012 — and they become quite similar when one notes that the haploid genomes sizes of *E. coli* and *Drosophila* differ by a factor of 50. Almost certainly, there is more dispensable (junk) DNA in a fly than in a microbe, so a linear scaling based on genome size is probably not entirely defensible. Still, the fairly narrow window of genomic deleterious mutation rates across taxa, when corrected for cell divisions per generation and genome size and for the unusually high amount of transposable element activity in *Drosophila,* is compelling.

How biased are these estimates by unknown variation in mutational effects? Keightley (1994) developed a maximum-likelihood technique for the analysis of mutation-accumulation experiments that assumes a Poisson distribution of mutational events and a gamma distribution of mutational effects. When applied to existing *Drosophila* results, his analyses suggest a highly L-shaped distribution of mutational effects, with most mutations being very close to neutral. However, a very broad range of distributions is compatible with the data, as are squared coefficients of variation of s ranging from $C_m = 0.25$ to approximately 30. Such results suggest that the genomic mutation rate to deleterious mutations in *Drosophila* could be as low as 0.8 and possibly as high as 20. The mean homozygous effect (s) of a new mutation could be as low as 0.002 or as high as 0.05. With this combination of mutation pressure and mutational effects, populations with effective sizes smaller than several hundreds of individuals are highly vulnerable to gradual mutational decay (Lynch and Gabriel 1990; Lande 1994; Lynch et al. 1995a,b).

Haldane (1935, 1947) and Crow (1993b) have pointed out that in many animal species the vast majority of mutations appear to arise in males rather than females, perhaps because most mutations arise during periods of cell division. In humans, for example, the mutation rate through males appears to be on the order of 10 times that through females. In women, the number of cell divisions between zygote and egg is approximately 24, regardless of maternal age. In men, however, approximately 36 cell divisions have occurred in the germ line by the time of puberty (age 13), and thereafter the number increases at a rate of about 23 per year. Thus, for a male of age 20, the number of cell divisions separating zygote and sperm cell is estimated to be $36 + (20-13) \cdot 23 = 200$, and by the age of 40, this has increased to 657. As reviewed by Crow (1993b), the observed increase in mutation rate with paternal age is even greater than that expected from this simple extrapolation from the increase in number of germline cell divisions. These results are of some relevance to the existing *Drosophila* experiments where, due to the experimental design, all mutations arose within males. However, the difference between male and female mutation rates in this species, as well as in the mouse, appears to be on the order of only twofold (Crow 1992).

One final issue is the extent to which spontaneous deleterious mutations interact epistatically. Evolutionary biologists are particularly interested in this issue because the equilibrium fitness load resulting from segregating deleterious mutations depends on the form of the fitness function, i.e., on the relationship between fitness and number of mutations. As noted in Chapter 10, when the effects of mutations at different loci act independently (with multiplicative effects on fitness), the equilibrium mean fitness, relative to that of a mutation-free individual, is simply e^{-U}, where U is the genomic deleterious mutation rate (Haldane 1937). If deleterious mutations interact synergistically, such that fitness declines at an accelerating rate with an increase in mutation number, the equilibrium load is reduced for the simple reason that the strength of selection increases with subsequent mutations (Kimura and Maruyama 1966, Milkman 1978, Crow and Kimura 1979, Kondrashov 1988, B. Charlesworth 1990). On the other hand, with diminishing-returns epistasis, the equilibrium load is increased because the strength of selection (per mutation) weakens as mutations accumulate.

Data on epistatic interactions among deleterious mutations are few. If such interactions do exist, they appear to be weak and synergistic (Simmons and Crow 1977, Crow and Simmons 1983). However, almost all of the power in this argument derives from a single observation of Mukai (1969) that mean fitness in his mutation-accumulation lines began to decline at an accelerating rate in later generations of his experiments. The curvilinearity that he observed was not strong, and was primarily a function of two late-generation assays, which are not independent.

Analysis of Natural Populations

Mutation-accumulation experiments of the sort described above are extremely labor intensive and require evolutionarily inert controls that are not easily maintained in most species. An alternative approach to characterizing the features of the deleterious mutation process is to evaluate various properties of the distribution of fitness in a natural population and to interpret them in light of their expected values under mutation-selection balance. Morton et al. (1956) and Mukai and Yamaguchi (1974) first suggested how the genomic deleterious mutation rate in a random-mating population might be estimated from the observed inbreeding depression for fitness, provided an estimate of the average degree of dominance were available, and Charlesworth et al. (1990) expanded their approach to obligately selfing populations. Both methods have been generalized by Deng and Lynch (1996b) to yield joint estimators of U, $\overline{h}$, and $\overline{s}$ for mutations affecting total fitness. Their method, which we review below, requires estimates of both the mean fitness and the genetic variance for fitness in inbred and outbred generations.

Recall from Chapter 10 that the equilibrium frequency of a deleterious allele under mutation-selection balance is $u/(hs)$, where u is the genic mutation rate, and hs is the fractional loss of fitness in heterozygotes. Using this result, and assuming that mutations are distributed among individuals in a Poisson fashion and interact multiplicatively to define total fitness (i.e., there is no epistasis), the mean fitness and genetic variance in fitness in a random-mating (outcrossing) population can be shown to be

$$\overline{W}_O = W_{max}\exp(-U) \tag{12.11a}$$

and

$$\sigma^2_W(O) = \overline{W}^2_O\{\exp[U(\overline{hs}\,)] - 1\} \tag{12.11b}$$

where W_{max} is the expected measure of fitness in a mutation-free individual, and $\overline{hs}$ denotes the mean effect of a mutation in the heterozygous state. Suppose now that progeny are produced via self-fertilization of a random sample of parental genotypes, and that multiple progeny are assayed per parent. The expected mean fitness and genetic variance in fitness among selfed families are

$$\overline{W}_S = W_{max} \exp\{-(U/4)[2 + (1/\widetilde{h})]\} \tag{12.11c}$$

$$\sigma^2_W(S) = \overline{W}^2_S\{\exp[\,(U/4)(\overline{s} + (1/4)[\,\overline{s/h}\,] + \overline{hs})] - 1\} \tag{12.11d}$$

where $\widetilde{h}$ is the harmonic mean dominance coefficient of new mutations, and $\overline{s/h}$ is the mean value of the ratio s/h for new mutations. These four equations can be rearranged to eliminate the parameter W_{max}, yielding a set of estimators for U, $\overline{s}$, and $\overline{h}$.

First, we define a set of observable measures,

$$\widehat{x} = \ln\left(\frac{\text{Var}(W_O)}{\overline{W}^2_O} + 1\right) \tag{12.12a}$$

$$\widehat{y} = \ln\left(\frac{\overline{W}_S}{\overline{W}_O}\right) \tag{12.12b}$$

$$\widehat{z} = \ln\left(\frac{\text{Var}(W_S)}{\overline{W}^2_S} + 1\right) \tag{12.12c}$$

where $\text{Var}(W_O)$ is the observed genetic variance in fitness in the parental generation and $\text{Var}(W_S)$ is the observed genetic variance in fitness among selfed families. A variety of methods for estimating genetic variances will be covered in the following chapters. Note that $\widehat{y}$, the natural logarithm of the ratio of fitnesses observed in selfed and outcrossed generations, is a measure of inbreeding depression, whereas $\widehat{x}$ and $\widehat{z}$ are, respectively, functions of the squared coefficients of genetic variation of fitness in the random-mating and selfed populations.

Given information on the means and genetic variances in fitness, $\widehat{x}$, $\widehat{y}$, and $\widehat{z}$ can be computed, and then estimates of the mutation parameters can be obtained from the solutions of Equations 12.11a–d,

$$\widehat{h} \simeq \frac{1}{8\sqrt{\widehat{z}/\widehat{x}} - 4}\left(C_h + \frac{2 - C^2_{s,h}}{1 + C_{s,h}}\right) \tag{12.13a}$$

$$\widehat{U} \simeq \frac{4\widehat{h}\widehat{y}}{2\widehat{h} - 1 - C_h} \tag{12.13b}$$

$$\widehat{s} \simeq \frac{\widehat{x}}{\widehat{U}\widehat{h}(1 + C_{s,h})} \tag{12.13c}$$

Here, $\widehat{h}$ and $\widehat{s}$ are estimates of the arithmetic mean properties of new mutations. Two new terms appearing in these formulae are functions of the distribution of mutational effects: $C_h = \sigma_h^2/\overline{h}^2$, the squared coefficient of variation of dominance coefficients for new mutations, and $C_{s,h} = \sigma_{s,h}/(\overline{h} \cdot \overline{s})$, the coefficient of covariation between s and h. Rarely will quantitative estimates of these properties be available. However, $C_{s,h}$ is expected to be negative (more deleterious mutations tending to be more recessive), and C_h is necessarily positive, although arguments presented in Deng and Lynch (1996b) suggest that it may be small. In the absence of information on these quantities, Equations 12.13a–c can be still be solved by setting $C_h = C_{s,h} = 0$. The estimators will then be biased — the average degree of dominance and the genomic mutation rate in a downward direction, and the average selection coefficient in an upward direction.

Thus, as in the case of the Bateman-Mukai estimators, the method of Deng and Lynch can yield estimates of one-sided bounds on the mutation parameters. Given these estimates, additional composite measures can also be approximated. For example, the long-term rate of decline in fitness that would result if natural selection were rendered ineffective (as in a very small population) is

$$\Delta M = \frac{\widehat{U}\widehat{s}}{2} \tag{12.14}$$

Results in Deng and Lynch (1996b) suggest that Equation 12.14 is an almost unbiased estimator because the downward bias in the estimate $\widehat{U}$ is balanced by the upward bias in the estimate $\widehat{s}$. Moreover, ΔM can be simply estimated by the quantity $\widehat{z}/2$, where $\widehat{z}$ is defined in Equation 12.12c.

Deng and Lynch (1996b) present a parallel set of equations, which provides the basis for parameter estimation by outcrossing ordinarily selfing populations. These estimators tend to be less biased by variance in mutational effects, apparently because most of the bias is caused by highly deleterious recessives, which are normally purged from regularly selfing populations (Chapter 10). Further information on the practical application of the technique, its power as a function of sample size, and its sensitivity to violations of the assumptions of the model is provided in Deng and Lynch (1996b).

Although the method of Deng and Lynch has not yet been applied to natural populations, a few estimates of U have been obtained from the observed improvement in fitness in outcrossed progeny of normally self-fertilizing plants. Such estimates make use of the estimator of Charlesworth et al. (1990),

$$U = \frac{\widehat{y}}{\widehat{h} - 0.5} \tag{12.15}$$

where $\widehat{y}$ is defined in Equation 12.12b. As noted above, this method requires an independent estimate of the degree of dominance, $\widehat{h}$. A variety of methods for estimating h were covered in Chapter 10, although it must be stressed that these estimate the average degree of dominance of segregating mutations. The latter is likely to be somewhat less than the degree of dominance of new mutations (the parameter that should be used in Equation 12.15), since selection removes more dominant mutations with greater efficiency. In their original survey of the literature, Charlesworth et al. (1990) obtained results that suggested that U must be at least 0.5 for individual fitness components. Johnston and Schoen (1995) obtained estimates of U for total fitness in four populations of the annual plant *Amsinckia* ranging from 0.24 to 0.87, and Charlesworth et al. (1994) obtained estimates of $U \simeq 0.8$ and 1.5 for two species of *Leavenworthia.* All of these estimates are in rough accord with the more direct results cited above for *Drosophila* and *Arabidopsis,* supporting the idea that the genomic deleterious mutation rate for total fitness is on the order of 0.2 to 2.0 per individual per generation.

The Persistence of New Mutations

Provided one has estimates for the standing amount of genetic variance for characters in natural populations, estimates of the mutational variance can be used to gain insight into the strength of selection operating against new mutations. The simplest way to view the problem is to assume that most new mutations are kept rare because of their deleterious nature, and to let hs (the average selection against heterozygous mutations) be the fraction of standing genetic variance that is removed each generation by selection. Then, letting $\sigma^2_G(t)$ denote the genetic variance at time t, the dynamics of genetic variance can be represented as $\sigma^2_G(t) = (1 - hs)\,\sigma^2_G(t-1) + \sigma^2_m$. At equilibrium, $\sigma^2_G = \sigma^2_m/(hs)$ (Barton 1990, Kondrashov and Turelli 1992). Thus, the ratio σ^2_m/σ^2_G is a measure of the strength of selection against new mutations, and its reciprocal approximates the mean number of generations that a new mutation persists in a population before being removed by selection (Crow 1992). An alternative and more general definition of σ^2_G/σ^2_m is the mean number of individuals affected by a mutation before it is eradicated (Li and Nei 1972). In an infinite population, the mean persistence time and the mean number of affected individuals are equivalent.

Alternative interpretations are required in the extreme case in which mutations are neutral. Under these circumstances, the equilibrium genetic variance is a consequence of a balance between random genetic drift and mutation and is equal to $2N_e\sigma^2_m$, as noted above. The mean persistence time is then $2N_e$ generations.

From estimates of σ^2_m and σ^2_G, Crow (1992, 1993b) has estimated the mean persistence of viability mutations in *Drosophila* to be approximately 40 generations, both for mild detrimentals and recessive lethals. These results are remarkably consistent with estimates of the selection intensity against heterozygous viability mutations cited above. For example, the reciprocal of hs for mildly deleterious mutations is approximately 45. Actual persistence times are a function of the total fitness effects of new mutations, whereas *Drosophila* mutation-accumulation experiments have typically focused only on the viability-to-adulthood component. Thus, the consistency of the two results suggests that the fitness consequences of new mutations may be largely expressed early in life.

A broader review of the data appears in Houle et al. (1996). Summarizing over a diversity of characters and species, all estimates of the mean persistence time are less than 1000 generations, and more than half are less than 100 generations. The mean persistence times for mutations affecting life-history traits are on the order of 50 generations, again suggesting an average intensity of selection against heterozygotes of about 2%. In *D. melanogaster,* the mean persistence times for fecundity and longevity are approximately 50 and 60 generations, not greatly different from the results for egg-to-adult viability. For morphological traits in this species, mean persistence times tend to be more on the order of 100 generations, much too low to be explained by a neutral model. For abdominal and sternopleural bristle numbers, they are on the order of 700 and 400 generations, consistent with the idea that there is weak selection against new bristle-number mutations as a consequence of their pleiotropic effects on fitness (Nuzhdin et al. 1995).

Such low persistence times raise fundamental questions about our understanding of the nature and relevance of standing pools of quantitative-genetic variation. Is such variation available for adaptive change when a population is exposed to a new selective regime? Or is much of it simply a consequence of the recurrent introduction of mutations with unconditionally deleterious side-effects on fitness? Under the former view, the ability of a population to respond to selection is directly proportional to the amount of additive genetic variance (Bulmer 1971, 1972, 1980; Lande 1975; Bürger et al. 1989; Houle 1989b). Under the latter view, genetic variation observed within populations is of little relevance to adaptive evolution, as the divergence of mean phenotypes among populations mostly involves rare mutations that have only weak pleiotropic side effects (Robertson 1967, Hill and Keightley 1988, Barton 1990, Keightley and Hill 1990, Kondrashov and Turelli 1992). Since all results from mutation-accumulation experiments support the contention that the vast majority of new mutations are deleterious, this latter hypothesis must be taken seriously.

13

Detecting Major Genes

In some cases, the bulk of character variation (either total or genetic) can be attributed to one or a few **major genes**. For a variety of reasons, it is of great interest to detect such genes. From a biological standpoint, the presence of major genes offers the potential for their isolation and genetic characterization, which in turn may be highly informative as to the underlying biological processes generating character variation. From a theoretical standpoint, several quantitative-genetic models (especially those dealing with selection response) assume a large number of loci of roughly equal (and small) effects. The validity of these models is severely compromised by the presence of major genes. If one or two major loci account for most of the genetic variation of a trait, essentially any problem of interest can be correctly modeled using standard machinery of one- and two-locus population genetics.

The observation of a continuous unimodal distribution of phenotypes is often taken as support for a large number of genes of roughly equal effect. It cannot be overstressed that this is an *assumption*. If environmental variation is sufficiently large relative to the effects of any individual gene or if major alleles are at sufficiently low frequency, the effects of segregating major genes can be completely obscured. The most powerful tests for the presence of major genes, which use information from linked molecular-marker loci, are examined in Chapters 14–16. This chapter illustrates how purely phenotypic data can be used to infer the presence or absence of major genes.

Our treatment starts with the simplest form of analysis, departure-from-normality tests applied to situations in which no genealogical information is available. We next consider several fairly simple tests that can be used when groups of known sibs are identified. We then introduce mixture models, wherein the phenotypic distribution is assumed to result from a weighted mixture of several underlying distributions (one for each major-locus genotype). Such models form the foundation for much of the discussion over the next four chapters. We conclude by examining **complex segregation analysis**, a collection of likelihood methods widely used in human genetics. The roots of many of the approaches discussed here stem from epidemiological genetics, a subject reviewed by Cavalli-Sforza and Bodmer (1971), Morton et al. (1983), and Weiss (1993). Additional discussions on the detection of major genes from phenotypic data can be found in reviews by Mayo (1989), Hill and Knott (1990), and Le Roy and Elsen (1992).

ELEMENTARY TESTS

While detection of major genes is not an easy task, it is facilitated considerably with experimental populations in which controlled breeding designs can be implemented. For example, a population segregating a major allele initially at low frequency is expected to show a significant increase in heritability after only a few generations of selection (Latter 1965, Frankham and Nurthen 1981), reflecting the increase in additive genetic variation as the rare major allele increases in frequency. Heritability can likewise show a sudden decrease if the selected major gene has a moderate to high initial frequency.

A related approach for detecting major genes using selection is the **select-and-backcross method** (reviewed by Wright 1952, 1968). The procedure is simple. Two populations differing in the character are crossed, with the largest F_1 individuals backcrossed to individuals from the parental population with the smaller mean character value. Continual selection for large character value followed by backcrossing to the smaller line removes genes (from the large line) of minor effect while retaining genes having large effects (Wright's **leading factors**). The ability to maintain a large character value in the face of continual backcrossing to the smaller line suggests the presence of major genes. Wright (1968) successfully used this approach to isolate leading factors for several traits in guinea pigs, and the Booroola gene in sheep (Examples 3 and 4 in Chapter 4) was detected in this manner (Piper and Bindon 1988). Natural populations that are unamenable to controlled breeding generally preclude such approaches, but as shown below, a random collection of individuals from such a population can sometimes be informative.

Departures from Normality

Consider a locus segregating a major allele and assume that the distribution of phenotypes for each of the major-locus genotypes is normal. The resulting distribution, a mixture of normals, is generally *not* normal. When a major gene is segregating, the phenotypic distribution can exhibit multimodality, skewness, and/or kurtosis. Figure 13.1 shows that when major allele frequencies are intermediate, the resulting phenotypic distribution is platykurtic, being more flatly peaked than a normal. The distribution becomes leptokurtic (more peaked than a normal) and skewed when alleles of large effect are at extreme frequencies, i.e., somewhat near zero or one (Hammond and James 1970, O'Donald 1971). When gene frequencies become very close to zero or one, the distribution again more closely approaches normality, as there is effectively only a single genotype in the population.

A variety of tests for normality have been developed, some of which were discussed in Chapter 11, but none of these is particularly powerful. As a

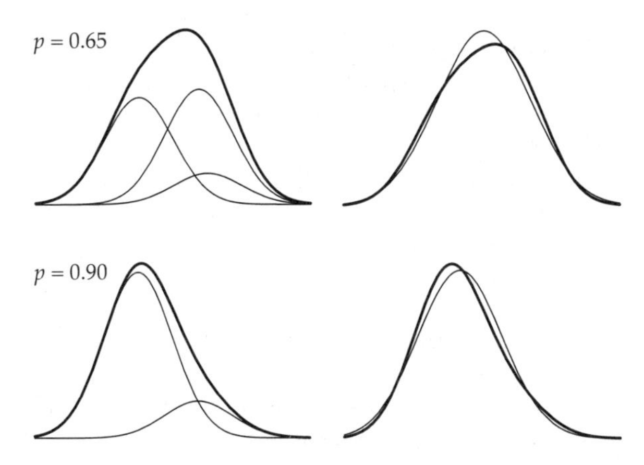

Figure 13.1 Phenotypic distributions resulting from a single diallelic major locus, where each genotype has normally distributed phenotypes with the same variance σ^2. Here the mean genotypic values are $\mu_{qq} = -\sigma$, $\mu_{Qq} = 0.75\sigma$, $\mu_{QQ} = \sigma$, and genotypes are in Hardy-Weinberg proportions, with the frequency of allele q being $p = 0.65$ (Top) and $p = 0.9$ (Bottom). The left side of each figure shows the three underlying distributions (thin lines) and the resulting mixture distribution (thick line). The right side shows how well the resulting mixture distribution fits a normal having the same mean and variance.

consequence, one must proceed with extreme caution in inferring the *absence* of major genes from a phenotypic distribution that is not significantly different from normal. As Figure 13.1 shows, the departure from normality can be rather small, requiring very large sample sizes for detection. Thoday and Thompson (1976) found that tests based on kurtosis have poor power for detecting major genes. In their simulations, a character with no environmental variation determined by four additive loci (each with two alleles at frequency 0.5 and relative effects of x, $2x$, $3x$, and $4x$) required a sample of roughly 750 individuals to detect nonnormality, while with three additive loci of equal effect, sample sizes in excess of 1000 were required. Conversely, when the distribution significantly departs from normality, one must be equally cautious in inferring the *presence* of a major gene without additional support. For example, a phenotypic distribution can display skew in the absence of major genes due to scale effects (Chapter 11). As we discuss shortly, mixture models provide a more sophisticated and powerful approach to major-gene detection, assessing whether the fit of the phenotypic distribution is consistent with a mixture of normals.

Tests Based on Sibship Variances

The use of collections of known relatives, rather than random individuals, greatly improves the power of tests for major genes. Several simple tests based on sibship comparisons have been proposed. The presence of major alleles increases the variance within sibships in which the alleles are segregating. For example, if Q and q denote alternative alleles at a major locus, the expected phenotypic variance among sibs from $Qq \times Qq$ parents is greater than for sibs from $qq \times qq$ or $QQ \times QQ$ parents, reflecting segregation within these families, while sibs from $QQ \times Qq$ and $qq \times Qq$ parents display intermediate levels of variance. A number of tests for unequal variances among sibships have been suggested, the simplest being **Bartlett's test** for homogeneity of variances (e.g., Sokal and Rohlf 1995). The difficulty with this approach is that nongenetic causes may underlie differences in sibship variances. Bartlett's test is also rather sensitive to departures from normality. As a way around these problems, Mérat (1968) proposed a test based on the notion that families with significantly elevated variances should also exhibit exceptional platykurtosis and possibly skewness if the inflation of the variance is caused by segregating alleles of large effect.

Fain (1978) proposed an alternative test that utilizes both the means and variances of sibships. The basis for this test is the observation that, for a character determined by many genes of roughly equal effect, there should be no relationship between the variance of the sibship and the phenotypic value of the parents (Pearson 1904, Penrose 1969, Felsenstein 1973, Stark 1976). If, however, the character is determined by a few genes of large effect, parents with the most extreme phenotypes are likely to be homozygotes, while parents of intermediate phenotypes are more likely to be heterozygotes. This relationship results in a roughly quadratic regression of offspring variance on midparental phenotypic value,

$$\text{Var}(z_i) = a + b_1\overline{z}_i + b_2\overline{z}_i^2 \tag{13.1}$$

where $\text{Var}(z_i)$ is the phenotypic variance within the ith sibship, and $\overline{z}_i$ is the midparental value for this sibship. Sibship means replace midparental values when the latter are unknown. A significant value of b_2 is taken as an indication of a major gene. Due to scaling effects (Chapter 11), the variance often increases linearly with the mean, so a significant b_1, by itself, does not necessarily indicate a major gene.

Example 1. Bucher et al. (1982) examined a large sample of families that were classified into groups based on cholesterol levels. One particular group (the High group) showed significant heterogeneity of within-sibship variance for cholesterol level ($P < 0.001$ using Bartlett's test of homogeneity of variances), while the other groups did not. Likewise, the regression of the within-sibship variance on the sibship means had a significant quadratic term for the High group, while the quadratic term was not significant in the other groups. These data suggest that a major gene was segregating in at least some of the families forming the High group (i.e., at least one parent was a heterozygote) but was not in the families forming the other groups. Thus, there are additional sources of variance beyond the major gene (such as environmental factors and / or additional polygenes) that contribute to the difference between groups.

Our second example of Fain's test is based on Mitchell-Olds and Bergelson's (1990) work on the annual plant *Impatiens capensis*. These authors found that the regression of within-sibship variance on sibship mean usually had significant linear (but not quadratic) terms, suggesting that scale effects are common. However, the regression for germination date also showed a significant quadratic term (Figure 13.2), suggesting the presence of a major gene influencing this character.

The power of different sibship variance tests has been considered by several investigators. MacCluer and Kammerer (1984) concluded that both Bartlett's and Fain's test have low power for detecting a major gene even when 100 nuclear families are used. However, they assumed a small number of sibs per family (3 to 5), as their interest was in human populations. On the positive side, they found that these tests are unlikely to give a false indication of a major gene when none is present (under the simplified assumption of normally distributed environmental effects). Mayo et al. (1980) note that Fain's test is compromised if heterozygotes have lower environmental variances than homozygotes, as this difference partially masks the increased genetic variance in sibships from heterozygous parents.

Major-gene Indices (MGI)

Karlin et al. (1979) show that, under polygenic inheritance with additive loci, offspring more closely resemble the average of their parents (the midparent) than either individual parent, whereas the converse is true when a major gene is segregating. Exploiting this relationship, they proposed a class of indices for indicating the presence of major genes,

$$\text{MGI}(a) = \frac{E[\,|z_o - (z_f + z_m)/2|^a\,]}{E[\,(\,|z_o - z_f|\,|z_o - z_m|)^{a/2}\,]} \tag{13.2}$$

where z_o is the phenotypic value of an offspring whose parents have values z_f and z_m, and a is a free parameter set by the investigator. Values of $a > 1$ accentuate large deviations between offspring and parents, while values of $a < 1$ accentuate small deviations. If the character is entirely determined by segregation of a major allele with no environmental variation, $\text{MGI}(a) > 1$ for all $a \geq 0$ (independent of the amount of dominance), while if the character is entirely determined by an infinite number of genes of equal effect with a normal distribution of environmental effects, $\text{MGI}(a) < 1$ for all $a \geq 0$. Karlin et al. recommend evaluation of the index at $a = 0.5$, 1, and 2, but this choice of a values is rather ad hoc. Improvements and embellishments on this test have been suggested by Famula (1986), Carmelli et al. (1979), and Karlin and Williams (1981), but Le Roy and Elsen (1992) found that both the Bartlett and Fain tests are more powerful than MGI-based tests.

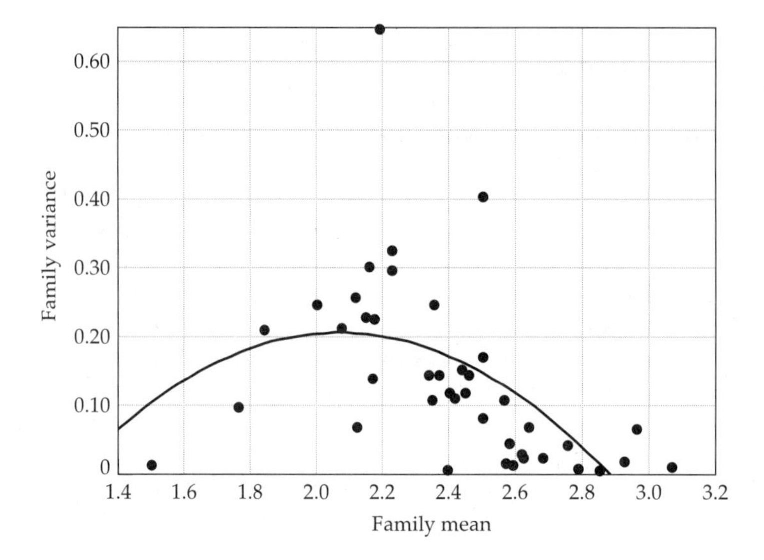

Figure 13.2 An example of Fain's test, plotting family variance as a function of family mean for germination date in *Impatiens capensis*. There is a highly significant quadratic term ($r^2 = 0.32$ vs. $r^2 = 0.18$ for a simple linear regression), suggestive of a segregating major gene for germination date in this population. (Data from Mitchell-Olds and Bergelson 1990, kindly provided by Tom Mitchell-Olds.)

Nonparametric Line-cross Tests

Collins (1967, 1968, 1973), Mode and Gasser (1972), and Birnbaum (1972) independently proposed simple line-cross tests to detect major genes. Suppose a character is completely determined by one locus with two alleles, Q and q. Each genotype has an associated distribution of environmental values, which may differ between genotypes, and the phenotypic distribution in the population is a weighted mixture of the distributions associated with the three underlying genotypes,

$$p(z) = p_{QQ}(z)\,\Pr(QQ) + p_{Qq}(z)\,\Pr(Qq) + p_{qq}(z)\,\Pr(qq) \tag{13.3}$$

where Pr(QQ) is the probability that a randomly chosen individual is QQ, and $p_{QQ}(z)$ is the distribution of environmental values for a QQ individual, with the other terms defined similarly.

Based on this simple relationship, the test proceeds as follows. Consider two parental populations, P_1 and P_2, fixed for different alleles (Q and q, respectively). If the outbreak of variation in the F_2 is entirely due to segregation at a single locus, then the F_2 distribution is specified exactly by Equation 13.3, with $p_{Qq}(z) = p_{F_1}(z)$, $p_{QQ}(z) = p_{P_1}(z)$, $p_{qq}(z) = p_{P_2}(z)$, $\Pr(QQ) = \Pr(qq) = 1/4$, and $\Pr(Qq) = 1/2$, giving

$$p_{F_2}(z) = \frac{p_{P_1}(z)}{4} + \frac{p_{F_1}(z)}{2} + \frac{p_{P_2}(z)}{4} \tag{13.4}$$

Standard goodness-of-fit criteria (such as a Kolmogorov-Smirnov test; Sokal and Rohlf 1995) are then used to compare the observed and expected distributions. Related tests have also been suggested by Elston (1981b) and Stolk et al. (1984), and these were used by Hagger et al. (1995) to reject the hypothesis of a single dominant gene to account for egg weight differences between two inbred lines of chickens.

This approach can, in theory, be extended to two or more major loci, although tests of more than two loci require an impractical number of crosses. If more than one locus is segregating, knowledge of the parental and F_1 distributions is generally not sufficient to predict the F_2 distribution. Additional information, such as phenotypic distributions of certain backcrosses (e.g., the distribution of B_1, $B_1 \times B_2$, etc.), is required to account for the new F_2 genotypes generated by segregation (Collins 1967, 1968, 1973).

Equation 13.3 provides our first formal introduction to **mixture models**, where the observed distribution is a weighted mixture of underlying distributions. The approach taken above is nonparametric in that it does not initially assume any particular form for the underlying distributions. While this is an advantage in some cases, it results in a reduction in power relative to parametric tests that incorporate the correct form of the underlying distributions. We now move to considerations of parametric mixture models that assume the underlying distributions to be normal. This class of models forms the basis for much of the analysis over the next four chapters.

MIXTURE MODELS

Given that rejection of normality, by itself, is not sufficient to imply a major gene, a more powerful approach is to test whether the phenotypic distribution is consistent with a mixing (or **commingling**) of two or more normals, as would be expected if the distribution of phenotypic values about each major genotype were normal (Tan and Chang 1972, Elston et al. 1974, Boerwinkle et al. 1986, Hoeschele 1988). While this is an improvement over previous methods, such a test is still fraught with a number of potential problems. Even if the phenotypic distribution is consistent with a mixture of normals, there are a variety of explanations other than major genes — for example, the population may be distributed over two or more significantly different environments. Thus, a slightly more powerful approach is to further specify the weights of the underlying normals.

If the distribution is consistent with three underlying normals with Hardy-Weinberg weights p^2, $2p(1-p)$, and $(1-p)^2$, where the allele frequency p is estimated from the data by maximum likelihood (see Example 2), then we may have more confidence in a major-locus interpretation. On the other hand, a lack of fit to this model does not necessarily exclude a major gene (e.g., the major gene may not be in Hardy-Weinberg equilibrium or the underlying phenotypic distributions for each major locus genotype may be nonnormal). In the absence of information from linked markers, the most powerful mixture-model tests for major genes use information from relatives to specify the weights of the underlying distributions. Before developing this **complex segregation analysis** approach, we first review some general features of mixture models.

The Distribution under a Mixture Model

Assume the distribution of interest results from a weighted mixture of several underlying distributions. If there are $i = 1, \cdots, n$ underlying distributions, $p_1(z), \cdots, p_n(z)$, each with frequency $\Pr(i)$, the resulting probability density of an observed variable z is given by a generalization of Equation 13.3,

$$p(z) = \sum_{i=1}^{n} \Pr(i) \cdot p_i(z)$$

It is usually assumed that the underlying distributions are normals, so this becomes

$$p(z) = \sum_{i=1}^{n} \Pr(i) \cdot \varphi(z, \mu_i, \sigma_i^2) \tag{13.5}$$

where

$$\varphi(z, \mu_i, \sigma_i^2) = \frac{1}{\sqrt{2\pi\sigma_i^2}} \exp\left[-\frac{(z-\mu_i)^2}{2\sigma_i^2} \right]$$

is the probability density function for a normally distributed random variable with mean μ_i and variance σ_i^2. Equation 13.5 has $3n - 1$ parameters to estimate: the $n - 1$ mixing proportions $\Pr(i)$, and the n means and variances of the underlying distributions. It is usually assumed that all the variances are equal, reducing the number of unknown parameters to $2n$. Various genetic hypotheses allow us to further specify and evaluate the structure of the mixing proportions.

Parameter Estimation

Parameters of mixture models are typically estimated by maximum likelihood procedures (Hasselblad 1966, Day 1969, Everitt and Hand 1981, Redner and Walker 1984, Titterington et al. 1985, McLachlan and Basford 1988), in which case Equation 13.5 gives the likelihood function $\ell(z)$ of the unknown parameters $\Pr(1), \cdots, \Pr(n)$, $\mu_1, \cdots, \mu_n$, $\sigma_1^2, \cdots, \sigma_n^2$ as a function of the observed value z. As just mentioned, typically one sets most (or all) of the variances equal to each other; one reason for equating variances is that, if they are all free to vary, there can be singularities in the likelihood function.

As an example of how likelihood functions are constructed, consider the situation for a random individual drawn from a population with a single segregating diallelic major locus. Indexing the three genotypes by i where $i = QQ$, Qq, and qq, and assuming that individuals with major-locus genotype i are normally distributed with mean μ_i and common variance σ^2, the resulting likelihood for the jth individual is

$$\ell(z_j) = \Pr(QQ)\, p_{QQ}(z_j) + \Pr(Qq)\, p_{Qq}(z_j) + \Pr(qq)\, p_{qq}(z_j) \tag{13.6a}$$

$$= \Pr(QQ)\, \varphi(z_j, \mu_{QQ}, \sigma^2) + \Pr(Qq)\, \varphi(z_j, \mu_{Qq}, \sigma^2) + \Pr(qq)\, \varphi(z_j, \mu_{qq}, \sigma^2)$$

where z_j is the character value in the focal individual. The likelihood for an individual is easily generalizable to more complicated genetic models. For example, if there are more than two alleles, or multiple loci, the likelihood has the form of Equation 13.6a with the sum now extending over all n_g multilocus genotypes. For n random (unrelated) individuals, denoting the observed phenotypic values by $\mathbf{z} = (z_1, z_2, \cdots, z_n)$, the overall likelihood is just the product of the n individual likelihoods,

$$\ell(\mathbf{z}) = \ell(z_1, z_2, \cdots, z_n) = \prod_{j=1}^{n} \ell(z_j) \tag{13.6b}$$

Assuming random mating, the Hardy-Weinberg principle describes the frequencies $\Pr(\cdot)$ of the major locus genotypes as a function of p, the frequency of one allele. This leaves five parameters to estimate — p, σ^2, μ_{QQ}, μ_{Qq}, and μ_{qq}.

Appendix 4 reviews the basic features of the maximum likelihood approach, and the reader may wish to consult this before continuing (additional features are developed in Chapter 27). Maximum likelihood estimates (MLEs) are those values of the unknown parameters that maximize the likelihood function when treating the observed data $\mathbf{z} = (z_1, \cdots, z_n)$ as fixed constants. Such estimates can be obtained by numerical maximization of the likelihood function (e.g., Gill et al. 1981, Fletcher 1987) or by other iterative approaches. In particular, expectation-maximization (EM) methods are both very powerful and very flexible, accommodating missing or incomplete data (Appendix 4). Sample variances and covariances of MLEs can either be obtained directly from the likelihood functions or via approximation methods (e.g., Meyer and Hill 1992). See Appendix 4 for details.

Hypothesis Testing

An important issue in tests for major genes is model fitting, i.e., evaluating whether the full model is needed, or if some subset of the model gives essentially the same fit. For example, we might initially assume a mixture of two normals with different means and common variance, so that the full model has parameters μ_1, μ_2, σ^2, and p. Is the fit using these four parameters significantly better than the fit assuming a single underlying normal with parameters μ and σ^2? For large sample sizes, the **likelihood ratio** (LR) statistic test for whether the full model provides a better fit than a particular subset of the model is

$$\Lambda(\mathbf{z}) = -2\ln\left[\frac{\widehat{\ell}_r(\mathbf{z})}{\widehat{\ell}(\mathbf{z})}\right] = -2\left\{\ln\left[\widehat{\ell}_r(\mathbf{z})\right] - \ln\left[\widehat{\ell}(\mathbf{z})\right]\right\} \tag{13.7}$$

where $\widehat{\ell}(\mathbf{z})$ is the likelihood function evaluated at the MLE for the full model, and $\widehat{\ell}_r(\mathbf{z})$ is the maximum of the likelihood function for the restricted model under which r parameters of the full model are assigned fixed values. Under appropriate conditions, the LR test statistic is approximately distributed as χ^2_r, i.e., as a χ^2 distribution with r degrees of freedom (Wald 1943); see Appendix 4.

The simplest restricted model assumes no mixture at all, so that the overall distribution is just a single normal distribution with unknown mean and variance. As shown in Appendix 4, the resulting likelihood is just the product of n identical normals with mean μ and variance σ^2. The MLEs in this case are the sample mean $\overline{z}$ and the uncorrected sample variance

$$\text{Var}(z) = \frac{1}{n}\sum_{j=1}^{n}(z_j - \overline{z})^2$$

Note that the ML estimate of the variance is slightly different from the unbiased variance estimator, which divides the sum of squares by $n-1$. Substituting the MLEs of σ^2 and μ gives the maximum value of this restricted likelihood as

$$\widehat{\ell}_r(\mathbf{z}) = \prod_{j=1}^{n}\left\{\frac{1}{\sqrt{2\pi\,\text{Var}(z)}}\exp\left[-\frac{(z_j - \overline{z})^2}{2\,\text{Var}(z)}\right]\right\}$$

Taking logarithms and recalling the definition of Var gives

$$-2\ln\left[\widehat{\ell}_r(\mathbf{z})\right] = n\cdot\left[\ln\text{Var}(z) + \ln 2\pi + 1\right] \tag{13.8}$$

Example 2. Consider the likelihood-ratio test statistic for whether a diallelic major gene (in Hardy-Weinberg frequencies, with the phenotypes for each major-locus genotype normally distributed with constant variance) provides a better fit of the data than a single normal distribution. Assume that the data consist of n (unrelated) individuals, randomly chosen from the population. From Equations 13.7 and 13.8, the likelihood-ratio test statistic is given by

$$\begin{aligned}\Lambda(\mathbf{z}) &= -2\left\{\ln\left[\widehat{\ell}_r(\mathbf{z})\right] - \ln\left[\widehat{\ell}(\mathbf{z})\right]\right\} \\ &= 2\ln\left[\widehat{\ell}(\mathbf{z})\right] + n\cdot\left[\ln\text{Var}(z) + \ln 2\pi + 1\right]\end{aligned}$$

where

$$\widehat{\ell}(\mathbf{z}) = \max\left[\prod_{j=1}^{n} \ell(z_j)\right]$$

the maximum being taken over all admissible values of p $(0 \leq p \leq 1)$, μ_{QQ}, μ_{Qq}, μ_{qq} $(-\infty < \mu < \infty)$, and σ^2 $(\sigma^2 \geq 0)$ and

$$\ell(z_j) = p^2 \cdot \varphi(z_j, \mu_{QQ}, \sigma^2) + 2p(1-p) \cdot \varphi(z_j, \mu_{Qq}, \sigma^2) + (1-p)^2 \cdot \varphi(z_j, \mu_{qq}, \sigma^2)$$

Since the full model has five unknown parameters while the reduced model has two (μ, σ^2), the test statistic Λ is approximately distributed as χ^2_3. Hence Λ values exceeding 7.82 and 11.4 indicate that a mixture of three normals provides a better fit at the 5% and 1% levels of significance, respectively, than a single normal.

Likelihood-ratio tests require that alternate hypotheses be **nested**, one model being a subset of the other (i.e., by fixing some parameters of one model we recover the second). If they are not, the large-sample distribution does not necessarily approach a χ^2. Likelihood functions involving nonnested hypotheses can be compared by using Akaike's (1974) **information content** (AIC),

$$\text{AIC} = -2\ln(\text{maximum likelihood}) + 2(\text{number of fitted parameters}) \tag{13.9}$$

The model with the smallest AIC is chosen as the most parsimonious. The AIC is a descriptive statistic only and not a formal hypothesis test, but it provides a useful measure for comparing rather different models. An alternative approach involves the use of resampling methods (Schork and Schork 1989, Churchill and Doerge 1994). Two such methods, permutation tests and bootstrapping, are discussed in Chapter 15.

How does one proceed if the underlying distributions of a mixture model are not normal? MacLean et al. (1976), Elston (1984), and Schork and Schork (1988) suggest that the Box-Cox (1964) power transform can be used to normalize each of the underlying distributions. Recall from Chapter 11 that this transformation is described by a single parameter λ, with $x = (z^\lambda - 1)/\lambda$ for $\lambda \neq 0$ and $x = \ln z$ for $\lambda = 0$. Hence, in the likelihood function, the ith underlying distribution uses $x_i = (z^{\lambda_i} - 1)/\lambda_i$ in place of z, allowing each underlying distribution to have a different transform. If we assume that an observed distribution is generated by k underlying distributions that can be transformed to normality by the appropriate Box-Cox transform, the resulting mixture model can be expressed as a mixture of k normals. The ith normalized underlying distribution has mean μ_i, variance σ_i^2, and transformation parameter λ_i, all of which can be estimated by standard ML methods. This approach offers a test for whether an observed skewed distribution results from a single naturally skewed distribution, from a mixture of underlying normals, or from a mixture of underlying distributions that can be transformed to normality. If the distribution results from a mixture of normals, a standard mixture model should give a significantly better fit than a single transformed distribution (where the transformation parameter λ is estimated from the data). Whether a mixture model provides a significantly better fit can be evaluated by a likelihood-ratio test. Likewise, the hypothesis that the underlying mixture distributions are themselves skewed can be tested by fitting the λ_i from the data and comparing this to a model with $\lambda_i = 1$ (no transformation).

COMPLEX SEGREGATION ANALYSIS

Starting with Elston and Stewart (1971) and Morton and MacLean (1974), human population geneticists have developed the method of complex segregation analysis, or CSA, to test between alternative modes of inheritance. As we will see in the next few sections, CSA extends the simple mixture model (Equation 13.6a) by using pedigree information to modify the mixture proportions. Assumptions about the mode of inheritance specify the mixture distribution weights, allowing likelihood-ratio tests for different models of transmission (a single major gene, no major gene but background polygenes, major gene plus background polygenes, and so forth). We illustrate some of these applications below, showing how known relationships among relatives define the transmission probabilities of both major genes and polygenes.

The extensive literature on complex segregation analysis is reviewed by Elston and Rao (1978), Boyle and Elston (1979), Elston (1980, 1981a, 1990a), and Morton et al. (1983). While the bulk of the literature deals with human populations, Le Roy et al. (1990) and Knott et al. (1991a,b) examine applications in animal breeding. Tourjee et al. (1995) give an application to plants, examining the genetic basis of flower color in *Gerbera jamesonii*. Similar likelihood methods for major loci have been developed for F_2 segregation in crosses between inbred lines (Tan and Chang 1972, Elston and Stewart 1973, Tan and D'Angelo 1979, Elston 1984, Janss and Van Der Werf 1992, Loisel et al. 1994, Changjian et al. 1994).

While complex segregation analysis is the most assumption-burdened test for detecting major genes, it is also the most powerful of the marker-free methods when the assumptions hold (MacCluer et al. 1983, MacCluer and Kammerer 1984). For example, the commingling tests discussed above (e.g., Example 2) that simply fit a mixture of normals and use no information from relatives can easily miss major genes that segregation analysis can detect (Kwon et al. 1990).

Complex segregation analysis assumes normality of the underlying distributions, which greatly simplifies the form of the likelihood function. If this assumption is violated, false detection of a major locus can occur (MacLean et al. 1975, Go et al. 1978, Morton 1984). As mentioned above, if the underlying distributions are suspected to be nonnormal, one strategy is to use a likelihood approach that incorporates a transformation parameter for each underlying distribution. Instead of individually transforming each underlying distribution (through the likelihood function), one could simply apply a single transform to the observed distribution. However, this approach can raise serious issues of interpretation (e.g., Asamoah et al. 1987). While the use of transformations is not a resolved issue, Demenais et al. (1986) suggest that tests that incorporate estimates of transmission probabilities (see Example 4 below) remove the need to transform the data before performing segregation analysis. A final complication is that when significant genotype $\times$ environment interaction is present, the power to detect a major gene is greatly reduced (Eaves 1984, Tiret et al. 1993).

The following sections illustrate some general approaches for constructing likelihood functions for full-sib families under increasingly more general genetic models. We start by assuming only a single diallelic major locus, and then consider separately common family effects and background segregation of polygenic loci. Likelihoods for more complicated pedigrees or other experimental designs (such as inbred-line crosses) follow using similar arguments. Before proceeding, an example will show the types of hypotheses that can be addressed and the types of parameters that need to be incorporated into the likelihood functions underlying complex segregation analysis.

Example 3. Morton and MacLean (1974) consider a model with both a segregating diallelic major gene (alleles Q and q) and a completely additive polygenic background. Conditioned on the genotype at the major locus, the distributions of phenotypic values are assumed to be normally distributed with means μ_{QQ}, μ_{Qq}, or μ_{qq}, and common variance $\sigma^2 = \sigma_E^2 + \sigma_A^2$ (the environmental variance plus the additive genetic variance contributed by the background polygenes). Assuming the major-locus genotypes are in Hardy-Weinberg proportions, this model is described by six parameters: p = frequency of Q, the means of each major-locus genotype (μ_{QQ}, μ_{Qq}, μ_{qq}), the environmental variance σ_E^2, and the genetic variance from the polygenic contribution σ_A^2. The resulting likelihood function (see Equations 13.11b and 13.22 below) is complex as it incorporates transmission of both the major alleles and polygenic background from parent to offspring, conditioning over all possible parental genotypes. The amount of support for various genetic hypotheses can be tested using likelihood ratios of appropriate subsets of the full model, as given in the following table:

	Model	Free Parameters	Restricted Parameters
1.	No genetic effects	μ, σ_E^2	$\mu_{QQ} = \mu_{Qq} = \mu_{qq} = \mu$ $p = 0$, $\sigma_A^2 = 0$
2.	Major gene, no background polygenes	μ_{QQ}, μ_{Qq}, μ_{qq}, p, σ_E^2	$\sigma_A^2 = 0$

Table continues

	Model (Continued)	Free Parameters	Restricted Parameters
3.	Background polygenes, no major gene	μ, σ_E^2, σ_A^2	$\mu_{QQ} = \mu_{Qq} = \mu_{qq} = \mu$ $p = 0$
4.	Full model: Major gene, background polygenes	μ_{QQ}, μ_{Qq}, μ_{qq}, p, σ_E^2, σ_A^2	None

For example, a test of support for a major gene is given by the likelihood ratio using model 1 (a single normal distribution) as the restricted model and model 2 as the full model. The resulting test statistic has $5 - 2 = 3$ degrees of freedom, with twice the log of the maximum of the restricted likelihood function given by Equation 13.8. If the major-gene model provides a significant improvement, model 4 (major gene plus polygenic background) can next be tested against the major-gene-only model (2), with the test statistic having $6 - 5 = 1$ degree of freedom. More complicated models are analyzed in a similar fashion.

Likelihood Functions Assuming a Single Major Gene

We start by computing the likelihood for a single individual, then proceed to an entire family, and finally to the collection of all families in our sample. Assume that a single diallelic locus underlies the character and consider the jth offspring from the ith family, o_{ij}, which has father f_i and mother m_i (for notational ease, in the following we use f, m, and o_j, reminding the reader that these, of course, change as we change families). Denote the phenotypic value of this offspring by z_{ij}. Index the major-locus genotypes by g where $g = 1$ for QQ, $g = 2$ for Qq, and $g = 3$ for qq, with g_f, g_m, and g_{o_j} denoting the genotypes of the parents (father and mother) and their jth offspring. Phenotypic values for each major-locus genotype are assumed to be normally distributed with means μ_g and common variance σ^2. Finally, let $\Pr(g_o \,|\, g_f,\, g_m)$ be the probability that an offspring has genotype g_o given that its parents have genotypes g_f and g_m.

Conditioned on the parental genotypes, the likelihood for the ijth offspring is

$$\ell(z_{ij} \,|\, g_f,\, g_m) = \sum_{g_o=1}^{3} \Pr(g_o \,|\, g_f,\, g_m) \cdot \varphi(z_{ij}, \mu_{g_o}, \sigma^2) \tag{13.10a}$$

This conditional likelihood is a mixture model with mixing proportions given by Mendelian segregation. For example, if the father and mother have major-locus genotypes QQ and Qq, then $g_f = 1$ and $g_m = 2$, and

$$\begin{aligned} \Pr(g_o = 3 \,|\, g_f = 1,\, g_m = 2) &= \Pr(qq \,|\, g_f = QQ,\, g_m = Qq) = 0 \\ \Pr(g_o = 2 \,|\, g_f = 1,\, g_m = 2) &= \Pr(Qq \,|\, g_f = QQ,\, g_m = Qq) = 1/2 \\ \Pr(g_o = 1 \,|\, g_f = 1,\, g_m = 2) &= \Pr(QQ \,|\, g_f = QQ,\, g_m = Qq) = 1/2 \end{aligned} \tag{13.10b}$$

so that with these parents Equation 13.10a reduces to

$$\ell(z_{ij}\,|\,QQ,Qq) = \frac{1}{2}\cdot\varphi(z_{ij},\mu_{QQ},\sigma^2) + \frac{1}{2}\cdot\varphi(z_{ij},\mu_{Qq},\sigma^2) \tag{13.10c}$$

Conditioned on parental genotype values, each offspring in a family is independent, implying that the likelihood for a full-sib family of n_i offspring is the product of individual likelihoods, giving the conditional likelihood for the ith family as

$$\ell(z_{i\cdot}\,|\,g_f,\,g_m) = \prod_{j=1}^{n_i}\ell(z_{ij}\,|\,g_f,\,g_m) \tag{13.11a}$$

Since we do not know the QTL genotypes of the parents, the unconditional likelihood for the ith family is obtained by summing over all nine possible pairs of parental genotypes,

$$\ell(z_{i\cdot}) = \sum_{g_f=1}^{3}\sum_{g_m=1}^{3}\ell(z_{i\cdot}\,|\,g_f,\,g_m)\,\Pr(g_f,\,g_m) \tag{13.11b}$$

Assuming the parents are chosen independently, $\Pr(g_f,\,g_m) = \Pr(g_f)\cdot\Pr(g_m)$. Further, if genotypes are in Hardy-Weinberg proportions, parental genotype frequencies are completely specified by the frequency p of allele Q, e.g.,

$$\Pr(g_f = 1,\,g_m = 1) = \Pr(g_f = QQ)\cdot\Pr(g_m = QQ) = p^2\cdot p^2$$
$$\Pr(g_f = 2,\,g_m = 1) = \Pr(g_f = Qq)\cdot\Pr(g_m = QQ) = 2p(1-p)\cdot p^2, \text{ etc.}$$

If there are $n_g > 3$ major-locus genotypes (either because of multiple alleles at the major locus or because of several major loci), the appropriate likelihood has sums ranging over the n_g genotypes, and the transmission probabilities are modified to account for the assumed model. Likewise, if the parental phenotypic values (z_f, z_m) are known, these can also be incorporated into the likelihood. Since $\ell(z\,|\,g) = \varphi(z,\mu_g,\sigma^2)$, the probability that the genotype is g_i given the phenotype is z is

$$\Pr(g_i\,|\,z) = \frac{\Pr(g_i)\;\varphi(z,\mu_{g_i},\sigma^2)}{\displaystyle\sum_{j=1}^{n_g}\Pr(g_j)\;\varphi(z,\mu_{g_j},\sigma^2)} = \frac{\Pr(g_i)\;\varphi(z,\mu_{g_i},\sigma^2)}{p(z)} \tag{13.12}$$

where $p(z)$ is the phenotypic density function for the entire population. Parental phenotypes are then incorporated by replacing $\Pr(g)$ by $\Pr(g\,|\,z)$. Equation 13.12 follows directly from Bayes' theorem (Equation 13.24), which will be discussed shortly.

Assuming different families are unrelated, the total likelihood is the product of the individual likelihoods from the n_f families,

$$\ell(\mathbf{z}) = \prod_{i=1}^{n_f} \ell(z_{i\cdot}) \tag{13.13}$$

where $\ell(z_{i\cdot})$ is given by Equation 13.11b. Although there are numerous summation and product indices in this likelihood, there are only five unknown parameters: the three genotypic means, the common variance σ^2, and the QTL allele frequency p.

While the most obvious test for a major gene compares the full model with the restricted model of a single underlying normal, Elston et al. (1975) suggest that a much more robust approach is to treat the transmission probabilities $\Pr(g_o \,|\, g_f, g_m)$ as unknown parameters and base hypothesis tests on these. Above, we specified the transmission probabilities based on Mendelian assumptions of inheritance (e.g., Equation 13.10b), but we can also treat them as parameters to be estimated. This is most conveniently done by considering τ_x, the probability that genotype x transmits a Q allele. For a diallelic locus, there are three τ values to estimate, one for each genotype. From the definition of τ, the transmission probabilities can be expressed as

$$\begin{aligned}
\Pr(qq \,|\, g_f, g_m) &= (1-\tau_{g_f})(1-\tau_{g_m}) \\
\Pr(Qq \,|\, g_f, g_m) &= \tau_{g_f}(1-\tau_{g_m}) + \tau_{g_m}(1-\tau_{g_f}) \\
\Pr(QQ \,|\, g_f, g_m) &= \tau_{g_f}\,\tau_{g_m}
\end{aligned} \tag{13.14}$$

For example, Equations 13.10b become

$$\begin{aligned}
\Pr(qq \,|\, g_f = QQ,\, g_m = Qq) &= (1-\tau_{QQ})(1-\tau_{Qq}) \\
\Pr(Qq \,|\, g_f = QQ,\, g_m = Qq) &= \tau_{QQ}(1-\tau_{Qq}) + \tau_{Qq}(1-\tau_{QQ}) \\
\Pr(QQ \,|\, g_f = QQ,\, g_m = Qq) &= \tau_{QQ}\,\tau_{Qq}
\end{aligned} \tag{13.15}$$

so that with these parents, Equation 13.10c becomes

$$\begin{aligned}
\ell(z_{ij} \,|\, QQ, Qq) = {} & \tau_{QQ}\,\tau_{Qq} \cdot \varphi(z_{ij}, \mu_{QQ}, \sigma^2) \\
& + [\,\tau_{QQ}(1-\tau_{Qq}) + \tau_{Qq}(1-\tau_{QQ})\,] \cdot \varphi(z_{ij}, \mu_{Qq}, \sigma^2) \\
& + (1-\tau_{QQ})(1-\tau_{Qq}) \cdot \varphi(z_{ij}, \mu_{qq}, \sigma^2)
\end{aligned}$$

Note that this likelihood reduces to Equation 13.10c using Mendelian segregation transmission probabilities ($\tau_{QQ} = 1$ and $\tau_{Qq} = 1/2$).

Elston et al. (1975) suggest that three criteria must be satisfied for acceptance of a major-gene hypothesis: (1) a significantly better overall fit of a mixture model compared with a single normal, (2) failure to reject the hypothesis of Mendelian segregation ($\tau_{QQ} = 1$, $\tau_{Qq} = 1/2$, $\tau_{qq} = 0$), and (3) rejection of the hypothesis of equal transmission for all genotypes ($\tau_{QQ} = \tau_{Qq} = \tau_{qq}$). Criterion (1) reduces false positives due to polygenic background loci, while criteria (2) and (3) offer some robustness against nonnormality of the underlying distributions and resemblance due to common environmental effects (Elston 1981a). While incorporation of transmission-probability criteria into likelihood models decreases the possibility of a false positive (Go et al. 1978, Goldin et al. 1981, Demenais et al. 1986), it does so at a cost of decreased power. Loss of power can be significant if the major gene is recessive (Borecki et al. 1995).

The fact that not all families are expected to be segregating the major gene has important consequences for the optimal number and size of families for detecting a major gene. Burns et al. (1984) showed that, for a fixed number of individuals, highest power is generally obtained by examining a moderate number of families of moderate size, as opposed to many small families or a few large families. If a small number of large families is chosen, we run the risk that none of the families are segregating the gene. Conversely, with a large number of small families, while some are likely to have the gene segregating, power for detecting a major gene is reduced due to the small sample size in each segregating family.

Example 4. As a demonstration of the utility of the Elston et al. (1975) criteria for accepting a major-gene hypothesis, we consider McGuffin and Huckle (1990) test for a genetic basis for attending medical school. The trait here is scored as a binary variable,

$$z = \begin{cases} 1 & \text{attending medical school} \\ 0 & \text{not attending medical school} \end{cases}$$

As is discussed below (Equation 13.28), complex segregation analysis can be easily modified to accommodate such binary characters. Of 249 students at the Wales College of Medicine, 13.4% had mothers/fathers who also attended medical school, a 61-fold increase in "risk" relative to the general population (0.2%). Taking μ as the population mean, the expected means at an underlying major locus can be modeled by using measures of additivity (a) and dominance (k), and the allele frequency p. General single-locus (a, k, p all estimated) and recessive ($k = 0$, a and p estimated) models were fitted and compared with a null model ($a = k = p = 0$). These single-locus models, which assumed Mendelian transmission probabilities ($\tau_1 = 1, \tau_2 = 1/2, \tau_3 = 0$), were then compared against two alternate transmission models — a generalized model where the three parameters (τ_1, τ_2, τ_3) were estimated from the data, and an equal transmission model ($\tau_1 = \tau_2 = \tau_3 = \tau$). The latter simply fits a mixture model to the data without allowing for Mendelian transmission. The resulting log likelihoods for these models were as follows:

Model	Parameters (in addition to μ)		Constant +
	Free	Fixed	$-$ 2 ln (likelihood)
Null	None	$a = k = p = 0$	283.60
Equal transmission	a, p, $\tau_1 = \tau_2 = \tau_3$	$k = 0$	283.60
General single-locus	a, k, p	$\tau_1 = 1, \tau_2 = 1/2, \tau_3 = 0$	120.14
Recessive	a, p	$k = 0, \tau_1 = 1, \tau_2 = 1/2, \tau_3 = 0$	120.14
General transmission	a, p, τ_1, τ_2, τ_3	$k = 0$	111.22

The general single-locus model gives a significantly better fit than the null (single underlying normal) model, with a likelihood-ratio test statistic of 283.60 $-$ 120.14 = 163.46 (three degrees of freedom). However, since the recessive model gives the same fit with fewer parameters, it is chosen as the standard for further analysis. The recessive model gives a significantly better fit than the equal transmission model (a mixture distribution not incorporating Mendelian segregation). Thus, criteria (1) and (3) for a major gene hold, since a mixture gives a better fit than a single normal, and the hypothesis of equal transmission (τ_1=τ_2=τ_3) is rejected. However, the general transmission model (τ_i estimated from the data) gives a significantly better fit than the Mendelian segregation hypothesis (τ_1 = 1, τ_2 = 1/2, τ_3 = 0), with a likelihood-ratio test statistic of 120.14 $-$111.22 = 8.92 with three degrees of freedom (P <0.03). Thus, these data fail the major-gene criterion (2), as the hypothesis of Mendelian segregation is rejected. Shared environmental effects, rather than major gene effects, likely account for this association between relatives.

Common-family Effects

Members of full-sib families usually share environmental effects, and likelihood functions accounting for these have been developed (Morton and MacLean 1974; Knott and Haley 1992a,b). Let the ith family have a common effect c_i, and assume that these effects are normally distributed among families with mean zero and variance σ_c^2. With this modification, the expected phenotypic value of an offspring with genotype g_o from family i is $\mu_{g_o} + c_i$. As before, we assume that the phenotypic values for each genotype (conditional on c_i) are normally distributed with variance σ^2, giving the conditional likelihood for the n_i offspring from this family as

$$\ell(z_{i\cdot} \,|\, g_f, g_m, c_i) = \prod_{j=1}^{n_i} \left[\sum_{g_{o_j}=1}^{3} \Pr(g_{o_j} \,|\, g_f, g_m) \cdot \varphi(z_{ij}, \mu_{g_{o_j}} + c_i, \sigma^2) \right] \qquad (13.16)$$

Averaging over all possible values of the common-family effect c_i gives

$$\ell(z_{i\cdot} \,|\, g_f,\, g_m) = \int_{-\infty}^{\infty} \ell(z_{i\cdot} \,|\, g_f,\, g_m,\, c) \cdot \varphi(c, 0, \sigma_c^2)\, dc \tag{13.17}$$

Finally, using the above expression for $\ell(z_{i\cdot} \,|\, g_f,\, g_m)$, averaging over all possible parental genotypes gives the unconditional likelihood for this family (Equation 13.11b). Assuming the QTL genotypes are in Hardy-Weinberg proportions, the unconditional likelihood has six unknown parameters: the three genotypic means, the allele frequency p, and the variances σ^2 and σ_c^2. Assuming the n_f families in our pedigree are unrelated, the total likelihood is the product of the individual family likelihoods (Equation 13.13).

The likelihood for the ith family under the restricted model assuming common-family effects, but no major genes, is

$$\begin{aligned}\ell(z_{i\cdot}) &= \int_{-\infty}^{\infty} \ell(z_{i\cdot} \,|\, c) \cdot \varphi(c, 0, \sigma_c^2)\, dc \\ &= \int_{-\infty}^{\infty} \left[\prod_{j=1}^{n_i} \varphi\left(z_{ij}, \mu + c, \sigma^2\right) \right] \cdot \varphi\left(c, 0, \sigma_c^2\right)\, dc \end{aligned} \tag{13.18}$$

A test for common-family effects but no major gene is given by the likelihood-ratio test using Equation 13.18 versus the likelihood function with $\sigma_c^2 = 0$. The latter is just the likelihood function assuming a single underlying normal, which has its maximum value given by Equation 13.8. Likewise, the likelihood-ratio test for a major gene but no common-family effects uses the full likelihood and a restricted likelihood assuming $\sigma_c^2 = 0$.

Polygenic Background

Our final modification assumes a background of segregating polygenes in addition to the major gene. The resulting likelihood functions are often called **mixed models** in the human genetics literature, although this is a different usage of the term from its standard linear-model interpretation (Chapter 26). A variety of such "mixed-model" likelihoods have been proposed for full-sib families (Elston and Stewart 1971, Morton and MacLean 1974, Ott 1979, Lalouel and Morton 1981, Lalouel et al. 1983, Demenais and Bonney 1989, Fernando et al. 1994, Stricker et al. 1995b).

We will consider the background polygenes to be completely additive, and assume that the background genetic value A is normally distributed with mean 0 and variance σ_A^2. The phenotypic value of an individual with major-locus genotype g and background polygenic value A is assumed to be normally distributed with mean $\mu_g + A$ and variance σ_E^2. Ignoring common-family effects, if the jth sib has background genetic value A_{o_j}, the conditional likelihood for the jth sib in the ith family becomes

$$\ell(z_{ij}\,|\,g_f,\,g_m,\,A_{o_j}) = \sum_{g_{o_j}=1}^{3} \Pr(g_{o_j}\,|\,g_f,\,g_m)\,\varphi(z_{ij}, \mu_{g_{o_j}} + A_{o_j}, \sigma_E^2) \tag{13.19}$$

The conditioning on offspring genetic value A_o is removed in two stages. First, A_o is removed by conditioning on parental polygenic values $(A_f,\ A_m)$,

$$\ell(z_{ij}\,|\,g_f,\,g_m,\,A_f,\,A_m) = \int_{-\infty}^{\infty} \ell(z_{ij}\,|\,g_f,\,g_m,A_{o_j})\,p(A_{o_j}\,|\,A_f,\,A_m)\,dA_{o_j} \tag{13.20}$$

The additive genetic value of an offspring is assumed to be normally distributed with mean $(A_f+A_m)/2$ and variance $\sigma_A^2/2$, so that the conditional density function is

$$p(A_o\,|\,A_f,\,A_m) = \varphi\left(A_o, \frac{A_f + A_m}{2}, \frac{\sigma_A^2}{2}\right) \tag{13.21}$$

as developed in Example 7 of Chapter 8. Second, averaging over all possible parental background polygenic values, A_f and A_m, gives a likelihood function for the ith family that is conditioned only on the major-locus genotypes of the parents,

$$\ell(z_{i\cdot}\,|\,g_f,\,g_m) =$$

$$\int_{-\infty}^{\infty}\int_{-\infty}^{\infty} \left[\prod_{j=1}^{n_i} \ell(z_{ij}\,|\,g_f,\,g_m,\,A_f,\,A_m)\right] \varphi(A_f, 0, \sigma_A^2)\,\varphi(A_m, 0, \sigma_A^2)\,dA_f\,dA_m \tag{13.22}$$

This expression assumes that parents are drawn at random, are unrelated, and not inbred, but these restrictions can be removed by averaging over an alternative joint distribution of A_f, A_m. Finally, Equation 13.22 is substituted into Equation 13.11b to obtain the unconditional likelihood for the entire family. The resulting likelihood has six unknown parameters: the three major-locus means, allele frequency p (assuming genotypes are in Hardy-Weinberg proportions), σ_E^2, and σ_A^2.

Under the restricted model of an additive polygenic background but no major-locus or common-family effects, the likelihood of an individual conditioned on the polygenic values of its parents is

$$\ell(z_{ij}\,|\,A_f,\,A_m) = \int_{-\infty}^{\infty} \varphi(z_{ij}, \mu + A_o, \sigma_E^2)\;\varphi\left(A_o, \frac{A_f + A_m}{2}, \frac{\sigma_A^2}{2}\right) dA_o \tag{13.23a}$$

giving the unconditional likelihood for the ith family as $\ell(z_{i\cdot}) =$

$$\int_{-\infty}^{\infty}\int_{-\infty}^{\infty} \left[\prod_{j=1}^{n_i} \ell(z_{ij}\,|\,A_f,\,A_m)\right] \varphi(A_f, 0, \sigma_A^2)\,\varphi(A_m, 0, \sigma_A^2)\,dA_f\,dA_m \tag{13.23b}$$

Incorporation of a common-family effect into either Equation 13.22 or 13.23 is straightforward and follows the logic leading to Equation 13.16.

The presence of multiple integrals in the common-family effect and polygenic likelihood functions usually means that considerable computing power is required to obtain the MLEs in even modest pedigrees. Knott et al. (1990, 1991a) provide excellent approximations for these Gaussian integrals, decreasing computational requirements. However, since multiple local maxima can occur on the likelihood surface, care must still be taken in numerically computing the global likelihood (Demenais et al. 1986, Borecki et al. 1995).

Other Extensions

Extensions allowing for multivariate traits (Blangero and Konigsberg 1991) and genotype $\times$ environment interaction (e.g., Blangero et al. 1990, Konigsberg et al. 1991) have also been developed. Moreover, as we show in Chapter 26, likelihood functions can be constructed to incorporate any number of fixed effects (e.g., effects due to age, sex, or specific environments) using the general linear mixed model. Removing such fixed effects prior to analysis increases the power of tests of oligogenic models. An alternative formulation for likelihoods in pedigrees has been suggested by Bonney (Bonney 1984, 1992; Bonney et al. 1989; Demenais and Bonney 1989). These **regressive models** have as their parameters correlations between relatives (i.e., the correlations between parent and offspring and between full sibs), rather than explicit genetic parameters to express these correlations.

Likelihoods for more general pedigrees follow using the same logic as above — conditioning on all possible genotypes and then averaging over these genotypes to obtain the unconditional likelihood. Several computer packages have been developed for complex segregation analysis in small pedigrees containing on the order of tens of individuals (Elston and Stewart 1971, Hasstedt and Cartwright 1979, Lalouel and Morton 1981, Elston et al. 1986, Lange et al. 1988), which are compared by MacCluer et al. (1983) and Konigsberg et al. (1989). However, while small pedigrees can be handled, the computational requirements for complex multigenerational pedigrees (which are common in human genetics) are extremely demanding, making their analysis by classical methods difficult except in special situations. Starting with Elston and Stewart (1971), a number of approaches and approximations have been suggested (Lange and Elston 1975; Cannings et al. 1976, 1978; Ott 1979; Lange and Boehnke 1983; Schork 1991, 1992; Goradia et al. 1992; Fernando et al. 1993; Stricker et al. 1995a). One exciting new approach is the use of intensive resampling methods, such as the Gibbs sampler (German and German 1984, Gelfand and Smith 1990). Here, one randomly samples a large number of the possible genotypes within a pedigree and uses the conditional likelihoods averaged over this set as an estimate of the unconditional likelihood (Thompson and Guo 1991; Guo and Thompson 1992, 1994; Thompson et al. 1993; Janss et al. 1995).

Ascertainment Bias

The preceding likelihood functions assume that families (or more generally, pedigrees) are chosen at random from the population as a whole. In many cases nonrandom sampling is much more efficient, as even a large random sample can miss pedigrees segregating a rare major gene. When pedigrees are not chosen at random, the likelihood function must be modified to account for how the observed pedigrees were sampled (**ascertained**). Pedigrees are ascertained through **probands**, individuals who cause a particular pedigree to enter the sample. As first noted by Weinberg (1927), failure to account for how probands are ascertained can bias the analysis.

To see the importance of correctly accounting for the sampling scheme, let $\ell(z_{i\cdot} \,|\, g_f, g_m)$ denote the likelihood function for the ith family, conditioned on the major-locus genotypes of the parents (Equation 13.11a). To obtain the unconditional likelihood, we compute the expectation of this likelihood over all possible parental genotypes giving Equation 13.11b,

$$\ell(z_{i\cdot}) = \sum_{g_f, g_m} \ell(z_{i\cdot} \,|\, g_f, g_m) \,\Pr(g_f,\, g_m)$$

In our previous treatment, we assumed that parents are chosen at random, so that (for a random-mating population) $\Pr(g_f,\, g_m)$ is entirely determined by p, the frequency of allele Q. However, if parents are *not* chosen at random, $\Pr(g_f,\, g_m)$ is no longer just a function of p, but also of how the parents were sampled.

Unfortunately, using an incorrect model of ascertainment can create as much bias as performing an analysis without considering ascertainment problems. Ascertainment correction is a very complicated subject, the details of which we will not pursue further. Some basic concepts are reviewed by Morton (1959), Cavalli-Sforza and Bodmer (1971, see their Appendix II), and Elston (1980, 1981a). Recent important papers include Cannings and Thompson (1977), Elston and Sobel (1979), Ewens and Shute (1986), Shute and Ewens (1988a,b), Hodge (1988), and Vieland and Hodge (1995). A nice discussion of some of the subtleties inherent in defining ascertainment schemes is given by Greenberg (1986).

Estimating Individual Genotypes

When a major locus is indicated, the investigator may wish to estimate the genotypes of particular individuals. For example, when the trait is determined by both a major gene and a polygenic background, extreme individuals in some families result from having extreme major-locus genotypes, while in other families they result from having extreme polygenic values. This is a particular problem with studies of human disease where pedigrees are gathered from very wide sampling on the basis of extreme phenotypes. Obviously, one would like to sort out these different causes before proceeding to detailed molecular analyses. For example, in a study of 70 pedigrees displaying high levels of blood cholesterol, Moll et al. (1984) found strong evidence for segregation of a major gene in only three cases, with only polygenic and/or environmental factors influencing cholesterol in the remaining pedigrees. Their strategy was to first fit a mixed (major gene plus background polygene) model to the data, and then use the estimated model parameters to predict the major-locus genotype of each parent, given observed parental and offspring phenotypic values.

The key to predicting the major-locus genotype is **Bayes' theorem**, which can be used to estimate the probability of each genotype given the phenotypic values and estimates of the major gene parameters obtained from segregation analysis (such as allele frequencies and genotypic means). Bayes' theorem is as follows: suppose there are n possible outcomes $(b_1, b_2, \cdots, b_n)$ of a random variable that we cannot observe. Given the observed outcome of a correlated variable A, what is the probability of b_j? From the definition of a conditional probability, $\Pr(b_j \mid A) = \Pr(b_j, A)/\Pr(A)$. We can decompose this further, by noting that $\Pr(b_j, A) = \Pr(b_j)\Pr(A \mid b_j)$ and $\Pr(A) = \sum_i^n \Pr(b_i)\, \Pr(A \mid b_i)$. Putting these together gives Bayes' theorem,

$$\Pr(b_j \mid A) = \frac{\Pr(b_j)\Pr(A \mid b_j)}{\Pr(A)} = \frac{\Pr(b_j)\Pr(A \mid b_j)}{\sum_{i=1}^{n} \Pr(b_i)\, \Pr(A \mid b_i)} \tag{13.24}$$

In particular, the probability that an individual with phenotypic value z has genotype j (for $1 \le j \le n$) is

$$\Pr(g_j \mid z) = \frac{\Pr(g_j)\Pr(z \mid g_j)}{\Pr(z)} = \frac{\Pr(g_j)\Pr(z \mid g_j)}{\sum_{i=1}^{n} \Pr(g_i)\, \Pr(z \mid g_i)}$$

Note that $\Pr(z)$ is simply the distribution of phenotypic values and the probabilities involving g_i can be computed using the ML estimates of the major gene parameters. More generally, the phenotypes of an individual's offspring and/or additional relatives can be used by replacing the single observation z with a vector $\mathbf{z}$ of phenotypes. Computing $\Pr(\mathbf{z} \mid g_i)$, the probability of the observed vector $\mathbf{z}$ of phenotypes, given that the individual of interest has genotype g_i, is straightforward but can be very tedious (Elston and Stewart 1971, Heuch and Li 1972, van Arendonk et al. 1989, Kinghorn et al. 1993).

ANALYSIS OF DISCRETE CHARACTERS

The likelihood functions developed above can be easily modified to accommodate complex segregation analysis of dichotomous (binary) characters, such as the presence/absence of a disease (e.g., Elston and Rao 1978). Define the **penetrance**

ψ_g of a genotype g as the probability that a random individual of that genotype displays the trait. Coding the character as

$$y = \begin{cases} 0 & \text{does not display the trait} \\ 1 & \text{displays the trait} \end{cases} \tag{13.25a}$$

gives the likelihood function for an individual with genotype g as

$$\ell(\, y \,|\, g\,) = (\psi_g)^y \, (1 - \psi_g)^{1-y} = \begin{cases} 1 - \psi_g & \text{for } y = 0 \\ \psi_g & \text{for } y = 1 \end{cases} \tag{13.25b}$$

More generally, if the character has n discrete states, and $\psi_{k,g}$ is the probability that an individual of genotype g has character state k, the likelihood function becomes

$$\ell(\, y \,|\, g\,) = \prod_{k=1}^{n} (\psi_{k,g})^{\delta(y,k)} \qquad \text{where} \qquad \delta(y,k) = \begin{cases} 1 & \text{if } y = k \\ 0 & \text{otherwise} \end{cases} \tag{13.26}$$

Thus, our treatment below of dichotomous characters easily extends to polychotomous traits.

Single-locus Penetrance Model

Assume that a single diallelic locus underlies a dichotomous trait, and denote the penetrances of genotypes QQ, Qq, and qq by ψ_1, ψ_2, and ψ_3, respectively. If allele Q has frequency p, under Hardy-Weinberg expectations the **population prevalence**, K, of the trait becomes

$$K = p^2 \cdot \psi_1 + 2p(1-p) \cdot \psi_2 + (1-p)^2 \cdot \psi_3 \tag{13.27}$$

As above, likelihood functions are constructed by standard conditioning arguments, using Equation 13.25b. For example, consider a collection of n full sibs from the ith family, with trait values $y_{i1,}, \cdots, y_{in}$ (taking values of zero or one). The likelihood for the jth sib from this family, conditioned on it having parental genotypes g_f and g_m, is

$$\ell(\, y_{ij} \,|\, g_f, g_m\,) = \sum_{g_o=1}^{3} \Pr(g_o \,|\, g_f, g_m) \; (\psi_{g_o})^{y_{ij}} \; (1 - \psi_{g_o})^{1-y_{ij}} \tag{13.28}$$

Using Equations 13.11a,b to average over all possible parental genotypes yields the unconditional likelihood for this family. The resulting likelihood has four unknown parameters: allele frequency p, and the gene effects as measured by the three penetrances ψ_i. Curtis and Stam (1995) note that this likelihood parameter space can be reduced when an estimate of the population prevalence K

is available, as this imposes the restriction given by Equation 13.27 on the four parameters.

Major Gene Plus a Polygenic Background

The penetrance approach can easily be extended to allow for both a major gene and a polygenic background. For a dichotomous character, the likelihood for an individual with major-locus genotype g and background polygenic value A is

$$\ell(y \mid g, A) = [\, \psi(g, A)\,]^{y} \; [\, 1 - \psi(g, A)\,]^{1-y} \tag{13.29a}$$

where $\psi(g, A)$ is the penetrance for an individual with this genotype. Substituting this likelihood into the previous mixed-model likelihoods (Equations 13.19–23) allows them to accommodate binary traits. For example, the likelihood $\ell(\, y \,|g_f, g_m, A_f, A_m\,)$ for a particular sib, conditioned on the major-locus genotypes and background polygenic values of its parents is

$$\sum_{i=1}^{3} \int_{-\infty}^{\infty} \ell(\, y \mid g_i, A\,) \cdot \Pr(\, g_i \mid g_f, g_m\,) \cdot \Pr(\, A \mid A_f, A_m\,)\, dA \tag{13.29b}$$

where $\Pr(\, A \mid A_f, A_m\,)$, given by Equation 13.21, is a function of the additive genetic variance σ_A^2 of the polygenic values. The conditioning on parental polygenic values is removed by integration, paralleling our development of the mixed-model segregation-analysis likelihood function (Equations 13.19–23). Likewise, common-family environmental effects can be incorporated using an analysis along the lines leading to Equations 13.16–18.

The penetrances $\psi(g, A)$ are usually modeled by assuming an underlying **liability model**. This approach, briefly introduced in Chapter 11 and more fully discussed in Chapter 25, assumes an underlying normal distribution of liability z. One trait value is displayed if the liability exceeds some threshold T, while the alternative trait value is displayed if liability lies below the threshold. In this case,

$$\psi(g, A) = \Pr(z > T \mid g, A\,) = \int_{T}^{\infty} \varphi(z, \mu_g + A, 1)\, dz \tag{13.30}$$

As is discussed in Chapter 25, the variance of the liability distribution can always be set equal to one. Defining $\Phi(x) = \Pr(U \leq x)$ as the cumulative distribution function for a unit normal U, the integral in Equation 13.30 can be written as

$$\int_{T-(\mu_g+A)}^{\infty} \varphi(z, 0, 1)\, dz = 1 - \int_{-\infty}^{T-(\mu_g+A)} \varphi(z, 0, 1)\, dz = 1 - \Phi(T - \mu_g - A\,)$$

Using the useful identity $\Phi(-x) = 1 - \Phi(x)$, Equation 13.30 thus becomes

$$\psi(g, A) = \Phi(\mu_g + A - T\,) \tag{13.31}$$

Segregation analysis proceeds by substituting this expression into Equation 13.29a. The resulting likelihood for the major-gene-only model has five parameters to estimate: the three major-locus means μ_g, allele frequency q, and the threshold value T. (Alternatively, one can set $T = 1$ and estimate the variance of the liability function.) Examples of this general approach applied to several different genetic models can be found in Thaller et al. (1996).

Finally, we note than alternative approach for computing penetrances that avoids having to evaluate the cumulative normal function involves the use of **logistic regressions**. The motivation for this approach is that the **logistic function**

$$f(x) = \frac{1}{1 + e^{-x}} \tag{13.32}$$

provides a reasonable approximation to the cumulative normal, with

$$\Phi(x) \simeq \frac{1}{1 + \exp(-\theta\, x)} = f(\theta x) \qquad \text{where} \qquad \theta = \frac{\pi}{\sqrt{3}} \tag{13.33}$$

(Liao 1994). Hence, we can model the penetrance as a logistic function. For example, the penetrance for the major gene and polygenic background becomes $\psi(g, A) = f(a + \alpha_g + \alpha_A)$ where α_g and α_A are the effects of the major-locus genotype and polygenic background, and a is a term that accounts (among other things) for the general prevalence of the trait. Likewise, one can incorporate other factors (such as the effect of sex, specific environments, or specific age groups) as extra terms in the argument of logistic function.

14

Principles of Marker-based Analysis

In classical Mendelian analysis, scorable genetic variants are used to map and characterize genes. This analysis requires that individual phenotypes be highly informative about the underlying genotypes. Quantitative genetics deals with the opposite setting wherein the phenotype provides very little information on the underlying genotype. Here the units of analysis are genetic variances rather than the underlying genes themselves. While this approach is sufficient in many settings, we ultimately would like to move away from this strictly statistical framework towards direct examination of effects of individual quantitative-trait loci (QTLs).

As discussed below, there are occasions when one can choose suitable **candidate loci** based on biological knowledge of the character, but in most situations this is not possible. However, QTLs can be assayed *indirectly* by using linked **marker loci**. This indirect approach has long been recognized, but until recently it has been regarded as of minor importance because of the lack of sufficient genetic markers. Thanks to modern molecular biology, this situation has now changed dramatically. The ability to detect genetic variation directly at the DNA level has resulted in an essentially endless supply of markers for any species of interest. It is now routine to use at least 50–200 molecular markers, and often many more, for fine-scale analysis. Not surprisingly, there has been an explosion in the use of marker-based methods in quantitative genetics.

This chapter introduces the basics of marker-based mapping, while the next two chapters consider specific refinements for inbred lines (Chapter 15) and outbred populations (Chapter 16). Our discussion is loosely organized by historical developments of marker-based methods. We start by considering classical approaches for quantifying the effects of entire chromosomes (or large chromosomal regions). We then turn to finer genetic resolution, examining a variety of topics concerning markers and genetic maps, including the use of markers to facilitate construction of nearly isogenic lines. We conclude by discussing tests for candidate loci and approaches for cloning individual QTLs.

CLASSICAL APPROACHES

In some species, the clever use of special biological features (e.g., balancer chromosomes and the lack of male recombination in *Drosophila*) allows one to assay the effects of whole chromosomes on characters of interest. Because such **chromosomal assays** can be done with just a few markers, they were the main method of marker-based trait analysis before the advent of molecular markers. Although restricted to a few species, such assays nonetheless have provided insight into the genetic architecture of a number of characters.

Chromosomal Assays

The idea here is straightforward — using genetic markers, a chromosome (or chromosomal segment) from one line is substituted into an otherwise standard genetic background (usually an inbred line to minimize background variance). *Drosophila* is highly amenable to such studies given its short generation time, few chromosomes, a well-characterized genetic map, the existence of balancer strains (e.g., Figure 5.7), and the lack of recombination in males. Application of these features allows intact chromosomes from one line to be substituted into another line. Chromosome substitution can also be done in wheat using certain genetic tricks that allow construction of individuals carrying an extra chromosome (e.g., Sears 1953, Law 1966, Snape et al. 1977, Law and Gale 1979, Choo 1983). The effects of large chromosomal segments have also been examined in maize through the use of reciprocal translocations (e.g., D. Robertson 1989) and in mice by using marked chromosomes (e.g., Kluge and Geldermann 1982).

In such investigations, given that each chromosome or chromosomal segment typically contains a significant fraction of the total genome, the considerations discussed in Chapter 9 apply, and it is best to speak of genetic *factors*, rather than individual genes. In addition to providing a very coarse estimate of the location of genetic factors, chromosomal assays provide estimates of the minimum number of underlying loci, the degree of dominance measured by the composite (aggregate) effects of loci on the chromosome/chromosomal segment, and the degree of epistasis as measured by interactions between assayed chromosomes.

With sufficiently large sample sizes, even factors of very small effect can be detected with this approach, and the types of characters that can be analyzed are limited only by the imagination of the investigator. For example, in *D. melanogaster*, Caligari and Mather (1988) used chromosomal assays to locate factors for aggression and competitive response to chromosomes 2 and 3. Sokolowski (1980) examined larval foraging behavior, mapping genetic factors for larval locomotor behavior to chromosome 2 and for feeding rates to both chromosomes 2 and 3. Luckinbill et al. (1988) localized factors deferring senescence to all major chromosomes, with chromosome 3 accounting for roughly 70% of the observed variation in females.

Estimating the additive, dominance, and epistasic composite effects associated with particular chromosomes is straightforward. For example, let X_B and X_b represent the X chromosomes from two different inbred strains. The composite amount of dominance is estimated by comparing the mean phenotypic values of X_{BB}, X_{Bb}, and X_{bb} females in a standardized genetic background. Nonadditive interactions (epistasis) between chromosomes can be detected in a similar manner by considering pairs of chromosomes (Robertson and Reeve 1953, Robertson 1954, Cooke and Mather 1962). Using these approaches, the genetic architectures for a number of characters have been examined in *Drosophila* (Table 14.1). Characters correlated with fitness tend to show epistasis and directional dominance, while those presumably weakly correlated with fitness do not. A serious limitation of these assays is the size of the unit of analysis — a chromosome or large chromosomal section can contain QTLs having effects in opposite directions, canceling the individual effects and leaving a composite effect close to zero. This is especially true of epistatic effects (Hastings 1986).

Natural chromosomal assays become possible when chromosomal rearrangements are segregating within a population. Because chromosomal inversions suppress recombination, alternative inversions can be treated as intact, nonrecombining units. This approach was used extensively by Dobzhansky in his studies of the fitnesses of natural inversions in *Drosophila* (Lewontin et al. 1981), and has also been applied to morphological characters. For example, Hasson et al. (1992) found that three inversions on chromosome 2 accounted for at least 25% of the total variance for thorax length in *Drosophila buzzatii* (also see Ruiz et al. 1991). Two of the inversions showed dominance, while a third showed apparent overdominance. Other studies have detected segregating rearrangements with effects on body size in the grasshopper *Moraba scurra* (White and Andrew 1960), on wing length in the seaweed fly *Coelopa frigida* (Butlin et al. 1982, Wilcockson et al. 1995), and on wing length in *D. melanogaster* (Herández et al. 1993).

Thoday's Method

Thoday (1961, 1979) introduced a considerable refinement for crude chromosomal assays. His method is of substantial interest for purely historical reasons, but it also points out potential limitations of the flanking-marker mapping methods discussed in subsequent chapters. Consider a hypothetical QTL between two linked markers, with allele Q fixed in one inbred line (the tester line, also fixed for alleles A and B at the flanking marker loci), and q in another line (fixed for markers a and b). Substituting an aqb chromosome into the tester line gives AQB/aqb heterozygotes, and these individuals are subsequently backcrossed to the tester line. Recombinant aB progeny are then scored for the trait of interest. Since these recombinant chromosomes are either aqB or aQB , a QTL with detectable effects is indicated by the presence of two distinct character classes (corresponding to QQ and Qq individuals). The presence of a linked QTL outside of the interval can also generate distinct classes, but this is not an issue if only one recombination event occurs per chromosome. The main limitation of Thoday's method is the difficulty of isolating single recombinant chromosomes and propagating them intact (without further recombination) in organisms other than *Drosophila*.

Table 14.1 The genetic architecture of various characters in *Drosophila* inferred by chromosomal assay.

Character	Dominance		Epistasis	Reference
	Present	Directional		
Fitness-related Characters				
Viability	Yes	Yes	Yes	Breese and Mather 1960
	—	—	Yes	Seager and Ayala 1982
	No	—	Weak	Ferrari 1987
Male mating activity	—	—	No	Kosuda 1993
Egg hatchability	Yes	Yes	Yes	Kearsey and Kojima 1967
Fecundity	Yes	Yes	—	Keller and Mitchell 1964
Egg-pupal survival	Yes	Yes	—	Keller and Mitchell 1964
Development time	Yes	Yes	—	Keller and Mitchell 1964
	No	—	Weak	Ferrari 1987
Progeny yield	Yes	Yes	Yes	Barnes 1966
Egg production				
D. melanogaster	Yes	Yes	—	Keller and Mitchell 1964
D. pseudoobscura	Yes	—	Yes	Kojima and Kelleher 1963
Physiological Characters				
DDT-resistance	Yes	No	Yes	Dapkus and Merrell 1977
ADH activity	Yes	Yes	Yes	McDonald and Ayala 1978
G6PD activity	—	—	Yes	Miyashita and
6PGD activity	—	—	Yes	Laurie-Ahlberg 1984
Morphological Characters				
Abdominal bristles	Weak	No	Weak	Keller and Mitchell 1962
	Weak	No	Weak	Breese and Mather 1957
Sternopleural bristles	Weak	No	Weak	Breese and Mather 1957
	Weak	No	No	J. Hill 1964
Thorax length	Yes	Yes	Yes	Robertson 1954
	Weak	No	Weak	Keller and Mitchell 1962
Wing length	Yes	Yes	Yes	Robertson 1954
	No	—	No	Keller and Mitchell 1962
Body weight				
D. melanogaster	No	—	Weak	Kearsey and Kojima 1967
D. pseudoobscura	Weak	No	Weak	Frahm and Kojima 1966

Source: Expanded from a shorter table in Kearsey and Kojima (1967).
Note: Unless otherwise indicated, all analyses are for *D. melanogaster*. Most of these results were obtained by comparing differences between selected lines. Results for many characters were obtained by both whole and sectional chromosomal assays. Directional dominance occurs when multiple assayed regions show the same direction of dominance (e.g., high values tending to be dominant over low values). Features not examined are denoted by —.

Example 1. A large sample of *Ab* recombinant chromosomes was generated from a cross between an *ab* line with negative character value and an *AB* tester line with mean zero. Each recombinant chromosome was used to create a line with a recombinant *Ab* chromosome and an *AB* tester chromosome in an otherwise genetically homogeneous background. Construction of such lines requires the use of balancer chromosomes, which prevent any further recombination from occurring (see Figure 5.7). The mean character values from the resulting lines formed two distinct clusters, one (comprising 43%) clustering around -1.5, the other clustering around 0. Hence, there appears to be a single factor (at our level of resolution), with the 0 mean class corresponding to Q/Q and the -1.5 class corresponding to Q/q, implying $\mu_{QQ} - \mu_{Qq} = 1.5$. The frequencies of these classes among *Ab* recombinant chromosomes imply that the *A–Q* recombination fraction is 43% of the *A–B* recombination fraction (assuming no interference).

McMillan and Robertson (1974) found that the power of Thoday's method is generally very poor. For high power, both the number of recombinant chromosomes (N) and the number of replicates (n) per recombinant chromosome need to be large. Even if a factor has a large effect, it will remain undetected unless the sample of recombinant chromosomes contains both factor alleles, which requires a sufficiently large N. To see this result, set the distance between the two markers to one, and let f $(0 \leq f \leq 1)$ denote the relative distance between factor and marker *A*, so that a fraction f of the *aB* recombinants are *aqB* and $1 - f$ are *aQB*. The probability that N recombinant (*aB*) chromosomes all contain the same QTL allele is $f^N + (1 - f)^N$. If f is close to zero or one, N has to be considerable for a sample to contain a sufficient number of both QTL genotypes for detection. The power for detection is maximized when the factor is exactly between the two markers. Graphs of power are given by McMillan and Robertson (1974).

Thoday's method can be extended to allow for multiple factors between two markers, but not without complications. As shown in Example 2, the estimation of map distances and effects of multiple QTLs between markers critically assumes that all QTL alleles in the line being examined have the same directional effect (McMillan and Robertson 1974). The following caveat will be a recurring theme as we continue to consider molecular-marker approaches: *these methods can be compromised if the line being examined contains linked QTL alleles of opposite effects.* Data from inbred lines show that such situations are not uncommon (Chapter 15).

Example 2. Suppose there are three QTLs between the marker loci *A* and *B*, as shown in the following figure. Assuming no double crossovers, let f_1, f_2, f_3, and $1 - (f_1 + f_2 + f_3)$, respectively, be the fractions of recombination between flanking markers that occur between *A* and the first QTL, between the first and second QTL, the second and third, and the third and B. Let Q_i denote alleles from the high line, which have effect H_i relative to alleles q_i from the low line.

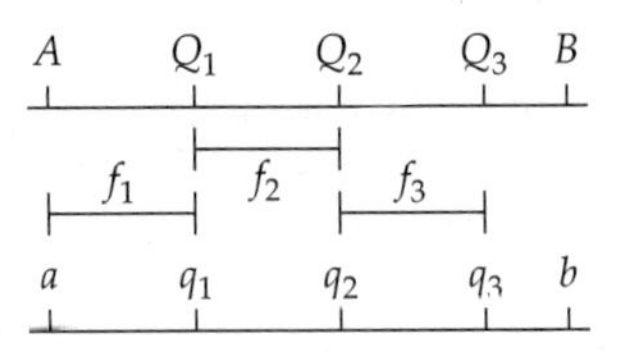

The expected frequencies and effects of *Ab* recombinant chromosomes are obtained as follows. As shown in the following figure, a fraction f_1 of *Ab* chromosomes have no high-line alleles, as the crossover occurred between A and Q_1. Likewise, a fraction f_2 contain high-line allele Q_1 with effect H_1. The last two classes follow similarly. Thoday's method estimates the H_i and f_i by making the *assumption* that all alleles from the high line increase character value (all $H_i > 0$) so that the mean of the largest phenotypic class corresponds to $H_1 + H_2 + H_3$, the mean of second largest class to $H_1 + H_2$, etc. This ordering allows us to uniquely estimate the H_i and f_i.

Fraction	Chromosome type	Character value
f_1	A q_1 q_2 q_3 b	0
f_2	A Q_1 q_2 q_3 b	H_1
f_3	A Q_1 Q_2 q_3 b	$H_1 + H_2$
$1 - f_1 - f_2 - f_3$	A Q_1 Q_2 Q_3 b	$H_1 + H_2 + H_3$

If some of the H_i differ in sign, this ordering breaks down, as illustrated in following example offered by McMillan and Robertson (1974). Setting the mean of the low line at 0, suppose that 23% of the recombinants have value 0, 7% have value 0.5, 41% have value 1.3, and the remaining 29% have value 3.7. Assuming that the effects of all Q alleles are positive, the second smallest mean corresponds to $H_1 = 0.5$. Likewise, the third smallest mean corresponds to $H_1 + H_2 = 1.3$, and hence $H_2 = 1.3 - 0.5 = 0.8$. Continuing in this fashion gives the estimated values shown below on the left. In actuality, McMillan and Robertson generated these data using a very different configuration than suggested by the estimate, as shown on the right.

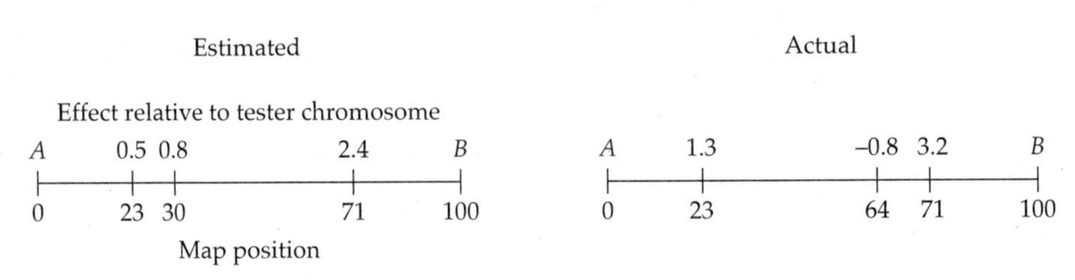

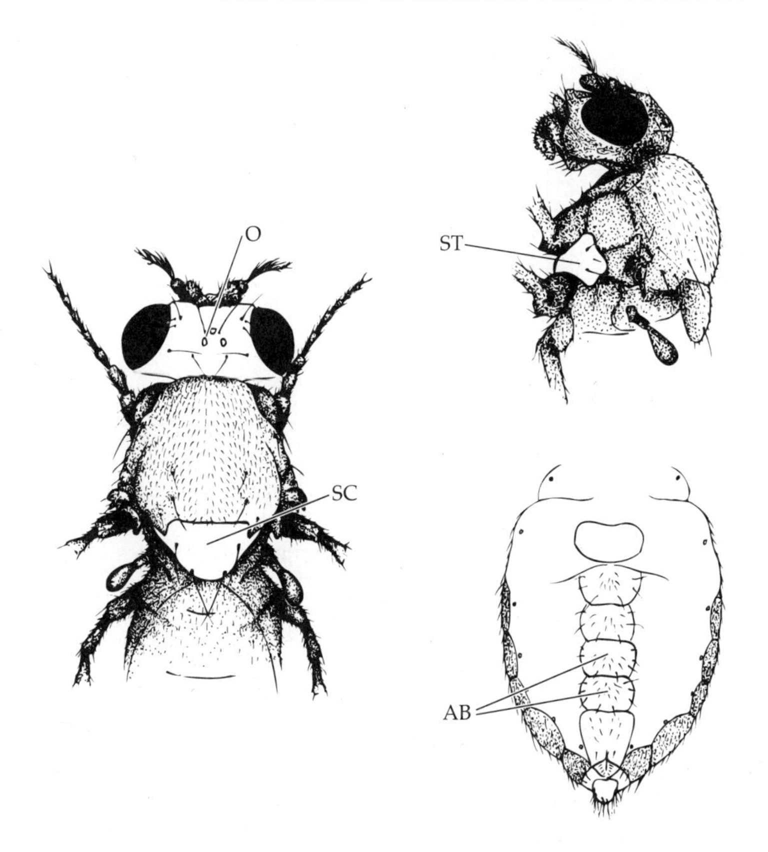

Figure 14.1 Location of head and thoracic bristles in adult *Drosophila*. The most frequently studied bristle systems in quantitative-genetic experiments are sternopleural bristles (ST) on the side of the thorax between the first and second legs, abdominal bristles on the underside of the abdomen (AB), and scutellar bristles (SC) on the upper side of the last thoracic region. Ocellar bristles (O) on the top of the head are also occasionally used.

Genetics of *Drosophila* Bristle Number

Drosophila possess a large number of bristles regularly distributed over the adult integument. As discussed in Chapter 12, a number of bristle systems, most notably sternopleural, abdominal, and scutellar bristles, have been widely used in quantitative-genetic studies (Figure 14.1).

The genetics of sternopleural bristle number have been intensively studied in selected lines using Thoday's method (reviewed in Spickett 1963; Thoday et al. 1964; Spickett and Thoday 1966; Davies and Workman 1971; Davies 1971; Thoday and Thompson 1976; Thoday 1979; Shrimpton and Robertson 1988a,b; Mackay

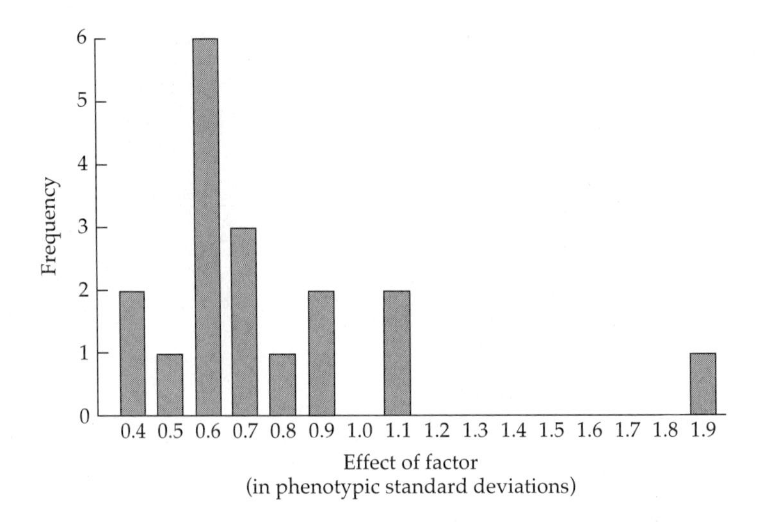

Figure 14.2 Distribution of the effects on sternopleural bristles of *Drosophila melanogaster* third chromosome factors isolated by Shrimpton and Robertson (1988b). This distribution is biased towards large factors, as effects under roughly 0.3 phenotypic standard deviations were not statistically detectable in this study.

1995, 1996). For example, Spickett and Thoday (1966) examined two selected lines (*vg4* and *vg6*), which showed increases of 15 and 20 bristles, respectively, relative to the standard Oregon R strain. For line *vg6*, 88% of this increase could be accounted for by five factors detectable via Thoday's approach, while three factors (each spanning a distance amounting to less than 2% recombination) accounted for 80% of the increase in line *vg4*. In more detailed analyses restricted to the third chromosome, Davies (1971) and Shrimpton and Robertson (1988a,b) isolated a minimum of 8 and 17 factors, respectively. Extrapolating the latter figure to the rest of the genome yields roughly 60 loci influencing sternopleural bristle number. The distribution of effects for the bristle factors isolated by Shrimpton and Robertson (Figure 14.2) is roughly consistent with results from P-element tagging (Chapter 12), suggesting that sternopleural bristle number is controlled by a few major genes supplemented by numerous genes of smaller effect. Evidence for epistasis was found in most studies.

Spickett (1963) examined the developmental aspects of the isolated factors from line *vg4* in further detail. One factor increased bristle number over the entire fly by increasing total fly cell number. An increase in cell number normally results in a larger fly, but Spickett found a second linked factor that decreased average cell size, resulting in a normal-sized fly with a larger number of cells and more bristles. Two other factors had more local effects. The first was linked to the previous two factors on chromosome 3 and increased the number of bristles in a specific region of the sternopleurite. The second was on chromosome 2 and modified development of a single large bristle to several smaller bristles in this region. Clearly, genetic changes in characters even as superficially simple as bristle number can result from modifications in a variety of developmental pathways.

Genetics of *Drosophila* Speciation

Considerable debate revolves around the question of whether speciation events are precipitated by a few genes of major effect (Templeton 1980, Carson and Templeton 1984) or by changes at many loci with individually small effects (Charlesworth et al. 1982, Barton and Charlesworth 1984). Most data on the number of genes involved in reproductive isolation come from assays of crosses between *Drosophila* species that show partial fertility

For example, *D. pseudoobscura / persimilis* hybrids produce normal females but sterile males. Dobzhansky (1936) backcrossed fertile F_1 hybrid females to marked males of both species, substituting single chromosomes from one species into an otherwise normal genetic background of the other. He found that testes size (correlated with male fertility) was affected by factors on all five chromosomes, with the strongest effect being on the X chromosome. Orr (1987) found that the reduction in sperm motility in hybrids between these species is largely due to incompatibilities between the X and Y chromosomes from different species. A finer analysis by Wu and Beckenbach (1983) suggested at least nine factors affecting hybrid male fertility (five X-linked, three autosomal, and one Y-linked).

Coyne (1984) also found that at least five factors are responsible for male sterility in *D. simulans/mauritiana* hybrids, with the main effect again being due to incompatibilities between the X and Y chromosomes of the different species (Figure 14.3). When small sections of the *D. simulans* X chromosome were introgressed into a *D. mauritiana* background by repeated backcrossing, Palopoli and Wu (1994) found four factors influencing male sterility, two of which displayed very strong epistasis. Given that they examined less than 20% of the X chromosome, this extrapolates to at least 40 X-linked factors (assuming an equal number of factors are fixed by both species) involved in reproductive isolation between these two species (Davis and Wu 1996). That the genetic units from the above studies are best viewed as factors, rather than single major genes, is shown by further work from Wu's lab. Introgression of small segments of the X from *D. mauritiana* and *D. sechellia* into *D. simulans* initially suggested that a 500 kilobase region harbored a single major sterility gene (Perez et al. 1993), but a more refined analysis suggested at least two epistatically interacting factors (Perez and Wu 1995).

An important caveat for all of these studies is that they measure only the *current* number of factors contributing to reproductive isolation, which can be much greater than the actual number involved in the original isolation event. Some indications of the potential magnitude of this bias can be seen by considering the classic **Dobzhansky-Muller model** of speciation, which assumes that speciation occurs through epistasis, with isolated populations fixing mutually incompatible

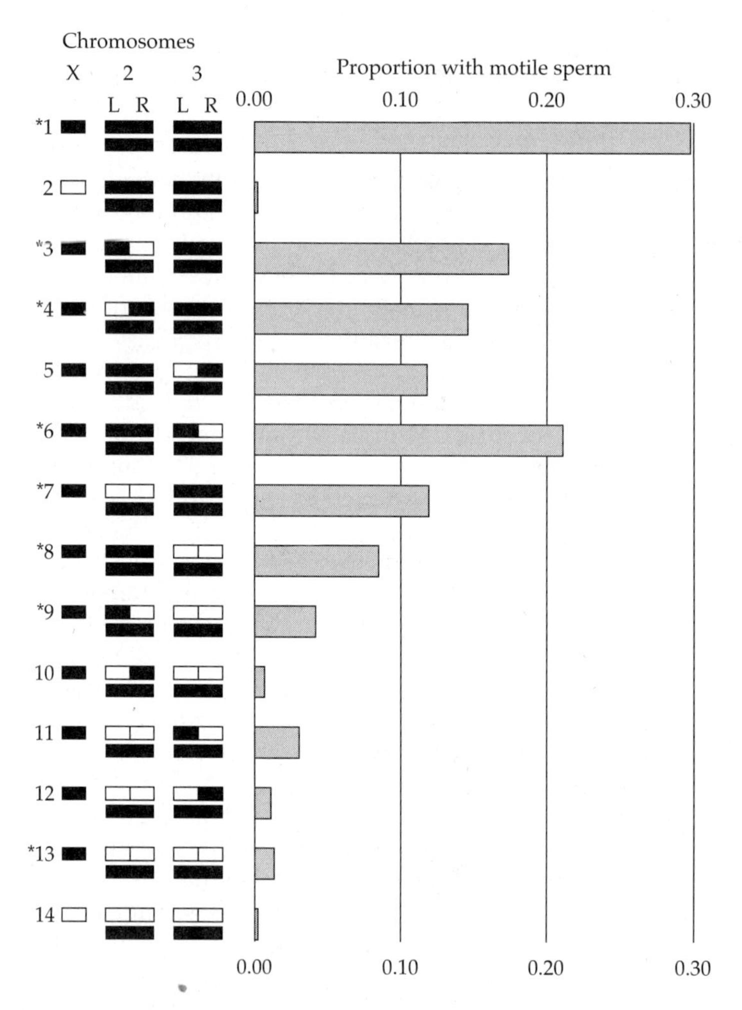

Figure 14.3 Locating male-sterility genes in *D. simulans/mauritiana* hybrids with chromosomal assays. Fertility is measured by the proportion of males with motile sperm, with chromosome segments from *D. simulans* being shown in black, and those from *D. mauritiana* in white. L and R refer to the left and right arms of chromosomes 2 and 3. All genotypes have a *D. simulans* Y chromosome. Asterisks denote genotypes that produced progeny when crossed to *D. simulans* females. (From Coyne 1984.)

alleles (Dobzhansky 1936, Muller 1939, Orr 1996). Their model assumes that a base population fixed for *aabb* is split into two isolated populations, with separate mutations, *A* and *B*, arising in each population. In the *bb* background, *A* has no serious negative fitness consequences, so fixation can occur by drift, giving an *AAbb* population. Likewise, the other population fixes allele *B* to yield *aaBB*, where again *B* has no serious fitness effects in the *aa* background. However, *A* and *B*, when together, are incompatible, such that an *AaBb* hybrid has reduced fitness relative to the ancestral population. A number of such incompatibilities are known in species hybrids (see Orr 1995 for a review), and they are related (at least in principle) to synthetic lethals (Chapter 10). Orr (1995) found that under the Dobzhansky-Muller model, the number of fixed incompatible loci separating the two populations should increase with (at least) the square of separation time, and that incompatibilities involving three or more loci evolve more readily than interactions involving pairs of loci (also see Cabot et al. 1994). Hence, complex interactions among "speciation genes" are expected to be the norm, rather than the exception. This introduces further complications into the already difficult problem of determining how many factors are actually involved in the generation of the initial isolation event that allows the accumulation of further incompatibility genes.

Despite these interpretative difficulties, some general conclusions emerge from these studies. First, in *Drosophila* there is a large X effect—factors with the largest effects on reproductive isolation and hybrid inviability are located on the X chromosome (Coyne and Orr 1989b, Coyne 1992). This effect is restricted to reproductive isolation, as no such X chromosome localization is seen for genes influencing morphological differences between *Drosophila* species (e.g., Coyne 1983, 1985; Liu et al. 1996). Consistent with this, Ford and Aquadro (1996) report evidence for selection on X-linked sites, but not autosomal sites, among three semispecies of the *Drosophila athabasca* complex showing reproductive isolation but no obvious morphological differentiation.

The second (more general) conclusion, first noted by Haldane (1922), is that when only one sex in a cross-species hybrid is sterile or inviable, it is usually the heterogametic sex (e.g., XY males in most animals, WZ females in birds and butterflies). The *Drosophila* data show that male sterility/inviability arises first (often rather quickly), with female sterility/inviablity often developing only after a considerable amount of additional time (Coyne and Orr 1989a, 1997). Much has been written about **Haldane's rule** and its possible causes (e.g., Charlesworth et al. 1987; Orr and Coyne 1989b; Read and Nee 1991; Coyne et al. 1991; Frank 1991; Coyne 1992; Wu and Davis 1993; Coyne and Orr 1993; Orr 1993a,b; Virdee 1993; Turelli and Orr 1995; Zeng 1996; Sawamura 1996; Orr and Turelli 1996; True et al. 1996). While a variety of mechanisms have been proposed (e.g., differences in the substitution rates on autosomes vs. sex chromosomes, higher mutation rates for male vs. female sterility), Turelli and Orr (1995, Orr and Turelli 1996) suggest that dominance provides an explanation for Haldane's rule. If alleles decreasing hybrid fitness are partly recessive, the cumulative effects of sex-linked genes are greater when hemizygous than when heterozygous, resulting in a more pronounced effect in heterogametic hybrids. This effect can be seen by noting that although XX hybrids carry approximately twice the number of deleterious X-linked genes as XY hybrids, if the expression of deleterious effects is less than 50% in heterozygotes, then the overall deleterious effects are smaller in XX hybrids. Finally, while most studies of Haldane's rule examine differences between populations, much can be learned by examining within-population variation (e.g., Wade et al. 1994).

MOLECULAR MARKERS

As mentioned in the introduction, the ability to score variation at the molecular level provides a huge increase in the number of available markers in any analysis. The first molecular markers used were allozymes, protein variants detected by differences in migration on starch gels in an electric field. Since the late 1960s, this class of markers has been extensively applied to a variety of population-genetic problems. However, as methods for evaluating variation directly at the DNA level became widely available during the mid-1980s, DNA-based markers largely replaced allozymes in mapping studies. Allozymic variants have the advantage of being relatively inexpensive to score in large numbers of individuals (although DNA markers are rapidly closing this price gap), but there is often insufficient protein variation for high-resolution mapping. On the other hand, there are effectively no limitations on either the genomic location or the number of DNA markers.

A wide variety of techniques can be used to measure DNA variation. Direct sequencing of DNA provides the ultimate measure of genetic variation, but much cruder (and quicker) scoring of variation is sufficient for most purposes. One of the simplest approaches is to **digest** DNA with a variety of **restriction enzymes**, each of which cuts the DNA at a specific sequence or **restriction site** (commonly four to six bases). When the digested DNA is run on a gel under an electric current, the fragments separate out according to size. As Figure 14.4 shows, a variety of mutational events can generate length variation. If we attempted to score the entire genome for fragment lengths, the result would be a complete (and uninformative) smear on the gel. Instead, individual bands are isolated from this smear by using labeled DNA **probes** that have base-pair complementarity to particular regions of the genome (often chosen at random). This approach is the basis for assays of **restriction fragment length polymorphisms** or RFLPs (Figure 14.5). Each RFLP probe generally scores a single-marker locus, and the marker alleles are **codominant**, as heterozygotes and homozygotes can be distinguished. The number of detectable RFLPs is impressive (Botstein et al. 1980; Doris-Keller et al. 1987; Beckmann and Soller 1983, 1986a,b; Soller and Beckmann 1988).

A particularly interesting RFLP approach involves the use of tissue-specific cDNA clones as the probes. (cDNAs are generated from the mRNAs of genes being expressed in that tissue.) This procedure can allow for enrichment of genes of

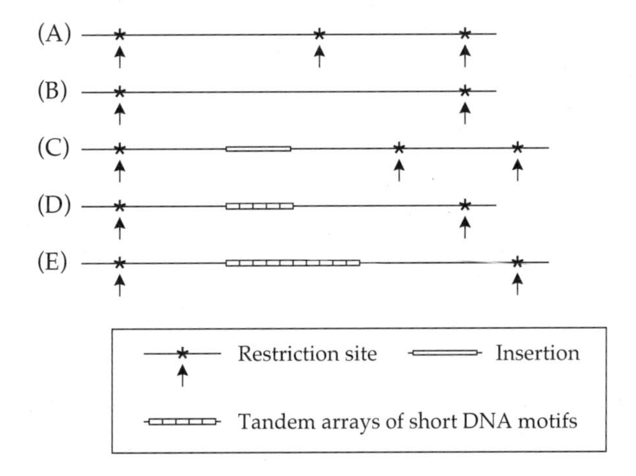

Figure 14.4 A variety of mutational events generate variation in restriction fragment lengths, scored by the distances between restriction sites. Single (or multiple) base-pair changes can create or destroy restriction sites, (A) vs. (B). Insertions of mobile genetic elements can create dramatic size differences (C). DNA sequences often exist as tandem arrays of short repeated sequences. Unequal crossing over and/or replication slippage creates variation in the size of such arrays, (D) vs. (E).

potential interest in the marker pool, enabling observed differences to be treated as potential candidate loci. For example, Kinzer at al. (1990) used 19 ripening-specific cDNA clones from the tomato (*Lycopersicon esculentum*) to detect polymorphisms in these loci between *L. esculentum* and a wild relative *L. pennellii*.

A rather different molecular marker approach uses short **primers** for DNA replication via the **polymerase chain reaction** (PCR) to delimit fragment sizes. A region flanked (in opposite orientation) by primer binding sequences that are sufficiently close together allows the PCR reaction to replicate this region, generating an amplified fragment. If primer binding sites are missing or are too far apart, the PCR reaction fails and no fragments are generated for that region. This procedure is the basis for **randomly amplified polymorphic DNAs** or RAPDs (Williams et al. 1990), sequence polymorphisms detected by using random short sequences as primers (Figure 14.5). RAPDs have an advantage over RFLPs in that a single probe (here, a particular random primer) can reveal several loci at once, each corresponding to different regions of the genome with appropriate primer sites. They also require smaller amounts of DNA. However, by their very nature, RAPDs are generally dominant, as they indicate presence/absence of a particular site, so marker genotypes can be ambiguous (see Figure 14.5). Ragot and Hoisington (1993) conclude that RAPDs are generally more time- /cost-effective in

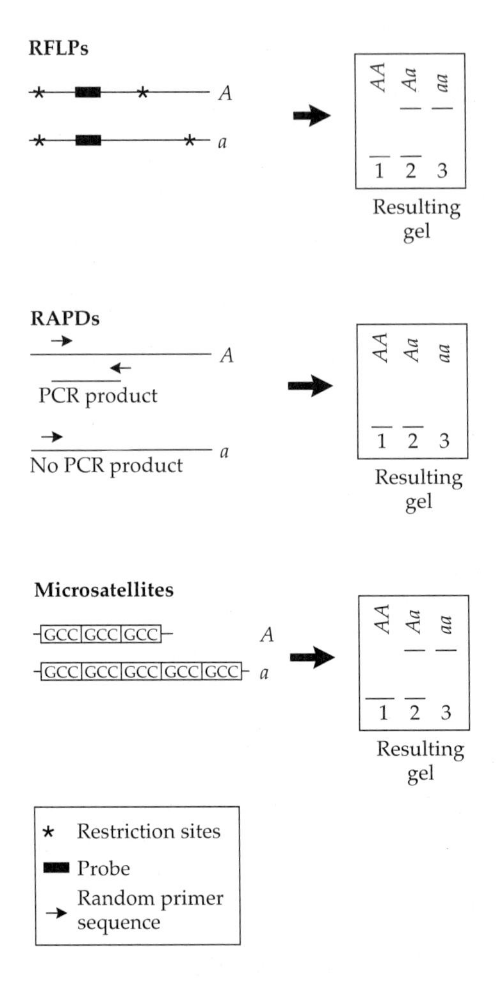

Figure 14.5 **Top**: RFLPs, restriction fragment length polymorphisms, variable lengths of restriction fragments in a particular region of the genome, are revealed by use of a probe for that region. This marker system is codominant (all marker genotypes can be distinguished), as an RFLP heterozygote shows fragments of both lengths. **Middle**: RAPDs, randomly amplified polymorphic DNAs, are revealed by the use of a random primer sequence. A successful PCR reaction requires sequences complementary to the primer in opposite orientation at sufficiently close distance. Running these products out on a gel, both *AA* and *Aa* exhibit a fragment and the site is scored as present, while the fragment does not amplify in *aa* individuals and the site is scored as absent. Hence, RAPDs are dominant markers. **Bottom**: Microsatellite DNAs exhibit variation in the array lengths of short sequences of tandemly repeated DNAs. Fragment length is scored, so that these markers are also codominant. Codominance occurs even when variation is scored via PCR with primers that flank the repeat region.

Table 14.2 Properties of the most commonly used classes of molecular marker loci.

	RFLP	RAPD	Microsatellite DNA
Number of alleles/locus	Few	Few	Many
Genomic abundance	High	Very high	Medium
Marker dominance	Codominant	Dominant	Codominant
Amount of DNA required	2–10 μg	10–25 ng	50–100 ng

Source: Rafalski and Tingey 1993.
Note: RFLP = restriction fragment length polymorphism, and RAPD = randomly amplified polymorphic DNA.

small studies where a modest number of individuals have to be genotyped, while RFLPs are better for larger studies.

Mapping in outbred populations is most efficient with marker loci having a large number of alleles (Chapter 16). In this case, **microsatellite DNAs**, short arrays of simple repeated sequences (Figure 14.5) have become the markers of choice. Microsatellites tend to be very highly polymorphic, a consequence of their high mutation rate to new alleles (changes in the number of repeats in the array). Since array length is scored, microsatellites are codominant, as heterozygotes show two different lengths and hence can be distinguished from homozygotes.

While the marker types listed in Table 14.2 are the most commonly used, other categories of markers can be very useful (reviewed in Rafalski and Tingey 1993). For example, several studies have used mobile genetic elements (such as retroviruses) as markers (e.g., Rise et al. 1991, Nuzhdin et al. 1993, Keightley and Bulfield 1993, Ebert et al. 1993, Long et al. 1995). Due to the high rates of movement of some transposable elements, individuals often differ in the presence or absence of elements at particular sites. Scoring sites for the presence or absence of elements yields dominant markers. Finally, important advances in marker scoring are offered by two recently developed methods: **representational difference analysis** or RDA (Lisitsyn et al. 1993, Lisitsyn 1995) and **genomic mismatch scanning** or GMS (Nelson et al. 1993). Both methods examine the entire genome, allowing one to isolate only those sequences that are shared by two populations (GMS) or those that differ between populations (RDA). Judicious use of these methods will very likely provide powerful approaches for the isolation of QTLs (Lander 1993, Aldhous 1994).

GENETIC MAPS

Genetic maps show the ordering of loci along a chromosome and the relative distances between them. As detailed throughout the remainder of this chapter, such maps are essential to the localization of QTLs. The lack of genetic markers historically prevented the construction of detailed maps in all but a few well studied species with short generation times. However, as predicted by Botstein et al. (1980), molecular markers have sparked an explosion of genetic maps in humans and economically important plants and animals. The theory for constructing genetic maps is now highly refined, and we only introduce a few important issues here. For more detailed reviews, the reader should consult Bailey (1961), Lalouel (1992), and the excellent text by Ott (1991).

Map Distances vs. Recombination Frequencies

Genetic map construction involves both the ordering of loci and the measurement of distance between them. Ideally, distances should be additive so that when new loci are added to the map, previously obtained distances do not need to be radically adjusted. Unfortunately, recombination frequencies are not additive and hence are inappropriate as distance measures. To illustrate, suppose that three loci are arranged in the order *A*, *B*, and *C* with recombination frequencies c_{AB}, c_{AC}, and c_{BC}. Each recombination frequency is the probability that an odd number of crossovers occurs between the markers, while $1 - c$ is the probability of an even number (including zero). There are two different ways to get an odd number of crossovers in the interval *A–C*: an odd number in *A–B* and an even number in *B–C*, or an even number in *A–B* and an odd number in *B–C*. If there is no **interference**, so that the presence of a crossover in one region has no effect on the frequency of crossovers in adjacent regions, these probabilities can be related as

$$c_{AC} = c_{AB}\,(1 - c_{BC}) + (1 - c_{AB})\,c_{BC} = c_{AB} + c_{BC} - 2c_{AB}\,c_{BC} \tag{14.1}$$

This is **Trow's formula** (Trow 1913). More generally, if the presence of a crossover in one region depresses the probability of a crossover in an adjacent region,

$$c_{AC} = c_{AB} + c_{BC} - 2(1 - \delta)c_{AB}\,c_{BC} \tag{14.2}$$

where the **interference parameter** δ ranges from zero if crossovers are independent (no interference) to one if the presence of a crossover in one region completely suppresses crossovers in adjacent regions (complete interference).

Thus, in the absence of very strong interference, recombination frequencies can only be considered to be additive if they are small enough that the product $2c_{AB}c_{BC}$ can be ignored. This is not surprising given that the recombination frequency measures only a part of all recombinant events (those that result in an odd number of crossovers). A map distance m, on the other hand, attempts to measure the total number of crossovers (both odd and even) between two markers. This is a naturally additive measure, as the number of crossovers between *A* and *C* equals the number of crossovers between *A* and *B* plus the number of crossovers between *B* and *C*.

A number of **mapping functions** attempt to predict the number of crossovers (m) from the observed recombination frequency (c). The simplest, derived by Haldane (1919), assumes that crossovers occur randomly and independently over the entire chromosome, i.e., no interference. Let $p(m,k)$ be the probability of k crossovers between two loci m map units apart. Under the assumptions of this model, Haldane showed that $p(m,k)$ follows a Poisson distribution, so that the observed fraction of gametes containing an odd number of crossovers is

$$c = \sum_{k=0}^{\infty} p(m, 2k+1) = e^{-m} \sum_{k=0}^{\infty} \frac{m^{2k+1}}{(2k+1)!} = \frac{1-e^{-2m}}{2} \tag{14.3}$$

where m is the expected number of crossovers. Rearranging, we obtain Haldane's mapping function, which yields the (Haldane) map distance m as a function of the observed recombination frequency c,

$$m = -\frac{\ln(1-2c)}{2} \tag{14.4}$$

For small c, $m \simeq c$, while for large m, c approaches 1/2. Map distance is usually reported in units of **Morgans** (after T. H. Morgan, who first postulated a chromosomal basis for the existence of linkage groups) or as **centiMorgans** (cM), where 100 cM = 1 Morgan. For example, a Haldane map distance of 20 cM ($m = 0.2$) corresponds to a recombination frequency of $c = (1 - e^{-0.2})/2 \simeq 0.16$.

Although Haldane's mapping function is still used frequently, several other functions allow for the possibility of crossover interference in adjacent sites (Bailey 1961, Felsenstein 1979, Karlin 1982, Pascoe and Morton 1987, Ott 1991, Zhao and Speed 1996). For example, geneticists often use Kosambi's (1944) mapping function, which allows for modest interference,

$$m = \frac{1}{4} \ln\left(\frac{1+2c}{1-2c}\right) \tag{14.5}$$

Figure 14.6 compares the Haldane and Kosambi mapping functions. For small c (< 0.15), both give $m \simeq c$. For large m, both approach $c = 1/2$. The appropriateness of a potential mapping function can be assessed by checking if the computed map distances depart significantly from additivity (e.g., Pascoe and Morton 1987).

There is no universal relationship between map distance and the actual physical distance between loci (Table 14.3). A centiMorgan can correspond to a span of between ten thousand to a million nucleotide base pairs (10 kb to 1,000 kb), depending on the species. Even within a chromosome, there can be rather dramatic differences. For example, crossovers are often suppressed in particular chromosomal regions (increasing the number of base pairs per cM), such as near the centromere and telomeres (chromosome ends), e.g., True et al. (1996). Further, the rate of recombination is under genetic control with some modifier genes having a general influence throughout the genome and others having a fine-scaled influence on specific chromosomal regions (Brooks 1988). Thus, there may be considerable variation among individuals and among populations in the strength of linkage. The recombination rate can also vary between the sexes. In mammals, for example, females usually have greater map distances than do males. In male *Drosophila*, crossing-over is generally entirely suppressed.

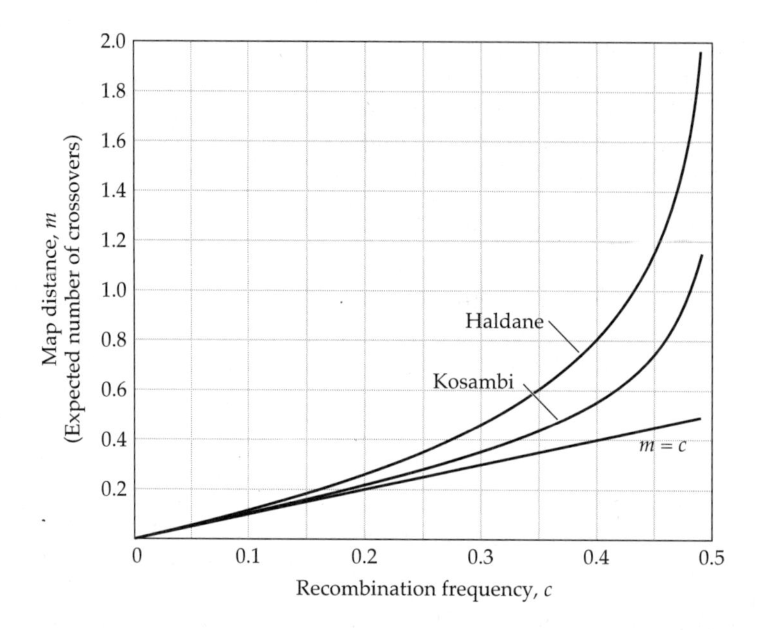

Figure 14.6 Comparison of the Haldane, Kosambi, and simple recombination-frequency ($m = c$) mapping functions, which translate an observed recombination frequency c into an estimate of the expected number of crossovers (the map distance m). The different functions make different assumptions about interference. Haldane's function assumes none, Kosambi's assumes a moderate amount, and the simple recombination-frequency model assumes complete interference ($\delta = 1$ in Equation 14.2). For a given c, the Haldane map distance is largest, and the recombination-frequency distance is the smallest. For $c < 0.15$, all three are extremely similar, while the Kosambi and simple function are close for $c < 0.25$.

Table 14.3 The relationship between map distance and physical distance.

Species	Haploid Genome Size (kilobase pairs)	Map Length (centiMorgans)	Kilobases/centiMorgan
Yeast	2.2×10^4	3,700	6
Neurospora	4.2×10^4	500	80
Arabidopsis	7.0×10^4	500	140
Drosophila	2.0×10^5	290	700
Tomato	7.2×10^5	1,400	510
Human	3.0×10^6	2,710	1,110
Corn	3.0×10^6	1,400	2,140

Source: Catchside 1977, Shields 1989.
Note: These data are rather crude approximations, averaged over the entire genome and do not account for the substantial variation in map-distance relationships that can exist within and between individuals.

How Many Markers Are Needed?

For most organisms, genetic maps are not yet available, and one must start with randomly obtained markers. As a rough approximation for the number of random polymorphic markers necessary for a desired saturation of the linkage map, assume a circular linkage map of total length L map units. In order to have a fraction p of all loci within m map units of some marker, the required number of randomly distributed markers is

$$n = \frac{\ln(1-p)}{\ln(1-2m/L)} \tag{14.6a}$$

(Lange and Boehnke 1982, Beckmann and Soller 1983). This rearranges to give the proportion of the genome within m map units of a marker as approximately

$$p \simeq 1 - \exp\left(-\frac{2mn}{L}\right) \tag{14.6b}$$

(Jacob et al. 1991). Assuming a circular genome map ignores the effect of chromosome ends, causing Equation 14.6a to underestimate n and Equation 14.6b to overestimate p. A more general expression, which accounts for linear chromosomes, is given by Bishop et al (1983). Assuming a haploid chromosome number C,

$$p = 1 - \frac{2C\left[(1-x)^{n+1} - (1-2x)^{n+1}\right]}{n+1} - (1-2x\,C)\,(1-2x)^n \tag{14.7}$$

where $x = m/L$.

An alternative consideration is the average distance between a locus and the nearest random marker. Martin et al. (1991) show that the expected distance of a gene from the closest of n random markers is

$$E(m) = \frac{L}{2(n+1)} \tag{14.8a}$$

with the upper 95% confidence interval for this distance given by

$$\frac{L}{2}\left(1 - 0.05^{1/n}\right) \tag{14.8b}$$

It is important to note that the above expressions refer to *randomly chosen* markers. It is much more efficient to use sets of equally spaced markers. If the entire genome is covered by marker loci equally spaced at m map units apart, no locus is more than $m/2$ from any marker, and the average distance of a locus from a marker (assuming a uniform distribution of loci along the chromosome) is $m/4$.

Example 3. The human linkage map is 33 Morgans long with haploid chromosome number $C = 23$. How many random markers are required to achieve a 90% probability that at least one marker is within 10 map units (centiMorgans) of a randomly chosen gene?

Here $L = 3300$ cM, $m = 10$ cM, and $p = 0.9$. The circular-chromosome approximation (Equation 14.6a) gives

$$n = \frac{\ln(1 - .9)}{\ln(1 - 2/330)} \simeq 379$$

Numerically solving Equation 14.7 with $p = 0.9$ gives $n = 404$. Hence, the effect of ignoring chromosome ends underestimates n by about 7%.

Suppose that 110 random markers are used. From Equation 14.8a, the expected distance of a particular gene from the nearest of these 110 markers is 14.9 cM, while Equation 14.8b gives the upper 95% confidence interval for this distance as 44.3 cM. In contrast, suppose these 110 markers are not random, but are instead chosen to be equally spaced at 30 cM apart. With this spacing, no locus is more than 15 cM from a marker, and all loci are, on average, 7.5 cM from a marker.

MARKER-TRAIT ASSOCIATIONS

We now turn our attention to the mapping of individual QTLs. A variety of experimental designs have been proposed, and we consider these in some detail in Chapters 15 (inbred line crosses) and 16 (outbred populations). Since those chapters focus on methods for estimating both QTL effects and map position, we defer further discussion of most statistical issues until then. The remainder of this chapter is concerned with general design strategies for mapping, and eventual cloning, of QTLs.

That QTLs, like normal genes, can be mapped (assigned to linkage groups) was first demonstrated by Payne (1918), who found that the X chromosome from selected lines of *Drosophila* contains multiple factors influencing scutellar bristle number. Over the next two decades, a number of workers found associations between Mendelian markers and quantitative traits in crosses between inbred lines. For example, Sax (1923) crossed two inbred bean lines (*Phaseolus vulgaris*) differing in seed pigment and weight, with the pigmented parents having heavier seeds than the nonpigmented parents. These crosses demonstrated that seed pigment is determined by a single locus with two alleles, *P* and *p*. Among F_2 segregants from this cross, *PP* and *Pp* seeds were, respectively, 4.3 ± 0.8 and 1.9 ± 0.6 centigrams heavier than *pp* seeds. Hence the *P* allele is linked to a factor (or factors) that act in an additive fashion on seed weight. In a similar manner, Lindstrom (1924) demonstrated linkage between a locus for fruit color and factors for fruit size in tomatoes, while Smith (1937) observed associations between corolla size differences and several independently segregating flower color genes in tobacco. Similar studies were reported in maize (Lindstrom 1931), peas (Rasmusson 1927), barley (Wexelesen 1933, 1934), and mice (Green 1931, 1933).

These studies raised the possibility that QTLs could be mapped with some precision, perhaps even allowing for the characterization of individual loci, given a population in linkage disequilibrium with a sufficient number of markers. Disequilibrium is a key feature, as it creates marker-trait associations, with different marker genotypes having different expected values for characters influenced by QTLs linked to these markers. This simple idea is the foundation for QTL mapping.

Creating populations with linked loci in disequilibrium is straightforward — crosses between inbred lines have this property, as do sibs and other collections of relatives. Other approaches are detailed in Chapters 15 and 16. For mapping purposes, crosses between inbred lines have the fewest complications. The progeny from such crosses display maximum disequilibrium and, being genetically identical, are equally informative as parents. Using such F_1 parents, a variety of populations (such as F_2s or backcrosses) can be generated.

A number of problems, which are elaborated on in Chapter 16, conspire to make QTL mapping considerably more difficult in outbred populations than with inbred-line crosses. Not all parents are guaranteed to be informative, as some may be segregating marker alleles, but not linked QTLs, or vice-versa. Further, the marker-QTL linkage phase can differ between different sets of relatives. For example, marker allele M might be associated with QTL allele q in one family, but with Q in another. Thus, marker-trait associations have to be assessed in each set of relatives, rather than (as with inbred-line crosses) averaged over all individuals.

Example 4. Likelihood functions for QTL mapping follow from standard mixture models (Chapter 13). Consider the simple backcross design, where two completely inbred lines (with marker/QTL genotypes *MMQQ* and *mmqq*) are crossed to form an F_1 (*MQ*/*mq*) which is then backcrossed to the *MMQQ* population. If c denotes the marker-QTL recombination frequency, then a fraction $(1-c)$ of M-bearing F_1 gametes contain Q, while c contain q. Likewise $(1-c)$ and c of m-bearing gametes contain q and Q, respectively. Because the gamete from the parental population is always *MQ*, the conditional probabilities of QTL genotypes given marker genotypes are

$$\Pr(QQ \,|\, MM) = \Pr(Qq \,|\, Mm) = 1 - c$$

$$\Pr(Qq \,|\, MM) = \Pr(QQ \,|\, Mm) = c$$

Hence, if z_j is the character value for individual j, the likelihood depends on the marker genotype,

$$\ell(z_j) = \begin{cases} (1-c)\cdot\varphi(z_j, \mu_{QQ}, \sigma^2) + c\cdot\varphi(z_j, \mu_{Qq}, \sigma^2), & \text{if marker} = MM \\ \\ c\cdot\varphi(z_j, \mu_{QQ}, \sigma^2) + (1-c)\cdot\varphi(z_j, \mu_{Qq}, \sigma^2), & \text{if marker} = Mm \end{cases}$$

As in Chapter 13, we have assumed that phenotypes are normally distributed with potentially different means for each genotype but a common variance, with $\varphi(z, \mu, \sigma^2)$ denoting the density function for a normal distribution with mean μ and variance σ^2. The total likelihood for n measured backcross individuals is the product of individual likelihoods, $\prod \ell(z_j)$. The maximum-likelihood estimates of the four model parameters (μ_{QQ}, μ_{Qq}, c, σ^2) are obtained by maximizing the likelihood with respect to these variables, treating the observed z_j as fixed constants.

Hypothesis testing follows by standard likelihood-ratio tests (Chapter 13, Appendix 4). The appropriate test for the presence of a QTL linked to the marker compares the likelihood under the assumed full model with the likelihood under a model of no QTL, in which each individual character value is assumed to be drawn from the same normal with unknown mean μ and variance σ^2. The maximum of this restricted likelihood is given by Equation 13.8. The resulting likelihood-ratio test has $4 - 2 = 2$ degrees of freedom (four free parameters in the full model; μ and σ^2 in the restricted model). Knott and Haley (1992a) show that this test is not biased by the presence of unlinked QTLs. It is, however, biased if one or more additional QTLs are linked to the marker under consideration.

An observed association between alleles at a polymorphic marker locus and the value of a quantitative trait can result either because of gametic phase disequilibrium between the marker locus and a QTL or because the marker itself has a pleiotropic effect on the trait. A simple test for pleiotropy versus disequilibrium is to examine a marker-trait association over several generations of random mating. An association due to pleiotropy will not decay over time, while one due to linkage alone will. A caveat is that if linkage is very tight, it can take many generations before any substantial decay is observed (Equation 5.13). If one suspects that the marker itself is the QTL, a variety of **candidate-locus** approaches (to be discussed below) can be used to test this hypothesis.

Example 5. Following crosses between inbred lines of barley, Powell et al. (1985a,b) found that two loci with recessive dwarfing alleles (*ert* and *denso*) were associated with other quantitative traits: *ert* with reduced seed weight and *denso* with reduced height. By crossing different lines to produce an F_2, it was found that the *ert* locus accounted for roughly 84% of the additive genetic variation in seed weight after both one and three generations of recombination, suggesting that the association between *ert* and seed weight is either due to pleiotropy or very tight linkage. In contrast, the *denso* locus accounted for 58% of the additive genetic variance for height in lines undergoing a single round of recombination but only 35% in lines undergoing three rounds. Hence, for the *denso* locus, much of the initial association was due to gametic phase disequilibrium.

Selective Genotyping and Progeny Testing

Many quantitative traits are less expensive to score than marker genotypes. In such cases, if our interest is in a single trait, it pays to first score a number of individuals for the trait and then genotype only a selected subset of these. Known as **selective genotyping**, this strategy can result in a large increase in power for the simple reason that much of the linkage information resides in individuals with extreme phenotypes (Lebowitz et al. 1987, Lander and Botstein 1989, Carey and Williamson 1991, Darvasi and Soller 1992). However, while selective genotyping offers increased power to *detect* a QTL, it also produces biased estimates of their effects. Unbiased estimates can be obtained by maximum likelihood, either using a likelihood function that directly accounts for the sampling bias (Darvasi and Soller 1992) or by treating unscored genotypes as missing values (Lander and Botstein 1989).

In those organisms where asexual propagation is possible, **progeny testing** (or **replicated progeny**) is another design strategy for increasing power (Cowen 1988, Simpson 1989, Lander and Botstein 1989, Soller and Beckmann 1990, Knapp and Bridges 1990). The idea here is to reduce the effects of environmental variance by asexually replicating each genotype, using the mean values of these replicated progeny in place of individual values. This is an efficient strategy in that if the heritability of the character is small, scoring only a few replicated progeny can result in a significant increase in power. Soller and Beckmann (1990) show that most of the increase in power occurs by measuring 10 or fewer replicated progeny. Besides offering some increase in power, progeny replication allows marker-trait associations to be examined across environments, providing a basis for estimating QTL $\times$ environment interactions.

Recombinant Inbred Lines (RILs)

When asexual propagation is not possible, a related method is to use **recombinant inbred lines** (RILs). These can be constructed for any organism by taking an F_1 line through multiple rounds of selfing (e.g., Burr and Burr 1991) or multiple generations of brother-sister mating (e.g., Bailey 1981). The resulting lines have essentially no within-line genetic variance (ignoring new mutation and small amounts of residual heterozygosity), whereas the genetic variance between lines is considerable, as each RIL represents a different multilocus genotype. A related approach, currently restricted to certain plants and the zebrafish (Streisinger et al. 1981), uses **doubled-haploid lines** (DHLs). Here haploid gametes from F_1 parents are chemically treated to double the chromosome number, instantly producing completely homozygous individuals (Jensen 1989, Knapp et al. 1990, Knapp 1991, Luo and Kearsey 1991).

Once the considerable work to generate and molecularly characterize a set of RILs or DHLs has been done, any character can be examined, and the previous marker information used to look for marker-trait associations. Only one

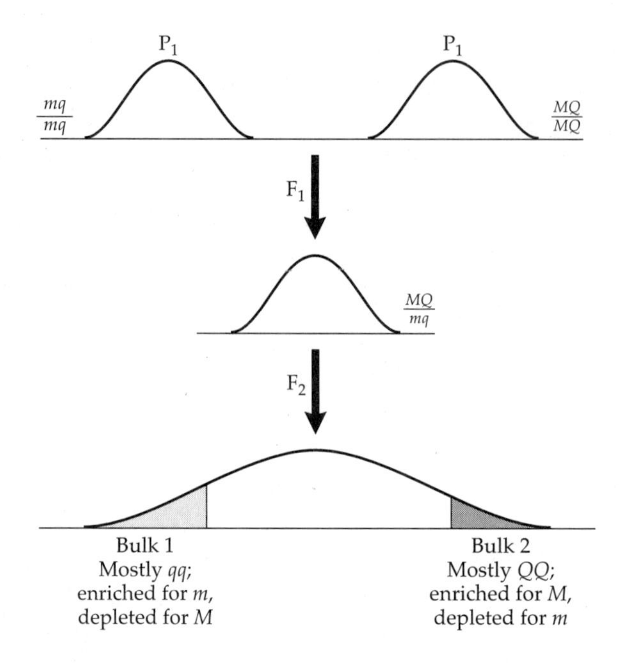

Figure 14.7 Bulked segregant analysis for mapping QTLs. Assume that a major QTL with alleles Q/q is tightly linked to a marker locus with alleles M/m. If we cross two inbred populations (*MMQQ* and *mmqq*) and then sample the extreme tails of the F_2 phenotype distribution, the lower tail is enriched for *qq*, and the upper tail is enriched for *QQ*. Since the marker is closely linked to the QTL, the lower tail is enriched for marker allele *m* and depleted for *M*, while the opposite is true in the upper tail.

laboratory needs to characterize the marker genotypes for the set of lines, while subsequent investigators (with these genotypes in hand) can look for marker-trait associations across the set of lines. Like asexual lineages, RILs/DHLs offer a particularly easy approach for measuring QTL $\times$ environment interactions, since the same lines can be raised over different sets of environments. For these reasons, behavioral geneticists are increasingly interested in using RILs for mapping behavioral QTLs in mice (e.g., Plomin et al. 1991, Belknap et al. 1993, Crabbe et al. 1994).

Bulked Segregant Analysis

A variant of selective genotyping that is especially powerful when using RILs is **bulked segregant analysis** or BSA (Arnheim et al. 1985, Michelmore et al. 1991). Here individuals are combined (**bulked** or **pooled**) into groups based on trait value (Figure 14.7). Marker alleles in linkage disequilibrium with QTL alleles are expected to have a nonrandom distribution across bulks, in the extreme having a particular marker allele present only in one bulk and the alternative allele present only in the other bulk. Unlinked markers (and marker alleles in linkage equilibrium with QTLs) are expected to be randomly distributed across bulks. Generally DNA from each bulk is screened en mass for a number of marker loci, and markers are scored as present/absent. This all/none scoring is not necessarily very stringent, as alleles with frequencies less than 5% to 10% are generally scored as absent. However, methods do exist for quantifying the relative frequencies of marker alleles in each pool (e.g., using optical densities of observed allelic bands), see Pacek et al. (1993), Khatib et al. (1994), and Darvasi and Soller (1994a).

Because bulked segregant analysis requires very high enrichment of the alternative QTL alleles in the tails of the trait distribution, it is not expected to work well for QTLs of small to modest effect. It also requires either inbred-line crosses or progeny from very large segregating families, as the method fails if groups with the marker and QTL alleles in different linkage phases are lumped. In spite of these limitations, BSA is so straightforward, allowing rapid screening of a very large number of markers, that it is reasonable to consider it in many cases, particularly with RILs.

Finally, we note that BSA can be used to find additional markers linked to a particular region (Giovannoni et al. 1991). Sorting individuals from a cross into separate pools based on alternate alleles at a marker enriches those pools for additional linked markers. Thus, bulking on a marker of interest provides for very efficient screening of a large number of markers, for example by using random PCR primers to search for linked RAPDs markers. Such an increase in marker density is required for finer mapping of QTL position.

Example 6. Yaghoobi et al. (1995) used bulked segregant analysis to map a major gene for root-knot nematode resistance in tomatoes. Forty-eight backcross individuals from a cross between resistant and susceptible strains (F_1 backcrossed to the susceptible parent) were scored, 25 of which were susceptible, and 23 resistant. This 1:1 segregation ratio is consistent with the hypothesis that a single dominant gene underlies resistance. To map this gene, two bulks of DNA, based on six resistant and six susceptible backcross plants were formed. Each bulk was screened for RAPDs using 520 different 10-base primers. Each primer gave on average about eight bands, resulting in 4,160 bands being scored in each pool, about 3% of which varied significantly between pools. These significant primers were then used to probe each of the original 48 backcross plants. One marker was present in 20 of 23 resistant plants and none (out of 25) of the susceptible plants, suggesting linkage to a major resistance gene.

Example 7. As a test of BSA, Mansur et al. (1993) examined four traits — maturity, plant height, lodging (a measure of plant structure), and seed yield — in 284 recombinant inbred lines generated by a cross between two soybean (*Glycine max*) cultivars. The four chosen traits were measured in each of the 284 RILs grown in two distinct environments (Minnesota and Chile) and the 20 highest- and lowest-performing lines, averaged over both environments, were selected for each trait. DNA was extracted from each of the extreme lines and bulked into a high and low sample for each of the traits. The resulting DNA was tested using radioactive probes for RFLP markers, and the amount of hybridization to each probe was quantified using a phosphoimager. The authors had previously used RFLPs to map a number of QTLs for these and several other traits using maximum-likelihood interval mapping (Chapter 15) in F_2 families (Mansur et al. 1993). The previous RFLP marker–QTL associations were confirmed, and one marker that showed marginal linkage to maturity and height under interval mapping showed very strong linkage to these traits as well as to lodging and yield.

QTL Mapping by Marker Changes in Populations under Selection

Selective genotyping amounts to a single-generation selection experiment, as individuals are chosen from the tails of the character distribution. This approach can be generalized by divergently selecting a base population for several generations and testing for significant changes in marker allele frequencies between up- and down-selected lines. Marker loci initially in linkage disequilibrium to a QTL will increase (decrease) in frequency as linked QTL alleles increase (decrease) due to selection.

Long-term selection creates two somewhat opposing forces for mapping. While each generation of selection increases the difference in QTL frequencies between divergently selected populations, recombination is expected to decrease the amount of linkage disequilibrium between markers and QTLs. Thus, unless the markers and QTLs are tightly linked, fairly rapid QTL allele-frequency changes (i.e., QTLs of modest to large effects) are required for a linked marker to show significant frequency changes (Lebowitz et al. 1987). Keightley and Bulfield (1993) show that the marker and QTL must be closer than 20 cM for significant marker allele-frequency changes to be likely.

Given the small population sizes of most selection experiments, drift alone is expected to change allele frequencies, motivating the need for tests for whether observed changes exceed those expected by chance events alone (Schaffer et al. 1977). The selection coefficient s on a QTL, which determines how quickly allele frequencies change, is a function of the QTL effect and the amount of selection on the character. Thus, s can provide information about the size of QTL effects. A maximum likelihood approach for estimating s was suggested by Keightley and Bulfield (1993) and Keightley et al. (1996). Here, computer simulations generate the distribution of allele-frequency change between two populations under divergent selection (assuming selection coefficient s) and drift (using an estimate of the effective population size). This procedure is repeated over a range of possible

s values, and the value giving the highest probability of the observed change is taken as the ML estimator of s.

Marker alleles associated with body size have been found in several studies of divergently selected lines of mice (Garnett and Falconer 1975, Simpson et al. 1982, Gray and Tait 1993, Keightley and Bulfield 1993, Keightley et al. 1996). For example, after 22 generations of selection, Keightley and Bulfield (1993) found significant changes at four of 16 assorted marker loci (two coat color, two allozyme, 12 proviral inserts). This same set of crosses was further examined using 124 microsatellite loci by Keightley et al. (1996), who found 11 QTLs with estimated effects ranging from 0.17 to 0.28 phenotypic standard deviations. In a study in maize, Stuber at al. (1980) found that 8 of 20 allozyme markers showed significant differences for yield in divergently selected lines starting from F_2 crosses between inbred lines. To test whether these associations might simply be due to chance, Stuber et al. (1982) independently selected on the markers, finding a response for yield in the direction predicted by the original marker-trait associations.

Example 8. Nuzhdin et al. (1993) used allele-frequency changes to detect markers associated with fitness in a cross of high and low fitness lines of *Drosophila melanogaster*. These authors used the presence / absence of the mobile genetic elements *mdg1* and *copia* as markers, finding 19 locations on chromosome 2 where the high and low lines differed in the presence / absence of inserted elements. An F_1 was backcrossed to the low line for three generations to generate a base population with high-line alleles at expected frequency 1/32. This base population was then allowed to reproduce, with marker frequencies sampled after 11, 13, and 17 generations of natural selection. The frequencies of nine high-line markers (all located in a region around the centromere of chromosome 2) showed significantly higher values than expected by drift, suggesting linkage to one (or more) QTLs increasing fitness. The associated selection coefficients estimated by the maximum likelihood approach of Keightley and Bulfield ranged from 0.3 to 0.7.

MARKER-BASED ANALYSIS USING NEARLY ISOGENIC LINES (NILs)

Nearly isogenic lines (NILs), inbred lines containing one or more small regions of DNA from a **donor** parent in an otherwise standard background, play an important and growing role in QTL mapping and cloning. NILs (also referred to as **cogenic strains**) are constructed by first crossing a donor parent to an inbred line (the **recurrent** parent) to form an F_1. The resulting offspring are then backcrossed to the recurrent parent for several generations. In species where selfing is allowed, the NIL is formed by at least one generation of selfing. Where selfing is not possible, after the backcrossing is complete, individuals are repeatedly sib mated to form the final inbred line. Note that NILs differ significantly from recombinant inbred lines (RILs). While both are inbred and can start from F_1 parents, RILs are generated by repeated inbreeding of the descendant lines and hence contain about 50% donor DNA, as opposed to the very small fraction of donor DNA expected in NILs.

While half of the F_1 genome contains donor parent DNA, the expected proportion $p(b)$ of the donor genome left following b generations of backcrossing and a final generation of selfing is

$$p(b) = (1/2)^{b+1} \tag{14.9}$$

Typically, five to seven generations of backcrossing are performed, giving the expected proportion of donor genome as 1.2% to 0.5%. This repeated backcrossing scheme is the basis for the Wehrhahn-Allard estimator of gene number, discussed in Chapter 9. When subjected to molecular-marker analysis, NILs allow for very powerful QTL analysis (Muehlbauer et al. 1988, Tanksley et al. 1989, Dudley 1993), as the following examples illustrate.

Example 9. Eshed and Zamir (1995) formed 50 NILs from a cross between the cultivated tomato (*Lycopersicon esculentum*) and a wild relative (*L. pennellii*). Based on analysis with 375 markers, each line contained a single RFLP-defined fragment from *L. pennellii*, averaging around 33 cM (or roughly 3% donor DNA). Comparing the mean value of each NIL against a standard, the authors found 23 QTLs for fruit soluble-solids content and 18 QTLs for fruit mass. By comparison, in an analysis of marker-trait associations in RILs, Goldman et al. (1995) found 7 and 13 QTLs, respectively. Eshed and Zamir subjected two regions to much finer scale mapping, using markers to select for recombinants within these regions. Upon finer analysis, a 55 cM region influencing fruit size was shown to contain at least three separate QTLs. A 37 cM region showing heterosis for soluble-solids yield was shown to result from associative overdominance. This latter region was subdivided by recombination into a partially dominant QTL that increases yield and a linked recessive that reduces yield.

Example 10. The simplest scheme for constructing NILs is to backcross without any selection (intentional or otherwise). A collection of such NILs is expected to be random with regard to the donor regions maintained. Most NILs, however, are produced by a select-and-backcross procedure, and hence are not random with regard to the retained donor DNA. This example shows how such an approach can be exploited for QTL mapping.

"Exotic" sorghum is tall and has a short-daylength requirement for flowering, while commercial sorghum cultivars are short (for easier harvesting) and daylength insensitive. Exotic strains are "converted" to commercial cultivars by first crossing to a standard short, daylength-neutral donor parent and continually backcrossing each generation to the exotic parent while selecting for short height and daylength neutrality. The resulting converted strains are expected to contain mostly exotic DNA, but are short and daylength neutral, due to the retention of donor QTL alleles influencing these traits. Hence, one approach to mapping QTLs for these traits is to search for regions that are retained in the converted derivatives.

This approach was used as a check of conventional QTL mapping by Lin et al. (1995), who examined 71 markers in nine exotic accessions of *Sorghum bicolor* and their converted derivatives. The authors compared the positions of retained donor regions in these accessions with the positions of QTLs for height and daylength sensitivity mapped in a previous experiment using the F_2 progeny in a cross of *S. bicolor* × *S. propinquum*. Seven of the nine QTLs (six for height, three for daylength) detected using the F_2 cross coincided with regions of donor parent DNA retained in the converted strains despite several generations of backcrossing. One donor region was seen in all nine derivatives, and in the F_2 crosses this region accounted for 55% of the variation in height and 86% of the variation in daylength sensitivity. In contrast, three converted regions (found in at least one line) did not coincide with mapped QTLs. Whether these regions indeed contain QTLs for height or daylength could be examined by crossing the converted strains with their parental exotics and looking for associations between the retained donor markers and trait values.

Example 11. Martin et al. (1991) used 144 random primers to generate RAPDs for comparing two tomato NILs, both derived from the same donor and recurrent parents. One NIL contained an introgressed segment for resistance to the bacterial pathogen *Pseudomonas syringae*, while the other lacked resistance. The random primers generated approximately 625 discrete products, seven of which differed between the two lines. Of the four chosen for further analysis, three were shown by cosegregation between the marker and pathogen resistance to be tightly linked to the resistance gene. The power of this approach can be seen by comparing the time for identification and confirmation of these three markers by this analysis (roughly one month) with the time required to construct a dense genetic map of tomatoes (over 2 years).

Marker-based Introgressions

Although NILs formed by selection have generally focused on phenotypic characters such as disease resistance or a small set of desirable characters, one can also select for the retention of specific donor marker alleles. Such marker selection allows the introgression of a specific region (such as one containing QTLs of

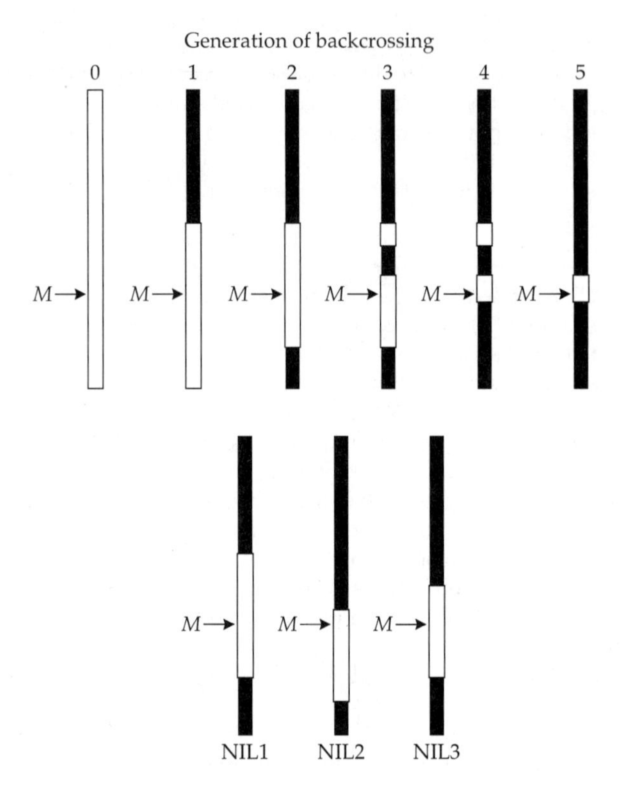

Figure 14.8 **Top:** Construction of a nearly isogenic line (NIL) containing a region of interest, obtained by selecting on a marker *M* while backcrossing to a standard stock. Since the final lines are constructed by selfing, we follow a single (haploid) chromosome, as the NIL has two identical copies of this chromosome. Clear regions are from the donor parent, solid from the recurrent parent. **Bottom:** If several lines are independently selected for the marker, the resulting NILs will contain different retained regions around *M*. With a dense marker map, crosses between such lines allow for finer mapping of the region around *M*.

modest to small effect) into an otherwise isogenic background (Young et al. 1988, Tanksley et al. 1989, Paterson et al. 1991, Dudley 1993). Independent NILs formed while selecting on the same marker are expected to differ in the regions retained around the marker (Figure 14.8), so crossing such NILs allows for finer scale mapping. This approach has been used to fine map QTLs (mainly resistance genes) in maize (Bentolila et al. 1991, Koester et al. 1993, Touzet et al. 1995), rice (Yu et al. 1991, Ronald et al. 1992), lettuce (Paran et al. 1991), wheat (Hartl et al. 1993; Schachermayr et al. 1994, 1995), barley (Schüller et al. 1992, Hinze et al. 1991), oats (Penner et al. 1993), and tomatoes (Osborn et al. 1987, Tanksley and Hewitt 1988,

Sarfatti et al. 1989, Paterson et al. 1990).

Selecting for the retention of a donor marker while backcrossing to the recurrent parent yields a NIL with a block of donor DNA surrounding the marker that can be rather large. One consequence of this is that linked undesirable genes can be dragged along with the marker, a phenomenon that plant breeders refer to as **linkage drag** (Brinkman and Frey 1977). Conversely, a single marker may remain associated with several favorable linked QTLs within the introgressed region. The amount of donor DNA present following marker-selected introgression has been examined by several authors (Bartlett and Haldane 1935, Hanson 1959a–d, Stam and Zeven 1981, Naveira and Barbadilla 1992). The donor DNA remaining in the NILs can either reside on the chromosome under marker selection or on the other (unmarked) chromosomes. We consider these two sources separately.

Suppose there are n chromosomes and a total genome length of $L = L_M + \sum L_i$, with L_M being the length of the marker chromosome and L_i being the length of the ith unmarked chromosome (all lengths in Morgans). Consider the DNA retained around the marker, and suppose that the selected marker is in the center of the chromosome. Letting $t = b + 1$, the proportion $p_M(b)$ of donor DNA on this marker chromosome after b generations of backcrossing and a final generation of selfing has expected value

$$E[\, p_M(b)\,] = 2\left(\frac{1 - e^{-tL_M/2}}{tL_M}\right) \tag{14.10a}$$

(Hanson 1959b), while the variance is

$$\sigma^2[\, p_M(b)\,] = \frac{2}{(tL_M)^2}\left[1 - \left(tL_M + e^{-tL_M/2}\right) e^{-tL_M/2}\right] \tag{14.10b}$$

(Naveira and Barbadilla 1992). For $tL_M >> 1$, these are approximately

$$E[\, p_M(b)\,] \simeq \frac{2}{tL_M}, \qquad \sigma^2[\, p_M(b)\,] \simeq \frac{2}{(tL_M)^2} \tag{14.10c}$$

Since these equations refer to the proportion of this chromosome retained, the corresponding chromosome lengths retained are given by $L_M \cdot p_M(b)$, yielding a mean length $\pm$ SD (in Morgans) of $2/t \pm \sqrt{2}/t$ for $t\, L_M >> 1$.

The restriction of the marker being in the center of the chromosome was removed by Stam and Zeven (1981), who showed that using Equations 14.10a–c for a randomly placed marker generally introduces only a small error. Of broader concern is that these equations consider only the segment around the marker, while the marked chromosome can also retain blocks of donor DNA elsewhere. Stam and Zeven show that these two effects (noncentrality of the marker and blocks of donor DNA retained outside of the marker) largely cancel, leaving Hanson's original result as a very good approximation.

Turning now to the unmarked chromosomes, let $p_{U_i}(b)$ denote the proportion of donor DNA on an unmarked chromosome of length L_i. The mean and variance are given by

$$E[\,p_{U_i}(b)\,] = \frac{1}{2^t} \tag{14.11a}$$

$$\sigma^2[\,p_{U_i}(b)\,] = 2\left(\frac{1 - e^{-tL_i/2}}{t\,L_i\,2^t}\right) - \frac{1}{2^{2t}} \tag{14.11b}$$

as shown by Stam and Zeven (1981). The expected total length of donor DNA retained over all unmarked chromosomes thus becomes

$$\sum_{i=1}^{n-1} L_i\,E[\,p_{U_i}(b)\,] = \frac{1}{2^t}\sum_{i=1}^{n-1} L_i = \frac{L - L_M}{2^t} \tag{14.12a}$$

with variance

$$\sum_{i=1}^{n-1} L_i^2 \cdot \sigma^2[\,p_{U_i}(b)\,] \tag{14.12b}$$

Example 12. Suppose there are $n = 15$ chromosomes, each of length 75 cM, and that we select for retention of a marker located in the middle of chromosome 1 during five generations of backcrossing and a generation of selfing. Applying Equation 14.10a, the expected fraction of donor DNA on chromosome 1 is

$$E[\,p_M(5)\,] = 2\left(\frac{1 - e^{-6\,(0.75/2)}}{0.75 \cdot 6}\right) \simeq 0.40$$

or, in units of chromosome length, $0.4 \cdot 75$ cM = 30 cM. Equation 14.10b gives the standard deviation of the expected proportion retained as 0.23, or 17.25 cM. Likewise, applying Equations 14.11a,b, the proportion of donor DNA on any nonmarker chromosome is 0.016 ± 0.077, or 1.2 ± 5.8 cM. The fraction of the total expected donor DNA in the NIL contributed by the marked chromosome is $30/(30 + 14 \cdot 1.2) = 0.64$.

Fewer and more generations of backcrossing give the following expected values:

	Lengths of Introgressed Segment (cM)		% from
b	Marker chromosome	All unmarked	Marked Chromosome
2	45.0 ± 19.1	131.25 ± 255.9	25.6
7	23.8 ± 14.8	4.10 ± 36.7	85.3
10	17.9 ± 12.0	0.51 ± 11.3	97.3
12	15.3 ± 10.5	0.13 ± 5.2	99.2
15	12.5 ± 8.7	0.02 ± 1.7	99.9

A NIL line created by selecting for retention of a marker contains significantly more donor DNA than a NIL created without any selection. The total map length in this example is 1,125 cM, giving the expected total length of donor DNA in an NIL formed without selection as $1125/2^{b+1}$, or 17.58, 4.39, and 0.55 cM after 5, 7, and 10 generations of backcrossing, respectively.

If donor and recurrent-parent DNAs are sufficiently different, recombination between donor and recurrent chromosomes can be greatly reduced, resulting in a much longer segment of donor DNA being retained than expected (e.g., Young and Tanksley 1989a, Paterson et al. 1990). In these cases, the above expressions may give rather serious underestimates. For example, Young and Tanksley (1989a) examined the length of donor DNA introgressed by selecting for a donor marker (resistance to tobacco mosaic virus) in a cross between two species of tomatoes.

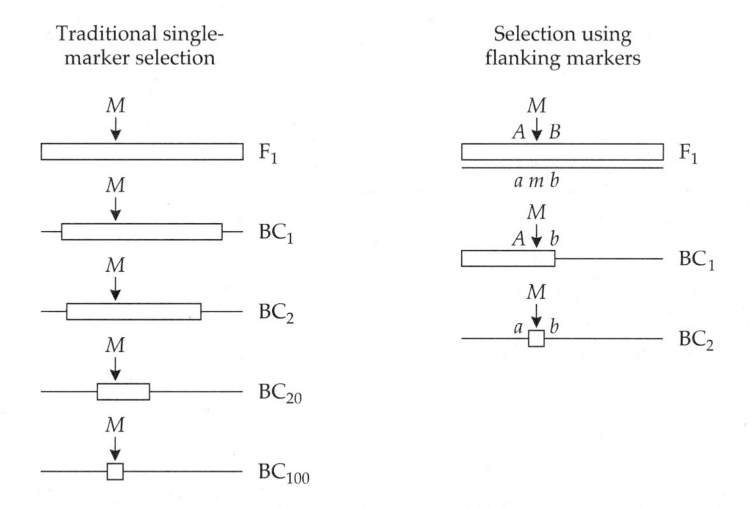

Figure 14.9 Selecting on flanking markers can greatly accelerate the elimination of donor DNA around a selected donor marker *M*. **Left:** Reduction of donor material around a single selected marker leads to an expected value of around 2 cM following 100 generations of selection and backcrossing to the recurrent parent. **Right:** Two generations of selection using a dense marker map can also reduce the introgressed segment to under 2 cM. First, choose two flanking markers (*a* and *b*) from the recurrent parent that are both within 1 cM of *M*, with the donor and recurrent genotypes being *AMB* and *amb*, respectively. In generation one, we select an individual that is either *AMb* or *aMB*. Using this individual as the parent for the second generation, we then select for the recombinant on the opposite side, giving an *aMb* individual with an introgressed region containing *M* that is less than 2 cM. (After Tanksley et al. 1989.)

Even after 11 generations of backcrossing, the length of the introgressed segment in one line was over 51 cM, as compared to the theoretical expectation of 16.7 ± 11.8 cM from Equation 14.10c.

Young and Tanksley (1989a,b) suggested that selection using multiple markers (instead of *single* markers) can greatly accelerate the formation of NILs with very short introgressed segments. Likewise, selection for recurrent markers on nontarget chromosomes increases the rate of removal of donor DNA on other chromosomes.

Equation 14.10c shows that 99 backcross generations are required to reduce the average length of a marker-selected introgressed region to 2 cM. However, with a dense marker map, such a region can be obtained in one or two generations by using markers that flank the region of interest (Figure 14.9). If we can screen a very large number of individuals, this can be done in a single generation by looking for a double recombinant. Using flanking markers 1 cM on either side of the region of interest, the probability of a double recombinant (assuming no interference) is 0.01^2, or about one out of every 10,000 progeny. If interference is significant, which is likely for tightly linked markers, far more progeny may need to be scored. This problem can be avoided by screening for two generations, with first-generation individuals showing a single recombinant (on either side of the marker) being used as parents for the second generation (Figure 14.9).

Example 13. Suppose the region of interest is flanked by markers, each at recombination frequency c from the locus of interest. To have a 95% probability of recovering a double recombinant in a single generation, one must examine roughly $3/c^2$ individuals, or 30,000 individuals for $c = 0.01$ (a 2 cM interval). This follows since the probability of no double recombinants among n sampled chromosomes is $(1-c^2)^n$, giving the probability of at least one double as $1 - (1-c^2)^n$. Setting this equal to 0.95 and solving for n gives $n = -\ln(0.05)/\ln(1-c^2) \simeq 3/c^2$. How much of a savings can the two-generation method offer?

If n_1 first-generation individuals are screened, the probability that at least one of them is a single recombinant is $1-(1-2c)^{n_1}$ (with $2c$ being the probability of recombination on either side of the marker, $1-2c$ is the probability that a chromosome is nonrecombinant with respect to the flanking markers). Using a single recombinant to form the next generation, the probability of a single recombination event on the opposite side of the region is $1-(1-c)^{n_2}$. The probability of obtaining a double recombinant after two generations is the probability of a single recombinant in the first generation times the probability of a single recombinant in the second. Thus, to have an overall probability of 95%, we need a probability of 97.5% for each of the two events, as $0.975^2 = 0.95$. The sample size required to have a 97.5% chance of a single recombinant (on either side of the region) in the first generation is obtained by solving $0.975 = 1 - (1-2c)^{n_1}$, yielding $n_1 \simeq 1.85/c$. Likewise, solving $0.975 = 1 - (1-c)^{n_2}$ gives $n_2 \simeq 3.7/c = 2\,n_1$. Hence, using this two-generation approach, to have a 95% probability of recovering a double recombinant, a total of $n_1 + n_2 = 5.55/c$ individuals must be scored. For $c = 0.01$, this is 555 individuals (185 in the first generation, 370 in the second), only two percent of the 30,000 required for the single-generation method.

FINE MAPPING OF MAJOR GENES USING POPULATION-LEVEL DISEQUILIBRIUM

In small populations, or in populations that have recently undergone a rapid expansion, the amount of disequilibrium between tightly linked markers generated by random drift may be sufficient to allow for very fine mapping of major genes using a random population sample. This approach is called **linkage disequilibrium** (LD) or **allelic association mapping** by human geneticists and is commonly applied to binary (presence/absence) traits. LD mapping can be applied to only a very restricted set of binary traits, as the assumption is that the trait has a very simple genetic basis, such that individuals displaying the trait can be traced to a single allele at one locus. Under this assumption, one tries to find markers that are associated with the allele by comparing the distribution of markers in individuals having the trait versus those lacking the trait. If the trait is influenced by multiple loci, or even multiple alleles at the same locus, marker associations will be obscured. Given its extreme sensitivity to such allelic heterogeneity, it is unlikely that LD mapping can be applied to QTLs of small to moderate effects. Nonetheless, this is an important method for mapping major genes.

The basic idea was suggested by Bodmer (1986, cited by Hill and Weir 1994), wherein once the rough (5–10 cM) position of a gene is found,

> "... it should be possible to saturate the relevant region with further polymorphic markers and look for the ones that have a population association with the trait, rather than increasing the numbers of families analyzed to search for closer linkage. This is the most efficient way of finding closely linked markers, since [human] family data are very inefficient at distinguishing between small recombination fractions such as 0.5% versus 5.0%."

Population-level disequilibrium can result from past admixture and/or from genetic drift. For neutral markers, the expected amount of disequilibrium in a stable population represents a balance between its creation by drift and its removal by recombination. For a randomly mating population in which the drift-recombination balance has been achieved, the expected squared correlation between the presence of two linked marker alleles is

$$E\left(R^2\right) = \frac{1}{1 + 4N_e\, c} \tag{14.13}$$

where c is the recombination frequency between loci and N_e is the effective population size (Hill and Robertson 1968). In principle, Equation 14.13 can be used in combination with estimates of N_e and the squared correlation R^2 between individuals with the trait and a particular marker allele to estimate c, the recombination frequency between the marker and QTL. However, while human geneticists have successfully found linked markers in population-level disequilibrium with major disease alleles (e.g., Kerem et al. 1989, Pritchard et al. 1991), it has proven difficult in most settings to use these markers to actually estimate the recombination frequency between the marker and major locus. One reason is that R^2 is expected to be quite small in large, stable populations for even very tightly linked markers. A second reason is the rather high variance associated with R^2. Hill and Weir (1994) show that this statistic is not very informative, as the resulting likelihood surface for ML estimates of c is very flat, reflecting this high variance. Empirical studies confirm this finding, showing that the amount of disequilibrium between very tightly linked human markers is highly variable (Jorde et al. 1994).

LD Mapping in Expanding Populations

The amount of linkage disequilibrium can be considerably greater than that predicted by Equation 14.13 if the population has not yet achieved a balance between drift and recombination. Such is the case in expanding populations, and geneticists have sought out isolated human populations that appear to have recently undergone such expansions for LD mapping of disease genes. The ideal situation occurs when most of the disease alleles descend from a single ancestral mutation, so that all current copies have maintained some of the ancestral marker haplotype. The age of the ancestral mutation from which all current copies derive can neither be too young nor too old. It must be sufficiently old that recombination has reduced the expected size of the retained region to one sufficiently small for fine mapping. However, if the mutation is too old, this region will be too small for analysis (as it will contain no observable linked markers). These conditions are often satisfied in expanding populations that can be traced back to a small number of recent founders. An example is the current Finnish population, which arose from a small founder group about 2000 years ago (reviewed in de la Chapelle 1993), and other such human populations are discussed by Jorde (1995). In a rapidly expanding population, all copies of a mutant disease allele often trace back to a single copy present in the founding population (or originating shortly after the population expansion began).

Motivated by the Luria-Delbrück theory for estimating mutation rates in exponentially growing populations of bacteria (Luria and Delbrück 1943, Sarkar 1991), Hästbacka et al. (1992) and Lehesjoki et al. (1993) have suggested that the same logic can be used to estimate recombination frequencies from population disequilibrium information. The simplest approach proceeds as follows. Suppose the disease allele was either present as a single copy (and hence associated with a single chromosomal haplotype) in the founder population or arose by mutation very shortly after the population was formed. Assume that there is no **allelic heterogeneity**, so that all disease-causing alleles in the population descend directly from the original mutation, and consider a marker locus tightly linked to the disease locus. The probability that a disease-bearing chromosome has not experienced recombination between the **disease susceptibility** (DS) gene and marker after t generations is just $(1-c)^t \simeq e^{-ct}$, where c is the marker-DS recombination frequency. Suppose the disease is predominantly associated with a particular haplotype, which presumably represents the ancestral haplotype on which the DS mutant arose. Equating the probability of no recombination to the observed proportion π of disease-bearing chromosomes with this predominant haplotype gives $\pi = (1-c)^t$, where t is the age of the mutation or the age of the founding population (whichever is more recent). Hence, one estimate of the recombination frequency is

$$c = 1 - \pi^{1/t} \tag{14.14}$$

Example 14. Hästbacka et al. (1992) examined the gene for diastrophic dysplasis (DTD), an autosomal recessive disease, in Finland. A total of 18 **multiplex** families (showing two or more affected individuals) allowed the gene to be localized to within 1.6 cM from a marker locus (*CSF1R*) using standard pedigree methods. To increase the resolution using pedigree methods requires significantly more multiplex families. Given the excellent public health system in Finland, however, it is likely that the investigators had already sampled most of the existing families. As a result, the authors turned to LD mapping.

While only multiplex families provide information under standard mapping procedures, this is not the case with LD mapping wherein single affected individuals can provide information. Using LD mapping thus allowed the sample size to increase by 59. A number of marker loci were examined, with the *CSF1R* locus showing the most striking correlation with DTD. The investigators were able to unambiguously determine the haplotypes of 152 DTD-bearing chromosomes and 123 normal chromosomes for the sampled individuals. Four alleles of the *CSF1R* marker gene were detected. The frequencies for these alleles among normal and DTD chromosomes were found to be:

	Chromosome type			
Allele	Normal		DTD	
1-1	4	3.3%	144	94.7%
1-2	28	22.7%	1	0.7%
2-1	7	5.7%	0	0%
2-2	84	68.3%	7	4.6%

Given that the majority of DTD-bearing chromosomes are associated with the rare 1-1 allele (present in only 3.3% of normal chromosomes), the authors suggested that all DTD-bearing chromosomes in the sample descended from a single ancestor carrying allele 1-1. Since 95% of all present DTD-bearing chromosomes are of this allele, $\pi = 0.95$. The current Finnish population traces back to around 2000 years to a small group of founders, which underwent around $t = 100$ generations of exponential growth. Using these estimates of π and t, Equation 14.14 gives an estimated recombination frequency between the *CSF1R* gene and the DTD gene as $c = 1 - (0.95)^{1/100} \simeq 0.00051$. Thus, the two genes are estimated to be separated by 0.05 cM, or about 50 kb (using the rough rule for humans that 1 cM = 10^6 bp). Subsequent cloning of this gene by Hästbacka et al. (1994) showed it be to 70 kb proximal to the *CSF1R* marker locus. Thus, LD mapping increased precision by about 34-fold over that possible using segregation within pedigrees (0.05 cM vs. 1.6 cM).

The DTD example is exceptional in that the disease allele was initially associated with a rare haplotype. Since the majority of current disease-bearing chromosomes have this haplotype, this suggests that the sample of disease alleles is unlikely to contain a significant fraction of new mutants. Lehesjoki et al. (1993) offer a modification when the predominant marker haplotype among disease-bearing chromosomes is also a common haplotype for normal chromosomes. Let p_a denote the frequency of the most common haplotype of disease-bearing chromosomes and p_n denote the frequency of this haplotype among normal chromosomes. Let π represent the fraction of disease-bearing chromosomes descended from the common ancestor that have not undergone any recombination between the marker and QTL, and α be the proportion of all disease alleles descending from the founder copy (as opposed to being new mutants). We can then divide disease-bearing chromosomes into two classes. First, a fraction $\alpha\,\pi$ of these chromosomes have a disease allele that both traces back to the founder copy (α) and does so through chromosomes that have not undergone recombination between the marker and disease loci (π). Second, a fraction $(1 - \alpha\,\pi)$ are either new mutants (which arose on some random chromosome) or a product of recombination between the founder copy and some random chromosome. In either event, the chance that such individuals have the predominant haplotype is just $(1 - \alpha\,\pi)\, p_n$. Putting these two results together gives

$$p_a = \alpha\,\pi + (1 - \alpha\,\pi)\, p_n \tag{14.15}$$

Rearranging Equation 14.15 and solving, we find that

$$\delta_p = \alpha\,\pi = \frac{p_a - p_n}{1 - p_n} \tag{14.16}$$

is a measure of the extent to which disease chromosomes are restricted to the predominant haplotype (Bengtsson and Thomsom 1981). Recalling Equation 14.14,

$$\pi = (1 - c)^t = \frac{\delta_p}{\alpha} \tag{14.17}$$

giving an estimate of the recombination frequency as

$$c = 1 - \left(\frac{\delta_p}{\alpha} \right)^{1/t} \tag{14.18}$$

Note that for the DTD data in Example 14, $p_n \simeq 0$, so that $\delta_p \simeq p_a$, and we assumed that all disease alleles descended from the founder copy ($\alpha = 1$), so that $c \simeq 1 - (p_a)^{1/t}$. Note also that if c is known, one can solve instead for t, the age of the mutation. Risch et al. (1995) used this approach to date the appearance of a mutation causing idiopathic torsion dystonia (ITD) disease in Ashkenazi Jews at about 350 years.

The fraction α of all disease alleles that are direct descendants from the ancestral copy (as opposed to being new mutants) is obtained as follows. Assuming that selection is weak on carriers (heterozygotes), mutants that have arisen after the initial appearance of the original DS allele in the founding population should comprise a fraction $1 - (1 - \mu)^t \simeq \mu\, t$ of all chromosomes, as $(1 - \mu)^t$ is the probability that no mutations have occurred. Here μ is the mutation rate at which new disease-causing alleles appear and t is the number of generations since the original mutation arose or since the founding of the population, whichever is more recent. If q is the frequency of disease alleles, then $\mu\, t/q$ is the expected fraction of disease-carrying chromosomes due to new mutants. Hence, the expected fraction of all disease-carrying chromosomes that descend from the original mutation is

$$\alpha = \left(1 - \frac{\mu\, t}{q} \right) \tag{14.19}$$

which, when applied to Equation 14.18, yields

$$c = 1 - \left(\frac{\delta_p}{1 - (\mu\, t/q)} \right)^{1/t} \tag{14.20}$$

Example 15. Sulisalo et al. (1994) examined the major gene for cartilage-hair hypoplasia (CHH), an autosomal recessive disease, in the Finnish population. As in Example 14, pedigree information allowed unambiguous determination the haplotypes associated with most CCH-bearing chromosomes. The authors observed that 85% of these contain a particular allele at marker *D95163*, while only 41% of non-CHH chromosomes carry this allele. Hence,

$$\delta_p = \frac{p_a - p_n}{1 - p_n} = \frac{0.85 - 0.41}{1 - 0.41} = 0.75$$

implying

$$c = 1 - \left(\frac{\delta_p}{\alpha}\right)^{1/t} = 1 - \left(\frac{0.75}{\alpha}\right)^{1/t}$$

To estimate the fraction α of all current CHH alleles that directly trace back to the ancestral copy, first note that the frequency of CHH alleles in Finland is estimated to be 0.0066. Assuming $t = 100$ and $\mu = 1 \times 10^{-5}$,

$$\alpha = \left(1 - \frac{\mu\, t}{q}\right) = \left(1 - \frac{100 \times 10^{-5}}{0.0066}\right) = 0.85$$

Thus, 85% of all present CHH alleles are estimated to be direct descendants of the founder copy, which implies $c = 0.12$ cM. Taking $\mu = 1 \times 10^{-6}$ gives $\alpha = 0.98$ and $c = 0.27$ cM. By contrast, traditional pedigree-based mapping was able to localize the CHH gene to only a 1.7 cM region.

LD mapping has been successfully applied to other disease genes segregating in the Finnish population, such as congenital nephrotic syndrome (Kestilä et al. 1994) and progressive myoclonus epilepsy (Lehesjoki et al. 1993). Kaplan et al. (1995) were able to apply LD mapping to the cystic fibrosis (CF) gene, using much more heterogeneous populations from Europe and elsewhere. The likely reason for success is that 70% of CF chromosomes worldwide appear to result from a single three-base deletion (Kerem et al. 1989), so that allelic heterogeneity is, at worst, a modest problem. Kaplan et al. (1995) found, however, that LD mapping was not very successful for Huntington's disease (multiple ancestral haplotypes) or Friedreich ataxia (high allelic heterogeneity) using European or North American populations.

Kaplan and Weir (1995) show that commonly used approximations for the confidence intervals for estimates given by Equations 14.14 and 14.18 are downwardly biased and can be very misleading. This is not the case with maximum likelihood approaches. Kaplan et al. (1995) provide likelihood functions incorporating two marker loci linked to the disease locus, allowing for statistical tests of gene ordering of the markers and the DS gene. Similar tests for ordering are presented in Risch et al. (1995). Further discussion of likelihood functions for disequilibrium mapping can be found in Terwilliger (1995).

CANDIDATE LOCI

In some cases, there may be sufficient physiological/biochemical information to suspect that certain known loci influence character expression. In human genetics, it is common practice to directly test for population-level associations between trait value and specific alleles at such **candidate loci**. For example, Boerwinkle and Sing (1987) showed that three common alleles at the human apolipoprotein E locus account for about 8% of the total variation in cholesterol levels. Interestingly, a particular allele of apolipoprotein E also appears to be a major determinant of Alzheimer's disease. The mean age of onset for homozygotes and heterozygotes for this allele is 68.4 ± 1.2 and 75.5 ± 1.0 years, respectively, while the mean age for individuals with no copies of this allele is 84.3 ± 1.3 (Corder et al. 1993). Other candidate loci for Alzheimer's disease are reviewed by Pericak-Vance and Haines (1995). As the next example shows, results using the candidate-locus approach can be rather unexpected.

Example 16. Winkelman and Hodgetts (1992) examined the growth hormone (GH) gene as a candidate locus for body weight in selected lines of mice. Molecular analysis disclosed the presence of an allele, GH^h, present in all four lines selected for increased weight (being fixed in three of these). An alternative allele, GH^c, was fixed in all five control lines. One of the up-selected lines was crossed to two separate control lines to create two different F_2 populations. The GH^h allele had a significant, but unexpected, effect on body weight in both F_2 populations, as it *decreased* weight. For one population, the genotypes $GH^hGH^h : GH^hGH^c : GH^cGH^c$ had 42-day weights of 29.2 : 30.2 : 31.4, while in the other population these respective weights were 34.6 : 34.9 : 38.8. Thus, the GH^h allele was additive in one of the F_2 backgrounds, but dominant in the other. The GH^h allele behaved rather differently once the F_2 populations were subjected to selection, again increasing in frequency in up-selected lines, and decreasing in frequency in down-selected lines. Winkelman and Hodgetts suggested that the association between the GH^h allele and increased weight was a product of epistastic interactions in mice selected for high weight.

How are candidate loci chosen? One obvious approach is to consider loci known from laboratory studies (such as knockout experiments) to have mutant alleles with major effects on the character of interest, as in natural populations such loci may also be segregating alleles with smaller effects (Chapter 12). Results from several QTL mapping experiments offer support for this approach (Chapter 15). In maize, for example, Beavis et al. (1991), Edwards et al. (1992), Veldboom et al. (1994), and Berke and Rocheford (1995) found that many QTLs for height map near the locations of known height mutants. Similar findings have been obtained with bristle number in *Drosophila* by Mackay and Langley (1990), Lai et al. (1994), and Long et al. (1995).

The Transmission/Disequilibrium Test

When considering genetic disorders, the frequency of a particular candidate (or marker) allele in affected (or **case**) individuals is often compared with the frequency of this allele in unaffected (or **control**) individuals. The problem with such **association studies** is that a disease-marker association can arise simply as a consequence of population structure, rather than as a consequence of linkage. Such **population stratification** occurs if the total sample consists of a number of divergent populations (e.g., different ethnic groups) which differ in both candidate-gene frequencies and incidences of the disease. Population structure can severely compromise tests of candidate gene associations, as the following example illustrates.

Example 17. Hanson et al. (1995) used segregation analysis (Chapter 13) to find evidence for a major gene for Type 2 diabetes mellitus segregating at high frequency in members of the Pima and Tohono O'odham tribes of southern Arizona. In an attempt to map this gene, Knowler et al. (1988) examined how the simple presence/absence of a particular haplotype, Gm^+, was associated with diabetes. Their sample showed the following associations:

Gm^+	Total subjects	% with Diabetes
Present	293	8%
Absent	4,627	29%

The resulting χ^2 value (61.6, 1 df) shows a highly significant negative association between the Gm^+ haplotype and diabetes, making it very tempting to suggest that this haplotype marks a candidate diabetes locus (either directly or by close linkage).

However, the presence/absence of this haplotype is also a very sensitive indicator of admixture with the Caucasian population. The frequency of Gm^+ is around 67% in Caucasians as compared to $< 1\%$ in full-heritage Pima and Tohono O'odham. When the authors restricted the analysis to such full-heritage adults (over age 35 to correct for age of onset), the association between haplotype and disease disappeared:

Gm^+	Total subjects	% with Diabetes
Present	17	59%
Absent	1,764	60%

Hence, the Gm^+ marker is a predictor of diabetes not because it is linked to genes influencing diabetes but rather because it serves as a predictor of whether individuals are from a specific subpopulation. Gm^+ individuals usually carry a significant fraction of genes of Caucasian extraction. Since a gene (or genes) increasing the risk of diabetes appears to be present at high frequency in individuals of full-blooded Pima/Tohono O'odham extraction, admixed individuals have a lower chance of carrying this gene (or genes).

The problem of population stratification can be overcome by employing tests that use family data, rather than data from unrelated individuals, to provide the case and control samples (Woolf 1955, Rubinstein et al. 1981, Falk and Rubinstein 1987, Terwilliger and Ott 1992, Spielman et al. 1993). This is done by considering the transmission (or lack thereof) of a parental marker allele to an affected offspring. Focusing on transmission within families controls for association generated entirely by population stratification and provides a direct test for linkage *provided* that a population-wide association between the marker and disease gene exists (Spielman et al. 1993, Ewens and Spielman 1995). This method can be applied to any family that has at least one affected offspring. The assumption of population-wide disequilibrium is required as one pools marker associations across all families.

The **transmission/disequilibrium test**, or TDT (Spielman et al. 1993), compares the number of times a marker allele is transmitted (T) versus not-transmitted (NT) from a marker heterozygote parent to affected offspring. Under the hypothesis of no linkage, these values should be equal, and the test statistic becomes

$$\chi^2_{td} = \frac{(T - NT)^2}{(T + NT)} \tag{14.21}$$

which follows a χ^2 distribution with one degree of freedom. This is also known as **McNemar's test** (Sokal and Rohlf, 1995). A number of related tests have also been proposed, and Schaid and Sommer (1994) provide a nice overview of their properties.

How are T and NT determined? Consider an M/m parent with three affected offspring. If two of those offspring received this parent's M allele, while the third received m, we score this as two transmitted M, one not-transmitted M. Conversely, if we are following marker m instead, this is scored as one transmitted m, two not-transmitted m. As the following example shows, each marker allele is examined separately under the TDT.

Example 18. Copeman et al. (1995) examined 21 microsatellite marker loci in 455 human families with Type 1 diabetes. One marker locus, *D2S152*, had three alleles, with one allele (denoted 228) showing a significant effect under the TDT. Parents heterozygous for this marker transmitted allele 228 to diabetic offspring 81 times, while transmitting alternative alleles only 45 times, giving

$$\chi^2 = \frac{(81 - 45)^2}{(81 + 45)} = 10.29$$

which has a corresponding P value of 0.001. As summarized below, the other two alleles (230 and 240) at this marker locus did not show a significant TD effect.

Allele	T	NT	χ^2	P
228	81	45	10.29	0.001
230	59	73	1.48	0.223
240	36	24	2.40	0.121

Hence, this marker is linked to a QTL influencing Type 1 diabetes, with allele 228 in (coupling) linkage disequilibrium with an allele that increases the risk for this disease.

Spielman et al. (1993) note that the TDT can give a false positive if the marker shows **segregation distortion**, where heterozygotes preferentially segregate one allele. This distortion can be controlled for by using a standard 2×2 contingency table χ^2 test, considering separately the transmission of a marker allele to affected and unaffected offspring,

	Transmitted	Not Transmitted
Affected Offspring	T_A	NT_A
Unaffected Offspring	T_U	NT_U

with resulting test statistic

$$\chi^2_{td} = \frac{(T_A - NT_A)^2}{(T_A + NT_A)} + \frac{(T_U - NT_U)^2}{(T_U + NT_U)} \tag{14.22}$$

controlling for any segregation distortion. A number of family-based methods for detecting marker-trait associations that do not require population-wide disequilibrium have been developed, and these are examined in some detail in Chapter 16.

Estimating Effects of Candidate Loci

While the estimation of genotype means for a candidate locus seems straightforward, there are several potential sources of bias. While it is possible in some settings to distinguish between the direct effects of a candidate locus and its indirect effects due to association with linked QTLs (e.g., Bovenhuis and Weller 1994), as noted above the presence of linkage disequilibrium greatly confounds the interpretation of candidate-locus means. We consider other, more subtle, sources of bias below.

Let z_{ij} denote the phenotypic value of the jth individual with candidate-locus genotype i ($1 \leq i \leq n_g$). If μ_i is the mean character value for genotype i, then the simplest model is

$$z_{ij} = \mu_i + \epsilon_{ij} \tag{14.23}$$

If individuals are sampled at random from a large homogeneous population, the residuals are uncorrelated and the μ_i values can be estimated simply by using $\overline{z}_i$. However, when relatives are present in the sample, one must take the residual covariance matrix into consideration to obtain a correct estimator. The residual error can be decomposed into a genetic component (G) due to segregation at loci other than the target QTL plus the general (E) and specific (e) environmental effects,

$$\epsilon_{ij} = G_{ij} + E_{ij} + e_{ij} \tag{14.24a}$$

Indexing two distinct individuals by j and k,

$$\sigma(\epsilon_j, \epsilon_k) = \sigma(G_j, G_k) + \sigma(E_j, E_k) \tag{14.24b}$$

as special environmental effects are not shared by different individuals (Chapter 6), and G and E are assumed to be uncorrelated. If the relationships among sampled individuals are known, the elements of the covariance matrix can be computed from Equation 14.24b, using the methods of Chapter 7. For example, if i and j are full-sibs, $\sigma(G_j, G_k) = (\sigma_A^2/2) + (\sigma_D^2/4)$, where the variances refer to the contribution of background polygenes. Estimates of the variance components associated with G and E can be obtained by a variety of methods introduced in Chapters 17–27. Using these to estimate the residual covariance matrix, genotypic means can be estimated by a GLS regression (Chapter 8). An example of the importance of correcting for common relatives is provided by Bentsen and Klemetsdal (1991), who examined the association between egg productivity and six different MHC haplotypes in chickens. The relative rankings of 5 of the 6 haplotype effects differed when comparing uncorrected estimates with those obtained by correcting for shared ancestry. More generally, when additional fixed factors (such as sex- or age-specific effects) are included in the model, BLUP estimation can be used (Boerwinkle et al. 1986, Cowan et al. 1990, Hoeschele and Meinert 1990, Bentsen and Klemetsdal 1991, Blangero et al. 1992, Kennedy et al. 1992). Chapter 26 examines this in some detail.

Many estimates of the amount of variability attributable to a candidate locus are upwardly biased, even for a random population sample. If q_i is the frequency of candidate-locus genotype i, then the variance contributed by this locus is

$$\sigma_L^2 = \sum_{i=1}^{n_g} q_i\,(\,\mu_i - \mu\,)^2 \qquad \text{where} \qquad \mu = \sum_{i=1}^{n_g} q_i\,\mu_i \tag{14.25}$$

Boerwinkle and Sing (1986) showed that the obvious estimator

$$s^2 = \sum_{i=1}^{n_g} \widehat{q}_i\,(\,\overline{z}_i - \overline{z}\,)^2 \tag{14.26a}$$

using the estimates $\widehat{q}_i$, $\overline{z}_i$, and $\overline{z}$ in place of their true values is biased because it includes the sampling error of these three estimates. To account for the sampling variances in $\overline{z}_i$ and $\overline{z}$, Boerwinkle and Sing suggest the improved estimator (for a collection of unrelated individuals),

$$s_L^2 = \sum_{i=1}^{n_g} \widehat{q}_i \left(\overline{z}_i - \overline{z} \right)^2 - \left(\frac{n_g - 1}{n} \right) \mathrm{Var}(e) \tag{14.26b}$$

where Var(e) is the estimate of the residual variance (within genotypes) and we assume n measured individuals per genotype. While an improvement over Equation 14.26a, this estimator still ignores the sampling variance introduced by using $\widehat{q}_i$ and hence still tends to overestimate the variance accounted for by the candidate locus. Resampling approaches for constructing confidence intervals of both σ_L^2 and its additive and dominance components have been suggested by Kaprio et al. (1991) and Zerba et al. (1996).

Templeton and Sing's Method: Using the Historical Information in Haplotypes

While one can use a single marker to examine whether genetic variation at a candidate locus is associated with character variation, studies typically involve several closely linked markers, often including several polymorphic sites within the gene itself. How should one best extract information from this set of markers? Obviously, it is more powerful to consider **haplotypes** (the multiple-locus genotypes associated with the region of interest) than single markers. One drawback with this approach is that there can be an enormous number of haplotypes, resulting in small sample sizes for each haplotype and a reduction in the power of tests for haplotype-trait associations. What is needed is a logical way to combine information from different haplotypes. In an interesting series of papers, Templeton and Sing (Templeton et al. 1987, 1988, 1992; Templeton 1995) suggest that this can be done by incorporating information on the inferred evolutionary relationships of the sampled haplotypes. We simply outline their basic idea here.

The sequence information in haplotypes can be used to construct a **cladogram** estimating how the different haplotypes are evolutionarily related to each other. With such a cladogram in hand, one can then consider how character value is a function of nested sets of cladogram members, for example by using a nested ANOVA (Chapter 18). The motivation behind this approach is that history is the main cause of linkage disequilibrium, with associations between loci decaying with time. Hence, the more closely related a set of haplotypes, the more disequilibrium they should display, and the more likely they are to share QTL alleles. Figure 14.10 shows an example of this approach, detecting associations between alcohol dehydrogenase (ADH) activity and haplotypes in a 13 kb region surrounding the ADH gene in *Drosophila*.

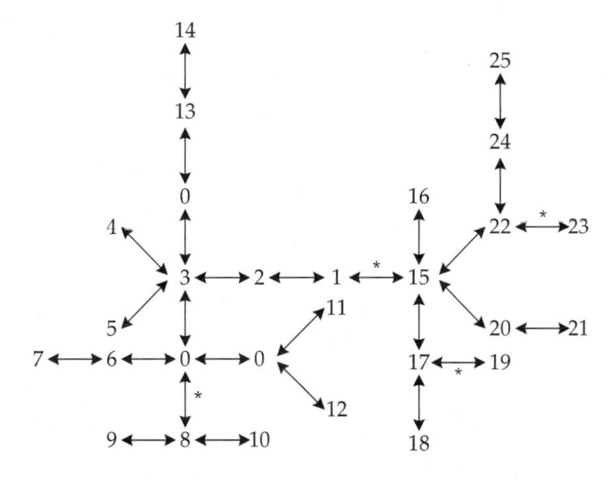

Figure 14.10 Inferred cladogram of haplotypes for a region surrounding the *Drosophila melanogaster* ADH gene. Numbers refer to different haplotypes, while "0" refers to intermediate haplotypes not present in the sample but required to interconnect the existing haplotypes. Distance between two haplotypes on the cladogram provides a measured of relatedness, so that haplotypes 8 and 9 are more closely related than 9 and 6 (one step vs. three steps). Using the hierarchical structure implied by this cladogram, a nested ANOVA was performed, with ANOVA levels set by the number of cladogram steps that haplotypes differ by. Four cladogram regions, denoted with asterisks, were found to be associated with significant phenotypic effects. (After Templeton et al. 1987.)

CLONING QTLs

Cloning a gene (transferring a DNA sequence containing the gene to a plasmid or other manipulable vector) allows the complete power of modern molecular biology to be used in the study of QTLs. With a clone in hand, one can sequence the gene, study its expression in different tissues and at different developmental times, and isolate homologous DNA sequences from other populations/species. More direct manipulation is also possible, such as the placement of modified copies of the gene back into the organism, or into different species, creating **transgenic** individuals. While a variety of molecular approaches can be used to clone a gene whose product is known (Maniatis et al. 1982), QTLs are of special interest precisely because their products are typically unknown. Two different cloning strategies have been suggested for genes with unknown products but discernible phenotypes: **transposon tagging** and **positional cloning**. We discuss these in turn. An interesting review of recent ideas for cloning QTLs is given by Tanksley et al. (1995).

Transposon Tagging

When a mobile genetic element (or **transposon**) inserts itself into or near a gene, it can disrupt expression, creating a visible mutation, and providing the basis for cloning by **transposon tagging** (Bingham et al. 1981). In several species, it is now possible to introduce transposons modified for high insertion rates into the germline either by genetic or micro-injection techniques. Using these elements as probes, standard molecular techniques can then be used to isolate any region of DNA within which an element has inserted.

Examples of the successful use of this procedure primarily involve genes with large effects: the tagging of the *white* eye-color locus in *D. melanogaster* with *P*-factors (Bingham et al. 1981), the *dilute* coat-color gene in the mouse with a murine leukemia virus (Rinchik et al. 1986), and a dwarfing gene in the mustard *Arabidopsis* with the *Agrobacterium* T-DNA plasmid (Feldmann et al. 1989), to name a few. For detection of QTLs by transposon tagging, the use of inbred lines is essential in order to reduce variation from segregation at other loci. Typically, one starts with a line with little or no genetic variation and then selects for new mutations affecting the character, which are then examined for indications that at least one scored element has moved. Soller and Beckmann (1987) discuss a variety of statistical considerations associated with cloning QTLs by such insertional mutagenesis techniques.

Given the need for completely inbred lines, is this approach really of general use? The answer is that a few well conceived studies with model systems may be of broad relevance to related species. A generalization appearing out of the morass of sequencing data is the relative conservation of gene sequences between widely divergent species. Hence, a QTL probe from a mouse might be used to detect its homologue in other mammals, a probe from *Arabidopsis* used for other plants, and a probe from *Drosophila* used for other insects. Having a clone in hand offers the possibility of looking at population variation at that locus, using techniques such as PCR (Saiki et al. 1985, Scharf et al. 1986, Engelke et al. 1988), or high-resolution restriction mapping (Kreitman and Aguade 1986). If variant alleles are detected in a natural population using a QTL probe, the candidate-gene approaches discussed above can be used estimate the effects of each allele on the character(s) of interest.

Positional Cloning and Comparative Mapping

In theory, if we can localize the position of a QTL to a sufficiently small region of DNA, it becomes possible to examine all of the genes in that region. For example, in humans the use of rare chromosome deletions or translocations that correlate with the presence of a disease have been very successfully exploited to delimit the region in which the disease gene resides (Collins 1992, 1995). This is the idea behind **positional cloning**, which can be done by brute force (sequencing

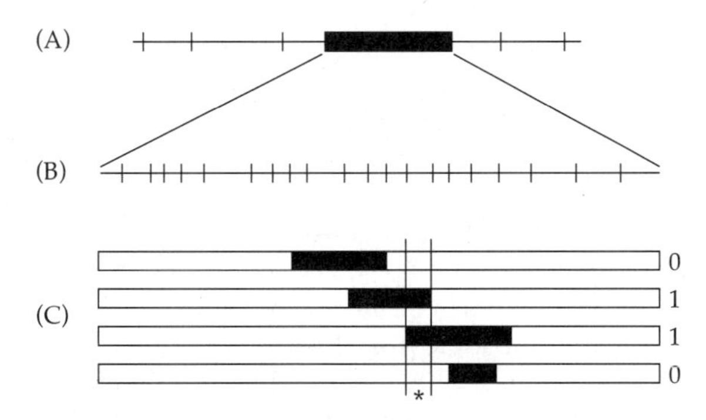

Figure 14.11 One approach to positional cloning of a QTL using a series of NILs. **(A)**: First the QTL (which increases character value by a scaled value of one) is localized to an interval (typically 30–50 cM). **(B)**: A much finer map of markers is then generated within this interval, and NILs are constructed by marker selection. **(C)**: Comparing this series of NILs, the QTL position is narrowed to the region indicated by *. If this region is not sufficiently small for further analysis, steps **(B)** and **(C)** are repeated until the desired size is obtained. This region is then cloned, and the genes within it are examined for potential clues as to which loci should be considered (or rejected) as candidates.

of the entire region) or by more clever methods. An example of the latter is **zoo blotting** — the use of sequences from related species to isolate only those parts of the region of interest that are conserved between species, as these likely represent genes. Another approach is to use schemes that directly isolate genes, such as exon amplification or **exon trapping** methods (Duyk et al. 1990, Buckler et al. 1991). The timing and pattern of expression of genes localized within a region can suggest exclusion or inclusion of others as candidate QTLs (generating **positional candidates**). As we become increasingly better at associating particular amino acid sequences with particular functional units (e.g., known DNA binding regions or specific catalytic sites), the hope is that we will be able to recognize candidate genes from unique features of the DNA sequence itself.

A serious limitation of positional cloning is the sheer physical distance involved in terms of number of DNA base pairs, even with very tight linkage. As noted above, a recombination fraction of 1% corresponds to roughly 10^5 to 10^7 base pairs (Table 14.3), which typically translates into 10 to 100 genes. Thus, the isolation of tightly linked markers flanking a QTL is just the initial step. Once such a region is believed to be isolated, it is important to actually confirm that it contains a QTL, for example by using markers to introgress the region into a standardized background and directly testing its effect. If the region has an affect on the character, the next step is to use NILs for ever-finer mapping (Figure

14.11), which may require the procurement of additional markers in the region of interest. As mentioned above, bulk segregant analysis provides one approach for obtaining such markers. When the localized region is sufficiently small, it can be sequenced and all open reading frames identified.

The amount of labor involved in positional cloning is far from trivial. Moreover, there is a practical limitation to constructing ever-finer NILs. Even if we have an infinitely dense marker map, the number of individuals that must be screened to isolate recombinants over shorter and shorter distances soon becomes prohibitively large. Using the logic from Example 13, the number of individuals required to have a 95% chance of at least one recombinant between markers c units apart is $\simeq 3/c$. Currently manageable amounts of DNA are around 10^4 to 10^5 bases. Taking the average value of 1 cM $\simeq 10^6$ DNA bases, this amounts to genetic map distances of 0.1 and 0.01 cM, or the scoring of 3,000 and 30,000 individuals, respectively. In spite of these potential problems, positional cloning has been successfully used in several model organisms, including *Arabidopsis* (e.g., Arondel et al. 1992), tomato (Martin et al. 1993), mice (e.g., Bultman et al. 1992, Copeland et al. 1993, Takahashi et al. 1994), and humans (e.g., Orkin 1986, Royer-Pokora et al. 1986, Rommens et al. 1989, Gessler et al. 1990).

A related approach for positional cloning follows from the observation that map position of homologous genes is partly conserved across species, providing an avenue by which results from one species can be used to map QTLs in a related species. For example, humans and mice show rather extensive conservation of gene order, with the average conserved linkage block in autosomes being around 8 to 10 cM (Nadeau 1989, Copeland et al. 1993). Thus, a marker tightly linked to a QTL in mouse has a good chance of also being tightly linked to the homologous gene (if it exists) in humans. The major grains (rice, wheat, maize) also show conservation of gene order within blocks on the order of 5 to 10 cM (Bennetzen and Freeling 1993). Similar observations have been made for the mustards *Brassica* and *Arabidopsis* (Teutonico and Osborn 1994), and for potatoes (*Solanum*) and tomatoes (*Lycopersicon*), both in the nightshade family (Bonierbale et al. 1988).

Such observations suggest that candidate regions for a QTL in one species are also candidate regions for QTLs in related species, an approach called **comparative mapping**. There are indications that this approach may be fruitful. Fatokun et al. (1992) found a major QTL for seed weight at similar positions in cowpea (*Vigna unguiculata*) and the distantly related mung bean (*V. radiata*). QTLs for several fruit characters have been found in similar positions in three species of tomatoes (Paterson et al. 1988, 1990; Goldman et al. 1995). The most striking examples come from the major grass grains: rice, maize, and sorghum (Lin et al. 1995, Pereira and Lee 1995, Paterson et al. 1995). Domestication of all three species in recent human history involved selection for three traits: increased seed size, reduced disarticulation of the mature inflorescence (seeds remaining together on the plant for easier harvesting), and daylength-insensitive flowering (increasing the growing season). Paterson et al. (1995) showed that a few QTLs, mapping to similar

15

Mapping and Characterizing QTLs: Inbred Line Crosses

Crosses between completely inbred lines offer an ideal setting for detecting and mapping QTLs by marker-trait associations, as all F_1s are genetically identical and show complete linkage disequilibrium for genes differing between lines. A number of designs have been proposed to exploit these features. While usually involving crop plants, QTL line-cross analysis has also been applied to a number of animal species, especially mice (reviewed by Frankel 1995). A particularly interesting example is the work of Hunt et al. (1995), who mapped QTLs for foraging behavior in honey bees using a design that exploited the haploid nature of male honey bees.

While the QTL mapping literature can seem very complex to the uninitiated, it is based on a few simple ideas. We start by reviewing the basic building blocks, first considering different line-cross populations useful for QTL mapping. The key element from which the formal theory of QTL mapping is constructed is the conditional probability of a particular QTL genotype given an observed marker genotype, and we examine this probability next. These probabilities allow a full development of the two principal methods for QTL detection and estimation — linear models (using differences in marker means) and maximum likelihood (using information from the entire marker-trait distribution), both of which are examined in some detail. We conclude our discussion of statistical issues by examining methods that accommodate multiple linked QTLs and by visiting the important issue of the sample sizes required to achieve a given power for detecting a QTL. Finally, we reward the persevering reader with a review of a number of interesting studies that have used inbred lines for QTL mapping.

FOUNDATIONS OF LINE-CROSS MAPPING

The idea behind using marker information to map and characterize QTLs is quite simple: by crossing two inbred lines, linkage disequilibrium is created between loci that differ between the lines, and this in turn creates associations between marker loci and linked segregating QTLs. A large number of experimental designs and statistical methodologies have been developed to exploit this information. Our attempt to make this field more accessible starts with an overview of some of the basic experimental designs and some of the key tools for the analysis of results from these designs.

Experimental Designs

The large number of possible designs can be categorized by the type of line-cross populations used for generating disequilibrium (e.g., F_2 vs. backcross populations) and the unit of marker analysis used (e.g., single markers vs. interval mapping). We consider these in turn.

Starting with two completely inbred parental lines, P_1 and P_2, a number of line-cross populations derived from the F_1 can be used for QTL mapping. The **F_2 design** examines marker-trait associations in the progeny from a cross (or selfing) of F_1s, while the **backcross design** examines marker-trait associations in the progeny formed by backcrossing the F_1 to one of the parental lines. While these are the most widely used designs, other line-cross populations can offer further advantages (and disadvantages). For example, the F_1 can be used to create recombinant inbred lines (RILs) and doubled haploid lines (DHLs), which allow marker-trait associations to be scored in a completely homozygous background and across multiple environments (Chapter 14). The F_2 design has an advantage over designs using backcross, RIL, or DHL populations, because it generates three genotypes at each marker locus, which allows the estimation of the degree of dominance associated with detected QTLs. Designs using an F_t population (formed by randomly mating F_1s for $t - 1$ generations) allow for even higher resolution of QTL map positions than do F_2s, albeit at the expense of decreased power of QTL detection. The properties of such **advanced intercross lines** (AILs) are discussed below.

More complex designs can be considered wherein individuals are genotyped in one population, while trait values are scored in a future population derived from the genotyped individuals. Fisch et al. (1996) present a general treatment for such designs. One example is the **$F_{2:3}$ design**, wherein F_2 individuals are genotyped and then selfed. The trait value associated with a genotyped individual is estimated by the mean value of the resulting F_3 family. Scoring the phenotype as the mean of several individuals (as opposed to measurement of a single individual) can offer increased power over a standard F_2 design by reducing the sampling variance.

Designs combining information from multiple crosses are expected to be more powerful than those involving a single cross, and as a result, designs using multiple line-cross populations are starting to be considered. Examples include diallel designs whose basic structure is examined in Chapter 20 (Rebaï and Goffinet 1993, Rebaï et al. 1994a), and Comstock and Robinson's (1952) classic Design III wherein the F_2 from two inbred lines is backcrossed to both parental lines (Cockerham and Zeng 1996). We will not consider these multiple-line designs further, but their continued development is clearly an important area for future work. Finally, several workers have considered designs involving crosses between lines that are not completely inbred, such as a cross of an outbred line to a completely inbred tester (Beckmann and Soller 1988, Dudley 1992, Haley et al. 1994).

Experimental designs are also classified by the unit of marker analysis chosen by the investigator. Marker-trait associations can be assessed using one-, two-, or multiple-locus marker genotypes. Under a **single-marker analysis**, the distribution of trait values is examined separately for each marker locus. Each marker-trait association test is performed independent of information from all other markers, so that a chromosome with n markers offers n separate single-marker tests. As discussed below, single-marker analysis is generally a good choice when the goal is simple *detection* of a QTL linked to a marker, rather than *estimation* of its position and effects. Under **interval mapping** (or **flanking-marker analysis**), a separate analysis is performed for each *pair* of adjacent marker loci. The use of such two-locus marker genotypes results in $n-1$ separate tests of marker-trait associations for a chromosome with n markers (one for each marker interval). Interval mapping offers a further increase in power of detection (albeit usually a slight one) and more precise estimates of QTL effects and position. Both single-marker and interval mapping approaches are biased when multiple QTLs are linked to the marker/interval being considered. Methods simultaneously using three or more marker loci attempt to reduce or remove such bias. **Composite interval mapping** (Zeng 1993, 1994; Jansen 1993b, 1994b; Jansen and Stam 1994) considers a marker interval plus a few other well-chosen single markers in each analysis, so that (as above) $n-1$ tests for interval-trait associations are performed on a chromosome with n markers. **Multipoint mapping** considers all of the linked markers on a chromosome simultaneously, resulting in a single analysis for each chromosome (Kearsey and Hyne 1994; Hyne and Kearsey 1995; Wu and Li 1994, 1996).

Conditional Probabilities of QTL Genotypes

The basic element upon which the formal theory of QTL mapping is built is the conditional probability that the QTL genotype is Q_k, given the observed marker genotype is M_j. From the definition of a conditional probability, this is

$$\Pr(Q_k \mid M_j) = \frac{\Pr(Q_k M_j)}{\Pr(M_j)} \tag{15.1}$$

The joint and marginal probabilities, $\Pr(Q_k M_j)$ and $\Pr(M_j)$, are functions of the experimental design and the linkage map (the position of the putative QTLs with respect to the marker loci). Computing these probabilities is a relatively simple matter of bookkeeping (see Example 1), but can get rather tedious as the number of markers and/or QTLs under consideration increases.

When computing joint probabilities involving more than two loci, one must also account for recombinational interference between loci (Chapter 14). Consider a single QTL flanked by two markers, M_1 and M_2. The gamete frequencies depend on three parameters: the recombination frequency c_{12} between markers, the recombination frequency c_1 between marker M_1 and the QTL, and the recombination frequency c_2 between the QTL and marker M_2. Under the assumption of no interference, $c_{12} = c_1 + c_2 - 2c_1c_2$, while $c_{12} = c_1 + c_2$ under complete interference (Chapter 14). When c_{12} is small, gamete frequencies are essentially identical under either interference assumption. Typically, c_{12} is assumed known, leaving two unknown recombination parameters (c_1 and c_2) under general assumptions about interference. In either case, there is only one parameter to estimate, as assuming complete interference $c_2 = c_{12} - c_1$, and assuming no interference $c_2 = (c_{12} - c_1)/(1 - 2c_1)$. Hence, for flanking-marker analysis, we restrict attention to the single recombination parameter c_1, the distance from marker locus M_1 to the QTL. When considering analysis of single-marker loci, for notational ease we drop the subscript, using c in place of c_1.

Conditional probabilities involving more than three linked loci are generally dealt with by first assuming an appropriate mapping function on which distances are additive (Chapter 14), and then translating these distances into recombination frequencies. When a large number of markers is considered, missing marker information can become a problem. Many individuals can be left with incomplete multilocus marker genotypes, excluding them from further analysis. Martínez and Curnow (1992) show how information from linked markers can be used to estimate the genotype at missing or incomplete (i.e., dominant) markers.

Example 1. Consider a single-marker analysis using the F_2 formed by crossing two inbred lines, $MMQQ \times mmqq$. If the recombination frequency between the marker locus and the QTL is c, the expected F_1 gamete frequencies are

$$\Pr(MQ) = \Pr(mq) = (1-c)/2, \qquad \Pr(Mq) = \Pr(mQ) = c/2$$

The probability that an F_2 individual is $MMQQ$ is $\Pr(MQ) \cdot \Pr(MQ) = [\,(1-c)/2\,]^2$. Likewise, $2 \cdot \Pr(MQ) \cdot \Pr(mQ) = 2(\,c/2\,)[\,(1-c)/2\,]$ is the probability of an $MmQQ$ individual, and so on. Since the probabilities of the marker genotypes MM, Mm, and mm are 1/4, 1/2, and 1/4, Equation 15.1 gives the F_2 conditional probabilities as

$$\Pr(QQ\,|\,MM\,) = (1-c)^2, \quad \Pr(Qq\,|\,MM\,) = 2c(1-c), \quad \Pr(qq\,|\,MM\,) = c^2$$

$$\Pr(QQ\,|\,Mm\,) = c(1-c), \; \Pr(Qq\,|\,Mm\,) = (1-c)^2 + c^2, \; \Pr(qq\,|\,Mm\,) = c(1-c)$$

$$\Pr(QQ\,|\,mm\,) = c^2, \quad \Pr(Qq\,|\,mm\,) = 2c(1-c), \quad \Pr(qq\,|\,mm\,) = (1-c)^2$$

This same logic extends to multiple marker loci. Suppose the QTL is flanked by two scored markers, and consider the F_2 in a cross of lines fixed for M_1QM_2 and m_1qm_2. What are the conditional probabilities of the three QTL genotypes when the marker genotype is $M_1M_1M_2M_2$? Since all F_1s are M_1QM_2/m_1qm_2, under the assumptions of no interference, the frequency of F_1 gametes involving M_1M_2 are

$$\Pr(M_1QM_2) = (1-c_1)(1-c_2)/2, \qquad \Pr(M_1qM_2) = c_1\,c_2/2$$

giving expected frequencies in the F_2 of $M_1M_1M_2M_2$ offspring as

$$\begin{aligned}
\Pr(M_1QM_2/M_1QM_2) &= [\,(1-c_1)(1-c_2)/2\,]^2 \\
\Pr(M_1QM_2/M_1qM_2) &= 2\,[\,(1-c_1)(1-c_2)/2\,]\,[\,c_1\,c_2/2\,] \\
\Pr(M_1qM_2/M_1qM_2) &= (\,c_1\,c_2/2\,)^2
\end{aligned}$$

where $c_2 = (c_{12} - c_1)/(1-2c_1)$. The overall frequency of $M_1M_1M_2M_2$ individuals, $\Pr(M_1M_1M_2M_2)$, is the sum of the three above terms, or $(1-c_{12})^2/4$. Substituting into Equation 15.1 gives

$$\begin{aligned}
\Pr(QQ\,|\,M_1M_1M_2M_2) &= \frac{(1-c_1)^2(1-c_2)^2}{(1-c_{12})^2} \\
\Pr(Qq\,|\,M_1M_1M_2M_2) &= \frac{2c_1c_2(1-c_1)(1-c_2)}{(1-c_{12})^2} \\
\Pr(qq\,|\,M_1M_1M_2M_2) &= \frac{c_1^2\,c_2^2}{(1-c_{12})^2}
\end{aligned} \tag{15.2}$$

Conditional probabilities for other marker genotypes are computed in a similar fashion. Since c_1c_2 is usually very small if c_{12} is moderate to small, essentially all $M_1M_1M_2M_2$ individuals are QQ. For example, assuming $c_1 = c_2 = c_{12}/2$ (the worst case), the conditional probabilities of an $M_1M_1M_2M_2$ individual being QQ are 0.96, 0.98, and 0.99 for $c_1 = c_2 = 0.25$, 0.2, and 0.1.

We now move on to the conditional probabilities for other single-marker line cross designs, starting with backcrosses. For a B_1 population, where the F_1 is backcrossed to P_1 (with genotype $MMQQ$), one parental gamete is always MQ. Hence, for a single-marker analysis, there are only two marker genotypes, MM and Mm. Using the frequencies for the four possible gametes (Example 1) of the F_1 parent gives the following conditional probabilities

$$\begin{aligned}
&\Pr(QQ\,|\,MM\,) = 1-c, &\quad &\Pr(Qq\,|\,MM\,) = c \\
&\Pr(QQ\,|\,Mm\,) = c, &\quad &\Pr(Qq\,|\,Mm\,) = 1-c
\end{aligned} \tag{15.3a}$$

Likewise, when backcrossing to the P_2 ($mmqq$), the two possible single-locus marker genotypes are Mm and mm, and the conditional probabilities become

$$\begin{aligned}
&\Pr(qq\,|\,mm\,) = 1-c, &\quad &\Pr(Qq\,|\,mm\,) = c \\
&\Pr(qq\,|\,Mm\,) = c, &\quad &\Pr(Qq\,|\,Mm\,) = 1-c
\end{aligned} \tag{15.3b}$$

For designs involving more than one generation of recombination, the single-generation recombination frequency c is simply replaced by a corrected frequency $\widetilde{c}$ that is a function of the particular design. We consider three such designs: advanced intercross lines (AILs), recombinant inbred lines (RILs), and double-haploid lines (DHLs).

Advanced intercross lines (Darvasi and Soller 1995) are obtained by crossing two inbred lines, but instead of stopping at the F_2, random mating proceeds for t generations, generating an F_t. In this case, unlike the strategy used to create RILs (Chapter 14), inbreeding is avoided by keeping the breeding population size large. As the result of the multiple rounds of recombination, markers in an F_t individual show an expansion of the genetic map relative to an F_2, with the expected frequency of a recombinant gamete in the F_t for a pair of loci at recombination fraction c being

$$\widetilde{c} = \frac{1-(1-c)^{t-2}(1-2c)}{2} \simeq \frac{t}{2}\,c \tag{15.4}$$

where the approximation holds for $ct << 1$ (Darvasi and Soller 1995, Liu et al. 1996). For example, if the marker-QTL recombination frequency is $c = 0.01$, only 1% of the F_2 gametes are recombinant (*Mq*, *mQ*), but this increases to 2.5% in an F_5 and 9.1% in an F_{20}. The conditional genotype probabilities for an F_t AIL are given by the F_2-design expressions in Example 1, with $\widetilde{c}$ substituted for c.

Recombinant inbred lines (RILs) also involve several generations of recombination, but here genotypes are fixed by inbreeding. Starting with a *MQ/mq* F_1 parent, there are only four possible genotypes in the resulting RILs—*MMQQ*, *MMqq*, *mmQQ*, and *mmqq*. The frequency of recombinant gametes (*Mq*, *mQ*) in RILs approaches a limiting value of $\widetilde{c} = 2c/(1+2c)$ for selfed lines and $\widetilde{c} = 4c/(1+6c)$ for lines formed by brother-sister mating (Haldane and Waddington 1931). Thus, the expected frequencies of genotypes in RILs are

Line genotype	Frequency
$MMQQ$, $mmqq$	$(1-\widetilde{c})/2$
$MMqq$, $mmQQ$	$\widetilde{c}/2$

While doubled-haploid lines (DHLs) also have only these four genotypes, they are formed by a single generation of meiosis, so that $\widetilde{c} = c$. Hence, among either RILs or DHLs, the conditional QTL probabilities are

$$\begin{aligned} \Pr(QQ\,|\,MM) &= 1-\widetilde{c}, \qquad & \Pr(qq\,|\,MM) &= \widetilde{c} \\ \Pr(QQ\,|\,mm) &= \widetilde{c}, \qquad & \Pr(qq\,|\,mm) &= 1-\widetilde{c} \end{aligned} \tag{15.5a}$$

where

$$\widetilde{c} = \begin{cases} c & \text{for DHLs} \\ 2c/(1+2c) & \text{for RILs formed by selfing} \\ 4c/(1+6c) & \text{for RILs formed by brother-sister mating} \end{cases} \tag{15.5b}$$

Expected Marker-class Means

With these conditional probabilities in hand, the expected trait values for the various marker genotypes follow immediately. Suppose there are N QTL genotypes, $Q_1, \cdots, Q_N$, where the mean of the kth QTL genotype is μ_{Q_k}. The mean value for marker genotype M_j is just

$$\mu_{M_j} = \sum_{k=1}^{N} \mu_{Q_k} \Pr(\, Q_k \,|\, M_j \,) \tag{15.6}$$

The QTL effects enter through the μ_{Q_k}, while the QTL positions enter through the conditional probabilities $\Pr(\, Q_k \,|\, M_j \,)$. Equation 15.6 is completely general, allowing for multilocus marker genotypes and multiple QTLs.

Example 2. Consider the single-marker F_2 design with a single QTL linked (at recombination frequency c) to the marker. Denote the QTL genotypic values by

$$\mu_{QQ} = \mu + 2a, \quad \mu_{Qq} = \mu + a(1+k), \quad \text{and} \quad \mu_{qq} = \mu$$

where a measures the additive value and k the degree of dominance. Applying the conditional probabilities developed in Example 1 to Equation 15.6, the mean values for the marker genotypes are

$$\begin{aligned}\mu_{MM} &= \mu + 2a(1-c)^2 + 2c(1-c)(1+k)a \\ \mu_{Mm} &= \mu + 2ac(1-c) + [1-2c(1-c)](1+k)a \\ \mu_{mm} &= \mu + 2ac^2 + 2c(1-c)(1+k)a\end{aligned}$$

If the marker and QTL are unlinked ($c = 1/2$), all markers have the same mean, $\mu + a[1 + (k/2)]$. Rearranging these equations gives

$$(\mu_{MM} - \mu_{mm})/2 = a(1-2c) = a^* \tag{15.7a}$$

$$\frac{\mu_{Mm} - (\mu_{MM} + \mu_{mm})/2}{(\mu_{MM} - \mu_{mm})/2} = k(1-2c) = k^* \tag{15.7b}$$

Hence, one strategy for detecting QTLs is to test for significant differences between the mean trait values associated with different marker genotypes. This is the basis for QTL detection via regression or ANOVA, which we generically refer to as **linear model** approaches.

This example shows that while contrasts of single-marker means can be used to estimate both a^* and k^*, these underestimate the magnitude of a and k by the (unknown) fraction $1 - 2c$. If the marker and QTL are tightly linked, this error is small, but it increases rather dramatically as c approaches 1/2. A small difference between marker-homozygote means is thus compatible with either a tightly linked QTL of small effect or a loosely linked QTL of large effect. As we will show shortly, when multilocus marker genotypes are considered, the use of appropriate combinations of marker means allows for separate estimates of QTL effect and position.

If there are N QTLs linked to the marker, the ith of which is at recombination frequency c_i from the marker and has associated additive and dominance effects a_i and k_i, then (from Edwards et al. 1987),

$$(\mu_{MM} - \mu_{mm})/2 = \sum_{i=1}^{N} a_i^* \tag{15.8a}$$

$$\frac{\mu_{Mm} - (\mu_{MM} + \mu_{mm})/2}{(\mu_{MM} - \mu_{mm})/2} = \sum_{i=1}^{N} a_i^* k_i^* \Big/ \sum_{i=1}^{N} a_i^* \tag{15.8b}$$

where $a_i^* = a_i(1 - 2c_i)$ and $k_i^* = k_i(1 - 2c_i)$. If some of the linked QTLs have effects of opposite sign, some cancellation occurs, reducing the marker-trait association. Moreover, with multiple linked QTLs, the degrees of dominance (k_i) are confounded with the homozygous effects (a_i).

Marker-class means for other designs follow by applying the appropriate conditional probabilities to Equation 15.6. For example, for the B_1 design, from Equation 15.3a,

$$\mu_{MM} - \mu_{Mm} = (\mu_{QQ} - \mu_{Qq})(1 - 2c) = a(1 - k)(1 - 2c) \tag{15.9a}$$

Thus, under a backcross design the scaled QTL effects are influenced strongly by the (unknown) degree of dominance k. If Q is completely dominant to q, $k = 1$, and there is no marker-QTL effect. Conversely, if q is dominant to Q, $k = -1$ and the scaled effect becomes $2a(1 - 2c)$, which is the same as under an F_2 design. Recalling Equation 15.3b, the reciprocal backcross ($B_2 = F_1 \times P_2$) yields a similar expression,

$$\mu_{Mm} - \mu_{mm} = (\mu_{Qq} - \mu_{qq})(1 - 2c) = a(1 + k)(1 - 2c) \tag{15.9b}$$

Note that the ratio of Equation 15.9a to 15.9b gives $(1-k)/(1+k)$, so that (provided only a single QTL is linked to the marker) an estimate of k can be obtained if one has access to *both* backcross populations.

The expressions developed in Example 2 for F_2 analysis hold for an F_t population, provided $\widetilde{c}$ (given by Equation 15.4) replaces c. For example, $\mu_{MM} - \mu_{mm} = 2a(1 - 2\,\widetilde{c})$, and so forth. Since $(1 - 2c) > (1 - 2\widetilde{c})$, AILs have smaller differences between marker means, and hence reduced power of QTL detection, relative to the F_2 design. Despite this, Darvasi and Soller (1995) advocate the use of AILs for fine-mapping of QTLs, as the expansion of the genetic map offers a higher precision of estimates of QTL position. We expand on this point below.

For RILs and DHLs, the recombination parameter $\widetilde{c}$ is given by Equation 15.5b, and from Equations 15.5a and 15.6 it follows that

$$\mu_{MM} = \mu_{QQ}\,(1 - \widetilde{c}) + \mu_{qq}\,\widetilde{c} \qquad \text{and} \qquad \mu_{mm} = \mu_{QQ}\,\widetilde{c} + \mu_{qq}\,(1 - \widetilde{c}) \tag{15.10a}$$

giving

$$\frac{\mu_{MM} - \mu_{mm}}{2} = a\,(1 - 2\,\widetilde{c}) = a^* \tag{15.10b}$$

again providing an estimate of a composite parameter of the QTL effect (a) and position (c). Because $\widetilde{c}$ is smallest in DHLs (see Equation 15.5b), the largest marker effect (and greatest power for QTL detection) occurs in this type of line, followed by selfed RILs, and finally by sib-mated RILs.

Finally, note that by considering two-locus (rather than single-locus) marker means, separate estimates of QTL effect and position can be obtained. Taking the genotype at two adjacent marker loci (M_1/m_1 and M_2/m_2) as the unit of analysis, consider the difference between the contrasting double homozygotes in an F_2. If the markers flank a QTL, then under the assumption of no interference, Equation 15.2 (and its analog for $m_1m_1m_2m_2$ probabilities) implies

$$\begin{aligned}\frac{\mu_{M_1M_1M_2M_2} - \mu_{m_1m_1m_2m_2}}{2} &= a\left(\frac{1 - c_1 - c_2}{1 - c_1 - c_2 + 2c_1\,c_2}\right)\\ &\simeq a\,(1 - 2c_1\,c_2)\end{aligned} \tag{15.11a}$$

where c_1 is the M_1-QTL recombination frequency. Equation 15.11a is essentially equal to a when the distance between flanking markers $c_{12} \le 0.20$, as here $(1 - 2c_1\,c_2) \ge 0.98$. Thus, recalling from Equation 15.7a that $\mu_{M_1M_1} - \mu_{m_1m_1} = 2a(1 - 2c_1)$, we can obtain estimates of the recombination frequencies by substituting Equation 15.11a for a and rearranging to give

$$\begin{aligned}c_1 &= \frac{1}{2}\left(1 - \frac{\mu_{M_1M_1} - \mu_{m_1m_1}}{2a}\right)\\ &\simeq \frac{1}{2}\left(1 - \frac{\mu_{M_1M_1} - \mu_{m_1m_1}}{\mu_{M_1M_1M_2M_2} - \mu_{m_1m_1m_2m_2}}\right)\end{aligned} \tag{15.11b}$$

Estimates for other flanking-marker designs are given by Knapp et al. (1990), Knapp and Bridges (1990), and Knapp (1991).

Marker Variances and Higher-order Moments

The same linkage disequilibrium that generates differences in the mean trait values of different marker genotypes can also create differences in the variance and higher moments (e.g., skewness and kurtosis). Such differences are not uncommon. In a cross of tomato species, for example, Weller et al. (1988) found significantly different variances for 28% (40 of 180) of the possible marker-trait associations, while 17% showed significant differences in skewness. In some instances these moments may be of more interest than the mean (Weller and Wyler 1992). For example, a reduction in the variance of flowering time shortens the harvesting window, and by reducing costs this may be more significant than changing mean harvesting time per se.

Table 15.1 The expected correlation between marker genotype (coded as $x = 1$ for *MM*, $x = -1$ for *mm*) and phenotypic value z can be used to estimate c.

Genotype	Freq.	x	z
MMQQ	$(1-\widetilde{c})/2$	1	z_{QQ}
MMqq	$\widetilde{c}/2$	1	z_{qq}
mmQQ	$\widetilde{c}/2$	−1	z_{QQ}
mmqq	$(1-\widetilde{c})/2$	−1	z_{qq}

$$\sigma(x,z) = E(x \cdot z) = a(1-2\widetilde{c})$$

$$\sigma^2(x) = E(x^2) = 1$$

$$\sigma^2(z) = E(z^2) = (1/2)[E(z_{QQ})^2 + \sigma_e^2] + (1/2)[E(z_{qq})^2 + \sigma_e^2] = a^2 + \sigma_e^2$$

Note: We assume that the difference in QTL means is $E(z_{QQ}) - E(z_{qq}) = 2a$ and that the phenotypic distributions conditioned on the QTL genotypes have common (within-line) variance σ_e^2. Since $\sigma(x,z)$ and $\sigma^2(z)$ are unchanged by a change in the mean of z (Chapter 3), we can arbitrarily set $E(z_{QQ}) = -E(z_{qq}) = a$. Coded this way, $E(x) = E(z) = 0$, simplifying calculation of $\sigma(x,z)$ and $\sigma^2(z)$.

Several workers have suggested the use of these higher-order moments for detection of a linked QTL and estimation of its effects (Zhuchenko et al. 1978, 1979; Korol et al. 1981, 1983, 1987; Ginzburg 1983; Asins and Carbonell 1988; Zhang et al. 1992). One difficulty with this approach is that variances (and higher moments) are estimated with far less precision than means, reducing both the power of detection and the accuracy of estimates. Another complication is that not all designs are capable of revealing significant changes in higher moments (e.g., Asins and Carbonell 1988).

RILs and DHLs provide one case where functions of higher-order moments (here, a correlation coefficient) may be of value. Here, single-locus marker information can be used to estimate the recombination frequency c (Hu et al. 1995). As shown in Table 15.1, coding the alternative marker homozygotes as $x = \pm 1$, the expected marker-trait correlation becomes

$$\rho = \frac{\sigma(z,x)}{\sigma(x)\,\sigma(z)} = \frac{a(1-2\widetilde{c})}{\sqrt{a^2+\sigma_e^2}} = \frac{1-2\widetilde{c}}{\sqrt{1+C^2}} \tag{15.12a}$$

where $C = \sigma_e/a$, with a being the QTL effect and σ_e^2 being the within-line variance (see Table 15.1). (The C term was neglected by Hu et al. 1995.) Rearranging and letting r be an estimate of ρ suggests the estimator

$$\widetilde{c} = \frac{1 - r\sqrt{1+C^2}}{2} \leq \frac{1-r}{2} \tag{15.12b}$$

While the value of C is unknown, by ignoring it one can obtain an upwardly biased estimate of c by first taking $\widetilde{c} = (1-r)/2$ and then using Equation 15.5b to translate this value of $\widetilde{c}$ into c. Rearranging Equation 15.10b,

$$a = \frac{\mu_{MM} - \mu_{mm}}{2(1 - 2\,\widetilde{c}\,)} \tag{15.12c}$$

which, upon using $0 \leq \widetilde{c} \leq (1-r)/2$, gives

$$\frac{\mu_{MM} - \mu_{mm}}{2} \leq a \leq \frac{\mu_{MM} - \mu_{mm}}{2r} \tag{15.12d}$$

Hence, the use of both the observed correlation r and the difference in marker means $(\overline{z}_{MM} - \overline{z}_{mm})$ allows the estimation of upper bounds for both c and a.

Overall Significance Level with Multiple Tests

The final statistical issue we need to introduce before exploring specific designs in detail is the problem of the proper significance level for an entire mapping experiment. In each mapping experiment, a large number of tests for marker-trait associations are typically performed. Thus, even if the significance level α (the probability of a false positive) for each test is set at a very small value, there is usually a high probability that the entire collection of tests (i.e., the entire mapping experiment) will show at least one false positive. If n independent tests with significance level α are conducted, the probability γ that at least one test shows a false positive is

$$\gamma = 1 - (1 - \alpha)^n \tag{15.13a}$$

Setting $\alpha = 0.01$ for each individual test, the probability of at least one false positive in 25 tests is 0.22, which increases to 0.633 for 100 tests, and is essentially one for 500 tests. The latter number of tests is not uncommon in QTL mapping studies. Hence, unless we use a very stringent significance value for each test, we run a very high risk of detecting false associations.

Suppose we wish to achieve an overall significance level γ for the *entire* experiment. With n *independent* tests, the standard **Bonferroni correction** for multiple comparisons, derived by rearranging Equation 15.13a, states that an overall significance level γ requires that each individual test be based on a significance level of

$$\alpha = 1 - (1 - \gamma)^{1/n} \simeq \frac{\gamma}{n} \tag{15.13b}$$

However, while this correction is appropriate for tests using unlinked markers (such as those on different chromosomes), tests involving linked markers are generally not independent.

A more robust approach for obtaining overall significance levels utilizes resampling procedures such as **permutation tests**, wherein the original analysis is replicated many times on data sets generated by appropriate reshuffling of the original data (Churchill and Doerge 1994, Doerge and Churchill 1996). Here, one randomly shuffles the observed trait values over individuals (marker genotypes), generating a sample with the original marker information but with trait values randomly assigned over genotypes. The test statistic is then computed on this new sample, and this procedure is repeated many times, generating an empirical distribution of the test under the hypothesis of no marker-trait associations. Churchill and Doerge suggest that 1,000 resamplings is sufficient for a significance level of 5%, but that 10,000 or more resamplings may be required to generate a stable critical value for the 1% level. By keeping the marker information for each individual together, this approach nicely accounts for missing markers, differences in marker densities, and any nonrandom segregation of marker alleles. (The latter is not uncommon in wide line crosses.)

QTL DETECTION AND ESTIMATION USING LINEAR MODELS

We now have all of the necessary machinery in place to consider particular estimation methods in greater detail. As noted above, the simplest test for a marker-trait association involves the comparison of the trait means of alternate marker genotypes. This is the basis for linear-model approaches for detecting QTLs. When only two genotypes are compared (such as with single-marker backcross-, RIL-, or DHL-designs), this can be accomplished with a simple *t* test (e.g., Sokal and Rohlf 1995). Most designs, however, involve more than two marker genotypes. For example, the single-marker F_2 design has three marker genotypes: *MM, Mm, mm*. In such cases, all marker genotypic means (or some subset of them) can be compared by using standard linear-model approaches, such as ANOVA or regression.

The simplest linear model considers the phenotypic value z_{ik} of the kth individual of marker genotype i as a mean value μ plus a marker effect b_i and a residual error e_{ik},

$$z_{ik} = \mu + b_i + e_{ik} \tag{15.14a}$$

This is a one-way ANOVA model (Chapter 18), with the presence of a linked QTL being indicated by a significant between-marker variance. Equivalently, we can express this model as a multiple regression, with the phenotypic value for individual j given by

$$z_j = \mu + \sum_{i=1}^{n} b_i\, x_{ij} + e_j \tag{15.14b}$$

where the x_{ij} are n indicator variables (one for each marker genotype),

$$x_{ij} = \begin{cases} 1 & \text{if individual } j \text{ has marker genotype } i, \\ 0 & \text{otherwise.} \end{cases}$$

The number of marker genotypes (n) in Equations 15.14a,b depend on both the number of marker loci and the type of design being used. With a single marker, $n = 2$ for a backcross, RIL, or DH design, while $n = 3$ for an F_2 design (using codominant markers). When two or more marker loci are simultaneously considered, b_i corresponds to the effect of a *multilocus* marker genotype, and n is the number of such genotypes considered in the analysis. In the regression framework, evidence of a linked QTL is provided by a significant r^2, which is the fraction of character variance accounted for by the marker genotypes (Chapter 8). Finally, as mentioned above, the presence of a linked QTL can cause different marker genotypes to have different trait variances. If the difference in variance between marker classes is substantial, the standard ANOVA assumption of variance homogeneity is violated and appropriate corrections are required for hypothesis testing (Asins and Carbonell 1988, Xu 1995).

Estimation of dominance requires information on all three genotypes at a marker locus, i.e., an F_2, F_t, or other design (such as *both* backcross populations). In these cases, dominance can be estimated using an appropriate function of the marker means (e.g., Equations 15.7b, 15.8b). Epistasis between QTLs can be modeled by including interaction terms. Here, an individual with genotype i at one marker locus and genotype k at a second is modeled as $z = \mu + a_i + b_k + d_{ik} + e$, where a and b denote the single-locus marker effects, and d is the interaction term due to epistasis between QTLs linked to those marker loci. In linear regression form this model becomes

$$z_j = \mu + \sum_i^{n_1} a_i \, x_{ij} + \sum_k^{n_2} b_k \, y_{kj} + \sum_i^{n_1} \sum_k^{n_2} d_{ik} \, x_{ij} \, y_{kj} + e_j \tag{15.14c}$$

where x_{ij} and y_{kj} are indicator variables for two different marker genotypes (with n_1 and n_2 genotypes, respectively). Significant a_i and/or b_k terms indicate significant effects at the individual marker loci, while significant d_{ik} terms indicate epistasis between the effects of the two markers.

Essentially the same approach can be used to look for genotype $\times$ environment interactions when markers are examined in several environments. Here the basic model for an individual with marker genotype i measured in the kth environment is $z = \mu + b_i + E_k + I_{ik} + e$, where a significant E_k indicates an environmental effect, while a significant I_{ik} implies a marker $\times$ environment interaction. For example, if the character has significant sex-specific effects, these can be incorporated by using the model $z = \mu + b_i + s_k + I_{ik} + e$ for an individual of marker genotype i and sex k. Here, a significant s_k implies a significant sex effect, while a significant I_{ik} implies a significant marker $\times$ sex interaction. Long et al. (1995) give an example of the utility of this approach, finding very significant sex-specific effects for bristle number in *Drosophila*.

Example 3. Edwards et al. (1987) examined two F_2 maize populations. Cross 1 consisted of 1776 individuals scored for 16 markers, while Cross 2 used a different set of parental lines and consisted of 1930 individuals scored for 20 markers. As the frequency distribution (below) shows, the detected marker effects (measured by the fraction r^2 of total F_2 phenotypic variance accounted for by each significant marker-trait association) were generally quite small.

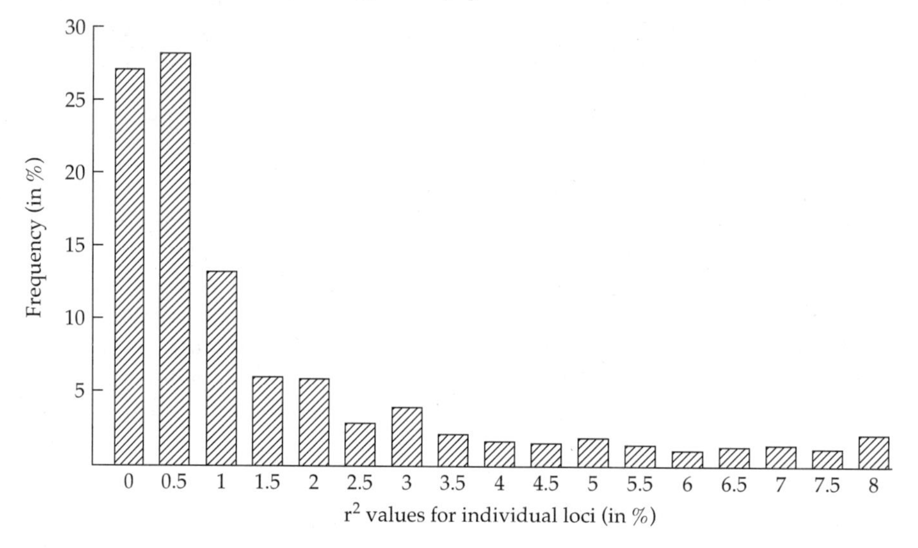

A total of 82 vegetative characters were examined, with 60% (Cross 1) and 64% (Cross 2) of all possible marker-trait combinations showing significant effects (at the $\alpha = 0.05$ level) using single-marker ANOVA. On average, each trait showed 10 (Cross 1) and 14 (Cross 2) significant marker associations. Dominance was common, while pairwise epistasis, as tested by incorporating a marker $\times$ marker interaction term into the linear model (Equation 15.14c), was rare.

The same two F_2 populations were used by Stuber et al. (1987) to examine 25 yield-related characters, with similar results. In that study, most marker-trait combinations were significant (66% and 72% at the $\alpha = 0.05$ level), and most marker effects were small (over half of the significant associations having r^2 values less than two percent). As a group, yield-related traits displayed more dominance than vegetative traits, but many yield traits were still largely additive.

A more recent study by Edwards et al. (1992) examined a subset of the vegetative characters in Cross 2, using a much larger number of markers (114 RFLPs). While only 187 F_2 individuals were scored, 15% of marker-trait associations were significant, and the overall results with respect to the distribution of effects were similar to those for the 1987 experiments.

QTL DETECTION AND ESTIMATION VIA MAXIMUM LIKELIHOOD

Maximum likelihood (ML) methods are especially popular in the QTL mapping literature. While linear models use only marker means, ML uses the full information from the marker-trait distribution and, as such, is expected to be more powerful. The tradeoff is that ML is computationally intensive, requiring rather special programs to solve the likelihood equations, while linear model analysis can be performed with almost any standard statistical package. Further, while modifying the basic model (such as adding extra factors) is rather trivial in the linear model framework, with ML new likelihood functions need to be constructed and solved for each variant of the original model. Although writing down a set of likelihood equations for the model of interest is relatively straightforward (e.g., Example 4), obtaining the ML estimates is much more difficult. One approach outlined in Appendix 4 is to use specialized algorithms, of which EM (expectation-maximization) methods have been successfully adapted to many of the mixture-model problems in QTL mapping (e.g., Lander and Botstein 1989; Carbonell and Gerig 1991; Luo and Kearsey 1992; van Ooijen 1992; Carbonell et al. 1992; Luo and Wolliams 1993; Jansen 1992, 1993a, 1994a, 1996; Jansen and Stam 1994). Alternatively, as we discuss later, a creative use of regressions can often provide excellent approximations to ML solutions. For the remainder of this chapter, we assume that the reader has recently read Chapter 13 and Appendix 4, which introduces much of the ML machinery used here.

Assuming that the distribution of phenotypes for an individual with QTL genotype Q_k is normal with mean μ_{Q_k} and variance σ^2, and following the logic of Chapter 13, the likelihood for an individual with phenotypic value z and marker genotype M_j becomes

$$\ell(z \mid M_j) = \sum_{k=1}^{N} \varphi(z, \mu_{Q_k}, \sigma^2) \Pr(Q_k \mid M_j) \tag{15.15}$$

where $\varphi(z, \mu_{Q_k}, \sigma^2)$ denotes the density function for a normal distribution with mean μ_{Q_k} and variance σ^2, and a total of N QTL genotypes is assumed. This likelihood is a mixture-model distribution (Chapter 13). The mixing proportions, $\Pr(Q_k \mid M_j)$, are functions of the genetic map (the position(s) of the QTL(s) with respect to the observed markers) and the experimental design, while the QTL effects enter only though the means μ_{Q_k} and variance σ^2 of the underlying distributions.

Example 4. Consider the single-marker F_2 design with a single QTL linked to the marker. Making the standard assumption that phenotypes are normally distributed about each QTL genotype, substitution of the F_2 conditional probabilities (Example 1) into Equation 15.15 gives the likelihood functions for the three different marker genotypes as

$$\ell(z \mid MM) = (1-c)^2\varphi(z,\mu_{QQ},\sigma^2) + 2c(1-c)\,\varphi(z,\mu_{Qq},\sigma^2) + c^2\varphi(z,\mu_{qq},\sigma^2)$$
$$\ell(z \mid Mm) = c(1-c)\varphi(z,\mu_{QQ},\sigma^2) + \left[\,(1-c)^2 + c^2\,\right]\varphi(z,\mu_{Qq},\sigma^2) + c(1-c)\varphi(z,\mu_{qq},\sigma^2)$$
$$\ell(z \mid mm) = c^2\varphi(z,\mu_{QQ},\sigma^2) + 2c(1-c)\,\varphi(z,\mu_{Qq},\sigma^2) + (1-c)^2\varphi(z,\mu_{qq},\sigma^2)$$

as obtained by Weller (1986). The total likelihood for n F_2 individuals is the product of the individual likelihoods,

$$\ell(\mathbf{z}) = \prod_{i=1}^{n} \ell(z_i \mid M_i)$$

While rather complex, the total likelihood is a function of just five parameters: the QTL position (c), the three QTL means ($\mu_{QQ}, \mu_{Qq}, \mu_{qq}$), and the common variance (σ^2).

As in the case of segregation analysis (Chapter 13), the likelihood equations can be modified to account for dichotomous (binary) and polychotomous (ordinal) characters through the use of logistic regressions and probit scales (Ghosh et al. 1993, Hackett and Weller 1995, Visscher et al. 1996a, Xu and Atchley 1996). Alternatively, one can simply ignore the discrete structure of the data, treating them as if they were continuous (e.g., coding alternative binary characters as 0/1) and applying ML. When flanking markers are used, this approach gives essentially the same power and precision as methods specifically designed for polychotomous traits (Hackett and Weller 1995, Visscher et al. 1996a), but when single markers are used, this approach can give estimates for QTL position that are rather seriously biased (Hackett and Weller 1995). An alternative approach for treating nonnormally distributed characters is given by Kruglyak and Lander (1995c), who develop a nonparametric interval mapping procedure.

Likelihood Maps

In the likelihood framework, tests of whether a QTL is linked to the marker(s) under consideration are based on the likelihood-ratio statistic,

$$\text{LR} = -2\ln\left[\frac{\max\ \ell_r(\mathbf{z})}{\max\ \ell(\mathbf{z})}\right]$$

where $\max\ \ell_r(\mathbf{z})$, given by Equation 13.8, is the maximum of the likelihood function under the null hypothesis of no segregating QTL (i.e., under the assumption

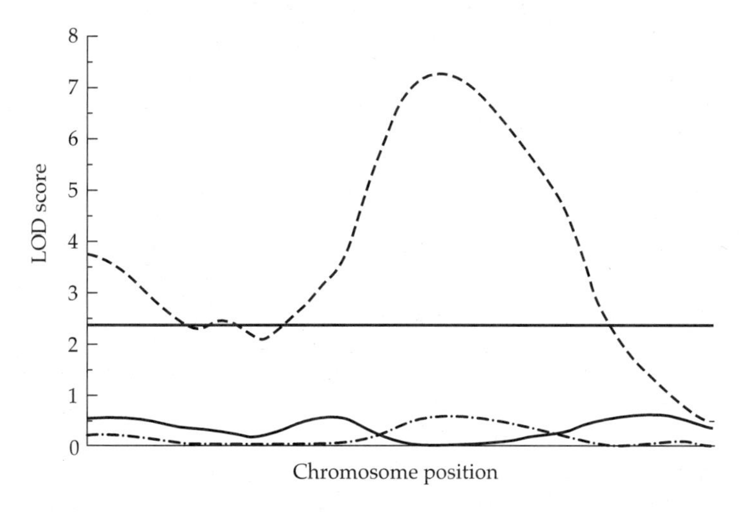

Figure 15.1 Likelihood map for QTL positions on chromosome 10 in a cross of two tomato species. Evidence for a QTL is provided when the likelihood function exceeds the significance threshold (indicated by the horizontal line). The upper dashed curve gives the LOD score for fruit pH as a function of map position, showing strong evidence of a QTL near the middle of the chromosome. The lower two curves (solid and broken) are for fruit weight and soluble-solid concentration, neither of which shows a significant QTL effect on this chromosome. (After Paterson et al. 1988.)

that the phenotypic distribution is a single normal). This test statistic is approximately χ^2-distributed, with the degrees of freedom given by the extra number of fitted parameters in the full model. For a model assuming a single QTL, most designs have five parameters in the full model (the three QTL means, the variance, and the QTL position), and two in the reduced model (the mean and variance), giving three degrees of freedom. Certain designs (such as a backcross, RIL, or DHL) involve situations where only two QTL means enter (e.g., *QQ* and *qq* for RILs/DHLs, *Qq* and *QQ* or *qq* for a backcross), and here the likelihood ratio has two degrees of freedom.

The amount of support for a QTL at a particular map position is often displayed graphically through the use of **likelihood maps** (Figures 15.1, 15.2), which plot the likelihood-ratio statistic (or a closely related quantity) as a function of map position of the putative QTL. For example, the value of the likelihood map at $c = 0.05$ gives the likelihood-ratio statistic that a QTL is at recombination fraction 0.05 from the marker vs. a model assuming no QTL. This approach for displaying the support for a QTL was introduced by Lander and Botstein (1989), who plotted the LOD (**likelihood of odds**) scores (Morton 1955b). The LOD score for a particular value of c is related to the likelihood-ratio test statistic (LR) by

$$\text{LOD}(c) = \log_{10}\left[\frac{\max\,\ell_r(\mathbf{z})}{\max\,\ell(\mathbf{z},c)}\right] = \frac{\text{LR}(c)}{2\,\ln 10} \simeq \frac{\text{LR}(c)}{4.61} \tag{15.16}$$

showing that the LOD score is simply a constant times the likelihood-ratio statistic. Here $\max\,\ell(\mathbf{z},c)$ denotes the maximum of the likelihood function given a QTL at recombination frequency c from the marker. Another variant is simply to plot $\max\,\ell(\mathbf{z},c)$ instead of the likelihood-ratio statistic, as the restricted likelihood, $\max\,\ell_r(\mathbf{z})$, is the same for each value of c.

The likelihood map projects the multidimensional likelihood surface (which is a function of the QTL means, variance, and map position) on to a single dimension, that of the map position, c. The ML estimate of c is that which yields the maximum value on the likelihood map, and the values for the QTL means and variance that maximize the likelihood given this value of c are the ML estimates for the QTL effects. Thus, in the likelihood framework, *detection* of a linked QTL and *estimation* of its position are coupled — if the likelihood ratio exceeds the critical threshold for that chromosome, it provides evidence for a linked QTL, whose position is estimated by the peak of the likelihood map. If the peak does not exceed this threshold, there is no evidence for a linked QTL.

Precision of ML Estimates of QTL Position

Since ML estimates are approximately normally distributed for large sample sizes, confidence intervals for QTL effects and position can be constructed using the sampling variances for the ML estimates (Appendix 4). Approximate confidence intervals are often constructed using the **one-LOD rule** (Figure 15.2), with the confidence interval being defined by all those values falling within one LOD score of the maximum value (Conneally et al. 1985, Lander and Botstein 1989). The motivation for such **one-LOD support intervals** follows from the fact that the large-sample distribution of the LR statistic follows a χ^2 distribution. If only one parameter in the likelihood function is allowed to vary, as when testing whether c equals a particular value (say the observed ML estimate), the LR statistic has one degree of freedom. Because a one-LOD change corresponds to an LR change of 4.61 (Equation 15.16), which for a χ^2 with one degree of freedom corresponds to a significance value of 0.04 (e.g., $\Pr(\chi_1^2 \geq 4.61) = 0.04$), it follows that one-LOD support intervals approximate 95% confidence intervals under the appropriate settings. However, the one-LOD rule often gives confidence intervals that are too short. Mangin et al. (1994a,b) show this to be the case for QTLs of small effect (one-LOD confidence intervals having between 60% and 95% probability of actually containing the QTL), and they develop an improved method for such cases. Simulation studies led van Ooijen (1992) to suggest that support intervals should be based on two-LOD differences in order to have a high probability of containing the QTL. A more rigorous approach to obtaining standard errors of both map

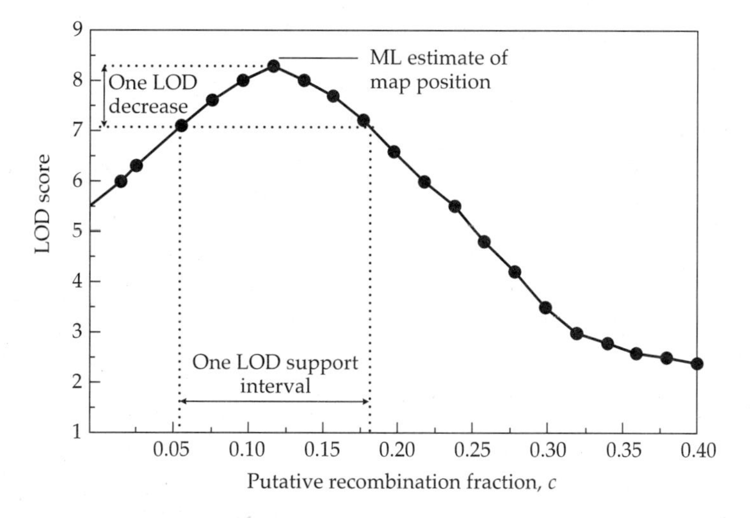

Figure 15.2 Hypothetical likelihood map for the marker-QTL recombination frequency c in a single-marker analysis. Points connected by straight lines are used to remind the reader that likelihood maps are computed by plotting the maximum of the likelihood function for each c value, usually done by considering steps of 0.01 to 0.05. A QTL is indicated if any part of the likelihood map exceeds a critical value. In such cases, the ML estimate for map position is the value of c giving the highest likelihood. Approximate confidence intervals for QTL position (one-LOD support intervals) are often constructed by including the set of all c values giving likelihoods within one LOD score of the maximum value.

position and QTL effects is to use the inverse of the Fisher information matrix associated with the likelihood function (Appendix 4), although this inverse is often considerably more difficult to compute than LOD support intervals.

Resampling methods provide a very robust procedure for constructing confidence intervals for QTL position, and Visscher et al. (1996b) suggest using a **bootstrap** approach (Efron 1979, 1982). Suppose the original data set consists of n individuals. A bootstrap sample is generated by drawing n values, *with replacement*, from the original data set. Such a sample will have some of the original values present multiple times and others not present at all. A series of N such samples are generated and an estimate (map position in this case) is computed for each, generating a distribution of estimates (the **empirical bootstrap distribution**). The resulting 95% bootstrap confidence interval has as its lower value the estimate corresponding to the 2.5% cumulative frequency point of the empirical bootstrap distribution, while the upper value is that corresponding to the upper 97.5% of the bootstrap distribution. Simulation studies by Visscher et al. show that this approach usually yields confidence intervals very close to the correct length when at least 200 bootstrap samples are used.

The length of the confidence interval is influenced by the number of individuals sampled, the effect of the QTL in question, and the marker density. Darvasi et al. (1993) show that precision is not significantly increased by increasing marker density beyond a certain point (around one marker every 5 to 10 cM). Given such a dense map, van Ooijen (1992) found that ML mapping using flanking markers with reasonable sample sizes (200–300 F_2 or backcross individuals) allows a QTL accounting for 5% of the total variance to be mapped to a 40 cM interval, while one accounting for 10% can be mapped to a 20 cM interval. Unfortunately, these interval sizes are distressingly large for cloning QTLs or even defining their positions to smaller intervals for RIL construction.

One strategy for increasing the precision of mapping is to use lines with expanded genetic maps, such as RILs or AILs. With these designs, estimates of the map position are in terms of the cumulative recombination frequency $\widetilde{c} = tc$, so that the confidence interval for c is reduced by a factor of $1/t$. For example, recombinant inbred lines have a two- to four-fold expansion of the map (Equation 15.5b), and hence reduce the length of the confidence interval for c by $1/2$ to $1/4$ relative to an F_2. Even more dramatic reductions are possible using advanced intercross lines. A sample size and marker density that yield a 20 cM confidence interval in an F_2 design give a 3.4 cM confidence interval for the same QTL in an F_{10} design. (This follows from Equation 15.4, which shows that a Haldane distance of 20 cM, corresponding to $c = 0.165$, translates into $\widetilde{c} = c/5 = 0.033$ and a Haldane distance of 3.4 cM with an F_{10} AIL.) Likewise, an F_{20} design would give a 1.7 cM confidence interval.

ML Interval Mapping

ML mapping with line crosses usually employs the genotypes of a pair of flanking markers as the unit of analysis. The likelihood functions for such **ML interval mapping** follow from Equation 15.15 using the appropriate conditional probabilities for QTL genotypes given the two-locus marker genotypes (Jensen 1989, Lander and Botstein 1989, Knapp et al. 1990, Carbonell et al. 1992, van Ooijen 1992, Korol et al. 1996). Example 5 shows the basic structure of the resulting likelihood functions. As with single-marker analysis, support for a QTL is evaluated with a likelihood map for the interval, with the peak of the likelihood map corresponding to the ML estimate of QTL position within that interval and its significance given by a likelihood-ratio test.

Example 5. Likelihood functions for interval mapping follow by substituting the appropriate conditional probabilities into Equation 15.15. For example, consider the F_2 formed by crossing two inbred lines. Assuming no interference, from

Equation 15.2 the likelihood for marker genotype $M_1M_1M_2M_2$ is

$$\ell(z \mid M_1M_1M_2M_2) = \left[\frac{(1-c_1)^2\,(1-c_2)^2}{(1-c_{12})^2}\right] \cdot \varphi(z, \mu_{QQ}, \sigma^2)$$

$$+ \left[\frac{2\,c_1\,c_2\,(1-c_1)\,(1-c_2)}{(1-c_{12})^2}\right] \cdot \varphi(z, \mu_{Qq}, \sigma^2)$$

$$+ \left[\frac{c_1^2\,c_2^2}{(1-c_{12})^2}\right] \cdot \varphi(z, \mu_{qq}, \sigma^2)$$

Likelihoods for the other eight flanking-marker genotypes follow similarly and can be found in Luo and Kearsey (1992), Carbonell et al. (1992), and van Ooijen (1992). Even though these likelihoods involve three recombination parameters (c_{12}, c_1, c_2), the distance between markers (c_{12}) is usually taken as known, and hence $c_2 = (c_{12} - c_1)/(1 - 2\,c_1)$ (assuming no interference) or $c_2 = c_{12} - c_1$ (complete interference). This leaves five parameters to estimate: three QTL means, the common variance σ^2, and the position c_1 of the putative QTL within the interval. Likelihoods for other designs follow using the appropriate conditional probabilities.

One of the first applications of ML interval mapping was performed by Paterson et al. (1988), who examined 237 backcross individuals in a cross between the tomato species *Lycopersicon esculentum* and *L. chmielewskii* for several fruit-related traits (Figure 15.1 gives the chromosome 10 likelihood maps for three traits). By using 68 markers (63 RFLP and 5 isozyme variants), 95% of the genome was within 20 cM of a marker. Six QTLs affecting fruit mass, four affecting concentration of soluble solids, and five affecting fruit pH were detected. A follow-up study (Paterson et al. 1990) using NILs (Chapter 14) detected additional QTLs. However, this finer mapping could not confirm the presence of one putative QTL that showed a highly significant peak on the likelihood map in the 1988 study, suggesting it was a false positive.

With ML interval mapping, the likelihood map for an entire chromosome is constructed by pasting together the likelihood maps for each successive interval. If the order of markers on a particular chromosome is $M_1 - M_2 - M_3 - \cdots - M_n$, the likelihood map for the $M_1 - M_2$ interval is constructed using only marker information from these two loci, the map for the $M_2 - M_3$ interval uses only information from M_2 and M_3, etc. The map resulting from joining the maps for each interval together is smooth; see Figures 15.1 and 15.3.

Given the multiple-test nature of these plots (since each map is actually multiple intervals), the appropriate threshold value for the collection of internal maps that constitutes the likelihood map for a chromosome is debatable. Knott and Haley (1992a) note that the total number of independent tests is bounded above by the number of intervals examined, but since these intervals are linked, they are not independent tests (Zeng 1993). The lower bound is set by the number of chromosomes examined, as these segregate independently. Hence, we first set a threshold level for each chromosome that ensures a desired genome-wide significance level for the entire collection of chromosomes. If C chromosomes are examined, Equation 15.13b implies that in order to obtain a genome-wide significance level γ, the significance level used to set thresholds for each chromosome is

$$1 - (1 - \gamma)^{1/C} \simeq \gamma/C \tag{15.17}$$

Rebaï et al. (1994b) suggest an improved approach that takes into account differences in chromosome lengths. Turning now to the significance values for intervals on a given chromosome, suppose the chromosome of interest has m intervals and we have set the chromosome-wide significance as γ_i. Simulation studies by Zeng (1994) suggest that if the number of markers is not too large, then, for large sample sizes, the critical value for each interval is approximately given by a χ^2_k value with significance γ_i/m. Here k is the number of free parameters in the likelihood-ratio test. More exact approximations assuming an infinitely dense map have been developed (Lander and Botstein 1989, Feingold et al. 1993), as have those for a finite number of markers (Zeng 1994, Rebaï et al. 1994b). Simulations by Doerge and Rebaï (1996) show that, dense marker methods (assuming a very large number of markers per chromosome) are conservative, with the probability of a test statistic exceeding the α-level threshold being less than α when no QTL is present.

Example 6. Suppose five chromosomes are used for ML-interval mapping in an F_2 design. Chromosomes 1 through 5 have 10, 5, 20, 30, and 40 markers, respectively. In order to achieve a genome-wide level of significance of $\gamma = 0.10$, what are the approximate critical values for each chromosome? Applying Equation 15.17, the overall level of significance for each chromosome is $1-(1-0.1)^{1/5} = 0.021$.

The critical values for each chromosome vary with the number of markers. For chromosome 1, the significance levels for each test become approximately $0.021/10 = 0.0021$. Recall that the degrees of freedom for the test of no QTLs in an F_2 design are $5-2 = 3$. Since $\Pr(\chi^2_3 > 14.71) = 0.0021$, this implies that the critical values for the likelihood ratios for chromosome 1 is 14.7. Similarly, the critical values for the remaining four chromosomes are 13.2, 16.2, 17.0, and 17.6.

An alternative approach to obtaining critical values is to use permutation tests to set the threshold levels (Churchill and Doerge 1994, Doerge and Churchill 1996). This resampling procedure has the advantage of being robust to the actual distribution of effects. Further, resampling is superior to analytical approximations for data with missing and incomplete marker information, as the permutation test, by keeping genotypes intact during reshuffling, automatically incorporates the special nature of each data set (Doerge and Rebaï 1996).

Finally, it should be mentioned that the null hypothesis usually assumed, that of no QTLs, may be misleading. Crossed lines are often chosen because they differ in traits of interest, so that there is certainly segregating genetic variance in the F_2 and other line-cross populations. Visscher and Haley (1996) note that if such background variance is present, it results in a more frequent rejection of the null hypothesis of no QTL than expected. They argue that the more appropriate null hypothesis should be that, taking the strain differences into account, the amount of genetic variance explained by a chromosome segment is that expected by chance, and they propose several tests of this hypothesis.

Approximating ML Interval Mapping by Haley-Knott Regressions

One problem with ML estimators is that they can be rather computationally demanding. Among other things, this limits the applicability of resampling methods, which require thousands of ML estimates to be computed per experiment. Fortunately, a simple regression procedure gives an excellent approximation of the likelihood map for ML interval mapping (Haley and Knott 1992, Martínez and Curnow 1992). This procedure greatly facilitates matters, as regressions are easily computed. Haley and Knott's (1992) idea is to express the regression coefficients as a function of the unknown QTL parameters. Using the Falconer parameterization for genotypic means,

$$\mu_{QQ} = \mu + a, \qquad \mu_{Qq} = \mu + d, \qquad \mu_{qq} = \mu - a \tag{15.18a}$$

this is done by considering the regression

$$z_j = \mu + a \cdot x(M_j) + d \cdot y(M_j) + e_j \tag{15.18b}$$

The variables x and y, which depend on both the flanking-marker genotype of the individual (M) and the assumed map position of the putative QTL, are obtained as follows. Taking the expectation of Equation 15.18b over all individuals with marker genotype M_i gives

$$\mu_{M_i} = \mu + a \cdot x(M_i) + d \cdot y(M_i) \tag{15.19a}$$

From Equation 15.6,

$$\begin{aligned} \mu_{M_i} &= (\mu + a)\Pr(QQ \,|\, M_i) + (\mu + d)\Pr(Qq \,|\, M_i) + (\mu - a)\Pr(qq \,|\, M_i) \\ &= \mu + a \cdot \big[\Pr(QQ \,|\, M_i) - \Pr(qq \,|\, M_i)\big] + d \cdot \Pr(Qq \,|\, M_i) \end{aligned} \tag{15.19b}$$

Equating like terms in Equations 15.19a and 15.19b gives

$$x(M_i) = \Pr(QQ \,|\, M_i\,) - \Pr(qq \,|\, M_i), \qquad y(M_i) = \Pr(Qq \,|\, M_i) \tag{15.20}$$

Thus, the x and y values are functions of the conditional QTL probabilities given the flanking-marker genotypes. For example, for the F_2 design with no interference, Equation 15.2 gives

$$x(M_1M_1M_2M_2) = \frac{(1-c_1)^2(1-c_2)^2 - c_1^2\,c_2^2}{(1-c_{12})^2}$$

$$y(M_1M_1M_2M_2) = \frac{2c_1c_2(1-c_1)(1-c_2)}{(1-c_{12})^2}$$

Haley and Knott give expressions for the eight other F_2 marker genotypes, and values for other designs easily follow when the appropriate conditional probabilities are employed. This regression approach was independently suggested by Martínez and Curnow (1992) for the analysis of backcross populations. These authors also detail how missing marker information can be accommodated (Martínez and Curnow 1994a).

By analogy with likelihood maps, the regression given by Equation 15.18b is computed for each c_1 value within the $M_1 - M_2$ interval, with that value giving the regression with the largest r^2 being taken as the estimate of QTL position. For each c_1 value, Equation 15.20 yields the set of x and y values, allowing μ, a, and d to be estimated by ordinary least-squares regression (Equation 8.33a),

$$\mathbf{b}_{c_1} = \begin{pmatrix} \widehat{\mu} \\ \widehat{a} \\ \widehat{d} \end{pmatrix} = \left(\mathbf{X}_{c_1}^T\,\mathbf{X}_{c_1}\right)^{-1}\mathbf{X}_{c_1}^T\mathbf{z} \tag{15.21}$$

where the ith row of the design matrix $\mathbf{X}_{c_1}$ is $(1, x(M_i, c_1), y(M_i, c_1)\,)$.

Haley and Knott show that r^2 plots for this regression are related to likelihood plots. Assuming that phenotypes are normally distributed about each QTL genotype, then if the QTL is completely linked to either marker ($c_1 = 0$ or $c_1 = c$), the residuals for the regression given by Equation 15.18b are normally distributed. In this case, the regression estimates are also ML estimates and the likelihood-ratio test can be expressed as

$$\text{LR} = n\,\ln\left(\frac{\text{SS}_T}{\text{SS}_E}\right) = -n\,\ln(1-r^2) \tag{15.22}$$

where SS_T and SS_E are the total and error (or residual) sums of squares associated with the regression (Equations A3.16a,c), with the second equality following from Equation A3.15. If the QTL is not completely linked to either marker, the distribution of residuals follows a mixture of normals, as some marker genotype classes will contain different QTL genotypes. However, Haley and Knott (1992) and Rebaï et al. (1995) show that the function given by Equation 15.22 gives extremely similar values to the true likelihood ratio. Haley and Knott suggest that the number of degrees of freedom appropriate for this test is the number of estimated QTL parameters plus an additional degree of freedom for map position c_1. Xu (1995) notes that this regression approach tends to overestimate the residual variance, and presents a correction. More generally, if the linear model has additional factors (accounting for, say, differences due to sex and age), the LR test is modified to become

$$\text{LR} = n \ln \left(\frac{\text{SS}_E(reduced)}{\text{SS}_E(full)} \right) \tag{15.23}$$

where the error sums of squares are now for the full model and the reduced model (the latter incorporating all factors but the QTL effects).

Example 7. Consider the following hypothetical data set: 10 F_2 individuals scored for flanking marker genotypes M_1/m_1 and M_2/m_2, separated by recombination frequency $c_{12} = 0.30$. The following marker genotypes and their associated character values are observed:

$M_1m_1M_2m_2$	$M_1M_1M_2M_2$	$M_1m_1M_2M_2$	$m_1m_1M_2m_2$	$M_1M_1M_2m_2$
3.9	5.6	3.7	3.9	5.3
$m_1m_1m_2m_2$	$M_1m_1M_2M_2$	$M_1M_1M_2M_2$	$M_1m_1M_2M_2$	$M_1m_1M_2m_2$
1.1	3.6	5.4	3.7	3.3

This yields the observation vector

$$\mathbf{z}^T = (3.9,\ 5.6,\ 3.7,\ 3.9,\ 5.3,\ 1.1,\ 3.6,\ 5.4,\ 3.7,\ 3.3)$$

Assuming no interference, $c_2 = (0.3 - c_1)/(1 - 2c_1)$. For each c_1 value ($0 \leq c_1 \leq 0.3$), a regression is fitted by first using Equation 15.20 to compute the elements of the design matrix for that value of c_1 and then using Equation 15.21 to obtain the regression coefficients. For example, consider three different QTL positions: $c_1 = 0$ (QTL at marker M_1), $c_1 = 0.15$ (QTL in the middle), and $c_1 = 0.3$ (QTL at marker M_2). The resulting regressions for these three c_1 values are

c_1	$\widehat{\mu}$	$\widehat{a}$	$\widehat{d}$	r^2
0.00	3.97	1.47	−0.33	0.730
0.15	3.70	1.89	−0.26	0.732
0.30	2.75	1.65	1.35	0.597

These regressions are obtained using the design matrices

$$\mathbf{X}_0 = \begin{pmatrix} 1 & 0 & 1 \\ 1 & 1 & 0 \\ 1 & 0 & 1 \\ 1 & -1 & 0 \\ 1 & 1 & 0 \\ 1 & -1 & 0 \\ 1 & 0 & 1 \\ 1 & 1 & 0 \\ 1 & 0 & 1 \\ 1 & 0 & 1 \end{pmatrix}, \quad \mathbf{X}_{0.15} = \begin{pmatrix} 1 & 0.00 & 0.85 \\ 1 & 0.91 & 0.09 \\ 1 & 0.35 & 0.60 \\ 1 & -0.56 & 0.40 \\ 1 & 0.56 & 0.40 \\ 1 & -0.91 & 0.09 \\ 1 & 0.35 & 0.60 \\ 1 & 0.91 & 0.09 \\ 1 & 0.35 & 0.60 \\ 1 & 0.00 & 0.85 \end{pmatrix}, \quad \mathbf{X}_{0.3} = \begin{pmatrix} 1 & 0 & 1 \\ 1 & 1 & 0 \\ 1 & 1 & 0 \\ 1 & 0 & 1 \\ 1 & 0 & 1 \\ 1 & -1 & 0 \\ 1 & 1 & 0 \\ 1 & 1 & 0 \\ 1 & 1 & 0 \\ 1 & 0 & 1 \end{pmatrix}$$

To complete the analysis, regressions are computed for the full range of c_1 values, generating the following plot of regression r^2 as a function of c_1.

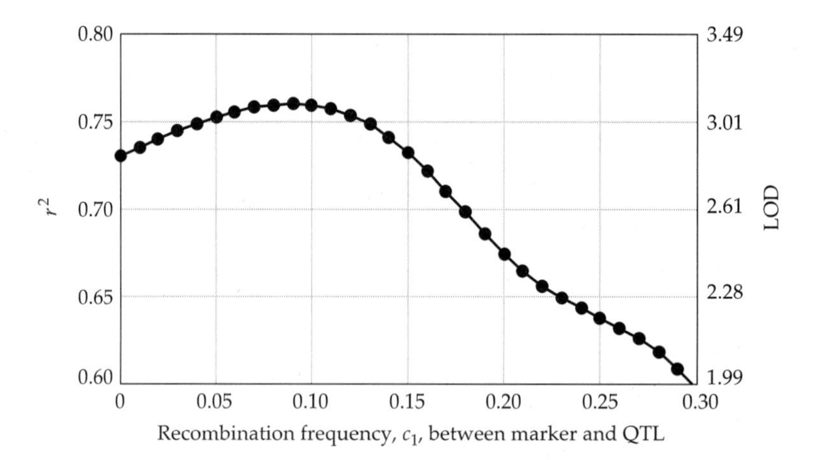

The maximum value of r^2 (0.76) occurs at $c_1 = 0.09$, and the associated regression coefficients are $\widehat{\mu} = 3.90$, $\widehat{a} = 1.76$ and $\widehat{d} = -0.46$. Hence, the data suggest that a QTL lies between these two markers at recombination fraction $c_1 = 0.09$ from marker locus M_1, with estimated genotypic means

$$\widehat{\mu}_{QQ} = \widehat{\mu} + \widehat{a} = 5.66, \quad \widehat{\mu}_{Qq} = \widehat{\mu} + \widehat{d} = 3.44, \quad \widehat{\mu}_{qq} = \widehat{\mu} - \widehat{a} = 2.14$$

Does this example show significant evidence of a QTL? From Equation 15.22, with $n = 10$ and $r^2 = 0.76$, the likelihood ratio (LR) becomes $-\ln(1-0.76^2) =$ 14.27. Note that only two QTL parameters are fitted (a and d) because the reduced model fits a mean μ. Hence, the critical value for the likelihood ratio is a χ^2 with three degrees of freedom (for a, d, c_1),

$$\Pr[\,\chi_3^2 > -n\ln(1-r^2)\,] = \Pr[\,\chi_3^2 > 14.27\,] = 0.003$$

showing that the QTL effect is indeed significant.

Approximate confidence intervals can be constructed by using those values giving scores within one LOD of the maximum value. We can translate r^2 values into LOD scores by using LOD = LR/4.61 = $-n\ \ln(1 - r^2)/4.61$. The MLE has $r^2 = 0.76$ and $n = 10$, for a LOD score of $-10\ \ln(1-0.76^2)/4.61 = 3.10$. Hence any c_1 value with a LOD score of 2.10 or greater is in the one-LOD support interval for QTL position. The resulting interval is $c_1 = 0$ to 0.28, so that although there is very strong evidence for a QTL, there is extreme uncertainty as to its position within the interval. This is not surprising given the very small sample size.

DEALING WITH MULTIPLE QTLs

All of the methods discussed so far are best characterized as **one-at-a-time** approaches for mapping QTLs, as they all assume a single QTL linked to the marker(s) of interest. While such methods can detect the presence of multiple QTLs (e.g., finding marker effects on a number of different chromosomes), they cannot discern whether significant effects at several linked markers / intervals are due to a common QTL or to several linked QTLs. The presence of multiple QTLs also introduces serious biases into estimates of QTL effects and positions derived from one-at-a-time approaches.

For example, while the presence of multiple (significant) peaks on a likelihood map for a given chromosome is generally taken as an indication of multiple QTLs, *such peaks do not necessarily correspond to the correct QTL positions* (Martínez and Curnow 1992, Haley and Knott 1992). Figure 15.3 gives an example of two linked QTLs embedded within four markers. Using a likelihood function that assumes only a single QTL, interval mapping correctly indicates likelihood peaks in the intervals flanked by $M_1 - M_2$ and $M_3 - M_4$. However, the resulting map also shows a much higher peak between $M_2 - M_3$, incorrectly suggesting the presence of a third QTL in this region. Some programs allow one to "fix" a QTL at the position corresponding to the highest peak of a multiply peaked map, and then search for a second linked QTL. As Figure 15.3 shows, this procedure can introduce serious bias. While specific tests for the presence of linked QTLs in adjacent intervals using sets of three overlapping markers have been suggested (Martínez and Curnow 1992, 1994b; Haley and Knott 1992), these are not without problems (Whittaker et al. 1996).

Unlinked QTLs also have an effect on one-at-a time methods (albeit not as dramatic), as segregation at such loci contributes to the phenotypic variance. Reducing or removing this segregation variance reduces the residual variance for the marker / interval under consideration, increasing the power for QTL detection and improving the precision of estimates. Example 8 provides a dramatic illustration.

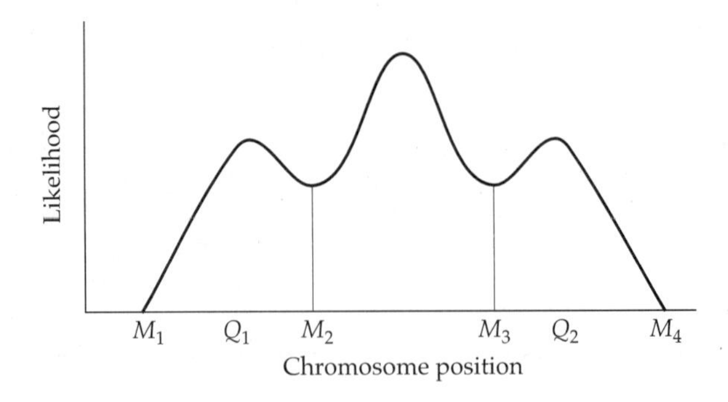

Figure 15.3 A false (or **ghost**) QTL generated by using a single-QTL likelihood function when two linked QTLs are actually present.

Most of the single-QTL methods developed above can be extended to multiple QTLs by considering additional marker loci and using conditional probabilities for multilocus genotypes. This approach has been used to develop explicit models for two or three linked QTLs (e.g., Knapp 1991, Haley and Knott 1992, Martínez and Curnow 1992, 1994b, Jansen 1996, Satagopan et al. 1996). We focus here on three particularly flexible regression-based approaches. The first is **marker-difference regression**, which considers all of the markers on one chromosome in a single analysis by using the regression of differences between the mean values of different genotypes. The second is **composite interval mapping**, which controls for both the effects of linked and unlinked QTLs by using the appropriate marker cofactors. Finally, Wright and Mowers (1994) and Whittaker et al. (1996) have shown how positional information for linked QTLs can be extracted from the regression coefficients of a standard multiple regression incorporating several linked markers.

Example 8. Lin et al. (1995) examined flowering date through ML interval mapping of 370 F_2 individuals from a cross between cultivated and exotic sorghum (*Sorghum bicolor* × *S. propinquum*). Only a single QTL for flowering date was detected, and this accounted for 85.7% of the total variance. The data were then adjusted to account for the effects of this major gene by using $(z - b_i)$ in place of the trait value z for an individual with genotype i at a marker linked to the major gene. Here, the b_i $(1 \leq i \leq 3)$ are the regression coefficients generated by a standard marker-trait regression using this marker locus. While the uncorrected F_2 phenotypic distribution was clearly bimodal, the adjusted data did not deviate from normality. Using the marker-adjusted data, two additional QTLs for flowering time were found (both unlinked to the original QTL), accounting for an additional 8.3% and 4.2% of the total variance. This example illustrates the potential importance of including additional marker information into the analysis when multiple QTLs are present. In this case, removing the effects of a major unlinked QTL reduced the residual variance sufficiently to enable detection of additional QTLs.

Marker-Difference Regression

Two groups (Kearsey and Hyne 1994, Hyne and Kearsey 1995, Wu and Li 1994, 1996) proposed a very simple, yet powerful, regression method that simultaneously considers all of the markers on a single chromosome. While the authors refer to this method as **marker regression** or **joint mapping**, we will use the more descriptive term **marker-difference regression**, or MDR, to emphasize that this approach is rather different from the regressions that we have considered up to this point. With MDR, each data point in the regression corresponds to a population mean value, rather than to values for single individuals (as in our previous regressions). While this data structure results in far fewer points in the regression, the use of means allows the inclusion of individuals missing some marker information and also allows the joint incorporation of information from several experiments.

The motivation for MDR follows from Equation 15.7a. We first present the method under the assumption of a single QTL to illustrate the main points before extending it to multiple QTLs. Suppose there are n linked markers on a chromosome containing a single QTL (with alleles Q and q). If the ith marker is at recombination frequency c_i from the QTL, the expected difference between marker homozygote means is

$$y_i = \mu(M_i M_i) - \mu(m_i m_i) = 2a(1 - 2c_i)$$

Thus, if we plot the differences y_i vs. $(1-2c_i)$ for each marker on the chromosome, the resulting n points are expected to fall on a straight line passing through the origin with slope $2a = \mu_{QQ} - \mu_{qq}$. Figure 15.4 illustrates this point, showing two regressions using the same set of marker differences but assuming two different locations for the QTL. The regression computed using the correct position of the QTL is linear, while that assuming the incorrect position is highly nonlinear. As with Haley-Knott regressions, one slides the position of a putative QTL along the chromosome, computing a regression at each point. The regression giving the best fit (i.e., the largest r^2) corresponds to the estimate of QTL position, and the slope of that regression divided by two provides an estimate of the QTL effect, a.

To formally develop this approach, suppose that there are n linked markers scored along a single chromosome, and consider the regression

$$y_i = \overline{z}(M_i M_i) - \overline{z}(m_i m_i) = \beta\, x_i + e_i \tag{15.24}$$

with the $x_i = 1 - 2c_i$ values obtained by fixing the QTL position and then computing c_i. Because the residuals are correlated and potentially heteroscedastic,

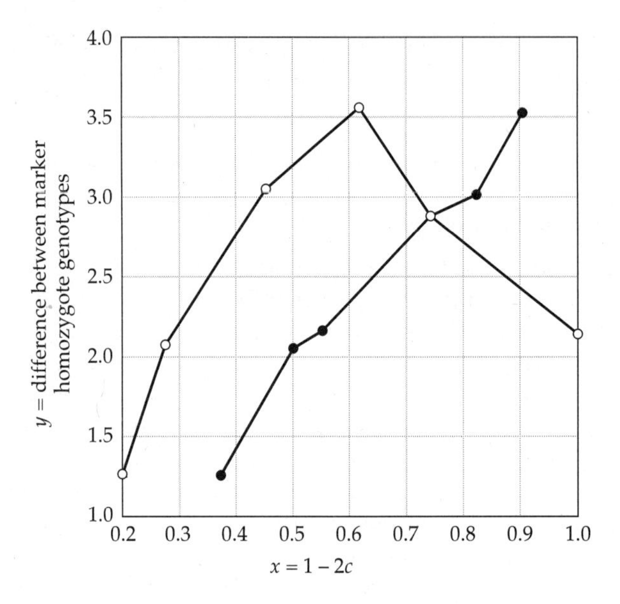

Figure 15.4 Marker-difference regression plot for the data given in Example 9. Open circles assume a QTL at map position 90 cM, closed circles a QTL at position 60 cM (the true position). Note that the relationship is linear when the correct position is used, but highly nonlinear under the incorrect position.

generalized least-squares regression (Chapter 8) must be used, with

$$\widehat{\beta} = \left(\mathbf{X}^T\mathbf{V}^{-1}\mathbf{X}\right)^{-1}\mathbf{X}^T\mathbf{V}^{-1}\mathbf{y} \tag{15.25a}$$

which has sample variance

$$\sigma^2(\widehat{\beta}) = \left(\mathbf{X}^T\mathbf{V}^{-1}\mathbf{X}\right)^{-1} \tag{15.25b}$$

where

$$\mathbf{y} = \begin{pmatrix} y_1 \\ \vdots \\ y_n \end{pmatrix}, \quad \mathbf{X} = \begin{pmatrix} 1 - 2c_1 \\ \vdots \\ 1 - 2c_n \end{pmatrix} \tag{15.25c}$$

and

$$\mathrm{V}_{ij} = \begin{cases} \dfrac{\mathrm{Var}(M_iM_i)}{n(M_iM_i)} + \dfrac{\mathrm{Var}(m_im_i)}{n(m_im_i)} & i = j \\ \\ (1 - 2c_{ij})\sqrt{\mathrm{V}_{ii}\;\mathrm{V}_{jj}} & i \neq j \end{cases} \tag{15.25d}$$

where $\mathrm{Var}(M_x)$ is the sample variance of $\overline{z}(M_x)$ and $n(M_x)$ is the sample size for marker class M_x (Wu and Li 1996).

Assuming normally distributed residuals, from Equation A3.11a the residual sum of squares,

$$\mathrm{SS}_E = \widehat{\mathbf{e}}^T \mathbf{V}^{-1} \widehat{\mathbf{e}} = (\mathbf{y} - \mathbf{X}\widehat{\beta})^T \mathbf{V}^{-1} (\mathbf{y} - \mathbf{X}\widehat{\beta}) \tag{15.26}$$

follows a χ^2 distribution with $n - 2$ degrees of freedom (n data points minus two estimated parameters, the QTL effect $\beta = 2a$ and the assumed position). The test for a significant QTL effect compares the SS_E for this regression with that for the regression assuming no marker effect ($y_i = \mu + e_i$). The SS_E for the reduced model is also χ^2-distributed, but with $n - 1$ degrees of freedom. Recalling (Appendix 5) the additivity property of the chi-square, the difference in residual sums of squares for these two models follows a χ^2_1 distribution under the null hypothesis. Hence, the regression is significant at the α level if SS_E(reduced model) – SS_E(QTL model) exceeds $\chi^2_1(\alpha)$, the α-level cutoff for a χ^2_1. Separate regressions are computed for each chromosome, so that to obtain a genome-wide level of significance γ, each chromosomal regression is tested with significance level $\alpha = 1-(1-\gamma)^{1/C} \simeq \gamma/C$, where C is the number of chromosomes examined.

Example 9. Consider the following hypothetical data (plotted in Figure 15.4) generated by assuming a single QTL with effect $a = 2.0$ at map position 60 cM along a chromosome containing six markers:

Marker Position (cM)	$\overline{z}(M_iM_i) - \overline{z}(m_im_i)$
10	1.26
25	2.06
50	3.04
65	3.54
75	2.90
90	2.15

We assume that the variance associated with each marker class is the same with $\mathrm{Var}(M_x) = 5$, and that 50 individuals of each marker class are scored, giving $\mathrm{V}_{ii} = 2 \cdot 5/50 = 0.2$. Using this and the c_{ij} values with Equation 15.25d fills out the rest of $\mathbf{V}$. For a MDR analysis, one computes a separate regression for each possible QTL position. Consider the regression for a QTL assumed to be at map position 50 cM. For the first marker, the QTL-marker map distance is 40 cM, which (assuming a Haldane map distance, Equation 14.3) translates into a recombination frequency of

$$c_1 = \frac{1 - e^{(-2 \cdot 0.4)}}{2} \simeq 0.275$$

giving $x_1 = (1-2\,c_1) = 0.45$, and the data point associated this marker becomes (0.45, 1.26). Computing the remaining data points and applying Equations 15.25

and 15.26 gives a regression with $SS_E = 11.03$. After this procedure is repeated for all positions along the chromosome, the resulting plot of SS_E vs. putative QTL position (shown below) exhibits a minimum value (0.43) at map position 61, and hence r^2 is maximized at this position (see Equation A3.15).

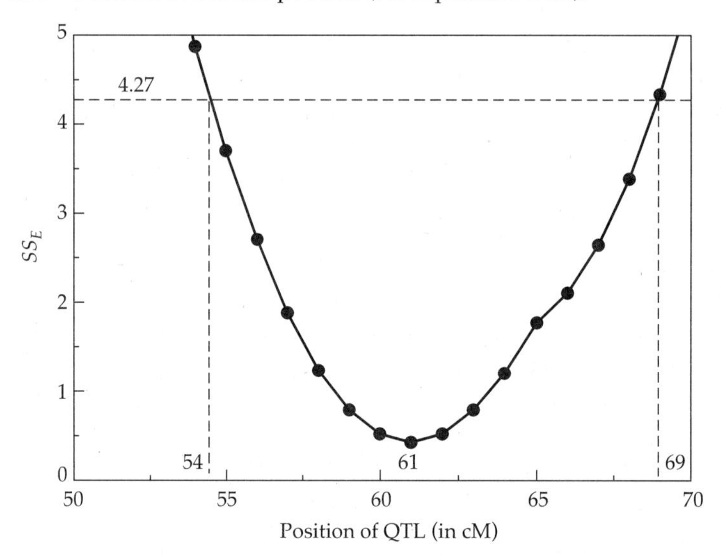

Whether the fit under the single-QTL model is a significant improvement over a model assuming no QTL can be assessed by comparing the error sum of squares of the QTL model ($SS_E = 0.43$) with the error sum of squares of the reduced (no QTL) model $y_i = \mu + e_i$. Since the QTL model fits an extra parameter, the difference in sums of squares follows a χ^2 distribution with one degree of freedom (one df) under the hypothesis of no QTL effect. For the reduced model, $SS_E = 19.16$, which is obtained by setting $\mathbf{X}$ equal to a vector of ones and applying Equation 15.25a. Hence, the QTL effect is highly significant as $\Pr(\chi_1^2 > 19.16 - 0.43\,) < 0.001$.

The adequacy of the single-QTL model can be assessed by noting that if this model is correct, SS_E follows a χ_4^2 distribution (there are six data points and two fitted parameters, for four df). Since $\Pr(\chi_4^2 > 0.43) = 0.99$, SS_E is not larger than expected by chance, suggesting that there is no need to consider additional QTLs.

Using the estimated map position, the resulting regression has slope 3.84, giving the estimated QTL effect as $\widehat{a} = 3.84/2 = 1.92$. From Equation 15.25b, we have $\sigma^2(2\,\widehat{a}\,) = \left(\mathbf{X}^T\mathbf{V}^{-1}\mathbf{X}\right)^{-1} = 0.16$, giving the standard error of $\widehat{a}$ as $\sqrt{0.16}/2 = 0.20$. Since SS_E follows a χ_1^2 distribution, the 95% confidence interval for QTL position contains those values giving regressions with SS_E not exceeding $\chi_1^2(0.05) = 3.84$ of the minimal SS_E value of 0.43 (i.e., SS_E values less than 4.27). This gives the confidence interval for the QTL position as 54 to 69 cM (see figure).

This approach easily extends to multiple QTLs. Recalling Equation 15.8a, if there are N linked QTLs, the jth of which is at recombination frequency c_{ji} from marker i, then (assuming no epistasis),

$$y_i = \mu(M_iM_i) - \mu(m_im_i) = 2a_1(1 - 2c_{1i}) + \cdots + 2a_N(1 - 2c_{Ni}) \qquad (15.27a)$$

This immediately suggests the multiple regression

$$y_i = \beta_1 \cdot x_{1i} + \cdots + \beta_N \cdot x_{Ni} + e_i \qquad (15.27b)$$

where $x_{ji} = (1 - 2c_{ji})$ and $\beta_j = 2a_j$. The estimates are still given by Equation 15.25a, with $\mathbf{y}$ and $\mathbf{V}$ being defined in the univariate case, and

$$\boldsymbol{\beta} = \begin{pmatrix} \beta_1 \\ \vdots \\ \beta_N \end{pmatrix} \quad \text{and} \quad \mathbf{X} = \begin{pmatrix} 1 - 2c_{11} & \cdots & 1 - 2c_{N1} \\ \vdots & \ddots & \vdots \\ 1 - 2c_{1n} & \cdots & 1 - 2c_{Nn} \end{pmatrix}$$

where N is the number of assumed QTLs, and n is the number of markers. As above, one computes the regression over the set of all possible QTL positions, with the estimates of QTL positions being given by the regression with the smallest SS_E value (or largest r^2). Each additional QTL reduces the degrees of freedom of SS_E by two (one for QTL effect, one for position). The test for whether adding another QTL significantly improves the fit compares the difference in the resulting two error sums of squares (for models assuming N versus $N - 1$ QTLs) with the appropriate critical value for a χ^2_2.

Interval Mapping with Marker Cofactors

The careful reader will note that marker-difference regression does not require knowledge of the multilocus marker genotypes of any individual, as all that enters into the analysis are the population means for each separate marker. An alternative approach for dealing with multiple QTLs that incorporates multilocus marker information from individuals is to modify standard interval mapping to include additional markers as cofactors in the analysis. Using the appropriate unlinked markers can partly account for the segregation variance generated by unlinked QTLs (Jansen 1992, 1993b; Zeng 1993, 1994), while the effects of linked QTLs can be reduced by including markers linked to the interval of interest (Stam 1991; Zeng 1993, 1994; Rodolphe and Lefort 1993). This general approach of adding marker cofactors to an otherwise standard interval analysis, often referred to as **composite interval mapping** or CIM, results in substantial increases in power to detect a QTL and in the precision of estimates of QTL position (Jansen 1993b, 1994a,b, 1996; Jansen and Stam 1994; Jansen et al. 1995; Zeng 1994; van Ooijen 1994; Utz and Melchinger 1994). Figure 15.5 shows a rather dramatic example of the improvement using CIM over interval analysis.

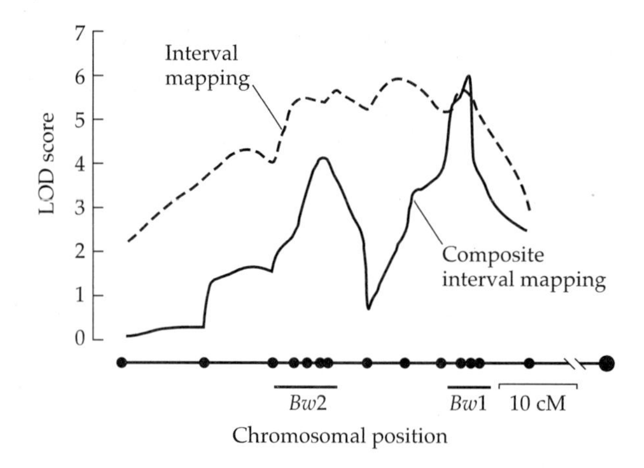

Figure 15.5 Likelihood plots for the X chromosome for QTLs influencing body weight in mice. The likelihood map under standard ML interval mapping (dashed line) shows a single very broad peak. Using the same data, the CIM likelihood map (solid lines) shows two distinct peaks. *Bw1* and *Bw2* denote the two putative body-weight QTLs, and the dots on the chromosome indicate the positions of the marker loci. (After Dragani et al. 1995.)

Suppose the interval of interest is flanked by markers i and $i+1$. One way to incorporate information from additional markers is to consider the sum over some collection of markers outside the interval of interest,

$$\sum_{k \neq i, i+1} b_k \cdot x_{kj} \tag{15.28a}$$

where k denotes a marker locus and j the individual being considered. Letting M_k and m_k denote alternative alleles at the kth marker, the values of the indicator variable x_{kj} depend on the marker genotype of j, with

$$x_{kj} = \begin{cases} 1 & \text{if individual } j \text{ has marker genotype } M_k M_k \\ 0 & \text{if individual } j \text{ has marker genotype } M_k m_k \\ -1 & \text{if individual } j \text{ has marker genotype } m_k m_k \end{cases} \tag{15.28b}$$

This is simply a convenient recoding of a regression of trait value on the number of M_k alleles. Hence, b_k is an estimate of the additive marker effect for locus k. For a backcross or RIL design, each marker has only two genotypes and the indicator variable takes on values 1 and -1. More generally, if there is considerable dominance, the effects of the kth marker locus can be more fully accounted for by considering a more complex regression with a term for each genotype, e.g.,

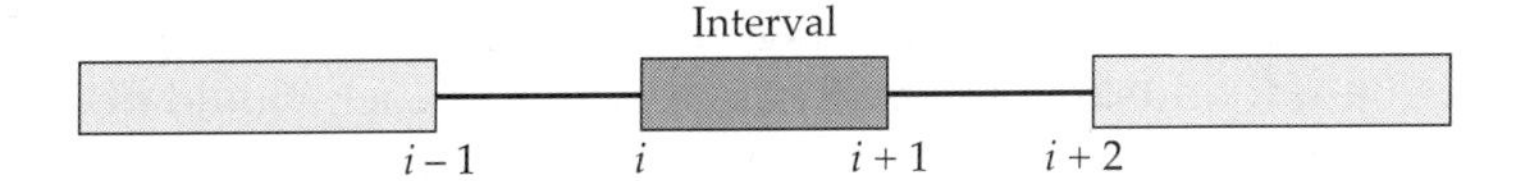

Figure 15.6 Suppose the interval being examined by CIM is between markers i and $i+1$. Addition of the adjacent markers $i-1$ and $i+2$ as cofactors absorbs the effects of any linked QTLs to the left of marker $i-1$ and to the right of marker $i+2$. Their inclusion, however, does not remove the effects of QTLs present in the two intervals, $(i-1, i)$ and $(i+1, i+2)$, flanking the interval of interest.

$b_{k1}x_{k1j} + b_{k2}x_{k2j} + b_{k3}x_{k3j}$, where the indicator variable x_{k1j} is one if j has marker genotype M_kM_k, else it is zero. The other two indicator variables for this marker locus are defined accordingly.

Composite interval mapping proceeds by adding this regression term to the model being considered. For example, upon adding marker cofactors, the Haley-Knott regression focusing on the interval bracketed by markers i and $i+1$ becomes

$$z_j = \left[\mu + a \cdot x(M_i) + d \cdot y(M_i) \right] + \sum_{k \neq i, i+1} b_k \cdot x_{kj} + e_j \tag{15.29}$$

Estimation of the QTL parameters (μ, a, d, c_i) for the interval proceeds as before, e.g., $x(M_i)$ and $y(M_i)$ are given by Equation 15.20 using marker loci i and $i+1$ as the flanking markers, with c_i being the putative QTL-marker i recombination frequency. For each c_i value in the interval, the regression given by Equation 15.29 is fitted (i.e., a, d, and the b_k), and (as before) the c_i value giving the regression with the largest r^2 is taken as the estimate of the QTL position. The significance of the interval can be tested by using Equation 15.23 to compare the full model (Equation 15.29) with the reduced model,

$$z_j = \mu + \sum_{k \neq i, i+1} b_k \cdot x_{kj} + e_j \tag{15.30}$$

which includes the marker cofactors but ignores the interval.

Just which markers should be added? While there is no single solution, the two markers directly flanking the interval being analyzed should always be included. Suppose the interval of interest is delimited by markers i and $i+1$ (Figure 15.6). Zeng (1994) showed that adding markers $i-1$ and $i+2$ as cofactors accounts for all linked QTLs to the left of marker $i-1$ and to the right of marker $i+2$. Thus, while these cofactors do not account for the effects of linked QTLs in the intervals immediately adjacent to the one of interest (i.e., the intervals $(i-1, i)$ and $(i+1, i+2)$ in Figure 15.6), they do account for all other linked QTLs.

The number of *unlinked* markers that should be used as cofactors is unclear, as inclusion of too many factors greatly reduces power (Zeng 1994). Jansen and

Stam (1994) recommend that the number of cofactors not exceed $2\sqrt{n}$, where n is the number of individuals in the analysis. A first approach would be to include all unlinked markers showing significant marker-trait associations (detected, for example, by standard single-marker regression). If several linked markers from a single chromosome all show significant effects, one might just use the marker having the largest effect. A related strategy, suggested by Jansen (1992, 1993b; Jansen and Stam 1994), is to first perform a multiple regression using all markers and then eliminate those that are not significant.

A multiple-trait extension of composite interval mapping given by Jiang and Zeng (1995) offers improved power for QTL detection and increased precision in estimation (relative to single-trait analysis) by incorporating the correlated error structure among traits (see also Ronin et al. 1995). Jiang and Zeng also develop likelihood-ratio tests for genotype $\times$ environment interaction and for tests of pleiotropy versus close linkage (one pleiotropic QTL vs. multiple linked QTLs each influencing separate characters).

Hypothesis testing and estimation for CIM follow by simple modifications of the appropriate results for interval mapping. Zeng (1993, 1994) showed that CIM test statistics for linked intervals are only weakly correlated, so that one can approximate each interval as an independent test. Zeng also found that the likelihood ratios within each interval are close to χ^2-distributed, so that an overall significance level of γ for an experiment examining m intervals can be obtained by equating the critical value within each interval to a χ^2 with significance level γ/m.

Resampling methods are easily extended to CIM. Doerge and Churchill (1996) suggest the following permutation test to account for multiple QTLs. A standard permutation test is first used to detect the marker with the greatest marker-trait association. Individuals are then divided (or **stratified**) according to their genotypes at this marker locus, and permutations are performed within each stratified group to generate new test statistics to find the next most significant QTL. This procedure is repeated until no significant effects are detected. Although permutation and bootstrap approaches are numerically intense, the rapid computation of solutions using Haley-Knott regressions makes these approaches feasible.

Finally, we note that other mapping approaches besides interval mapping can be improved by considering marker cofactors. For example, we can enhance the power of marker-difference regression by including unlinked markers to reduce the residual variance from unlinked QTLs. Since MDR uses the mean values for each marker, the individual data must be adjusted first to remove the effects from unlinked QTLs. Suppose n markers (unlinked to the chromosome of interest) are chosen because they show significant effects. The marker-adjusted value z_j^* of the original trait value of individual j, z_j, is given by

$$z_j^* = z_j - \left(\sum_{k=1}^{n} b_k \cdot x_{kj} \right) \tag{15.31}$$

and a MDR analysis is then performed using these adjusted values.

Detecting Multiple Linked QTLs Using Standard Marker-Trait Regressions

Consider the standard multiple regression of trait value on the single-locus genotypes at each of n markers,

$$z_j = \mu + \sum_{k=1}^{n} b_k \cdot x_{kj} + e_j \tag{15.32}$$

where j indexes the individual being considered, and the x_{kj} are given by Equation 15.28b. A rather remarkable finding, due to Wright and Mowers (1994) and Whittaker et al. (1996), is that the regression coefficients b_k for adjacent markers provide information on whether these markers flank a QTL. Further, the b_k can be used in many cases to obtain direct estimates of QTL effect and position.

When a QTL is **isolated** — an interval contains a single QTL and both flanking intervals are free of QTLs — the regression coefficients for the two markers immediately flanking the QTL depend only on this QTL and are not influenced by other linked QTLs (Stam 1991, Zeng 1993). A consequence of this finding is that markers flanking a QTL have regression coefficients of the same sign, while markers not adjacent to a QTL (i.e., there is at least one marker in the regression between the marker of interest and the nearest QTL) have expected regression coefficients of zero. Hence, one can simply scan the regression coefficients to see which intervals show support for a QTL (see Example 10).

Whittaker et al. (1996) further show, for an isolated additive QTL, that the regression coefficients for the flanking markers can be directly used to estimate QTL effect and position. Suppose the markers i and $i+1$ flank an isolated QTL. Whittaker et al. found that for an F_2 population, the estimated distance from marker i to the QTL is

$$c_i = \frac{1}{2}\left[1 - \sqrt{1 - \frac{4\, b_{i+1}\theta_i\,(1-\theta_i)}{b_{i+1} + b_i\,(1-2\theta_i)}}\;\right] \tag{15.33a}$$

where $\theta_i = c_{i,i+1}$ is the distance between the markers. Likewise, an estimate of the QTL's additive effect a, independent of amount of dominance at this QTL, is given by

$$a^2 = \frac{[\,b_i + (1-2\theta_i)\,b_{i+1}\,] \cdot [\,b_{i+1} + (1-2\theta_i)\,b_i\,]}{1-2\theta_i} \tag{15.33b}$$

where both b_i and b_{i+1} have the same sign as a.

Example 10. Whittaker et al. (1996) used a simulation study to generate 2000 F_2 progeny in a setting with three chromosomes, each with five markers evenly spaced at 25 cM (implying $c \simeq 0.2$ under Haldane's mapping function). QTLs were placed in the intervals flanked by markers (1, 2), (4, 5), (7, 8), (13, 14), and (14, 15). The multiple regression involving all 15 markers (Equation 15.32) had associated regression coefficients of:

Marker	1	2	3	4	5
b_i	−0.2996	−0.1422	−0.0221	0.2209	0.1956

Marker	6	7	8	9	10
b_i	−0.0189	−0.1922	−0.2404	0.0100	−0.0108

Marker	11	12	13	14	15
b_i	−0.0254	0.0371	0.3019	0.2644	0.3370

Looking for pairs of adjacent regression coefficients that have the same sign and are both significantly different from zero (as judged using standard regression tests, not shown) suggests evidence for QTLs in the intervals (1, 2), (4, 5), (7, 8), (13, 14), and (14, 15). The regression using just these nine markers had essentially the same SS_E as the full regression using all 15 markers, suggesting that none of the omitted markers are adjacent to QTLs (or they are adjacent to multiple linked QTLs whose effects cancel). However, removal of any one of the nine markers results in a regression with a significantly greater error sum of squares, supporting the hypothesis that all of these markers are adjacent to QTLs. Using these nine markers only, the new regression coefficients become

Marker	1	2	4	5
b_i	−0.2975	−0.1323	0.2296	0.1962

Marker	7	8
b_i	−0.2407	−0.2377

Marker	13	14	15
b_i	0.3145	0.2640	0.3355

Since the QTLs in intervals (1, 2), (4, 5), and (7, 8) appear to be isolated (no evidence for QTLs in adjacent intervals), Equations 15.33a,b can be used to estimate their effects and positions. For the QTL in the interval flanked by markers 1 and 2,

$$c_1 = \frac{1}{2}\left[1 - \sqrt{1 - \frac{4\,(-0.1323)\cdot 0.2\,(1-0.2)}{(-0.1323) + (-0.2975)(1 - 2\cdot 0.2)}}\,\right] = 0.074$$

and the estimate of the squared effect of the QTL is

$$a_1^2 = \frac{[(-0.2975) + (1 - 2 \cdot 0.2)(-0.1323)][(-0.1323) + (1 - 2 \cdot 0.2)(-0.2975)]}{1 - 2 \cdot 0.2}$$
$$= (0.442)^2$$

implying $a_1 = -0.442$ (since the regression coefficients $b_1, b_2 < 0$). Similarly, the estimates for the QTL in the interval (4, 5) are $c_4 = 0.105$ and $a_4 = 0.440$, while for the QTL in (7, 8), we find $c_7 = 0.112$ and $a_7 = -0.494$. The estimated values were rather close to the true values used in the simulations ($a_4 = -a_7 = -a_1 = 0.447$, $c_1 = 0.07$, $c_4 = 0.11$, and $c_7 = 0.11$).

Whittaker et al. (1996) make a final important point that applies to all multiple-QTL methods. Unless a QTL is isolated — it is the only QTL in a particular interval and the flanking intervals lack QTLs — these methods cannot separate out the effects of multiple linked QTLs. In particular, if an interval contains multiple QTLs, we cannot estimate their effects and positions (or even the correct number of QTLs), a point stressed by McMillan and Robertson (1974). (See Example 2 from Chapter 14.) While one obvious solution is simply to increase the marker density to the point where each QTL is indeed isolated, any increase in the marker density must be accompanied by a sufficient increase in sample size to ensure that a sufficient number of recombination events have occurred between adjacent markers.

SAMPLES SIZES REQUIRED FOR QTL DETECTION

Before investing the time and expense in a QTL mapping experiment, it is critical to have an understanding of the sample sizes required for the detection of QTLs of specified effects. The probability of a significant marker-trait association is increased by increasing the difference between means and/or by decreasing the within-marker class (or residual) variance. Increasing the sample size reduces the within-class variance, while changing the experimental design can increase the difference between means. The residual variance can also often be decreased by adding explanatory factors to the model, such as sex- or age-effects.

The following discussion of sample size is restricted to single-marker *t* tests using the F_2 or backcross designs, where a QTL is indicated if the means for two alternative marker genotypes are significantly different. Using the theory of power calculations (reviewed in Appendix 5), simple expressions can be obtained for these designs. The broad utility of the results developed below is that both theoretical (Simpson 1989, 1992; Haley and Knott 1992; Darvasi et al. 1993; Rebaï et al. 1995) and empirical (e.g., Stuber at el. 1992, deVicente and Tanksley 1993,

Nodari et al. 1993, Damerval et al. 1994, Champux et al. 1995, Kennard and Harvey 1995) studies show that t tests and more elaborate flanking-marker methods have very similar power for detection, especially when adjacent markers are no farther than 20 cM apart. Hence, the sample size expressions developed below provide a baseline for most designs.

We start by considering the t test for an F_2 design, where the presence of a linked QTL is indicated when $\overline{z}_{MM} - \overline{z}_{mm}$ is significantly different from zero. Suppose that the marker is completely linked to a single QTL with additive value a, in which case $E(\overline{z}_{MM} - \overline{z}_{mm}) = 2a$. Assuming that the distribution of phenotypes about each QTL genotype has constant variance σ_e^2, then if the numbers of *MM* and *mm* individuals measured are n_1 and n_2,

$$\sigma^2\left(\overline{z}_{MM} - \overline{z}_{mm}\right) = \sigma^2\left(\overline{z}_{MM}\right) + \sigma^2\left(\overline{z}_{mm}\right) = \left(\frac{1}{n_1} + \frac{1}{n_2}\right)\sigma_e^2 \tag{15.34}$$

If n total F_2 individuals are scored, we expect only one in four to be a particular marker homozygote, giving $n_1 = n_2 = n/4$ and the expected variance $8\,\sigma_e^2/n$. If $r_{F_2}^2$ denotes the fraction of the total F_2 phenotypic variance $[\sigma_z^2(F_2)\,]$ due to segregation at the QTL, then $\sigma_e^2 = (1 - r_{F_2}^2)\,\sigma_z^2(F_2)$. Hence, if n is reasonably large, the observed difference in marker means is approximately normally distributed, with

$$\overline{z}_{MM} - \overline{z}_{mm} \sim N\left[2a,\ 8(1 - r_{F_2}^2)\,\sigma_z^2(F_2)/n\right] \tag{15.35a}$$

Under the null hypothesis of no QTL, this difference is distributed as a normal with mean zero and variance $8\sigma_z^2(F_2)/n$.

Using the machinery developed in Appendix 5, the sample size required to have probability $1-\beta$ of detecting a QTL using a test with an α level of significance becomes

$$n_{F_2} = \frac{8(1 - r_{F_2}^2)}{\delta_{F_2}^2}\left(\frac{z_{(1-[\alpha/2])}}{\sqrt{1 - r_{F_2}^2}} + z_{(1-\beta)}\right)^2 \tag{15.35b}$$

where $z_{(p)}$ satisfies $\Pr(\,U \leq z_{(p)}) = p$ with $U \sim N(0,1)$, and

$$\delta_{F_2} = \frac{\mu_{QQ} - \mu_{qq}}{\sigma_z(F_2)} = \frac{2a}{\sigma_z(F_2)} \tag{15.36a}$$

is the difference in QTL means in units of F_2 phenotypic standard deviations. The variance contributed by F_2 segregation at this locus is $\sigma_Q^2(F_2) = a^2(2 + k^2)/4$, where k is the dominance coefficient, implying

$$r_{F_2}^2 = \frac{\sigma_Q^2(F_2)}{\sigma_z^2(F_2)} = \frac{a^2(2 + k^2)/4}{\sigma_z^2(F_2)} = \frac{\delta_{F_2}^2(2 + k^2)}{16} \tag{15.36b}$$

which for a completely additive QTL ($k = 0$) is $r^2_{F_2} = \delta^2_{F_2}/8$. Using Equation 15.36b, we can alternatively express the required sample size in terms of the fraction of variation accounted for by the QTL,

$$n_{F_2} = \left(\frac{1 - r^2_{F_2}}{r^2_{F_2}} \right) \left(\frac{z_{(1-[\alpha/2])}}{\sqrt{1 - r^2_{F_2}}} + z_{(1-\beta)} \right)^2 [1 + (k^2/2)] \tag{15.37}$$

Example 11. What sample sizes are required to detect a completely linked QTL using a test with $\alpha = 0.05$ and $\beta = 0.1$ (i.e., a 5% probability of a false positive and a 10% probability of missing a true association)? From normal tables, Pr(U < 1.96) = 0.975 and Pr(U < 1.28) = 0.9, so that $z_{(1-[\alpha/2])} = z_{(0.975)} = 1.96$ and $z_{(1-\beta)} = z_{(0.9)} = 1.28$. Substituting these into Equation 15.37 gives the following sample sizes for a completely additive ($k = 0$) and a completely dominant or completely recessive ($k \pm 1$) QTL whose segregation accounts for r^2 of the total F_2 variance:

r^2	0.5	0.3	0.1	0.05	0.01
Additive QTL	16	31	101	206	1046
Dominant QTL	25	46	151	309	1568

Note that the presence of dominance can significantly inflate the required F_2 sample size.

Turning now to the backcross designs, consider $B_1 = F_1 \times P_1$ (i.e., $MQ/mq \times MQ/MQ$). Here $n_1 = n_2 = n/2$, while $\mu_{QQ} - \mu_{Qq} = a(1-k)$, giving

$$\overline{z}_{MM} - \overline{z}_{Mm} \sim N\left[a(1-k),\ 4(1 - r^2_{B_1})\,\sigma^2_z(B_1)/n\right] \tag{15.38a}$$

Using the same logic as above, the required sample size is found to be

$$n_{B_1} = \left(\frac{1 - r^2_{B_1}}{r^2_{B_1}} \right) \left(\frac{z_{(1-[\alpha/2])}}{\sqrt{1 - r^2_{B_1}}} + z_{(1-\beta)} \right)^2 \tag{15.38b}$$

with

$$r^2_{B_1} = \frac{\delta^2_{B_1}}{4}, \qquad \text{where} \qquad \delta_{B_1} = \frac{a(1-k)}{\sigma_z(B_1)} \tag{15.38c}$$

For the B_2 population, the results are similar, except that the comparison is now $\overline{z}_{Mm} - \overline{z}_{mm}$ and $-k$ replaces k in the above expressions. Comparing the F_2 and the backcross design (for small to modest r^2), the ratio of samples sizes to achieve the same power is approximately

$$\frac{n_{B_1}}{n_{F_2}} \simeq \left[\frac{2}{(1-k)^2} \right] \left[\frac{\sigma_z(B_1)}{\sigma_z(F_2)} \right]^2 \tag{15.39}$$

Thus, if the backcross and F_2 phenotypic variances are the same, the backcross design requires twice as many individuals as an F_2 for a completely additive QTL ($k = 0$). When dominance is present, depending on its direction relative to the backcross population used, the backcross design can require more than twice as many individuals as an F_2 ($k > 0$ for B_1, $k < 0$ for B_2) or fewer individuals than the F_2. (If $k = -1$, the required sample size for the B_1 is only half of that for an F_2 design.) A further complication is that the phenotypic variance is generally rather different in the F_2 and backcross populations due to changes in the variance from background QTLs. In the F_2 population, all QTL alleles have frequency 1/2, which gives maximum additive variance (provided all QTLs are additive). In a backcross, the allele frequency is 1/4, and additive genetic variance is often reduced significantly relative to that in the F_2. Thus, if background QTLs contribute significantly to the character, the backcross can show a reduced variance and more power.

If the QTL is not completely linked to the marker, two corrections are required for the above expressions. First, the difference in means for F_2 homozygous marker genotypes now estimates $2a(1-2c)$. A more subtle correction is that the variance about the marker means increases when $c \neq 0$, as the phenotypic distribution for each marker class is now a mixture of distributions with different means. In spite of these complications, to a very good approximation the sample sizes required for a specific power of QTL detection are given by $n_0/(1-2c)^2$, where n_0 is the required sample size under complete linkage (Soller et al. 1976, Soller and Genizi 1978). Thus, the power to detect a linked QTL falls off as $(1-2c)^2$ decreases, being very weak when $c > 0.2$ (25 cM under the Haldane map).

Example 12. Suppose we wish to have a 90% chance of detecting (using a test with $\alpha = 0.05$) a QTL whose segregation accounts for 10% the total F_2 variance. Further assume that all of the genetic variation at this locus is additive. From Example 11, 101 individuals are required to detect this QTL using a completely linked marker. With a marker at recombination frequency c from the QTL, $n = 101/(1-2c)^2$, giving sample sizes of 281, 158, and 125 for c = 0.2, 0.1, 0.05, respectively.

One can increase the power to detect a linked QTL either by increasing the number of markers (which decreases c and hence increases the difference between marker means) or by increasing the number of individuals genotyped (which decreases the sampling variance). To see the relative importance of each, note from Equation 15.35a that the t statistic has approximate expected value

$$E\left[\frac{\mu_{MM}-\mu_{mm}}{\sigma(\overline{z}_{MM}-\overline{z}_{mm})}\right] \simeq \sqrt{n}\,(1-2c)\left[\frac{a}{\sqrt{2}\,(1-r^2_{F_2})\,\sigma_z(F_2)}\right] \tag{15.40}$$

The term in brackets is fixed for a given QTL, so that the test statistic scales with the square root of the sample size. Increasing the number of markers results in an increase in the test statistic, but there is a point of diminishing returns when markers are already closely spaced. For example, for $c = 0.2$ (corresponding to markers spaced 50 cM apart), moving to an infinitely dense map ($c = 0$) requires that only 36% as many individuals be scored to give the same power. However, for markers spaced at $c = 0.1$ and 0.05, these percentages become 81% and 90%.

Darvasi and Soller (1994b) show under rather general conditions that the spacing of markers giving the highest chance of detecting a QTL, given the constraint of scoring a fixed total number of marker genotypes (marker loci $\times$ individuals) is 20 to 30 cM. Here each QTL is no further than 10 to 15 cM (and on average is within 5 to 7.5 cM) from any marker. Thus, for markers spaced 10 cM or closer, there is really little point in further increasing the marker density when the goal is simple detection of a linked QTL. However, increasing marker density does become important if the goal is a highly precise estimate of QTL position or the dissection of a cluster of tightly linked QTLs.

Example 13. As mentioned in Example 3, Edwards et al. (1987, 1992) examined the same cross of two maize strains with two different designs. The 1987 design used 1,776 F_2 individuals and 17 markers, while the 1992 design used 187 F_2 individuals and 114 markers. The two designs represent a tradeoff between increased marker density (1992 design) and increased sample size (1987 design), as both examined a somewhat similar number of total marker genotypes (1776 $\times$ 17 = 30,200 vs. 187 $\times$ 114 = 21,300). Comparisons of c values in the two studies is problematic, given that only a fraction of the genome was covered in the 1987 study (about 40% of the genome was within 20 cM of a marker), while under the 1992 design most of the genome was 5 to 10 cM from a marker. Choosing $c = 0.25$ (1987 design) and $c = 0.08$ (1992 design), from Equation 15.40 the expected ratio of t statistics becomes

$$\frac{\sqrt{1776}\,(1-2\cdot 0.25)}{\sqrt{187}\,(1-2\cdot 0.08)} = 1.8$$

showing that (for these c values) the 1987 design had greater power.

ML interval mapping is expected to be somewhat more powerful than the simple single-marker t test, so the above results can be considered as upper bounds for the required sample size, although they are not greatly exaggerated. For example, the power of ML interval mapping to detect QTLs has been examined by several authors (Lander and Botstein 1989, van Ooijen 1992, Carbonell et al. 1993, Darvasi et al. 1993), who conclude that with a reasonable density of markers (one every 20 cM), 250 F_2 individuals are sufficient to detect a QTL whose segregation accounts for at least 5% of the F_2 variation. How does this compare with the required sample size for a t test? Since markers spaced at 20 cM intervals imply that a marker is within 10 cM from the QTL, using the result for $r^2 = 0.05$ from Example 11 gives the required sample size for a t test as $206/(1 - 2 \cdot 0.1)^2 = 263$. Hence, the above t test guidelines are also reasonable for ML interval mapping.

Power under Selective Genotyping

The idea behind selective genotyping is that scoring characters is often much less expensive than scoring markers. Hence, if n individuals are scored and genotyped in a normal design, there may be merit in scoring a larger number of individuals $n_z > n$ for the trait value, and then choosing a subset $n_g \leq n$ of these for genotyping. Typically, the uppermost and lowermost fractions (p) of scored individuals are genotyped, giving $n_g = 2p\,n_z \leq n$. Darvasi and Soller (1992) show that selective genotyping by scoring n_z individuals and genotyping $n_g = 2\,p\,n_z$ gives the same power as an analysis genotyping all n individuals, when

$$n_z = \frac{n}{2p + 2\,z_{(1-p)}\,\varphi(z_{(1-p)})} \tag{15.41}$$

Here, $\varphi(z_{(1-p)})$ is the unit normal density function evaluated at $z_{(1-p)}$, where $\Pr(U > z_{(1-p)}) = p$ with $U \sim N(0,1)$. Figure 15.7 plots the ratio n_z/n as a function of p. For example, selective genotyping using the uppermost and lowermost 10% ($p = 0.1$) of the population requires that $n_z = 1.54\,n$ individuals be phenotyped but only $n_g = 2 \cdot 0.1 \cdot n_z = 0.3\,n$ be genotyped. Since a decrease in p reduces the number of individuals that must be genotyped but increases the number that must be scored for the trait, the optimal p value depends on the relative costs of phenotyping and genotyping each individual (Darvasi and Soller 1992).

Power and Repeatability of Mapping Experiments

Even under designs where power is low, if the number of QTLs is large, it is likely that at least a few will be detected. In such cases of low power, the contributions of detected QTLs can be significantly (often *very* significantly) overestimated. Such a scenario, wherein we detect a small number of QTLs that appear to account for a significant fraction of the total character variation, can lead to the false

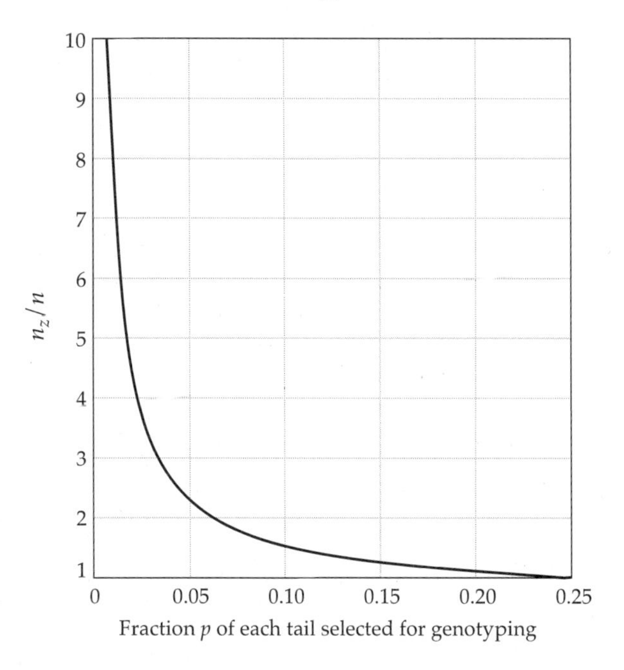

Figure 15.7 Under selective genotyping, n_z individuals are scored for the trait value, with the uppermost and lowermost fraction p of these being genotyped, giving $n_g = 2\,p\,n_z$. Here we plot, as a function of p, the number of individuals scored for the trait (n_z) that yields the same power as scoring and genotyping all individuals in a population of size n.

conclusion that character variation is largely determined by a few QTLs of major effect (Beavis 1994, Utz and Melchinger 1994).

As shown in Figure 15.8, the lower the power, the more the effects of a detected QTL are overestimated. For example, a QTL accounting for 0.75% of the total F_2 variation has only a 3% chance of being detected with 100 F_2 progeny with markers spaced at 20 cM. However, for cases in which such a QTL is detected, the average estimated total variance it accounts for is 15.8%, a 19-fold overestimate of the correct value. With 1,000 F_2 progeny, the probability of detecting such a QTL increases to 25%, and each detected QTL on average accounts for approximately 1.5% of the total variance, only a twofold overestimate. Further, these are the *average* values for the estimates. As shown in Figure 15.9, the distribution of observed effects is skewed, with a few loci having large estimated effects, and the rest small to modest effects. Such distributions of effects, commonplace in QTL mapping studies, have usually been taken as being representative of the true distribution of effects. Beavis's simulation studies show that they can be spuriously generated by a set of loci with equal effects.

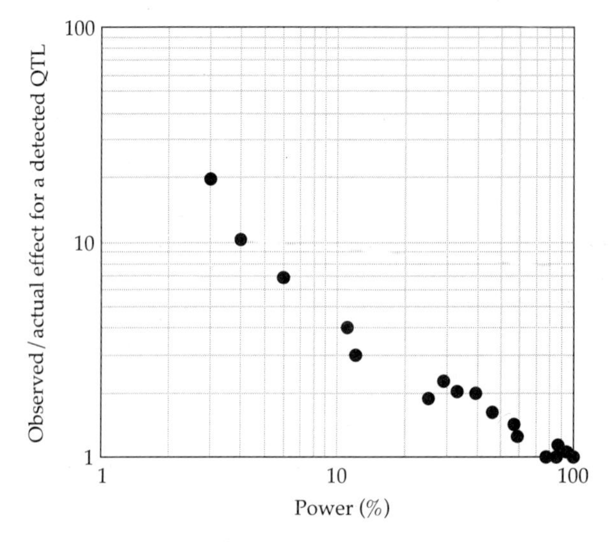

Figure 15.8 Relationship between the probability (power) of detecting a QTL and the amount by which the estimated effect of a *detected* QTL overestimates it actual value. (Based on results from a simulation study of Beavis 1994.)

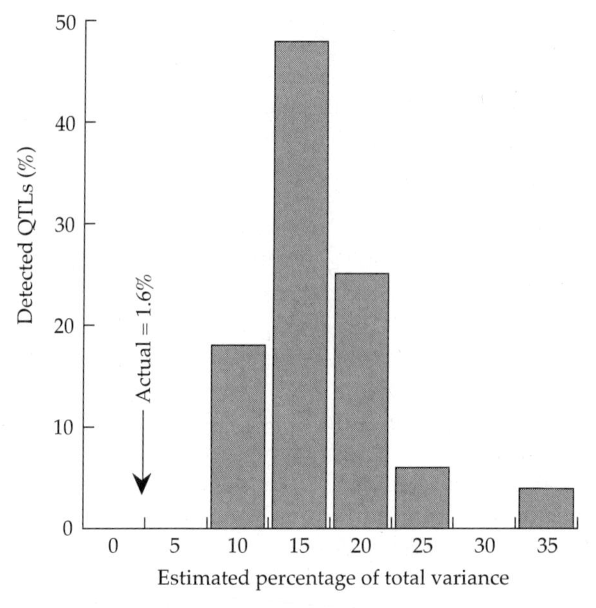

Figure 15.9 Distribution of the estimated effects of detected QTLs. Here 40 QTLs, each accounting for 1.58% of the variance, are assumed. Using 100 F_2 individuals, only 4% of such loci were detected. The average estimated fraction of total variation fraction accounted for by each detected QTLs was 16.3%, with the distribution of estimates skewed towards larger values. (From Beavis 1994.)

SELECTED APPLICATIONS

We close by examining some selected applications of QTL mapping using inbred lines, followed by a summary of the conclusions that can be drawn from these studies to date.

The Nature of Transgressive Segregation

QTL mapping experiments provide insight into the nature of **transgression** (or **transgressive segregation**), whereby some F_2 individuals show more extreme character values than are seen in either parental line. One explanation for such outliers is nonadditive gene action, i.e., epistasis and/or overdominance. Alternatively, transgressive segregation could be caused by the parental lines being fixed for sets of alleles having opposite effects, e.g., one line fixed for $+-/+-$, the other $-+/-+$, which would generate more extreme genotypes in the F_2 than observed in either parent (e.g., $++/++$ and $--/--$). This latter explanation is the one supported by most QTL studies.

For example, Li et al. (1995) observed transgressive segregation for heading date in the cross of *Lemont* and *Teqing* strains of rice (*Oryza sativa*). Using 113 markers (with an average spacing 19 cM) and 2,418 F_4 lines, three regions that together account for 77% of the phenotypic variance in heading date were mapped. While the difference in average heading date between parental strains was just 6 days, one region from *Lemont* decreased heading date by 8 days, while another from *Teqing* decreased it by 7 days. Hence, these lines were fixed for alternative alleles at major gene loci, resulting in effects that largely canceled.

Transgressive segregation was also observed in 8 of 11 traits measured in a large F_2 population from a cross of *Lycopersicon esculentum* (cultivated tomato) and *L. pennellii*, its wild Peruvian relative (deVicente and Tanksley 1993). Of the 74 QTLs detected for these 11 traits, 36% showed alleles having effects on the character that were opposite from parental-line differences (alleles reducing a trait being found in parents from the large line, and vice versa). Pairwise epistasis was ruled out as a major cause for the observed transgressive segregation, as the number of significant epistatic associations did not exceed that expected by chance. However, overdominance (or associative overdominance) contributed in a few cases, with marker heterozygote means being more extreme than those for marker homozygotes. Likewise, Weller (1987) and Weller et al. (1988) observed that around 25% of the significant marker-QTL relationships in their tomato crosses were opposite in sign from the parental differences. A similar study based on a cross of two phenotypically similar cultivars of soybeans also noted that transgressive segregation due to complementary QTL alleles was quite common (Mansur et al. 1993).

Transgressive segregation has also been found when lines resistant for certain insect pests or plant pathogens have been crossed to sensitive lines. For example, of seven detected maize QTLs conferring increased resistance to the European corn borer in a resistant × sensitive cross, five came from the resistant parent, while two came from the sensitive parent (Schön et al. 1993). Dirlewanger et al. (1994) similarly found that a sensitive pea line carried a resistance allele for *Ascochyta* fungal blight that was not present in a more resistant line.

Transgressive segregation has important evolutionary implications. Lewontin and Birch (1966) suggested that interspecies and wide-population hybrids can result in rapid adaptation to new environments. If transgressive segregation in population crosses is the rule rather than the exception, then the hybrids from such crosses possess the genetic variability to extend, perhaps considerably, the phenotypic range of a trait relative to either parental population. At a minimum, it is clear that mean phenotypic differences between lines are often very poor predictors of underlying genetic differences.

QTLs Involved in Reproductive Isolation in *Mimulus*

Bradshaw et al. (1995) examined the genetic basis of floral differences between sibling species of monkey flower, *Mimulus lewisii* and *M. cardinalis*. Although the ranges of these species overlap and laboratory F_1 hybrids are completely interfertile, hybrid plants are not found in nature. Presumably, this is due to nonoverlap of pollinators. *Mimulus lewisii* shows characters typical of bumblebee-pollinated plants: pink flowers with yellow nectar guides, a wide corolla, small volume of highly concentrated nectar, and short anthers and stigma. *Mimulus cardinalis*, on the other hand, shows a typical suite of hummingbird-pollinated characters: red petals lacking nectar guides, a narrow tubular corolla, high nectar volumes, and long anthers and stigma.

Using 93 F_2 plants and 159 markers, a number of QTLs for these characters were detected by ML interval mapping. As shown in Table 15.3, four of the characters appear to each have a QTL accounting for over 50% of the total F_2 variance, while all other characters had a QTL accounting for at least 25% of the total variance. Hence, it appears that the bulk of the differences in pollination characters (and hence reproductive isolation) can be accounted for by one or two loci for each character. However, with these small sample sizes, some caution is in order, given our previous comments about overestimation of QTL effects when power is low.

QTLs Involved in Protein Regulation

Quantitative-genetic approaches are often thought to be restricted to phenotypic characters such as body weight, height, or some measure of shape. However, they apply equally well to molecular characters. Damerval et al. (1994) analyzed the spot volumes of 72 anonymous proteins (from a specific seed tissue in maize) separated by high-resolution 2-D polyacrylamide gel electrophoresis. Genes controlling protein volume are, by definition, **regulatory genes** influencing the amount of that protein. Sixty F_2 individuals were scored with 76 RFLP markers, and both

Table 15.3 Number of detected QTLs influencing pollination characters involved in reproductive isolation between *Mimulus cardinalis* and *Mimulus lewisii* and their estimated individual effects (measured by % of variance explained).

	Number of QTLs	% Phenotypic Variance ($r^2 \times 100$)
Pollinator attraction characters		
Petal anthocyanins	2	33.5, 21.5
Petal carotenoids	1	88.3
Corolla width	3	68.7, 33.0, 25.7
Petal width	3	42.4, 41.2, 25.2
Pollinator reward		
Nectar volume	2	53.1, 48.9
Nectar concentration	2	28.5, 23.9
Pollination efficiency		
Stamen length	4	27.7, 27.5, 21.3, 18.7
Pistil length	2	51.9, 43.9

Source: Bradshaw et al. 1995.
Note: Due to sampling error, the sum of individual r^2 values exceeds 100% in a few cases.

ML-interval mapping and single-marker ANOVA detected a total of 70 QTLs affecting 46 of the 72 proteins. Of these 46 proteins, 25 were influenced by two or more QTLs (up to a maximum of five). Of the 70 detected QTLs, 33 showed strict additivity, while the remaining 37 showed at least some dominance. The amount of variation in protein volume accounted for by a single QTL ranged from 16% (the lower detection limit for this sample size) to 67%, and the cumulative variation accounted for by all detected QTLs for each protein ranged from 37% to 90%. Perhaps the most striking observation was the presence of significant epistasis. Four proteins had QTLs that were only detected through epistasis (their single-locus effects were not significant). In all, 14% of the 72 proteins showed detectable epistasis (Figure 15.10).

QTLs in the Illinois Long-term Selection Lines of Maize

In 1896, C. Hopkins initiated a set of maize lines selected for high and low oil and high and low protein content (Hopkins 1899). Selection on these lines continues today, and results from this remarkable study after 76 and 90 generations of selection have been summarized by Dudley (1977) and Dudley and Lambert (1992). The smooth and continuous long-term response in these lines suggests that a number of genes of relatively small effect underlie the differences. Crosses of the divergently selected lines have been used in three QTL mapping studies.

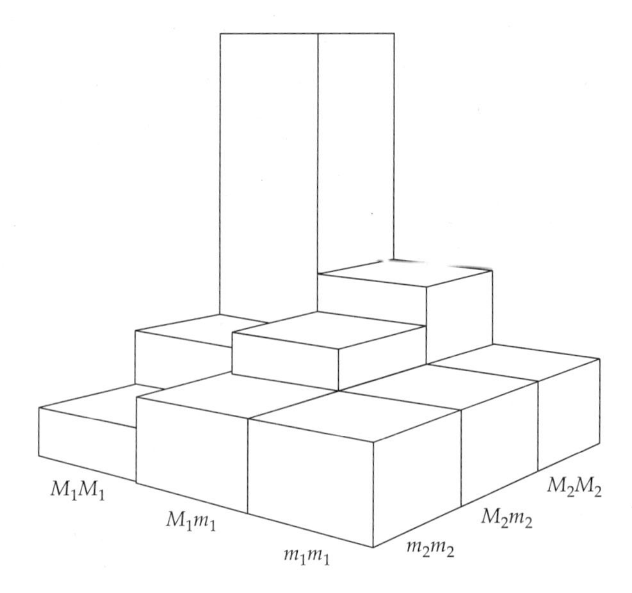

Figure 15.10 An example of epistasis for QTLs influencing protein volume in maize. Height indicates the amount of protein volume for each of the genotypes. In this case, the $M_1M_1M_2M_2$ marker genotype had the greatest effect on protein volume, while $M_1M_1m_2m_2$ had the smallest. (After Damerval et al. 1994.)

Goldman et al. (1993, 1994) crossed the selected (76-generation) high- and low-protein lines and then examined 100 F_3 families (formed by selfing F_2s) using 100 markers spanning the maize genome at an average spacing of about 20 cM. Using single-marker ANOVA, 22 markers on 10 chromosome arms were significantly associated with protein concentration, 19 markers on nine arms were associated with starch concentration, 26 on 13 arms with oil concentration, and 18 on 10 arms with kernel weight. Many of the marker-trait associations extended across clusters of linked markers. In these cases, single-marker ANOVA cannot distinguish between several linked markers all detecting the same linked QTL or multiple QTLs, and methods using multiple linked markers would be more illuminating, although these were not used.

A multiple regression involving only six (unlinked) markers accounted for 65% of the variation in protein concentration, and this increased to 84% when five significant pairwise epistatic interactions between these markers were incorporated into the regression (using Equation 15.14c). Seven markers accounted for 66% of the variation in starch, increasing to 78% when one significant pairwise epistatic interaction was included. Four marker loci accounted for 43% of the variation in oil concentration, while six markers accounted for 47% of the variation in kernel weight. These last two values are similar to what is seen in maize in other QTL mapping experiments, but the values for protein and starch seem rather high. In particular, it is very surprising that so few loci could account for such a significant fraction of the differences, especially given the long-term continuous and gradual change in the lines. One possible explanation is low power resulting in overestimation of QTL effects. Alternatively, if these values are indeed correct, selection response may have occurred by successive fixation of a series of new alleles at each of the major loci. Berke and Rocheford (1995) examined a cross of two other variant lines from this experiment (High Oil with Low Oil), and found similar results, with six loci accounting for 58% of the genetic variation in oil concentration and seven markers accounting for 56% of the variation in starch concentration.

QTLs Involved in the Differences Between Maize and Teosinte

Maize and teosinte are dramatically different (see Figure 5.2), to the point that they were originally placed in separate genera. Hybrids, however, are fully interfertile and maize is believed to have resulted from domestication of teosinte (Beadle 1980, Doebley 1992). In an elegant series of papers, Doebley and colleagues (Doebley et al. 1990, 1994, 1995a, 1997; Doebley and Stec 1991, 1993; Dorweiler et al. 1993) have begun to characterize the genes involved in these dramatic differences.

Maize and teosinte have major differences in plant architecture (Figure 5.2, Table 15.4). Teosinte has multiple long lateral branches, topped with male inflorescences (tassels). In maize, these branches are very short and topped with ears. These differences in plant architecture can be quantified by considering four characters: internode length on lateral branches (small in maize, long in teosinte), the number of branches (none to few in maize, many in teosinte), percentage of lateral branches topped with tassels as opposed to ears (mostly tassels in teosinte, ears in maize), and the number of secondary ears on each lateral branch (few in maize, many in teosinte). Table 15.4 gives the mean values for these characters in maize and teosinte.

Differences in the structure of the female inflorescence (the ear) are even more dramatic (Figure 5.2, Table 15.4). The teosinte ear has 5–10 cupulate fruitcases arranged in pairs. Each of these has a single spikelet that gives rise to a kernel, resulting in 10–20 kernels per teosinte ear. Each mature fruitcase is covered by a hardened outer glume that seals in the kernel, making harvesting very difficult. In contrast, the maize ear is composed of 100 or more cupules (arranged in multiple rows rather than pairs of rows), each cupule containing two spikelets, leading to two kernels per spikelet. These changes result in the maize ear having an order of magnitude more kernels than the teosinte ear. The maize outer glume is soft, so the kernels remain exposed for easy harvesting. Finally, while the teosinte ear easily disarticulates (to scatter seeds), kernels on the maize ear stay intact, further facilitating the harvesting of kernels.

QTLs were mapped in two different crosses, each involving a different primitive maize race and a different subspecies of teosinte (Doebley et al. 1990; Doebley and Stec 1991, 1993). As shown in Table 15.4, a few QTLs of major effect account

Table 15.4 Character differences between maize and teosinte (primitive maize race Reventado × *Zea mays parviglumis*).

	Means		QTLs			
	Maize	Teos.	N	Max	Min	r^2
Plant Architectural Characters						
Lateral branch internode length (LBIL)	0.7	21.9	5	0.45	0.05	0.63
Number of branches (LIBN)	0.0	5.8	4	0.24	0.04	0.42
% male primary lateral inflorescences (STAM)	0.0	97	5	0.23	0.05	0.52
No. secondary ears/lateral branch (PROL)	1.0	8.4	7	0.25	0.04	0.63
Ear Characters						
Cupules along a single rank (CUPR)	37.4	5.3	6	0.25	0.04	0.61
Disarticulation: 1 = none, 10 = full (DISA)	1.0	10.0	6	0.42	0.04	0.60
Glume score: 1 = soft, 10 = hard (GLUM)	1.0	10.0	2	0.41	0.08	0.75
% cupules with only one spikelet (PEDS)	0.0	100	5	0.25	0.08	0.69
Number of rows of cupules (RANK)	5.6	2.0	6	0.36	0.05	0.87

Source: From Doebley and Stec 1993.
Note: Listed are mean character values (Means), the number of detected QTLs (N), the r^2 value for the largest (Max) and smallest (Min) detected QTLs, and the total r^2 for a model containing all detected QTLs. Locations for the QTLs detected in this cross are plotted as the upper bars in Figure 15.11.

for most of the differences between characters. These QTLs are mostly in very similar positions in the two crosses (Figure 15.11), with both sets of crosses showing five regions of the maize genome that account for most of the differences.

Such results are consistent with Beadle's hypothesis of five major genes accounting for the difference between maize and teosinte (Beadle 1939). Beadle arrived at this figure by examining 50,000 F_2 maize × teosinte offspring, finding the frequency of all-maize or all-teosinte phenotypes to be $\simeq 1/500$. If n genes are involved, the expected F_2 frequency of either parental genotype is $(1/4)^n + (1/4)^n$. Setting this equal to $1/500$ and solving gives $n = 5$.

Focusing on the five major regions, marker-selected NILs (Chapter 14) were constructed to further characterize the QTLs. Doebley's first target was a QTL on chromosome 4, which accounted for 50% of the variance in glume score. A small maize segment containing this region was introgressed into a teosinte background by three generations of backcrossing and selection for flanking markers (Dorweiler et al. 1993). When NILs with the introgressed teosinte region were backcrossed to the maize recurrent parent, the resulting F_2 progeny showed two discrete classes for glume score, as would be expected with a single major gene. The putative gene was named *tga1*, for *teosinte glume architecture 1*. A similar analysis of two other regions, QTL-3L and QTL-1L, showed strong epistasis for a number of key traits separating maize and teosinte (Doebley et al. 1995a; see Example 1 from Chapter 5). By using marker-selected introgressed lines,

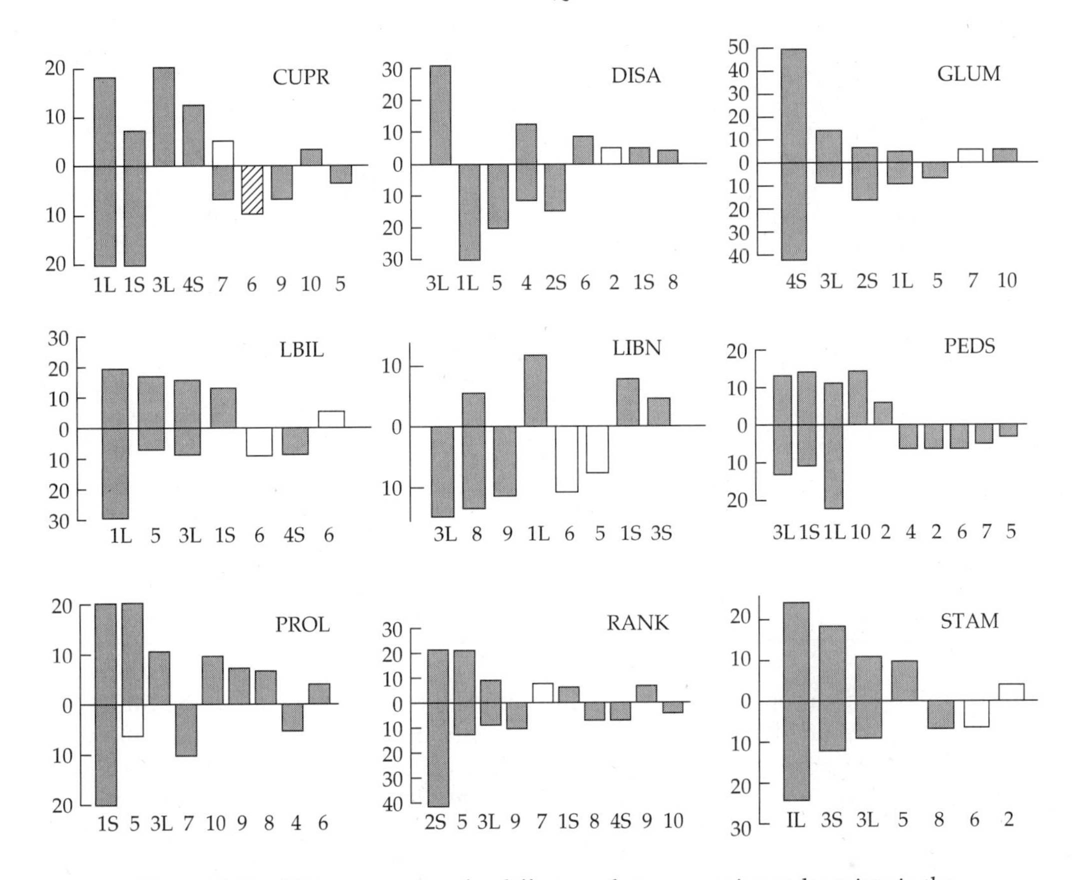

Figure 15.11 QTLs accounting for differences between maize and teosinte in the nine characters listed in Table 15.4. Each bar represents a detected QTL, with bar height indicating its r^2 value. Acronyms for characters are listed in Table 15.4. Chromosome position (chromosome number and short vs. long arm) is indicated under each bar. The upper bars refer to QTLs detected in a cross between a primitive maize strain and one subspecies of teosinte (*Z. m. parviglumis*), while the lower bars are for a cross involving a different primitive maize strain and another teosinte subspecies (*Z. m. mexicana*). Aligned bars denote QTLs mapping to very similar positions in both crosses. White bars indicate QTL effects in the opposite direction from parental phenotype, while the cross-hatched bar indicates apparent overdominance. (From Doebley and Stec 1993.)

QTL-1L was shown by complementation tests to be the locus *teosinte branched 1* (*tb1*). In maize, mutants at this locus result in teosinte-like features for inflorescence sex (tassels, not ears) and number (many instead of one), and length of lateral branches (long, not short). Doebley found that the joint effects of *tb1* and

Table 15.5 QTLs influencing age-specific weight and age-specific growth rates in mice, measured at weekly intervals.

	Age-specific weight (weeks)									
	1	2	3	4	5	6	7	8	9	10
No. of QTLs	7	10	16	13	15	15	14	14	16	17
Total r^2	0.29	0.30	0.56	0.52	0.59	0.63	0.64	0.56	0.67	0.76

	Age-specific growth			
	Early	Middle	6-week	Late
No. of QTLs	11	12	14	12
Total r^2	0.39	0.51	0.54	0.38

Source: From Cheverud et al. 1996.
Note: Early, Middle, and Late correspond to growth from 1 to 3 weeks, growth from 3 to 6 weeks, and growth from 6 to 10 weeks, respectively, while 6-week refers to growth from 1 to 6 weeks.

QTL-3L, by themselves, are sufficient to account for essentially all of the differences in plant architecture between teosinte and maize. Further, these two loci result in substantial differences in ear architecture. Hence, there is direct evidence that just a few genes can account for a very significant amount of the dramatic differences between teosinte and maize.

QTLs for Age-specific Growth in Mice

Cheverud et al. (1996) examined weight and growth using 535 F_2 mice from a cross between two inbred lines differing in size. A total of 75 microsatellite markers were used, generating 55 intervals that averaged around 28 cM in length. ML-interval mapping was used to examine age-specific weight (at ages 1 through 10 weeks) and age-specific growth rate. As Table 15.5 shows, considerable numbers of QTLs were found for all characters. All detected QTLs had small effects, with the largest accounting for around 10% of the F_2 phenotypic variance, while the average (detected) effect was around 4% of the F_2 variance. Note also from Table 15.5 that as age increases, so does the number of QTLs for weight and the total r^2 value of these detected loci. Early vs. late weight and growth showed different genetic architectures. First, largely distinct sets of QTLs are involved (Figure 15.12). Second, dominance was found to be much more important in early weight and growth than in late weight and growth.

Summary of QTL Mapping Experiments

QTL mapping using inbred-line crosses has been widely applied in many other

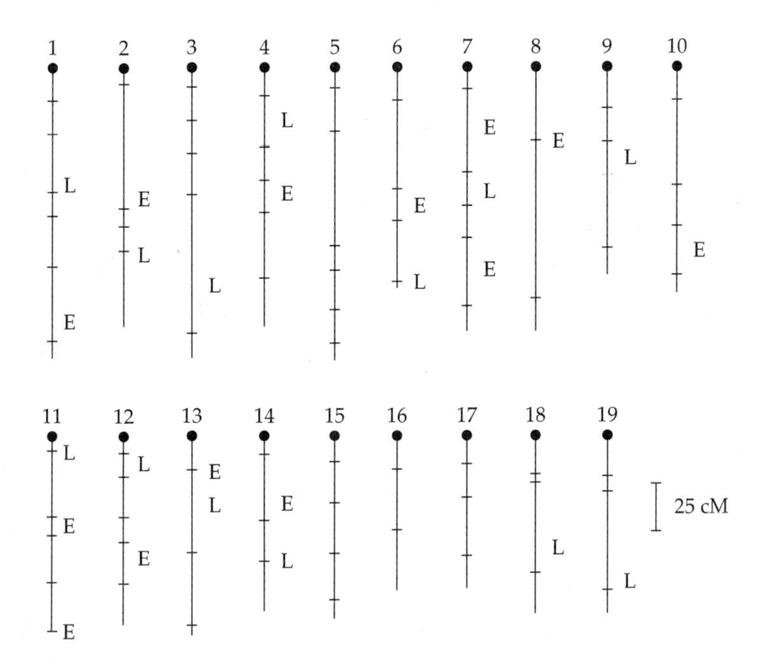

Figure 15.12 Locations for early (E) and late (L) growth QTLs on the 19 mouse autosomes. The marker locations are indicated by hatch marks. (From Cheverud et al. 1996.)

species of plants (mostly crops) and a few other animal species. The basic conclusion from these studies is that experiments using modest numbers of individuals (100–200) and markers (20–100) generally detect QTLs. In a survey we performed on 52 experiments covering a total of 222 traits, almost half (45%) of all traits had a QTL accounting for at least 20% of the total phenotypic variance (Figure 15.13). Figure 15.14 shows that there is little (if any) correlation between the number of detected QTLs and the total percentage of variation they explain. Most studies (84%) found the total contribution from all detected QTLs to be at least 20% of the total variance, and for a third of the traits it was at least 50%. In spite of these values, we emphasize the fact that the effects of *detected* QTLs can be severely overestimated, especially when the power to detect them is low (Figures 15.8, 15.9).

A second conclusion to be drawn from the existing data is that dominance is common. Epistatic interactions, on the other hand, appear to be fairly rare, although there are notable exceptions (e.g., Damerval et al. 1994, Doebley et al. 1995a, Lark et al. 1995, Long et al. 1995, Eshed and Zamir 1996, Cockerham and Zeng 1996). This general lack of epistasis may not reflect biological

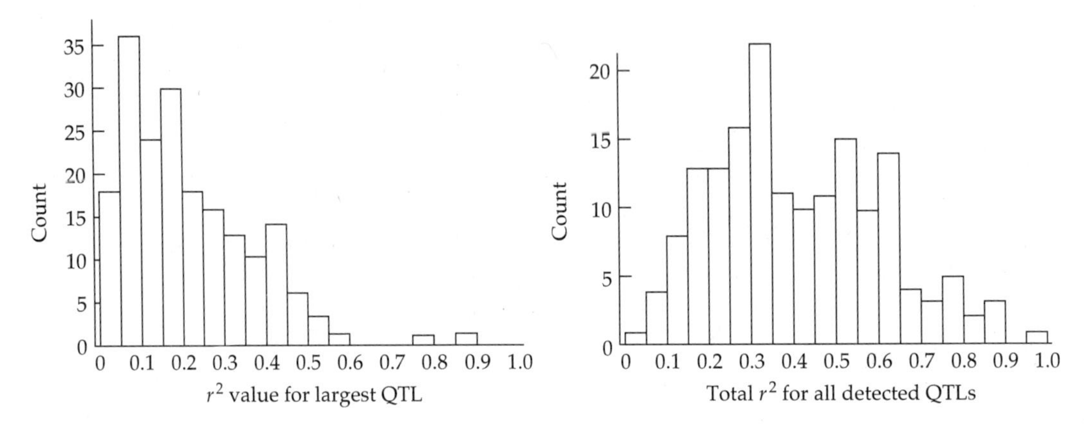

Figure 15.13 Summary of results from 52 QTL mapping experiments using inbred-line crosses (mainly crop plants), examining a total of 222 traits. **Left:** Distribution of r^2 values for the QTL of largest effect. **Right:** Distribution of the total effects accounted for by all detected QTLs.

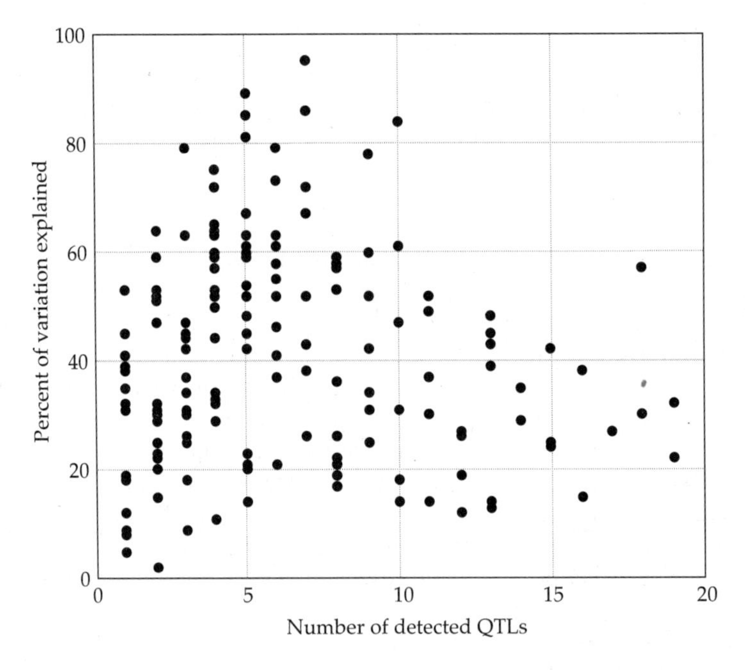

Figure 15.14 Joint distribution of the total percent of variation attributable to detected QTLs and QTL number for 52 experiments (covering 222 traits) from Figure 15.13.

reality, as several factors complicate its detection (Gallais and Rives 1993). Since most epistasis tests only examine markers showing significant single-locus effects, the results are very likely biased towards loci showing reduced epistasis. Smaller sample sizes for each multilocus genotype also reduce the power for detecting epistasis. It is perhaps noteworthy that some of the strongest evidence for epistasis comes from experiments using RILs, which control for background effects outside of the regions of interest.

Another issue that remains unresolved is the frequency of genotype × environment interactions involving detected QTLs. Two different approaches have been used to study this interaction: the consistency of marker-trait associations across environments and ANOVA methods incorporating specific terms for marker × environment interactions. The former measure is very crude, simply asking whether a marker-trait association is detected in all scored environments. If it is, this is generally taken as evidence of no G×E, while detection of an association in only some of the environments is often taken as evidence for G×E. However, a QTL can have a significant effect in all environments even in the presence of very significant G×E interaction. Likewise, low power of detection can result in a QTL being detected in only some of the replicates of an experiment, even when its effects are identical across environments. Consistent with this expectation, Koester et al. (1993) found that QTLs with small effects are less likely to be detected across environments than are QTLs with large effects. By explicitly testing for marker × environment interactions, ANOVA methods provide a more sensitive measure of G × E effects. As Table 15.6 shows, the conclusions from studies using either method are mixed.

One final caveat is necessary with respect to results from inbred-line crosses. Most parental populations are chosen because of their wide difference in traits of interest, so the relevance of these results to within-population variation remains unclear. Indeed, marker loci in these inbred-line crosses often show strong segregation distortion, suggesting very significant genetic divergence between parental lines (e.g., Vallejos and Tanksley 1983, Edwards et al. 1987, Paterson et al. 1988, Bonierbale et al. 1988, Doebley and Stec 1991, Schön et al. 1993). Direct methods for mapping QTLs responsible for within-population variation are developed in the next chapter.

Table 15.6 Selected studies examining G × E interaction in detected QTLs.

Organism/trait	Reference
Tomatos	
3 fruit characters in California (2 locations) and Israel. 4/29 QTLs detected in all three environments, 10/29 in two, 15/29 in one.	Paterson et al. 1991

Table 15.6 (Continued).

Organism/trait	Reference
Maize	
Flowering time, height in 3 North Carolina locations. Marker-trait associations displaying largest effects are generally constant over environments; markers with less significant associations are not as constant over environments.	Koester et al. 1993
11 yield-related characters in 4 North Carolina locations. Of 70 detected QTLs, 21% detected in all four locations, 34% in 2 or 3 locations, 44% in only one location.	Ragot et al. 1995
Yield in two locations in Northern Italy. Most detected QTLs consistent across environments.	Ajmone-Marsan et al. 1995
ANOVA analysis of 7 traits in 4–6 locations (4 in North Carolina, 1 in Iowa, 1 in Illinois). Little evidence of G×E in 4 traits, (yield, ear height, plant height, leaf area). Strong G×E in 3 others (days to tassel, grain moisture, ear number).	Stuber et al. 1992 Cockerham and Zeng 1996
ANOVA analysis of 7 characters over 2 different years. 6/28 significant marker-trait associations for starch concentration showed G×E, 6/16 for protein concentration, 12/16 for anthesis date, 7/14 for ear weight, 11/18 for height, 3/27 for kernel weight, and 9/31 for oil concentration.	Berke and Rocheford 1995
Corn borer resistance, height in 2 Iowa locations. All 10 detected QTLs (7 resistance, 3 height) gave very similar LOD maps across environments, although only 4/7 height QTLs were significant in both.	Schön et al. 1993
Gray Leaf Spot resistance in three environments. 22/33 significant marker-trait associations found in a single environment, 9/33 in two, 2/33 in all three. Averaged over all environments, only 20 of these marker-trait associations were significant.	Bubeck et al. 1993
Rapeseed (*Brassica napus*)	
Flowering time in 3 different vernalization treatments. LOD scores very similar over all three treatments.	Ferreira et al. 1995

Table 15.6 (Continued).

Organism/trait	Reference
Peas	
Node number in field and greenhouse locations. 3 QTLs detected only in greenhouse, 3 only in field, 1 in both.	Dirlewanger et al. 1994
Arabidopsis thaliana	
Flowering time in 6 vernalization/photoperiod treatments. Four of 12 QTLs detected by composite interval mapping showed QTL × E interactions.	Jansen et al. 1995

16

Mapping and Characterizing QTLs: Outbred Populations

For many species, especially humans, the luxury of manipulatable inbred lines is out of the question, in which case we must resort to outbred populations for QTL mapping. Outbred populations are also of interest for reasons other than experimental constraints. For example, QTLs detected by inbred-line crosses usually represent fixed differences *between* lines, or even different species, and the relevance of these results to QTLs segregating *within* populations remains unclear. This fundamental distinction that inbred-line crosses detect QTLs responsible for between-population differences while outbred populations detect QTLs responsible for within-population variation makes these complementary, rather than competing, approaches.

Using within-population variation, as opposed to fixed differences between populations, results in significantly decreased power for QTL detection. With inbred lines, all F_1 parents have identical genotypes (including the same linkage phase), so all individuals are informative, and linkage disequilibrium is maximized. Further, QTL effects are expressed as means with inbred lines (the average value of each QTL genotype), but as genetic variances with outbred populations. Since variances are estimated with much less precision than means, estimates from outbred populations are expected to be inherently less precise. All of these complications conspire to make QTL mapping in outbred populations a difficult, but not impossible, enterprise.

As with inbred lines, a variety of designs have been proposed for obtaining samples with the linkage disequilibrium required for QTL mapping. Although crosses between outbred lines can sometimes generate the required amounts of disequilibrium, collections of relatives are generally relied upon. We start our discussion by reviewing the types of outbred parents that are informative for QTL mapping and then consider analyses using collections of relatives, starting with ANOVA and maximum likelihood methods for family data. We conclude by examining approaches that use comparisons between pairs of relatives, as such comparisons are widely applied in mapping human disease genes.

MEASURES OF INFORMATIVENESS

The major difference between QTL analysis using inbred-line crosses vs. outbred populations is that while the parents in the former are genetically uniform, parents in the latter are genetically variable. This distinction has several consequences. First, only a fraction of the parents from an outbred population are **informative**. For a parent to provide linkage information, it must be heterozygous at both a marker *and* a linked QTL, as only in this situation can a marker-trait association be generated in the progeny. Only a fraction of random parents from an outbred population are such double heterozygotes. With inbred lines, F_1's are heterozygous at all loci that differ between the crossed lines, so that all parents are fully informative. Second, there are only two alleles segregating at any locus in an inbred-line cross design, while outbred populations can be segregating any number of alleles. Finally, in an outbred population, individuals can differ in marker-QTL linkage phase, so that an M-bearing gamete might be associated with QTL allele Q in one parent, and with q in another. Thus, with outbred populations, marker-trait associations must be examined *separately* for each parent. With inbred-line crosses, all F_1 parents have identical genotypes (including linkage phase), so one can simply average marker-trait associations over all offspring, regardless of their parents.

Before considering the variety of QTL mapping methods for outbred populations, some comments on the probability that an outbred family is informative are in order. A parent is **marker-informative** if it is a marker heterozygote, **QTL-informative** if it is a QTL heterozygote, and simply **informative** if it is both. Unless *both* the marker and QTL are highly polymorphic, most parents will not be informative. Given the need to maximize the fraction of marker-informative parents, classes of marker loci successfully used with inbred lines may not be optimal for outbred populations. For example, RFLPs are widely used in inbred lines, but these markers are typically diallelic and hence have modest polymorphism (at best). Microsatellite marker loci (Chapter 14), on the other hand, are highly polymorphic and hence much more likely to yield marker-informative individuals.

As shown in Table 16.1, there are three kinds of marker-informative crosses. With a highly polymorphic marker, it may be possible to examine marker-trait associations for both parents. With a **fully informative family** ($M_iM_j \times M_kM_\ell$) all parental alleles can be distinguished, and both parents can be examined by comparing the trait values in M_i- vs. M_j- offspring and M_k- vs. $M_\ell-$ offspring. With a **backcross family** ($M_iM_j \times M_kM_k$), only the heterozygous parent can be examined for marker-trait associations. Finally, with an **intercross family** ($M_iM_j \times M_iM_j$), homozygous offspring (M_iM_i, M_jM_j) are unambiguous as to the origin of parental alleles, while heterozygotes are ambiguous, because allele M_i (M_j) could have come from either parent.

In designing experiments, it is useful to estimate the fraction of families

Table 16.1 Types of marker-informative matings.

Fully informative: $M_i M_j \times M_k M_\ell$

Parents are different marker heterozygotes.
All offspring are informative in distinguishing alternative alleles from both parents.

Backcross: $M_i M_j \times M_k M_k$

One parent is a marker heterozygote, the other a marker homozygote.
All offspring informative in distinguishing heterozygous parent's alternative alleles.

Intercross: $M_i M_j \times M_i M_j$

Both parents are the same marker heterozygote.
Only homozygous offspring informative in distinguishing alternative parental alleles.

Note: Here i, j, k, and ℓ index different marker alleles.

expected to be marker-informative. One measure of this is the **polymorphism information content** (Botstein et al. 1980), or PIC, of the marker locus,

$$\text{PIC} = 1 - \sum_{i=1}^{n} p_i^2 - \sum_{i=1}^{n-1} \sum_{j=i+1}^{n} 2\, p_i^2\, p_j^2 \leq \frac{(n-1)^2(n+1)}{n^3} \tag{16.1}$$

which is the probability that one parent is a marker heterozygote and its mate has a *different* genotype (i.e., a backcross or fully informative family, but excluding intercross families). In this case, we can distinguish between the alternative marker alleles of the first parent in all offspring from this cross. The upper bound (given by the right hand side of Equation 16.1) occurs when all marker alleles are equally frequent, $p_i = 1/n$. The **proportion of fully informative matings** (PFIM), where marker alleles from *both* parents can be distinguished in all offspring, is given by

$$\begin{aligned} \text{PFIM} &= \sum_{i=1}^{n-1} \sum_{j=i+1}^{n} 2\, p_i\, p_j \left[\left(\sum_{k=1}^{n-1} \sum_{\ell=k+1}^{n} 2\, p_k\, p_\ell \right) - 2\, p_i\, p_j \right] \\ &\leq \frac{(n-1)\,(n-2)\,(n+1)}{n^3} \end{aligned} \tag{16.2}$$

where i, j, k, and ℓ denote *different* marker alleles (Götz and Ollivier 1992). Again, the upper bound occurs when all alleles are equally frequent. Figure 16.1 plots the maximum values of both of these measures as a function of the number of marker alleles.

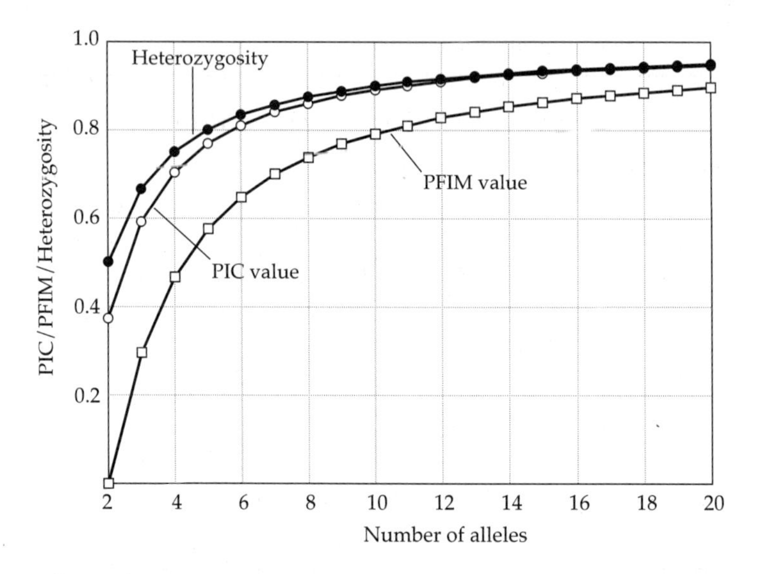

Figure 16.1 The relationship between three measures of marker information: heterozygosity, PIC, and PFIM. Plotted are the maximum values for each measure as a function of the number of marker alleles under the assumption of an even allele-frequency distribution ($p_i = 1/n$).

Of course, a marker-informative parent is not necessarily also QTL-informative. For a QTL with n_Q alleles, the ith of which has frequency q_i, the probability that a parent is a QTL heterozygote (in a random mating population) is

$$\left(1 - \sum_{i=1}^{n_Q} q_i^2\right) \leq \left(1 - \frac{1}{n_Q}\right) \tag{16.3}$$

For a population where parents are in linkage equilibrium, the probability that (at least) one parent is a double (QTL/marker) heterozygote and that alternative marker alleles from this parent can unambiguously be distinguished in all offspring is

$$\Pr(\text{at least one parent fully informative}) = \text{PIC} \times \left(1 - \sum_{i=1}^{n_Q} q_i^2\right) \tag{16.4a}$$

while the probability that both parents satisfy this condition is

$$\Pr(\text{both parents fully informative}) = \text{PFIM} \times \left(1 - \sum_{i=1}^{n_Q} q_i^2\right)^2 \tag{16.4b}$$

Thus, under fairly general settings, we expect most parents to be uninformative about a particular QTL / marker combination.

Example 1. Consider a marker locus segregating five alleles in a population. Here,

$$\text{PIC} \leq \frac{4^2 \cdot 6}{5^3} = 0.768 \qquad \text{and} \qquad \text{PFIM} \leq \frac{4 \cdot 3 \cdot 6}{5^3} = 0.576$$

Hence, in order to have 100 families that are marker-informative for at least one parent, 100 / 0.768 = 130 randomly drawn families must be examined. If one requires these families to be informative for both parents, then 100 / 0.576 = 174 families must be examined. These are the *best-case* scenarios for this marker, occurring if all five alleles have equal frequency.

Now suppose that the marker is linked to a QTL with three alleles whose frequencies are 0.5, 0.3, and 0.2. The probability that an individual is a QTL heterozygote is $1 - (0.5^2 + 0.3^2 + 0.2^2) \simeq 0.62$. Thus, to have 100 families with one or both parents fully informative requires sampling (at least) $130/0.62 \simeq 210$ and $174/0.62^2 \simeq 453$ random families, respectively.

SIB ANALYSIS: LINEAR MODELS

One can use family data to search for QTLs by comparing offspring carrying alternative marker alleles from the same parent. There are a number of ways to accomplish this apparently simple task, and our treatment first considers collections of relatives, such as entire half- or full-sib families.

Sibship-based methods for detecting linked QTLs historically developed along two different fronts. Animal breeders typically rely on a few large half-sib families, each resulting from a single sire (often via artificial insemination). Human geneticists, on the other hand, rely on a number of small full-sib families. These different designs not only reflect the biological limitations present in different species but also mirror concerns about the tradeoff between number of families (increasing the chance of sampling at least one QTL-informative sibship) and number of sibs per family (increasing the power to detect marker differences, *provided* the family is QTL-informative). Neither design is necessarily optimal, since the highest power (for a fixed sample size) occurs when a single large *informative* sibship is used (A. Hill 1975, Weller et al. 1990, Luo 1993). Unfortunately, although we can gauge if a particular parent is marker-informative, we have no way of knowing beforehand whether that parent is also a QTL heterozygote.

A variety of linear model approaches for dealing with sibship data have been proposed. We start by considering the simplest case of a single half-sib family. Care is required when combining information from several half-sibships, and we examine this issue next. Finally, we consider how half-sib analyses are extended to full sibs. Upon completion of our discussion of linear models, we will provide a broad overview of the maximum likelihood approach to sib analysis.

A Single Half-sib Family

Suppose a single sire is crossed to n unrelated dams and a single offspring from each mating is measured. Ideally, this sire should be informative at a number of marker loci covering a significant fraction of the genome (as can be accomplished by using a large number of highly informative markers). While we have much less control over the choice of QTL-informative sires, if a major gene is involved, a pilot segregation analysis with a number of candidate sires can be used to estimate the probability that a particular sire is a QTL heterozygote (e.g., Equation 13.24), and those sires with the highest such probabilities can be used. One might also select those sires showing the greatest offspring variance (the idea being that this variance increases with the number of heterozygous QTLs), but within-family variance can be a poor indicator of sire QTL heterozygosity.

Let M_1 and M_2 denote the alternative sire alleles at a heterozygous marker locus, and let z_{ij} denote the phenotype of the jth individual with sire marker allele i. The simplest linear model for examining whether this marker is linked to a QTL (heterozygous in the sire) is

$$z_{ij} = \mu + m_i + e_{ij} \tag{16.5}$$

where the marker effect m_i is the expected deviation from the mean trait value μ, given that an offspring has sire allele M_i. Since there are only two marker classes, a t test can be used to determine if offspring with M_1 vs. M_2 sire alleles have significantly different means,

$$t = \frac{\overline{z}_{M_1} - \overline{z}_{M_2}}{\sqrt{\text{Var}(\overline{z}_{M_1}) + \text{Var}(\overline{z}_{M_2})}} = \frac{\text{MC}}{\text{SE}} \tag{16.6}$$

Here, MC denotes the **marker contrast** between alternative sire alleles, and SE is the standard error of the marker contrast. A significant t value implies that the marker is linked to a QTL that is heterozygous in the sire. Typically, only those offspring whose sire alleles can be determined are included in this analysis. If the marker locus is not very polymorphic, many offspring will be uninformative. Dentine and Cowan (1990) present a GLS regression method that allows uninformative individuals to be included in the analysis.

Finally, consider the expected value of the marker contrast, given the sire genotype. Suppose the sire is M_1Q_i/M_2Q_j and that c is the recombination frequency between the marker and the QTL. The conditional probabilities for alternate QTL alleles in gametes carrying a known sire marker allele are

$$\Pr(\,Q_i\,|\,M_1\,) = \Pr(\,Q_j\,|\,M_2\,) = 1-c\,, \qquad \Pr(\,Q_i\,|\,M_1\,) = \Pr(\,Q_j\,|\,M_2\,) = c \tag{16.7}$$

Suppose there are n_Q QTL alleles ($Q_1, \cdots, Q_{n_Q}$; with frequencies $q_1, \cdots, q_{n_Q}$). Following Chapter 4, decompose the mean value of Q_iQ_j as $\mu_{Q_iQ_j} = \mu + \alpha_i + \alpha_j + \delta_{ij}$, where the deviation from the population mean is determined by the average effects of both alleles ($\alpha_i + \alpha_j$) plus the dominance deviation (δ_{ij}). Assuming random mating, the dam contributes an allele at random, giving the expected value in offspring with sire marker M_1 as

$$
\begin{aligned}
E\left[\,\mu_{M_1} \,\middle|\, \text{sire} = M_1Q_i/M_2Q_j\,\right] &= (1-c)\sum_{k=1}^{n_Q} \mu_{Q_iQ_k}\, q_k + c\sum_{k=1}^{n_Q} \mu_{Q_jQ_k}\, q_k \\
&= \mu + (1-c)\,\alpha_i + c\,\alpha_j
\end{aligned} \tag{16.8}
$$

which follows since $\sum \alpha_k\, q_k = \sum \delta_{jk}\, q_k = 0$ (Chapter 4). The expected value for the marker contrast from this sire becomes

$$
E\left[\,\mu_{M_1} - \mu_{M_2} \,\middle|\, \text{sire} = M_1Q_i/M_2Q_j\,\right] = (1-2c)\,(\alpha_i - \alpha_j) \tag{16.9}
$$

Thus, assuming a single QTL linked to the marker, the expected marker contrast for a given sire is the difference of the average effects of his alleles at this QTL, scaled by the QTL-marker recombination frequency.

While a marker contrast may be significantly different from zero *within* a sibship, the expected value of this contrast *across* a random collection of sibships is zero for a population in linkage equilibrium, as $E(\alpha_x) = 0$. Thus, one cannot simply combine contrasts across families, as these are expected to cancel. On the other hand, *squared* marker contrasts have expected values that are a function of the additive variance of the linked QTL. To see this relationship, assume that each sire marker allele has sample size $n/2$ and that the within-marker (residual) variance is the same for both marker alleles. Under these assumptions, $\sigma^2(\overline{z}_{M_1}) = \sigma^2(\overline{z}_{M_2}) = \sigma^2_e/(n/2)$, implying that the variance of the observed marker contrast about its expected value within each family is

$$
\sigma^2\,(\text{MC}) = E\left(\text{SE}^2\right) = \sigma^2(\overline{z}_{M_1}) + \sigma^2(\overline{z}_{M_2}) = \frac{4\sigma^2_e}{n} \tag{16.10a}
$$

Combining this result with Equation 16.9 gives the expected value of a squared marker contrast as

$$
\begin{aligned}
E\left(\text{MC}^2\right) &= E\left[\,(\,\overline{z}_{M_1} - \overline{z}_{M_2}\,)^2\,\right] = E\left[\,(\,\mu_{M_1} - \mu_{M_2})^2\,\right] + \sigma^2(\overline{z}_{M_1}) + \sigma^2(\overline{z}_{M_2}) \\
&= (1-2c)^2\, E\left[\,(\alpha_i - \alpha_j)^2\,\right] + \frac{4\,\sigma^2_e}{n} \\
&= (1-2c)^2\sigma^2_A + \frac{4\,\sigma^2_e}{n}
\end{aligned} \tag{16.10b}
$$

where the last step follows from $E[\,(\alpha_i - \alpha_j)^2\,] = 2E(\,\alpha_i^2\,) = \sigma^2_A$ (Equation 4.23b).

Several Half-sib Families

Equation 16.10b immediately suggests that one approach for combining information from N (unrelated) half-sib families is to use the sum of the squared t statistics (Equation 16.6) for each (informative) marker contrast,

$$T = \sum_{i=1}^{N} t_i^2 = \sum_{i=1}^{N} \frac{\text{MC}_i^2}{\text{SE}_i^2} \tag{16.11a}$$

where from Equations 16.10a and 16.10b,

$$E(\,T\,) \simeq N\,\frac{E(\,\text{MC}^2\,)}{E(\,\text{SE}^2\,)} = N\left[\,1 + (1-2c)^2 \frac{n}{4}\,\frac{\sigma_A^2}{\sigma_e^2}\,\right] \tag{16.11b}$$

(We stress that σ_A^2 is the additive genetic variance accounted for by the QTL, not the trait as a whole.) A value of T significantly greater than N suggests linkage to a QTL. If the sample size within each marker class is sufficiently large, so that the observed value of SE^2 can be taken to be very close to the population value of $\sigma^2(\text{MC})$, then the distribution of T is approximately χ^2 with N degrees of freedom.

Rearranging Equation 16.11b suggests the following method-of-moments estimate of $(1-2c)^2\sigma_A^2$,

$$\frac{4\,\text{Var}(e)}{n}\left(\frac{T}{N} - 1\right) \tag{16.12}$$

where $\text{Var}(e)$ is an estimate of σ_e^2 (the within-marker class variance). Aspects of this approach are further examined by Geldermann (1975), Weller et al. (1990), and van der Beck et al. (1995).

A closely related, and more widely used, approach is a nested ANOVA (Neimann-Sørensen and Robertson 1961, Lowry and Schultz 1959), which considers marker effects nested within each sibship. We only briefly consider the properties of nested ANOVA here, as these are extensively discussed in Chapter 18 (where effects of dams are nested within those for sires for the different purpose of estimating genetic variance components summed over all loci). The basic linear model is

$$z_{ijk} = \mu + s_i + m_{ij} + e_{ijk} \tag{16.13}$$

where z_{ijk} denotes the phenotype of the kth individual of marker genotype j from sibship i, s_i is the effect of sire i, m_{ij} is the effect of marker genotype j in sibship i, and e_{ijk} is the within-marker, within-sibship residual. It is assumed that s, m, and e have means equal to zero, are uncorrelated, and are normally distributed with variances σ_s^2 (the between-sire variance), σ_m^2 (the between-marker, within-sibship variance), and σ_e^2 (the residual or within-marker, within-sibship variance). A significant marker variance indicates linkage to a segregating QTL, and is tested by using the statistic

$$F = \frac{\text{MS}_m}{\text{MS}_e} \tag{16.14}$$

where the mean squares are given in Table 18.3. Assuming normality, Equation 16.14 follows an F distribution under the null hypothesis that $\sigma_m^2 = 0$. Assuming a balanced design with N sires, each with $n/2$ half-sibs in each marker class, Equation 16.14 has N and $N(n-2)$ degrees of freedom. Chapter 18 discusses the appropriate degrees of freedom under an unbalanced design.

Example 2. Three studies have attempted to map QTLs influencing milk production characters in dairy cows with nested ANOVA, using electrophoretically scored markers to examine half-sibships. The pioneering study of Neimann-Sørensen and Robertson (1961) used six markers and six production characters in Red Danish cattle, with 123 sires and 12 to 13 daughters per sire. None of the 36 possible marker-trait associations were significant at the 1% level, while three were significant at the 5% level. Four production characters in Holsteins (14 markers) and Guernseys (15 markers) were examined by Gonyon et al. (1987) and Haenlein et al. (1987), respectively. In Holsteins, the average number of sires was 181, with an average of 5.3 daughters per sire. Of 56 possible marker-character combinations, six were significant at the 1% level, and another three were significant at the 5% level. The sample sizes were smaller in Guernseys, with an average of 88 sires and 5.4 daughters per sire. None of the resulting 64 marker-character combinations were significant at the 1% level, while three were significant at the 5% level.

Taking into account the problem of multiple comparisons, the Red Danish data do not show convincing evidence of marker effects, as the probability of finding 3 (or more) of 36 tests significant at the 5% level is 0.10. The same is true for the Guernsey data, as the probability of 3 (or more) out of 64 tests being significant at the 5% level is 0.40. On the other hand, at least some of the Holstein associations are likely to be significant, even after accounting for multiple comparisons.

The connection between sums of squared marker contrasts used in t tests (Equation 16.11) and nested ANOVAs can be seen by considering the marker mean squares, MS_m, which can be expressed as a function of the marker contrasts,

$$\text{MS}_m = \frac{1}{N}\sum_{i=1}^{N}\frac{n}{2}\left[(\overline{z}_{i1}-\overline{z}_{i\cdot})^2+(\overline{z}_{i2}-\overline{z}_{i\cdot})^2\right] = \frac{n}{4N}\sum_{i=1}^{N}\text{MC}_i^2 \tag{16.15}$$

The first equality follows from Table 18.3, while the second follows by noting that the mean phenotype of progeny of the ith sire is $\overline{z}_{i\cdot} = (\overline{z}_{i1}+\overline{z}_{i2})/2$, where $\overline{z}_{ij}$ is the mean trait value for marker j in family i.

Again referring to Table 18.3, we see that for a balanced design the mean squares have expected values of

$$E(\,\text{MS}_m\,) = \sigma_e^2 + n\sigma_m^2 \qquad \text{and} \qquad E(\,\text{MS}_e\,) = \sigma_e^2 \tag{16.16}$$

Table 16.2 Sample sizes required to have a 90% chance of detecting a linked QTL using ANOVA with significance level $\alpha = 0.05$.

	$\Delta = 0.10$		$\Delta = 0.05$		$\Delta = 0.01$	
N	n	N_T	n	N_T	n	N_T
5	150	1,505	290	2,905	1450	14,505
20	40	1,620	110	4,420	540	21,620
100	22	4,500	40	8,100	200	40,100
200	14	5,800	28	11,400	132	53,000

Note: The QTL effect is measured by $\Delta = (1-2c)^2\sigma_A^2/\sigma_e^2$, where c is the marker-QTL recombination frequency. Here σ_e^2 is the residual variance, i.e., the variance within half-sib families not accounted for by the marker effects. N and n denote the number of sires and the number of offspring per sire, assuming $n/2$ offspring per marker class per sire. $N_T = N\,(2n+1)$ is the total number of genotyped individuals (one for each sire, dam, and offspring). All sires are assumed to be marker informative, and we assume that N is sufficiently large that at least one family is QTL-informative. The number $N(\gamma)$ of random sires required to have probability γ that at least one of them is QTL-informative (for a QTL with n_Q alleles with frequencies $q_1, \cdots, q_{n_Q}$) satisfies Pr(none of the N sires are informative) $= \left(\sum_{i=1}^{n_Q} q_i^2\right)^{N(\gamma)} = 1-\gamma$. Solving gives $N(\gamma) = \ln(1-\gamma)/\ln\left(\sum_{i=1}^{n_Q} q_i^2\right) \geq \gamma/\ln(n_Q)$, where the last step follows from $\sum_{i=1}^{n_Q} q_i^2 \geq 1/n_Q$ and $\ln(1-\gamma) \simeq -\gamma$.

with Equations 16.10b and 16.15 gives,

$$\sigma_m^2 = \frac{E(\text{MS}_m) - E(\text{MS}_e)}{n} = (1-2c)^2\,\frac{\sigma_A^2}{4} \tag{16.17}$$

Thus, an estimate of the QTL effect (measured by its additive variance, scaled by the distance between QTL and marker), can be obtained from the observed mean squares.

One immediate drawback of measuring a QTL effect by its variance in an outbred population is that even a completely linked QTL with a large effect can nonetheless have a small σ_m^2. Consider a strictly additive diallelic QTL with allele frequency p, where $\sigma_A^2 = 2a^2p(1-p)$. Even if a is large, the additive genetic variance can still be quite small if the QTL allele frequencies are near zero or one. An alternative way of visualizing this relationship is to note that the probability of a QTL-informative sire is $2p(1-p)$. If this is small, even if a is large, σ_A^2 will be small, as most families will not be informative. In those rare informative families, however, the between-marker effect is large. Combining this

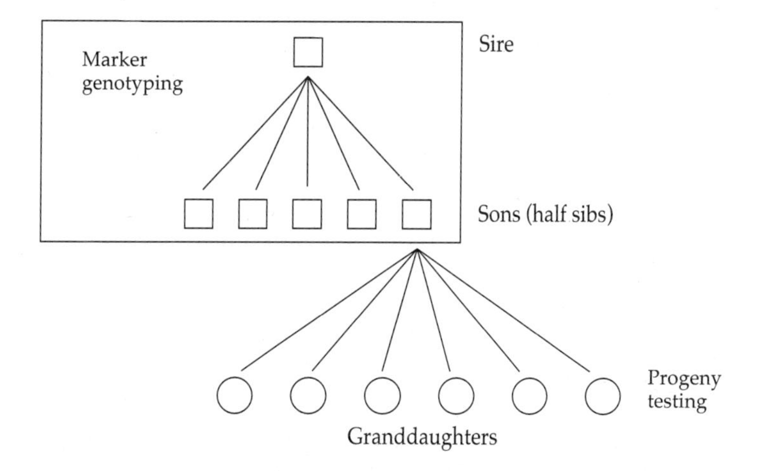

Figure 16.2 The granddaughter design of Weller et al. (1990). Here, each sire produces a number of half-sib sons that are scored for the marker genotypes. The character value for each son is determined by progeny testing, with the trait value being scored in a large number of daughters (again half-sibs) from each son. More general designs allowing for both full- and half-sib families are examined by van der Beck et al. (1995).

Contrast this to the situation with inbred-line crosses, where the QTL effect estimates $2a$, since here all families are informative, rather that the fraction $2p(1-p)$ seen an outbred population.

POWER OF NESTED ANOVA DESIGNS

The power of nested ANOVA designs to detect QTLs in a collection of half-sib families has been examined by Soller and Genizi (1978), Weller et al. (1990), Luo (1993), and van der Beck et al. (1995). Details of the power calculations are given in Appendix 5, and some results are presented in Table 16.2. Inspection of this table shows that increasing the number of sibs per family (n) is more efficient than increasing the number of families (N), as the total number of individuals that must be genotyped (N_T) is smaller for small values of N. These power calculations assume that N is sufficiently large so as to include at least one QTL-informative family.

One approach for increasing power is the **granddaughter design** (Weller et al. 1990), under which each sire produces a number of sons that are genotyped for sire marker alleles (Figure 16.2), and the trait values for each son are taken to be the mean value of the traits in offspring from the son (rather than the direct measures of the son itself). This design was developed for milk-production characters in dairy cows, where the offspring are granddaughters of the original sires. The linear model for this design is

$$z_{ijk\ell} = \mu + g_i + m_{ij} + s_{ijk} + e_{ijk\ell} \tag{16.18}$$

where g_i is the effect of grandsire i, m_{ij} is the effect of marker allele j $(= 1, 2)$ from the ith sire, s_{ijk} is the effect of son k carrying marker allele j from sire i, and $e_{ijk\ell}$ is the residual for the ℓth offspring of this son. Sire marker-allele effects are halved by considering granddaughters (as opposed to daughters), as there is only a 50% chance that the grandsire allele is passed from its son onto its granddaughter. However, this reduction in the expected marker contrast is usually more than countered by the smaller standard error associated with each contrast due to the large number of offspring used to estimate trait value. Van der Beck et al. (1995) examine the power of a number of three-generation designs, concluding that they generally offer more power than two-generation (parent-offspring) designs.

Finally, Mackinnon and Georges (1992) show that another factor that can decrease power is selection, which has the effect of reducing the between-marker constrasts. Even a small amount of selection acting on the trait of interest can have a large effect on power.

A Single Full-sib Family

Large half-sib families are not feasible for many species, and one must instead consider full-sib families, which are often much smaller (as is the case in humans). Some domesticated animals such as pigs and poultry can have full-sib families of at least modest size, while others (such as certain trees) can have huge full-sib families.

With a single full-sib family, one can again use a simple one-way ANOVA to detect QTLs (Equation 16.5). The only modification required is that the number of offspring marker genotypes ranges from two to four, depending on the marker genotypes of the parents (Table 16.1). With a backcross family ($M_iM_j \times M_kM_k$), the resulting ANOVA has two marker genotypes (M_iM_k and M_jM_k), corresponding to the alternate alleles from the informative parent (and hence the marker contrast $\overline{z}_{M_i} - \overline{z}_{M_j}$). With a fully informative family ($M_iM_j \times M_kM_\ell$), all genotypes are informative as to parental alleles, and the resulting ANOVA considers all four marker genotypes (giving rise to two marker contrasts, $\overline{z}_{M_i} - \overline{z}_{M_j}$ and $\overline{z}_{M_k} - \overline{z}_{M_\ell}$, one for each parent). Finally, while an intercross family ($M_iM_j \times M_iM_j$) has three offspring marker genotypes, heterozygotes provide no information on parental marker alleles and are usually excluded. The ANOVA in this case considers only M_iM_i and M_jM_j offspring (leading to the marker contrast $\overline{z}_{M_iM_i} - \overline{z}_{M_jM_j}$). Obviously, as one considers different markers, a family that is fully informative for one marker may be a backcross family for another, and noninformative for a third.

With a large half-sib family, the effect of a QTL allele from the common parent (typically the sire) is averaged over a large number of random QTL alleles from the other parents, and hence only its additive effect enters into the expected values of marker contrasts. Within a full-sib family, however, each QTL allele from an informative parent is averaged over at most two QTL alleles from the other parent, which results in dominance entering into the expected marker contrasts. To see this result, denote the genotypes of the parents by $M_1Q_1\,/\,M_2Q_2$ and $M_3Q_3\,/\,M_4Q_4$. This notation simply indicates that the marker allele M_i is associated with QTL allele Q_i, and does not indicate the actual state of the allele itself. We also allow the possibility that two or more alleles are the same, so that (for example), $M_1 = M_4$ and $Q_1 = Q_2$.

Consider the gametes from the first parent. Among M_1-bearing gametes, $(1 - c)$ contain Q_1 while the rest (c) contain Q_2. Likewise, of the M_3-bearing gametes, $(1 - c)$ contain Q_3 while the rest (c) contain Q_4. The resulting expected genotypic value for M_1M_3 offspring becomes

$$\mu_{M_1M_3} = (1 - c)^2\,\mu_{Q_1Q_3} + c(1 - c)\,\mu_{Q_1Q_4} + c(1 - c)\,\mu_{Q_2Q_3} + c^2\,\mu_{Q_2Q_4} \quad (16.19a)$$

The other three marker means follow similarly. These values can be used to compute the average genotypic value of offspring (within this family) carrying a particular parental marker allele. For example, for M_1,

$$\begin{aligned} \mu_{M_1} &= \frac{\mu_{M_1M_3} + \mu_{M_1M_4}}{2} \\ &= (1 - c)\left(\frac{\mu_{Q_1Q_3} + \mu_{Q_1Q_4}}{2}\right) + c\left(\frac{\mu_{Q_2Q_3} + \mu_{Q_2Q_4}}{2}\right) \end{aligned} \quad (16.19b)$$

Using these results, the expected value of the contrast between offspring bearing alternative marker alleles from the first parent becomes

$$\begin{aligned} \mu_{M_1} - \mu_{M_2} &= (1 - 2c)\left[\frac{\mu_{Q_1Q_3} + \mu_{Q_1Q_4}}{2} - \frac{\mu_{Q_2Q_3} + \mu_{Q_2Q_4}}{2}\right] \\ &= (1 - 2c)\left[(\alpha_1 - \alpha_2) + \left(\frac{\delta_{13} + \delta_{14}}{2} - \frac{\delta_{23} + \delta_{24}}{2}\right)\right] \end{aligned} \quad (16.20)$$

where we have again decomposed the QTL genotype as $\mu_{Q_iQ_j} = \mu + \alpha_i + \alpha_j + \delta_{ij}$. As expected, if this parent is a QTL homozygote ($Q_1 = Q_2$), the contrast has expected value zero. Comparing this with the similar contrast in a half-sib design (Equation 16.9) shows that the two are identical in the absence of dominance $\delta_{ij} = 0$. When dominance is present, the contrast in a full-sib design can be larger or smaller than that which would occur in a half-sib design, depending on the dominance relationships between the QTL alleles of the focal individual and those of its mate.

For an intercross family (both parents are the same marker heterozygote, say, M_1M_2), the contrast of interest is between the marker homozygotes, and (noting that $M_1 = M_3$ and $M_2 = M_4$) Equation 16.19a gives

$$\begin{aligned} \mu_{M_1M_1} - \mu_{M_2M_2} &= (1 - 2c)\,[\,\mu_{Q_1Q_3} - \mu_{Q_2Q_4}\,] \\ &= (1 - 2c)\,[\,(\alpha_1 + \alpha_3) - (\alpha_2 + \alpha_4) + (\delta_{13} - \delta_{24})\,] \end{aligned} \quad (16.21)$$

Note that even though the parents share marker alleles (so that $M_1 = M_3$ and $M_2 = M_4$), they need not share the corresponding QTL alleles (so that Q_1 and Q_3 are potentially different, as are Q_2 and Q_4).

Several Full-sib Families

As with half-sibs, the contrasts noted above are conditional on the parental genotypes, and have expected value zero when considered over a random collection of parents. However, either scaled squared marker contrasts or nested ANOVA can be used to combine results from a collection of full-sib families into an estimate of the genetic variance associated with the marker alleles. The squared marker contrasts used in both backcross and fully informative families have expected values (under a balanced design of $n/2$ sibs per marker class) of

$$\begin{aligned} E(\mathrm{MC}^2) &= E[(\overline{z}_{M_1} - \overline{z}_{M_2})^2] \\ &= (1-2c)^2 \left[2\,E(\alpha^2) + 4\,(1/2)^2 E(\delta^2) \right] + \frac{4\,\sigma_e^2}{n} \\ &= (1-2c)^2 \left(\sigma_A^2 + \sigma_D^2 \right) + \frac{4\,\sigma_e^2}{n} \end{aligned} \tag{16.22}$$

In the absence of dominance ($\sigma_D^2 = 0$), this equation reduces to the value for half-sibs (Equation 16.10b).

Intercross families use a slightly different marker contrast,

$$\mathrm{MC}_I = \overline{z}_{M_1 M_1} - \overline{z}_{M_2 M_2}$$

Since the sample sizes for the two marker genotypes are now $n_{M_i M_i} = n/4$, the sampling variance of the marker contrasts is doubled (with $E(\mathrm{SE}_I^2) = 8\sigma_e^2/n$), giving the expected value of the squared marker contrast as

$$\begin{aligned} E(\mathrm{MC}_I^2) &= (1-2c)^2 \left[4\,E(\alpha^2) + 2E(\delta^2) \right] + \frac{8\,\sigma_e^2}{n} \\ &= 2\left[(1-2c)^2 \left(\sigma_A^2 + \sigma_D^2 \right) + \frac{4\,\sigma_e^2}{n} \right] \end{aligned} \tag{16.23}$$

This is twice the expected value for the contrasts for backcross and fully informative families, exactly compensating for the doubling of sampling variance. Hence, for all informative contrasts we have

$$E(t^2) \simeq \frac{E(\mathrm{MC}^2)}{E(\mathrm{SE}^2)} = 1 + \frac{n}{4}\,(1-2c)^2 \left(\frac{\sigma_A^2 + \sigma_D^2}{\sigma_e^2} \right) \tag{16.24}$$

As with Equation 16.11, a test statistic T is obtained as the sum of the t^2 values over all N informative contrasts, with T significantly greater than N being taken as evidence for a linked QTL.

As in the case of half-sib analysis, nested ANOVA is often used in place of sums of squared marker constrasts. The ANOVA model for full-sib families becomes

$$z_{ijk} = \mu + f_i + m_{ij} + e_{ijk}$$

where f_i, the effect of the ith full-sib family, replaces the half-sib family (or sire) effect s_i in the half-sib design, and the number of marker effects ranges from two to four, depending on the type of family. Again, a significant between-marker within-sibship variance (σ^2_m) indicates linkage to a segregating QTL, and this is tested using the statistic $F = \text{MS}_m/\text{MS}_e$. The use of nested ANOVAs for full-sib families was developed by Jayakar (1970) and A. Hill (1975), while Soller and Genizi (1978), Luo (1993), Knott (1994), van der Beck (1995), and Muranty (1996) examine the power of these designs.

If one ignores the marker heteroyzgotes in intercross families and considers the marker contrasts for all informative parents, then using the same logic that leads to Equation 16.15 shows that the contribution to MS_m is $(n/4)\text{MC}^2$ for each informative contrast in a backcross or fully informative family and $(n/8)\text{MC}^2_I$ for an intercross marker contrast. Recalling Equations 16.22 and 16.23, the expected value of MS_m becomes

$$E(\,\text{MS}_m\,) = n(1-2c)^2\left(\frac{\sigma^2_A}{4} + \frac{\sigma^2_D}{4}\right) + \sigma^2_e \tag{16.25}$$

so that the variance associated with marker alleles is

$$\sigma^2_m = (1-2c)^2\left(\frac{\sigma^2_A}{4} + \frac{\sigma^2_D}{4}\right) \tag{16.26a}$$

This can be estimated by using

$$\text{Var}(m) = \frac{\text{MS}_m - \text{MS}_e}{n} \tag{16.26b}$$

Thus, when dominance ($\sigma^2_D > 0$) is present, full-sib designs are expected to be more powerful for QTL detection than half-sib designs, as σ^2_m is greater. Even in the absence of dominance, full-sib designs are generally more powerful, as power depends on the number of informative marker contrasts (one for each backcross and intercross family, two for each fully informative family). In the extreme case where all N families in a design are fully marker informative, a full-sib design has $2N$ contrasts vs. N for a half-sib design with the same number of families.

SIB ANALYSIS: MAXIMUM LIKELIHOOD

ML estimation of QTL position and effects using a set of relatives from an outbred population proceeds as with inbred-line crosses — one constructs an appropriate likelihood function for the assumed model, obtains the parameter values that

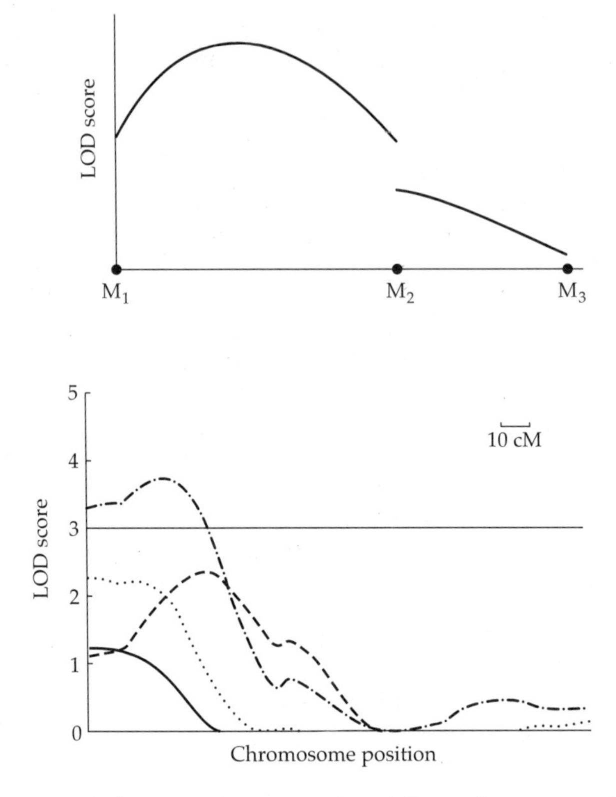

Figure 16.3 Top: When adjacent marker loci differ in the amount of heterozygosity, interval mapping in an outbred population gives a likelihood map with discontinuities at some of the marker loci (here, at the second marker). **Bottom:** Likelihood maps that simultaneously use all markers on a chromosome (instead of two at a time, as with interval mapping) have no such discontinuities. This likelihood map considers bovine chromosome U5(10) and examines four milk production traits in the granddaughters of a single sire. Only fat yield (denoted by – · – ·) showed a significant LOD score (exceeding the critical value of three), indicating a putative QTL on this chromosome. (After Georges et al. 1995.)

maximize this function, and tests alternative hypotheses with likelihood ratios. One subtle difference for outbred parents is that likelihood maps displaying the amount of support for a putative QTL often show discontinuities between adjacent markers (Knott and Haley 1992b); see Figure 16.3. Such discontinuities arise when adjacent loci differ in the fraction of informative individuals in the sample because of between-locus variance in the marker allele frequencies (a situation that does not arise with inbred-line crosses). As Figure 16.3 shows, these discontinuities can be avoided by constructing likelihood functions that use all of the marker information on a chromosome (Georges et al. 1995, Kruglyak and Lander 1995b).

The major complication in applying ML methods to outbred populations is that the resulting likelihood functions are much more complex than for inbred-line crosses due to the need to consider all possible genotypes of measured pedigree members. For example, for an inbred-line cross, all F_1 parents have the same genotype, say MQ/mq. In contrast, in an outbred population, for a diallelic QTL there are 16 possible genotypes for a pair of parents with marker genotype M/m, as each parent could be QM/Qm, QM/qm, qM/Qm, or qM/qm. The situation is even more complex when more than two alleles are segregating at the QTL or when one considers two (or more) markers simultaneously. For a pedigree with even a modest number of parents, the number of possible genotypes that must be considered is enormous.

LeRoy and Elsen (1995) show for a single marker that likelihood methods can be considerably more powerful for detecting a QTL in a sib design than linear model-based approaches under loose linkage. However, with multiple markers, Knott et al. (1996) find little difference between linear model and ML approaches. Both studies find these two approaches have the same power when the marker and QTL are completely linked. One likely explanation for the apparent discrepancy under loose linkage is that while Le Roy and Elsen assume only a single QTL is segregating, they do not incorporate a factor into their likelihood function to account for segregation at unlinked sites. Hence, as the marker-QTL recombination frequency approaches 1/2, their likelihood function approaches that for classical segregation analysis (i.e., no marker information, see Chapter 13). Thus, as the marker-QTL linkage becomes looser, the power of the linear model decreases to the significance level of the test, while the ML method approaches the power of classical segregation analysis. A likelihood approach which included a factor for account for between-family differences due to segregation of unlinked background polygenes should give similar power to linear model based approach except when the QTL effect is very large (S. Knott and C. Haley, pers. comm.).

Constructing Likelihood Functions

Likelihood functions for detecting QTLs using marker information in sets of sibs follow by suitable modifications of those developed for complex segregation analysis (Chapter 13). Recall that to construct the latter, the likelihood for an individual with trait value z_o is computed by first obtaining the likelihood conditional on the QTL genotypes of its parents (Q_s for the sire and Q_d for the dam),

$$\ell(\, z_o \,|\, Q_s, Q_d) = \sum_{Q_o} \varphi(z_o, \mu_{Q_o}, \sigma^2\,)\, p(Q_o \,|\, Q_s, Q_d)$$

where we make the standard assumption that, conditional on QTL genotype, phenotypes are normally distributed with common variance σ^2. This conditioning is removed by averaging over all possible parental values,

$$\ell(\, z_o\,) = \sum_{Q_s} \sum_{Q_d} \ell(\, z_o \,|\, Q_s, Q_d)\, p(Q_s)\, p(Q_d)$$

When marker data are present, the likelihood for an individual with marker genotype M is obtained by averaging over the probability of each QTL genotype given the marker genotypes,

$$\ell(\, z_o \,|\, M_o, M_s, M_d, Q_s, Q_d) = \sum_{Q_o} \varphi(z_o, \mu_{Q_o}, \sigma^2\,)\, p(Q_o \,|\, M_o, M_s, M_d, Q_s, Q_d) \tag{16.27a}$$

where the transmission probabilities $p(Q_o \,|\, M_o, M_s, M_d, Q_s, Q_d)$ are functions of the marker genotypes of the offspring (M_o) and its parents (M_s, M_d). Averaging over all possible parental genotypes gives

$$\ell(\, z_o \,|\, M_o, M_s, M_d) = \sum_{Q_s}\sum_{Q_d} \ell(\, z_o \,|\, M_o, M_s, M_d, Q_s, Q_d)\, p(Q_s|\, M_s)\, p(Q_d|\, M_d) \tag{16.27b}$$

In a similar fashion, the likelihood function for any pedigree can (in principle) be computed by suitably conditioning on the genotypes of parents and other relatives.

Likelihood functions for half- and full-sib families have been developed for single markers (Risch 1984, Demenais et al. 1988, Weller 1990, Haley 1991, Amos 1994, Le Roy and Elsen 1995, Mackinnon and Weller 1995, Elsen et al. 1997), interval mapping (Knott and Haley 1992b), and multipoint mapping (Georges et al. 1995, Uimari et al. 1996, Knott et al. 1996). A framework for developing likelihood functions that allows for multiple QTLs is given by Stricker et al. (1995b). Hoeschele and VanRaden (1993a,b) and Uimari et al. (1996) discuss Bayesian modifications of the likelihood functions to allow for the incorporation of prior information, such as the number of chromosomes (e.g., Smith 1953, 1959).

Example 3. Consider Knott and Haley's (1992b) development of a likelihood function for interval mapping in a full-sib family with a segregating diallelic QTL (with alleles Q and q, and Q allele frequency equal to p) flanked by two marker loci, A and B. The transmission probabilities required by Equation 16.27a follow from arguments identical to those used in Chapter 15 and depend on QTL-marker distances and assumptions about interference. For example, if both parents have genotype $AQB\,/\,aqb$, where c_1 is the A–Q distance and c_2 the Q–B distance, then (assuming no interference) from Example 1 in Chapter 15,

$$\Pr\left(o = AQB/AqB \,|\, s, d = AQB/aqb\right) = 2\left[\frac{(1-c_1)(1-c_2)}{2}\right]\left(\frac{c_1\, c_2}{2}\right)$$

The recombination frequency c between these markers is assumed known, in which case c_2 follows as a function of c_1 and c (Chapter 15). For a given pair of parental genotypes, many of these parent-offspring conditional probabilities are zero, as usually only a subset of all possible offspring genotypes can occur for a given pair of parental genotypes.

The likelihood for all n offspring in a family, conditional on parental genotypes, is

$$\ell(\,\mathbf{z}\,|\,\mathbf{M_o}, M_s, M_d, Q_s, Q_d) = \prod_{j=1}^{n} \ell(\,z_{o_j}\,|\,M_{o_j}, M_s, M_d, Q_s, Q_d) \quad (16.28a)$$

where $\mathbf{M_o}$ denotes the vector of the n offspring marker genotypes. As before, the conditioning on the sire and dam QTL genotype (including phase) is removed by averaging over all possible genotypes, giving the unconditional likelihood for the observed markers as

$$\ell(\,\mathbf{z}\,|\,\mathbf{M_o}, M_s, M_d) =$$

$$\sum_{Q_s}\sum_{Q_d} \ell(\,\mathbf{z}\,|\,\mathbf{M_o}, M_s, M_d, Q_s, Q_d) \cdot \Pr(Q_s\,|\,M_s) \cdot \Pr(Q_d\,|\,M_d) \quad (16.28b)$$

where the sum for each parent is over all QTL genotypes and all QTL-marker phases. In $\Pr(Q_x\,|\,M_x)$, Q_x denotes both the possible QTL genotype and phase with respect to the observed marker M_x. For an individual of known marker genotype, say $A_iA_jB_kB_\ell$, there are two possible marker phases ($A_iB_k\,/\,A_jB_\ell$ and $A_jB_k\,/\,A_iB_\ell$), each of which has four possible QTL genotypes (Q/Q, Q/q, q/Q, q/q), for a total of eight possible marker-QTL phases for each parent. Assuming Hardy-Weinberg proportions hold and that the base population is in linkage equilibrium (so that both marker-QTL phases are equally frequent), the eight possible marker-QTL genotypes for an $A_iB_k\,/\,A_jB_\ell$ individual have expected frequencies as follows:

Marker-QTL Genotype		$\Pr(Q_x\,|\,A_iA_jB_kB_\ell)$
$A_iQ\,B_k\,/\,A_jQ\,B_\ell$,	$A_jQ\,B_k\,/\,A_iQ\,B_\ell$	$p^2/2$
$A_iQ\,B_k\,/\,A_jq\,B_\ell$,	$A_jQ\,B_k\,/\,A_iq\,B_\ell$	$p(1-p)/2$
$A_iq\,B_k\,/\,A_jQ\,B_\ell$,	$A_jq\,B_k\,/\,A_iQ\,B_\ell$	$p(1-p)/2$
$A_iq\,B_k\,/\,A_jq\,B_\ell$,	$A_jq\,B_k\,/\,A_iq\,B_\ell$	$(1-p)^2/2$

The likelihood can be suitably modified to account for common-family effects (Knott and Haley 1992b) and background (unlinked) polygenes by using arguments identical to those leading to Equations 13.17 and 13.22. The full likelihood over N unrelated families is just the product of the individual family likelihoods. Equation 16.28b has six parameters to estimate (three QTL means, σ^2, c_1, and p), giving a likelihood-ratio test with four degrees of freedom when compared to the null likelihood of no segregating QTL (Equation 13.8).

The properties of ML estimates using relatives from outbred populations have not been as extensively examined as have those for line-cross populations. Mackinnon and Weller (1995) examined the accuracy of single-marker ML estimation under the half-sib design, finding that while ML estimates of QTL effects tend to be reasonably accurate, the same is not the case for estimates of QTL position, which have very flat likelihood surfaces. They also observed that estimates of QTL parameters tend to be correlated, so that a poor estimate in one parameter (e.g., map position) can have a detrimental effect on other estimates (e.g., QTL effects). Interval mapping is expected to have a more peaked likelihood surface for QTL position (and a resulting smaller variance), so these concerns may be less important for analysis using multiple markers.

While marker-based likelihood methods have greater power than segregation analysis (Chapter 13) for detecting QTLs (Goldin et al. 1984, Demenais et al. 1988, Knott and Haley 1992b, Le Roy and Elsen 1995), their power is still poor compared to inbred-line-cross designs. Knott and Haley (1992b) show that the use of flanking markers increases power, as does incorporation of a within-family effect into the likelihood function. Choosing highly polymorphic marker loci further increases power by increasing the frequency of informative sibships. Even with these modifications, however, power remains marginal — with markers at 20 map unit intervals, a QTL whose segregation accounts for 1/3 of the total variance is barely detectable when 250 full-sib families of four individuals each are used. Further, as with segregation analysis, there is a tradeoff between increasing the number of families versus increasing family size. For a fixed total number of progeny, one must sample a sufficient number of families (to ensure that at least one of them is informative) and a sufficient number of individuals within each family (to have power within an informative family).

MAXIMUM LIKELIHOOD OVER GENERAL PEDIGREES: VARIANCE COMPONENTS

The above likelihood functions explicitly model the transmission of QTL genotypes from parent to offspring, requiring estimation of QTL allele frequencies and genotype means (as well as assumptions about the number of segregating alleles). While this approach can be extended to multigenerational pedigrees, the number of possible combinations of genotypes for individuals in the entire pedigree increases exponentially with the number of pedigree members, and solving the resulting likelihood functions becomes increasingly more difficult. An alternative is to construct likelihood functions using the **variance components** associated with a QTL (or linked group of QTLs) in a genetic region of interest, rather than explicitly modeling all of the underlying genetic details. This approach allows for very general and complex pedigrees. The basic idea is to use marker information to compute the fraction of a genetic region of interest that is **identical by descent** between two individuals. Recall that two alleles are identical by descent, or **ibd**, if we can trace them back to a single copy in a common ancestor (Chapter 7).

Goldgar (1990) was the first to suggest this approach for QTL mapping. We sketch the basic ideas here and refer the reader to Chapter 27 for a more detailed discussion of the estimation of variance components in a likelihood framework. Consider the simplest case, in which the genetic variance is additive for the QTLs in the region of interest as well as for background QTLs unlinked to this region. Under Goldgar's model, an individual's phenotypic value is decomposed as

$$z_i = \mu + A_i + A_i^* + e_i \tag{16.29}$$

where μ is the population mean, A is the contribution from the chromosomal interval being examined, A^* is the contribution from QTLs outside this interval, and e is the residual. The random effects A, A^*, and e are assumed to be normally distributed with mean zero and variances σ^2_A, $\sigma^2_{A^*}$, and σ^2_e. Here σ^2_A and $\sigma^2_{A^*}$ correspond to the additive variances associated with the chromosomal region of interest and background QTLs in the remaining genome, respectively. We assume that none of these background QTLs are linked to the chromosome region of interest so that A and A^* are uncorrelated, and we further assume that the residual e is uncorrelated with A and A^*. Under these assumptions, the phenotypic variance is $\sigma^2_A + \sigma^2_{A^*} + \sigma^2_e$.

Assuming no shared environmental effects, from Chapter 7 the phenotypic covariance between two individuals is

$$\sigma(z_i, z_j) = R_{ij}\,\sigma^2_A + 2\Theta_{ij}\,\sigma^2_{A^*} \tag{16.30}$$

where R_{ij} is the fraction of the chromosomal region shared ibd between individuals i and j, and $2\Theta_{ij}$ is twice Wright's coefficient of coancestry (i.e., $2\Theta_{ij} = 1/2$ for full sibs). Note that R_{ij} is a random variable whose expected value is $2\Theta_{ij}$. Goldgar (1990) and Guo (1994a,b) discuss the estimation of R_{ij} from flanking marker information. For the special case of a nuclear family, Kruglyak and Lander (1995b) discuss the estimation of R_{ij} at any point along a chromosome using all marker information (multipoint mapping). Approaches such as genomic mismatch scanning (Chapter 14) can also be used with a very dense genetic map to estimate R_{ij} with high precision. For a vector **z** of observations on n individuals, the associated covariance matrix **V** can be expressed as contributions from the region of interest, from background QTLs, and from residual effects,

$$\mathbf{V} = \mathbf{R}\,\sigma^2_A + \mathbf{A}\,\sigma^2_{A^*} + \mathbf{I}\,\sigma^2_e \tag{16.31a}$$

where **I** is the $n \times n$ identity matrix, and **R** and **A** are matrices of known constants,

$$\mathbf{R}_{ij} = \begin{cases} 1 & \text{for } i = j \\ \widehat{R}_{ij} & \text{for } i \neq j \end{cases}, \qquad \mathbf{A}_{ij} = \begin{cases} 1 & \text{for } i = j \\ 2\Theta_{ij} & \text{for } i \neq j \end{cases} \tag{16.31b}$$

The elements of $\mathbf{R}$ contain the estimates of ibd status for the region of interest based on marker information (see below), while the elements of $\mathbf{A}$ are given by the pedigree structure.

The resulting likelihood is a multivariate normal with mean vector $\boldsymbol{\mu}$ (all of whose elements are μ) and variance-covariance matrix $\mathbf{V}$. From Equation 8.24,

$$\ell(\,\mathbf{z}\,|\,\mu, \sigma_A^2, \sigma_{A^*}^2, \sigma_e^2\,) = \frac{1}{\sqrt{(2\pi)^n\,|\mathbf{V}|}}\,\exp\left[-\frac{1}{2}(\mathbf{z}-\boldsymbol{\mu})^T\,\mathbf{V}^{-1}\,(\mathbf{z}-\boldsymbol{\mu})\right] \qquad (16.32)$$

This likelihood has four unknown parameters (μ, σ_A^2, $\sigma_{A^*}^2$, and σ_e^2). A significant σ_A^2 indicates the presence of at least one QTL in the interval being considered, while a significant $\sigma_{A^*}^2$ implies background genetic variance contributed from QTLs outside the focal interval. Both of these hypotheses can be tested by likelihood-ratio tests. Using the methodology of Chapters 26 and 27, the simple model given by Equation 16.29 can easily be extended to allow for other genetic effects (such as dominance), multiple chromosomal regions (Schork 1993), shared environmental effects, and any number of fixed effects (such as differences between the sexes, different known environmental effects, and so forth). Incorporation of such additional effects can increase power by reducing the residual variance.

Estimating QTL Position

Goldgar's original approach makes no assumptions about how many QTLs a tested chromosomal segment contains. As such, it simply provides a test for detecting the presence of one or more QTLs in a given region, along with providing a composite estimate of their effects as measured by the genetic variance attributable to that region. However, if we are willing to assume that only a single QTL resides in the region, Goldgar's method can be modified to estimate QTL position (Amos 1994, Xu and Atchley 1995). The idea, as used in many previous tests, is that QTL position enters through estimates of R_{ij}. For example, Amos (1994) shows that for a single marker linked to a QTL at recombination frequency c,

$$\widehat{R}_{ij} = \begin{cases} 1/2 + (1-2c)^2\,(\pi_{ij} - 1/2) & \text{full sibs} \\ 1/4 + (1-2c)^2\,(\pi_{ij} - 1/4) & \text{half sibs} \end{cases} \qquad (16.33)$$

where $\widehat{R}_{ij}$ is the estimate of the fraction of QTL alleles ibd between individuals i and j, and π_{ij} is the proportion of ibd marker alleles (the estimation of which is discussed below). Values for other relatives are given by Amos, allowing most pedigree structures to be easily incorporated. One then proceeds by treating c as a further parameter to estimate, displaying support for a QTL at any particular position by a standard likelihood map. Xu and Atchley (1995) consider multiple markers and QTLs, while Kruglyak and Lander (1995b) present a complete multipoint approach for nuclear families.

THE HASEMAN-ELSTON REGRESSION

Starting with Haseman and Elston (1972), human geneticists have developed a number of methods for detecting QTLs using pairs of relatives as the unit of analysis. The idea is to consider the number of alleles identical by descent (ibd) between individuals for a given marker. If a QTL is linked to the marker, pairs sharing ibd marker alleles should also tend to share ibd QTL alleles and hence are expected to be more similar than pairs not sharing ibd marker alleles. This fairly simple idea is the basis for a large number of relative-pair methods (often referred to as **allele sharing** methods), which we now examine in some detail. We first consider Haseman and Elston's original regression method (and its recent extensions), which deals with continuous characters, and finish by discussing a number of closely related methods that have been developed for dichotomous characters, the latter being motivated by the desire to map human disease genes.

Derivation of the Haseman-Elston Regression

If a marker locus is linked to a QTL, the difference in character values between two relatives is expected to decrease as they share more marker alleles ibd. This is the basis for the Haseman-Elston regression (1972, generalized by Amos and Elston 1989, Amos et al. 1990, Elston 1990b), which regresses the squared difference in character value between a pair of relatives on the proportion of ibd marker alleles for that pair.

To formally develop the Haseman-Elston (H-E) test, consider a QTL with no dominance. Suppose there are n pairs of relatives (all of the same type, such as all full- or all half-sibs), and denote the character values for the jth pair by z_{1j} and z_{2j}. The model assumed is

$$z_{ij} = \mu + A_{ij} + e_{ij} \tag{16.34}$$

where A is the QTL effect (with mean zero and additive genetic variance σ^2_A), and e is the residual value. The latter includes all effects not accounted for by the QTL of interest, including environmental effects and contributions from other QTLs unlinked to the marker. The difference in residual values for the jth pair, $e_j = e_{1j} - e_{2j}$, is assumed to have mean zero, variance σ^2_e, and to be uncorrelated with $(A_{1j} - A_{2j})$. Note that the difference between pairs has the especially nice feature of accounting for shared environmental factors, as effects common to both relatives (such as family effects) subtract out. The squared difference for each pair $Y_j = (z_{1j} - z_{2j})^2$ has expected value

$$\begin{aligned} E(Y_j) &= E\left[\,(A_{1j} - A_{2j} + e_{1j} - e_{2j})^2\,\right] \\ &= E\left[\,(A_{1j} - A_{2j})^2\,\right] + \sigma^2_e \\ &= 2\,\left[\sigma^2_A - \sigma(A_{1j}, A_{2j})\,\right] + \sigma^2_e \end{aligned} \tag{16.35}$$

From Chapter 7, $\sigma(A_{1j}, A_{2j}) = \sigma_A^2 \cdot \pi_{jt}$, where π_{jt} (=0, 1/2, or 1) is the proportion of alleles ibd at the QTL. (In what follows, we use the subscripts t and m to distinguish trait vs. marker loci.) Hence, the expectation of Y conditional on the proportion of QTL alleles ibd is

$$E(Y_j \mid \pi_{jt}) = (2\sigma_A^2 + \sigma_e^2) - (2\sigma_A^2) \cdot \pi_{jt} \tag{16.36}$$

As expected, if $\sigma_A^2 > 0$, the regression of Y_j on the proportion π_{jt} of *trait* alleles ibd has a negative slope.

While we do not know π_{jt}, we can estimate it using π_{jm}, the proportion of alleles ibd at a linked marker locus. (We detail shortly how to estimate π_{jm} using the observed marker information.) Conditioning on the number (i = 0, 1, or 2) of marker alleles ibd gives

$$E(Y_j \mid \pi_{jm}) = \sum_{i=0}^{2} E\left(Y_j \mid \pi_{jt} = \frac{i}{2} \right) \cdot \Pr\left(\pi_{jt} = \frac{i}{2} \,\middle|\, \pi_{jm} \right) \tag{16.37}$$

The second term in the summation is the conditional probability for the proportion of QTL alleles ibd given the fraction of marker alleles ibd. These are functions of the type of relatives being considered and the recombination fraction c between the marker and QTL, and are obtained in a similar fashion to the conditional probabilities obtained in Example 1 in Chapter 15. Values of $\Pr(\pi_t \mid \pi_m)$ for many common types of relative pairs are given by Bishop and Williamson (1990) and Risch (1990b). Substituting these into Equation 16.37 gives the regression of the squared difference Y_j on the proportion π_{jm} of *marker* alleles ibd as

$$E(Y_j \mid \pi_{jm}) = \alpha + \beta \cdot \pi_{jm} \tag{16.38a}$$

where the slope β and intercept α depend on the type of relatives and the recombination fraction c. For full sibs, substituting the appropriate values of $\Pr(\pi_{jt} \mid \pi_{jm})$ into Equation 16.37 gives (Haseman and Elston 1972),

$$\beta = -2(1-2c)^2\,\sigma_A^2, \qquad \text{and} \qquad \alpha = 2[\,1 - 2(1-c)c\,]\,\sigma_A^2 + \sigma_e^2 \tag{16.38b}$$

A significant negative slope provides evidence of a QTL linked to the marker, with the power of this test scaling with $(1-2c)^2$ and σ_A^2. The expected slopes for other pairs of relatives are similarly found to be

$$\beta = \begin{cases} -2(1-2c)\,\sigma_A^2 & \text{grandparent–grandchild;} \\ -2(1-2c)^2\,\sigma_A^2 & \text{half-sibs;} \\ -2(1-2c)^2(1-c)\,\sigma_A^2 & \text{avuncular (aunt/uncle–nephew/niece).} \end{cases} \tag{16.39}$$

Thus, the Haseman-Elston test is quite simple: for n pairs of the same type of relatives, one regresses the squared difference of each pair on the fraction of alleles ibd at the marker locus. A significant negative slope for the resulting regression indicates linkage to a QTL. This is a one-sided test, as the null hypothesis (no linkage) is $\beta = 0$ versus the alternative $\beta < 0$.

There are several caveats with this general approach. First, different types of relatives cannot be mixed in the standard H-E test, requiring separate regressions for each type of relative pair. This procedure can be avoided by modifying the test by using an appropriately weighted multiple regression (Olson and Wijsman 1993, Olson 1994). A related issue is that (even when using the same type of relatives), the residuals of the H-E regression are heteroscedastic, as the variance in phenotype varies with the number of QTL alleles ibd. A weighted regression can account for this problem, and results in a slight increase in power (Amos et al. 1989). Second, parents and their offspring share *exactly* one allele ibd and hence cannot be used to estimate this regression, as there is no variability in the predictor variable (Elston 1990b). Finally, QTL position (c) and effect (σ_A^2) are confounded and cannot be separately estimated from the regression slope β. Thus, in its simplest form, the H-E method is a *detection* test rather than an *estimation* procedure. This conclusion is not surprising, given that the H-E method is closely related to the single-marker linear model. As we discuss shortly, estimation of c and σ_A^2 is possible by extending the H-E regression by using ibd status of two (or more) linked marker loci to estimate π_{jt}.

Turning to a QTL with dominance, the resulting model becomes

$$z_{ij} = \mu + A_{ij} + D_{ij} + e_{ij} \tag{16.40}$$

Dominance only causes a departure from the additive model when both QTL alleles are ibd (such as can occur with full sibs) and hence does not influence most types of relative pairs. Define the indicator variables $f_{1,jm}$ and $f_{2,jm}$, where $f_{i,jm} = 1$ if the jth pair share i marker alleles ibd, else it is zero, so that $\pi_{jm} = f_{2,jm} + f_{1,jm}/2$. For full sibs, Blackwelder and Elston (1982) show that

$$E(\,Y_j \mid f_{1,jm},\, f_{2,jm}) = \alpha + \beta \cdot \pi_{jm} + \gamma \cdot f_{1,jm} \tag{16.41a}$$

where

$$\beta = -2(1-2c)^2 \cdot (\sigma_A^2 + \sigma_D^2) \qquad \text{and} \qquad \gamma = (1-2c)^4 \cdot \sigma_D^2 \tag{16.41b}$$

For full sibs with known parental marker genotypes, Amos et al. (1989) show that ignoring the γ term in Equation 16.41a and simply using $\alpha + \beta\pi_{jm}$ yields an unbiased estimator of β. Hence, if our goal is simply QTL detection, we can ignore the effects of dominance with full-sib analysis.

Table 16.3 Probability of various pairs of marker genotypes in relatives as a function of the number of genes ibd between these relatives.

Type	Marker pairs	Pr(M | i=0)	Pr(M | i=1)	Pr(M | i=2)
I	$M_jM_j,\ M_jM_j$	p_j^4	p_j^3	p_j^2
II	$M_jM_j,\ M_kM_k$	$2\,p_j^2\,p_k^2$	0	0
III	$M_jM_j,\ M_jM_k$	$4\,p_j^3\,p_k$	$2\,p_j^2\,p_k$	0
IV	$M_jM_j,\ M_kM_\ell$	$4\,p_j^2\,p_k\,p_\ell$	0	0
V	$M_jM_k,\ M_jM_k$	$4\,p_j^2\,p_k^2$	$p_j\,p_k\,(p_j+p_k)$	$2\,p_j\,p_k$
VI	$M_jM_k,\ M_jM_\ell$	$8\,p_j^2\,p_k\,p_\ell$	$2\,p_j\,p_k\,p_\ell$	0
VII	$M_jM_k,\ M_\ell M_m$	$8\,p_j\,p_k\,p_\ell\,p_m$	0	0

Source: Elston 1990b.
Note: Haseman and Elston (1972) note that there are seven different types of marker pairs. Of these, three share no alleles ibd (Types II, IV, and VII), two share either zero or one alleles (Types III and VI), and two (Types I, V) can share zero, one, or two alleles ibd. Note that j, k, ℓ and m index *different* alleles.

Estimating the Number of Marker Genes ibd

An obvious question is how to determine the number of alleles that are ibd at a given marker. If both parents are *different* marker heterozygotes (i.e., the family is fully informative), ibd status can be determined with certainty in all of their offspring. However, other relative pairs are problematic, as ibd status must be estimated rather than determined with certainty. In the absence of pedigree information, Haseman and Elston (1972) and Elston (1990b) suggested the following general estimator of π_{jm} for any pair of relatives,

$$p_{jm} = \frac{f_2 \cdot \Pr(M \mid i=2) + (f_1/2) \cdot \Pr(M \mid i=1)}{f_2 \cdot \Pr(M \mid i=2) + f_1 \cdot \Pr(M \mid i=1) + f_0 \cdot \Pr(M \mid i=0)} \tag{16.42}$$

where f_i is the prior probability that the pair of relatives share i genes ibd, and the $\Pr(M \mid i)$ are the probabilities of the observed pair M of marker genotypes given that the pair shares i alleles ibd (Table 16.3). While easy to compute, one must exercise caution in the application of Equation 16.42, as it is rather sensitive to errors in the estimated marker allele frequencies (Babron et al. 1993). Risch (1990c) develops alternative ML estimators for p_{jm} based on these probabilities, while Hu et al. (1995) show how to modify Equation 16.42 to accommodate sex-linked markers.

The probability $\Pr(M \mid i)$ of observing a given pair M of marker genotypes in two relatives given that they share i marker alleles ibd is computed as follows. If the two relatives share no alleles ibd, the frequencies of marker genotype pairs are given by the probability of drawing two such individuals at random. For example, the probability the pair has genotypes M_jM_j and M_jM_k is $2 \cdot p_j^2 \cdot 2\, p_j\, p_k = 4\, p_j^3\, p_k$. If one allele is ibd between the pair of relatives, the shared allele is chosen at random (say M_j with probability p_j) while the remaining allele in each relative is chosen at random. For the previous marker pair, the probability becomes $p_j \cdot (2p_jp_k) = 2\, p_j^2\, p_k$. If both alleles are ibd, marker genotypes are identical and the probability of observing the pair equals the probability of observing one such genotype, i.e., p_i^2 for $(M_iM_i,\ M_iM_i)$, $2p_i\, p_j$ for $(M_iM_j,\ M_iM_j)$. Using this logic, values for the seven different classes of paired marker genotypes can be computed (Table 16.3).

Example 4. Suppose a marker has four alleles with frequencies $p_1 = 0.5$, $p_2 = 0.25, p_3 = 0.15$, and $p_4 = 0.1$. The marker genotypes for three independent pairs of full sibs were found to be: Pair 1 = $(M_1M_1,\ M_1M_1)$; Pair 2 = $(M_3M_4,\ M_3M_4)$; Pair 3 = $(M_4M_4,\ M_4M_4)$. Recall from Chapter 7 that for full sibs, $f_0 = f_2 = 1/4$, $f_1 = 1/2$. Applying Equation 16.42 and using Table 16.3, the estimated proportions of ibd marker alleles for these three pairs are

$$p_{1m} = \frac{(1/4)\, p_1^2 + [(1/2)/2]\, p_1^3}{(1/4)\, p_1^2 + (1/2)\, p_1^3 + (1/4)\, p_1^4} \simeq 0.67$$

$$p_{2m} = \frac{(1/4)\, 2\, p_3\, p_4 + [(1/2)/2]\, p_3\, p_4\, (p_3 + p_4)}{(1/4) 2\, p_3\, p_4 + (1/2)\, p_3\, p_4\, (p_3 + p_4) + (1/4)\, 4\, p_3^2\, p_4^2} \simeq 0.88$$

$$p_{3m} = \frac{(1/4)\, p_4^2 + [(1/2)/2]\, p_4^3}{(1/4)\, p_4^2 + (1/2)\, p_4^3 + (1/4)\, p_4^4} \simeq 0.91$$

These values are substituted in place of π_{jm} when computing the H-E regression. By comparing the first and third pairs, it can be seen that rarer alleles are more informative as to ibd status.

Power and Improvements

A number of authors have examined the power of the Haseman-Elston test (Robertson 1973; Blackwelder and Elston 1974, 1982; Amos and Elston 1989; Götz and Ollivier 1992), and large-sample approximations for power have been developed by Blackwelder and Elston (1982) and Amos and Elston (1989). The general conclusion from these studies is that the standard Haseman-Elston test has poor power in many settings. For example, Amos and Elston (1989) found that 320 full-sib pairs are required to have a 90% chance of detecting (at significance level

$\alpha = 0.05$) a major gene with a heritability of 0.5 (hence accounting for 50% of the phenotypic variance) when the marker and major gene are completely linked. This number increases to 778 sib pairs if the marker and major gene are separated by recombination fraction $c = 0.1$. The sample size increases as more distant relative pairs are considered, requiring 1,264 grandparent-grandchild pairs; 1,980 half-sib pairs; and 2,445 avuncular pairs. Since these power calculations assume that the fraction π_m of marker alleles ibd is known exactly, in reality, even more pairs are required as marker-uninformative pairs are discarded or π_m is estimated for such pairs.

While these numbers look bleak, the situation is not necessarily as bad with large sibships. In a sibship of size s, there are $s(s-1)/2$ pairwise comparisons, so that single families of 5, 10, and 25 sibs offer 10, 45, and 325 sib-pairs. The question remains as to whether all pairwise comparisons within a family are independent. There is some dispute on this point, with Hodge (1984) showing that the number of independent sib pairs approaches $(3/2)s - 2$ for $s > 4$. However, simulation studies have suggested that all pairwise combinations of sibs within a family can often be treated as independent pairs (Blackwelder and Elston 1982, 1985; Amos et al. 1989; Götz and Ollivier 1992; but see Daly and Lander 1996). Wilson and Elston (1993) show that while all pairs can be used, for small sample sizes the appropriate degrees of freedom for a t test for the regression is given by $(\sum_i [s_i - 1]) - 2$, rather than $(\sum_i s_i[s_i - 1]/2) - 2$, where s_i is the number of sibs in family i. Thus, the H-E test may not be unreasonable for modest sample sizes, provided one can obtain a few informative families with a large number of sibs.

Significant power gains are possible through the use of selective genotyping, wherein pairs are chosen based on extreme trait values in at least one member (Boehnke and Moll 1989, Carey and Williamson 1991, Cardon and Fulker 1994). For example, for a locus with associated heritability $h^2 = 0.3$ at recombination frequency $c = 0.1$ from the marker, simulation studies by Cardon and Fulker (1994) found that the probability of detecting this locus using Haseman-Elston interval mapping (see below) and 250 sib-pairs is 0.28. Under selective genotyping, where the pairs are now chosen by having (at least) one sib with an extreme character value, the power increases to 0.57, 0.65, and 0.84 using 250 pairs with one sib in the uppermost 10%, 5%, and 1%, respectively, of the population.

Interval Mapping by a Modified Haseman-Elston Regression

Fulker and Cardon (1994) proposed an interval mapping modification that allows separate estimates of QTL effect (σ_A^2) and position (c) by considering two linked markers simultaneously. Recall Equation 16.36,

$$E(Y_j \mid \pi_{jt}) = E[\,(z_{1j} - z_{2j})^2 \mid \pi_{jt}\,] = (2\sigma_A^2 + \sigma_e^2) - (2\,\sigma_A^2) \cdot \pi_{jt}$$

Fulker and Cardon show that the proportion π_{jt} of alleles ibd at the trait locus can be estimated by using a regression on $\pi_{jm}^{(1)}$ and $\pi_{jm}^{(2)}$, the proportion of alleles ibd at two marker loci flanking the QTL, where

$$\widehat{\pi}_{jt} = \frac{1 - \beta_1 - \beta_2}{2} + \beta_1 \, \pi_{jm}^{(1)} + \beta_2 \, \pi_{jm}^{(2)} \tag{16.43}$$

and

$$\beta_1 = \frac{(1-2c_1)^2 - (1-2c_2)^2(1-2c)^2}{1-(1-2c)^4}, \qquad \beta_2 = \frac{(1-2c_2)^2 - (1-2c_1)^2(1-2c)^2}{1-(1-2c)^4}$$

Here, as elsewhere, c_1 is the recombination frequency between the first marker and the QTL, c_2 between the QTL and second marker, and c between markers (assumed known). As in Chapter 15, c_2 follows directly from c_1 and c under various assumptions about interference (e.g., with complete interference $c_2 = c - c_1$), leaving the β_i as functions of only one unknown parameter, c_1. When the markers are reasonably close ($c < 0.2$), these coefficients simplify to $\beta_{m,1} \simeq c_2/c$ and $\beta_{m,2} \simeq c_1/c$. Substituting into Equation 16.36,

$$E(Y_j) = \alpha - (2\,\sigma_A^2)\left(\beta_1 \, \pi_{jm}^{(1)} + \beta_2 \, \pi_{jm}^{(2)}\right) \tag{16.44}$$

where

$$\alpha = \left[\sigma_A^2 \, (1 + \beta_1 + \beta_2) + \sigma_e^2\right]$$

Following the approach of Haley and Knott (1992), Fulker and Cardon compute this regression for each value of c_1 ($0 \leq c_1 \leq c$), with that value giving the regression with the largest r^2 taken to be the estimate of map position, and $2\sigma_A^2$ estimated from the associated regression coefficients. This method offers increased power for QTL detection relative to the standard Haseman-Elston method (Fulker and Cardon 1994, Cardon and Fulker 1994). This increase is small when the single-marker power is either very poor or very high, but can be considerable when single-marker power is modest. As with the single-marker regression, selective genotyping further increases power (Cardon and Fulker 1994).

Example 5. Cardon et al. (1994) used this interval mapping test to examine whether QTLs influencing reading disability are linked to markers on human chromosome 6. Previous work had suggested linkage to this chromosome. For more refined mapping, these authors used a set of five highly informative markers. In the following figure, support for QTL position is given by plotting the regression t statistic as a function of putative QTL position on chromosome 6.

The number of alleles for each marker locus ranged from 9–13 with associated heterozygosities of 0.6–0.9. Two independent sets of relatives were examined, a collection of 358 siblings from 19 families, and an independent collection of 50 dizygotic twins. Selective genotyping was used, with both sets of relatives chosen by the presence of one sib in each pair who scored very poorly on standardized reading tests (following suitable correction for other environmental factors). Interval mapping on both relative sets shows strong (and independent) support for a QTL near the markers *D6S105* and *TNFB*. Support for a second QTL at 70 cM is suggested from the twin data, but this is not confirmed by the siblings data set. (Figure after Cardon et al. 1994.)

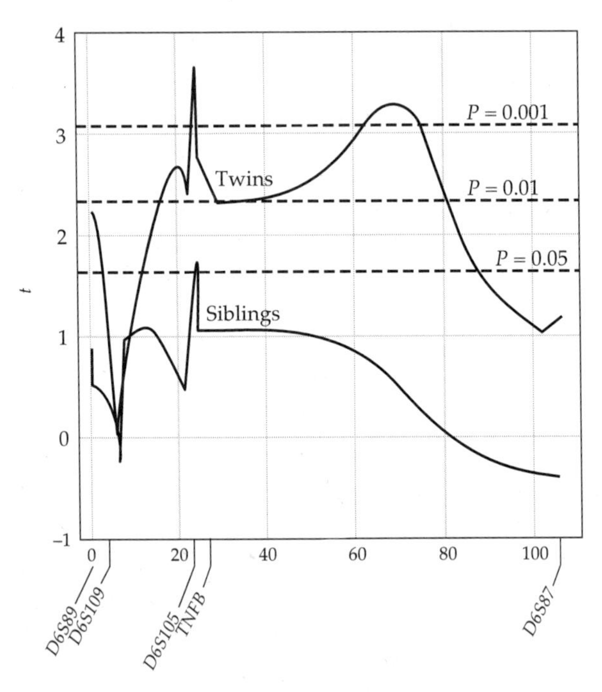

Standard (i.e., two-marker) interval mapping can be extended to **multipoint interval mapping** by considering all linked markers on a chromosome (Fulker et al. 1995, Kruglyak and Lander 1995b). In the approach of Fulker et al., the two-marker regression estimate of the QTL ibd status (Equation 16.43) is replaced by a multiple regression that incorporates all n linked markers on a chromosome,

$$\widehat{\pi}_{jt} = \alpha + \sum_{k=1}^{n} \beta_k \, \pi_{jm}^{(k)} \tag{16.45}$$

where the fraction $\widehat{\pi}_{jt}$ of ibd QTL alleles for the jth pair is predicted using the fractions $\pi_{jm}^{(k)}$ of ibd alleles at each of the n marker loci on the chromosome containing the putative QTL. The β_k are obtained from standard GLS regressions (Chapter

8), where the vector $\boldsymbol{\beta}^T = (\beta_1, \cdots, \beta_n)$ satisfies $\boldsymbol{\beta} = \mathbf{V}^{-1}\mathbf{C}$, where

$$V_{ik} = \sigma\left(\pi_m^{(i)}, \pi_m^{(k)}\right) = \begin{cases} \text{Var}\left(\pi_m^{(i)}\right) & \text{for } i = k \\ 8\,(1 - 2c_{ik})\,\text{Var}\left(\pi_m^{(i)}\right)\text{Var}\left(\pi_m^{(k)}\right) & \text{for } i \neq k \end{cases} \tag{16.46a}$$

and

$$C_i = \sigma\left(\pi_t, \pi_m^{(i)}\right) = (1 - 2c_i)\,\text{Var}\left(\pi_m^{(i)}\right) \tag{16.46b}$$

Here $\text{Var}(\pi_m^{(i)})$ is the observed sample variance for ibd status at the ith marker (i.e., the sample variance of the $\pi_{jm}^{(i)}$ over all included pairs), c_{ik} is the distance between markers i and k (assumed known), and c_i is the (unknown) distance between the QTL and marker i.

Again the regression is computed by first assuming a particular position for the QTL along the chromosome, computing $\widehat{\pi}_{jt}$ using Equation 16.45, and substituting this value into the standard Haseman-Elston regression (Equation 16.36). This process is repeated for all possible positions of the QTL along the chromosome, with the estimate of QTL position taken as that which gives the regression with the largest r^2 value. (This is equivalent to the regression giving the largest t statistic.) Fulker et al. show that this method offers high precision for mapping and (unlike two-locus interval mapping) accounts for differences in the information content of adjacent markers. When only two markers are used, Fulker et al. show that QTL position is biased towards the more informative marker. Multipoint interval mapping avoids this problem by incorporating all the marker information along the entire chromosome. Since the multipoint regression is easily computed, there is little reason to prefer simple interval mapping (Equation 16.43) over multipoint mapping (Equation 16.45).

An improved multipoint estimator is given by Kruglyak and Lander (1995b). While Equation 16.45 uses marker information to estimate the *mean* value of π_{jt}, Kruglyak and Lander use all the marker information to provide an ML estimate at each point along a chromosome of the actual *distribution* of ibd values (i.e., the probabilities at any particular chromosomal position that the pair shares zero, one, or two ibd alleles). An EM method (Appendix 4) is then used to compute the regression using this distribution, resulting in greater power than in an analysis one based simply on the estimated mean of π_{jt}. As before, regressions are computed at each point along a chromosome, generating a t statistic plot for the regression significance as a function of putative QTL position.

MAPPING DICHOTOMOUS CHARACTERS

Finally, we turn our attention to mapping and characterizing QTLs that underlie dichotomous characters. This problem has received a tremendous amount of attention from human geneticists, whose ultimate goal is the isolation of **disease**

susceptibility (DS) genes — QTLs that influence human disease. When a disease can be treated as a continuous character by measuring the appropriate underlying continuous variables, such as blood pressure or cholesterol level, the methods discussed in previous sections of this chapter can be applied. However, for many diseases, the appropriate underlying physiological variables are unresolved, and the observed phenotype is naturally dichotomous, with individuals being scored as either normal or affected. Even when the appropriate underlying variables are known, individuals are often still placed into discrete categories, such as high vs. normal blood pressure. We frame our discussion in terms of normal vs. affected character states, although our treatment is general.

If the disease has a simple Mendelian basis, wherein a single gene underlies the disease and all individuals of a given genotype have the same character state, standard mapping approaches for simple Mendelian genes can be used (reviewed by Ott 1991, Terwilliger and Ott 1994). However, most human diseases have a more complex genetic basis, with no single gene showing perfect segregation with the disease. Even when genetic causes are attributable to a single locus, lack of perfect association can occur for a variety of reasons (Lander and Schork 1994). There can be **incomplete penetrance** (not all susceptible genotypes show the disease), **phenocopies** (individuals display the disease for environmental, rather than genetic, reasons) and **genetic heterogeneity** (multiple alleles at a given locus have different effects on disease susceptibility). Likewise, the disease could have a polygenic basis. All of these factors confound a simple Mendelian analysis, requiring that other methods be used.

Starting with Penrose (1935, 1954b), human geneticists have developed a variety of methods for mapping DS genes, and these are reviewed by Weiss (1993), Terwilliger and Ott (1994), Lander and Schork (1994), and Weeks and Lathrop (1995). One approach is to modify the likelihood functions constructed for segregation analysis of dichotomous characters (Equation 13.29) to incorporate marker information by using Equation 16.27. For a diallelic QTL, the resulting likelihood function has five unknown parameters: QTL position (c), allele frequency (p), and effects (measured by the penetrances ψ_1, ψ_2, ψ_3 of the three QTL genotypes). Likelihood-ratio tests for linkage follow by comparing the restricted model assuming $c = 1/2$ (no linkage) against the full model where c is estimated from the data. Instead of attempting to maximize the entire likelihood over a complex pedigree structure, human geneticists usually further reduce the parameter space by assuming p is known, then performing the likelihood-ratio test for linkage under one or more reduced models where at least one of the penetrances is assigned a fixed value. Typically, it is assumed that the DS locus is either completely dominant or fully recessive, so that if D and N denote the disease and normal alleles at the DS locus, the two models generally considered are

$$\text{Dominance model:} \qquad \psi_{DD} = \psi_{DN} = \psi \qquad \psi_{NN} = \psi_p$$

$$\text{Recessive model:} \qquad \psi_{DD} = \psi \qquad \psi_{DN} = \psi_{NN} = \psi_p$$

where the two parameters to estimate by ML are ψ, the penetrance of a susceptible genotype, and ψ_p, the probability of a phenocopy. Frequently, the model is reduced further by preassigning a value to ψ_p (most often being set to zero). A serious limitation of this approach is that it is very model-dependent. Using the wrong model can potentially result in reports of false linkage or (more likely) failure to detect a linked QTL (Clerget-Darpoux et al. 1986, Curtis and Stam 1995). Further, likelihood methods can be computationally very intense, especially in complex pedigrees.

Recurrent and Relative Risks for Pairs of Relatives

Relative-pair approaches, on the other hand, are computationally very simple and require no assumptions about the underlying genetic model. Not surprisingly, these methods are very popular, and they form the bulk of our discussion. Such tests are based on the common theme of **allele sharing** — if a marker is linked to a QTL, two relatives with the same character state should, on average, share more ibd marker alleles than expected by chance.

Before examining these methods, a brief introduction to measures of risk is in order. Human geneticists have devised a number of measures for the risk that an individual is affected, given that a relative is, and these can be expressed as a function of the genetic parameters of the DS locus. Assigning values to the alternative states of the character of 0 (if unaffected) or 1 (if affected) yields several useful identities. Under this coding, the character follows a binomial distribution with mean K (the population prevalence) and variance $K(1-K)$. If the character has a genetic basis, relatives of an affected individual should have a probability greater than K of also being affected. Letting z_1 and z_2 denote the character states of two relatives, this probability is quantified by the **recurrence risk**,

$$K_R = \Pr(\, z_2 = 1 \,|\, z_1 = 1\,) \tag{16.47a}$$

which is the conditional probability that a relative of relationship R (say, a full sib) to an affected individual is also affected. Since $\Pr(A, B) = \Pr(A \,|\, B) \cdot \Pr(B)$, it follows that the probability that both relatives are affected is $K_R \cdot K$. An alternative measure is the **relative risk**, λ_R, the increase in risk given that one relative is affected,

$$\lambda_R = \frac{K_R}{K} \tag{16.47b}$$

For example, the population prevalence for type 1 diabetes is 0.4% ($K = 0.004$), while the probability that a sib has diabetes given it has an affected sib is 6% ($K_s = 0.06$), giving the relative risk in sibs as $\lambda_S = 0.06/0.004 = 15$ (Davies et al. 1994).

James' identity (1971) relates the recurrence risk to the population prevalence and the covariance between relatives,

$$K_R = K + \frac{\text{Cov}(z_1, z_2)}{K} \tag{16.48a}$$

while rearranging gives Risch's (1990a) expression,

$$\lambda_R - 1 = \frac{\text{Cov}(z_1, z_2)}{K^2} \tag{16.48b}$$

James' identity is completely general and is obtained as follows. Because the z_i are zero-one indicator variables, $z_1 \cdot z_2 = 0$ unless $z_1 = z_2 = 1$, and hence

$$\text{Cov}(z_1, z_2) + K^2 = E(z_1\, z_2) = 1 \cdot \Pr(z_1 = 1,\ z_2 = 1) = K_R \cdot K$$

The first step follows by recalling $E(z_1\, z_2) = \text{Cov}(z_1, z_2) + E(z_1)\, E(z_2)$ and $E(z_i) = K$.

Ignoring epistasis and shared environmental effects,

$$\text{Cov}(z_1, z_2) = 2\Theta_{12}\, \sigma_A^2 + \Delta_{12}\, \sigma_D^2$$

where σ_A^2 and σ_D^2 are the additive and dominance variances associated with this character (measured on the 0/1 scale), Θ_{12} is the coefficient of coancestry, and Δ_{12} is the coefficient of fraternity for these two individuals (Chapter 7). Hence,

$$K_R = K + \frac{2\Theta_R\, \sigma_A^2 + \Delta_R\, \sigma_D^2}{K} \tag{16.49a}$$

Genetic variances can be estimated using the observed recurrent risks for different sets of relatives (Suarez et al. 1978, Olson 1995). For example, assuming only additive and dominance genetic variance is present and rearranging Equation 16.49a gives

$$\sigma_A^2 = 2K(K_O - K) \qquad \text{and} \qquad \sigma_D^2 = 4K(K_S - K_O) \tag{16.49b}$$

where K_O and K_S denote, respectively, the recurrent risk for parent-offspring and full-sib pairs. Likewise, the total genetic variance σ_G^2, even in the presence of epistasis, is

$$\sigma_G^2 = K(K_{MZ} - K) \tag{16.49c}$$

where K_{MZ} is the recurrent risk for monozygotic (identical) twins.

Finally, the rate of decrease of $\lambda_R - 1$ in sets of ever-distant relatives provides a test for epistasis (Risch 1990a). To see this, consider sets of relatives where $\Delta_R = 0$, so that dominance terms can be ignored. In this case Equation 16.48b can be generalized to

$$\lambda_R - 1 = \frac{1}{K^2}\left[\, 2\Theta_R\, \sigma_A^2 + (2\Theta_R)^2 \sigma_{AA}^2 + (2\Theta_R)^3 \sigma_{AAA}^2 + \cdots \right] \tag{16.49d}$$

In the absence of epistasis, $\lambda_R - 1$ decreases by a factor of two when comparing relatives whose coefficient of coancestry also differs by a factor of two (e.g., half sibs vs. first cousins). A greater than twofold decrease in $\lambda_R - 1$ implies epistasis, and, of course, multiple loci. Risch used this approach to provide evidence for multiple interacting loci for schizophrenia.

Affected Sib-pair Tests

When dealing with a dichotomous character, pairs of relatives can be classified into three groups: pairs where both are normal, **singly affected** pairs with one affected and one normal member, and **doubly affected** pairs. The first and last pairs are also called **concordant**, while pairs that differ are called **discordant**. The motivation behind relative-pair tests is that if a marker is linked to a QTL influencing the trait, concordant and discordant pairs should have different distributions for the number of ibd marker alleles.

In addition to being much more robust than ML methods for dichotomous characters, relative-pair tests also have the advantage of selective genotyping in that pairs are usually chosen so that at least one member is affected. The pairs of relatives considered are usually full sibs, and a number of variants of these **affected sib-pair**, or ASP, methods have been proposed (e.g., Day and Simons 1976; de Vries et al. 1976; Green and Woodrow 1977; Suarez 1978; Suarez et al. 1978; Fishman et al. 1978; Suarez and Eerdewegh 1984; Badner et al. 1984; Hopper et al. 1984; Thompson 1986; Lange 1986a,b; Risch 1987, 1990a–c; Faraway 1993; Commenges 1994; Curtis and Stam 1994; Kruglyak and Lander 1995b). Most of these are detection tests, rather than estimation procedures, as they cannot provide independent estimates of QTL effect and position. While our attention focuses on full-sib pairs, this basic approach can easily be applied to any pair of relatives, *provided* there is variability in the number of ibd alleles. (This excludes parent-offspring pairs, as these share exactly one allele ibd.) Most affected sib-pair tests have the basic structure of comparing the observed ibd frequencies (or some statistic based on them) of doubly affected pairs with either their expected values under no linkage or with the corresponding values in singly affected pairs. There are many possible tests based on this idea and most, it seems, have made their way into the literature. We consider three here.

Among those n_i pairs with i affected members ($i = 0, 1, 2$), let p_{ij} denote the frequency of such pairs with j ibd marker alleles ($j = 0, 1, 2$). From the binomial distribution, the estimator $\widehat{p}_{ij}$ has mean p_{ij} and variance $p_{ij}\,(1 - p_{ij})/n_i$. One ASP test is based on $\widehat{p}_{22}$, the observed frequency of doubly affected pairs that have two marker alleles ibd. Under the assumption of no linkage, $\widehat{p}_{22}$ has mean 1/4 (as full sibs have a 25% chance of sharing both alleles ibd) and variance $(1/4)(1-1/4)/n_2 = 3/(16\,n_2)$, suggesting the test

$$T_2 = \frac{\widehat{p}_{22} - 1/4}{\sqrt{\dfrac{3}{16\,n_2}}} \tag{16.50a}$$

For a large number of doubly affected pairs, T_2 is approximately distributed as a unit normal under the null hypothesis of no linkage. This test is one-sided, as $p_{22} > 1/4$ under linkage.

An alternative approach is to consider statistics that employ the mean number of ibd marker alleles, $p_{i1} + 2\,p_{i2}$. Under the hypothesis of no linkage, this has expected value $1\cdot(1/2)+2\cdot(1/4) = 1$ and variance $[\,1^2\cdot(1/2)+2^2\cdot(1/4)\,]-1^2 = 1/2$. For doubly affected pairs, the test statistic becomes

$$T_m = \sqrt{2\,n_2}\,\left(\widehat{p}_{21} + 2\,\widehat{p}_{22} - 1\,\right) \tag{16.50b}$$

which again for large samples is approximately distributed as a unit normal and is a one-sided test, as $p_{21} + 2\,p_{22} > 1$ under linkage.

Finally, maximum likelihood-based goodness-of-fit tests can be used (Risch 1990b,c). In keeping with the tradition of human geneticists, ML-based tests usually report LOD (likelihood of odds) scores in place of the closely related likelihood ratio (LR). (Recall from Equation 15.16 that 1 LR = 4.61 LOD.) Here the data are n_{20}, n_{21}, and n_{22}, the number of doubly affected sibs sharing zero, one, or two marker alleles ibd, with the unknown parameters to estimate being the population frequencies of these classes (p_{20}, p_{21}, p_{22}). The MLEs for these population frequencies are given by $\widehat{p}_{2i} = n_{2i}/n_2$. The LOD score for the test of no linkage becomes

$$MLS = \log_{10}\left[\prod_{i=0}^{2}\left(\frac{\widehat{p}_{2i}}{\pi_{2i}}\right)^{n_{2i}}\right] = \sum_{i=0}^{2} n_{2i}\,\log_{10}\left(\frac{\widehat{p}_{2i}}{\pi_{2i}}\right) \tag{16.51}$$

where π_{2i} is the probability that the pair of doubly affected sibs shares i alleles ibd in the absence of linkage to a QTL. (For full sibs, $\pi_{20} = \pi_{22} = 1/4$, $\pi_{21} = 1/2$.) The test statistic given by Equation 16.51 is referred to as the **maximum LOD score**, or MLS, with a score exceeding three being taken as significant evidence for linkage (Risch 1990b, Morton 1955b).

Example 6. An alternative formulation for the MLS test is to consider each informative parent separately, simply scoring whether or not a doubly affected sib pair shares a marker allele from this parent. This approach generates 0 (match, both affected sibs share the allele) or 1 (no match) ibd data. Under the null hypothesis of no linkage, each state (0 or 1) has probability 1/2, and the MLS test statistic becomes

$$MLS = (1 - n_1)\log_{10}\left(\frac{1-\widehat{p}_1}{1/2}\right) + n_1\log_{10}\left(\frac{\widehat{p}_1}{1/2}\right)$$

where n_1 and p_1 are, respectively, the number and frequency of sibs sharing the parental allele. This method has the advantage that sibs informative for only one parental marker can still be used. Using this approach, Davies et al. (1994) did a genome-wide search for markers linked to DS genes influencing human type 1 diabetes. Among doubly affected sibs, one marker on chromosome 6, *D6S273*, had 92 pairs sharing parental alleles and 31 pairs not sharing parental alleles. A second marker on the opposite end of this chromosome, *D6S415*, had 74 pairs sharing parental alleles and 60 not sharing alleles. The MLS scores for these two markers are

$$MLS(D6S273) = 31 \cdot \log_{10}\left(\frac{2 \cdot 31}{123}\right) + 92 \cdot \log_{10}\left(\frac{2 \cdot 92}{123}\right) = 6.87$$

$$MLS(D6S415) = 60 \cdot \log_{10}\left(\frac{2 \cdot 60}{134}\right) + 74 \cdot \log_{10}\left(\frac{2 \cdot 74}{134}\right) = 0.32$$

Thus, the first marker shows significant evidence of linkage, while the second does not. Translating these LOD scores into LR values (the latter being distributed as a χ^2 with one degree of freedom) gives LR = $4.61 \cdot 6.87 = 31.6$ ($P < 0.001$) for *D6S273* and LR = $4.61 \cdot 0.32 = 1.47$ ($P = 0.2$) for *D6S415*.

Power of ASP Tests and Related Issues

The power of different sib-pair tests has been examined by several authors (Suarez et al. 1978, Blackwelder and Elston 1985, Goldin and Gershon 1988, Risch 1990a–c, Schaid and Nick 1990, Sribney and Swift 1992, Goldin and Weeks 1993, Green and Shah 1993, Shah and Green 1994), who conclude that most of the power resides in doubly affected pairs. Blackwelder and Elston (1985) and Schaid and Nick (1990) compared the performance of tests based on the mean number of ibd alleles with those based on a single ibd frequency class (e.g., Equation 16.50b vs. 16.50a). If dominance is modest or linkage is loose, T_m is a more powerful test than T_2. However, with strong dominance and tight linkage, the test based on T_2 is slightly more powerful. Sribney and Swift (1992), noting that assortative mating and multiple underlying loci can inflate the number of required sib-pairs to achieve a given power, suggest that the use of discordant sib-trios (two normal and one affected, or vice versa) can offer higher power in these cases.

The ibd distribution for a marker locus linked to a DS gene has been solved for full sibs by Suarez et al. (1978) and for more general pairs of relatives by Risch (1990b). These expressions are useful for both power calculations (Appendix 5; Figure 16.4) and for showing how the DS parameters influence the ibd distribution. Since many tests use only doubly affected full-sib pairs, we restrict attention to this case. It is convenient to express the ibd distribution as the expected frequency under no linkage plus a deviation due to linkage,

$$p_{20} = \frac{1}{4} - d_{20}, \quad p_{21} = \frac{1}{2} - d_{21}, \quad p_{22} = \frac{1}{4} + d_{22}, \tag{16.52a}$$

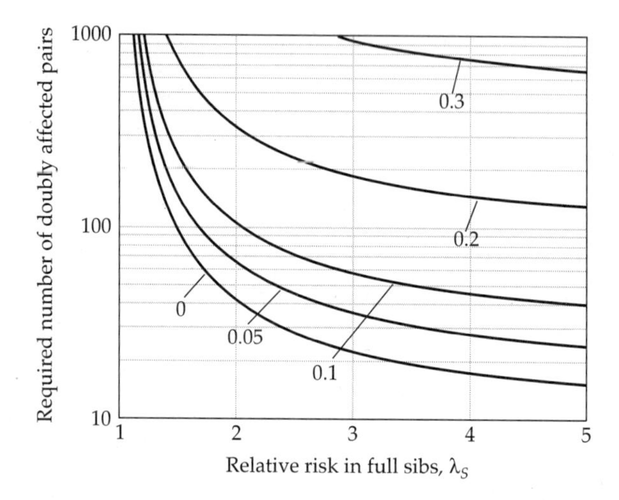

Figure 16.4 Sample size required to detect a QTL using the T_m test (Equation 16.50b). Each curve represents a different value for the recombination fraction c between the QTL and linked marker, and plots the number n_2 of doubly affected sib-pairs required to detect (with 75% power at the 5% significance level) a linked QTL with no dominance variance as a function of λ_S for that QTL. If dominance is present, the required n_2 are below the plotted values, so these curves serve as upper bounds on the desirable sample size regardless of the amount of dominance. See Appendix 5 for these power calculations.

Here, as above, p_{2i} is the probability that a doubly affected pair shares i ibd marker alleles. If c is the recombination fraction between marker and QTL, these deviations are given by

$$d_{20} = \Gamma + [\,\Gamma \cdot \Delta \cdot 4c(1-c)\,] \tag{16.52b}$$
$$d_{21} = 2\Gamma \cdot \Delta \cdot (1-2c)^2 \tag{16.52c}$$
$$d_{22} = \Gamma + \{\,\Gamma \cdot \Delta \cdot 2[\,(1-c)^2 + c^2]\,\} \tag{16.52d}$$

with

$$\Gamma = \frac{(1-2c)^2}{4}\left(1 - \frac{1}{\lambda_S}\right) \qquad \text{and} \qquad \Delta = \frac{\lambda_S - \lambda_O}{\lambda_S - 1}$$

where λ_S and λ_O are the relative risks in full-sib and parent-offspring pairs attributable to the DS locus. If there is no dominance at the DS locus ($\lambda_S = \lambda_O$ and hence $\Delta = 0$), then $d_{21} = 0$, so this class provides no information on linkage. Further, in this case the magnitude of the deviations for p_{20} and p_{22} are identical, and hence each class provides the same amount of linkage information. When dominance is present, $d_{22} > d_{20}$, implying that the p_{22} class provides the most linkage information. Equation 16.52a assumes that a single linked locus underlies the disease. When multiple loci are involved, the marker ibd distribution is no longer a simple function of the population λ_R values, and can be rather involved (Risch 1990a,b).

The power using different types of relative pairs has been examined by Risch (1990b,c) and Bishop and Williamson (1990). For diseases with large λ_S values, grandparent-grandchild pairs offer the highest power of detection when tightly linked markers are available. For diseases with smaller λ_S values and/or less tightly linked markers, sibs offer the most power.

An important issue related to power is the verification of previous linkage claims. Such verification has been especially problematic in studies claiming associations between markers and certain psychiatric disorders (reviewed in Gershon and Cloninger 1994, Risch and Botstein 1996). An analysis of this problem was presented by Suarez et al. (1994), who found that when the character is determined by several QTLs of roughly similar effects, the number of families required to detect the *first* marker-trait association is much smaller than the number of independent families required to *replicate* the initial claim (also see Bell and Lathrop 1996). In effect, the waiting time (measured as number of sib pairs that must be sampled) required for replication of a claim is far greater than the waiting time to detect an association. This result occurs because if the trait is determined by n QTLs of equal effect, and if p is the probability of detecting a given QTL (i.e., the power of the design), then there are n chances for detecting significant marker effects (and hence a power of approximately np), but only one chance of replicating the initial observation (with probability p). This waiting time is especially a problem if a number of rare genes of large effect are involved, as affected individuals in one set of families may arise from one gene, while a second set of families involves another QTL and hence does not replicate the original finding.

Genomic Scanning

Given the increasing density of the human genetic map, **genomic scans** for diseases have become popular. In their simplest form, the investigator computes an ASP test statistic (such as the MLS value) at each of a number of markers, scanning the genome for markers showing linkage to DS genes. Examples include scans for type 1 (Davies et al. 1994) and type 2 (Hanis et al. 1996) diabetes and for multiple sclerosis (Ebers et al. 1996). Lander and Kruglyak's (1995) suggested guidelines (for humans) on the appropriate significance level for such tests have generated much discussion (Witte et al. 1996, Curtis 1996, Lander and Kruglyak 1996, Kruglyak 1996).

A more sophisticated version of genomic scanning has been developed by Kruglyak and Lander (1995b), who used a multipoint method (simultaneously incorporating information from all linked markers) to estimate the ibd distribution at each point along a chromosome. This estimated distribution is then used to compute some ASP statistic for that point. For example, for arbitrary chromosomal map position c, Kruglyak and Lander obtain estimates of $\pi_{20}(c)$, $\pi_{21}(c)$, and $\pi_{22}(c)$ — the ibd values at this position in doubly affected sib pairs — which can then be used to compute any ASP statistics, such as MLS. This method allows for a genomic scan of all chromosomal positions, not just those points corresponding to marker loci. As with likelihood maps, estimates of the locations of DS genes are given by local maxima of the test statistic plot, provided they exceed the significance threshold. With the putative DS gene position in hand, estimates of its effect can be obtained by using the estimated ibd distribution at that position. Recall that the effect of a QTL can be expressed in terms of the relative risks in full-sib and parent-offspring pairs (λ_S and λ_O). First noting that estimates of d_{20} and d_{21} can be obtained from the estimated ibd values, $\widehat{d}_{20} = 1/4 - \widehat{\pi}_{20}$ and $\widehat{d}_{21} = 1/2 - \widehat{\pi}_{21}$, substitution of these into Equations 16.52b,c (with $c = 0$) and rearranging gives

$$\widehat{\lambda}_S = \frac{1}{1 - 4\,\widehat{d}_{20}} = \frac{1}{4\,\widehat{\pi}_{20}} \qquad \text{and} \qquad \widehat{\lambda}_O = \frac{1 - 2\,\widehat{d}_{21}}{1 - 4\,\widehat{d}_{20}} = \frac{\widehat{\pi}_{21}}{2\,\widehat{\pi}_{20}} \tag{16.53a}$$

If desired, these can be translated into estimates of genetic variances using Equations 16.47b and 16.49b. Kruglyak and Lander (1995a) found that 200 affected pairs are sufficient to localize a gene with effect $\lambda_S = \lambda_O = 5$ to within 1 cM. Localization of a gene with $\lambda_S = 2$ with the same precision requires 700 pairs.

Exclusion Mapping and Information Content Mapping

While the support for a QTL is often presented by likelihood (or related) plots, two other types of plots provide useful mapping information. **Exclusion mapping** displays the support for the hypothesis that a QTL is *absent* at a particular position. With a standard LOD score plot, one tests the data against the null hypothesis of no linked QTL. In contrast, under exclusion mapping likelihood-ratio test statistics are still plotted, but now the null hypothesis is that the position contains a QTL of specified (or greater) effect. An **exclusion LOD score** of -2 or less is taken as sufficient evidence to *exclude* the presence of a gene of this effect or larger segregating within the sample population (Figure 16.5). Using this approach, Ebers et al. (1996) were able to exclude 88 percent of the genome of their sampled population from containing a DS gene for multiple sclerosis of effect $\lambda_S \geq 3.0$. Likewise, Hyer et al. (1991) and Hanis et al. (1996) were able to exclude regions from containing genes influencing diabetes. Further details can be found in Hyer et al. (1991), Risch (1993), and Kruglyak and Lander (1995b). One major caveat is that even a highly significant exclusion only refers to the sample data. It is possible that a DS gene of major effect is segregating in the population, but has been missed by the sampling scheme.

Another useful type of plot is an **information content map**, which graphs the amount of marker information at each point along a chromosome (Figure 16.6; Kruglyak and Lander 1995b). Here, one plots the r^2 value for a model that predicts

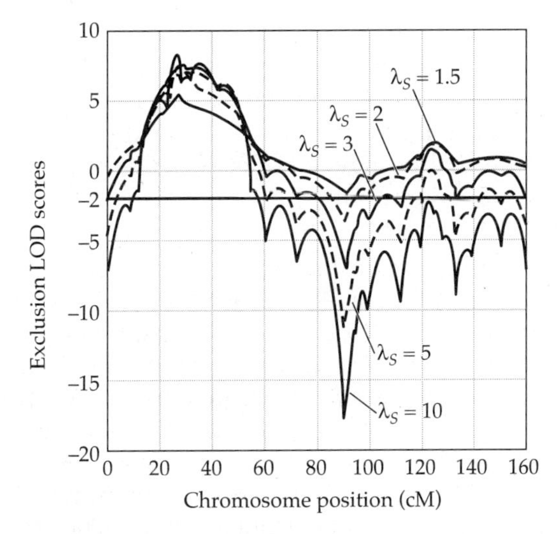

Figure 16.5 LOD scores for exclusion mapping, under the assumption of a DS gene with effect λ_S (and no dominance, so $\lambda_O = \lambda_S$) for a scan of human chromosome 6 for genes influencing diabetes susceptibility (Davies et al. 1994). A LOD score below -2 provides sufficient evidence that a DS gene of specified effect (or greater) is absent from this region. For example, a DS gene with $\lambda_S \geq 5$ can be excluded from chromosome positions 70 – 120 cM and 130 – 140 cM. (From Kruglyak and Lander 1995b.)

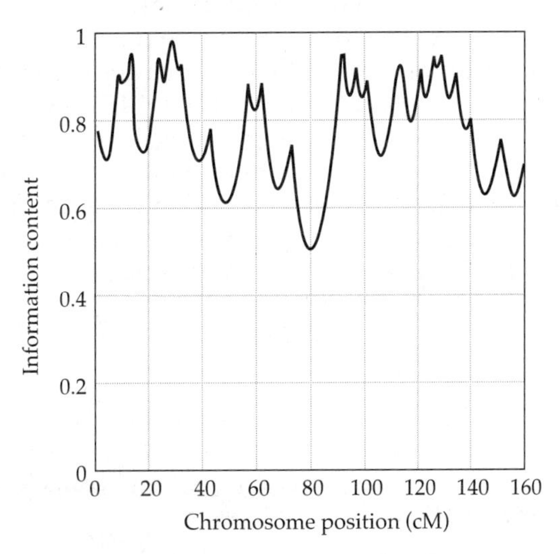

Figure 16.6 Information content mapping of a chromosome 6 diabetes scan (Davies et al. 1994) shows those regions where increasing marker density will provide more linkage information (e.g., 70 – 90 cM and 140 – 160 cM). Information content is highest at the markers, falling off in the intervals between markers. (From Kruglyak and Lander 1995b.)

the ibd distribution for each point along the chromosome. Regions with high r^2 values imply that the ibd distribution is estimated with little error, so additional markers are not needed to predict ibd status. In contrast, regions with smaller r^2 values would benefit by increasing marker density, as this would improve the ability to predict the ibd status in regions between markers.

Affected Pedigree Member Tests

In many cases, the investigator may have a number of pedigrees, each containing several affected members. While an affected sib-pair analysis could be used, there is potentially more information to be gained by considering *all* affected relatives, not just those that happen to be sibs (or some other group of like relatives). This is the motivation behind an **affected pedigree member**, or APM, analysis wherein all pairs of doubly affected relatives within each pedigree are examined in a single analysis. One immediate obstacle to such an analysis is ascertaining the ibd status for marker alleles within each pair of relatives. At a minimum, this requires genotyping all individuals within a pedigree, not just those who are affected. Further, for late-onset diseases (such as Alzheimer's), it is often impossible to genotype the parents, and hence estimates of ibd status of sib-pairs can be equivocal, at best. Instead of working with ibd status, Lange (1986a,b) suggested using the **identity by state** (ibs) status among marker alleles. While ibs status is trivial to ascertain, ibs-based tests are expected to have lower power unless the marker locus is sufficiently polymorphic that ibs status is a strong indicator of ibd status.

While Lange's initial suggestion was to use ibs status for pairs of affected sibs, Weeks and Lange (1988) generalized this approach to consider all pairs of affected members within a pedigree, and a number of extensions have been proposed (Weeks and Lange 1992, Ward 1993, Schroeder et al. 1994, Brown et al. 1994, Davis et al. 1996). The APM test of Weeks and Lange is developed as follows. Consider a pair of affected relatives, i and j, which have marker genotypes $M_{i1}M_{i2}$ and $M_{j1}M_{j2}$. If the marker is linked to a DS gene, doubly affected relatives should have more similar marker genotypes than expected by chance alone. The simplest measure of the shared marker alleles between relatives is to consider

$$Z_{ij} = \frac{1}{4}\sum_{k=1}^{2}\sum_{\ell=1}^{2} \delta(M_{ik}, M_{j\ell}) \quad \text{where} \quad \delta(x,y) = \begin{cases} 1 & \text{if } x = y \\ 0 & \text{otherwise} \end{cases} \tag{16.54}$$

which is just the probability that a randomly chosen marker allele from relative i is identical in state to a random marker allele in relative j. Note that Z_{ij} is trivial to compute for *any* pair of affected relatives. Weeks and Lange's test statistic for a pedigree with n (genotyped) affected relatives is

$$T = \frac{Z - E(Z)}{\sigma(Z)} \qquad \text{where} \qquad Z = \sum_{i<j} Z_{ij} \tag{16.55a}$$

the sum being taken over all $n(n-1)/2$ pairs of doubly affected relatives. For sufficiently large sample sizes, T is approximately distributed as a unit normal, where $E(Z)$ and $\sigma^2(Z)$ are computed from pedigree information; see Weeks and Lange for details. Simulations can be used to compute confidence levels (e.g., Davis et al. 1996).

Since the measure Z_{ij} given by Equation 16.54 weights all genotypes equally, Weeks and Lange suggest that a more reasonable measure weights matching ibs alleles by some function of their allele frequencies,

$$Z_{ij} = \frac{1}{4} \sum_{k=1}^{2} \sum_{\ell=1}^{2} \delta(M_{ik}, M_{j\ell}) f(p_{M_{ik}}) \tag{16.55b}$$

where $p_{M_{ik}}$ is the frequency of allele M_{ik}. The motivation for this modification is that a match involving a rare allele in distant relatives should be weighted more than a match involving a frequent allele (e.g., Example 4). Weeks and Lange suggest three candidate weight functions, $f(p) = 1$ (no weight), $1/\sqrt{p}$, and $1/p$. The first places no weight on allele frequencies and recovers Equation 16.55a. It is most applicable if all alleles have roughly equal frequencies. The last two functions give greater weight to rarer alleles. While $f = 1/p$ places the most weight on rare alleles, Weeks and Lange note that it often creates severe nonnormality in the data. They offer $f = 1/\sqrt{p}$ as a compromise function weighting rare alleles but preserving normality. One criticism of this method is that the weighting functions are rather ad hoc, and it is possible that some of the functions may give significant results while others do not, making the interpretation of linkage somewhat problematic. One solution might be to simply use weights that minimize the sampling variance of the Z_{ij}, which is a standard procedure. A second criticism is that ibs-based tests are sensitive to estimates of the marker-allele frequencies and can give false positives (indications of a linked QTL when none is present) when incorrect frequencies are used (Babron et al. 1993, Van Eerdewegh et al. 1993).

III

Estimation Procedures

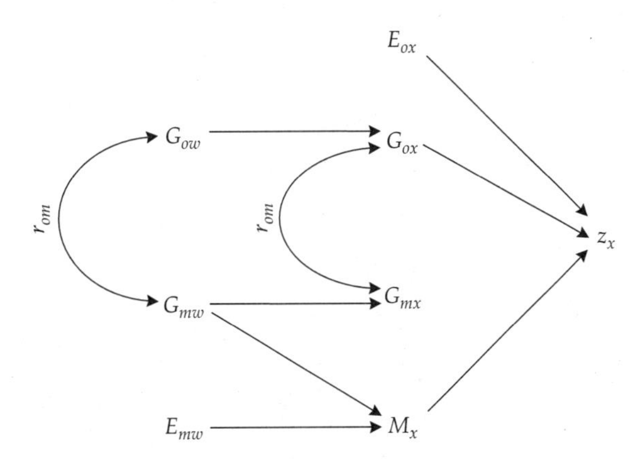

17

Parent-Offspring Regression

As we will see in upcoming chapters, the number of techniques for estimating the components of variance for quantitative traits is quite large. In choosing among the alternatives, two issues arise. First, consideration has to be given to the kinds of relatives that should be analyzed. There are often practical limitations to this problem. Certain kinds of relationships are observed more readily in some species than in others, and some types of phenotypic covariances between relatives are more likely to approximate desired quantities than others. Second, prior to performing an actual analysis, attention should be given to the experimental design. The degree of precision that can be achieved in a quantitative-genetic survey is a function of the number of individuals that are measured, and the way in which effort is allocated to numbers of families versus numbers of individuals within families.

One of the most commonly used methods for estimating heritabilities is the regression of offspring phenotypes on those of their parents, and there are many reasons for this. First, for many species, associations of parents and offspring are the most easily identified relationships in the field. Second, the essential computations are based on least-squares regression, the statistical properties of which are well known (Chapter 3). Third, we have seen in Chapter 7 that neither dominance nor linkage influences the covariance between parents and offspring. Fourth, parent-offspring regression is the only simple method for heritability estimation that is unbiased by selection on the parents. Finally, and perhaps most importantly, the desire to obtain a heritability estimate usually stems from a specific interest in the resemblance between parent and offspring phenotypes, so it is natural that this resemblance should be measured directly.

It is often the case, particularly in natural populations, that the identity of only one parent of an individual can be established with certainty. By collecting seeds, for example, one can be virtually certain of maternity in plants. But since the dispersal of pollen by insects or wind is highly unpredictable, it is often impossible to establish paternity without an elaborate analysis of molecular markers. In other cases, a character may be expressed in only one of the sexes, e.g., clutch size. When the only measurable parent is the mother, care must be taken to ensure that the maternal-progeny covariance is not inflated by maternal effects (Chapter 23). For the time being, we will assume that such effects are not a problem. In addition, we will assume that genotype $\times$ environment interaction and covariance are of negligible importance; such complications are taken up in Chapter 22.

ESTIMATION PROCEDURES

Balanced Data

It is a rather exceptional circumstance when one has the same amount of data from all families, but to simplify discussion, we will start with the assumption that only a single offspring and a single parent are observed in each family. The appropriate linear model for such an analysis is

$$z_{oi} = \alpha + \beta_{op} z_{pi} + e_i$$

where z_{oi} and z_{pi} represent the offspring and parent phenotypes for the ith family, $\alpha = \mu_o - \beta_{op}\mu_p$ is the intercept, β_{op} is the regression coefficient, and e_i is the residual deviation from the regression. In statistical terms, from Chapter 3 we know that the least-squares regression coefficient, $b_{op} = \text{Cov}(z_o, z_p)/\text{Var}(z_p)$, provides an estimate of β_{op}. If there are no environmental causes of resemblance between parents and offspring, we then have (from Chapter 7),

$$E(b_{op}) = \frac{\sigma(z_o, z_p)}{\sigma^2(z_p)} \simeq \frac{(\sigma_A^2/2) + (\sigma_{AA}^2/4) + (\sigma_{AAA}^2/8) + \cdots}{\sigma_z^2} \tag{17.1}$$

Thus, under the stated assumptions, a simple (possibly upwardly biased) estimate of $h^2 = \sigma_A^2/\sigma_z^2$ is twice the (single) parent-offspring regression, $2b_{op}$.

Greater precision is possible when both parents can be measured, as one can then regress offspring phenotypes on the mean phenotypes of their parents (also known as the **midparent values**). Our model is now slightly altered to become

$$z_{oi} = \alpha + \beta_{o\overline{p}} \left(\frac{z_{mi} + z_{fi}}{2} \right) + e_i$$

where z_{mi} and z_{fi} refer to the phenotypes of mothers and fathers. The least-squares slope of the midparent-offspring regression, $b_{o\overline{p}}$, is a direct estimate of the heritability. To obtain this result, let us assume that the phenotypic variance is the same in both sexes and in both generations and that the resemblance between relatives is independent of their sex. We then have

$$\begin{aligned} b_{o\overline{p}} &= \frac{\text{Cov}[z_o, (z_m + z_f)/2]}{\text{Var}[(z_m + z_f)/2]} \\ &= \frac{[\text{Cov}(z_o, z_m) + \text{Cov}(z_o, z_f)]/2}{[\text{Var}(z) + \text{Var}(z)]/4} \\ &= \frac{2\text{Cov}(z_o, z_p)}{\text{Var}(z)} = 2b_{op} \end{aligned} \tag{17.2}$$

In obtaining this result, we have assumed that there is no assortative mating, i.e., $\text{Cov}(z_m, z_f) = 0$. Referring to Equation 17.1, we see that $b_{o\bar{p}} \simeq \sigma_A^2/\sigma_z^2$, ignoring terms involving epistasis.

Finally, consider the situation when multiple (n) offspring are measured in each family. The expected phenotypic covariance of a parent i and the average of its $j = 1, \cdots, n$ offspring may be written $\sigma[(\sum_{j=1}^{n} z_{oij}/n), z_p]$. Under the assumptions of the previous paragraph, all n of the covariance terms contained in this expression have the same expected value, reducing it to $n\sigma(z_o, z_p)/n = \sigma(z_o, z_p)$, which is the same as the expectation for single offspring. Thus, provided family sizes are equal, the interpretation of a parent-offspring regression is the same whether individual offspring data or the progeny means are used in the analysis.

The results for the multiple-offspring regression help clarify why heritabilities are usually estimated from regression rather than correlation coefficients. Assuming equal phenotypic variances in the two generations, the correlation between single offspring and single parents, r_{op}, is identical to the regression coefficient b_{op} (see Equation 3.15b). The single offspring-midparent regression, $r_{o\bar{p}}$, is equal to $b_{o\bar{p}}/\sqrt{2}$, a simple transformation. However, with multiple offspring per family, the variance of offspring family means is a function of both the family size and the heritability itself (see next section), rendering the interpretation of the correlation coefficient difficult. Such problems do not arise with the regression coefficient, which does not involve the use of the offspring family variance. The following section provides a broader coverage of the issues that arise when multiple offspring are assayed per family.

Unequal Family Sizes

When there is significant variation in family size, one is confronted with the problem of how to weight the information from families of different sizes. With a goal of minimizing the sampling error of the heritability estimate, it is logical that families of larger size should be given more weight in a regression since their mean phenotype estimates are more accurate. Should one simply weight each family by the number of offspring measured, as would be the case if one were to regress each individual on its parent's phenotype? The answer is no — the appropriate weights are less than proportional to the actual family sizes. Once one has measured a very large number of offspring from a family, very little improvement in the precision of the family mean will be obtained by making additional measurements.

In one of the first applications of weighted least-squares regression (Chapter 8), Kempthorne and Tandon (1953; see also Bohren et al. 1961) showed that the appropriate weights are proportional to the inverse of the residual sampling variances of family means about the parent-offspring regression. Although more sophisticated maximum-likelihood approaches now exist for the analysis of populations with arbitrary family structures (Chapter 27), these will not be particularly transparent to the reader at this point in the book, so we consider the Kempthorne-

Tandon derivation in some detail. To obtain their result, we require the use of the **intraclass correlation,** here defined to be the phenotypic correlation between sibs,

$$t = \frac{\text{Cov}(S)}{\text{Var}(z)} \tag{17.3}$$

where $\text{Cov}(S)$ denotes the phenotypic covariance of sibs. The intraclass correlation estimates the fraction of the total phenotypic variance attributable to factors causing resemblance between members of the same sib family. It follows that $(1-t)$ estimates the fraction of the phenotypic variance due to differences among individuals of the same family. Stated in another way, $(1-t)\text{Var}(z)$ and $t\cdot\text{Var}(z)$ are estimates of the within- and among-family components of phenotypic variance.

Letting $\overline{z}_{oi}$ be the mean phenotype of offspring from the ith family, the linear model becomes

$$\overline{z}_{oi} = \alpha + \beta_{op} z_{pi} + e_i$$

To perform a weighted regression, we need expressions for the variance of the residual errors around the regression (the e_i) as a function of family size. The residual variance is the sum of two components: (1) the variance of the "true" family mean deviations from the regression, and (2) the sampling variance of the estimated family means around their expectations. The first of these components is independent of family size.

For a family of size n, it follows from above that the second component of the residual variance is simply $(1-t)\text{Var}(z)/n$. The first component is easily obtained by process of elimination. The variance of the true offspring family means is estimated by $t\,\text{Var}(z)$, and from this we have to subtract the variance accounted for by the regression. For a single-parent regression, Equation 3.17 gives the regression variance as $r_{op}^2\text{Var}(z) = b_{op}^2\text{Var}(z)$. (Under the assumption of equal parent and offspring variances, Equation 3.15b implies the regression and correlation coefficients are the same for the single parent-offspring regression.) For a midparent analysis, the variance due to regression is $r_{o\overline{p}}^2\text{Var}(z) = b_{o\overline{p}}^2\text{Var}(z)/2$. This follows from Equation 3.15b, as

$$b_{o\overline{p}}^2 = r_{o\overline{p}}^2 \frac{\text{Var}(z_o)}{\text{Var}(z_{\overline{p}})} = 2\,r_{o\overline{p}}^2$$

Thus, the variance of "true" family means from the regression is $(t-B)\text{Var}(z)$, where $B = b_{op}^2$ or $b_{o\overline{p}}^2/2$ depending on whether one or both parents are used. Summing up, we obtain the expression for the conditional variance of the ith family mean,

$$\text{Var}(e_i) = \left(t - B + \frac{1-t}{n_i}\right)\text{Var}(z) \tag{17.4}$$

The sampling variance of the parent-offspring regression coefficient is minimized by weighting the contribution of different families by the reciprocal of this

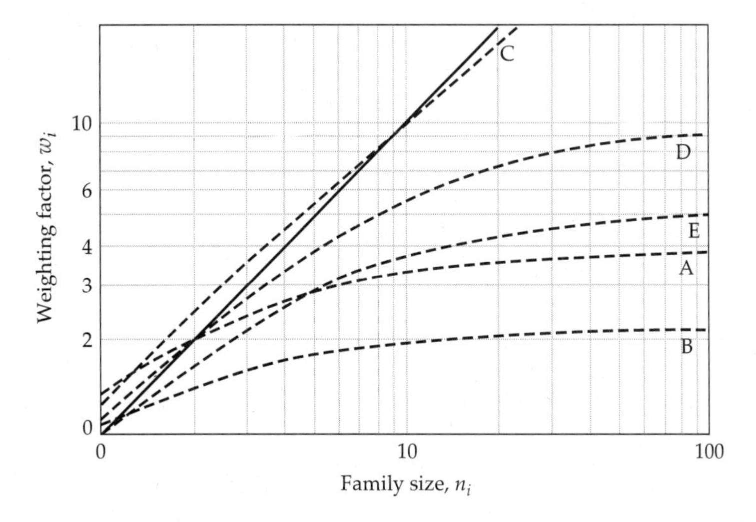

Figure 17.1 The least-squares weights for families with n_i offspring in populations with values of t and B equal respectively to: (A) 0.5, 0.25; (B) 0.5, 0.05; (C) 0.2, 0.18; (D) 0.2, 0.1; and (E) 0.2, 0.01. The solid line is the uncorrected weighting, i.e., simple family size, n_i. Dashed lines are solutions to Equation 17.5a.

quantity (see Example 11, Chapter 8). Since $\text{Var}(z)$ is a constant factor, it can be dropped from the analysis. Thus, the weight for the ith family is

$$w_i = \frac{n_i}{n_i(t - B) + (1 - t)} \tag{17.5a}$$

and, from Equation 8.36b, the weighted least-squares regression coefficient is

$$b = \frac{\sum_{i=1}^{N} w_i(\overline{z}_{oi} - \overline{z}_o)(z_{pi} - \overline{z}_p)}{\sum_{i=1}^{N} w_i(z_{pi} - \overline{z}_p)^2} \tag{17.5b}$$

where

$$\overline{z}_p = \sum_{i=1}^{N} w_i z_{pi} \Big/ \sum_{i=1}^{N} w_i \qquad \text{and} \qquad \overline{z}_o = \sum_{i=1}^{N} w_i \overline{z}_{oi} \Big/ \sum_{i=1}^{N} w_i \tag{17.5c}$$

are the weighted mean phenotypes for the parent and offspring generations. (For a midparent regression, $z_{\overline{p}i}$ needs to be substituted for z_{pi}.)

Equation 17.5a shows that as n_i becomes large, the weight w_i approaches the limiting value $(t - B)^{-1}$, i.e., once the family size is very large, very little is gained by measuring additional progeny. This asymptotic value is reached more rapidly when t is large, because in that case, only a few offspring are sufficient to give an accurate estimate of the family mean. The diminishing returns of large family size can be seen especially clearly by considering case (B) in Figure 17.1 in the context of a single-parent regression. Here, since $B - b_{op}^2 = 0.05$, the heritability is moderate ($h^2 \simeq 2\sqrt{B} = 0.45$). Assuming that the families consist of full-sibs, $t = 0.5$ implies that there is considerable resemblance between sibs due to factors other than additive genetic variance (since the correlation would be expected to be approximately $h^2/2$ on the basis of additive genetic covariance alone). Under these circumstances, families with 10 measured progeny should only be given twice as much weight as families with single offspring.

A practical issue that arises in applying the weighted regression technique is that the weighting factor, w_i, is a function of both t and B. Although t can be calculated directly as the correlation between sib phenotypes, B is a function of the regression coefficient, precisely the quantity that we want to estimate. Resolution of this difficulty is relatively straightforward. A preliminary estimate of B can be obtained from the slope of an unweighted regression analysis. This B is then substituted into Equation 17.5a to generate some preliminary weights. The new regression coefficient generated by Equation 17.5b is then compared with the initial unweighted estimate, and, if the values are the same, the computation is over. If they are different, the second estimate of B is used to generate new weights and a third regression estimate. The entire procedure is repeated until satisfactory convergence has been attained, which usually requires only a few iterations.

Standardization of Data from the Different Sexes

Often, the mean (and / or the variance) of traits differs between males and females. This can result in different estimates of h^2 depending upon which of the sexes is utilized in a parent-offspring regression, since the denominator of the regression coefficient is the phenotypic variance. The problem is sometimes resolved as a scaling issue by using standardized variables, precisely the approach used in Example 5 from Chapter 7 in the analysis of human stature data. For each individual, the observed value minus the mean for that sex is divided by the sex-specific standard deviation. Such a transformation equalizes the phenotypic means and variances across the sexes, to 0 and 1, respectively.

Sex-specific corrections do not always equilibrate the parent-offspring regressions involving the four son-daughter and mother-father combinations. Real sex-specific differences in genetic components of variance may occur, for example, due to variation associated with sex chromosomes or to sex-limited expression of specific genes (Chapter 24).

PRECISION OF ESTIMATES

As in all attempts to estimate parameters, it is always desirable to ascertain the degree of precision of heritability estimates. Since the statistical properties of least-squares regression are well known, this is relatively easy to do with parent-offspring analysis. Provided the data have been measured or transformed so that the joint distribution of parent and offspring phenotypes is bivariate normal, the sampling variance of the single parent-single offspring regression is, from Equation A1.20a, approximately

$$\text{Var}(b_{op}) \simeq \frac{(1 - r_{op}^2)\text{Var}(z_o)}{N\,\text{Var}(z_p)} \tag{17.6}$$

where N is the number of parent-offspring pairs. This expression reduces to $(1 - r^2)/N$ when the phenotypic variances in the two generations are equal.

Equation 17.6 also applies to regressions involving midparents if $\text{Var}(\overline{z}_p) = \text{Var}(z_p)/2$ is substituted for $\text{Var}(z_p)$ and $r_{o\overline{p}}$ for r_{op}, and it applies to regressions involving multiple progeny when $\text{Var}(z_o) \cdot [t + (1-t)/n]$ is substituted for $\text{Var}(z_o)$. For unequal family sizes, Kempthorne and Tandon (1953) show that, when the convergent regression coefficient has been attained,

$$\text{Var}(b) \simeq \frac{\text{Var}(z_o)}{\sum_{i=1}^{N} w_i (z_{pi} - \overline{z}_p)^2} \tag{17.7}$$

Provided the joint distribution of offspring and parent phenotypes is bivariate normal, the sampling distribution of a regression coefficient is also normal. The standard error of b can then be used to construct a confidence interval for the heritability estimate. Provided the number of families $N > 15$ (which is generally necessary for any reasonable degree of precision), the 95% confidence interval for a regression coefficient is approximately $b \pm 2\text{SE}(b)$, where $\text{SE}(b) = \sqrt{\text{Var}(b)}$. For a midparent-offspring regression, the confidence interval for the slope is also the confidence interval for h^2. For a regression involving single parents, the confidence interval for h^2 is twice that of the regression coefficient.

OPTIMUM EXPERIMENTAL DESIGN

Prior to embarking on a long-term, labor-intensive study, it is important to consider how the sampling variance of the parent-offspring regression coefficient might be minimized. Given the constraint of being able to measure a certain number of individuals, the primary question is, How should one's resources be allocated to measuring numbers of families versus numbers of offspring/family? Klein et al. (1973) and Klein (1974) present a useful series of tables outlining the expected standard errors of parent-offspring regressions under various experimental designs.

Latter and Robertson (1960) developed a general procedure for determining the optimal design, showing how the solution depends upon the nature of the constraints on the investigator. We first consider the situation when the investigator is simply limited by the total number of offspring that can be measured (T). If progeny from N families are measured, then the number of progeny measured per family (n) must satisfy $T = Nn$. A general expression for the sampling variance of a regression coefficient has already been given in Equation 17.6. The numerator of that expression is the residual variance around the regression, which was defined in another manner in Equation 17.4. Making the appropriate substitutions in Equation 17.6,

$$\text{Var}(b_{op}) \simeq \frac{n(t - b_{op}^2) + (1 - t)}{Nn} \tag{17.8a}$$

for a single-parent regression, and

$$\text{Var}(b_{o\overline{p}}) \simeq \frac{2[n(t - b_{o\overline{p}}^2/2) + (1 - t)]}{Nn} \tag{17.8b}$$

for a midparent regression. In both cases, since Nn is taken to be the constant T, it is clear that the sampling variance of the regression coefficient is minimized by measuring just a single offspring ($n = 1$) from $N = T$ families.

Now suppose there is a baseline cost to evaluating a family, irrespective of family size. In a natural setting, for example, a certain amount of effort may be necessary to locate a known parent and offspring. In a laboratory setting, a certain amount of time may be necessary for the basic setup and maintenance of a family. Let the limiting resource be the T total time units available for the study, and let τ be the baseline time required for evaluating a family. If one scales the unit of time to be that required for the processing of a single individual in excess of the baseline investment for the family, then $\tau + n$ is the time that it takes to process a family of n offspring, and the number of families that can be processed is $N = T/(\tau + n)$. The optimal family size $\widehat{n}$ can be computed by substituting $N = T/(\tau + n)$ into Equations 17.8a,b, setting the derivative of Var(b) with respect to n equal to zero, and solving for $\widehat{n}$. In both cases,

$$\widehat{n} = \left[\frac{\tau(1 - t)}{t - B}\right]^{1/2} \tag{17.9}$$

Recall that both t and B are functions of h^2. Thus, with this slightly different and perhaps more realistic constraint, the optimal experimental design depends upon the value of the very quantity that we wish to solve for. This obviously reduces the general utility of Equation 17.9. However, an educated guess can sometimes be made, based upon past experience or information in the literature, as to the approximate value of h^2. On average, an estimate of $\widehat{n}$ based upon this information ought to be better than a blind guess.

As an example of the use of Equation 17.9, consider the special case in which the character of interest has a purely additive genetic basis and there are no

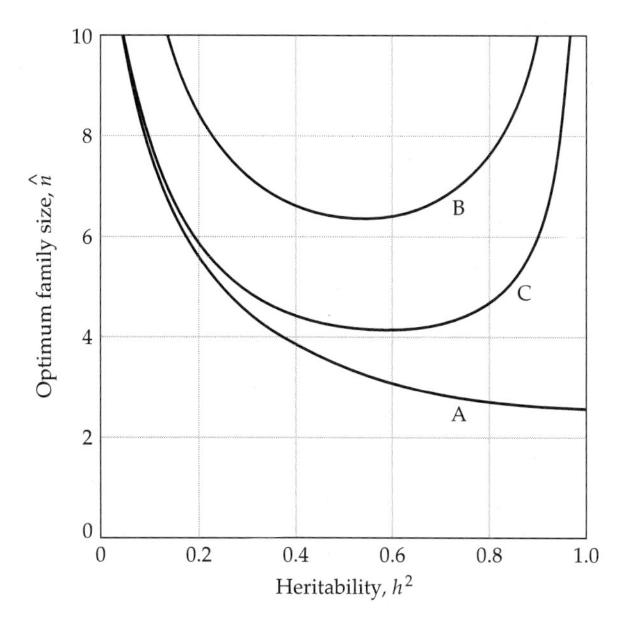

Figure 17.2 Family sizes that minimize the sampling variance of h^2 for the case in which the cost of securing a family is three times that of obtaining measures from individual progeny. (A) Single parents, full-sib families; (B) single parents, half-sib families; (C) midparents, full-sib families.

environmental effects causing resemblance between sibs. Then, for full- and half-sib families respectively, t is $h^2/2$ and $h^2/4$. For single-parent and midparent regressions, B is $h^4/4$ and $h^4/2$. Substituting the appropriate values into Equation 17.9, we obtain optimal designs for the three kinds of parent-offspring regressions:

$$\text{Single parents, full-sib families:} \qquad \widehat{n} = \left[\frac{2\tau}{h^2}\right]^{1/2} \tag{17.10a}$$

$$\text{Single parents, half-sib families:} \qquad \widehat{n} = \left[\frac{\tau(4-h^2)}{h^2(1-h^2)}\right]^{1/2} \tag{17.10b}$$

$$\text{Midparents, full-sib families:} \qquad \widehat{n} = \left[\frac{\tau(2-h^2)}{h^2(1-h^2)}\right]^{1/2} \tag{17.10c}$$

The relationship of $\widehat{n}$ to h^2 is given in Figure 17.2 for these three kinds of experimental designs for the special case in which $\tau = 3$ (i.e., the cost of securing a new family is three times that required for measuring an individual in an established family). Note that the optimal family size is a complex function of the type of family evaluated as well as of the heritability, but that $\widehat{n}$ is never less than 2 in this particular example.

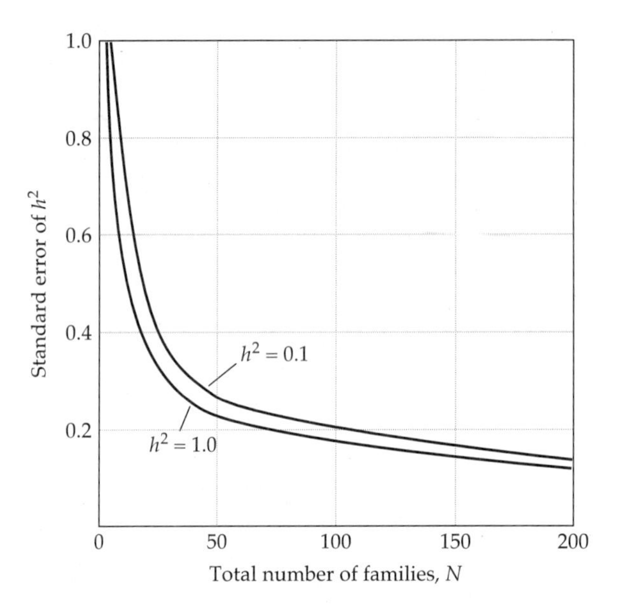

Figure 17.3 Expected standard errors of heritability estimates obtained from regressions of single offspring on single parents. The functions are defined by Equation 17.11a.

By use of the above equations, some insight can be gained into the magnitude of the standard errors of heritability that would arise under an optimal design. Consider the situation where the optimal design has been determined to be $n = 1, N = T$. Substituting $n = 1$ into Equations 17.8a,b, and again assuming the covariance between sibs to be solely due to additive genetic variance, we find:

$$\text{Single-parent regression:} \qquad \text{SE}(h^2) = \left(\frac{4 - h^4}{N} \right)^{1/2} \tag{17.11a}$$

$$\text{Midparent regression:} \qquad \text{SE}(h^2) = \left(\frac{2 - h^4}{N} \right)^{1/2} \tag{17.11b}$$

Three important results can be noted from these expressions. First, since the range of possible values for h^2 is 0 to 1, the standard error of h^2 is nearly independent of h^2. For a single-parent regression, it is approximately equal to $2/\sqrt{N}$. This is illustrated in Figure 17.3 where the standard error is plotted as a function of N for extreme values of $h^2 = 0.1$ and 1.0. Second, compared to a single-parent regression, a midparent regression yields a 30 to 40% improvement in precision. Thus, it is generally well worth gathering data on both parents if possible. Third, unless the heritability is quite high, the detection of statistically significant heritabilities by parent-offspring regression can require large sample sizes. If our estimate of h^2 were 1.0, we would require a standard error of about 0.5 to say with 95% confidence that the true heritability is significantly greater than zero. In the case of a single-parent regression, this would require the measurement of only about 12 parent-offspring pairs. As the heritability declines, however, the sample sizes required for the demonstration of significance rapidly increases. If $h^2 = 0.1$, ~ 1600 parent-offspring pairs are required to obtain a standard error of 0.05.

While the sample size required for a given standard error is straightforward, a more rigorous approach is to compute the actual **power** of the experimental design. The power is the probability that a test statistic will be significant, given the sample size and some assumed true values for the unknown parameters. Power calculations are examined in detail in Appendix 5. Suppose the true $h^2 = 1$, and we test the hypothesis of a significant regression using a test with significance level $\alpha = 0.05$. Taking 12 parent-offspring pairs (as suggested above), the probability of a significant regression coefficient is only 0.53. If the sample size is doubled to 24, the probability of a significant regression increases to 0.82, while 38 parent-offspring pairs are required in order to have a 90% probability. Likewise, for $h^2 = 0.1$, taking 1600 parent-offspring pairs (as suggested by the standard error approach) gives only a 64% chance of the resulting regression being significant. A sample size of 3500 pairs is required to have a 90% chance that the regression is significant.

Hill (1990) has suggested that a further reduction in the standard error of a parent-offspring regression is obtainable if most of the effort is applied to families of parents with phenotypes far from the population mean. The increase in efficiency is a simple consequence of the fact that parents with phenotypes near the mean provide little information on the slope of the regression. A special application of this idea is covered in the following section.

Assortative Mating

Reeve (1961) and Hill (1970) have suggested the use of assortative mating to improve the accuracy of heritability estimates derived from midparent-offspring regressions. The rationale for this approach is that it increases the variance of midparent values from $\sigma^2(z_p)/2$ to $(1+\rho_z)\sigma^2(z_p)/2$, where ρ_z is the phenotypic correlation between mates. Since the variance of a regression coefficient is inversely proportional to the variance of the explanatory variable (Equation 17.6), assortative mating should reduce the sampling variance of $b_{o\overline{p}}$ by a factor of $(1+\rho_z)^{-1}$, e.g., by 50% with full assortative mating.

As noted in Chapter 7 (Table 7.4), assortative mating increases the additive-genetic covariance between parents and offspring from $\sigma_A^2/2$ to $(1+\rho_z)\sigma_A^2/2$. Thus, since both the parent-offspring covariance and the midparent variance are increased by the same factor, assortative mating does not alter the expected value of the midparent-offspring regression. This result is strictly true only in the absence of nonadditive gene action.

With nonadditive gene action, some caution is needed with this approach, as assortative mating can bias the regression coefficient. Although it was suggested in Chapter 7 that dominance has a negligible effect on the covariance of assortatively mated parents and their progeny, this condition requires that the variance of the character is influenced by a large number of loci, each with minor effects. If that is not the case, assortative mating can cause considerable covariance between the nonadditive effects in parents and offspring, as well as between the additive effects in parents and nonadditive effects in progeny (and vice versa) (Wright 1952). Gimelfarb (1985) has shown that under certain circumstances, assortative mating can cause a more than twofold inflation in the slope of a parent-offspring regression, particularly if h^2 is small. Thus, unless one has prior knowledge that nonadditive sources of variance and major alleles are unimportant, assortative mating should probably be avoided in heritability estimation.

ESTIMATION OF HERITABILITY IN NATURAL POPULATIONS

Because it determines the potential response to natural selection, the genetic variation that exists for quantitative characters in natural populations is of fundamental interest to evolutionary biologists. Unfortunately, for many species, it is nearly impossible to carry out a quantitative-genetic analysis in the wild. Many individuals may die before expressing the character of interest, and in mobile animals, a large fraction of the population may be capable of avoiding capture. It is also extremely difficult to identify parentage with certainty in the field, although the situation is improving with the development of new molecular-marker methods.

Intensive efforts with banded bird populations have led to numerous parent-offspring analyses of body size, morphology, and clutch size. Although no individual study is immune to criticism (Hailman 1986, Boag and van Noordwijk 1987), the results are certainly compatible with the idea that significant amounts of additive genetic variation exist for such traits (Table 17.1). For body size (as indexed by tarsus length) and bill morphology, heritabilities are often on the order of 0.5 or greater. The regression coefficients are often independent of the sexes and retain a high level even when progeny are cross-fostered by unrelated mothers, suggesting that postnatal maternal effects are of relatively minor importance (Chapter 23).

For species that cannot be tracked in the field, the investigator has no choice but to remove a segment of the population to the laboratory. Such an approach is of concern since the heritabilities of traits may be as much a function of the environment as of population-genetic structure. For example, the magnitude of environmental variance is likely to differ significantly between artificial and natural settings. Ruiz et al. (1991) obtained phenotypic variances for adult body size

Table 17.1 Heritability estimates ($\pm$ SE) for natural populations of birds obtained by parent-offspring regression.

Species	Mother-Daughter	Father-Son	Reference
Clutch size			
Anser caerulescens	0.61 ± 0.19	—	Findlay and Cooke 1983
Ficedula albicollis	0.32 ± 0.14	—	Gustafsson 1986
Geospiza fortis	-0.17 ± 0.12	—	Gibbs 1988
Parus major	0.48 ± 0.10	—	Perrins and Jones 1974
	0.37 ± 0.12	—	van Noordwijk et al. 1981
Sturnus vulgaris	0.34 ± 0.08	—	Flux and Flux 1982
Tarsus length			
Ficedula albicollis	0.50 ± 0.22	0.43 ± 0.14	Gustafsson 1986
	0.65 ± 0.07	0.53 ± 0.04	Merilä and Gustafsson 1993
*Ficedula hypoleuca**	0.50 ± 0.22	—	Alatalo and Lundberg 1986
Geospiza fortis	0.38 ± 0.30	0.46 ± 0.31	Boag and Grant 1978
Geospiza scandens	0.94 ± 0.39	1.26 ± 0.36	Boag 1983
*Melospiza melodia**	1.12 ± 0.37	0.90 ± 0.33	Smith and Dhondt 1980
	0.34 ± 0.15	0.37 ± 0.18	Smith and Zach 1979
*Parus caeruleus**	0.78 ± 0.26	0.62 ± 0.28	Dhondt 1982
Bill length			
Ficedula albicollis	0.43 ± 0.12	0.37 ± 0.12	Gustafsson 1986
	0.43 ± 0.07	0.44 ± 0.06	Merilä and Gustafsson 1993
Geospiza fortis	1.09 ± 0.42	1.06 ± 0.27	Boag and Grant 1978
Geospiza scandens	-0.18 ± 0.70	0.44 ± 0.41	Boag 1983
*Melospiza melodia**	0.22 ± 0.34	0.59 ± 0.24	Smith and Dhondt 1980

Note: * indicates that the progeny were cross-fostered to minimize postnatal maternal effects. *Anser caerulescens* is the lesser snow goose, *Geospiza* sps. are Darwin's finches, *Ficedula* sps. are flycatchers, *Melospiza melodia* is the song sparrow, *Parus* sps. are tits, and *Sturnus vulgaris* is the starling.

in natural *Drosophila* populations that were seven to nine times larger than those observed in lab-reared derivative populations. In addition, if genotype $\times$ environment interaction is important, the relative rankings and dispersion of genotypic values may be altered by lab rearing. Depending upon the magnitudes and directions of all of these effects, heritability estimates extracted from manipulated populations may be either upwardly or downwardly biased with respect to the wild. Mitchell-Olds and Rutledge (1986) give a useful overview of the salient issues in plant studies. Weigensberg and Roff (1996) examined 22 cases where both laboratory and natural estimates of heritabilities are available. The correlation between measures was significant, with $r = 0.6$, and while laboratory heritabilities tended to be larger than field estimates, the difference was not significant.

With some species, a possible compromise is to remove adults from the field, mate them, and assay their progeny in the artificial setting (Highton 1960, Underhill 1969, Coyne and Beechum 1987, Prout and Barker 1989). Riska et al. (1989) have shown that a lower bound, h^2_{min}, to the heritability in the field can be estimated by regressing the phenotypes of lab-reared progeny on their field-reared parents. Let the regression coefficient involving wild midparents and lab-reared offspring be $b'_{o\overline{p}}$, the phenotypic variance of the natural population be $\text{Var}_n(z)$, and the additive genetic variance in the laboratory environment (obtained either from the covariance of lab-reared sibs or of lab-reared parents and offspring) be $\text{Var}_l(A)$. Then,

$$h^2_{min} = (b'_{o\overline{p}})^2 \frac{\text{Var}_n(z)}{\text{Var}_l(A)} = \left[\frac{\text{Cov}_{l,n}(A)}{\text{Var}_n(z)}\right]^2 \frac{\text{Var}_n(z)}{\text{Var}_l(A)} \tag{17.12}$$

where $\text{Cov}_{l,n}(A)$ is the additive genetic covariance between the trait as expressed in the wild and in the lab. (For an analysis involving single parents, $(2b'_{op})^2$ needs to be substituted for $(b'_{o\overline{p}})^2$ in Equation 17.12.) To see that this provides a lower bound, define

$$\gamma = \frac{\text{Cov}_{l,n}(A)}{\sqrt{\text{Var}_n(A)\text{Var}_l(A)}}$$

to be the additive genetic correlation between environments (Chapter 21). The expected value of h^2_{min} is then $\gamma^2 h^2_n$, which is necessarily $\leq h^2_n$, the heritability in the wild. h^2_{min} is an unbiased estimate of h^2_n only if the genetic correlation across environments is equal to one.

LINEARITY OF THE PARENT-OFFSPRING REGRESSION

We have been operating under the assumption that the true relationship between parent and offspring phenotypes is linear, and indeed, when such data are plotted, there is normally little evidence of nonlinearity (Figure 17.4). There are good statistical reasons for this, including the central limit theorem (Chapter 2) — when multiple independent factors jointly influence the expression of a character, the pairwise distribution of phenotypes in relatives will approach bivariate normality in a randomly mating population, insuring a linear regression (Chapter 8). Such conditions will be approximated as the number of freely recombining loci increases. For purely additive loci, the expected regression is always linear, and dominance is unlikely to cause significant nonlinearity unless the character is strongly influenced by a few rare recessive alleles, all with effects in the same direction (Bulmer 1980, Gimelfarb 1986). Bulmer (1976, 1980) has shown that linkage influences the residual variance but not the linearity of the parent-offspring regression, provided that the loci are in gametic phase equilibrium.

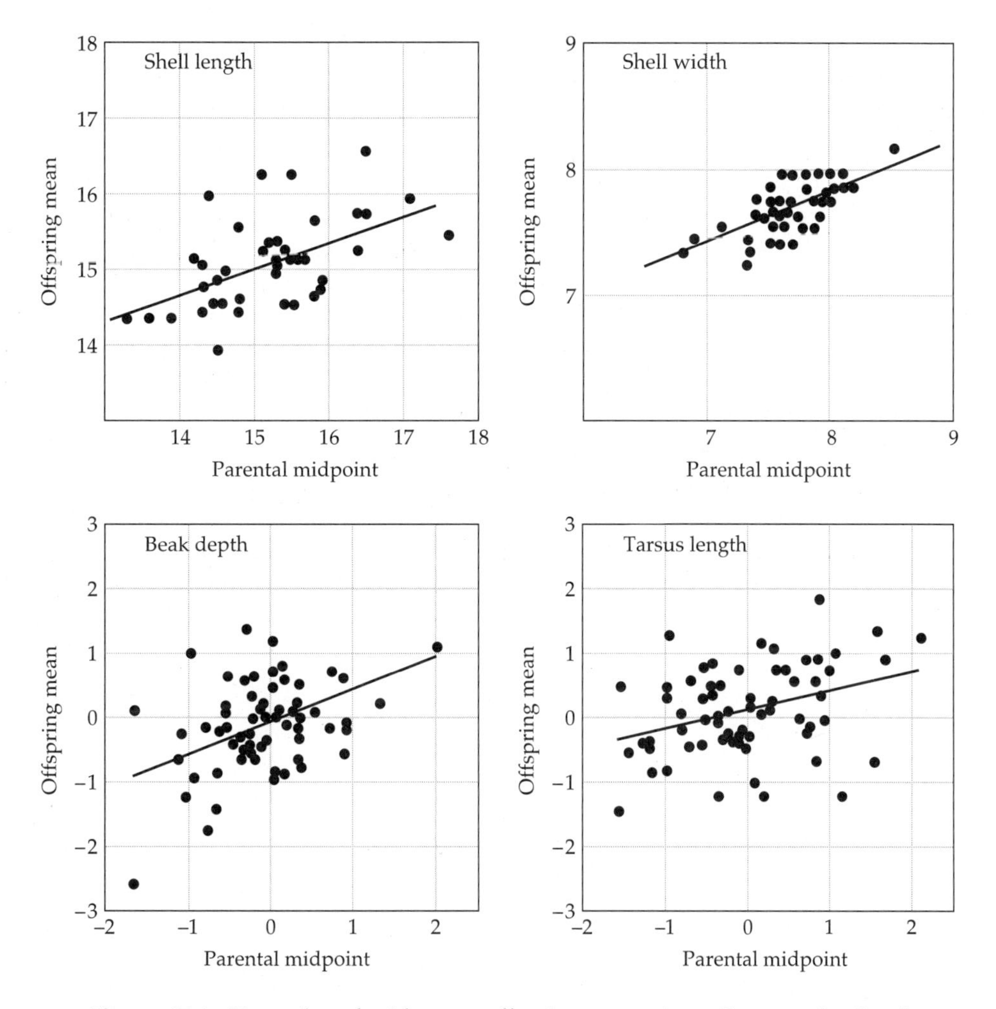

Figure 17.4 Examples of midparent-offspring regressions. **Top panels:** Land snail *Partula taeniata* (Murray and Clarke 1968). **Bottom panels:** Song sparrow *Melospiza melodia* (Smith and Zach 1979).

A few observations of nonlinear parent-offspring regressions have been reported (Nishida 1972, Meyer and Enfield 1975, Gimelfarb and Willis 1994), and some theoretical explanations can be offered to explain them:

1. If extended out to the most extreme genotypic values, where individuals are homozygous for all "positive" or all "negative" alleles, the regression must ultimately become nonlinear. Any further increase or decrease in parental phenotype could only be due to environmental effects and would

not be reflected in the offspring phenotype. In theory, such limits can be approached under strong directional selection, but they are expected to be uncommon in natural populations due to the rarity of extreme multilocus homozygotes.

2. Nishida and Abe (1974) and Robertson (1977a) have pointed out that linearity of a parent-offspring regression requires that the distributions of genetic and environmental effects be of the same form. That is, even if the underlying distribution of genotypic values is normal, the regression will not be strictly linear unless the environmental deviations are also normally distributed. Nevertheless, if numerous environmental factors influence the expression of a trait, the central limit theorem will again ensure that this source of nonlinearity will not be great.

3. Robertson (1977a) has shown that nonlinearity may arise if the variance of environmental deviations is a function of the genotypic value. Suppose, for example, that highly "positive" genotypic values were associated with exceptionally high variance for environmental effects. Such a condition would tend to reduce the correspondence (and hence the regression) between parental phenotype and genotype on the high end of the scale.

18

Sib Analysis

Situations in which one is unable to acquire information from both parent and offspring generations are common. For example, when the character of interest is age-specific, it can be impractical to study phenotypic correlations with ancestors if the generation time exceeds a few years. This is a common problem in forest genetics. For species with nonoverlapping generations, it is impossible to find adults and their offspring in the field at the same time. Moreover, many animal species lay their eggs externally and then abandon them, making it difficult to ascertain parentage. In all of these cases, the analysis of contemporary (or **collateral**) relatives, sibs in particular, provides an attractive alternative to parent-offspring regression in estimating quantitative-genetic parameters.

There are essentially three types of sib analyses: those employing half-sib families, those employing full-sib families, and those combining both. Each of these family structures permit one to partition the total phenotypic variance into within- and among-family components, both of which can be interpreted in terms of covariances between relatives. The branch of statistics called **analysis of variance** (ANOVA) is designed to deal with these kinds of data.

In addition to outlining the practical aspects of sib analysis and its underlying assumptions, this chapter will serve the function of introducing the fundamental principles underlying analysis of variance. This discussion will help set the stage for understanding more advanced applications of ANOVA that appear in the next several chapters. We start with the simple half-sib model, initially assuming a balanced design, and show how observed within- and among-family components of variance can be related to the underlying causal components of variance discussed in Chapter 7. Complications that arise with unbalanced data sets are then described. The second half of the chapter considers the nested full-sib, half-sib design. A major advantage of this method over the half-sib design is its ability to provide insight into the potential significance of dominance and/or shared environmental effects.

In describing both methods, we follow a similar organization, initially covering issues involving model definition and parameter interpretation, then considering methods of parameter estimation and hypothesis testing, and finally discussing optimal designs for minimizing the sampling error in statistical analysis. We assume throughout that parents have been sampled randomly from the population and randomly mated, so that the simple causal interpretations of covariances between sibs given in Chapter 7 can be used. For highly unbalanced designs (unequal family sizes) and situations involving nonrandom mating and selection, maximum-likelihood procedures have been developed as an alternative to ANOVA in parameter estimation. We defer discussion of these more advanced methods to Chapter 27. Since both methods rely on the same model and interpretation of the variance components and both yield the same estimates with balanced designs, the basic discussion in this chapter should still be useful to even the most ardent defenders of ML.

HALF-SIB ANALYSIS

The utility of half-sib analysis stems from the close relationship between the additive genetic variance and the covariance between half sibs (Chapter 7). Under random mating and free recombination, four times the genetic covariance between half sibs is $\sigma_A^2 + (\sigma_{AA}^2/4) + (\sigma_{AAA}^2/16) + \cdots$, which is approximately equal to twice the genetic covariance between parents and offspring, $\sigma_A^2 + (\sigma_{AA}^2/2) + (\sigma_{AAA}^2/4) + \cdots$, provided that the components of epistatic genetic variance are small relative to σ_A^2. Thus, if epistasis is of minor importance and if common environmental effects do not contribute to the phenotypic resemblance of half sibs, $4\sigma(\text{HS})$ provides an estimate of σ_A^2. (Here $\sigma(\text{HS})$ denotes the expected covariance between a pair of half sibs. Similar notation is used below for other types of sibs.)

The potential for common environmental effects is the main drawback of any sib analysis, and special precautions should always be taken to minimize the problem. If the variance due to common environmental effects is of the order of σ_A^2, then $4\sigma(\text{HS})$ can greatly exceed σ_A^2. The most reliable way to minimize this problem is to exclusively employ paternal half-sib families in order to eliminate common maternal effects. A typical paternal half-sib design involves the random mating of each of N males to n different females and evaluation of a single offspring from each female (Figure 18.1). Under this design, all of the progeny of a given male are half sibs, unrelated to progeny of other males.

There are two logical ways to analyze paternal half-sib data. One could simply perform a regression of half sib on half sib using all possible combinations within a family as entries (assuming a balanced design). Problems arise with this approach, since pairs of points within families are not independent. Karlin et al. (1981) suggest several regression methods involving different weightings for family size, but the statistical justification for these has not been established.

The traditional approach to analyzing half-sib data is the **one-way analysis of variance**, based on the linear model

$$z_{ij} = \mu + s_i + e_{ij} \tag{18.1}$$

where z_{ij} is the phenotype of the jth offspring of the ith father, s_i is the effect of the

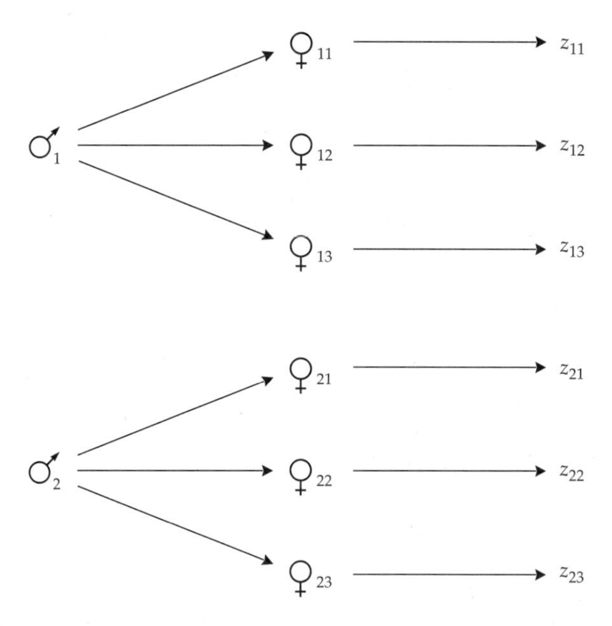

Figure 18.1 A paternal half-sib mating design. Each male is mated to several unique (unrelated) females, and a single offspring from each female is assayed.

ith father (the **sire effect**), and e_{ij} is the residual error resulting from segregation, dominance, genetic variance among mothers, and environmental variance. Stated another way, e_{ij} is the deviation of the phenotype of the ijth individual from the expected value for the ith family. As deviations from the linear model, the e_{ij} have expectations equal to zero. We further assume that the e_{ij} are uncorrelated with each other and have common variance σ_e^2, the **within-family variance**. The N sires are assumed to be a random sample of the entire population so that $E(s_i) = 0$. The variance among sire effects (the **among-family variance**) is denoted by σ_s^2.

A basic assumption of linear models underlying ANOVA is that the random factors are uncorrelated with each other. As first recognized by Fisher in his classical 1918 paper, this leads to a key feature — *the analysis of variance partitions the total phenotypic variance into the sum of the variances from each of the contributing factors.* For example, for the half-sib model, the critical assumption is that the residual deviations are uncorrelated with the sire effects, i.e., $\sigma(s_i, e_{ij}) = E(s_i e_{ij}) = 0$. Thus, the total phenotypic variance equals the variance due to sires plus the residual variance,

$$\sigma_z^2 = \sigma_s^2 + \sigma_e^2 \tag{18.2}$$

A second relationship that proves to be very useful is that *the phenotypic covariance between members of the same group equals the variance among groups.* For the model given in Equation 18.1, this can be shown quite simply. Members of the same group (paternal half sibs) share sire effects, but have independent residual deviations, so

$$\begin{aligned}\sigma(\text{PHS}) &= \sigma(\, z_{ij}, z_{ik}\,) \\ &= \sigma[\,(\mu + s_i + e_{ij}), (\mu + s_i + e_{ik})\,] \\ &= \sigma(\, s_i, s_i\,) + \sigma(\, s_i, e_{ik}\,) + \sigma(\, e_{ij}, s_i\,) + \sigma(\, e_{ij}, e_{ik}\,) \\ &= \sigma_s^2 \end{aligned} \tag{18.3}$$

Thus, the covariance between paternal half sibs equals the variance among sire effects. This is a particularly useful identity since, as we will see below, ANOVA provides a simple means of estimating σ_s^2. Hence, for the ideal case in which a character has no epistatic variance, the additive genetic variance can be estimated as four times the among-family variance, i.e., $4\sigma_s^2 = \sigma_A^2$.

The pure half-sib design employs the simplest possible linear model. However, the general logic just outlined applies to the estimation of variance components in all linear models, including those employed in subsequent chapters. Thus, the steps we have just taken are worth summarizing. First, the linear model is written down. Second, with the assumptions of the model made explicit, an expression for the total phenotypic variance is written in terms of components. Third, the components of variance associated with the model are expressed as covariances between specific classes of relatives. Fourth, using the mechanistic interpretations of phenotypic covariances between relatives outlined in Chapter 7, the observable variance components are used to partition the phenotypic variance into its causal sources. We now demonstrate the practical utility of this approach by showing how ANOVA generates estimates of the within- and among-family components of variance from phenotypic data.

One-way Analysis of Variance

ANOVA uses **sums of squares,** the derivation of which we outline below. Throughout, we use SS to denote an observed sum of squares, and E(SS) to denote its expected value. As we will see shortly, scaled sums of squares, known as **mean squares,** are used to estimate variance components. We use the parallel notation, MS and E(MS), to denote observed and expected mean squares.

Consider the balanced design in which n half sibs are assayed from each of N males, so that there are a total $T = Nn$ individuals in the analysis. The quantity

$$\text{SS}_T = \sum_{i=1}^{N}\sum_{j=1}^{n}(z_{ij} - \overline{z})^2 \tag{18.4}$$

defines the observed **total sum of squares** around the grand mean $\overline{z}$. ANOVA partitions SS_T into components describing variation among the s_i (i.e., among families) and among the e_{ij} within families. This partitioning is readily accomplished by expanding around the observed family means, $\overline{z}_i = \sum_{j=1}^{n} z_{ij}/n$,

$$\begin{aligned} SS_T &= \sum_{i=1}^{N}\sum_{j=1}^{n}[(z_{ij}-\overline{z}_i)+(\overline{z}_i-\overline{z})]^2 \\ &= \sum_{i=1}^{N}\sum_{j=1}^{n}[(z_{ij}-\overline{z}_i)^2+2(z_{ij}-\overline{z}_i)(\overline{z}_i-\overline{z})+(\overline{z}_i-\overline{z})^2] \end{aligned} \tag{18.5}$$

The middle term of this expression is equal to zero, since by the definition of a mean, $\sum_{j=1}^{n}(z_{ij}-\overline{z_i}) = 0$. The third term may be written as $n\sum_{i=1}^{N}(\overline{z}_i-\overline{z})^2$ since it does not contain j. Thus, the total sum of squares is partitioned into an among- and a within-family component,

$$\begin{aligned} SS_T &= n\sum_{i=1}^{N}(\overline{z}_i-\overline{z})^2+\sum_{i=1}^{N}\sum_{j=1}^{n}(z_{ij}-\overline{z}_i)^2 \\ &= SS_s + SS_e \end{aligned} \tag{18.6}$$

The within-family sum of squares (SS_e) is simply the sum of the squared deviations of individual measures from their observed family means, while the among-family sum of squares (SS_s) is the sum (over all progeny) of the squared deviations of observed family means from the grand mean.

Assuming that the parents are a random sample of the population at large, the sums of squares can be used to obtain unbiased estimates of the within- and among-family components of variance in the following way. We note first that the expected within-family sum of squares is

$$E(SS_e) = \sum_{i=1}^{N} E\left[\sum_{j=1}^{n}(z_{ij}-\overline{z}_i)^2\right] = N(n-1)\sigma_e^2 \tag{18.7a}$$

This result follows from the fact that $\sum_{j=1}^{n}(z_{ij}-\overline{z}_i)^2/(n-1)$ is an unbiased estimate of the variance among sibs in the ith family (Chapter 2) and from our assumption that the variance within each family is equal to σ_e^2.

For the among-family sum of squares, similar reasoning leads to

$$E(SS_s) = nE\left[\sum_{i=1}^{N}(\overline{z}_i-\overline{z})^2\right] = n(N-1)\sigma^2(\overline{z}_i) \tag{18.7b}$$

where $\sigma^2(\overline{z}_i)$ is the expected variance of the observed family means, here (with a balanced design) assumed to be the same for all families. Further simplification of this expression is possible. The variance of observed family means is a function of the variance of the true family means, the $(\mu+s_i)$, as well as of their sampling error, the $\overline{e}_i = \overline{z}_i - (\mu + s_i)$. Thus, assuming that the measurement error is independent of the family mean,

$$\sigma^2(\overline{z}_i) = \sigma^2(\mu + s_i) + \sigma^2(\overline{e}_i) \tag{18.8}$$

Since μ is a constant, the first term of this expression is the among-family variance, σ_s^2, while the second is the expected sampling variance of a mean, σ_e^2/n (Chapter 2). Substituting into Equation 18.7b,

$$E(\text{SS}_s) = (N - 1)(\sigma_e^2 + n\sigma_s^2) \tag{18.9}$$

Finally, rearranging Equations 18.7a and 18.9, the variance components can be expressed in terms of the expected sums of squares,

$$\sigma_e^2 = \frac{E(\text{SS}_e)}{N(n-1)} \tag{18.10a}$$

$$\sigma_s^2 = \frac{1}{n}\left[\frac{E(\text{SS}_s)}{N-1} - \frac{E(\text{SS}_e)}{N(n-1)}\right] \tag{18.10b}$$

Note that the sums of squares in these expressions are divided by constants. Such weighted sums of squares are the mean squares (MS) referred to above, and the quantities in their denominators are the associated degrees of freedom (df). For the half-sib model,

$$\text{MS}_s = \frac{\text{SS}_s}{N-1} \tag{18.11a}$$

$$\text{MS}_e = \frac{\text{SS}_e}{N(n-1)} \tag{18.11b}$$

are the observed among- and within-family mean squares. Substitution of observed mean squares for their expectations in Equations 18.10a,b yields the following unbiased estimators of σ_s^2, σ_e^2, and σ_z^2,

$$\text{Var}(s) = \frac{\text{MS}_s - \text{MS}_e}{n} \tag{18.12a}$$

$$\text{Var}(e) = \text{MS}_e \tag{18.12b}$$

$$\text{Var}(z) = \text{Var}(s) + \text{Var}(e) \tag{18.12c}$$

A summary of the steps for obtaining the observed mean squares, generalized to allow for unequal family sizes, is given in Table 18.1. This general procedure of estimating variance components from observed mean squares is an example of the **method of moments**, as the unknown variances can be expressed in terms of observable moments (here, the mean squares).

Table 18.1 Summary of a one-way ANOVA involving N independent families, the ith of which contains n_i individuals.

Factor	df	SS	MS	E(MS)
Among-families	$N-1$	$\text{SS}_s = \sum_{i=1}^{N} n_i(\overline{z}_i - \overline{z})^2$	$\text{SS}_s/(N-1)$	$\sigma_e^2 + n_0\sigma_s^2$
Within-families	$T-N$	$\text{SS}_e = \sum_{i=1}^{N}\sum_{j=1}^{n_i}(z_{ij} - \overline{z}_i)^2$	$\text{SS}_e/(T-N)$	σ_e^2
Total	$T-1$	$\text{SS}_T = \sum_{i=1}^{N}\sum_{j=1}^{n_i}(z_{ij} - \overline{z})^2$	$\text{SS}_T/(T-1)$	σ_z^2

Note: The total sample size is $T = \sum_{i=1}^{N} n_i$, and $n_0 = [T - (\sum n_i^2/T)]/(N-1)$, which reduces to n with equal family sizes. Degrees of freedom are denoted by df, observed sums of squares by SS, and expected mean squares by E(MS).

The quantity

$$t_{\text{PHS}} = \frac{\text{Var}(s)}{\text{Var}(z)} \tag{18.13}$$

is the intraclass correlation (Fisher 1918, 1925), discussed previously in Chapter 17. It provides an estimate of the fraction of the phenotypic variance attributable to differences among sires. Recalling from above that $\sigma_s^2 = \sigma(\text{PHS}) \simeq \sigma_A^2/4$, the paternal half-sib ANOVA estimator of the heritability is

$$h^2 \simeq 4t_{\text{PHS}} \tag{18.14}$$

This expression again assumes that contributions from epistatic genetic variance are small.

Ratios of quantities estimated with sampling error are usually biased with respect to their parametric values (Appendix 1), and this is true for the intraclass correlation (Ponzoni and James 1978, Wang et al. 1991). Letting τ be the parametric value, the downward bias is approximately

$$\Delta_t = \tau - E(t_{\text{PHS}}) = \frac{2\tau(1-\tau)[(n-1)\tau + 1]}{nN} \tag{18.15}$$

In principle, correction for this bias can be made by substituting the observed t_{PHS} for τ in the preceding expression and adding the estimated bias Δ_t to t_{PHS} prior to estimating the heritability with Equation 18.14. The bias in t_{PHS} can be considerable if N is very small (less than 20), but for larger designs it is generally no more than a few percent.

Hypothesis Testing

In obtaining the variance-component estimators, Equations 18.12a,b, we made no assumptions as to how the data or their underlying components (s_i and e_{ij}) were distributed, other than the constraint that they are independent of each other. This distribution-free condition illustrates a useful feature of ANOVA that is not shared by many other estimation procedures — it yields variance-component estimates that are unbiased with respect to the true parametric values (although, as just noted, nonlinear functions, such as ratios, of these estimates will be biased). Unfortunately, this distribution-free property does not extend to the estimation of confidence intervals for the variance components, nor to most traditional methods of hypothesis testing.

Most conventional hypothesis tests involving ANOVA assume normality and homogeneity of error variances. Thus, prior to embarking on an analysis of variance, an attempt should always be made to ensure that the observed data are on an appropriate scale of measurement (Chapter 11). It should be realized, however, that normality of the observed data does not guarantee normality of the distributions of the underlying factors s_i and e_{ij}.

Assuming that adequate normalization has been accomplished, standard theoretical results can be used to test the hypothesis that the among-family component of variance, and hence the heritability, is significantly greater than zero. We accomplish this by recalling from Appendix 5 that when normally distributed variables with mean zero and variance one (unit normals) are squared, they follow a χ^2 distribution. Dividing an observed sum of squares (SS) by its associated E(MS) transforms the SS into a sum of squared unit normals, which is χ^2-distributed with the associated degrees of freedom. For the one-way ANOVA, from Equations 18.7a, 18.9, and 18.11, the expected mean squares are

$$E(\text{MS}_s) = \sigma_e^2 + n\sigma_s^2 \tag{18.16a}$$

$$E(\text{MS}_e) = \sigma_e^2 \tag{18.16b}$$

Thus,

$$\frac{\text{SS}_s}{\sigma_e^2 + n\sigma_s^2} \sim \chi^2_{N-1} \tag{18.17a}$$

$$\frac{\text{SS}_e}{\sigma_e^2} \sim \chi^2_{T-N} \tag{18.17b}$$

Recall also that the ratio of two χ^2-distributed variables, each divided by its respective degrees of freedom, follows an F distribution (Appendix 5).

Now notice that if $\sigma_s^2 = 0$, the denominators of Equations 18.17a and 18.17b are the same, in which case their ratio is simply SS_s/SS_e. Recalling that $\text{SS}_x/\text{df}_x = \text{MS}_x$,

$$F = \frac{\text{MS}_s}{\text{MS}_e} \tag{18.18}$$

provides a test of the hypothesis that $E(\text{MS}_s) = E(\text{MS}_e)$, or equivalently that $\sigma_s^2 = 0$. If $\sigma_s^2 > 0$, we expect the ratio of observed mean squares to be greater than one. However, it needs to be significantly larger than one if we are to be confident in our conclusion that $\text{MS}_s > \text{MS}_e$ did not occur just by chance. An explicit test of the null hypothesis of no sire effects is made by referring to standard F-distribution tables and comparing the observed value of F with the critical values associated with $(N-1)$ and $(T-N)$ degrees of freedom.

Sampling Variance and Standard Errors

In the analysis of heritability, a case can be made that hypothesis testing is of little biological relevance. Because polygenic mutation continually introduces genetic variation into populations, the heritabilities of essentially all characters must be nonzero, and the only real issue is their absolute magnitude. If an F-test signals nonsignificance, it most likely is a simple consequence of inadequate sample size.

Standard errors provide rough guides to the accuracy of variance-component estimates, and to estimate them, we require the sampling variances of the observed mean squares. Under the assumptions of normality and balanced design, a useful (and general) result is that the observed mean squares extracted from an analysis of variance are distributed independently with expected sampling variance

$$\sigma^2(\text{MS}_x) \simeq \frac{2E(\text{MS}_x)^2}{\text{df}_x + 2} \tag{18.19}$$

This fundamental relationship has been used in many contexts in quantitative genetics to derive expressions for variances and covariances of variance components extracted from ANOVA (Tukey 1956, 1957; Smith 1956; Bulmer 1957, 1980; Scheffé 1959). Searle et al. (1992) provide a particularly lucid overview of its utility.

Since the variance-component estimators, Equations 18.12a–c, are linear functions of the observed mean squares, the rules for obtaining variances and covariances of linear functions (Chapter 3 and Appendix 1) can be used in conjunction with Equation 18.19 to obtain the large-sample approximations

$$\text{Var}[\,\text{Var}(e)\,] = \text{Var}(\text{MS}_e) \simeq \frac{2(\text{MS}_e)^2}{T - N + 2} \tag{18.20a}$$

$$\begin{aligned} \text{Var}[\,\text{Var}(s)\,] &= \text{Var}\left[\frac{\text{MS}_s - \text{MS}_e}{n}\right] \\ &\simeq \frac{2}{n^2}\left(\frac{(\text{MS}_s)^2}{N+1} + \frac{(\text{MS}_e)^2}{T-N+2}\right) \end{aligned} \tag{18.20b}$$

$$\begin{aligned} \text{Cov}[\,\text{Var}(s), \text{Var}(e)\,] &= \text{Cov}\left[\left(\frac{\text{MS}_s - \text{MS}_e}{n}\right), \text{MS}_e\right] = -\frac{\text{Var}(\text{MS}_e)}{n} \\ &\simeq -\frac{2(\text{MS}_e)^2}{n(T-N+2)} \end{aligned} \tag{18.20c}$$

$$\text{Var}[\,\text{Var}(z)\,] = \text{Var}[\,\text{Var}(e)\,] + 2\text{Cov}[\,\text{Var}(s), \text{Var}(e)\,] + \text{Var}[\,\text{Var}(s)\,] \tag{18.20d}$$

The standard errors of the estimated within-family, among-family, and total phenotypic variance estimates are obtained by substituting observed mean squares into Equations 18.20a,c,d and taking square roots. Since the accuracy of the resultant standard errors depends on the accuracy of the observed mean squares, the standard errors are not very reliable if the degrees of freedom are small. Hence, the reference to "large-sample" estimators.

Using the techniques in Appendix 1, Osborne and Paterson (1952) showed that the large-sample variance of the intraclass correlation from a balanced one-way ANOVA is

$$\text{Var}(t) \simeq \frac{2(1-t)^2[1+(n-1)t]^2}{Nn(n-1)} \tag{18.21}$$

The standard error of h^2 derived by half-sib analysis is estimated by $4\sqrt{\text{Var}(t)}$. Again, the accuracy of this expression increases with the number of families in an analysis.

Confidence Intervals

Under the assumption of normality, approximate confidence intervals for variance-component estimates can be obtained from the expected distributions of the sums of squares (Harville and Fenech 1985, Searle et al. 1992). For the within-family variance, recalling the distribution of SS_e/σ_e^2 given in Equation 18.17b, the lower and upper values associated with the $100(1-\alpha)\%$ confidence level are simply

$$\frac{\text{SS}_e}{\chi^2_{(T-N),(\alpha/2)}} < \text{Var}(e) < \frac{\text{SS}_e}{\chi^2_{(T-N),(1-\alpha/2)}} \tag{18.22}$$

where $\chi^2_{(T-N),(\alpha/2)}$ and $\chi^2_{(T-N),(1-\alpha/2)}$ are the upper and lower χ^2 values associated with α given $(T-N)$ degrees of freedom. For example, for a 95% confidence interval (2.5% error on each side of the estimate), $\chi^2_{(T-N),0.025}$ is the point at which the probability of obtaining a higher χ^2_{T-N} by chance is 0.025 and $\chi^2_{(T-N),0.975}$ is the point at which the probability of obtaining a higher value is 0.975. These values can be found in tabular form in most elementary statistics texts.

For the among-family variance, the lower and upper confidence limits associated with the $100(1-\alpha)\%$ level are given by

$$\frac{\text{MS}_e}{n}\left[\frac{F}{F_{(N-1),\infty,(\alpha/2)}} - 1 - \left(\frac{F_{(N-1),(T-N),(\alpha/2)}}{F_{(N-1),\infty,(\alpha/2)}} - 1\right)\left(\frac{F_{(N-1),(T-N),(\alpha/2)}}{F}\right)\right] \tag{18.23a}$$

and

$$\frac{\text{MS}_e}{n}\left[F \cdot F_{\infty,(N-1),(\alpha/2)} - 1 + \left(1 - \frac{F_{\infty,(N-1),(\alpha/2)}}{F_{(T-N),(N-1),(\alpha/2)}}\right)\left(\frac{1}{F_{(T-N),(N-1),(\alpha/2)}}\right)\right] \tag{18.23b}$$

respectively, where the unsubscripted F is the ratio of observed mean squares defined by Equation 18.18, and the F values subscripted by their degrees of freedom are the critical values associated with $\alpha/2$. These values are also obtainable from standard tables.

Assuming normality of the underlying data, the $100(1-\alpha)\%$ confidence interval for the heritability is given by

$$4\left[\frac{(F/F_U)-1}{(F/F_U)+n-1}\right] < h^2 < 4\left[\frac{(F/F_L)-1}{(F/F_L)+n-1}\right] \tag{18.24}$$

(Scheffé 1959, Graybill 1961, Williams 1962). In these expressions, F is again the ratio of observed mean squares defined by Equation 18.18, and F_U and F_L are the upper and lower F values associated with $(\alpha/2)$ at $(N-1)$, $(T-N)$ degrees of freedom. Specifically, $F_U = F_{(N-1),(T-N),(\alpha/2)}$, whereas $F_L = 1/(F_{(T-N),(N-1),(\alpha/2)})$. (See Example 2 for an application of this equation.)

Although somewhat complicated, the preceding expressions are general, provided the data are normally distributed with homogeneous variance (i.e., $s \sim \text{N}(0,\sigma_s^2)$ and $e \sim \text{N}(0,\sigma_e^2)$ for all families). However, most confidence intervals reported in the literature are approximated by a simpler route. The usual procedure is to assume that the degrees of freedom are large enough that parameter estimates are approximately normally distributed. Then, symmetrical confidence intervals can be computed more simply from the standard errors, e.g., 95% confidence intervals are obtained by multiplying the standard error by 1.96 (Chapter 2). The degree to which this approach can yield biased confidence intervals will be illustrated in Example 2.

Negative Estimates of Heritability

Because ANOVA yields unbiased estimates of variance components, sampling error associated with the mean squares sometimes causes the estimated among-family variance to be negative when σ_s^2 is small. Since this, in turn, results in a negative estimate of h^2, to avoid embarrassment, investigators often either report negative variance component estimates as zero or do not report them at all. In individual studies, such treatment is reasonable in that the parametric value of a variance component cannot be negative, and at best, a negative estimate means that the results are unreliable. However, the censoring of negative variance component estimates from the literature can only have the cumulative effect of upwardly biasing the reported pool of values. Since the probability of obtaining

Table 18.2 Probabilities (to the nearest 0.01) of obtaining negative heritability estimates from a balanced half-sib analysis with N families, each containing n offspring.

	$h^2 = 0.19$		$h^2 = 0.36$		$h^2 = 0.80$	
N	$n = 5$	$n = 25$	$n = 5$	$n = 25$	$n = 5$	$n = 25$
5	0.46	0.22	0.38	0.11	0.22	0.03
10	0.38	0.09	0.27	0.02	0.10	0.00
25	0.27	0.00	0.11	0.00	0.00	0.00
50	0.18	0.00	0.05	0.00	0.00	0.00

Source: Searle et al. 1992.

negative estimates of σ_s^2 can be rather high when sample sizes are small (which is often the case in studies of natural populations), it seems prudent to report the actual values in published studies.

Algebraic considerations dictate that the lower limit to the sampling distribution of $h^2 = 4t_{\text{PHS}}$ is $-4/(n-1)$, which is quite negative for small n. Under the assumption of normality, it is straightforward to derive the probability of sampling a negative value of h^2 (Leone and Nelson 1966, Gill and Jensen 1968, Searle et al. 1992, Wang et al. 1992). This is equivalent to the probability of obtaining $\text{MS}_s < \text{MS}_e$, which is the same as the probability that an F-distributed variable is less than one. From Searle et al. (1992), for a paternal half-sib analysis,

$$\Pr(h^2 < 0) = \Pr\left[F_{(T-N),(N-1)} > 1 + \frac{nh^2}{4 - h^2}\right] \tag{18.25}$$

where $F_{(T-N),(N-1)}$ denotes an F-distributed variable with $(T - N), (N - 1)$ degrees of freedom. Thus, the probability of obtaining a negative heritability estimate depends on the true parametric value of h^2, as well as on the sampling design (N and n). If the true heritability exceeds 0.8 or so, there is very little chance of obtaining a negative estimate if $Nn = 100$ or more individuals are assayed. However, if the true heritability is on the order of 0.25 or less, the probability of obtaining a negative estimate can exceed 0.1 if fewer than 200 to 300 observations are made (Table 18.2). Similarly, we note that if the true value of h^2 is quite high, there is a possibility that its estimate derived by half-sib analysis will exceed the upper limit of 1.0 (Prabhakaran and Jain 1987).

Optimal Experimental Design

Robertson (1959a) first addressed the problem of optimal allocation of effort in the half-sib design, the objective being to minimize the sampling variance of the intraclass correlation (defined in Equation 18.21). Assuming that the constraint on the investigator is the total number of individuals that can be measured, $T = Nn$, the optimal family size is $\widehat{n} \simeq 1/\tau_{\text{PHS}} \simeq 4/h^2$. Since h^2 has an upper limit of one, this result clearly indicates that half-sib experiments with family sizes smaller than four should be avoided. Substituting $1/\tau_{\text{PHS}}$ for n in Equation 18.21, the expected standard error of the intraclass correlation under an optimal design is

$$\begin{aligned} \text{SE}(t_{\text{PHS}}) &\simeq 2(1-\tau_{\text{PHS}})\sqrt{\frac{2\tau_{\text{PHS}}}{T}} \\ &\simeq \left(1-\frac{h^2}{4}\right)\sqrt{\frac{2h^2}{T}} \end{aligned} \tag{18.26}$$

Four times this value is the smallest $\text{SE}(h^2)$ that one can reasonably hope to attain in a half-sib design involving T individuals.

As in the case of parent-offspring analysis, the optimal experimental design depends upon the heritability of the trait, but some general recommendations can still be made. Figure 18.2 shows the relationship between $\text{SE}(h^2)$ and n under two fairly extreme values of h^2 and for two values of $T = Nn$. If h^2 is relatively large there is a pronounced increase in $\text{SE}(h^2)$ as the family size exceeds the optimal value. However, when h^2 is small, $\text{SE}(h^2)$ is very insensitive to the design, provided that n is greater than approximately 20. Thus, with no information on h^2 prior to analysis, the most broadly satisfactory recommendation that can be made is to use family sizes on the order of 5 to 20.

Power is also a serious consideration for any experimental design, and this is discussed in detail in Appendix 5. In particular, Table A5.1 gives the power under the optimal half-sib design for various values of T and h^2.

Example 1. For situations in which the investigator has the option of performing a parent-offspring or a half-sib analysis, an obvious question is, Which protocol will give a more precise estimate of h^2?

Some insight into this matter can be gained by comparing the results immediately above with those obtained in the previous chapter. Let us assume that the major constraint on the investigator is the total number of individuals (T) that can be measured in the offspring generation (i.e., we will ignore the additional work that is required to measure the parents in a parent-offspring analysis). Assuming an optimal half-sib design, with $n = 1/\tau_{\text{PHS}}$, from Equation 18.26, the minimum standard error of h^2 under a half-sib design is $4\sqrt{2h^2/T}$. We previously found for an optimal single parent-offspring design (one offspring measured per parent) that $\text{SE}(h^2) \simeq 2\sqrt{1/T}$. The ratio of these two values is $\sqrt{8h^2}$. Thus, an optimally designed half-sib experiment will give a better estimate of h^2 than an optimally designed parent-offspring analysis if $h^2 < 1/8$ (provided that all of the above assumptions hold). Otherwise, the parent-offspring regression is preferable.

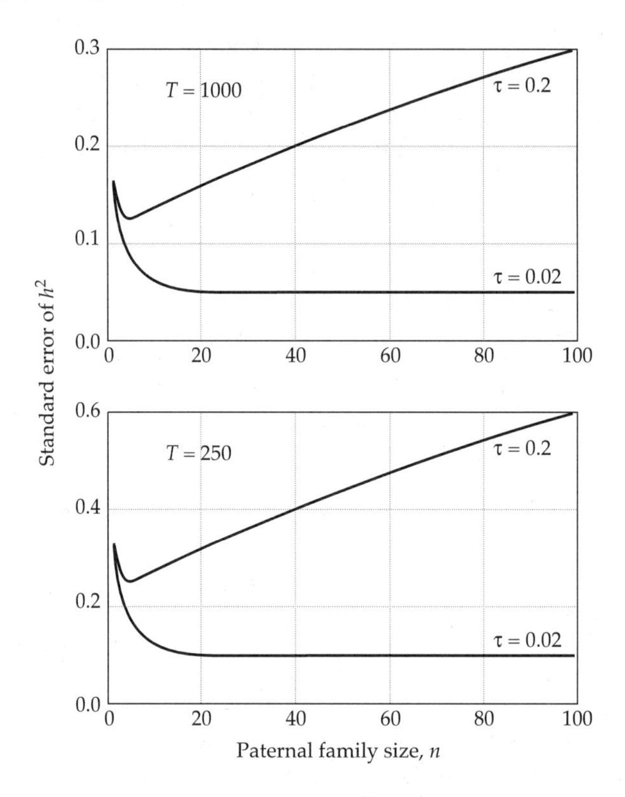

Figure 18.2 Expected standard errors of h^2 estimates derived from paternal half-sib analyses involving different family sizes (n) and two extreme intraclass correlations $(\tau \simeq h^2/4)$. T is the total number of individuals assayed, such that T/n is the number of half-sib families. The solutions are obtained from Equation 18.21.

Unbalanced Data

Accidental losses or natural mortality almost always cause inequities in family sizes in sib analyses. With unbalanced data, estimates of variance components can still be obtained by the method of moments, but this requires that the definitions of the expected mean squares first be modified appropriately (Table 18.1). All aspects of the unbalanced one-way ANOVA are identical to those outlined for the balanced design, except for the expected among-family mean square, which is no longer $(\sigma_e^2 + n\sigma_s^2)$, but $(\sigma_e^2 + n_0\sigma_s^2)$, where n_0 is a function of the sire-specific family sizes (Table 18.1). Thus, in obtaining estimates of the variance components by the method of moments, we still use Equations 18.12a,b, substituting n_0 for n.

Provided the data are normally distributed, the sums of squares obtained from an unbalanced one-way ANOVA are still independent, and expressions for the sampling variances and covariance of the variance components analogous to

Equations 18.20a–d are obtainable,

$$\text{Var}[\,\text{Var}(e)\,] \simeq \frac{2[\,\text{Var}(e)\,]^2}{T-N+2} \tag{18.27a}$$

$$\text{Var}[\,\text{Var}(s)\,] \simeq \frac{2}{n_0(N+1)}\left\{\frac{(T-1)[\,\text{Var}(e)\,]^2}{n_0(T-N)} + 2\text{Var}(e)\text{Var}(s) + \frac{\sum n_i^2 + (\sum n_i^2/T)^2 - 2\sum n_i^3/T}{n_0(N-1)}[\,\text{Var}(s)\,]^2\right\} \tag{18.27b}$$

$$\text{Cov}[\,\text{Var}(s), \text{Var}(e)\,] \simeq -\frac{2[\,\text{Var}(e)\,]^2}{n_0(T-N+2)} \tag{18.27c}$$

$$\text{Var}[\,\text{Var}(z)\,] \simeq \text{Var}[\,\text{Var}(e)\,] + 2\text{Cov}[\,\text{Var}(s), \text{Var}(e)\,] + \text{Var}[\,\text{Var}(s)\,] \tag{18.27d}$$

where the summations in Equation 18.27b are over sires. (These expressions contain corrections to results given in Searle et al. 1992).

Comparison of Equation 18.27a with 18.20a shows that the sampling variance of the within-family component of variance is unaffected by lack of balance. In fact, lack of balance does not alter the fact that the within-family sum of squares is χ^2-distributed. Thus, Equation 18.22 can still be used to obtain confidence intervals for the within-family component of variance. Unfortunately, the situation is not so simple with the among-family statistics. If $\sigma_s^2 = 0$, the ratio $F = \text{MS}_s/\text{MS}_e$ still has an F distribution with $(N-1)$ and $(T-N)$ degrees of freedom, so even with an unbalanced design, the ratio of mean squares provides a basis for testing the null hypothesis that $\sigma_s^2 = 0$. Searle et al. (1992, pp. 76–78) outline procedures for estimating confidence intervals for the among-family variance component and for t_{PHS}, but these procedures are quite complicated.

In recent years, maximum likelihood procedures have been developed as an alternative to ANOVA approaches for variance-component estimation. As a consequence of their relative insensitivity to unbalanced designs, these methods have been embraced widely by animal breeders. Unlike ANOVA, maximum likelihood techniques assume normality in *both* the estimation of parameters and hypothesis testing. A broad overview of the use of maximum likelihood methods in quantitative genetics is given in Chapters 26 and 27. For historical completeness, we note that Smith (1956) long ago introduced a weighted ANOVA procedure for unbalanced data that is closely related to maximum likelihood estimation. For the computation of the among-family sum of squares, he proposed that the family means be weighted by the inverse of their sampling variance. As in the case of weighted regression (discussed in the preceding chapter), the weights turn out to be a function of the variance components to be estimated, so an iterative solution is used in the estimation of the variance components. The weights proposed in Smith's (1956) paper are identical to those used in the maximum likelihood solution to the one-factor model (Searle et al. 1992).

Example 2. In order to illustrate the application of some of the above techniques, we now consider a specific study, an investigation of the quantitative genetics of chemical, antiherbivore defenses in the wild parsnip *Pastinaca sativa* (Berenbaum et al. 1986). This example will also highlight some of the interpretative issues that arise in a half-sib analysis.

In self-incompatible plants such as *Pastinaca*, it is likely that the ovules produced by an individual are fertilized by pollen from many different sources. If true, this would ensure that a sample of the seeds produced by a particular female will represent a half-sib family to a good approximation. Berenbaum et al. made this assumption in seeking to evaluate whether sufficient genetic variation for chemical defenses exists for *Pastinaca* to respond evolutionarily to herbivore pressures. We will consider only one of the characters studied — xanthotoxin concentration in the seeds.

From each of $N = 20$ wild plants, 10 seeds were randomly collected and grown in a greenhouse. Seed was then collected from the mature plants and analyzed chemically. Prior to analysis, some deaths occurred, necessitating an ANOVA with unbalanced data. At the final analysis, n_i was 10 in 5 cases; 9 in 8 cases; 8 in 4 cases; and 7, 6, and 4 each in one case, giving $T = 171$. Using the definition in Table 18.1, the weighted family size is $n_0 = 8.55$. The within- and among-family mean squares were $\text{MS}_e = 0.0370$ and $\text{MS}_s = 0.1156$, respectively. Substituting these three values into Equations 18.12a–c, we obtain the variance estimates, $\text{Var}(s) = 0.0092$, $\text{Var}(e) = 0.0370$, and $\text{Var}(z) = 0.0462$. The intraclass correlation is $t = \text{Var}(s)/\text{Var}(z) = 0.20$. The heritability is estimated to be $h^2 \simeq 4t = 0.80$, and using Equation 18.21 (with n_0 replacing n), the standard error of h^2 is approximately

$$\text{SE}(h^2) = 4\text{SE}(t) \simeq 4(1-t)[1+(n_0-1)t]\sqrt{2/[Nn_0(n_0-1)]} = 0.32$$

The ratio of observed mean squares is $F = 3.12$. From standard tables for the F distribution, with 19 and 151 degrees of freedom, there is only a 0.01 probability of drawing a value higher than 2.03 by chance. Thus, the among-family component of variance, and hence the heritability estimate, is highly significant.

As noted above, the estimation of confidence intervals for h^2 is rather complicated when family sizes are unequal. However, with only small inequities in family sizes, Equation 18.24 should be adequate for the purposes of this example. Five quantities, MS_e, MS_s, n_0, F_U, and F_L are required to solve this equation. For a 95% confidence interval, we obtain from an F-distribution table at the 0.025 level, $F_L = F_{19,151,0.025} = 1.80$ and $F_H = 1/F_{151,19,0.025} = 0.46$. Making the appropriate substitutions in Equation 18.24, we find the 95% confidence interval for h^2 to be 0.32 to 1.61. The confidence limits are clearly asymmetrical, and the upper limit substantially exceeds the maximum possible value for h^2.

Two factors could have resulted in an overestimation of h^2 in this study. First, since the authors employed *maternal* half-sib families, common maternal effects may have contributed to the phenotypic covariance between sibs. Second, when seeds are taken from naturally pollinated plants, the possibility that a significant proportion of sibs share the same father cannot be ruled out. If, in fact, many members of families are full sibs, then the heritability estimate will be inflated by use of a half-sib model. This inflation would occur because in estimating h^2 we have multiplied the intraclass correlation by four to get the hypothetical value $\text{Var}(A)$ in the numerator (ignoring epistatic and common-environment sources of variance). If the families consisted of full sibs with genetic covariance $\text{Var}(A)/2 + \text{Var}(D)/4$, multiplying by four would give the quantity $2\text{Var}(A) + \text{Var}(D)$ in the numerator of h^2. Thus, in the extreme case in which family members are all full sibs but mistakenly treated as half sibs, the heritability estimate can be inflated by a factor greater than two, due not only to the fact that $\text{Var}(A)$ is counted twice but also because of the inclusion of dominance genetic variance (and in this study, perhaps epistatic and common-environment variance as well). Jackson (1983) provides an in-depth analysis of the consequences of the contamination of a half-sib analysis with varying proportions of full sibs. Both sorts of problems might have been avoided by constructing paternal half-sib families by hand-pollination of bagged inflorescences (Mitchell-Olds 1986).

These potential difficulties may not have been of major importance in the *Pastinaca* study. A second estimate of $h^2 (\pm \text{SE})$ obtained by parent-offspring regression is 0.65 ± 0.21, which is lower than that obtained by sib analysis, but not significantly so. Nevertheless, there is need for caution here, too. Although dominance genetic variance does not contribute to the resemblance between parents and offspring, genetically based maternal effects can (Chapter 23). The authors found substantial heritabilities for several other defense compounds in seeds and in leaves.

Resampling Procedures

To avoid the interpretative pitfalls that can arise with hypothesis tests involving nonnormal and unbalanced data, several computer-based resampling procedures have been developed that make no assumptions about the form of the distribution of the data or the structure of experimental design (Miller 1968, 1974; Efron 1982; Milliken and Johnson 1984; Wu 1986; Little and Rubin 1987; Manly 1991; Crowley 1992). All of these techniques assume that the sample data provide a reasonably good representation of the distribution in the entire population. The data are then used to generate sampling distributions of desired statistics. Three basic approaches are used:

1. The **jackknife** procedure iteratively deletes one unit of the data set, each time using the truncated data to obtain a set of parameter estimates. For

the one-way ANOVA, a different paternal half-sib family is deleted in each analysis, and from the resultant N sets of parameter estimates, one obtains a mean estimate and a standard error for each variance component, heritability, and so on.

2. The **bootstrap** procedure repeatedly draws random samples from the original data set with replacement. With reasonably large sample sizes, the number of ways the data set can be sampled is effectively infinite, and usually a thousand or more analyses are performed to arrive at stable average values for the parameter estimates and their standard errors. Confidence intervals are constructed from the cumulative distribution of the individual estimates. For the one-way ANOVA, bootstrapping would be done over families, as our interest is in the among-family variance.

3. **Permutation tests** randomize the individual data with respect to families, while keeping the overall data structure (number of families and progeny per family) constant. Again, an essentially unlimited number of data sets can be constructed in this way, and from a large number of them, the distribution of the estimated among-family variance can be established under the null hypothesis that $\sigma_s^2 = 0$. This distribution is then used to evaluate the probability of obtaining by chance an estimate of σ_s^2 with a value as extreme as that found with the original data set. Mitchell-Olds (1986) used this approach in a sib analysis of life-history variation in the annual plant *Impatiens capensis* to test for significant heritabilities; later, the delete-one jackknife was applied to the same data (Mitchell-Olds and Bergelson 1990). The two types of analyses led to similar, although not identical, conclusions.

FULL-SIB ANALYSIS

The statistical methodology outlined above for half sibs also applies to full-sib analysis, provided the interpretations of the within- and among-family components of variance are modified appropriately. For the special case in which there are no dominance effects and no common environmental effects, twice the intraclass correlation coefficient for full sibs provides an estimate of the narrow-sense heritability. However, since the significance of these two sources of variance is generally unknown in advance of a quantitative-genetic analysis, it is best to avoid the exclusive use of full sibs to estimate h^2.

On the other hand, a *supplementary* full-sib analysis can be valuable precisely because it provides information on the relative significance of the components of variance associated with dominance and common environmental effects. Such an analysis is only a small step beyond the paternal half-sib design. As before, N randomly selected fathers (sires) are each mated to several different females (dams), but now several (rather than one) offspring are assayed

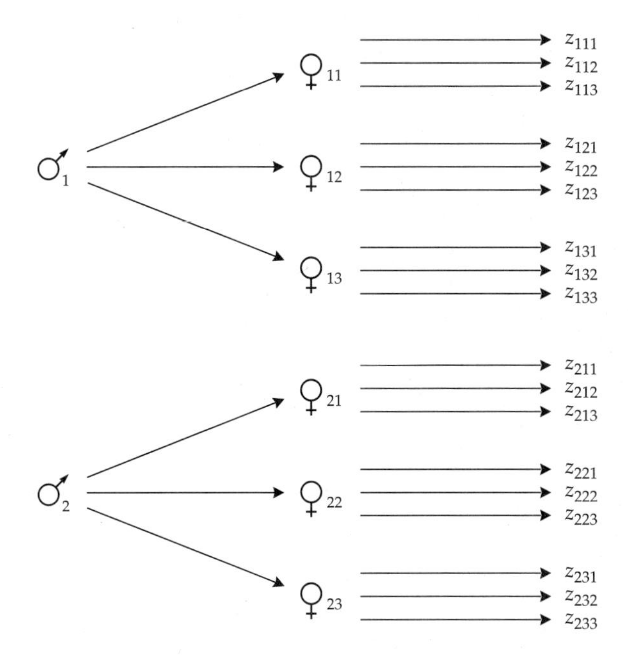

Figure 18.3 A nested full-sib, half-sib mating design. Each male is mated to several unique (unrelated) females, from each of which several offspring are assayed.

per dam (Figure 18.3). Under this design, all offspring of a given female are full sibs, while as before, progeny of different females that share the same mate are paternal half sibs.

The linear model for this **nested design** is

$$z_{ijk} = \mu + s_i + d_{ij} + e_{ijk} \tag{18.28a}$$

where z_{ijk} is the phenotype of the *k*th offspring from the family of the *i*th sire and *j*th dam, s_i is the effect of the *i*th sire, d_{ij} is the effect of the *j*th dam mated to the *i*th sire, and e_{ijk} is the residual deviation. As usual, under the assumption that individuals are random members of the same population, the s_i, d_{ij}, and e_{ijk} are defined to be independent random variables with expectations equal to zero. It then follows that the total phenotypic variance is

$$\sigma_z^2 = \sigma_s^2 + \sigma_d^2 + \sigma_e^2 \tag{18.28b}$$

where σ_s^2 is the variance among sires, σ_d^2 is the variance among dams within sires, and σ_e^2 is the variance within full-sib families.

The next step is to relate the observable components of variance to covariances between relatives. First, we note that the total phenotypic variance can be partitioned into two components, the variance within and among full-sib families. Following the rule that the variance among groups is equivalent to the covariance of members within groups, we know that the variance among full-sib families equals the phenotypic covariance of full sibs, $\sigma(\text{FS})$. Thus, the variance within full-sib families (the residual variance in the model) is simply

$$\sigma_e^2 = \sigma_z^2 - \sigma(\text{FS}) \tag{18.29a}$$

Similarly, as already shown in Equation 18.3, the variance among sires is equivalent to the covariance of individuals with the same father but different mothers, i.e., the covariance of paternal half sibs,

$$\sigma_s^2 = \sigma(\text{PHS}) \tag{18.29b}$$

Since the three components of variance must sum to σ_z^2,

$$\begin{aligned} \sigma_d^2 &= \sigma_z^2 - \sigma_s^2 - \sigma_e^2 \\ &= \sigma(\text{FS}) - \sigma(\text{PHS}) \end{aligned} \tag{18.29c}$$

Finally, we express the observable components of variance in terms of their underlying causal sources. Recalling the genetic/environmental interpretations of $\sigma(\text{PHS})$ and $\sigma(\text{FS})$, and ignoring sources of genetic variance beyond two-locus epistasis,

$$\sigma_s^2 \simeq \frac{\sigma_A^2}{4} + \frac{\sigma_{AA}^2}{16} \tag{18.30a}$$

$$\sigma_d^2 \simeq \frac{\sigma_A^2}{4} + \frac{\sigma_D^2}{4} + \frac{3\sigma_{AA}^2}{16} + \frac{\sigma_{AD}^2}{8} + \frac{\sigma_{DD}^2}{16} + \sigma_{E_c}^2 \tag{18.30b}$$

$$\sigma_e^2 \simeq \frac{\sigma_A^2}{2} + \frac{3\sigma_D^2}{4} + \frac{3\sigma_{AA}^2}{4} + \frac{7\sigma_{AD}^2}{8} + \frac{15\sigma_{DD}^2}{16} + \sigma_{E_s}^2 \tag{18.30c}$$

where $\sigma_{E_c}^2$ is the environmental component of variance due to common (maternal) environmental effects, and $\sigma_{E_s}^2$ is that due to individual (special) environmental effects.

An obvious problem with the preceding set of equations is that they are overdetermined — there are seven causal sources of variance, but only three observable variance components. The usual approach in sib analysis is to *assume* that the epistatic sources of variance are of negligible significance, so that

$$\sigma_s^2 \simeq \frac{\sigma_A^2}{4} \tag{18.31a}$$

$$\sigma_d^2 \simeq \frac{\sigma_A^2}{4} + \frac{\sigma_D^2}{4} + \sigma_{E_c}^2 \tag{18.31b}$$

$$\sigma_e^2 \simeq \frac{\sigma_A^2}{2} + \frac{3\sigma_D^2}{4} + \sigma_{E_s}^2 \tag{18.31c}$$

Under this assumption, as in half-sib analysis, four times the sire component of variance provides an estimate of the additive genetic variance. The difference $\sigma_d^2 - \sigma_s^2$ provides an indication as to whether dominance (and/or maternal effects) makes a significant contribution to the total phenotypic variance. However, the relative importance of these two factors cannot be determined without information on the resemblance between other types of relatives.

Nested Analysis of Variance

As with one-way ANOVA, estimates of the variance components can be obtained by the method of moments, i.e., by partitioning the total observed sum of squares into components, writing the expected mean squares as linear functions of the variance components, equating the observed mean squares to their expectations, and solving for the variances. Let $\overline{z}_{ij}$ be the mean phenotype of full-sib family ij, $\overline{z}_i$ be the mean phenotype of progeny of sire i, and $\overline{z}$ be the grand mean of the z_{ijk}. The total sum of squared deviations of the z_{ijk} from $\overline{z}$ can be partitioned into components describing deviations of observed sire means from the grand mean, deviations of the full-sib family means from their sire group means, and deviations of individual measures from their full-sib family means (Table 18.3).

The variance-component estimators are given by,

$$\text{Var}(s) = \frac{\text{MS}_s - \text{MS}_e - (k_2/k_1)(\text{MS}_d - \text{MS}_e)}{k_3} \tag{18.32a}$$

$$\text{Var}(d) = \frac{\text{MS}_d - \text{MS}_e}{k_1} \tag{18.32b}$$

$$\text{Var}(e) = \text{MS}_e \tag{18.32c}$$

where k_1, k_2, and k_3 are functions of the experimental design (equal to n, n, and Mn under a completely balanced design, where n is the number of offspring per full-sib family, and M is the number of dams per sire). Table 18.3 gives the general expressions for these quantities.

By analogy with Equation 18.13, the intraclass correlations for paternal half sibs and full sibs are

$$t_{\text{PHS}} = \frac{\text{Cov(PHS)}}{\text{Var}(z)} = \frac{\text{Var}(s)}{\text{Var}(z)} \tag{18.33a}$$

$$t_{\text{FS}} = \frac{\text{Cov(FS)}}{\text{Var}(z)} = \frac{\text{Var}(s) + \text{Var}(d)}{\text{Var}(z)} \tag{18.33b}$$

As in the half-sib design, $4t_{\text{PHS}}$ provides the best estimate of h^2 since it is not inflated by dominance and/or maternal effects. If, however, Var(s) and Var(d) are found to be approximately equal, then dominance and maternal effects can be ruled out as significant causal sources of covariance. In that case, the average

Table 18.3 Summary of a nested analysis of variance involving N sires, M_i dams within the ith sire, and n_{ij} offspring within the ijth full-sib family.

Factor	df	Sums of Squares	MS	E(MS)
Sires	$N-1$	$\sum_{i=1}^{N}\sum_{j=1}^{M_i} n_{ij}(\overline{z}_i - \overline{z})^2$	SS_s/df_s	$\sigma_e^2 + k_2\sigma_d^2 + k_3\sigma_s^2$
Dams (sires)	$N(\overline{M}-1)$	$\sum_{i=1}^{N}\sum_{j=1}^{M_i} n_{ij}(\overline{z}_{ij} - \overline{z}_i)^2$	SS_d/df_d	$\sigma_e^2 + k_1\sigma_d^2$
Sibs (dams)	$T - N\overline{M}$	$\sum_{i=1}^{N}\sum_{j=1}^{M_i}\sum_{k=1}^{n_{ij}} (z_{ijk} - \overline{z}_{ij})^2$	SS_e/df_e	σ_e^2
Total	$T-1$	$\sum_{i=1}^{N}\sum_{j=1}^{M_i}\sum_{k=1}^{n_{ij}} (z_{ijk} - \overline{z})^2$		

$$k_1 = \frac{1}{N(\overline{M}-1)}\left(T - \sum_{i=1}^{N}\frac{\sum_j^{M_i} n_{ij}^2}{n_i}\right)$$

$$k_2 = \frac{1}{N-1}\left(\sum_{i=1}^{N}\frac{\sum_j^{M_i} n_{ij}^2}{n_i} - \frac{\sum_i^N \sum_j^{M_i} n_{ij}^2}{T}\right)$$

$$k_3 = \frac{1}{N-1}\left(T - \frac{\sum_i^N n_i^2}{N}\right)$$

Note: T is the total number of individuals in the experiment, $\overline{M}$ is the mean number of dams/sire, and n_i is the total number of offspring of the ith sire. MS denotes an observed mean square, E(MS) denotes its expected value, and df denotes degrees of freedom.

of Var(s) and Var(d) provides an estimate of $\sigma_A^2/4$, and when multiplied by $4/\text{Var}(z)$ provides an estimate of h^2. Ideally, such an average should weight the two variance-component estimates by the inverse of their sampling variances (Grossman and Norton 1981).

Hypothesis Testing

Under the assumption of normality and balanced design, standard F ratios can be used to test for significant variation associated with sires and dams. In each case, the numerator of the F ratio is the observed mean square at the level containing the factor of interest, and the denominator is the observed mean square at the next lower level (which incorporates all factors except the one of interest). The test statistic for evaluating whether there is significant variance associated with sires is the F ratio MS_s/MS_d, since under the null hypothesis of $\sigma_s^2 = 0$, the expected value of the numerator is equal to that of the denominator (see Table 18.3). Similarly, the test statistic for significant dam effects is the ratio MS_d/MS_e, as the expected value of the numerator is again equal to that of the denominator under the null hypothesis of $\sigma_d^2 = 0$ (Table 18.3).

With unbalanced designs, hypothesis testing with F ratios becomes more difficult under the nested model. If the data are normally distributed, the logic developed above tells us that

$$F_{N(\overline{M}-1),(T-N\overline{M})} = \frac{\text{MS}_d}{\text{MS}_e} \tag{18.34a}$$

can still be employed as a test for significant dam effects, since the numerator and denominator have identical expectations under the null hypothesis of $\sigma_d^2 = 0$. However, since the coefficients (k_1 and k_2) associated with σ_d^2 in the mean squares associated with dams and sires are unequal in an unbalanced design (Table 18.3), the numerator and denominator of MS_s/MS_d no longer have equal expectations under the null hypothesis $\sigma_s^2 = 0$. Nevertheless, a linear function of the mean squares can be constructed for the numerator that does fulfill this requirement, leading to the test statistic

$$F_{r,N(\overline{M}-1)} = \frac{k_1\text{MS}_s + (k_2 - k_1)\text{MS}_e}{k_2\text{MS}_d} \tag{18.34b}$$

The main problem with this test statistic is the unknown degrees of freedom for the numerator, r. A general solution to this problem was developed by Satterthwaite (1946). Consider a linear function of m observed mean squares

$$Q = c_1\text{MS}_1 + c_2\text{MS}_2 + \cdots + c_m\text{MS}_m \tag{18.35a}$$

Satterthwaite showed that $rQ/E(Q)$ is approximately χ^2-distributed with degrees of freedom equal to

$$r = \frac{Q^2}{\displaystyle\sum_{i=1}^{m} \frac{(c_i\text{MS}_i)^2}{\text{df}_i}} \tag{18.35b}$$

Thus, for example, for the numerator of Equation 18.34b, the degrees of freedom is estimated by

$$r = \frac{Q^2}{\dfrac{(c_s\text{MS}_s)^2}{N-1} + \dfrac{(c_e\text{MS}_e)^2}{T-N\overline{M}}} \tag{18.35c}$$

where $c_s = k_1/k_2$, $c_e = (k_2 - k_1)/k_2$, and $Q = c_s\text{MS}_s + c_e\text{MS}_e$. This estimate of r is really only a first-order approximation, as Satterthwaite's derivation assumes that the observed mean squares in the function Q are independently distributed, a condition that is not strictly true with an unbalanced design.

Recalling that the variance associated with sires is an estimate of $\sigma^2_A/4$, the F ratio defined by Equation 18.34b with the numerator degrees of freedom defined by Equation 18.35c provides a test for significant additive genetic variance. Provided that both the sire and dam components of variance are significant, the next question is whether the latter is significantly greater than the former. From Equation 18.31b, it can be seen that a test of the null hypothesis $\sigma^2_s = \sigma^2_d$ is equivalent to a test for no significant dominance and/or common-environmental effects. This test also requires the construction of a linear function of mean squares whose expectation is equal to the expectation of MS_d under the null hypothesis. The appropriate F ratio is

$$F_{r,N(\overline{M}-1)} = \frac{c_s\text{MS}_s + c_e\text{MS}_e}{\text{MS}_d} \tag{18.36}$$

where $c_s = k_1/(k_2+k_3)$ and $c_e = (k_2+k_3-k_1)/(k_2+k_3)$. The numerator degrees of freedom is approximated by Equation 18.35c, with $Q = c_s\text{MS}_s + c_e\text{MS}_e$.

Resampling procedures provide an alternative to F ratios for testing for the significance of variance components under the nested design. The jackknife, with deletion around sire families, has been shown to provide a relatively robust approach for testing for significance of the sire component of variance (Arvesen and Schmitz 1970, Knapp and Bridges 1988, Mitchell-Olds and Bergelson 1990). Presumably, the jackknife or the bootstrap can also be used to test the hypothesis that $\sigma^2_s = \sigma^2_d$, by referring to the sampling distribution of $\text{Var}(d) - \text{Var}(s)$.

Sampling Error

Under a balanced design (with N sires, M dams per sire, and n progeny per dam), the large-sample variance for t_{PHS} and t_{FS} can be obtained from formulae provided by Osborne and Paterson (1952),

$$\begin{aligned}\text{Var}(t_{\text{PHS}}) \simeq\ & \frac{2\{(1-t_{\text{PHS}})(\phi+Mnt_{\text{PHS}})\}^2}{M^2(N-1)n^2} \\ & + \frac{2\{[1+(M-1)t_{\text{PHS}}]\phi\}^2}{M^2N(M-1)n^2} + \frac{2(n-1)[t_{\text{PHS}}(1-t_{\text{FS}})]^2}{NMn^2}\end{aligned} \tag{18.37a}$$

$$\begin{aligned}\text{Var}(t_{\text{FS}}) \simeq\ & \frac{2\{t'[\phi+Mnt_{\text{PHS}}]\}^2}{M^2(N-1)n^2} \\ & + \frac{2\{[M-(M-1)t']\phi\}^2}{M^2N(M-1)n^2} + \frac{2\{(1-t_{\text{FS}})[1+(n-1)t']\}^2}{MN(n-1)n^2}\end{aligned} \tag{18.37b}$$

where $t' = t_{\text{FS}} - t_{\text{PHS}}$, and $\phi = 1 - t_{\text{FS}} + nt'$. The standard error of $h^2 = 4t_{\text{PHS}}$ is $4\sqrt{\text{Var}(t_{\text{PHS}})}$.

A more general procedure for estimating $\text{Var}(t_{\text{PHS}})$, which allows for unbalanced designs, is to use the large-sample estimator for the variance of a ratio given in Appendix 1. Such a computation requires estimates of the sampling variances and covariances of $\text{Var}(s)$, $\text{Var}(d)$, and $\text{Var}(e)$, expressions for which are given in Hammond and Nicholas (1972) and Searle et al. (1992). If one is willing to assume normality and to ignore the sampling variance of $\text{Var}(z)$ and the sampling covariance between $\text{Var}(z)$, $\text{Var}(s)$, and $\text{Var}(d)$, some fairly simple and conservative (upwardly biased) estimates are possible,

$$\text{Var}(t_{\text{PHS}}) \simeq \frac{\text{Var}(\text{MS}_s) + (k_2/k_1^2)\text{Var}(\text{MS}_d) + [1-(k_2/k_1)]^2\text{Var}(\text{MS}_e)}{[k_3\text{Var}(z)]^2} \tag{18.38a}$$

$$\text{Var}(t_{\text{FS}}) \simeq \frac{\text{Var}(\text{MS}_s) + k_3^2[\phi^2\text{Var}(\text{MS}_d) + (1+\phi)^2\text{Var}(\text{MS}_e)]}{[k_3\text{Var}(z)]^2} \tag{18.38b}$$

with

$$\text{Var}(\text{MS}_x) = \frac{2(\text{MS}_x)^2}{\text{df}_x + 2}, \qquad \phi = \frac{(k_2/k_3) - 1}{k_1}$$

and k_1, k_2, and k_3 as defined in Table 18.3 (Dickerson 1969).

Graybill et al. (1956), Broemeling (1969), and Graybill and Wang (1979) provide expressions for the confidence limits of t_{PHS} and t_{FS} for the special case of a balanced design with normally distributed effects. These expressions are not necessarily very robust to violations of the assumptions of balance and normality.

Optimal Design

In the now familiar fashion, the optimal design for a nested analysis of variance is defined to be the combination of N, M, and n, subject to some constraint, that minimizes the sampling variance of the intraclass correlation of interest. Since $4t_{\text{PHS}}$ provides the most reliable estimate of h^2, it will generally be most desirable to minimize $\text{Var}(t_{\text{PHS}})$ as defined in Equation 18.37a, but the solution is quite complicated, as it depends upon both t_{PHS} and t_{FS}. Some feeling for the best design and the sensitivity of $\text{Var}(t_{\text{PHS}})$ to nonoptimal designs can be achieved by substituting different values for the design parameters (N, M, n) and for the possible values of t_{PHS} and t_{FS}.

Robertson (1959a) has shown that when dominance and common environmental effects are absent, the preferred design for estimating t_{PHS} is to use full-sib families of only single individuals, i.e., to rely on the pure half-sib analysis outlined in the previous section. If on the other hand, one desires approximately equal precision in the estimates of t_{PHS} and t_{FS}, it is advisable to allocate at least 3 to 4 females/male and to maintain full-sib families of $\sim 1/(2t_{\text{PHS}})$ (but no less than 2) progeny/female.

Bridges and Knapp (1987) performed simulation studies to evaluate the probability of obtaining negative estimates for σ_A^2 and σ_D^2 under the nested design.

Assuming no epistatic or common environmental effects, with designs of moderate size, the probability of obtaining a negative estimate of the additive genetic variance is usually on the order of only a few percent. However, the probability of obtaining a negative estimate of σ_D^2 is typically about an order of magnitude higher. Thus, although the nested design is often relied on as a means for detecting dominance, it is not particularly powerful in this regard.

Example 3. To evaluate the causal sources of variation in developmental rate in the flour beetle *Tribolium castaneum,* Dawson (1965) estimated the covariances between several types of relatives. Here we focus on the results from a nested design in which 30 males were each mated to three different females, with the goal of assaying 10 progeny per female. Some mortality among the dams and the offspring induced slight inequalities in family sizes, resulting in $k_1 = k_2 = 9.1$ and $k_3 = 25.7$. Following the layout in Table 18.3, the degrees of freedom and observed and expected mean squares for the nested ANOVA are:

Factor	df	Mean squares	E(MS)
Sires	29	5.949	$\sigma_e^2 + 9.1\sigma_d^2 + 25.7\sigma_s^2$
Dams within sires	56	3.925	$\sigma_e^2 + 9.1\sigma_d^2$
Sibs within dams	695	1.314	σ_e^2

From Equations 18.32a–c, the estimated variance components for sires, dams within sires, and sibs within dams are $\text{Var}(s) = 0.079$, $\text{Var}(d) = 0.288$, and $\text{Var}(e) = 1.314$. From Equations 18.33a,b, the intraclass correlations for paternal half sibs and full sibs are $t_{\text{PHS}} = 0.047$ and $t_{\text{FS}} = 0.218$. If all of the resemblance between relatives were due to additive genetic variance, we would expect $t_{\text{FS}} \simeq 2t_{\text{PHS}}$. The fact that t_{FS} is nearly five times t_{PHS} immediately suggests that dominance and/or common maternal effects may be contributing to the covariance between full sibs.

How much confidence can we have in these intraclass correlations? To evaluate the significance of the sire component of variance, we compute $F = 5.949/3.925 = 1.52$. Using an F-distribution table, for 29 and 56 degrees of freedom, we find that there is a 5% chance of observing an F as large as 1.68 by chance. Thus, the hypothesis that $\sigma_s^2 = 0$ cannot be rejected at this level. On the other hand, for the dam component of variance, $F = 3.925/1.314 = 2.99$, which is well above the critical 0.1% value for 56 and 695 degrees of freedom (1.70), implying that a significant fraction of the total variation in developmental rate is attributable to dams.

Approximate standard errors can be obtained for the two intraclass correlations using Equations 18.38a,b. After the appropriate substitutions, we find $\text{Var}(t_{\text{PHS}}) \simeq 0.0098$ and $\text{Var}(t_{\text{FS}}) \simeq 0.0127$. Taking square roots, we arrive at the standard errors SE(t_{PHS}) $\simeq 0.099$ and SE(t_{FS}) $\simeq 0.113$. As noted in the text, Equations 18.38a,b generally yield conservative (upwardly biased) estimates of the standard errors. When the more precise Equations 18.37a,b are used, we obtain SE(t_{PHS}) $\simeq 0.038$ and SE(t_{FS}) $\simeq 0.037$. When compared with the estimates t_{PHS} and t_{FS}, these results are consistent with our conclusion that the dam component of variance is much more significant than the sire component. The heritability of developmental rate is estimated as four times t_{PHS}, and its standard error is four times SE(t_{PHS}). Thus, Dawson's results yield the estimate $h^2 \simeq 0.19$, but with a standard error of 0.15.

Were these the only available data, we would have difficulty accepting the idea that developmental rate in *Tribolium castaneum* exhibits heritable (additive genetic) variation. However, additional data shed further light on the issue. Using twice the regression between parents and offspring, Dawson estimated the heritability to be 0.12 with paternal data and 0.30 with maternal data. He also performed a diallel analysis (Chapter 20), which generated both paternal and maternal half sibs. In this case, he estimated the heritability to be 0.13 with paternal half-sib data (similar to the nested design result of 0.19) and 0.41 with maternal half-sib data.

Thus, we see that the heritability estimates based on paternal half sibs and on the paternal-offspring regression are quite consistent, averaging about 0.15. The fact that maternal half sibs are much more similar than paternal half sibs, and that offspring are much more similar to mothers than to fathers, suggests that maternal effects are a significant source of variation. In the absence of significant epistatic sources of variation, the covariance between maternal half sibs is $\sigma^2_A/4 + \sigma^2_{E_c}$. Thus, multiplying the maternal half-sib correlation by four inflates the heritability estimate by $4\sigma^2_{E_c}/\sigma^2_z$. This suggests that the fraction of the phenotypic variance that is due to variation in maternal effects is approximately $(0.41 - 0.15)/4 = 0.06$. Finally, we recall that twice the intraclass correlation between full sibs (0.436) actually estimates $[\sigma^2_A + (\sigma^2_D/2) + 2\sigma^2_{E_c}]/\sigma^2_z$. Thus, with our previous results, the contribution to the phenotypic variance from dominance can be estimated as $2[0.436 - 0.15 - (2 \times 0.06)] \simeq 0.33$. In summary, assuming that epistasis is of negligible importance, these results suggest that the variance in developmental rate is approximately partitioned as: 15% additive genetic variance, 33% dominance genetic variance, 7% maternal-effects variance, and 45% unobserved causes (presumably special environmental effects).

19

Twins and Clones

Starting with Galton (1875), the analysis of twins has been a major focal point for research in human quantitative genetics (Bulmer 1970, Rowe 1994). Since approximately one in 100 humans are twins, substantial data bases can often be acquired through hospital records or through advertisements. One of the utilities of twin research arises from the fact that two types of twins are possible. **Monozygotic twins**, which are derived from the fragmentation of a single embryo, are genetically identical. **Dizygotic twins** are genetically equivalent to full sibs since they are derived from two eggs fertilized by different sperm. Thus, for characters that are genetically variable, a greater amount of phenotypic resemblance is expected within pairs of monozygotic twins than dizygotic twins. This is almost always observed.

When certain conditions regarding additivity of gene action and sources of environmental variation are met, some simple and powerful conclusions can be derived from twin analysis. However, as we have noted many times before, it is often impossible to verify whether all of the assumptions of a quantitative-genetic model are met. Consequently, any conclusions drawn from twin research should be carefully qualified, a practice that not all previous investigators have adhered to. The analysis of twin scores for psychological and intelligence tests has frequently led to the suggestion that human behavior and intelligence are largely genetically determined. Racial and societal overtones associated with such conclusions have generated some of the most bitter scientific debate of this century (Lewontin et al. 1984). This debate has been good for quantitative genetics because it has forced a meticulous consideration of the assumptions of traditional models. As a result, the relatively simple twin models that were relied upon until about 1970 are gradually being replaced by more elaborate models, particularly with respect to sources of environmental variation.

The initial focus of this chapter is on the classical approach to heritability analysis using combined data from monozygotic and dizygotic twins. We then describe the additional power that arises when one is fortunate enough to have data on twins raised apart or on the offspring of twins. Finally, we show how the basic principles of twin analysis can be extended to the analysis of phenotypic variation in asexual populations.

Table 19.1 Summary of a one-way ANOVA involving N independent pairs of twins.

Source of Variance	df	Sums of Squares	Observed Mean Squares	E(MS)
Among pairs	$N-1$	$\text{SS}_b = 2\sum_{i=1}^{N}(\overline{z}_i - \overline{z})^2$	$\text{SS}_b/(N-1)$	$\sigma_e^2 + 2\sigma_b^2$
Within pairs	N	$\text{SS}_e = \sum_{i=1}^{N}\sum_{j=1}^{2}(z_{ij} - \overline{z}_i)^2$	SS_e/N	σ_e^2
Total	$T-1$	$\text{SS}_T = \sum_{i=1}^{N}\sum_{j=1}^{2}(z_{ij} - \overline{z})^2$	$\text{SS}_T/(T-1)$	σ_z^2

Note: This general description applies to both monozygotic and dizygotic analyses. The total sample size is $T = 2N$, z_{ij} is the observed phenotype of the jth member of the ith pair of twins, $\overline{z}_i$ is the mean phenotype of the ith pair, and E(MS) denotes the expected mean squares.

THE CLASSICAL APPROACH

The most commonly used methodology for twin analysis relies on the simple one-way analysis of variance (Kempthorne and Osborne 1961, Haseman and Elston 1970, Christian et al. 1974, Kang et al. 1974, Eaves et al. 1978, Martin et al. 1978). Separate analyses are performed for monozygotic and dizygotic twins using the layout given in Table 19.1. Since twin groups always consist of two individuals, this is one of the few designs in quantitative-genetic analysis that is always perfectly balanced.

For both types of twins, we start with the simple linear model

$$z_{ij} = \mu + b_i + e_{ij} \tag{19.1}$$

where z_{ij} is the phenotype of the jth member ($j = 1$ or 2) of the ith pair, b_i is the effect of the ith pair, and e_{ij} is the residual error resulting from segregation (in the case of dizygotic twins) and environmental variance. As usual, we assume that the e_{ij} have zero expectations, are uncorrelated with each other, and have common variance within each family, σ_e^2. The twin pairs within each analysis are also assumed to be a random sample of the entire population so that $E(b_i) = 0$. For monozygotic and dizygotic twin analyses respectively, we denote the variance among pairs by $\sigma_b^2(\text{MZ})$ and $\sigma_b^2(\text{DZ})$ and the variance within pairs by $\sigma_e^2(\text{MZ})$ and $\sigma_e^2(\text{DZ})$.

As in the case of sib analysis (Chapter 18), estimates of the within- and among-pair components of variance are obtained by setting the observed mean squares equal to their expectations and solving. The observed variance components can, in turn, be used to derive inferences about causal sources of phenotypic variance. From Equation 19.1, it can be seen that under the assumption of independent residuals, the expected among-pair variance is equivalent to the covariance of members of the same pair, $\sigma(z_{i1}, z_{i2}) = \sigma(b_i, b_i) = \sigma_b^2$. In causal terms, for monozygotic twins, this is equivalent to

$$\sigma_b^2(\mathrm{MZ}) = \sigma_G^2 + \sigma_{E_c}^2(\mathrm{MZ}) \tag{19.2a}$$

where σ_G^2 is the total genetic variance, and $\sigma_{E_c}^2(\mathrm{MZ})$ is the variance due to common familial environment. The genetic covariance between monozygotic twins is equal to the total genetic variance because of the complete genetic identity of the individuals involved. For dizygotic twins,

$$\sigma_b^2(\mathrm{DZ}) = \frac{1}{2}\sigma_A^2 + \frac{1}{4}\sigma_D^2 + \frac{1}{4}\sigma_{AA}^2 + \cdots + \sigma_{E_c}^2(\mathrm{DZ}) \tag{19.2b}$$

Separate notation is needed for the variance due to common familial environment because it is not necessarily the same in the two types of twins. Since $\sigma_z^2 = \sigma_b^2 + \sigma_e^2$, it follows that the remainder of the phenotypic variance is in the within-pair component. Therefore,

$$\sigma_e^2(\mathrm{MZ}) = \sigma_{E_s}^2(\mathrm{MZ}) \tag{19.3a}$$

and

$$\sigma_e^2(\mathrm{DZ}) = \frac{1}{2}\sigma_A^2 + \frac{3}{4}\sigma_D^2 + \frac{3}{4}\sigma_{AA}^2 + \cdots + \sigma_{E_s}^2(\mathrm{DZ}) \tag{19.3b}$$

where the terms $\sigma_{E_s}^2(\mathrm{MZ})$ and $\sigma_{E_s}^2(\mathrm{DZ})$ refer to the residual environmental variance, i.e., environmental variance not attributable to common familial effects (Chapter 6).

All of the procedures outlined in Chapter 18 for the one-way analysis of variance of sibs apply to twin analysis. From Table 19.1, it can be seen that the among-pair variance is estimated by

$$\mathrm{Var}(b) = \frac{\mathrm{MS}_b - \mathrm{MS}_e}{2} \tag{19.4a}$$

Taking the sampling variance of the mean squares to be approximately $2(\mathrm{MS})^2/N$ (from Equation 18.19), and recalling that the mean squares are distributed independently under a balanced design, the large-sample variance of $\mathrm{Var}(b)$ is

$$\mathrm{Var}[\,\mathrm{Var}(b)\,] \simeq \frac{\mathrm{Var}(\mathrm{MS}_b) + \mathrm{Var}(\mathrm{MS}_e)}{4} = \frac{(\mathrm{MS}_b)^2 + (\mathrm{MS}_e)^2}{2N} \tag{19.4b}$$

The within-pair variance is approximated by

$$\mathrm{Var}(e) = \mathrm{MS}_e \tag{19.5a}$$

and its large-sample variance is

$$\text{Var}[\,\text{Var}(e)\,] \simeq 2(\text{MS}_e)^2/N \tag{19.5b}$$

Combining the estimators for σ_b^2 and σ_e^2, we obtain

$$\text{Var}(z) = \frac{\text{MS}_b + \text{MS}_e}{2} \tag{19.6}$$

as the estimator of σ_z^2. Its large-sample variance estimator is identical to that of Var(b). Confidence intervals for the within- and among-pair components of variance can be obtained by using Equations 18.22 and 18.23.

Recalling the arguments laid out in Chapter 18, the ratio of mean squares, $F = \text{MS}_b/\text{MS}_e$, provides a test statistic for evaluating the significance of the among-pair component of variance. The null hypothesis of no pair effects is tested by referring to standard F-distribution tables and comparing the observed F with the 5% or 1% critical values associated with $(N-1)$ and N degrees of freedom.

Heritability Estimation

An upper limit to the broad-sense heritability, based only on monozygotic-twin data, is the intraclass correlation (the fraction of the total variance that is attributable to differences among pairs)

$$H^2 = t_{\text{MZ}} = \frac{\text{MS}_b(\text{MZ}) - \text{MS}_e(\text{MZ})}{\text{MS}_b(\text{MZ}) + \text{MS}_e(\text{MZ})} \tag{19.7a}$$

From Equation 18.21, the standard error of this estimator is approximately

$$\text{SE}(H^2) \simeq \frac{1 - t_{\text{MZ}}^2}{\sqrt{N}} \tag{19.7b}$$

Under conditions of normality, the $100(1-\alpha)\%$ confidence interval for H^2 is given by

$$\left[\frac{(F/F_U) - 1}{(F/F_U) + 1}\right] < H^2 < \left[\frac{(F/F_L) - 1}{(F/F_L) + 1}\right] \tag{19.7c}$$

where F is the ratio of observed mean squares, and $F_U = F_{(N-1),N,(\alpha/2)}$ and $F_L = 1/[F_{N,(N-1),(\alpha/2)}]$ are the upper and lower F values associated with $(\alpha/2)$ (Searle et al. 1992).

Example 1. Reed et al. (1975) performed an analysis of total fingerprint ridge count for $N = 260$ pairs of monozygotic twins, obtaining the observed mean squares $\text{MS}_b = 3619.2$ and $\text{MS}_e = 82.9$. From Equations 19.7a,b, we obtain the estimate $H^2 = 0.955$ and the standard error $\text{SE}(H^2) = 0.005$. With an F ratio of 43.66, this heritability estimate is highly significant. To obtain the 99% confidence interval for H^2, we first find $F_U = F_{259,260,0.005} \simeq 1.4$ and $F_L = 1/F_{260,259,0.005} \simeq 0.7$. Substituting these values into Equation 19.7c, the lower and upper confidence limits are found to be 0.937 and 0.984.

As an estimator of the broad-sense heritability, the intraclass correlation will be inflated by the presence of any common environmental effects, $\sigma^2_{E_c}(\text{MZ})$. This can be checked by comparing the correlation between twins raised together with that between twins raised apart (by one or more adoptive parents). The difference between the two correlations provides a measure of the fraction of the phenotypic variance that is due to common-environment effects. Generally, intraclass correlations for twins raised together do exceed those for twins raised apart, but the difference is usually only on the order of a few percent (Table 19.2). Thus, in humans, a large fraction of the variation in size, physiology, and mental ability appears to have a genetic basis, and not much appears to be associated with shared environment (except in the case of IQ). There is still some need for caution here, however, since it is unlikely that the foster homes of twins are ever perfectly randomized, and postnatal separation does not eliminate prenatal maternal effects, which may be important in mammals (Chapter 23). In humans, an alternative means of testing for the importance of shared environment is to consider the correlation between unrelated adoptees raised by the same foster parents. For IQ, this correlation is approximately 0.34 (Rowe 1994), which is slightly higher than the difference between correlations for monozygotic twins raised together and apart (Table 19.2).

Some success at eliminating the contribution of common environmental effects in twin analysis has also been obtained by the following approximation

$$\widetilde{H}^2 \simeq \frac{2[\text{MS}_e(\text{DZ}) - \text{MS}_e(\text{MZ})]}{\text{Var}(z)} \tag{19.8}$$

where $\text{Var}(z)$ is the average of the estimates of total variance obtained for monozygotic and dizygotic types of twins. Provided the variance due to special environmental effects is comparable in the two types of twins, and some other assumptions to be discussed below are met, $\widetilde{H}^2$ has an expectation equal to approximately $[\sigma^2_A + (3\sigma^2_D/2) + (3\sigma^2_{AA}/2) + \cdots]/\sigma^2_z$. Thus, Equation 19.8 still overestimates the broad-sense heritability if the relative magnitude of dominance and/or epistatic variance is large. Nevertheless, a reasonable test of the hypothesis $\widetilde{H}^2 = 0$ is provided by the F statistic $\text{MS}_e(\text{DZ})/\text{MS}_e(\text{MZ})$. Estimates of broad-sense heritability using Equation 19.8, like those using Equation 19.7a, indicate that variation in human size attributes has a large genetic component (Table 19.3).

Table 19.2 Intraclass correlations for monozygotic twins raised together (t_{MZT}) and raised apart (t_{MZA}), both obtained by use of Equation 19.7a.

Character	t_{MZT}	t_{MZA}	$t_{\mathrm{MZT}} - t_{\mathrm{MZA}}$
Fingerprint ridge count	0.96	0.97	−0.01
Height	0.93	0.86	0.07
Weight	0.83	0.73	0.10
Blood pressure	0.70	0.64	0.06
Heart rate	0.54	0.49	0.05
IQ	0.88	0.69	0.19

Source: Bouchard et al. 1990.

Table 19.3 Broad-sense heritability estimates for size-related characters in humans, obtained by use of Equation 19.8.

Character	$\mathrm{MS}_e(\mathrm{DZ})$	$\mathrm{MS}_e(\mathrm{MZ})$	$\mathrm{Var}(z)$	$\widetilde{H}^2$
Height	1620.3	195.4	3031.7	0.94
Arm span	2132.0	317.7	3944.1	0.92
Middle finger length	11.9	1.4	22.3	0.94
Foot length	58.5	10.9	105.8	0.90
Chest circumference	1098.8	423.7	1776.6	0.76
Head breadth	14.9	4.2	25.5	0.84

Source: Clark 1956.
Note: Data are from 44 pairs of monozygotic twins and 37 pairs of dizygotic twins of the same sex in Michigan high schools and junior high schools.

Because monozygotic and dizygotic twins develop in different types of placental environments, one might question the assumptions that $\sigma^2_{E_s}(\mathrm{MZ}) = \sigma^2_{E_s}(\mathrm{DZ})$ and $\sigma^2_{E_c}(\mathrm{MZ}) = \sigma^2_{E_c}(\mathrm{DZ})$. Mammalian embryos are surrounded by two types of fetal membranes: an amnion (or birth sac), which is continuous with the body wall of the embryo and forms a fluid-filled chamber, and a more external chorion (placenta) through which exchange of nutrients, gases, and wastes takes place. Human dizygotic twins always have separate membranes, but only about a third of monozygotic twins do (Nance 1979). Thus, due to the greater sharing of prenatal environments, monozygotic twins may exhibit higher similarity due to shared environmental effects and deviate less because of specific effects. The fact that dichorionic and monochorionic monozygotic twins differ significantly in their within-pair mean squares for plasma cholesterol (Christian et al. 1976) and fingerprint patterns (Reed et al. 1978) is consistent with this idea.

More remarkably, embryo transplant experiments with highly inbred lines of mice have shown that monozygotic twins separated at the eight-cell stage are significantly more similar to each other than are dizygotic twins (separate since conception) treated in the same manner (Gärtner and Baunack 1981). In this study, dizygotic twins are no more different genetically than monozygotic twins since the parents are highly homozygous. Thus, these results suggest that by the third cell division, zygotes can be modified by common environmental effects that have substantial impact on the adult phenotype.

A number of other important assumptions underly twin analysis (Price 1950; Kempthorne and Osborne 1961; Nance 1976, 1979; Eaves et al. 1978; Martin et al. 1978; Rowe 1994). For example, one might question whether twins are a random subset of the gene pool of the general population. Since it is the population at large about which one normally wants to make inferences, there is reason for concern here. The dizygotic twinning rate differs significantly among races of humans as well as among individuals (Bulmer 1970). In addition, dizygotic twins are more frequently produced by older mothers, for whom congenital malformations are also more common. An additional concern is whether the environmental component of variance may be exceptionally high or low in twins relative to singletons. Parents may treat twins differently than singleton offspring. Because of their contemporaneity, twins may also experience an exceptional level of sib competition (or cooperation).

One or more of the above-mentioned problems would be suggested if the phenotypic variances for monozygotic twins and dizygotic twins were found to be significantly different. This can be tested by the ratio

$$F = \frac{\text{MS}_b(\text{DZ}) + \text{MS}_e(\text{DZ})}{\text{MS}_b(\text{MZ}) + \text{MS}_e(\text{MZ})} \tag{19.9a}$$

Using Satterthwaite's (1946) method (Chapter 18), the degrees of freedom associated with the numerator are approximately

$$df = \frac{N[\,\text{MS}_b(\text{DZ}) + \text{MS}_e(\text{DZ})\,]^2}{\text{MS}_b^2(\text{DZ}) + \text{MS}_e^2(\text{DZ})} \tag{19.9b}$$

and a parallel definition applies to the denominator.

THE MONOZYGOTIC-TWIN HALF-SIB METHOD

Nance and Corey (1976) introduced a clever method of twin analysis that eliminates many of the uncertainties of the classical method. Children produced by two monozygotic twins are cousins socially, but genetically they are related as half sibs (Figure 19.1). Moreover, unlike their monozygotic-twin parents, such

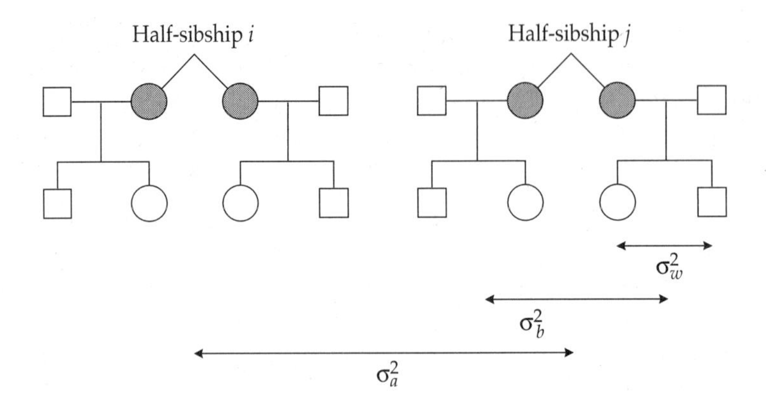

Figure 19.1 The monozygotic-twin half-sib design. Each pedigree contains two full-sib families derived from a pair of monozygotic twins, in this case the two sets of females denoted by the shaded circles.

half sibs are not raised in a common home. When data are available for the progeny of many pairs of such twins, a nested analysis of variance can be performed, analogous to that described previously for the full-sib, half-sib design (Table 18.3). The linear model to be analyzed is

$$z_{ijk} = \mu + a_i + b_{ij} + e_{ijk}$$

where z_{ijk} is the phenotype of the kth offspring of the jth member ($j = 1$ or 2) of the ith twin pair, a_i is the effect of the ith pair, b_{ij} is the effect of the jth family within the ith pair (resulting from genetic differences between the unrelated parents and from common environmental effects), and e_{ijk} is the residual deviation resulting from segregration and special environmental effects. As usual under the assumption that individuals are random members of the population, a_i, b_{ij}, and e_{ijk} are defined to have means equal to zero and to be uncorrelated with each other. σ_a^2, the variance between monozygotic-twin half-sibships, is equivalent to the covariance among half-sibs living in different homes. σ_b^2 is the variance between full-sibships within half-sibships, and σ_e^2 is the variance within full-sibships. Estimates of these variance components are obtainable from the observed mean squares of a nested ANOVA in the manner described in Chapter 18. Their causal components are summarized in Table 19.4.

The **monozygotic-twin half-sib** (MTHS) method has several major advantages over classical twin analysis. First, since it is performed on normal singleton children, the potential problem of environmental effects specific to twin phenotypes is eliminated. Second, the MTHS method removes the necessity of relying on data from both monozygotic and dizygotic twins. Third, when separate analyses are performed on female and male twins, it becomes possible to estimate the

Table 19.4 Coefficients for the causal components of variance that contribute to the expected variance components σ_a^2, σ_b^2, and σ_e^2 extracted from the nested ANOVA in a monozygotic-twin half-sib design.

	σ_A^2	σ_D^2	σ_{AA}^2	σ_{AD}^2	σ_{DD}^2	$\sigma_{G_m}^2$	$\sigma_{E_c'}^2$	$\sigma_{E_c}^2$	$\sigma_{E_s}^2$
Female twins (maternal half sibships)									
σ_a^2	1/4	0	1/16	0	0	1	1	0	0
σ_b^2	1/4	1/4	3/16	1/8	1/16	0	0	1	0
Male twins (paternal half sibships)									
σ_a^2	1/4	0	1/16	0	0	0	1	0	0
σ_b^2	1/4	1/4	3/16	1/8	1/16	1	0	1	0
General									
σ_e^2	1/2	3/4	3/4	7/8	5/16	0	0	0	1

genetic variance for maternal effects, $\sigma_{G_m}^2$. As described in Table 19.4, the causal components of σ_a^2 are identical for maternal and paternal half-sibships except that $\sigma_{G_m}^2$ makes no contribution in the latter case (since the "half sibs" have different maternal genotypes). The opposite holds for σ_b^2. Thus, $\sigma_b^2(PHS) - \sigma_b^2(MHS)$ and $\sigma_a^2(MHS) - \sigma_a^2(PHS)$ both provide estimates of $\sigma_{G_m}^2$. Using this relationship, Nance (1979) was able to show that genetic maternal effects are responsible for about 22% of the variance in human height. (A more in-depth discussion of the analysis of maternal effects is provided in Chapter 23, where it is shown that $\sigma_{G_m}^2$ actually contains several subsidiary components.)

The variance due to common environmental effects is partitioned into two components in Table 19.4. $\sigma_{E_c'}^2$ is equivalent to the covariance between half sibs due to common environmental effects. Even though the half sibs do not live in the same home, $\sigma_{E_c'}^2$ may be nonzero in the MTHS design if monozygotic twins raise their offspring in a similar manner due to common cultural inheritance. (Although it is not shown in the table, $\sigma_{E_c'}^2$ could also vary between paternal and maternal half-sibships). $\sigma_{E_c}^2$ is the covariance between full sibs due to common maternal environment, in excess of $\sigma_{E_c'}^2$.

In all, there are three sources of environmental variance as well as several sources of genetic variance in Table 19.4, but only five equations. Thus, not all of the parameters can be estimated with the nested design. However, when data are also recorded on the twins themselves, six other types of relationships can be evaluated (Table 19.5). Estimates of the variance components can then be obtained in the usual way by the solution of simultaneous equations, although it should be noted that the observed covariances are not all independent since they share common individuals.

Table 19.5 Additional covariances between relatives that can be observed with a MTHS design, and their causal components, when the parents are measured.

	Variance Component								
Covariance	σ^2_A	σ^2_D	σ^2_{AA}	σ^2_{AD}	σ^2_{DD}	$\sigma^2_{G_m}$	$\sigma^2_{E'_c}$	$\sigma^2_{E_c}$	$\sigma^2_{E_s}$
MZ twins	1	1	1	1	1	1	0	1	0
Mother-offspring	1/2	0	1/4	0	0	1/2	b_1	0	0
Father-offspring	1/2	0	1/4	0	0	0	b_2	0	0
Twin aunt-offspring	1/2	0	1/4	0	0	1/2	b_3	0	0
Twin uncle-offspring	1/2	0	1/4	0	0	0	b_4	0	0
Variance within MZ twin pairs	0	0	0	0	0	0	0	0	1

Note: To account for the possibility that cultural inheritance across generations is a source of resemblance between relatives, arbitrary coefficients $(b_1, \ldots, b_4)$ have been denoted for $\sigma^2_{E'_c}$; these may or may not be equal.

Nance and Corey (1976) and Haley et al. (1981) have pointed out that when separate analyses are performed on male and female progeny, the additive and dominance components of variance due to sex-linked loci can be determined with the MTHS design. For example, when an analysis is restricted to male progeny, in maternal half-sibships σ^2_a contains additive genetic variance associated with the X chromosome since the male progeny within the sibship derive their X chromosomes from identical maternal genotypes, whereas σ^2_b contains no variance associated with the X chromosome. On the other hand, in paternal half-sibships σ^2_b contains additive genetic variance associated with the X chromosome since the male progeny in related half-sib families derive their X chromosomes from different mothers, whereas σ^2_a contains no variance associated with the X chromosome since the paternal X chromosome is not inherited in the male offspring. (See Chapter 24 for further discussion of methods for estimating genetic variance due to sex-linked loci.)

An even more elaborate analysis than the preceding one has been outlined by (Haley and Last 1981), incorporating both dizygotic and monozygotic twins. The progeny of different dizygotic twins are genetically equivalent to first cousins.

Example 2. Fingerprint traits have long been employed in studies of polygenic inheritance in man (Holt 1968). They are easily assayed, and many permanent records of them exist for deceased and living individuals. The broad-sense heritabilities of dermatoglyphic traits tend to be close to 1.0, so standard errors of genetic variance estimates are relatively low. Fingerprints do not change with age, and they are not subject to assortative mating or to common postnatal environmental effects.

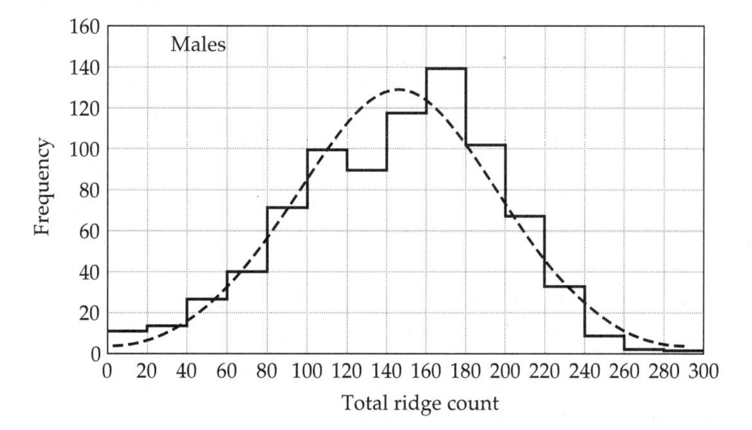

In order to illustrate some of the concepts of this chapter and their broader utility, we will consider some of the results for total dermal ridge count (TDRC), the sum of the ridges in the central cores of all ten fingertips. The above figure (from Holt 1955) gives the observed distribution for a sample of 825 British males and the fitted normal curve. The phenotype distribution for this character is approximately normal, and there is a wide range of variation. Correlations between a variety of relatives from several independent studies (with sample sizes ranging from 100 to 700) are recorded below. The results are highly consistent across studies.

Relationship	r	Reference
Monozygotic twins	0.96	Lamy et al. 1957
	0.95	Holt 1968
	0.96	Reed et al. 1975
Dizygotic twins	0.45	Lamy et al. 1957
	0.49	Holt 1968
	0.54	Reed et al. 1975
Full sibs	0.50	Holt 1968
	0.43	Mi and Rashad 1975
	0.46	Reed et al. 1979
Half sibs	0.16	Reed et al. 1979
	0.16	Nance 1979
Mother-offspring	0.41	Mi and Rashad 1975
	0.40	Reed et al. 1979
	0.39	Matsuda 1973
Father-offspring	0.41	Mi and Rashad 1975
	0.40	Reed et al. 1979
	0.48	Matsuda 1973

TDRC appears to be almost completely genetically determined with a broad-sense heritability very close to 0.96. The correlations between ordinary full sibs and between dizygotic twins are essentially the same (averaging 0.46 and 0.49 respectively). In the absence of dominance and maternal effects, this would also be the expected correlation between parent and offspring. However, the latter appears to be very close to 0.42. Since the average mother-offspring correlation is less than the father-offspring regression, maternal effects appear to be unimportant. Thus, since $\sigma_D^2/4$ contributes to the covariance between full sibs but not between parent and offspring, $(0.48 - 0.42) \times 4 = 0.24$ provides an estimate of the proportion of the total phenotypic variance that is attributable to dominance (ignoring epistatic terms involving dominance).

Epistasis may also play an important role in the expression of TDRC (Nance 1979). The genetic covariance for parent and offspring is $(\sigma_A^2/2) + (\sigma_{AA}^2/4) + \cdots$, while that for half sibs is $(\sigma_A^2/4) + (\sigma_{AA}^2/16) + \cdots$. Thus, in the presence of epistatic genetic variance, the phenotypic correlation r_{PO} is expected to be greater than $2r_{HS}$. Using the estimates for r_{HS} in the table, which are from analyses of the offspring of monozygotic twins, $r_{PO} - 2r_{HS} \simeq 0.10$, provides an estimate of $[(\sigma_{AA}^2/8) + \cdots]/\sigma_z^2$. This result suggests that approximately $8 \times 10\% = 80\%$ of the phenotypic variance is attributable to epistatic genetic variance, leaving little room for simple additive genetic variance. This estimate may be substantially inflated by sampling error. However, Heath et al. (1984) provide convincing evidence that about 40% of the variance of another trait, total finger pattern intensity, results from additive $\times$ additive epistatic interactions.

CLONAL ANALYSIS

Many species of microorganisms, plants, and animals have an obligately asexual mode of reproduction. For such organisms, the narrow-sense heritability is an unmeasurable and meaningless concept. The relevant measure of genetic variability is the broad-sense heritability $H^2 = \sigma_G^2/\sigma_T^2$, the ratio of the among-clone component of variance to the total variance.

The simplest approach to estimating H^2 in an asexual population is to perform a one-way ANOVA. By randomly choosing a group of mothers and assaying n progeny from each of them, a within- and among-clone mean square is obtained (see Table 18.1). Under the assumption of no shared maternal environmental effects, the former is an estimate of the environmental variance due to special effects, $\sigma_{E_s}^2$, while the expected value of the latter is $\sigma_{E_s}^2 + n\sigma_G^2$ under a balanced design. It follows that $(\text{MS}_b - \text{MS}_e)/n$ is an estimate of the total genetic variance, σ_G^2. Breeders have used this approach to estimate broad-sense heritabilities in plants that can be clonally propagated (Burton and DeVane 1953, Keller and Likens 1955).

Some species complexes contain both asexual and sexual individuals, or

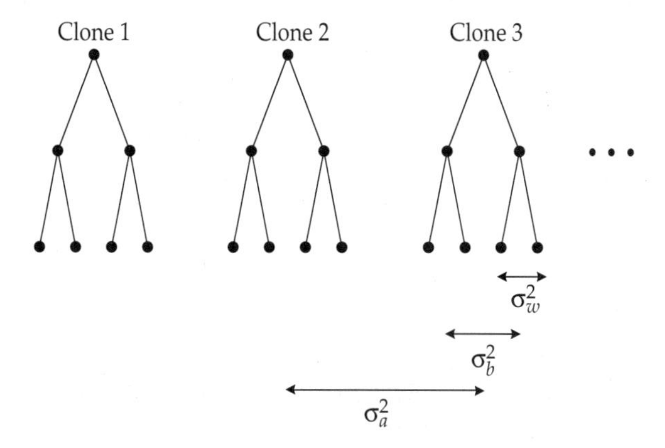

Figure 19.2 A nested design for the quantitative-genetic analysis of a population of clones. In this example, four descendants are measured for each clone, pairs of which have descended from different sublines.

allow for the possibility of artificial clonal propagation. This suggests a simple technique for estimating the broad-sense heritability for the sexual component of the population. The phenotypic variance within clones, σ_e^2, can only be due to environmental causes. Thus, letting σ_z^2 be the phenotypic variance of the sexual population, $(\sigma_z^2 - \sigma_e^2)/\sigma_z^2$ provides an estimate of H^2 for the sexual component of the population. A potential weakness of this approach is the assumption that the environmental variance of both types of individuals is the same. The genotypes of asexual and sexual individuals may respond differently to the environment, and this may be the reason why Browne et al. (1984) obtained some highly negative estimates of H^2 when they applied this technique to the brine shrimp *Artemia*.

A more general limitation of the above approaches to estimating the broad-sense heritability is that maternal effects are not factored out. As noted above, even in the absence of any genetic variation, significant differences may arise between families as a result of maternal-line influences. This problem can be resolved through the use of a nested analysis of variance (Lynch 1985). Prior to analysis, N clonal lines are each split into M sublines, each of which is maintained for one or more generations (Figure 19.2). In the final analysis, each subline is replicated n times. The data are then subjected to an ordinary nested analysis of variance, again as described in Chapter 18. σ_e^2 is the environmental variance within an immediate family, while σ_b^2 (the variance among sublines within clones) is the residual environmental variance. If the sublines are separated for a single generation, σ_b^2 is equivalent to the maternal-effects variance. For sublines separated for two generations, σ_b^2 is the sum of the variance resulting from maternal and grandmaternal effects, and for a separation of three generations it includes the great-grandmaternal-effects variance as well. The among-clone component of variance, σ_a^2, provides an estimate of the total genetic variance (plus any residual variance associated with ancestral effects).

Example 3. An application of nested ANOVA to a clonal population of the microcrustacean *Daphnia pulex* is given in the table below (data provided by K. Spitze). A group of 77 clones from a single pond were grown in the laboratory under controlled food and temperature conditions. Prior to analysis, each clone was split into two sublines for a generation, and two offspring from each subline (within each clone) were measured daily for growth in the first three instars. (There is a decline in clone number between instars because of mortality.)

	df	Sums of Squares	Mean Squares	Expected MS	Variance Estimates ($\times 10^4$)
Instar 1					
Clone	76	0.2580	0.00340	$\sigma_e^2 + 2\sigma_b^2 + 4\sigma_a^2$	$\text{Var}(a) = 4.8^{**}$
Sublines (clone)	77	0.1152	0.00150	$\sigma_e^2 + 2\sigma_b^2$	$\text{Var}(b) = 5.2^{**}$
Replicates	154	0.0700	0.00045	σ_e^2	$\text{Var}(e) = 4.5$
Instar 2					
Clone	52	0.4595	0.00884	$\sigma_e^2 + 2\sigma_b^2 + 4\sigma_a^2$	$\text{Var}(a) = 9.4^{*}$
Sublines (clone)	53	0.2682	0.00506	$\sigma_e^2 + 2\sigma_b^2$	$\text{Var}(b) = 14.6^{**}$
Replicates	106	0.2270	0.00214	σ_e^2	$\text{Var}(e) = 21.4$
Instar 3					
Clone	35	0.4754	0.01358	$\sigma_e^2 + 2\sigma_b^2 + 4\sigma_a^2$	$\text{Var}(a) = 20.1^{*}$
Sublines (clone)	36	0.1998	0.00555	$\sigma_e^2 + 2\sigma_b^2$	$\text{Var}(b) = -2.7$
Replicates	72	0.4390	0.00610	σ_e^2	$\text{Var}(e) = 61.0$

Estimates of the three variance components are obtained by setting the observed mean squares equal to their expectations and solving. Their significance, as determined by an F test of the ratio of adjacent mean squares, is denoted by $*$ and $**$, respectively, for the 0.05 and 0.01 levels. Note that the variance among sublines is highly significant in the first two instars, suggesting the presence of substantial maternal effects. Such effects appear to be dissipated by the following instar.

The broad-sense heritability is estimated by

$$H^2 = \frac{\text{Var}(a)}{\text{Var}(a) + \text{Var}(b) + \text{Var}(e)}$$

which gives values of 0.33, 0.21, and 0.26 for the three instars. All of these estimates are significant as revealed by F tests for the significance of the among-clone variance estimates, $\text{Var}(a)$.

Had the parental lines not been taken through the subline generation, the maternal-effects variance, $\text{Var}(b)$, would have been confounded with the genetic variance, $\text{Var}(a)$. The broad-sense heritability from a one-way analysis of variance would then have been equivalent to

$$H^2 = \frac{\text{Var}(a) + \text{Var}(b)}{\text{Var}(a) + \text{Var}(b) + \text{Var}(e)}$$

which yields estimates of 0.69, 0.54, and 0.23. Thus, failure to account for maternal effects in this study would have resulted in heritability estimates inflated by a factor of two for the first two instars.

20

Cross-Classified Designs

The theoretical underpinnings of the one-way and nested analyses of variance were described in Chapter 18 in the context of experimental designs involving full and half sibs. A third type of ANOVA is the **factorial**, or **cross-classified, design** with interaction. The focus of this chapter is on the simplest application of this model — the two-way classification in which the two factors are groups of mothers and fathers mated multiply to each other. We confine most of our focus to two extreme situations — the parental categories being either completely inbred lines or individuals extracted from a random-mating base population. The latter type of analysis requires that females can be mated to multiple males, and that offspring with known paternity can be recovered reliably. With inbred lines, replicate members of the same line are genetically identical, and it is not essential that individuals be multiply mated. Both approaches are used widely in the analysis of plant populations.

Many different two-factor designs have been used in quantitative genetics. In some cases, different sets of genotypes serve as paternal and maternal sources of gametes, whereas in others (**diallels**) the paternal and maternal sets of genotypes are identical. All possible crosses are not always assayed — reciprocal crosses are sometimes excluded, as are crosses within categories (selfed crosses). In some analyses, the parents are assumed to represent a random sample of the population about which inferences are to be made (and hence are treated as random effects), whereas in other situations they are the only genotypes of interest (fixed effects). Finally, different investigators often apply rather different linear models to data sets derived from the same type of experimental design.

Although the diversity of approaches to cross-classified analysis sometimes borders on the bewildering, some unifying features and distinct advantages emerge. First, all of the approaches rely on an ability to generate multiple types of sib relationships, e.g., full sibs, paternal and maternal half sibs, reciprocal and nonreciprocal sibs (a shared parent of reciprocal sibs serves as the father of one and the mother of the other), and selfed and nonselfed families. Since different causal factors contribute differentially to the resemblance between different types of relatives, this increase in the number of relationships that can be observed in a single experiment expands the number of variance components that can be estimated beyond what is possible in parent-offspring and nested sib analyses. Second, since the performance of individual genotypes is assayed in multiple familial backgrounds, it is possible to evaluate individual breeding values as well as relative performances of specific crosses. In agriculture, such information plays a central role in the development of economically valuable breeds, including elite hybrids. Third, for the same amount of effort, cross-classified designs often yield more precise estimates of variance components than can be achieved with other methods of analysis. Fourth, cross-classified analysis provides a means of estimating the average degree of dominance of alleles underlying quantitative traits.

In this chapter, we outline several of the linear models that have been employed in cross-classified analysis, showing how the components of variance associated with each model can be interpreted in terms of covariances between relatives, and hence in terms of causal sources normally associated with the resemblance between relatives. In our overview of the estimation procedures, we confine our attention to balanced designs analyzed in an ANOVA framework. With two-way ANOVA, the computational complexities that arise with unbalanced data are much more formidable than those outlined in Chapter 18 (Searle et al. 1992), and the maximum likelihood methods covered in Chapters 26 and 27 provide powerful and elegant alternative means of estimating variance components. Although ANOVA continues to dominate the landscape of two-factor analysis, this is rapidly changing as efficient and user-friendly computer programs for ML estimation are becoming available.

NORTH CAROLINA DESIGN II

Comstock and Robinson (1948, 1952) devised a series of experimental designs for estimating quantitative-genetic parameters for situations in which one or both sexes can be mated multiply. Their protocols became known as the **North Carolina designs** as they were employed extensively in breeding programs in that state. The nested full-sib, half-sib analysis covered in Chapter 18 is equivalent to **North Carolina Design I. Design II**, the subject of this section, involves all possible crosses between two sets of individuals — a group of $i = 1, \ldots, N_s$ sires and an independent group of $j = 1, \ldots, N_d$ dams (Figure 20.1). This design has proven especially applicable to plants that produce multiple flowers. Since pollen is usually produced in abundance, for species with separate sexes, the size of an experiment is generally limited by the number of flowers that female plants produce. For hermaphroditic species, the amount of effort that is required to emasculate flowers can become limiting. In principle, Design II can also be applied to animals, but when females are mated multiply, care must be taken to control for sperm storage and the possible effects of maternal age on progeny phenotypes.

Under Design II, the linear model for the phenotype of the kth offspring of the $i \times j$ mating can be expressed as

$$z_{ijk} = \mu + s_i + d_j + I_{ij} + e_{ijk} \tag{20.1a}$$

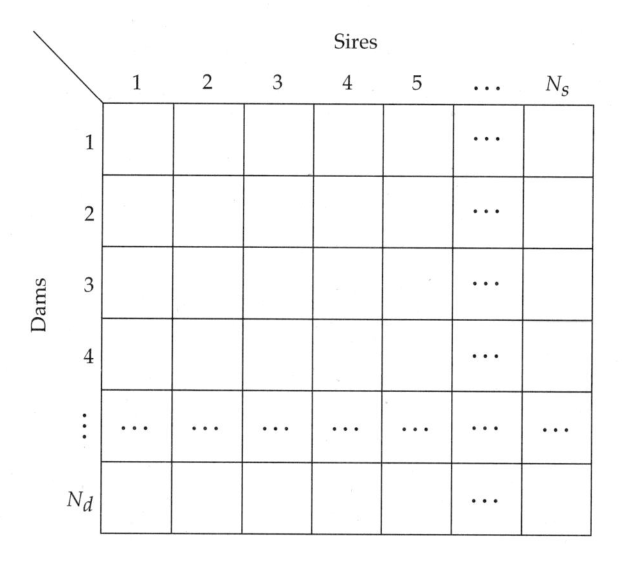

Figure 20.1 Mating scheme for North Carolina Design II. Each cell contains n observations.

where μ is the mean phenotype in the population, s_i and d_j are the additive effects (breeding values) of the ith sire and jth dam, I_{ij} is the nonadditive (interaction) effect due to the combination of genes from parents i and j, and e_{ijk} is the deviation of the observed phenotype of the kth offspring of parents i and j from the model's prediction. Note that this model is identical in form to that given for the nested design (Equation 18.28a), except that the effect of nested dam in the previous model, d_{ij}, has been replaced with the sum of an additive and an interaction effect, $d_i + I_{ij}$.

In the following, we assume that the parents are sampled randomly from some random-mating population about which inferences are to be made. Under this random-effects interpretation, we are usually more interested in estimating population parameters than the attributes of specific individuals. All of the effects in Equation 20.1a are independent, have zero expectations, and have variances respectively equal to σ_d^2, σ_s^2, σ_I^2, and σ_e^2. The total phenotypic variance is simply the sum of these four components,

$$\sigma_z^2 = \sigma_d^2 + \sigma_s^2 + \sigma_I^2 + \sigma_e^2 \tag{20.1b}$$

The effects in this model are defined as

$$s_i = \mu_{i\cdot} - \mu \tag{20.2a}$$

$$d_j = \mu_{\cdot j} - \mu \tag{20.2b}$$

$$I_{ij} = \mu_{ij} - \mu - s_i - d_j \tag{20.2c}$$

$$e_{ijk} = z_{ijk} - \mu - s_i - d_j - I_{ij} \tag{20.2d}$$

Table 20.1 Summary of a two-way analysis of variance with interaction, assuming a perfectly balanced design.

Factor	df	Sums of Squares	E(MS)
Sires	$N_s - 1$	$nN_d \sum_i (\bar{z}_{i\cdot} - \bar{z})^2$	$\sigma_e^2 + n\sigma_I^2 + nN_d\sigma_s^2$
Dams	$N_d - 1$	$nN_s \sum_j (\bar{z}_{\cdot j} - \bar{z})^2$	$\sigma_e^2 + n\sigma_I^2 + nN_s\sigma_d^2$
Interaction	$(N_d - 1)(N_s - 1)$	$n \sum_{i,j} (\bar{z}_{ij} - \bar{z}_{i\cdot} - \bar{z}_{\cdot j} + \bar{z})^2$	$\sigma_e^2 + n\sigma_I^2$
Error	$N_s N_d (n - 1)$	$\sum_{i,j,k} (z_{ijk} - \bar{z}_{ij})^2$	σ_e^2

$$\mathrm{Var}(e) = \mathrm{MS}_e \qquad \mathrm{Var}[\,\mathrm{Var}(e)\,] = \frac{2(\mathrm{MS}_e)^2}{\mathrm{df}_e + 2}$$

$$\mathrm{Var}(I\,) = \frac{\mathrm{MS}_I - \mathrm{MS}_e}{n} \qquad \mathrm{Var}[\,\mathrm{Var}(I\,)\,] = \frac{2}{n^2}\left[\frac{(\mathrm{MS}_I)^2}{\mathrm{df}_I + 2} + \frac{(\mathrm{MS}_e)^2}{\mathrm{df}_e + 2}\right]$$

$$\mathrm{Var}(d\,) = \frac{\mathrm{MS}_d - \mathrm{MS}_I}{nN_s} \qquad \mathrm{Var}[\,\mathrm{Var}(d\,)\,] = \frac{2}{(nN_s)^2}\left[\frac{(\mathrm{MS}_d)^2}{\mathrm{df}_d + 2} + \frac{(\mathrm{MS}_I)^2}{\mathrm{df}_I + 2}\right]$$

$$\mathrm{Var}(s) = \frac{\mathrm{MS}_s - \mathrm{MS}_I}{nN_d} \qquad \mathrm{Var}[\,\mathrm{Var}(s)\,] = \frac{2}{(nN_d)^2}\left[\frac{(\mathrm{MS}_s)^2}{\mathrm{df}_s + 2} + \frac{(\mathrm{MS}_I)^2}{\mathrm{df}_I + 2}\right]$$

Note: The lower-left half of the table gives estimators for the variance components σ_e^2, σ_I^2, σ_d^2, and σ_s^2. Large-sample variance expressions (derived under the assumption of normality by using Equation 18.19) appear to their right; the square roots of these provide standard errors. N_s and N_d are the numbers of sires and dams, and n is the number of progeny measured per family. MS_d, MS_s, MS_I, and MS_e are, respectively, the observed mean squares for dams, sires, interactions, and errors; these are obtained by dividing the respective observed sums of squares by their associated degrees of freedom (df).

where $\mu_{i\cdot}$ and $\mu_{\cdot j}$ are, respectively, the expected phenotypes of offspring of the ith father and the jth mother, and μ_{ij} is the expected phenotype of progeny in the full-sib family of parents i and j.

The method-of-moments procedures encountered in previous chapters provide a straightforward means to partitioning the total variance into its components. First, using approaches analogous to those developed in Chapter 18, the total sum of squared deviations of individual observations from the grand mean are partitioned into quantities relating to paternal and maternal factors, paternal $\times$ maternal interaction, and residual error (Table 20.1). Second, the expected mean squares, E(MS), are expressed as linear functions of the variance components σ_d^2, σ_s^2, σ_I^2, and σ_e^2. Finally, estimates of the variance components are obtained by equating the observed mean squares to their expectations and solving the system of four linear equations.

Design II has two particularly useful features. First, the variance components σ_s^2 and σ_d^2 are respectively equivalent to the covariances between paternal and maternal half sibs. This equivalency can be verified by referring to Equation 20.1a and recalling that only like terms (those with identical indices) are correlated,

$$\sigma(\text{PHS}) = \sigma(z_{ijk}, z_{ij'k'}) = \sigma^2(s_i) = \sigma_s^2 \tag{20.3a}$$

$$\sigma(\text{MHS}) = \sigma(z_{ijk}, z_{i'jk'}) = \sigma^2(d_j) = \sigma_d^2 \tag{20.3b}$$

Provided variation at sex-linked loci is of negligible importance and both sexes are equally inbred, both covariances should be equivalent with respect to inherited nuclear genes. However, the covariance between maternal half sibs will be inflated by any genetic or environmental maternal effects. Thus, the difference between Var(d) and Var(s) provides an estimate of the variance due to maternal effects. Second, the interaction variance, Var(I), is equivalent to Cov(FS) – Cov(PHS) – Cov(MHS), where Cov(FS) is the covariance between full sibs. This can also be seen by noting

$$\begin{aligned}\sigma(\text{FS}) &= \sigma(z_{ijk}, z_{ijk'}) = \sigma^2(d_i) + \sigma^2(s_j) + \sigma^2(I_{ij}) \\ &= \sigma_d^2 + \sigma_s^2 + \sigma_I^2\end{aligned} \tag{20.3c}$$

Using the expressions for the covariances between relatives given in Chapter 7, the observable variance components can be expressed in terms of hypothetical underlying causal factors,

$$\sigma_d^2 = \sigma(\text{MHS}) \simeq \frac{\sigma_A^2}{4} + \frac{\sigma_{AA}^2}{16} + \sigma_{G_m}^2 + \sigma_{E_c}^2 \tag{20.4a}$$

$$\sigma_s^2 = \sigma(\text{PHS}) \simeq \frac{\sigma_A^2}{4} + \frac{\sigma_{AA}^2}{16} \tag{20.4b}$$

$$\sigma_I^2 = \sigma(\text{FS}) - \sigma(\text{PHS}) - \sigma(\text{MHS}) \simeq \frac{\sigma_D^2}{4} + \frac{\sigma_{AA}^2}{8} + \frac{\sigma_{AD}^2}{8} + \frac{\sigma_{DD}^2}{16} \tag{20.4c}$$

where $\sigma_{G_m}^2$ is the variance due to genetic maternal effects and $\sigma_{E_c}^2$ is the variance due to environmental maternal effects. Since the maternal-effects variance contributes to both the full-sib and maternal half-sib covariances, it cancels out in Var(I), as does the additive genetic variance, leaving only variance due to non-additive nuclear gene action. Thus, an advantage of the factorial Design II over the nested Design I (Chapter 18) is that the former allows for separate estimates of variance due to maternal effects $(\sigma_{G_m}^2 + \sigma_{E_c}^2)$ and that due to dominance (σ_D^2). Equations 20.4a–c ignore all epistatic interactions involving more than two loci.

In practical applications of Design II, several $N_s \times N_d$ sets (**blocks**) of independent parents are usually analyzed simultaneously to allow the sampling of a large number of genotypes (see Example 1). The general layout of Table 20.1 still applies except that the sums of squares (SS) and degrees of freedom are computed individually for each individual block and then summed prior to estimating the mean squares for the whole experiment. The expectations for the mean squares are the same as in Table 20.1.

Example 1. Dawson (1965) set up 43 blocks of 2 male $\times$ 2 female factorial experiments ($N_s = N_d = 2$) for the flour beetle *Tribolium castaneum* and monitored development time in the progeny. (For each block, the offspring of the four crosses $s_1 \times d_1$, $s_1 \times d_2$, $s_2 \times d_1$, and $s_2 \times d_2$ were examined.) The ANOVA table follows. Note from Table 20.1 that for each 2×2 factorial, the degrees of freedom for sires, dams, and interactions are all equal to one. The within-family sample size varied slightly around 8, but to maintain compatibility with the layout for a balanced design, it is treated as a constant $n = 8$ here, with little effect on the final results. Thus, the error degrees of freedom for each factorial is $N_s N_d(n-1) = 2 \times 2 \times 7 = 28$. At all levels, the total degrees of freedom are obtained by multiplying those for individual factorials by the number of blocks (43).

Factor	df	SS	MS	E(MS)	Estimates (SE)
Sires	43	257.3	5.98	$\sigma_e^2 + 8\sigma_I^2 + 16\sigma_s^2$	$\text{Var}(s) = 0.073\ (0.101)$
Dams	43	362.6	8.43	$\sigma_e^2 + 8\sigma_I^2 + 16\sigma_d^2$	$\text{Var}(d) = 0.226\ (0.128)$
Interaction	43	207.3	4.82	$\sigma_e^2 + 8\sigma_I^2$	$\text{Var}(I) = 0.370\ (0.127)$
Error	1,204	2,539.4	1.86	σ_e^2	$\text{Var}(e) = 1.860\ (0.076)$
Total					$\text{Var}(z) = 2.529$

The variance component estimates given in the above table are obtained by equating the observed mean squares to their expectations. Under the assumption of normality, the standard errors of these estimates (SE) are obtained as the square roots of the expressions in the bottom right of Table 20.1.

The following hypotheses can be evaluated by use of F tests:

$$\sigma_s^2 = 0 \qquad F_{43,43} = \text{MS}_s/\text{MS}_I = 1.240\ (NS)$$
$$\sigma_d^2 = 0 \qquad F_{43,43} = \text{MS}_d/\text{MS}_I = 1.749\ (P < 0.05)$$
$$\sigma_I^2 = 0 \qquad F_{43,1204} = \text{MS}_I/\text{MS}_e = 2.592\ (P < 0.001)$$
$$\sigma_d^2 = \sigma_s^2 \qquad F_{43,43} = \text{MS}_d/\text{MS}_s = 1.409\ (NS)$$

where the subscripts on F denote the degrees of freedom associated with the test.

As noted in Chapter 18, the general procedure in determining the numerators and denominators for these test statistics is to use terms whose expectations are identical under the null hypothesis. Thus, each of the first three hypotheses is evaluated by dividing the respective observed mean square by the mean square whose expectation is identical except for the absence of the variance component of interest. For example, to test for significant sire effects, the sire mean square, whose expectation is $\sigma_e^2 + n\sigma_I^2 + nN_d\sigma_s^2$, is divided by the interaction mean square, whose expectation is $\sigma_e^2 + n\sigma_I^2$. To test whether the two variance components σ_s^2 and σ_d^2 are equal, a ratio of mean squares differing only in the components of interest is constructed, i.e., here we are evaluating whether $E(\text{MS}_s) = E(\text{MS}_d)$.

Although the component of variance associated with sires is not significant, those associated with dams and interactions are, the latter highly so. As noted from Equation 20.4c, the presence of significant interaction variance implies the existence of nonadditive genetic variance. Assuming epistatic genetic variance is of negligible importance, the dominance genetic variance is estimated by $4\text{Var}(I) = 1.480$, accounting for 59% of the observed phenotypic variance. The difference between the variance associated with dams and sires, 0.153, provides an estimate of the maternal-effects variance. This accounts for another 7% of the phenotypic variance, although failure to reject the hypothesis $\sigma_d^2 = \sigma_s^2$ implies that this result is not statistically significant. Finally, four times the sire component of variance, although again not significant, provides an estimate of the additive genetic variance, 0.292, and accounts for 12% of the phenotypic variance.

These results, although not significant in all respects, are qualitatively similar to those obtained by parent-offspring and sib analyses (Example 3, Chapter 18), which suggested that 15%, 29%, and 7% of the phenotypic variance is attributable to additive genetic, nonadditive genetic, and maternal effects, respectively. The relatively consistent results obtained by different approaches lends confidence to the conclusion that there is a large amount of genetic variance for development rate in *Tribolium,* most of which is nonadditive.

The Average Degree of Dominance

Comstock and Robinson (1952) were particularly interested in the situation in which the frequencies of all genes at segregating loci are equal to one-half, as occurs when two inbred lines have been mated randomly to form an F_2 generation, the members of which are then utilized in a factorial design. Recalling our scheme for representing the three genotypic values at a locus (0 for B_1B_1, $(1+k)a$ for B_1B_2, and $2a$ for B_2B_2), it can be seen from Equations 4.12a,b that when $p = q = 0.5$, the additive genetic variance in the F_2 and later generations is $\sum a_i^2/2$, while the dominance genetic variance is $\sum(k_i a_i)^2/4$, the summations being over loci. Thus, under the assumptions of equal gene frequencies, no epistasis, and no gametic phase disequilibria, twice the ratio of dominance to additive genetic variance provides an estimate of the average value of k^2, each locus being weighted by the square of the magnitude of the homozygous effect on the character. In terms of the observable components of variance, this weighted mean value of k^2 is estimated by

$$\widetilde{D} = \frac{2\text{Var}(I)}{\text{Var}(s)} \tag{20.5}$$

A value of $\widetilde{D}$ equal to zero implies that there is no dominance, whereas $0 < \widetilde{D} < 1$ implies partial dominance, $\widetilde{D} = 1$ complete dominance, and $\widetilde{D} > 1$ overdominance. Strictly speaking, this technique does not reveal the direction of dominance, since the sign of k is eliminated by squaring, but a direct examination of the data (to see whether family means tend to resemble higher vs. lower performing parents) can resolve that issue.

Epistasis and gametic phase disequilibria are potential sources of bias in the estimate $\widetilde{D}$. For example, additive × additive epistatic variance contributes $\sigma^2_{AA}/4$ to $2\text{Var}(I)$ but only $\sigma^2_{AA}/16$ to $\text{Var}(s)$, so epistasis will always cause an upward bias in the estimate of $\overline{k^2}$. The direction and magnitude of bias caused by gametic phase disequilibrium depends upon the linkage phase between constituent loci, as can be seen by reference to Equations 5.16a,b. The situation is quite complex, but the bias in $\widetilde{D}$ is again most likely to be in the upward direction. If genes with like effects are coupled (positives with positives, negatives with negatives), then both σ^2_A and σ^2_D will be biased in the same direction (although not necessarily to the same extent), and the bias in their ratio may not be great. However, if genes are in repulsion disequilibrium (alleles with negative effects tending to be associated with alleles with positive effects), the additive genetic variance is reduced while the dominance genetic variance is increased. This will inflate the estimated $\overline{k^2}$ relative to its true value, and can sometimes lead to the false impression of overdominance (even when none of the individual loci is more than partially dominant).

Such **associative overdominance** (Chapter 10) can be common in the early generations of a cross between lines and is detectable as a downward trend in $\widetilde{D}$ in successive generations of mating. Moll et al. (1965) demonstrated such a trend in two crosses between inbred lines of maize by analyzing morphological characters in the F_2 and later generations of random mating. The general pattern was for the additive genetic variance to remain constant across generations, while the interaction variance declined, leading to a reduction in $\widetilde{D}$. In the case of total grain yield, $\widetilde{D}$ was initially greater than one in both crosses, but declined to near one in one cross and to 0.77 in the other. Additional estimates of $\widetilde{D}$ for grain yield, derived after several generations of random mating, are 0.79 (Gardner 1963) and 0.71 (Hallauer and Miranda 1981). Since $\overline{k^2} = \overline{k}^2 + \sigma^2_k$, these results imply that $\overline{k} \leq \sqrt{0.75}$. Such results are consistent with the idea that "hybrid vigor" for grain yield in crosses of inbred lines of maize is due to linked favorable genes exhibiting partial to complete dominance (see Chapter 10). For plant height, ear height, and ear number, the estimates of $\widetilde{D}$ are lower, averaging 0.16, 0.12, and 0.23, respectively, in late-generation analyses (Hallauer and Miranda 1981). For these characters, the unfavorable alleles appear to be only slightly recessive.

The Cockerham-Weir Model

Cockerham and Weir (1977b) generalized Design II to incorporate reciprocal crosses, which are possible with hermaphrodites and with inbred lines. The advantage of this modification over the Comstock-Robinson approach is that an explicit partitioning of nuclear and extranuclear effects is possible.

Under the Cockerham-Weir model, the phenotype of the kth offspring of the cross between i as father and j as mother is represented as

$$z_{ijk} = \mu + n_i + n_j + t_{ij} + p_i + m_j + k_{ij} + e_{ijk} \tag{20.6a}$$

where n_i and n_j are the additive nuclear contributions of parents i and j, t_{ij} is the nonadditive interaction of the nuclear contributions, m_j and p_i are the maternal and paternal extranuclear effects of dam j and sire i, and k_{ij} is the sum of all nuclear-extranuclear and extranuclear-extranuclear interactions. In the analysis of this model, we make the usual assumptions that the effects are distributed independently with zero means. In addition, we assume that the variances of additive nuclear effects through mothers and fathers are identical, and that the reciprocal dominance effects, t_{ij} and t_{ji}, are equal. On the other hand, the reciprocal effects k_{ij} and k_{ji} are not necessarily equal, since the cytoplasmic elements contributed by sires and dams will generally be different. For example, in animals, the mitochondrial genome is usually maternally inherited, and in plants, one parent usually contributes the chloroplast genome. The total phenotypic variance is

$$\sigma_z^2 = 2\sigma_n^2 + \sigma_t^2 + \sigma_m^2 + \sigma_p^2 + \sigma_k^2 + \sigma_e^2 \tag{20.6b}$$

The connection between the Cockerham-Weir and Comstock-Robinson models can be seen by setting $d_j = n_j + m_j$, $s_i = n_i + p_i$, and $I_{ij} = t_{ij} + k_{ij}$ in Equation 20.1a. It follows that

$$\sigma_d^2 = \sigma_n^2 + \sigma_m^2 \tag{20.7a}$$

$$\sigma_s^2 = \sigma_n^2 + \sigma_p^2 \tag{20.7b}$$

$$\sigma_I^2 = \sigma_t^2 + \sigma_k^2 \tag{20.7c}$$

These equivalencies reveal two hidden assumptions in the Comstock-Robinson model — an absence of paternal extranuclear effects ($\sigma_p^2 = 0$), and an absence of any interaction effects involving extranuclear factors ($\sigma_k^2 = 0$).

Table 20.2 Relationships between the variance components from the Cockerham-Weir and Comstock-Robinson models, and their descriptions in terms of covariances between relatives.

Cockerham-Weir	Comstock-Robinson	Covariance between Sibs
σ_n^2	$\sigma_{d,s}$	$\sigma(\text{RHS})$
σ_p^2	$\sigma_s^2 - \sigma_{d,s}$	$\sigma(\text{PHS}) - \sigma(\text{RHS})$
σ_m^2	$\sigma_d^2 - \sigma_{d,s}$	$\sigma(\text{MHS}) - \sigma(\text{RHS})$
σ_t^2	σ_{I_d,I_s}	$\sigma(\text{RFS}) - 2\sigma(\text{RHS})$
σ_k^2	$\sigma_I^2 - \sigma_{I_d,I_s}$	$\sigma(\text{FS}) + 2\sigma(\text{RHS})$ $-\sigma(\text{MHS}) - \sigma(\text{PHS}) - \sigma(\text{RFS})$

Note: PHS, MHS, and FS denote conventional paternal half sibs, maternal half sibs, and full sibs, while RHS and RFS refer to reciprocal half sibs and reciprocal full sibs. Reciprocal half sibs share one parent, which is the father of one and the mother of the other. With reciprocal full sibs, the father of one individual is the mother of the other, and vice versa.

Despite the differences between the two models, the variance components of the Cockerham-Weir model can be estimated by an extension of the two-factor ANOVA described above. Recall that under Design II without reciprocals, it is possible to extract three components of variance: σ_s^2, σ_d^2, and σ_I^2. With reciprocals, it is also possible to estimate the covariance of maternal and paternal contributions (d_i and s_i) from the same parent, $\sigma_{d,s}$, as well as the covariance of interaction effects involving father i and mother j and vice versa, σ_{I_d,I_s}. These have expectations

$$\sigma_{d,s} = \sigma(n_i + m_i, n_i + p_i) = \sigma_n^2 \tag{20.7d}$$

$$\sigma_{I_d,I_s} = \sigma(t_{ij} + k_{ij}, t_{ji} + k_{ji}) = \sigma_t^2 \tag{20.7e}$$

Thus, by use of Equations 20.7a–e, the five variance components for the Cockerham-Weir model can be expressed in terms of those for the Comstock-Robinson model (Table 20.2).

The complete layout for the observed and expected mean squares and cross-products for Design II with reciprocals is given in Table 20.3, the top part of which is a simple extension of Table 20.1. Estimates of the Comstock-Robinson variance-covariance components, and hence of the Cockerham-Weir parameters, are obtainable by the method of moments (equating the observed mean squares and cross products to their expectations). Interpretations of both sets of variance-covariance components in terms of covariances between relatives are summarized in Table 20.2.

Table 20.3 Summary of the analysis of variance of a factorial mating design involving reciprocal crosses between two distinct sets of parents, for a perfectly balanced design.

Factor	df	Sums of Squares	E(MS)
Dams	N_1+N_2-2	$nN_2\sum_i^{N_1}(\overline{z}_{i\cdot}-\overline{z}_1)^2 + nN_1\sum_j^{N_2}(\overline{z}_{j\cdot}-\overline{z}_2)^2$	$\sigma_e^2+n\sigma_I^2+nN'\sigma_d^2$
Sires	N_1+N_2-2	$nN_2\sum_i^{N_1}(\overline{z}_{\cdot i}-\overline{z}_2)^2 + nN_1\sum_j^{N_2}(\overline{z}_{\cdot j}-\overline{z}_1)^2$	$\sigma_e^2+n\sigma_I^2+nN'\sigma_s^2$
Interact.	$2(N_1-1)(N_2-1)$	$n\sum_{i,j}(\overline{z}_{ij}-\overline{z}_{i\cdot}-\overline{z}_{\cdot j}+\overline{z}_1)^2 + (\overline{z}_{ji}-\overline{z}_{\cdot i}-\overline{z}_{j\cdot}+\overline{z}_2)^2$	$\sigma_e^2+n\sigma_I^2$
Error	$2N_1N_2(n-1)$	$\sum_{i,j,k}(z_{ijk}-\overline{z}_{ij})^2 + (z_{jik}-\overline{z}_{ji})^2$	σ_e^2

Factor	df	Sums of Cross Products	E(MCP)
d, s	N_1+N_2-2	$nN_2\sum_i^{N_1}(\overline{z}_{i\cdot}-\overline{z}_1)(\overline{z}_{\cdot i}-\overline{z}_2) + nN_1\sum_j^{N_2}(\overline{z}_{j\cdot}-\overline{z}_2)(\overline{z}_{\cdot j}-\overline{z}_1)$	$n\sigma_{I_d,I_s}+nN'\sigma_{d,s}$
I_d, I_s	$(N_1-1)(N_2-1)$	$n\sum_{i,j}(\overline{z}_{ij}-\overline{z}_{i\cdot}-\overline{z}_{\cdot j}+\overline{z}_1) \times (\overline{z}_{ji}-\overline{z}_{\cdot i}-\overline{z}_{j\cdot}+\overline{z}_2)$	$n\sigma_{I_d,I_s}$

$$\text{Var}(e) = \text{MS}_e \qquad \text{Var}[\text{Var}(e)] = \frac{2(\text{MS}_e)^2}{\text{df}_e+2}$$

$$\text{Var}(I) = \frac{\text{MS}_I-\text{MS}_e}{n} \qquad \text{Var}[\text{Var}(I)] = \frac{2}{n^2}\left[\frac{(\text{MS}_I)^2}{\text{df}_I+2}+\frac{(\text{MS}_e)^2}{\text{df}_e+2}\right]$$

$$\text{Var}(d) = \frac{\text{MS}_d-\text{MS}_I}{nN'} \qquad \text{Var}[\text{Var}(d)] = \frac{2}{(nN')^2}\left[\frac{(\text{MS}_d)^2}{\text{df}_d+2}+\frac{(\text{MS}_I)^2}{\text{df}_I+2}\right]$$

$$\text{Var}(s) = \frac{\text{MS}_s-\text{MS}_I}{nN'} \qquad \text{Var}[\text{Var}(s)] = \frac{2}{(nN')^2}\left[\frac{(\text{MS}_s)^2}{\text{df}_s+2}+\frac{(\text{MS}_I)^2}{\text{df}_I+2}\right]$$

$$\text{Cov}(I_d,I_s) = \frac{\text{MCP}_{I_d,I_s}}{n} \qquad \text{Var}[\text{Cov}(I_d,I_s)] = \frac{2(\text{MCP}_{I_d,I_s})^2}{n^2(\text{df}_{I_d,I_s}+2)}$$

$$\text{Cov}(d,s) = \frac{\text{MCP}_{d,s}-\text{MCP}_{I_d,I_s}}{nN'}$$

$$\text{Var}[\text{Cov}(d,s)] = \frac{2}{(nN')^2}\left[\frac{(\text{MCP}_{d,s})^2}{\text{df}_{d,s}+2}+\frac{(\text{MCP}_{I_d,I_s})^2}{\text{df}_{I_d,I_s}+2}\right]$$

Note: $i=1,\ldots,N_1$ and $j=1,\ldots,N_2$ denote the numbers of parents in the two sets, and

n is the number of progeny per mating. $\overline{z}_1$ is the mean phenotype observed with set 1 as dams, and $\overline{z}_2$ with set 2 as dams. $\overline{z}_{i\cdot}$ is the mean phenotype observed for all progeny from dam i, and $\overline{z}_{\cdot i}$ for all progeny from sire i. $N' = 2N_1N_2/(N_1 + N_2)$. MS and MCP denote observed mean squares and cross-products, obtained by dividing the observed sums of squares and cross-products by their associated degrees of freedom (df).

As an example of how the Comstock-Robinson and Cockerham-Weir models can be applied to the same set of data, consider the following results, derived from an unpublished experiment of Terumi Mukai (previously discussed by Cockerham and Weir 1977b).

Example 2. By using marked chromosomes with crossover suppressors, Mukai constructed 14 lines of *Drosophila melanogaster* carrying unique second chromosomes extracted from a natural population. Presumably, the lines varied randomly with respect to other chromosomes and extranuclear factors. A 7×7 factorial experiment ($N_1 = N_2 = 7$) was performed on these lines to evaluate the relative viabilities of various chromosomal heterozygotes. Each of the 49 crosses was done reciprocally and in duplicate ($n = 2$). A portion of the data, along with the marginal means, is given below, with one set of parents (1 to 7) denoted vertically and the other set (8 to 14) horizontally. Each cell contains four estimates, the top two being maternal $\times$ paternal replicates, and the bottom two the reciprocals; each of the assays involves a large number (unknown to the authors) of flies. $\overline{z}_{i\cdot}$ is the marginal mean of progeny with chromosome i inherited through mothers, and $\overline{z}_{\cdot i}$ through fathers.

	8		9			14		$\overline{z}_{i\cdot}$	$\overline{z}_{\cdot i}$
1	0.63	0.95	0.77	1.09	......	0.92	1.17	0.89	
	1.12	0.84	0.73	0.83	......	0.93	0.97		1.11
2	1.11	0.95	0.62	0.75	......	1.16	0.95	0.98	
	0.77	1.07	0.66	1.09	......	0.87	0.72		0.96
	...	...	...	...	......	...	...		
	...	...	...	...	......	...	...		
7	1.17	1.19	0.82	0.73	......	1.12	1.04	1.05	
	0.87	1.03	0.76	0.89	......	1.31	0.86		0.93
$\overline{z}_{\cdot j}$	1.02		0.76		......	1.13			
$\overline{z}_{j\cdot}$		1.06		0.82	......		1.02		

From the raw data, the sums of squares and cross products and the Comstock-Robinson variance-covariance components were calculated following the procedures in Table 20.3.

Factor	df	MS or MCP	Estimate (SE)
Dams	12	0.2213	$\text{Var}(d) = 0.0090\ (0.0061)$
Sires	12	0.3107	$\text{Var}(s) = 0.0154\ (0.0085)$
Interaction	72	0.0956	$\text{Var}(I) = 0.0388\ (0.0113)$
Error	98	0.0180	$\text{Var}(e) = 0.0180\ (0.0025)$
Recip. Main	12	0.0633	$\text{Cov}(d, s) = 0.0012\ (0.0019)$
Recip. Interaction	36	0.0466	$\text{Cov}(I_d, I_s) = 0.0233\ (0.0053)$

Several specific hypotheses about the mode of gene action can be tested with F ratios (given in the following table), and these lead to the conclusion that σ_k^2, $(\sigma_p^2 + \sigma_m^2)$, and σ_t^2 are all significantly greater than zero ($P < 0.01$, 0.01, and 0.001, respectively), while the additive component of variance associated with nuclear genes (σ_n^2) is not. The hypothesis that extranuclear maternal and paternal effects are equally variable $(\sigma_m^2 = \sigma_p^2)$ cannot be rejected.

Hypothesis	Test Statistic	Degrees of Freedom
$\sigma_k^2 = 0$	$\dfrac{\text{MS}_I - \text{MCP}_{I_d,I_s}}{\text{MS}_e} = 2.72$	$(N_1 - 1)(N_2 - 1)$, $2N_1N_2(n-1)$
$\sigma_p^2 + \sigma_m^2 = 0$	$\dfrac{(\text{MS}_s + \text{MS}_d)/2 - \text{MCP}_{d,s}}{\text{MS}_I - \text{MCP}_{I_d,I_s}} = 4.15$	$N_1 + N_2 - 2$, $(N_1 - 1)(N_2 - 1)$
$\sigma_t^2 = 0$	$\dfrac{\text{MS}_I + \text{MCP}_{I_d,I_s}}{\text{MS}_I - \text{MCP}_{I_d,I_s}} = 2.91$	$(N_1 - 1)(N_2 - 1)$, $(N_1 - 1)(N_2 - 1)$
$\sigma_n^2 = 0$	$\dfrac{\text{MCP}_{d,s}}{\text{MCP}_{I_d,I_s}} = 1.36$	$N_1 + N_2 - 2$, $(N_1 - 1)(N_2 - 1)$
$\sigma_m^2 = \sigma_p^2$	$\dfrac{\text{MS}_s}{\text{MS}_d} = 1.40$	$N_1 + N_2 - 2$, $N_1 + N_2 - 2$

Note that the rather unusual appearance of the F-ratio expressions in the preceding table is due to the translation of the mean squares and cross products between models. All of the ratios do, in fact, satisfy the desired property that the numerator and denominator have equal expectations under the null hypothesis. For example, using the expressions in Tables 20.2 and 20.3, it can be shown that the expected values of the numerator and denominator of the statistic testing $\sigma_k^2 = 0$ are $\sigma_e^2 + n\sigma_k^2$ and σ_e^2.

Using the relationships in Table 20.2, the variance components given above for the Comstock-Robinson model (and their standard errors) can be transformed into those for the Cockerham-Weir model:

$$\begin{aligned}
\text{Var}(n) &= 0.0012\ (0.0019) \\
\text{Var}(p) &= 0.0142\ (0.0087) \\
\text{Var}(m) &= 0.0078\ (0.0064) \\
\text{Var}(t) &= 0.0233\ (0.0053) \\
\text{Var}(k) &= 0.0155\ (0.0125)
\end{aligned}$$

Almost all of the genetic variance is attributable to extranuclear effects (σ_p^2, σ_m^2, and σ_k^2) and to dominance (σ_t^2).

DIALLELS

In the types of experiments just discussed, two sets of distinct genotypes are crossed (i.e., the genotypes on the horizontal and vertical axes of Figure 20.1 are different). Diallel experiments, first introduced by Schmidt (1919), utilize the same set of parents on both axes, with the same individuals (or inbred lines) serving as both male and female parents. With N parents, there are N^2 potential crosses in such an experiment, but depending upon whether reciprocals and/or within-line crosses are included, four types of diallel analysis are possible. In addition, depending on whether the parental genotypes are viewed as fixed or random effects, there are two approaches to the analysis of data. As described above, a random-effects analysis applies when the parental genotypes are taken to be random with respect to a base population about which genetic inferences are to be made. The primary goal in this case is the estimation of variance components. Under a fixed-effects interpretation, the parents are the only genotypes of immediate interest. This approach is often used in plant breeding programs where the goal is to estimate the average effects of specific lines and to identify higher yielding combinations of parents. Griffing (1956) provides a lucid outline of the eight types of analysis, and additional reviews are provided by Cockerham (1963), Hinkelmann (1976), Baker (1978), Hallauer and Miranda (1981), Wright (1985), and Christie and Shattuck (1992). We will simply give an overview of some of the more commonly used approaches.

Pooled Reciprocals, No Self Crosses

Assuming there are no maternal or paternal effects (aside from direct inheritance) and no significant sex-linked effects, reciprocal crosses are expected to yield

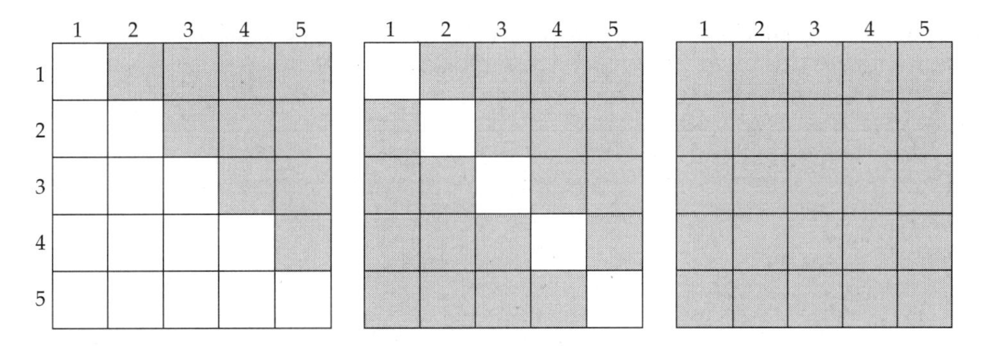

Figure 20.2 Mating scheme for an **incomplete diallel** with no reciprocal crosses and no crosses within the parental lines (left); for an incomplete diallel with reciprocals but no selfed (i.e., within parental line) crosses (center); and for a **full diallel** (right). Unshaded cells denote crosses that are not made.

equivalent progeny phenotype distributions. The simplest type of diallel analysis makes this assumption (either pooling progeny from reciprocal crosses or performing crosses in only one direction) and excludes crosses within parental types (Figure 20.2, left). The model to be analyzed is

$$z_{ijk} = \mu + g_i + g_j + s_{ij} + e_{ijk} \tag{20.8a}$$

where g_i and g_j are the **general combining abilities** (GCAs) of parents i and j, and $s_{ij} = s_{ji}$ is the **specific combining ability** (SCA) of $i \times j$ matings. GCA describes the average performance of a parent in hybrid combination with other genotypes. It is equivalent to the breeding value of an individual. SCA describes the degree to which specific parental combinations lead to deviations in progeny phenotypes from expectations based on average parental performance. These definitions were first elucidated by Sprague and Tatum (1942).

Equation 20.8a is functionally identical to the model employed in Design II without reciprocals, except that here it is assumed that the sire and dam effects are drawn from the same distribution. In the following discussion of this model, we make the usual assumptions that the effects (g, s, and e) are independently distributed with zero means. Under a random-effects interpretation, the variances of the three effects, g, s, and e, are denoted as σ^2_{GCA}, σ^2_{SCA}, and σ^2_e, and the total phenotypic variance is

$$\sigma^2_z = 2\sigma^2_{GCA} + \sigma^2_{SCA} + \sigma^2_e \tag{20.8b}$$

The GCA variance is equivalent to the covariance between half sibs. This equivalency can be seen from Equation 20.8a, since under the assumption that different

Table 20.4 Summary of the analysis of variance for an incomplete diallel (no reciprocal or within-line crosses) for an otherwise balanced design, involving N parental genotypes and n progeny per full-sib family.

Factor	Degrees of Freedom	Sums of Squares	E(MS)
GCA	$N-1$	$\dfrac{n(N-1)^2}{N-2}\sum_i(\bar{z}_{i\cdot}-\bar{z})^2$	$\sigma^2_e + n\sigma^2_{SCA} + n(N-2)\sigma^2_{GCA}$
SCA	$\dfrac{N(N-3)}{2}$	$n\sum_{i<j}(\bar{z}_{ij}-\bar{z})^2 - \text{SS}_{GCA}$	$\sigma^2_e + n\sigma^2_{SCA}$
Error	$(n-1)\left[\dfrac{N(N-1)}{2}-1\right]$	$\sum_{i<j}\sum_k(z_{ijk}-\bar{z}_{ij})^2$	σ^2_e

Note: z_{ijk} is the phenotype of the kth offspring from cross $i \times j$, $\bar{z}_{ij}$ is the mean phenotype of offspring from cross $i \times j$, and $\bar{z}_{i.}$ is the mean phenotype of offspring from parent i averaged over all mates. SS_{GCA} denotes the sum of squares associated with general combining ability.

types of effects are uncorrelated in the same individuals, $\sigma(\text{HS}) = \sigma(z_{ijk}, z_{ij'k'}) = \sigma^2(g_i) = \sigma^2_{GCA}$. The SCA variance is equivalent to the covariance between full sibs minus twice σ^2_{GCA}, since $\sigma(\text{FS}) = \sigma(z_{ijk}, z_{ijk'}) = 2\sigma^2(g_i) + \sigma^2(s_{ij}) = 2\sigma^2_{GCA} + \sigma^2_{SCA}$.

Diallel analysis is usually as concerned with the estimation of the g_i and s_{ij} values for the experimental lines as with the estimation of population components of variance. From Griffing (1956), the least-squares estimators of the effects and their standard errors are:

$$\begin{aligned} \widehat{g}_i &= \frac{N-1}{N-2}(\bar{z}_{i\cdot} - \bar{z}) \qquad & \text{SE}(\widehat{g}_i) &= \sqrt{\frac{(N-1)\text{Var}(e)}{nN(N-2)}} \\ \widehat{s}_{ij} &= \bar{z}_{ij} - \widehat{g}_i - \widehat{g}_j - \bar{z} \qquad & \text{SE}(\widehat{s}_{ij}) &= \sqrt{\frac{(N-3)\text{Var}(e)}{n(N-1)}} \end{aligned} \tag{20.9}$$

where N is the number of parental lines, n is the number of progeny per full-sib family, and Var(e) is an estimate of the variance within full-sib families. With a fixed-effects interpretation, this concludes the analysis.

The layout of the analysis of variance under a random-effects interpretation, in which interest is in estimating the variances of effects, is given in Table 20.4. Note that, unlike other two-way ANOVA tables outlined in this chapter, there is only one line for main effects. This is a consequence of our assumption that the distributions of paternal and maternal contributions to progeny are identical. If the experiment is replicated in multiple blocks of N parents, the degrees of freedom and sums of squares are summed over experiments prior to computation of the mean squares, but the definitions of the expected mean squares do not change. Under the usual assumptions of normality, significance of the GCA and SCA components of variance can be evaluated by use of the F ratios $\text{MS}_{GCA}/\text{MS}_{SCA}$ and $\text{MS}_{SCA}/\text{MS}_e$, respectively.

Diallels are frequently performed on completely inbred parental lines. Assuming such lines are extracted randomly from a random-mating base population and maintained without selection for the character under analysis, they will be in approximate gametic phase equilibrium between loci (each line representing a random gamete), and Hardy-Weinberg equilibria within loci will be restored by a single generation of random mating. There will be no genetic variance within families, so σ_e^2 is simply the variance due to special environmental effects ($\sigma_{E_s}^2$), and from Equation 20.8b, $2\sigma_{GCA}^2 + \sigma_{SCA}^2$ is the total genetic variance. When the parents are completely inbred, the covariance between half sibs is equivalent to a parent-offspring covariance in a random-mating population since as a consequence of both copies in a parent being identical, the half sibs have exactly one gene identical-by-descent at each locus. Therefore, when completely inbred lines are used,

$$\sigma_{GCA}^2 = \frac{\sigma_A^2}{2} + \frac{\sigma_{AA}^2}{4} + \cdots \tag{20.10a}$$

Subtracting twice this quantity from the total genetic variance, and ignoring all epistatic genetic variance except that due to additive × additive effects,

$$\sigma_{SCA}^2 = \sigma_D^2 + \frac{\sigma_{AA}^2}{2} + \cdots \tag{20.10b}$$

If, on the other hand, the parents are members of a random-mating population, the within-family variance contains the genetic variance expected to be segregating within full-sib families as well as the environmental variance

$$\sigma_e^2 = \frac{\sigma_A^2}{2} + \frac{3\sigma_D^2}{4} + \frac{3\sigma_{AA}^2}{4} + \cdots + \sigma_{E_s}^2 \tag{20.11a}$$

The GCA variance, again equivalent to the covariance of half sibs, becomes,

$$\sigma_{GCA}^2 = \frac{\sigma_A^2}{4} + \frac{\sigma_{AA}^2}{16} + \cdots \tag{20.11b}$$

Finally, the variance associated with specific combining ability is

$$\sigma_{SCA}^2 = \sigma(\text{FS}) - 2\sigma_{GCA}^2 = \frac{\sigma_D^2}{4} + \frac{\sigma_{AA}^2}{8} + \cdots \tag{20.11c}$$

Example 3. The following example concerns a diallel between $N = 6$ lines of the oyster, *Crassostrea gigas*. By stripping eggs and sperm from the hermaphroditic adults, Lannan (1980) performed all possible $6(6-1)/2 = 15$ crosses (excluding reciprocals) and recorded an index of female gonadal development in the progeny. Both the number of parental lines and the number of offspring per full-sib family ($n = 9$) are very small, so this example will serve simply to illustrate how the observed mean squares lead to parameter estimates by the method of moments, with no attention being given to the significance of the resultant estimates. The results of the analysis of variance are given in the following table:

Factor	df	Mean Squares	E(MS)	Estimates
GCA	5	1957.14	$\sigma_e^2 + 9\sigma_{SCA}^2 + 36\sigma_{GCA}^2$	$\text{Var}(GCA) = 54.08$
SCA	9	10.13	$\sigma_e^2 + 9\sigma_{SCA}^2$	$\text{Var}(SCA) = -5.24$
Error	112	57.31	σ_e^2	$\text{Var}(e) = 57.31$

The remaining analysis is somewhat complicated by the fact that the parental lines had been inbred by full-sib mating to $f = 0.375$. To simplify matters, we will assume that all of the genetic variance is due to additive effects, an assumption that seems justified based on the observation that the SCA variance estimate is negative. Recall from Equation 7.5a that provided the parents are unrelated and inbred to the same degree, the coefficient of coancestry (and the additive genetic covariance) of two individuals is inflated by the factor $(1 + f)$ relative to its expectation in a random-mating population. Thus, for this experiment, the expected GCA variance, which is equivalent to the covariance between half sibs, is $(1+0.375)\sigma_A^2/4$. Setting this equal to the observed Var(GCA), the additive genetic variance in the random-mating source population is estimated to be 157.32. This estimate nearly equals the total phenotypic variance $(2\sigma_{GCA}^2+\sigma_{SCA}^2+\sigma_e^2)$ revealed by the analysis, suggesting the possibility that the vast majority of the phenotypic variance has an additive genetic basis.

Reciprocals, No Self Crosses

A statistical model for the incomplete diallel with reciprocals was first proposed by Yates (1947). Under this design, all $N \times N$ crosses are employed, except those on the diagonals (Figure 20.2, center). As in the case of Design II with reciprocals, this type of diallel yields reciprocal full and half sibs, in addition to paternal half-sib, maternal half-sib, and full-sib families, providing an opportunity to estimate variance components beyond those possible in a design without

Table 20.5 Summary of the analysis of variance for a diallel mating design involving reciprocal crosses between lines but no crosses within lines.

Factor	df	Sums of Squares	Expected Mean Squares
GCA	$N-1$	$\frac{n(N-1)^2}{2(N-2)} \sum_i (\bar{z}_{i\cdot} - \bar{z})^2 + (\bar{z}_{\cdot i} - \bar{z})^2$	$\sigma_e^2 + n\sigma_k^2 + 2n\sigma_t^2 + \frac{n(N-2)}{2} (\sigma_m^2 + \sigma_p^2 + 4\sigma_n^2)$
SCA	$\frac{N}{2}(N-3)$	$\frac{n}{2} \sum_{i<j} (\bar{z}_{ij} - \bar{z})^2 + (\bar{z}_{ji} - \bar{z})^2 - \text{SS}_{GCA}$	$\sigma_e^2 + n\sigma_k^2 + 2n\sigma_t^2$
RGCA	$N-1$	$\frac{n(N-1)}{2N} \sum_i (\bar{z}_{i\cdot} - \bar{z}_{\cdot i})^2$	$\sigma_e^2 + n\sigma_k^2 + \frac{nN}{2}(\sigma_m^2 + \sigma_p^2)$
RSCA	$\frac{N-1}{2}(N-2)$	$\frac{n}{2} \sum_{i<j} (\bar{z}_{ij} - \bar{z}_{ji})^2 - \text{SS}_{RGCA}$	$\sigma_e^2 + n\sigma_k^2$
Error	$N(N-1)(n-1)$	$\sum_{i \neq j} \sum_k (z_{ijk} - \bar{z}_{ij})^2$	σ_e^2

Note: N is the number of parental lines, and n is the number of progeny evaluated per cross. $\bar{z}_{i\cdot}$ is the mean phenotype observed for all progeny from mother i, and $\bar{z}_{\cdot i}$ for all progeny from father i. $\bar{z}_{ij}$ is the mean phenotype of progeny of the mating between the ith dam and the jth sire, and $\bar{z}_{ji}$ is the mean of the reciprocal cross. z_{ijk} is the observed phenotype of the kth offspring of the $i \times j$ cross. RGCA and RSCA refer respectively to reciprocal general and specific combining abilities. The expected mean squares are given in terms of variance components associated with the Cockerham-Weir model.

reciprocals. Our description will be in terms of the Cockerham-Weir (1977b) model, Equation 20.6a, described above in the context of Design II. The analysis of variance, given in Table 20.5, contains mean squares for **reciprocal general** (RGCA) and **specific** (RSCA) **combining abilities**, in addition to the GCA and SCA terms just described. These are functions of the five causal components of genetic variance associated with the Cockerham-Weir model (see Table 20.5).

With this design, the variances associated with maternal and paternal extranuclear effects, σ_m^2 and σ_p^2, always appear together as a sum in the expected mean squares, so the method of moments cannot generate separate estimates of them. To accomplish that, Cockerham and Weir (1977b) proposed the use of the symmetrical products,

$$T_m = \frac{\sum_{i \neq j} [(N-1)\bar{z}_{i.}^2 - \bar{z}_{ij}^2]}{N(N-1)(N-2)} \tag{20.12a}$$

$$T_p = \frac{\sum_{i \neq j} [(N-1)\bar{z}_{.i}^2 - \bar{z}_{ij}^2]}{N(N-1)(N-2)} \tag{20.12b}$$

the expectations of which are

$$E(T_m) = \mu^2 + \sigma_n^2 + \sigma_m^2 \tag{20.13a}$$

$$E(T_p) = \mu^2 + \sigma_n^2 + \sigma_p^2 \tag{20.13b}$$

The difference $T_m - T_p$ provides an estimate of $\sigma_m^2 - \sigma_p^2$, which when combined with the estimate of $(\sigma_m^2 + \sigma_p^2)$ obtained from the analysis of variance, allows the separation of σ_m^2 and σ_p^2.

From the standpoint of the estimation of individual effects, the Cockerham-Weir model has some limitations under this design. n_i and m_i, and n_i and p_i are confounded, so only $(2n_i + m_i + p_i)$ and $(m_i - p_i)$ are estimable. An alternative approach is the **reciprocal-effects model** of Griffing (1956),

$$z_{ijk} = \mu + g_i + g_j + s_{ij} + r_{ij} + e_{ijk} \tag{20.14}$$

which is a simple extension of Equation 20.8a. Here, g_i and s_{ij} are equivalent to n_i and t_{ij} in the Cockerham-Weir model, whereas $r_{ij} = p_i + m_j + k_{ij}$ is a composite estimate of the extranuclear effects. The constraint $r_{ij} = -r_{ji}$ arises naturally from the model definition of $\mu + g_i + g_j + s_{ij}$ as the mean phenotype of offspring from $i \times j$ and $j \times i$ matings. The least-squares estimators and standard errors of the effects in Griffing's (1956) model are:

$$\begin{aligned}
\widehat{g}_i &= \frac{1}{2N(N-2)} \left[N \sum_{j \neq i} (\bar{z}_{ij} + \bar{z}_{ji}) - 2 \sum_{j \neq k} \bar{z}_{jk} \right] & \text{SE}(\widehat{g}_i) &= \sqrt{\frac{(N-1)\text{Var}(e)}{2Nn(N-2)}} \\
\widehat{s}_{ij} &= \frac{\bar{z}_{ij} + \bar{z}_{ji}}{2} - \widehat{g}_i - \widehat{g}_j - \frac{1}{N(N-1)} \sum_{k \neq l} \bar{z}_{kl} & \text{SE}(\widehat{s}_{ij}) &= \sqrt{\frac{(N-3)\text{Var}(e)}{2n(N-1)}} \\
\widehat{r}_{ij} &= \frac{\bar{z}_{ij} - \bar{z}_{ji}}{2} & \text{SE}(\widehat{r}_{ij}) &= \sqrt{\frac{\text{Var}(e)}{2n}}
\end{aligned} \tag{20.15}$$

Still another analytical approach to diallels with reciprocals assumes an absence of paternal, nuclear-extranuclear, and extranuclear-extranuclear effects (i.e., $p_i = k_{ij} = 0$). This reduces the Cockerham-Weir model to

$$z_{ijk} = \mu + n_i + n_j + t_{ij} + m_i + e_{ijk} \tag{20.16}$$

Details of the analysis involving this model can be found in Bulmer (1980).

Example 4. Thomas-Orillard and Jeune (1985) performed all possible reciprocal crosses between six ($N = 6$) strains of *D. melanogaster* (three French and three African) and examined the female progeny for number of ovarioles. ANOVA was performed on means of 50 measures in each of two replicate blocks ($n = 2$). The ANOVA table follows, and from it, the estimated variance components for the Cockerham-Weir model are obtained by setting the observed mean squares equal to their expectations (defined in Table 20.5). We illustrate the computation of variance components primarily for heuristic purposes, as the lines involved in this experiment can hardly be viewed as a random sample of the species.

Source	df	Mean Squares	Estimates
GCA	5	59.649	$\text{Var}(n) = 3.182$
SCA	9	9.593	$\text{Var}(t) = 1.616$
RGCA	5	1.853	$\text{Var}(m) + \text{Var}(p) = -0.213$
RSCA	10	3.129	$\text{Var}(k) = 0.999$
Error	30	1.130	

From the results provided in the original analysis, T_m and T_p are found to be 1,718.958 and 1,720.854, respectively. Setting these equal to their expectations and subtracting, $\text{Var}(m) - \text{Var}(p) = -1.896$, which when combined with the results in the table leads to the estimates $\text{Var}(m) = -1.054$ and $\text{Var}(p) = 0.841$. Using F ratios, the following hypotheses can be tested:

Hypothesis	Test Statistic	Degrees of Freedom
$\sigma_n^2 = 0$	$\dfrac{\text{MS}_{GCA} + [(N-2)\text{MS}_{RSCA}/N]}{\text{MS}_{SCA} + [(N-2)\text{MS}_{RGCA}/N]} = 5.70$	See remarks below.
$\sigma_k^2 = 0$	$\dfrac{\text{MS}_{RSCA}}{\text{MS}_e} = 2.77$	$(N-1)(N-2)/2$, $N(N-1)(n-1)$
$\sigma_m^2 + \sigma_p^2 = 0$	$\dfrac{\text{MS}_{RGCA}}{\text{MS}_{RSCA}} = 0.59$	$(N-1)$, $(N-1)(N-2)/2$
$\sigma_t^2 = 0$	$\dfrac{\text{MS}_{SCA}}{\text{MS}_{RSCA}} = 3.07$	$N(N-3)/2$, $(N-1)(N-2)/2$

(Note that the degrees of freedom for the test of $\sigma_n^2 = 0$ are not a simple function of N and n, due to the fact that the numerator and denominator of the test statistic are sums of observed mean squares. For situations like this, the degrees of freedom can be approximated using the method of Satterthwaite (1946), described in Chapter 18. For this particular example, the approximate degrees of freedom are 5 and 11).

The variance of nuclear additive effects (n), nuclear nonadditive effects (t), and extranuclear interaction effects (k) are all significant (at the 0.01, 0.05, and 0.05 levels, respectively), the latter arising despite the fact that the summed variance of the extranuclear effects $(\sigma_m^2 + \sigma_p^2)$ is nonsignificant. Nearly half of the observed genetic variance is attributable to nonadditive effects, $\sigma_t^2 + \sigma_k^2$.

Complete Diallels

The main thing to be gained by performing a complete diallel, which includes crosses within lines, is the information gathered on the effects of inbreeding. We will not elaborate on these issues to any great extent, since the consequences of inbreeding have already been reviewed extensively in Chapter 10. Obviously, if parents are completely inbred (or very nearly so), there is nothing gained by considering crosses within lines. Various parameterizations of the linear model for the complete diallel have been given by Eberhart and Gardner (1966), Griffing (1956), and Wearden (1964). Further discussion on these models can be found in Morley-Jones (1965), Gardner and Eberhart (1966), Walters and Morton (1978), and Carbonell et al. (1983).

The interpretation of the variance components extracted from a complete-diallel analysis can be problematical. Depending on whether reciprocal crosses have been included or not, $1/N$ or $2/(N+1)$ of the individuals in the diallel table will be inbred, and this will inflate the homozygosity in comparison to the ancestral base population. Thus, if the reference of interest is a random-mating base population, analyses that exclude selfed families should be employed. A useful application of the complete diallel in the estimation of the average degree of dominance is given below.

Partial Diallels

In the types of diallels described above, the number of families required for a balanced design increases with the square of the number of parents. This requirement can impose a serious constraint on the number of parental genotypes that can be analyzed simultaneously. To alleviate this problem, many modifications, collectively referred to as **partial diallels**, have been suggested. The general features of all of these modifications, two of which are illustrated in Figure 20.3, are that individuals are mated to only a subset of the parental lines and that these subsets are partially overlapping.

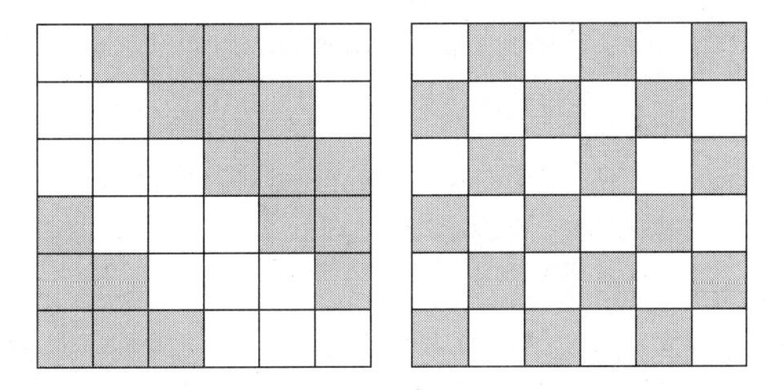

Figure 20.3 Two possible structures for partial diallels. The reciprocal crosses are sometimes excluded.

Since partial diallels generate all of the sib relationships discussed in the previous sections, they can be employed fully in the estimation of causal components of variance. However, due to the structure of the data, the design of the analysis of variance is rather complicated, and the construction of standard errors and hypothesis tests by this route is not straightforward. The methods to be discussed in Chapters 26 and 27 appear to be ideally suited to this type of analysis. Technical details regarding the optimal design and analysis of partial diallels can be found in Gilbert (1958), Kempthorne and Curnow (1961), Curnow (1963), Fyfe and Gilbert (1963), Hinkelmann (1975), and Ceranka et al. (1987).

HAYMAN-JINKS ANALYSIS

Our preceding discussion has mostly focused on the estimation of properties of individuals and populations. Diallel analysis can also provide insight into the mode of gene action. In particular, for the special case in which there are only two alleles per segregating locus (as when all parental lines are derived from a cross between two inbred lines), useful inferences about the degree of dominance can be made from the distribution of mean phenotypes in a complete diallel. The essential relationships were first pointed out by Hayman (1954) and Jinks (1954) for the case in which the parental lines are completely inbred. These were generalized later to arbitrary parental genotypes by Dickinson and Jinks (1956). A broad overview of the theory is given in Mather and Jinks (1982).

Let $\overline{z}_i$ be the mean phenotype of the ith parent (often estimated by a single individual, unless the genotype can be cloned), $\overline{z}_{ii}$ be the mean phenotype of its offspring obtained by selfing, and $\overline{z}_{ij}$ be the mean phenotype of progeny from the $i \times j$ and $j \times i$ crosses (i.e., the pooled means of reciprocal crosses). Further let $\overline{z}_{i.}$ be the (marginal) mean phenotype of all families having i as the mother (including the $i \times i$ cross), and let $\overline{z}_{.i}$ denote the same type of marginal mean indexed on father i. From these phenotypic measures, we can estimate several phenotypic variances and covariances that are useful in evaluating the dominance properties of segregating alleles:

1. $\sigma^2_{P_1}$, the variance among the actual parents of the diallel $(\overline{z}_1, \overline{z}_2, \ldots, \overline{z}_N)$, and $\sigma^2_{P_2}$, the variance of the family means obtained by selfing the parents $(\overline{z}_{11}, \overline{z}_{22}, \ldots, \overline{z}_{NN})$. These two variance estimates are unlikely to be equal — the genetic contributions to $\sigma^2_{P_1}$ and $\sigma^2_{P_2}$ will be identical only if the parents are completely inbred, and the sampling variances contributing to both terms will usually be unequal since the P_1 and P_2 phenotypes are normally based on single and multiple measures, respectively.

2. $\sigma^2_{\overline{r}}$, the variance of marginal means, is the average of the variance among the N paternal means $(\overline{z}_{\cdot 1}, \overline{z}_{\cdot 2}, \ldots, \overline{z}_{\cdot N})$, and the variance among the N maternal means $(\overline{z}_{1\cdot}, \overline{z}_{2\cdot}, \ldots, \overline{z}_{N\cdot})$.

3. $\overline{\sigma}^2_r$, the average variance of family means around their array means, is obtained by averaging the N variance estimates for means within rows and their N complementary estimates for means within columns of the diallel table. For example, for row i, $\sigma^2_r(i)$ is based on the means $(\overline{z}_{i1}, \overline{z}_{i2}, \ldots, \overline{z}_{iN})$. This computation is made for all rows and then for all columns, and the $2N$ variance estimates are averaged to give an estimate of $\overline{\sigma}^2_r$.

4. $\overline{\sigma}_{P_1,r}$, the covariance of family means in arrays with the phenotypes of their nonrecurrent parents, is also obtained by averaging over rows and columns. For example, for the ith row, $\sigma_{P_1,r}(i)$ is the covariance of the elements in the vectors $(\overline{z}_1, \overline{z}_2, \ldots, \overline{z}_N)$ and $(\overline{z}_{i1}, \overline{z}_{i2}, \ldots, \overline{z}_{iN})$. This computation is made for all rows and then for all columns, and the $2N$ covariance estimates are averaged to give an estimate of $\overline{\sigma}_{P_1,r}$. $\overline{\sigma}_{P_2,r}$ is obtained in the same manner as $\overline{\sigma}_{P_1,r}$, but through the use of P_2 rather than P_1 measures.

5. σ_{P_1,P_2}, the covariance of parents and their selfed family means, involves the vectors $(\overline{z}_1, \overline{z}_2, \ldots, \overline{z}_N)$ and $(\overline{z}_{11}, \overline{z}_{22}, \ldots, \overline{z}_{NN})$.

Expectations of all of the preceding quantities can be expressed in terms of gene frequencies and homozygous and dominance effects. The final expressions are summarized most easily by using the composite parameters defined in Table 20.6,

$$\sigma^2_{P_1} = D + H_3 - 4G_2 + \sigma^2(\overline{P}_1) \tag{20.17a}$$

$$\sigma^2_{P_2} = D + \frac{H_3}{4} - 2G_2 + \sigma^2(\overline{P}_2) \tag{20.17b}$$

$$\overline{\sigma}^2_r = \frac{D}{4} + H_1 - G_1 + \frac{(N-1)\sigma^2(\overline{F}_1) + \sigma^2(\overline{P}_2)}{N} \tag{20.17c}$$

Table 20.6 Composite genetic parameters used in the Hayman-Jinks analysis. Summations are over loci, each indexed by l.

		$p_l = q_l = 0.5$	
	Arbitrary Gene Frequencies	$f = 0$	$f = 1$
D	$2\sum p_l q_l(1+f)a_l^2$	$\sum a_l^2/2$	$\sum a_l^2$
H_1	$\frac{1}{2}\sum p_l q_l[1-2p_l q_l(1-f)](1+f)(a_l k_l)^2$	$\sum(a_l k_l)^2/16$	$\sum(a_l k_l)^2/4$
H_2	$\sum[p_l q_l(1+f)a_l k_l]^2$	$\sum(a_l k_l)^2/16$	$\sum(a_l k_l)^2/4$
H_3	$2\sum p_l q_l[1-2p_l q_l(1-f)](1-f)(a_l k_l)^2$	$\sum(a_l k_l)^2/4$	0
H_4	$2\sum(1-f)(1+f)(p_l q_l a_l k_l)^2$	$\sum(a_l k_l)^2/8$	0
G_1	$\sum p_l q_l(p_l - q_l)(1+f)a_l^2 k_l$	0	0
G_2	$\sum p_l q_l(p_l - q_l)(1-f)a_l^2 k_l$	0	0

Note: f refers to the level of inbreeding in the parents. The columns labeled $f = 0$ and $f = 1$ refer to the special cases in which the two genes per heterozygous locus are in equal frequencies $p_l = q_l = 0.5$, as would be the case when all parents are descended from a cross between two inbred lines.

$$\sigma^2_{\overline{r}} = \frac{D}{4} + H_1 - H_2 - G_1 \tag{20.17d}$$

$$\overline{\sigma}_{P_1,r} = \frac{D}{2} + \frac{H_3}{2} - H_4 - G_1 - G_2 \tag{20.17e}$$

$$\overline{\sigma}_{P_2,r} = \frac{D}{2} + \frac{H_3}{4} + \frac{H_4}{2} - G_1 - \frac{G_2}{2} + \frac{\sigma^2(\overline{P}_2)}{N} \tag{20.17f}$$

$$\sigma_{P_1,P_2} = D + \frac{H_3}{2} - 3G_2 \tag{20.17g}$$

The terms $\sigma^2(\overline{P}_1)$, $\sigma^2(\overline{P}_2)$, and $\sigma^2(\overline{F}_1)$ are the expected sampling variances of P_1, P_2, and F_1 family means, estimates of which can be obtained from single plots by dividing the within-family variance by the sample size, or from variances among replicate plot means. In the following, we assume that the contributions of sampling variance to the preceding expressions have been eliminated (e.g., by subtracting the sampling variance of the mean), so that the quantities of interest have a purely genetic interpretation (solely a function of the quantities in Table 20.6).

Letting the appropriately subscripted V and C denote estimates of the preceding variances (Equations 20.17a–d) and covariances (Equations 20.17e–g) after removal of the sampling variance contributions, estimators for the composite genetic parameters in Table 20.6 are found by rearranging Equations 20.17a-g to give

$$\widehat{D} = V_{P_1} + 4V_{P_2} - 4C_{P_1,P_2} \tag{20.18a}$$

$$\widehat{H}_1 = \frac{V_{P_1}}{4} + V_{P_2} + \overline{V}_r + \overline{C}_{P_1,r} - 2\overline{C}_{P_2,r} - C_{P_1,P_2} \tag{20.18b}$$

$$\widehat{H}_2 = \overline{V}_r - V_{\overline{r}} \tag{20.18c}$$

$$\widehat{H}_3 = 4V_{P_1} + 4V_{P_2} - 8C_{P_1,P_2} \tag{20.18d}$$

$$\widehat{H}_4 = V_{P_1} - 2\overline{C}_{P_1,r} + 2\overline{C}_{P_2,r} - C_{P_1,P_2} \tag{20.18e}$$

$$\widehat{G}_1 = \frac{V_{P_1}}{2} + 2V_{P_2} + \overline{C}_{P_1,r} - 2\overline{C}_{P_2,r} - 2C_{P_1,P_2} \tag{20.18f}$$

$$\widehat{G}_2 = V_{P_1} + 2V_{P_2} - 3C_{P_1,P_2} \tag{20.18g}$$

When parents are completely inbred, these expressions can be simplified by letting $V_{P_1} = V_{P_2} = C_{P_1,P_2}$, and $\overline{C}_{P_1,r} = \overline{C}_{P_2,r}$.

As shown in Table 20.6, all seven of these composite parameters are defined by gene frequencies and gene effects. Functions of these quantities can yield considerable insight into the mechanistic basis of genetic variation. For example, the quantity $2(2H_1 + H_4)/D$ is equal to $\sum p_l q_l a_l^2 k_l^2 / \sum p_l q_l a_l^2$. Thus,

$$\widetilde{D}' = \frac{2(2\widehat{H}_1 + \widehat{H}_4)}{\widehat{D}} \tag{20.19}$$

provides a weighted estimate of the average value of k^2, with the weighting in favor of loci with large effects $|a|$ and/or high heterozygosities. For the special case in which $p_l = q_l = 0.5$, the weighting involves only the squared homozygous effects, since $2(2H_1 + H_4)/D = \sum a_l^2 k_l^2 / \sum a_l^2$. In that case, $\widetilde{D}'$ is identical to the quantity estimated by Comstock and Robinson's (1952) $\widetilde{D}$, Equation 20.5.

When the parental lines $(i = 1, \ldots, N)$ are completely inbred, there are three ways in which further insight into the dominance properties of quantitative-trait loci can be acquired from the relationship between the expected array covariances, $\sigma_{P_2,r}(i)$, and variances, $\sigma_r^2(i)$. First, an additional weighted estimate of the average degree of dominance can be acquired in the following manner. If the environmental contributions to $\sigma_{P_2,r}(i)$ and $\sigma_r^2(i)$ are removed, the expected values of these quantities can be expressed in terms of the frequencies and effects of alleles in the entire collection of lines,

$$\sigma_r^2(i) = \sum p_l q_l a_l^2 (1 - \delta_{il} k_l)^2 \tag{20.20a}$$

$$\sigma_{P_2,r}(i) = 2 \sum p_l q_l a_l^2 (1 - \delta_{il} k_l) \tag{20.20b}$$

where the summation is over all variable loci, indexed by l, and δ_{il} equals +1 or −1 depending on whether the line is fixed for the dominant or recessive allele at

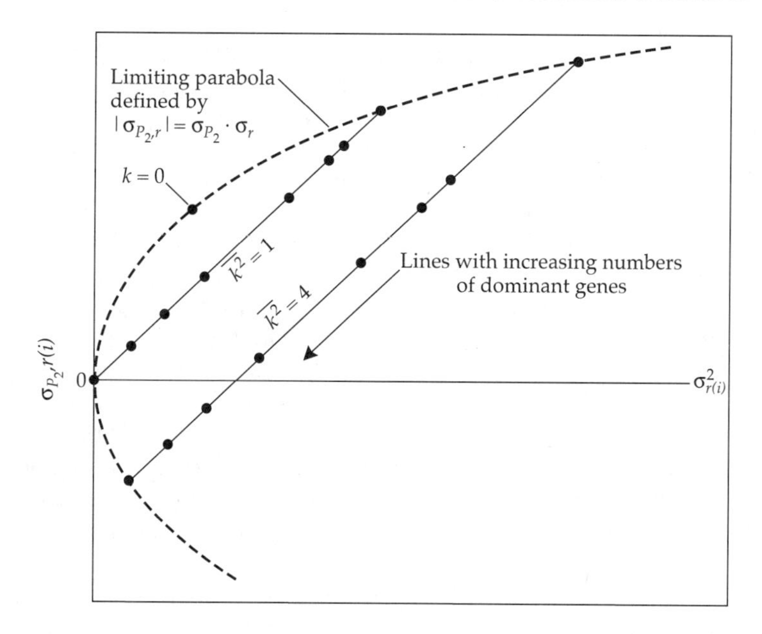

Figure 20.4 Expected relationship between $\sigma_{P_2,r}(i)$, the covariance of family means in array i with the means of their nonrecurrent parents, and $\sigma_r^2(i)$, the variance of family means in array i. The parental lines are assumed to be completely inbred. Prior to actual regression, the environmental contribution to these two parameter estimates needs to be removed. The data points for each parental line are then expected to fall along a straight line with slope equal to one. The intercept depends upon the squared degree of dominance, with a positive intercept denoting partial dominance ($\overline{k^2} < 1$), a zero intercept complete dominance ($\overline{k^2} = 1$), and a negative intercept overdominance ($\overline{k^2} > 1$).

the lth locus. These two quantities are related as

$$\sigma_{P_2,r}(i) = \sigma_r^2(i) + \sum p_l q_l a_l^2 (1 - k_l^2) \tag{20.21}$$

Thus, provided the environmental contributions to the observed statistics $C_{P_2,r}(i)$ and $V_r(i)$ have been removed, the regression of $C_{P_2,r}(i)$ on $V_r(i)$ has an expected slope equal to one and an expected intercept, α_r, equal to $\sum p_l q_l a_l^2(1 - k_l^2)$. Note also that when the parents are completely inbred, as we are assuming, $\sigma^2_{P_2} = 4\sum p_l q_l a_l^2$. Hence, an average value of k^2 (again, using the $p_l q_l a_l^2$ as weights) is obtained from using the estimated intercept $\widehat{\alpha}_r$,

$$\widetilde{D}'' = \frac{V_{P_2} - 4\widehat{\alpha}_r}{V_{P_2}} \tag{20.22}$$

Second, inferences can be made regarding the relative numbers of dominant alleles carried in different inbred lines. The geometric relationship between the

$\sigma_{P_2,r}(i)$ and their corresponding $\sigma_r^2(i)$ is shown in Figure 20.4. Due to sampling error, the data points from an actual analysis will not all fall exactly on a straight line, but the regression coefficient is not expected to be significantly different from one unless the parental lines are harboring significant heterozygosity or there is significant epistatic variance. Since the absolute value of a correlation coefficient cannot exceed one, $|\sigma_{P_2,r}(i)| \leq \sigma_{P_2} \cdot \sigma_r(i)$, so the data are expected to be confined to a parabola defined by σ_{P_2}. Any parental line that is homozygous for all of the dominant alleles will produce an array with minimal variance and covariance, yielding a point on the parabola close to the origin. (In the case of complete dominance, such a parental line will exhibit no genetic variance among its F_1 families and will fall at the origin.) On the other hand, a line fixed for all of the recessive alleles yields the maximum array variance and covariance, and falls on the point where the regression line intersects the upper end of the parabola. Thus, in general, the relative positions of the points on the regression line indicate the relative numbers of dominant genes in the parental lines. For the special case of additivity, all of the observations are expected to fall at a single point on the limiting parabola.

Third, elaborating further on the pattern just noted for pure parental lines, the expected sum of the covariance and variance involving the ith array is $\sigma_{P_2,r}(i) + \sigma_r^2(i) = [(3D/4) + H_1] - 4\sum \delta_{il} p_l q_l a_l^2 k_l$. This quantity is expected to increase linearly with the number of recessive alleles in the parental line, i.e., with the number of negative δ_{il}. Thus, by regressing the index $\widetilde{R} = C_{P_2,r}(i) + V_r(i)$ on $\overline{z}_{ii}$, it is possible to evaluate whether the number of dominant alleles in a parental line is correlated with its mean phenotype. Such a comparison was made for mean flowering time in *Nicotiana rustica* by Mather and Jinks (1982) (Figure 20.5). In general, later flowering lines had higher values of $\widetilde{R}$, suggesting that they carried more recessive genes. However, the line with the earliest flowering time had an intermediate value of $\widetilde{R}$ (point 8 in Figure 20.5), suggesting that not all of the alleles for early flowering are dominant.

NORTH CAROLINA DESIGN III AND THE TRIPLE TEST CROSS

As still another means of estimating the degree of dominance, Comstock and Robinson (1948, 1952) proposed a line-cross technique (**North Carolina Design III**) involving a synthetic F_2 population constructed from two inbred parental lines. Random members of the F_2 generation are backcrossed to each of the parental lines, and the average performances of the backcross families are evaluated in a series of replicate plots (Figure 20.6). Let $\overline{z}_{1ij}$ and $\overline{z}_{2ij}$ denote the mean phenotypes of progeny derived from the ith F_2 individual backcrossed to the two parent lines in the jth plot. Further, let $S_{ij} = \overline{z}_{1ij} + \overline{z}_{2ij}$ and $\Delta_{ij} = \overline{z}_{1ij} - \overline{z}_{2ij}$ denote the sums and differences of mean phenotypes involving the jth backcross replicates of the ith F_2 individual. Using the values of S_{ij} and Δ_{ij} as units of

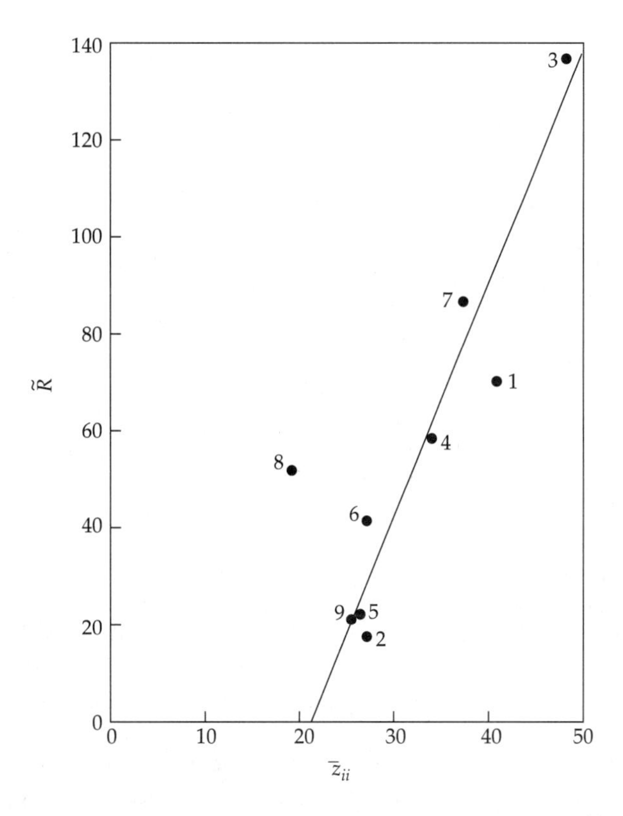

Figure 20.5 Estimates of the index of number of recessive genes ($\widetilde{R}$) versus the mean phenotype of inbred lines ($\overline{z}_{ii}$) for flowering time (days after July 1) in nine lines of tobacco *Nicotiana rustica* subjected to a complete diallel. (From Mather and Jinks 1982.)

observation, one-way ANOVA can be used to estimate the variances of the family sums and differences.

In the absence of epistasis and gametic phase disequilibrium, these variance components have very simple definitions. The variance among sums, $\sigma^2(S)$, is equivalent to the additive genetic variance in the F_2 generation, which as noted above, is simply $\sum_\ell (a_\ell)^2/2$ when gene frequencies are equal (as they are in the synthetic population). The variance among differences, $\sigma^2(\Delta)$, is equivalent to twice the dominance genetic variance in the F_2 generation, $\sum_\ell (a_\ell k_\ell)^2/2$.

The restriction of this technique to situations in which gene frequencies are equal is rather limiting. However, where the technique applies, a major advantage is that it provides estimates of the additive and dominance components of variance with nearly equal precision, unlike the situation with other multiple mating

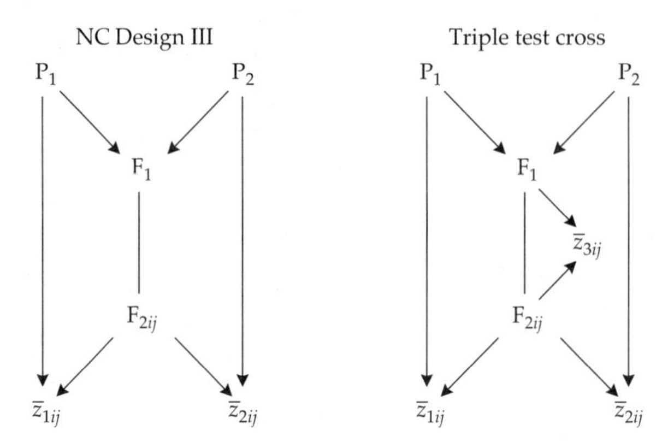

Figure 20.6 Crossing schemes for North Carolina Design III and the triple test cross.

designs. In addition, the ratio $\text{Var}(\Delta)/\text{Var}(S)$ provides a weighted estimate of the squared degree of dominance, having an expectation identical to that given above for Design II, Equation 20.5, provided the assumptions of gametic phase equilibrium and no epistatic interactions are met. As noted above, violations of either of these assumptions are likely to result in inflated estimates of the degree of dominance.

Kearsey and Jinks (1968) suggested a **triple test cross** that can shed light on the likely significance of epistatic sources of variance. Their approach is very similar to Design III, except that individuals are backcrossed to the F_1 derivatives of the parental lines as well as to the parental lines themselves (Figure 20.6). The logic behind the test is that F_1 individuals produce recombinant gametes, whose average associated gene expression will deviate from that of the mean of the parental line gametes if epistatic interactions are significant. Letting $\overline{z}_{3ij}$ denote the mean phenotype of progeny from a cross between the F_1 and the ith F_2 individual in the jth plot, the relevant unit of observation becomes $\overline{z}_{1ij} + \overline{z}_{2ij} - 2\overline{z}_{3ij}$, which has zero expectation in the absence of epistasis (see Table 9.1) and may be positive or negative otherwise. The test for significant epistatic variance again uses a one-way ANOVA, in this case to evaluate whether the variance among the observed family values of $\overline{z}_{1i} + \overline{z}_{2i} - 2\overline{z}_{3i}$ is greater than that expected from sampling error. The triple test cross is capable of detecting epistasis involving the entire sample of loci for which the tester lines differ. In the absence of significant epistatic variance, one can more justifiably proceed to estimate the average degree of dominance via the Comstock-Robinson approaches.

With the triple test cross, each member of the assay population is mated to only three tester lines, so a larger sample of the population can be accommodated than is usually possible in other factorial designs. Additional information on the method can be found in Jinks and Perkins (1970) and Mather and Jinks (1982).

SOME CLOSING STATISTICAL CONSIDERATIONS

This chapter has served to illustrate the multiplicity of experimental designs and biological interpretations that become possible when individuals of both sexes can be mated multiply. Cross-classified designs are advantageous because they generate several kinds of relationships, allowing the estimation of multiple components of variance. They also offer a significant advantage with respect to statistical power. Although only a few studies have attempted to evaluate the relative precision of variance-component estimates procured by different mating schemes, the evidence is compelling that, for the same amount of effort, diallels yield more precise estimates of both the additive and dominance components of variance than NC Design II (Pederson 1972, Namkoong and Roberds 1974, Bridges and Knapp 1987). Moreover, for a fixed number of crosses, partial diallels, because they sample over more genotypes, appear to be superior to full diallels. Of all the methods that generate both full and half sibs, NC Design I (the nested sib analysis) is clearly the least desirable from the standpoint of statistical power. Thus, the substantial reliance of plant breeders on cross-classified designs is well justified.

Accompanying the increased statistical power of cross-classified designs is a substantial increase in statistical complexity. The major issues have been largely invisible in this chapter because our introduction to estimation procedures and hypothesis testing has focused on the ideal situation in which the data are balanced. In this optimal situation, the properties of ANOVA are both well behaved and well understood — there is a unique partitioning of the sums of squares, the expected mean squares have straightforward definitions, estimates of the variance components are unbiased, and provided the data are distributed normally, F ratios provide a simple means for hypothesis testing and construction of confidence intervals.

On the other hand, with unbalanced data, there are numerous ways to partition the sums of squares, and the derivation of expressions for the expected mean squares, necessary for application of the method of moments, is quite complicated (Henderson 1953). To this day, there is no consensus among statisticians as to the optimal approach to two-way ANOVA with unbalanced data, and as a consequence, most standard statistical packages compute several (usually four) types of sums of squares, leaving it to the practitioner to decide which ones to rely on. The methods vary in the way they factor out the various model effects in computing the sum of squares associated with each individual factor, and as a consequence, can sometimes yield rather different results. In none of the approaches are the observed mean squares necessarily independent, and the expressions for their expectations generally do not yield to definitions of simple ratios whose numerators and denominators are equal under a null hypothesis. Both issues raise significant problems in hypothesis testing.

So as not to overshadow the salient biological and conceptual issues of cross-classified analysis with statistical details of estimation procedures, we have chosen to leave further investigation of these issues to the reader. An explicit introduction into some of the issues has already been provided in Chapter 16 in the context of one-way and nested ANOVA, and Milliken and Johnson (1984) and Searle et al. (1992) provide lucid and comprehensive overviews for the two-factor model. The problems are not trivial and they are formidable algebraically, but not insurmountable. Many of them can be avoided entirely, provided the model effects are normally distributed, by using the maximum likelihood procedures outlined in Chapters 26 and 27.

21

Correlations Between Characters

The phenotypic values of different traits in the same individual are often found to be correlated. In humans, for example, individuals that are tall also tend to have large feet. Such **phenotypic correlations** can arise from two causes. First, the expression of two characters may be modified by the same environmental factors operating within individuals. Some environmental factors may influence both characters in the same direction, e.g., variation in resource availability during development may influence the growth of all organs. Others may have opposite effects, as when an environmental cue to initiate the allocation of resources to reproduction causes a curtailment of growth. The joint influences of all such factors determine whether a within-individual **environmental correlation** will exist between the traits. Such a correlation cannot be assumed to be a species-specific constant. Just as the magnitude of the environmental variance for a trait can depend on the nature of the environment in which a population is assayed (Chapter 6), so can the covariance of environmental deviations for two traits.

Second, **genetic correlations** between characters can arise by two mechanisms. As a result of complex biochemical, developmental, and regulatory pathways, a single gene will almost always influence multiple traits, a phenomenon known as **pleiotropy** (Wright 1968). The direction of pleiotropy may differ among genetic factors. Thus, at least in principle, strong pleiotropy need not result in a strong genetic correlation between characters if the pleiotropic effects from different loci cancel each other. A second possible source of genetic correlation is gametic phase disequilibrium between genes affecting different characters, i.e., the tendency of genes with like effects on two characters to be positively or negatively associated in the same individuals. Since the pleiotropic effects of genes may be influenced by their genetic background, and since the degree of gametic phase disequilibrium will be a function of the past history of populations, care should also be taken in extrapolating estimates of the genetic correlation across populations or across time.

A knowledge of the mechanisms underlying the correlations between different traits is fundamental to understanding the degree of integration of the phenotype and to resolving the constraints imposed on evolutionary processes. Depending on their sign, genetic correlations between two characters can either facilitate or impede adaptive evolution. A conflict arises when two negatively genetically correlated traits are both selected in the same direction, as the selective advance of each character tends to pull the other character in the opposite direction. A perfect genetic correlation between two traits is equivalent to an absolute evolutionary constraint, since no change in either character can occur without a parallel change in the other.

Deciphering the relative contributions of environmental and genetic factors to phenotypic correlations is one of the most powerful and revealing applications of quantitative genetics. However, as we will see below, the estimation of genetic correlations is also an extremely demanding enterprise, requiring substantially larger sample sizes than are necessary in univariate analysis. For many purposes, a simple knowledge of the sign of a genetic correlation can be very revealing, and with the appropriate effort, is certainly achievable. However, as in the case of heritability analysis, completely unambiguous estimates of the causal sources of covariance are not usually possible, e.g., it is generally not possible to obtain a completely unbiased estimate of the genetic correlation due to additive genetic factors, any more than it is possible to obtain an absolutely clean estimate of σ_A^2 by heritability analysis.

In this chapter, we first show how additive genetic covariances between traits can be approximated by a simple extension of the methods already developed for variance-component analysis of single traits. Combining these covariances with estimates of the additive genetic variances of the traits provides a basis for estimating the genetic correlation. Subtracting the genetic components of variance and covariance at the genetic level from those at the phenotypic level yields the environmental components, which can then be used to estimate the environmental correlation. Following a consideration of issues associated with statistical analysis and hypothesis testing, we present several examples of the impact that genetic correlation analysis is having on evolutionary thinking.

THEORETICAL COMPOSITION OF THE GENETIC COVARIANCE

Assuming the contribution from gametic phase disequilibrium to be negligible, Mode and Robinson (1959) showed how the genetic covariance can be subdivided into various components. Their work is a straightforward extension of the results of Cockerham (1954) and Kempthorne (1954) for single characters. In order to simplify the presentation, recall the procedure used in Chapter 5 for decomposing the total genotypic value of a trait influenced by two loci. For characters 1 and 2, we have

$$\begin{aligned} G_1 = \mu_{G,1} &+ [\alpha_{i,1} + \alpha_{j,1} + \alpha_{k,1} + \alpha_{l,1}] + [\delta_{ij,1} + \delta_{kl,1}] \\ &+ [(\alpha\alpha)_{ik,1} + (\alpha\alpha)_{il,1} + (\alpha\alpha)_{jk,1} + (\alpha\alpha)_{jl,1}] \\ &+ [(\alpha\delta)_{ikl,1} + (\alpha\delta)_{jkl,1} + (\alpha\delta)_{ijk,1} + (\alpha\delta)_{ijl,1}] + (\delta\delta)_{ijkl,1} + \cdots \end{aligned} \tag{21.1a}$$

$$\begin{aligned} G_2 = \mu_{G,2} &+ [\alpha_{i,2} + \alpha_{j,2} + \alpha_{k,2} + \alpha_{l,2}] + [\delta_{ij,2} + \delta_{kl,2}] \\ &+ [(\alpha\alpha)_{ik,2} + (\alpha\alpha)_{il,2} + (\alpha\alpha)_{jk,2} + (\alpha\alpha)_{jl,2}] \\ &+ [(\alpha\delta)_{ikl,2} + (\alpha\delta)_{jkl,2} + (\alpha\delta)_{ijk,2} + (\alpha\delta)_{ijl,2}] + (\delta\delta)_{ijkl,2} + \cdots \end{aligned} \tag{21.1b}$$

As in Chapter 5, α and δ denote additive and dominance effects, i and j denote alleles at locus 1, and k and l denote alleles at locus 2. The genotypic values of each trait are composed of six components: the mean genotypic value for the population, the total deviation from this mean due to the additive effects of the four alleles, the additional deviations due to the dominance effects at the two loci, and additive $\times$ additive, additive $\times$ dominance, and dominance $\times$ dominance epistatic effects. Although they are written in terms of only two loci, the above expressions can be generalized to any number of loci.

Recalling that unlike terms (those having different subscripts) in Equations 21.1a,b are uncorrelated under random mating and gametic phase equilibrium (Chapter 5), the genetic variances for each trait can be written as

$$\sigma_G^2(1) = \sigma_A^2(1) + \sigma_D^2(1) + \sigma_{AA}^2(1) + \sigma_{AD}^2(1) + \sigma_{DD}^2(1) + \cdots \tag{21.2a}$$

$$\sigma_G^2(2) = \sigma_A^2(2) + \sigma_D^2(2) + \sigma_{AA}^2(2) + \sigma_{AD}^2(2) + \sigma_{DD}^2(2) + \cdots \tag{21.2b}$$

The genetic covariance can also be partitioned into components. Using Equations 21.1a,b and noting again that unlike terms are uncorrelated,

$$\sigma_G(1,2) = \sigma_A(1,2) + \sigma_D(1,2) + \sigma_{AA}(1,2) + \sigma_{AD}(1,2) + \sigma_{DD}(1,2) + \cdots \tag{21.3}$$

where, for example,

$$\sigma_A(1,2) = \sigma(\alpha_{i,1}, \alpha_{i,2}) + \sigma(\alpha_{j,1}, \alpha_{j,2}) + \sigma(\alpha_{k,1}, \alpha_{k,2}) + \sigma(\alpha_{l,1}, \alpha_{l,2}) + \cdots$$

is the additive genetic covariance between characters 1 and 2. The second component of Equation 21.3 is attributable to the covariance of dominance effects, and so on for the epistatic components of variance. Note that when the two traits are the same, Equation 21.3 reduces to 21.2a.

In the past several chapters, numerous methods for the estimation of variance components were covered. All of these methods are based on the same principle — that the expected phenotypic covariances between various kinds of relatives can be expressed as linear functions of causal components of genetic, and in some cases environmental, variance. These principles extend naturally to the estimation of causal components of phenotypic covariance, except that instead of comparing the same trait in two relatives, two different characters are compared — one in each relative. The expected phenotypic covariance of character 1 in individual x and character 2 in individual y follows naturally from the formulations of Cockerham (1954) and Kempthorne (1954):

$$\begin{aligned}\sigma_G(1_x, 2_y) = 2\Theta_{xy}\sigma_A(1,2) + \Delta_{xy}\sigma_D(1,2) + (2\Theta_{xy})^2\sigma_{AA}(1,2) \\ + 2\Theta_{xy}\Delta_{xy}\sigma_{AD}(1,2) + \Delta_{xy}^2\sigma_{DD}(1,2) + \cdots\end{aligned} \tag{21.4}$$

where Θ_{xy} is the coefficient of coancestry, and Δ_{xy} is the coefficient of fraternity of x and y (both defined as in Chapter 7). Note that when $x = y$, then $2\Theta_{xy} = \Delta_{xy} =$

1, and Equation 21.4 reduces to the total genetic covariance given in Equation 21.3.

ESTIMATION OF THE GENETIC CORRELATION

All of the regression and ANOVA techniques for estimating components of variance reviewed in the last four chapters extend readily to the decomposition of the covariance between two traits, as first recognized by Hazel (1943). Three of the most frequently used approaches will be covered here.

Pairwise Comparison of Relatives

We start with a method that is both conceptually and computationally simple, requiring only that data are available for pairs of relatives. Suppose, for example, that measures of traits 1 and 2 have been obtained for both midparents (denoted by x) and offspring means (denoted by y). Four types of phenotypic covariances can then be computed: trait 1 in midparents and offspring, trait 2 in midparents and offspring, trait 1 in midparents and 2 in offspring, and vice versa. The first two of these relate to the genetic variances of the traits, the second two to the genetic covariance between the traits. Ignoring possible contributions from common environmental effects, their expected values are respectively:

$$\sigma(z_{1x}, z_{1y}) = \frac{\sigma_A^2(1)}{2} + \frac{\sigma_{AA}^2(1)}{4} + \cdots \tag{21.5a}$$

$$\sigma(z_{2x}, z_{2y}) = \frac{\sigma_A^2(2)}{2} + \frac{\sigma_{AA}^2(2)}{4} + \cdots \tag{21.5b}$$

$$\sigma(z_{1x}, z_{2y}) = \frac{\sigma_A(1,2)}{2} + \frac{\sigma_{AA}(1,2)}{4} + \cdots \tag{21.5c}$$

$$\sigma(z_{2x}, z_{1y}) = \frac{\sigma_A(1,2)}{2} + \frac{\sigma_{AA}(1,2)}{4} + \cdots \tag{21.5d}$$

These expressions are arrived at by use of Equation 21.4, after substituting the midparent-offspring measures of relatedness, $\Theta_{xy} = 1/4$ and $\Delta_{xy} = 0$. Assuming negligible epistatic effects, the sum of the "cross-covariances," $\sigma(z_{1x}, z_{2y}) + \sigma(z_{2x}, z_{1y})$, is equal to the additive genetic covariance. Each additive genetic variance is equivalent to twice the respective within-character covariance, i.e., $\sigma_A^2(1) = 2\sigma(z_{1x}, z_{1y})$. Thus, an approximation of the additive genetic correlation based on midparent-offspring analysis is

$$\rho_A \simeq \frac{\sigma(z_{1x}, z_{2y}) + \sigma(z_{2x}, z_{1y})}{2\sqrt{\sigma(z_{1x}, z_{1y}) \cdot \sigma(z_{2x}, z_{2y})}} \tag{21.6a}$$

Substitution of observed for expected covariances yields the estimate r_A. An alternative to Equation 21.6b involves the geometric (rather than arithmetic) mean covariance,

$$\rho_A \simeq \sqrt{\frac{\sigma(z_{1x}, z_{2y}) \cdot \sigma(z_{2x}, z_{1y})}{\sigma(z_{1x}, z_{1y}) \cdot \sigma(z_{2x}, z_{2y})}} \tag{21.6b}$$

However, there are two reasons why this estimator is less desirable than Equation 21.6a. First, since geometric means are always less than arithmetic means, Equation 21.6b will tend to yield biased correlation estimates that are closer to zero than those generated with Equation 21.6a. Second, if one estimate of the genetic covariance is negative and the other positive, Equation 21.6b is undefined.

Equation 21.6a is general in that, as a first approximation, it applies to any set of relatives with constant degrees of relationship. For example, x and y could represent the two members of a pair of dizygotic twins. Alternatively, x might represent the mean of several members of a sib group and y that of the remaining (nonoverlapping) members. The sib groups can even consist of mixtures of half and full sibs, as is often the case in wild-caught gravid females. This generality follows from the fact that the coefficient Θ_{xy} in the expressions $\sigma(z_{1x}, z_{1y})$, $\sigma(z_{2x}, z_{2y})$, $\sigma(z_{1x}, z_{2y})$, and $\sigma(z_{2x}, z_{1y})$ is always the same, and therefore cancels out in Equations 21.6a,b.

Nonetheless, the above formulation has the same kinds of uncertainties that we have encountered in estimators of heritability. First, only in the absence of nonadditive genetic variance and common environmental effects does Equation 21.6a reduce exactly to the additive genetic correlation, $\sigma_A(1,2)/[\,\sigma_A(1)\sigma_A(2)\,]$. Bias from dominance can be eliminated entirely by using relatives with $\Delta_{xy} = 0$ (such as parent-offspring or half-sib pairs). The presence of epistatic genetic variance and/or common environmental effects in x and y will inflate the estimates of the additive genetic variances, but since covariances can be positive or negative, the same complications may bias estimates of the additive genetic covariance in either direction. Thus, the directional effect of confounding factors on estimates of the additive genetic correlation remains uncertain in most cases. A second undesirable feature of Equation 21.6a is that because it is not actually a product-moment correlation, it can sometimes yield estimates that are outside of the true range of ± 1, especially when sample sizes are small.

Nested Analysis of Variance and Covariance

The nested full-sib, half-sib design provides an alternative approach to estimating genetic correlations. Recall that with this design, several different females are mated to each sire, and a nested analysis of variance yields estimates of the additive genetic variances (Chapter 18). A parallel analysis can also provide an estimate of the additive genetic covariance. Analysis of covariance is identical in form to analysis of variance except that the former employs mean cross-products of the deviations of traits 1 and 2 rather than mean squared deviations of individual traits (Table 21.1). A lucid overview of the procedure is given by Grossman and Gall (1968).

Table 21.1 Summary of a nested analysis of covariance involving N sires, M_i dams within the ith sire, and n_{ij} offspring within the ijth full-sib family.

Factor	df	Sums of Cross-products	E(MCP)
Sires	$N-1$	$\sum_i^N \sum_j^{M_i} n_{ij}(\overline{z}_{1i}-\overline{z}_1)(\overline{z}_{2i}-\overline{z}_2)$	$\sigma_e(1,2)+k_2\sigma_d(1,2)$ $+k_3\sigma_s(1,2)$
Dams (sires)	$N(\overline{M}-1)$	$\sum_i^N \sum_j^{M_i} n_{ij}(\overline{z}_{1ij}-\overline{z}_{1i})(\overline{z}_{2ij}-\overline{z}_{2i})$	$\sigma_e(1,2)+k_1\sigma_d(1,2)$
Sibs (dams)	$T-N\overline{M}$	$\sum_i^N \sum_j^{M_i} \sum_k^{n_{ij}} (z_{1ijk}-\overline{z}_{1ij})(z_{2ijk}-\overline{z}_{2ij})$	$\sigma_e(1,2)$
Total	$T-1$	$\sum_i^N \sum_j^{M_i} \sum_k^{n_{ij}} (z_{1ijk}-\overline{z}_x)(z_{2ijk}-\overline{z}_y)$	$\sigma_z(1,2)$

$$\sigma_s(1,2) \simeq \frac{\sigma_A(1,2)}{4} + \frac{\sigma_{AA}(1,2)}{16}$$

$$\sigma_d(1,2) \simeq \frac{\sigma_A(1,2)}{4} + \frac{\sigma_D(1,2)}{4} + \frac{3\sigma_{AA}(1,2)}{16} + \frac{\sigma_{AD}(1,2)}{8} + \frac{\sigma_{DD}(1,2)}{16}$$

$$\sigma_e^2(1,2) \simeq \frac{\sigma_A(1,2)}{2} + \frac{3\sigma_D(1,2)}{4} + \frac{3\sigma_{AA}(1,2)}{4} + \frac{7\sigma_{AD}(1,2)}{8} + \frac{15\sigma_{DD}(1,2)}{16} + \sigma_{E_s}(1,2)$$

Note: T is the total number of individuals in the experiment, and $\overline{M}$ the mean number of dams/sire. For character 1, z_{1ijk} is the observed phenotype of the *k*th offspring of the *j*th dam mated to the *i*th sire, $\overline{z}_{1ij}$ is the mean phenotype of the *ij*th full-sib family, $\overline{z}_{1i}$ is the mean phenotype of all progeny of the *i*th sire, and $\overline{z}_1$ is the mean phenotype of all individuals. Similar notation is used for character 2. MCP denotes a mean cross-product, obtained by dividing a sum of cross-products by its respective degrees of freedom. The coefficients k_1, k_2, and k_3 are defined in Table 18.3.

Although most nested analyses are not perfectly balanced, it is usually preferable to use only individuals for which measures of both characters are available in the analysis, so that estimates of both the variances and the covariance are based on the same sample. Obviously, when a large number of individuals are missing one measure, this has the unfortunate side-effect of making the variance estimates less accurate than they would be otherwise, so in extreme cases it may be preferable to use all of the data.

The nested analysis of variance (covariance) yields nine mean squares (cross-products) — at the sire, dam, and progeny levels, for the variance of character 1, the variance of character 2, and the covariance of 1 and 2. From the observed mean squares and cross-products, and the standard expressions for their expectations (Table 21.1), estimates for the sire, dam, and within-family components of variance and covariance can then be extracted by the method of moments (equating mean squares with their expectations and solving). Recall from Chapter 18 that in a univariate analysis, the sire component of variance is equivalent to the covariance of paternal half sibs, thereby providing an estimate of one-fourth the additive genetic variance for the trait, assuming sources of epistatic variance are of negligible significance. Similarly, in an analysis of covariance, the sire component provides an estimate of one-fourth the additive genetic covariance between the two traits. Thus, letting $\text{Var}(s_1), \text{Var}(s_2)$, and $\text{Cov}(s_1, s_2)$ denote estimates of the sire components of variance and covariance, the genetic correlation is estimated by

$$r_A = \frac{\text{Cov}(s_1, s_2)}{\sqrt{\text{Var}(s_1)\text{Var}(s_2)}} \tag{21.7}$$

As discussed in the previous section, this measure of the genetic correlation can only be taken to be an approximation, since additive epistatic interactions are potentially included in the estimates of the genetic variances and covariance. However, the reliance on paternal half sibs should minimize the complications that can arise from common-environment effects. Like regression analysis, analysis of variance can also yield estimates of the genetic correlation that are outside the range of true possibilities (−1,1). The likelihood of this situation happening can be substantial, as Hill and Thompson (1978) have shown for one-way (nonnested) ANOVA of sib families. For example, if the intraclass correlations for both traits are $t = 0.0625$ (assuming only additive genetic variance, this implies $h^2 = 0.25$ or 0.125, depending on whether families consist of half or full sibs), and 160 families of 5 sibs are analyzed, the probability of obtaining a genetic correlation or heritability out of bounds is only about 0.06. However, if $t = 0.025$ ($h^2 = 0.10$ or 0.05), with the same sample sizes, the probability of obtaining an unrealistic estimate is nearly 0.5.

In principle, as in the case of components of variance, the various components of a genetic covariance can be extracted by comparison of the cross-covariances between different types of relatives, e.g., full vs. half sibs in the nested design. It would then be possible to procure estimates of genetic correlations due to dominance and various forms of epistasis in addition to additive effects. However, previous chapters have amply demonstrated the difficulties in accomplishing such partitioning with components of genetic variance with any reasonable degree of accuracy. Since the sampling variance of cross-covariances is also very high (see below), there appears to be little hope of a further dissection of the genetic correlation unless the number of families sampled is in the thousands.

Regression of Family Means

Because of technical difficulties in acquiring genetic correlation estimates and in testing for their significance, a number of investigators have opted to use the correlation between family mean phenotypes as a surrogate for estimating ρ_A. Here one simply regresses the family mean phenotype of character 1 on that of character 2, using the *same* individuals to compute each mean. The rationale for such an approach is that as the size of a family increases, the sampling error of the mean becomes diminishingly small, leaving the family mean phenotype as an estimate of the family mean genotypic value. However, if the heritability of either trait is low, this approach can yield misleading results because the variances and covariances of family means will be biased estimates of the additive genetic expectations. Although we do not advocate this approach, we elaborate on it somewhat to illustrate the interpretative difficulties that can arise.

Suppose each family consists of a group of n individuals, all related with coefficient of coancestry Θ (for example, a group of paternal half-sibs), and let $\overline{z}_{1i}$ be the mean phenotype of the ith family and z_{1ij} be the phenotype of the jth member of that family. Assuming that all of the resemblance between relatives is a consequence of additive gene action (in particular, that there are no shared environmental effects), the expected variance among family means for character 1 can then be expressed as

$$\sigma^2(\overline{z}_{1i}) = \frac{1}{n^2}\sigma^2\left(\sum_{j=1}^{n} z_{1ij}\right) = \frac{1}{n}\sigma^2(z_{1ij}) + \frac{n(n-1)}{n^2}\sigma(z_{1ij}, z_{1ik})$$

$$= \frac{1}{n}[\,\sigma_z^2(1) - 2\Theta\sigma_A^2(1)\,] + 2\Theta\sigma_A^2(1) \tag{21.8a}$$

where $\sigma_z^2(1)$ is the phenotypic variance of the trait. The same logic gives the expected covariance between family means for traits 1 and 2 as

$$\sigma(\overline{z}_{1i}, \overline{z}_{2i}) = \frac{1}{n}[\,\sigma_z(1,2) - 2\Theta\sigma_A(1,2)\,] + 2\Theta\sigma_A(1,2) \tag{21.8b}$$

After some algebraic rearrangement, and continuing to ignore all sources of variation and covariation except those due to additive genetic effects, the correlation of family means is found to be

$$\rho_{\overline{z}} \simeq \frac{\sigma(\overline{z}_{1,i}, \overline{z}_{2,i})}{\sigma(\overline{z}_{1,i}) \cdot \sigma(\overline{z}_{2,i})} = \rho_A \left[\frac{\phi h_1 h_2 + (\rho_z/\rho_A)}{\sqrt{(\phi h_1^2 + 1)(\phi h_2^2 + 1)}} \right] \tag{21.9}$$

where $\phi = 2\Theta(n-1)$, $h_1^2 = \sigma_A^2(1)/\sigma_z^2(1)$ and $h_2^2 = \sigma_A^2(2)/\sigma_z^2(2)$ are the heritabilities of the traits, and ρ_z is the phenotypic correlation. The quantity in brackets defines the factor by which the regression of family means deviates from the desired value ρ_A. The amount of bias can be seen to depend on ϕ, h_1^2, h_2^2, and on the ratio of phenotypic to genetic correlations.

In the extreme case in which the heritabilities of both traits are equal to one, the genetic and phenotypic correlations are equal (as there is no environmental variance), and the correlation of family means provides an unbiased estimate of the genetic correlation, i.e., $\rho_{\overline{z}} = \rho_A$. However, such concordance is unlikely to arise under any other circumstances. More generally, in order for the correlation between family means to closely approximate the genetic correlation, both ϕh_1^2 and ϕh_2^2 must be substantially larger than one and than $|\rho_z/\rho_A|$. Even if the two traits have moderately high heritabilities, the first condition requires large family sizes. Consider, for example, the situation in which both traits have heritabilities of 0.5, 50 paternal half sibs ($2\Theta = 0.25$) are sampled per family, and the phenotypic correlation is five times larger than the genetic correlation. Substituting for the quantities in the brackets, we find that the correlation of family means inflates the estimate of the genetic correlation by a factor of 1.8.

A perceived advantage of the family-mean approach is that $\rho_{\overline{z}}$ is a true product-moment correlation. Thus, unlike the other estimators described above, the correlation among family means cannot exceed ± 1, and its significance can be evaluated in a straightforward manner using standard tables of critical values for the sample correlation coefficient. However, the actual utility of this property seems questionable, given the uncertainty of what $\rho_{\overline{z}}$ actually measures.

A more reasonable path to estimating the genetic correlation from family means involves a combination of univariate ANOVA with covariance analysis, as follows. If each family is divided into two independent groups, one used to estimate the mean of character 1 and the other for character 2, the expected covariance between means is simply $2\Theta\sigma_A(1,2)$. Assuming common family environment is not a significant source of variation, there is no bias from environmental covariance because the two groups being compared contain different individuals. Combining the resultant estimate of $\sigma_A(1,2)$ with estimates of $\sigma_A^2(1)$ and $\sigma_A^2(2)$ obtained by ANOVA provides a basis for a relatively unbiased estimate of ρ_A.

COMPONENTS OF THE PHENOTYPIC CORRELATION

As noted in the introduction, phenotypic covariance between two traits arises from both genetic and environmental causes. We have just seen how the basic machinery for estimating heritabilities can be extended to the estimation of genetic correlations. However, the environmental correlation, ρ_E, can only be calculated *directly* under a very special set of circumstances. If a collection of genetically homogeneous individuals (either a highly inbred line or a single clone) is used, so there is no genetic variance among individuals within the group, the phenotypic and environmental correlations are equivalent.

Ordinarily, a more circuitous route to the estimation of ρ_E is necessary. The usual approach is to estimate the phenotypic and genetic correlations first, and to extract the environmental correlation from the algebraic relationship between the three quantities. The phenotypic correlation between two traits is easily acquired, as it is simply the correlation of the measures of the two traits in the same individual,

$$\rho_z = \frac{\sigma_z(1,2)}{\sigma_z(1) \cdot \sigma_z(2)} \tag{21.10}$$

where $\sigma_z(1)$ and $\sigma_z(2)$ are the phenotypic standard deviations of the two traits and $\sigma_z(1,2)$ is the covariance of traits 1 and 2 in the same individual. The relationship between the three types of correlations is fairly easily derived. Noting that covariances are additive (Chapter 3) and assuming that all covariance is due to additive genetic and special environmental effects, $\sigma_z(1,2) = \sigma_A(1,2) + \sigma_E(1,2)$, where

$$\sigma_A(1,2) = \rho_A \sigma_z(1) h_1 \sigma_z(2) h_2$$
$$\sigma_E(1,2) = \rho_E \left(\sigma_z(1)\sqrt{1-h_1^2}\right)\left(\sigma_z(2)\sqrt{1-h_2^2}\right)$$

Equation 21.10 then expands to

$$\rho_z = h_1 h_2 \rho_A + \rho_E \sqrt{(1-h_1^2)(1-h_2^2)} \tag{21.11}$$

rearrangement of which leads to

$$\rho_E = \frac{\rho_z - h_1 h_2 \rho_A}{\sqrt{(1-h_1^2)(1-h_2^2)}} \tag{21.12}$$

(An alternative derivation based on path analysis is provided in Appendix 2.) Estimates of the environmental correlation are obtained by substituting observed for expected quantities in this expression.

The derivation of Equation 21.12 assumes zero covariance between the genetic value of trait 1 in individual x and the environmental deviation of trait 2 in relative y, and vice versa, a reasonable assumption if maternal effects can be ruled out. This problem aside, it should also be emphasized that because Equations 21.6, 21.7, and 21.9 only provide approximations of the additive genetic correlation, Equation 21.12 yields only an approximation of the environmental correlation. All of the nonadditive genetic variance and covariance that is not included in the estimation of ρ_A will contribute to ρ_E. For example, when the genetic covariance is estimated by twice the covariance of offspring and midparents, the actual composition of the excess "environmental" covariance is

$$\sigma_z(1,2) - \left[\sigma_A(1,2) + \frac{\sigma_{AA}(1,2)}{2}\right] = \sigma_E(1,2) + \sigma_D(1,2) + \frac{\sigma_{AA}(1,2)}{2} + \sigma_{AD}(1,2) + \cdots$$

Since any of the covariance terms can be positive or negative, estimates of environmental correlations can be biased in either direction by nonadditive gene action.

Phenotypic Correlations as Surrogate Estimates of Genetic Correlations

Because of the inherent difficulties in estimating additive genetic correlations, it is of great interest to know if these have a strong tendency to reflect the more easily acquired phenotypic correlations in magnitude and/or sign. If this were true, phenotypic correlations would provide useful, and more accessible, insight into the directionality of constraints on multivariate evolution. Moreover, because phenotypic correlations can normally be estimated with a high degree of accuracy, while genetic correlations usually have very large standard errors, if the parametric values of genetic and phenotypic correlations tended to be equal, then estimates of the phenotypic correlation could more closely approximate the true genetic correlation than the genetic correlation estimate itself.

It has been noticed that estimates for genetic correlations tend to slightly exceed phenotypic correlations in absolute magnitude (Searle 1961, Kohn and Atchley 1988, Koots et al. 1994). But a broader analysis indicates that this may be due to biases that arise with small sample sizes. When the "effective number of families," Nh_1h_2, where N is the actual number of families, exceeds 50 or so, the average difference between the two types of correlation becomes negligible (Cheverud 1988).

Broad surveys of the literature have led Cheverud (1988, 1995) and Roff (1995, 1996) to the conclusion that ρ_z and ρ_A not only normally have the same sign, but are also of the same magnitude (Figure 21.1). The pattern appears to be particularly clear for morphological (as opposed to life-history) characters (Roff 1996, Simons and Roff 1996). Few others have been bold enough to make the assertion that $\rho_z \simeq \rho_A$, and Willis et al. (1991) point out several reasons why the generality of such a statement should be treated with caution. Certainly, it is still an open question as to whether environmental factors that jointly influence two traits operate through the same biochemical/developmental pathways as pleiotropic genetic factors, and cases do exist in which the estimates r_z and r_A differ in sign (Mousseau and Roff 1987) and magnitude (Hébert et al. 1994). Nevertheless, the similarities between existing estimates of genetic and phenotypic correlations are striking. Because the latter is a function of the former, some correspondence is expected just on the basis of sampling error, but it seems unlikely that this accounts for the entire pattern. Further in-depth study of this fundamentally important problem is certainly in order.

STATISTICAL ISSUES

As with any parameter estimates, it is useful to have measures of the sampling variances and/or confidence intervals of the phenotypic, genetic, and environ-

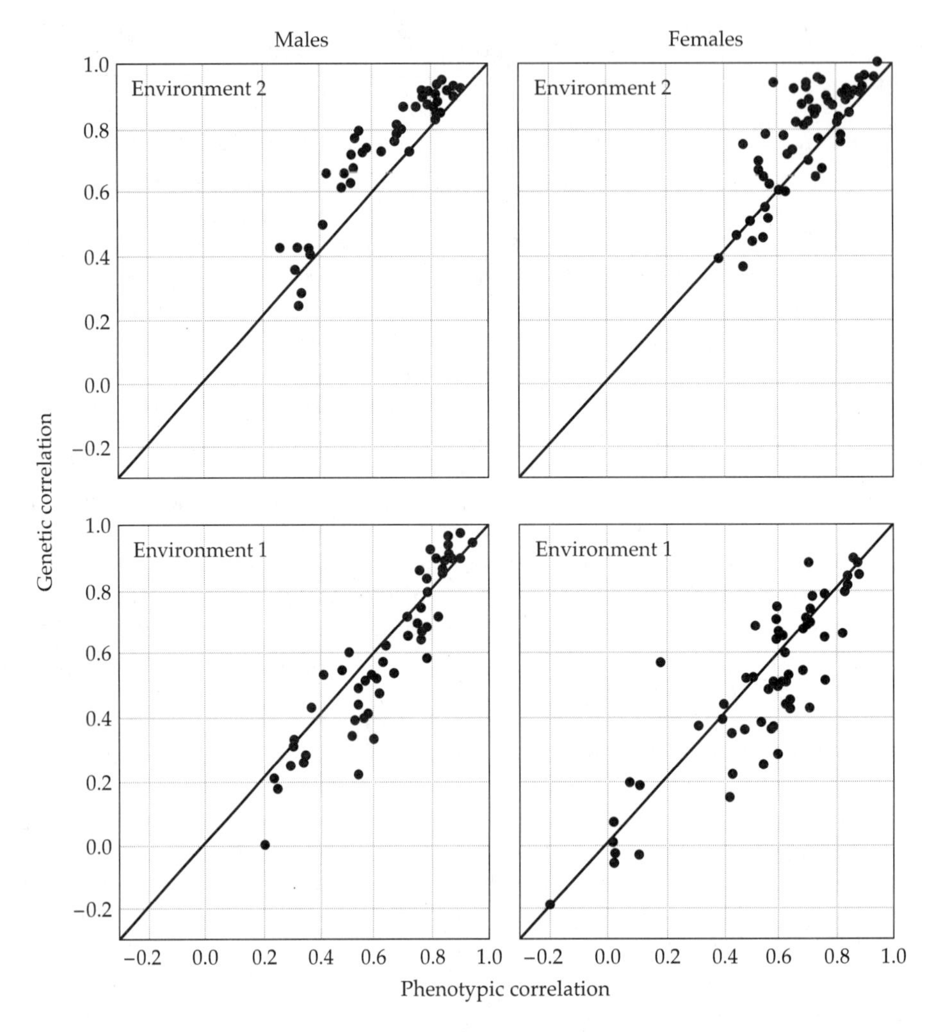

Figure 21.1 Comparison of genetic vs. phenotypic correlations for pairs of morphological traits in the sand cricket, *Gryllus firmus*. Results are given for the two sexes, raised in two environments that varied in temperature and photoperiod. (From Roff 1995.)

mental correlations to aid in their interpretation. For the genetic and environmental correlations, the problems here are considerable. The estimators for ρ_A and ρ_E generally utilize a combination of results from different applications of ANOVA and/or regression analysis, all employing data on the same individuals. The sampling properties of functions of statistics derived from nonindependent analyses are poorly understood.

Hypothesis Tests

Testing for a significant phenotypic correlation is straightforward, as it is a conventional product-moment correlation, which can be evaluated against critical values widely available in tables in statistics texts. For genetic correlations, the simplest option for testing the hypothesis that $\rho_A = 0$ is based on the principle that significant genetic covariance between two traits implies a significant genetic correlation. With the pairwise-comparison method, the significance of the regression of character 2 in set y on character 1 in set x, and vice versa, can be evaluated by the standard test for a regression slope.

Methods for evaluating the significance of an environmental correlation are less well developed. However, with the nearly universal availability of high-speed computers, a simple procedure enables simultaneous tests for all three types of correlation (r_z, r_A, and r_E). Bootstrapping and jackknifing over families (Chapter 18) are being relied upon increasingly to generate empirically derived sampling distributions of the desired statistics (Dorn and Mitchell-Olds 1991, Brodie 1993, Paulsen 1994, Roff and Preziosi 1994). For each quasisample of the data set, the estimates r_z and r_A are computed directly, and then r_E is obtained by use of Equation 21.11. After randomly generating numerous sets of such estimates, one can construct confidence intervals for the three parameters from their observed sampling distributions. With the paired-comparison method, an alternative procedure is to randomize members of the set y with respect to those of set x, evaluating the probability (under the null hypothesis of no correlation) of obtaining a correlation coefficient as extreme as that observed with the true data set. A similar approach can be applied to the nested design, by randomizing full-sib families with respect to sires and dams.

Finally, we note that studies of genetic correlation usually involve the simultaneous analysis of several characters, not just two, resulting in tables of correlations between all possible pairs of characters. Care then needs to be taken so as not to overinterpret the significance levels attached to single correlations. For example, suppose that one were studying a set of N traits, none of which are actually correlated. The probability that none of the $N(N-1)/2$ observed correlations is significant (at level α) is $(1-\alpha)^{N(N-1)/2}$. The probability that at least one correlation would appear, by chance, to be significant at the level α or smaller is one minus this quantity. This probability can be substantial — if $N = 7$, there is a 0.19 probability that at least one of the 21 correlations will spuriously appear to be significant at the $P = 0.01$ level, and a 0.66 probability at the $\alpha = 0.05$ level. Adjusting significance tests to account for multiple comparisons is straightforward when the different tests involve independent data (Rice 1989; see Chapter 14). However, in the analysis of correlated characters, the nonindependence of data renders conventional multiple-comparison procedures invalid, and to our knowledge no satisfactory solution to the problem exists.

Standard Errors

By Taylor expansion (Appendix 1), Reeve (1955) first obtained an expression for the approximate large-sample variance of r_A for the single parent-offspring regression, and Hammond and Nicholas (1972) subsequently generalized this to include all types of parent-offspring combinations:

$$\mathrm{Var}(r_A) \simeq \frac{1}{N}\left[\frac{(1-r_A^2)^2}{2} + \frac{A(n+k)(1-r_A^2)(B-Cr_z r_A)}{4k} + \frac{2A(Br_A - Cr_z)^2}{k} + \frac{A[C^2(1-r_A^2)(1-r_z^2)-(B/2)+(Cr_A r_z/2)]}{k}\right] \tag{21.13}$$

where $A = 1$ or 2 for regressions involving midparents or single parents (respectively), $B = [(1/h_1^2)+(1/h_2^2)]/2$, $C = 1/(h_1 h_2)$, N is the number of families, n is the number of offspring/dam in each family, and k is the number of offspring/sire in each family. For midparent-offspring regressions, $n = k$ is the family size, whereas in the regression of paternal half-sibs on fathers, $n = 1$ and k is the number of dams/sire. Reeve (1955) provides an expression for n for use in unbalanced designs, and VanVleck and Henderson (1961) present an equation for the case in which only a single character is measured in each individual (such as character 1 in parents and character 2 in offspring).

The preceding formula serves as a useful guide in choosing an adequate experimental design. If the constraint on the investigator is the total number of individuals that can be measured, $T = Nk$, then it can be seen that $\mathrm{Var}(r_A)$ is minimized by maximizing the number of families (N). (This approach minimizes the first term in the equation, while the remaining terms, whose denominators are $T = Nk$, are unaffected.) Thus, the optimal design for estimating a genetic correlation is to measure a single offspring from as many families as possible. This recommendation is similar to that for heritability estimation based on parent-offspring regression (Chapter 17).

VanVleck and Henderson (1961) and Brown (1969) used simulation studies to evaluate the degree of accuracy in using Equation 21.13 to estimate the standard error of a genetic correlation by substituting observed for expected quantities. For $N < 100$, they found that estimates from Equation 21.13 are biased downwards and may underestimate the true sampling variance by as much as an order of magnitude with appreciable frequency. However, if N is on the order of 1,000, and the true value of r_A is not very nearly ± 1.0, Equation 21.13 is quite accurate.

VanVleck and Henderson (1961) and Brown (1969) also examined the sampling distribution of r_A. Provided gene action is additive, Equation 21.6a appears to yield an unbiased estimate of r_A, and provided the true value ρ_A is not very nearly ± 1.0, the sampling distribution of r_A approximates normality as sample sizes become large. Thus, when N is large, $2\sqrt{\mathrm{Var}(r_A)}$ usually can be taken as

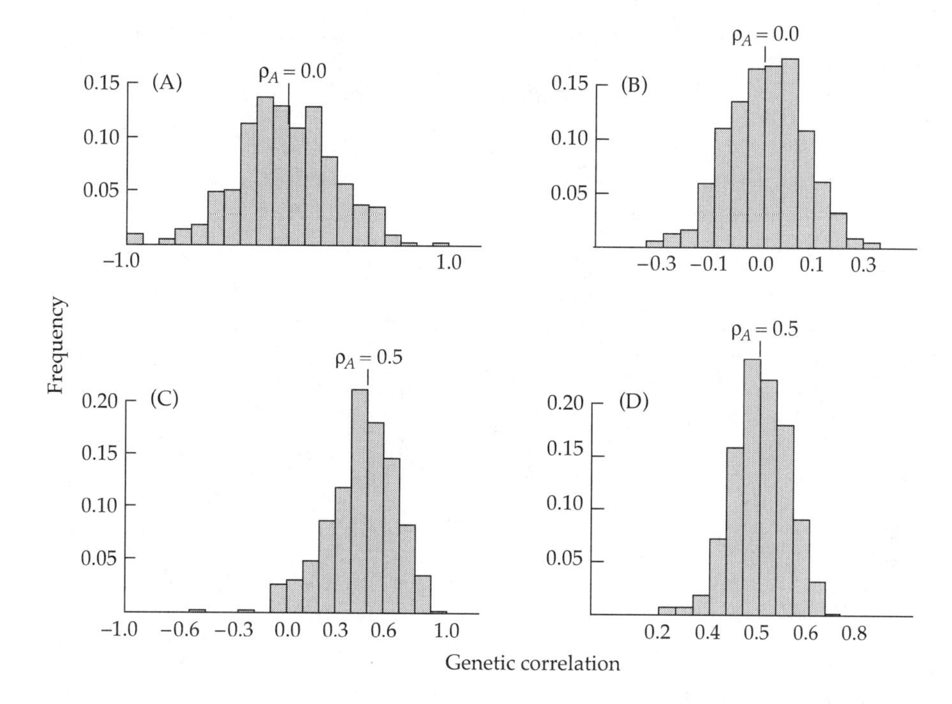

Figure 21.2 Sampling distributions of the additive genetic correlation estimated with regression of single offspring on N single parents, using simulated data sets. In all cases, both characters have heritabilities equal to 0.4. Values of the underlying parameters are: (A) $\rho_z = 0.0$, $\rho_A = 0.0$, $N = 200$; (B) $\rho_z = 0.0$, $\rho_A = 0.0$, $N = 1000$; (C) $\rho_z = 0.5$, $\rho_A = 0.5$, $N = 200$; (D) $\rho_z = 0.5$, $\rho_A = 0.5$, $N = 1000$. (From Brown 1969.)

an estimate of the 95% confidence interval for r_A. It is clear from Figure 21.2 that unless the number of families analyzed is on the order of 1,000 (for single parent-single offspring regressions), the standard error of r_A will be quite large. Certainly, experiments that involve fewer than several hundred individuals should be avoided if at all possible (see also Klein 1974).

Several attempts have also been made to obtain expressions for the large-sample variance of correlation coefficients obtained from nested full-sib, half-sib analyses (Mode and Robinson 1959, Robertson 1959b, Tallis 1959, Scheinberg 1966, Abe 1969, Grossman 1970, Hammond and Nicholas 1972, Grossman and Norton 1974). The algebra is quite tedious, and a number of the early papers contain errors or are rather restrictive in their applicability. The most general expression, derived by Hammond and Nicholas (1972), is rather complex, but can be summarized as

$$\text{Var}(r_i) \simeq 2\left(\frac{r_i}{e}\right)^2 \left(\frac{a^2 S}{\text{df}_s + 2} + \frac{b^2 D}{\text{df}_d + 2} + \frac{c^2 W}{\text{df}_w + 2}\right) \tag{21.14a}$$

where i denotes z, A, or E depending on which type of correlation is being considered, and df_s, df_d, and df_w refer to the degrees of freedom at the sire, dam, and progeny levels. The terms S, D, and W are of the form

$$\left(\frac{Z_1}{V_1} - \frac{Z_{1,2}}{C_{1,2}}\right)^2 + \left(\frac{Z_2}{V_2} - \frac{Z_{1,2}}{C_{1,2}}\right)^2 + 2\left(\frac{Z_{1,2}}{V_1} - \frac{Z_2}{C_{1,2}}\right)\left(\frac{Z_{1,2}}{V_2} - \frac{Z_1}{C_{1,2}}\right) \tag{21.14b}$$

where Z_1, Z_2, and $Z_{1,2}$ refer to the mean squares and cross-products of characters 1 and 2 at the level of sires (when calculating S), dams (when calculating D), and replicates (when calculating W). V_1, V_2, and $C_{1,2}$ denote the estimates of the variances and covariances at the phenotypic, genetic, and environmental levels, depending upon which correlation is being dealt with; these estimates are derived by the usual route of the method of moments. Finally, the constants a, b, c, and e depend upon the nature of the correlation as follows:

	a	b	c	e
Phenotypic	k_1	$k_3 - k_2$	$k_2 - k_1 + k_3(k_1 - 1)$	$k_1 k_3$
Genetic	1	$-k_2/k_1$	$(k_2 - k_1)/k_1$	$k_3/4$
Environmental	-2	$2k_2/k_1$	$[(k_1 - k_2)/k_1] + k_3$	k_3

with the k_i coefficients being defined in Table 18.3.

Robertson (1959b) and Tallis (1959) have considered the optimal design for estimating the genetic correlation from a nested analysis of variance and covariance, concluding that the design that minimizes the sampling variance of the heritabilities also applies to the genetic correlations. Thus, the recommendations of Chapter 18 may be referred to.

The important message of this section is that attaining a reasonable degree of confidence in any study of genetic correlation requires a very substantial data base. Often, with sample sizes less than a few hundred individuals, the strongest statement that can be made is whether the correlation is significantly positive or negative. The conventional approach in regression analysis, and the one that we focused on in the preceding paragraphs, is to take $\rho = 0$ as the null hypothesis. For genetic studies concerned with constraints on the evolutionary process, an alternative is to let $\rho_A = \pm 1$ be the null hypothesis and to evaluate it against the observed data using resampling procedures.

Bias Due to Selection

Selection in the parental generation on the characters of interest or any other characters correlated with them can lead to biased estimates of the genetic correlation by altering the variances and covariances relative to the expectations prior to selection (Van Vleck 1968, Robertson 1977b, Meyer and Thompson 1984). In principle, this problem can be significant in studies of wild populations exposed to natural selection. Here we consider how serious the bias can be and how it might be corrected for.

Utilizing an early result of Pearson (1903), Lande and Price (1989) showed how the bias can be eliminated under the assumption that the joint distribution of the characters in parents and offspring is multivariate normal in the absence of selection. Under those conditions, the partial regression coefficients in a multiple regression of offspring on parent characters are unaffected by selection, provided all of the characters under selection are actually included in the analysis. Letting **C** denote the matrix of covariances between characters in unselected offspring and parents, and **P** be the phenotypic variance-covariance matrix in the unselected parents, then Pearson's result implies

$$\mathbf{C}\mathbf{P}^{-1} = \mathbf{C}_s\mathbf{P}_s^{-1} \tag{21.15a}$$

with the subscript s denoting matrices after selection. (Recall from Chapter 8 that partial regression coefficients are obtained as the product of the covariance matrix involving predictor and response variables and the inverse of the variance-covariance matrix for the predictor variables.) Rearranging, the matrix of covariances between unselected parents and offspring can be expressed as

$$\mathbf{C} = \mathbf{C}_s\mathbf{P}_s^{-1}\mathbf{P} \tag{21.15b}$$

The genetic correlations that we wish to estimate are $\rho_A(i,j) = C_{ij}/\sqrt{C_{ii}C_{jj}}$, where C_{ij} denotes the element in row i and column j of **C**.

The above relationship shows that the observed covariances in $\mathbf{C}_s$ can be transformed into the desired elements of **C** if the phenotypic variance-covariance matrices before (**P**) and after ($\mathbf{P}_s$) selection are known. The latter is what we observe from the sampled parents. Lande and Price (1989) suggest that the elements of **P** might be obtained from the unselected offspring (of the selected parents) since a single generation of selection rarely causes a significant alteration in the phenotypic covariance structure of populations. Obviously, this approach is possible only if the forces of selection operating on the parents can be removed from the offspring, and no new ones are added, and if the sources of environmental variation contributing to **P** in the offspring generation can be kept the same as those in the parental generation. Such conditions may be difficult to achieve in many empirical settings.

As a simple example (from Lande and Price 1989) of the magnitude of bias in the genetic correlation that can be induced by selection, suppose that the variance of the first of two characters under investigation has been reduced by a fraction k by selection such that the phenotypic variance in the observed (selected) parents is $\sigma_s^2(1) = (1-k)\sigma_z^2(1)$, where $\sigma_z^2(1)$ is the phenotypic variance of character 1 before selection. Assume further that selection did not operate on any other correlated traits. Since the regression of character 2 on 1 accounts for a fraction ρ_z^2 of the variance of character 2 (Equation 3.17), a change in the variance of character 1

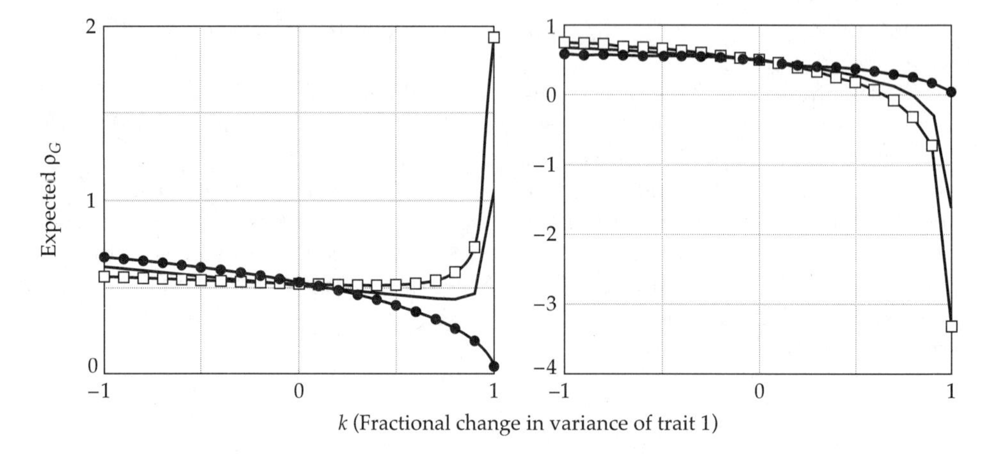

Figure 21.3 Bias in the expected genetic correlation between two characters when the phenotypic variance of character 1 in the parents has been modified to $(1-k)\sigma_z^2(1)$ but has been unaccounted for in the analysis. Solid circles: correlation based on covariance of offspring character 2 on parent character 1; open squares: correlation based on covariance of offspring character 1 on parent character 2; solid line: average. Left: $h_1^2 = 0.3,\ h_2^2 = 0.7$. Right: $h_1^2 = 0.7,\ h_2^2 = 0.3$. The true value ρ_A is 0.5 in both examples. (From Lande and Price 1989.)

equal to $-k\sigma_z^2(1)$ must induce a change in the variance of character 2 equal to $-k\rho_z^2\sigma_z^2(2)$. Therefore, the phenotypic variance for character 2 in the parents after selection is $\sigma_s^2(2) = (1-k\rho_z^2)\sigma_z^2(2)$. From Pearson's result, we know that selection does not change the regression coefficient, so if selection reduces the phenotypic variance of character 1 by the factor k, it must reduce the phenotypic covariance by the same factor, i.e., $\sigma_s(1,2) = (1-k)\sigma_z(1,2)$. Substituting these quantities into $\mathbf{P}_s$ and solving $\mathbf{C}_s = \mathbf{C}\mathbf{P}^{-1}\mathbf{P}_s$, we obtain the expected covariances between parents (p) and offspring (o) after selection,

$$\sigma_s(z_{1o}, z_{1p}) = \frac{\sigma_A^2(1)}{2}(1-k) \tag{21.16a}$$

$$\sigma_s(z_{2o}, z_{2p}) = \frac{\sigma_A^2(2)}{2}\left(1 - \frac{k\rho_A\rho_z h_1}{h_2}\right) \tag{21.16b}$$

$$\sigma_s(z_{1o}, z_{2p}) = \frac{\sigma_A(1,2)}{2}\left(1 - \frac{k\rho_z h_1}{\rho_A h_2}\right) \tag{21.16c}$$

$$\sigma_s(z_{2o}, z_{1p}) = \frac{\sigma_A(1,2)}{2}(1-k) \tag{21.16d}$$

assuming that additive effects are the only source of covariance between relatives.

These expressions illustrate several important points about estimates of genetic correlations in selected populations (Figure 21.3). First, selection that causes a change in the phenotypic variance of the parents will usually induce changes in all four of the covariances between parents and offspring.

Second, selection causes the two covariances between characters 1 and 2 to be unequal, i.e., $\sigma_s(z_{1o}, z_{2p}) \neq \sigma_s(z_{1p}, z_{2o})$. In some cases, especially when the heritability of the unselected trait is relatively low, the two measures of genetic covariance may actually differ in sign; this requires that $\rho_z h_1/(\rho_A h_2) > 1/k$. In principle, this property could provide a way of assessing whether selection has caused a significant bias in the parent sample, provided maternal effects can be ruled out. However, in light of the difficulties in procuring accurate estimates of covariances, such a test is not likely to be very powerful. In general, averaging the two types of covariances (Equations 21.16c and d) does not improve the situation much, since both are often biased in the same direction (Figure 21.3).

Third, in extreme cases, the *expected* value of the genetic correlation, as defined by Equation 21.6a, can exceed its theoretical limits of ± 1 (Figure 21.3). Combining this problem with the substantial sampling variance of genetic correlations, wildly unrealistic genetic correlations are possible with selected populations.

If any generalization can be drawn from these results, it is that the direction and magnitude of selection bias on the estimation of ρ_A is difficult to assess in the absence of prior information on the composite parameter $k\rho_z h_1/h_2$ and on ρ_A itself. (The situation is worse, of course, if selection is acting on both traits and/or on additional correlated traits.) Clearly, in situations where selection is likely to be a problem, application of Equation 21.15b prior to analysis is highly desirable.

The most feasible way to eliminate selection bias is to assay the study population in a highly protective environment, but this approach raises another fundamental issue. If the characters under investigation are sensitive to genotype $\times$ environment interaction (Chapter 22), then a change in environment may induce a real shift in ρ_A, so that one is no longer estimating the correlation of interest. Simons and Roff (1996) found that patterns of genetic correlations among morphological characters in crickets are essentially the same when estimated in constant laboratory vs. variable field conditions, although the correspondence among correlations involving life-history traits is less pronounced. On the other hand, Gebhardt and Stearns (1988) found that the genetic correlation between development time and weight at eclosion in *Drosophila mercatorum* changed sign from one environment to another. This issue is of special concern when one is most interested in quantifying genetic constraints in harsh environments, where selective mortality may be quite high and genetic constraints may play their most important role. Clearly, more work is needed on the degree to which genetic correlations (and covariances) respond to environmental changes.

APPLICATIONS

These warnings are not meant to be totally discouraging. Studies that are appropriately designed and meet with the appropriate precautions can yield substantial insight into the constraints on the evolution of multivariate phenotypes. The following examples will provide a feeling for the diversity of problems that can be evaluated with a genetic covariance analysis.

Genetic Basis of Population Differentiation

Ecologists are well aware that different populations of the same species often exhibit rather different diets. To a large extent this may simply reflect shifts in the relative availabilities of prey types in different areas. An alternative possibility is the existence of genetic differences in feeding behavior. Arnold (1981a–c) studied this issue in garter snakes (*Thamnophis*). In California, coastal populations of this snake are primarily terrestrial predators of slugs, while inland populations prey more exclusively on fish and amphibians. A dietary shift between these two areas is clearly necessary, since slugs are absent from inland habitats. Arnold made several observations that were consistent with genetic differences in the feeding habits of coastal and inland snakes. For example, about 75% of naive, newborn snakes from the coast would attack slugs in laboratory experiments, while only about 35% of the inland snakes would do so. The slug-refusing individuals were quite persistent in their decision, starving to death unless alternate prey were offered.

A standard laboratory test was devised to evaluate whether the divergence in prey preference was due to genetic differences in chemoreceptive responses. Cotton swabs were either rubbed on different prey or soaked in their extracts and presented to naive, newborn snakes. The number of tongue flicks/minute was then taken to be a measure of chemoreceptive response. As expected, the coastal population was much more receptive to slugs (Table 21.2). On the other hand, the inland population was not significantly more responsive to fish and amphibian odors than was the coastal population. Furthermore, the coastal snakes exhibited a much stronger response to leeches than did the inland snakes. This latter point was surprising, since leeches are unknown in the diets of coastal snakes.

Some insight into these results was provided by a genetic analysis of full-sib families obtained from field-collected gravid females (19 females with a total of 211 young in the inland population and 20 females with 102 young from the coast). Because of the full-sib design, the genetic variances and covariances may be biased by the presence of dominance, but common environmental effects were ruled out on the basis of prior experiments. Estimates for the heritabilities of chemoreceptive responses are given in Table 21.2 and for the genetic correlations in Table 21.3.

Table 21.2 A comparison of chemoreceptive responses to prey odors by newborn garter snakes, *Thamnophis elegans,* from coastal and inland California.

	Mean Tongue-flick Rate			Heritability	
	Coast	Inland	Difference	Coast	Inland
Slugs	30.9	4.3	1.72	0.2	0.2
Leeches	11.8	2.5	1.24	0.6	0.3
Salamanders	9.0	7.4	0.21	0.4	0.2
Frogs and tadpoles	30.5	25.8	0.17	0.4	0.3
Control	1.4	1.6	–0.26	0.0	0.1

Source: Arnold (1981a,c).
Note: The difference between population mean phenotypes is in units of standard deviations of ln-transformed values.

The estimated genetic correlation between responses to leeches and slugs is 0.9 for both populations. Thus, any evolutionary change in one of these responses is expected to cause a similar shift in the other through pleiotropy. This result helps explain the increase in receptivity to leeches for the coastal population, which has evolved in the direction of slug specialization. Arnold further suggests that the dichotomy between the two populations may be magnified by an evolutionary reduction in receptivity to leeches in the inland population. There is no positive selection for slug predation in this population because there are no slugs. There are leeches, however, and their consumption may be deleterious, since they often pass through the snake's gut alive, causing some damage in the process.

Table 21.3 A comparison of the genetic correlations for chemoreceptive responses to prey odors in coastal (above diagonal) and inland (below diagonal) populations of *Thamnophis elegans*.

	Sl	Le	Sa	Fr	Co
Slugs	—	0.9	1.0	0.9	0.0
Leeches	0.9	—	0.8	0.9	0.2
Salamanders	0.5	0.8	—	0.9	0.6
Frogs and tadpoles	0.6	0.4	0.2	—	0.2
Controls	-0.4	0.0	0.3	-0.3	—

Source: Arnold (1981b).
Note: Standard errors of the estimates are on the order of 0.3. Sl = slugs, Le = leeches, Sa = salamanders, Fr = frogs and tadpoles, and Co = control.

If the hypothesized selective pressures due to slugs and leeches are correct, then the similarities in responses to salamanders and frogs in the two populations can also be clarified. In the coastal population, chemoreceptive responses to slugs, salamanders, and frogs are almost perfectly genetically correlated so the relatively strong response to vertebrates may be largely a pleiotropic effect of selection for predation on slugs. The responses to leeches, salamanders, and frogs have lower, but still positive, genetic correlations in the inland population. Thus, an antagonism would exist between positive selection for predation on vertebrates and selection for avoidance of leeches.

This kind of reasoning would not have been reached had Arnold relied solely on phenotypic correlations. The phenotypic correlations between the various chemoreceptive responses were uniformly low in both populations, ≤ 0.3 in all but one case.

The Homogeneity of Genetic Covariance Matrices Among Species

As noted in the previous example, characters evolve in response to natural selection as a direct consequence of the forces of selection operating on the characters themselves and as an indirect consequence of selection operating on all genetically correlated traits (see Chapter 8). Thus, any attempt to project the long-term consequences of selection on specific characters is highly dependent on the degree of constancy of the genetic covariances over time. Such constancy is also required if much progress is to be made in retrospective evaluations of the evolutionary forces that may be responsible for observed changes in the fossil record (Reyment 1991). One approach to evaluating the stability of the genetic covariance matrix is to perform temporal surveys of genetic variances and covariances in individual populations. But such comparisons cannot usually be made on very long time scales. An alternative is to compare the genetic covariance structure of isolated populations or species. Similarity in this case would be consistent with long-term stability.

Lofsvold (1986) used the latter approach in an analysis of skull morphology in the white-footed mouse (*Peromyscus leucopus*) and two subspecies of the deer mouse (*P. maniculatus bairdii* and *P. m. nebrascensis*). The specimens were obtained from a museum collection of preserved skulls obtained from full-sib families of wild-caught individuals bred in the laboratory of L. R. Dice in the 1930s. Fifteen cranial characters were measured with calipers. After adjusting for sexual dimorphism (Chapter 24), the additive genetic variances and covariances were estimated from regressions of offspring means on paternal phenotype. Lofsvold used three approaches to compare the genetic covariance matrices. Two of these are explained relatively easily, while the third requires a rather advanced understanding of multivariate statistics and will not be considered here.

The first approach was to treat the corresponding elements of two genetic covariance matrices as paired observations and compute the ordinary correlation coefficient, r_M, between them. An r_M equal to 1.0 would then indicate perfect proportionality between the two matrices, while $r_M = 0.0$ would indicate a complete lack of correspondence. This is not a test of the absolute equality of two matrices, since $r_M = 1.0$ would arise if one matrix were simply a product of the second and a constant. Moreover, since the statistical distribution of r_M is unknown, it is not possible to attach any degree of confidence to r_M.

Lofsvold's second approach eliminates these difficulties, but in a rather arbitrary fashion. An index of similarity, γ, was computed by taking the sum of the cross-products of the corresponding elements of the two genetic covariance matrices being compared. The off-diagonal elements of one matrix were then randomly rearranged by rows, and a new index computed for the randomized matrix and the other, unaltered, matrix. The randomization procedure was repeated many times, yielding an empirical distribution of γ for randomly constructed matrices. The significance of the observed γ was then determined from the cumulative distribution of the randomly generated values of γ. For example, if a randomly generated γ greater than the observed γ arose at a frequency less than 5%, the hypothesis that the observed similarity is no greater than that expected by chance was rejected at the 0.05 level. Although this and other procedures involving the permutation of matrix elements have been used frequently to compare variance/covariance structures, the statistical and biological justification for their use has not been established.

The three different approaches taken in this study yielded essentially the same conclusions. While the genetic covariance structures for the two *P. maniculatus* subspecies were similar (proportionally), comparisons between *P. leucopus* and *P. maniculatus* indicated pronounced differences. Thus, for this genus, the assumption of constant genetic covariance structure cannot be extended beyond the species level. The potential response of one *Peromyscus* species to selection cannot be extrapolated from information on the genetic covariance of another.

Several other attempts have been made to test the hypothesis that closely related populations or taxa have similar genetic covariance and/or correlation matrices (Cheverud 1988, 1989; Kohn and Atchley 1988; Cheverud et al. 1989; Venable and Búrquez 1990; Wilkinson et al. 1990; Spitze et al. 1991; Platenkamp and Shaw 1992; Brodie 1993; Paulsen 1994), most of them failing to detect significant differences, probably because of low statistical power. Many statistics beyond those mentioned above have been employed in these studies, e.g., the sum of squared differences between like elements, the difference between matrix determinants or between dominant eigenvalues, and the correlation between the elements of the leading eigenvectors. The statistical properties of most of these tests are poorly understood, if not completely unknown (Cowley and Atchley 1992, Shaw 1992). Consequently, most investigators apply several different techniques to their data in hopes that a consistent message will emerge, and that has usually been the case. However, such results may be a bit misleading, since the different methods are clearly not independent.

Prior to the application of any test for covariance matrix similarity, a fundamental issue that needs to be dealt with is measurement scale. If there is an association between means and variances, two taxa can exhibit different covariance structures simply because they differ in mean phenotypes. Methods for dealing with this type of problem have been covered in Chapter 11. An equally serious issue is the fact that different characters are often measured on fundamentally different scales (e.g., length vs. volume). Without some kind of transformation, the contributions of different characters to matrix similarity indices will depend on the different measurement scales. The problem cannot be eliminated by simply standardizing all characters to have equal variances, since that would be contrary to the goal of the analysis. Spitze et al. (1991) suggest that each scale of measurement be transformed such that the variance of all characters measured on that scale averages to one across populations. This approach has the effect of standardizing different scales with respect to each other, while preserving differences in variance among characters within and between populations.

Resampling procedures (Chapter 16) seem to provide a reasonable alternative means for the comparison of covariance matrices (Spitze et al. 1991). Once the data have been transformed, the information from both populations can be pooled into one synthetic population. Then two quasipopulations can be constructed by randomly selecting families from the synthetic population, and allocating them such that the two quasipopulations have the same number of families as the true population samples. This procedure is repeated a thousand or so times, and the similarity index of interest is computed for each pair of quasipopulations, generating a null distribution of the statistic against which the observed value for the true populations is tested. One can then evaluate whether the observed value is significantly greater than (or less than) what would be obtained by drawing two samples from a common population. In addition to making no assumptions about the distribution of phenotypes, this bootstrap procedure has the advantage of applying to any similarity index.

If one is willing to assume that the characters under consideration have a multivariate normal distribution, a maximum likelihood method is available for testing the hypothesis that any element (or group of elements) of the covariance matrices differ between two populations (Shaw 1991). Unfortunately, simulations have shown the power of this test to be quite low. For example, considering only the univariate case, if the additive genetic variances in two populations differ by a factor of 2.5, a nested sib analysis involving 100 sires / population, 3 dams / sire, and 3 offspring / dam would detect the difference at the 0.05 level of significance only about 50% of the time. With the same design but only 40 sires, the difference would go undetected 80% of the time. Such results should not be too surprising. We have seen repeatedly that the procurement of accurate variance / covariance estimates demands large sample sizes; detecting a significant difference between two separate estimates can only be more difficult. It remains to be seen whether

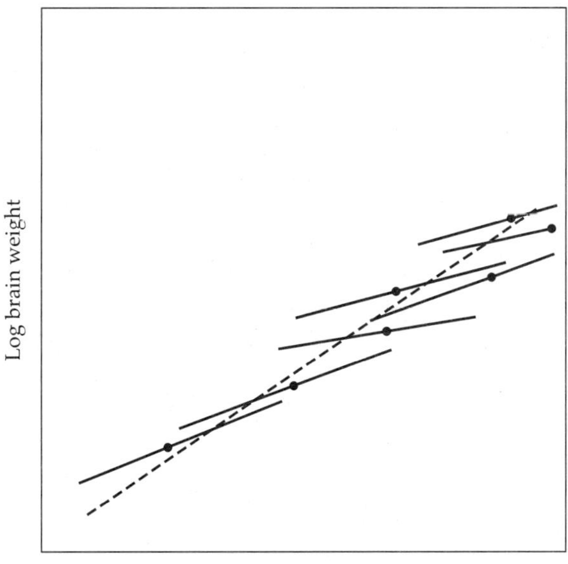

Figure 21.4 Intraspecific lines of allometric growth for brain:body size vs. the interspecific pattern. (Solid lines give the regression, and solid dots the means, for individual species; the dashed line is the regression of the species means.)

the power of the bootstrap and other randomization tests exceeds that of the maximum likelihood procedure.

Evolutionary Allometry

As noted in Chapter 11, the scaling of anatomical features to body size (allometry) has long been recognized as an important contributor to shape differences between species. Because of our unusually large heads, a great deal of attention has been focused upon the relationship of brain size to body size. When log-transformed adult brain weights are regressed on log-transformed adult body weights for members of the same species, a linear relationship is observed with the slope generally on the order of 0.2 to 0.4. Usually, this also applies to the mean phenotypes of different species in the same genus. However, when adults of distantly related species (e.g., different genera within an order) are compared, a higher slope of about 0.6 is obtained (Figure 21.4).

The different slopes at different taxonomic levels has long been a perplexing problem (Lande 1979, 1985). Drawing from extensive laboratory work, Riska and Atchley (1985) suggested an attractive hypothesis to account for these differences. Using a nested sib design with cross-fostering (to factor out maternal effects; Chapter 23), they analyzed approximately 500 laboratory rats and 1,500

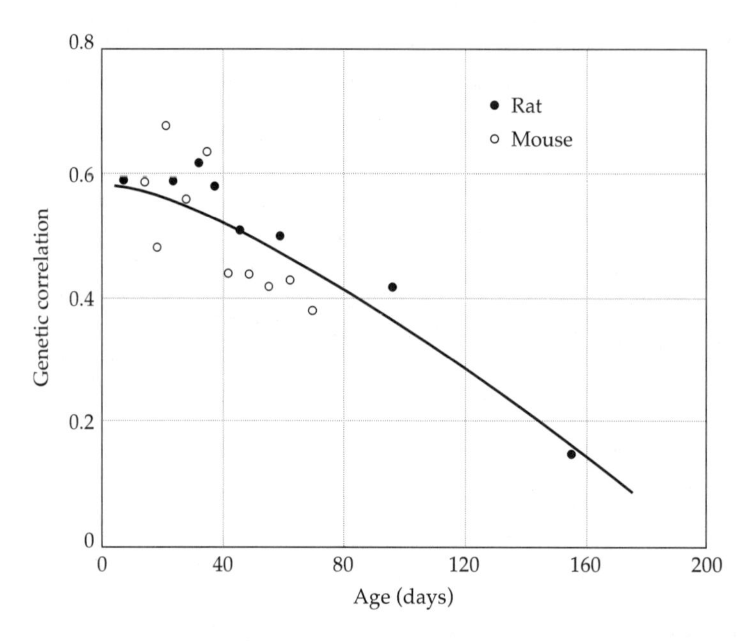

Figure 21.5 Reduction with age of the genetic correlation between final adult brain and age-specific body weights. (From Riska and Atchley 1985.)

mice (Atchley and Rutledge 1980, Atchley et al. 1984). The experimental design provided estimates of the genetic correlations between brain size, body size, and body growth during different age intervals. In both species, the correlation between eventual adult brain weight and age-specific body weight declines with age (Figure 21.5). This reduction was found to be due to an increasingly negative correlation between brain weight and growth increment late in life. Thus, the positive genetic correlation between brain and body size is primarily a function of genes that are active in prenatal and early postnatal growth.

Riska and Atchley pointed out that mammalian growth can be partitioned roughly into an early phase in which cell numbers of most organs are increasing and a late phase in which cell numbers are nearly constant but cell sizes are increasing in different organs to different degrees. Thus, genes that influence early growth have general pleiotropic effects on the size of most organs, while those operating later in life have more specific targets. In artificial selection experiments for body size in mice, a large proportion of the response is due to changes in cell size (Falconer et al. 1978), indicating that late-acting genes with relatively mild pleiotropic effects are selected upon. Based on the pattern in Figure 21.5, such selection would be expected to yield a relatively mild change in brain weight, as observed in intraspecific studies. In contrast, body size differences between distantly related species are due almost entirely to variation in cell numbers (Raff and Kaufman 1983), suggesting that diversification at this taxonomic level is largely a consequence of evolutionary changes in the early phase of growth where the correlated response of brain size to selection on body size would be expected to be strong.

These results provide a good example of how the genetic analysis of correlated characters can lead to a mechanistic understanding of developmental pattern. Several other studies have used this approach to understand the degree to which the component parts of multivariate phenotypes are free to evolve. For example, in a study of wing color patterns in two species of butterflies, after correcting for size differences among individuals, Paulsen (1994) found extremely high genetic correlations between various wing venation measures, between various "eye-spot" diameters, and between various "eye-spot" positions. The lack of correlation between characters in different trait sets implies that the three sets are free to evolve independently in response to natural selection, and helps explain the pathways by which butterfly wing color diversification takes place. In a similar study with *Drosophila,* Cowley and Atchley (1990) found that traits derived from the same imaginal disc during development are more closely correlated genetically than are those from different discs. In wild radish *(Raphanus raphanistrum),* genetic correlations are much higher within than between functionally related groups of characters (flowers vs. leaves) (Connor and Via 1993). Similarly, in studies of primate cranial morphology, genetic correlations are consistently higher among functionally related traits (cranial vault vs. oral cavity) than among unrelated traits Cheverud (1982, 1989, 1995). In all of these cases, it is reasonable to hypothesize that the observed patterns of correlation are a consequence of pleiotropy, i.e., of functionally similar traits sharing the same developmental pathways.

Consider, however, Brodie's (1989, 1993) study of the garter snake *Thamnophis ordinoides,* which revealed a strong genetic correlation between color pattern and antipredator behavior. Although not impossible, the coupling of these two traits via the pleiotropic effects of gene action seems implausible. An alternative hypothesis is that selection for adaptive combinations of color pattern and behavior lead to the build-up of gametic phase disequilibria among pairs of polymorphic loci. A simple test of this hypothesis would be to randomly mate the snakes for several generations and maintain them under relaxed selection. If the association between behavior and color pattern were a consequence of gametic phase disequilibrium, the genetic correlation should decline over time.

Evolution of Life-history Characters

A widespread belief in evolutionary ecology is that negative genetic correlations between fitness characters are the rule in natural populations. Indirect and direct evidence of such tradeoffs have indeed been recorded frequently (see reviews in Reznick 1985, Partridge and Harvey 1985, Bell and Koufopanou 1986, Scheiner et al. 1989), but a number of clear cases of positive genetic correlations have also been reported (e.g., Giesel and Zettler 1980, Hegmann and Dingle 1982, Mitchell-Olds 1986, Rausher and Simms 1989, Spitze et al. 1991). In a broad review of the literature, Roff (1996) found that genetic correlations between fitness characters tend to be lower than those between morphological characters. However, there is a broad degree of overlap in the distributions, and the majority of life-history correlations are still positive.

Positive genetic correlations between fitness characters can be artifactual (Rose 1984, Service and Rose 1985, Clark 1987) — a consequence of using inbred lines, some of which suffer from inbreeding depression more than others, or of performing assays in a novel laboratory environment to which populations are not adapted. But not all of the data seem to be explained so easily.

The usual argument for the negative correlation hypothesis is that alleles that simultaneously improve several traits tend to be advanced rapidly by selection, while those with several negative effects tend to be eliminated (Falconer and Mackay 1996, Rose 1982). Such a sorting process is expected to leave circulating a pool of alleles with favorable effects on some fitness traits but unfavorable effects on others, i.e., a set of alleles with equivalent effects on total fitness. Curtsinger et al. (1994) have cast doubt upon this seductively simple hypothesis, pointing out that the conditions for the maintenance of stable polymorphisms by **antagonistic pleiotropy** are quite restrictive. However, their argument is not entirely satisfying, since they only considered the maintenance of variation by balancing selection, ignoring the recurrent introduction of new alleles by mutation. As noted in Chapter 12, polygenic mutation introduces variance for quantitative characters at a high enough rate that substantial genetic variance can be maintained by a balance with purifying selection, and this is likely to be true for genetic covariance as well.

van Noordwijk and de Jong (1986) and Houle (1991) have shown how the sign of a genetic correlation between fitness characters can depend on the pleiotropic properties of mutations. If no genetic variation exists for the ability to acquire resources, then there will necessarily be a genetic tradeoff in the amount of resources that can be allocated to two competing processes. Suppose, however, that genetic variation exists for acquisition ability so that some individuals acquire more total resources and therefore are able to allocate more to both characters. A positive genetic correlation between the two characters would then be possible. As noted above, selection would be expected to eliminate such variation by fixing favorable genes, but if deleterious mutation continuously generated individuals with low acquisition abilities, a positive genetic correlation between fitness characters would be maintained by selection-mutation balance. Given that most mutations tend to be deleterious (Chapter 12), such a situation is not out of the realm of possibility, assuming that many more mutations influence acquisition than allocation of resources. Resolution of these issues will require a deeper understanding of the pleiotropic effects of mutations than is currently available.

22

Genotype $\times$ Environment Interaction

To this point, we have been conceptualizing the phenotype of an individual as the sum of independent genetic and environmental contributions, with both factors having essentially continuous distributions and the environmental effect being a random deviate with expectation zero. As noted in Chapter 6, more precise statements can sometimes be made about the environmental contribution when individuals are distributed over a discrete set of environments. For example, two different locations or years may present very different growth conditions for a crop variety. Such a difference can be recorded as a macroenvironmental effect, and no further complications are introduced provided all varieties of interest respond in the same way. However, things are not so simple if varieties differ in their response to environmental change, i.e., if there is genotype $\times$ environment interaction.

Issues concerning genotype $\times$ environment interaction arise in many contexts in quantitative genetics. Genotype $\times$ environment interaction is a major concern in attempts to develop economically important breeds of plants and animals with wide geographical utility. It is also of considerable concern in quantitative-genetic investigations of natural populations when broad inferences are made from studies of individuals raised in simple, often novel, laboratory environments. Finally, genotype $\times$ environment interaction is of substantial importance in genetic epidemiological studies that show certain genotypes to have elevated sensitivities to environmental risk factors.

As a point of departure, recall that a general description of the phenotype of an individual of genotype *j* living in the *i*th environmental setting is

$$z_{ijk} = \mu + G_j + I_{ij} + E_{g,i} + E_{s,ijk} \tag{22.1}$$

where μ is the average phenotype over all genotypes and environments, G_j is the effect of the *j*th genotype averaged over all environments, $E_{g,i}$ is the average (general) effect of the *i*th environment on all genotypes, I_{ij} is the interaction effect between genotype j and environment i, and $E_{s,ijk}$ is the special environmental effect (residual deviation) for the *k*th replicate of genotype *j* in environment *i*. The effect of the *i*th environment is interpreted as the difference between the mean phenotype of all genotypes grown in environment *i* and the grand mean, i.e., as $E_{g,i} = \mu_i - \mu$. Similarly, the *j*th genotypic value is defined to be the difference between the mean phenotype of genotype j over all environments and the grand mean, $G_j = \mu_j - \mu$, and the *ij*th genotype $\times$ environment interaction is $I_{ij} = \mu_{ij} - \mu - G_j - E_{g,i} = \mu_{ij} - \mu_j - \mu_i + \mu$.

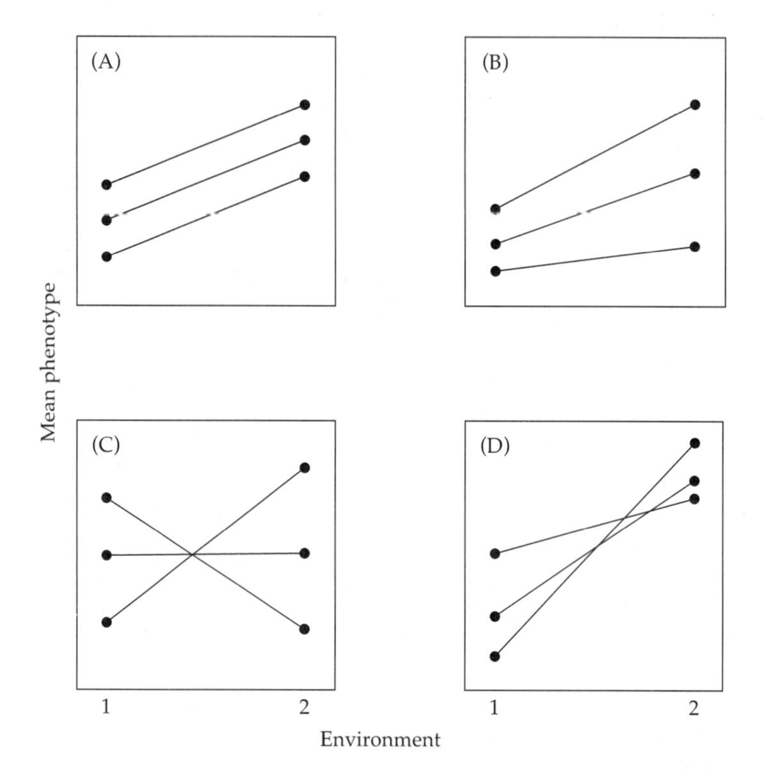

Figure 22.1 Reaction norms for three genotypes in response to two environments. (A) No genotype × environment interaction. (B) Genotype × environment interaction is due entirely to a change in scale. (C) Genotype × environment interaction is due to a change in ranking. (D) There is a change of scale as well as a change in ranking.

If $I_{ij} = 0$ for all ij, then all genotypes respond to macroenvironmental change in a parallel fashion, as shown in Figure 22.1A for the case of two environments. A nonparallel response to environmental change implies genotype × environment interaction with respect to phenotypic expression. Such interaction can come about in two not necessarily exclusive ways: (1) a change of scale, such that higher-ranking genotypes in one environment react more (or less) strongly to conditions in the second environment (Figure 22.1B), and (2) a change of ranking (Figure 22.1C). With multiple genotypes in multiple environments, many more complex patterns are possible.

The function relating mean phenotypic response of a genotype to a change in the environment is called a **reaction norm** (Woltereck 1909, Schmalhausen 1949). Like any other character, reaction norms can evolve in response to environmental pressures. For example, in Figure 22.1D, selection for the high-performing genotype in environment 2 leads to the evolution of a steep reaction norm (i.e., high variance in phenotypic expression across environments), while selection for the high-performing genotype in environment 1 leads to a population dominated by individuals with low environmental sensitivity. If selection occurred in both environments, and genotypes were distributed randomly across environments, the reaction norm giving the highest average performance would come to dominate.

This chapter is concerned with reaction norm analysis, i.e., with the statistical methodology for quantifying differences among genotypes in response to environmental change. Two classes of traits are of interest in this regard. **Labile** traits are those for which phenotypic expression can adjust rapidly within individuals, through physiological and/or behavioral means, to changes in the environment. Examples include behavioral changes in the presence vs. absence of predators, competitors, and/or mates. **Nonlabile** traits are those for which phenotypic expression becomes fixed during some sensitive period of development, e.g., age at first reproduction or adult size in an arthropod with determinate growth. Both types of traits can exhibit genotype × environment interaction. But from a practical standpoint, there is an important distinction between labile and nonlabile traits. The entire reaction norm of a labile trait can be determined at the individual level by scoring the same individual in a number of environments, but with nonlabile traits, different points on the reaction norm need to be assayed in different individuals. Many traits, of course, have both labile and nonlabile components. For example, the behavior of an individual can often be conditioned by prior experiences, and the metabolic rates of individuals can depend both on current environmental conditions and on conditions that existed during development.

The statistical methods relevant to reaction norm analysis largely follow from previous chapters. We start by considering the special, but common, case in which multiple genotypic groups are assayed in two environments. With this sort of experimental design, the concept of genetic correlation (Chapter 21) can be applied to the problem of genotype × environment interaction. We then show how a fuller resolution of the issues can be achieved by assaying genotypes over a more complete spectrum of environments. With multiple environments, two-way analysis of variance (Chapter 20) becomes useful, the two factors being genotypes and environments. As in all of the methods introduced in previous chapters, maximum likelihood provides an alternative framework for analyzing genotype × environment interaction (Platenkamp and Shaw 1992), an issue that we defer until Chapter 27. We conclude the chapter by considering some recent theoretical developments that may help guide future interpretations of reaction norm analyses.

Before proceeding, we raise the caveat that the literature on genotype × environment interaction analysis is littered with disagreements and inconsistencies as to how parameter estimates are to be interpreted. Much of the controversy concerns differences in opinions as to whether environmental treatments should be considered to be random vs. fixed effects. Treating the environments as fixed implies concern only with the particular set of environments considered (such as particular climates for cultivars). Treating the environments as random implies that the sampled environments represent a random sample from a larger universe of environments. The confusion goes beyond this issue, since there is more than one way to parameterize a linear model containing fixed effects. Moreover, some applications of genotype × environment interaction even depart from the basic linear model given above, treating the interaction term as a function of the product of the genotypic value and environmental effect, $I_{ij} = G_j \cdot E_{g,i}$ (Gauch 1988, Gimelfarb 1994, Piepho 1995). We have chosen to focus on the more general model, Equation 22.1, and in the following pages we attempt to sort out the relevant issues regarding fixed and random effects. Before analyzing any experiment on genotype × environment interaction, the practitioner will need to give careful consideration to the merits of alternative approaches.

Reviews on genotype × environment interaction from various applied perspectives can be found in Dickerson (1962), Comstock and Moll (1963), Bradshaw (1965), Pani and Lasley (1972), Freeman (1973), J. Hill (1975), Barlow (1981), Simmonds (1981), Hohenboken (1985), Schlichting (1986), Becker and Léon (1988), Wahlsten (1990), and Kang and Gauch (1995). The overwhelming message from the extensive empirical literature on varieties of crops and domesticated animals is that genotype × environment interaction is extremely common. However, there are still surprisingly few data on its significance in natural populations.

GENETIC CORRELATION ACROSS TWO ENVIRONMENTS

Falconer (1952) had the useful insight that the same character measured in two different environments can be treated as two different traits. Genotype × environment interaction can then be detected from the genetic correlation between the two traits. If the family genetic effects do not change across environments or if they are related such that performance of any genotype in environment 2 is proportional to that in environment 1 (Figures 22.1A,B), the genetic correlation of family members across environments is equal to one. The null hypothesis of no significant genotype × environment interaction is rejected whenever the genetic correlation across environments is significantly less than one.

To achieve a more explicit understanding of the relationship between the genetic correlation across environments and the amount of genotype × environment interaction, a clear description of the linear model is required. Throughout this chapter, we use a liberal definition of "genotype," letting j denote an average member of the jth genetic group, where the group may consist of genetically identical individuals (e.g., members of a clone or an inbred line) or of members of the same family (e.g., paternal half sibs). From Equation 22.1, the phenotypes of two individuals (k and l) of genetic group j, each in a different environment (1 and 2), can be represented as

$$z_{1jk} = \mu + G_j + I_{1j} + E_1 + e_{1jk} \tag{22.2a}$$

$$z_{2jl} = \mu + G_j + I_{2j} + E_2 + e_{2jl} \tag{22.2b}$$

Here, to simplify notation, we drop the subscript g on the general environmental effect E_g. We also use e rather than E_s to denote the residual effect, since the former potentially contains some genetic as well as environmental sources of variation when the "genotype" is a collection of related, but not identical, individuals.

Using the definitions given in the introduction, this model can be simplified in two ways. First, since $E_1 = \mu_1 - \mu$, $E_2 = \mu_2 - \mu$, and $\mu = (\mu_1 + \mu_2)/2$, it follows that $E_1 = -E_2$. Thus, the macroenvironmental effects can be denoted by $\pm E$. Second, since $I_{1j} = \mu_{1j} - \mu - G_j - E$, $I_{2j} = \mu_{2j} - \mu - G_j + E$, and $G_j = [(\mu_{1j} + \mu_{2j})/2] - \mu$, it follows that $I_{1j} = -I_{2j}$, so the interaction effects for specific genetic groups in the alternate environments can be denoted as $\pm I_j$. With these modifications, the preceding equations become

$$z_{1jk} = \mu + G_j + I_j + E + e_{1jk} \tag{22.3a}$$

$$z_{2jl} = \mu + G_j - I_j - E + e_{2jl} \tag{22.3b}$$

The fundamental properties of this linear model can be summarized as follows: (1) μ is the grand mean of all genotypes across both environments; (2) the genetic (family) effects are assumed to be random variables with expectation zero and variance σ_G^2; (3) for a given environment, the interaction effects are random variables with expectation zero and variance σ_I^2; (4) within genetic groups, the mean interaction effect is constrained to be zero; (5) genotypic values and interaction effects may be correlated with covariance $\sigma_{G,I}$ (Figure 22.1B provides an example of such covariance, genotypes with higher values of G having higher values of I); (6) the average value of the two environmental effects is zero; and (7) the residual deviations within environments have expectation zero and variance σ_e^2 for all genetic groups.

Within each environment, the genotypic effects are $G_{1j} = G_j + I_j$ and $G_{2j} = G_j - I_j$. Thus, the variances among genotypic group (typically family) effects within each environment and the covariance across environments are

$$\sigma_G^2(1) = \sigma^2(G_j + I_j) = \sigma_G^2 + \sigma_I^2 + 2\sigma_{G,I} \tag{22.4a}$$

$$\sigma_G^2(2) = \sigma^2(G_j - I_j) = \sigma_G^2 + \sigma_I^2 - 2\sigma_{G,I} \tag{22.4b}$$

$$\sigma_G(1,2) = \sigma[(G_j + I_j), (G_j - I_j)] = \sigma_G^2 - \sigma_I^2 \tag{22.4c}$$

We emphasize that here and below, $\sigma_G^2(1)$, $\sigma_G^2(2)$, and $\sigma_G(1,2)$ refer to variances and covariances of family (genotypic group) means, not to genetic variances at the levels of individuals.

Three distinctive features emerge from these expressions. First, within any environment, the portion of the genotypic variance that is responsive to environmental change (σ_I^2) is confounded with the variance of genotypic means across environments (σ_G^2). That is, in a single-environment setting, the contributions of σ_G^2 and σ_I^2 to the genetic variance cannot be isolated (since they appear as a sum). Second, the genetic variance differs between the two environments whenever there is a covariance between mean effects and interaction effects of the genetic groups ($\sigma_{G,I} \neq 0$). In the context of Figure 22.1, this requires that the slopes and elevations of genotypic reaction norms be correlated. Third, from the variances of family means within environments and the covariance of family means across environments, the genetic variance and interaction variance can be isolated, as

$$\sigma_G^2 = \frac{1}{2}\left[\sigma_G(1,2) + \frac{\sigma_G^2(1) + \sigma_G^2(2)}{2}\right] \tag{22.5a}$$

$$\sigma_I^2 = \frac{1}{2}\left[\frac{\sigma_G^2(1) + \sigma_G^2(2)}{2} - \sigma_G(1,2)\right] \tag{22.5b}$$

$$\sigma_{G,I} = \frac{\sigma_G^2(1) - \sigma_G^2(2)}{4} \tag{22.5c}$$

Now recall the rule that an among-family variance is equivalent to the phenotypic covariance of individuals within families (Chapter 18). Thus, additive genetic variance contributes to $\sigma_G^2(1)$ and $\sigma_G^2(2)$ in the amounts $2\Theta\sigma_A^2(1)$ and $2\Theta\sigma_A^2(2)$, where Θ is the coefficient of coancestry among individuals within families (or more generally, in the group of individuals chosen). Similarly, additive genetic covariance contributes to $\sigma_G(1,2)$ in the amount $2\Theta\sigma_A(1,2)$. Thus, since all three contributions are proportional to 2Θ, ignoring nonadditive genetic effects, the additive genetic correlation across environments is defined to be

$$\rho_\times = \frac{\sigma_A(1,2)}{\sigma_A(1)\,\sigma_A(2)} = \frac{\sigma_G(1,2)}{\sigma_G(1)\,\sigma_G(2)} \tag{22.6}$$

This expression shows that the genetic correlation across environments can be estimated directly using measures of the among-family variances, $\sigma_G^2(1)$ and $\sigma_G^2(2)$, and covariance, $\sigma_G(1,2)$.

Substituting Equations 22.4a–c, $\rho_\times$ can be expressed in causal terms as

$$\rho_\times = \frac{\sigma_G^2 - \sigma_I^2}{\sqrt{(\sigma_G^2 + \sigma_I^2)^2 - 4\sigma_{G,I}^2}} \tag{22.7a}$$

This expression helps clarify a number of issues. First, in the absence of genotype $\times$ environment interaction, $\sigma_I^2 = \sigma_{G,I} = 0$, and $\rho_\times = 1$. Second, if there is a perfect negative correlation between rankings of genetic groups in the two environments such that all families have the same mean performance across environments ($\sigma_G^2 = 0$), then $\sigma_{G,I}$ must also be zero, and $\rho_\times = -1$. Third, from Equations 22.4a,b it can be seen that when the among-family variances in the two environments are the same [i.e., $\sigma_G^2(1) = \sigma_G^2(2)$], then $\sigma_{G,I} = 0$, reducing Equation 22.7a to

$$\rho_\times = \frac{\sigma_G^2 - \sigma_I^2}{\sigma_G^2 + \sigma_I^2} \tag{22.7b}$$

and implying that genotype × environment interaction reduces the genetic correlation below +1. If, on the other hand, $\sigma_G^2(1) \neq \sigma_G^2(2)$, $\rho_\times$ can equal one even when there is significant interaction variance. This requires only that the average effects of genotypes be correlated perfectly with their interaction effects, i.e., $\sigma_{G,I} = \sigma_G\,\sigma_I$, or equivalently, $I_j = b \cdot G_j$, where b is a constant independent of genotype.

In summary, a genetic correlation across environments significantly less than one cannot exist in the absence of genotype × environment interaction. On the other hand, a genetic correlation across environments equal to one need not imply an absence of genotype × environment interaction, although when $\rho_\times = 1$, any genotype × environment interaction will most likely be a simple scale effect. In practical applications, such an effect can be verified by simply transforming the data prior to analysis so that the among-family variances in both environments are equal.

Estimation Procedures

To obtain an estimate of the genetic correlation across environments, $r_\times$, estimates of the among-family variances are required for both environments, as is an estimate of the covariance of family means across environments. The simplest way of procuring such estimates is to split the families, assaying for each family one set of individuals in one environment and another set in the second environment. Methods for estimating the among-family components of variance within single environments, $\sigma_G^2(1)$ and $\sigma_G^2(2)$, have been presented in previous chapters (e.g., one-way ANOVA), so we will consider that issue no further. Via (1984) reviews several procedures for estimating $\sigma_G(1,2)$, the most straightforward of which is the simple computation of the covariance of family means in the different environments. In all such analyses, care should be taken to avoid the use of families containing individuals that share family environmental effects, i.e., individuals that share the same mothers. After the among-family variances and covariances have been estimated, the estimate $r_\times$ is obtained by substituting observed values for their expectations in Equation 22.6.

As in the case of other genetic correlation estimators encountered in Chapter 21, this indirect way of estimating $\rho_\times$ is not equivalent to the computation of a product-moment correlation, and as a consequence, sampling error can cause estimates to exceed the natural boundaries of ± 1. For purposes of hypothesis testing, however, a product-moment correlation, $r(\overline{z}_{1j}, \overline{z}_{2j})$, can be obtained from the simple regression of family means (i.e., $\overline{z}_{1j}$ on $\overline{z}_{2j}$). The significance level associated with such a regression provides a conservative test as to whether $r_\times$ is significantly different from zero, since the correlation coefficient will be biased towards zero by the inclusion of residual variance in the denominator (see below).

Based on the arguments in the preceding section, a more interesting question in the analysis of genotype $\times$ environment interaction is whether the correlation across environments is significantly different from $+1$. For this purpose, the jackknife and bootstrap procedures discussed in the previous chapter may be exploited profitably. If the true genetic correlation across environments were $+1$, one would expect the distribution of estimates of $\rho_\times$ derived from bootstrap samples to significantly overlap $+1$.

A number of investigators have used the correlation of observed family means, $r(\overline{z}_{1j}, \overline{z}_{2j})$, as an estimate of the genetic correlation across environments. As noted above and in Chapter 21, the absolute value of any such estimate will be biased downwardly relative to $\rho_\times$ since the variances of family means will be inflated by contributions from measurement error and environmental variance. Thus, although this procedure does not affect the sign of the correlation (since the covariance is still estimated in the way we just outlined), it does artifactually create a tendency towards falsely concluding that genotype $\times$ environment interaction is present. A slight modification of Equation 21.9 shows the magnitude of the bias,

$$\rho(\overline{z}_{1j}, \overline{z}_{2j}) \simeq \rho_\times \left[\frac{2\Theta n h_1 h_2}{\sqrt{(\phi h_1^2 + 1)(\phi h_2^2 + 1)}} \right] \tag{22.8}$$

where $\phi = 2\Theta(n-1)$, and $h_1^2 = \sigma_A^2(1)/\sigma_z^2(1)$ and $h_2^2 = \sigma_A^2(2)/\sigma_z^2(2)$ are the heritabilities of the trait as expressed in the two environments. Note that with moderately large sample sizes ($n \geq 10$ in each environment) and assuming the heritabilities in both environments are approximately the same, the quantity in brackets is approximately $\phi h^2/(\phi h^2 + 1)$. Thus, the correlation between family means can substantially underestimate $\rho_\times$ unless $\phi h^2 >> 1$. That is, unless the heritabilities in both environments are quite high, large family sizes $[n >> 1/(2\Theta h^2)]$ are essential to reduce the sampling bias of the correlation of family means to a reasonable level. The following empirical example demonstrates the misleading nature of correlations of family means.

Example 1. In an attempt to determine whether cowpea weevils (*Callosobruchus maculatus*) exhibit variance in their relative performances on alternative hosts, Fox (1993) raised members of full-sib families ($\Theta = 0.25$ and $n = 5$) on both cowpea and azuki beans. The correlations between family means (and their standard errors obtained by bootstrapping over families) are given for development time and three morphological characters in the following table. (Data are for females only.) Note that the family-mean correlations $r(\overline{z}_{1j}, \overline{z}_{2j})$ for all four characters are well over three standard errors below $+1$, suggesting significant genotype $\times$

environment interaction.

Character	$r(\overline{z}_{1j}, \overline{z}_{2j})$	h^2	$\dfrac{2\Theta n h^2}{\phi h^2 + 1}$	$r_\times$
Development time	0.24 (0.16)	0.68	0.72	0.33 (0.22)
Pronotum width	0.30 (0.14)	0.50	0.62	0.48 (0.25)
Elytron length	0.33 (0.14)	0.48	0.61	0.54 (0.23)
Emergence weight	0.36 (0.17)	0.30	0.46	0.78 (0.37)

The author provides data that yield the average heritability estimates given in the table. Using these as well as the values of Θ and n given above, the quantity in the large brackets in Equation 22.8 can be estimated for each character. These bias estimates, which appear in the fourth column of the table, give the expected degree to which the correlation between family means underestimates the true correlation across environments; a value of one implies there is no bias. Division of the correlations between family means by their respective biases yields the estimates of $\rho_\times$ given in the final column. The corrected standard errors reported in the table were obtained in the same manner, i.e., by dividing the previous standard errors by the bias. However, this procedure must lead to the standard errors being downwardly biased since it does not account for sampling error associated with the heritability estimates.

All of the estimates of the correlation across environments ($r_\times$) are substantially higher than those for the correlation between family means, $[r(\overline{z}_{1j}, \overline{z}_{2j})]$. That for emergence weight is clearly not significantly different from $+1$, while those for pronotum width and elytron length are of questionable significance. Thus, the correlation between family means substantially inflates the confidence that should be attached to the hypothesis of genotype × environment interaction.

Several other studies have used correlations of family means across environments to test for local adaptation of insect pests to different host plants. Full-sib families of leaf miners and pea aphids have been found to have significantly negative genetic correlations for fitness-related characters when raised on different crops (Via 1984, 1991), suggesting the potential for evolutionary specialization to local host plants. In this case, it is unlikely that a reanalysis of the data would lead to a change in interpretation; since the correlation between family means yields a downwardly biased estimate of the *absolute* value of $\rho_\times$, the true values of the genetic correlations are probably more negative than Via's estimates. Thus, Via's study populations appear to exhibit a real genetic tradeoff in performance on alternative hosts. On the other hand, several other studies have found positive correlations between family mean performances on different hosts (Rausher 1984, Futuyma and Phillipi 1987, Takano et al. 1987, Karowe 1990). Although such results imply an absence of tradeoffs, as in the case of Fox's analysis, they are less clear with respect to genotype × environment interaction (see also Fry 1993).

TWO-WAY ANALYSIS OF VARIANCE

Two-way ANOVA provides an approach for detecting genotype $\times$ environment interaction when data are available for more than two environments. Applications date back to Sprague and Federer (1951). The sums of squares are computed in the same manner as in a diallel (Table 20.1), the two factors now being genotypes (or families) and environments, rather than fathers and mothers. However, as pointed out by Ayres and Thomas (1990) and Fry (1992), much confusion exists in the literature regarding the interpretation of the mean squares (and the variance components extracted from them) in genotype $\times$ environment analysis.

We will assume throughout that the genetic groups under analysis are sampled randomly from some larger population about which inferences are to be made and are hence random effects. The main issue then concerns the treatment of the macroenvironmental effects. If the environments in which the genotypes are assayed are regarded as random with respect to the possible set of environments in which the study species is located naturally, then a random-effects interpretation is appropriate. This would be the case, for example, if genotypes of a plant species were replicated in plots randomly distributed over the native habitat (Stratton 1994). If, on the other hand, the environments are selected for a particular reason and/or are the only ones of interest, as in the assay of crop varieties at future production sites, they should be regarded as fixed effects. In this case, a mixed-model interpretation (random genotypes, fixed environments) is required.

We start with the random-effects model, as the basic machinery has already been introduced in Chapter 20. As we will see below, because of its focus on two specific environments, Falconer's approach, described by Equations 22.2a,b, is equivalent to a mixed-model analysis, with the environmental effects being interpreted as fixed factors. To distinguish that model from the random-effects model, a slight change in notation is required. We denote terms from the random-effects model with tildes,

$$z_{ijk} = \mu + \widetilde{G}_j + \widetilde{I}_{ij} + \widetilde{E}_i + e_{ijk} \tag{22.9}$$

The properties of this model are identical to those encountered in Chapter 20 — the model components $\widetilde{G}_j$, $\widetilde{I}_{ij}$, $\widetilde{E}_j$, and e_{ijk} all have expected values equal to zero, are distributed independently, and have variances respectively denoted by $\sigma^2_{\widetilde{G}}$, $\sigma^2_{\widetilde{I}}$, $\sigma^2_{\widetilde{E}}$, and σ^2_e. A simple change of terms from Table 20.1 yields the expressions for the expected mean squares given in Table 22.1.

Properties of the random-effects model are widely agreed upon by statisticians, but things are more complicated under the mixed model. Under one interpretation, all of the properties given above for the random-effects model hold, except for the definition of the environmental effects as fixed constants rather than random variables. In this case, the expressions for the expected mean

Table 22.1 Alternative interpretations of the expected mean squares for two-way analysis of variance of genotype × environment interaction under a random-effects vs. a mixed (genotypes random, environments fixed) model.

		Expected Mean Squares	
Factor	Degrees of Freedom	Random Effects	Mixed Model
Environment	$N_E - 1$	$\sigma_e^2 + n\sigma_{\widetilde{I}}^2 + nN_G\sigma_{\widetilde{E}}^2$	$\sigma_e^2 + \dfrac{nN_E}{(N_E-1)}\sigma_I^2 + \dfrac{nN_G}{N_E-1}\sum_{i=1}^{N_E} E_i^2$
Genetic	$N_G - 1$	$\sigma_e^2 + n\sigma_{\widetilde{I}}^2 + nN_E\sigma_{\widetilde{G}}^2$	$\sigma_e^2 + nN_E\sigma_{\widetilde{G}}^2$
G × E	$(N_E - 1)(N_G - 1)$	$\sigma_e^2 + n\sigma_{\widetilde{I}}^2$	$\sigma_e^2 + \dfrac{nN_E}{(N_E-1)}\sigma_I^2$
Error	$N_E N_G(n-1)$	σ_e^2	σ_e^2

Note: The sums of squares are computed in an identical manner under both models, and are defined in Table 22.2. A completely balanced design is assumed, with n members of N_G genetic groups being assayed in each of N_E environments.

squares for genotypes, interaction effects, and residual deviations are identical to those for the random-effects model. A second interpretation is that because the macroenvironmental effects are assumed to be fixed effects, their mean value should be constrained to be zero. Given this constraint, it is argued that there should be a partial restriction on the interaction effects, such that $\sum_{i=1}^{N_E} I_{ij} = 0$, i.e., within each genetic group, the interaction effects should also have mean zero across environments. This is the interpretation that we made in introducing the causal determinants of Falconer's correlation across environments, where we let the interaction effect of genotype j be $+I_j$ in one environment and $-I_j$ in the other, and where we let the two environmental effects be $\pm E$. To maintain continuity, we adhere to this latter interpretation of the mixed model throughout this chapter. A good general introduction to the issues distinguishing the two interpretations of the mixed model can be found in Searle et al. (1992, pp. 123–127). Itoh and Yamada (1990) provide an overview of the two approaches in genotype × environment interaction analysis.

For the remainder of our discussion on the mixed model, we return to Equation 22.1, emphasizing again that the only changes in assumptions from the random-effects model (Equation 22.8) are that the observed macroenvironmental effects are assumed to be constants with mean zero (under the random-effects model, the expected value of the mean macroenvironmental effect is zero, but the sample mean is not so constrained), and that the interaction effects within genotypes are constrained to have mean zero. As in the random-effects model, G_j is still a random variable with expectation zero, but we now denote its variance as σ^2_G (as opposed to $\sigma^2_{\widetilde{G}}$). The interaction effects are random variables *across* genotypes with zero means and variance σ^2_I. However, the summation restriction on the interaction effects causes them to be nonindependent *within* genotypes, the average covariance within genotypes being $-\sigma^2_I/(N_E - 1)$. For example, with Falconer's model, where $N_E = 2$, $\sigma(I_{1j}, I_{2j}) = \sigma(+I_j, -I_j) = -\sigma^2_I$. This negative covariance causes the sample-wide expectation of the interaction variance to be $[N_E/(N_E - 1)]\sigma^2_I$, rather than σ^2_I (which is the expectation in the absence of the zero-sum restriction), leading to the unusual appearance of the interaction-variance contribution to the expected mean squares in Table 22.1.

Note that despite these subtle technical differences, there is a simple and close connection between the variance components underlying the random-effects and mixed models, as can be seen by comparing the expressions for the expected mean squares in Table 22.1,

$$\sigma^2_I = \frac{N_E - 1}{N_E}\,\sigma^2_{\widetilde{I}} \tag{22.10a}$$

$$\sigma^2_G = \sigma^2_{\widetilde{G}} + \frac{\sigma^2_{\widetilde{I}}}{N_E} \tag{22.10b}$$

A simple way of understanding these relationships is to recall (from Chapter 2) that, with a sample size of N, the average squared deviation of an observed variable from an observed mean is a downwardly biased estimate of the population variance by a factor of $(N-1)/N$. In effect, by defining the mean interaction effects within genotypes to be zero, the mixed model ignores this sampling bias. On the other hand, the sampling variance of mean interaction effects within genotypes, $(\sigma^2_{\widetilde{I}}/N_E)$, ignored by the mixed model, inflates the genotypic variance. The net effect is that the sum of components of variance associated with genotypes and interaction effects is identical in both models, i.e., $\sigma^2_G + \sigma^2_I = \sigma^2_{\widetilde{G}} + \sigma^2_{\widetilde{I}}$.

Still another way of understanding the main difference between the random- and fixed-effects models provides some help in interpreting the meaning of the variance components. Under the mixed model, σ^2_G is a measure of the variance among marginal mean genotypic values (i.e., genotypic means averaged over all sampled environments), including that caused by sampling of the interaction effects. This equivalence can be seen from Equation 22.10b, which shows σ^2_G to be the sum of the variance of random genotypic values $(\widetilde{G}_j)$ and the variance of a mean interaction effect based on a sample size of N_E. On the other hand, despite its notation as a variance, $\sigma^2_{\widetilde{G}}$ can be seen from Equation 22.9 to be equivalent to the covariance between family members raised in different environments (Hocking 1985, Fry 1992). This result follows from the standard rule, encountered in previous chapters, that an among-group variance component extracted by ANOVA is equivalent to the covariance of individuals within groups. Thus, in the context of genotype × environment interaction, $\sigma^2_{\tilde{G}}$ can actually take on negative values if, for example, there is a strong tendency for the reaction norms of different genetic groups to cross. As a variance of marginal means, σ^2_G is constrained to be positive, as are $\sigma^2_{\tilde{I}}$ and σ^2_I (although estimates of them can be negative).

With the simple translations given by Equations 22.10a,b, and the fact that both models utilize exactly the same observed mean squares, it is relatively easy to move from one sort of analysis to the other. The variance components underlying the two models are strictly identical only in the absence of genotype × environment interaction ($\sigma^2_I = \sigma^2_{\tilde{I}} = 0$), and they can be quite different when N_E is only two. Note, however, from Equations 22.10a,b, that as the number of environments (N_E) becomes large, $\sigma^2_{\tilde{G}}$ converges on σ^2_G (as does $\sigma^2_{\tilde{I}}$ on σ^2_I), and the likelihood of a negative $\sigma^2_{\tilde{G}}$ becomes diminishingly small.

As can be seen from the structure of the expected mean squares in Table 22.1, under either model the hypothesis of no significant interaction variance can be evaluated by use of the F ratio MS_I/MS_e. With the random-effects model, the test statistic for significant genotype effects is MS_G/MS_I, whereas with the mixed model, it is MS_G/MS_e. This difference follows from the requirement that, under the null hypothesis of no significant effects, both mean squares have the same expected values.

Care must be taken in using the variance components extracted from ANOVA to interpret the heritability of a trait. Under either model, assuming purely additive gene effects, the average additive genetic variance within an environment is $(\sigma^2_{\tilde{G}} + \sigma^2_{\tilde{I}})/(2\Theta)$, where Θ is the coefficient of coancestry within genetic groups in the analysis. Thus, if the population were to be confined to a single environment, the *expected* heritability would be

$$h^2 = \frac{\sigma^2_{\tilde{G}} + \sigma^2_{\tilde{I}}}{2\Theta(\sigma^2_e + \sigma^2_{\tilde{G}} + \sigma^2_{\tilde{I}})} \tag{22.11a}$$

If, on the other hand, the population is viewed as being distributed randomly over heterogeneous macroenvironments, then the interaction variance does not contribute to the resemblance between relatives (assuming family members develop in different macroenvironments), and the total phenotypic variance needs to include that due to macroenvironmental effects. Thus, the relevant measure of heritability becomes

$$h^2 = \frac{\sigma^2_{\tilde{G}}}{2\Theta(\sigma^2_e + \sigma^2_{\tilde{G}} + \sigma^2_{\tilde{I}} + \sigma^2_{\tilde{E}})} \tag{22.11b}$$

Example 2. A highly informative, although labor-intensive, approach to ascertaining the degree of genotype × environment interaction in nature is to transplant clonal replicates to various positions in a natural landscape. In one of the few experiments of this type, Stratton (1994) assayed the performance of three genotypes of the asexual annual plant *Erigeron annuus* at 630 locations in an 0.5 hectare old field. As can be seen in the following contour map of total seed production, average performance varied dramatically across the landscape.

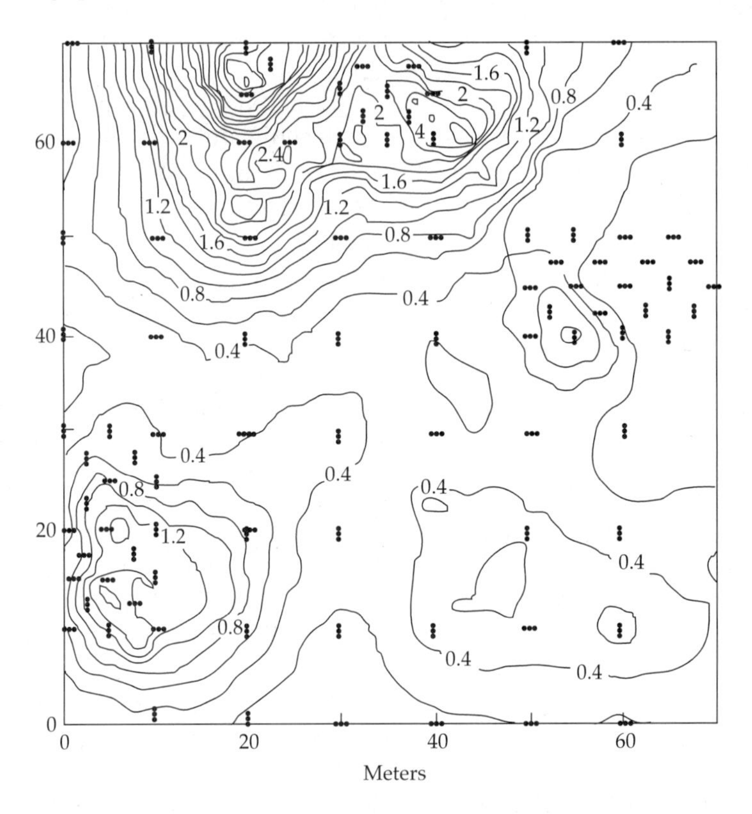

Stratton distributed his plants over the landscape in a hierarchical design: sections separated by distances > 25 m, clusters within sections separated by 3 to 25 m, plots within clusters separated by 0.5 to 1.0 m, and subplots within plots with individuals separated by 10 cm. Significant genotype $\times$ environment interaction was detected in this study, but 78% of it was at the level of subplots, i.e., at a spatial scale of 10 cm or less. Seed dispersal experiments indicated that almost all seeds disperse much farther from their parents than 10 cm, implying that the relevant measure of heritability for this population is given by Equation 22.11b. With only three clones in the experiment, the results can hardly be viewed as giving reliable heritability estimates, and we simply point them out for heuristic purposes. The vast majority of the variance for lifetime fitness (70%) appears at the level of replicate plants within subplots. Such variance is potentially a consequence of microenvironmental variation occurring at scales less than 10 cm and of developmental noise. The phenotypic variance attributable to differences among all higher levels of spatial scale is 22%, that attributable to genotype × environment interaction is 8%, and that associated with genotypes is only 0.3%. Thus, the broadsense heritability for total fitness in this population (0.003) is essentially zero.

Relationship to Falconer's Correlation Across Environments

There is a close relationship between the ANOVA approach to genotype × environment interaction and Falconer's genetic correlation across environments (Robertson 1959b; Dickerson 1962; Yamada 1962, Yamada et al. 1988). Here we show the connection between Falconer's fixed-effects model and the random-effects interpretation (Equation 22.9). For the time being, we assume that the total genetic variance is the same in different environments, so that Falconer's correlation is given by Equation 22.7b, which is expressed in terms of the same variance components that we have used in the mixed model. Substituting from Equations 22.10a,b, the equivalent expression in terms of the random-effects model is

$$\rho_\times = \frac{\sigma^2_{\widetilde{G}} + [(2 - N_E)/N_E]\sigma^2_{\widetilde{I}}}{\sigma^2_{\widetilde{G}} + \sigma^2_{\widetilde{I}}} \tag{22.12a}$$

which reduces to

$$\rho_\times = \frac{\sigma^2_{\widetilde{G}}}{\sigma^2_{\widetilde{G}} + \sigma^2_{\widetilde{I}}} \tag{22.12b}$$

when there are only two environments. Written in this manner, for analyses involving only two environments, $\rho_\times$ can be interpreted as an intraclass correlation of genotypic values in the different environments. Thus, a third way of computing the genetic correlation across two environments is to perform a conventional two-way ANOVA on the data, computing the variance components by the method of moments, and substituting the appropriate values into either Equation 22.7b or 22.12b. One advantage of the ANOVA approach is that the F ratio, MS_I/MS_e, provides an explicit test of the hypothesis of no genotype × environment interaction.

Equation 22.7b, or equivalently Equation 22.12a, can also be used in experiments employing multiple ($N_E > 2$) environments to estimate the average degree of genetic correlation across all pairs of environments. Following the logic developed above, $\rho_\times$ can be negative, but geometric constraints make the chance of this very small when N_E is large.

A slight complication arises when the among-family variance differs among environments, as this causes the interaction variance extracted from two-way ANOVA to be inflated by scale effects. The resultant downward bias in $\rho_\times$ can be corrected by subtracting from the denominator of Equation 22.12a (or 22.7b) the variance of the environment-specific genetic standard deviations (Robertson 1959b, Dickerson 1962, Itoh and Yamada 1990, Muir et al. 1992). To apply this correction, one-way ANOVAs need to be performed on the data for each of the N_E environments. In each case, the among-family component of variance is obtained by the method of moments. Square roots are then taken of each of the N_E estimates, and the variance of the among-family standard deviations is subtracted from the denominator of the preliminary estimate of $\rho_\times$ to provide a scale-independent estimate of the correlation across environments.

FURTHER CHARACTERIZATION OF INTERACTION EFFECTS

The methods outlined in the previous sections of this chapter serve merely to test whether a significant amount of genotype $\times$ environment interaction exists within a population and to provide quantitative estimates of the amount of phenotypic variance associated with interaction effects. An absence of interaction variance implies that the reaction norms of different genotypes are essentially parallel, in which case the population mean reaction norm provides a good indication of the pattern of response to environmental change. On the other hand, the presence of significant interaction variance raises numerous questions. Can the interaction effects be broken down further into biologically interpretable components? To what extent are some genotypes more sensitive to environmental change than others? Is there a correlation between a genotype's stability of phenotypic expression across environments and its average performance?

If the environments in which genotypes are assayed are characterized for different chemical, physical, or biological properties, it is possible to partition the interaction effects into components depending on these features. That is, the interaction effect of genetic group *j* in environment *i* could be described as

$$I_{ij} = \sum_{l=1}^{m} \beta_{jl} x_{il} + \epsilon_{ij} \tag{22.13a}$$

where x_{il} is the measure of the lth feature of the ith environment, β_{jl} is the partial regression coefficient of the *j*th genotype's interaction effects (I_{ij}) on the x_{il}, and ϵ_{ij} is the deviation of I_{ij} from the regression prediction. Although this approach can provide insight into the special features of the environment that contribute to interaction effects, it has the disadvantage of demanding data on numerous aspects of the environment, few or none of which may actually be of direct relevance to the study organism.

Joint-regression Analysis

An alternative approach to Equation 22.13a is to let the mean performance of all genotypes serve as a bioassay of the overall suitability of different environments, allowing each genotype's interaction effects to be expressed as a function of average population performance,

$$I_{ij} = \beta_j(\mu_i - \mu) + \epsilon_{ij} \tag{22.13b}$$

where μ_i is the mean phenotype of all genotypes in environment i. This expression has the advantage that all of the complex (and perhaps unobservable) features of the environment are integrated into a single measure, the average environmental effect $E_i = \mu_i - \mu$. Recalling Equation 22.1, and substituting observed for expected values, this definition of I_{ij} leads to the relationship

$$\overline{z}_{ij} = \overline{z}_{\cdot j} + (1 + B_j)\widehat{E}_i + \epsilon_{ij} \tag{22.14}$$

where $\widehat{E}_i = (\overline{z}_{i\cdot} - \overline{z}_{\cdot\cdot})$. Thus, a regression of environment-specific mean phenotypes for the jth genetic group on the general environmental effects has an intercept equal to $\overline{z}_{\cdot j}$ and a slope equal to $(1 + B_j)$, where B_j is an estimate of β_j.

This idea of partitioning genotype-specific interactions into a component explained by mean population performance and a residual component was first suggested by Yates and Cochran (1938). Now known as **joint-regression analysis,** it was largely neglected until Finlay and Wilkinson (1963) applied it to varieties of barley. Since then it has been used widely in the analysis of crop cultivars. Statistical aspects of this method are covered in Eberhart and Russell (1966), Perkins and Jinks (1968), A. J. Wright (1971; 1976a,b), and Freeman (1973). J. Hill (1975) gives an overview of early applications. For an extension that includes regression on genotypic means (G_j) as well as on the cross-product G_jE_i (see A. J. Wright 1971).

Application of Equation 22.14 to all genotypes in a genotype × environment analysis provides a basis for ranking the genotypes with respect to their responsiveness to environmental change. $B_j = 0$ implies that the average response of genotype j to the environment is the same as the mean response of the population of genotypes in the analysis. $B_j > 0$ implies a stronger than average response, whereas $-1 < B_j < 0$ implies a weaker than average response. If $B_j = -1$, the genotype's mean performance is completely uncorrelated with that of the population mean, and $B_j < -1$ implies that the genotype tends to respond to environmental change in a direction contrary to the average genotype in the population.

It is important to note that B_j is only a measure of a genotype's *tendency* to respond to environmental change in a manner parallel to the population mean. Two genotypes with identical B coefficients may in fact have rather different sets of I_{ij}. For example, the first genotype might perform above average in environments where the second genotype is below average and vice-versa, or the residual variance around the regression $(\sigma^2_{\epsilon j})$ might differ between the genotypes. Thus, $\sigma^2_{\epsilon j}$ provides additional information about a genotype's response to the environment, a larger value of $\sigma^2_{\epsilon j}$ implying that the genotype's performance is only weakly correlated with the overall pattern in the population.

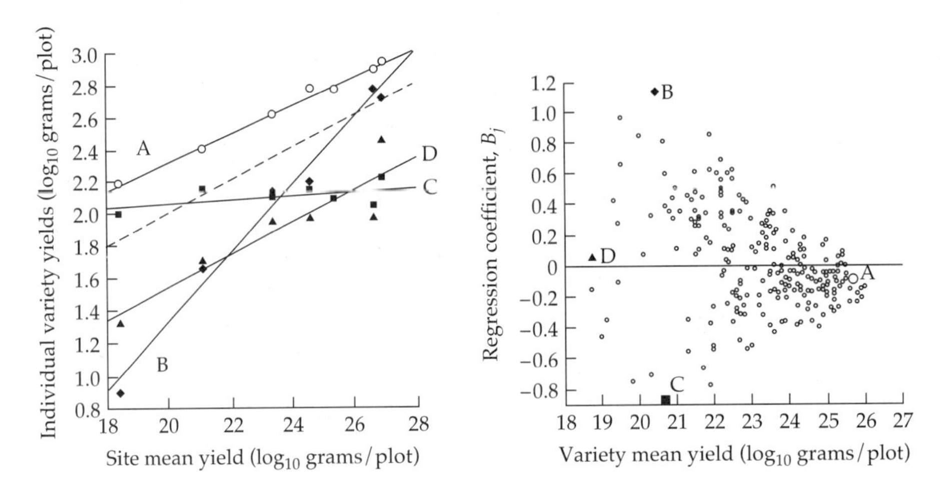

Figure 22.2 **Left:** The reaction norms for yield $(\overline{z}_{ij})$ for four varieties of barley $(j = A, B, C, D)$ as a function of site-specific mean yields for all varieties $(\overline{z}_{i\cdot})$, where i is an index of the sites. The dashed line denotes the average population performance. **Right:** The relationship between the response to environmental change (B_j) and mean performance $(\overline{z}_{\cdot j})$ for 277 barley varieties. (From Finlay and Wilkinson 1963.)

In summary, joint-regression analysis can be used to obtain estimates of three properties of each genetic group in a genotype $\times$ environment analysis: a measure of mean performance $(\widehat{G}_j = \overline{z}_{\cdot j} - \overline{z}_{\cdot\cdot})$, a measure of relative responsiveness to environmental change (B_j), and a measure of consistency of response relative to the linear model $(\sigma^2_{\epsilon j})$.

Finlay and Wilkinson (1963) evaluated the yields of 277 varieties of barley, obtained throughout the world, in seven environments. Figure 22.2 (left) illustrates the joint-regression analyses for four of these varieties on the environment-specific means $\overline{z}_{i\cdot}$. Variety A responds to environmental change in much the same way as average members of the sample $(B_j = -0.10)$, but produces above average yields in all environments. On the other hand, variety C is nearly unresponsive to environmental change $(B_j = -0.86)$ and has a lower than average yield in benign environments, but a higher than average yield in harsh environments. The joint distribution of genotypic mean performances $(\overline{z}_j)$ and environmental sensitivity (B_j) for all 277 lines (Figure 22.2, right) shows that: (1) all lines exhibit the same directional response to the environment (all B_j are > -1), and (2) no high performing lines have very high or very low responsiveness to the environment (as $\overline{z}_{\cdot j}$ becomes high, the range of B_j becomes quite narrow, with a mean B only slightly less than zero). Conversely, low performing lines display a very wide range of responses. Thus, although $\overline{z}_{\cdot j}$ and B_j are essentially uncorrelated in this study, they are certainly not independent.

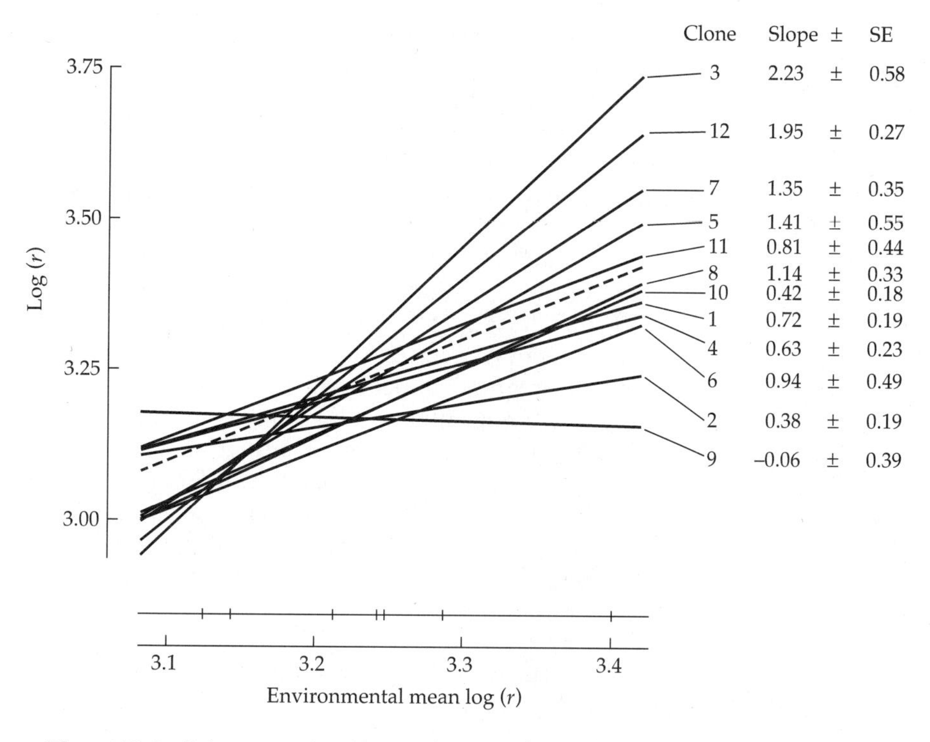

Figure 22.3 Joint regressions for 12 clones of the green alga *Chlamydomonas reinhardtii.* All of the clones were derived from a single cross between two parental strains, and assayed in eight environments varying in nitrogen, phosphorus, and carbon availability. The mean phenotype, log(r), is the logarithm of the rate of exponential growth at low density. The dashed line represents the common regression with a slope equal to one. Note that the variance in performance among clones increases with increasing environmental value. The clonal performances converge to similar values in intermediate environments and then diverge as environmental conditions become progressively worse, i.e., there is a tendency for the joint-regression lines to cross. (From Bell 1991; see also Bell 1990.)

Application of joint-regression analysis to other types of organisms has often yielded a positive correlation between mean performance ($\overline{z}_{\cdot j}$) and B_j (Eberhart and Russell 1966, Perkins and Jinks 1968, Fripp and Caten 1973, Jinks and Connolly 1973, J. Hill 1975, Zuberi and Gale 1976, Jinks and Pooni 1982, Garbutt and Zangerl 1983). The existence of a such a genetic correlation between environmental sensitivity and mean performance can have profound evolutionary consequences, as can be seen by referring to Figure 22.3. For example, if directional selection operates in a positive direction in an environment that magnifies the expression of the trait, a steep reaction norm (high phenotypic plasticity) will evolve as a correlated response. A subsequent change of the environment toward poorer conditions would then result in a substantial reduction in the mean phenotype, to levels lower than would be exhibited by the genotype best adapted to the poor environment (but poorly adapted to favorable environments). Simmonds (1981) reviews evidence that improvement of agricultural practices combined with selection in the improved environments has led to the inadvertent selection of genotypes whose phenotypic expression is highly responsive to environmental change.

The usual approach to estimating the β_j has been to simply perform a least-squares regression of the $\overline{z}_{ij}$ on the $\overline{z}_{i\cdot}$, although this can lead to slightly biased estimates of the regression coefficient β_j. Due to the fact that $(\overline{z}_{i\cdot} - \overline{z}_{\cdot\cdot})$ is only an estimate of the true environmental effect E_i, the variance of the latter, which appears in the denominator of the regression coefficient B_j, is inflated by sampling variance. This source of bias does not influence the rankings of the genotypes with respect to B_j, since the covariance is unaffected (Hardwick and Wood 1972). However, somewhat more problematical is the fact that $\overline{z}_{i\cdot}$ contains a contribution from $\overline{z}_{ij}$. This contribution inflates the correlation between the two measures, to a degree that differs from genotype to genotype depending on their contributions to the $\overline{z}_{i\cdot}$, and it can be of some significance when the number of genetic groups is small.

A modification suggested by A. J. Wright (1976b) eliminates the latter problem by regressing the jth genotypic means on an index of the environment based on the means of the other $(N_G - 1)$ genotypes,

$$B_j = \left(\frac{N_G - 1}{N_G - 2}\right)\left(\frac{C_j}{\overline{C}} - 1\right) \tag{22.15a}$$

where

$$C_j = \text{Cov}(\overline{z}_{ij}, \overline{z}_{i\cdot}) - \frac{\text{Var}(\overline{z}_{ij})}{N_G} \tag{22.15b}$$

with $\text{Var}(\overline{z}_{ij})$ being the variance of the environment-specific mean phenotypes for genotype j, $\text{Cov}(\overline{z}_{ij}, \overline{z}_{i\cdot})$ being the covariance of the performance of the jth genotype on estimates of the environmental effects based on the remaining $(N_G - 1)$ genotypes, and $\overline{C}$ being the mean of the N_G estimates of the C_j. The significance of each regression can be tested by evaluating the correlation

$$r_j = B_j\sqrt{\frac{N_G\,\overline{C}}{(N_G - 1)\text{Var}(\overline{z}_{ij})}} \tag{22.15c}$$

against its critical value with $(N_E - 2)$ degrees of freedom.

Table 22.2 Partitioning of the interaction sum of squares by joint-regression analysis, under the random-effects model with a completely balanced design.

Factor	Degrees of Freedom	Sums of Squares	Expected MS
Environment (E)	$N_E - 1$	$nN_G \sum_i (\overline{z}_i - \overline{z})^2$	$\sigma_e^2 + n\sigma_{\widetilde{I}}^2 + nN_G\sigma_{\widetilde{E}}^2$
Genotype (G)	$N_G - 1$	$nN_E \sum_j (\overline{z}_j - \overline{z})^2$	$\sigma_e^2 + n\sigma_{\widetilde{I}}^2 + nN_E\sigma_{\widetilde{G}}^2$
Interaction (I)	$(N_E - 1)(N_G - 1)$	$n \sum_{i,j} (\overline{z}_{ij} - \overline{z}_i - \overline{z}_j + \overline{z})^2$	$\sigma_e^2 + n\sigma_{\widetilde{I}}^2$
Regression (B)	$N_G - 1$	$\mathrm{SS}_E \cdot \sum_i B_j^2 / N_G$	$n(\sigma_{\widetilde{I}}^2 - \sigma_\epsilon^2)$
Residual (ϵ)	$(N_E - 2)(N_G - 1)$	$n \sum_{i,j} \epsilon_{ij}^2$	$\sigma_e^2 + n\sigma_\epsilon^2$
Error (e)	$N_E N_G (n - 1)$	$\sum_{i,j,k} (z_{ijk} - \overline{z}_{ij})^2$	σ_e^2

Note: SS_E and $E(\mathrm{MS}_I)$ are, respectively, the error sum of squares and the expected interaction mean square.

In effect, joint-regression analysis partitions the interaction sum of squares in a two-way ANOVA into two components — one due to the heterogeneity of regressions (variance of the B_j), and one due to still unexplained interaction variation (variance of the ϵ_{ij}) (Table 22.2). This partitioning follows directly from Equation 22.13b and the fact that under a least-squares approach the residual deviations (ϵ_{ij}) are uncorrelated with the environmental effects. There is, of course, no reason for pursuing joint regression if the total interaction variance is not significant, but if the estimate of σ_I^2 is significant, then one or both of its components must be as well.

The variance associated with heterogeneity of regressions can be tested for significance by use of the F ratio of the total interaction mean square and the residual interaction mean square, $\mathrm{MS}_I/\mathrm{MS}_\epsilon$, as under the null hypothesis $\sigma_{\widetilde{I}}^2 = \sigma_\epsilon^2$ and the expected mean squares are identical. The residual interaction effects can be tested for significance $(\sigma_\epsilon^2 > 0)$ using the ratio $\mathrm{MS}_\epsilon/\mathrm{MS}_e$. If the regression mean square alone is significant, then within the limits of sampling error all of the interaction effects are predicted by linear regressions on the environmental values. If only the residual mean square term is significant, then there is no general linear relationship between the interaction effects and environmental values, and the joint-regression analysis has offered no further insight into the basis of the interaction effects.

Example 3. The following table summarizes the results from the experiment of Finlay and Wilkinson (1963) mentioned above (Figure 22.2). Each of the $N_G = 277$ barley varieties was grown in $N_E = 7$ environments in $n = 3$ replicates. By use of F tests, the authors found all of the sources of variance to be significant. Our purpose is to quantify their relative contributions to the total phenotypic variance using the relationships given in Table 22.2.

Source of Variation	df	Mean Squares	Variance Estimates
Environments	6	125.5803	0.1510
Genotypes	276	0.5618	0.0235
Genotype × Environment	1,656	0.0616	0.0160
Regressions	276	0.2227	0.0130
Deviations from regressions	1,380	0.0294	0.0030
Error	3,878	0.0205	0.0205

Equating the observed mean squares with their expected values (Table 22.2), by the method of moments, we obtain the total interaction variance as $\text{Var}(I) = N_E(\text{MS}_I - \text{MS}_e)/[n(N_E - 1)] = 0.0160$, using the relationship in Equation 22.10a. Similarly, the variance due to deviations around the regressions is estimated by $\text{Var}(\epsilon) = (\text{MS}_\epsilon - \text{MS}_e)/n = 0.0030$. The remaining interaction variance, 0.016 − 0.003 = 0.013, is explained by the joint-regression analysis. The components of variance due to the main effects are estimated in a similar way. Since 67% (0.1510/0.2270) of the total variation is explained by macroenvironmental differences ($\sigma^2_{\widetilde{E}}$), and 10% (0.0205/0.2270) by microenvironmental variation (σ^2_e), variation among growing sites has a substantial impact on the performance of most varieties. Of the variation attributable to genotypes (0.0235 + 0.0160), 40% involves genotype × environment interaction. Only about 1% of the total phenotypic variance is due to deviations around the regressions, while about 6% is accounted for by the regressions.

Testing for Cross-over Interaction

For cases in which significant genotype × environment interaction is revealed by ANOVA or other means, a closer look at the data often reveals that the estimated norms of reaction have a wide variety of shapes, parallel in some ranges of environments, diverging in others, and intersecting on occasion (Figure 22.4).

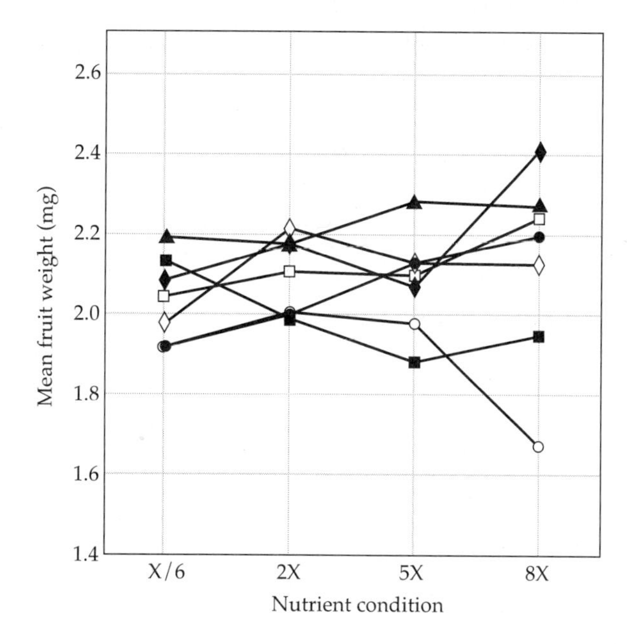

Figure 22.4 Reaction norms for mean fruit weight (mg) for seven genotypes extracted from a single population of the herbaceous annual plant *Polygonum persicaria*. The genotypes were propagated clonally and grown with replication under four nutrient conditions. Two-way ANOVA revealed significant genotype × environment interaction for this trait. (From Sultan and Bazzaz 1993.)

The latter condition, a **cross-over interaction,** is of particular interest as it cannot be eliminated by a change in scale. Provided they are not artifacts of sampling error, cross-over interactions imply that different genotypes are adapted to different environments, a mechanism that can lead to the maintenance of genetic variance in populations inhabiting heterogeneous environments (Gillespie and Turelli 1989). In agriculture, strong cross-over interactions between alternative cultivars or breeds suggests the need for developing locally adapted strains.

There are several ways to test for the significance of observed cross-over interactions, two of which are outlined in Baker (1988). Here we mention only the test developed by Gail and Simon (1985), which considers the performances of two genotypic groups, 1 and 2, over all N_E environmental states. For each environment, the difference between genotypes is computed, and the differences are grouped into positive ones ($\overline{z}_{i1}$ exceeding $\overline{z}_{i2}$) and negative ones ($\overline{z}_{i2}$ exceeding $\overline{z}_{i1}$). Two test statistics are then constructed, one for each group, from the sum of squared deviations standardized by their respective sampling variances,

$$Q^{+} = \sum_{i=1}^{N_E} \frac{(\overline{z}_{i1} - \overline{z}_{i2})^2\, \delta_i}{\text{Var}(\overline{z}_{i1}) + \text{Var}(\overline{z}_{i2})} \tag{22.16a}$$

$$Q^{-} = \sum_{i=1}^{N_E} \frac{(\overline{z}_{i2} - \overline{z}_{i1})^2\, \delta_i}{\text{Var}(\overline{z}_{i1}) + \text{Var}(\overline{z}_{i2})} \tag{22.16b}$$

where $\delta_i = 1$ when the difference within parentheses is positive and $\delta_i = 0$ otherwise. To test for the significance of cross-over interaction between the two genotypic groups, the minimum of the two statistics is compared with critical values given in Table 1 of Gail and Simon (1985). Large values of the final test statistic imply that cross-overs occur more frequently than can be expected by chance, given the observed sampling variances of the means.

Example 4. Baker (1988) reports differences in mean grain yield (g/m^2) between two cultivars of wheat grown under nine different conditions (combinations of sites and years). On five occasions, the first cultivar outperformed the second: $\overline{z}_{i1} - \overline{z}_{i2} = 20, 17, 84, 22,$ and 84. On four occasions, the second cultivar outperformed the first: $\overline{z}_{i1} - \overline{z}_{i2} = -12,\ -43,\ -78,$ and -21. The standard errors of the differences between means were fairly similar under all nine conditions, averaging 15.3. We will use the square of this value, 235.1, as a general estimate of $[\text{Var}(\overline{z}_{i1}) + \text{Var}(\overline{z}_{i2})]$. Squaring the differences between means, and substituting into Equations 22.16a,b, $Q^{+} = 65.0$ and $Q^{-} = 36.2$. The minimum of these values substantially exceeds the 0.001 critical value (19.9) for nine environments given in Table 1 of Gail and Simon (1985). Thus, it can be concluded that significant cross-over interactions exist between the two cultivars.

Concepts of Stability and Plasticity

Definitions for the stability of phenotypic expression are nearly as numerous as their applications. In general, measures of **phenotypic stability** attempt to quantify a genotype's tendency to exhibit constant phenotypic expression in different environments. In contrast, **phenotypic plasticity** refers to the relative responsiveness of a genotype's outward appearance (or behavior) to environmental change. Both concepts play a central role in evolutionary ecology where, for example, fundamental questions exist as to how genotypes persist in temporally and/or spatially variable environments by modulating phenotypic expression developmentally. In agriculture, the patterns by which different breeds respond to temperature, moisture, and nutrient availability are primary determinants of their desirability in different geographic/economic settings. For subsistence farmers, stability of a cultivar's performance across a broad range of environmental conditions is usually essential, whereas large-scale farms that can afford the luxury of fertilization and irrigation generally view a strong phenotypic response to optimal growth conditions to be most desirable.

Reviews on the historical development, rationale, and use of the various indices of phenotypic stability can be found in Becker (1981), Skrøppa (1984), Lin et al. (1986), Wescott (1986), Nassar and Hühn (1987), Becker and Léon (1988), and Muir et al. (1992). In the following paragraphs, we provide a brief synopsis on four of the more frequently utilized indices, briefly mentioning their pros and cons. In doing so, we refer to the basic linear model (Equation 22.1) and its joint-regression extension (Equation 22.13b),

$$\begin{aligned} z_{ijk} &= \mu + G_j + I_{ij} + E_i + e_{ijk} \\ &= \mu + G_j + \beta_j(\mu_i - \mu) + \epsilon_{ij} + E_i + e_{ijk} \end{aligned}$$

First, as noted above, the average responsiveness of different genotypes to environmental change is often evaluated by assaying their performance in multiple environments. By performing one-way analyses of variance on each of the $j = 1, \ldots, N_G$ genotypic groups, genotype-specific estimates of the among-environment component of variance, $\sigma_E^2(j)$, can be extracted from the mean squares in the usual manner. Such measures estimate the variance of genotype-specific values of $(I_{ij} + E_i)$ over the set of assayed environments, providing a basis for ranking genotypes with respect to average sensitivity to macroenvironmental change.

Second, from the type of analysis just mentioned, the within-environment components of variance, $\sigma_e^2(j)$, provide a measure of the relative sensitivity of individual genotypes to microenvironmental change within macroenvironmental settings. A shortcoming of this measure of developmental stability is that, with genetically variable groups, it contains genetic variance due to segregation as well as environmental variance due to microenvironmental effects. Although little work has been done on the subject, it clearly would be of interest to know whether genotypes with high sensitivity to macroenvironmental change also have relatively high levels of developmental instability, i.e., whether the $\sigma_E^2(j)$ and the $\sigma_e^2(j)$ tend to be correlated.

Third, Wricke (1962) proposed using the contribution of each genotype to the interaction sum of squares of the two-way ANOVA as a measure of "ecovalence." According to Wricke's definition, a high ecovalence implies that a genotype's performance in a specific environment $(\overline{z}_{ij})$ is poorly predicted by the overall genotypic and environmental means ($\overline{z}_{\cdot j}$ and $\overline{z}_{i\cdot}$). Recalling that $\widehat{I}_{ij} = \overline{z}_{ij} - \overline{z}_{i\cdot} - \overline{z}_{\cdot j} + \overline{z}_{\cdot\cdot}$, under the random-effects model

$$\text{Var}_j(\widehat{I}\,) = \frac{\sum_{i=1}^{N_E} (\widehat{I}_{ij} - \overline{I}_j)^2}{N_E - 1} \tag{22.17}$$

provides an estimate of $\sigma_I^2(j) + [\sigma_e^2(j)/n_j]$, where $\overline{I}_j = \sum_{i=1}^{N_E} \widehat{I}_{ij}/N_E$ is the mean interaction effect for genotype j, and n_j is the average sample size of genotype j within each of the N_E treatments. The term $\sigma_e^2(j)/n_j$ is a measure of the sampling variance of the $\overline{z}_{ij}$. If the average sample sizes and the residual variances are the same for different genotypes, then $[\sigma_e^2(j)/n_j]$ is a constant, and the genotypic values of $\text{Var}_j(\widehat{I})$ provide an unbiased basis for ranking with respect to stability of interaction effects. If, however, the sampling variances are heterogeneous, the genotype-specific estimates $\text{Var}_e(j)/n_j$ should be subtracted from the estimated $\text{Var}_j(\widehat{I})$ before comparing genotypes. Further details on these matters can be found in Shukla (1972, 1982), Muir et al. (1992), and Piepho (1994).

Fourth, as noted above, the joint-regression coefficient β_j is a measure of the relationship between the interaction effects (I_{ij}) of a genotype and the environmental values defined by the mean performance of the entire population. Taken alone, β_j provides only a weak understanding of the phenotypic response to environmental change, as a lack of close correspondence between the behavior of an individual genotype and that of the population at large need not imply anything about the genotype's actual variation across environments. Moreover, as a regression coefficient, β_j provides no information about the goodness-of-fit of a joint regression. As noted above, information on the latter is provided by $\sigma_\epsilon^2(j)$, the residual variance around the regression.

Finally, we note that whereas $\sigma_E^2(j)$ and $\sigma_e^2(j)$ are intrinsic properties of the genotype, estimable without regard to other members of the population, all of the other measures of genotypic stability mentioned above can only be estimated when data are available on other genotypes. Such measures indicate only the degree to which a genotype's average response to environmental change reflects the population pattern. Thus, although low values of $\sigma_I^2(j)$ or near-zero estimates of β_j imply that a genotype's phenotypic expression is predictable from data on other members of the population, they convey no information on a genotype's intrinsic phenotypic stability across environments.

Additional Issues

More complicated experimental designs for analyzing genotype $\times$ environment interaction are frequently employed, and details on their analysis can be found in most statistics texts concerned with linear models. For example, the nested full-sib, half-sib design can be embedded in a two-way ANOVA by raising replicate members of each full-sib family in different environments (Pani and Lasley 1972), and diallel designs can be extended in a similar manner (Cockerham 1963). In principle, these approaches can allow the partitioning of the interaction variance into components associated with additive and dominance genetic effects. If more than one environmental factor is employed simultaneously in an analysis (e.g., family members might be assayed in all possible combinations of several temperature and light treatments, or at several sites in several years), it becomes possible to test for the existence of higher-order (e.g., three-way) interactions (Gordon et al. 1972, Bell 1991). Genotype × environment analysis can also be extended to the analysis of two traits. All of the approaches that we have described for estimating the genetic correlation of the same trait across environments can be used to estimate the correlation between different traits across environments (Aastveit and Aastveit 1993).

THE QUANTITATIVE GENETICS OF GENOTYPE × ENVIRONMENT INTERACTIONS

Very little empirical work has been done on the evolutionary properties of reaction norms, despite their fundamental significance in issues such as the evolution of specialization vs. generalization, the costs and advantages of developmental homeostasis, and the extent to which genotype × environment interaction can maintain genetic variation in natural populations. It is still an open question as to whether phenotypic plasticity evolves in response to selection for plasticity genes *per se* or whether it is simply a by-product of selection favoring the expression of specific phenotypes in different environments (Scheiner and Lyman 1991, Scheiner 1993, Schlichting and Pigliucci 1993, Via 1993, de Jong 1995). Although many types of phenotypic plasticity are certainly adaptive, it is also still an open question as to whether many phenotypic responses to environmental changes are simply pathological consequences of the inability of genotypes to cope with an altered environment. Nonoptimal reaction norms can be historical artifacts (a consequence of past selection), but they can also be expected if an absolute genetic constraint (such as the complete lack of additive genetic variance for a reaction norm feature) is present.

As a potential guide to future empirical research, we close this chapter by considering how environment-dependent expression of genotypes can be related to conventional quantitative-genetic models. The methodological approaches presented above give a somewhat distorted view of the situation, since by necessity experiments usually employ discrete environments, although the states of the specific treatments are generally taken from an underlying continuous scale. A more realistic view might be to treat the expression of each allele as a continuous function of the underlying environmental determinants. An interesting start in this direction has been made by de Jong (1990). Consider a character influenced by a single locus with two alleles, and let the allelic effects be linear functions of an environmental variable E,

$$\alpha_1 = a_1 + c_1 E \tag{22.18a}$$

$$\alpha_2 = a_2 + c_2 E \tag{22.18b}$$

If $c_i \neq 0$, the expression of the ith allele is environment dependent. Under the assumption of additive gene action, the response (reaction norm) of a genotype to a change in environment is

$$R_{ij} = \alpha_i + \alpha_j = (a_i + a_j) + (c_i + c_j)E \tag{22.19}$$

For a polygenic trait, the reaction norm is obtained by summing the R_{ij} over all loci.

Now recall from Chapter 4 that under an additive model the average effect of allelic substitution in a randomly mating population is $\alpha = \alpha_1 - \alpha_2$, and that the additive genetic variance at the locus is $\sigma_A^2 = 2pq\alpha^2$, where p and q are the frequencies of the two alleles. Substituting from above,

$$\sigma_A^2 = 2pq[(a_1 - a_2) + (c_1 - c_2)E]^2 \tag{22.20a}$$

describes the genetic variance as a function of the environmental parameter. Expanding Equation 22.20a, the genetic variance can be expressed as a function of three terms, the variances and covariance of intercepts and slopes,

$$\sigma_A^2 = \sigma_a^2 + 2E\sigma_{a,c} + E^2\sigma_c^2 \tag{22.20b}$$

Unless the expressions of the two alleles respond to the environment in exactly the same way (in which case $\sigma_c^2 = \sigma_{a,c} = 0$ because $c_1 = c_2$) the genetic variance must change with the environment. Moreover, there will be a point on the environmental gradient at which the allelic functions cross, $E^* = (a_1 - a_2)/(c_2 - c_1)$. When the environment is at this state, the effect of allelic substitution is zero, and as a consequence so is the genetic variance. With a polygenic character, it seems unlikely that E^* would ever be the same for every locus, a requirement for the total genetic variance to be eliminated completely, but there are many ways in which σ_A^2 might vary with E. Thus, de Jong's model, and variants of it, provide a simple starting point for understanding how components of genetic variance might change with the state of the environment. Ward (1994), for example, reviews data that suggest that heritabilities tend to increase when traits are expressed in more extreme environments. Although such changes could result from the expression of different sets of genes in different environments, de Jong's model makes it clear that this need not be the case.

Now consider the genetic covariance for the trait as expressed in two different environments, E_j and E_k,

$$\sigma_{Aj,Ak} = 2pq[(a_1 - a_2) + (c_1 - c_2)E_j][(a_1 - a_2) + (c_1 - c_2)E_k] \tag{22.21}$$

This expression again shows that unless $c_1 = c_2$, the covariance across environments will depend on both E_j and E_k. If either E_j or E_k is equal to E^*, the covariance will be zero. If both are greater or both are less than E^*, the covariance will be positive, but if the two environmental states straddle the critical value, the covariance will be negative. Thus, this relatively simple model also provides a mechanistic explanation for how the genetic correlation across environments might change sign depending upon which environments are employed in an analysis.

de Jong (1990) has generalized these results to an arbitrary number of loci in gametic phase disequilibrium. Further extension to allow for nonlinear regressions of allelic effects on E and / or nonadditive interactions (dominance and epistasis) is relatively straightforward, although tedious, and a start in this direction has been made by Ward (1994) and de Jong (1995).

A related, but rather different, approach to the analysis of reaction norm genetics has been suggested by Gomulkiewicz and Kirkpatrick (1992). Instead of focusing on the detailed genetics of individual loci, they view an individual as possessing a reaction norm function with underlying genetic and environmental components. As in de Jong's model, the genotypic reaction norm function describes the expected phenotypic response over the entire environmental gradient E, but in the Gomulkiewicz-Kirkpatrick model, no assumptions are made about the form of the function, e.g., one genotype might possess a linear reaction norm, and another a nonlinear reaction norm. The additive genetic variance structure for reaction norm properties is encapsulated in a genetic variance-covariance function, a three-dimensional surface, the height of which describes the additive genetic covariance of performance between all possible points on the environmental gradient, i.e., the surface of $\sigma(A_{E_x}, A_{E_y})$ for all (x, y). Two conditions on this surface can lead to an absolute constraint on the evolution of the reaction norm — additive genetic variances equal to zero, or additive genetic correlations equal to ± 1.

Gomulkiewicz and Kirkpatrick (1992) discuss how an interpolation between the elements of a genetic covariance matrix, obtained by assaying performances of family members at discrete points along the environmental continuum, can yield an approximation to the continuous reaction norm function. Due to the large sampling errors that are normally associated with estimates of genetic covariances (Chapter 21), such extrapolations can be expected to be rather crude unless sample sizes are enormous. However, this does not detract from the conceptual value of the approaches of both de Jong and Gomulkiewicz and Kirkpatrick as means of elucidating the critical issues underlying the evolution of reaction norms.

23

Maternal Effects

From the standpoint of genetic analysis, we have already had three main encounters with maternal effects: the nested sib design (Chapter 18), the monozygotic twin half-sib method (Chapter 19), and the factorial designs with reciprocal crosses (Chapter 20). However, none of these techniques provides a full resolution to the issue of quantifying variance associated with maternal effects. For example, under the nested design, the degree by which the dam component of variance exceeds the sire component is a function of dominance and epistatic genetic variance as well as of the variance due to common maternal environment. Thus, with this sort of analysis, excess variance associated with dams cannot be taken as definitive evidence of maternal effects variance. Factorial designs with reciprocal crosses remove this ambiguity, but their application is restricted to inbred lines and populations of hermaphrodites.

This chapter expands our discussion of maternal effects by presenting additional methods for their analysis. We start by extending our previous descriptions of the causal sources of phenotypic variance and covariance between relatives (Chapter 7) to allow for maternal effects in a wide variety of contexts. Although we restrict attention to situations in which epistatic sources of maternal genetic variation are of negligible importance, it will be seen that there are still a large number of ways in which maternal effects contribute to the phenotypic variance in populations. We then consider several empirical approaches for acquiring estimates of the causal components of variance (or functions of them). Examples throughout the chapter will emphasize the significance of maternal effects in organisms ranging from mammals with extended maternal care to seed plants provisioning their young with endosperm, showing how the failure to account for such effects can result in gross misunderstandings about the mode of phenotypic inheritance. We close the chapter by demonstrating how the models and methods for analyzing maternal effects can be extended to indirect effects of other relatives, including fathers, grandparents, and sibs.

Early theoretical work on the subject of maternal effects was done by Dickerson (1947), Koch and Clark (1955), and Kempthorne (1955), and later generalized by Willham (1963, 1972), whose approach we initially adhere to (Figure 23.1). Letting w denote the mother of individual x, the phenotypic value of a character of x can be viewed as the sum of two components. The first component is a function of the direct expression of x's genotype and special environmental effects, $z_{ox} = \mu + G_{ox} + E_{ox}$, where o denotes a direct effect, and μ is the

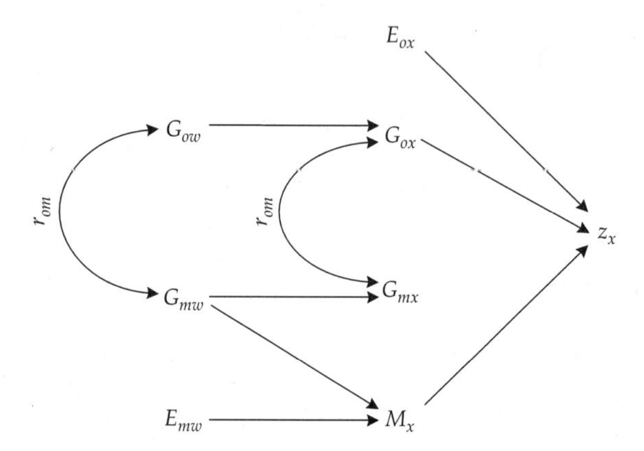

Figure 23.1 Path diagram (Appendix 2) representing the determination of the phenotype z_x of an individual x by direct genetic effects G_{ox}, direct environmental effects, E_{ox}, and maternal effects M_x. The mother of x is denoted by w; r_{om} is the genetic correlation between direct and maternal effects. Assuming x is female, its genetic maternal effect, G_{mx}, is expressed in its offspring.

population mean phenotype. The second component, the maternal effect, is an indirect effect of the maternal phenotype, and it too can have genetic and environmental components, $M_x = G_{mw} + E_{mw}$. In what follows, it is important to keep in mind that a maternal effect is a composite function of many possible aspects of the maternal phenotype, none of which may be the character being evaluated in the progeny. Although they may have a genetic component, maternal effects are an environmental source of variance from the standpoint of the offspring. Thus, M_x, a property of the mother, is expressed in the offspring. Summing up, the phenotype of individual x is represented by

$$z_x = z_{ox} + M_x = \mu + (G_{ox} + E_{ox}) + (G_{mw} + E_{mw}) \tag{23.1}$$

Throughout, we assume that the direct and indirect genetic effects, G_{ox} and G_{mw}, are random variables, potentially containing additive and dominance effects, but exhibiting no epistasis. A genetic covariance, which we denote by σ_{G_o,G_m}, may exist between the two effects, as for example when genes with direct effects on maternal body size also affect characters that influence the provisioning of offspring. A fuller description of the genetic variances and covariances is given in the following section. The environmental effects, E_{ox} and E_{mw}, are also assumed to be random deviates with expectations zero, and they too may exhibit a covariance, which we denote by σ_{E_o,E_m}. However, we will assume that the genetic and environmental effects are independently distributed with no interactions between them.

COMPONENTS OF VARIANCE AND COVARIANCE

Because the addition of maternal effects doubles the number of factors in the linear model for the phenotype, it greatly increases the number of causal components that potentially contribute to the resemblance between relatives. Consider the phenotypic covariance between two individuals — x with mother w, and y with mother z. Assuming no genotype $\times$ environment interaction, there are still eight causal components of covariance,

$$\sigma_z(x,y) = \sigma_{G_o}(x,y) + \sigma_{G_o,G_m}(x,z) + \sigma_{G_o,G_m}(y,w) + \sigma_{G_m}(w,z) + \sigma_{E_o}(x,y) + \sigma_{E_o,E_m}(x,z) + \sigma_{E_o,E_m}(y,w) + \sigma_{E_m}(w,z) \quad (23.2)$$

We first consider the four sources of environmental covariance given in the second line of this formula:

1. $\sigma_{E_o}(x,y)$ is the covariance of the direct effect of the environment on individuals x and y. It will generally be zero, except when x and y are the same individual, in which case it is part of the environmental variance, $\sigma^2_{E_o}$.

2. $\sigma_{E_o,E_m}(x,z)$ is the covariance of the direct effect of the environment on individual x with the environmental contribution of the maternal effect transmitted by y's mother. It may be of significance in a number of situations. For example, if x is the mother of y, then $x = z$, and $\sigma_{E_o,E_m}(x,z) = \sigma_{E_o,E_m}(x,x) = \sigma_{E_o,E_m}$, the covariance between the direct environmental effect on a female and the maternal environmental effect that she contributes to her offspring. As another example, if x and y share the same mother z, $\sigma_{E_o,E_m}(x,z)$ will be unequal to zero if the direct effect of the environment on sib x influences the mother's treatment of sib y. Because of symmetry with $\sigma_{E_o,E_m}(x,z)$, the preceding arguments also apply to $\sigma_{E_o,E_m}(y,w)$.

3. $\sigma_{E_m}(w,z)$ is the covariance of the environmental component of the maternal effects on x and y. If x and y are the same individual, then $w = z$ and $\sigma_{E_m}(w,z) = \sigma^2_{E_m}$, the variance of environmental maternal effects. For other relationships involving direct ancestors, the situation is less clear-cut, but a simple approach suggested by Falconer (1965a) can be utilized as a first approximation. Here it is assumed that the environmental maternal effect transmitted by an individual x consists of two components: a residual fraction (b) of the maternal effect from x's mother (bE_{mw}), and a unique contribution (E'_{mx}), so $E_{mx} = bE_{mw} + E'_{mx}$. Suppose, for example, that x is the mother of y, then $x = z$, and $\sigma_{E_m}(w,z) = \sigma_{E_m}(w,x) = \sigma[E_{mw}, (bE_{mw} + E'_{mx})] = b\sigma^2_{E_m}$. In the case of maternal sibs, $w = z$, and if the unique components of E_{mw} transmitted to each offspring are independent, we would expect that $\sigma_{E_m}(w,z) = b^2\sigma^2_{E_m}$. However,

through sib competition and/or cooperation, the maternal effect transmitted to x may depend on that transmitted to y. Thus, a more general approach is to define the covariance of environmental maternal effects on maternal sibs as $(b^2+c)\sigma^2_{E_m}$, where $c\sigma^2_{E_m}$ is the covariance of unique maternal environmental effects dispensed on sibs. The quantity $(b^2+c)\sigma^2_{E_m}$ is equivalent to the variance due to common maternal environment used in previous chapters.

To obtain a description of the genetic components of covariance between relatives, we will adhere to the usual assumptions of gametic phase equilibrium and random mating. Recalling Equation 7.12, the genetic covariance due to direct effects is

$$\sigma_{G_o}(x,y) = 2\Theta_{xy}\sigma^2_{A_o} + \Delta_{xy}\sigma^2_{D_o} \tag{23.3a}$$

where Θ_{xy} and Δ_{xy} are the coefficients of coancestry and fraternity for individuals x and y, and $\sigma^2_{A_o}$ and $\sigma^2_{D_o}$ are the additive and dominance components of variance involving direct genetic effects. The next two genetic covariance terms in Equation 23.2 can be expanded by noting that they involve the direct genetic effects of one individual and the maternal genetic effects contributed by the mother of the second individual,

$$\sigma_{G_o,G_m}(x,z) = 2\Theta_{xz}\sigma_{A_o,A_m} + \Delta_{xz}\sigma_{D_o,D_m} \tag{23.3b}$$

$$\sigma_{G_o,G_m}(y,w) = 2\Theta_{yw}\sigma_{A_o,A_m} + \Delta_{yw}\sigma_{D_o,D_m} \tag{23.3c}$$

The final genetic term in Equation 23.2 describes the covariance of the maternal genetic effects contributed by the two mothers,

$$\sigma_{G_m}(w,z) = 2\Theta_{wz}\sigma^2_{A_m} + \Delta_{wz}\sigma^2_{D_m} \tag{23.3d}$$

Table 23.1 summarizes the coefficients for the different causal components of variance and covariance that contribute to the phenotypic covariances between several types of relatives. Expressions for many other kinds of relatives are given by Willham (1963, 1972), Eisen (1967), and Thompson (1976).

Some interesting and counterintuitive relationships can be seen in Table 23.1. For example, the presence of maternal effects causes the expected phenotypic covariance between father and offspring to be $(\sigma^2_{A_o}/2) + (\sigma_{A_o,A_m}/4)$ rather than $(\sigma^2_{A_o}/2)$. Why should the resemblance between father and offspring be influenced by maternal effects genes? The answer resides in the paternal grandmother's genes. These fully determine the maternal effect on the father's phenotype, and also comprise 25% of the offspring's genome, where their direct effects are expressed. Thus, a genetic correlation between the direct and maternal effects of genes modifies the phenotypic covariance between a male and his descendants. Note further that because σ_{A_o,A_m} is a covariance, it may be negative.

Table 23.1 Coefficients for the components of covariance and variance contributing to the resemblance between relatives in a model that includes maternal effects.

	A_o	D_o	A_o, A_m	D_o, D_m	A_m	D_m	E_o	E_o, E_m	E_m
Father–offspring	1/2	0	1/4	0	0	0	0	0	0
Mother–offspring	1/2	0	5/4	1	1/2	0	0	1	b
Full sibs	1/2	1/4	1	0	1	1	0	2	$b^2 + c$
Reciprocal full sibs	1/2	1/4	1	0	0	0	0	0	0
Paternal half sibs	1/4	0	0	0	0	0	0	0	0
Maternal half sibs	1/4	0	1	0	1	1	0	2	$b^2 + c$
Reciprocal half sibs	1/4	0	1/2	0	0	0	0	0	0
MGM–grandchild	1/4	0	5/8	0	1/4	0	0	0	b
PGM–grandchild	1/4	0	1/8	0	0	0	0	0	0
MGF–grandchild	1/4	0	5/8	0	1/4	0	0	0	0
PGF–grandchild	1/4	0	1/8	0	0	0	0	0	0

Note: It is assumed that epistatic sources of genetic variance are absent. b and c are constants. MGM and PGM stand for maternal and paternal grandmothers, MGF and PGF for maternal and paternal grandfathers.

Indeed, if σ_{A_o,A_m} were sufficiently negative, the regression of offspring on father could be negative. Similar arguments apply to the mother-offspring regression. Such negative regressions have been observed on occasion. For example, Gibbs (1988) observed significant negative mother-offspring regressions for clutch size in Darwin's medium ground finches.

Example 1. Consider the genetic covariance between mother y and offspring x. Since maternal effects are assumed present, we also need to consider the mothers z and w of y and x, respectively. Here $y = w$, whereas z represents individual x's maternal grandmother (as shown in the left of the accompanying figure on the next page).

For this set of relationships, $2\Theta_{xy} = 1/2$, $2\Theta_{xz} = 1/4$, $2\Theta_{wz} = 1/2$, and since w and y represent the same individual, $2\Theta_{yw} = 2\Theta_{yy} = 1$. Because an individual inherits only one gene from each parent, $\Delta_{xy} = \Delta_{xz} = \Delta_{wz} = 0$, but again, since w is y, $\Delta_{yw} = \Delta_{yy} = 1$. Making the appropriate substitutions in

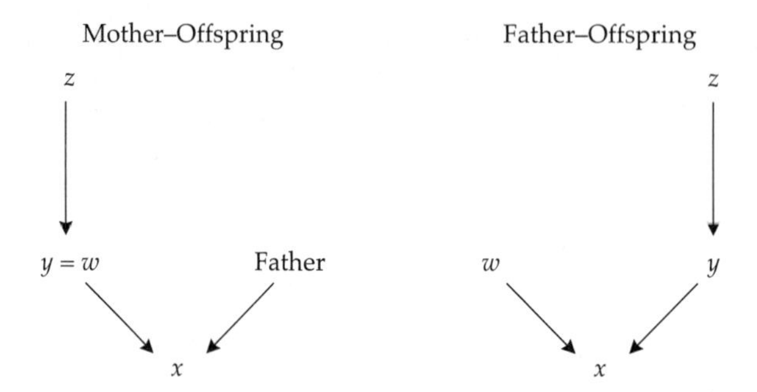

Equations 23.3a-d, the genetic covariance between mother and offspring is found to be

$$\sigma_G(M,O) = \frac{\sigma^2_{A_o}}{2} + \frac{5\sigma_{A_o,A_m}}{4} + \sigma_{D_o,D_m} + \frac{\sigma^2_{A_m}}{2}$$

From arguments presented above, the environmental covariance between mother and offspring can be expressed as

$$\sigma_E(M,O) = \sigma_{E_o,E_m} + b\sigma^2_{E_m}$$

On the other hand, a father-offspring relationship can be represented by letting the offspring be x, the father be y, and the paternal grandmother be z (as shown in the right of the accompanying figure). We then have the coefficients $2\Theta_{xy} = 1/2$, $2\Theta_{xz} = 1/4$, and since the father and mother of x are assumed to be unrelated, $2\Theta_{yw} = 2\Theta_{wz} = 0$. All of the Δ coefficients are also zero in this case. Summing up the terms,

$$\sigma_G(F,O) = \frac{\sigma^2_{A_o}}{2} + \frac{\sigma_{A_o,A_m}}{4}$$

As noted above, the second term represents the covariance between the maternal effect expressed in the father (via z) and the direct effects of the genes of z, one-quarter of which are transmitted to x. Assuming the father is derived from a different maternal lineage than its mate, the environmental covariance between father and offspring is zero.

Thus, the expected difference between the mother-offspring and father-offspring covariances is

$$(\sigma_{A_o,A_m} + \sigma_{D_o,D_m} + \sigma_{E_o,E_m}) + \frac{\sigma^2_{A_m}}{2} + b\sigma^2_{E_m}$$

The first term (in parentheses) is the covariance between all of the direct effects on a mother's phenotype and the maternal effect she contributes to her offspring's phenotype. The second term is half the additive genetic variance for maternal effects; it arises because a mother transmits half the genes that determine her maternal effect to her progeny. The third term is the fraction of the environmental maternal effect on the mother that is transmitted to her offspring (through, for example, physiological effects or cultural inheritance).

Let the difference between the regressions of offspring on mother and offspring on father be m. Since the final two terms in the preceding equation are necessarily positive, one would ordinarily expect m to be positive. However, the opposite has sometimes been observed. For example, Falconer (1965a) found the difference between regressions to be -0.13 for litter size in mice, and Janssen et al. (1988) obtained a difference of approximately -0.5 for age at maturity in springtails. Such a pronounced reduction of the mother-offspring regression relative to that for father-offspring provides a strong indication that the covariance between direct and maternal effects is negative, i.e., that genes whose direct effects cause an increase in the expression of the trait have an antagonistic effect on the trait's expression through their maternal effects.

Cytoplasmic Transmission

Throughout this book, including the preceding section, we have been assuming that all of the genes contributing to phenotypic variation reside in the nuclear genome. However, since critical metabolic functions are carried out by products of genes contained in cytoplasmic organelles (mitochondria and chloroplasts), it seems prudent to keep in mind the possibility that some variation in the expression of quantitative traits may owe its origin to variation among organelle lineages. Organelle genomes are almost always inherited uniparentally, usually through the mother, and except in the case of plant mitochondria, usually with little or no recombination. Thus, from the standpoint of quantitative genetics, an organelle genome can be treated as a single haploid locus.

Extension of the usual expressions of genetic variance and covariance to include cytoplasmic inheritance is straightforward (Beavis et al. 1987). Here we assume that all of the variation due to nuclear genes is due to direct additive and dominance (nonepistatic) effects, and that all cytoplasmic genes are effectively inherited as a single linkage group through mothers. In addition, we assume that the nuclear genes are in gametic phase equilibrium with respect to each other and with respect to the cytoplasmic genomes. The total genetic variance among individuals is then

$$\sigma^2_G = \sigma^2_A + \sigma^2_D + \sigma^2_C + \sigma^2_{AC} + \sigma^2_{DC} \tag{23.4}$$

where σ^2_A and σ^2_D are the familiar components of genetic variance due to the additive and dominance effects of nuclear genes, σ^2_C is the variance of additive effects of the cytoplasmic genome, and σ^2_{AC} and σ^2_{DC} are the variances of the interaction (epistatic) effects between cytoplasmic and nuclear (additive and dominance, respectively) gene effects. Note that because of their haploid nature, organelle genomes do not have dominance effects.

Table 23.2 Coefficients for the components needed for the expression describing the expected phenotypic covariance between relatives in a model that includes cytoplasmic gene expression.

	κ_{xy}	σ^2_C	σ^2_{AC}	σ^2_{DC}
Father–offspring	0	0	0	0
Mother–offspring	1	1	1/2	0
Full sibs	1	1	1/2	1/4
Reciprocal full sibs	0	0	0	0
Paternal half sibs	0	0	0	0
Maternal half sibs	1	1	1/4	0
Reciprocal half sibs	0	0	0	0
Maternal grandmother–grandchild	1	1	1/4	0
Paternal grandmother–grandchild	0	0	0	0
Maternal grandfather–grandchild	0	0	0	0
Paternal grandfather–grandchild	0	0	0	0

Note: The last three columns give the coefficients that precede the three terms in Equation 23.5.

The genetic covariance between relatives due to cytoplasmic effects is

$$\sigma_{G_c}(x,y) = \kappa_{xy}\sigma^2_C + 2\Theta_{xy}\kappa_{xy}\sigma^2_{AC} + \Delta_{xy}\kappa_{xy}\sigma^2_{DC} \tag{23.5}$$

where κ_{xy} is the probability that individuals x and y share organelle genomes that are identical by descent. Although there are multiple copies of organelles in individuals, generally all are identical. Thus, κ_{xy} is equal to one when x and y are members of the same maternal lineage (assuming maternal inheritance), and zero otherwise. Thus, in a random-mating population, unless the direct path from x to y contains only females, κ_{xy} is equal to zero.

Table 23.2 gives the coefficients for the components of genetic variance associated with cytoplasmic genes for some commonly observed relationships. A complete expression for the genetic covariance between a specific group of relatives is obtained by adding the three terms in this table to those outlined in Table 23.1. This inclusion obviously makes an already complicated case even more complex, raising the question as to whether it is even possible to separate the effects associated with organelles (and/or nuclear $\times$ cytoplasmic interaction) from those associated with maternal effects.

Such a partitioning would be possible if offspring could be separated from their parents and from each other at an early enough stage of development that none of the sources of maternal effects variance described in the preceding section contributed to their phenotypic similarities. Such a situation may be extremely difficult to accomplish with most organisms, but methods of cross-fostering and/or embryo transplantation (below) may prove useful in some cases. In any event, assuming that the resemblance due to maternal effects (other than those involving the cytoplasm) can be eliminated, some simple relationships emerge using the coefficients in Tables 23.1 and 23.2. For example,

$$\sigma^2_C = 2\sigma(MHS) - \sigma(M, O) \tag{23.6a}$$

$$\sigma^2_{AC} = 2[2\sigma(M, O) - \sigma(F, O) - 2\sigma(MHS)] \tag{23.6b}$$

$$\sigma^2_{DC} = 4[\sigma(FS) - \sigma(RFS) - \sigma(M, O) + \sigma(F, O)] \tag{23.6c}$$

A few attempts have been made to quantify the significance of cytoplasmic gene differences at the phenotypic level. A large survey of dairy cattle, spanning 10 generations, suggested that about 3% of the variance in milk production is attributable to the maternal mitochondrial lineage (Bell et al. 1985). In a study of two strains of tobacco *(Nicotiana tabacum)*, with identical nuclear but different mitochondrial genomes, large (5 to 30%) differences were found in germination rate, growth rate, and age at first flowering (Pollak 1991). On the other hand, Forbes and Allendorf (1991) were unable to detect any morphological differences between mitochondrial haplotypes in a hybrid swarm of trout, despite the rather high (2%) nucleotide divergence between the two mitochondrial types.

Postpollination Reproductive Traits in Plants

Quantitative-genetic analyses of seeds and their component parts (ovules, endosperm, seed coats) are perhaps more numerous than those of any other characters. In agronomy, large-scale studies on the genetic properties of grain yield have long been driven by economic interests. In evolutionary ecology, studies on the genetics of seed architecture and maturation have been stimulated by interest in reproductive strategies and parent-offspring conflict. Remarkably, almost all studies of seed properties have been performed as though such traits are properties of the maternal genotype, like any other nonreproductive diploid tissue. This, however, is not the case. Three genetically distinct tissues contribute to the expression of various seed properties: (1) the seed coat is a direct product of the diploid maternal genotype, (2) the endosperm is triploid, having two doses of maternal genes and one of paternal genes, and (3) the embryo is a diploid product of the paternal and maternal gametes. To complicate matters further, the paternal genomes that contribute to endosperm and embryo are derived from different gametes.

The main issues have been addressed by Shaw and Waser (1994), and we encourage those with interests in the subject to read their paper; see also Huidong

(1988) and Foolad and Jones (1992). Here we just present a simple example to point out why the issues are nontrivial. Consider the additive variance for direct genetic effects on a character. The usual expectation is that the covariance between full sibs is $\sigma^2_{A_o}/2$, while that between half sibs (either maternal, paternal, or reciprocal) is $\sigma^2_{A_o}/4$. Suppose, however, that the character is an attribute of the seed that is largely determined by the properties of the endosperm. One problem that arises immediately is that the tissues being compared (the endosperm) are not related to the same degree as the plants that bear them, for the simple reason that endosperms acquire additional haploid complements of genes from the pollen donors (which, due to segregation, are almost certainly different from each other). For example, if two seeds within the same maternal plant are products of fertilization by the same male, their endosperms are derived from the same maternal diploid genome, and half of their paternally derived genes are identical by descent. If the two seeds are products of a reciprocal full-sib mating, then one of them has two complements of the first parent's genome and one complement from the second parent, while the situation for the second seed is reversed.

The net effect of all of these different degrees of gene sharing is that different types of half-sibs no longer have the same additive genetic covariance, nor do true full sibs and reciprocal full sibs. Similar arguments apply to the dominance component of variance, and the matter is complicated further by the likelihood that the expression of genes in endosperm may depend on whether they are derived from maternal vs. paternal sources. Add to these concerns the fact that other genomes, distinct from the endosperm, contribute to the properties of a seed, and it seems rather doubtful whether clean estimates of the causal components of variance of seed properties can ever be procured by the usual method of comparing resemblances of different types of relatives. On the other hand, so little empirical work exists on the issues that it is difficult to evaluate whether any or all of the potential complications pointed out by Shaw and Waser (1994) have serious practical implications. Applying a North Carolina II design (Chapter 20) to a population of wild radish (*Raphanus sativus*), Nakamura and Stanton (1989) were able to show that although the pollen donor contributed to the phenotype of the endosperm and the embryo, it never accounted for more than 2% of the phenotypic variance through either of these routes. This result was true even for embryos that were grown on artificial medium after being removed from their seed coats and endosperm. Such results suggest that the classical treatment of seed phenotypes as a property of the maternal genotype may not be so unreasonable. However, more studies of this nature are needed to clarify the issues.

CROSS-FOSTERING EXPERIMENTS

For species in which it is possible to transplant progeny to surrogate mothers, a cross-fostering experiment can reveal whether maternal effects contribute to the resemblance between relatives. In the absence of maternal effects subsequent to the transplantation event, unrelated individuals that are raised by the same mother should exhibit zero phenotypic covariance with each other as well as with their foster mother. Cross-fostering is expected to decrease the phenotypic covariance of parents and their true offspring if maternal effects are significant. The phenotypic covariance between mother and her fostered offspring (FO, raised by a nonrelative) is identical to the father-offspring covariance,

$$\sigma_z(M, FO) = \frac{\sigma^2_{A_o}}{2} + \frac{\sigma_{A_o,A_m}}{4} \tag{23.7a}$$

whereas the phenotypic covariance between foster mother and unrelated foster child is

$$\sigma_z(FM, FO) = (\sigma_{A_o,A_m} + \sigma_{D_o,D_m} + \sigma_{E_o,E_m}) + \frac{\sigma^2_{A_m}}{2} + b\sigma^2_{E_m} \tag{23.7b}$$

These expressions clarify several things. First, note that the two equations sum to the usual covariance between mother and offspring, $\sigma(M, O)$ (Example 1) — the first equation is the phenotypic covariance of a mother and her offspring, exclusive of transmitted maternal effects, whereas the second equation is the covariance of a mother and her offspring, exclusive of direct effects. Second, Equation 23.7b is identical to the difference between the mother-offspring and father-offspring covariance (Example 1). Thus, the regression of foster offspring on foster parent provides a second means of estimating the composite parameter m introduced in Example 1. Third, we see again that since the different types of covariances between mothers and offspring contain covariances between direct and maternal effects, they can take on negative values.

Cross-fostering experiments have been used frequently with natural populations of banded birds to evaluate the validity of heritability estimates derived from parent-offspring regressions. Regressions on true parents typically lead to high heritability estimates (0.5 to 1.0) for morphological traits such as beak dimensions, tarsus and wing length, and body weight (Table 17.1). In order to counter the criticism that such estimates are inflated by maternal effects, some investigators have exchanged eggs of incubating females. Such experiments only control for the effects of maternal investment subsequent to egg laying. Generally, the regressions employing foster parents have not been significantly different from zero, and the regressions of fostered offspring on true parents have remained high, suggesting that variance due to maternal effects is in fact negligible (Figure 23.2). However, exceptions do exist. In great tits (*Parus major*), significant positive regressions on foster parents have been observed for body weight when growth conditions are poor (Gebhardt-Henrich and van Noordwijk 1991).

Although the preceding approach provides a simple way to test for the presence of maternal effects, it does not allow any further separation of terms such as those involving the covariance between direct and maternal effects. A more

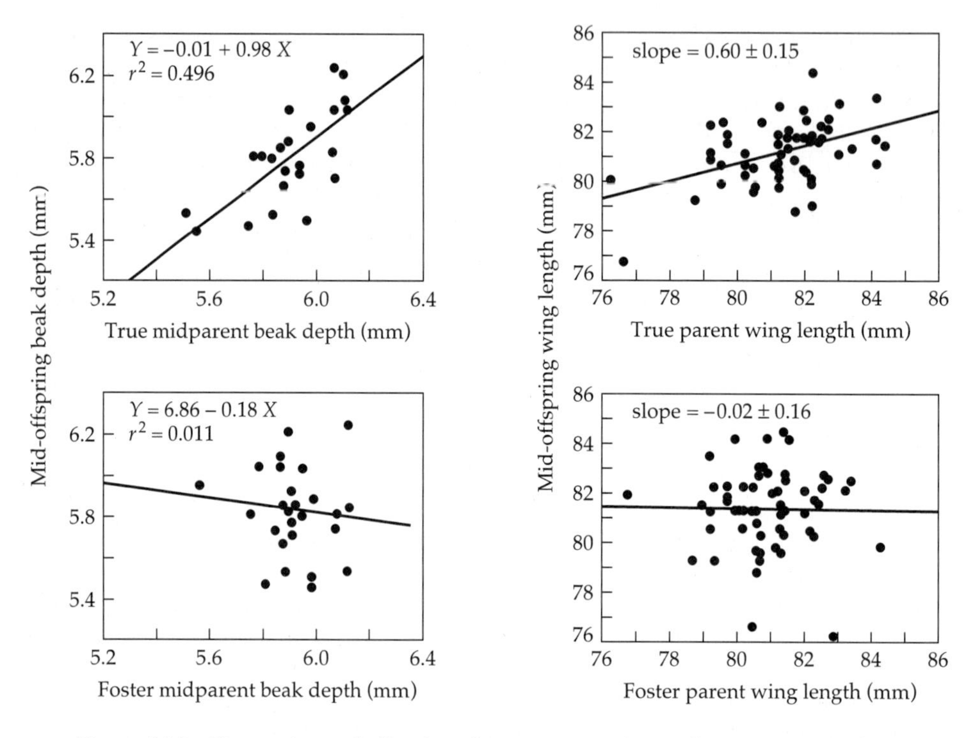

Figure 23.2 Regressions of offspring phenotypes on those of true parents and on those of foster parents. Data on the left are for beak depth in a natural population of song sparrows (Smith and Dhondt 1980). Those on the right are for wing length in a population of the collared flycatcher (Gustafsson and Merilä 1994). In both cases, the true parent-offspring regression is significant, while the foster-parent regression is not.

powerful approach is to incorporate cross-fostering into a full-sib design (Rutledge et al. 1972). Rather than completely exchanging offspring between mothers, pairs of unrelated mothers giving birth on the same day are forced to exchange half of their progeny. Such cross-fostering gives rise to a replicated 2×2 factorial design, with each exchange creating four situations: offspring of mother A raised by A, offspring of A raised by mother B, offspring of B raised by A, and offspring of B raised by B. The rationale of this approach is that maternal effects contribute to the covariance among sibs raised by the same mother (whether it is the true mother, or the foster mother), but not to the covariance among sibs raised by different mothers.

The progeny phenotypes in such an experiment can be described by the following linear model,

$$z_{ijkl} = \mu + P_i + M_{ij} + N_{ik} + I_{ijk} + e_{ijkl} \tag{23.8}$$

Table 23.3 Interpretation of the expected mean squares for a replicated two-way analysis of variance of a cross-fostering experiment, assuming a random-effects model and a completely balanced design.

Factor	df	Sums of Squares	Expected Mean Squares
Pairs	$N_p - 1$	$4n\sum_i (\overline{z}_i - \overline{z})^2$	$\sigma_e^2 + n\sigma_I^2 + 2n\sigma_M^2 + 2n\sigma_N^2 + 4n\sigma_P^2$
Mothers	N_p	$2n\sum_{i,j} (\overline{z}_{ij} - \overline{z}_i)^2$	$\sigma_e^2 + n\sigma_I^2 + 2n\sigma_M^2$
Nurses	N_p	$2n\sum_{i,k} (\overline{z}_{ik} - \overline{z}_i)^2$	$\sigma_e^2 + n\sigma_I^2 + 2n\sigma_N^2$
M × N	N_p	$n\sum_{i,j,k} (\overline{z}_{ijk} - \overline{z}_{ij} - \overline{z}_{ik} + \overline{z}_i)^2$	$\sigma_e^2 + n\sigma_I^2$
Error	$4N_p(n-1)$	$\sum_{i,j,k,l} (z_{ijkl} - \overline{z}_{ijk.})^2$	σ_e^2

Note: $\overline{z}$ is the mean phenotype over all families, $\overline{z}_i$ is the mean phenotype of progeny in the *i*th pair, $\overline{z}_{ij}$ is the mean phenotype of progeny of the *j*th mother within the *i*th pair, $\overline{z}_{ik}$ is the mean phenotype of offspring raised by the *k*th nurse within the *i*th pair, and $\overline{z}_{ijk}$ is the mean phenotype of progeny of the *j*th mother raised by the *k*th nurse within the *i*th pair.

where P_i is the average effect of the *i*th cross-fostered pair, M_{ij} is the direct effect of the *j*th (genetic) mother within the *i*th pair ($j = 1$ or 2), N_{ik} is the effect of the *k*th (unrelated) nurse within the *i*th pair ($k = 1$ or 2), I_{ijk} is the $M \times N$ interaction within the *i*th pair, and e_{ijkl} is the residual error for the *l*th offspring of the *j*th mother raised by the *k*th nurse within the *i*th pair. All of the effects in the model are assumed to be independently distributed with zero expectations. The complete layout of the analysis of variance for such an experiment is given in Table 23.3, where we assume a balanced design with N_p cross-fostering pairs, each involving n full-sibs within a particular mother-nurse grouping (i.e., each mother has $2n$ offspring, n of which are cross-fostered).

We next consider how the observable variance components for mothers, nurses, and mother × nurse interactions relate to the causal components of variance and covariance outlined above. As we have now encountered the general definitions of the variance components in the two-way ANOVA in two preceding chapters (20 and 22), it is not necessary to go into great detail. The variance component due to mothers, σ_M^2, is equivalent to the covariance between full sibs raised by different nurses. (This can be seen by use of Equation 23.8, noting that within the *i*th pair, full sibs that are raised by different nurses share only the term

M_{ij}.) The variance due to nurses, σ_N^2, is the covariance of unrelated individuals raised by the same nurse. The variance due to mother × nurse interaction, σ_I^2, is the covariance of full sibs raised by their mother minus $(\sigma_M^2 + \sigma_N^2)$.

Using these definitions and the coefficients described in Table 23.1, the observable components of variance can be expressed in causal terms,

$$\sigma_M^2 = \frac{\sigma_{A_o}^2}{2} + \frac{\sigma_{D_o}^2}{4} \tag{23.9a}$$

$$\sigma_N^2 = \sigma_{A_m}^2 + \sigma_{D_m}^2 + (b^2 + c)\sigma_{E_m}^2 \tag{23.9b}$$

$$\sigma_I^2 = \sigma_{A_o,A_m} + 2\sigma_{E_o,E_m} \tag{23.9c}$$

The sum of these three components is equal to the expected phenotypic covariance among full sibs raised by their mother. An attractive feature of the cross-fostering design is that it allows a clean partitioning of the causal sources of phenotypic resemblance into components due to the variance of direct effects, variance of maternal effects, and covariance of direct and maternal effects. Note that σ_M^2 is influenced only by the direct effects of genes, since maternal effects do not contribute to the resemblance of individuals raised by unrelated females. (However, any maternal effects that are expressed prior to the cross-fostering (e.g., prenatal effects) will contribute to σ_M^2. On the other hand, σ_N^2 is influenced only by maternal effects, since the individuals concerned are unrelated. Finally, σ_I^2 defines the remaining contribution to the covariance among full sibs raised by the same mother, the covariance of direct and maternal effects.

Without observations on additional kinds of relatives, further decomposition of these quantities into their subsidiary (additive, dominance, and environmental) components is not possible. Riska et al. (1985) suggest how additional information can be extracted from a cross-fostering experiment when phenotypic data are available for sires and dams as well as their offspring. In this case, separate estimates of $\sigma_{A_o}^2$, $\sigma_{D_o}^2$, and $\sigma_{A_m}^2$ can be acquired, although $\sigma_{D_m}^2 + (b^2 + c)\sigma_{E_m}^2$ and $\sigma_{A_o,A_m} + 2\sigma_{E_o,E_m}$ still appear as composite terms.

Body Weight in Mice

The cross-fostering design has been used extensively to evaluate the sources of variance for body weight in laboratory populations and domesticated species of mammals. Here we consider the results from an experiment with an outbred laboratory mouse strain (ICR), in which Rutledge et al. (1972) mated a large number of virgin females to unrelated males. Pairs of unrelated females that released litters within a 12-hour period were treated as cross-foster groups. Their litters were standardized to four males and four females, and then half of each sex were exchanged randomly between mothers. Twenty-eight such pairs were constructed. For identification purposes, all offspring were toe-clipped. Weaning was enforced at 21 days, and subsequently all offspring were weighed to the nearest gram at 3 to 7 day intervals. The results of the analysis of variance for each time interval

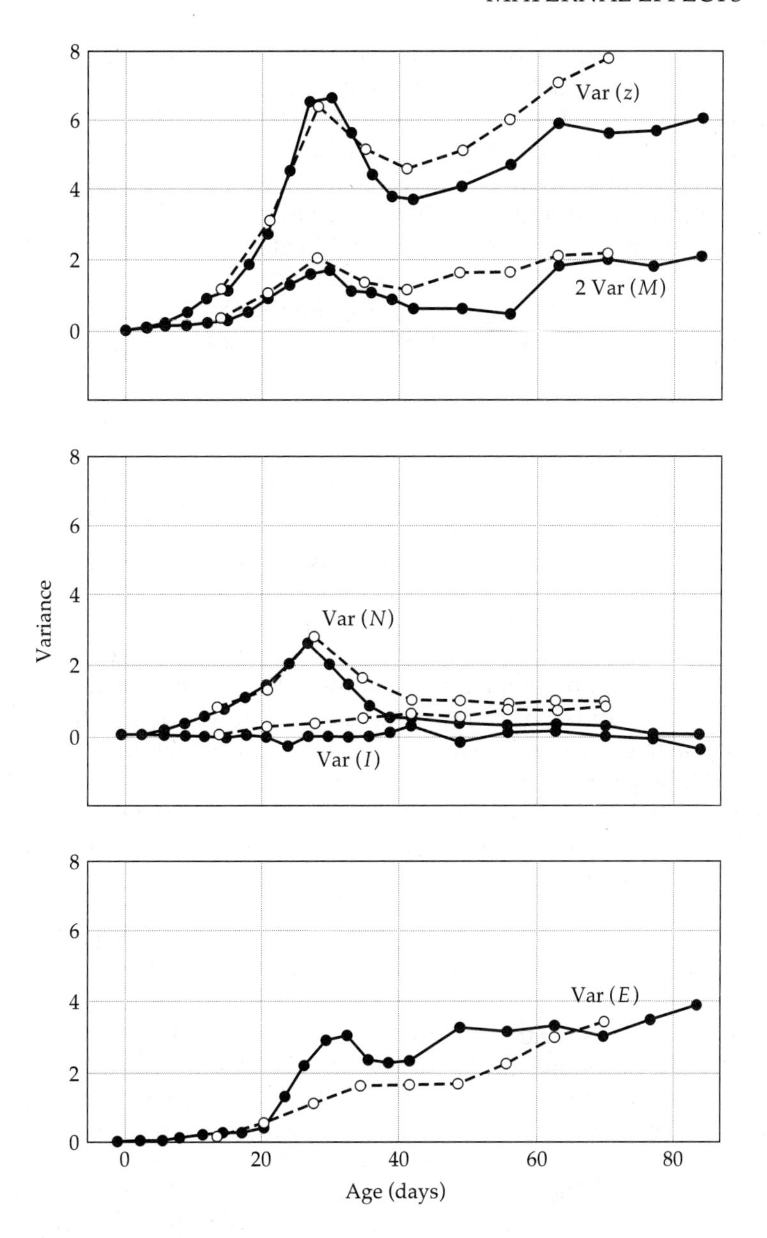

Figure 23.3 Ontogenetic changes in components of variance for body size in mice, determined from cross-fostering experiments. Solid lines are for data from Rutledge et al. (1972); dashed lines are for data from Riska et al. (1984, 1985).

are outlined in Figure 23.3. Also illustrated in this figure are results obtained from a second, much larger (345 cross-fostering pairs) experiment performed on the same strain of mice by Riska et al. (1984, 1985). Despite the 10-year lapse between these studies, the results are essentially the same.

The phenotypic variance for weight in these mice, Var(z) in Figure 23.3, reaches a maximum at approximately 4 weeks, declines until approximately 7 weeks, at which point sexual maturity is attained, and then exhibits a monotonic increase. Thus, early differences in growth rates cause an initial divergence in size. The reduced variance in size near the time of maturity is caused by "compensatory" (Monteiro and Falconer 1966, Atchley 1984) or "targeted" (Riska et al. 1984) growth. If we let the weight of an individual at time t be equal to the sum of the weight at time $t-1$ and a growth increment, so that $W_t = W_{t-1} + \Delta_W$, then the variance in weight at time t may be written as

$$\sigma^2(W_t) = \sigma^2(W_{t-1}) + 2\sigma(W_{t-1}, \Delta_W) + \sigma^2(\Delta_W)$$

Since $\sigma^2(\Delta_W)$ is necessarily positive, a reduction in the variance of weight can only arise if the covariance between weight and growth rate, $\sigma(W_{t-1}, \Delta_W)$, is sufficiently negative to offset $\sigma^2(\Delta_W)$. The mechanism for convergent growth appears to be size-dependent initiation of sexual maturity accompanied by a reduction in growth rate (Monteiro and Falconer 1966). Different individuals reach the critical size at different times. Thus, in the interval of 4 to 8 weeks, small mice continue to grow rapidly while larger individuals that have attained sexual maturity exhibit a pronounced reduction in growth.

The peak in phenotypic variance at 4 weeks, 1 week after weaning, can be seen to be due largely to maternal effects — the majority of the phenotypic variance up to this point is attributable to Var(N). At 12 days, Var(N)/Var(z) attains a maximum of ~ 0.7, but even at 70 days, 7 weeks after weaning, significant maternal effects on body size are still detectable. On the other hand, the covariance between direct and maternal effects [as revealed by Var(I)] appears to be very small, but slightly positive. (Recall from Chapter 20 that interaction "variances" are really estimates of covariances and hence can be negative.) Since body weight in mice exhibits negligible dominance effects (Atchley 1984), 2 Var(M) is a good estimator of the additive genetic variance for direct effects on body weight, and this is roughly constant following weaning. The final rise in phenotypic variance following maturation is almost entirely due to a steady increase in the component of variance containing special environmental effects, Var(E).

The conclusions from these studies appear to be broadly generalizable. Working with the same or different strains of mice (Monteiro and Falconer 1966, Hanrahan and Eisen 1973, Cheverud et al. 1983) and rats (Atchley and Rutledge 1980), other authors have obtained essentially the same results. All of these studies have considered only postnatal maternal effects, since progeny were cross-fostered after birth. However, recent embryo-transplant experiments have demonstrated the presence of significant prenatal effects that persist for up to two months after birth

(Cowley et al. 1989, Cowley 1991, Pomp et al. 1989). Excellent reviews on cross-fostering experiments in mice, rats, and other species may be found in Legates (1972), Cheverud et al. (1983), and Atchley (1984). Reviews on maternal effects, not restricted to cross-fostering designs, are also available for cattle (Koch 1972, Shi et al. 1993), sheep (Bradford 1972), and swine (Robison 1972).

EISEN'S APPROACH

For situations in which cross-fostering is unfeasible, estimates of the maternal-effects variance and covariance components can be acquired by the method of moments, provided that measures of phenotypic covariance can be obtained for enough relationships (Eisen 1967). The causal components of variance are estimated in the usual way, by setting the observed phenotypic covariances equal to their expectations, and solving the set of linear equations for estimates of the underlying causal components. For example, from Table 23.1, the father-offspring covariance minus twice the covariance between paternal half-sibs provides an estimate of $\sigma_{A_o,A_m}/4$. A complete solution for all nine of the causal components of variance and covariance in Table 23.1 requires observations on at least nine types of relatives, and hence, data of a multigenerational nature. Data sets of a smaller scope can nevertheless be revealing. We present the following example simply to illustrate one powerful design for applying Eisen's (1967) method.

Bondari's Experiment

Starting with 331 males of the flour beetle *Tribolium castaneum*, each of which was mated to two unrelated females, Bondari et al. (1978) developed three types of mating structures to estimate the causal components of variance for pupal weight (Figure 23.4). As shown in Table 23.4, each design provides the basis for a nested analysis of variance, details of which have been covered in Chapter 18.

Design I. The primary purpose for this design, which is simply the full-sib, half-sib nested design outlined in Chapter 18, is to obtain an estimate of $\sigma^2(A_o)$. Recalling the results of Chapter 18, the variance among sires is formally equivalent to the covariance among paternal half-sibs, which we know has the expectation $\sigma^2_{A_o}/4$ (Table 23.1).

The design utilized all $N = 331$ males, each mated to two random females $(M = 2)$, with $n = 2$ pupae being weighed within each full-sib family. Setting the observed mean squares in Table 23.4 equal to their expectations and solving, we obtain estimates of the three hierarchical components of variance: $\text{Var}(a) = 3{,}384$, $\text{Var}(b) = 7{,}514$, and $\text{Var}(e) = 25{,}852$, respectively referring to sires, dams within sires, and offspring within dams. Thus, the total phenotypic variance is estimated to be $\text{Var}(z) = 36{,}750$, and the additive genetic variance involving direct effects is estimated by $\text{Var}(A_o) = 4\text{Var}(a) = 13{,}536$.

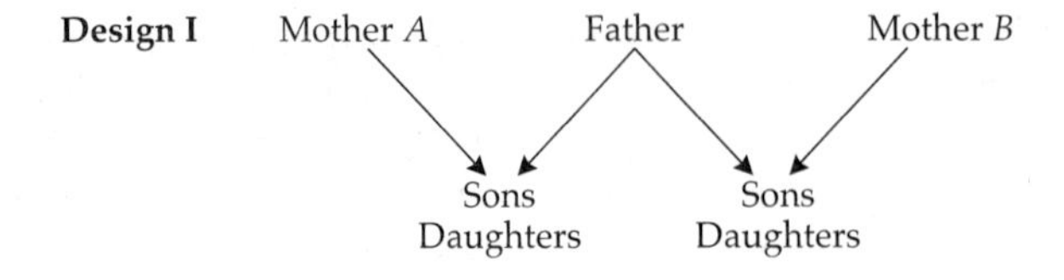

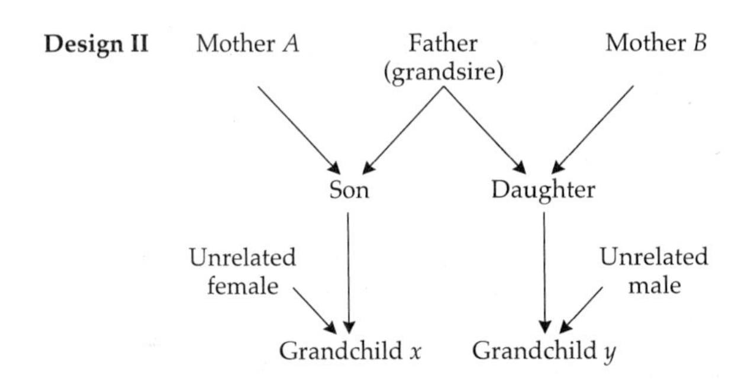

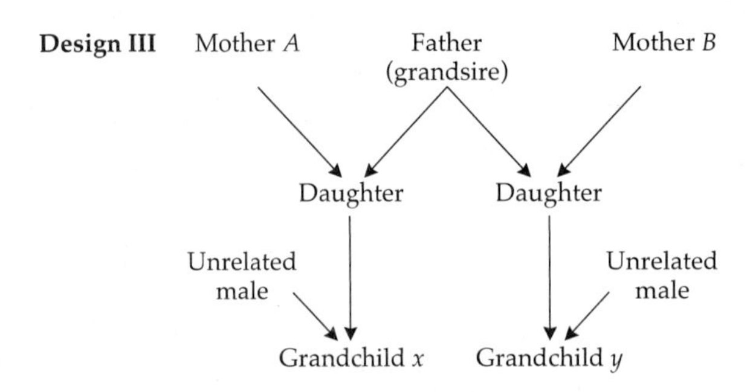

Figure 23.4 Pedigree structure for families employed in the three experimental designs of Bondari et al. (1978).

Design II. This experiment, which spans three generations, provides a means for estimating the additive genetic covariance between direct and maternal effects, σ_{A_o,A_m}. A fraction ($N = 208$) of the fathers in Design I served as the grandsires in Design II. For each grandsire, a single son was taken from the progeny of one mate and crossed to an unrelated female to produce a full-sib family; a single daughter was taken from the progeny of the second mate and crossed to an unrelated male to produce a second full-sib family. The members of the third-generation families served as the experimental units.

As in Design I, the results can be analyzed by nested ANOVA, in this case with σ_a^2 being the variance associated with grandsires. Note in Figure 23.4 that

Table 23.4 Summary of nested analyses of variance involving the three experimental designs employed by Bondari et al. (1978).

Factor	Degrees of Freedom	Mean Squares	Expected Mean Squares
Design I			
Sires	$(N_a - 1) = 330$	54,418	$\sigma_e^2 + n\sigma_b^2 + N_b n\sigma_a^2$
Dams (sires)	$N_a(N_b - 1) = 331$	40,880	$\sigma_e^2 + n\sigma_b^2$
Sibs (dams)	$N_a N_b(n - 1) = 330$	25,852	σ_e^2
	$\sigma_a^2 = \sigma(\text{PHS}) = \dfrac{\sigma_{A_o}^2}{4}$		
Design II			
Grandsires	$(N_a - 1) = 207$	48,499	$\sigma_e^2 + n\sigma_b^2 + N_b n\sigma_a^2$
Parents (grandsires)	$N_a(N_b - 1) = 208$	46,672	$\sigma_e^2 + n\sigma_b^2$
Sibs (parents)	$N_a N_b(n - 1) = 416$	23,867	σ_e^2
	$\sigma_a^2 = \sigma(\text{HFC, II}) = \dfrac{\sigma_{A_o}^2}{16} + \dfrac{\sigma_{A_o,A_m}}{8}$		
Design III			
Grandsires	$(N_a - 1) = 122$	48,336	$\sigma_e^2 + n\sigma_b^2 + N_b n\sigma_a^2$
Parents (grandsires)	$N_a(N_b - 1) = 123$	38,495	$\sigma_e^2 + n\sigma_b^2$
Sibs (parents)	$N_a N_b(n - 1) = 246$	23,310	σ_e^2
	$\sigma_a^2 = \sigma(\text{HFC, III}) = \dfrac{\sigma_{A_o}^2}{16} + \dfrac{\sigma_{A_o,A_m}}{4} + \dfrac{\sigma_{A_m}^2}{4}$		

Note: For each of three designs, σ_a^2, σ_b^2, and σ_e^2 denote the three observable hierarchical components of variance, and N_a, N_b, and n denote the nested sample sizes for the three hierarchical levels. The design was completely balanced. Further details on the analysis of nested data can be found in Chapter 18.

grandchild *x* has a father that is a half-sib of the mother of grandchild *y*. Thus, *x* and *y* are half-first cousins. From the rule that an among-group variance is equivalent to the covariance between members in the same group, σ_a^2 is equivalent to the covariance between half-first cousins related through parents of the opposite sex. The expected phenotypic covariance for such a relationship is Var(HFC,II) = $(\sigma_{A_o}^2/16) + (\sigma_{A_o,A_m}/8)$.

Again substituting observed for expected mean squares in Table 23.4 and solving, we obtain the estimate $\text{Var}(a) = \text{Var}(\text{HFC, II}) = 457$. Recalling from Design I that $\text{Var}(A_o) = 13{,}536$, and substituting into the expression for $\sigma^2(\text{HFC, II})$, we estimate the covariance between direct and maternal effects to be $\text{Cov}(A_o, A_m) = -3{,}112$. Thus, the data suggest that there is a negative genetic correlation between direct and maternal effects for pupal weight.

Design III. This experiment provides an estimate of the variance of additive genetic maternal effects, $\sigma^2_{A_m}$. The design is identical in form to Design II, except that the grandchildren x and y have mothers that are half-sibs and fathers that are unrelated. Thus, x and y are still half-first cousins, but because they are related through their mothers, they are influenced by common maternal-effect genes. The expression for the covariance between these types of relatives, which is a function of $\sigma^2_{A_o}$, σ_{A_o,A_m}, and $\sigma^2_{A_m}$, is given in Table 23.4.

The remaining $N = 123$ fathers from Design I served as grandsires in this experiment. Again substituting observed for expected mean squares in Table 23.4, we find $\text{Var}(a) = 2{,}460$. Equating this grandsire variance component to $\sigma(\text{HFC, III})$, and substituting the previous estimates of $\sigma^2(A_o)$ and $\sigma(A_o, A_m)$ into the expression for $\sigma(\text{HFC, III})$, yields $\text{Var}(A_m) = 9{,}568$.

All three experiments yielded similar estimates of the phenotypic variance, the average of which is $\text{Var}(z) = 35{,}280$. Thus, from the results of Design I, an estimate of the heritability of pupal weight, unbiased by maternal effects, is $h^2 = 13{,}536/35{,}280 = 0.38$. An estimate of the additive genetic correlation between direct and maternal effects is $\text{Cov}(A_o, A_m)/\sqrt{\text{Var}(A_o)\text{Var}(A_m)} = -0.27$. Thus, in this species, genes that increase pupal weight through their direct effects decrease it through maternal effects. Although the authors did not pursue it, this multigenerational experiment could have yielded estimates of the covariance between several other types of relatives, and hence of additional causal sources of variance.

FALCONER'S APPROACH

In all of the procedures discussed above, the maternal effect was treated as a general feature of the mother, with no specific character in the mother being identified as contributing to the effect. An alternative approach is to identify explicitly one or more maternal characters that are likely to be the source of the maternal effect, and to consider how these modify the expression of other characters in offspring. Falconer (1965a) introduced a simple model in which a single maternal character affects its own expression, e.g., maternal body size influencing the size of offspring via maternal care effects. Under this model, an individual's phenotype is described in the usual way, with the addition of a third term describing the maternal effect,

$$z_i = A_i + m\, z_{i1} + E_i \tag{23.10}$$

Here z_{i1} represents the phenotype of individual i's mother (the 1 denoting 1 generation back), and m, the maternal effect coefficient, is defined as the partial regression of offspring phenotype on maternal phenotype, holding the genetic contribution constant. (We will see shortly that m is just the difference between maternal-offspring and paternal-offspring regressions, as previously denoted). The remainder of the phenotype is assumed to be determined by an additive genetic effect A_i and an independently distributed residual deviation E_i.

By extension, the phenotype of the mother can be written as a function of her mother, i.e., $z_{i1} = A_{i1} + mz_{i2} + E_{i1}$, and that is the case for the phenotype z_{i2} of i's grandmother, as well as for all more remote members of i's maternal lineage. Consequently, the phenotype of i can be expressed by the infinite series,

$$z_i = \sum_{t=0}^{\infty} m^t (A_{it} + E_{it}) \tag{23.11}$$

where t denotes the number of generations back in the maternal lineage.

This simple model yields some interesting features. Assuming that the absolute value of m is less than one, which is necessary for the phenotypic variance to equilibrate, the covariance between mother and offspring is

$$\begin{aligned}\sigma(M,O) &= \sigma[z_{i1}, (A_i + mz_{i1} + E_i)] = \sigma(z_{i1}, A_i) + m\sigma_z^2 \\ &= \sigma\left(\sum_{t=0}^{\infty} m^t A_{i(t+1)}, A_i\right) + m\sigma_z^2 \\ &= \sigma\left(\sum_{t=0}^{\infty} \frac{m^t}{2^{t+1}} A_i, A_i\right) + m\sigma_z^2 \\ &= \frac{\sigma_A^2}{2-m} + m\sigma_z^2 \end{aligned} \tag{23.12a}$$

The covariance between paternal and offspring phenotypes is found in the same manner, noting that there is no covariance between the maternal effect mz_{i1} and the father's phenotype,

$$\sigma(F,O) = \frac{\sigma_A^2}{2-m} \tag{23.12b}$$

Using Equation 23.11, it is also possible to compute the equilibrium phenotypic variance,

$$\sigma_z^2 = \frac{(2+m)\sigma_A^2 + (2-m)\sigma_E^2}{(2-m)(1-m^2)} \tag{23.12c}$$

(Falconer 1965a).

The preceding derivations help clarify the meaning of Falconer's maternal effect coefficient m, and also suggest ways to estimate it. First, m is seen to be the difference between the two parent-offspring regressions

$$m = \frac{\sigma(M,O) - \sigma(F,O)}{\sigma_z^2} \tag{23.13}$$

Thus, Falconer's m is identical to the m that we defined in Example 1. In terms of the covariance components of Wilham's model, Equation 23.1,

$$m = \frac{(\sigma_{A_o,A_m} + \sigma_{D_o,D_m} + \sigma_{E_o,E_m}) + \dfrac{\sigma^2_{A_m}}{2} + b\sigma^2_{E_m}}{\sigma^2_z} \tag{23.14}$$

Second, from Equation 23.7b, m is also identical to the regression of foster child on foster mother. Third, by setting $\sigma^2_A = 0$ in Equation 23.12a, it can be seen that m is equivalent to the regression of offspring on mother in a group of genetically uniform individuals. Thus, for species that can be propagated vegetatively, m can be estimated as the average within-clone regression of offspring on mother. All three interpretations of m are consistent with its definition as the regression of offspring on mother above that expected on the basis of gene transmission.

For some organisms, m can be estimated by experimentally manipulating the expression of the maternal character and monitoring the phenotypes of offspring. For example, by adding and subtracting eggs from clutches of the collared flycatcher (*Ficedula albicollis*), Schluter and Gustafsson (1993) obtained an estimate of $m = -0.25$ (Figure 23.5), i.e., the addition of an egg to a mother's clutch causes the average clutch size of her daughters to decline by 0.25 eggs.

Restricted to a single character, Falconer's model may seem a bit abstract, but it is readily extended to multiple characters (Kirkpatrick and Lande 1989, Lande and Price 1989). In multivariate terms, Equation 23.11 generalizes to

$$\mathbf{z} = \mathbf{a} + \mathbf{M}\mathbf{z}_1 + \mathbf{e} \tag{23.15}$$

where $\mathbf{M}$ is a matrix of maternal effect coefficients, with the element m_{ij} defining the strength of the maternal effect of character j in the mother on character i in the progeny. $\mathbf{z}$, $\mathbf{a}$, and $\mathbf{e}$ are, respectively, vectors of the phenotypic values, additive genetic values, and special environmental effects on the traits in the individual, and $\mathbf{z}_1$ is the vector of phenotypic values in the mother. Note that, in general, $\mathbf{M}$ is unlikely to be symmetric. From Kirkpatrick and Lande (1989), the multivariate analogs of Equations 23.12a–c are

$$\mathbf{C}^m = \frac{1}{2}\mathbf{G}\left(\mathbf{I} - \frac{1}{2}\mathbf{M}^T\right)^{-1} + \mathbf{MP} \tag{23.16a}$$

$$\mathbf{C}^f = \frac{1}{2}\mathbf{G}\left(\mathbf{I} - \frac{1}{2}\mathbf{M}^T\right)^{-1} \tag{23.16b}$$

$$\mathbf{P} = \mathbf{G} + \mathbf{E} + \mathbf{MPM}^T + \mathbf{M}(\mathbf{C}^f)^T + \mathbf{C}^f\mathbf{M}^T \tag{23.16c}$$

where $\mathbf{I}$ is the identity matrix, $\mathbf{C}^m$ and $\mathbf{C}^f$ are the parent-offspring phenotypic covariance matrices (m denoting mothers, f denoting fathers) for the different

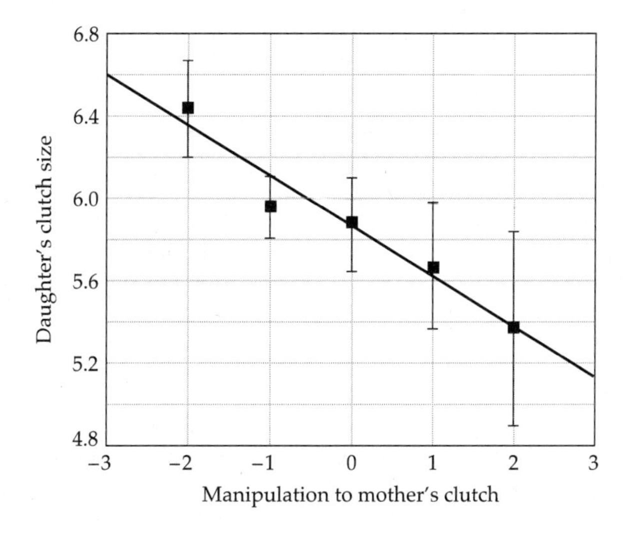

Figure 23.5 The consequences of experimental modifications of maternal clutch size for the mean clutch size of resultant daughters in their first year of breeding. Data are for the collared flycatcher (*Ficedula albicollis*). (From Schluter and Gustafsson 1993.)

traits, **G** is the matrix of additive genetic variances and covariances for the traits, **P** is the phenotypic covariance matrix, and T denotes a transpose traits, **G** is the matrix of additive genetic variances and covariances for the traits, **P** is the phenotypic covariance matrix, and T denotes a transpose.

From the difference between Equations 23.16a,b, the matrix of maternal effects is found to be

$$\mathbf{M} = (\mathbf{C}^m - \mathbf{C}^f)\mathbf{P}^{-1} \tag{23.17}$$

This expression provides a means of obtaining unbiased estimates of the m_{ij}, provided that all of the maternal characters influencing the characters under study are included in the analysis. (That, of course, may be a formidable task, requiring a deeper understanding of the biology of the system than is usually available). Once **M** has been obtained, the genetic covariance matrix of direct effects can be obtained by rearrangement of Equation 23.16b,

$$\mathbf{G} = 2\mathbf{C}^f \left(\mathbf{I} - \frac{1}{2}\mathbf{M}^T\right) \tag{23.18}$$

Since **M** contains n^2 coefficients, where n is the number of traits under consideration, it seems unlikely that any of them could be estimated very accurately if n is large, so this general procedure is difficult to apply to most practical problems. An alternative approach is to restrict the number of nonzero elements in **M** based on one's intuition about the biology of the character of interest, as shown in the following example.

Example 2. Here we consider a two-character situation (from Lande and Price 1989) in which only one of the four possible maternal effect coefficients is nonzero. Adult size (character 1) of a mother has a direct maternal effect on her offspring's size at birth (character 2) so that $m_{21} = m \neq 0$, but no direct effect on the offspring's size at maturity $(m_{11} = 0)$. Furthermore, the mother's size at birth has no maternal influence on her offspring's size at birth $(m_{22} = 0)$ or maturity $(m_{12} = 0)$. Thus, all of the elements of $\mathbf{M}$ are zero but m_{21}. If there are no other maternal characters influencing size at birth or maturity, and the other assumptions of the Kirkpatrick-Lande model are met, then unbiased definitions of the expected values of the parent-offspring covariance matrices are given by Equations 23.16a,b, which reduce to

$$\mathbf{C}^m = \begin{pmatrix} \sigma^2_{A_1}/2 & \sigma_{A_1,A_2}/2 + m_{21}\sigma^2_{A_1}/4 \\ \sigma_{A_1,A_2}/2 + m_{21}\sigma^2_{z_1} & \sigma^2_{A_2}/2 + m_{21}\sigma_{A_1,A_2}/4 + m_{21}\sigma_{z_1,z_2} \end{pmatrix}$$

$$\mathbf{C}^f = \begin{pmatrix} \sigma^2_{A_1}/2 & \sigma_{A_1,A_2}/2 + m_{21}\sigma^2_{A_1}/4 \\ \sigma_{A_1,A_2}/2 & \sigma^2_{A_2}/2 + m_{21}\sigma_{A_1,A_2}/4 \end{pmatrix}$$

where A_1 denotes the direct additive effect on adult size, and A_2 the direct additive effect on size at birth.

Provided that observed values of the elements of the two parent-offspring covariance matrices, $\mathbf{C}^m$ and $\mathbf{C}^f$, are available along with estimates of the phenotypic variance of the maternal trait, $\sigma^2_{z_1}$, and the phenotypic covariance between the two traits, σ_{z_1,z_2}, estimates of the genetic variances and covariance and of the maternal effect coefficient can be acquired by equating the observed elements of $\mathbf{C}^m$ and $\mathbf{C}^f$ to their expectations. For example, letting subscripts denote the rows and columns of matrix elements,

$$\begin{aligned} \sigma^2_{A_1} &= C^m_{11} + C^f_{11} \\ \sigma_{A_1,A_2} &= 2C^f_{21} \\ \sigma^2_{A_2} &= C^m_{22} + C^f_{22} - m_{21}[C^f_{21} + \sigma_{z_1,z_2}] \\ m_{21} &= \frac{2(C^m_{12} + C^f_{12} - 2C^m_{21})}{C^m_{11} + C^f_{11} - 4\sigma^2_{z_1}} \end{aligned}$$

Applying this model to weight data for Darwin's finches (Price and Grant 1985) and great tits (van Noordwijk 1984), Lande and Price (1989) obtained estimates of $m_{21} = 0.6$ and 0.3 respectively. Assuming the model is valid, these results suggest that the maternal contribution to hatchling body size associated with maternal adult size can be quite substantial in birds with maternal care.

EXTENSION TO OTHER TYPES OF RELATIVES

With its focus on maternal effects, this entire chapter has concentrated on one particular way in which an individual can modify the phenotype of another by means other than direct inheritance. Extension of these ideas to effects from other types of intrafamilial interactions, such as paternal effects (in species with paternal care) or sib effects (in species where sibs compete or cooperate for resources) is relatively straightforward. Consider an individual x living in a typical nuclear family with social interactions. At various stages of development, the individual's phenotype may be influenced by the direct expression of its own genotype and environmental effects (z_{ox}), by maternal (M_x) and/or paternal (P_x) effects, by effects from sibs with which it is raised (S_x), and later in life, by indirect effects of its mate (H_x) and its progeny (R_x). Thus, Equation 23.1, which contains only maternal effects, can be expanded to include these other contributions,

$$z_x = z_{ox} + M_x + P_x + S_x + H_x + R_x \tag{23.19}$$

(Lynch 1987). Many types of behavioral interactions can contribute to the additional intrafamilial effects. For example, a sib effect can be positive in the case of sib cooperation or negative in the case of sib competition. The behavior and/or physiological condition of an interacting mate can have effects on an individual's phenotype. An offspring effect can be negative, as when juveniles impose high energetic demands for parental care, or positive, as when older progeny assist their parents.

As in the case of maternal effects, each of the new terms in Equation 23.19 can be treated as a sum of components due to additive genetic effects and residual deviations: $P_x = G_{px} + E_{px}$, $S_x = G_{sx} + E_{sx}$, $H_x = G_{hx} + E_{hx}$, and $R_x = G_{rx} + E_{rx}$. With 12 effects contributing to Equation 23.19, and the possibility that some of them may be correlated, expressions for the genetic variance and covariance between relatives are quite complicated with the complete model. Rather than give a complete description, we provide two examples of the use of simplified versions of the model.

Example 3. Consider the case in which the only intrafamilial effects on the phenotype derive from fathers and mothers. Equation 23.19 then reduces to

$$z_x = z_{ox} + M_x + P_x$$

Such a model should provide a reasonable description of prereproductive traits of organisms, provided that interactions with sibs during development are either nonexistent or have negligible phenotypic consequences. How does this model alter our interpretations of the covariances between relatives compared to the maternal-effects model?

First, we note that the phenotypic covariance between relatives x and y can be written

$$\begin{aligned}\sigma_z(x,y) = \sigma(z_{ox}, z_{oy}) &+ [\sigma(z_{ox}, M_y) + \sigma(z_{oy}, M_x)] \\ &+ \sigma(M_x, M_y) + [\sigma(z_{ox}, P_y) + \sigma(z_{oy}, P_x)] \\ &+ [\sigma(M_x, P_y) + \sigma(M_y, P_x)] + \sigma(P_x, P_y)\end{aligned}$$

Thus, six types of factors can contribute to the resemblance between relatives: variances associated with direct, maternal, and paternal effects, and covariances between direct and maternal effects, between direct and paternal effects, and between maternal and paternal effects.

The first four terms of this equation (i.e., those involving direct and / or maternal effects) have all been described above (Table 23.1). The next two terms, which describe the covariance between direct and paternal effects, can be evaluated by extending the procedures used earlier in the chapter. For example, for the *genetic* covariance of direct and paternal effects, we let s denote the father of x and t denote the father of y. Modifying Equations 23.3a,b, we then obtain

$$\sigma_{G_o,G_p}(x,t) + \sigma_{G_o,G_p}(y,s) = 2(\Theta_{xt} + \Theta_{ys})\sigma_{A_o,A_p} + (\Delta_{xt} + \Delta_{ys})\sigma_{D_o,D_p}$$

For most relationships, simple expressions for the contributions of σ_{A_o,A_p} and σ_{D_o,D_p} to the phenotypic covariance can be obtained by reversing the sexes in Table 23.1. For example, when $y = s$ is the father of x, then $\Theta_{xt} = 1/8$, $\Theta_{ys} = 1/2$, $\Delta_{xt} = 0$, and $\Delta_{ys} = 1$, yielding

$$\sigma_{G_o,G_p}(x,t) + \sigma_{G_o,G_p}(y,s) = \frac{5\sigma_{A_o,A_p}}{4} + \sigma_{D_o,D_p}$$

This equation has the same structure as the covariance between direct and maternal effects when y is the mother of x (Table 23.1). Similar procedures show that the covariance of direct and paternal effects contributes $\sigma_{A_o,A_p}/4$ to the mother-offspring covariance, which compares with the contribution of the covariance of direct and maternal effects to the father-offspring covariance, $\sigma_{A_o,A_m}/4$. Comparable analogies can be used to deduce the contribution of $\sigma^2_{A_p}$ to phenotypic resemblance.

Finally, we consider the genetic covariance between maternal and paternal effects. Here we let s and w denote the father and mother of x and t and z denote the father and mother of y, which leads to

$$\sigma_{G_m,G_p}(w,t)+\sigma_{G_m,G_p}(z,s)=2(\Theta_{wt}+\Theta_{zs})\sigma_{A_m,A_p}+(\Delta_{wt}+\Delta_{zs})\sigma_{D_m,D_p}$$

For all of the relationships given in Table 23.1, except reciprocal sibs, this covariance is equal to zero because the dams are unrelated to the sires. In the case of reciprocal full sibs, $w = t$ and $s = z$, so the covariance is $2(\sigma_{A_m,A_p} + \sigma_{D_m,D_p})$. For reciprocal half-sibs, the father of x is the mother of y, but the other parents are unrelated. This reduces the covariance to $(\sigma_{A_m,A_p} + \sigma_{D_m,D_p})$.

From these results, it is clear that by adding terms in $\sigma^2_{A_p}$, $\sigma^2_{D_p}$, σ_{A_o,A_p}, and σ_{D_o,D_p}, the presence of genetic paternal effects can substantially complicate the expressions for the covariance between relatives beyond those described in Table 23.1. We leave it to the reader to work out the additional contributions from environmental paternal effects. The addition of more terms to a model describing the resemblance between relatives can only magnify the difficulties in achieving clean estimates of causal components of variance. Nevertheless, some of the empirical procedures for detecting maternal effects described in this chapter suggest ways in which paternal effects might be detected. For example, for species in which it is possible to cross-foster with respect to sire, a simple means of testing for paternal effects would be to estimate the foster child-foster father covariance. Factorial designs incorporating reciprocal crosses can also be useful in this regard (Chapter 20).

Example 4. For organisms that invest substantially in postnatal parental care, it is conceivable that progeny, through their demands on parental resources, can affect aspects of a parent's phenotype. A simple way to test for such effects is to consider pairs of unrelated parents mated to the same individual:

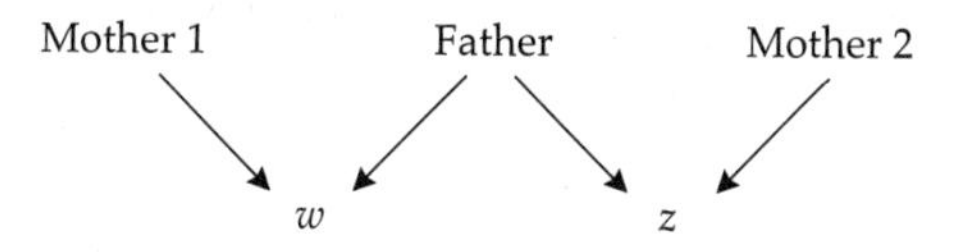

In this case, even under the complete model (Equation 23.19), the paired parents would share no direct, maternal, paternal, or sib effects. However, their offspring would be related as half-sibs. Denoting the offspring of the two individuals by w and z, and noting $2\Theta_{wz} = 1/4$, the expected (postreproductive) phenotypic covariance between unrelated parents mated to the same individual is

$$\sigma[(G_{rw}+E_{rw}),(G_{rz}+E_{rz})]=2\Theta_{wz}\sigma^2_{G_r}=\frac{\sigma^2_{G_r}}{4}$$

where G_{rw} and E_{rw} denote the genetic and environmental effects of progeny w on the phenotype of the maternal parent, and we assume that environmental progeny effects are uncorrelated

Thus, four times the phenotypic covariance between unrelated individuals sharing half-sib progeny provides an estimate of the genetic variance associated with offspring effects, $\sigma^2_{G_r}$. Additional ideas on the analysis of offspring (fetal) effects can be found in Bar-Anan et al. (1976) and Van Vleck (1978), who present evidence that a negative genetic correlation exists between direct genetic and fetal effects operating on calving ability in cattle.

Current thought on the evolution of social behavior is heavily dominated by two theoretical concepts. Hamilton's (1964) idea of inclusive fitness led to the conclusion that a behavior that is costly to an individual's fitness will nonetheless evolve if it increases a relative's fitness enough that the cost:benefit ratio is less than $2\Theta_{xy}$. Trivers (1974) popularized the notion that there is a genetic conflict between parents and offspring such that offspring are selected to acquire as many resources from their parents as possible, whereas parents are selected for their total reproductive output (see also Haig 1993). Almost all applications of these ideas assume that behavioral characters expressed in different social contexts are free to evolve independently. However, since individuals play overlapping roles of offspring, sib, parent, and mate at various stages in their lives, it seems likely that behavioral attributes expressed in these different roles are correlated genetically. If this is the case, the direction of evolution of a species' behavioral repertoire can be radically altered relative to conventional expectations (Cheverud 1984, Lynch 1987). For example, offspring behavior that elicits more parental care cannot be expected to evolve if the traits underlying such behavior are strongly negatively correlated with characters that influence fitness later in life, e.g., parenting ability. A quantitative-genetic perspective may prove useful in future attempts to decipher the genetic constraints on the evolution of social behavior.

24

Sex Linkage and Sexual Dimorphism

In sexual species, males and females often differ with respect to the mean and variance of traits. In addition, the relative contribution of the different components of variance may differ between the sexes. There are two reasons for this. First, in the case of sex-linked genes, females have two possible genes at each locus, while males have only one (assuming, as we will below, that males are the heterogametic sex). Second, the expression of a gene may vary depending upon the sexual background within which it is found.

Although very little empirical work has been focused on these problems from a quantitative-genetic perspective, the theory, which involves a straightforward extension of Kempthorne's (1954) linear model, is fairly well developed. As the entire subject is relevant to the problem of sexual dimorphism and sexual selection, we examine the general principles in some detail. These were first presented by Bohidar (1964) and later clarified by James (1973) and Grossman and Eisen (1989). The two issues mentioned above, sex linkage and sex-specific expression, will be considered separately first and then combined in general form.

SEX-LINKED LOCI AND DOSAGE COMPENSATION

In the case of a specific sex-linked locus, the genotypic value of a male with allele i can be written

$$G'_{Mi} = \mu_M + \alpha'_{Mi} \tag{24.1a}$$

where μ_M is the mean phenotype of males, and α'_{Mi} is the average effect of the ith allele measured as a deviation from μ_M, with a prime distinguishing a sex-linked locus from an autosomal locus. Since the expectation of α'_{Mi} is zero, the total genetic variance due to this locus in males is simply $\sigma^2(A'_M) = \sum p_i(\alpha'_{Mi})^2$, where p_i is the frequency of the ith allele at the locus. Since males contain only one gene per sex-linked locus, there is no dominance genetic variance at such loci.

The situation in females is slightly more complicated since they have two alleles at each sex-linked locus. In this case, we write the genotypic value as

$$G'_{Fij} = \mu_F + \alpha'_{Fi} + \alpha'_{Fj} + \delta'_{Fij} \tag{24.1b}$$

where μ_F is the mean phenotype of females, α'_{Fi} and α'_{Fj} are the average effects of the ith and jth alleles as expressed in females, and δ'_{Fij} is the dominance effect of the *ij*th genotype. It follows that the genetic variance due to a sex-linked locus in females is $\sigma^2(\alpha'_{Fi}) + \sigma^2(\alpha'_{Fj}) + \sigma^2(\delta'_{Fij}) = \sigma^2(A'_F) + \sigma^2(D'_F)$, the sum of the additive and dominance components of variance.

From Equations 24.1a,b, it can be seen that depending on the degree to which the additive effects of genes differ between the sexes and on the degree of dominance in females, the genetic variance associated with sex-linked loci is unlikely to be the same in the two sexes. For example, if $\alpha'_{Mi} = \alpha'_{Fi}$ for all alleles, and if there are no dominance effects in females, then $\sigma^2(A'_F) = 2\sigma^2(A'_M)$, i.e., the additive genetic variance due to the sex-linked locus in males would be only half that in females. However, α'_{Mi} is often unequal to α'_{Fi}. For example, in females of many placental mammals, most of one of the X chromosomes is inactivated randomly in different cell lineages early in development (reviewed by Migeon 1994). In such species, females are functionally haploid mosaics at sex-linked loci. A familiar and most dramatic single-locus example of such mosaicism is that of the "tortoise-shell" cat. In cats, coat color genes lie on the X chromosome. If the two alleles in a female have different effects (black and yellow in the case of the "tortoise-shell" genotype), a mottled coat pattern results from the inactivation of different alleles in different somatic regions.

The adjustment of the total activity of sex-linked loci to achieve equality in the two sexes is known as **dosage compensation** (Muller 1932). For the compensatory mechanism noted above (complete inactivation of random alleles in females), a more appropriate expression for the genotypic value of females might be

$$G'_{Fij} = \mu_F + \frac{1}{2}\alpha'_{Mi} + \frac{1}{2}\alpha'_{Mj} + \delta'_{Fij} \tag{24.1c}$$

This relationship implies an additive genetic variance of $\sigma^2(\alpha'_{Mi})/4 + \sigma^2(\alpha'_{Mj})/4 = \sigma^2(\alpha'_M)/2$, i.e., half that in males. Although we include a term for dominance in the above expression, it is unclear whether dominance occurs in the normal sense when only a single allele is active in each cell.

A second type of dosage compensation occurs when the same X chromosome is inactivated in all cells of a female. In this case, there is clearly no dominance effect, and our expression now becomes

$$G'_{Fij} = \mu_F + \alpha'_{Fi} \tag{24.1d}$$

so the additive genetic variance is simply $\sigma^2(\alpha'_{Fi})$. This type of dosage compensation appears to be close to the situation in kangaroos and other marsupials, where the paternally derived X chromosome is inactivated in all cells (Cooper 1971). Note that even in this case, the sex-linked variation will only be equal in the two sexes if $\alpha'_{Fi} = \alpha'_{Mi}$.

The mechanism of sex-linked dosage compensation in *Drosophila,* and presumably other insects, is somewhat different from that in mammals in that complete X chromosome inactivation does not occur. Instead, both X chromosomes are active in females, transcribing at about half the rate of those in males (Lucchesi 1978). Such dosage compensation has been demonstrated for a polygenic trait, abdominal bristle number, in *Drosophila* by Frankham (1977).

Using standard techniques involving chromosomal stocks with visible markers and cross-over suppressors, Frankham (1977) constructed 17 lines of *Drosophila melanogaster* with identical homozygous autosomal backgrounds, cytoplasmic backgrounds, and Y chromosomes. The only difference between the lines was their X chromosome. A simple one-way analysis of variance was then performed on the lines. Previous work had shown that the expression of bristle number genes is sex modified, counts in males being about 0.8 times those in females. Thus, in order to minimize the potential effects of scale differences on the variance, all measurements were log transformed (Chapter 11). The within-line component of variance provided an estimate of the environmental variance, since the lines were completely homozygous within the limits of the marker-inversion technique, and the among-line component estimated the genetic variance due to loci on the X chromosome. Separate analyses of males and females resulted in essentially identical estimates of X-linked genetic variance (both 0.00023).

Provided the scaling phenomenon has been accounted for properly, these results may have arisen in two ways. On average, only one of the two bristle number genes at each sex-linked locus may be active in females, in which case Equation 24.1d would be the appropriate model. Alternatively, the expression of both alleles may be suppressed at sex-linked loci in females, in which case a model of the form of Equation 24.1c would be required.

The first possibility was ruled out by further work. In one of the first successful attempts to identify a polygene, Frankham et al. (1980) verified that the sex-linked ribosomal RNA gene cluster is a major source of quantitative-genetic variation for bristle number in *Drosophila.* Complete inactivation of a gene with major effects in females would cause females to be much more similar to one parent (the one whose descendent X remained active) than the other, whereas partial inactivation of both would result in an intermediate phenotype. Frankham (1977) performed reciprocal crosses between his pure lines, and found that the phenotypic means of the reciprocal F_1s were, in fact, intermediate to those of the parentals and not significantly different from each other. Thus, the data suggest that the expression of sex-linked genes for bristle number is partially suppressed in females.

The mammalian (mouse / human) and *Drosophila* forms of dosage compensation outlined above are unlikely to be the only ones utilized by different species, but very little work has been done with other organisms. For the loci that have been examined in birds (Cock 1964) and Lepidoptera (Johnson and Turner 1979), groups where the females are heterogametic, there seems to be no dosage compensation. Two species of crickets have been examined, one having a mammalian form and the other a *Drosophila* form of compensation (Rao and Arora 1979, Rao and Ali 1982). Although the outcome of dosage compensation in the nematode *Caenorhabditis elegans* is essentially the same as in *Drosophila,* it is achieved by a very different mechanism (Parkhurst and Meneely 1994, Kelley and Kuroda 1995). In the following, we will treat Equations 24.1a,b as our general models for X-linked expression. Although we will not pursue them, models for the expression of Y-linked traits are very straightforward, in that females need not be considered and the male model is simply that of a haploid locus.

SEX-MODIFIED EXPRESSION OF AN AUTOSOMAL LOCUS

Just as the expressed effect of a sex-linked gene may vary between the sexes, so may the expression of any autosomal gene. For a single autosomal locus, we denote the genotypic values of females and males respectively as

$$G_{Fij} = \mu_F + \alpha_{Fi} + \alpha_{Fj} + \delta_{Fij} \tag{24.2a}$$

$$G_{Mij} = \mu_M + \alpha_{Mi} + \alpha_{Mj} + \delta_{Mij} \tag{24.2b}$$

The autosomal additive genetic variances of the two sexes are defined in the usual manner to be

$$\sigma^2(A_F) = \sigma^2(\alpha_{Fi}) + \sigma^2(\alpha_{Fj}) = 2\sum p_i\alpha_{Fi}^2 = 2\sigma^2(\alpha_F) \tag{24.3a}$$

$$\sigma^2(A_M) = \sigma^2(\alpha_{Mi}) + \sigma^2(\alpha_{Mj}) = 2\sum p_i\alpha_{Mi}^2 = 2\sigma^2(\alpha_M) \tag{24.3b}$$

and the dominance components are

$$\sigma^2(D_F) = \sum p_{ij}\delta_{Fij}^2 = \sigma^2(\delta_F) \tag{24.4a}$$

$$\sigma^2(D_M) = \sum p_{ij}\delta_{Mij}^2 = \sigma^2(\delta_M) \tag{24.4b}$$

where p_{ij} is the frequency of the *ij*th genotype.

Gametic Imprinting

Before proceeding, we mention a rather different type of sex modification of gene action. Over the past few years, several studies, mostly in transgenic mice, have demonstrated the phenomenon of **gametic imprinting,** whereby certain genes carry a memory of their gametic origin and are expressed differently in progeny depending upon whether they are paternally or maternally inherited (Solter 1988; Hall 1990; Barlow 1994, 1995). Under this type of sex modification, the expression of a gene depends not just on the gender of the zygote within which it is found, but also on the sex from which it was inherited. Some imprinted genes are totally inactivated, whereas others have suppressed activities in specific tissues. The interesting aspect of this phenomenon is that the imprint on a gene is only transiently heritable — it is totally erased in a single generation if it passes through the nonimprinting sex. For example, if a gene inactivated by paternal imprinting is inherited by an offspring, it will be reactivated after it is transferred to the next generation through a daughter's gamete, while remaining inactivate when passed through a son. The mechanisms of imprinting appear to involve DNA methylation, but are otherwise poorly understood. Essentially nothing is known of its importance in the expression of quantitative characters.

However, a simple example illustrates the potential significance of gametic imprinting in the interpretation of quantitative-genetic analyses. Consider a single locus for which imprinting occurs in male gametes and erasure of imprinting occurs in female gametes. This implies that every individual has one imprinted (+) allele and one nonimprinted (−) allele. Thus, if an A_+/a_- individual is a male, it will produce A_+ and a_+ gametes, and if it is a female, it will produce A_- and a_- gametes. In both cases, there has been a change in state in one of the two alleles. This implies that there is a 50% chance that an allele inherited by an offspring will have experienced a change in state relative to that expressed in the parent (compared to a zero probability for a locus not experiencing imprinting). Now consider the similarity of gene expression for sibs. In this case, regardless of whether imprinting occurs, there is a 50% chance that the alleles derived by sibs from the same heterozygous parent are identical in expression (in the sibs). Thus, the net result of gametic imprinting is to reduce the expected phenotypic covariance between parents and offspring relative to that between sibs. A difference between the parent-offspring covariance and twice the covariance between paternal half sibs that is significantly less than zero would be compatible with the hypothesis of significant gametic-imprinting effects on the expression of a quantitative trait.

EXTENSION TO MULTIPLE LOCI AND THE COVARIANCE BETWEEN RELATIVES

As first shown by Bohidar (1964), it is a relatively straightforward procedure to extend the general model of Kempthorne (1954) to describe polygenic situations in which sex linkage and sex modification of gene expression are involved. The only modification of the description of the phenotypic resemblance between relatives is the need to consider the sexes of the individuals involved. To simplify discussion, we ignore sources of genetic variation due to maternal effects (Chapter 23).

Let us first examine relationships involving only males. Three groups of factors must be considered. First, for the autosomal loci, we must account for the additive, dominance, and various epistatic effects. Second, for the sex-linked loci, there are no terms involving dominance since only a single gene is active at each locus, but additive and epistatic effects involving additive interactions must be considered. Third, the possibility of epistatic interactions between autosomal and sex-linked loci must be recognized. Under random mating, as shown by Kempthorne (1954) and discussed previously (Chapter 5), the covariances between different terms in the expression for a genotypic value are all zero, so that the genotypic variance for males may be written

$$\begin{aligned}\sigma^2(G_M) = \sigma^2(A_M) + \sigma^2(D_M) + \sigma^2(A_M A_M) + \sigma^2(A_M D_M) + \cdots \\ + \sigma^2(A'_M) + \sigma^2(A'_M A'_M) + \cdots \\ + \sigma^2(A_M A'_M) + \sigma^2(A'_M D_M) + \cdots \end{aligned} \tag{24.5a}$$

where primed and unprimed elements refer to sex-linked and autosomal loci, respectively. Variances with two terms within parentheses denote epistatic effects. For example, $\sigma^2(A_M A'_M)$ is the variance due to additive $\times$ additive interactions between autosomal and sex-linked loci in males.

Under the usual assumptions of random mating and free recombination, the genotypic covariance between two males may be expressed as a sum of variance terms each weighted by an appropriate measure of relatedness. Letting x and y represent the two males, then

$$\begin{aligned}\sigma(G_{Mx}, G_{My}) = 2\Theta_{xy}\sigma^2(A_M) + \Delta_{xy}\sigma^2(D_M) + \phi_{xy}\sigma^2(A'_M) \\ + \sum (2\Theta_{xy})^\alpha \Delta_{xy}^\beta \phi_{xy}^\lambda \sigma^2[A_M^\alpha D_M^\beta (A'_M)^\lambda] \end{aligned} \tag{24.5b}$$

Except for the introduction of the new term ϕ_{xy}, this expression is nearly identical in structure to Kempthorne's equation for autosomal loci (Chapter 7). ϕ_{xy} is the probability that a sex-linked gene in male x is identical by descent with that at the same locus in male y. Unlike Θ_{xy}, ϕ_{xy} is not preceded by a 2 because sex-linked loci exist in a haploid state in males. The final summation collects all of the covariance due to epistatic effects; it involves all terms for which $\alpha + \beta + \lambda \geq 2$, with α, β, and λ representing, respectively, the number of autosomal additive, autosomal dominance, and sex-linked additive effects in the epistatic interaction.

Expressions for some common male-male relationships are outlined in Table 24.1. The respective values of $2\Theta_{xy}$ and Δ_{xy} should be familiar by now. ϕ_{xy} is zero for a father-son relationship, since the son always obtains its X chromosome from its mother. Similarly, $\phi_{xy} = 0$ for paternal half sibs, since the X chromosome of each half sib comes from a different mother. On the other hand, $\phi_{xy} = 1/2$ for full brothers and maternal half brothers, since they share the same mother, who contributes to each of them one of her two X chromosomes. Finally, one of the X chromosomes of a mother must have come from her father. Thus, there is a 50% chance that the X chromosome contained in a male is identical by descent with that of his maternal grandfather. The table shows that if genetic variance exists at sex-linked loci, the covariance between grandfather and grandson will be greatest in the case of a maternal grandfather.

Table 24.1 Coefficients needed for the expressions describing the expected phenotypic covariance between relatives in a model that includes sex-linkage and sex-dependent gene expression.

Male-male relationships	$2\Theta_{xy}$	Δ_{xy}	ϕ_{xy}	
Father–son	1/2	0	1/4	
Full brothers	1/2	1/4	1/2	
Paternal half brothers	1/4	0	0	
Maternal half brothers	1/4	0	1/2	
Paternal grandfather–grandson	1/4	0	0	
Maternal grandfather–grandson	1/4	0	1/2	
Monozygotic twins	1	1	1	
Female-female relationships	$2\Theta_{xy}$	Δ_{xy}	$2\Theta'_{xy}$	Δ'_{xy}
Mother–daughter	1/2	0	1/2	0
Full sisters	1/2	1/4	3/4	1/2
Paternal half sisters	1/4	0	1/2	0
Maternal half sisters	1/4	0	1/4	0
Paternal grandmother–granddaughter	1/4	0	1/2	0
Maternal grandmother–granddaughter	1/4	0	1/4	0
Monozygotic twins	1	1	1	1
Male-female relationships	$2\Theta_{xy}$	Δ_{xy}	γ_{xy}	
Father–daughter	1/2	0	1	
Mother–son	1/2	0	1	
Full brother and sister	1/2	1/4	1/2	
Paternal half brother and sister	1/4	0	0	
Maternal half brother and sister	1/4	0	1/2	
Paternal grandfather–granddaughter	1/4	0	0	
Maternal grandfather–granddaughter	1/4	0	1/2	
Paternal grandmother–grandson	1/4	0	0	
Maternal grandmother–grandson	1/4	0	1/2	

Note: Coefficients for other types of relationships can be found in Grossman and Eisen (1989).

The genetic variance of females can be described in a similar manner, except that we must also account for dominance interactions at the sex-linked loci,

$$\begin{aligned}\sigma(G_F) = &\sigma^2(A_F) + \sigma^2(D_F) + \sigma^2(A_F A_F) + \sigma^2(A_F D_F) + \cdots \\ &\sigma^2(A'_F) + \sigma^2(D'_F) + \sigma^2(A'_F A'_F) + \sigma^2(A'_F D'_F) + \cdots \\ &\sigma^2(A_F A'_F) + \sigma^2(A'_F D_F) + \sigma^2(A_F D'_F) + \sigma^2(D_F D'_F) + \cdots \end{aligned} \tag{24.6a}$$

The genetic covariance between two females is then

$$\sigma(G_{Fx}, G_{Fy}) = 2\Theta_{xy}\sigma^2(A_F) + \Delta_{xy}\sigma^2(D_F) + 2\Theta'_{xy}\sigma^2(A'_F) + \Delta'_{xy}\sigma^2(D'_F)$$
$$+ \sum (2\Theta_{xy})^\alpha \Delta^\beta_{xy} (2\Theta'_{xy})^\lambda (\Delta'_{xy})^\delta \sigma^2[A^\alpha_F D^\beta_F (A'_F)^\lambda (D'_F)^\delta] \quad (24.6b)$$

where Θ'_{xy} and Δ'_{xy} are the coefficients of coancestry and fraternity for sex-linked loci in female-female relationships. Again, the summation compiles all of the genetic covariance due to epistatic interactions such that $\alpha + \beta + \lambda + \delta \geq 2$, where α, β, λ, and δ, respectively, represent the number of autosomal additive, autosomal dominance, sex-linked additive, and sex-linked dominance effects in an epistatic interaction.

Note that the coefficients of identity for female-female relationships often differ for autosomal and sex-linked loci. Consider, for example, full sisters. We already know that $2\Theta_{xy} = 1/2$ and $\Delta_{xy} = 1/4$ in this case. However, for a sex-linked locus, $2\Theta'_{xy} = 3/4$ and $\Delta'_{xy} = 1/2$. This result can be seen as follows. If single X chromosomes are drawn randomly from two full sisters, both may be paternal in origin with probability $1/4$, in which case they must be identical by descent since a father has only one X chromosome. Both may be maternal in origin with probability $1/4$, in which case there is a $1/2$ probability of identity by descent. If one is paternal and the other maternal, the two X-linked alleles cannot be identical by descent. Therefore, $2\Theta_{xy} = 2[(1)(1/4) + (1/2)(1/4) + (0)(1/2)] = 3/4$. Both full sisters obtained the same X chromosome from their father and have a 50% chance of obtaining the same X from their mother. Therefore, $\Delta'_{xy} = 1/2$, which contrasts with $\Delta_{xy} = 1/4$ for autosomal loci.

Finally, we consider the slightly more complicated case of genetic covariance between members of different sexes. In this case, a new coefficient, γ_{xy}, must be introduced to denote the probability that a sex-linked gene in male y is identical by descent with that in a female relative x. In addition, new notation needs to be introduced to describe the covariance between gene effects as expressed in the different sexes — $\sigma(A_F, A_M)$ and $\sigma(D_F, D_M)$ for additive and dominance effects at autosomal loci, and $\sigma(A'_F, A'_M)$ for sex-linked additive effects. For epistasis, we will use terms of the form $\sigma(\ldots_{FM}, \ldots_{FM}, \ldots)$, where, for example, $\sigma(A'_{FM} D_{FM})$ refers to the covariance across the sexes of epistatic effects involving sex-linked additive and autosomal dominance interaction. These modifications lead to the following expression for the male-female covariance

$$\sigma(G_{Fx}, G_{My}) = 2\Theta_{xy}\sigma(A_F, A_M) + \Delta_{xy}\sigma(D_F, D_M) + 2\gamma_{xy}\sigma(A'_F, A'_M)$$
$$+ \sum (2\Theta_{xy})^\alpha \Delta^\beta_{xy} (2\gamma_{xy})^\lambda \sigma[A^\alpha_{FM} D^\beta_{FM} (A'_{FM})^\lambda] \quad (24.7)$$

Note that there are no terms involving sex-linked dominance effects (D') since they do not exist in males.

We will not pursue the methodological details of estimating components of genetic variance and covariance in the presence of sex linkage and sex modification any further, except by means of example. It should be clear from arguments in previous chapters and from the expressions in Table 24.1 that through the estimation of covariances between appropriate sets of relatives, most of the causal components of genetic variance can be extracted by the method of moments, assuming that the epistatic components are of negligible importance. Any of the methods described in previous chapters can be used for these purposes, the only new distinction being the need to analyze males and females separately. Further information on methodology employing sib analyses may be found in Bohidar (1964) and Eisen and Legates (1966). Risch (1979) gives results for populations undergoing assortative mating.

From estimates of the additive genetic covariances, it is possible to estimate the genetic correlation across the sexes for autosomal and/or sex-linked loci, defined respectively as

$$\rho_{FM}(A) = \frac{\sigma(A_F, A_M)}{\sigma(A_F)\,\sigma(A_M)} \tag{24.8a}$$

$$\rho_{FM}(A') = \frac{\sigma(A'_F, A'_M)}{\sigma(A'_F)\,\sigma(A'_M)} \tag{24.8b}$$

High values for these correlations suggest a high degree of overlap in the sets of genes expressed in the different sexes. All of the techniques described in Chapter 22 for estimating Falconer's genetic correlation across environments are relevant here, as the two sexes can be treated as fixed effects. (They are essentially two genetic environments within which gene expression occurs.)

Example 1. In one of the few attempts to evaluate the contribution of sex-linked loci and sex-modified gene expression to patterns of phenotypic variation, Cowley et al. (1986) and Cowley and Atchley (1988) performed a nested sib analysis on various morphological traits in *Drosophila melanogaster.* The experiment was quite large, involving analyses on 1482 flies of each sex, sampled from 988 full-sib families and 494 paternal half-sib families. By scoring phenotypes in both male and female progeny, six estimates of covariance between relatives were obtained — full sibs and paternal half sibs, within each sex and between the sexes.

Under the assumptions that the autosomal additive genetic variance and the variance due to common environment was the same in both sexes, and that there was no significant dominance genetic variance, the causal relationships outlined in Table 24.1 were used to obtain estimates of $\sigma^2(A)$, $\sigma^2(A'_M)$, $\sigma^2(A'_F)$, $\sigma(A_M, A_F)$, and $\sigma(A'_M, A'_F)$.

The results in the following table are averages for groups of adult characters derived from different imaginal disks (the embryonic precursors of adult tissues). The heritabilities are defined as $[\text{Var}(A) + \text{Var}(A'_x)]/\text{Var}(z_x)$, where $\text{Var}(A'_x)$ and $\text{Var}(z_x)$are, respectively, the sex-linked additive genetic variance and the total phenotypic variance in the xth sex. The genetic correlations across the sexes include both the autosomal and sex-linked components of additive genetic covariance and variance, i.e., they are estimates of

$$\rho_{FM}(A + A') = \frac{\sigma(A_F, A_M) + \sigma(A'_F, A'_M)}{\sqrt{[\sigma^2(A) + \sigma^2(A'_F)][\sigma^2(A) + \sigma^2(A'_M)]}}$$

	Heritability		$\text{Var}(A'_x)/\text{Var}(z_x)$		
Disc	Males	Females	Males	Females	$r_{FM}(A + A')$
Labial	0.38	0.36	0.12	0.05	0.67
Clypeo-labral	0.66	0.65	0.20	0.18	0.84
Eye/antenna	0.43	0.44	0.07	0.08	0.91
Wing	0.54	0.44	0.08	0.00	0.76
Mesothoracic leg	0.44	0.42	0.12	0.07	0.86

The estimated heritabilities in males average only a few percent greater than those in females. Averaged over all characters, sex-linked loci account for about 12% of the phenotypic variance among males and 8% among females. The total additive genetic correlations across the sexes are quite high, suggesting that an average of approximately 81% of the additive variance in the two sexes is caused by shared genes.

VARIATION FOR SEXUAL DIMORPHISM

In the previous sections, we recognized the fact that many characters are sexually dimorphic. However, we did not consider whether the dimorphism itself is evolutionarily labile, as opposed to being a fixed difference that inevitably results from physiological and/or hormonal differences between males and females. Sexual dimorphisms are of interest to evolutionary ecologists for two reasons. First, dimorphisms in foraging strategies and/or morphologies are potentially selectively advantageous because they reduce competition between the sexes for food; a highly dimorphic pair of parents may provide an exceptionally broad base of food for their offspring, thereby enhancing their fitness. Second, sexual dimorphisms frequently have been attributed to sexual selection. When individuals exercise mate choice, they exert a selective pressure on the characters in the opposite sex that are the criteria for choice.

Robertson (1959b) first pointed out that the existence of genetic variation for sexual dimorphism requires that the correlation between the effects of genes in males and females must be less than one. Strictly speaking, this criterion holds only when the genetic variance is equal in the two sexes (analogous to the situation with genotype $\times$ environment interaction, Chapter 22). For an allele i at an autosomal locus, $(\alpha_{Mi} - \alpha_{Fi})$ provides a measure of the dimorphic effect. The variance of this quantity is $\sigma^2(\alpha_M) - 2\sigma(\alpha_M, \alpha_F) + \sigma^2(\alpha_F) = \sigma^2(\alpha_M) + \sigma^2(\alpha_F) - 2\rho_{FM}[\sigma(\alpha_F)\sigma(\alpha_M)]$, where ρ_{FM} is the correlation between α_{Mi} and α_{Fi}. If the two sexes have the same variance, $\sigma^2(\alpha_{Mi} - \alpha_{Fi}) = 2\sigma^2(\alpha)[1 - \rho_{FM}]$, which is zero only when $\rho_{FM} = 1$. By summing over all loci, we may define the total additive genetic variance for sexual dimorphism associated with autosomal loci to be

$$\sigma^2_{M-F}(A) = \sigma^2(A_M) + \sigma^2(A_F) - 2\sigma(A_F, A_M) \tag{24.9}$$

all three of the components on the right having been defined in the previous section. Thus, the additive genetic variance for sexual dimorphism can be calculated from estimates of the covariance between relatives that yield estimates of the components $\sigma^2(A_M)$, $\sigma^2(A_F)$, and $\sigma(A_F, A_M)$.

Example 2. Under the assumptions of no dominance or epistasis and no contribution from sex-linked loci, Eisen and Legates (1966) estimated the components of Equation 24.9 by performing a nested sib analysis on a laboratory strain of random-bred mice. As in the preceding example, the design was quite large, involving approximately 100 sires, 200 dams, and 800 male and 800 female offspring. As can be seen in the second and third columns of the following table, significant sexual dimorphism exists for body weight in the mouse, and this becomes more pronounced as the animals mature. Significant additive genetic variation exists in both of the sexes, but the additive effects of the genes in the two sexes are correlated nearly perfectly. Consequently, the estimated additive genetic variance for sexual dimorphism, $V(A_{M-F})$, is not significantly different from zero, although it approaches significance at 8 weeks. Similar conclusions were reached by Hanrahan and Eisen (1973) in a later, even larger study with the same strain.

	Mean Weight (g)						
Age	Male	Female	$V(A_M)$	$V(A_F)$	$C(A_F, A_M)$	$r_{FM}(A)$	$V(A_{M-F})$
3	11.4 (0.1)	11.1 (0.1)	0.04 (1.1)	0.02 (0.9)	0.02 (0.9)	0.64 (0.4)	0.02 (1.0)
6	30.0 (0.1)	25.4 (0.1)	4.52 (2.0)	1.40 (0.9)	2.27 (1.0)	0.90 (0.1)	1.38 (1.0)
8	34.0 (0.1)	27.8 (0.1)	5.76 (2.3)	1.39 (1.0)	2.25 (1.0)	0.80 (0.1)	2.65 (1.4)

Numbers in parentheses are standard errors of the estimates, ages are in weeks, and V and C denote Var and Cov, respectively.

In addition to the results in the previous example, there have been a number of reports of genetic correlations across the sexes that are not significantly different from one: pupal weight in *Tribolium* (Enfield et al. 1966), weight and fleece characteristics in domesticated sheep (Vesely and Robison 1971), height in humans (Rogers and Mukherjee 1992), bill color in zebra finches (Price and Burley 1993), and morphological characters in *Drosophila* (Example 1). In all of these cases, although a sexual dimorphism has clearly evolved for the character under study, the further evolution of dimorphism is tightly constrained — selection on either of the sexes would cause an almost perfectly correlated response in the opposite sex.

Some exceptions to this pattern of $r_{FM}(A) = 1$ have emerged. For example, Møller (1993) obtained an additive genetic correlation of tail length in male and female barn swallows (*Hirundo rustica*) of 0.55 ± 0.16. Meagher (1992) used correlations of family means to estimate genetic correlations across the sexes for a diversity of characters in the dioecious plant *Silene latifolia*. This approach causes a downward bias in the absolute value of estimates (Chapter 21), so it is difficult to evaluate the significance of some of his results. On the other hand, Meagher found significantly *negative* correlations across the sexes for some pairs of reproductive traits. Since the true correlations were probably even more negative, this suggests negative pleiotropic effects of some genes on fitness as expressed in males vs. females. The following example provides a clear-cut case of heritability for sexual dimorphism.

Example 3. Simmons and Ward (1991) estimated the four possible parent-offspring regressions for hind tibia length (a measure of body size) in a population of dung flies (*Scathophaga stercoraria*) raised in a common laboratory environment. Prior to analysis, male and female data were transformed to have equal variances. The regressions of sons on fathers ($b_{sf} = 0.24$) and daughters on mothers ($b_{dm} = 0.21$) were much larger than those for sons on mothers ($b_{sm} = 0.11$) and daughters on fathers ($b_{df} = 0.05$). Assuming the resemblance between relatives is due entirely to autosomal additive gene effects, the expected value of the regressions across the sexes are $\sigma(A_F, A_M)/2\sigma_z^2$, whereas the expectations for the same-sex regressions are $E(b_{sf}) = \sigma^2(A_M)/2\sigma_z^2$ and $E(b_{dm}) = \sigma^2(A_F)/2\sigma_z^2$. Therefore, an estimate of the genetic correlation across the sexes is

$$r_{FM}(A) = \frac{b_{sm} + b_{df}}{2\sqrt{b_{sf} \cdot b_{dm}}} = 0.36$$

The fact that $r_{FM}(A)$ is much less than one implies the existence of substantial genetic variation for sexual dimorphism in body size in this species. The character studied did exhibit sexual dimorphism, with females being $\sim 80\%$ of the size of males.

25

Threshold Characters

While the analytical procedures discussed in previous chapters are applicable to essentially all continuously distributed traits, the states of many characters fall into discrete categories. Important "all-or-none" or **dichotomous** characters include survivorship and the expression of congenital malformations. **Polychotomous** traits, meristic traits that can be partitioned into more than two discrete classes, include numbers of reproductive events, numbers of vertebrae or other skeletal parts, and so forth. Although the expression of some discrete traits, such as gender, may be a consequence of the expression of a single segregating factor, multiple loci are often involved. Thus, general genetic models for discrete phenotypic states need to be consistent with an underlying multifactorial basis for the character of interest.

The incidence of a character among the relatives of individuals expressing a trait relative to the incidence in the entire population is generally referred to as the **relative recurrence risk** (Chapter 16). For a character with no heritable basis, we expect this ratio to be equal to one. As the relative recurrence risk increases, it becomes more plausible that the variance of the character has a genetic basis. But how can such information be translated into a more conventional estimate of heritability?

A possible solution to this problem was first offered by Wright (1934c,d) in a study on digit number in guinea pigs. While guinea pigs normally have only three hind toes, four-toed (polydactylous) individuals occasionally appear in laboratory stocks, and the incidence of the trait can be increased to 100% by selection and inbreeding (Castle 1906). Through a carefully designed breeding program with initially homozygous strains, Wright rejected the hypothesis that polydactyly has a simple genetic basis. The most compelling evidence came from the observation that when different strains that bred true for the three-toed condition were mated to the same pure four-toed strain, very different proportions of four-toed progeny were obtained. These results and other oddities were shown to be consistent with a model in which the development of a fourth toe depends on the level of a continuously distributed underlying trait. Wright suggested that the four-toed condition would only arise if the total contribution of genes and environment to the underlying trait exceeded a certain threshold (Figure 25.1).

Attributes that are categorical on an outward (observed) scale but believed to be continuous on an underlying (unobserved) scale are known as **threshold** (Wright 1934c,d) or **quasi-continuous** (Grüneberg 1952) characters. The threshold

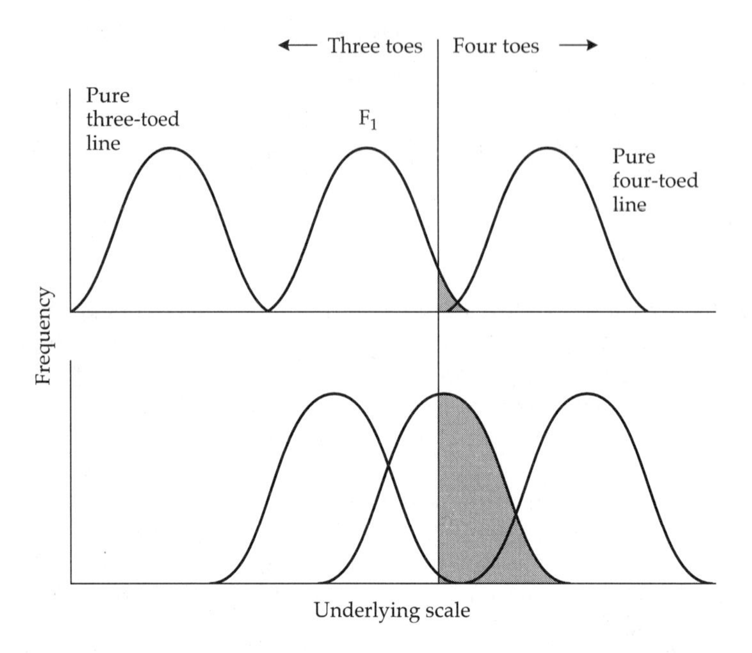

Figure 25.1 Wright's (1934c,d) explanation for how the incidence of a dichotomous character can vary in crosses between parents of two types. Individuals from both types of parental strains are assumed to be normally distributed with respect to some underlying determinant of the dichotomous trait. However, the three-toed line in the lower panel is assumed to have a mean phenotype on the underlying scale that is closer to the threshold than that in the upper panel. The mean phenotype of the F_1 progeny is assumed to be intermediate between that of the parents, and progeny whose underlying measure exceeds the threshold exhibit the four-toed condition (shaded areas).

model assumes a stepwise **risk function** for phenotypes on the underlying scale (Figure 25.2). All individuals with underlying phenotypic values above the threshold exhibit the trait; all those below it do not. A stepwise risk function on the underlying phenotypic scale implies a sigmoid risk function on the underlying genotypic scale, provided the environmental deviations on this scale are normally distributed (Smith 1971, Curnow 1972, Mendell and Elston 1974). Individuals with genotypic values above the threshold are at less than 100% risk because some of them have underlying phenotypes below the threshold. Likewise, some individuals with genotypic values below the threshold have phenotypic values above it. The greater the environmental contribution to the variance of the trait, the more gradual the risk function on the genotypic scale (Figure 25.2).

Since the nature of the underlying trait is almost always unknown, the interpretation of categorical data with a threshold model may appear to require an extreme act of faith. However, there are a number of ways to test the general validity

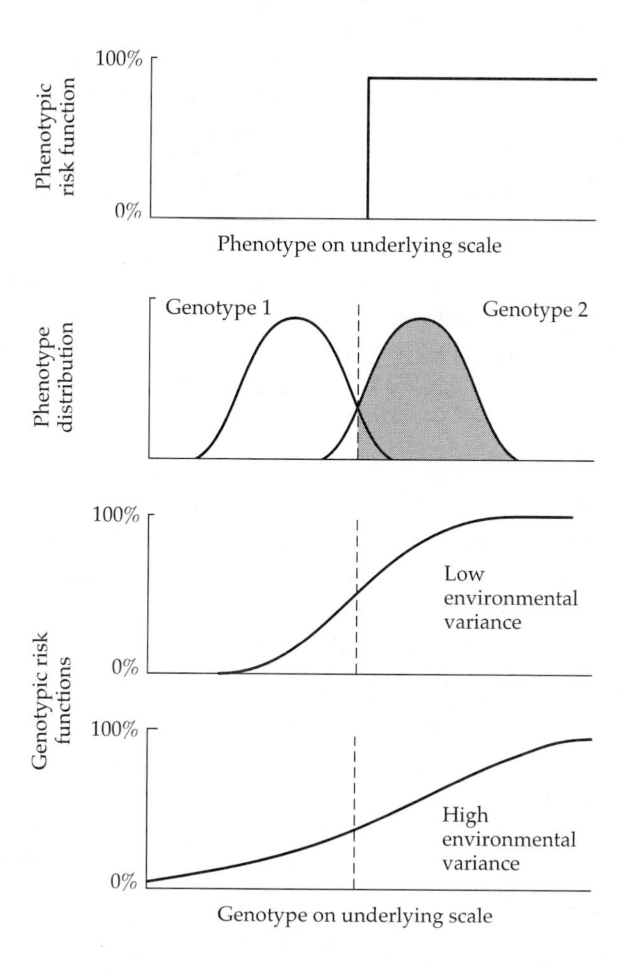

Figure 25.2 The relationship between phenotypic and genotypic risk functions and the environmental component of variance for the trait on the underlying scale. The stepwise phenotypic risk function is given in the upper panel. The next lower panel shows conditional phenotypic distributions for two genotypes; individuals in the portions of the distributions to the right of the threshold exhibit the trait, and their incidence represents the risk for their respective genotype. The bottom two panels plot genotypic risk as a function of genotypic value, where the latter is simply the mean of the genotype-specific conditional distribution. As the conditional genotype distributions become narrower (i.e., environmental variances become smaller), the genotypic risk function converges on the phenotypic risk function, whereas high environmental variance induces a flat genotypic risk function.

of the model. In Chapter 11, for example, we showed how threshold developmental maps provide plausible models for phenomena ranging from canalization to genetic assimilation. Moreover, in certain situations, the underlying determinant of character expression may be revealed through careful experimentation. For example, Alberch and Gale (1985) suggested that the number of primordial cells in the amphibian limb bud determines the digital structure. Through the use of mitotic inhibitors and promoters, they showed that below a certain threshold number of cells, complete loss of a digit may occur. A similar mechanism may have been operating in Wright's guinea pigs. Finally, as will be demonstrated below, the validity of a threshold model can be evaluated through the comparison of phenotypes of various types of relatives.

Threshold models are used extensively in genetic counseling to predict the risk of congenital malformations and psychiatric disorders in offspring of affected relatives (Carter 1965, 1969; James 1971; Smith 1971; Gershon et al. 1976). General mathematical/statistical reviews of the theory have been written by Curnow and Smith (1975) and Gianola (1982). Evolutionary problems that have recently been investigated with threshold models include limb loss in tetrapods (Lande 1978), environmental sex determination (Bulmer and Bull 1982), and mating preference (O'Donald and Majerus 1985).

HERITABILITY ON THE UNDERLYING SCALE

If the phenotype distribution on the underlying scale is treated as a standard normal (with mean = 0 and variance = 1), a relatively simple approach is available for estimating heritability on the underlying scale. The technique was independently developed by Crittenden (1961) and Falconer (1965b), both of whom saw an analogy between the phenotypes of affected parents and the response of a population to truncation selection. Both authors were concerned with the inheritance of genetic disorders in humans. For that reason, Falconer (1965b) called the underlying scale **liability** and the affected individuals **propositi**. For reference, we refer to the grand mean for liability in the base population as $\mu_p = 0$ and the mean liability of propositi as μ_w (Figure 25.3). For the most part, for simplicity, we assume the propositi to be parents and their relatives to be offspring. However, the following analysis can be readily extended to any degree of relatedness.

As for any quantitative trait, the unobserved underlying character (liability) can be treated as the sum of a genotypic value and an environmental deviation, and assuming that it is normally distributed, the regression for liability in different sets of relatives will be linear. Further assuming that there is no change in the mean environmental contribution to liability between generations and no selection, the mean liability among all offspring will equal that for the base population (μ_p). This value provides one point on the expected parent-offspring regression. Now suppose that affected parents (with mean liability μ_w) produce offspring with

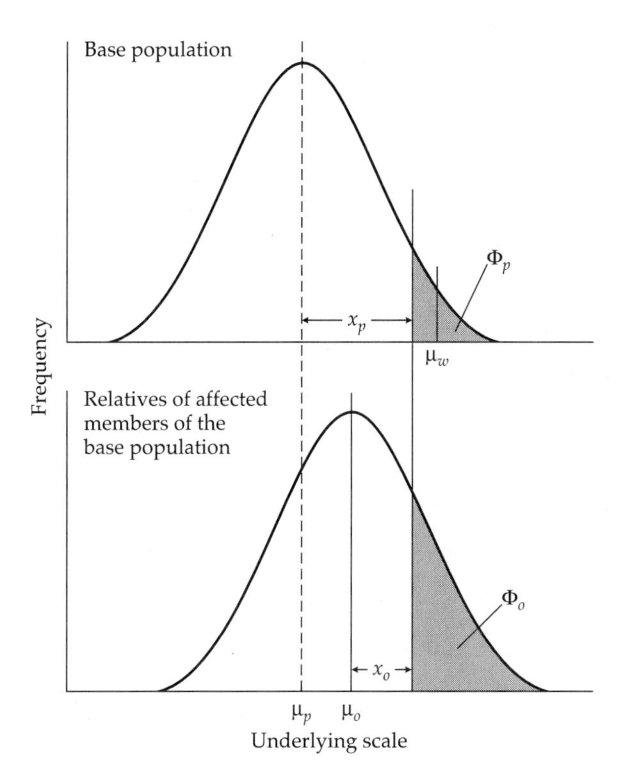

Figure 25.3 Upper panel: The phenotype distribution of a parental population on the underlying scale, assumed to be normal. The dashed line denotes the parental mean phenotype, μ_p, whereas the solid line denotes the threshold above which the character is expressed. A fraction Φ_p of the population exhibits the trait, and the mean phenotype of such individuals is μ_w. **Lower panel:** The phenotype distribution for a class of relatives of affected parents, e.g., progeny. Due to the nonzero heritability of the trait, this conditional distribution is shifted to the right ($\mu_o > \mu_p$) and shows a higher incidence Φ_o.

expected liability μ_o. These values provide a second point on the regression. It follows that the slope of the parent-offspring regression is $\beta_{op} = (\mu_o - \mu_p)/(\mu_w - \mu_p)$, which reduces further to $\beta_{op} = \mu_o/\mu_w$ since we have scaled the distribution of liability in the base population such that $\mu_p = 0$. This definition of β_{op} has the usual interpretation of a regression between parents and offspring. That is, $2\beta_{op} = [\sigma_A^2 + (\sigma_{AA}^2/2) + \cdots]/\sigma_z^2$.

Thus, the problem of estimating heritability on the underlying scale reduces to obtaining estimates of the mean liabilities in affected parents and their offspring, μ_w and μ_o. Neither of these is directly observable, but from the properties of a normal distribution, estimates of them are obtainable from the incidence of the disorder in the population (Φ_p) and the proportion of offspring of propositi that are affected (Φ_o). First, we note that the estimated mean $\overline{z}_o$ can also be written as $(x_p - x_o)$, where x_p and x_o are the estimated distances of the threshold from the mean liability in the two samples (Figure 25.3). Given the observations Φ_p and Φ_o, values for x_p and x_o are readily obtained from a table of the standard normal distribution (Chapter 2). These values are in units of phenotypic standard deviations on the underlying scale of liability. Second, the mean liability of affected members of the base population, $\overline{z}_w$, can be determined by use of the equation for the mean of the tail of a normal distribution. From Equation 2.15, $\overline{z}_w = p(x_p)/\Phi_p$, where $p(x_p) = (2\pi)^{-1/2}\exp(-x_p^2/2)$ is the height of the standardized normal distribution at the threshold in the base population. Thus, the estimated parent-offspring regression on the underlying scale is

$$b_{op} = \frac{\overline{z}_o}{\overline{z}_w} = \frac{(x_p - x_o)\Phi_p}{p(x_p)} \tag{25.1a}$$

Note that this expression generalizes to any set of relatives (r) by writing

$$b_{rp} = \frac{(x_p - x_r)\Phi_p}{p(x_p)} \tag{25.1b}$$

The large-sample variance estimate for b_{op}, derived by Taylor expansion, is

$$\text{Var}(b_{op}) \simeq \left[\frac{1}{\overline{z}_w} - b_{op}(\overline{z}_w - x_p)\right]^2 \left[\frac{\Phi_p(1-\Phi_p)}{N_p\, p^2(x_p)}\right] + \frac{\Phi_o(1-\Phi_o)}{\overline{z}_w^2\, N_o\, p^2(x_o)} \tag{25.1c}$$

where N_p and N_o are, respectively, the sample sizes for total individuals in the parental generation and offspring of affected individuals.

Because of its elegance and simplicity, the Crittenden-Falconer technique has been utilized widely. However, the reader should be aware of two assumptions made in the preceding derivation. First, in relying on a linear model, we assumed implicitly that the distribution of liability is normal in both the affected parents and their offspring. Clearly, this is not true for the affected parents, which are a truncated sample from a normal distribution (Figure 25.3), nor is it likely to be exactly true for their offspring. The second simplifying assumption is that both the base population and the relatives of propositi have unit variance for liability. However, if the character is heritable, the variance of liability in offspring of affected parents will be less than one since they represent only a subset of the population.

Although Crittenden (1961) and Falconer (1965b) were aware of these difficulties, they left their solution to later investigators. Utilizing statistical theory that had been developed much earlier by Pearson (1900) and Everitt (1910), Edwards (1969) and Smith (1970) showed how the preceding problems can be eliminated by considering all possible pairs of parents and offspring rather than just affected

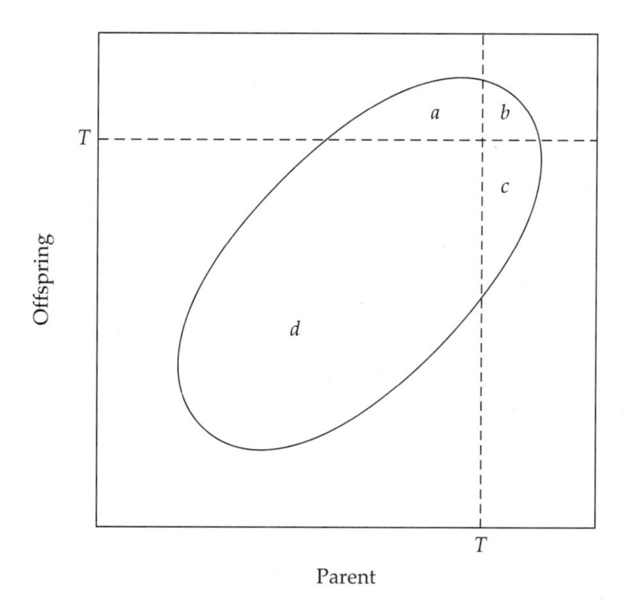

Figure 25.4 The bivariate distribution of parent and offspring phenotypes on the underlying scale. The threshold value for the character is indicated by the dashed lines. The region denoted d marks combinations of parents and offspring that are unaffected, b denotes affected parents and offspring, a denotes unaffected parents and affected offspring, and c denotes affected parents and unaffected offspring. In the ideal case with no selection, the portions of the curves denoted a and c should be equal.

parents. This allows the entire joint distribution of parents and offspring to be partitioned into four types of pairs: both affected, both unaffected, parents affected but offspring unaffected, and vice versa (Figure 25.4). Given this information, the exact phenotypic correlation between parents and offspring on the underlying standard bivariate normal scale can be extracted from tables provided by Pearson (1900) and Everitt (1910) or from integral equations derived by Curnow (1972). Both procedures are a bit tedious, but a useful approximation has been given by Edwards (1969),

$$r_{op} = b_{op} \simeq h^2/2 = \frac{0.57 \ln k}{-\ln(\Phi_p) - 0.44 \ln k - 0.18} \tag{25.2}$$

where $k = \Phi_o/\Phi_p$. An analytical approximation that accounts for the change in variance across generations but not for the nonnormality of the affected parents was derived by Reich et al. (1972),

$$r_{op} = \frac{x_p - x_o \left\{1 - (x_p^2 - x_o^2)[1 - (x_p/\overline{z}_w)]\right\}^{1/2}}{\overline{z}_w + x_o^2(\overline{z}_w - x_p)} \tag{25.3}$$

where again $\overline{z}_w = p(x_p)/\Phi_p$.

Table 25.1 Estimates of the heritability of three human congenital disorders obtained by the methods of Crittenden-Falconer (Equation 25.1a), Edwards (Equation 25.2), and Reich et al. (Equation 25.3), as described in the text.

	Incidence					Heritability		
Disease	Φ_p	Φ_r	x_p	x_r	$p(x_p)$	(25.1a)	(25.2)	(25.3)
Harelip	0.001		3.090		0.003			
MZ		0.500		0.000		1.03	0.99	1.03
1st degree		0.040		1.751		0.89	0.90	0.86
2nd degree		0.007		2.457		0.84	0.81	0.83
3rd degree		0.003		2.748		0.91	0.86	0.90
Club Foot	0.001		3.090		0.003			
MZ		0.325		0.454		0.88	0.88	0.86
1st degree		0.021		2.034		0.70	0.70	0.68
2nd degree		0.006		2.512		0.77	0.74	0.75
3rd degree		0.002		2.878		0.57	0.53	0.56
Schizophrenia	0.010		2.326		0.027			
MZ		0.443		0.143		0.81	0.92	0.83
1st degree		0.077		1.426		0.67	0.75	0.74
2nd degree		0.027		1.927		0.59	0.63	0.63
3rd degree		0.016		2.144		0.54	0.56	0.56

Note: Incidence data are from Carter (1965, 1969) and McGue et al. (1983). Φ_p and Φ_r are, respectively, the incidences of the disorders in the population and in relatives of affected individuals. MZ denotes monozygotic twins; first-degree relatives include parent-offspring and full-sibs; second-degree relatives include aunt (uncle)-niece (nephew); and third-degree relatives are first cousins. Each of the methods gives an estimate of the regression between relatives for inferred phenotypes on the underlying scale. The heritabilities are computed by dividing the estimated regression by twice the coefficient of coancestry (1, 1/2, 1/4, and 1/8 for monozygotic twins, and first-, second-, and third-degree relatives, respectively).

The degree of inaccuracy that results from the use of the Crittenden-Falconer equation can be seen in Table 25.1, where the heritabilities of three human disorders are calculated using Equations 25.1–3. For each attribute, data are available for several degrees of relationship and sample sizes are large (several hundreds to thousands). There is no strong evidence for dominance genetic variance for these traits, so different types of relatives of the same degree have been pooled.

Three important observations can be gleaned from this table. First, for all three traits, the incidence in affected relatives (Φ_r) is substantially higher than that in the population at large (Φ_p). Thus, regardless of the model or the class of relatives employed in the computation of h^2, it is clear that the variable expression of these traits has a genetic basis. Second, the high estimates of h^2, all in excess of 0.5, arise despite the fact that the incidences of the disorders in the population are very low. Thus, the incidence of a trait in a population provides no information about its heritability. Third, the equation of Reich et al. (1972) produces results that are essentially the same as those of Edwards (1969) formula, indicating that the nonnormality correction is of negligible significance. More remarkable is the excellent agreement between the results of the Crittenden-Falconer model and those from the more exact treatments. In no case do the estimates differ by more than 10%.

Less clear is why all three approaches yield lower heritability estimates with increasing distance of relationship. One possibility is that significant sources of epistatic variance contribute to the expression of these traits. All components of epistatic genetic variance are completely confounded with the additive genetic variance in the case of monozygotic twins (Chapter 19), i.e., the heritability estimate derived for monozygotic twins is more appropriately described as a broad-sense heritability. However, epistatic components of variance make diminishingly smaller contributions to the covariance between relatives as the coefficient of coancestry declines (Chapter 7; also see Equation 16.49d). Another potential explanation is that shared environmental effects contribute disproportionately to the resemblance between close relatives. Still a third possibility is that relationships in humans (other than twins) are sometimes less than expected due to uncertain paternity.

Example 1. For the case in which individuals can be clonally replicated, a simple method exists for estimating the broad-sense heritability on the underlying scale. Suppose that n individuals are scored for the character in each of N clones, and let n_i be the number of affected individuals in the ith clone. The incidence of the trait in the population is then $\Phi_p = \sum n_i/(Nn)$. Within a clone exhibiting affected individuals, the incidence of affected relatives is estimated by $(n_i - 1)/(n-1)$, since an affected individual has $(n-1)$ sibs, $(n_i - 1)$ of which are also affected. Thus, an estimate of the incidence among relatives is $\Phi_r = \sum(n_i - 1)/[N_a(n-1)]$, where N_a is the number of clones with affected individuals, and the summation is over such clones. The broad-sense heritability can be estimated by using Φ_r in place of Φ_o in the solution of Equations 25.1a, 25.2, or 25.3. The regression coefficient provides an estimate of h^2 since $2\Theta = 1$ for clonemates.

As in all attempts to estimate heritability from the resemblance between relatives, an important assumption of the above procedures is that selection does not alter the relative incidence of the trait in different pairs of relatives (i.e., the values of a, b, c, and d in Figure 25.4) prior to their assessment. Selection may be a serious source of bias for certain types of relatives and characters. In the case of human genetic disorders, for example, affected individuals may fail to reproduce for physiological reasons or, as a response to genetic counseling, may differentially abort affected fetuses. Comparisons of full sibs do not necessarily provide a solution to such problems. For example, individuals whose first offspring is affected may tend to decide against having future offspring, in which case they would be eliminated from the analysis. With appropriate medical records, correction might be made for such bias, but it would not be a trivial task.

MULTIPLE THRESHOLDS

In principle, the model developed above can be extended to any number of thresholds. Indeed, in his original analyses of digit number in guinea pigs, Wright (1934c,d) actually considered three classes of individuals: those with the usual three toes, those with an incomplete fourth toe, and those with four complete digits. The computations become rather lengthy for more than three character states, but a two-threshold model, which we will focus upon here, is fairly straightforward.

There are two very useful attributes of multiple threshold models. First, they provide a means for evaluating the relative means and variances of liability in different populations. Second, by providing several estimates of the regression between relatives, they provide an internal check on the consistency of the model.

With a three-character state model, there are two thresholds (Figure 25.5). We will refer to these as T_1 and T_2, the **wide** and **narrow thresholds**, respectively. With this notation, individuals are classified as normal (character state 1, with liability less than T_1), **wide** (character states 2 and 3, with liability exceeding T_1), or **narrow** (character state 3, with liability exceeding T_2). From above, we know that x_{p2}, the distance of T_2 from the mean in standard deviations, can be extracted from tables if the incidence of character state 3 is known. Similarly, x_{p1} is obtainable from the total incidence of character states 2 and 3. The absolute distance between the two thresholds, $(x_{p2} - x_{p1})\sigma$, may then be defined as one threshold unit, implying that the standard deviation in threshold units is $\sigma = 1/(x_{p2} - x_{p1})$. The mean, measured as the deviation from the wide threshold, is then $\overline{z}_p = -x_{p1}\sigma$. Thus, if different populations can be assumed to have their thresholds located at the same position on the phenotypic scale for liability, this simple approach allows a comparison of the means and standard deviations of their underlying phenotypic distributions (in threshold units).

The two-threshold model provides four routes to estimating the heritability of liability (Reich et al. 1972). First, by ignoring one of the thresholds, the standard single-threshold model can be applied to the data. In this case, individuals having character states 2 and 3 can be treated as the same (normal vs. wide), or individuals with character states 1 and 2 can be aggregated (narrow vs. not narrow). Either way, the population is divided into two classes of individuals.

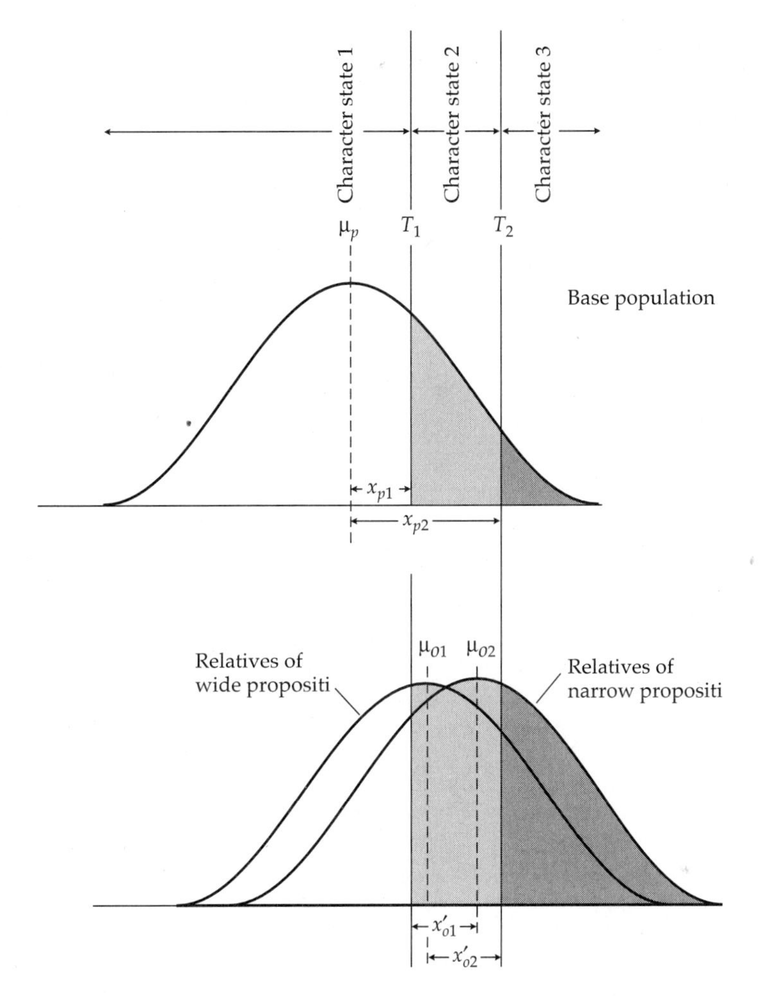

Figure 25.5 Upper panel: The phenotype distribution of a parental population on the underlying scale, assumed to be normal, with two thresholds determining the expression of three character states. The dashed line denotes the parental mean phenotype, μ_p, whereas the two solid lines denote the thresholds. In the lower panel, the distribution on the left refers to offspring from parents above the wide threshold T_1, whereas that on the right refers to offspring from parents above the narrow threshold T_2.

Two estimates of heritability can also be obtained from information on "cross prevalence." Consider, for example, the propositi to be parents with character state 3. From their incidence, we obtain an estimate of x_{p2}. The affected offspring of these parents can be scored as all individuals exhibiting character states 2 or 3. The normal deviate for these individuals, x'_{o1} (Figure 25.5), refers to the widely affected offspring of narrowly affected parents. Similarly, x'_{o2} is obtained from the narrowly affected offspring of widely affected (character state 2 or 3) parents (as the distance of the mean of these offspring from the upper threshold). From this information, two additional estimators of the parent-offspring regression are:

$$b_{op} = \frac{x_{p2} - x'_{o2}\sqrt{1 - (x_{p2}^2 - x_{o2}'^2)[1 - (x_{p1}/a_1)]}}{a_1 + x_{o2}'^2(a_1 - x_{p1})} \tag{25.4a}$$

$$b_{op} = \frac{x_{p1} - x'_{o1}\sqrt{1 - (x_{p1}^2 - x_{o1}'^2)[1 - (x_{p2}/a_2)]}}{a_2 + x_{o1}'^2(a_2 - x_{p2})} \tag{25.4b}$$

where $a_1 = p(x_{p1})/\Phi_{p1}$ and $a_2 = p(x_{p2})/\Phi_{p2}$. If the assumptions of the threshold model have been met, then all four estimates of b_{op} should be similar.

Example 2. The application of the two-threshold model can be illustrated with data on the incidence of diabetes in the Edinburgh population (Smith et al. 1972). Considerable variation exists for the age of onset of diabetes, and the authors wanted to know whether early-onset vs. late-onset diabetes simply represent different levels of liability. Through interviews of diabetics attending a clinic, information on the incidence of the disease in first-degree relatives was obtained. The affected individuals were partitioned into those first exhibiting the disease before and after age 25 years. Thus, the narrowly affected individuals were those affected between ages 0 and 25. A large survey estimated the incidence of narrowly affected individuals to be $\Phi_{p2} = 0.0006$, whereas the total (or wide) incidence of the disease was $\Phi_{p1} = 0.0039$. From a table of the standard normal distribution, $x_{p2} = 3.24$, $x_{p1} = 2.66$, $a_2 = 3.51$, and $a_1 = 2.97$. Thus, the wide and narrow thresholds are approximately 2.7 and 3.2 standard deviations above the mean on the liability scale, and the distance between the two thresholds is $(x_{p1} - x_{p2}) = 0.58$ standard deviations.

Data on the incidence of late- and early-onset diabetes in first degree relatives are given in the table at the top of the following page. If early-onset patients represent a subset of the population with more extreme genetic values for liability than late-onset patients, the total incidence of the disease should be higher in the relatives of the former than the latter. The data show this to be true — first-degree relatives of early- and late-onset patients have total incidences of 0.0520 and 0.0312, respectively.

Converting the incidences of disease in relatives to their respective x values, the four possible heritability estimates (twice the regression coefficients) range from 0.46 to 0.76. From Equation 25.1b, the standard errors of h^2 based on narrow propositi and relatives and on wide propositi and relatives are found to be approximately 0.10 and 0.03, respectively. Thus, with one possible exception, the four estimates are in approximate agreement, leading to an overall estimate of $h^2 = 0.56$ for the liability.

Propositi	Relatives	Incidence Among Relatives:		b_{op}	Equation
Narrow	Narrow	0.0205	$x_{o2} = 2.044$	0.38	(25.3)
	Wide	0.0315	$x'_{o1} = 1.859$	0.23	(25.4a)
Wide	Narrow	0.0045	$x'_{o2} = 2.612$	0.24	(25.4b)
	Wide	0.0267	$x_{o1} = 1.932$	0.27	(25.3)

GENETIC CORRELATIONS AMONG THRESHOLD TRAITS

The various applications of genetic correlations encountered in previous chapters are useful in many contexts with threshold traits. For example, in insects with wing dimorphisms, wing development is often highly dependent on photoperiod or other environmental conditions (Roff 1986, 1994). Salamanders in the genus *Ambystoma* may develop cannibalistic morphs or exhibit paedomorphosis under specific environmental conditions. Many species of zooplankton are known to modify their morphologies in the presence of predators (Havel 1987). Situations like these raise questions as to whether the genotypes that respond to one set of environmental cues are the same as those that respond to a second set of stimuli.

As noted in Chapter 24, another application of the genetic correlation concerns the expression of characters in the different sexes. For threshold characters, it has often been noted that when the incidence of affected individuals differs between the sexes, the sex with lower incidence has a higher frequency of affected relatives (Table 25.2). One potential explanation for such a reversal in frequencies is that affected members of the sex with the lower incidence tend to have a higher liability due to a displacement of the threshold to the right of the mean phenotype on the underlying scale.

When separate data are available for the two sexes, four separate regressions can be calculated using Equations 25.1a, 25.2, or 25.3: one for males only (b_{MM}),

Table 25.2 Incidence of pyloric stenosis in a human population.

Propositi	Incidence in Population	Incidence Among First-degree Relatives: Male	Female
Male	0.005	0.050	0.022
Female	0.001	0.171	0.066

Source: Carter 1961.

one for females only (b_{FF}), one for male propositi and female relatives (b_{FM}), and vice versa (b_{MF}). Ignoring nonadditive sources of variance and sex linkage, and assuming an absence of common environmental effects, the male-male and female-female regression coefficients will have expected values

$$\beta_{MM} = \frac{2\Theta\sigma^2(A_M)}{\sigma^2(z_M)} \tag{25.5a}$$

$$\beta_{FF} = \frac{2\Theta\sigma^2(A_F)}{\sigma^2(z_F)} \tag{25.5b}$$

where Θ is the coefficient of coancestry for the relatives, $\sigma^2(A_M)$ and $\sigma^2(z_M)$ are the additive genetic variance and phenotypic variance for males, and $\sigma^2(A_F)$ and $\sigma^2(z_F)$ are those for females. The regressions of males on females and females on males have expectations

$$\beta_{MF} = \frac{2\Theta\sigma^2(A_M, A_F)}{\sigma^2(z_F)} \tag{25.5c}$$

$$\beta_{FM} = \frac{2\Theta\sigma^2(A_M, A_F)}{\sigma^2(z_M)} \tag{25.5d}$$

where $\sigma^2(A_M, A_F)$ is the additive genetic covariance between the sexes. The genetic correlation across the sexes can be estimated as

$$r_{FM}(A) = \sqrt{\frac{b_{MF} \cdot b_{FM}}{b_{MM} \cdot b_{FF}}} \tag{25.6a}$$

The preceding expressions are completely generalizable. For example, instead of pertaining to the expression of a trait in male and female relatives in the same environment, they may refer to male relatives in two different environments, to female relatives in different environments, or to male and female relatives in different environments. Thus, a formula of the form of Equation 23.6a can be used to estimate the genetic correlation across environments for the expression of a dichotomous trait. Moreover, the same approach can be used to obtain the genetic correlation between two different dichotomous traits, e.g., survivorship and the presence/absence of a morphological trait. For two characters x and y,

$$r_{xy}(A) = \sqrt{\frac{b_{xy} \cdot b_{yx}}{b_{xx} \cdot b_{yy}}} \tag{25.6b}$$

Finally, it is relatively straightforward to compute the genetic correlation between a dichotomous and a continuously distributed character. Denoting two such traits as d and c, Equation 25.6b still applies. The regression for the dichotomous character (b_{dd}) can be obtained by the methods discussed earlier in this chapter, while the regression for the continuously distributed trait (b_{cc}) can be obtained by conventional methods outlined in previous chapters, e.g., parent-offspring regression. The regression involving the two traits, b_{cd}, can be obtained by dividing the deviation of the mean of the continuously distributed trait in relatives of affected individuals from the mean in the population at large ($\overline{z}_{cr} - \overline{z}_{cp}$) by the mean of the affected individuals for the dichotomous trait on the underlying scale ($\overline{z}_{dr} = x_{dp} - x_{do}$), with a similar definition applying to b_{dc}.

Example 3. The application of the preceding ideas will be illustrated with a familiar dichotomy in humans — handedness. The data base consists of responses to questionnaires distributed to college undergraduates and service recruits in Scotland (Annett 1973). The incidences of left-handedness in males and females are 0.118 and 0.114, so at least on the outward scale the two sexes have essentially identical phenotype distributions. In the following analyses, we assume that the variances for both sexes are also equal on the underlying scale. The table summarizes the incidences of left-handedness in brothers and sisters of male and female propositi and the associated probit (x) scores. The regression coefficients are computed by use of Equation 25.3.

Propositi	Relatives	Incidence in Relatives	x	$2b$	Expectation
Males	Brothers	0.143	1.067	0.15	h_M^2
	Sisters	0.114	1.208	−0.02	$\rho_{FM} h_M h_F$
Females	Brothers	0.135	1.103	0.12	$\rho_{FM}(A) h_M h_F$
	Sisters	0.156	1.010	0.24	h_F^2

Averaging over the two sexes, the heritability estimate for liability to left-handedness is approximately 0.20. By use of Equation 25.6b, we obtain $r_{FM}(A) = 0.26$. Thus, the genetic correlation across the sexes appears to be small. In light of these results and the fact that the study population consisted of full sibs, a reasonable interpretation is that handedness is primarily a chance event of development, with genetics playing a minor role, and perhaps a small contribution of the variance coming from common environmental effects.

HERITABILITY ON THE OBSERVED SCALE

The question often arises as to why heritabilities of dichotomous characters are measured on an unobservable underlying scale rather than on the directly observed scale. The latter approach was actually taken by most early investigators

(Lush et al. 1948, Robertson and Lerner 1949, Dempster and Lerner 1950, VanVleck 1971). Assuming additivity, a simple expression for heritability on the observed scale can be obtained as follows. Suppose the character of interest is survivorship to a certain age. Then all nonsurvivors can be scored as 0 and all survivors as 1. The frequency of survivors in the population is simply the incidence Φ_p, and the phenotypic variance is the familiar variance of a binomial distribution,

$$\sigma_z^2 = \Phi_p(1 - \Phi_p) \tag{25.7}$$

It follows that the heritability on the observed scale is

$$h_o^2 = \frac{\sigma^2(A_o)}{\Phi_p(1 - \Phi_p)} \tag{25.8a}$$

where the subscript o denotes the observed scale, and $\sigma^2(A_o)$ is the additive genetic variance on the observed scale. In an appendix to Dempster and Lerner (1950), Robertson showed that $\sigma^2(A_o) \simeq [p(x_p)]^2 h^2$, where h^2 is the heritability on the unobserved scale of liability. Thus, the relationship between the heritabilities on the underlying and observed scales is

$$h_o^2 = h^2 \left\{ \frac{[p(x_p)]^2}{\Phi_p(1 - \Phi_p)} \right\} \tag{25.8b}$$

Examination of this formula reveals several undesirable properties of heritability estimates on the observed scale:

1. h_o^2 is a function of the incidence in the population (Figure 25.6). With constant phenotypic variance on the liability scale, h_o^2 changes with a shift in mean liability because this induces a change in Φ_p.

2. The maximum possible value of h_o^2 is $2/\pi \simeq 0.64$, which arises when $h^2 = 1$, $\Phi_p = 0.5$, and $p(x_p) = (2\pi)^{-1/2}$. This implies that a substantial proportion of the genetic variance on the observed scale is nonadditive even if all of the genetic variance on the liability scale is additive. The reason for this relationship has been outlined in Figure 25.2 — the regression of risk on liability of genotypes is necessarily nonlinear since the probability of expressing the trait is bounded between 0 and 100%.

3. For this same reason, genotypic values and environmental deviations are not independent on the observed scale.

Sometimes it is more practical to initially calculate heritability on the observed scale and then compute h^2 indirectly by rearranging Equation 25.8b. Any of the procedures described in earlier chapters can be used for this purpose, applying the analysis to the $0, 1$ variables. One problem with this approach, pointed

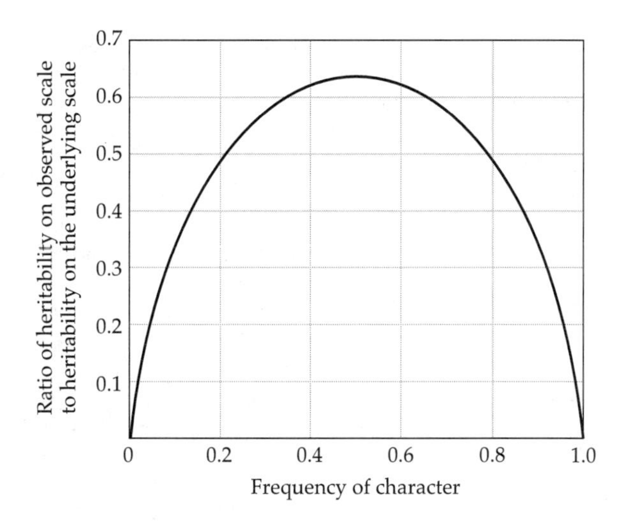

Figure 25.6 The ratio of heritabilities on observable (h_o^2) and underlying (h^2) scales as a function of the incidence of expression in the population (Φ_p).

out by VanVleck (1971), is that the relationship between h_o^2 and h^2 depends on the incidence Φ_p in a nonlinear fashion. Consequently, the substitution of *sample estimates* of Φ_p into Equation 25.8b results in a biased conversion between the two heritabilities.

Example 4. In map turtles, the average sex ratio (% males) of clutches is closely coupled with the temperature at which the eggs are incubated. A rather tight threshold exists at 29.2°C. Below 28°C, virtually all eggs develop into males, while above 30°C clutches are entirely female (see the accompanying figure below). In effect, the response curve is a phenotypic risk function if temperature is viewed as the underlying scale of liability.

Bull et al. (1982) were interested in determining the extent to which variation in offspring sex at the intermediate temperature was due to genetic variation among females. Twenty gravid females were collected in the field and induced to lay eggs in the laboratory. Ten eggs from each mother were then randomly distributed in an incubator maintained at 29.2 – 29.3°C. The offspring were sexed upon hatching and scored as 0 if female and 1 if male. A one-way ANOVA was then performed on the full-sib family data.

The additive genetic variance on the observed scale was estimated to be 0.13 from the among-family component of variance. As an estimate of $\sigma^2(A_o)$, this among-family variance could be inflated by dominance genetic variance for the sex determination mechanisms, but the influence of common-family environment should have been eliminated by the random design. The total proportion of male hatchlings was 0.41. Heritability on the observed scale is therefore approximately $0.13/[0.41(1 - 0.41)] = 0.54$. Using $\Phi_p = 0.41$, $x_p = 0.228$ and $p(x_p) = 0.389$, from Equation 25.8b, heritability on the underlying scale is estimated to be $h^2 \simeq (0.54)(0.41)(1 - 0.41)/(0.389)^2 = 0.86$.

The authors note that the above computation assumes a constant incubation temperature for all females. In the field, however, different females will inevitably place their eggs in areas of somewhat different temperatures. Let the phenotypic variance in liability at a constant temperature (as in the laboratory experiment) be σ_x^2. In the field, the phenotypic variance is $\sigma_x^2 + \sigma_T^2$, where σ_T^2 is the additional variance in liability resulting from a variable environment. Thus, the heritability in the field is $h^2[\sigma_x^2/(\sigma_x^2 + \sigma_T^2)]$. For the study population, rough estimates of σ_x^2 and σ_T^2 were 0.09 and 1.0, reducing the expected heritability in the field to only 0.06.

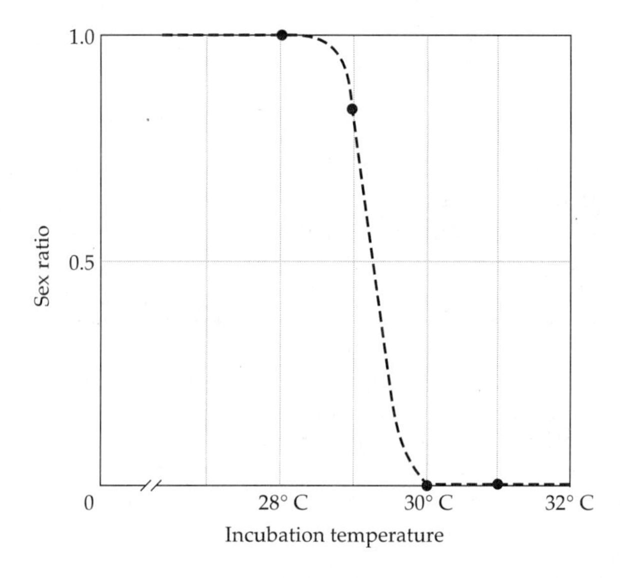

26

Estimation of Breeding Values

Most of the analytical methods encountered in earlier chapters of this book have assumed a data base derived from a setting with a regular design and fairly constant sample sizes in different genetic groups. Such situations are often approachable in laboratory or greenhouse populations, but many natural populations and agricultural species present the investigator with highly unbalanced family sizes and fragmentary data from numerous kinds of relationships. Data sets from which individuals have been eliminated by natural and/or artificial selection may deviate substantially from the base population about which one wishes to make inferences. Further culling of the data to accommodate conventional statistical techniques such as ANOVA, even if nonselective, still leads to an inefficient use of information. The goal of the next two chapters is to present a general overview of a family of statistical approaches that allows the efficient estimation of quantitative-genetic parameters under arbitrary settings, including those involving extended pedigrees, unequal family sizes, assortative mating, and selection.

All of the methods to be covered below are based on the general mixed model, which embraces the vast majority of estimation problems encountered in previous chapters. This chapter introduces best linear unbiased prediction (BLUP), a general method for predicting random effects (such as breeding values and maternal effects), while Chapter 27 is concerned with the estimation of genetic variances by restricted maximum likelihood (REML). These two methods are related in that BLUP assumes that the appropriate variance components are known, while REML procedures estimate variance components in an iterative fashion from BLUP estimates of random effects. Although the basic properties of these techniques have been known for decades, because of their computational demands, their practical application is a fairly recent phenomenon. BLUP is now by far the dominant methodology for estimating breeding values.

After a brief introduction to the general mixed model, we will develop expressions for BLUEs (best linear unbiased estimators) of fixed effects and for BLUPs of random effects under the assumption that variances are known in the base population. The remainder of the chapter considers several specific applications of BLUP, starting with the estimation of the breeding value of a single trait under a strictly additive model and then considering more advanced issues, including the estimation of dominance values and maternal effects and methods for dealing with repeated records and multiple traits.

There is a huge and sophisticated literature on BLUP methodology, detailed reviews of which can be found in Henderson (1977a, 1984a, 1988a), Schaeffer (1991), Kennedy (1991), Searle et al. (1992), and Mrode (1996). Our goal is simply to introduce the general framework and provide some specific examples that may increase the accessibility and attractiveness of the method to nonspecialists. BLUP is primarily used for the identification of individuals with maximum genetic merit in selection programs and for monitoring actual selection response. However, the method is very general and has been applied to a wide variety of additional problems ranging from the prediction of line-cross means (Henderson 1977c, 1984a) to the estimation of QTL effects (Kennedy et al. 1992) to the estimation of unusual genetic transmission properties such as maternal and cytoplasmic effects (e.g., Southwood et al. 1989; Zhu and Weir 1994a,b) and parental imprinting (Schaeffer et al. 1989, Tier and Sölkner 1993). Because the next two chapters rely very heavily on matrix algebra, before proceeding, the reader will likely benefit from reviewing Chapter 8, especially the sections on the multivariate normal and the general linear model.

THE GENERAL MIXED MODEL

Consider a column vector $\mathbf{y}$ containing the phenotypic values for a trait measured in n individuals. We assume that these observations are described adequately by a linear model with a $p \times 1$ vector of fixed effects ($\boldsymbol{\beta}$) and a $q \times 1$ vector of random effects ($\mathbf{u}$). The first element of the vector $\boldsymbol{\beta}$ is typically the population mean, and other factors included may be gender, location, year of birth, experimental treatment, and so on. The elements of the vector $\mathbf{u}$ of random effects are usually genetic effects such as additive genetic values. In matrix form,

$$\mathbf{y} = \mathbf{X}\boldsymbol{\beta} + \mathbf{Z}\mathbf{u} + \mathbf{e} \tag{26.1}$$

where $\mathbf{X}$ and $\mathbf{Z}$ are respectively $n \times p$ and $n \times q$ **incidence matrices** ($\mathbf{X}$ is also called the **design matrix**), and $\mathbf{e}$ is the $n \times 1$ column vector of residual deviations assumed to be distributed independently of the random genetic effects. Usually, all of the elements of the incidence matrices are equal to 0 or 1, depending upon whether the relevant effect contributes to the individual's phenotype. Because this model jointly accounts for fixed and random effects, it is generally referred to as a **mixed model** (Eisenhart 1947). Analysis of Equation 26.1 forms the basis for the remainder of this chapter and the next.

Example 1. Suppose that three sires are chosen at random from a population, and each mated to a randomly chosen dam. Two offspring from each mating are evaluated, some in environment 1 and some in environment 2. Let y_{ijk} denote the phenotypic value of the kth offspring of sire i in environment j. The model is then

$$y_{ijk} = \beta_j + u_i + e_{ijk}$$

This model has three random effects (u_1, u_2, u_3), which measure the contribution from each sire, and two fixed effects (β_1, β_2), which describe the influence of the two environments. The model assumes an absence of sire $\times$ environment interaction.

As noted above, a total of six offspring were measured. One offspring of sire 1 was assigned to environment 1 and had phenotypic value $y_{1,1,1} = 9$, while the second offspring was assigned to environment 2 and had phenotypic value $y_{1,2,1} = 12$. The two offspring of sire 2 were both assigned to environment 1 and had values of $y_{2,1,1} = 11$ and $y_{2,1,2} = 6$. One offspring of sire 3 was assigned to environment 1 and had phenotypic value $y_{3,1,1} = 7$, while the second offspring was assigned to environment 2 and had phenotypic value $y_{3,2,1} = 14$. The resulting vector of observations can be written as

$$\mathbf{y} = \begin{pmatrix} y_{1,1,1} \\ y_{1,2,1} \\ y_{2,1,1} \\ y_{2,1,2} \\ y_{3,1,1} \\ y_{3,2,1} \end{pmatrix} = \begin{pmatrix} 9 \\ 12 \\ 11 \\ 6 \\ 7 \\ 14 \end{pmatrix}$$

giving the mixed model as

$$\mathbf{y} = \mathbf{X}\boldsymbol{\beta} + \mathbf{Z}\mathbf{u} + \mathbf{e}$$

where the incidence matrices for fixed and random effects and the vectors of these effects are respectively

$$\mathbf{X} = \begin{pmatrix} 1 & 0 \\ 0 & 1 \\ 1 & 0 \\ 1 & 0 \\ 1 & 0 \\ 0 & 1 \end{pmatrix}, \qquad \mathbf{Z} = \begin{pmatrix} 1 & 0 & 0 \\ 1 & 0 & 0 \\ 0 & 1 & 0 \\ 0 & 1 & 0 \\ 0 & 0 & 1 \\ 0 & 0 & 1 \end{pmatrix}, \qquad \boldsymbol{\beta} = \begin{pmatrix} \beta_1 \\ \beta_2 \end{pmatrix}, \qquad \mathbf{u} = \begin{pmatrix} u_1 \\ u_2 \\ u_3 \end{pmatrix}$$

Now consider the means and variances of the component vectors of the mixed model. Since $E(\mathbf{u}) = E(\mathbf{e}) = \mathbf{0}$ by definition, $E(\mathbf{y}) = \mathbf{X}\boldsymbol{\beta}$. Denote the $(n \times n)$ covariance matrix for the vector $\mathbf{e}$ of residual errors by $\mathbf{R}$ and the $(q \times q)$ covariance matrix for the vector $\mathbf{u}$ of random genetic effects by $\mathbf{G}$. Excluding the difference among individuals due to fixed effects, from Equation 8.21b and the assumption that $\mathbf{u}$ and $\mathbf{e}$ are uncorrelated, the covariance matrix for the vector of observations $\mathbf{y}$ is

$$\mathbf{V} = \mathbf{Z}\mathbf{G}\mathbf{Z}^T + \mathbf{R} \tag{26.2}$$

The first term accounts for the contribution from random genetic effects, while the second accounts for the variance due to residual effects. We will generally assume that residual errors have constant variance and are uncorrelated, so that $\mathbf{R}$ is a diagonal matrix, with $\mathbf{R} = \sigma_E^2\,\mathbf{I}$.

We are now in a position to contrast the mixed model and the general linear model. Under the general linear model (Chapter 8),

$$\mathbf{y} = \mathbf{X}\boldsymbol{\beta} + \mathbf{e}^* \qquad \text{where} \quad \mathbf{e}^* \sim (\mathbf{0}, \mathbf{V}) \quad \text{implying } \ \mathbf{y} \sim (\mathbf{X}\boldsymbol{\beta}, \mathbf{V})$$

where the notation $\sim\ (a, b)$ means that the random variable has mean a and variance b. On the other hand, the mixed model partitions the vector of residual effects into two components, with $\mathbf{e}^* = \mathbf{Z}\mathbf{u} + \mathbf{e}$, giving

$$\mathbf{y} = \mathbf{X}\boldsymbol{\beta} + \mathbf{Z}\mathbf{u} + \mathbf{e} \qquad \text{where} \quad \mathbf{u} \sim (\mathbf{0}, \mathbf{G}) \ \text{ and } \ \mathbf{e} \sim (\mathbf{0}, \mathbf{R})$$

$$\text{implying } \ \mathbf{y} \sim (\mathbf{X}\boldsymbol{\beta}, \mathbf{V}) = (\mathbf{X}\boldsymbol{\beta}, \mathbf{Z}\mathbf{G}\mathbf{Z}^T + \mathbf{R})$$

When analyzed in the appropriate way, both formulations yield the same estimate of the vector of fixed effects $\boldsymbol{\beta}$, while the mixed-model formulation further allows estimates of the vector of random effects $\mathbf{u}$.

For the mixed model, we observe $\mathbf{y}$, $\mathbf{X}$, and $\mathbf{Z}$, while $\boldsymbol{\beta}$, $\mathbf{u}$, $\mathbf{R}$, and $\mathbf{G}$ are generally unknown. Thus, mixed-model analysis involves two complementary estimation issues: (1) estimation of the vectors of fixed and random effects, $\boldsymbol{\beta}$ and $\mathbf{u}$, and (2) estimation of the covariance matrices $\mathbf{G}$ and $\mathbf{R}$. These covariance matrices are generally assumed to be functions of a few unknown variance components. For the remainder of this chapter, we consider estimators of $\boldsymbol{\beta}$ and $\mathbf{u}$ under the assumption that $\mathbf{y}$, $\mathbf{X}$, $\mathbf{Z}$, $\mathbf{G}$, and $\mathbf{R}$ are all known. Estimation of the variance components (and hence $\mathbf{R}$ and $\mathbf{G}$) from $\mathbf{y}$, $\mathbf{X}$, and $\mathbf{Z}$ is the subject of the next chapter.

Estimating Fixed Effects and Predicting Random Effects

As outlined in the preceding chapters, the primary goal of a quantitative-genetic analysis is often solely to estimate variance components. However, there are also numerous situations in which inferences about fixed effects (such as the effect of a particular environment or year) and / or random effects (such as the breeding value of a particular individual) are the central motivation. Inferences about fixed effects have come to be called **estimates,** whereas those that concern random effects are known as **predictions.** Procedures for obtaining such estimators and predictors have been developed using a variety of approaches, such as likelihood theory (Appendix 4). The most widely used procedures are BLUE and BLUP, referring respectively to **best linear unbiased estimator** and **best linear unbiased predictor.** They are *best* in the sense that they minimize the sampling variance, *linear* in the sense that they are linear functions of the observed phenotypes $\mathbf{y}$, and *unbiased* in the sense that $E[\,\text{BLUE}(\boldsymbol{\beta})\,] = \boldsymbol{\beta}$ and $E[\,\text{BLUP}(\mathbf{u})\,] = \mathbf{u}$.

For the mixed model given by Equation 26.1, the BLUE of $\boldsymbol{\beta}$ is

$$\widehat{\boldsymbol{\beta}} = \left(\mathbf{X}^T\mathbf{V}^{-1}\mathbf{X}\right)^{-1}\mathbf{X}^T\mathbf{V}^{-1}\mathbf{y} \tag{26.3}$$

with $\mathbf{V}$ as given by Equation 26.2. Notice that this is just the generalized least-squares (GLS) estimator discussed in Chapter 8. Henderson (1963) showed that the BLUP of $\mathbf{u}$ is

$$\widehat{\mathbf{u}} = \mathbf{G}\mathbf{Z}^T\mathbf{V}^{-1}\left(\mathbf{y} - \mathbf{X}\widehat{\boldsymbol{\beta}}\right) \tag{26.4}$$

which is equivalent to the conditional expectation of $\mathbf{u}$ given $\mathbf{y}$ under the assumption of multivariate normality (cf. Equation 8.27). As noted above, the practical application of both of these expressions requires that the variance components be known. Thus, prior to a BLUP analysis, the variance components need to be estimated by ANOVA or REML.

Example 2. What are the BLUP values for the sire effects (u_1, u_2, u_3) in Example 1? In order to proceed, we require the covariance matrices for sire effects and errors. We assume that the residual variances within both environments are the same (σ_E^2), so $\mathbf{R} = \sigma_E^2\,\mathbf{I}$, where $\mathbf{I}$ is the 6×6 identity matrix. Assuming that all three sires are unrelated and drawn from the same population, $\mathbf{G} = \sigma_S^2\,\mathbf{I}$, where $\mathbf{I}$ is the 3×3 identity matrix and σ_S^2 is the variance of sire effects. Assuming only additive genetic variance, the sire effects (breeding values) are half the sires' additive genetic values. Thus, since the sires are sampled randomly from an outbred base population, $\sigma_S^2 = \sigma_A^2/4$, where σ_A^2 is the additive genetic variance. Assuming that $\sigma_A^2 = 8$ and $\sigma_E^2 = 6$, the covariance matrix $\mathbf{V}$ for the vector of observations $\mathbf{y}$ is given by $\mathbf{Z}\mathbf{G}\mathbf{Z}^T + \mathbf{R}$, or

$$\mathbf{V} = \frac{8}{4}\begin{pmatrix} 1 & 0 & 0 \\ 1 & 0 & 0 \\ 0 & 1 & 0 \\ 0 & 1 & 0 \\ 0 & 0 & 1 \\ 0 & 0 & 1 \end{pmatrix}\begin{pmatrix} 1 & 0 & 0 \\ 0 & 1 & 0 \\ 0 & 0 & 1 \end{pmatrix}\begin{pmatrix} 1 & 1 & 0 & 0 & 0 & 0 \\ 0 & 0 & 1 & 1 & 0 & 0 \\ 0 & 0 & 0 & 0 & 1 & 1 \end{pmatrix} + 6\begin{pmatrix} 1 & 0 & 0 & 0 & 0 & 0 \\ 0 & 1 & 0 & 0 & 0 & 0 \\ 0 & 0 & 1 & 0 & 0 & 0 \\ 0 & 0 & 0 & 1 & 0 & 0 \\ 0 & 0 & 0 & 0 & 1 & 0 \\ 0 & 0 & 0 & 0 & 0 & 1 \end{pmatrix}$$

$$= \begin{pmatrix} 8 & 2 & 0 & 0 & 0 & 0 \\ 2 & 8 & 0 & 0 & 0 & 0 \\ 0 & 0 & 8 & 2 & 0 & 0 \\ 0 & 0 & 2 & 8 & 0 & 0 \\ 0 & 0 & 0 & 0 & 8 & 2 \\ 0 & 0 & 0 & 0 & 2 & 8 \end{pmatrix} \quad \text{giving} \quad \mathbf{V}^{-1} = \frac{1}{30} \cdot \begin{pmatrix} 4 & -1 & 0 & 0 & 0 & 0 \\ -1 & 4 & 0 & 0 & 0 & 0 \\ 0 & 0 & 4 & -1 & 0 & 0 \\ 0 & 0 & -1 & 4 & 0 & 0 \\ 0 & 0 & 0 & 0 & 4 & -1 \\ 0 & 0 & 0 & 0 & -1 & 4 \end{pmatrix}$$

Using this result, a few simple matrix calculations give

$$\widehat{\boldsymbol{\beta}} = \begin{pmatrix} \widehat{\beta}_1 \\ \widehat{\beta}_2 \end{pmatrix} = \left(\mathbf{X}^T\mathbf{V}^{-1}\mathbf{X}\right)^{-1}\mathbf{X}^T\mathbf{V}^{-1}\mathbf{y} = \frac{1}{18}\begin{pmatrix} 148 \\ 235 \end{pmatrix}$$

and

$$\widehat{\mathbf{u}} = \begin{pmatrix} \widehat{u}_1 \\ \widehat{u}_2 \\ \widehat{u}_3 \end{pmatrix} = \mathbf{G}\mathbf{Z}^T\mathbf{V}^{-1}\left(\mathbf{y} - \mathbf{X}\widehat{\boldsymbol{\beta}}\right) = \frac{1}{18}\begin{pmatrix} -1 \\ 2 \\ -1 \end{pmatrix}$$

Example 3. As mentioned in Chapter 13, the effects of different genotypes at a single QTL are often estimated by ordinary least squares (OLS), using the model

$$y_{ij} = g_i + e_{ij}$$

where y_{ij} is the observed phenotype of the jth individual of genotype i, g_i is the mean genotypic value for the ith genotype at the locus of interest, and e_{ij} is a residual deviation assumed to be independently distributed among individuals. While this model may be reasonable for a random collection of individuals from a large population, when some sampled individuals are relatives, the sharing of alleles at other loci influencing the trait will induce correlations between residuals. If this is the case, OLS analysis can produce biased estimates of the QTL effects. When one of the QTL genotypes is very rare, as is often the case, the sampled individuals may be intentionally selected from the same pedigree, so the problem of bias is not trivial.

Use of a mixed model provides a means for accounting for associations among background QTLs in a way that eliminates bias in estimates of QTL effects. If the relatives in question share only additive effects (as in a pedigree with no full sibs or double first cousins, or when there is no nonadditive gene action), the correlations among residuals are accounted for by the additive genetic relationship matrix $\mathbf{A}$, where A_{ij} is twice the coefficient of coancestry, $2\Theta_{ij}$. When sibs are included and dominance is present at background QTLs, both $\mathbf{A}$ and a dominance relationship matrix (see below) are required.

Here we assume that all of the background genetic effects are additive, in which case the simplest mixed model can be applied,

$$y_{ij} = g_i + a_{ij} + e_{ij}$$

with the contribution from the different single-locus genotypes (g_i) being treated as fixed effects. The additive genetic background effects (a_{ij}) and the residual environmental deviations (e_{ij}) are treated as random effects, both with expected values equal to zero, and with respective variances σ_A^2 and σ_E^2. Note that σ_A^2 is the background additive genetic variance for the trait in excess of that caused by the QTL.

In matrix form,

$$\mathbf{y} = \mathbf{Xg} + \mathbf{Za} + \mathbf{e}$$

If there is a single observation for each individual, as we assume below, then $\mathbf{Z} = \mathbf{I}$ and the covariance matrix for the vector of observations $(\mathbf{y})$ is

$$\mathbf{V} = \sigma_A^2\,\mathbf{A} + \sigma_E^2\,\mathbf{I}$$

Thus, the covariance between the residual errors of two individuals (i and j) is just $2\Theta_{ij}\sigma_A^2$, while the variance of individual errors is $\sigma_A^2 + \sigma_E^2$. The error in using OLS to estimate single gene effects is that $\mathbf{A}$ is assumed to equal an identity matrix, so that $\mathbf{V}$ is incorrectly assumed to be a diagonal matrix.

From Equation 26.3, the estimates of the QTL means are given by

$$\widehat{\mathbf{g}} = \left(\mathbf{X}^T\mathbf{V}^{-1}\mathbf{X}\right)^{-1}\mathbf{X}^T\mathbf{V}^{-1}\mathbf{y}$$

Kennedy et al. (1992) showed that mixed-model estimates of QTL effects are much more reliable than OLS estimates, especially in small selected populations. Building on this approach, several authors (Hoeschele 1988, Hofer and Kennedy 1993, Kinghorn et al. 1993) have proposed BLUP-based segregation analysis for estimating the effects of an unknown major gene. Here the elements in the design matrix $\mathbf{X}$ associated with g_i are probabilistic estimates for the major-locus genotypes of each individual.

Note that the solution of Equations 26.3 and 26.4 requires the inverse of the covariance matrix $\mathbf{V}$. In the preceding example, $\mathbf{V}^{-1}$ was not particularly difficult to obtain. However, when $\mathbf{y}$ contains many thousands of observations, as is commonly the case in cattle breeding, the computation of $\mathbf{V}^{-1}$ can be quite difficult. As a way around this problem, Henderson (1950, 1963, 1973, 1984a) offered a more compact method for jointly obtaining $\widehat{\boldsymbol{\beta}}$ and $\widehat{\mathbf{u}}$ in the form of his **mixed-model equations** (MME),

$$\begin{pmatrix} \mathbf{X}^T\mathbf{R}^{-1}\mathbf{X} & \mathbf{X}^T\mathbf{R}^{-1}\mathbf{Z} \\ \mathbf{Z}^T\mathbf{R}^{-1}\mathbf{X} & \mathbf{Z}^T\mathbf{R}^{-1}\mathbf{Z}+\mathbf{G}^{-1} \end{pmatrix} \begin{pmatrix} \widehat{\boldsymbol{\beta}} \\ \widehat{\mathbf{u}} \end{pmatrix} = \begin{pmatrix} \mathbf{X}^T\mathbf{R}^{-1}\mathbf{y} \\ \mathbf{Z}^T\mathbf{R}^{-1}\mathbf{y} \end{pmatrix} \tag{26.5}$$

While these expressions may look considerably more complicated than Equations 26.3 and 26.4, $\mathbf{R}^{-1}$ and $\mathbf{G}^{-1}$ are trivial to obtain if $\mathbf{R}$ and $\mathbf{G}$ are diagonal, and hence the submatrices in Equation 26.5 are much easier to compute than $\mathbf{V}^{-1}$. A second advantage of Equation 26.5 can be seen by considering the dimensionality of the matrix on the left. Recalling that $\mathbf{X}$ and $\mathbf{Z}$ are $n \times p$ and $n \times q$ respectively, $\mathbf{X}^T\mathbf{R}^{-1}\mathbf{X}$ is $p \times p$, $\mathbf{X}^T\mathbf{R}^{-1}\mathbf{Z}$ is $p \times q$, and $\mathbf{Z}^T\mathbf{R}^{-1}\mathbf{Z}+\mathbf{G}^{-1}$ is $q \times q$. Thus, the matrix that needs to be inverted to obtain the solution for $\widehat{\boldsymbol{\beta}}$ and $\widehat{\mathbf{u}}$ is of order $(p+q) \times (p+q)$, which is usually considerably less than the dimensionality of $\mathbf{V}$ (an $n \times n$ matrix).

Although there are several ways to derive the mixed-model equations (Robinson 1991), Henderson (1950) originally obtained them by assuming that the covariance matrices $\mathbf{G}$ and $\mathbf{R}$ are known and that the densities of the vectors $\mathbf{u}$ and $\mathbf{e}$ are each multivariate normal with no correlations between them. Equation 26.5 then yields the maximum likelihood estimates of the fixed and random effects. Henderson (1963) later showed that the mixed-model equations do not actually depend on normality, and that $\widehat{\boldsymbol{\beta}}$ and $\widehat{\mathbf{u}}$ are BLUE and BLUP, respectively, under general conditions provided the variances are known.

Example 4. Using the values from Examples 1 and 2, we find that

$$\mathbf{X}^T\mathbf{R}^{-1}\mathbf{X} = \frac{1}{6}\begin{pmatrix} 4 & 0 \\ 0 & 2 \end{pmatrix}, \qquad \mathbf{X}^T\mathbf{R}^{-1}\mathbf{Z} = \left(\mathbf{Z}^T\mathbf{R}^{-1}\mathbf{X}\right)^T = \frac{1}{6}\begin{pmatrix} 1 & 2 & 1 \\ 1 & 0 & 1 \end{pmatrix}$$

$$\mathbf{G}^{-1} + \mathbf{Z}^T\mathbf{R}^{-1}\mathbf{Z} = \frac{5}{6}\begin{pmatrix} 1 & 0 & 0 \\ 0 & 1 & 0 \\ 0 & 0 & 1 \end{pmatrix}, \quad \mathbf{X}^T\mathbf{R}^{-1}\mathbf{y} = \frac{1}{6}\begin{pmatrix} 33 \\ 26 \end{pmatrix}, \quad \mathbf{Z}^T\mathbf{R}^{-1}\mathbf{y} = \frac{1}{6}\begin{pmatrix} 21 \\ 17 \\ 21 \end{pmatrix}$$

Thus, after factoring out 1/6 from both sides, the mixed-model equations for these data become

$$\begin{pmatrix} 4 & 0 & 1 & 2 & 1 \\ 0 & 2 & 1 & 0 & 1 \\ 1 & 1 & 5 & 0 & 0 \\ 2 & 0 & 0 & 5 & 0 \\ 1 & 1 & 0 & 0 & 5 \end{pmatrix} \begin{pmatrix} \widehat{\beta}_1 \\ \widehat{\beta}_2 \\ \widehat{u}_1 \\ \widehat{u}_2 \\ \widehat{u}_3 \end{pmatrix} = \begin{pmatrix} 33 \\ 26 \\ 21 \\ 17 \\ 21 \end{pmatrix}$$

Taking the inverse gives the solution

$$\begin{pmatrix} \widehat{\beta}_1 \\ \widehat{\beta}_2 \\ \widehat{u}_1 \\ \widehat{u}_2 \\ \widehat{u}_3 \end{pmatrix} = \frac{1}{270} \begin{pmatrix} 100 & 25 & -25 & -40 & -25 \\ 25 & 175 & -40 & -10 & -40 \\ -25 & -40 & 67 & 10 & 13 \\ -40 & -10 & 10 & 70 & 10 \\ -25 & -40 & 13 & 10 & 67 \end{pmatrix} \begin{pmatrix} 33 \\ 26 \\ 21 \\ 17 \\ 21 \end{pmatrix} = \frac{1}{18} \begin{pmatrix} 148 \\ 235 \\ -1 \\ 2 \\ -1 \end{pmatrix}$$

which is identical to the results obtained in Example 2.

Although the method of predicting random effects using BLUP methodology was first discussed by Henderson (1949, 1950), the expression "best linear unbiased predictor" was apparently first used by Goldberger (1962), with the acronym BLUP due to Henderson (1973). In a relatively short time, BLUP has become the method of choice for estimating the breeding values of individuals from field records of large and complex pedigrees. For BLUPs to be the best unbiased estimates, the appropriate genetic variances must be known without error. Kackar and Harville (1981) show that BLUP estimates remain unbiased when estimates of genetic variances are used in place of actual values (as is usually the case), although they are not guaranteed to be the best of all unbiased linear estimators.

Estimability of Fixed Effects

It is sometimes impossible to obtain unique BLUE estimates for all of the fixed factors in a model. Suppose, for example, that

$$\boldsymbol{\beta} = \begin{pmatrix} \beta_1 \\ \beta_2 \\ \beta_3 \end{pmatrix} \quad \text{with} \quad \mathbf{X} = \begin{pmatrix} 1 & 1 & 0 \\ 1 & 1 & 0 \\ 0 & 0 & 1 \end{pmatrix}$$

Here, factors 1 and 2 are completely confounded, as they contribute equally to all individuals, so unique estimates of β_1 and β_2 cannot be acquired. Generally, when two or more columns of $\mathbf{X}$ are not independent, it is still possible to obtain unique BLUEs for certain linear combinations of $\boldsymbol{\beta}$ through the use of **generalized inverses** (Appendix 3). With the preceding design matrix $\mathbf{X}$, the solution is simple — by combining the two factors into a single new factor, $\beta_1 + \beta_2$, the new model becomes

$$\boldsymbol{\beta}_* = \begin{pmatrix} \beta_1 + \beta_2 \\ \beta_3 \end{pmatrix} \quad \text{with} \quad \mathbf{X}_* = \begin{pmatrix} 1 & 0 \\ 1 & 0 \\ 0 & 1 \end{pmatrix}$$

Since the columns of $\boldsymbol{\beta}_*$ are now independent, a unique solution exists for $\mathbf{X}_*^T\mathbf{V}^{-1}\mathbf{X}_*$, and from Equation 26.3, the two BLUEs of the fixed effects are given by

$$\widehat{\boldsymbol{\beta}}_* = \left(\mathbf{X}_*^T\mathbf{V}^{-1}\mathbf{X}_*\right)^{-1}\mathbf{X}_*^T\mathbf{V}^{-1}\mathbf{y}$$

Situations in which linear combinations of fixed effects are required commonly arise when a very large number of fixed factors are included in the model, as occurs in large breeding programs involving multiple environments (such as different herds and different years.) Henderson (1984a) provides an extended discussion of the issues. Throughout the remainder of the book, we assume that $\boldsymbol{\beta}$ is estimable, either immediately or after an appropriate transformation. Appendix 3 discusses how to determine which combinations of effects are estimable when singular matrices exist.

Standard Errors

A relatively straightforward extension of Henderson's mixed-model equations provides estimates of the standard errors of the fixed and random effects. Let the inverse of the leftmost matrix in Equation 26.5 be

$$\begin{pmatrix} \mathbf{X}^T\mathbf{R}^{-1}\mathbf{X} & \mathbf{X}^T\mathbf{R}^{-1}\mathbf{Z} \\ \mathbf{Z}^T\mathbf{R}^{-1}\mathbf{X} & \mathbf{Z}^T\mathbf{R}^{-1}\mathbf{Z}+\mathbf{G}^{-1} \end{pmatrix}^{-1} = \begin{pmatrix} \mathbf{C}_{11} & \mathbf{C}_{12} \\ \mathbf{C}_{12}^T & \mathbf{C}_{22} \end{pmatrix} \tag{26.6}$$

where $\mathbf{C}_{11}$, $\mathbf{C}_{12}$, and $\mathbf{C}_{22}$ are, respectively, $p \times p$, $p \times q$, and $q \times q$ submatrices. Using this notation, Henderson (1975) showed that the sampling covariance matrix for the BLUE of $\boldsymbol{\beta}$ is given by

$$\sigma(\widehat{\boldsymbol{\beta}}) = \mathbf{C}_{11} \tag{26.7a}$$

that the sampling covariance matrix of the prediction errors ($\widehat{\mathbf{u}} - \mathbf{u}$) is given by

$$\sigma(\,\widehat{\mathbf{u}} - \mathbf{u}\,) = \mathbf{C}_{22} \tag{26.7b}$$

and that the sampling covariance of estimated effects and prediction errors is given by

$$\sigma(\widehat{\boldsymbol{\beta}}, \widehat{\mathbf{u}} - \mathbf{u}\,) = \mathbf{C}_{12} \tag{26.7c}$$

(We consider $\widehat{\mathbf{u}} - \mathbf{u}$ rather than $\widehat{\mathbf{u}}$ as the latter includes variance from both the prediction error and the random effects $\mathbf{u}$ themselves.) The standard errors of the fixed and random effects are obtained, respectively, as the square roots of the diagonal elements of $\mathbf{C}_{11}$ and $\mathbf{C}_{22}$. For very large animal breeding designs where the inverse of the MME matrix may be difficult to compute, Meyer (1989a) presents methods for approximating the diagonal elements of the inverse of this matrix (and hence the standard errors).

Example 5. Consider the mixed-model equation from Example 4. Here for the fixed factors β_1, β_2 and the random effects u_1, u_2, u_3, the inverse of the coefficient matrix is

$$\left(\begin{array}{cc|ccc} 4 & 0 & 1 & 2 & 1 \\ 0 & 2 & 1 & 0 & 1 \\ \hline 1 & 1 & 5 & 0 & 0 \\ 2 & 0 & 0 & 5 & 0 \\ 1 & 1 & 0 & 0 & 5 \end{array}\right)^{-1} = \frac{1}{270}\left(\begin{array}{cc|ccc} 100 & 25 & -25 & -40 & -25 \\ 25 & 175 & -40 & -10 & -40 \\ \hline -25 & -40 & 67 & 10 & 13 \\ -40 & -10 & 10 & 70 & 10 \\ -25 & -40 & 13 & 10 & 67 \end{array}\right)$$

Hence,

$$\mathbf{C}_{11} = \frac{1}{270}\begin{pmatrix} 100 & 25 \\ 25 & 175 \end{pmatrix} \quad \text{and} \quad \mathbf{C}_{22} = \frac{1}{270}\begin{pmatrix} 67 & 10 & 13 \\ 10 & 70 & 10 \\ 13 & 10 & 67 \end{pmatrix}$$

so that, for example,

$$\sigma^2(\widehat{\beta_1}) = \frac{100}{270}, \quad \sigma^2(\widehat{\beta_2}) = \frac{175}{270}, \quad \sigma(\widehat{\beta_1}, \widehat{\beta_2}) = \frac{25}{270}$$

and, likewise,

$$\sigma^2(\widehat{u_2} - u_2) = \frac{70}{270}, \quad \sigma(\widehat{u_1} - u_1, \widehat{u_3} - u_3) = \frac{13}{270}, \quad \text{and so on.}$$

MODELS FOR THE ESTIMATION OF BREEDING VALUES

While the general mixed model (Equation 26.1) forms the fundamental framework for BLUP analysis, there are numerous ways in which this model can be formulated. Three specific variants of the model provide the basis for most attempts to estimate breeding values. So-called **animal models** estimate the breeding values of each measured individual, while **gametic models** describe the breeding values of measured individuals in terms of parental contributions. The **reduced animal model** combines aspects of both the animal and gametic models in specific applications in which parental breeding values are the only ones of interest. In the following sections, we show how each of these models can be readily adapted to the mixed-model equations.

The Animal Model

Assuming only a single fixed factor (the population mean) under the simplest animal model, the observation for individual i is expressed as

$$y_i = \mu + a_i + e_i \tag{26.8}$$

where a_i is the additive genetic value of individual i. With k individuals, the model can be expressed as in Equation 26.1 with

$$\mathbf{X} = \begin{pmatrix} 1 \\ 1 \\ \vdots \\ 1 \end{pmatrix}, \qquad \boldsymbol{\beta} = \mu, \qquad \mathbf{u} = \begin{pmatrix} a_1 \\ a_2 \\ \vdots \\ a_k \end{pmatrix}$$

The matrix $\mathbf{G}$ describing the covariances among the random effects (here the breeding values) follows from standard results for the covariances between relatives. From Equation 7.12, the additive genetic covariance between two relatives i and j is given by $2\Theta_{ij}\sigma_A^2$, i.e., by twice the coefficient of coancestry times the additive genetic variance in the base population. Hence, under the animal model, $\mathbf{G} = \sigma_A^2\,\mathbf{A}$, where the **additive genetic** (or **numerator**) **relationship matrix A** has elements $A_{ij} = 2\Theta_{ij}$.

The covariance matrix $\mathbf{R}$ for the vector of residual errors requires a little more care. The standard assumption is that $\mathbf{R} = \sigma_E^2\,\mathbf{I}$, so that the residual error for each observation has the same variance σ_E^2 and is uncorrelated with all other residual errors. There are many ways in which this assumption can fail. For example, if the character displays any dominance and i and j are full sibs, $\sigma(e_i, e_j) = \sigma_D^2/4$. Shared environmental effects can also cause correlations between residual effects. These complications will be considered below in some detail, but for now we assume that the residual errors have the simple covariance structure $\mathbf{R} = \sigma_E^2\,\mathbf{I}$, implying $\mathbf{R}^{-1} = \sigma_E^{-2}\,\mathbf{I}$.

Since $\mathbf{G}^{-1} = \sigma_A^{-2}\,\mathbf{A}^{-1}$, the mixed-model equations (Equation 26.5) for the animal model reduce to

$$\begin{pmatrix} \mathbf{X}^T\mathbf{X} & \mathbf{X}^T\mathbf{Z} \\ \mathbf{Z}^T\mathbf{X} & \mathbf{Z}^T\mathbf{Z} + \lambda\,\mathbf{A}^{-1} \end{pmatrix} \begin{pmatrix} \widehat{\boldsymbol{\beta}} \\ \widehat{\mathbf{u}} \end{pmatrix} = \begin{pmatrix} \mathbf{X}^T\mathbf{y} \\ \mathbf{Z}^T\mathbf{y} \end{pmatrix} \tag{26.9a}$$

where $\lambda = \sigma_E^2/\sigma_A^2 = (1 - h^2)/h^2$ under the assumption of additive gene action. Since the only fixed factor is the mean μ (so that $\boldsymbol{\beta} = \mu$ and $\mathbf{X} = \mathbf{1}$, a vector of ones) and each individual has only a single observation (so that $\mathbf{Z} = \mathbf{I}$), with n individuals, Equation 26.9a reduces to

$$\begin{pmatrix} n & \mathbf{1}^T \\ \mathbf{1} & \mathbf{I} + \lambda\,\mathbf{A}^{-1} \end{pmatrix} \begin{pmatrix} \widehat{\mu} \\ \widehat{\mathbf{u}} \end{pmatrix} = \begin{pmatrix} \sum^n y_i \\ \mathbf{y} \end{pmatrix} \tag{26.9b}$$

Example 6. Consider the pedigree of individuals given in the figure below, where each individual has a single measurement and the only fixed factor is the mean.

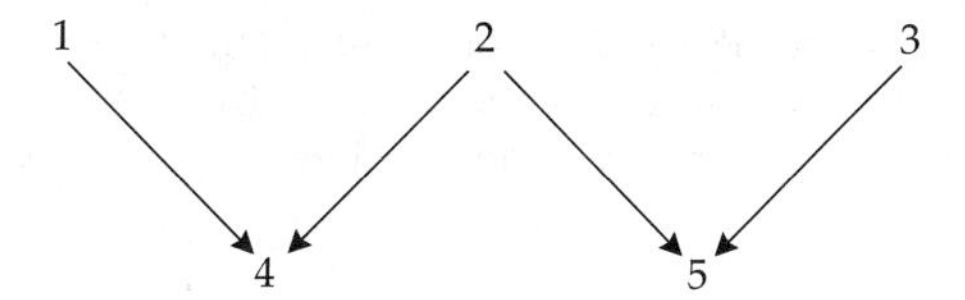

With the vector of observations,

$$\mathbf{y} = \begin{pmatrix} y_1 \\ y_2 \\ y_3 \\ y_4 \\ y_5 \end{pmatrix} = \begin{pmatrix} 7 \\ 9 \\ 10 \\ 6 \\ 9 \end{pmatrix}$$

we can use Equation 26.9b with $\widehat{\mathbf{u}}^T = (\widehat{a}_1,\ \widehat{a}_2,\ \widehat{a}_3,\ \widehat{a}_4,\ \widehat{a}_5)$. Assuming that individuals 1, 2, and 3 are unrelated and not inbred, the relationship matrix becomes

$$\mathbf{A} = \begin{pmatrix} 1 & 0 & 0 & 1/2 & 0 \\ 0 & 1 & 0 & 1/2 & 1/2 \\ 0 & 0 & 1 & 0 & 1/2 \\ 1/2 & 1/2 & 0 & 1 & 1/4 \\ 0 & 1/2 & 1/2 & 1/4 & 1 \end{pmatrix}$$

Suppose it is known that $\sigma_E^2 = \sigma_A^2$, so that $\lambda = 1$. Then,

$$\mathbf{I} + \lambda\,\mathbf{A}^{-1} = \begin{pmatrix} 5/2 & 1/2 & 0 & -1 & 0 \\ 1/2 & 3 & 1/2 & -1 & -1 \\ 0 & 1/2 & 5/2 & 0 & -1 \\ -1 & -1 & 0 & 3 & 0 \\ 0 & -1 & -1 & 0 & 3 \end{pmatrix}$$

Since $n = 5$ and $\sum y_i = 41$, Equation 26.9b gives the mixed-model equations for these data as

$$\begin{pmatrix} 5 & 1 & 1 & 1 & 1 & 1 \\ 1 & 5/2 & 1/2 & 0 & -1 & 0 \\ 1 & 1/2 & 3 & 1/2 & -1 & -1 \\ 1 & 0 & 1/2 & 5/2 & 0 & -1 \\ 1 & -1 & -1 & 0 & 3 & 0 \\ 1 & 0 & -1 & -1 & 0 & 3 \end{pmatrix} \begin{pmatrix} \widehat{\mu} \\ \widehat{a}_1 \\ \widehat{a}_2 \\ \widehat{a}_3 \\ \widehat{a}_4 \\ \widehat{a}_5 \end{pmatrix} = \begin{pmatrix} 41 \\ 7 \\ 9 \\ 10 \\ 6 \\ 9 \end{pmatrix}$$

the solutions of which are

$$\widehat{\mu} = \frac{440}{53} \simeq 8.302, \qquad \begin{pmatrix} \widehat{a}_1 \\ \widehat{a}_2 \\ \widehat{a}_3 \\ \widehat{a}_4 \\ \widehat{a}_5 \end{pmatrix} = \begin{pmatrix} -662/689 \\ 4/53 \\ 610/689 \\ -732/689 \\ 381/689 \end{pmatrix} \simeq \begin{pmatrix} -0.961 \\ 0.076 \\ 0.885 \\ -1.062 \\ 0.553 \end{pmatrix}$$

Note that the average breeding value in the base population (individuals 1, 2, and 3) is zero (as expected for a random sample of the population). This is no longer the case once we leave the base population, unless all base-population individuals contribute equally to progeny production.

The Gametic Model

The gametic model is often used when parental breeding values are of more concern than offspring values, as when one is attempting to estimate the breeding value of bulls from large arrays of descendants. In this model the additive genetic value of each offspring is expressed in terms of its parents' breeding values. In particular, letting a_{si} and a_{di} be the breeding values for individual i's sire and dam, we can express i's breeding value as

$$a_i = \left(\frac{a_{si}}{2} + \frac{a_{di}}{2}\right) + e_{ai} \tag{26.10a}$$

the sum of its predicted value (the average of parental values) and a random deviation e_{ai} resulting from Mendelian segregation. Thus, we can rewrite the simplest animal model (Equation 26.8) as

$$\begin{aligned} y_i &= \mu + a_i + e_i \\ &= \mu + \left(\frac{a_{si}}{2} + \frac{a_{di}}{2}\right) + (e_{ai} + e_i) \end{aligned} \tag{26.10b}$$

The **sire model** used in Example 1 is a variation of the gametic model wherein the dam contribution is ignored (and hence incorporated into the error term).

The residual error in the gametic model, $e_{ai} + e_i$, contains a genetic component e_{ai} (the segregation error) in addition to an uncorrelated environmental component e_i. As in the animal model, the e_i are assumed to be uncorrelated with common variance σ_E^2 so that their covariance matrix is $\sigma_E^2\,\mathbf{I}$. For noninbred parents, the variance of the segregation error is $\sigma_A^2/2$, while if the parents are inbred,

$$\sigma^2(e_{ai}) = \left(1 - \frac{f_{si} + f_{di}}{2}\right)\frac{\sigma_A^2}{2} = (1 - \overline{f}_i)\,\frac{\sigma_A^2}{2} \tag{26.11a}$$

where f denotes the inbreeding coefficient and $\overline{f}_i$ is the average amount of inbreeding for both parents (Dempfle 1990). The inbreeding coefficient can be obtained directly from the additive genetic relationship matrix of the parents ($\mathbf{A}$). Since $A_{ii} = 2\Theta_{ii} = 1 + f_i$ (Equation 7.3), we have

$$\overline{f}_i = \frac{A_{si,si} + A_{di,di}}{2} - 1 \tag{26.11b}$$

Because they are random deviations around expectations, the segregation errors for all individuals are uncorrelated. Hence, the covariance matrix for the residual errors becomes

$$\mathbf{R} = \sigma_E^2\,\mathbf{I} + \sigma_A^2 \begin{pmatrix} (1-\overline{f}_1)/2 & \cdots & 0 \\ \vdots & \ddots & \vdots \\ 0 & \cdots & (1-\overline{f}_k)/2 \end{pmatrix} = \sigma_E^2\,\mathbf{W} \tag{26.12}$$

where $\mathbf{W}$ is a diagonal matrix with diagonal elements equal to $1 + (1-\overline{f}_i)/(2\lambda)$, with $\lambda = \sigma_E^2/\sigma_A^2$ (as in the animal model). Note that $\sigma_E^2 W_{ii}$ is the total within-family variance (the within-family segregation variance plus the within-family environmental variance).

The Reduced Animal Model

Quaas and Pollak (1980) combined features of both the animal and gametic model to obtain a **reduced animal model** (RAM) for large pedigrees that contain only parents and their offspring (i.e., ignoring individuals in the third generation and beyond). Suppose k parents and a total l of their offspring are each measured once. In the reduced animal model, parents are treated as in the full animal model, $y_i = \mu + a_i + e_i$, while their offspring are described by the gametic model. Only the k parental additive genetic values are estimated, so that $\mathbf{u}^T = (a_1, \cdots, a_k)^T$. Partitioning the vector of observations $\mathbf{y}$ into a $k \times 1$ vector of parental observations ($\mathbf{y}_p$) and an $l \times 1$ vector of offspring observations ($\mathbf{y}_o$), the reduced animal model can be expressed using the general mixed model (Equation 26.1) by letting

$$\mathbf{y} = \begin{pmatrix} \mathbf{y}_p \\ \mathbf{y}_o \end{pmatrix}, \qquad \mathbf{Z} = \begin{pmatrix} \mathbf{I}_k \\ \mathbf{Z}^* \end{pmatrix}, \qquad \mathbf{e} = \begin{pmatrix} \mathbf{e}_p \\ \mathbf{e}_o \end{pmatrix} \tag{26.13a}$$

giving

$$\mathbf{y} = \begin{pmatrix} \mathbf{y}_p \\ \mathbf{y}_o \end{pmatrix} = \mathbf{X}\boldsymbol{\beta} + \begin{pmatrix} \mathbf{I}_k \\ \mathbf{Z}^* \end{pmatrix} \mathbf{u} + \begin{pmatrix} \mathbf{e}_p \\ \mathbf{e}_o \end{pmatrix} \tag{26.13b}$$

where $\mathbf{I}_k$ is the $k \times k$ identity matrix, and $\mathbf{Z}^*$ is the $l \times k$ incidence matrix, which records the parents of each offspring. Here, in any particular row of $\mathbf{Z}^*$, the two elements corresponding to the parents of that individual are set equal to 1/2 while all other elements are zero. $\mathbf{Z}^*\mathbf{u}$ is then a vector of the expected breeding values of each offspring as given by the average of the parental breeding values. The genetic covariance matrix is $\mathbf{G} = \sigma_A^2\,\mathbf{A}$, where $\mathbf{A}$ is the $k \times k$ relationship matrix for the parents, while the $(k+l) \times (k+l)$ covariance matrix for the vector of residuals becomes

$$\mathbf{R} = \sigma_E^2 \begin{pmatrix} \mathbf{I}_k & \mathbf{0} \\ \mathbf{0} & \mathbf{W} \end{pmatrix} \tag{26.14a}$$

implying

$$\mathbf{R}^{-1} = \sigma_E^{-2} \begin{pmatrix} \mathbf{I}_k & \mathbf{0} \\ \mathbf{0} & \mathbf{W}^{-1} \end{pmatrix} \tag{26.14b}$$

with the offspring submatrix **W** defined as above (Equation 16.12). Substituting these results into Equation 26.5 and solving gives the BLUE for the vector of fixed effects and the BLUP for the vector of parental breeding values.

Quaas and Pollak (1980) show that the offspring breeding values can be obtained from the BLUE estimates of fixed effects and the BLUP estimates of parental breeding values. When estimates of the breeding values of both parents are available ($\widehat{a}_{si}, \widehat{a}_{di}$), the estimate of i's breeding value given its character value y_i is

$$\widehat{a}_i = \frac{1}{2}\left[\widehat{a}_{si} + \widehat{a}_{di}\right] + \left[\frac{\sigma_A^2(1-\overline{f}_i)/2}{\sigma_E^2 + \sigma_A^2(1-\overline{f}_i)/2}\right]\left[y_i - \widehat{\mu} - \frac{1}{2}\left(\widehat{a}_{si} + \widehat{a}_{di}\right)\right] \tag{26.15}$$

The first term is the predicted breeding value based only on the parental breeding values, while the product of the last two terms provides an estimate of the segregational deviation of the actual breeding value from its expectation. This additional contribution to the predicted offspring breeding value takes advantage of the information implicit in the offspring's phenotype — the first term in parentheses is the ratio of the within-family additive genetic variance to the total within-family variance (the **within-family heritability**). Equation 26.15 pertains to a model with only a single fixed effect. When multiple fixed effects are present, the appropriate BLUE estimates are also subtracted from each observation.

At first sight, the reduced animal model seems much more complicated than the animal model, so why might it be preferred? The reason is related to computational efficiency. For very large data sets, solutions to the reduced animal model require only inverses of matrices of the order of the number of parents, while the full animal model requires inverses of the order of the total number of measured individuals.

Example 7. We now reconsider Example 6, modeling the data with the reduced animal model. Here, the only genetic parameters to estimate are the three parental breeding values. The resulting model has $\boldsymbol{\beta}$ and $\mathbf{X}$ as in Example 6, but

$$\mathbf{y} = \begin{pmatrix} \mathbf{y}_p \\ \mathbf{y}_o \end{pmatrix} = \begin{pmatrix} y_1 \\ y_2 \\ y_3 \\ \cdots \\ y_4 \\ y_5 \end{pmatrix}, \quad \mathbf{u} = \begin{pmatrix} a_1 \\ a_2 \\ a_3 \end{pmatrix}, \quad \mathbf{Z}^* = \begin{pmatrix} 1/2 & 1/2 & 0 \\ 0 & 1/2 & 1/2 \end{pmatrix}$$

giving

$$\mathbf{Z} = \begin{pmatrix} \mathbf{I} \\ \mathbf{Z}^* \end{pmatrix} = \begin{pmatrix} 1 & 0 & 0 \\ 0 & 1 & 0 \\ 0 & 0 & 1 \\ 1/2 & 1/2 & 0 \\ 0 & 1/2 & 1/2 \end{pmatrix}$$

Since the three parents were assumed to be unrelated and noninbred, $\mathbf{G} = \sigma_A^2\,\mathbf{A} = (\sigma_E^2/\lambda)\,\mathbf{A}$, where

$$\mathbf{A} = \mathbf{A}^{-1} = \begin{pmatrix} 1 & 0 & 0 \\ 0 & 1 & 0 \\ 0 & 0 & 1 \end{pmatrix}$$

Note that by restricting attention to just the parents, the relationship matrix $\mathbf{A}$ is much simpler than that in Example 6 where parents and offspring are considered jointly. Since the parents are assumed to be noninbred,

$$\mathbf{W} = \left(1 + \frac{1}{2\lambda}\right)\begin{pmatrix} 1 & 0 \\ 0 & 1 \end{pmatrix} = \left(1 + \frac{1}{2\lambda}\right)\mathbf{I}$$

Because we assumed $\lambda = 1$, $\mathbf{R}$ is diagonal with elements $\sigma_E^2 \times (1, 1, 1, 3/2, 3/2)$, so that $\mathbf{R}^{-1}$ is diagonal, with elements $\sigma_E^{-2} \times (1, 1, 1, 2/3, 2/3)$. To obtain the mixed-model equations (ignoring the factor σ_E^{-2} common to all equations), we first obtain

$$\mathbf{X}^T\mathbf{R}^{-1}\mathbf{X} = \frac{13}{3}, \quad \mathbf{Z}^T\mathbf{R}^{-1}\mathbf{Z} = \frac{1}{6}\begin{pmatrix} 7 & 1 & 0 \\ 1 & 8 & 1 \\ 0 & 1 & 7 \end{pmatrix}, \quad \mathbf{X}^T\mathbf{R}^{-1}\mathbf{y} = 36,$$

$$(\mathbf{X}^T\mathbf{R}^{-1}\mathbf{Z})^T = \mathbf{Z}^T\mathbf{R}^{-1}\mathbf{X} = \frac{1}{3}\begin{pmatrix} 4 \\ 5 \\ 4 \end{pmatrix}, \quad \mathbf{Z}^T\mathbf{R}^{-1}\mathbf{y} = \begin{pmatrix} 9 \\ 14 \\ 13 \end{pmatrix}$$

$$\mathbf{Z}^T\mathbf{R}^{-1}\mathbf{Z} + \mathbf{G}^{-1} = \frac{1}{6}\begin{pmatrix} 13 & 1 & 0 \\ 1 & 14 & 1 \\ 0 & 1 & 13 \end{pmatrix}$$

giving the final form of the mixed-model Equation 26.5 as

$$\begin{pmatrix} 13/3 & 4/3 & 5/3 & 4/3 \\ 4/3 & 13/6 & 1/6 & 0 \\ 5/3 & 1/6 & 14/6 & 1/6 \\ 4/3 & 0 & 1/6 & 13/6 \end{pmatrix}\begin{pmatrix} \widehat{\mu} \\ \widehat{a}_1 \\ \widehat{a}_2 \\ \widehat{a}_3 \end{pmatrix} = \begin{pmatrix} 36 \\ 9 \\ 14 \\ 13 \end{pmatrix}$$

which has solutions

$$\widehat{\mu} = \frac{440}{53}, \qquad \begin{pmatrix} \widehat{a}_1 \\ \widehat{a}_2 \\ \widehat{a}_3 \end{pmatrix} = \begin{pmatrix} -662/689 \\ 4/53 \\ 610/689 \end{pmatrix}$$

To obtain the estimates of the offspring breeding values by use of Equation 26.15, first note that neither offspring has inbred parents and that $\lambda = 1$, so that the within-family heritability is

$$\frac{(1 - \overline{f}_i)/(2\lambda)}{1 + (1 - \overline{f}_i)/(2\lambda)} = \frac{(1-0)/2}{1 + (1-0)/2} = \frac{1}{3}$$

Hence,

$$\widehat{a}_i = \frac{1}{2}(\widehat{a}_{si} + \widehat{a}_{di}) + \left(\frac{1}{3}\right)\left[y_i - \frac{440}{53} - \frac{1}{2}(\widehat{a}_{si} + \widehat{a}_{di})\right]$$

giving the BLUPs for the offspring breeding values as

$$\widehat{a}_4 = \frac{1}{2}\left(-\frac{662}{689} + \frac{4}{53}\right) + \left(\frac{1}{3}\right)\left[6 - \frac{440}{53} - \frac{1}{2}\left(-\frac{662}{689} + \frac{4}{53}\right)\right] = -\frac{732}{689}$$

$$\widehat{a}_5 = \frac{1}{2}\left(\frac{4}{53} + \frac{610}{689}\right) + \left(\frac{1}{3}\right)\left[6 - \frac{440}{53} - \frac{1}{2}\left(\frac{4}{53} + \frac{610}{689}\right)\right] = \frac{381}{689}$$

These estimates are identical to those obtained from the full animal model (Example 6). This result is expected, as the RAM and full models are **equivalent** (Appendix 3).

In closing this section, we emphasize several aspects of BLUP (as applied to the animal, gametic, and reduced animal models) that highlight its exceptional degree of flexibility. First, because the relationship matrix **A** records the flow of genetic information through the pedigree, BLUP provides unbiased estimates of breeding values even in populations under selection, provided the individuals upon which selection operated are included in the analysis (Kennedy and Sorensen 1988). For further information on this subject, including modifications that need to be made when measurements are unavailable for some selected individuals, see Henderson (1975, 1990), Goffinet (1983), Gianola et al. (1988), and Fernando and Gianola (1990). Second, the breeding value estimates from different generations reflect changes in mean phenotypes and additive genetic variances resulting from selection and/or random genetic drift and inbreeding (recall Example 6). Third, because breeding value estimates are conditional expectations given the entire vector of observations ($\mathbf{y}$), the evaluations of individuals are based on weighted information from all measured relatives. Individual evaluations are adjusted for the breeding values of their progeny and mates, thereby accounting for any assortative mating, and as noted in Equation 26.15, the deviation of the individual's own phenotype from its conditional expectation provides an additional bit of information. Further adjustments to the BLUPs can be made for individuals with relatives other than parents and progeny, see Mrode (1996) for examples.

SIMPLE RULES FOR COMPUTING A AND $\mathbf{A}^{-1}$

While the identity-coefficient methods introduced in Chapter 7 can be used to compute each of the individual elements of **A**, this approach can be rather tedious for large pedigrees. However, as pointed out by Henderson (1976), certain features of the pattern of gene flow through pedigrees can be exploited to greatly facilitate the computation of **A**. As noted above, $A_{ii} = 1 + f_i$. In addition, $A_{ij} = 2\Theta_{ij} = 0$ if *i* and *j* are unrelated individuals. Building on these relationships, the following helpful rules outlined by Emik and Terrill (1949) can be used to obtain the elements of **A** for an arbitrary pedigree.

Order the individuals so that parents precede their offspring, and let the first b noninbred and unrelated individuals comprise the base population. With these individuals forming our starting point, the upper-left $b \times b$ submatrix of **A** is an identity matrix. This submatrix is expanded iteratively, one row and one column at a time, until the entire **A** matrix is filled out. If individual *i* has parents indexed by *g* and h, then its diagonal element is

$$A_{ii} = 1 + f_i = 1 + \Theta_{gh} = 1 + \frac{A_{gh}}{2} \tag{26.16a}$$

For a pair of individuals *i* and *j*, with $j < i$,

$$A_{ij} = A_{ji} = \Theta_{jg} + \Theta_{jh} = \frac{A_{jg} + A_{jh}}{2} \tag{26.16b}$$

If a parent is unknown, it is assumed to be noninbred and unrelated to any other measured individual (except, of course, its known descendants), so that if k indexes an unmeasured parent, we assume $A_{kk} = 1$ and $A_{ik} = 0$ (for $i \neq k$ where i indexes any individual except known descendants of k).

Example 8. Consider the additive genetic relationship matrix **A** for the five measured individuals in the pedigree given in the following figure.

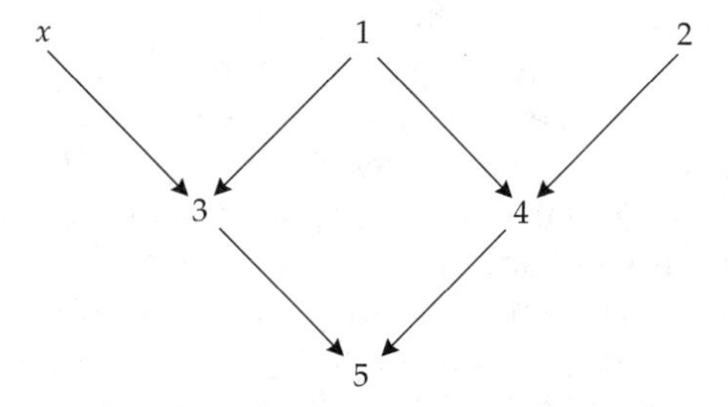

Here, measured individuals 1 and 2 as well as the unmeasured (and potentially unknown) individual *x* are assumed to be unrelated members of a noninbred base population. Since 1 and 2 are unrelated and noninbred, $A_{11} = A_{22} = 1$ and $A_{12} = A_{21} = 0$. Likewise, $A_{xx} = 1$, $A_{x1} = A_{x2} = 0$. Now consider individual 3, whose parents are 1 and *x*. Applying Equations 26.16a,b, we have $A_{33} = 1 + A_{x1}/2 = 1$ and

$$A_{13} = A_{31} = \frac{A_{1x} + A_{11}}{2} = \frac{1}{2}, \quad A_{23} = A_{32} = \frac{A_{2x} + A_{21}}{2} = 0$$

Individual 4 has parents 1 and 2, giving $A_{44} = 1 + A_{12}/2 = 1$ and

$$A_{14} = A_{41} = \frac{A_{11} + A_{12}}{2} = \frac{1}{2}, \quad A_{24} = A_{42} = \frac{A_{21} + A_{22}}{2} = \frac{1}{2},$$

and

$$A_{34} = A_{43} = \frac{A_{31} + A_{32}}{2} = \frac{1}{4}$$

Turning to individual 5, whose parents are 3 and 4, $A_{55} = 1 + A_{34}/2 = 1 + 1/8 = 1.25$, and

$$A_{15} = A_{51} = \frac{A_{13} + A_{14}}{2} = \frac{1}{2}, \quad A_{25} = A_{52} = \frac{A_{23} + A_{24}}{2} = \frac{1}{4}$$

$$A_{35} = A_{53} = \frac{A_{33} + A_{34}}{2} = \frac{5}{8}, \quad A_{45} = A_{54} = \frac{A_{43} + A_{44}}{2} = \frac{5}{8}$$

Thus,

$$\mathbf{A} = \begin{pmatrix} 1 & 0 & 1/2 & 1/2 & 1/2 \\ 0 & 1 & 0 & 1/2 & 1/4 \\ 1/2 & 0 & 1 & 1/4 & 5/8 \\ 1/2 & 1/2 & 1/4 & 1 & 5/8 \\ 1/2 & 1/2 & 5/8 & 5/8 & 9/8 \end{pmatrix}$$

In many natural settings, and some breeding situations, only one of the parents (generally, the mother) is known with certainty. One approach to dealing with such uncertainty is simply to assume that the unknown sire is unrelated to any measured individual in the base population (as we did above). However, when all potential sires have been measured, an **average relationship matrix** can be computed by assigning all potental sires equal weight (Henderson 1988b). For example, if the sire of individual i is one of k potential males, then each of these is assumed to be the sire with probability $1/k$, and the entry for each potential sire in the ith row and column of $\mathbf{A}$ becomes $1/(2k)$. In principle, molecular-marker information may be used to assign more accurate probabilities of paternity. Konigsberg and Cheverud (1992) conclude that when more than a few individuals are potential sires, either method (ignoring the sire or replacing the values in $\mathbf{A}$ for each potential sire by a probabilistic estimate) gives similar power. However, when the sire can be limited to a few potential individuals, the use of the average relationship matrix is a more powerful approach.

While the preceding rules provide a simple means for acquiring $\mathbf{A}$, a more fundamental problem remains — it is the inverse of $\mathbf{A}$, not $\mathbf{A}$ itself, that enters into the mixed-model equations. With very large numbers of individuals in a pedigree, the procurement of $\mathbf{A}^{-1}$ can be extremely demanding computationally.

Thus, considerable attention has been given to the development of shortcuts to compute the elements of $\mathbf{A}^{-1}$.

A remarkable result, due again to Henderson (1976), is that in a noninbred population, one can directly write down the inverse of $\mathbf{A}$ by a simple method without having to compute $\mathbf{A}$ itself. For n individuals, order n operations are required to obtain $\mathbf{A}^{-1}$ by Henderson's method, while order n^2 and n^3 operations are required to first obtain $\mathbf{A}$ and then invert it by normal methods. The critical feature of Henderson's method is the recognition that the relationship matrix can be expressed as the product

$$\mathbf{A} = \mathbf{TDT}^T \tag{26.17a}$$

and that its inverse is given by

$$\mathbf{A}^{-1} = (\mathbf{T}^{-1})^T\mathbf{D}^{-1}\mathbf{T}^{-1} \tag{26.17b}$$

$\mathbf{D}$ is a diagonal matrix, the elements of which are proportional to the variances associated with Mendelian (segregational) sampling conditional on the parents. These elements are easily acquired in a noninbred population, for in this case D_{ii} is equal to 0.5, 0.75, or 1.0 when both, one, or none of individual *i*'s parents are included in the matrix $\mathbf{A}$. Because $\mathbf{D}$ is diagonal, its inverse is also diagonal with elements equal to either 2, 4/3, or 1.

$\mathbf{T}$ is a lower triangular matrix, the elements of which trace the flow of genes through the sample. The elements above the diagonal are all equal to zero, while the diagonal elements are all equal to one. The element in the *j*th row in the column below the *i*th diagonal are defined to be the fraction of genes of individual *j* that are expected in individual *i*; for nonrelatives, these elements equal zero, and in the absence of inbreeding, the elements involving relatives are equal to 1/2, 1/4, and 1/8 for first-, second-, and third-degree relatives, etc. The inverse of $\mathbf{T}$ is even easier to obtain than $\mathbf{T}$ itself. $\mathbf{T}^{-1}$ is also a lower triangular matrix with zeros above the diagonal and ones on the diagonal. Below the diagonal in the *j*th row, all elements are equal to zero, except for those corresponding to the column of *j*'s known parents, which are set equal to -0.5.

Using these simple rules, all of the elements of $\mathbf{D}^{-1}$ and $\mathbf{T}^{-1}$ can be rapidly acquired for any pedigree, allowing the efficient computation of $\mathbf{A}^{-1}$. This eliminates the need to invert $\mathbf{A}$ by conventional methods. Quaas (1976), Tier (1990), and Mrode (1996) present extensions of these techniques that allow for inbreeding. These important developments have allowed BLUP methodology to be practically applied to extremely large and complex data sets in animal breeding (in particular, cattle) where the number of individuals in a pedigree can approach one million. Alternative solutions to mixed-model equations with very high dimensions are discussed by Schaeffer and Kennedy (1986), Misztal et al. (1987), Misztal and Gianola (1987), Groeneveld and Kovac (1990a), and Ducrocq (1992).

Example 9. Recall the pedigree given in Example 6, where the inverse of the relationship matrix was

$$\mathbf{A}^{-1} = \begin{pmatrix} 3/2 & 1/2 & 0 & -1 & 0 \\ 1/2 & 2 & 1/2 & -1 & -1 \\ 0 & 1/2 & 3/2 & 0 & -1 \\ -1 & -1 & 0 & 2 & 0 \\ 0 & -1 & -1 & 0 & 2 \end{pmatrix}$$

This can also be obtained by applying Equation 26.17b. Because the parents of individuals 1, 2, and 3 are unknown, while both parents of individuals 4 and 5 are contained within the observed pedigree, $\mathbf{D}$ is diagonal with elements (1.0, 1.0, 1.0, 0.5, 0.5). The inverse of $\mathbf{D}$ is therefore diagonal with elements (1.0, 1.0, 1.0, 2.0, 2.0). Following the rules outlined above, the gene-flow matrix is

$$\mathbf{T} = \begin{pmatrix} 1 & 0 & 0 & 0 & 0 \\ 0 & 1 & 0 & 0 & 0 \\ 0 & 0 & 1 & 0 & 0 \\ 0.5 & 0.5 & 0 & 1 & 0 \\ 0 & 0.5 & 0.5 & 0 & 1 \end{pmatrix}$$

Note that within a given column, below the diagonal, the elements equal to 0.5 denote parent-offspring relationships. Taking the inverse directly,

$$\mathbf{T}^{-1} = \begin{pmatrix} 1 & 0 & 0 & 0 & 0 \\ 0 & 1 & 0 & 0 & 0 \\ 0 & 0 & 1 & 0 & 0 \\ -0.5 & -0.5 & 0 & 1 & 0 \\ 0 & -0.5 & -0.5 & 0 & 1 \end{pmatrix}$$

$\mathbf{T}^{-1}$ can also be obtained by the shorter route — below the diagonal, within each individual's row, each known parent is denoted with a -0.5. Using Equation 26.17b, the inverse of $\mathbf{A}$ (given above) is recovered.

Allowing for Mutation when Computing A

While we have been using $\mathbf{G} = \sigma^2_A\,\mathbf{A}$ as the covariance matrix for the vector of additive effects, this relationship ignores the effects of mutation. Such an approximation is reasonable in many settings, but for multiple-generation pedigrees, the effects of mutation can become rather significant. Wray (1990) showed that t generations of mutation are accounted for by using

$$\mathbf{G} = \sigma^2_A \left(\mathbf{A} + \frac{\sigma^2_{m0}}{\sigma^2_A} \sum_{k=1}^{t} \mathbf{A}_k \right) \tag{26.18}$$

where σ^2_{m0} is the mutational rate of input of new additive genetic variance (Chapter 12). Here $\mathbf{A}$ is the standard relationship matrix ignoring mutation, while $\mathbf{A}_k$ is the relationship matrix computed under the assumption that ancestors born in generations 0 to $k-1$ are ignored. Hence, $\sigma^2_A\,\mathbf{A}$ accounts for the additive genetic variation present in the base population, while $\sigma^2_m\,\mathbf{A}_k$ accounts for the variation generated by mutation in generation k. Equation 26.18 is useful in both estimating the mutational variance using REML and in the analysis of long-term selection experiments where input to the additive genetic variance from new mutations is likely to be important (Mackay et al. 1992b, 1995).

JOINT ESTIMATION OF SEVERAL VECTORS OF RANDOM EFFECTS

The mixed-model equations can be easily extended to situations where two (or more) vectors of random effects are of interest, as for example, in the estimation of both additive and dominance values or in the estimation of breeding values and maternal effects. With two vectors of random effects ($\mathbf{u}_1$ and $\mathbf{u}_2$) uncorrelated with each other, the mixed model becomes

$$\mathbf{y} = \mathbf{X}\boldsymbol{\beta} + \mathbf{Z}_1\mathbf{u}_1 + \mathbf{Z}_2\mathbf{u}_2 + \mathbf{e} \tag{26.19a}$$

The vectors of random effects can have different dimensions (q_1 for $\mathbf{u}_1$, q_2 for $\mathbf{u}_2$), so with n individuals in the vector $\mathbf{y}$, the incidence matrix $\mathbf{Z}_i$ is $n \times q_i$ (for $i = 1, 2$). Letting $\mathbf{R}$ be the $n \times n$ covariance matrix for the vector of residual errors $\mathbf{e}$, and $\mathbf{G}_i$ be the $q_i \times q_i$ covariance matrix for $\mathbf{u}_i$, the MMEs become

$$\begin{pmatrix} \mathbf{X}^T\mathbf{R}^{-1}\mathbf{X} & \mathbf{X}^T\mathbf{R}^{-1}\mathbf{Z}_1 & \mathbf{X}^T\mathbf{R}^{-1}\mathbf{Z}_2 \\ \mathbf{Z}_1^T\mathbf{R}^{-1}\mathbf{X} & \mathbf{Z}_1^T\mathbf{R}^{-1}\mathbf{Z}_1 + \mathbf{G}_1^{-1} & \mathbf{Z}_1^T\mathbf{R}^{-1}\mathbf{Z}_2 \\ \mathbf{Z}_2^T\mathbf{R}^{-1}\mathbf{X} & \mathbf{Z}_2^T\mathbf{R}^{-1}\mathbf{Z}_1 & \mathbf{Z}_2^T\mathbf{R}^{-1}\mathbf{Z}_2 + \mathbf{G}_2^{-1} \end{pmatrix} \begin{pmatrix} \widehat{\boldsymbol{\beta}} \\ \widehat{\mathbf{u}}_1 \\ \widehat{\mathbf{u}}_2 \end{pmatrix} = \begin{pmatrix} \mathbf{X}^T\mathbf{R}^{-1}\mathbf{y} \\ \mathbf{Z}_1^T\mathbf{R}^{-1}\mathbf{y} \\ \mathbf{Z}_2^T\mathbf{R}^{-1}\mathbf{y} \end{pmatrix} \tag{26.19b}$$

Equation 26.19b can be extended in an obvious fashion to incorporate additional uncorrelated vectors of random effects. The following sections outline a few common applications of this extension of the mixed model.

BLUP Estimates of Dominance Values

In animal breeding practices, most attention is generally given to the estimation of additive genetic values (breeding values) of individuals, since nonadditive effects are not transmitted to offspring in noninbreeding populations. However, in certain situations, the total genetic merit (additive plus nonadditive genetic values) can be exploited, as for example, when individuals can be cloned or full-sib families can be reliably reconsitituted from frozen embryos.

Here we let $\mathbf{u}_1 = \mathbf{a}$ and $\mathbf{u}_2 = \mathbf{d}$ represent vectors of additive and dominance genetic values, and we assume an outbred base population, as inbreeding produces a correlation between additive and dominance values. To expand from the purely additive model, the main modification that we require is a means for converting pedigree information into the covariance matrix of dominance effects. This idea has already been covered in Chapter 7, where we found that the covariance of dominance values for two individuals *i* and *j* is the product of the dominance genetic variance and the coefficient of fraternity, $\sigma_D^2 \Delta_{ij}$. From Equation 7.7, the latter is given by

$$\Delta_{ij} = \Theta_{gk}\,\Theta_{hl} + \Theta_{gl}\,\Theta_{hk} \tag{26.20a}$$

where *i*'s parents are indexed by *g* and *h* and *j*'s by *k* and *l*, and as above, Θ is the coefficient of coancestry. The **dominance genetic relationship matrix D** contains these coefficients of fraternity for the measured individuals. Recalling that the elements of $\mathbf{A}$ are $2\Theta_{ij}$, the off-diagonal elements of $\mathbf{D}$ can be computed as

$$D_{ij} = \frac{A_{gk}\,A_{hl} + A_{gl}\,A_{hk}}{4} \tag{26.20b}$$

whereas the diagonal elements are all $D_{ii} = 1$.

Now we can develop the BLUP estimates for the vectors of additive and dominance genetic values. Assuming that only a single observation is made for each individual, $\mathbf{Z}_1 = \mathbf{Z}_2 = \mathbf{I}$, giving the mixed model

$$\mathbf{y} = \mathbf{X}\boldsymbol{\beta} + \mathbf{a} + \mathbf{d} + \mathbf{e}$$

The covariance matrices for the random effects are $\mathbf{G}_1 = \boldsymbol{\sigma}(\mathbf{a},\mathbf{a}) = \sigma_A^2\,\mathbf{A}$ and $\mathbf{G}_2 = \boldsymbol{\sigma}(\mathbf{d},\mathbf{d}) = \sigma_D^2\,\mathbf{D}$, and the covariance matrix for residual environmental deviations is $\mathbf{R} = \boldsymbol{\sigma}(\mathbf{e},\mathbf{e}) = \sigma_E^2\,\mathbf{I}$. Assuming that $\mathbf{a}$, $\mathbf{d}$, and $\mathbf{e}$ are uncorrelated, Equation 26.19b reduces to

$$\begin{pmatrix} \mathbf{X}^T\mathbf{X} & \mathbf{X}^T & \mathbf{X}^T \\ \mathbf{X} & \mathbf{I} + \lambda_A \mathbf{A}^{-1} & \mathbf{I} \\ \mathbf{X} & \mathbf{I} & \mathbf{I} + \lambda_D \mathbf{D}^{-1} \end{pmatrix} \begin{pmatrix} \widehat{\boldsymbol{\beta}} \\ \widehat{\mathbf{a}} \\ \widehat{\mathbf{d}} \end{pmatrix} = \begin{pmatrix} \mathbf{X}^T\mathbf{y} \\ \mathbf{y} \\ \mathbf{y} \end{pmatrix} \tag{26.21}$$

where we have multiplied through by σ_E^2 to simplify the equations, $\lambda_A = \sigma_E^2/\sigma_A^2$, and $\lambda_D = \sigma_E^2/\sigma_D^2$.

Henderson (1984a) noted an interesting relationship between the BLUP estimates of additive and dominance values that further simplifies the MMEs. Subtracting the second row from the last in Equation 26.21 and rearranging gives

$$\mathbf{A}^{-1}\widehat{\mathbf{a}}/\sigma_A^2 = \mathbf{D}^{-1}\widehat{\mathbf{d}}/\sigma_D^2 \tag{26.22}$$

implying that

$$\widehat{\mathbf{d}} = (\sigma_D^2/\sigma_A^2)\mathbf{D}\mathbf{A}^{-1}\widehat{\mathbf{a}} \tag{26.23}$$

Substituting this expression into Equation 26.21 reduces the mixed-model equations to

$$\begin{pmatrix} \mathbf{X}^T\mathbf{X} & \mathbf{X}^T[\,\mathbf{I} + (\sigma_D^2/\sigma_A^2)\,\mathbf{D}\mathbf{A}^{-1}\,] \\ \mathbf{X} & \mathbf{I} + (\sigma_D^2/\sigma_A^2)\,\mathbf{D}\mathbf{A}^{-1} + (\sigma_E^2/\sigma_A^2)\,\mathbf{A}^{-1} \end{pmatrix} \begin{pmatrix} \widehat{\boldsymbol{\beta}} \\ \widehat{\mathbf{a}} \end{pmatrix} = \begin{pmatrix} \mathbf{X}^T\mathbf{y} \\ \mathbf{y} \end{pmatrix} \tag{26.24}$$

with $\widehat{\mathbf{d}}$ obtained by using Equation 26.23.

The above analysis provides insight into an important issue that applies to the purely additive model. In the presence of nonadditive gene action, application of the conventional expression ($\mathbf{y} = \mathbf{X}\boldsymbol{\beta} + \mathbf{a} + \mathbf{e}$) can yield biased estimates of the additive genetic value. The reason for this bias is that the standard applications of the additive model assume that the residual errors are uncorrelated with constant variance, i.e., $\mathbf{R} = \sigma_E^2\,\mathbf{I}$. If dominance genetic variance is present and pairs of individuals with nonzero coefficients of fraternity (such as full sibs and/or double first cousins) are included in the analysis, this simple residual variance structure is no longer correct. Assuming that the environmental values of different individuals are independent and uncorrelated with genetic values, the true residual covariance matrix has the form $\mathbf{R} = \sigma_E^2\,\mathbf{I} + \sigma_D^2\,\mathbf{D}$ under the animal model, so unless $\mathbf{D}$ is zero, using $\sigma_E^2\,\mathbf{I}$ for $\mathbf{R}$ biases the breeding value estimates. The correct residual covariance structure requires that $\mathbf{d}$ be treated as a separate random effect, and even then we assume an absence of higher-order (epistatic) effects.

The above approaches can be extended in a straightforward fashion to the estimation of epistatic effects, provided these are distributed independently of the additive and dominance effects (as expected in a random-mating population) (Henderson 1977a, 1985a,b; Mrode 1996). Smith and Mäki-Tanila (1990) and Uimari and Kennedy (1990) discuss further modifications that are required in inbred populations with nonadditive gene action. However, even for the simplest model with dominance in a random-mating population, rigorous practical applications of BLUP are limited due to difficulties in obtaining accurate estimates of the dominance genetic variance, which are required for the solution of Equation 26.24. We leave it to the interested reader to pursue further issues involving nonadditive genetic values via the references cited above.

Repeated Records

Another situation in which correlations are expected among residual errors arises when multiple observations are made on individuals, a common procedure used to reduce measurement error. Here, assuming dominance is of negligible importance, the residual error can be described as $p + e$, where p is the "permanent" environmental effect common to all observations on the same individual, and e is the residual error between observations of the same individual due, for example, to measurement error and changes in some environmental factors. Recall from Chapter 6 that the repeatability of a character (r) is defined to be the correlation between different measurements in the same individual. If j and k denote different observations on the same individual i, the covariance between repeated measurements is

$$\begin{aligned}\sigma(y_{ij}, y_{ik}) = r\sigma_y^2 &= \sigma(a_i + p_i + e_{ij}, a_i + p_i + e_{ik}) \\ &= \sigma_A^2 + \sigma_P^2\end{aligned}$$

which follows from the assumption that the residual errors for the same individual are uncorrelated. Assuming purely additive gene action, $\sigma_A^2 = h^2\sigma_y^2$ of the covariance is due to genetic effects, leaving $(r - h^2)\sigma_y^2 = \sigma_P^2$ as the covariance due to permanent environmental effects.

One approach to estimating breeding values when repeated measures are contained in the data set is to continue to apply the simple animal model (Equation 26.8), with suitable $\mathbf{Z}$ to accommodate multiple records, modifying the residual covariance matrix $\mathbf{R}$ such that

$$R_{jk} = \begin{cases} (1 - h^2)\sigma_y^2 & j = k \text{ (the same measurement in an individual)} \\ (r - h^2)\sigma_y^2 & j \text{ and } k \text{ are repeated measures} \\ 0 & j \text{ and } k \text{ are measures on different individuals.} \end{cases} \tag{26.25}$$

Since the resulting covariance matrix is not diagonal, it is not always easily inverted, a potentially serious complication for extremely large data sets.

An alternative approach follows the same rationale as the model incorporating dominance, i.e., explicitly accounting for shared environmental effects (rather than incorporating them into the residual error structure) by introducing a new random factor into the model, such that

$$y_{ij} = \mu + a_i + p_i + e_{ij} \tag{26.26}$$

(Henderson 1977a). With this structure, all residual effects are again uncorrelated with common variance $\sigma_E^2 = (1 - r)\sigma_y^2$.

Suppose a total of k measurements are made on n individuals (such a balanced design is not essential). We can write this in the general mixed-model framework as

$$\mathbf{y} = \mathbf{X}\boldsymbol{\beta} + \mathbf{Z}(\mathbf{a} + \mathbf{p}) + \mathbf{e} = \mathbf{X}\boldsymbol{\beta} + \mathbf{Z}\mathbf{a} + \mathbf{Z}\mathbf{p} + \mathbf{e} \tag{26.27}$$

where $\mathbf{a}^T = (a_1, \cdots, a_n)^T$ and $\mathbf{p}^T = (p_1, \cdots, p_n)^T$, and each row of the $nk \times n$ incidence matrix $\mathbf{Z}$ has all zeros except a one at the position corresponding to the individual measured. This model has the resulting covariance matrices

$$\sigma(\mathbf{a}, \mathbf{a}) = h^2\sigma_y^2\,\mathbf{A}, \qquad \sigma(\mathbf{p}, \mathbf{p}) = (r - h^2)\,\sigma_y^2\,\mathbf{I}, \qquad \sigma(\mathbf{e}, \mathbf{e}) = (1 - r)\,\sigma_y^2\,\mathbf{I}$$

with **a**, **p**, and **e** being assumed to be uncorrelated. Applying Equation 26.19b and removing the common factor σ_y^2 from all expressions, the mixed-model equations become

$$\begin{pmatrix} \mathbf{X}^T\mathbf{X} & \mathbf{X}^T\mathbf{Z} & \mathbf{X}^T\mathbf{Z} \\ \mathbf{Z}^T\mathbf{X} & \mathbf{Z}^T\mathbf{Z} + \lambda_A\,\mathbf{A}^{-1} & \mathbf{Z}^T\mathbf{Z} \\ \mathbf{Z}^T\mathbf{X} & \mathbf{Z}^T\mathbf{Z} & \mathbf{Z}^T\mathbf{Z} + \lambda_P\,\mathbf{I} \end{pmatrix} \begin{pmatrix} \widehat{\boldsymbol{\beta}} \\ \widehat{\mathbf{a}} \\ \widehat{\mathbf{p}} \end{pmatrix} = \begin{pmatrix} \mathbf{X}^T\mathbf{y} \\ \mathbf{Z}^T\mathbf{y} \\ \mathbf{Z}^T\mathbf{y} \end{pmatrix} \tag{26.28a}$$

where

$$\lambda_A = \frac{\sigma_E^2}{\sigma_A^2} = \frac{1-r}{h^2}, \qquad \lambda_P = \frac{\sigma_E^2}{\sigma_P^2} = \frac{1-r}{r-h^2} \tag{26.28b}$$

Example 10. To compare the two different methods for dealing with repeated records, suppose three unrelated and noninbred individuals are measured, with two observations on individual one ($y_1 = 7$, $y_2 = 8$), three observations on individual two ($y_3 = 6$, $y_4 = 6$, $y_5 = 5$), and one observation on individual three ($y_6 = 9$). Assume that the only fixed factor is the mean and that the character has heritability $h^2 = 0.4$ and repeatability $r = 0.5$, giving $1 - h^2 = 0.6$ and $r - h^2 = 0.1$. For either formulation, we have

$$\mathbf{y} = \begin{pmatrix} 7 \\ 8 \\ 6 \\ 6 \\ 5 \\ 9 \end{pmatrix}, \quad \boldsymbol{\beta} = (\,\mu\,), \quad \mathbf{X} = \begin{pmatrix} 1 \\ 1 \\ 1 \\ 1 \\ 1 \\ 1 \end{pmatrix}, \quad \mathbf{Z} = \begin{pmatrix} 1 & 0 & 0 \\ 1 & 0 & 0 \\ 0 & 1 & 0 \\ 0 & 1 & 0 \\ 0 & 1 & 0 \\ 0 & 0 & 1 \end{pmatrix}$$

Since all three individuals are assumed to be unrelated, $\mathbf{A} = \mathbf{I}$.

To apply the permanent-effects model $y_i = \mu + a_i + p_i + e_i$, note that

$$\lambda_A = \frac{1-r}{h^2} = \frac{1-0.5}{0.4} = 1.25, \qquad \lambda_P = \frac{1-r}{r-h^2} = \frac{1-0.5}{0.5-0.4} = 5,$$

$$\mathbf{X}^T\mathbf{X} = 6, \qquad \mathbf{X}^T\mathbf{y} = 41,$$

$$\mathbf{Z}^T\mathbf{X} = (\mathbf{X}^T\mathbf{Z})^T = \begin{pmatrix} 2 \\ 3 \\ 1 \end{pmatrix}, \qquad \mathbf{Z}^T\mathbf{y} = \begin{pmatrix} 15 \\ 17 \\ 9 \end{pmatrix}, \qquad \mathbf{Z}^T\mathbf{Z} = \begin{pmatrix} 2 & 0 & 0 \\ 0 & 3 & 0 \\ 0 & 0 & 1 \end{pmatrix},$$

$$\mathbf{Z}^T\mathbf{Z} + \lambda_A\,\mathbf{A}^{-1} = \begin{pmatrix} 3.25 & 0 & 0 \\ 0 & 4.25 & 0 \\ 0 & 0 & 2.25 \end{pmatrix}, \qquad \mathbf{Z}^T\mathbf{Z} + \lambda_P\,\mathbf{I} = \begin{pmatrix} 7 & 0 & 0 \\ 0 & 8 & 0 \\ 0 & 0 & 6 \end{pmatrix}$$

giving the MMEs (Equation 26.28a) as

$$\begin{pmatrix} 6 & 2 & 3 & 1 & 2 & 3 & 1 \\ 2 & 3.25 & 0 & 0 & 2 & 0 & 0 \\ 3 & 0 & 4.25 & 0 & 0 & 3 & 0 \\ 1 & 0 & 0 & 2.25 & 0 & 0 & 1 \\ 2 & 2 & 0 & 0 & 7 & 0 & 0 \\ 3 & 0 & 3 & 0 & 0 & 8 & 0 \\ 1 & 0 & 0 & 1 & 0 & 0 & 6 \end{pmatrix} \begin{pmatrix} \widehat{\mu} \\ \widehat{a}_1 \\ \widehat{a}_2 \\ \widehat{a}_3 \\ \widehat{p}_1 \\ \widehat{p}_2 \\ \widehat{p}_3 \end{pmatrix} = \begin{pmatrix} 41 \\ 15 \\ 17 \\ 9 \\ 15 \\ 17 \\ 9 \end{pmatrix}$$

which has solutions

$$\widehat{\mu} \simeq 7.174, \qquad \begin{pmatrix} \widehat{a}_1 \\ \widehat{a}_2 \\ \widehat{a}_3 \end{pmatrix} \simeq \begin{pmatrix} 0.174 \\ -0.904 \\ 0.730 \end{pmatrix}, \qquad \begin{pmatrix} \widehat{p}_1 \\ \widehat{p}_2 \\ \widehat{p}_3 \end{pmatrix} \simeq \begin{pmatrix} 0.043 \\ -0.226 \\ 0.183 \end{pmatrix}$$

Conversely, applying the simple animal model $y_i = \mu + a_i + e_i$, from Equation 26.25 the covariance matrix for the residual errors becomes

$$\mathbf{R} = \sigma_y^2 \begin{pmatrix} 0.5 & 0.1 & 0 & 0 & 0 & 0 \\ 0.1 & 0.5 & 0 & 0 & 0 & 0 \\ 0 & 0 & 0.5 & 0.1 & 0.1 & 0 \\ 0 & 0 & 0.1 & 0.5 & 0.1 & 0 \\ 0 & 0 & 0.1 & 0.1 & 0.5 & 0 \\ 0 & 0 & 0 & 0 & 0 & 0.5 \end{pmatrix}$$

Likewise, $\mathbf{G} = h^2\sigma_y^2\,\mathbf{I}$, and hence $\mathbf{G}^{-1} = (h^2\sigma_y^2)^{-1}\,\mathbf{I}$. Removing the factor σ_y^2 common to all expressions gives

$$\mathbf{X}^T\mathbf{R}^{-1}\mathbf{X} \simeq 8.27, \qquad \mathbf{Z}^T\mathbf{R}^{-1}\mathbf{X} = (\mathbf{X}^T\mathbf{R}^{-1}\mathbf{Z})^T \simeq \begin{pmatrix} 2.86 \\ 3.75 \\ 1.67 \end{pmatrix}$$

$$\mathbf{Z}^T\mathbf{R}^{-1}\mathbf{Z} + \mathbf{G}^{-1} = \mathbf{Z}^T\mathbf{R}^{-1}\mathbf{Z} + \frac{1}{h^2}\,\mathbf{I} \simeq \begin{pmatrix} 5.36 & 0 & 0 \\ 0 & 6.25 & 0 \\ 0 & 0 & 4.17 \end{pmatrix}$$

$$\mathbf{X}^T\mathbf{R}^{-1}\mathbf{y} = 57.68, \qquad \mathbf{Z}^T\mathbf{R}^{-1}\mathbf{y} \simeq \begin{pmatrix} 21.43 \\ 21.25 \\ 15 \end{pmatrix}$$

Substituting into Equation 26.5 gives the MMEs

$$\begin{pmatrix} 8.27 & 2.86 & 3.75 & 1.67 \\ 2.86 & 5.36 & 0 & 0 \\ 3.75 & 0 & 6.25 & 0 \\ 1.67 & 0 & 0 & 4.17 \end{pmatrix} \begin{pmatrix} \widehat{\mu} \\ \widehat{a}_1 \\ \widehat{a}_2 \\ \widehat{a}_3 \end{pmatrix} = \begin{pmatrix} 57.68 \\ 21.43 \\ 21.25 \\ 15 \end{pmatrix}$$

which gives the same estimates as obtained with the permanent-effects model.

Maternal Effects

As discussed in detail in Chapter 23, for species with maternal care, the phenotype of an individual can depend on both genetic and environmental components of maternal effects, in addition to the individual's direct additive genetic value and a random environmental deviation. As first pointed out by Quaas and Pollak (1981), Willham's (1963) model for maternal effects is readily incorporated into a mixed-model framework, allowing the estimation of breeding values as well as maternal genetic and environmental effects. Here we simply consider the animal-model framework, although application of the reduced animal model is also straightforward (Mrode 1996). The structure of the model is very similar to that provided above for dominance and for repeated measures, the main difference being the addition of a third vector of random effects.

Letting $\boldsymbol{\beta}$ denote the vector of fixed effects, $\mathbf{a}$ the vector of breeding (direct additive genetic) values, $\mathbf{m}$ the vector of genetic maternal effects, and $\mathbf{c}$ the vector of environmental maternal effects, the mixed-model becomes

$$\mathbf{y} = \mathbf{X}\boldsymbol{\beta} + \mathbf{Z}_1\mathbf{a} + \mathbf{Z}_2\mathbf{m} + \mathbf{Z}_3\mathbf{c} + \mathbf{e} \tag{26.29}$$

where $\mathbf{Z}_1$, $\mathbf{Z}_2$, and $\mathbf{Z}_3$ are, respectively, the incidence matrices for random breeding values, genetic maternal effects, and environmental maternal effects. Assuming there are n individuals in the observation vector $\mathbf{y}$ (each with a single measure) and m mothers in the analysis (not all of which may have been observed), the vector of breeding values $\mathbf{a}$ contains q elements, where q is the total number of observations plus the number of unobserved mothers. The vectors of genetic maternal effects and environmental maternal effects are also $q \times 1$, while that for residual deviations is $n \times 1$. Assuming a single observation for each individual, $\mathbf{Z}_1$ has dimensionality $n \times q$ with all elements in each row being equal to zero except for a one in the column denoting the individual. $\mathbf{Z}_2$ also has dimensionality $n \times q$, and again consists entirely of zeros, with a one in each row denoting the individual's mother. Finally, $\mathbf{Z}_3$ has dimensionality $n \times q$, with all elements being equal to zero, except a single one in each row denoting the individual's mother. (Note that with this latter treatment, we assume that environmental maternal effects are not transmitted across generations. With cross-generational transmission, as in Falconer's model (Chapter 23), elements in $\mathbf{Z}_3$ involving maternal grandmothers and others in the maternal line may take on nonzero values.)

The breeding values, genetic and environmental maternal effects, and residual environmental effects are assumed to be random with respective variances $\sigma^2_{A_o}$, $\sigma^2_{A_m}$, $\sigma^2_{E_m}$, and $\sigma^2_{E_s}$. All effects are also assumed to be distributed independently except for the direct and maternal genetic effects, which may have covariance σ_{A_o,A_m} due to pleiotropic gene effects. Letting $\mathbf{A}$ be the $q \times q$ relationship matrix, the covariance matrix of breeding values is $\sigma^2_{A_o}\mathbf{A}$, the covariance matrix of genetic maternal effects is $\sigma^2_{A_m}\,\mathbf{A}$, and the covariance matrix of direct and maternal genetic effects is $\sigma_{A_o,A_m}\,\mathbf{A}$.

The BLUEs and BLUPs for this model are the solutions to the following mixed-model equations:

$$\begin{pmatrix} \mathbf{X}^T\mathbf{X} & \mathbf{X}^T\mathbf{Z}_1 & \mathbf{X}^T\mathbf{Z}_2 & \mathbf{X}^T\mathbf{Z}_3 \\ \mathbf{Z}_1^T\mathbf{X} & \mathbf{Z}_1^T\mathbf{Z}_1 + \lambda_1\,\mathbf{A}^{-1} & \mathbf{Z}_1^T\mathbf{Z}_2 + \lambda_2\,\mathbf{A}^{-1} & \mathbf{Z}_1^T\mathbf{Z}_3 \\ \mathbf{Z}_2^T\mathbf{X} & \mathbf{Z}_2^T\mathbf{Z}_1 + \lambda_2\,\mathbf{A}^{-1} & \mathbf{Z}_2^T\mathbf{Z}_2 + \lambda_3\,\mathbf{A}^{-1} & \mathbf{Z}_2^T\mathbf{Z}_3 \\ \mathbf{Z}_3^T\mathbf{X} & \mathbf{Z}_3^T\mathbf{Z}_1 & \mathbf{Z}_3^T\mathbf{Z}_2 & \mathbf{Z}_3^T\mathbf{Z}_3 + \lambda_4\,\mathbf{I} \end{pmatrix} \begin{pmatrix} \widehat{\boldsymbol\beta} \\ \widehat{\mathbf{a}} \\ \widehat{\mathbf{m}} \\ \widehat{\mathbf{c}} \end{pmatrix}$$

$$= \begin{pmatrix} \mathbf{X}^T\mathbf{y} \\ \mathbf{Z}_1^T\mathbf{y} \\ \mathbf{Z}_2^T\mathbf{y} \\ \mathbf{Z}_3^T\mathbf{y} \end{pmatrix} \tag{26.30}$$

Letting the covariance matrix of direct and maternal genetic effects be

$$\mathbf{g} = \begin{pmatrix} \sigma^2_{A_o} & \sigma_{A_o,A_m} \\ \sigma_{A_o,A_m} & \sigma^2_{A_m} \end{pmatrix}$$

and its inverse be

$$\mathbf{g}^{-1} = \begin{pmatrix} g^{11} & g^{12} \\ g^{21} & g^{22} \end{pmatrix}$$

then

$$\begin{pmatrix} \lambda_1 & \lambda_2 \\ \lambda_2 & \lambda_3 \end{pmatrix} = \sigma^2_{E_s} \begin{pmatrix} g^{11} & g^{12} \\ g^{21} & g^{22} \end{pmatrix}$$

and $\lambda_4 = \sigma^2_{E_s}/\sigma^2_C$.

Multiple Traits

In principle, in the estimation of breeding values for multiple traits, one can simply perform univariate BLUP analyses on each individual trait, but this does not necessarily make efficient use of the available data. When characters are correlated, measurements on each trait provide some information on the breeding values of the other correlated traits. Multivariate BLUP takes this information directly into account by computing breeding values as conditional expectations given the measurements of all traits in all relatives. On the other hand, limitations of multivariate BLUP include the need for accurate estimates of the genetic and environmental covariances among traits and its high computational demands. We have already considered the difficulties in procuring reliable estimates of covariances among traits (Chapter 22), and we return to the issue of computational speed at the end of this section.

To illustrate the basic principles of multivariate BLUP, we follow the approach of Henderson and Quaas (1976), restricting our attention to the simplest of situations in which each of k traits has been measured once in each of n individuals. (For modifications necessary with missing data, see Henderson and Quaas 1976 and Mrode 1996.) The $(nk) \times 1$ dimensional column vector of observations is denoted by the stack of univariate vectors,

$$\mathbf{y} = \begin{pmatrix} \mathbf{y}_1 \\ \mathbf{y}_2 \\ \vdots \\ \mathbf{y}_k \end{pmatrix}$$

where the ith element of the column vector $\mathbf{y}_j$ corresponds to the observation of character j in the ith individual.

Here we assume that each trait follows the animal model

$$\mathbf{y}_j = \mathbf{X}_j \boldsymbol{\beta}_j + \mathbf{Z}_j \mathbf{a}_j + \mathbf{e}_j$$

where there are q_j fixed effects associated with character j so that $\mathbf{X}_j$ and $\boldsymbol{\beta}_j$ have, respectively, dimensionality $n \times q_j$ and $q_j \times 1$ for each trait. Assuming there is a single measurement for each character in each individual, $\mathbf{Z}_j = \mathbf{I}$, and the mixed model can then be written as

$$\begin{pmatrix} \mathbf{y}_1 \\ \mathbf{y}_2 \\ \vdots \\ \mathbf{y}_k \end{pmatrix} = \begin{pmatrix} \mathbf{X}_1 & \mathbf{0} & \cdots & \mathbf{0} \\ \mathbf{0} & \mathbf{X}_2 & \cdots & \mathbf{0} \\ \vdots & \vdots & \ddots & \vdots \\ \mathbf{0} & \mathbf{0} & \cdots & \mathbf{X}_k \end{pmatrix} \begin{pmatrix} \boldsymbol{\beta}_1 \\ \boldsymbol{\beta}_2 \\ \vdots \\ \boldsymbol{\beta}_k \end{pmatrix} + \begin{pmatrix} \mathbf{a}_1 \\ \mathbf{a}_2 \\ \vdots \\ \mathbf{a}_k \end{pmatrix} + \begin{pmatrix} \mathbf{e}_1 \\ \mathbf{e}_2 \\ \vdots \\ \mathbf{e}_k \end{pmatrix} \tag{26.31}$$

To obtain the mixed-model equations for the total vectors of fixed effects and breeding values, it only remains to specify the $(nk) \times (nk)$ covariance matrices $\mathbf{R}$ and $\mathbf{G}$ associated with the total vector $\mathbf{e}^T = (\mathbf{e}_1^T, \cdots, \mathbf{e}_k^T)$ of residual errors and the total vector $\mathbf{a}^T = (\mathbf{a}_1^T, \cdots, \mathbf{a}_k^T)$ of random effects.

First, consider the vector of environmental effects, $\mathbf{R}$. While residual deviations for the same character measured in different individuals can often be assumed to be uncorrelated, this is not necessarily the case for different characters measured in the same individual, which can exhibit an environmental correlation (Chapter 21). The covariance matrix between $\mathbf{e}_i$ and $\mathbf{e}_j$ can be written as $\boldsymbol{\sigma}(\mathbf{e}_i, \mathbf{e}_j) = \epsilon_{ij}\, \mathbf{I}$, where $\epsilon_{ij} = \sigma_E^2(i,j)$ is the environmental covariance between traits i and j as expressed in the same individual. The resulting $(nk) \times (nk)$ covariance matrix for the total error vector $\mathbf{e} = (\mathbf{e}_1^T, \cdots, \mathbf{e}_k^T)$ becomes

$$\begin{pmatrix} \sigma(\mathbf{e}_1, \mathbf{e}_1) & \sigma(\mathbf{e}_1, \mathbf{e}_2) & \cdots & \sigma(\mathbf{e}_1, \mathbf{e}_k) \\ \sigma(\mathbf{e}_2, \mathbf{e}_1) & \sigma(\mathbf{e}_2, \mathbf{e}_2) & \cdots & \sigma(\mathbf{e}_2, \mathbf{e}_k) \\ \vdots & \vdots & \ddots & \vdots \\ \sigma(\mathbf{e}_k, \mathbf{e}_1) & \sigma(\mathbf{e}_k, \mathbf{e}_2) & \cdots & \sigma(\mathbf{e}_k, \mathbf{e}_k) \end{pmatrix} = \begin{pmatrix} \mathbf{I}\,\epsilon_{11} & \mathbf{I}\,\epsilon_{12} & \cdots & \mathbf{I}\,\epsilon_{1k} \\ \mathbf{I}\,\epsilon_{21} & \mathbf{I}\,\epsilon_{22} & \cdots & \mathbf{I}\,\epsilon_{2k} \\ \vdots & \vdots & \ddots & \vdots \\ \mathbf{I}\,\epsilon_{k1} & \mathbf{I}\,\epsilon_{k2} & \cdots & \mathbf{I}\,\epsilon_{kk} \end{pmatrix} \tag{26.32}$$

An extremely convenient notation for $\mathbf{R}$ utilizes the **Kronecker product**. For a $k \times l$ matrix $\mathbf{A}$ and a $m \times n$ matrix $\mathbf{B}$,

$$\mathbf{A} = \begin{pmatrix} a_{11} & a_{12} & \cdots & a_{1l} \\ a_{21} & a_{22} & \cdots & a_{2l} \\ \vdots & \vdots & \ddots & \vdots \\ a_{k1} & a_{k2} & \cdots & a_{kl} \end{pmatrix} \quad \text{and} \quad \mathbf{B} = \begin{pmatrix} b_{11} & b_{12} & \cdots & b_{1n} \\ b_{21} & b_{22} & \cdots & b_{2n} \\ \vdots & \vdots & \ddots & \vdots \\ b_{m1} & b_{m2} & \cdots & b_{mn} \end{pmatrix}$$

the Kronecker product of $\mathbf{A}$ and $\mathbf{B}$, denoted $\mathbf{A} \otimes \mathbf{B}$, is the $(km) \times (ln)$ matrix

$$\mathbf{A} \otimes \mathbf{B} = \begin{pmatrix} \mathbf{B}\,a_{11} & \mathbf{B}\,a_{12} & \cdots & \mathbf{B}\,a_{1l} \\ \mathbf{B}\,a_{21} & \mathbf{B}\,a_{22} & \cdots & \mathbf{B}\,a_{2l} \\ \vdots & \vdots & \ddots & \vdots \\ \mathbf{B}\,a_{k1} & \mathbf{B}\,a_{k2} & \cdots & \mathbf{B}\,a_{kl} \end{pmatrix}$$

where each element is itself a matrix (of order $m \times n$) with

$$\mathbf{B}\,a_{ij} = \begin{pmatrix} a_{ij}\,b_{11} & a_{ij}\,b_{12} & \cdots & a_{ij}\,b_{1n} \\ a_{ij}\,b_{21} & a_{ij}\,b_{22} & \cdots & a_{ij}\,b_{2n} \\ \vdots & \vdots & \ddots & \vdots \\ a_{ij}\,b_{m1} & a_{ij}\,b_{m2} & \cdots & a_{ij}\,b_{mn} \end{pmatrix}$$

An especially useful feature of Kronecker products (indeed, our primary reason for using this notation) is that for two square nonsingular matrices $\mathbf{A}$ and $\mathbf{B}$,

$$(\mathbf{A} \otimes \mathbf{B})^{-1} = \mathbf{A}^{-1} \otimes \mathbf{B}^{-1} \tag{26.33}$$

Let $\mathbf{E}$ be the $k \times k$ covariance matrix of within-individual environmental effects, with the ijth element being $\sigma_E(i, j) = \epsilon_{ij}$. In Kronecker product notation, the covariance matrix $\mathbf{R}$ for the total vector of errors $\mathbf{e}^T = (\mathbf{e}_1^T, \cdots, \mathbf{e}_k^T)$ is

$$\mathbf{R} = \mathbf{E} \otimes \mathbf{I} \tag{26.34}$$

giving

$$\mathbf{R}^{-1} = \mathbf{E}^{-1} \otimes \mathbf{I}^{-1} = \begin{pmatrix} \mathbf{I}\,\epsilon^{11} & \mathbf{I}\,\epsilon^{12} & \cdots & \mathbf{I}\,\epsilon^{1k} \\ \mathbf{I}\,\epsilon^{21} & \mathbf{I}\,\epsilon^{22} & \cdots & \mathbf{I}\,\epsilon^{2k} \\ \vdots & \vdots & \ddots & \vdots \\ \mathbf{I}\,\epsilon^{k1} & \mathbf{I}\,\epsilon^{k2} & \cdots & \mathbf{I}\,\epsilon^{kk} \end{pmatrix} \tag{26.35}$$

where ϵ^{ij} denotes the ijth element of $\mathbf{E}^{-1}$. Thus, although $\mathbf{R}$ is $(nk) \times (nk)$, its inverse can be computed from the inverse of a much smaller $k \times k$ matrix.

The same argument can be used to obtain the covariance matrix $\mathbf{G}$ of the additive effects and its inverse. Let $\mathbf{C}$ be the $k \times k$ matrix of additive genetic covariances, with $c_{ij} = \sigma_A(i, j)$ being the additive genetic covariance between characters i and j within an individual. The covariance between the additive genetic value of character i in individual l and the additive genetic value of character j in individual m is (from Chapter 21) the additive genetic covariance between characters i and j times twice the coefficient of coancestry $(2\Theta_{lm})$ between l and m. In terms of the relationship matrix, this is $c_{ij} A_{lm}$. Thus, the covariance of $\mathbf{a}_i$ and $\mathbf{a}_j$ is $c_{ij}\,\mathbf{A}$, and the resulting $(nk) \times (nk)$ matrix $\mathbf{G}$ for the total vector $\mathbf{a}^T = (\mathbf{a}_1^T, \cdots, \mathbf{a}_k^T)$ of estimated breeding values becomes

$$\mathbf{G} = \begin{pmatrix} \mathbf{A}\, c_{11} & \mathbf{A}\, c_{12} & \cdots & \mathbf{A}\, c_{1k} \\ \mathbf{A}\, c_{21} & \mathbf{A}\, c_{22} & \cdots & \mathbf{A}\, c_{2k} \\ \vdots & \vdots & \ddots & \vdots \\ \mathbf{A}\, c_{k1} & \mathbf{A}\, c_{k2} & \cdots & \mathbf{A}\, c_{kk} \end{pmatrix} = \mathbf{C} \otimes \mathbf{A} \tag{26.36}$$

where $\mathbf{A}$ (as before) is the $n \times n$ relationship matrix. Hence,

$$\mathbf{G}^{-1} = \begin{pmatrix} \mathbf{A}^{-1} c^{11} & \mathbf{A}^{-1} c^{12} & \cdots & \mathbf{A}^{-1} c^{1k} \\ \mathbf{A}^{-1} c^{21} & \mathbf{A}^{-1} c^{22} & \cdots & \mathbf{A}^{-1} c^{2k} \\ \vdots & \vdots & \ddots & \vdots \\ \mathbf{A}^{-1} c^{k1} & \mathbf{A}^{-1} c^{k2} & \cdots & \mathbf{A}^{-1} c^{kk} \end{pmatrix} = \mathbf{C}^{-1} \otimes \mathbf{A}^{-1} \tag{26.37}$$

where c^{ij} is the ijth element of $\mathbf{C}^{-1}$.

We can now substitute directly into the MMEs using these expressions for $\mathbf{R}^{-1}$ and $\mathbf{G}^{-1}$. Recalling that we assumed $\mathbf{Z} = \mathbf{I}$, Equation 26.5 becomes

$$\begin{pmatrix} \mathbf{X}^T(\mathbf{E}^{-1} \otimes \mathbf{I})\mathbf{X} & \mathbf{X}^T(\mathbf{E}^{-1} \otimes \mathbf{I}) \\ (\mathbf{E}^{-1} \otimes \mathbf{I})\mathbf{X} & (\mathbf{E}^{-1} \otimes \mathbf{I}) + (\mathbf{C}^{-1} \otimes \mathbf{A}^{-1}) \end{pmatrix} \begin{pmatrix} \widehat{\boldsymbol{\beta}} \\ \widehat{\mathbf{a}} \end{pmatrix} = \begin{pmatrix} \mathbf{X}^T(\mathbf{E}^{-1} \otimes \mathbf{I})\mathbf{y} \\ (\mathbf{E}^{-1} \otimes \mathbf{I})\mathbf{y} \end{pmatrix} \tag{26.38}$$

where

$$\mathbf{X} = \begin{pmatrix} \mathbf{X}_1 & 0 & \cdots & 0 \\ 0 & \mathbf{X}_2 & \cdots & 0 \\ \vdots & \vdots & \ddots & \vdots \\ 0 & 0 & \cdots & \mathbf{X}_k \end{pmatrix}, \quad \boldsymbol{\beta} = \begin{pmatrix} \boldsymbol{\beta}_1 \\ \boldsymbol{\beta}_2 \\ \vdots \\ \boldsymbol{\beta}_k \end{pmatrix}, \quad \mathbf{a} = \begin{pmatrix} \mathbf{a}_1 \\ \mathbf{a}_2 \\ \vdots \\ \mathbf{a}_k \end{pmatrix}, \quad \mathbf{y} = \begin{pmatrix} \mathbf{y}_1 \\ \mathbf{y}_2 \\ \vdots \\ \mathbf{y}_k \end{pmatrix}$$

with $\boldsymbol{\beta}$ having dimensionality $(\sum^k q_i) \times 1$ and $\mathbf{a}$ and $\mathbf{y}$ having dimensionality $(nk) \times 1$.

Since the dimensionality of the multivariate MME coefficient matrix is at least $(nk) \times (nk)$, solving the mixed-model equations with more than two or three traits can be rather tortuous. However, multivariate BLUP can be greatly streamlined by constructing a canonical transformation for the characters being considered (Thompson 1977, Ducrocq and Besbes 1993). The idea here is to transform the vector of correlated traits into a new vector of uncorrelated variables. The transformed variables can then be analyzed by standard univariate BLUP analyses, with the breeding values of the transformed variables being subsequently transformed back to the original scale of measurement.

The canonical transformation is accomplished by use of a transformation matrix $\mathbf{Q}$. Letting $\mathbf{y}_i$ be the vector of observations (on the n traits) for individual i, then the transformed variables are given by $\mathbf{y}_i^* = \mathbf{Q}\mathbf{y}_i$. The matrix $\mathbf{Q}$ is chosen to satisfy the conditions that the residual covariance matrix of transformed variables is the identity matrix

$$\mathbf{QEQ}^T = \mathbf{I}$$

and that the covariance matrix of transformed variables is a diagonal matrix

$$\mathbf{QCQ}^T = \mathbf{W}$$

(Anderson 1984, Mrode 1996). Letting $\widehat{\mathbf{a}}_i^*$ be the vector of breeding values estimated on the transformed scale for the *i*th individual, these can then be converted back to the original scale of measurement by

$$\widehat{\mathbf{a}}_i = \mathbf{Q}^{-1}\widehat{\mathbf{a}}_i^*$$

See Árnason (1982) for an example, and Meyer (1985), Schaeffer (1986), Jensen and Mao (1988), Itoh and Iwaisaki (1990), and Mrode (1996) for general treatments.

27

Variance-Component Estimation with Complex Pedigrees

In the numerous forms of analysis of variance (ANOVA) discussed in previous chapters, variance components were estimated by equating observed mean squares to expressions describing their expected values, these being functions of the variance components. ANOVA has the nice feature that the estimators for the variance components are unbiased regardless of whether the data are normally distributed, but it also has two significant limitations. First, field observations often yield records on a variety of relatives, such as offspring, parents, or sibs, that cannot be analyzed jointly with ANOVA. Second, ANOVA estimates of variance components require that sample sizes be well balanced, with the number of observations for each set of conditions being essentially equal. In field situations, individuals are often lost, and even the most carefully crafted balanced design can quickly collapse into an extremely unbalanced one. Although modifications to the ANOVA sums of squares have been proposed to account for unbalanced data (Henderson 1953, Searle et al. 1992), their sampling properties are poorly understood.

Unlike ANOVA estimators, maximum likelihood (ML) and restricted maximum likelihood (REML) estimators do not place any special demands on the design or balance of data. Such estimates are ideal for the unbalanced designs that arise in quantitative genetics, as they can be obtained readily for any arbitrary pedigree of individuals. Since many aspects of ML and REML estimation are quite difficult technically, the detailed mathematics can obscure the general power and flexibility of the methods. Therefore, our main concern is to make the theory more accessible to the nonspecialist, and as a consequence, we are not as thorough in our coverage of the literature as in previous chapters. Also, unlike elsewhere in this book, we occasionally rely upon mathematical machinery (such as matrix derivatives) that is not fully developed here (see Appendix 3 for an introduction). This chapter is mathematically difficult in places, and the reader will do well to review some of the advanced topics in Chapter 8 (such as the multivariate normal and expectations of quadratic products) and Appendix 4.

We start at a relatively elementary level, providing a simple example to show how ML and REML procedures can be used to estimate variance components and how these estimates differ. We then develop the ML and REML equations for variance-component estimation under the general mixed model (introduced in Chapter 26). Extension of these methods to multiple traits, wherein full covariance matrices, rather than single variance components, must be estimated, are then reviewed. We conclude our coverage of ML/REML by examining a number of computational methods for solving the ML/REML equations.

ML/REML methods provide a powerful approach to estimating variance components in populations with complex but known pedigrees. In studies of natural populations, however, the relationships between individuals are often uncertain. We close the book with a brief discussion of a new and conceptually simple procedure that yields estimates of variance components using relatedness estimates indirectly inferred from information on molecular markers. This exciting development is of potentially great utility for the quantitative-genetic analysis of natural populations in undisturbed settings.

While our focus is largely on the estimation of additive genetic and environmental variances, we remind the reader that ML/REML analysis can be applied to a wide variety of issues (as was the case with BLUP), including those involving the estimation of nonadditive genetic variances (Henderson 1985b), mutational variances (Wray 1990), genetic covariances across environments (Platenkamp and Shaw 1992), and maternal and cytoplasmic genetic variances (Southwood et al. 1989).

ML VERSUS REML ESTIMATES OF VARIANCE COMPONENTS

Although algebraically tedious, maximum likelihood (ML) is conceptually very simple. It was introduced to variance component-estimation by Hartley and Rao (1967). For a specified model, such as Equation 26.1, and a specified form for the joint distribution of the elements of $\mathbf{y}$, ML estimates the parameters of the distribution that maximize the likelihood of the observed data. This distribution is almost always assumed to be multivariate normal. An advantage of ML estimators is their efficiency — they simultaneously utilize all of the available data and account for any nonindependence.

One drawback with variance-component estimation via the usual maximum likelihood approach is that all fixed effects are assumed to be known without error. This is rarely true in practice, and as a consequence, ML estimators yield biased estimates of variance components. Most notably (as we show below), estimates of the residual variance tend to be downwardly biased. This bias occurs because the observed deviations of individual phenotypic values from an estimated population mean tend to be smaller than their deviations from the true (parametric) mean. Such bias can become quite large when a model contains numerous fixed effects, particularly when sample sizes are small.

Unlike ML estimators, restricted maximum likelihood (REML) estimators maximize only the portion of the likelihood that does not depend on the fixed effects. In this sense, REML is a *restricted* version of ML. The elimination of bias by REML is analogous to the removal of bias that arises in the estimate of a variance component when the mean squared deviation is divided by the degrees of freedom instead of by the sample size (Chapter 2, and below). REML does not always eliminate all of the bias in parameter estimation, since many methods for obtaining REML estimates cannot return negative estimates of a variance component. However, this source of bias also exists with ML, so REML is clearly the preferred method for analyzing large data sets with complex structure. In the ideal case of a completely balanced design, REML yields estimates of variance components that are identical to those obtained by classical analysis of variance. Since it was first introduced to breeders by Patterson and Thompson (1971), many thorough references to REML, its justification, and its various applications have been published (Harville 1977; Ott 1979; Henderson 1984b, 1986; Gianola and Fernando 1986; Little and Rubin 1987; Robinson 1987; Searle 1987; Shaw 1987; Searle et al. 1992).

A Simple Example of ML versus REML

In an attempt to make the distinction between ML and REML likelihood equations as simple and transparent as possible, we start with a useful pedagogical connection between ML and REML noticed by Foulley (1993), confining our attention to a very simple application — the estimation of the mean and variance of a set of independent observations. In this case, the mixed model reduces to

$$\mathbf{y} = \mathbf{1}\mu + \mathbf{e} \tag{27.1}$$

where μ is the population mean (the fixed effect), $\mathbf{1}$ is a $n \times 1$ column vector of ones (equivalent to the design matrix $\mathbf{X}$ in Equation 26.1), and the covariance matrix of residuals about the mean is assumed to be $\mathbf{R} = \sigma^2\mathbf{I}$.

What are the ML estimates of μ and σ^2 based on the n sampled individuals? Assuming the phenotypes are independent of each other and normally distributed, the probability density of the data $\mathbf{y}$ conditional on the parametric mean and variance is the product of the n univariate normal densities,

$$\begin{aligned} p(\mathbf{y}\,|\,\mu,\sigma^2) &= \prod_{i=1}^{n} p(y_i\,|\,\mu,\sigma^2) \\ &= (2\pi)^{-n/2}(\sigma^2)^{-n/2}\exp\left[-\sum_{i=1}^{n}\frac{(y_i-\mu)^2}{2\sigma^2}\right] \end{aligned} \tag{27.2}$$

where y_i is the phenotypic value of the ith individual. Taking the natural logarithm of the expression on the right, the log-likelihood (Appendix 4) for the observed data set is

$$L(\mathbf{y}\,|\,\mu,\sigma^2) = -\frac{n}{2}\left[\ln(2\pi) + \ln(\sigma^2) + \frac{1}{n\sigma^2}\sum_{i=1}^{n}(\,y_i - \mu\,)^2\right] \tag{27.3a}$$

Although this is the logarithm of the likelihood of the data given the moments of the normal distribution (μ and σ^2), it can also be viewed as the log-likelihood of the parameter estimates, $L(\mu, \sigma^2 \mid \mathbf{y})$, treating the y_i as constants and μ and σ^2 as variables. To obtain estimates of these two distributional parameters, we need at least two observable statistics. Letting

$$\overline{y} = \frac{1}{n}\sum_{i=1}^{n} y_i \qquad \text{and} \qquad V = \frac{1}{n}\sum_{i=1}^{n}(y_i - \overline{y})^2$$

we have

$$\begin{aligned}\sum_{i=1}^{n}(y_i - \mu)^2 &= \sum_{i=1}^{n}(y_i - \overline{y} + \overline{y} - \mu)^2 \\ &= \sum_{i=1}^{n}(y_i - \overline{y})^2 + \sum_{i=1}^{n}(\overline{y} - \mu)^2 + 2(\overline{y} - \mu)\sum_{i=1}^{n}(y_i - \overline{y}) \\ &= n\left[V + (\overline{y} - \mu)^2\right] \end{aligned} \qquad (27.3b)$$

Substituting this final expression into Equation 27.3a, the log-likelihood can be expressed as

$$L(\mu, \sigma^2 \mid \mathbf{y}) = -\frac{n}{2}\left[\ln(2\pi) + \ln(\sigma^2) + \frac{V + (\overline{y} - \mu)^2}{\sigma^2}\right] \qquad (27.3c)$$

Differentiating with respect to μ and σ^2 yields

$$\frac{\partial L(\mu, \sigma^2 | \mathbf{y})}{\partial \mu} = \frac{n(\overline{y} - \mu)}{\sigma^2} \qquad (27.4a)$$

$$\frac{\partial L(\mu, \sigma^2 | \mathbf{y})}{\partial \sigma^2} = -\frac{n}{2\sigma^2}\left[1 - \frac{V + (\overline{y} - \mu)^2}{\sigma^2}\right] \qquad (27.4b)$$

By setting these equations equal to zero and solving, we obtain estimators for the population mean and variance that maximize the likelihood function given the observed data $\mathbf{y}$. From Equation 27.4a, we obtain an estimator for the mean that is completely independent of the variance,

$$\widehat{\mu} = \overline{y} \qquad (27.5a)$$

where $\widehat{\ }$ denotes an estimate. This shows that the standard definition of a sample mean is, in fact, the ML estimate of the parametric value. Unfortunately, the solution to Equation 27.4b,

$$\widehat{\sigma}^2 = V + (\overline{y} - \mu)^2 \qquad (27.5b)$$

is not independent of the estimated mean, $\overline{y}$, unless the estimated mean happens to coincide perfectly with the true mean μ. The maximum likelihood estimator of σ^2 is obtained by assuming that the mean is, in fact, estimated without error, yielding

$$\widehat{\sigma}^2 = V \tag{27.5c}$$

Since the term ignored in Equation 27.5b is necessarily positive, Equation 27.5c gives a downwardly biased estimate of the true variance σ^2.

REML removes this bias by accounting for the error in the estimation of μ. From Equation 27.5b, the expected amount by which $\widehat{\sigma}^2$ underestimates σ^2 is the expected value of $(\overline{y} - \mu)^2$, which is simply the sampling variance of the mean, σ^2/n. Thus, an improved estimator is

$$\widehat{\sigma}^2 = V + E[(\overline{y} - \mu)^2] = V + \frac{\sigma^2}{n} \tag{27.5d}$$

We cannot, of course, know exactly what this bias is because we do not know σ^2 with certainty (indeed, we are trying to estimate it). However, the bias is estimable because we have a preliminary estimate of σ^2, the maximum likelihood estimate V. Thus, starting with the initial estimate of $\widehat{\sigma}^2(0) = V$, a second improved estimate of the variance is

$$\widehat{\sigma}^2(1) = V + \frac{\widehat{\sigma}^2(0)}{n} = V + \frac{V}{n}$$

However, just as this changes the estimate of the variance, it also changes the estimate of $(\overline{y} - \mu)^2$. Hence, a third estimate of σ^2 would be

$$\widehat{\sigma}^2(2) = V + \frac{\widehat{\sigma}^2(1)}{n} = V + \frac{V + (V/n)}{n}$$

This sequence suggests an iterative approach for estimating the variance,

$$\widehat{\sigma}^2(t+1) = V + \frac{\widehat{\sigma}^2(t)}{n} \tag{27.6a}$$

The final (stable) solution to this equation, $\widehat{\sigma}^2$, is obtained by setting $\widehat{\sigma}^2(t+1) = \widehat{\sigma}^2(t)$, yielding

$$\widehat{\sigma}^2 = \frac{n}{n-1}V = \frac{\sum_{i=1}^{n}(y_i - \overline{y})^2}{n-1} \tag{27.6b}$$

which is the unbiased estimator of the variance that we normally use (Chapter 2).

To obtain a solution for this particular example, iteration of Equation 27.6a is not really necessary. However, with models containing multiple fixed effects in the form of the vector $\mathbf{u}$, closed solutions such as Equation 27.6b are not usually possible, particularly in complex pedigree analyses involving unbalanced data.

In those cases, as we will see below, iterative procedures can still yield solutions that are asymptotically unbiased.

Note that the REML estimators given by Equations 27.5a and 27.6b were derived under the assumption of normality. That these same solutions can be acquired without reference to any particular distribution (Chapter 2) provides some evidence that REML estimators may often be fairly robust to violations of the normality assumption.

ML ESTIMATES OF VARIANCE COMPONENTS IN THE GENERAL MIXED MODEL

In light of the fundamental role that the mixed model plays in quantitative genetics, we attempt in this section to give a clear step-by-step development of the maximum likelihood procedures, following the same steps that were used above for the simple model ($\mathbf{y} = \mathbf{1}\,\mu + \mathbf{e}$). Although REML is preferred over ML as a method of analysis, we start with ML, since REML estimation can be expressed as an ML problem by a simple linear transform.

We start with the general mixed model (Equation 26.1), $\mathbf{y} = \mathbf{X}\boldsymbol{\beta} + \mathbf{Z}\mathbf{u} + \mathbf{e}$, and we assume that $\mathbf{u} \sim \text{MVN}(\mathbf{0}, \mathbf{G})$ and $\mathbf{e} \sim \text{MVN}(\mathbf{0}, \mathbf{R})$. Under this model, $\mathbf{y}$ is also multivariate normal, with mean $\mathbf{X}\boldsymbol{\beta}$ and variance-covariance matrix $\mathbf{V} = \mathbf{Z}\mathbf{G}\mathbf{Z}^T + \mathbf{R}$. Recalling the form of the multivariate normal distribution (Equation 8.24), the probability density of the data $\mathbf{y}$, analogous to that in Equation 27.2, is

$$p(\mathbf{y}\,|\,\mathbf{X}\boldsymbol{\beta}, \mathbf{V}) = (2\pi)^{-n/2} |\,\mathbf{V}\,|^{-1/2} \exp\left[-\frac{1}{2}(\mathbf{y} - \mathbf{X}\boldsymbol{\beta})^T \mathbf{V}^{-1} (\mathbf{y} - \mathbf{X}\boldsymbol{\beta})\right] \qquad (27.7a)$$

The next step, analogous to Equation 27.3a, is to take the natural logarithm of the expression on the right of Equation 27.7a. This yields the log-likelihood of $\boldsymbol{\beta}$ and $\mathbf{V}$ given the observed data $(\mathbf{X}, \mathbf{y})$ as

$$L(\boldsymbol{\beta}, \mathbf{V}\,|\,\mathbf{X}, \mathbf{y}) = -\frac{n}{2}\ln(2\pi) - \frac{1}{2}\ln|\,\mathbf{V}\,| - \frac{1}{2}(\mathbf{y} - \mathbf{X}\boldsymbol{\beta})^T \mathbf{V}^{-1} (\mathbf{y} - \mathbf{X}\boldsymbol{\beta}) \qquad (27.7b)$$

The following discussion considers $\mathbf{u} = \mathbf{a}$ to be the vector of additive genetic (breeding) values. The variance components that we are trying to estimate are embedded within $\mathbf{G}$ and $\mathbf{R}$, and we assume that $\mathbf{G} = \sigma^2_A\,\mathbf{A}$, where $\mathbf{A}$ is the additive genetic relationship matrix, and that $\mathbf{R} = \sigma^2_E\,\mathbf{I}$, i.e., the residual deviations of different individuals are independent and homoscedastic.

This approach extends readily to the estimation of additional variance components by using the generalized model

$$\mathbf{y} = \mathbf{X}\boldsymbol{\beta} + \sum_{i=1}^{m} \mathbf{Z}_i \mathbf{u}_i + \mathbf{e} \qquad (27.8a)$$

where the m vectors of random effects ($\mathbf{u}_i$) are assumed to be uncorrelated, with $\mathbf{u}_i \sim \text{MVN}(\mathbf{0}, \sigma_i^2\,\mathbf{B}_i)$ and $\mathbf{B}_i$ being a matrix of known constants. This more general model can incorporate estimates of dominance and other nonadditive variances, and maternal effects variances, to name a few (see Chapter 26). The log-likelihood is still given by Equation 27.7b, but now the covariance matrix $\mathbf{V}$ consists of $m+1$ (unknown) variances,

$$\mathbf{V} = \sum_{i=1}^{m} \sigma_i^2\,\mathbf{Z}_i\,\mathbf{B}_i\,\mathbf{Z}_i^T + \sigma_E^2\,\mathbf{I} \tag{27.8b}$$

We now move on to the partial derivatives of the log-likelihood required for the derivation of the ML estimators. Consider first the derivative with respect to the vector of fixed effects, $\boldsymbol{\beta}$. This derivative involves only the final term of Equation 27.7b, and its procurement is facilitated by using a general result for matrix derivatives. Applying Equation A3.25d,

$$\frac{\partial\,[(\mathbf{y}-\mathbf{X}\boldsymbol{\beta})^T\mathbf{V}^{-1}(\mathbf{y}-\mathbf{X}\boldsymbol{\beta})]}{\partial\,\boldsymbol{\beta}} = -2\mathbf{X}^T\mathbf{V}^{-1}(\mathbf{y}-\mathbf{X}\boldsymbol{\beta}) \tag{27.9}$$

which yields

$$\frac{\partial\,L(\boldsymbol{\beta},\mathbf{V}\,|\,\mathbf{X},\mathbf{y})}{\partial\,\boldsymbol{\beta}} = \mathbf{X}^T\mathbf{V}^{-1}(\mathbf{y}-\mathbf{X}\boldsymbol{\beta}) \tag{27.10}$$

Obtaining the partial derivatives with respect to the variances σ_A^2 and σ_E^2 involves two other general results from matrix theory (Searle 1982, pp. 335–336). If $\mathbf{M}$ is a square matrix whose elements are functions of a scalar variable x, then

$$\frac{\partial\,\ln|\,\mathbf{M}\,|}{\partial\,x} = \text{tr}\left(\mathbf{M}^{-1}\frac{\partial\,\mathbf{M}}{\partial\,x}\right) \tag{27.11a}$$

$$\frac{\partial\,\mathbf{M}^{-1}}{\partial\,x} = -\mathbf{M}^{-1}\frac{\partial\,\mathbf{M}}{\partial\,x}\mathbf{M}^{-1} \tag{27.11b}$$

where tr, the **trace**, denotes the sum of the diagonal elements of a square matrix (Chapter 8). The trace operator appears frequently in this chapter, and the following properties will prove useful

$$\text{tr}\,(a\,\mathbf{A}) = a\,\text{tr}\,(\mathbf{A}) \tag{27.12a}$$

$$\text{tr}\,(\,\mathbf{I}_n\,) = n \tag{27.12b}$$

$$\text{tr}\,(\mathbf{B}_{n\times m}\mathbf{A}_{m\times n}) = \text{tr}\,(\mathbf{A}_{m\times n}\mathbf{B}_{n\times m}) \tag{27.12c}$$

$$\text{tr}\,(\mathbf{A}+\mathbf{C}) = \text{tr}\,(\mathbf{A}) + \text{tr}\,(\mathbf{C}) \tag{27.12d}$$

where $\mathbf{I}_n$ is the $n \times n$ identity matrix.

Recall that prior to the differentiation of Equation 27.3a, we rewrote the sum of squared deviations of observed mean phenotypes from the population mean in terms of $(y_i - \mu)$ and $\overline{y} - \mu$. Performing the analogous changes in matrix form, we find that

$$(\mathbf{y} - \mathbf{X}\boldsymbol{\beta})^T\mathbf{V}^{-1}(\mathbf{y} - \mathbf{X}\boldsymbol{\beta}) = (\mathbf{y} - \mathbf{X}\widehat{\boldsymbol{\beta}})^T\mathbf{V}^{-1}(\mathbf{y} - \mathbf{X}\widehat{\boldsymbol{\beta}}) + (\widehat{\boldsymbol{\beta}} - \boldsymbol{\beta})^T\mathbf{X}^T\mathbf{V}^{-1}\mathbf{X}(\widehat{\boldsymbol{\beta}} - \boldsymbol{\beta}) \tag{27.13}$$

where $\widehat{\boldsymbol{\beta}}$ is the estimate of $\boldsymbol{\beta}$. (This step is not really necessary here, but its incorporation will allow us to see the bias in ML estimates of the variance components, as it did in the previous section.)

Moving now to the derivatives with respect to the variance components, we first assume the simple case of only two unknown variances, typically σ_E^2 and σ_A^2. Writing $\mathbf{V}$ in terms of these two components, we have $\mathbf{V} = \sigma_A^2\,\mathbf{ZAZ}^T + \sigma_E^2\,\mathbf{I}$. Using the notation of σ_i^2 to denote the variance component being estimated, we have

$$\frac{\partial\,\mathbf{V}}{\partial\,\sigma_i^2} = \mathbf{V}_i = \begin{cases} \mathbf{I} & \text{when } \sigma_i^2 = \sigma_E^2 \\ \mathbf{ZAZ}^T & \text{when } \sigma_i^2 = \sigma_A^2 \end{cases} \tag{27.14a}$$

Substituting Equation 27.13 into Equation 27.7b, using Equations 27.11a,b, and letting σ_i^2 denote either σ_A^2 or σ_E^2, we obtain the general equation

$$\frac{\partial L(\boldsymbol{\beta}, \mathbf{V}\,|\,\mathbf{X}, \mathbf{y})}{\partial\sigma_i^2} = -\frac{1}{2}\text{tr}(\mathbf{V}^{-1}\mathbf{V}_i) + \frac{1}{2}(\mathbf{y} - \mathbf{X}\widehat{\boldsymbol{\beta}})^T\mathbf{V}^{-1}\mathbf{V}_i\mathbf{V}^{-1}(\mathbf{y} - \mathbf{X}\widehat{\boldsymbol{\beta}}) + \frac{1}{2}(\widehat{\boldsymbol{\beta}} - \boldsymbol{\beta})^T\mathbf{X}^T\mathbf{V}^{-1}\mathbf{V}_i\mathbf{V}^{-1}\mathbf{X}(\widehat{\boldsymbol{\beta}} - \boldsymbol{\beta}) \tag{27.14b}$$

where $\mathbf{V}_i$ is given by Equation 27.14a. Equations 27.10 and 27.14b are directly analogous to Equations 27.4a,b derived above. Note that $\mathbf{V}_i$ is a fixed matrix of known constants, whereas $\mathbf{V} = \sigma_A^2\,\mathbf{ZAZ}^T + \sigma_E^2\,\mathbf{I}$ is a function of the variance-component estimates. More generally, with m random effects plus a residual error (Equation 27.8a), Equation 27.14b holds for each of the $m+1$ variance components with

$$\frac{\partial\,\mathbf{V}}{\partial\,\sigma_i^2} = \mathbf{V}_i = \begin{cases} \mathbf{I} & \text{when } \sigma_i^2 = \sigma_E^2 \\ \mathbf{Z}_i\mathbf{B}_i\mathbf{Z}_i^T & \text{otherwise} \end{cases} \tag{27.15}$$

The maximum likelihood (ML) estimators are obtained by setting Equations 27.10 and 27.14b equal to zero and solving. Using Equation 27.10 alone, a little rearranging gives the ML estimate of the vector of fixed effects as

$$\widehat{\boldsymbol{\beta}} = (\mathbf{X}^T\widehat{\mathbf{V}}^{-1}\mathbf{X})^{-1}\mathbf{X}^T\widehat{\mathbf{V}}^{-1}\mathbf{y} \tag{27.16}$$

Note that this is the BLUE (best linear unbiased estimator) of $\boldsymbol{\beta}$ obtained in the previous chapter (Equation 26.3). The ML estimators for the variance components are obtained by setting $\widehat{\boldsymbol{\beta}} = \boldsymbol{\beta}$ in Equation 27.14b, rendering the last term equal to zero. Rearranging, we obtain

$$\text{tr}(\widehat{\mathbf{V}}^{-1}\mathbf{V}_i) = (\mathbf{y} - \mathbf{X}\widehat{\boldsymbol{\beta}})^T\widehat{\mathbf{V}}^{-1}\mathbf{V}_i\widehat{\mathbf{V}}^{-1}(\mathbf{y} - \mathbf{X}\widehat{\boldsymbol{\beta}}) \tag{27.17a}$$

This equation can be simplified by using the matrix

$$\mathbf{P} = \mathbf{V}^{-1} - \mathbf{V}^{-1}\mathbf{X}(\mathbf{X}^T\mathbf{V}^{-1}\mathbf{X})^{-1}\mathbf{X}^T\mathbf{V}^{-1} \tag{27.17b}$$

which will appear frequently throughout the rest of the chapter. In particular, we have the very useful result that

$$\mathbf{P}\mathbf{y} = \mathbf{V}^{-1}\mathbf{y} - \mathbf{V}^{-1}\mathbf{X}(\mathbf{X}^T\mathbf{V}^{-1}\mathbf{X})^{-1}\mathbf{X}^T\mathbf{V}^{-1}\mathbf{y} = \mathbf{V}^{-1}(\mathbf{y} - \mathbf{X}\widehat{\boldsymbol{\beta}}) \tag{27.17c}$$

Using this identity, Equation 27.17a can be more compactly written as

$$\text{tr}(\widehat{\mathbf{V}}^{-1}\mathbf{V}_i) = \mathbf{y}^T\widehat{\mathbf{P}}\mathbf{V}_i\widehat{\mathbf{P}}\mathbf{y} \tag{27.17d}$$

where we use the notation $\widehat{\mathbf{P}}$ to remind the reader that $\mathbf{P}$, being a function of $\mathbf{V}$, depends on the variance components that we are trying to estimate. Although it may not be immediately apparent, Equation 27.17d is directly analogous to Equation 27.5c. The variance estimates that we wish to obtain, $\widehat{\sigma}^2_A$ and $\widehat{\sigma}^2_E$, are contained on both sides of Equation 27.17d, embedded in the inverted variance-covariance matrix $\widehat{\mathbf{V}}^{-1}$ that appears in $\mathbf{P}$.

In summary, the ML estimates satisfy the solutions to Equation 27.16 (for the fixed effects) and the set of equations for the variance components (Equation 27.17d). For the additive model assumed above, the two variance equations are

$$\text{tr}(\widehat{\mathbf{V}}^{-1}) = \mathbf{y}^T\widehat{\mathbf{P}}\widehat{\mathbf{P}}\mathbf{y} \qquad \text{for } \sigma^2_E \tag{27.18a}$$

$$\text{tr}(\widehat{\mathbf{V}}^{-1}\mathbf{Z}\mathbf{A}\mathbf{Z}^T) = \mathbf{y}^T\widehat{\mathbf{P}}\mathbf{Z}\mathbf{A}\mathbf{Z}^T\widehat{\mathbf{P}}\mathbf{y} \qquad \text{for } \sigma^2_A \tag{27.18b}$$

More generally, with m random effects plus a residual (Equation 27.8a), the set of $m+1$ ML equations for the variances of random effects is

$$\text{tr}(\widehat{\mathbf{V}}^{-1}) = \mathbf{y}^T\widehat{\mathbf{P}}\widehat{\mathbf{P}}\mathbf{y} \qquad \text{for } \sigma^2_E \tag{27.19a}$$

$$\text{tr}(\widehat{\mathbf{V}}^{-1}\mathbf{Z}_i\mathbf{B}_i\mathbf{Z}_i^T) = \mathbf{y}^T\widehat{\mathbf{P}}\mathbf{Z}_i\mathbf{B}_i\mathbf{Z}_i^T\widehat{\mathbf{P}}\mathbf{y} \qquad \text{for } \sigma^2_i,\ 1 \le i \le m \tag{27.19b}$$

where $\widehat{\mathbf{P}}$ now uses

$$\widehat{\mathbf{V}} = \sum_{i=1}^{m} \widehat{\sigma}^2_i\, \mathbf{Z}_i\, \mathbf{B}_i\, \mathbf{Z}_i^T + \widehat{\sigma}^2_E\,\mathbf{I} \tag{27.19c}$$

These solutions have two troublesome properties. First, unlike our simple example at the start of this chapter where there was a closed form estimator for the fixed effect μ, the ML vector of fixed effects $\widehat{\boldsymbol{\beta}}$ is a function of the variance-covariance matrix $\widehat{\mathbf{V}}$, which in turn contains the variance components that we wish to estimate. Second, because these solutions involve the inverse of $\widehat{\mathbf{V}}$, they are nonlinear functions of the variance components. As a consequence, there is no simple one-step solution. ML estimation of $\boldsymbol{\beta}$, σ_A^2, and σ_E^2 requires an iterative procedure, several steps of which are described below.

Example 1. Consider the simple animal model, $\mathbf{y} = \mathbf{X}\boldsymbol{\beta} + \mathbf{a} + \mathbf{e}$, where there is only one observation per individual ($\mathbf{Z} = \mathbf{I}$), and we assume $\mathbf{a} \sim \text{MVN}(\mathbf{0}, \sigma_A^2\,\mathbf{A})$ and $\mathbf{e} \sim \text{MVN}(\mathbf{0}, \sigma_E^2\,\mathbf{I})$. In this case, the ML equations become

$$\widehat{\boldsymbol{\beta}} = (\mathbf{X}^T\widehat{\mathbf{V}}^{-1}\mathbf{X})^{-1}\mathbf{X}^T\widehat{\mathbf{V}}^{-1}\mathbf{y}$$

$$\text{tr}(\widehat{\mathbf{V}}^{-1}) = \mathbf{y}^T\widehat{\mathbf{P}}\widehat{\mathbf{P}}\mathbf{y}$$

$$\text{tr}(\widehat{\mathbf{V}}^{-1}\mathbf{A}) = \mathbf{y}^T\widehat{\mathbf{P}}\mathbf{A}\widehat{\mathbf{P}}\mathbf{y}$$

where

$$\widehat{\mathbf{V}} = \widehat{\sigma}_A^2\,\mathbf{A} + \widehat{\sigma}_E^2\,\mathbf{I}$$

and $\widehat{\mathbf{P}}$ is obtained by substituting $\widehat{\mathbf{V}}$ into Equation 27.17b.

If we further allow for dominance, the model becomes modified to $\mathbf{y} = \mathbf{X}\boldsymbol{\beta} + \mathbf{a} + \mathbf{d} + \mathbf{e}$. Assuming $\mathbf{a} \sim \text{MVN}(\mathbf{0}, \sigma_A^2\,\mathbf{A})$, $\mathbf{d} \sim \text{MVN}(\mathbf{0}, \sigma_D^2\,\mathbf{D})$, and $\mathbf{e} \sim \text{MVN}(\mathbf{0}, \sigma_E^2\,\mathbf{I})$, the ML equations now become

$$\widehat{\boldsymbol{\beta}} = (\mathbf{X}^T\widehat{\mathbf{V}}^{-1}\mathbf{X})^{-1}\mathbf{X}^T\widehat{\mathbf{V}}^{-1}\mathbf{y}$$

$$\text{tr}(\widehat{\mathbf{V}}^{-1}) = \mathbf{y}^T\widehat{\mathbf{P}}\widehat{\mathbf{P}}\mathbf{y}$$

$$\text{tr}(\widehat{\mathbf{V}}^{-1}\mathbf{A}) = \mathbf{y}^T\widehat{\mathbf{P}}\mathbf{A}\widehat{\mathbf{P}}\mathbf{y}$$

$$\text{tr}(\widehat{\mathbf{V}}^{-1}\mathbf{D}) = \mathbf{y}^T\widehat{\mathbf{P}}\mathbf{D}\widehat{\mathbf{P}}\mathbf{y}$$

where $\widehat{\mathbf{P}}$ is a function of

$$\widehat{\mathbf{V}} = \widehat{\sigma}_A^2\,\mathbf{A} + \widehat{\sigma}_D^2\,\mathbf{D} + \widehat{\sigma}_E^2\,\mathbf{I}$$

Standard Errors of ML Estimates

Recall from the theory of maximum likelihood (Appendix 4) that standard errors of ML estimates can be obtained from the appropriate elements of the inverse of the Fisher information matrix ($\mathbf{F}$) involving the vector of parameters being estimated ($\boldsymbol{\Theta}$). The elements of $\mathbf{F}$ are functions of the second derivatives of the log-likelihood function, evaluated by substituting ML estimates of the parameters,

$$\mathbf{F}_{ij} = -E\left(\frac{\partial^2 L}{\partial\,\theta_i\,\partial\,\theta_j}\right) \simeq -\left.\frac{\partial^2 L}{\partial\,\theta_i\,\partial\,\theta_j}\right|_{\boldsymbol{\Theta} = \widehat{\boldsymbol{\Theta}}} \tag{27.20}$$

The sampling variance of the ML estimate of the parameter θ_i is approximated by F_{ii}^{-1} (the ith diagonal element of $\mathbf{F}^{-1}$), while the sampling covariance between the ML estimates of θ_i and θ_j is approximated by F_{ij}^{-1}.

Computing the partials for the mixed model gives the information matrix for the ML estimates of $\boldsymbol{\beta}$ and $\boldsymbol{\sigma}^2$ (the vector of variance-component estimates) as

$$\mathbf{F} = \begin{pmatrix} \mathbf{X}^T\mathbf{V}^{-1}\mathbf{X} & \mathbf{0} \\ \mathbf{0} & \mathbf{S} \end{pmatrix} \tag{27.21}$$

where

$$S_{ij} = \frac{1}{2}\,\mathrm{tr}\left(\mathbf{V}^{-1}\mathbf{V}_i\mathbf{V}^{-1}\mathbf{V}_j\right) \tag{27.22}$$

with $\mathbf{V}_i$ given by Equation 27.15 (Searle et al. 1992). Inverting gives

$$\mathbf{F}^{-1} = \begin{pmatrix} \left(\mathbf{X}^T\mathbf{V}^{-1}\mathbf{X}\right)^{-1} & \mathbf{0} \\ \mathbf{0} & \mathbf{S}^{-1} \end{pmatrix} \tag{27.23}$$

Hence,

$$\sigma(\beta_i, \beta_j) = \left(\mathbf{X}^T\mathbf{V}^{-1}\mathbf{X}\right)^{-1}_{ij}, \qquad \sigma(\sigma_i^2, \sigma_j^2) = \left(\mathbf{S}^{-1}\right)_{ij} \tag{27.24}$$

The ML estimates for fixed effects are uncorrelated with those for variance components, i.e., $\sigma(\beta_i, \sigma_j^2) = 0$.

Example 2. For the simple model with dominance (Example 1), the Fisher information submatrix $\mathbf{S}$ dealing with the ML variance estimates $(\sigma_A^2, \sigma_D^2, \sigma_E^2)$ is

$$\mathbf{S} = \frac{1}{2}\begin{pmatrix} \mathrm{tr}(\mathbf{V}^{-1}\mathbf{A}\mathbf{V}^{-1}\mathbf{A}) & \mathrm{tr}(\mathbf{V}^{-1}\mathbf{A}\mathbf{V}^{-1}\mathbf{D}) & \mathrm{tr}(\mathbf{V}^{-1}\mathbf{A}\mathbf{V}^{-1}) \\ \mathrm{tr}(\mathbf{V}^{-1}\mathbf{A}\mathbf{V}^{-1}\mathbf{D}) & \mathrm{tr}(\mathbf{V}^{-1}\mathbf{D}\mathbf{V}^{-1}\mathbf{D}) & \mathrm{tr}(\mathbf{V}^{-1}\mathbf{D}\mathbf{V}^{-1}) \\ \mathrm{tr}(\mathbf{V}^{-1}\mathbf{A}\mathbf{V}^{-1}) & \mathrm{tr}(\mathbf{V}^{-1}\mathbf{D}\mathbf{V}^{-1}) & \mathrm{tr}(\mathbf{V}^{-1}\mathbf{V}^{-1}) \end{pmatrix}$$

where $\mathbf{V}$ is as given in Example 1.

RESTRICTED MAXIMUM LIKELIHOOD

REML is based on a linear transformation of the observation vector $\mathbf{y}$ that removes the fixed effects from the model. The simplest way to see how this is done is to imagine a transformation matrix $\mathbf{K}$ associated with the design matrix $\mathbf{X}$ for the model under consideration such that

$$\mathbf{KX} = \mathbf{0} \tag{27.25}$$

Applying this transformation matrix to the mixed model yields

$$\begin{aligned} \mathbf{y}^* = \mathbf{Ky} &= \mathbf{K}(\mathbf{X}\boldsymbol{\beta} + \mathbf{Za} + \mathbf{e}) \\ &= \mathbf{KZa} + \mathbf{Ke} \end{aligned} \tag{27.26a}$$

The linear contrasts $\mathbf{y}^*$ are equivalent to residual deviations from the estimated fixed effects, akin to using $y_i^* = y_i - \overline{y}$ in the introductory example used at the start of this chapter. REML estimates of variance components are equivalent to ML estimates of the transformed variables. Thus, we can use the ML solutions outlined above by making the following substitutions:

$$\mathbf{Ky} \text{ for } \mathbf{y}, \qquad \mathbf{KX} = \mathbf{0} \text{ for } \mathbf{X}, \qquad \mathbf{KZ} \text{ for } \mathbf{Z}, \qquad \mathbf{KVK}^T \text{ for } \mathbf{V} \tag{27.26b}$$

While REML appears to require the additional task of finding a matrix $\mathbf{K}$ that satisfies Equation 27.25, the REML equations can actually be expressed directly in terms of $\mathbf{V}$, $\mathbf{y}$, and $\mathbf{P}$. This result follows from the very useful identity, proven in Searle et al. (1992), that $\mathbf{K}$ satisfies

$$\mathbf{P} = \mathbf{K}^T(\mathbf{KVK}^T)^{-1}\mathbf{K} \tag{27.27a}$$

Noting that

$$(\mathbf{y}^*)^T(\mathbf{V}^*)^{-1}\mathbf{y}^* = (\mathbf{y}^T\mathbf{K}^T)(\mathbf{KVK}^T)^{-1}(\mathbf{Ky}) = \mathbf{y}^T\mathbf{Py} \tag{27.27b}$$

and substituting the expressions given as 27.26b into Equation 27.17a, after some rearrangement, the ML equations yield the REML estimators,

$$\text{tr}(\widehat{\mathbf{P}}) = \mathbf{y}^T\widehat{\mathbf{P}}\widehat{\mathbf{P}}\mathbf{y} \qquad \text{for } \sigma_E^2 \tag{27.28a}$$

$$\text{tr}(\widehat{\mathbf{P}}\mathbf{ZAZ}^T) = \mathbf{y}^T\widehat{\mathbf{P}}\mathbf{ZAZ}^T\widehat{\mathbf{P}}\mathbf{y} \qquad \text{for } \sigma_A^2 \tag{27.28b}$$

Note that REML does not return estimates of $\boldsymbol{\beta}$, since the fixed effects are removed by setting $\boldsymbol{\beta}^* = \mathbf{0}$.

Since the transformation $\mathbf{y}^* = \mathbf{Ky}$ satisfying Equation 27.25 solely depends on the design matrix, this general approach still holds with m uncorrelated random vectors. In this case, Equation 27.8a expands to

$$\mathbf{y}^* = \sum_{i=1}^{m} \mathbf{K}\,\mathbf{Z}_i\,\mathbf{u}_i + \mathbf{Ke} \tag{27.29}$$

and the REML equations for the $m+1$ variance components become

$$\text{tr}(\widehat{\mathbf{P}}) = \mathbf{y}^T\widehat{\mathbf{P}}\widehat{\mathbf{P}}\mathbf{y} \qquad \text{for } \sigma_E^2 \tag{27.30a}$$

$$\text{tr}(\widehat{\mathbf{P}}\,\mathbf{Z}_i\,\mathbf{B}_i\,\mathbf{Z}_i^T) = \mathbf{y}^T\,\widehat{\mathbf{P}}\,\mathbf{Z}_i\,\mathbf{B}_i\,\mathbf{Z}_i^T\,\widehat{\mathbf{P}}\mathbf{y} \qquad \text{for } \sigma_i^2,\ 1 \le i \le m \tag{27.30b}$$

where $\widehat{\mathbf{P}}$ is now a function of $\widehat{\mathbf{V}} = \sum \widehat{\sigma}_i^2\,\mathbf{Z}_i\,\mathbf{B}_i\,\mathbf{Z}_i^T + \widehat{\sigma}_E^2\,\mathbf{I}$.

With REML, the information matrix contains only items corresponding to variance-component estimates, so $\mathbf{F} = \mathbf{S}$, where

$$S_{ij} = \frac{1}{2}\,\text{tr}(\,\mathbf{P}\mathbf{V}_i\mathbf{P}\mathbf{V}_j\,) \tag{27.31a}$$

with $\mathbf{V}_i$ given by Equation 27.15. Estimates of the sampling variances and covariances of the variance-component estimates are obtained from the inverse of the matrix $\mathbf{S}$, as described above.

Example 3. The REML variance-component estimates for the single-records dominance model of Example 2, $\mathbf{y} = \mathbf{X}\boldsymbol{\beta} + \mathbf{a} + \mathbf{d} + \mathbf{e}$, satisfy

$$\begin{aligned} \text{tr}(\widehat{\mathbf{P}}) &= \mathbf{y}^T\widehat{\mathbf{P}}\widehat{\mathbf{P}}\mathbf{y} && \text{for } \sigma_E^2 \\ \text{tr}(\widehat{\mathbf{P}}\mathbf{A}) &= \mathbf{y}^T\widehat{\mathbf{P}}\mathbf{A}\widehat{\mathbf{P}}\mathbf{y} && \text{for } \sigma_A^2 \\ \text{tr}(\widehat{\mathbf{P}}\mathbf{D}) &= \mathbf{y}^T\widehat{\mathbf{P}}\mathbf{D}\widehat{\mathbf{P}}\mathbf{y} && \text{for } \sigma_D^2 \end{aligned}$$

where $\widehat{\mathbf{P}}$ is defined as in Equation 27.17b with $\mathbf{V} = \widehat{\sigma}_A^2\,\mathbf{A} + \widehat{\sigma}_D^2\,\mathbf{D} + \widehat{\sigma}_E^2\,\mathbf{I}$. For purposes of estimating sampling variances and covariances of these estimates, the information matrix is given by

$$\mathbf{S} = \frac{1}{2}\begin{pmatrix} \text{tr}(\,\mathbf{PAPA}\,) & \text{tr}(\,\mathbf{PAPD}\,) & \text{tr}(\,\mathbf{PAP}\,) \\ \text{tr}(\,\mathbf{PAPD}\,) & \text{tr}(\,\mathbf{PDPD}\,) & \text{tr}(\,\mathbf{PDP}\,) \\ \text{tr}(\,\mathbf{PAP}\,) & \text{tr}(\,\mathbf{PDP}\,) & \text{tr}(\,\mathbf{PP}\,) \end{pmatrix}$$

When the estimate of $\mathbf{P}$ is inserted into this matrix, the standard errors of the variance-component estimates are obtained as the square roots of the diagonal elements of $\mathbf{S}^{-1}$, and the covariance between $\widehat{\sigma}_i^2$ and $\widehat{\sigma}_j^2$ is given by S_{ij}^{-1}.

Multivariate Analysis

When multiple characters are measured in individuals, the most complete analysis includes all characters, even if our interest is only in a subset of them. Aside from the information provided on the genetic covariances among traits, multivariate analysis can improve the accuracy of variance-component estimates for single traits for the simple reason that correlated characters provide information about each other. Balancing these advantages are computational difficulties that increase with the number of characters under consideration.

Extension of univariate REML to the analysis of multiple traits is straightforward. Suppose n characters are of interest. From our development of multivariate BLUP in the previous chapter, the variance components for the simple additive model now become the elements of the $n \times n$ covariance matrices $\mathbf{C}$ and $\mathbf{E}$ of additive genetic and environmental effects, whose elements are, respectively, denoted by $\sigma_A(i,j)$ and $\sigma_E(i,j)$. Thus, instead of two variances, we are now faced with the estimation of $n(n+1)$ variance-covariance elements. From Chapter 26, for the simple animal model $\mathbf{G} = \mathbf{C} \otimes \mathbf{A}$ and $\mathbf{R} = \mathbf{E} \otimes \mathbf{I}$, where $\otimes$ denotes the Kronecker product, and $\mathbf{A}$ is the relationship matrix. The log-likelihood function for multiple characters is given by Equation 27.7b, with the design matrix $\mathbf{X}$ now constructed as in Equation 26.31, and

$$\mathbf{V} = \mathbf{G} + \mathbf{R} = \mathbf{C} \otimes \mathbf{A} + \mathbf{E} \otimes \mathbf{I}$$

While the set of $n(n+1)$ REML equations can be obtained by differentiating the log-likelihood with respect to each variance / covariance component, the computational demands of multivariate analysis increase rapidly with the number of characters. However, as we saw for multivariate BLUP analysis in Chapter 26, canonical transformation provides an elegant way of reducing an n-dimensional multivariate analysis to n one-dimensional analyses. An additional complexity that arises when this approach is applied to REML is that the transformation matrix $\mathbf{Q}$ is a function of $\mathbf{C}$ and $\mathbf{E}$, the matrices that we are trying to estimate. As will be seen below, this can generally be accommodated by the iterative procedures that are routinely employed in REML analysis, by starting with some initial estimate of $\mathbf{Q}$, computing $\mathbf{C}$ and $\mathbf{E}$ by univariate analyses based on this $\mathbf{Q}$, and then using these new estimates to compute a new $\mathbf{Q}$. Further details can be found in Meyer (1985), Schaeffer (1986), Taylor et al. (1985), Jensen and Mao (1988), and Thompson and Hill (1990).

ML/REML Estimation in Populations under Selection

Selection changes the additive genetic variance of a character by generating gametic phase disequilibria and changing allele frequencies. Thus, genetic variance estimates generated from selected individuals can be quite different from those in the unselected base population from which they descended. Under the infinitesimal model (which assumes the character to be determined by a very large number of loci, each with small effect), changes in the additive genetic variance are entirely due to gametic phase disequilibria. Thus, under this model, once selection is stopped, recombination decays the disequilibria away, returning the additive genetic variance to the level present before selection. Hence, to the extent that this model is approximated in nature, interest is usually in the additive genetic variance in the unselected base population rather than that observed within a set of selected individuals.

Under certain experimental settings, REML variance-component estimates have the unique feature of being uninfluenced by selection. In particular, if the base population consists of unrelated, unselected, and noninbred individuals and phenotypic data are available for all selected and unselected individuals, then REML yields essentially unbiased estimates of the additive genetic variance in the base population (Henderson 1949, Henderson et al. 1959, Curnow 1961, Thompson 1973, Rothschild et al. 1979, Sorensen and Kennedy 1984, Gianola and Fernando 1986, Gianola et al. 1988, Fernando and Gianola 1990). On the other hand, van der Werf and colleagues (van der Werf 1990, van der Werf and de Boer 1990, van der Werf and Thompson 1992) show that when the base population consists of previously selected individuals, REML provides no protection from biased estimates of the additive genetic variance in the population prior to selection, even if the entire pedigree of individuals back to the base population is included. Likewise, if selection acts on a suite of unmeasured characters that are correlated with characters included in the model, REML can generate biased estimates of the variances and covariances of the measured characters (Schaeffer and Song 1978).

Given that the conditions under which REML yields unbiased estimates are likely to be violated in most natural populations, why should such estimates be used? One reason is that, even though imperfect, likelihood methods always at least partially account for biases introduced by selection, in part because the additive genetic relationship matrix $\mathbf{A}$ corrects for the pattern of flow of genetic information from generation to generation (Sorensen and Kennedy 1984). Other variance-component estimators, such as those derived from ANOVA, make no such correction.

SOLVING THE ML/REML EQUATIONS

Because the equations for the ML/REML solutions are highly nonlinear, closed analytical solutions are only available in very special cases (e.g., certain completely balanced designs). In principle, the solutions can be obtained by performing an exhaustive grid search — computing the log-likelihood of the data at each point on a grid covering the entire range of parameter space, and letting the solution be defined by the point on the grid giving the largest log-likelihood. However, this procedure is impractical under ML if $\boldsymbol{\beta}$ contains more than a few elements, since each element of β adds to the dimensionality of the search. Under REML, the dimensionality of parameter space can be greatly reduced, but the likelihood function is considerably more complicated to compute. Thus, simple grid searches are rarely used by themselves, although they are sometimes used in conjunction with other methods that restrict the search to one or a few dimensions.

A wide variety of iterative techniques for solving ML / REML equations have been proposed based on various modifications of two basic approaches: the Newton-Raphson algorithm and the EM algorithm. Both procedures start with preliminary estimates of the parameters (obtained, for example, by ordinary least-squares analysis), and using information on the slope of the likelihood surface, these estimates are then moved in a direction that increases the log-likelihood of the data. The revised estimates are subsequently modified in an iterative fashion, until a satisfactory degree of convergence on a final set of estimates has been achieved. With these types of approaches, the search for ML / REML solutions avoids spending huge amounts of computational time in regions of low likelihood. Such hill-climbing methods are not guaranteed to converge on the global maximum of the likelihood function, but potential problems with secondary peaks in the likelihood surface can be investigated through the use of different starting values.

Our review of numerical methods for obtaining solutions to the ML / REML equations is intentionally brief, focusing only on the general principles. All of the methods are very intensive computationally when large pedigrees are involved, as they usually require the inversion of large matrices at each step. Detailed reviews of this highly technical area appear in Meyer (1989b), Harville and Callanan (1990), and Searle et al. (1992).

Derivative-based Methods

The Newton-Raphson (NR) algorithm, a standard method for numerically solving coupled sets of nonlinear equations, has been used extensively to solve ML / REML equations (Harville 1977, Jennrich and Sampson 1976, Searle et al. 1992). Specific applications to genetic variance-component estimation include Lange et al. (1977) for ML estimates of additive and dominance variances for single characters and Meyer (1983, 1985) for REML estimates of the additive genetic covariance matrix for multiple characters. We confine our discussion of Newton-Raphson iteration to REML estimates, as applications to ML follow in a similar fashion.

The Newton-Raphson method obtains the REML estimate of the vector of parameters Θ by starting with some initial value $\Theta^{(0)}$ and then iterating to convergence to a final solution by using

$$\Theta^{(k+1)} = \Theta^{(k)} - \left(\mathbf{H}^{(k)}\right)^{-1} \left.\frac{\partial L}{\partial \Theta}\right|_{\Theta^{(k)}} \tag{27.32}$$

where $\partial L / \partial \Theta$ is a column vector of the partials of the log-likelihood function with respect to each parameter evaluated at the estimate $\Theta^{(k)}$, and $\mathbf{H}$ is the **Hessian**

matrix of all second-order partial derivatives of the log-likelihood L with respect to the variance components. $\mathbf{H}^{-1}$ and $\partial L/\partial \boldsymbol{\Theta}$ respectively provide measures of the curvature and the slope (and directionality) of the likelihood surface, given the current estimates. Their product gives a projected degree of movement of the vector $\boldsymbol{\Theta}$ towards an improved set of values to be used in the next iteration.

Consider again the mixed model with m random factors plus a residual,

$$\mathbf{y} = \mathbf{X}\boldsymbol{\beta} + \sum_{i=1}^{m} \mathbf{Z}_i \mathbf{u}_i + \mathbf{e}$$

where $\mathbf{u}_i \sim \text{MVN}(\mathbf{0}, \sigma_i^2\, \mathbf{B}_i)$ for $1 \leq i \leq m$, and $\mathbf{B}_i$ is a square symmetric $n_i \times n_i$ matrix of known constants. The residuals are also assumed to be multivariate normal with $\mathbf{e} \sim \text{MVN}(\mathbf{0}, \sigma_E^2\, \mathbf{I})$. Since $\mathbf{y}$ is the sum of multivariate normals, it is also multivariate normal with $\mathbf{y} \sim \text{MVN}(\mathbf{X}\boldsymbol{\beta}, \mathbf{V})$, where

$$\mathbf{V} = \sum_{i=1}^{m} \sigma_i^2\, \mathbf{Z}_i\, \mathbf{B}_i\, \mathbf{Z}_i^T + \sigma_E^2\, \mathbf{I}$$

Under REML, $\boldsymbol{\Theta} = (\sigma_1^2,\, \sigma_2^2,\, \cdots,\, \sigma_E^2)^T$ and Equations 27.14b and 27.27b give the elements of $\partial L/\partial \boldsymbol{\Theta}$ as

$$\left.\frac{\partial L}{\partial \sigma_i^2}\right|_{\boldsymbol{\Theta}^{(k)}} = -\frac{1}{2}\,\text{tr}\left(\mathbf{P}^{(k)}\mathbf{V}_i\right) + \frac{1}{2}\,\mathbf{y}^T\mathbf{P}^{(k)}\mathbf{V}_i\mathbf{P}^{(k)}\mathbf{y} \tag{27.33}$$

where $\mathbf{V}_i$ is given by Equation 27.15 and $\mathbf{P}^{(k)}$ is calculated from Equation 27.17b using the current variance-component estimates in $\boldsymbol{\Theta}^{(k)}$. Searle et al. (1992) give the elements of $\mathbf{H}$ for REML as

$$\mathbf{H}_{ij}^{(k)} = \frac{\partial^2 L}{\partial \sigma_i^2\, \partial \sigma_j^2} = \frac{1}{2}\,\text{tr}\left(\mathbf{P}^{(k)}\,\mathbf{V}_i\,\mathbf{P}^{(k)}\,\mathbf{V}_j\right) - \mathbf{y}^T\mathbf{P}^{(k)}\,\mathbf{V}_i\,\mathbf{P}^{(k)}\,\mathbf{V}_j\,\mathbf{P}^{(k)}\,\mathbf{y} \tag{27.34}$$

where again the partials are evaluated using $\boldsymbol{\Theta}^{(k)}$.

A common variant of the Newton-Raphson algorithm is **Fisher's scoring method**, which replaces the inverse of the Hessian matrix in Equation 27.32 by its expected value, which after allowing for a change in sign, turns out to be defined by the inverse of Fisher's information matrix, $-\mathbf{F}^{-1}$ (Equation 27.20). This reduces the iterative equation to

$$\boldsymbol{\Theta}^{(k+1)} = \boldsymbol{\Theta}^{(k)} + \left(\mathbf{F}^{(k)}\right)^{-1} \left.\frac{\partial L}{\partial \boldsymbol{\Theta}}\right|_{\boldsymbol{\Theta}^{(k)}} \tag{27.35a}$$

with

$$F_{ij}^{(k)} = \frac{1}{2}\,\text{tr}(\,\mathbf{P}^{(k)}\,\mathbf{V}_i\,\mathbf{P}^{(k)}\,\mathbf{V}_j\,) \tag{27.35b}$$

There are several motivations for employing this modification. First, as noted above, the inverse of the information matrix, when evaluated at the REML values, estimates the standard errors for these estimates. Second, $\mathbf{F}$ is easier to compute than $\mathbf{H}^{-1}$ (compare Equations 27.34 and 27.35b). Finally, Fisher's scoring method appears to be slightly more robust to initial values than strict Newton-Raphson iteration (Jennrich and Sampson 1976).

Example 4. Again consider the simple animal model with a single observation per individual, $\mathbf{y} = \mathbf{X}\boldsymbol{\beta} + \mathbf{a} + \mathbf{e}$. For REML estimates, letting

$$\boldsymbol{\Theta}^{(k)} = \begin{pmatrix} (\sigma_A^2)^{(k)} \\ (\sigma_E^2)^{(k)} \end{pmatrix} \quad \text{gives} \quad \frac{\partial L}{\partial \boldsymbol{\Theta}} = \frac{1}{2} \begin{pmatrix} -\,\text{tr}(\,\mathbf{P}\,) + \mathbf{y}^T\mathbf{P}\mathbf{P}\mathbf{y} \\ -\,\text{tr}(\,\mathbf{P}\mathbf{A}\,) + \mathbf{y}^T\mathbf{P}\mathbf{A}\mathbf{P}\mathbf{y} \end{pmatrix}$$

Note that $\mathbf{P}$ is a function of the current variance-component estimates, with

$$\mathbf{P} = (\mathbf{V}^{-1})^{(k)} - (\mathbf{V}^{-1})^{(k)}\mathbf{X}(\mathbf{X}^T(\mathbf{V}^{-1})^{(k)}\mathbf{X})^{-1}\mathbf{X}^T(\mathbf{V}^{-1})^{(k)}$$

where

$$\mathbf{V}^{(k)} = (\sigma_A^2)^{(k)}\,\mathbf{A} + (\sigma_E^2)^{(k)}\,\mathbf{I}$$

with $\mathbf{A}$ being the relationship matrix for the inviduals being measured. Likewise, from Equation 27.34 the Hessian matrix $\mathbf{H}$ is given by

$$\left.\frac{\partial^2 L}{\partial \boldsymbol{\Theta}^2}\right|_{\boldsymbol{\Theta}^{(k)}} = \frac{1}{2} \begin{pmatrix} \text{tr}(\,\mathbf{PP}) - 2\mathbf{y}^T\mathbf{PPPy} & \text{tr}(\,\mathbf{PAP}\,) - 2\mathbf{y}^T\mathbf{PAPPy} \\ \text{tr}(\,\mathbf{PAP}\,) - 2\mathbf{y}^T\mathbf{PAPPy} & \text{tr}(\,\mathbf{PAPA}\,) - 2\mathbf{y}^T\mathbf{PAPAPy} \end{pmatrix}$$

and the Fisher information matrix by

$$\mathbf{F} = -E\left(\frac{\partial^2 L}{\partial \boldsymbol{\Theta}^2}\right) = \frac{1}{2} \begin{pmatrix} \text{tr}(\,\mathbf{PP}) & \text{tr}(\,\mathbf{PAP}\,) \\ \text{tr}(\,\mathbf{PPA}\,) & \text{tr}(\,\mathbf{PAPA}\,) \end{pmatrix}$$

Note that $\mathbf{P}$ is really indexed by k since it depends on the current estimates of the unknown variance components, $\widehat{\sigma}_A^2$ and $\widehat{\sigma}_E^2$.

EM Methods

The idea behind the EM (**expectation/maximization**) algorithm for variance-component analysis is that if we knew the values of the random effects, we could estimate the variances in a simple fashion directly from them. Focusing on the general mixed model defined by Equation 27.8a, the variances of the random and residual effects are defined respectively to be

$$\sigma_i^2 = \frac{E[\,\mathbf{u}_i^T \mathbf{B}_i^{-1} \mathbf{u}_i\,]}{n_i} \tag{27.36a}$$

$$\sigma_E^2 = \frac{E[\,\mathbf{e}_i^T \mathbf{e}_i\,]}{n} \tag{27.36b}$$

where n and n_i are, respectively, the number of elements in $\mathbf{e}$ and $\mathbf{u}_i$. Equation 27.36a follows from Equation 8.22, which, since $E[\mathbf{u}_i] = \mathbf{0}$, reduces to

$$E[\,\mathbf{u}_i^T \mathbf{B}_i^{-1} \mathbf{u}_i\,] = \text{tr}(\mathbf{B}_i^{-1} \sigma_i^2 \mathbf{B}_i) = \sigma_i^2 \text{tr}(\mathbf{I}_{n_i}) = n_i\, \sigma_i^2$$

The last two steps follow from Equations 27.12a and 27.12b, respectively. Equation 27.36b follows in a similar fashion. In actuality, of course, we only know $\mathbf{y}$, not the underlying vectors of random effects $(\mathbf{u}_i)$ or residual deviations $(\mathbf{e})$.

Underlying the EM algorithm is the idea, discussed in Chapter 26, that the information in $\mathbf{y}$ provides a basis for making predictions about the elements of $\mathbf{u}_i$ and $\mathbf{e}$. In the context of variance-component analysis, we need to go a step beyond BLUP estimation of $\mathbf{u}_i$ and $\mathbf{e}$, as it is actually the quadratic products of $\mathbf{u}_i$ and $\mathbf{e}$ in the numerators of Equations 27.36a,b that we need to predict. Here, in the interest of clarity and space, we skip over a number of steps to the final solution (see Searle et al. 1992, pp. 297–304 for a complete derivation). Searle et al. show that the conditional distribution of $\mathbf{u}$ given the observed $\mathbf{y}$ is MVN, with

$$E[\mathbf{u}_i \,|\, \mathbf{y}] = \sigma_i^2\, \mathbf{Z}_i^T \mathbf{V}^{-1} (\mathbf{y} - \mathbf{X}\boldsymbol{\beta}) = \sigma_i^2\, \mathbf{Z}_i^T \mathbf{P} \mathbf{y}$$

and

$$\sigma^2(\mathbf{u}_i \,|\, \mathbf{y}) = \sigma_i^2\, \mathbf{I}_{n_i} - \sigma_i^4\, \mathbf{Z}_i^T \mathbf{V}^{-1} \mathbf{Z}_i$$

Substituting into Equation 8.22, after some simplication, the expectation of the quadratic product in Equation 27.36a, conditional on the observed $\mathbf{y}$, becomes

$$E[\,\mathbf{u}_i^T \mathbf{B}_i^{-1} \mathbf{u}_i \,|\, \mathbf{y}\,] = n_i \sigma_i^2 + \sigma_i^4\, [\,\mathbf{y}^T\, \mathbf{P} \mathbf{V}_i \mathbf{P} \mathbf{y} - \text{tr}(\,\mathbf{P}\mathbf{V}_i\,)\,] \tag{27.37a}$$

where $\mathbf{V}_i = \mathbf{Z}_i\, \mathbf{B}_i\, \mathbf{Z}_i^T$ as given by Equation 27.15. Similar logic gives

$$E[\,\mathbf{e}^T \mathbf{e} \,|\, \mathbf{y}\,] = n\sigma_E^2 + \sigma_E^4\, [\,\mathbf{y}^T\, \mathbf{P}\mathbf{P}\mathbf{y} - \text{tr}(\,\mathbf{P}\,)\,] \tag{27.37b}$$

These expressions define expected quadratic values, conditional on the particular set of observations $\mathbf{y}$, under the assumption that the true variance components are known. The astute reader will immediately notice that our problem has hardly been solved, since we are trying to estimate the variance components.

The EM algorithm (Dempster et al. 1977) attempts to circumvent this problem by starting with some initial estimates of the variance components, and then substituting these as well as $\mathbf{y}$ into Equations 27.37a,b to obtain estimates of the quadratic products. These latter estimates are then substituted into Equations 27.36a,b to obtain improved estimates of the variance components, and then the entire process is repeated again and again until satisfactory convergence has been achieved. Defining the quantities estimated by Equations 27.37a,b in the kth iteration as $\widehat{q}_i^{(k)}$, and $\widehat{q}_E^{(k)}$, the EM algorithm can be summarized as follows: (1) the E step computes the *expected* quadratic products conditional upon $\mathbf{y}$, $\widehat{q}_i^{(k)}$, and $\widehat{q}_E^{(k)}$, and (2) the M step substitutes these conditional expectations into the *maximum* likelihood estimators (Equations 27.36a,b) to generate the next round of REML variance-component estimates, $(\widehat{\sigma}_E^2)^{(k+1)}$ and $(\widehat{\sigma}_i^2)^{(k+1)}$, which are then applied to the next E step. The final REML estimates are achieved when $(\widehat{\sigma}_E^2)^{(k)} \simeq (\widehat{\sigma}_E^2)^{(k+1)}$ and $(\widehat{\sigma}_i^2)^{(k)} \simeq (\widehat{\sigma}_i^2)^{(k+1)}$.

That estimates obtained via the EM algorithm do indeed correspond to the REML solutions can be seen by recalling the REML Equations 27.30a,b. Upon convergence of the EM algorithm, the terms in brackets on the right sides of Equations 27.37a,b must be equal to zero, which is equivalent to the REML solutions. When convergence is reached, the estimates of the variance components are used to obtain the final estimate of $\widehat{\mathbf{V}}$, and the vector of fixed effects is then estimated by

$$\widehat{\boldsymbol{\beta}} = (\mathbf{X}^T \widehat{\mathbf{V}}^{-1} \mathbf{X})^{-1} \mathbf{X}^T \widehat{\mathbf{V}}^{-1} \mathbf{y}$$

In general, solutions via the EM algorithm can take considerably more iterations to converge than those via Newton-Raphson iteration, especially when heritabilities are low. Moreover, as with the Newton-Raphson algorithm, the EM algorithm is by no means guaranteed to converge on the REML solution; it sometimes generates multiple solutions for different starting conditions (Groeneveld and Kovac 1990b). Such problems can result from multiple peaks in the likelihood surface. Since the EM method in essence uses the first derivatives of the likelihood function to adjust the variance-component estimates (compare Equation 27.33 and the terms in brackets in Equations 27.37a,b), it can get stuck on inflection points in the likelihood surface as well. Rounding errors can also compromise the iterative solutions (Boichard et al. 1992). As in the case of derivative-based methods, many of these problems can be minimized by performing multiple analyses from different starting points.

Example 5. Consider again the animal model with dominance and a single record per individual, $\mathbf{y} = \mathbf{X}\boldsymbol{\beta} + \mathbf{a} + \mathbf{d} + \mathbf{e}$. The EM equations for the REML estimates of σ_A^2, σ_D^2 , and σ_E^2 are

$$(\widehat{\sigma}_A^2)^{(k+1)} = (\widehat{\sigma}_A^2)^{(k)} + \frac{(\widehat{\sigma}_A^4)^{(k)}}{n} \cdot \left\{ \mathbf{y}^T \mathbf{P}^{(k)} \mathbf{A} \mathbf{P}^{(k)} \mathbf{y} - \text{tr}[\mathbf{P}^{(k)} \mathbf{A}] \right\}$$

$$(\widehat{\sigma}_D^2)^{(k+1)} = (\widehat{\sigma}_D^2)^{(k)} + \frac{(\widehat{\sigma}_D^4)^{(k)}}{n} \left\{ \mathbf{y}^T \mathbf{P}^{(k)} \mathbf{D} \mathbf{P}^{(k)} \mathbf{y} - \text{tr}[\mathbf{P}^{(k)} \mathbf{D}] \right\}$$

$$(\widehat{\sigma}_E^2)^{(k+1)} = (\widehat{\sigma}_E^2)^{(k)} + \frac{(\widehat{\sigma}_E^4)^{(k)}}{n} \left\{ \mathbf{y}^T \mathbf{P}^{(k)} \mathbf{P}^{(k)} \mathbf{y} - \text{tr}[\mathbf{P}^{(k)}] \right\}$$

where $\mathbf{P}^{(k)}$ is defined by Equation 27.17b using $\mathbf{V}^{(k)}$ for $\mathbf{V}$ where

$$\mathbf{V}^{(k)} = (\widehat{\sigma}_A^2)^{(k)} \mathbf{A} + (\widehat{\sigma}_D^2)^{(k)} \mathbf{D} + (\widehat{\sigma}_E^2)^{(k)} \mathbf{I}$$

Additional Approaches

Aside from its technical complexities, one of the major limitations of the EM algorithm for variance-component estimation is the huge computational demand imposed by the need to invert the **V** matrix each iteration. Thus, several attempts have been made to develop EM-like algorithms that circumvent the inversion of **V**. For example, Smith and Graser (1986) and Graser et al. (1987) propose a method wherein $\lambda = \sigma_A^2/\sigma_E^2$ is assumed to be fixed and then, conditional on this λ, an ML estimate of σ_E^2 is obtained. With this estimate of σ_E^2 in hand, a search is then performed to obtain a new maximum likelihood estimate of λ, and the method is repeated until the estimates of λ and σ_E^2 stabilize. Meyer (1991) extends this method to multiple characters by performing a grid search over a larger parameter space.

An alternative method was developed by Thompson and Shaw (1990, 1992) for both univariate and multivariate applications of the animal model. Conventional application of the EM algorithm does the equivalent of estimating the breeding values of each individual conditional on the entire set of observations. The key to the Thompson-Shaw method is the computation of expected breeding values, which are taken to be conditional only on local pedigrees (the individual, its parents and offspring, and its mate) and on the variance-component estimates from the previous iteration. With this approach, the **V** matrix that needs to be inverted for each individual contains only the members of the local pedigree, and this submatrix is diagonal since, within a pedigree containing only parents and offspring, breeding values differ only because of random segregation. Additional simplifications of the EM equations are presented by Thompson and Shaw.

A MOLECULAR-MARKER-BASED METHOD FOR INFERRING VARIANCE COMPONENTS

One of the greatest technical limitations of all methods for estimating genetic variance components is their requirement for pairs of individuals of known relatedness. Except in humans, some animals in zoological parks, and some domesticated species, relationships of free-ranging individuals are generally unknown, and even in the best situations, paternity is often uncertain. Thus, almost all quantitative-genetic analyses are performed in artificial settings where the investigator has control of matings. When such settings are imposed on progeny of individuals derived from natural populations, uncertainty always remains as to the relevance of the observed results to the field situation, because changes in the environment can induce changes in variance components. As discussed in Chapter 17, some attempts have been made to circumvent this difficulty by comparing the phenotypes of wild-caught parents to those of lab-reared progeny, but such approaches can only be applied to species that can be readily raised in controlled settings, and even in these cases, the results can be biased. The ideal setting for any quantitative-genetic analysis of a natural population is the noninvasive procurement of phenotypic information from random individuals of known relatedness.

In principle, the absence of direct observations on relationships can be overcome by utilizing information recorded from molecular markers. Several methods have been suggested for the estimation of pairwise values of the coefficients of coancestry (Θ_{ij}) and fraternity (Δ_{ij}) from information on shared alleles at codominant marker loci (Thompson 1975, Lynch 1988c, Queller and Goodnight 1989, Ritland 1996a). These estimators are not necessarily very efficient unless large numbers of polymorphic loci are assayed, but most of them do provide unbiased estimates. Ritland (1996b) made the clever leap of showing how estimates of pairwise relatedness can be combined with estimates of pairwise phenotypic similarity to generate estimates of variance components in undisturbed natural populations. Ritland's method, the fundamentals of which we briefly outline below, is conceptually very simple.

Recall that the basic premise underlying all conventional methods for estimating the additive genetic variance of a trait is the fact that, for a character with a purely additive-genetic basis, the phenotypic covariance between relatives (i and j) has expected value $2\Theta_{ij}\sigma_A^2$. Define the **phenotypic similarity** of two individuals with phenotypes z_i and z_j to be

$$s_{ij} = (z_i - \bar{z})(z_j - \bar{z}) \tag{27.38}$$

where $\bar{z}$ is the mean phenotype in the population. Since this expression is in the form of a phenotypic covariance, under the purely additive model (assuming random mating and no shared environmental effects, and an accurate estimate of $\bar{z}$), the expected value of s_{ij} is simply $2\Theta_{ij}\sigma_A^2$. Thus, with a collection of individuals, the observed phenotypic similarity can be written in the form of a linear model,

$$s_{ij} = 2\widehat{\Theta}_{ij}\sigma_A^2 + e_{ij} \tag{27.39}$$

where $\widehat{\Theta}_{ij}$ is the estimated value of Θ_{ij} for the two individuals, and e_{ij} is the residual deviation of the observed similarity from its expectation. This expression assumes that the marker loci are in gametic phase equilibrium with the loci underlying the quantitative trait.

Equation 27.39 suggests that an estimate of the narrow-sense heritability, σ_A^2/σ_z^2, can be procured by regressing pairwise measures of phenotypic similarity on estimates of the coefficient of coancestry (with half the slope providing the estimate of σ_A^2, and the observed phenotypic variance in the population $\text{Var}(z)$ providing the estimate of σ_z^2). This idea is closely related to the logic underlying the Haseman-Elston regression in QTL analysis (Chapter 16).

Because the $\widehat{\Theta}_{ij}$ are only estimates (and often rather inaccurate ones), a conventional least-squares analysis would lead to downwardly biased estimates of σ_A^2 as a consequence of the inflated estimate of the variance of relatedness. Ritland (1996b) outlines a method that provides an estimate of σ_Θ^2, the actual variance of relatedness, which excludes the sampling variance resulting from the use of a finite number of marker loci. Letting $\text{Var}(\Theta)$ be the estimated actual variance of relatedness and $\text{Cov}(s, \widehat{\Theta})$ be the covariance of phenotypic similarity and estimated relatedness, the heritability can be estimated by

$$\widehat{h}^2 = \frac{\text{Cov}(s, \widehat{\Theta})}{2\text{Var}(\Theta)\text{Var}(z)} \tag{27.40}$$

under the assumptions of the ideal additive model.

This general strategy can be easily extended to the estimation of genetic correlations among traits. For two characters (x and y), the analog of phenotypic similarity is

$$s_{xy,ij} = \frac{(z_{x,i} - \bar{z}_x)(z_{y,j} - \bar{z}_y) + (z_{y,i} - \bar{z}_y)(z_{x,j} - \bar{z}_x)}{2} \tag{27.41}$$

which has the same form as a phenotypic covariance between traits. An estimate of the additive genetic covariance can be obtained by the regression of the $s_{xy,ij}$ on $\widehat{\Theta}_{ij}$. Letting the regression slopes involving the $s_{xy,ij}$, $s_{xx,ij}$, and $s_{yy,ij}$, be b_{xy}, b_x, and b_y, respectively, the additive genetic correlation is then estimated by

$$r_A = \frac{b_{xy}}{\sqrt{b_x b_y}} \tag{27.42}$$

A useful feature of this approach is that, unlike the situation with heritability estimation, a corrected estimate of the variance of relatedness is not required.

Although the three raw regression coefficients are biased, the proportional bias is identical for all three and cancels out in Equation 27.42.

Ritland's method provides an exciting potential framework for the quantitative-genetic analysis of natural populations, especially for species that are difficult to perform controlled matings on and/or to raise in the lab. In preliminary applications with two populations of the yellow monkeyflower (*Mimulus guttatus*), with 300 individuals assayed per population at 10 polymorphic loci, significant heritabilities and strong positive genetic correlations were obtained for a variety of characters associated with fitness (Ritland and Ritland 1996). Interestingly, lab-based estimates of h^2 using individuals of known relatedness were often substantially lower than those obtained in the field with the marker-based technique. This suggests that heritabilities in the wild are not always depressed, contrary to the conventional wisdom.

A number of important technical aspects of this technique remain to be explored. First, there is the practical issue of the spatial scale on which to sample individuals. Successful application of the technique requires the presence of adequate variance of actual relatedness among pairs of sampled individuals. With too large an average distance between sampled individuals, nearly all individuals will be essentially unrelated, and the marker-based approach will have no power. The scale beyond which this becomes important will depend on the average dispersal distances of individuals in the species under consideration. For this reason, the technique will presumably prove much more useful with sedentary plants than with mobile animals. On the other hand, with too small a spatial scale of sampling, some phenotypic similarity is likely to arise from the sharing of common environments. Ritland (1996b) outlines how the inclusion of geographic distance between individuals in the model can allow for the factoring out of phenotypic covariance due to shared environments.

Second, as outlined above, the model ignores the contribution of nonadditive gene action to phenotypic similarity. However, Ritland (1996b) shows how the model can be readily extended to the joint estimation of additive and dominance genetic variances, using the simple idea that in the presence of dominance, the genetic covariance between individuals is $2\Theta_{ij}\sigma_A^2 + \Delta_{ij}\sigma_D^2$. Applications of the model with dominance involve the regression of phenotypic similarity on estimates of both Θ_{ij} and Δ_{ij}.

Third, the marker-based approach to variance-component analysis raises a number of basic statistical issues. The linear modeling approach taken by Ritland (1996b) would appear to be only one of several alternative estimation methods. In the context of all of the other methods outlined for complex-pedigree analysis in this chapter, Ritland's method is equivalent to partitioning all pairs of measured individuals into discrete classes based on Θ_{ij}, and then simply regressing phenotypic similarity on Θ, ignoring the nonindependence of the data (the very complexity that REML is designed to deal with). However, rather than transforming the phenotypic data to pairwise measures of similarity prior to analysis, one could conceivably work directly with the individual data in a REML-like framework, as described above. In this case, the elements of the relationship matrix **A** would be estimates rather than actual measures of relatedness. This alternative approach would partially account for the nonindependence of data, which might in turn lead to more efficient estimators for the variance components. On the other hand, it is unclear how sampling variance of relatedness, assumed to be zero in conventional REML analysis, would influence the parameter estimates. The central point is that there seem to be underlying similarities between the statistical issues raised with Ritland's method and with REML analysis as conventionally applied by animal breeders.

Finally, it is unclear how sensitive the results from Ritland's method are to the presence of linkage between QTLs and marker loci. Even if all loci are in gametic phase equilibrium in the survey population, unless they are unlinked, they will be nonindependent in restricted regions of the pedigree (Chapters 14–16). This suggests that the residual errors in estimates of Θ_{ij} will not be independent of those in s_{ij} if the marker loci are linked to QTLs influencing the trait. An interesting avenue for future investigation is whether the joint distribution of molecular-marker and phenotype information can be used to partition the total genetic variance into components associated with linked vs. unlinked QTLs.

IV

Appendices

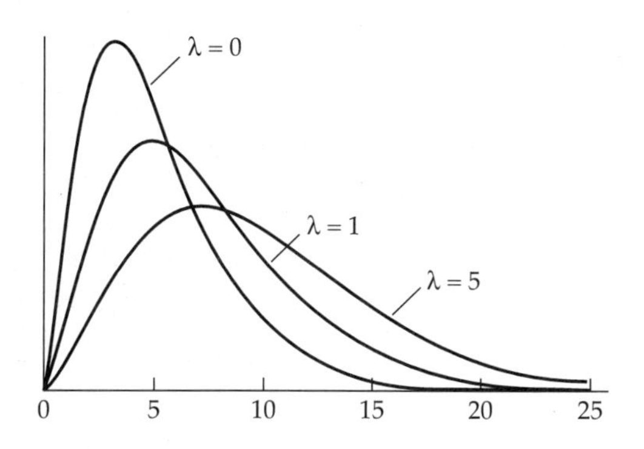

Appendix 1

Expectations, Variances, and Covariances of Compound Variables

Many situations arise in quantitative genetics in which a measurement or a statistic is a complicated function of one or more variables. This raises many serious difficulties that are at the heart of statistical analysis. We are interested ultimately in the distribution and accuracy of the observed statistic, but at best we know something about the distribution of its component variables, and even that is known with error.

As an example of these problems, consider the regression coefficient $\beta_{yx} = \sigma(x, y)/\sigma^2(x)$, which is a ratio of a covariance and a variance. With finite sample sizes, the observed statistics $\text{Cov}(x, y)$ and $\text{Var}(x)$ are only estimates of $\sigma(x, y)$ and $\sigma^2(x)$. Thus, in order to have some confidence in the estimate b_{yx} of β_{yx}, we need to know something about the accuracy of $\text{Cov}(x, y)$ and $\text{Var}(x)$. Moreover, we need to know how sampling variance of $\text{Cov}(x, y)$ and $\text{Var}(x)$ translates into sampling variance of b_{yx}, i.e., we need to know the sampling variance of a ratio. The problem is exacerbated because the same individuals are typically used to estimate $\text{Cov}(x, y)$ and $\text{Var}(x)$, which causes these two estimates to be correlated.

Fortunately, there is a fairly general approach (often referred to as the **delta method**) for dealing with all of these issues. This appendix provides a brief overview of the theory underlying the delta method and uses it to derive a number of fundamental results that will prove helpful throughout the book.

THE DELTA METHOD

Consider an arbitrary expression f, which is a function of x. Performing a Taylor series expansion around an arbitrary constant c,

$$f = f(c) + (x - c)\frac{\partial f(c)}{\partial x} + (x - c)^2 \frac{\partial^2 f(c)}{2\,\partial x^2} + (x - c)^3 \frac{\partial^3 f(c)}{3 \cdot 2\,\partial x^3} + \cdots \tag{A1.1}$$

where $f(c)$ refers to the function evaluated at $x = c$, and the partial derivatives are first evaluated with respect to x, after which c is substituted for x.

Example 1. Consider the function $f = \sqrt{x}$. Here, $\partial f / \partial x = 1/(2\sqrt{x})$ and $\partial^2 f / \partial x^2 = -1/(4x^{3/2})$, giving the second-order Taylor series about c as

$$f = \sqrt{x} \simeq \sqrt{c} + \frac{x-c}{2\sqrt{c}} - \frac{(x-c)^2}{8c^{3/2}} + \cdots$$

Expectations of Complex Variables

As our first example of the utility of the Taylor expansion, consider the case where x is a random variable and we wish to determine the expected value of the function f averaged over all x. Generally speaking, *the mean value of a function is only equal to the function evaluated at the mean of x in the special cases in which the function is linear in x or x is a constant*. Hence, we cannot just directly substitute the sample mean when trying to evaluate the mean of some function of the data. However, we can get around this problem by expanding f about the mean of x, using Equation A1.1 with $c = \mu_x$, and then taking the expectation,

$$E(f) = E\left[f(\mu_x) + (x-\mu_x)\frac{\partial f(\mu_x)}{\partial x} + (x-\mu_x)^2 \frac{\partial^2 f(\mu_x)}{2\partial x^2} + \cdots \right]$$

$$= f(\mu_x) + E(x-\mu_x)\frac{\partial f(\mu_x)}{\partial x} + E[(x-\mu_x)^2]\frac{\partial^2 f(\mu_x)}{2\partial x^2} + \cdots \quad \text{(A1.2)}$$

The last step follows since the derivative terms, evaluated at μ_x, are really just constants. Further simplification is possible since, by the definition of a mean, $E(x - \mu_x) = 0$, and $E[(x-\mu_x)^2]$ is the expected variance of x, $\sigma^2(x)$. Thus, ignoring third- and higher-order terms,

$$E(f) \simeq f(\mu_x) + \sigma^2(x)\,\frac{\partial^2 f(\mu_x)}{2\partial x^2} \quad \text{(A1.3)}$$

This relationship shows that the expected value of f is a function of both the mean and the variance of x if f is a nonlinear function of x. From Example 1, if $f = \sqrt{x}$, then

$$E(f) \simeq \sqrt{\mu_x} - \frac{\sigma^2(x)}{8\,\mu_x^{3/2}}$$

This same approach can be used to derive expressions for the expectations of functions that depend on more than a single variable. In this case, f must be expanded around the means of each of the component variables. With two component variables, for example, an expansion around μ_x and μ_y leads (after some algebra) to

$$E(f) = f(\mu_x, \mu_y) + \sigma^2(x)\,\frac{\partial^2 f(\mu_x, \mu_y)}{2\partial x^2} + \sigma(x,y)\,\frac{\partial^2 f(\mu_x, \mu_y)}{\partial x\,\partial y} + \sigma^2(y)\,\frac{\partial^2 f(\mu_x, \mu_y)}{2\partial y^2} + \cdots \tag{A1.4a}$$

When f is a function of n variables, $(x_1,\ x_2,\ \ldots,\ x_n)$, ignoring partials higher than the second order, this generalizes to

$$E(f) \simeq f(\mu_{x_1}, \mu_{x_2}, \cdots, \mu_{x_n}) + \frac{1}{2}\sum_{i=1}^{n} \sigma^2(x_i)\frac{\partial^2 f}{\partial x_i^2} + \sum_{i=1}^{n}\sum_{j>i}^{n} \sigma(x_i, x_j)\frac{\partial^2 f}{\partial x_i \partial x_j} \tag{A1.4b}$$

Example 2. By definition, the variance, $\sigma^2(x)$, is equal to the average value of $(x_i - \mu_x)^2 = x_i^2 - \mu_x^2$ in a population, where x_i denotes the measure for the ith individual. The population mean, μ_x, is generally not known with certainty, but rather is estimated by $\overline{x}$. Thus, it is tempting to estimate the variance by using the mean value of $x_i^2 - \overline{x}^2$. Here we show that this quantity gives a (slightly) downwardly biased estimate of $\sigma^2(x)$, and that there is a simple solution to the problem.

Expanding $x_i^2 - \overline{x}^2$, we wish to know whether the expected value of the function

$$f_i = x_i^2 - \left(\frac{\sum_{j=1}^{n} x_j}{n}\right)^2$$

is equal to $\sigma^2(x)$. Under the assumption that the population has been sampled randomly, $\mu_{x_i} = \mu_{x_j} = \mu_x$, $\sigma^2(x_i) = \sigma^2(x_j) = \sigma^2(x)$, and $\sigma(x_i, x_j) = 0$ for all i, j. Equation A1.4b then reduces to

$$E(f_i) = f_i(\mu_x) + \frac{\sigma^2(x)}{2}\left(\frac{\partial^2 f_i}{\partial x_i^2} + \sum_{j \neq i} \frac{\partial^2 f_i}{\partial x_j^2}\right) \tag{A1.5a}$$

The required partial derivatives are

$$\frac{\partial f_i}{\partial x_i} = 2x_i - \frac{2\sum_{j=1}^{n} x_j}{n^2} \qquad \frac{\partial^2 f_i}{\partial x_i^2} = 2\left(1 - \frac{1}{n^2}\right)$$

$$\frac{\partial f_i}{\partial x_j} = -\frac{2\sum_{k=1}^{n} x_k}{n^2} \qquad \frac{\partial^2 f_i}{\partial x_j^2} = -\frac{2}{n^2}$$

Since all higher-order partials are equal to zero, Equation A1.5a is exact in this case. In addition, substitution of μ_x for x_i and all x_j, shows that $f_i(\mu_x) = 0$. Substituting the partial derivatives into Equation A1.5a gives

$$E(x_i^2 - \overline{x}^2) = \sigma^2(x) \cdot \left(\frac{n-1}{n}\right)$$

This shows that the mean value of $x_i^2 - \overline{x}^2$ gives a slightly downwardly biased estimate of the population variance, $\sigma^2(x)$. The problem is eliminated by simply multiplying the mean value of $x_i^2 - \overline{x}^2$ by $n/(n-1)$. Thus, an unbiased estimate of $\sigma^2(x)$ is

$$\text{Var}(x) = \left(\frac{n}{n-1}\right)(\overline{x^2} - \overline{x}^2) \tag{A1.5b}$$

$$= \frac{1}{n-1}\sum_{i=1}^{n}(x_i - \overline{x})^2 \tag{A1.5c}$$

Variances of Complex Variables

The preceding approach can also be used to obtain an expression for the variance of a function. Again expanding around $c = \mu_x$, and substituting for f from Equation A1.1 and $E(f)$ from Equation A1.2,

$$\sigma_f^2 = E\left\{[f - E(f)]^2\right\}$$

$$= E\left\{\left[\left(f(\mu_x) + (x - \mu_x)\frac{\partial f(\mu_x)}{\partial x} + \cdots\right) - \left(f(\mu_x) + \sigma^2(x)\frac{\partial^2 f(\mu_x)}{2\partial x^2} + \cdots\right)\right]^2\right\}$$

$$= E\left\{\left[(x - \mu_x)\frac{\partial f(\mu_x)}{\partial x} + [(x - \mu_x)^2 - \sigma^2(x)]\frac{\partial^2 f(\mu_x)}{2\partial x^2} + \cdots\right]^2\right\} \tag{A1.6}$$

Ignoring all but the two lowest-order terms, and noting that $E(x - \mu_x) = 0$,

$$\sigma_f^2 \simeq E\left[(x-\mu_x)^2\right]\left[\frac{\partial f(\mu_x)}{\partial x}\right]^2 + 2E\left[(x-\mu_x)^3\right]\left[\frac{\partial f(\mu_x)}{\partial x}\right]\left[\frac{\partial^2 f(\mu_x)}{2\partial x^2}\right]$$

$$- 2E(x-\mu_x)\sigma^2(x)\left[\frac{\partial f(\mu_x)}{\partial x}\right]\left[\frac{\partial^2 f(\mu_x)}{2\partial x^2}\right] + E\left[(x-\mu_x)^4\right]\left[\frac{\partial^2 f(\mu_x)}{2\partial x^2}\right]^2$$

$$- 2E\left[(x-\mu_x)^2\right]\sigma^2(x)\left[\frac{\partial^2 f(\mu_x)}{2\partial x^2}\right]^2 + \sigma^4(x)\left[\frac{\partial^2 f(\mu_x)}{2\partial x^2}\right]^2$$

$$= \sigma^2(x)\left[\frac{\partial f(\mu_x)}{\partial x}\right]^2 + 2\mu_{3x}\left[\frac{\partial f(\mu_x)}{\partial x}\right]\left[\frac{\partial^2 f(\mu_x)}{2\partial x^2}\right]$$

$$+ \left[\mu_{4x} - \sigma^4(x)\right]\left[\frac{\partial^2 f(\mu_x)}{2\partial x^2}\right]^2 \tag{A1.7a}$$

where $\mu_{3x} = E[(x-\mu_x)^3]$ and $\mu_{4x} = E[(x-\mu_x)^4]$ are the third and fourth moments about the mean of x. When f is a function of two variables, Equation A1.6a expands to

$$\sigma_f^2 = E\Bigg\{\Bigg[(x-\mu_x)\frac{\partial f(\mu_x,\mu_y)}{\partial x} + (y-\mu_y)\frac{\partial f(\mu_x,\mu_y)}{\partial y}$$

$$+ \left[(x-\mu_x)^2 - \sigma^2(x)\right]\frac{\partial^2 f(\mu_x,\mu_y)}{2\partial x^2} + \left[(x-\mu_x)(y-\mu_y) - \sigma(x,y)\right]\frac{\partial^2 f(\mu_x,\mu_y)}{\partial x\,\partial y}$$

$$+ \left[(y-\mu_y)^2 - \sigma^2(y)\right]\frac{\partial^2 f(\mu_x,\mu_y)}{2\partial y^2} + \cdots \Bigg]^2\Bigg\} \tag{A1.7b}$$

where $\sigma(x, y)$ is the covariance of x and y. An approximation often used in place of Equation A1.7b is obtained by ignoring all but the first-order terms. Then, if f is a function of n variables,

$$\sigma_f^2 \simeq \sum_{i=1}^{n}\sum_{j=1}^{n} \sigma(x_i, x_j)\left(\frac{\partial f}{\partial x_i}\right)\left(\frac{\partial f}{\partial x_j}\right) \tag{A1.7c}$$

where $\sigma(x_i, x_j)$ is a variance if $i = j$ and a covariance otherwise, and the partial derivatives are evaluated at the expectations for all underlying variables.

Example 3. Imagine sampling a population many times independently for n individuals, each time estimating the mean. Can we express the variance of the sample means, $\sigma^2(\overline{x})$, in terms of the variance of individual measures, $\sigma^2(x)$?

Recall that the definition of a sample estimate of the mean is

$$\overline{x} = \frac{x_1 + x_2 + \cdots + x_n}{n}$$

Under the assumption that the population is sampled randomly, there is no covariance between the measures of different individuals, so that Equation A1.7c reduces to

$$\sigma_f^2 \simeq \sum_{i=1}^{n} \sigma^2(x_i) \left(\frac{\partial f}{\partial x_i} \right)^2 \tag{A1.8a}$$

where in this example $f = \overline{x}$. The partial derivative of $\overline{x}$ with respect to each individual measure x_i is simply $1/n$, and assuming a homogeneous population, $\sigma^2(x_i) = \sigma^2(x)$ for all i. Thus, substituting into Equation A1.8a, the sampling variance of a mean is

$$\sigma^2(\overline{x}) = \sum_{i=1}^{n} \frac{\sigma^2(x_i)}{n^2} = \frac{\sigma^2(x)}{n} \tag{A1.8b}$$

i.e., the sampling variance of a mean is equal to the variance of individual measures divided by the sample size.

The practical utility of this expression might be questioned since the parameter $\sigma^2(x)$ is something that we can only estimate. However, recall from Example 2 that an unbiased estimator of $\sigma^2(x)$ is $\text{Var}(x) = n(\overline{x^2} - \overline{x}^2)/(n-1)$. It follows that

$$\text{Var}(\overline{x}) = \frac{\text{Var}(x)}{n} \tag{A1.8c}$$

is an unbiased estimator of $\sigma^2(\overline{x})$. The square root of $\text{Var}(\overline{x})$ is known as the **standard error of the mean.**

The practice of substituting an observed (and, ideally, unbiased) statistic for a population parameter in sampling-variance equations is widely used to obtain approximate sampling variances of statistics. Since the accuracy of formulations employing such estimates increases with sample size, these formulations are usually referred to as **large-sample variance** expressions, their square roots yielding **standard errors** of the statistic. It is often possible to use the standard error to construct a **confidence interval** around the estimate, such that the parametric value is encompassed within the confidence limits with a certain probability. However, the construction of confidence intervals for complex functions is generally very difficult not only because of the approximate nature of the variance estimates but also because the forms of distributions of complex functions are usually unknown.

As pointed out in Chapter 2, if it is reasonable to assume that the statistic of interest is approximately normally distributed, then the estimate plus or minus twice the square root of the sampling variance provides an estimate of the 95% confidence interval, i.e., of the interval within which the true parametric value lies with approximately 95% probability. For variables that are not normally distributed, a more general statement can be made. By **Chebyshev's theorem,** if f and μ_f are observed and parametric values of a function, the probability that μ_f is in the range $f \pm k\sigma(f)$ is at least $1 - (1/k^2)$. For example, the probability that a parametric value lies within two (true) standard deviations of its estimate is at least 3/4 regardless of the sampling distribution of the estimator. The probability increases to 95% if $k \simeq 4.5$. In other words, ± 4.5 times the standard error of an estimate provides a very conservative estimate of the 95% confidence interval.

Covariances of Complex Variables

The above procedures can also be used to evaluate the covariance between two compound functions f and g determined by common variables $(x_1, \ldots, x_n)$. Again ignoring all but the first-order terms,

$$\sigma(f,g) = E\{[f - E(f)][g - E(g)]\} \simeq \sum_{i=1}^{n}\sum_{j=1}^{n} \sigma(x_i, x_j)\left(\frac{\partial f}{\partial x_i}\right)\left(\frac{\partial g}{\partial x_j}\right) \tag{A1.9}$$

Example 4. Consider two linear functions of the set of variables $(x_1, \cdots, x_n)$,

$$f = \alpha_1 x_1 + \alpha_2 x_2 + \cdots + \beta_n x_n$$

$$g = \beta_1 x_1 + \beta_2 x_2 + \cdots + \beta_n x_n$$

What is the covariance between f and g? The partial derivatives with respect to the x_i variables are $\partial f/\partial x_i = \alpha_i$ and $\partial g/\partial x_i = \beta_i$. Substituting into Equation A1.9,

$$\sigma(f,g) = \sum_{i=1}^{n} \alpha_i \beta_i \sigma^2(x_i) + \sum_{i=1}^{n}\sum_{j \neq i}^{n} \alpha_i \beta_j \sigma(x_i, x_j)$$

This is an exact result, because here all of the higher-order partial derivatives are equal to zero.

VARIANCES OF VARIANCES AND COVARIANCES

Since quantitative genetics is often concerned with quantifying components of variance and covariance, it is useful to have expressions for the variance of these.

The general rules set forth in the previous sections are readily applied to these issues, since variances and covariances are linear functions of sums of variables and their squares and cross-products. The rest of this appendix examines some specific applications of these general approximation procedures, largely by example.

Example 5. Here we consider the sampling variance of a variance estimate, i.e., $\sigma^2[\text{Var}(x)]$. This quantity can be thought of as the expected variance that would arise among variance estimates obtained from a large number of independent samples from the same population. With n observations $(x_1, x_2, \cdots, x_n)$,

$$f = \text{Var}(x) = \frac{n}{n-1}\left[\frac{\sum_{i=1}^{n} x_i^2}{n} - \left(\frac{\sum_{i=1}^{n} x_i}{n}\right)^2\right]$$

Provided that the individual measures of x are obtained from random members of the same population, the covariance between all measures is zero, and the variance associated with each measure is the same, $\sigma^2(x)$. Thus, each of the n observed variables makes an identical contribution to the variance of $\text{Var}(x)$. The partial derivatives with respect to measure x_i, evaluated at its mean, are

$$\frac{\partial f}{\partial x_i} = \frac{n}{n-1}\left(\frac{2x_i}{n} - \frac{2\sum_{j=1}^{n} x_j}{n^2}\right)\Bigg|_{x_i=\mu_x} = \frac{n}{n-1}\left(\frac{2\mu_x}{n} - \frac{2n\mu_x}{n^2}\right) = 0$$

$$\frac{\partial^2 f}{\partial x_i^2} = \frac{n}{n-1}\left(\frac{2}{n} - \frac{2}{n^2}\right) = \frac{2}{n}$$

Substituting into Equation A1.7a, the variance of $\text{Var}(x)$ caused by variation in the ith measure is $[\mu_{4x} - \sigma^4(x)]/n^2$, and summing over all n measures,

$$\sigma^2[\text{Var}(x)] = \frac{\mu_{4x} - \sigma^4(x)}{n} \tag{A1.10a}$$

When x is normally distributed, $\mu_{4x} = 3\sigma^4(x)$ (Kendall and Stuart 1977), giving

$$\sigma^2[\text{Var}(x)] = \frac{2\sigma^4(x)}{n} \tag{A1.10b}$$

Equations A1.10a,b are exact expressions for the sampling variance of a variance because all partial derivatives higher than second order are equal to zero.

Our final problem is to modify Equation A1.10b in such a way that an unbiased estimate of the sampling variance of the variance can be obtained from the sample statistic $\text{Var}(x)$. It is tempting to simply substitute $[\text{Var}(x)]^2$ for $\sigma^4(x)$, but this is not quite correct. Recalling the definition of a variance, $E(z^2) = \sigma^2(z) + \mu_z^2$, and substituting $\text{Var}(x)$ for z, and $\sigma^2(x)$ for μ_z, we find that

$$E\{[\text{Var}(x)]^2\} = \sigma^2[\text{Var}(x)] + [\sigma^2(x)]^2$$

Substituting for $\sigma^2[\text{Var}(x)]$ from Equation A1.10b,

$$E\{[\text{Var}(x)]^2\} = \left(1 + \frac{2}{n}\right)\sigma^4(x)$$

which shows that the quantity $n\,[\text{Var}(x)]^2/(n+2)$, rather than $[\text{Var}(x)]^2$, provides an unbiased estimate of $\sigma^4(x)$. Thus, an unbiased estimate of the sampling variance of a variance is given by

$$\text{Var}[\text{Var}(x)] = \frac{2[\text{Var}(x)]^2}{n+2} \tag{A1.10c}$$

General information on the variances and covariances of moments can be found in Chapter 10 of Kendall and Stuart (1977). Letting $m_r = n^{-1}\sum(x_i - \overline{x})^r$ and $\mu_r = E[(x - \mu_x)^r]$ represent observations and expectations for the *r*th moment about the mean, the following approximations apply to the sampling variances and covariances,

$$\sigma^2(m_r) \simeq \frac{1}{n}\left(\mu_{2r} - \mu_r^2 + r^2\mu_2 \cdot \mu_{r-1}^2 - 2r\mu_{r-1} \cdot \mu_{r+1}\right) \tag{A1.11}$$

$$\sigma(m_r, m_q) \simeq \frac{1}{n}\Big(\mu_{r+q} - \mu_r \cdot \mu_q + r \cdot q \cdot \mu_2 \cdot \mu_{q-1}$$

$$- r \cdot \mu_{r-1} \cdot \mu_{q+1} - q \cdot \mu_{r+1} \cdot \mu_{q-1}\Big) \tag{A1.12}$$

where $\mu_0 = \mu_1 = 0$. Ideally, in the application of any of these formulae, unbiased estimates of the moments (μ_r) should be employed (as illustrated in Examples 3 and 5). A few other useful results from Kendall and Stuart (1977) follow.

The covariance of an observed mean and a moment about that mean is

$$\sigma(\overline{x}, m_r) = \frac{1}{n}(\mu_{r+1} - r \cdot \mu_2 \cdot \mu_{r-1}) \tag{A1.13a}$$

Thus, an unbiased estimate of the covariance between estimates of the mean and the variance (estimated from the same sample) is

$$\text{Cov}[\,\overline{x}, \text{Var}(x)\,] = \frac{\text{Skw}(x)}{n} \tag{A1.13b}$$

where $\text{Skw}(x)$, given as Equation 2.7 in the text, is an unbiased estimate of the third moment. For symmetrical distributions, such as the normal, all odd moments have expectations equal to zero, in which case the covariance between mean and variance is zero.

The following results apply to the second-order moments of a bivariate normal distribution. The variance of a covariance estimate is

$$\sigma^2\,[\,\text{Cov}(x,y)\,] = \frac{\sigma^2(x)\,\sigma^2(y) + [\,\sigma(x,y)\,]^2}{n} \tag{A1.14}$$

The covariance of variance and covariance estimates sharing a common variable is

$$\sigma\,[\,\text{Cov}(x,y), \text{Var}(x)\,] = \frac{2\sigma^2(x)\,\sigma(x,y)}{n} \tag{A1.15}$$

The covariance of the variances for two variables is

$$\sigma[\,\text{Var}(x), \text{Var}(y)\,] = \frac{2\,[\,\sigma(x,y)\,]^2}{n} \tag{A1.16}$$

All three of these expressions can be used to obtain large-sample variance and covariance estimates by substituting, on the right, observed moments for their expectations. Such expressions are very slightly upwardly biased, by factors of $(n+1)/n$ or $(n+2)/n$, but because the estimates of variances and covariances of higher-order moments are highly unreliable unless the sample size is large, this distinction is trivial for most practical purposes.

Expressions for the variances and covariances of moments about the origin can be found in Kendall and Stuart (1977, p. 244), and for the variances and covariances of other bivariate moments in Kendall and Stuart (1977, p. 250).

Example 6. What is the expected sampling variance of the directional selection differential S? In Chapter 3, it is shown that for any character whose phenotype is denoted by z, S is equivalent to the covariance between z and relative fitness w, i.e., $S = \sigma(z, w)$. Applying Equation A1.14, we obtain

$$\sigma^2(S) = \frac{\sigma^2(z)\sigma^2(w) + [\,\sigma(z,w)\,]^2}{n}$$

Note that Equation A1.14 assumes z and w are bivariate normally distributed, implying that we assume that fitness has some optimal value and falls off (roughly quadratically) around this optimun.

Some insight into the relative magnitude of the sampling variance of S can be acquired by considering the coefficient of sampling variation,

$$\text{CV}(S) = \frac{\sigma(S)}{E(S)}$$

where $E(S) = \sigma(z, w)$. Letting ρ be the correlation between phenotype and relative fitness, this reduces to

$$\text{CV}(S) = \frac{1}{\rho}\left(\frac{1+\rho^2}{n}\right)^{1/2}$$

The minimum value of $\text{CV}(S)$ arises when the character is the sole determinant of fitness, i.e., $\rho = 1$, in which case $\text{CV}(S) = \sqrt{2/n}$. This shows that unless sample sizes are fairly high, the standard error of S relative to its expected value can be quite high — for $n = 50,\ 100$, and 250, the CVs are, respectively, $\geq 0.20,\ 0.14$, and 0.09.

EXPECTATIONS AND VARIANCES OF PRODUCTS

Consider the product $f = uv$, where u and/or v may be variables or functions of variables. Here $\partial f/\partial v = u$, $\partial f/\partial u = v$, $\partial^2 f/\partial u\, \partial v = 1$, and all other partial derivatives are zero. By including all terms involving these three nonzero partials in the application of Equation A1.4a, we obtain the expectation

$$E(uv) = \mu_u \mu_v + \sigma(u, v) \tag{A1.17}$$

where μ_u and μ_v are, respectively, the expected values of u and v. This expression also follows directly from the definition of a covariance, $\sigma(u, v) = E(uv) - \mu_u \mu_v$.

Applying Equation A1.7b,

$$\sigma^2(uv) = E\left\{\left[(u-\mu_u)\mu_v + (v-\mu_v)\mu_u + (v-\mu_v)(u-\mu_u) - \sigma(u,v)\right]^2\right\}$$

After a little algebra, this becomes

$$\begin{aligned}\sigma^2(uv) &= \mu_v^2\sigma^2(u) + \mu_u^2\sigma^2(v) + E[(v-\mu_v)^2(u-\mu_u)^2] - [\sigma(u,v)]^2 \\ &\quad + 2\Big(\mu_u\mu_v\sigma(u,v) + \mu_u E[(v-\mu_v)^2(u-\mu_u)] \\ &\quad + \mu_v E[(v-\mu_v)(u-\mu_u)^2]\Big)\end{aligned} \tag{A1.18a}$$

If u and v are bivariate normally distributed, then from Kendall and Stuart (1977, p. 85),

$$E[(v-\mu_v)^2(u-\mu_u)^2] = \sigma^2(u)\sigma^2(v) + 2\,[\sigma(u,v)]^2$$

and

$$E[(v-\mu_v)^2(u-\mu_u)] = E[(v-\mu_v)(u-\mu_u)^2] = 0$$

in which case

$$\sigma^2(uv) = \mu_v^2\sigma^2(u) + \mu_u^2\sigma^2(v) + [\,\sigma(u,v)\,]^2 + 2\mu_u\mu_v\sigma(u,v) + \sigma^2(u)\sigma^2(v) \quad \text{(A1.18b)}$$

When u and v are independent, this equation reduces further to

$$\sigma^2(uv) = \mu_v^2\sigma^2(u) + \mu_u^2\sigma^2(v) + \sigma^2(u)\sigma^2(v) \quad \text{(A1.18c)}$$

A useful compendium of facts regarding the variances and covariances of products of random variables can be found in Bohrnstedt and Goldberger (1969).

EXPECTATIONS AND VARIANCES OF RATIOS

Many of the statistics utilized in quantitative genetics are ratios of moments. These include the coefficient of variation, and regression and correlation coefficients. Thus, it is useful to have a general expression for the variance of a ratio, and the delta method yields a good approximation. Letting $f = u/v$, the first-order partials become $\partial f/\partial u = v^{-1}$ and $\partial f/\partial v = -u/v^2$, giving $\partial^2 f/\partial u^2 = 0$, $\partial^2 f/\partial v^2 = 2u/v^3$, and $\partial^2 f/\partial u\partial v = \partial^2 f/\partial v\partial u = -1/v^2$. Evaluating these second-order partials at μ_u and μ_v (again, the expected values of u and v), Equation A1.4a gives

$$E\left(\frac{u}{v}\right) \simeq \frac{\mu_u}{\mu_v}\left(1 + \frac{\sigma^2(v)}{\mu_v^2} - \frac{\sigma(u,v)}{\mu_u\mu_v}\right) \quad \text{(A1.19a)}$$

Likewise, from Equation A1.7c,

$$\begin{aligned}\sigma^2(u/v) &\simeq \sigma^2(u)\left(\frac{1}{\mu_v}\right)^2 + \sigma^2(v)\left(\frac{\mu_u}{\mu_v^2}\right)^2 - 2\sigma(u,v)\left(\frac{1}{\mu_v}\right)\left(-\frac{\mu_u}{\mu_v^2}\right)\\ &= \left(\frac{\mu_u}{\mu_v}\right)^2\left(\frac{\sigma^2(u)}{\mu_u^2} - \frac{2\sigma(u,v)}{\mu_u\mu_v} + \frac{\sigma^2(v)}{\mu_v^2}\right) \quad \text{(A1.19b)}\end{aligned}$$

Both Equations A1.19a and A1.19b are approximations since $\partial f^2/\partial v^2 \neq 0$.

Sampling Variances of Regression and Correlation Coefficients

The least-squares regression coefficient is given by $b = u/v$, where $u = \text{Cov}(x,y)$ and $v = \text{Var}(x)$. To apply Equation A1.19b, we need to know μ_u, μ_v, $\sigma^2(u)$, $\sigma^2(v)$, and $\sigma(u,v)$. Since $\text{Var}(x)$ and $\text{Cov}(x,y)$ are unbiased estimators of the variance and covariance, $\mu_u = \sigma(x,y)$ and $\mu_v = \sigma^2(x)$. Under the assumption that x and y are bivariate normally distributed, we can also use the above results to obtain the variances and covariance of u and v: from Equation A1.14, $\sigma^2(u) =$

$[\sigma^2(x)\sigma^2(y) + \sigma^2(x,y)]/n$; from Equation A1.10b, $\sigma^2(v) = 2\sigma^4(x)/n$; and from Equation A1.15, $\sigma(u,v) = 2\sigma^2(x)\sigma(x,y)/n$.

Substituting these expressions into Equation A1.19b, we obtain (after some algebra) a result first obtained by Pearson (1896)

$$\sigma^2(b) \simeq \frac{\sigma^2(y)(1-\rho^2)}{n\sigma^2(x)} \tag{A1.20a}$$

where $\rho = \sigma(x,y)/[\sigma(x)\sigma(y)]$ is the correlation coefficient. With much additional algebra, it can also be shown that the sampling variance of a correlation coefficient, $r = \text{Cov}(x,y)/[\text{Var}(x)\text{Var}(y)]^{1/2}$, is

$$\sigma^2(r) \simeq \frac{(1-\rho^2)^2}{n} \tag{A1.20b}$$

Both of these expressions are strictly valid only under the assumption of bivariate normality (Kendall and Stuart 1977; p. 250). In practice, large-sample variances for b and r are estimated by substituting observed for expected variances and covariances in Equations A1.20a,b, and by using $(n - 2)$ in place of n in the denominator, although this latter modification is usually of trivial importance.

Example 7. Dickerson (1969) argued that because the numerator (the covariance) of a regression coefficient is estimated with much lower accuracy than the denominator (the variance), the sampling variance of the latter can be safely ignored in computing the standard error of a regression coefficient. Under this assumption, the approximate sampling variance of a regression coefficient is simply $\sigma^2[\text{Cov}(x,y)]/\sigma^4(x)$, which upon substitution for $\sigma^2[\text{Cov}(x,y)]$ becomes

$$\sigma^2(b) \simeq \frac{\sigma^2(y)(1+\rho^2)}{n\sigma^2(x)}$$

Comparing this to Equation A1.20a, it can be seen that ignoring the sampling variance in the denominator leads to a conservative estimate of the standard error of b, i.e., to a standard error that is upwardly biased. The ratio of the standard errors resulting from both expressions is $[(1-\rho^2)/(1+\rho^2)]^{1/2}$. For $\rho = 0.5$, the ratio is 0.77.

Sampling Variance of a Coefficient of Variation

As our final example of finding an approximate (large-sample) sampling variance of a ratio, we consider $\sigma^2[\,\text{CV}(x)\,]$, where the coefficient of variation $\text{CV}(x) =$

$\text{SD}(x)/\overline{x}$. This example shows how the approximations developed above can be successively used for even rather complex functions. As a first step, consider the expectation and variance of SD, the sample estimate of standard deviation. We start with the variance, $\sigma^2[\,\text{SD}(x)\,] = \sigma^2[\sqrt{\text{Var}(x)}\,]$. Applying Equation A1.7c (with $n = 1$ variable), we have

$$\sigma^2(\sqrt{v}\,) \simeq \sigma^2(v) \left(\frac{\partial \sqrt{v}}{\partial v} \right)^2 \bigg|_{\mu_v} = \frac{\sigma^2(v)}{4\,\mu_v} \tag{A1.21}$$

Recalling that $\sigma^2[\,\text{Var}(x)\,] = [(\mu_{4x} - \sigma^4(x)]/n$, and letting $v = \text{Var}(x)$, then $\mu_v = \sigma^2(x)$ and Equation A1.21 gives

$$\sigma^2[\,\text{SD}(x)\,] = \sigma^2[\,\sqrt{\text{Var}(x)}\,] \simeq \frac{\sigma^2[\,\text{Var}(x)\,]}{4\sigma^2(x)} = \frac{\mu_{4x} - \sigma^4(x)}{4n\sigma^2(x)} \tag{A1.22}$$

To compute the approximate expected value of the SD, we apply Equation A1.3,

$$E(\sqrt{v}\,) \simeq \sqrt{\mu_v} - \sigma^2(v)\,\frac{\mu_v^{-3/2}}{8} \tag{A1.23a}$$

Letting $v = \text{Var}(x)$, $\mu_v = \sigma^2(x)$ and $\sigma^2[\,\text{Var}(x)\,] = [\mu_{4x} - \sigma^4(x)]/n$,

$$\mu_{\text{SD}} = E[\,\text{SD}(x)\,] \simeq \sigma(x) + \frac{\sigma(x)}{8n}\left[1 - \frac{\mu_{4x}}{\sigma^4(x)}\right] \tag{A1.23b}$$

Next, consider $\sigma[\,\overline{x},\ \text{SD}(x)\,]$, the covariance between the standard deviation and the sample mean. Since we know $\sigma[\,\overline{x},\ \text{Var}(x)\,] \simeq \mu_{3x}/n$ (Equation A1.13b), we first use Equation A1.9 to obtain an approximation for $\sigma[\,\overline{x},\ \text{SD}(x)\,]$ as a function of $\overline{x}$ and $\text{Var}(x)$. Letting $f = u$ and $g = \sqrt{v}$, Equation A1.9 gives

$$\sigma(\,u, \sqrt{v}\,) \simeq \sigma^2(u) \left(\frac{\partial u}{\partial u}\right) \left(\frac{\partial \sqrt{v}}{\partial u}\right) + \sigma^2(v) \left(\frac{\partial \sqrt{v}}{\partial v}\right) \left(\frac{\partial u}{\partial v}\right)$$
$$+\, \sigma(u, v) \left[\left(\frac{\partial u}{\partial u}\right) \left(\frac{\partial \sqrt{v}}{\partial v}\right) + \left(\frac{\partial u}{\partial v}\right) \left(\frac{\partial \sqrt{v}}{\partial u}\right) \right] \tag{A1.24}$$

For our particular case,

$$\frac{\partial u}{\partial v} = \frac{\partial\,\overline{x}}{\partial\,\text{Var}(x)} = 0 \qquad \text{and} \qquad \frac{\partial\,\sqrt{v}}{\partial\,u} = \frac{\partial\,\text{SD}(x)}{\partial\,\overline{x}} = 0$$

so three of the terms in Equation A1.24 are zero, leaving

$$\sigma(\,u, \sqrt{v}\,) = \sigma(u, v)/(2\sqrt{\mu_v})$$

Hence, from Equation A1.13b,

$$\sigma[\,\overline{x},\ \mathrm{SD}(x)\,] \simeq \frac{\sigma[\,\overline{x},\ \mathrm{Var}(x)\,]}{2\sigma(x)} \simeq \frac{\mu_{3x}}{2n\sigma(x)} \tag{A1.25}$$

Everthing is now in place for applying Equation A1.19b to obtain $\sigma^2(\mathrm{SD}(x)/\,\overline{x}\,)$. Let $u = \mathrm{SD}(x) = [\mathrm{Var}(x)]^{1/2}$ and $v = \overline{x}$. Hence $\mu_v = \mu_x$, $\sigma^2(v) = \sigma^2(\overline{x}) = \sigma^2(x)/n$ (Example 3), and Equations A1.22 and A1.25 provide expressions for $\sigma^2(u)$ and $\sigma(u,v)$. Substituting these into Equation A1.19b gives

$$\sigma^2[\,\mathrm{CV}(x)\,] \simeq \frac{1}{n}\left(\frac{\mu_{\mathrm{SD}}}{\mu_x}\right)^2 \left(\frac{\mu_{4x} - \sigma^4(x)}{4\sigma^2(x)\,\mu_{\mathrm{SD}}^2} - \frac{\mu_{3x}}{\mu_x\,\sigma(x)\,\mu_{\mathrm{SD}}} + \frac{\sigma^2(x)}{\mu_x^2}\right) \tag{A1.26a}$$

where μ_{SD} is given by Equation A1.23b. If x is normally distributed, $\mu_{3x} = 0$, $\mu_{4x} = 3\sigma^4(x)$, $\mu_{\mathrm{SD}} = \sigma(x)[\,1 - 1/(\,4n\,)\,]$ and further simplification is possible. Ignoring terms of order n^2 and higher, then

$$\sigma^2[\,\mathrm{CV}(x)\,] \simeq \frac{1}{2n}\left[\frac{\sigma(x)}{\mu_x}\right]^2 \left\{1 + 2\left(\frac{\sigma(x)}{\mu_x}\right)^2\right\} \tag{A1.26b}$$

in which case an estimate for the large-sample variance of the CV is simply

$$\mathrm{Var}[\,\mathrm{CV}(x)\,] \simeq \frac{[\,\mathrm{CV}(x)\,]^2}{2n}\,\{1 + 2[\,\mathrm{CV}(x)\,]^2\} \tag{A1.26c}$$

Appendix 2

Path Analysis

Wright (1921a) developed the method of path analysis as a means of interpreting the correlation between two variables in terms of hypothetical paths of causation between them. Initially, he was interested in the relative importance of general and specific growth factors for the variation of bones sizes in small mammals (Wright 1918), but he quickly realized the broad utility of his new technique and later applied it to many problems in genetics, agricultural economics, physiology, and ecology. It is surprising that, despite the method's wide use in the social sciences, animal and plant breeding, and genetic epidemiology, it has never been very popular among evolutionary theorists. The major exception is its general use in estimating degrees of relatedness and inbreeding (Chapter 7).

Provided the underlying assumptions are kept in mind, path analysis provides an extremely powerful, and conceptually, simple tool. Many of the fundamental principles of quantitative genetics can be derived by its use. Exceptionally lucid accounts of the theory and applications are given by Li (1975) and Pedhazur (1982). Only the major results are highlighted in the next few pages.

The purpose of path analysis is the quantification of the relative contributions of causal sources of variance and covariance *once a certain network of interrelated variables has been accepted*. It is not a technique for identifying the actual sources of causality, which can only come from careful experimentation. In response to periodic abuses and criticism of the technique, Wright (1932, 1934e, 1968, 1983, 1984) repeatedly emphasized this point.

UNIVARIATE ANALYSIS

Through visual display, path analysis can greatly facilitate the analysis of a complex problem. Consider, for example, a system of four measurable variables, one (y) dependent and three (z_1, z_2, and z_3) of potential explanatory value. Such a system can be displayed in the form of a **path diagram** (Figure A2.1). In this diagram, a single-headed arrow denotes a direct path from an explanatory variable to y, implying a cause-and-effect relationship. The connections between the explanatory variables are represented by double-headed arrows. It is assumed that y is a linear function of the z_i. Finally, unless y is known to be completely

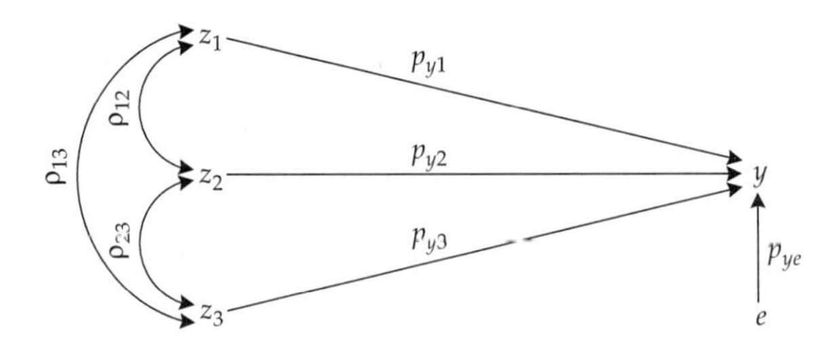

Figure A2.1 Path diagram for the variable y in terms of three explanatory variables (z_1, z_2, and z_3) and a residual error (e). p denotes a path coefficient and ρ a correlation.

determined by the observed explanatory variables, an arrow is also drawn from the independent residual term, e.

Figure A2.1 is only one of several possible path diagrams for a four-variable system. It is of general interest in that it represents a multiple regression (Chapter 8). A great deal of information is contained in this diagram. For example, it can easily be seen that z_1 potentially influences y in three ways: directly by the path $z_1 \rightarrow y$ and indirectly by the paths $z_1 \leftrightarrow z_2 \rightarrow y$ and $z_1 \leftrightarrow z_3 \rightarrow y$. Similarly, z_2 influences y through paths $z_2 \rightarrow y$, $z_2 \leftrightarrow z_1 \rightarrow y$, and $z_2 \leftrightarrow z_3 \rightarrow y$, and z_3 influences y through $z_3 \rightarrow y$, $z_3 \leftrightarrow z_2 \rightarrow y$, and $z_3 \leftrightarrow z_1 \rightarrow y$. The independent residual term operates only through path $e \rightarrow y$.

The labels on the double-headed arrows are the simple correlation coefficients (ρ) between the two denoted variables, while the quantities along single-headed arrows are **path coefficients** (p). The path coefficients for this diagram are standardized partial regression coefficients. If, prior to a multiple regression, each of the variables (y, z_1, z_2, and z_3) are standardized by subtracting the mean and dividing by the standard deviation so that the transformed variables all have zero means and unit variances, the subsequent partial regression coefficients equal the path coefficients, as can be seen in the following manner.

We start with the general linear model

$$y = \alpha + \beta_1 z_1 + \beta_2 z_2 + \cdots + \beta_n z_n + e \tag{A2.1a}$$

where $\beta_1, \beta_2, \cdots, \beta_n$ are partial regression coefficients. Subtracting $\overline{y}$ from the left and its equivalent $(\alpha + \beta_1 \overline{z}_1 + \beta_2 \overline{z}_2 + \cdots + \beta_n \overline{z}_n + \overline{e})$ from the right,

$$y - \overline{y} = \beta_1(z_1 - \overline{z}_1) + \beta_2(z_2 - \overline{z}_2) + \cdots + \beta_n(z_n - \overline{z}_n) + (e - \overline{e}) \tag{A2.1b}$$

Squaring this expression and taking expectations, we obtain a general expression for the variance of y,

$$\sigma^2(y) = \sum_{i=1}^{n} (\beta_i)^2 \sigma^2(z_i) + 2 \sum_{i=1}^{n} \sum_{j>i}^{n} \beta_i \beta_j \sigma(z_i, z_j) + \sigma_e^2 \tag{A2.2}$$

The residual variable is uncorrelated with the remaining variables under a least-squares analysis (Chapter 8), so covariance terms involving e do not appear in Equation A2.2. Dividing all terms in Equation A2.2 by $\sigma^2(y)$, recalling that $\sigma(z_i, z_j) = \rho_{ij}\sigma(z_i)\sigma(z_j)$ where ρ_{ij} is the correlation between variables i and j, and defining

$$p_{yi} = \beta_i \left[\frac{\sigma(z_i)}{\sigma(y)}\right] \tag{A2.3a}$$

$$p_{ye} = \frac{\sigma(e)}{\sigma(y)} \tag{A2.3b}$$

we obtain one of the fundamental equations of path analysis,

$$1 = \sum_{i=1}^{n} p_{yi}^2 + 2\sum_{i=1}^{n}\sum_{j>i}^{n} p_{yi}\rho_{ij}p_{yj} + p_{ye}^2 \tag{A2.4}$$

This expression, known as the **equation of complete determination,** is a simple extension of the multiple regression equation. The p_{yi} are called **path coefficients**, and from Equation A2.3a can be seen to be standardized partial regression coefficients. Thus, the path coefficients are directly obtainable by multiplying partial regression coefficients by ratios of observed standard deviations. Path coefficient p_{yi} may be interpreted as the change in y in standard deviations caused by a change in z_i in standard deviations when all other background variables are held constant.

Equation A2.4 greatly expands the utility of multiple regression by explicitly partitioning the variance of y into proportional contributions from all of the direct and indirect paths of influence. The contribution from each path is a simple product of the correlation coefficients (for each double-headed arrow) and path coefficients (for each single-headed arrow) along a loop between variables. Thus, the contributions of the direct paths from z_1, z_2, z_3 and e to y are p_{y1}^2, p_{y2}^2, p_{y3}^2, and p_{ye}^2, respectively. The contributions from the indirect paths $y \leftarrow z_1 \leftrightarrow z_2 \rightarrow y$, $y \leftarrow z_2 \leftrightarrow z_3 \rightarrow y$, and $y \leftarrow z_1 \leftrightarrow z_3 \rightarrow y$ are $2p_{y1}\rho_{12}p_{y2}$, $2p_{y2}\rho_{23}p_{y3}$, and $2p_{y1}\rho_{13}p_{y3}$. Each of the indirect paths is counted twice since they influence y in both directions. For example, the contribution of the path $y \leftarrow z_1 \leftrightarrow z_2 \rightarrow y$ to the variance of y is the same as path $y \leftarrow z_2 \leftrightarrow z_1 \rightarrow y$.

It is important to note that in computing the joint influence of two explanatory variables on y, only the direct correlation between the two variables is considered. Hence, $y \leftarrow z_1 \leftrightarrow z_2 \leftrightarrow z_3 \rightarrow y$ is not a contributing path in Figure A2.1. The entire correlation between z_1 and z_3 is contained in ρ_{13}. Thus, the general rules of path analysis are that *there is only one two-headed arrow in any path,* and that the *arrows change direction only once in a path.* Note also that, unlike correlation coefficients, path coefficients need not have absolute values less than one. Moreover, the contributions from indirect paths may be negative. The only constraint on Equation A2.4 is that the total contributions sum to one.

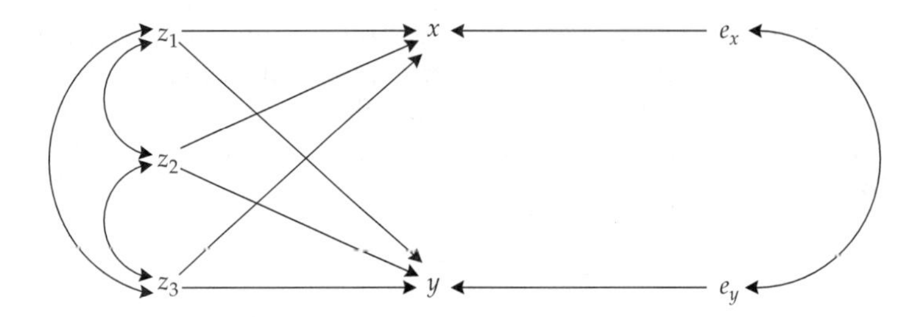

Figure A2.2 Path diagram for a system in which two dependent variables (x and y) are jointly influenced by three explanatory variables (z_1, z_2, and z_3) and residuals (e_x and e_y), which may be correlated.

BIVARIATE ANALYSIS

An exceedingly useful property of path analysis is its ability to quantify the degree of association between two variables in terms of one or more mutually shared explanatory variables. Figure A2.2 is identical in form to Figure A2.1 except that z_1, z_2, and z_3 are now causal determinants of two characters, x and y. There are 10 distinct pathways connecting x and y: the direct paths $x \leftarrow z_1 \rightarrow y$, $x \leftarrow z_2 \rightarrow y$, and $x \leftarrow z_3 \rightarrow y$, and the indirect paths $x \leftarrow z_1 \leftrightarrow z_2 \rightarrow y$, $x \leftarrow z_2 \leftrightarrow z_1 \rightarrow y$, $x \leftarrow z_1 \leftrightarrow z_3 \rightarrow y$, $x \leftarrow z_3 \leftrightarrow z_1 \rightarrow y$, $x \leftarrow z_2 \leftrightarrow z_3 \rightarrow y$, $x \leftarrow z_3 \leftrightarrow z_2 \rightarrow y$, and $x \leftarrow e_x \leftrightarrow e_y \rightarrow y$. The correlation between x and y is simply the sum of the products of path coefficients and correlation coefficients along these paths:

$$\rho_{xy} = p_{x1}p_{y1} + p_{x2}p_{y2} + p_{x3}p_{y3} + p_{x1}\rho_{12}p_{y2} + p_{x2}\rho_{12}p_{y1} + p_{x1}\rho_{13}p_{y3} + p_{x3}\rho_{13}p_{y1} + p_{x2}\rho_{23}p_{y3} + p_{x3}\rho_{23}p_{y2} + p_{xe}e_{xy}p_{ye}$$

where e_{xy} represents the correlation between the residual terms e_x and e_y. This expression may be generalized to define the correlation between any two variables with n common causal sources of variation,

$$\rho_{xy} = \left(\sum_{i=1}^{n} \sum_{j=1}^{n} p_{xi}\rho_{ij}p_{yj} \right) + p_{xe}e_{xy}p_{ye} \tag{A2.5}$$

For terms in which $i = j$, ρ_{ij} is set equal to one. Just as Equation A2.4 partitions a variance into components, Equation A2.5 partitions a correlation into a series of paths through shared explanatory variables.

APPLICATIONS

Path analysis can be very useful in quantitative genetics since explicit statements can often be made about causality and sometimes about additivity. Of the following examples, the first two illustrate how path analysis can be used to derive some fundamental relationships concerning phenotypic correlations. The third example considers an empirical problem.

Phenotypic Correlation Between Parents and Offspring

Much of the methodology of quantitative genetics relies on the comparison of phenotypic measures in close relatives. A common application involves the regression of offspring on parent, which was one of Wright's (1921a) earliest uses of path analysis. Defining an individual's phenotype (z) to be the sum of its genotypic value (G) and an environmental deviation (E), the phenotypes of a father (f) and mother (m) may be written

$$z_f = G_f + E_f$$
$$z_m = G_m + E_m$$

Provided the parents are not sibs, their environmental deviations may be treated as independent random variables with respect to G. If, however, mates select each other on the basis of phenotypes, a correlation may exist between z_f and z_m. We will denote this correlation by a. Under additive gene action, the genotypic value of an offspring is equal to the sum of the gametic contributions from its father (H_f) and mother (H_m),

$$G_o = H_p + H_m$$

An unambiguous path diagram can be constructed for such a familial structure (Figure A2.3). There are four path coefficients: h from genotypic to phenotypic values, e from environmental effects to phenotypic values, g from genotypic value to gametic value, and s from gametic value to genotypic value. These are assumed to be constant across generations.

We now consider the correlation between the phenotype of a parent and that of its offspring, ρ_{op}. Here we focus on the father-offspring correlation, although in this example, identical results arise for mother-offspring analysis. Regardless of which parent is considered, there are two paths connecting it to its offspring. The first results from the direct gametic contribution that a parent makes to its offspring; i. e., $z_f \leftarrow G_f \rightarrow H_f \rightarrow G_o \rightarrow z_o$. The contribution of this path to ρ_{of} is the product of four path coefficients, $hgsh$. The second path, which only exists under assortative mating, is an indirect route through a mate's gamete, i.e., $z_f \leftrightarrow z_m \leftarrow G_m \rightarrow H_m \rightarrow G_o \rightarrow z_o$. Its contribution to ρ_{of} is $ahgsh$. Summing up,

$$\rho_{of} = h^2 gs(1 + a) \tag{A2.6}$$

A further simplification of this expression, which eliminates the coefficients g and s, is possible. The genotypic value G_o is determined by the two direct paths $H_f \rightarrow G_o$ and $H_m \rightarrow G_o$, each of which contributes a proportion s^2 to the variance of G_o, and by the indirect paths $G_o \leftarrow H_f \leftrightarrow H_m \rightarrow G_o$ and $G_o \leftarrow H_m \leftrightarrow H_f \rightarrow G_o$. The correlation between H_f and H_m is determined by the single

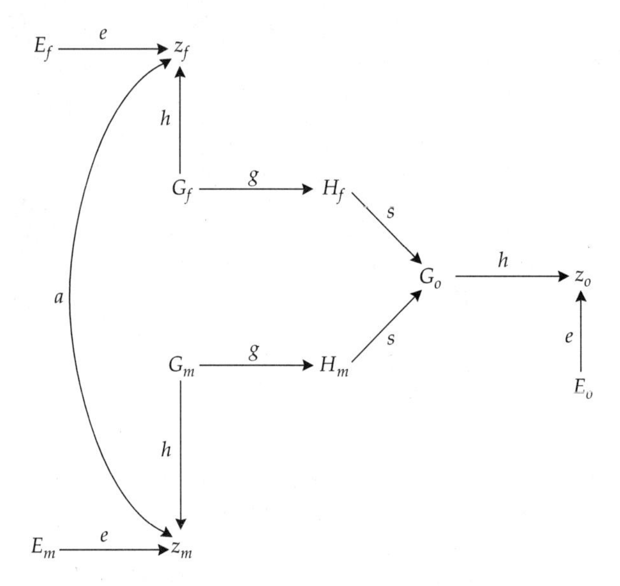

Figure A2.3 Wright's (1921a) path diagram, slightly modified, for the phenotypes of parents and offspring. Variables are defined in the text.

path $H_f \leftarrow G_f \leftarrow z_f \leftrightarrow z_m \rightarrow G_m \rightarrow H_m$ and is equal to h^2g^2a. Therefore, each indirect path makes a proportional contribution of $(hgs)^2a$ to the variance of G_o. The equation of complete determination for G_o is then

$$1 = 2s^2 + 2(hgs)^2a$$

which upon rearrangement yields

$$s = [2(1 + h^2g^2a)]^{-1/2} \tag{A2.7}$$

Wright (1921a) pointed out that, provided the path coefficients remain constant across generations, the correlation between a genotype and a gamete that it produces (G_f and H_f) will be the same as the correlation between a genotype and a gamete that produced it (G_o and H_f). It can be seen directly from the path diagram that the first of these correlations is simply g. There are two paths connecting G_o and H_f, ($H_f \rightarrow G_o$ and $H_f \leftarrow G_f \rightarrow z_f \leftrightarrow z_m \leftarrow G_m \rightarrow H_m \rightarrow G_o$), however, so their correlation is $s + ghahgs$. Equating this expression to g,

$$g = s(1 + h^2g^2a) \tag{A2.8}$$

Multiplying Equations A2.7 and A2.8 together, it can be seen that

$$gs = s^2(1 + h^2g^2a) = 0.5$$

Thus, Equation A2.6 simplifies to

$$\rho_{of} = h^2 \left(\frac{1+a}{2} \right) \tag{A2.9}$$

which in the absence of assortative mating ($a = 0$), reduces to

$$\rho_{of} = \frac{h^2}{2} \tag{A2.10}$$

Mate selection on the basis of phenotypes influences the resemblance between parents and offspring in a particularly simple manner. Perfect disassortative mating ($a = -1$) completely eliminates the correlation between parent and offspring phenotypes, while perfect assortative mating ($a = +1$) doubles it.

Quantitative geneticists have long referred to the fraction of phenotypic variance that is additive genetic in basis as the narrow-sense heritability and abbreviated it as h^2. The use of this notation, particularly the square, may seem puzzling. Returning to Figure A2.3, the origin of h^2 can now be seen to be a historical tribute to Wright's (1921a) path diagram. Under the additive model of gene action, the equation of complete determination for an individual's phenotype is

$$h^2 + e^2 = 1$$

where h^2 is the proportion of the phenotypic variance due to the direct path from the genotypic value.

Correlations Between Characters

Path analysis can also be used to describe the correlation between two different characters, x and y, in the same individual. Here we denote the two phenotypes as

$$z_x = G_x + E_x$$
$$z_y = G_y + E_y$$

where, as usual, E_x has a mean of zero and is independent of G_x, and the same properties apply to E_y and G_y. The path diagram joining the two traits is drawn in its most general form in Figure A2.4. The path $G_x \leftrightarrow G_y$ indicates the possibility of a correlation between genotypic values of the two traits, owing to their expression being mutually determined by shared genes. Correlation between the environmental effects on the two traits is denoted by ρ_e, while those between the genotypic value of one trait and the environmental deviation of the other are indicated by $\rho_{xe,yg}$ and $\rho_{xg,ye}$. The phenotypic correlation between

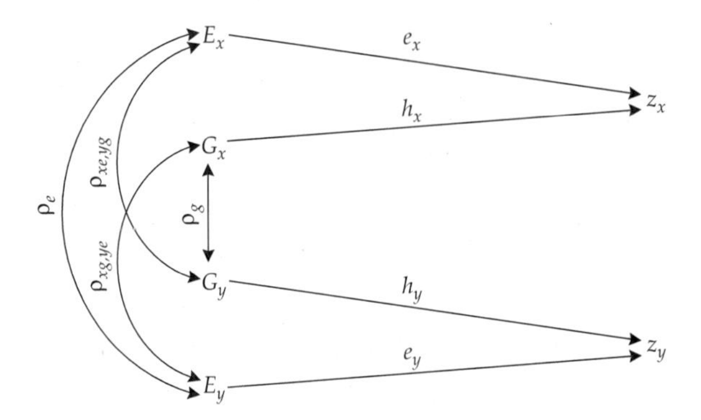

Figure A2.4 Path diagram for the phenotypic correlation between two characters (z_x and z_y) in terms of genetic values (G_x and G_y) and environmental deviations (E_x and E_y).

characters x and y, ρ_{xy}, derives from four possible paths: $z_x \leftarrow G_x \leftrightarrow G_y \rightarrow z_y$, $z_x \leftarrow E_x \leftrightarrow E_y \rightarrow z_y$, $z_x \leftarrow E_x \leftrightarrow G_y \rightarrow z_y$, and $z_x \leftarrow G_x \leftrightarrow E_y \rightarrow z_y$. Summing the appropriate products of path and correlation coefficients,

$$\rho_{xy} = h_x \rho_g h_y + e_x \rho_e e_y + e_x \rho_{xe,yg} h_y + e_y \rho_{xg,ye} h_x \tag{A2.11}$$

Note that there are only two arrows pointing to z_x and that these come from variables that are uncorrelated (as E_x and G_x are not connected by any paths). The same is true for z_y. Thus, by the equation of complete determination,

$$h_x^2 + e_x^2 = 1$$
$$h_y^2 + e_y^2 = 1$$

Rearranging and substituting $e_x = \sqrt{1 - h_x^2}$ and $e_y = \sqrt{1 - h_y^2}$ into Equation A2.11,

$$\begin{aligned} \rho_{xy} = h_x h_y \rho_g &+ \rho_e \sqrt{(1 - h_x^2)(1 - h_y^2)} \\ &+ \rho_{xe,yg} h_y \sqrt{1 - h_x^2} + \rho_{xg,ye} h_x \sqrt{1 - h_y^2} \end{aligned} \tag{A2.12}$$

Thus, the phenotypic correlation between two traits is entirely described in terms of correlations between components of the traits and their heritabilities. Frequently in quantitative-genetic applications, the correlations between E_x and G_y and between E_y and G_x are zero. In this case, Equation A2.12 reduces to

$$\rho_{xy} = h_x h_y \rho_g + \rho_e \sqrt{(1 - h_x^2)(1 - h_y^2)} \tag{A2.13}$$

Growth Analysis

Biological features in which a whole can be considered to be the sum of several individual parts are numerous. For example, the total diet of a predator often consists of several prey species, and the total seed set by a plant can be partitioned into contributions from various flowers. The problem considered here is the size of an individual (or character) at time t, z_t. This can be expressed simply as the sum of the initial size, z_1, and an arbitrary number (n) of subsequent growth increments, z_2 to z_n,

$$z_t = z_1 + z_2 + \cdots + z_n \tag{A2.14}$$

With this type of model, all of the terms on the right necessarily sum to z_t, so there is no residual error term. Moreover, all of the partial regression coefficients (the β_i coefficients on the z_i in the multiple regression equation) are equal to one. Returning to Equations A2.1a and A2.2, it can be seen that the equation of complete determination for z_t reduces to

$$1 = \frac{1}{\sigma^2(z_t)} \left[\sum_{i=1}^{n} \sigma^2(z_i) + 2 \sum_{i=1}^{n} \sum_{j \geq i}^{n} \sigma(z_i, z_j) \right] \tag{A2.15}$$

Thus, the elements of the variance-covariance matrix for the growth components in Equation A2.14 provide a complete description of the direct and indirect contributions to the variance of size at time t.

As an example of the application of Equation A2.15, the growth dynamics of a population of feral pigeons will be examined. One hundred birds were weighed at regular intervals from shortly after birth to fledging. The path analysis here will consider the weight at day 26 as a function of an initial weight at day 2 plus four subsequent six-day growth increments (days 2–8, 8–14, 14–20, and 20–26). Each growth increment is the difference between adjacent weighings, so letting w_t be the weight on day t, Equation A2.14 becomes

$$\begin{aligned} w_{26} &= w_2 + (w_8 - w_2) + (w_{14} - w_8) + (w_{20} - w_{14}) + (w_{26} - w_{20}) \\ &= w_2 + \Delta w_2 + \Delta w_8 + \Delta w_{14} + \Delta w_{20} \end{aligned}$$

The path diagram (Figure A2.5) illustrates that there are five direct paths and ten indirect paths to w_{26} (each of which must be counted twice). The proportional contributions of these paths are directly obtainable from the variances and covariances of the observed variables and are summarized in Table A2.1. Recalling that for this model all $\beta_i = 1$, the path coefficients can be computed using Equation A2.3a. For example, the contribution of direct path $w_2 \to w_{26}$ is simply $[p(w_2, w_{26})]^2 = \sigma^2(w_2)/\sigma^2(w_{26})$, while the total contribution from the indirect paths $w_2 \leftrightarrow \Delta w_2 \to w_{26}$ and $\Delta w_2 \leftrightarrow w_2 \to w_{26}$ is $2\sigma(w_2, \Delta w_2)/\sigma^2(w_{26})$.

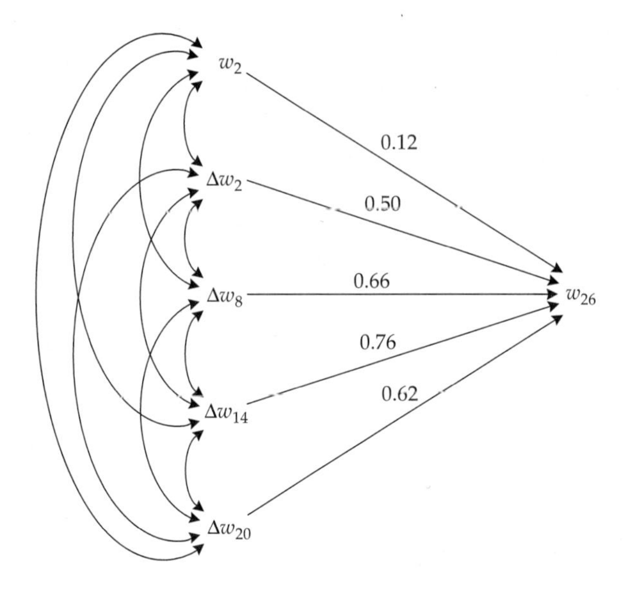

Figure A2.5 Path diagram for weight at day 26 as a function of five growth components for a population of feral pigeons. Numbers on the single-headed arrows are the path coefficients. The correlations between explanatory variables are given in Table A2.1.

Several aspects of the growth properties of this population are revealed by the path analysis. First, very little of the variation in size at age 26 is accounted for by the size at birth, i.e., $(p_{w_2,w_{26}})^2 = 0.014$. Most of it arises from variation in the post-natal growth rates. Second, all of the indirect paths make negative or

Table A2.1 Correlations (above diagonal) and path contributions (diagonal and below) of growth components to weight on day 26 for a sample of 100 feral pigeons.

	w_2	Δw_2	Δw_8	Δw_{14}	Δw_{20}
w_2	0.014	0.232*	0.100	−0.069	−0.186
Δw_2	0.027	0.255	−0.014	−0.316**	−0.045
Δw_8	0.015	−0.010	0.437	−0.157	−0.096
Δw_{14}	−0.012	−0.244	−0.159	0.584	−0.167
Δw_{20}	−0.027	−0.028	−0.079	−0.158	0.385

Source: D. Droge, unpubl. data

Note: * and ** denote correlations that are significance at the 5% and 1% levels. The diagonal elements (direct contributions to the variance of w_{26}) are simply the squares of the respectively path coefficients given in Figure A2.5. The below-diagonal elements denote the contributions resulting from correlations between characters x and y, and are obtained as $2p_x \rho_{xy} p_y$.

negligibly positive contributions to w_{26}. The sum of the direct (diagonal elements of Table A2.1) and indirect (below-diagonal elements) paths are 1.675 and -0.675 respectively. The contribution from the path involving Δw_2 and Δw_{14} ($p = -0.244$) is particularly pronounced because of the highly significant negative correlation between Δw_2 and Δw_{14} ($\rho = -\ 0.316$). On the other hand, while w_2 and Δw_2 are significantly positively correlated ($\rho = 0.232$), their indirect contribution to w_{26} is very small because of the small path coefficient from w_2 to w_{26} ($p_{w_2,w_{26}} = 0.12$). The preponderance of negative correlations between growth components is indicative of compensatory growth. Individuals that experience early periods of relatively rapid growth generally also experience subsequent periods of slowed growth. More details on this method of growth analysis, as well as estimators for the sampling variance of path coefficients, may be found in Lynch (1988d).

Appendix 3

Further Topics in Matrix Algebra and Linear Models

This appendix builds on Chapter 8, presenting additional results from matrix algebra and linear models. We start by introducing two useful matrix transforms, generalized inverses (for solving singular systems of equations) and the square root of a matrix (for obtaining a set of uncorrelated variables). These results are then used for a formal derivation of several properities of generalized least-squares (GLS) estimators. We next examine how linear model sums of squares can be written as quadratic forms and how these sums of squares are used in formal hypothesis testing. We conclude with two additional topics, equivalent linear models (which allow calculations for one model to be performed on a potentially much simpler model) and a brief introduction to matrix derivatives.

GENERALIZED INVERSES AND SOLUTIONS TO SINGULAR SYSTEMS OF EQUATIONS

Linear systems of equations are ubiquitous in quantitative genetics and we have presented solutions for such systems by assuming that the appropriate matrices are nonsingular, and hence can be inverted. However, in the real world of large, complex, and unbalanced designs, the existence of an inverse is by no means guaranteed. Consider the solution of the matrix equation $\mathbf{y} = \mathbf{A}\mathbf{x}$ for the unknown vector $\mathbf{x}$. If $\mathbf{A}$ is a square and nonsingular, then $\mathbf{x} = \mathbf{A}^{-1}\mathbf{y}$ is the unique solution. However, what happens if $\mathbf{A}$ is singular or is nonsquare? In this case either the system has no solution and is said to be **inconsistent** or else there are an infinite number of solutions. An example of an inconsistent system is

$$x_1 + x_2 = 1$$
$$x_1 + x_2 = 2$$

which cannot be satisfied by any $(x_1,\, x_2)$. Likewise, a system with an infinite number of solutions is

$$x_1 + x_2 = 1$$
$$x_1 + x_2 = 1$$

which has a line of solutions of the form $x_2 = 1 - x_1$ for arbitrary x_1. While these two simple systems can be solved by inspection, a more systematic approach is required for arbitary systems. This is provided by using **generalized inverses**.

Generalized Inverses

Suppose a matrix $\mathbf{A}^-$ exists such that

$$\mathbf{A}\mathbf{A}^-\mathbf{A} = \mathbf{A} \tag{A3.1}$$

where $\mathbf{A}$ is $p \times q$ and $\mathbf{A}^-$ is $q \times p$. Premultiplying both sides of the equation $\mathbf{Ax} = \mathbf{y}$ by $\mathbf{AA}^-$ gives

$$\mathbf{AA}^-\mathbf{Ax} = \mathbf{Ax} = \mathbf{AA}^-\mathbf{y}$$

and hence

$$\mathbf{A}(\mathbf{x} - \mathbf{A}^-\mathbf{y}) = \mathbf{0}$$

implying that, if the system is consistent, a solution is

$$\mathbf{x} = \mathbf{A}^-\mathbf{y} \tag{A3.2}$$

Given the analogy with the inverse of a nonsingular square matrix, a matrix $\mathbf{A}^-$ satisfying Equation A3.1 is called a generalized inverse (also **g-inverse, conditional inverse**) of $\mathbf{A}$. Unless $\mathbf{A}$ is nonsingular, Equation A3.1 does not define a unique matrix, so we refer to $\mathbf{A}^-$ as *a* generalized inverse instead of *the* generalized inverse. A unique generalized inverse, the **Moore-Penrose inverse**, can be obtained by imposing three additional conditions: $\mathbf{A}^-\mathbf{AA}^- = \mathbf{A}^-$, $(\mathbf{AA}^-)^T = \mathbf{AA}^-$, and $(\mathbf{A}^-\mathbf{A})^T = \mathbf{A}^-\mathbf{A}$. However, for our purposes any $\mathbf{A}^-$ satisfying Equation A3.1 is sufficient. Methods for computing generalized inverses are found in Henderson (1984a). More detailed treatment of the properties of generalized inverses are given by Dhrymes (1978), Searle (1982), Pringle and Rayner (1971), and Rao and Mitra (1971), and we summarize some of these results below.

Consistency and Solutions to Consistent Systems

When dealing with linear models for complex designs, it is not immediately clear if the resulting OLS/GLS equations have solutions. Generalized inverses provide a check of consistency, and hence of whether a system of equations has any solutions. A linear system $\mathbf{Ax} = \mathbf{y}$ is consistent if and only if

$$\mathbf{AA}^-\mathbf{y} = \mathbf{y} \tag{A3.3}$$

Given a consistent system, all solutions have the form

$$\mathbf{x} = \mathbf{A}^-\mathbf{y} + (\mathbf{I} - \mathbf{A}^-\mathbf{A})\mathbf{c} \tag{A3.4}$$

where $\mathbf{c}$ is an arbitrary $q \times 1$ column vector. For example, taking $\mathbf{c} = \mathbf{0}$ recovers Equation A3.2, while if $\mathbf{A}^{-1}$ exists, then $\mathbf{I} - \mathbf{A}^{-1}\mathbf{A} = \mathbf{0}$ and the solution $x = \mathbf{A}^{-1}\mathbf{y}$

is unique. To see that any expression of the form of Equation A3.4 is a solution, note that

$$\begin{aligned}\mathbf{Ax} &= \mathbf{A}(\mathbf{A}^{-}\mathbf{y} + (\mathbf{I} - \mathbf{A}^{-}\mathbf{A})\mathbf{c}) \\ &= \mathbf{AA}^{-}\mathbf{y} + (\mathbf{A} - \mathbf{AA}^{-}\mathbf{A})\mathbf{c} = \mathbf{y} + (\mathbf{A} - \mathbf{A})\mathbf{c} \\ &= \mathbf{y}\end{aligned}$$

which follows from Equations A3.3 and A3.1, respectively.

Example 1. Consider the following system of equations

$$\begin{aligned}x_1 + 2x_2 + 3x_3 &= 5 \\ 2x_1 + x_2 + 2x_3 &= 6\end{aligned}$$

which can be written in matrix form as $\mathbf{Ax} = \mathbf{y}$, with

$$\mathbf{A} = \begin{pmatrix} 1 & 2 & 3 \\ 2 & 1 & 2 \end{pmatrix}, \qquad \mathbf{x} = \begin{pmatrix} x_1 \\ x_2 \\ x_3 \end{pmatrix}, \qquad \mathbf{y} = \begin{pmatrix} 5 \\ 6 \end{pmatrix}$$

The matrix

$$\mathbf{A}^{-} = \begin{pmatrix} -11/26 & 9/13 \\ 4/13 & -3/13 \\ 7/26 & -1/13 \end{pmatrix}$$

satisfies $\mathbf{AA}^{-}\mathbf{A} = \mathbf{A}$ and thus is a generalized inverse of $\mathbf{A}$. Matrix multiplication shows that $\mathbf{AA}^{-} = \mathbf{I}$, implying $\mathbf{AA}^{-}\mathbf{y} = \mathbf{y}$. Thus, Equation A3.3 is satisfied and this system of equations is consistent for any $\mathbf{y}$. One solution is $\mathbf{x} = \mathbf{A}^{-}\mathbf{y}$, or

$$\begin{pmatrix} x_1 \\ x_2 \\ x_3 \end{pmatrix} = \begin{pmatrix} -11/26 & 9/13 \\ 4/13 & -3/13 \\ 7/26 & -1/13 \end{pmatrix} \begin{pmatrix} 5 \\ 6 \end{pmatrix} = \frac{1}{26} \begin{pmatrix} 53 \\ 4 \\ 23 \end{pmatrix}$$

More generally, since

$$\mathbf{I} - \mathbf{A}^{-}\mathbf{A} = \begin{pmatrix} 1/26 & 2/13 & -3/26 \\ 2/13 & 8/13 & -6/13 \\ -3/26 & -6/13 & 9/26 \end{pmatrix}$$

then from Equation A3.4, any solution to this system of equations has the form

$$\begin{pmatrix} x_1 \\ x_2 \\ x_3 \end{pmatrix} = \frac{1}{26} \begin{pmatrix} 53 \\ 4 \\ 23 \end{pmatrix} + \begin{pmatrix} 1/26 & 2/13 & -3/26 \\ 2/13 & 8/13 & -6/13 \\ -3/26 & -6/13 & 9/26 \end{pmatrix} \begin{pmatrix} c_1 \\ c_2 \\ c_3 \end{pmatrix}$$

which reduces to

$$\begin{pmatrix} x_1 \\ x_2 \\ x_3 \end{pmatrix} = \frac{1}{26}\begin{pmatrix} 53 \\ 4 \\ 23 \end{pmatrix} + c \cdot \begin{pmatrix} 1 \\ 4 \\ -3 \end{pmatrix}$$

where c is an arbitrary constant. Substitution shows this to be a solution.

Although an infinite number of solutions exists when $\mathbf{A}$ is singular, particular linear combinations (or **contrasts**) of the elements of $\mathbf{x}$ may have unique values. For example, consider the system $x_1 + x_2 = 1$. Here there are an infinite number of solutions for (x_1, x_2), but only a single solution, 1, for the contrast $x_1 + x_2$.

Consider some linear combination $\mathbf{b}^T\mathbf{x}$. If the vector of constants $\mathbf{b}$ satisfies

$$\mathbf{b}^T\mathbf{A}^-\mathbf{A} = \mathbf{b}^T \tag{A3.5a}$$

then $\mathbf{b}^T\mathbf{x}$ has a unique solution given by

$$\mathbf{b}^T\mathbf{x} = \mathbf{b}^T\mathbf{A}^-\mathbf{y} \tag{A3.5b}$$

To see this, note that Equation A3.4 gives the general solution as

$$\begin{aligned} \mathbf{b}^T\mathbf{x} &= \mathbf{b}^T(\mathbf{A}^-\mathbf{y} + [\mathbf{I} - \mathbf{A}^-\mathbf{A}]\mathbf{c}) \\ &= \mathbf{b}^T\mathbf{A}^-\mathbf{y} + (\mathbf{b}^T\mathbf{I} - \mathbf{b}^T\mathbf{A}^-\mathbf{A})\mathbf{c} \\ &= \mathbf{b}^T\mathbf{A}^-\mathbf{y} + (\mathbf{b}^T - \mathbf{b}^T)\mathbf{c} \\ &= \mathbf{b}^T\mathbf{A}^-\mathbf{y} \end{aligned}$$

which is independent of the arbitrary vector $\mathbf{c}$. Likewise, a vector of contrasts $\mathbf{Bx}$ has a unique solution $\mathbf{BA}^-\mathbf{y}$, provided $\mathbf{B}$ satisfies $\mathbf{BA}^-\mathbf{A} = \mathbf{B}$

Example 2. Consider the system of equations from Example 1. Is there a unique solution for the two linear contrasts $c_1 = x_2 - 4x_1$ and $c_2 = x_3 + 3x_1$? In matrix form,

$$\begin{pmatrix} c_1 \\ c_2 \end{pmatrix} = \begin{pmatrix} x_2 - 4x_1 \\ x_3 + 3x_1 \end{pmatrix} = \mathbf{Bx}$$

where

$$\mathbf{B} = \begin{pmatrix} -4 & 1 & 0 \\ 3 & 0 & 1 \end{pmatrix} \quad \text{and} \quad \mathbf{x} = \begin{pmatrix} x_1 \\ x_2 \\ x_3 \end{pmatrix}$$

Using the generalized inverse for $\mathbf{A}$ from Example 1, matrix multiplication shows that

$$\mathbf{BA}^{-}\mathbf{A} = \begin{pmatrix} -4 & 1 & 0 \\ 3 & 0 & 1 \end{pmatrix} = \mathbf{B}$$

Hence, the matrix version of Equation A3.5b gives the unique solution for this vector of contrasts as

$$\begin{pmatrix} c_1 \\ c_2 \end{pmatrix} = \mathbf{BA}^{-}\mathbf{y} = \begin{pmatrix} -4 & 1 & 0 \\ 3 & 0 & 1 \end{pmatrix} \begin{pmatrix} -11/26 & 9/13 \\ 4/13 & -3/13 \\ 7/26 & -1/13 \end{pmatrix} \begin{pmatrix} 5 \\ 6 \end{pmatrix} = \begin{pmatrix} -8 \\ 7 \end{pmatrix}$$

To see that this solution is indeed unique, note that we can rearrange the contrast equations to obtain $x_2 = c_1 + 4x_1$ and $x_3 = c_2 - 3x_1$. Substituting into the original set of equations (Example 1),

$$x_1 + 2x_2 + 3x_3 = x_1 + 2(c_1 + 4x_1) + 3(c_2 - 3x_1) = 2c_1 + 3c_2 = 5$$
$$2x_1 + x_2 + 2x_3 = 2x_1 + (c_1 + 4x_1) + 2(c_2 - 3x_1) = c_1 + 2c_2 = 6$$

so that the original set of three equations and three unknowns reduces to a two equation-two unknown system. In matrix form this is

$$\begin{pmatrix} 2 & 3 \\ 1 & 2 \end{pmatrix} \begin{pmatrix} c_1 \\ c_2 \end{pmatrix} = \begin{pmatrix} 5 \\ 6 \end{pmatrix}$$

Since the coefficient matrix is invertible, there is a unique solution for this pair of contrasts ($c_1 = -8$ and $c_2 = 7$).

Estimability of Fixed Factors

The above results have implications for the estimation of (fixed) factors in the general linear model, $\mathbf{y} = \mathbf{X}\boldsymbol{\beta} + \mathbf{e}$. Recall that the OLS solution for a vector $\boldsymbol{\beta}$ of fixed effects is $\widehat{\boldsymbol{\beta}} = (\mathbf{X}^T\mathbf{X})^{-1}\mathbf{X}^T\mathbf{y}$ (Chapters 8, 26). If the design matrix $\mathbf{X}$ has **full column rank** (all columns of $\mathbf{X}$ are independent), $(\mathbf{X}^T\mathbf{X})^{-1}$ exists and the OLS solution for $\boldsymbol{\beta}$ is unique. However, when $(\mathbf{X}^T\mathbf{X})$ is singular (and hence does not have a unique inverse), it is not possible to obtain unique OLS estimates for all the fixed factors in a model. For example, suppose β_1 indicates a sex effect and β_2 indicates the effect of a particular diet. If the design is such that all females use this diet, we do not have separate information on both sex and diet effects and hence can only estimate $\beta_1 + \beta_2$ rather than being able to estimate both β_1 and β_2 separately.

A linear combination of factors $\mathbf{b}^T\boldsymbol{\beta}$ is said to be **estimable** for a given design matrix $\mathbf{X}$ if there exists some column vector $\mathbf{a}$ that satisfies

$$E(\mathbf{a}^T\mathbf{y}) = \mathbf{b}^T\boldsymbol{\beta} \tag{A3.6a}$$

Estimability thus implies that there is some linear combination $\mathbf{a}^T\mathbf{y}$ of the original data whose expected value equals the desired linear combination of factors. Since $E(\mathbf{y}) = \mathbf{X}\boldsymbol{\beta}$, this definition implies that $\mathbf{b}^T\boldsymbol{\beta}$ is estimable if there exists a column vector $\mathbf{a}$ that satisfies $E(\mathbf{a}^T\mathbf{y}) = \mathbf{a}^T\mathbf{X}\boldsymbol{\beta} = \mathbf{b}^T\boldsymbol{\beta}$, implying $(\mathbf{a}^T\mathbf{X} - \mathbf{b}^T)\boldsymbol{\beta} = \mathbf{0}$, or

$$\mathbf{X}^T\mathbf{a} = \mathbf{b} \tag{A3.6b}$$

An alternative (and equivalent) condition is that $\mathbf{b}$ satisfies

$$\mathbf{b}^T(\mathbf{X}^T\mathbf{X})^-(\mathbf{X}^T\mathbf{X}) = \mathbf{b}^T \tag{A3.6c}$$

Henderson (1984a) gives other equivalent conditions. Equation A3.6c implies that if $\mathbf{X}^T\mathbf{X}$ is nonsingular, all linear combinations of $\boldsymbol{\beta}$ are estimable. Note that Equation A3.6c is identical to the condition given by Equation A3.5a (taking $\mathbf{A} = \mathbf{X}^T\mathbf{X}$), implying that these solutions are also unique estimates. If estimable, the OLS solution of the vector $\mathbf{b}^T\boldsymbol{\beta}$ given by

$$\text{OLS}(\mathbf{b}^T\boldsymbol{\beta}) = \mathbf{b}^T\left(\mathbf{X}^T\mathbf{X}\right)^-\mathbf{X}^T\mathbf{y} \tag{A3.6d}$$

is unique and independent of which generalized inverse is actually used.

Example 3. Consider the linear model $\mathbf{y} = \mathbf{X}\boldsymbol{\beta}$, where

$$\boldsymbol{\beta} = \begin{pmatrix} \beta_1 \\ \beta_2 \\ \beta_3 \end{pmatrix} \quad \text{and} \quad \mathbf{X} = \begin{pmatrix} 1 & 1 & 0 \\ 1 & 1 & 0 \\ 0 & 0 & 1 \end{pmatrix}, \quad \text{giving} \quad \mathbf{X}^T\mathbf{X} = \begin{pmatrix} 2 & 2 & 0 \\ 2 & 2 & 0 \\ 0 & 0 & 1 \end{pmatrix}$$

Note that $\mathbf{X}^T\mathbf{X}$ is singular, so we cannot obtain unique estimates of all three parameters. For this design matrix, are β_3, $\beta_1+\beta_2$, and β_1 estimable? These three combinations correspond to vectors of $\mathbf{b}^T = (0,0,1)$, $(1,1,0)$, and $(1,0,0)$, respectively. For the first two $\mathbf{b}$ vectors, we can find a vector $\mathbf{a}$ that satisfies $\mathbf{X}^T\mathbf{a} = \mathbf{b}$, viz.,

$$\mathbf{X}^T\begin{pmatrix} 0 \\ 0 \\ 1 \end{pmatrix} = \begin{pmatrix} 0 \\ 0 \\ 1 \end{pmatrix} \quad \text{and} \quad \mathbf{X}^T\begin{pmatrix} 1/2 \\ 1/2 \\ 0 \end{pmatrix} = \begin{pmatrix} 1 \\ 1 \\ 0 \end{pmatrix}$$

so that, from Equation A3.6b, these two linear combinations, β_3 and $(\beta_1+\beta_2)$, are estimable. However, since

$$\mathbf{X}^T\begin{pmatrix} a_1 \\ a_2 \\ a_3 \end{pmatrix} = \begin{pmatrix} a_1+a_2 \\ a_1+a_2 \\ a_3 \end{pmatrix} \neq \begin{pmatrix} 1 \\ 0 \\ 0 \end{pmatrix}$$

β_1 is not estimable as $a_1 + a_2$ cannot simultaneously equal zero and one, and hence there exists no vector $\mathbf{a}$ that satisfies $\mathbf{X}^T\mathbf{a} = \mathbf{b}$ for this particular $\mathbf{X}$ and $\mathbf{b}$.

THE SQUARE ROOT OF A MATRIX

The concept of the square root of a symmetric nonsingular matrix provides another useful matrix tool for the analysis of linear models. In particular, using the square root of the covariance matrix transforms a vector of correlated variables into a new vector of variables with covariance matrix $\mathbf{I}$, implying that the transformed variables are uncorrelated with unit variance.

Consider a symmetric nonsingular matrix $\mathbf{V}$ and define $\mathbf{V}^{1/2}$ as the matrix satisfying

$$\mathbf{V}^{1/2}\mathbf{V}^{1/2} = \mathbf{V} \tag{A3.7a}$$

In effect, $\mathbf{V}^{1/2}$ is the square root of a matrix, in that, when squared, we recover $\mathbf{V}$. Denoting the inverse of $\mathbf{V}^{1/2}$ as $\mathbf{V}^{-1/2}$, we have the following properties

$$\mathbf{V}^{-1/2}\mathbf{V}^{1/2} = \mathbf{I}, \quad \mathbf{V}^{-1/2}\mathbf{V}^{-1/2} = \mathbf{V}^{-1}, \quad \text{and} \quad \mathbf{V}^{-1/2}\mathbf{V} = \mathbf{V}^{1/2} \tag{A3.7b}$$

Likewise, both $\mathbf{V}^{1/2}$ and its inverse are symmetric.

Suppose the random vector $\mathbf{y}$ has covariance matrix $\mathbf{V}$ and consider the new vector $\mathbf{z} = \mathbf{V}^{-1/2}\mathbf{y}$. Recalling Equation 8.21b, the resulting covariance matrix for $\mathbf{z}$ becomes

$$\text{Var}(\mathbf{z}) = \mathbf{V}^{-1/2}\text{Var}(\mathbf{y})\mathbf{V}^{-1/2} = \mathbf{V}^{-1/2}\mathbf{V}\mathbf{V}^{-1/2} = \mathbf{I} \tag{A3.8}$$

Thus, the transformed variables have unit variance and are uncorrelated. Suppose $\mathbf{y}$ is an $n \times 1$ column vector with $\mathbf{y} \sim \text{MVN}(\boldsymbol{\mu}, \mathbf{V})$. It follows that

$$\mathbf{z} = \mathbf{V}^{-1/2}(\mathbf{y} - \boldsymbol{\mu}) \sim \text{MVN}(\mathbf{0}, \mathbf{I})$$

so that $z_i \sim \text{N}(0, 1)$, and hence the transformed variables are independent unit normals. Thus,

$$\begin{aligned}(\mathbf{y} - \boldsymbol{\mu})^T\mathbf{V}^{-1}(\mathbf{y} - \boldsymbol{\mu}) &= (\mathbf{y} - \boldsymbol{\mu})^T\mathbf{V}^{-1/2}\mathbf{V}^{-1/2}(\mathbf{y} - \boldsymbol{\mu}) \\ &= \mathbf{z}^T\mathbf{z} \\ &= \sum_{i=1}^{n} z_i^2 \sim \chi_n^2 \end{aligned} \tag{A3.9}$$

The last step follows by recalling that the sum of n squared unit normal random variables follows a χ^2 distribution with n degrees of freedom (Appendix 5). Thus when $\mathbf{y}$ is multivariate normal, the quadratic form $(\mathbf{y}-\boldsymbol{\mu})^T\mathbf{V}^{-1}(\mathbf{y}-\boldsymbol{\mu})$ follows a χ^2 distribution. As we will see shortly, Equation A3.9 is the basis for goodness-of-fit tests of linear models.

DERIVATION OF THE GLS ESTIMATORS

One important application of the square root of a matrix is that it allows us to obtain generalized least-squares (GLS) estimators from ordinary least-squares (OLS) estimators. Suppose the linear model is

$$\mathbf{y} = \mathbf{X}\boldsymbol{\beta} + \mathbf{e} \qquad \text{with } \mathbf{e} \sim (0, \mathbf{R}\,\sigma_e^2)$$

Premultiplying both sides by $\mathbf{R}^{-1/2}$ gives

$$\mathbf{z} = \mathbf{Z}\boldsymbol{\beta} + \mathbf{f} \qquad \text{with } \mathbf{f} \sim (0, \mathbf{I}\,\sigma_e^2)$$

where

$$\mathbf{z} = \mathbf{R}^{-1/2}\mathbf{y}, \qquad \mathbf{Z} = \mathbf{R}^{-1/2}\mathbf{X}, \qquad \mathbf{f} = \mathbf{R}^{-1/2}\mathbf{e}$$

OLS can be applied to this model since the transformed residuals are uncorrelated and homoscedastic. Thus, GLS estimates are obtained from the OLS solution by substituting

$$\mathbf{z} = \mathbf{R}^{-1/2}\mathbf{y} \text{ for } \mathbf{y}, \qquad \mathbf{Z} = \mathbf{R}^{-1/2}\mathbf{X} \text{ for } \mathbf{X}, \qquad \mathbf{f} = \mathbf{R}^{-1/2}\mathbf{e} \text{ for } \mathbf{e} \tag{A3.10}$$

Substituting into the OLS solutions (Equation 8.33a) gives the GLS estimate of $\boldsymbol{\beta}$ as

$$\begin{aligned}\widehat{\boldsymbol{\beta}} &= \left((\mathbf{X}^T\mathbf{R}^{-1/2})(\mathbf{R}^{-1/2}\mathbf{X}) \right)^{-1} (\mathbf{X}^T\mathbf{R}^{-1/2})(\mathbf{R}^{-1/2}\mathbf{y}) \\ &= \left(\mathbf{X}^T\mathbf{R}^{-1}\mathbf{X}\right)^{-1} \mathbf{X}^T\mathbf{R}^{-1}\mathbf{y}\end{aligned}$$

Likewise, substituting into the OLS covariance expression (Equation 8.33b) gives the resulting covariance matrix for the GLS estimates as

$$\text{Var}(\widehat{\boldsymbol{\beta}}) = \left(\mathbf{X}^T\mathbf{R}^{-1}\mathbf{X}\right)^{-1} \sigma_e^2$$

If the residuals follow a multivariate normal distribution, $\mathbf{e} \sim \text{MVN}(\mathbf{0}, \mathbf{V})$, and $\mathbf{y} = \mathbf{X}\boldsymbol{\beta} + \mathbf{e}$ is indeed the correct model, then $\mathbf{y} - \widehat{\mathbf{y}} \sim \text{MVN}(\mathbf{0}, \mathbf{V})$ and it follows from Equation A3.9 that

$$(\mathbf{y} - \widehat{\mathbf{y}})^T\, \mathbf{V}^{-1}\, (\mathbf{y} - \widehat{\mathbf{y}}) \sim \chi^2 \tag{A3.11a}$$

The degrees of freedom for the χ^2 distribution equal the number of observations minus the number of estimated parameters. Equation A3.11a provides a χ^2 test for the goodness-of-fit of a particular linear model. If **V** is a diagonal matrix, then

$$(\mathbf{y} - \widehat{\mathbf{y}})^T\, \mathbf{V}^{-1}\, (\mathbf{y} - \widehat{\mathbf{y}}) = \sum_{i=1}^{n} \frac{(y_i - \widehat{y}_i)^2}{V_{ii}} \sim \chi^2 \tag{A3.11b}$$

Similar modifications extend a number of other OLS results into GLS results (Table A3.1).

Table A3.1 Summary of useful results for the general linear model, $\mathbf{y} = \mathbf{X}\boldsymbol{\beta} + \mathbf{e}$, under ordinary least-squares (OLS) and generalized least-squares (GLS) assumptions about the distribution of residuals.

	OLS	GLS
Assumed distribution of residuals	$\mathbf{e} \sim (0, \sigma_e^2\,\mathbf{I})$	$\mathbf{e} \sim (0, \mathbf{V})$
Least-squares estimator of $\boldsymbol{\beta}$	$\widehat{\boldsymbol{\beta}} = (\mathbf{X}^T\mathbf{X})^{-1}\mathbf{X}^T\mathbf{y}$	$\widehat{\boldsymbol{\beta}} = (\mathbf{X}^T\mathbf{V}^{-1}\mathbf{X})^{-1}\mathbf{X}^T\mathbf{V}^{-1}\mathbf{y}$
Var$(\widehat{\boldsymbol{\beta}})$	$(\mathbf{X}^T\mathbf{X})^{-1}\sigma_e^2$	$(\mathbf{X}^T\mathbf{V}^{-1}\mathbf{X})^{-1}$
Predicted values, $\widehat{\mathbf{y}} = \mathbf{X}\widehat{\boldsymbol{\beta}}$	$\mathbf{X}(\mathbf{X}^T\mathbf{X})^{-1}\mathbf{X}^T\mathbf{y}$	$\mathbf{X}(\mathbf{X}^T\mathbf{V}^{-1}\mathbf{X})^{-1}\mathbf{X}^T\mathbf{V}^{-1}\mathbf{y}$
Var$(\widehat{\mathbf{y}})$	$\mathbf{X}(\mathbf{X}^T\mathbf{X})^{-1}\mathbf{X}^T\sigma_e^2$	$\mathbf{X}(\mathbf{X}^T\mathbf{V}^{-1}\mathbf{X})^{-1}\mathbf{X}^T$
Chi-square goodness-of-fit statistic (assuming $\mathbf{e} \sim$ MVN)	$\chi^2 = \sum_{i=1}^{n} \dfrac{(y_i - \widehat{y}_i)^2}{\sigma_e^2}$	$\chi^2 = (\mathbf{y} - \widehat{\mathbf{y}})^T\mathbf{V}^{-1}(\mathbf{y} - \widehat{\mathbf{y}})$

QUADRATIC FORMS AND SUMS OF SQUARES

The analysis of linear models relies very heavily on sums of squares, which can be expressed in matrix notation as quadratic forms. To introduce the reader to the machinery used to work with sums of squares, we first present expressions for the mean and variance of a quadratic form, and then express linear model sums of squares as quadratic forms.

Moments of Quadratic Forms

When $\mathbf{x}$ is a vector of random variables, the quadratic form $\mathbf{x}^T\mathbf{A}\mathbf{x}$ is a scalar random variable. If $\mathbf{x}$ has mean $\boldsymbol{\mu}$ and (nonsingular) covariance matrix $\mathbf{V}$, Equation 8.22 gives the expected value of this quadratic form as

$$E(\mathbf{x}^T\mathbf{A}\mathbf{x}) = \text{tr}(\mathbf{A}\mathbf{V}) + \boldsymbol{\mu}^T\mathbf{A}\boldsymbol{\mu} \tag{A3.12a}$$

where the **trace** of a square matrix, $\text{tr}(\mathbf{M}) = \sum M_{ii}$, is the sum of its diagonal elements. Further, if $\mathbf{x} \sim \text{MVN}(\boldsymbol{\mu}, \mathbf{V})$, then as shown in Searle (1971), the variance of the quadratic form has a fairly simple form,

$$\sigma^2(\mathbf{x}^T\mathbf{A}\mathbf{x}) = 2\,\text{tr}\,(\mathbf{A}\mathbf{V}\mathbf{A}\mathbf{V}) + 4\boldsymbol{\mu}^T\mathbf{A}\mathbf{V}\mathbf{A}\boldsymbol{\mu} \tag{A3.12b}$$

The Sample Variance Expressed as a Quadratic Form

As an introduction to expressing sums of squares as quadratic forms, consider the sample variance for n observations,

$$\text{Var}(x) = \frac{1}{n-1} \sum_{i=1}^{n} (x_i - \overline{x})^2$$

Define the **unit matrix** $\mathbf{J}_{n \times k}$ as an $n \times k$ matrix in which every element is unity, e.g.,

$$\mathbf{J}_{n\times 1} = \left.\begin{pmatrix} 1 \\ \vdots \\ 1 \end{pmatrix}\right\} n, \qquad \mathbf{J}_{2\times 3} = \begin{pmatrix} 1 & 1 & 1 \\ 1 & 1 & 1 \end{pmatrix}$$

Likewise, define the matrix

$$\mathbf{N} = \frac{1}{n-1}\left(\mathbf{I} - \frac{1}{n}\mathbf{J}\right) = \frac{1}{n-1}\begin{pmatrix} 1-1/n & -1/n & \cdots & -1/n \\ -1/n & 1-1/n & \cdots & -1/n \\ \vdots & \vdots & \ddots & \vdots \\ -1/n & -1/n & \cdots & 1-1/n \end{pmatrix} \tag{A3.13a}$$

where $\mathbf{J}$ is $n \times n$. Noting that

$$\mathbf{Nx} = \frac{1}{n-1}\left(\mathbf{x} - \frac{1}{n}\mathbf{Jx}\right) = \frac{1}{n-1}\begin{pmatrix} x_1 - \overline{x} \\ \vdots \\ x_n - \overline{x} \end{pmatrix} \tag{A3.13b}$$

it follows that

$$\mathbf{x}^T\mathbf{Nx} = \text{Var}(x) \tag{A3.14a}$$

To see this, observe that

$$\begin{aligned} \mathbf{x}^T\mathbf{Nx} &= \frac{1}{n-1}\begin{pmatrix} x_1 & \cdots & x_n \end{pmatrix}\begin{pmatrix} x_1 - \overline{x} \\ \vdots \\ x_n - \overline{x} \end{pmatrix} \\ &= \frac{1}{n-1}\sum_{i=1}^{n} x_i(x_i - \overline{x}) = \frac{1}{n-1}\left(\sum_{i=1}^{n} x_i^2 - \overline{x}\sum_{i=1}^{n} x_i\right) \\ &= \frac{1}{n-1}\sum_{i=1}^{n}(x_i - \overline{x})^2 = \text{Var}(x) \end{aligned} \tag{A3.14b}$$

Example 4. Since we have expressed $\text{Var}(x)$ as a quadratic form, we can use Equation A3.12a to compute its expected value and Equation A3.12b (under the assumption of normality) to compute its sampling variance. If $\mathbf{x} \sim (\boldsymbol{\mu}, \mathbf{V})$, the expected value of $\text{Var}(x)$ is

$$E[\,\text{Var}(x)\,] = E(\mathbf{x}^T\mathbf{N}\mathbf{x}) = \text{tr}(\mathbf{N}\mathbf{V}) + \boldsymbol{\mu}^T\mathbf{N}\boldsymbol{\mu}$$

To compute this expression, first note from Equation A3.14b that

$$\boldsymbol{\mu}^T\mathbf{N}\boldsymbol{\mu} = \frac{1}{n-1}\sum_{i=1}^{n}(\,\mu_i - \overline{\mu}\,)^2$$

Likewise, from Equation A3.13b

$$\mathbf{N}\mathbf{V} = \frac{\mathbf{V}}{n-1} - \frac{\mathbf{J}\mathbf{V}}{n(n-1)}$$

which has diagonal elements

$$(\mathbf{N}\mathbf{V})_{ii} = \frac{1}{n-1}\left(\sigma^2(z_i) - \frac{\sum_j \sigma(z_i, z_j)}{n}\right)$$

After some simplification, we have

$$\text{tr}(\mathbf{N}\mathbf{V}) = \sum_{i=1}^{n}(\mathbf{N}\mathbf{V})_{ii} = \frac{1}{n}\sum_{i=1}^{n}\sigma^2(z_i) - \frac{2}{n(n-1)}\sum_{i<j}\sigma(z_i, z_j)$$

Putting these results together gives

$$E[\text{Var}(x)] = \frac{1}{n}\sum_{i=1}^{n}\sigma^2(z_i) - \frac{2}{n(n-1)}\sum_{i<j}\sigma(z_i, z_j) + \frac{1}{n-1}\sum_{i=1}^{n}(\,\mu_i - \overline{\mu}\,)^2$$

where $\overline{\mu} = \sum \mu_i/n$. In the simple situation where all observations have the same mean and variance ($\mu_i = \mu$, $\sigma^2(z_i) = \sigma^2$) and are uncorrelated, this reduces to

$$E[\,\text{Var}(x)\,] = \sigma^2$$

Turning now to the sample variance of $\text{Var}(x)$, if we are willing to assume that $\mathbf{x}$ is multivariate normal, then from Equation A3.12b,

$$\sigma^2[\text{Var}(x)] = \sigma^2(\mathbf{x}^T\mathbf{N}\mathbf{x}) = 2\,\text{tr}\,[\mathbf{N}\mathbf{V}\mathbf{N}\mathbf{V}] + 4\boldsymbol{\mu}^T\mathbf{N}\mathbf{V}\mathbf{N}\boldsymbol{\mu}$$

If, for example, $\mathbf{V} = \sigma^2\,\mathbf{I}$ (the x_i are uncorrelated with common variance), then

$$\mathbf{NVNV} = \sigma^4\,\mathbf{NN} = \frac{\sigma^4}{(n-1)^2}\left(\mathbf{I} - \frac{1}{n}\,\mathbf{J}_{n\times n}\right)\left(\mathbf{I} - \frac{1}{n}\,\mathbf{J}_{n\times n}\right)$$

$$= \frac{\sigma^4}{(n-1)^2}\left(\mathbf{I} - \frac{2}{n}\,\mathbf{J}_{n\times n} + n^{-2}\,\mathbf{J}_{n\times n}\,\mathbf{J}_{n\times n}\right)$$

The *ij*th element in $\mathbf{J}_{n\times n}\,\mathbf{J}_{n\times n}$ is n, giving $\mathbf{J}^2_{n\times n} = n\,\mathbf{J}_{n\times n}$. Hence, the *i*th diagonal element of $\mathbf{NVNV}$ is

$$\frac{\sigma^4}{(n-1)^2}\left(1 - \frac{2}{n} + n^{-2}n\right) = \frac{\sigma^4}{n(n-1)}$$

giving $\text{tr}(\mathbf{NVNV}) = \sigma^4/(n-1)$. When all of the means are equal, it follows that $\mathbf{N}\boldsymbol{\mu} = \mathbf{0}$ and the second term in Equation A3.12b vanishes, giving

$$\sigma^2[\,\text{Var}(x)\,] = \frac{2\sigma^4}{n-1}$$

Sums of Squares Expressed as Quadratic Forms

In the same fashion that we decomposed total variance into genetic and phenotypic components (Chapters 3–7), we can decompose the total variance of a response vector $\mathbf{y}$ into the variance accounted for by the linear model and the remaining (error or residual) variance. This is typically done by considering the sums of squares, with the **total sum of squares** (SS_T) being the sum of two components, the **error** (or **residual) sum of squares** (SS_E) and the **model sum of squares** (SS_M),

$$SS_T = SS_M + SS_E$$

The total sum of squares measures the total variability in the data, while the model sum of squares measures the amount of variation accounted for by the linear model. As noted in our discussions of univariate regression in Chapter 3, the fraction of total variance explained by a linear model is given by the **coefficient of determination,**

$$r^2 = \frac{SS_M}{SS_T} = 1 - \frac{SS_E}{SS_T} \tag{A3.15}$$

The sums of squares have different forms under OLS and GLS. Under OLS, the residuals are assumed to be independent with common variance σ_e^2. In this case, each observation/residual is weighted equally, and the total sum of squares is simply

$$SS_T = \sum_{i=1}^{n}(\,y_i - \overline{y}\,)^2$$

Sums of squares can be expressed as a quadratic form of the vector of observations $\mathbf{y}$, allowing the use of Equations 3A.12a,b to obtain their expectations and variances. Recalling Equation A3.14b and A3.13a,

$$\mathrm{SS_T} = \mathbf{y}^T \left(\mathbf{I} - \frac{1}{n}\mathbf{J} \right) \mathbf{y} \tag{A3.16a}$$

where $\mathbf{J}$ is $n \times n$.

Now consider the error sum of squares

$$\mathrm{SS_E} = \sum_{i=1}^{n} (\, y_i - \widehat{y}_i \,)^2 = \sum_{i=1}^{n} \widehat{e}_i^{\,2}$$

Since $\widehat{\mathbf{e}} = \mathbf{y} - \widehat{\mathbf{y}}$ and $\widehat{\mathbf{y}} = \mathbf{Xb} = \mathbf{X}\left(\mathbf{X}^T\mathbf{X}\right)^{-1}\mathbf{X}^T\mathbf{y}$, we have

$$\mathrm{SS_E} = \widehat{\mathbf{e}}^T\widehat{\mathbf{e}}, \quad \text{where} \quad \widehat{\mathbf{e}} = \left[\mathbf{I} - \mathbf{X}\left(\mathbf{X}^T\mathbf{X}\right)^{-1}\mathbf{X}^T \right] \mathbf{y} \tag{A3.16b}$$

Expanding this expression and noting that $\mathbf{X}^T\mathbf{X}\left(\mathbf{X}^T\mathbf{X}\right)^{-1} = \mathbf{I}$, this simplifies to

$$\mathrm{SS_E} = \mathbf{y}^T \left[\mathbf{I} - \mathbf{X}\left(\mathbf{X}^T\mathbf{X}\right)^{-1}\mathbf{X}^T \right] \mathbf{y} \tag{A3.16c}$$

Finally, the model sum of squares is the difference between the total and error sums of squares,

$$\mathrm{SS_M} = \mathrm{SS_T} - \mathrm{SS_E} = \mathbf{y}^T \left[\mathbf{X}\left(\mathbf{X}^T\mathbf{X}\right)^{-1}\mathbf{X}^T - \frac{1}{n}\mathbf{J} \right] \mathbf{y} \tag{A3.16d}$$

Note that

$$\mathrm{SS_M} = \sum_{i=1}^{n} (\, \widehat{y}_i - \overline{y} \,)^2$$

so that (for OLS) the model sum of squares is the sum of squared deviations of the predicted values from the overall mean.

The sums of squares under generalized least-squares (GLS) are slightly different, as we have to correct for heteroscedasticity and/or the lack of independence among the residuals. Assume that the residuals have covariance matrix $\sigma_e^2\,\mathbf{R}$. From Equation A3.10, $\mathbf{y}$ is replaced by $\mathbf{R}^{-1/2}\mathbf{y}$ and $\mathbf{X}$ is replaced by $\mathbf{R}^{-1/2}\mathbf{X}$ in the above OLS expressions for sums of squares. Hence, the total sum of squares for GLS becomes

$$\begin{aligned} \mathrm{SS_T} &= \mathbf{y}^T\mathbf{R}^{-1/2} \left(\mathbf{I} - \frac{1}{n}\mathbf{J} \right) \mathbf{R}^{-1/2}\mathbf{y} \\ &= \mathbf{y}^T \left[\mathbf{R}^{-1} - \frac{1}{n}\mathbf{R}^{-1/2}\mathbf{J}\mathbf{R}^{-1/2} \right] \mathbf{y} \end{aligned} \tag{A3.17a}$$

Likewise, the error sum of squares becomes

$$\mathrm{SS_E} = \widehat{\mathbf{e}}^T\mathbf{R}^{-1}\widehat{\mathbf{e}}$$

$$= \mathbf{y}^T\left[\mathbf{R}^{-1} - \mathbf{R}^{-1}\mathbf{X}\left(\mathbf{X}^T\mathbf{R}^{-1}\mathbf{X}\right)^{-1}\mathbf{X}^T\mathbf{R}^{-1}\right]\mathbf{y} \tag{A3.17b}$$

and the model sum of squares becomes

$$\mathrm{SS_M} = \mathbf{y}^T\left[\mathbf{R}^{-1}\mathbf{X}\left(\mathbf{X}^T\mathbf{R}^{-1}\mathbf{X}\right)^{-1}\mathbf{X}^T\mathbf{R}^{-1} - \frac{1}{n}\mathbf{R}^{-1/2}\mathbf{J}\mathbf{R}^{-1/2}\right]\mathbf{y} \tag{A3.17c}$$

TESTING HYPOTHESES ABOUT LINEAR MODELS

Since sums of squares are very closely related to the variances accounted for by the various components of a particular linear model, it should not be surprising that hypothesis testing is based on the sums of squares. Such hypothesis tests can be quite involved, especially if we are evaluating the various components of a complex model. Here we consider the simplest case of testing the fit of the total model to the data.

If the residuals are multivariate-normally distributed with

$$\mathbf{e} \sim \mathrm{MVN}(\mathbf{0}, \sigma_e^2\,\mathbf{I}) \quad \text{for OLS;} \qquad \mathbf{e} \sim \mathrm{MVN}(\mathbf{0}, \sigma_e^2\,\mathbf{R}) \quad \text{for GLS}$$

then (recalling Equation A3.11a and A3.17b), $\mathrm{SS_E}/\sigma_e^2$ is the sum of squared unit normals and hence is χ^2-distributed. In particular, with n observations and p estimated parameters,

$$\frac{\mathrm{SS_E}}{\sigma_e^2} \sim \chi^2_{n-p} \tag{A3.18}$$

as a degree of freedom is lost for each estimated model parameter.

Suppose we have n observations and wish to compare two linear models, a **full model** fitting p parameters and a **reduced model** which uses only a subset ($q < p$) of the parameters in the full model. Do the additional $p - q$ fitted parameters provide a significant increase in the amount of variation accounted for by the model? Let $\mathrm{SS_{E_f}}$ and $\mathrm{SS_{E_r}}$ denote the appropriate (OLS or GLS) error sums of squares for the full and reduced models. Under the null hypothesis (that the full model provides the same fit as the reduced model), the difference in error sums of squares ($\mathrm{SS_{E_r}} - \mathrm{SS_{E_f}}$) is distributed as constant (σ_e^2) times a χ^2_{p-q}. Likewise, from Equation A3.18, $\mathrm{SS_{E_f}} \sim \sigma_e^2\,\chi^2_{n-p}$. Recalling the definition of the F distribution (Appendix 5), it follows that

$$\frac{(\mathrm{SS_{E_r}} - \mathrm{SS_{E_f}})\,/\,(p-q)}{\mathrm{SS_{E_f}}/\,(n-p)} = \left(\frac{n-p}{p-q}\right)\left(\frac{\mathrm{SS_{E_r}}}{\mathrm{SS_{E_f}}} - 1\right) \tag{A3.19}$$

is distributed as $F_{p-q,n-p}$ under the null hypothesis of no improved fit.

For example, we can ask if a particular linear model accounts for a significant fraction of the variation in y by considering that model versus the simplest reduced model $y_i = \mu + e_i$. It is easily seen that the least-squares solution for μ is $\overline{y}$ for OLS and the weighted mean for GLS, giving $SS_{E_r} = SS_T$. Since the number of parameters in the reduced model is $q = 1$, the test for whether a particular linear model accounts for a significant amount of the variation is

$$\left(\frac{n-p}{p-1}\right)\left(\frac{SS_T}{SS_{E_f}} - 1\right) = \left(\frac{n-p}{p-1}\right)\left(\frac{r^2}{1-r^2}\right) \qquad (A3.20)$$

where r^2 is the coefficient of determination for the full model (Equation A3.15). This test statistic follows an $F_{p-1,n-p}$ distribution.

EQUIVALENT LINEAR MODELS

Two linear models are said to be **equivalent** if they have the same mean vector $E(\mathbf{y})$ and covariance matrix $\sigma(\mathbf{y},\mathbf{y})$. The utility of equivalent models is that the parameters of one model can always be expressed as linear combinations of the parameters of any equivalent model. Hence, by choosing an appropriate equivalent model, one can often greatly simplify computations. An example of this approach is the reduced animal model of Quaas and Pollak (1980) discussed in Chapter 26. Likewise, Equation 26.23, for estimating the BLUP values of dominance effects as a function of estimated breeding values, also follows from using equivalent models. Additional examples from BLUP are given by Henderson (1985c). Our purpose here is to briefly introduce the use and construction of equivalent models.

Consider two different mixed linear models, both using the same vector $\mathbf{y}$ of observations but with different assumed vectors of fixed ($\boldsymbol{\beta}$ vs. $\boldsymbol{\beta}_*$) and random ($\mathbf{u}$ and $\mathbf{e}$ vs. $\mathbf{u}_*$ and $\mathbf{e}_*$) effects. Model 1 is

$$\mathbf{y} = \mathbf{X}\boldsymbol{\beta} + \mathbf{Z}\mathbf{u} + \mathbf{e}, \quad \text{where} \quad \mathbf{u} \sim (\mathbf{0}, \mathbf{G}) \quad \text{and} \quad \mathbf{e} \sim (\mathbf{0}, \mathbf{R})$$

while model 2 is

$$\mathbf{y} = \mathbf{X}_*\boldsymbol{\beta}_* + \mathbf{Z}_*\mathbf{u}_* + \mathbf{e}_*, \quad \text{where} \quad \mathbf{u}_* \sim (\mathbf{0}, \mathbf{G}_*) \quad \text{and} \quad \mathbf{e}_* \sim (\mathbf{0}, \mathbf{R}_*)$$

Recalling our treatment of general mixed linear models (Chapter 26), Equation 26.2 implies that for model 1,

$$\mathbf{y} \sim (\mathbf{X}\boldsymbol{\beta}, \mathbf{V}), \quad \text{where} \quad \mathbf{V} = \mathbf{Z}\mathbf{G}\mathbf{Z}^T + \mathbf{R}$$

while for model 2,

$$\mathbf{y} \sim (\mathbf{X}_*\boldsymbol{\beta}_*, \mathbf{V}_*), \quad \text{where} \quad \mathbf{V}_* = \mathbf{Z}_*\mathbf{G}_*\mathbf{Z}_*^T + \mathbf{R}_*$$

Thus, these two models are equivalent if

$$\mathbf{X}\boldsymbol{\beta} = \mathbf{X}_*\boldsymbol{\beta}_* \tag{A3.21a}$$

and $\mathbf{V} = \mathbf{V}_*$, or

$$\mathbf{Z}\mathbf{G}\mathbf{Z}^T + \mathbf{R} = \mathbf{Z}_*\mathbf{G}_*\mathbf{Z}_*^T + \mathbf{R}_* \tag{A3.21b}$$

Equations A3.21a,b provide the framework for constructing equivalent models, and hence obtaining models that are potentially easier to analyze. Consider the situation where our interest is in the prediction of random effects and we wish to obtain an equivalent model that considers the same fixed effects but uses a different vector of random effects. (For example, instead of considering a vector of both parental and offspring breeding values, we might simply consider the vector of parental breeding values, using the parental estimates to subsequently estimate the breeding values in their offspring.) If the original model is

$$\mathbf{y} = \mathbf{X}\boldsymbol{\beta} + \mathbf{Z}\mathbf{u} + \mathbf{e}, \qquad \text{where} \quad \mathbf{u} \sim (0, \mathbf{G}), \quad \text{and} \quad \mathbf{e} \sim (0, \mathbf{R})$$

an equivalent model using *any* vector of random effects $\mathbf{u}_* \sim (0, \mathbf{G}_*)$ is given by

$$\mathbf{y} = \mathbf{X}\boldsymbol{\beta} + \mathbf{Z}_*\mathbf{u}_* + \mathbf{e}_*, \qquad \text{where} \quad \mathbf{u}_* \sim (0, \mathbf{G}_*), \quad \text{and} \quad \mathbf{e}_* \sim (0, \mathbf{R}_*)$$

Since for these models to be equivalent, we require that $\mathbf{V} = \mathbf{V}_*$, it immediately follows from Equation A3.21b that the covariance matrix for the vector of new residual values, $\mathbf{e}_*$, is given by

$$\mathbf{R}_* = \mathbf{R} + \mathbf{Z}\mathbf{G}\mathbf{Z}^T - \mathbf{Z}_*\mathbf{G}_*\mathbf{Z}_*^T \tag{A3.22}$$

Given an estimate of $\mathbf{u}_*$, an estimate of $\mathbf{u}$ can be directly obtained, as parameters of a linear model can always be expressed as linear combinations of the parameters of any equivalent model. In this case, given the BLUP estimate ($\widehat{\mathbf{u}}_*$) of $\mathbf{u}_*$, the BLUP estimate of $\mathbf{u}$ is given by

$$\widehat{\mathbf{u}} = \mathbf{C}\mathbf{G}^{-1}\widehat{\mathbf{u}}_* \tag{A3.23}$$

where $\mathbf{C}$ is the covariance matrix between $\mathbf{u}_*$ and $\mathbf{u}$, and $\mathbf{G}$ is the covariance matrix associated with $\mathbf{u}$ (Henderson 1977b). This is just the linear regression of $\mathbf{u}_*$ on $\mathbf{u}$ (see Equation 8.27). Note that the vectors $\mathbf{u}_*$ and $\mathbf{u}$ can have different dimensionality, so that if $\mathbf{u}_*$ is $r \times 1$ and $\mathbf{u}$ is $q \times 1$, then $\mathbf{C}$ is an $r \times q$ matrix with $C_{ij} = \sigma(u_{*i}, u_j)$.

DERIVATIVES OF VECTORS AND MATRICES

Our final special topic in matrix algebra concerns the derivatives of vector- and matrix-valued functions, which we use rather extensively in Chapter 27. We present a few simple results here, and the reader is referred to Morrison (1976), Graham (1981), and Searle (1982) for more details. Consider first the simplest function of vector $\mathbf{x}$, namely the product of $\mathbf{x}$ and either a vector ($\mathbf{a}$) or matrix ($\mathbf{A}$) of constants. The derivatives of these functions with respect to the vector $\mathbf{x}$ become

$$\frac{\partial\, \mathbf{a}^T\mathbf{x}}{\partial\, \mathbf{x}} = \frac{\partial\, \mathbf{x}^T\mathbf{a}}{\partial\, \mathbf{x}} = \mathbf{a} \tag{A3.24a}$$

$$\frac{\partial\, \mathbf{A}\mathbf{x}}{\partial\, \mathbf{x}} = \mathbf{A}^T \tag{A3.24b}$$

Turning to quadratic forms, if $\mathbf{A}$ is symmetric, then

$$\frac{\partial\, \mathbf{x}^T\mathbf{A}\mathbf{x}}{\partial\, \mathbf{x}} = 2\mathbf{A}\mathbf{x} \tag{A3.25a}$$

Three useful identities involving quadratic forms follow

$$\frac{\partial\, (\mathbf{a}-\mathbf{x})^T\mathbf{A}(\mathbf{a}-\mathbf{x})}{\partial\, \mathbf{x}} = -2\mathbf{A}(\mathbf{a}-\mathbf{x}) \tag{A3.25b}$$

$$\frac{\partial\, (\mathbf{a}-\mathbf{B}\mathbf{x})^T(\mathbf{a}-\mathbf{B}\mathbf{x})}{\partial\, \mathbf{x}} = -2\mathbf{B}^T(\mathbf{a}-\mathbf{B}\mathbf{x}) \tag{A3.25c}$$

$$\frac{\partial\, (\mathbf{a}-\mathbf{B}\mathbf{x})^T\mathbf{A}(\mathbf{a}-\mathbf{B}\mathbf{x})}{\partial\, \mathbf{x}} = -2\mathbf{B}^T\mathbf{A}(\mathbf{a}-\mathbf{B}\mathbf{x}) \tag{A3.25d}$$

Example 5. The OLS solution for a linear model is the value of $\boldsymbol{\beta}$ that minimizes the residual sum of squares given $\mathbf{y}$ and $\mathbf{X}$. In matrix form,

$$\sum_{i=1}^{n} e_i^2 = \mathbf{e}^T\mathbf{e} = (\mathbf{y}-\mathbf{X}\boldsymbol{\beta})^T(\mathbf{y}-\mathbf{x}\boldsymbol{\beta})$$

Taking the derivative with respect to $\boldsymbol{\beta}$ and using Equation A3.25c (with $\mathbf{a} = \mathbf{y}$, $\mathbf{B} = \mathbf{X}$, and $\mathbf{x} = \boldsymbol{\beta}$) gives

$$\frac{\partial\, \mathbf{e}^T\mathbf{e}}{\partial\, \boldsymbol{\beta}} = \frac{(\mathbf{y}-\mathbf{X}\boldsymbol{\beta})^T(\mathbf{y}-\mathbf{x}\boldsymbol{\beta})}{\partial\, \boldsymbol{\beta}} = -2\mathbf{X}^T\,(\mathbf{y}-\mathbf{X}\boldsymbol{\beta})$$

Setting this equal to zero gives $\mathbf{X}^T\mathbf{X}\beta = \mathbf{X}^T\mathbf{y}$, or

$$\beta = \left(\mathbf{X}^T\mathbf{X}\right)^{-1}\mathbf{X}^T\mathbf{y}$$

If $\mathbf{X}^T\mathbf{X}$ is singular, a generalized inverse is used instead.

Appendix 4

Maximum Likelihood Estimation and Likelihood-ratio Tests

The method of maximum likelihood (ML), introduced by Fisher (1921), is widely used in human and quantitative genetics and we draw upon this approach throughout the book, especially in Chapters 13–16 (mixture distributions) and 26–27 (variance component estimation). Weir (1996) gives a useful introduction with genetic applications, while Kendall and Stuart (1979) and Edwards (1992) provide more detailed treatments.

LIKELIHOOD, SUPPORT, AND SCORE FUNCTIONS

The basic idea underlying ML is quite simple. Usually, when specifying a probability density function (say, a normal with unknown mean μ and unit variance), we treat the pdf as a function of z (the value of the random variable) with the distribution parameters $\boldsymbol{\Theta}$ assumed to be known. (While much of our discussion is in terms of a vector $\boldsymbol{\Theta}$, we use θ to indicate results for a single parameter.) With maximum likelihood estimation, we reverse the roles of the observed value and the distribution parameters by asking: Given a vector of observations $\mathbf{z}$, what can we say about $\boldsymbol{\Theta}$? To specify this alternative interpretation, the density function is denoted as $\ell(\boldsymbol{\Theta}\,|\,\mathbf{z})$, the **likelihood** of $\boldsymbol{\Theta}$ given the observed vector of data $\mathbf{z}$. This defines a **likelihood surface**, as $\ell(\boldsymbol{\Theta}\,|\,\mathbf{z})$ assigns a value to each possible point in the $\boldsymbol{\Theta}$-parameter space given the observed data $\mathbf{z}$. The **maximum likelihood estimate** (MLE) of the unknown parameters, $\widehat{\boldsymbol{\Theta}}$, is the value of $\boldsymbol{\Theta}$ corresponding to the maximum of $\ell(\boldsymbol{\Theta}\,|\,\mathbf{z})$, i.e., the MLE is the value of $\boldsymbol{\Theta}$ that is "most likely" to have produced the data $\mathbf{z}$. It is usually easier to find the maximum of a likelihood function by first taking its log and working with the resulting **log-likelihood**

$$L(\boldsymbol{\Theta}\,|\,\mathbf{z}) = \ln\left[\,\ell(\boldsymbol{\Theta}\,|\,\mathbf{z})\,\right] \tag{A4.1}$$

L is also referred to as the **support**. Since the natural log is a monotonic function, $\ell(\boldsymbol{\Theta})$ has the same maxima as $\ln\left[\ell(\boldsymbol{\Theta})\right]$, so that the maximum of L also corresponds to the maximum of the likelihood function. The **score** S of a likelihood function is the first derivative of L with respect to the likelihood parameters, with $S(\theta) = \partial L(\theta)/\partial\theta$ for a single parameter likelihood function, and

$$\mathbf{S}(\boldsymbol{\Theta}) = \frac{\partial L(\boldsymbol{\Theta})}{\partial \boldsymbol{\Theta}} = \begin{pmatrix} \partial L(\boldsymbol{\Theta})/\partial \Theta_1 \\ \vdots \\ \partial L(\boldsymbol{\Theta})/\partial \Theta_n \end{pmatrix} \tag{A4.2}$$

for a vector of n parameters. From elementary calculus it follows that the score evaluated at the MLE is zero, $\mathbf{S}(\widehat{\boldsymbol{\Theta}}) = \mathbf{0}$. This provides one approach for obtaining MLEs.

Example 1. Suppose n values, $z_1 \cdots z_n$, are sampled independently from an underlying normal with unknown mean μ and unit variance ($\sigma^2 = 1$). Letting $\mathbf{z} = (z_1, z_2, \cdots, z_n)$, what is the MLE for μ given $\mathbf{z}$? Since the observations are independent, the resulting probability density function for $\mathbf{z}$ is the product of n normal density functions,

$$\begin{aligned} p(\mathbf{z}, \mu) &= \prod_{i=1}^{n} (2\pi)^{-1/2} \exp\left[-(z_i - \mu)^2/2\right] \\ &= (2\pi)^{-n/2} \exp\left[-\sum_{i=1}^{n} (z_i - \mu)^2/2\right] \end{aligned} \tag{A4.3}$$

The log-likelihood (or support) becomes

$$L(\mu \,|\, \mathbf{z}) = \ln\left[\ell(\mu \,|\, \mathbf{z})\right] = -\left(\frac{n}{2}\right) \ln(2\pi) - \frac{1}{2} \sum_{i=1}^{n} (z_i - \mu)^2 \tag{A4.4}$$

which has the score function

$$S(\mu) = \frac{\partial L(\mu \,|\, \mathbf{z})}{\partial \mu} = \sum_{i=1}^{n} (z_i - \mu) = n(\overline{z} - \mu) \tag{A4.5}$$

Setting the score equal to zero and solving gives the MLE, $\widehat{\mu} = \overline{z}$.

Large-sample Properties of MLEs

MLEs have several important features when the sample size is large:

1. **Consistency**: As the sample size increases, the MLE converges to the true parameter value, e.g., $\widehat{\boldsymbol{\Theta}} \to \boldsymbol{\Theta}$.

2. **Invariance**: If $f(\Theta)$ is a function of the unknown parameters of the distribution, then the MLE of $f(\Theta)$ is $f(\widehat{\Theta})$, i.e., the MLE of a function of the parameters is simply that function evaluated at the MLE. For example, the MLE of $\sqrt{\theta} = (\,\widehat{\theta}\,)^{1/2}$.

3. **Asymptotic normality and efficiency**: As the sample size increases, the sampling distribution of the MLE converges to a normal and (generally) no other estimation procedure has a smaller variance. Hence, for sufficiently large sample sizes, estimates obtained via maximum likelihood typically have the smallest confidence intervals.

4. **Variance**: For large sample sizes, the variance of an MLE (assuming a single unknown parameter) is approximately the negative of the reciprocal of the second derivative of the log-likelihood function evaluated at the MLE $\widehat{\theta}$,

$$\sigma^2(\,\widehat{\theta}\,) \simeq -\left(\frac{\partial^2\, L(\theta\,|\,\mathbf{z})}{\partial\,\theta^2} \bigg|_{\theta=\widehat{\theta}} \right)^{-1} \tag{A4.6}$$

This is just the reciprocal of the curvature of the log-likelihood surface at the MLE. The flatter the likelihood surface around its maximum value (the MLE), the larger the variance; the steeper the surface, the smaller the variance. The minus sign appears because the second derivative is negative (downward curvature) at the maximum of the likelihood function.

Example 2. What is the large-sample variance of the MLE for μ from Example 1?

$$\frac{\partial^2\, L(\mu\,|\,\mathbf{z})}{\partial\mu^2} = \frac{\partial S(\mu\,|\,\mathbf{z})}{\partial\mu} = \frac{\partial\left(\sum_{i=1}^{n}(z_i-\mu)\right)}{\partial\mu} = -n$$

Applying Equation A4.6,

$$\sigma^2\,(\,\widehat{\mu}\,) \simeq \frac{1}{n}$$

Using the asymptotic normality of MLEs, the approximate distribution of the MLE is $\widehat{\mu} \sim \mathrm{N}(\mu, n^{-1})$, and the resulting 95 percent confidence interval for μ is $\widehat{\mu} \pm 1.96/\sqrt{n}$.

The Fisher Information Matrix

When estimating a vector of parameters, Equation A4.6 can be generalized by using the **Hessian matrix, H**, the matrix of second partials of the log-likelihood, whose ijth element is given by

$$\mathbf{H}_{ij} = \frac{\partial^2 \, L(\boldsymbol{\Theta} \,|\, \mathbf{z})}{\partial \, \Theta_i \, \partial \, \Theta_j} \tag{A4.7a}$$

$\mathbf{H}(\boldsymbol{\Theta}_o)$ refers to the Hessian matrix evaluated at the point $\boldsymbol{\Theta}_o$ and provides a measure of the local curvature of L around that point. The **Fisher information matrix** (**F**), the negative of expected value of the Hessian matrix for L,

$$\mathbf{F}(\boldsymbol{\Theta}) = -E\,[\,\mathbf{H}(\boldsymbol{\Theta})\,] \tag{A4.7b}$$

provides a measure of the multidimensional curvature of the log-likelihood surface. Alternately, **F** can be computed as the expected value of the outer product of the score vector,

$$\mathbf{F}(\boldsymbol{\Theta}) = E\left[\mathbf{S}(\boldsymbol{\Theta})\,\mathbf{S}(\boldsymbol{\Theta})^T\,\right] \tag{A4.7c}$$

The covariance matrix for the MLEs is simply the inverse of the information matrix, with

$$\sigma\left(\widehat{\Theta}_i, \widehat{\Theta}_j\right) = \left[\mathbf{F}(\boldsymbol{\Theta})^{-1}\right]_{ij} \tag{A4.7d}$$

As in the univariate case, if the likelihood surface is highly curved (very peaked) around the MLE, then the standard errors (being the inverse of the local curvature) are small, while if the likelihood is very flat, the sampling variance is large. For large sample sizes, **F** is often approximated by the Hessian matrix evaluated at the MLE,

$$\mathbf{F}(\boldsymbol{\Theta}) \simeq -\mathbf{H}(\widehat{\boldsymbol{\Theta}}) \tag{A4.7e}$$

Example 3. Suppose n values are sampled independently from a normal with unknown mean and variance. What are the MLEs and their sampling variances? Here $\boldsymbol{\Theta} = (\mu, \sigma)^T$. Noting that $\sum_{i=1}^{n}(z_i - \mu)^2 = n(\,\overline{z^2} - 2\,\overline{z}\,\mu + \mu^2)$, the same logic leading to Equation A4.3 shows that the log-likelihood function is

$$L(\mu, \sigma^2 \,|\, \mathbf{z}) = -\left(\frac{n}{2}\right)\ln(2\pi) - \left(\frac{n}{2}\right)\ln(\sigma^2) - \frac{n\,(\,\overline{z^2} - 2\,\overline{z}\,\mu + \mu^2\,)}{2\,\sigma^2} \tag{A4.8a}$$

Taking derivatives, the score vector becomes

$$\mathbf{S}(\boldsymbol{\Theta}) = \begin{pmatrix} \partial L(\boldsymbol{\Theta})/\partial\mu \\ \\ \partial L(\boldsymbol{\Theta})/\partial\sigma^2 \end{pmatrix} = \left(\frac{n}{\sigma^2}\right) \begin{pmatrix} \overline{z} - \mu \\ \\ \dfrac{\overline{z^2} - 2\,\overline{z}\,\mu + \mu^2}{2\,\sigma^2} - \dfrac{1}{2} \end{pmatrix} \tag{A4.8b}$$

Solving $\mathbf{S}(\widehat{\boldsymbol{\Theta}}) = \mathbf{0}$ gives the MLEs as

$$\widehat{\boldsymbol{\Theta}} = \begin{pmatrix} \widehat{\mu} \\ \\ \widehat{\sigma}^2 \end{pmatrix} = \begin{pmatrix} \overline{z} \\ \\ \overline{z^2} - \overline{z}^2 \end{pmatrix} \tag{A4.8c}$$

As the first step towards computing the Hessian and Fisher matrices, the second partials are found to be

$$\frac{\partial L^2}{(\partial\mu)^2} = -\frac{n}{\sigma^2}, \qquad \frac{\partial L^2}{\partial\mu\,\partial\sigma^2} = -\frac{n(\overline{z}-\mu)}{\sigma^4} \tag{A4.8d}$$

$$\frac{\partial L^2}{(\partial\sigma^2)^2} = \frac{n}{2\,\sigma^4}\left(1 - \frac{2\,(\overline{z^2} - 2\,\overline{z}\,\mu + \mu^2\,)}{\sigma^2}\right) \tag{A4.8e}$$

Since $E(\,\overline{z}\,) = \mu$, the first two derivatives have expected values of $-n/\sigma^2$ and 0. Likewise, since $E(\,\overline{z^2}\,) = \mu^2 + \sigma^2$, the expected value of Equation A4.8e becomes

$$E\left(\frac{\partial L^2}{(\partial\sigma^2)^2}\right) = \frac{n}{2\,\sigma^4}\left(1 - \frac{2\,(\mu^2 + \sigma^2 - 2\,\mu^2 + \mu^2\,)}{\sigma^2}\right) = -\frac{n}{2\,\sigma^4}$$

With the above results, the Fisher matrix becomes

$$\mathbf{F} = -E(\mathbf{H}) = \begin{pmatrix} \dfrac{n}{\sigma^2} & 0 \\ 0 & \dfrac{n}{2\,\sigma^4} \end{pmatrix}$$

Alternatively, evaluating the derivatives at the MLE, $\widehat{\boldsymbol{\Theta}} = (\,\overline{z},\,\widehat{\sigma}^{\,2})^T$, Equation A4.8d gives values of $-n/\widehat{\sigma}^{\,2}$ and 0, while Equation A4.8e gives $-n/(2\,\widehat{\sigma}^{\,4})$, so that the value of the Hessian matrix evaluated at the MLE becomes

$$\mathbf{H}(\widehat{\boldsymbol{\Theta}}) = -\begin{pmatrix} \dfrac{n}{\widehat{\sigma}^{\,2}} & 0 \\ 0 & \dfrac{n}{2\,\widehat{\sigma}^{\,4}} \end{pmatrix}$$

Applying Equation A4.7d gives the large-sample variances and covariance for the MLEs as

$$\sigma^2(\,\widehat{\mu}\,) = \sigma^2/n \simeq \widehat{\sigma}^{\,2}/n, \quad \sigma^2(\widehat{\sigma}^{\,2}) = 2\,\sigma^4/n \simeq 2\,\widehat{\sigma}^{\,4}/n, \quad \sigma(\,\widehat{\mu},\widehat{\sigma}^{\,2}) = 0$$

LIKELIHOOD-RATIO TESTS

Maximum likelihood provides for extremely convenient tests of hypotheses in the form of **likelihood-ratio**, or LR, tests (reviewed in Chapter 24 of Kendall and Stuart 1979) that examine whether a reduced model provides the same fit as a full model. The likelihood-ratio test statistic is given by

$$LR = 2\ln\left(\frac{\ell(\widehat{\boldsymbol{\Theta}}\,|\,\mathbf{z})}{\ell(\widehat{\boldsymbol{\Theta}}_r\,|\,\mathbf{z})}\right) = -2\ln\left(\frac{\ell(\widehat{\boldsymbol{\Theta}}_r\,|\,\mathbf{z})}{\ell(\widehat{\boldsymbol{\Theta}}\,|\,\mathbf{z})}\right) = -2\left[\,L(\widehat{\boldsymbol{\Theta}}_r\,|\,\mathbf{z}) - L(\widehat{\boldsymbol{\Theta}}\,|\,\mathbf{z})\right] \tag{A4.9}$$

where $\ell(\widehat{\boldsymbol{\Theta}}\,|\,\mathbf{z})$ is the likelihood evaluated at the MLE and $\ell(\widehat{\boldsymbol{\Theta}}_r\,|\,\mathbf{z})$ is the maximum of the likelihood function, subject to the restriction that r parameters unconstrained in the full likelihood analysis are assigned fixed values. For sufficiently large sample size, *the LR test statistic is χ^2_r-distributed,* a χ^2 with r degrees of freedom (Wald 1943).

Example 4. Suppose we wish to test the hypothesis that $\mu = 0$ in Example 1. Here the MLE is $\widehat{\mu} = \overline{z}$ and the LR test statistic becomes

$$-2\ln\left(\frac{\ell(0\,|\,\mathbf{z})}{\ell(\,\widehat{\mu}\,|\,\mathbf{z})}\right) = -2\ln\left(\frac{(2\pi)^{-n/2}\exp\left(-\sum_{i=1}^{n}(z_i-0)^2/2\right)}{(2\pi)^{-n/2}\exp\left(-\sum_{i=1}^{n}(z_i-\overline{z})^2/2\right)}\right)$$

$$= \sum_{i=1}^{n}\left[\,z_i^2 - (z_i-\overline{z})^2\,\right] = n\,\overline{z}^{\,2}$$

This test statistic is distributed as a χ^2 with one degree of freedom, as one parameter (μ) was assigned a fixed value in the reduced model. Since Prob($\chi^2_1 > 3.84$) = 0.05, the hypothesis $\mu = 0$ is rejected at the 5% level if the test statistic exceeds 3.84.

Now suppose we wish to test this hypothesis under the conditions of Example 3, where the variance is also unknown and hence must also be estimated. Here the MLEs for the full model are given by Equation A4.8c. Substituting $\mu = 0$ into Equation A4.8b gives the score function for the restricted model as

$$\frac{\partial L(\sigma^2)}{\partial\sigma^2} = \frac{n}{\sigma^2}\left(\frac{\overline{z^2}}{2\,\sigma^2} - \frac{1}{2}\right)$$

giving the MLE for σ^2 under this restriction as $\widehat{\sigma}_r^{\,2} = \overline{z^2}$. Substituting the MLEs into the likelihood functions, and once again using the identity $\sum(z_i-\mu)^2 = n(\,\overline{z^2} - 2\,\overline{z}\,\mu + \mu^2)$ gives the LR test statistic as

$$-2\ln\left(\frac{\ell(\,0,\widehat{\sigma}_r^{\,2}\,|\,\mathbf{z}\,)}{\ell(\,\widehat{\mu},\widehat{\sigma}^{\,2}\,|\,\mathbf{z}\,)}\right)$$

$$= -2\ln\left(\frac{(\,\overline{z^2}\,)^{-n/2}\cdot\exp\left[\,-n\,\overline{z^2}\,/\,(\,2\,\overline{z^2}\,)\,\right]}{(\,\overline{z^2} - \overline{z}^{\,2}\,)^{-n/2}\cdot\exp\left[\,-n\,(\,\overline{z^2} - \overline{z}^{\,2})\,/\,2\,(\,\overline{z^2} - \overline{z}^{\,2})\,\right]}\right)$$

$$= -n\,\ln\left(1 - \frac{(\,\overline{z}\,)^2}{\overline{z^2}}\right)$$

Again, for large samples this follows a χ_1^2 distribution as the value of one parameter is assigned a fixed value.

The *G*-test

A common likelihood-ratio based test is the ***G*-test** for goodness of fit. Consider n observations that have been apportioned into a set of N different categories, and denote these by the vector $\mathbf{n} = (n_1, n_2, \cdots, n_N)$. Likewise, let p_i represent the true population frequency of the ith category and let $\mathbf{p} = (p_1, p_2, \cdots, p_N)$. From the multinomial distribution, the likelihood of $\mathbf{p}$ given the observations $\mathbf{n}$ is

$$\ell(\mathbf{p} \,|\, \mathbf{n}) = k\, p_1^{n_1}\, p_2^{n_2} \,\cdots\, p_N^{n_N} \tag{A4.10a}$$

where k is the appropriate multinomial coefficient (which is independent of the p_i). It can be shown that the values of p_i that maximize Equation A4.10a (and hence are the MLE's) are $\widehat{p_i} = n_i/n$. This gives the value of the maximum of the likelihood function as

$$\ell(\widehat{\mathbf{p}} \,|\, \mathbf{n}) = k \left(\frac{n_1}{n}\right)^{n_1} \left(\frac{n_2}{n}\right)^{n_2} \cdots \left(\frac{n_N}{n}\right)^{n_N} \tag{A4.10b}$$

In order to test whether the observed data are consistent with a specified vector $\mathbf{q}$ of population frequencies, we need the value of the likelihood function under this constraint. Denoting the expected value for the number of individuals in category i by $\widehat{n}_i = q_i n$, we can write $q_i = \widehat{n}_i/n$. Substitution into Equation A4.10a gives the likelihood under $\mathbf{q}$ as

$$\ell(\mathbf{q} \,|\, \mathbf{n}) = k \left(\frac{\widehat{n}_1}{n}\right)^{n_1} \left(\frac{\widehat{n}_2}{n}\right)^{n_2} \cdots \left(\frac{\widehat{n}_N}{n}\right)^{n_N} \tag{A4.10c}$$

Applying Equation A4.9 yields the likelihood-ratio test (in this case, it is also called the *G*-test, for goodness of fit) that the observed data are consistent with $\mathbf{q}$,

$$G = -2\ln\left(\frac{\ell(\mathbf{q} \,|\, \mathbf{n})}{\ell(\widehat{\mathbf{p}} \,|\, \mathbf{n})}\right) = -2\sum_{i=1}^{N} n_i \ln\left(\frac{\widehat{n}_i}{n_i}\right) = -2\sum_{i=1}^{N} n_i \ln\left(\frac{q_i}{\widehat{p}_i}\right) \tag{A4.11}$$

Since the N frequencies sum to one, there are $N-1$ unconstrained parameters in the full likelihood, implying that G is asymptotically distributed as a χ_{N-1}^2 random variable. Since large sample sizes are required to give the likelihood-ratio test a χ^2 distribution, caution should be exercised in employing this test whenever any expected quantity is less than five (e.g., any $q_i < 5/n$), a problem that can sometimes be avoided by pooling cells. Sokal and Rohlf (1995) provide a thorough overview of these and other matters.

Likelihood-ratio Tests for the General Linear Model

As a final example of likelihood-ratio tests, consider the general linear model (Chapters 8, 26, 27), $\mathbf{y} = \mathbf{X}\boldsymbol{\beta} + \mathbf{e}$, where we assume that the $n \times 1$ vector of residual errors $\mathbf{e}$ is multivariate normal, with mean vector zero and covariance matrix $\mathbf{V}$, i.e., $\mathbf{e} \sim \text{MVN}(\mathbf{0}, \mathbf{V})$. From Equation 8.24, the density function for $\mathbf{e}$ is

$$(2\pi)^{-n/2} |\,\mathbf{V}\,|^{-1/2} \exp\left(-\frac{1}{2}\mathbf{e}^T\mathbf{V}^{-1}\mathbf{e}\right)$$

Writing the vector of residuals as $\mathbf{e} = \mathbf{y} - \mathbf{X}\boldsymbol{\beta}$ gives the resulting likelihood for $\boldsymbol{\beta}$ and $\mathbf{V}$, conditional on the observed data $(\mathbf{y}, \mathbf{X})$, as

$$\ell(\boldsymbol{\beta}, \mathbf{V} \,|\, \mathbf{y}, \mathbf{X}) = (2\pi)^{-n/2} |\,\mathbf{V}\,|^{-1/2} \exp\left(-\frac{1}{2}(\mathbf{y} - \mathbf{X}\boldsymbol{\beta})^T\mathbf{V}^{-1}(\mathbf{y} - \mathbf{X}\boldsymbol{\beta})\right)$$

which has log-likelihood

$$L = \ln \ell = -\frac{n}{2}\ln(2\pi) - \frac{1}{2}\ln|\,\mathbf{V}\,| - \frac{1}{2}(\mathbf{y} - \mathbf{X}\boldsymbol{\beta})^T\mathbf{V}^{-1}(\mathbf{y} - \mathbf{X}\boldsymbol{\beta}) \tag{A4.12}$$

Here $\boldsymbol{\beta}$ is a vector of fixed effects and the matrix $\mathbf{V}$ is a function of k variance components, with $\mathbf{V} = \sum_{i=1}^{k} \mathbf{R}_i\, \sigma_i^2$ where the $\mathbf{R}_i$ are matrices of known constants. Thus, the parameters to be estimated are the vector $\boldsymbol{\beta}$ of fixed effects and the k variances, σ_i^2.

Suppose we wish to compare the relative fit of two models that assume the same covariance structure (i.e., the same $\mathbf{V}$), but have different vectors of fixed effects, a vector $\boldsymbol{\beta}_f$ for the full model vs. a vector $\boldsymbol{\beta}_r$ for the reduced model that assumes fewer factors. The resulting likelihood-ratio test statistic is

$$\begin{aligned} LR &= -2\left[\,L(\widehat{\boldsymbol{\beta}}_r \,|\, \mathbf{y}, \mathbf{X}_r) - L(\widehat{\boldsymbol{\beta}}_f \,|\, \mathbf{y}, \mathbf{X}_f)\right] \\ &= \left[(\mathbf{y} - \widehat{\mathbf{y}}_r)^T\widehat{\mathbf{V}}^{-1}(\mathbf{y} - \widehat{\mathbf{y}}_r) - (\mathbf{y} - \widehat{\mathbf{y}}_f)^T\widehat{\mathbf{V}}^{-1}(\mathbf{y} - \widehat{\mathbf{y}}_f)\right] \end{aligned} \tag{A4.13}$$

where $\widehat{\mathbf{y}}_f = \mathbf{X}_f\widehat{\boldsymbol{\beta}}_f$ and $\widehat{\mathbf{y}}_r = \mathbf{X}_r\widehat{\boldsymbol{\beta}}_r$ are the predicted means under the full and reduced models. For large sample sizes, this test statistic follows a χ^2 distribution with $n_f - n_r$ degrees of freedom, where n_f and n_r are the degrees of freedom for the full and reduced models, respectively.

Example 5. Suppose the y_i values are the means of n different populations, e.g., data from a series of populations being used in a line-cross analysis (Chapter 9). Assuming the means are independent but with potentially different variances (due to differences in sample sizes, among other things), $\mathbf{V}$ is a diagonal matrix whose ith element is the variance of the ith mean. Denoting the variance of the ith mean by Var(y_i), then recalling Equation A3.11c, the quadratic product in the LR test reduces to

$$(\mathbf{y}-\widehat{\mathbf{y}})^T\,\widehat{\mathbf{V}}^{-1}(\mathbf{y}-\widehat{\mathbf{y}}) = \sum_{i=1}^{n} \frac{(y_i - \widehat{y}_i)^2}{\text{Var}(y_i)}$$

Hence, the likelihood-ratio test statistic for comparing a full model with a reduced model assuming fewer effects is given by

$$\sum_{i=1}^{n} \frac{[\,y_i - \widehat{y}_i(r)\,]^2}{\text{Var}(y_i)} - \sum_{i=1}^{n} \frac{[\,y_i - \widehat{y}_i(f)\,]^2}{\text{Var}(y_i)} \tag{A4.14}$$

which is just the difference in the χ^2 values for the fit of the full and reduced models. This test follows a χ^2 distribution with degrees of freedom given by the difference in degrees of freedom for the full and reduced models.

ITERATIVE METHODS FOR SOLVING ML EQUATIONS

While ML estimation and hypothesis testing with likelihood ratios is conceptually straightforward, in practice it can be quite difficult to accomplish due to the complexities associated with having to find the maximum of the likelihood function. Ideally, closed-form solutions to the MLEs can be obtained by deriving the score vector, setting it equal to zero, and solving. However, in many cases this is impractical and numerical approaches must be used. In very simple cases with one or two parameters, a brute force approach relying upon a **grid search** can be used, where one computes a one- or two-dimensional plot of the likelihood surface as a function of the unknown parameters. With more than two variables, this is impractical and a variety of iterative methods have been suggested as alternatives. We discuss two of these here, Newton-Raphson and EM methods (Chapter 27 discusses these methods further in the context of variance-component estimation). A potential problem with all iterative methods is that they may not converge to the true MLEs if the likelihood surface contains several local maxima. Iterative methods require an initial starting value, and a poor choice can result in the iteration converging to a solution that is a local, but not a global, maximum. Hence, when applying iterative methods, several starting points should be used.

Newton-Raphson Methods

Recall from elementary calculus that one can approximate a function $f(x)$ by expanding it in a power series around a point x_o,

$$f(x) \simeq f(x_0) + (x - x_o) f'(x_o)$$

This suggests one approach for finding roots of the equation $f(x) = 0$. Given some initial guess x_0, an improved value is obtained by solving

$$f(x) = 0 \simeq f(x_0) + (x - x_o)f'(x_o)$$

for x, or

$$x \simeq x_0 - \frac{f(x_0)}{f'(x_0)} \tag{A4.15a}$$

Noting that the score function is zero at the MLE [$S(\widehat{\theta}) = 0$], this suggests one approach for obtaining an iterative solution of the MLE. Applying Equation A4.15a to the score function, so that $f = S$ and $f' = \partial S(\theta)/\partial\theta = \partial L^2(\theta)/\partial^2\theta$, an updated estimate, $\widehat{\theta}^{(k+1)}$, of a current estimate $\widehat{\theta}^{(k)}$ is given by

$$\widehat{\theta}^{(k+1)} = \widehat{\theta}^{(k)} - \left(\frac{\partial L^2(\theta)}{\partial^2\theta} \bigg|_{\theta=\widehat{\theta}^{(k)}} \right)^{-1} S[\widehat{\theta}^{(k)}] \tag{A4.15b}$$

which is interated until $|\widehat{\theta}^{(k+1)} - \widehat{\theta}^{(k)}|$ is sufficiently small to declare convergence. This is the **Newton-Raphson** method, a member of a class of **quadratic methods**. Such methods involve second partial derivatives of the likelihood function and have a quadratic convergence rate. The same logic when applied to a multivariate Taylor series implies that a vector $\widehat{\boldsymbol{\Theta}}^{(k)}$ of current estimates is updated by using

$$\widehat{\boldsymbol{\Theta}}^{(k+1)} = \widehat{\boldsymbol{\Theta}}^{(k)} - \mathbf{H}^{-1}(\widehat{\boldsymbol{\Theta}}^{(k)})\,\mathbf{S}(\widehat{\boldsymbol{\Theta}}^{(k)}) \tag{A4.16}$$

where **S** and **H** are the vector of scores and the Hessian matrix, respectively, both evaluated at the current estimate.

One variant of this approach is **Fisher's scoring**, where the Hessian matrix **H** is replaced by its expected value, the negative of Fisher's information matrix **F** (Equation A4.7b),

$$\widehat{\boldsymbol{\Theta}}^{(k+1)} = \widehat{\boldsymbol{\Theta}}^{(k)} + \mathbf{F}^{-1}(\widehat{\boldsymbol{\Theta}}^{(k)})\,\mathbf{S}(\widehat{\boldsymbol{\Theta}}^{(k)}) \tag{A4.17}$$

One advantage of Fisher's scoring is that **F** is usually of a simpler form than **H**, often containing elements equal to zero that are non-zero in **H**. This can make **F** easier to compute and invert (e.g., compare Equations 27.34 and 27.35b). Further, Fisher's scoring appears to be more robust to poor initial starting choices than the strict Newton-Raphson method (Jennrich and Sampson 1986). In addition to the advantage of quadratic convergence, both Newton-Raphson and Fisher's scoring yield the covariance matrix of MLE estimates from **H** (or **F**) using the final interation values of $\boldsymbol{\Theta}$ and applying Equation A4.7. Additional quadratic methods are discussed by Kennedy and Gentle (1980).

Expectation-maximization (EM) Methods

Newton-Raphson and related methods require the first and second derivatives of the likelihood function, which can be difficult to obtain and/or computationally demanding (e.g., requiring the repeated inversion of large matrices). An alternative strategy is to use expectation-maximization (EM) methods, which were introduced by Dempster et al. (1977) as a very general iterative approach for data sets with missing (or incomplete) data. The idea is that, in many cases, if we had more information about certain observations, MLEs are easily obtained. For example, if observations are drawn from a mixture distribution (Chapter 13), obtaining the MLEs for the means and variances of the underlying distributions is trival *provided* we know from which distribution each individual observation is drawn. Thus the original data set is treated as incomplete data, missing additional information (e.g., for a mixture model, which distribution a specific observation is drawn from). Using a current estimate of the unknown parameter values, the expected value of the incomplete data is computed (e.g., for a mixture model, the category identity of each individual is estimated). This is the **expectation**, or **E step**. The result is a set of likelihood equations that are considerably easier to solve than the full likelihood (the **maximization**, or **M step**). The new estimates obtained from the M step are then used to update the expected values, and this approach is iterated until convergence. The EM method refers to a general class of approaches, and there can be several EM versions for solving the same problem.

While EM methods often have fairly simple forms and hence are easy to program, they can be extremely slow to converge to a solution. EM methods offer computational advantages over Newton-Raphson methods, as they do not have to compute second derivatives of the likelihood function and they do not directly evaluate the full likelihood function. However, this is a disadvantage in terms of constructing confidence intervals and LR tests, as other approaches must be used to obtain the standard errors of the MLEs and to compute the likelihoods needed for LR tests. Chapter 27 discusses an EM method for computing unknown variance components in linear models, while our focus here is on the other broad class of likelihood models used throughout this book, mixture models (introduced in Chapter 13).

EM for Mixture Model Likelihoods

Mixture models naturally appear in a number of quantitative-genetic settings, wherein the observed distribution is really a weighted sum of a number of underlying distributions. For example, when a major diallelic locus is segregating in a population, the phenotypic distribution is a weighted sum of the three distributions representing each major locus genotype (Chapter 13). The general likelihood function for a single observation z from the kinds of mixture models considered in this book has the form

$$\ell(\boldsymbol{\Theta} \mid z) = \sum_{k=1}^{N} \pi_k \cdot \varphi(z, \mu_k, \sigma^2) \tag{A4.18a}$$

where the distribution is assumed to result from N underlying normals, the kth of which has frequency π_k, mean μ_k, and common variance σ^2. We assume that the number N of underlying distributions is known and wish to estimate the $2N \times 1$ vector $\boldsymbol{\Theta}$ of the N means, the common variance, and the $N-1$ independent mixing proportions (the π_k). With n individuals independently drawn from this distribution, the full likelihood is

$$\ell(\boldsymbol{\Theta} \mid \mathbf{z}) = \prod_{i=1}^{n} \ell(\boldsymbol{\Theta} \mid z_i) \tag{A4.18b}$$

While appearing rather simple, the full likelihood function is complicated to work with analytically, and numerical approaches are usually employed.

When we observe a particular value, we don't know which underlying distribution (or category) it was drawn from. If we knew the category identity for each observation, the ML solutions for the mean and variance of the underlying distributions are easily computed. For example, if a single diallelic QTL is segregating, if we could determine whether individuals had QTL genotype *QQ*, *Qq*, or *qq*, then the mean for each genotype and the common variance could easily be estimated. This is the basis of the EM method. We start with some initial guess as to the category identity of each observation, which then allows us to easily compute an ML estimate of the means and variance of the underlying distribution. This guess is in the form of a weight vector for each individual, whose k element, $w(k \mid z)$, is the probability that an individual has the kth QTL genotype given they have trait value z. Using these mean and variance estimates, updated weights can be computed using **Bayes' theorem** (Equation 13.24) for conditional probabilities. Since $w(k \mid z) = \Pr(k \mid z)$, applying Bayes' theorem gives

$$w(k \mid z) = \frac{\Pr(k) \cdot \Pr(z \mid k)}{\Pr(z)} = \frac{\pi_k \cdot \varphi(z, \mu_k, \sigma^2)}{\Pr(z)} = \frac{\pi_k \cdot \varphi(z, \mu_k, \sigma^2)}{\sum_{j=1}^{N} \pi_j \cdot \varphi(z, \mu_j, \sigma^2)} \tag{A4.19}$$

These updated weights are then used to obtain new estimates of the category-specific means and the variance, and this procedure is repeated until convergence. Formally, this EM approach proceeds as follows (Aitkin and Wilson 1980):

(1) **Initial step**. Choose initial starting values for the MLEs of the variance $\widehat{\sigma}^{2\,(0)}$ and the vectors of mixture proportions and means,

$$\widehat{\boldsymbol{\pi}}^{(0)} = (\widehat{\pi}_1^{(0)}, \cdots, \widehat{\pi}_N^{(0)}), \qquad \widehat{\boldsymbol{\mu}}^{(0)} = (\widehat{\mu}_1^{(0)}, \cdots, \widehat{\mu}_N^{(0)}) \tag{A4.20}$$

(2) **E-step**. Define the weight $w^{(1)}(k \,|\, z_i)$ as the probability that observation z_i is drawn from distribution k given the current estimates $\widehat{\sigma}^{\,2\,(0)}$, $\widehat{\boldsymbol{\pi}}^{\,(0)}$, and $\widehat{\boldsymbol{\mu}}^{\,(0)}$. From Bayes' theorem,

$$w^{(1)}(k \,|\, z_i) = \frac{\widehat{\pi}_k^{(0)} \cdot \varphi(z_i, \widehat{\mu}_k^{(0)}, \widehat{\sigma}^{\,2\,(0)})}{\displaystyle\sum_{j=1}^{N} \widehat{\pi}_j^{(0)} \cdot \varphi(z_i, \widehat{\mu}_j^{(0)}, \widehat{\sigma}^{\,2\,(0)})} \tag{A4.21}$$

where $\varphi(z_i, \widehat{\mu}_k^{(0)}, \widehat{\sigma}^{\,2\,(0)})$ is the normal distribution evaluated at the value z_i using mean $\widehat{\mu}_k^{(0)}$ and variance $\widehat{\sigma}^{\,2\,(0)}$.

(3) **M-step**. Assuming these weights are correct, the updated estimates of the MLEs are obtained as follows:

(a) **Mixing proportions**: Given by the average probability of being in category k,

$$\widehat{\pi}_k^{(1)} = \overline{w}_k^{(1)} = \frac{1}{n}\sum_{i=1}^{n} w^{(1)}(k \,|\, z_i) \tag{A4.22a}$$

(b) **Means**: Given by the weighted average of the observations,

$$\widehat{\mu}_k^{(1)} = \frac{1}{n}\sum_{i=1}^{n} z_i \left(\frac{w^{(1)}(k \,|\, z_i)}{\overline{w}_k^{(1)}} \right) \tag{A4.22b}$$

(c) **Variance**: Given by the weighted variance of the observations,

$$\widehat{\sigma}^{\,2\,(1)} = \frac{1}{n}\sum_{i=1}^{n}\sum_{k=1}^{N} \left(z_i - \widehat{\mu}_k^{(1)} \right)^2 w^{(1)}(k \,|\, z_i) \tag{A4.22c}$$

These updated estimates are then used to compute new weights, and the whole procedure continues until the interations converge.

EM Modifications for QTL Mapping

One important application of mixture models involves the use of marker data to map QTLs (Chapters 14–16). Here estimates of the category identity are influenced not only by an individual's trait value z but also by its marker genotype value m. For example, suppose a single diallelic QTL (with alleles *Q*, *q*) is linked to a marker, and an inbred-line cross is used in an attempt to map and characterize this QTL. In this case, the likelihood for an individual with marker genotype m and trait value z is

$$\ell(\boldsymbol{\Theta} \,|\, z, m) = \sum_{k=1}^{3} \pi_k(m) \cdot \varphi(z, \mu_k, \sigma^2) \tag{A4.23}$$

where

$$\pi_k(m) = \Pr(Q_k \,|\, m) \qquad \text{and} \qquad \Theta = (\mu_{QQ},\, \mu_{Qq},\, \mu_{qq},\, \sigma^2)$$

where Q_k denotes the kth QTL genotype. The mixing proportions $\pi_k(m)$ are functions of the marker genotype, the QTL position (generally given by the marker-QTL recombination frequency distance c), and the particular design used (see Example 6 below).

Support for the presence of a QTL is usually displayed using **likelihood maps** (Chapters 15–16), which plot the maximum value of the likelihood function over all possible values of c, with the c value giving the largest value being taken as the MLE for QTL position. For a given value of c, say c_0, the EM method is used to obtain the MLE for Θ under the restriction that a QTL is at map position c_0. Here, the weights are again given by Bayes' theorem (Equation A4.21), where the weight for an individual with trait value z and marker genotype m assuming QTL position (c) is

$$\begin{aligned} \Pr(Q_k \,|\, m, z, c) &= \frac{\Pr(Q_k \,|\, m, c) \cdot \Pr(z \,|\, Q_k)}{\Pr(z \,|\, m, c)} \\ &= \frac{\Pr(Q_k \,|\, m, c) \cdot \varphi(z, \mu_{Q_k}, \sigma^2)}{\sum_{j=1}^{N} \Pr(Q_j \,|\, m, c) \cdot \varphi(z, \mu_{Q_j}, \sigma^2)} \end{aligned} \tag{A4.24}$$

Using these weights, updated estimates of the means and variance are obtained as above, and these are substituted back into Equation A4.24 to obtain new weights. This procedure is continued until the iterations converge. Ranging through all possible values of c and plotting the resulting maximum of the likelihood function for each c value thus generates a likelihood map for c.

Example 6. As an example of accounting for missing marker information, consider QTL mapping in an F_2 design from an inbred-line cross (Chapter 15) using dominant markers (such as RAPDs). Suppose marker allele M is dominant to allele m, so that the observed marker genotypes are mm and $M-$, the later consisting of the genotypes MM and Mm. Since in the F_2, $\Pr(M-) = \Pr(MM) + \Pr(Mm) = 3/4$,

$$\Pr(MM \,|\, M-) = \frac{\Pr(MM)}{\Pr(M-)} = \frac{(1/4)}{(3/4)} = 1/3$$

and likewise $\Pr(Mm \,|\, M-) = 2/3$. The conditional probability $\Pr(QQ \,|\, M-)$ that the QTL genotype is QQ given the marker genotype is $M-$ becomes

$$\begin{aligned}\Pr(QQ\,|\,M-) &= \Pr(QQ\,|\,MM)\Pr(MM\,|\,M-) \\ &\quad + \Pr(QQ\,|\,Mm)\Pr(Mm\,|\,M-) \\ &= \frac{1}{3}\Pr(QQ\,|\,MM) + \frac{2}{3}\Pr(QQ\,|\,Mm)\end{aligned}$$

From Example 1 from Chapter 15, for the F_2 design $\Pr(QQ\,|\,MM) = (1-c)^2$ and $\Pr(QQ\,|\,Mm) = c(1-c)$, giving

$$\Pr(QQ\,|\,M-) = \frac{1}{3}(1-c)^2 + \frac{2}{3}c(1-c) = \frac{1-c^2}{3}$$

Similarly, it can be shown that

$$\Pr(Qq\,|\,M-) = \frac{2(1-c+c^2)}{3} \quad \text{and} \quad \Pr(qq\,|\,M-) = \frac{c(2-c)}{3}$$

Thus, for a given c value, these mixing proportions are fixed constants. Similar logic gives the values for individuals with marker genotype **mm**.

To obtain the weights for the EM method, first index the three QTL genotypes by $k = 1, 2, 3$ for QQ, Qq, and qq. Given a current estimate of the three QTL means $\widehat{\mu}_k^{(t)}$ and variance $\widehat{\sigma}^{2\,(t)}$, the updated weights are obtained from Equation A4.24. For example, the updated weight that an individual is genotype QQ given it has trait value z and marker genotype $M-$ is

$$w^{(t+1)}(1\,|\,z,\, M-, c) = \frac{\Pr(QQ\,|\,M-,c)\cdot\Pr(z\,|\,QQ)}{\Pr(z\,|\,M-,c)}$$

$$\begin{aligned}&= \Pr(QQ\,|\,M-,c)\cdot\Pr(z\,|\,QQ)\Big/\Big[\Pr(QQ\,|\,M-,c)\cdot\Pr(z\,|\,QQ) \\ &\qquad + \Pr(Qq\,|\,M-,c)\cdot\Pr(z\,|\,Qq) + \Pr(qq\,|\,M-,c)\cdot\Pr(z\,|\,qq)\Big] \\ &= (1-c)^2\cdot\varphi(z,\widehat{\mu}_1^{(t)},\widehat{\sigma}^{2\,(t)})\Big/\Big[(1-c)^2\cdot\varphi(z,\widehat{\mu}_1^{(t)},\widehat{\sigma}^{2\,(t)}) \\ &\qquad + 2(1-c+c^2)\cdot\varphi(z,\widehat{\mu}_2^{(t)},\widehat{\sigma}^{2\,(t)}) + c(2-c)\cdot\varphi(z,\widehat{\mu}_3^{(t)},\widehat{\sigma}^{2\,(t)})\Big]\end{aligned}$$

The probabilities of the two other QTL genotypes follow similarly, as do the weights for individuals with marker genotype mm. Using these updated weights, new estimates of the means and variance are obtained from Equations A4.22b, c.

Appendix 5

Computing the Power of Statistical Tests

There are two types of errors that one can make when performing a statistical test. A **false positive** (a **Type I error**) occurs when the null hypothesis is rejected when in fact it is correct. We control for this by setting the significance level α of a test (the probability of a false positive) to be small. The other source of error is a **false negative** (a **Type II error**), *failing* to reject the null hypothesis when in fact it is false. The **power** of a test is defined to be the probability that the null hypothesis is rejected when it is indeed false. Hence if β is the probability of a false negative, the power is $1 - \beta$ (Figure A5.1).

Before embarking on a potentially very costly experiment, the investigator would like to be certain that the design ensures sufficiently high power given the objectives of the proposed study. Indeed, many experiments that report negative results may in fact have significant biological effects that are swamped out by the high sampling error generated by insufficient sample size. Power depends not

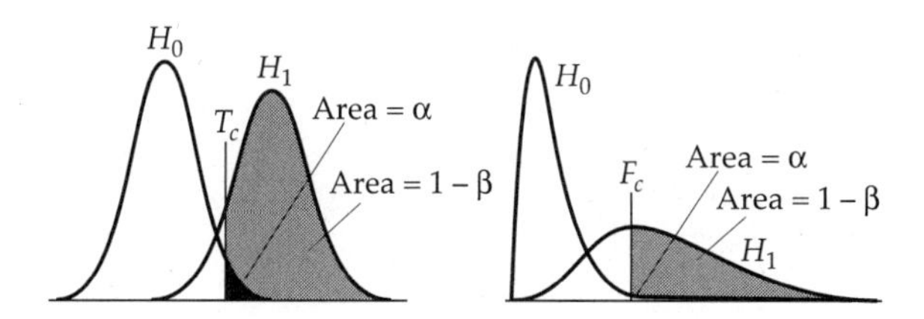

Figure A5.1 Power $(1-\beta)$ and significance (α) of a normally (**left**) and F-distributed test statistics (**right**). The distributions of the test statistic under the null hypothesis (H_0) and at a particular parameter value under the alternative hypothesis (H_1) are shown. Under the null hypothesis, the probability that the test statistic exceeds a critical value (T_c, F_c) is α, and we define this as the significance of the test. The power of the test for a particular value of the alternative hypothesis is the probability that the test statistic exceeds the critical value, and we denote this by $1 - \beta$. Hence, β is the probability that the test statistic is below the critical value (a false negative).

only on sample size and the actual values of the unknown distribution parameters being estimated, but also on the assumed level of significance α of the test. Here we consider some of the basics of computing the power of a given design for tests whose statistics are normally and F distributed. Both appear often throughout this book.

POWER OF NORMALLY DISTRIBUTED TEST STATISTICS

Assume the test statistic T is normally distributed under the null hypothesis, with $T \sim \text{N}(\mu_0, \sigma_0^2)$. For power calculations, assume the null hypothesis is false and that the test statistic actually has mean μ_1 and variance σ_1^2 but remains normally distributed, so that $T \sim \text{N}(\mu_1, \sigma_1^2)$. To obtain expressions for the required sample size for a given power, it is convenient to first write the sample variance as a function of sample size n, with $\sigma_i^2 = f_i^2/n$ for $i = 0, 1$. (Often $\sigma_0^2 = \sigma_1^2$, in which case $f_0 = f_1$.) Finally, let $z_{(\alpha)}$ satisfy

$$\Pr(U \leq z_{(\alpha)}) = \alpha \tag{A5.1a}$$

where U is a unit normal random variable. For example, $\Pr(U \leq 1.65) = 0.95$, so that $z_{(0.95)} = 1.65$. Two identities involving z_α will prove useful in the following discussions. First, note from Figure A5.2 that

$$\Pr(U > z_{(1-\alpha)}) = \alpha \tag{A5.1b}$$

so that, for example, $\Pr(U > 1.65) = 0.05$. Second, from the symmetry of the normal distribution it can be shown that

$$z_{(\alpha)} = -z_{(1-\alpha)} \tag{A5.1c}$$

We now have the all the necessary definitions in hand to consider the power of normally distributed tests. Hypothesis tests generally fall into two categories: **one-sided** and **two-sided**, and we examine these in turn before considering specific applications.

One-sided Tests

Some hypotheses are naturally one-sided. For example, we may wish to test whether heritability h^2 is significantly different from zero. In this case, true values of $h^2 < 0$ cannot occur, so that the alternative to the null hypothesis $h^2 = 0$ is $h^2 > 0$.

Consider the null hypothesis that the test statistic has mean $\mu = \mu_o$ versus the alternative hypothesis that $\mu = \mu_1$, where (for example) $\mu_1 > \mu_0$. Suppose that

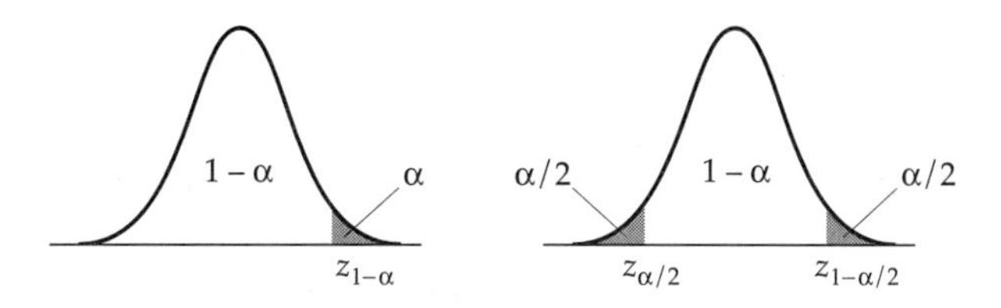

Figure A5.2 Areas under the normal curve for one-sided (left) and two-sided (right) tests of significance at level α.

under the null hypothesis, the test statistic $T \sim \mathrm{N}(\mu_o, \sigma_o^2)$, so that $(T - \mu_o)/\sigma_o$ is distributed as a unit normal, implying from Equation A5.1b that

$$\Pr\left(\frac{T-\mu_o}{\sigma_o} > z_{(1-\alpha)}\right) = \Pr\left(T > \mu_o + \sigma_o\, z_{(1-\alpha)}\right) = \alpha \tag{A5.2a}$$

Thus, comparing the observed statistic T with the α-level critical value, $T_c(\alpha)$, where

$$T_c(\alpha) = \mu_o + \sigma_o\, z_{(1-\alpha)} \tag{A5.2b}$$

gives the one-sided test of significance with probability α of a false positive. We reject the null hypothesis when $T > T_c(\alpha)$.

Suppose that the alternative hypothesis is correct, so that $T \sim \mathrm{N}(\mu_1, \sigma^2)$. What is the probability that the resulting test statistic is significant (and hence a false negative is avoided)? Since it is now $(T - \mu_1)/\sigma_1$ that follows a unit normal, the probability that the test statistic exceeds the critical value $T_c(\alpha)$ given by Equation A5.2b is

$$\begin{aligned}\Pr\left(T > T_c(\alpha)\right) &= \Pr\left(\frac{T-\mu_1}{\sigma_1} > \frac{T_c(\alpha)-\mu_1}{\sigma_1}\right) \\ &= \Pr\left(U > \frac{\mu_o-\mu_1}{\sigma_1} + z_{(1-\alpha)}\,\frac{\sigma_o}{\sigma_1}\right)\end{aligned} \tag{A5.3}$$

where U is a unit normal random variable. Noting that $\Pr(U > z_{(\beta)}) = 1 - \beta$ (Equation A5.1b), from Equation A5.3 it immediately follows that for the test to have power $1 - \beta$ requires

$$\frac{\mu_o-\mu_1}{\sigma_1} + z_{(1-\alpha)}\,\frac{\sigma_o}{\sigma_1} = z_{(\beta)}$$

Using our notation for writing the variance of the test statistic as a function f divided by the sample size n (i.e., $\sigma_i^2 = f_i^2/n$), this becomes

$$\sqrt{n}\,\frac{(\mu_o-\mu_1)}{f_1} + z_{(1-\alpha)}\,\frac{f_o}{f_1} = z_{(\beta)}$$

Since $-z_{(\beta)} = z_{(1-\beta)}$ (Equation A5.1c), this rearranges to give

$$n = \frac{f_1^2}{(\mu_1 - \mu_o)^2} \left(z_{(1-\alpha)} \frac{f_o}{f_1} + z_{(1-\beta)} \right)^2 \tag{A5.4a}$$

$$= \left(\frac{z_{(1-\beta)}\, f_1 + z_{(1-\alpha)}\, f_0}{\mu_1 - \mu_0} \right)^2 \tag{A5.4b}$$

$$\simeq \frac{f_1^2}{(\mu_1 - \mu_o)^2} \left(z_{(1-\alpha)} + z_{(1-\beta)} \right)^2 \tag{A5.4c}$$

Equation A5.4c holds when the sample variances under the null hypothesis and true parameter values are approximately equal ($f_0 \simeq f_1$), as is often the case.

Two-sided Tests

Other hypotheses are naturally **two-sided**. Here, the null hypothesis is $\mu = \mu_o$ versus the alternative of $\mu \neq \mu_o$. A glance at Figure A5.2 shows that under the null hypothesis,

$$\Pr\left(z_{(\alpha/2)} \leq \frac{T - \mu_o}{\sigma_o} \leq z_{(1-\alpha/2)} \right) = 1 - \alpha \tag{A5.5a}$$

Again recalling Equation A5.1c, this rearranges to become

$$\Pr\left(-\sigma_o\, z_{(1-\alpha/2)} \leq T - \mu_o \leq \sigma_o\, z_{(1-\alpha/2)} \right) = 1 - \alpha$$

giving

$$| T - \mu_o | > \sigma_o\, z_{(1-\alpha/2)} \tag{A5.5b}$$

as a two-sided test of significance with probability α of a false positive under the null hypothesis. Under the alternative hypothesis that $T \sim \mathrm{N}(\mu_1, \sigma_1^2)$, the power of this test is

$$\Pr[T - \mu_0 < -\sigma_o\, z_{(1-\alpha/2)}] + \Pr[T - \mu_0 > \sigma_o\, z_{(1-\alpha/2)}]$$

Since $(T - \mu_1)/\sigma_1 \sim U$, this is equivalent to

$$\Pr\left(U < \frac{\mu_o - \mu_1}{\sigma_1} - \frac{\sigma_o}{\sigma_1} z_{(1-\alpha/2)} \right) + \Pr\left(U > \frac{\mu_o - \mu_1}{\sigma_1} + \frac{\sigma_o}{\sigma_1} z_{(1-\alpha/2)} \right) \tag{A5.6}$$

Using the same logic leading to Equation A5.4, the sample size n required to give power $1 - \beta$ is found to be

$$n = \frac{f_1^2}{(\mu_o - \mu_1)^2} \left(z_{(1-\alpha/2)} \frac{f_o}{f_1} + z_{(1-\beta)} \right)^2 \tag{A5.7a}$$

$$\simeq \frac{f_1^2}{(\mu_o - \mu_1)^2} \left(z_{(1-\alpha/2)} + z_{(1-\beta)} \right)^2 \tag{A5.7b}$$

Thus, the two-sided expressions are identical to the one-sided expressions with α replaced by $\alpha/2$.

Example 1. One approach for detecting QTLs using a marker locus (with alleles M and m) in inbred-line crosses is to use the test statistic $T = \overline{z}_{MM} - \overline{z}_{mm}$, which compares the mean trait values of alternate marker homozygotes (Chapter 15). This leads to a simple t-test.

How powerful is this method for detecting QTLs? Assuming the marker is completely linked to the QTL, the mean of T is μ_1 = 2a, the difference between the means of alternate QTL homozygotes. As derived in Equation 15.34, $\sigma^2(T) = \sigma_1^2 = 8(1-r^2)\sigma_z^2/n$, where r^2 is the fraction of the total of the total character variance accounted for by this QTL and n is the total number of offspring sampled (only 1/4 of the individuals are a particular homozygote, so that the expected sample size for each homozygote marker class is $n/4$). Under the null hypothesis of no linked QTL, $\mu_0 = 0$ and $\sigma_0^2 = 8\sigma_z^2/n$. This test is two-sided, as a significantly positive or negative T indicates a linked QTL. Substituting into Equation A5.7a with $f_1^2 = 8(1-r^2)\sigma_z^2$ and $f_0^2 = 8\sigma_z^2$ gives the sample size required to have power $1-\beta$ in a test with significant level α

$$n = \left(\frac{\sqrt{8(1-r^2)\sigma_z^2}}{0-2a}\right)^2 \left(\sqrt{\frac{8\sigma_z^2}{8(1-r^2)\sigma_z^2}}\, z_{(1-\alpha/2)} + z_{(1-\beta)}\right)^2$$

$$= \left(2(1-r^2)\,\frac{\sigma_z^2}{a^2}\right)\left(\frac{z_{(1-\alpha/2)}}{\sqrt{1-r^2}} + z_{(1-\beta)}\right)^2 \tag{A5.8}$$

One popular approach when considering possible experimental designs is to compute the sample size required to give the expected confidence interval a preset length, so that the interval given by $\mu_1 \pm z_\alpha\, f_1/\sqrt{n}$ does not include μ_0. Here z_α is the appropriate value to give an α-level one- or two-sided test (depending on the hypothesis being tested). Solving for n,

$$(\mu_1 - \mu_o)^2 \geq \frac{z_\alpha^2\, f_1^2}{n}, \qquad \text{or} \qquad n \geq \frac{z_\alpha^2\, f_1^2}{(\mu_1 - \mu_o)^2}$$

This approach is often used in place of a direct power calculation, but just what is its actual power? From Equations A5.4c and A5.7b, this expression yields the n which gives a test whose power satisfies $z_{(1-\beta)} = 0$ or $\beta = 0.5$. Thus, designs using this sample size have only 50% power, i.e., they are expected to detect an effect only half the time.

Applications: Parent-offspring Regressions

As is detailed in Chapters 7 and 17, the slope b_{op} of the parent-offspring regression provides information on genetic variance components. What is the power of such parent-offspring regressions? We first note that N data points are used to compute the regression. For a single-parent regression, N is the number of parents, while for a midparent regression, N is the number of pairs of parents (we distinguish between these by indexing the regressions by p and $\overline{p}$, respectively). For each parental point, the offspring value could be that for a single offspring ($n = 1$) or the mean value of n offspring (indexed by o and $\overline{o}$, respectively).

Since the increase in power with sample size results from a reduction in the sampling variance, it is again useful to write the sampling variance of b_{op} as some function f divided by the number N of families, $\sigma^2(b_{op}) = f_{op}^2/N$. From Equation 17.6

$$f_{op}^2 = (1 - r_{op}^2)\,\frac{\sigma^2(z_o)}{\sigma^2(z_p)} \tag{A5.9}$$

where r_{op} is the parent-offspring correlation. The sampling variance under the null hypothesis that $b_{op} = 0$ follows by setting $r_{op} = 0$. If the parent-offspring regression has true slope b_{op}, what sample size (measured by number of parents) is required for a test to have power $1 - \beta$? Since this is a one-sided test ($b_{op} = 0$ vs. $b_{op} > 0$), substituting Equation A5.9 into Equation A5.4c (and hence assuming that the variances under null and alternative hypotheses are essentially equal, i.e., $\sigma_0^2 \simeq \sigma_1^2$) gives

$$N = \left[z_{(1-\alpha)} + z_{(1-\beta)}\right]^2\,\frac{(1 - r_{op}^2)}{b_{op}^2}\,\frac{\sigma^2(z_o)}{\sigma^2(z_p)} \tag{A5.10}$$

This expression, which assumes single parents and single offspring, can easily be modified to account for midparents and/or multiple offspring per family. With n measured offspring per family, $\sigma^2(z_{\overline{o}}) = \sigma^2(z_o)[t + (1-t)/n]$ replaces $\sigma^2(z_o)$, where t is the correlation between full sibs. Likewise, for a midparent-offspring regression, $\sigma^2(z_{\overline{p}}) = \sigma^2(z_p)/2$ replaces $\sigma^2(z_p)$, and $r_{o\overline{p}}$ and $b_{o\overline{p}}$ replace r_{op} and b_{op}. Chapter 17 discusses these modifications in more detail

---

Example 2. Suppose 200 (single) parent-offspring pairs are measured ($N = 200$, $n = 1$) for a character with $h^2 = 0.2$ and no epistasis or maternal effects. Assuming a test with $\alpha = 0.05$, what is the power of this design to detect this regression as being significant? Here $z_{(1-0.05)} = 1.65$ and from Chapter 7,

$$b_{op} = r_{op} = h^2/2 = 0.1$$

Assuming parent and offspring have equal variance, $\sigma^2(z_o) = \sigma^2(z_p)$, Equation A5.9 gives

$$\sigma^2(b_{op}) = \frac{f^2}{N} = \frac{1 - r_{op}^2}{N} = \frac{1 - 0.1^2}{200} = 0.00495$$

while under the null hypothesis of zero slope, $r_{op} = 0$ and hence $\sigma^2(b_{op}) = 1/200 = 0.005$. Equation A5.3 gives the power as

$$\Pr\left(U > \frac{\mu_o - \mu_1}{\sigma_1} + z_{(1-\alpha)}\,\frac{\sigma_o}{\sigma_1}\right) = \Pr\left(U > \frac{(0-0.1)}{\sqrt{0.00495}} + 1.65\,\frac{\sqrt{0.005}}{\sqrt{0.00495}}\right)$$
$$= \Pr(\,U > 0.236\,) = 0.40$$

The false-negative rate for this design is $\beta = 0.60$, so that the majority of time the observed slope will not be judged to be significantly greater than zero.

Measuring only a single offspring per family (n = 1), how many families must be used to have 90% probability that the observed slope is significantly positive (using a test with α = 0.05)? To make use of Equation A5.4, first note that here β = 0.1, and from unit normal tables,

$$Pr(U \leq 1.28) = 0.9, \quad \text{giving} \quad z_{(1-\beta)} = 1.28$$

hence

$$\left(z_{(1-\alpha)} + z_{(1-\beta)}\right)^2 = (1.65 + 1.28)^2 = 8.58$$

Substituting this result into Equation A5.10 gives the required N as

$$N = 8.58\,\frac{(1 - r_{op}^2)}{b_{op}^2}\,\frac{\sigma^2(z_o)}{\sigma^2(z_p)}$$

For a single-parent regression, $b_{op} = r_{op} = h^2/2 = 0.1$, and the required number of single-parent families is

$$N = 8.58\,\frac{(1 - 0.1^2)}{0.1^2} = 850$$

For a midparent-offspring regression, $b_{o\overline{p}} = h^2 = 0.2$, $r_{o\overline{p}}^2 = b_{o\overline{p}}^2/2 = 0.02$, and

$$\frac{\sigma^2(z_o)}{\sigma^2(z_{\overline{p}})} = \frac{\sigma^2(z_o)}{\sigma^2(z_p)/2} = 2$$

giving the required number of two-parent families as

$$N = \frac{8.58 \cdot 2\,(1 - 0.02)}{0.2^2} = 420$$

Applications: QTL Detection Tests Using Doubly Affected Sib Pairs

Chapter 16 examines the use of doubly affected full sib pairs to detect QTLs influencing a binary character (typically disease presence/absence, so that both sibs in each pair display the disease). If the marker is linked to a QTL influencing the character, we expect the pair members to share more marker alleles than expected by chance. A number of tests based on the number of ibd (identical by descent) marker alleles shared between pair members have been proposed for detecting a linked QTL. Letting p_{2i} and $\widehat{p}_{2i}$ denote the true and estimated fractions of doubly affected pairs sharing $i = 0, 1, 2$ ibd marker alleles (the leading 2 in the subscript indicates we are restricting attention to pairs where both sibs are affected), we consider two such tests here. One test statistic, $T_2 = \widehat{p}_{22}$, is based on the fraction of doubly affected pairs sharing 2 ibd alleles. A second test statistic, $T_m = \widehat{p}_{21}/2 + \widehat{p}_{22}$, corresponds to the mean number of ibd marker alleles shared by a doubly affected pair. The true ibd frequencies can be expressed as deviations from the values expected with no linked QTL,

$$p_{20} = \frac{1}{4} - d_{20}, \qquad p_{21} = \frac{1}{2} - d_{21}, \qquad p_{22} = \frac{1}{4} + d_{22}$$

Under the null hypotheses (the marker is unlinked to a QTL), the d_{2i} are zero, while Equation 16.52 gives their values as functions of the QTL effects and distance from the marker if a QTL is linked to the marker. Note from Equation 16.52 that $d_{2i} \geq 0$ under linkage.

We examine the power of the test based on $T_2 = \widehat{p}_{22}$ first. This test statistic has mean p_{22} and variance $p_{22}(1 - p_{22})/n$, where n is the number of doubly affected pairs. Expressing the variance of the test statistic as $\sigma^2(T) = f^2/n$, shows that here $f^2 = p_{22}(1 - p_{22})$. Under the null hypothesis of no linked QTL, $p_{22} = 1/4$, giving the mean and variance as $\mu_o = 1/4$ and $f_0^2 = (1/4)(3/4) = 3/16$. When a linked QTL is present, so that $p_{22} = 1/4 + d_{22}$, the mean and variance become

$$E(\widehat{p}_{22}) = \frac{1}{4} + d_{22}, \quad \text{and} \quad f_1^2 = p_{22}(1 - p_{22}) = (1/4 + d_{22})(3/4 - d_{22}) \quad \text{(A5.11a)}$$

Hence, $(\mu_1 - \mu_0)^2 = d_{22}^2$. The T_2 test is one-sided, as $p_{22} > 1/4$ under linkage to a QTL, and from Equation A5.4b, the required number of doubly affected pairs to have power $1 - \beta$ using a test with significance α is

$$n_{T_2} = \left(\frac{z_{(1-\beta)}\sqrt{(1/4 + d_{22})(3/4 - d_{22})} + z_{(1-\alpha)}\sqrt{3/16}}{d_{22}} \right)^2 \quad \text{(A5.11b)}$$

In a similar fashion, the expected value for the test statistic T_m is

$$\mu_1 = p_{21} + 2\,p_{22} = \left(\frac{1}{2} - d_{21}\right) + 2\left(\frac{1}{4} + d_{22}\right) = 1 + 2\,d_{22} - d_{21} \quad \text{(A5.12a)}$$

To compute the variance $\sigma_1^2 = f_1^2/n$, let x denote the fraction of ibd alleles in a randomly chosen individual. Taking expectations gives

$$f_1^2 = E(x^2) - \mu_x^2 = \left[1^2 \cdot \Pr(x=1) + 2^2 \cdot \Pr(x=2)\right] - \mu_x^2$$

$$= \left[1^2\left(\frac{1}{2} - d_{21}\right) + 2^2\left(\frac{1}{4} + d_{22}\right)\right] - (1 + 2\,d_{22} - d_{21})^2$$

$$= \frac{1}{2} + d_{21} - (2\,d_{22} - d_{21})^2 \tag{A5.12b}$$

Under the null hypothesis of no linked QTLs, $d_{21} = d_{22} = 0$, giving $\mu_0 = 1/2$ and $f_0^2 = 1/2$. Again, this is a one-sided test, as $T_m > 1/2$ for a marker linked to a QTL. Applying Equation A5.4b gives the required number of doubly affected pairs as

$$n_{T_m} = \left(\frac{z_{(1-\beta)}\sqrt{1/2 + d_{21} - (2d_{22} - d_{21})^2} + z_{(1-\alpha)}\sqrt{1/2}}{2d_{22} - d_{21}}\right)^2 \tag{A5.12c}$$

POWER OF *F*-RATIO TESTS

The analysis of variance (ANOVA) is widely used throughout this book, e.g., in estimating variance components under balanced experimental designs (Chapters 18–24) and in detecting QTLs via marker-trait associations (Chapters 14–16). ANOVA designs are typically based upon sums of squares (SS_x) and their associated mean squares ($\text{MS}_x = \text{SS}_x/n_x$, with n_x the associated degrees of freedom). ANOVA test statistics are generally given by ratios of mean squares, $F = \text{MS}_x/\text{MS}_y$. Under normality assumptions, sums of squares are χ^2-distributed, while under the null hypothesis F is distributed as a (central) F-ratio distribution with n_x (numerator) and n_y (denominator) degrees of freedom, and we denote this as $F \sim F_{n_x,n_y}$. Thus, a test of level α for the hypothesis of no additional effects on level x (compared to level y) is whether $F > F_{n_x,n_y,[1-\alpha]}$, where $F_{n_x,n_y,[\alpha]}$ satisfies

$$\Pr\left[F_{n_x,n_y} \le F_{n_x,n_y,[\alpha]}\right] = \alpha \tag{A5.13a}$$

Two useful identities related to this definition are

$$\Pr\left[F_{n_x,n_y} > F_{n_x,n_y,[1-\alpha]}\right] = \alpha \tag{A5.13b}$$

and

$$F_{n_x,n_y,[\alpha]} = \frac{1}{F_{n_y,n_x,[1-\alpha]}} \tag{A5.13c}$$

The power of the test is the probability that the test statistic F exceeds this critical value,

$$\Pr\left[\, F > F_{n_x,n_y,[1-\alpha]} \,\right] \tag{A5.13d}$$

In order to compute this probability, we require the distribution of the test statistic F under the alternative hypothesis. Here, some sums of squares follow a noncentral χ^2 distribution, and the resulting F statistic involving these follows a noncentral F distribution. Hence, before considering power, we describe the properties of central and noncentral χ^2 and F distributions, proofs of which can be found in standard texts such as Scheffé (1959), Searle (1971), and Johnson and Kotz (1970b).

Central and Noncentral χ^2 Distributions

The χ^2 distribution arises from sums of squared, normally distributed, random variables — if $x_i \sim \mathrm{N}(0,1)$, then $u = \sum_{i=1}^{n} x_i^2 \sim \chi^2_n$, a **central** χ^2 distribution with n degrees of freedom. It follows that the sum of two χ^2 random variables is also χ^2 distributed, so that if $u \sim \chi^2_n$ and $v \sim \chi^2_m$, then

$$u + v \sim \chi^2_{(n+m)} \tag{A5.14a}$$

Two other useful results are that if $x_i \sim \mathrm{N}(0, \sigma^2)$, then

$$\sum_{i=1}^{n} x_i^2 \sim \sigma^2 \cdot \chi^2_n \tag{A5.14b}$$

and for $\overline{x} = n^{-1}\sum_{i=1}^{n} x_i$,

$$\sum_{i=1}^{n} (x_i - \overline{x}\,)^2 \sim \sigma^2 \cdot \chi^2_{(n-1)} \tag{A5.14c}$$

In this last case, subtraction of the mean causes the loss of one degree of freedom.

A **noncentral** χ^2 arises when the random variables being considered have nonzero means. In particular, if $x_i \sim \mathrm{N}(\mu_i, 1)$, then $u = \sum_{i=1}^{n} x_i^2$ follows a noncentral χ^2 distribution with n degrees of freedom and **noncentrality parameter**

$$\lambda = \sum_{i=1}^{n} \mu_i^2 \tag{A5.15a}$$

and we write $u \sim \chi^2_{n,\lambda}$. As shown in Figure A5.3, increasing the noncentrality parameter λ shifts the distribution to the right. This is also seen by considering the mean and variance of u,

$$E(u) = n + \lambda \qquad \text{and} \qquad \sigma^2(u) = 2(n + 2\lambda) \tag{A5.15b}$$

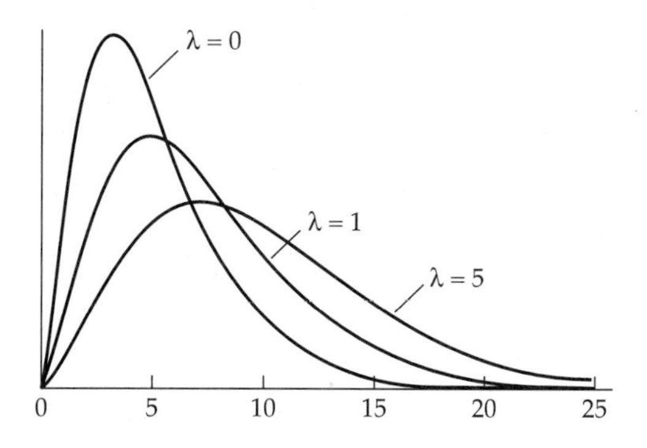

Figure A5.3 The probability distribution function for a noncentral χ^2. As the noncentrality parameter λ increases, the distribution is pulled to the right. We plot here a χ^2 random variable with $n = 5$ degrees of freedom and noncentrality parameters $\lambda = 0$ (a central χ^2), 1, and 5.

It follows directly from the definition that sums of noncentral χ^2 variables also follows a noncentral χ^2 distribution, so that if $u \sim \chi^2_{n,\,\lambda_1}$ and $v \sim \chi^2_{m,\,\lambda_2}$, then

$$(u + v) \sim \chi^2_{(n+m),(\lambda_1+\lambda_2)} \tag{A5.15c}$$

Finally, Equations A5.14b,c can be generalized to noncentral χ^2 random variables as follows. Suppose $x_i \sim \text{N}(\mu_i, \sigma^2)$, then

$$\sum_{i=1}^{n} x_i^2 \sim \sigma^2 \cdot \chi^2_{n,\lambda} \qquad \text{where} \qquad \lambda = \sum_{i=1}^{n} \mu_i^2 \tag{A5.15d}$$

and

$$\sum_{i=1}^{n} (\, x_i - \overline{x}\,)^2 \sim \sigma^2 \cdot \chi^2_{(n-1),\lambda} \qquad \text{where} \qquad \lambda = \sum_{i=1}^{n} \frac{\mu_i^2}{\sigma^2} \tag{A5.15e}$$

Central and Noncentral *F* Distributions

The ratio of two χ^2-distributed variables leads to the F distribution. In particular, if $u \sim \chi^2_n$ and $v \sim \chi^2_m$, then the ratio of these two χ^2 variables divided by their respective degrees of freedom follows a **central *F* distribution** with **numerator** and **denominator degrees of freedom** n and m (respectively), i.e., $(u/n)/(v/m) \sim F_{n,m}$. Since

$$\lim_{m\to\infty} F_{n,m} \to \frac{\chi^2_n}{n}$$

the F distribution can be approximated by a χ^2_n when the denominator degrees of freedom is large.

The **noncentral *F* distribution** results when the numerator χ^2 variable is noncentral. If $u \sim \chi^2_{n,\lambda}$ and $v \sim \chi^2_m$, then $F = (u/n)/(v/m)$ follows a noncentral F distributed with noncentrality parameter λ, and we write $F \sim F_{n,m,\lambda}$. As with the noncentral χ^2, increasing λ shifts the distribution further to the right. Again, this is seen in the mean and variance, with

$$E(F) = \frac{m}{m-2}\left(1 + \frac{2\lambda}{n}\right) \tag{A5.16a}$$

$$\sigma^2(F) = 2\left(\frac{m}{n}\right)^2 \left[\frac{(n+m)^2 + (n+2\lambda)(m-2)}{(m-2)^2(m-4)}\right] \tag{A5.16b}$$

Various mathematical and statistical packages provide routines for computing cumulative probabilities of noncentral χ^2 and F random variables, and a number of approximations have been suggested (e.g., Patnaik 1949, Severo and Zelen 1960, Tiku 1965). Winer et al. (1991) offer one such approximation based on the unit normal U, with the probability that a noncentral F-distributed random variable exceeds a value F_o being approximately

$$\Pr(\, F_{n,m,\lambda} > F_o\,) \simeq \Pr(\, U > z_o\,) \tag{A5.17a}$$

where

$$z_o = \frac{\sqrt{(2m-1)\,B} - \sqrt{2(n+\lambda) - A}}{\sqrt{A+B}}, \qquad A = \frac{n+2\lambda}{n+\lambda}, \qquad B = \frac{n}{m}\,F_o \tag{A5.17b}$$

From this general expression follow simplified approximations for the special cases of central F and noncentral χ^2 variables. Setting $\lambda = 0$ gives an approximation for the central F distribution as

$$\Pr(F_{n,m} > F_o) \simeq \Pr(\,U > \widetilde{z}_o\,), \quad \text{with} \quad \widetilde{z}_o = \frac{\sqrt{(2m-1)\,B} - \sqrt{2n-1}}{\sqrt{1+B}} \tag{A5.17c}$$

Likewise, taking the limit as $m \to \infty$ offers an approximation for the noncentral χ^2, since $\chi^2_{n,\lambda} \sim n \cdot F_{n,\infty,\lambda}$. Taking the limit of Equation A5.17c as $m \to \infty$ gives the probability that a noncentral χ^2 exceeds a value C_o as approximately

$$\Pr(\,\chi^2_{n,\lambda} > C_o\,) \simeq \Pr(\,U > \widetilde{z}_o\,), \quad \text{with} \quad \widetilde{z}_o = \frac{\sqrt{2C_o} - \sqrt{2(n+\lambda) - A}}{\sqrt{A}} \tag{A5.17d}$$

Power of Fixed-effects ANOVA Designs

We now have the necessary machinery in hand to perform power calculations for ANOVA designs. We consider fixed-effects designs first, using the simple one-way ANOVA to illustrate the basic approach. Letting y_{ij} denote the value of the

jth individual with treatment i, the model is $y_{ij} = \mu + \tau_i + e_{ij}$. We assume N fixed factors (the treatment effects $\tau_1, \cdots, \tau_N$); that the residuals are independent and normally distributed, with $e_{ij} \sim \text{N}(0, \sigma_e^2)$; and that the design is balanced with n observations for each treatment. As discussed in Chapter 18, the resulting treatment (or model) and error sums of squares are given by

$$\text{SS}_t = n \sum_{i=1}^{N} (\overline{y}_{i\cdot} - \overline{y}_{\cdot\cdot})^2, \qquad \text{SS}_e = \sum_{j=1}^{n} \sum_{i=1}^{N} (y_{ij} - \overline{y}_{i\cdot})^2 \tag{A5.18a}$$

where $\overline{y}_{i\cdot}$ is the sample mean for factor i and $\overline{y}_{\cdot\cdot}$ is the overall sample mean. As we will see shortly, these sums of squares have degrees of freedom $N-1$ and $N(n-1)$, respectively. The test for a significant treatment effect (at least one $\tau_i \neq 0$) is the F-ratio statistic

$$F = \frac{\text{MS}_t}{\text{MS}_e} = \frac{\text{SS}_t/(N-1)}{\text{SS}_e/N(n-1)} \tag{A5.18b}$$

The distribution of F depends on the distribution of the sums of squares. Since $\overline{y}_{i\cdot} \sim \text{N}(\tau_i, \sigma_e^2/n)$, Equation A5.15e gives

$$\text{SS}_t \sim n \cdot (\sigma_e^2/n) \cdot \chi^2_{N-1,\lambda} \quad \text{where} \quad \lambda = \sum_{i=1}^{N} \frac{\tau_i^2}{\sigma_e^2/n} \tag{A5.19}$$

The noncentrality parameter λ can also be expressed in terms of amount of the total variance attributable to model (treatment) effects, σ_τ^2. Since $\overline{\tau} = 0$ (by construction), it follows that

$$\sigma_\tau^2 = \frac{1}{N-1} \sum_{i=1}^{N} \tau_i^2 \tag{A5.20a}$$

and hence,

$$\lambda = n\,(N-1)\,\frac{\sigma_\tau^2}{\sigma_e^2} \tag{A5.20b}$$

Finally, noting that $y_{ij} - \overline{y}_{i\cdot} = e_{ij} - \overline{e}_{i\cdot}$ and that $e_{ij} \sim \text{N}(0, \sigma_e^2)$, we have

$$\text{SS}_e = \sum_{i=1}^{N} \left[\sum_{j=1}^{n} (e_{ij} - \overline{e}_{i\cdot})^2 \right] \sim \sum_{i=1}^{N} \sigma_e^2 \cdot \chi^2_{n-1} \sim \sigma_e^2 \cdot \chi^2_{N(n-1)} \tag{A5.21}$$

where the last two identities follow from Equations A5.14c and A5.14a, respectively. Thus the F-ratio test statistic is distributed as

$$F \sim \left(\frac{\sigma_e^2}{\sigma_e^2} \right) \left[\frac{\chi^2_{N-1,\lambda}/(N-1)}{\chi^2_{N(n-1)}/N(n-1)} \right] \sim F_{N-1,N(n-1),\lambda} \tag{A5.22}$$

where the noncentrality parameter is given by Equation A5.20b.

Under the hypothesis of no treatment effects, $\lambda = 0$ and $F \sim F_{(N-1),N(n-1)}$. A test with significance level α is thus given by whether $f > F_{(N-1),N(n-1),[1-\alpha]}$. When at least one treatment effect is nonzero, the test statistic f follows a noncentral F distribution, and the power of the test is given by

$$\Pr\left[F_{N-1,N(n-1),\lambda} > F_{N-1,N(n-1),[1-\alpha]}\right] \tag{A5.23}$$

To find the sample size required to give this test power $1-\beta$, we equate Equation A5.23 to $1-\beta$ and solve for n given the fixed values of α, β, N, and $\sum \tau_i^2/\sigma_e^2$ (or, equivalently, σ_τ^2/σ_e^2).

Example 3. Consider a fixed-effect design with four factors ($N=4$), and suppose that $\sigma_\tau^2/\sigma_e^2 = 1/3$, so that the treatment effects account for 25% of the total variance ($\sigma_\tau^2 + \sigma_e^2$). If each treatment has a sample size of $n=5$, what is the power to detect a significant treatment effect using a test of $\alpha = 0.05$? Here $N(n-1) = 16$, and we find from F distribution tables that

$$\Pr(F_{3,16} > 3.24) = 0.05$$

so that the critical value for the test is $F_{3,16,[0.95]} = 3.24$. Likewise, the noncentrality parameter is

$$\lambda = (N-1)\,n\,(\sigma_\tau^2/\sigma_e^2) = 3 \cdot 5 \cdot (1/3) = 5$$

and from noncentral F tables, the power is found to be

$$\Pr(F_{3,16,5} > 3.24) = 0.353$$

In the absence of noncentral F tables or programs, one could instead use the normal approximation given by Equation A5.17b. Just how good is this? Here $F_o = 3.24$, and substituting the degrees of freedom and $\lambda = 5$ yields $A = 1.625$ and $B = 0.6075$, giving

$$z_o = \frac{\sqrt{31 \cdot 0.6075} - \sqrt{2(3+5) - 1.625}}{\sqrt{1.625 + 0.6075}} = .3669$$

and hence

$$\Pr(F_{3,16,5} > 3.24) \simeq \Pr(U > 0.3669) = 0.357$$

showing that this approximation works quite well.

What sample size n is required to give this test 90% power? Here $\alpha = 0.05$, $N = 4$, and $\sigma_\tau^2/\sigma_e^2 = 1/3$ are fixed values. For $n = 15$, we find that $N(n-1) = 56$,

$\lambda = 15$, and (from central *F* tables) $F_{3,56,[0.95]} = 2.769$. Substituting these values into Equation A5.23 and using noncentral *F* tables gives

$$\Pr(F_{3,56,15} > 2.769) = 0.8957$$

showing that a test with this sample size has a power of 89.6%. Increasing n to 16, $N(n-1) = 60$, $\lambda = 16$, and $F_{3,60,[0.95]} = 2.758$, giving

$$\Pr(F_{3,60,16} > 2.758) = 0.9167$$

for a power of 91.7%.

Application: Power of QTL Mapping in Half-sib Families

As an example of analysis of a more complex design, consider the nested ANOVA design for mapping QTLs in half-sib families (Chapter 16). In the QTL half-sib design, each sire heterozygous for a marker locus is crossed to n dams and a single offspring is scored from each mating. Evidence of a QTL linked to the marker is indicated if the mean trait values for alternative sire alleles at this marker locus are significantly different. Since a different dam is used for each mating, all offspring from the same sire are half-sibs. The model here is

$$y_{ijk} = \mu + s_i + m_{ij} + e_{ijk} \tag{A5.24}$$

where y_{ijk} denotes the kth offspring of marker genotype j (with $j = 1$ for offspring carrying sire allele 1, $j = 2$ for sire allele 2) from sire i. We assume a completely balanced design, with N sires, each of which have n offspring equally distributed over the two alternative genotypes (the alternative sire marker alleles). The residuals are assumed independent with $e_{ijk} \sim \text{N}(0, \sigma_e^2)$ and the test statistic for a significant marker effect is $F \sim \text{MS}_m/\text{MS}_e$. To compute the power of this test, we need to obtain the distributions of sums of squares. Starting with the error sum of squares (see Chapter 18 for details),

$$\begin{aligned}\text{SS}_e &= \sum_{i=1}^{N}\sum_{j=1}^{2}\sum_{k=1}^{n/2}(y_{ijk} - \overline{y}_{ij\cdot})^2 = \sum_{i=1}^{N}\sum_{j=1}^{2}\sum_{k=1}^{n/2}(e_{ijk} - \overline{e}_{ij\cdot})^2 \\ &\sim \sum_{i=1}^{N}\sum_{j=1}^{2}\sigma_e^2 \cdot \chi^2_{(n/2-1)} = \sigma_e^2 \cdot \chi^2_{2N(n/2-1)}\end{aligned} \tag{A5.25}$$

The last two identities follow from Equations A5.14c and A5.14a, respectively. Hence, there are $N(n-2)$ degrees of freedom associated with SS_e. Turning to the marker sum of squares,

$$\text{SS}_m = \sum_{i=1}^{N}\sum_{j=1}^{2}\sum_{k=1}^{n/2}(\overline{y}_{ij\cdot} - \overline{y}_{i\cdot\cdot})^2 = \left(\frac{n}{2}\right)\sum_{i=1}^{N}\sum_{j=1}^{2}(\overline{y}_{ij\cdot} - \overline{y}_{i\cdot\cdot})^2 \tag{A5.26a}$$

Since $\sigma^2(\,\overline{e}_{ij\cdot}) = \sigma_e^2/(n/2)$, it follows that $\overline{y}_{ij\cdot} \sim \mathrm{N}(m_{ij}, 2\sigma_e^2/n\,)$. Applying Equation A5.15e gives

$$\sum_{j=1}^{2}(\,\overline{y}_{ij\cdot} - \overline{y}_{i\cdot\cdot})^2 = \frac{2\,\sigma_e^2}{n}\cdot\chi^2_{1,\lambda_i} \quad \text{where} \quad \lambda_i = \frac{n}{2\,\sigma_e^2}\cdot\left(m_{i1}^2 + m_{i2}^2\right) \qquad \text{(A5.26b)}$$

Substituting this into Equation A5.26a and applying Equation A5.15c gives

$$\mathrm{SS}_m \sim \sigma_e^2\cdot\chi^2_{N,\lambda} \qquad \text{where} \quad \lambda = \frac{n}{2\,\sigma_e^2}\sum_{i=1}^{N}(\,m_{i1}^2 + m_{i2}^2) \qquad \text{(A5.26c)}$$

Hence the marker sum of squares follows a noncentral χ^2 distribution with N degrees of freedom and noncentrality parameter λ. It thus remains to compute λ as a function of the unknown QTL parameters.

If the ith sire has marker-QTL genotype $Q_j M_1/Q_\ell M_2$, then m_{i1} is the expected within-sire deviation for offspring carrying sire marker allele 1 given the sire has this genotype, or

$$m_{i1} = E\,(\,\overline{y}_{i1} - \overline{y}_i\,|\,\mathrm{sire}_i = Q_j M_1/Q_\ell M_2)$$

Recall from Equation 16.8, that

$$E\,(\,\overline{y}_{i1}\,|\,\mathrm{sire}_i = Q_j M_1/Q_\ell M_2) = \mu + c\,\alpha_\ell + (1-c)\,\alpha_j$$

while $E\,(\,\overline{y}_i\,|\,\mathrm{sire}_i = Q_j/Q_\ell) = \mu + (\alpha_j + \alpha_\ell)/2$. Putting these together gives

$$m_{i1} = \mu + c\,\alpha_\ell + (1-c)\,\alpha_j - \left[\mu + \left(\frac{\alpha_j + \alpha_\ell}{2}\right)\right] = (0.5 - c)\,(\alpha_j - \alpha_\ell)$$

where α_j is the average effect of allele Q_j and c is the marker-QTL recombination frequency. Since $E(m_i) = 0$, it follows that $m_{i1} = -m_{i2}$ and hence

$$m_{i1}^2 + m_{i2}^2 = 2(0.5 - c)^2(\alpha_\ell - \alpha_j)^2$$

Taking the expected value over all possible sire QTL genotoypes,

$$E(m_{i1}^2 + m_{i2}^2) = 2(0.5 - c)^2 E\left[\,(\alpha_\ell - \alpha_j)^2\,\right] = 2(0.5 - c)^2\sigma_A^2$$

giving the noncentrality parameter for the marker sum of squares as

$$\lambda = N\cdot n\left[\frac{(1-2c)^2}{4}\right]\left(\frac{\sigma_A^2}{\sigma_e^2}\right) \qquad \text{(A5.26d)}$$

Thus, under the hypothesis of no linkage ($c = 0.5$), the test statistic is

$$F = \frac{\text{MS}_m}{\text{MS}_e} \sim F_{N,N(n-2)}$$

Hence, a sire marker effect is declared to be significant at the α-level if $F > F_{N,N(n-2),[1-\alpha]}$. The power of this test is the probability that the test statistic F exceeds this critical value,

$$\Pr\left[\, F_{N,N(n-2),\lambda} > F_{N,N(n-2),[1-\alpha]} \,\right] \tag{A5.27}$$

Example 4. Suppose we have a completely linked QTL ($c = 0$) with $\sigma_A^2/\sigma_e^2 = 0.1$. With $N = 20$ sires and $n = 35$ sibs/sire, what is the power of this test to detect this QTL using a significance level of $\alpha = 0.05$? Here $N(n-2) = 560$, and from central F tables, we find $F_{20,560,[0.95]} = 1.589$. Since $\lambda = 20{\cdot}36{\cdot}\ 0.1/4 = 15$, the power is

$$\Pr(F_{20,560,15} > 1.589) = 0.60$$

How many sibs are required per sire to have 90% power? Solving with increasing n values, we find that for $n = 53$, $N(n-2) = 1020$, $F_{20,1020,[0.95]} = 1.581$, and $\lambda = 20{\cdot}36{\cdot}\ 0.1/4 = 26.5$ for a power of

$$\Pr(F_{20,1020,26.5} > 1.581) = 0.90$$

Power of a Random-effects ANOVA Design

Under a random-effects model, the simple one-way ANOVA model becomes $y_{ij} = \mu + t_i + e_{ij}$, where t is now a random variable drawn from a normal distribution with variance σ_t^2, so that $t_i \sim \text{N}(0, \sigma_t^2)$. Residuals are assumed independent with $e_{ij} \sim \text{N}(0, \sigma_e^2)$ as before. Assuming a balanced design with n replicates for each of the N t_i values, the treatment and error sums of squares are also given by Equation A5.18a. As with the fixed-effects model, $\text{SS}_e \sim \sigma_e^2 \cdot \chi^2_{N(n-1)}$. However, the distribution of the treatment sum of squares remains a central χ^2, even when $\sigma_t^2 > 0$. To see this, note that $\overline{y}_{i\cdot} \sim \text{N}(0, \sigma^2)$ with $\sigma^2 = \sigma_t^2 + \sigma_e^2/n$. Applying Equation A5.14c, the same logic leading to Equation A5.19 now gives

$$\text{SS}_t \sim n \cdot \sigma^2 \cdot \chi^2_{N-1} = (n\,\sigma_t^2 + \sigma_e^2) \cdot \chi^2_{N-1} \tag{A5.28}$$

Hence, the F-ratio statistic is distributed as

$$F = \frac{\text{SS}_t/(N-1)}{\text{SS}_e/N(n-1)} \sim \frac{(n\,\sigma_t^2 + \sigma_e^2) \cdot \chi^2_{N-1}/(N-1)}{\sigma_e^2 \cdot \chi^2_{N(n-1)}/N(n-1)}$$

$$\sim \left(1 + n\,\frac{\sigma_t^2}{\sigma_e^2}\right) F_{(N-1),N(n-1)} \tag{A5.29}$$

Note that the term in parentheses is just $E(\text{MS}_t)/E(\text{MS}_e)$. Hence

$$\frac{F}{1+n\,(\sigma_t^2/\sigma_e^2)} \sim F_{N-1,N(n-1)} \tag{A5.30a}$$

giving the power of this test as

$$\Pr\left[F_{N-1,N(n-1)} > \frac{F_{N-1,N(n-1),[1-\alpha]}}{1+n\,(\sigma_t^2/\sigma_e^2)} \right] \tag{A5.30b}$$

Since, by definition (Equation A5.13b),

$$\Pr\left[F_{N-1,N(n-1)} > F_{N-1,N(n-1),[\beta]} \right] = 1-\beta$$

to have power $1-\beta$ requires that

$$F_{N-1,N(n-1),[1-\alpha]} = \left(1+n\,\frac{\sigma_t^2}{\sigma_e^2}\right) F_{N-1,N(n-1),[\beta]} \tag{A5.31}$$

Under more general random-effect models, if the test statistic is $F = \text{MS}_x/\text{MS}_y$, then

$$\frac{F}{R} \sim F_{n_x,n_y}, \qquad \text{where} \qquad R = \frac{E(\text{MS}_x)}{E(\text{MS}_y)} \tag{A5.32a}$$

and to obtain power $1-\beta$ requires choosing sample sizes such that

$$F_{n_x,n_y,[1-\alpha]} = \left[\frac{E(\text{MS}_x)}{E(\text{MS}_y)}\right] F_{n_x,n_y,[\beta]} \tag{A5.32b}$$

Example 5. Consider a random-effects model with design parameters similar to Example 3, with $\sigma_t^2/\sigma_e^2 = 1/3$, $N = 4$, and $n = 5$. What is the power of this design for a test with $\alpha = 0.05$? Here $F_{3,16,[0.95]} = 3.24$ and $1+n\,(\sigma_t^2/\sigma_e^2) = 1+5/3 = 8/3$. From Equation A5.30b, the power becomes

$$\Pr\left(F_{3,16} > \frac{3.24}{8/3} = 1.215\right) = 0.103$$

What value of n is required to give 90% power? With increasing values of n, we find that for $n = 38$, $N(n-1) = 148$, $1+n\,(\sigma_t^2/\sigma_e^2) = 13.67$, and $F_{3,148,[0.95]} = 2.66$, giving

$$\Pr\left(F_{3,148} > \frac{2.66}{13.67} = 0.195\right) = 0.90$$

Application: Power of the Half-sib Design for Variance Estimation

As an example of computing the power of random-effects ANOVA designs for estimating variance components, consider the standard half-sib design for estimating additive genetic variance (Chapter 18). Assuming a balanced design, N sires are each mated to n unique and unrelated females. The linear model (Equation 18.1) is $y_{ij} = \mu + s_i + e_{ij}$, where $s \sim (0, \sigma_s^2)$ and $e \sim (0, \sigma_e^2)$. Assume further that s and e are normally distributed. Then, from Equation A5.28, the sire effect sum of squares, SS_s, follows a χ^2 distribution, with

$$\text{SS}_s \sim n \cdot \sigma^2 \cdot \chi^2_{N-1} = (n\,\sigma_s^2 + \sigma_e^2) \cdot \chi^2_{N-1}$$

Likewise, from Equation A5.29, the resulting test statistic F for a significant sire variance follows an F distribution, with

$$F = \frac{\text{SS}_s/(N-1)}{\text{SS}_e/N(n-1)} \sim \left(1 + n\,\frac{\sigma_s^2}{\sigma_e^2}\right) F_{(N-1),N(n-1)} \tag{A5.33a}$$

Expressing the sire variance in terms of additive genetic and environmental variance components (ignoring epistasis),

$$\sigma_s^2 = \sigma_A^2/4 = (h^2/4)\sigma_z^2$$

Since $\sigma_z = \sigma_s^2 + \sigma_e^2$, it immediately follows that

$$\sigma_e^2 = \sigma_z^2 - \sigma_s^2 = \sigma_z^2(1 - h^2/4)$$

Substituting these into Equation A5.33a gives

$$F \sim \left(1 + \frac{n\,h^2}{4 - h^2}\right) F_{(N-1),N(n-1)} \tag{A5.33b}$$

From Equation A5.30b, the power of this test is

$$\Pr\left[\, F_{N-1,N(n-1)} > \frac{F_{N-1,N(n-1),[1-\alpha]}}{1 + n\,h^2/(4 - h^2)} \,\right] \tag{A5.33c}$$

Example 6. Suppose $h^2 = 0.25$, so that the sire variance is $\sigma_s^2 = h^2/4 = 0.0625$. What is the probability of detecting a significant sire variance (with a test of $\alpha = 0.05$) using 10 sires, each mated to 5 dams? Here $N = 10$ and $n = 5$, and from F tables, the critical value of the test is $F_{N-1,N(n-1),[1-\alpha]} = F_{9,40,[0.95]} = 2.12$. Applying Equation A5.33c, the power of this design to detect a sire variance at least this large is

$$\Pr\left[\, F_{9,40} > \frac{2.12}{1 + 5 \cdot 0.25/(4 - 0.25)} = 1.59 \,\right] = 0.15$$

Hence the power is very poor, with the test indicating a significant sire effect only 15 percent of the time.

If we keep the number of dams/sire constant at 5, how many sires are required to give this test 90% power? Trial and error shows that with N = 255 sires, the critical value becomes $F_{N-1,N(n-1),[1-\alpha]} = F_{254,1020,[0.95]} = 1.17$, and Equation A5.33c gives

$$\Pr\left[F_{254,1020} > \frac{1.17}{1+5\cdot 0.25/(4-0.25)} = 0.88\right] = 0.90$$

Note from the form of Equation A5.33c that the number of dams/sire, n, appears to assert a more important role than N. Keeping N constant at 10, how many dams/sire are required to have 90% power? Again trial and error shows for n = 47 that $F_{N-1,N(n-1),[1-\alpha]} = F_{9,460,[0.95]} = 1.90$, and Equation A5.33c gives

$$\Pr\left[F_{9,460} > \frac{1.90}{1+47\cdot 0.25/(4-0.25)} = 0.46\right] = 0.90$$

Finally, recall from Chapter 18 that if a total of T individuals are measured, the optimal number of sires per dam is given by $n = 4/h^4$. Hence $N = T/n = Th^2/4$, and the power under the optimal design for T individuals is

$$\Pr\left[F_{(\{Th^2/4\}-1),T(1-h^2/4)} > \frac{F_{(\{Th^2/4\}-1),T(1-h^2/4),[1-\alpha]}}{1+4/(4-h^2)}\right] \tag{A5.34}$$

Table A5.1 gives the power under the optimal design for various values of T and h^2. Note that with modest, but reasonable, heritabilities ($h^2 = 0.25$), at least 250 individuals must be used to have a reasonable chance of the half-sib design showing a significant sire variance.

Example 7. For Example 6, where $h^2 = 0.25$ and $T = 50$, the optimal number of dams/sire is $4/h^2 = 16$. Rounding up to approach $T = 50$, we take $N = 3$ and $n = 17$ for $T = 51$. Under this design, the power is computed to be 0.234. For $N = 2$ and $N = 25$ ($T = 50$) the power is 0.224, while for $N = 4$ and $n = 12$ ($T = 48$) the power is 0.211. As expected, the optimal design ($n = 17$) does indeed show the largest power.

Table A5.1 The power to detect a significant heritability under the optimal sampling scheme for a half-sib design as a function of h^2 and the total number T of individuals measured.

	h^2				
T	0.1	0.25	0.5	0.75	1.0
50	0.12	0.22	0.36	0.44	0.59
100	0.22	0.36	0.58	0.68	0.82
250	0.35	0.65	0.87	0.95	0.99
500	0.57	0.86	0.99	1.00	1.00
750	0.70	0.95	1.00	1.00	1.00
1000	0.80	0.99	1.00	1.00	1.00

Note: For fixed T, the optimal number of dams per sire is $n = 4/h^2$, or 40, 16, 8, 5, and 4 under the above respective heritabilities. Power is computed using the design closest to these optimal values, given the constraint that the total number of individuals in the balanced design is T. For example, for $h^2 = 0.1$, $n = 40$ is the optimal value. However, since design must have $N \geq 2$, for $T = 50$, power is computed using $N = 2$ and $n = 25$.

Literature Cited

The numbers in brackets following each reference denote the chapters in which the reference is cited.

Aastveit, A. H., and K. Aastveit. 1993. Effects of genotype-environment interactions on genetic correlations. *Theor. Appl. Genet.* 86: 1007–1013. [22]

Abe, T. 1969. On the sampling variances of the genetic correlation estimates from analysis of variance and covariance. *Jpn. Poultry Sci.* 6: 209–214. [21]

Adams, M. W., and D. B. Shank. 1959. The relationship of heterozygosity to homeostasis in maize hybrids. *Genetics* 44: 777–786. [6]

Ågren, J., and D. W. Schemske. 1993. Outcrossing rate and inbreeding depression in two annual monoecious herbs, *Begonia hirsuta* and *B. semiovata*. *Evolution* 47: 125–135. [10]

Aitchison, J., and J. A. C. Brown. 1966. *The log-normal distribution*. Cambridge Univ. Press, Cambridge, UK. [11]

Aitken, A. C. 1935. On least squares and linear combination of observations. *Proc. Royal Soc. Edinburgh A* 55: 42–47. [8]

Aitken, M., and G. T. Wilson. 1980. Mixture models, outliers, and the EM algorithm. *Technometrics* 22: 325–331. [A4]

Ajmone-Marsan, P., G. Monfredini, W. F. Ludwig, A. E. Melchinger, P. Franceschini, G. Pagnotto, and M. Motto. 1995. In an elite cross of maize a major quantitative trait locus controls one-fourth of the genetic variation for grain yield. *Theor. Appl. Genet.* 90: 415–424. [15]

Akaike, H. 1974. A new look at the statistical model identification. *IEEE Trans. on Automatic Control, AC-19.* 716–723. [13]

Alatalo, R., and A. Lundberg. 1986. Heritability and selection on tarsus length in the pied flycatcher *(Ficedula hypoleuca)*. *Evolution* 40: 574–583. [17]

Alberch, P., and E. A. Gale. 1985. A developmental analysis of an evolutionary trend: digital reduction in amphibians. *Evolution* 39: 8–23. [25]

Albrecht, G. H. 1978. Some comments on the use of ratios. *Syst. Zool.* 27: 67–71. [11]

Aldhous, P. 1994. Fast tracks to disease genes. *Science* 265: 2008–2010. [14]

Amos, C. I. 1994. Robust variance-components approach for assessing genetic linkage in pedigrees. *Am. J. Hum. Genet.* 54: 535–543. [16]

Amos, C. I., and R. C. Elston. 1989. Robust methods for the detection of genetic linkage for quantitative data from pedigrees. *Genet. Epidem.* 6: 349–360 (Correction 6: 727). [16]

Amos, C. I., and R. C. Elston. 1989. Robust methods for the detection of genetic linkage for quantitative data from pedigrees. *Genet. Epidem.* 6: 349–360 (Correction 6: 727). [16]

Amos, C. I., R. C. Elston, A. F. Wilson, and J. E. Bailey-Wilson. 1989. A more powerful robust sib-pair test of linkage for quantitative traits. *Genet. Epidem.* 6: 435–449. [16]

Amos, C. I., R. C. Elston, G. E. Bonney, B. J. B. Keats, and G. S. Berenson. 1990. A multivariate method for detecting genetic linkage, with application to a pedigree with adverse lipoprotein phenotype. *Am. J. Hum. Genet.* 47: 247–254. [16]

Anderson, T. W. 1984. *An introduction to multivariate statistical analysis.* 2nd Ed. John Wiley & Sons, NY. [8,26]

Anderson, V. L., and O. Kempthorne. 1954. A model for the study of quantitative inheritance. *Genetics* 39: 883–898. [10]

Andersson, D. I., and D. Hughes. 1996. Muller's ratchet decreases fitness of a DNA-based microbe. *Proc. Natl. Acad. Sci. USA* 93: 906–907. [12]

Angus, R. A., and R. J. Schultz. 1983. Meristic variation in homozygous and heterozygous fish. *Copeia* 1983: 287–299. [6]

Annett, M. 1973. Handedness in families. *Ann. Hum. Genet.* 37: 93–105. [25]

Antonovics, J. 1968. Evolution in closely adjacent populations. V. Evolution of self-fertility. *Heredity* 23: 219–238. [10]

Árnason, T. 1982. Prediction of breeding values for multiple traits in small non-random mating (horse) populations. *Acta Agric. Scand.* 32: 171–176. [26]

Arnheim, N., C. Strange, and H. Erlich. 1985. Use of pooled DNA samples to detect linkage disequilibrium of polymorphic restriction fragments and human disease: studies of the HLA class II loci. *Proc. Natl. Acad. Sci. USA* 82: 6970–6974. [14]

Arnold, S. J. 1981a. Behavioral variation in natural populations. I. Phenotypic, genetic and environmental correlations between chemoreceptive responses to prey in the garter snake, *Thamnophis elegans*. *Evolution* 35: 489– 509. [21]

Arnold, S. J. 1981b. Behavioral variation in natural populations. II. Inheritance of a feeding response in crosses between geographic races of the garter snake, *Thamnophis elegans*. *Evolution* 35: 510–515. [21]

Arnold, S. J. 1981c. The microevolution of feeding behavior. *In* A. Kamil and T. Sargent (eds.), *Foraging behavior: ecological, ethological and psychological approaches*, pp. 409–453. Garland Press, NY. [21]

Arondel, V., B. Lemieux, I. Hwang, S. Gibson, H. M. Goodman, and C. R. Somerville. 1992. Map-based cloning of a gene controlling omega-3 fatty acid desaturation in *Arabidopsis*. *Science* 258: 1353–1355. [14]

Arvesen, J. N., and T. H. Schmitz. 1970. Robust procedures for variance component problems using the jackknife. *Biometrics* 26: 677–686. [18]

Asamoah, A., A. F. Wilson, R. C. Elston, E. Dalferes Jr., and G. S. Berenson. 1987. Segregation and linkage analysis of dopamine-β-hydrozylase activity in a six-generation pedigree. *Am. J. Med. Genet.* 27: 613–621. [13]

Ashby, E., and E. Wangermann. 1954. The effects of meristem aging on the morphology and behavior of fronds in *Lemna minor*. *Ann. N.Y. Acad. Sci.* 57: 476–483. [6]

Asins, M. J., and E. A. Carbonell. 1988. Detection of linkage between restriction fragment length polymorphism markers and quantitative traits. *Theor. Appl. Genet.* 76: 623–626. [15]

Atchley, W. R. 1984. Ontogeny, timing of development, and genetic variance-covariance structure. *Am. Nat.* 123: 519–540. [23]

Atchley, W. R., and D. Anderson. 1978. Ratios and the statistical analysis of biological data. *Syst. Zool.* 27: 71–78. [11]

Atchley, W. R., and J. J. Rutledge. 1980. Genetic components of size and shape. I. Dynamics of components of phenotypic variability and covariability during ontogeny in the laboratory rat. *Evolution* 34: 1161–1173. [21,23]

Atchley, W. R., C. T. Gaskins, and D. Anderson. 1976. Statistical properties of ratios. I. Empirical results. *Syst. Zool.* 25: 137–148. [11]

Atchley, W. R., B. Riska, L. A. Kohn, A. A. Plummer, and J. J. Rutledge. 1984. A quantitative genetic analysis of brain and body size associations, their origin and ontogeny: data from mice. *Evolution* 38: 1165–1179. [21]

Atkinson, A. C. 1982. Regression diagnostics, transformations and constructed variables. *J. Royal Stat. Soc. B*: 44: 1–36. [11]

Avery, P. J., and W. G. Hill. 1977. Variability in genetic parameters among small populations. *Genet. Res.* 29: 193–213. [5]

Avery, P. J., and W. G. Hill. 1979. Variance in quantitative traits due to linked dominant genes and variance in heterozygosity in small populations. *Genetics* 91: 817–844. [5]

Ayres, M. P., and D. L. Thomas. 1990. Alternative formulations of the mixed-model ANOVA applied to quantitative genetics. *Evolution* 44: 221–226. [22]

Babron, M.-C., M. Martinez, C. Bonaïti-Pellié, and F. Clerget-Darpoux. 1993. Linkage detection by the affected-pedigree-member method: what is really tested? *Genet. Epidem.* 10: 389–394. [16]

Bachmann, K., K. L. Chambers, and H. J. Price. 1985. Genome size and natural selection: observations and experiments in plants. *In* T. Cavalier-Smith (ed.), *The evolution of genome size*, pp. 267–276. John Wiley & Sons, NY. [12]

Bader, R. S. 1965. Fluctuating asymmetry in the dentition of the house mouse. *Growth* 29: 291–300. [6]

Bader, R. S., and J. S. Hall. 1960. Osteometric variation and function in bats. *Evolution* 14: 8–17. [11]

Badner, J. A., A. Chakravarti, and D. K. Wagener. 1984. A test of nonrandom segregation. *Genet. Epidem.* 1: 329–340. [16]

Bailey, D. W. 1959. Rates of subline divergence in highly inbred strains of mice. *J. Heredity* 50: 26–30. [12]

Bailey, D. W. 1981. Strategic uses of recombinant inbred and cogenic strains in behavior genetics research. *In* E. S. Gershon, S. Matthysse, X. O. Breakefield, and R. D. Ciaranello (eds.), *Genetic research strategies for psychobiology and psychiatry*, pp. 189–198. Plenum, NY. [14]

Bailey, N. T. J. 1961. *The mathematical theory of genetic linkage*. Clarendon Press, Oxford, UK. [14]

Bailit, H. L., P. L. Workman, J. D. Niswander, and C. J. MacLean. 1970. Dental asymmetry as an indicator of genetic and environmental conditions in human populations. *Hum. Biol.* 42: 626–638. [6]

Baker, A. J. 1980. Morphometric differentiation in New Zealand populations of the house sparrow (*Passer domesticus*). *Evolution* 34: 638–653. [11]

Baker, R. J. 1978. Issues in diallel analysis. *Crop Sci.* 18: 533–536. [20]

Baker, R. J. 1988. Tests for crossover genotype-environment interactions. *Can. J. Plant Sci.* 68: 405–410. [22]

Baker, W. K., and E. A. Kaeding. 1981. Linkage disequilibrium at the alpha- esterase loci in a population of *Drosophila melanogaster* from Utah. *Am. Nat.* 117: 804–809. [5]

Bamshad, M., M. H. Crawford, D. O'Rourke, and L. B. Jorde. 1994. Biochemical heterozygosity and morphologic variation in a colony of *Papio hamadryas* baboons. *Evolution* 48: 1211–1221. [6]

Bar-Anan, R., M. Soller, and J. C. Bowman. 1976. Genetic and environmental factors affecting the incidence of difficult calving and prenatal calf mortality in Israeli-Friesian dairy herds. *Anim. Prod.* 22: 299–310. [23]

Barden, H. S. 1980. Fluctuating dental asymmetry: a measure of developmental asymmetry in Down syndrome. *Am. J. Phys. Anthropol.* 52: 169–173. [6]

Barker, J. S. F. 1979. Inter-locus interactions: a review of experimental evidence. *Theor. Pop. Biol.* 16: 323–346. [5]

Barlow, D. P. 1994. Imprinting: a gamete's point of view. *Trends Genet.* 10: 194–199. [24]

Barlow, D. P. 1995. Gametic imprinting in mammals. *Science* 270: 1610–1613. [24]

Barlow, R. 1981. Experimental evidence for interaction between heterosis and environment in animals. *Anim. Breed. Abst.* 49: 715–737. [10,22]

Barnes, B. 1966. Environment and selection in *Drosophila melanogaster*. Ph. D. thesis. Birmingham Univ., Birmingham, UK. [14]

Barrai, I., L. L. Cavalli-Sforza, and M. Mainardi. 1964. Testing a model of dominant inheritance for metric traits in man. *Heredity* 19: 651–668. [10]

Barrett, S. C. H., and D. Charlesworth. 1991. Effects of a change in the level of inbreeding on the genetic load. *Nature* 352: 522–524. [10]

Bartlett, M. S., and J. B. S. Haldane. 1935. The theory of inbreeding with forced heterozygosity. *J. Genet.* 31: 327–340. [14]

Barton, N. H. 1990. Pleiotropic models of quantitative variation. *Genetics* 124: 773–782. [12]

Barton, N. H., and B. Charlesworth. 1984. Genetic revolutions, founder events and speciation. *Ann. Rev. Ecol. Syst.* 15: 133–164. [14]

Barton, N. H., and G. M. Hewitt. 1981. Hybrid zones and speciation. *In* W. R. Atchley and D. S. Woodruff (eds.), *Evolution and speciation: essays in honor of M. J. D. White,* pp. 109–145. Cambridge Univ. Press, Cambridge, UK. [9]

Barton, N. H., and M. Turelli. 1989. Evolutionary quantitative genetics: how little do we know? *Ann. Rev. Genetics* 23: 337–370. [1]

Bateman, A. J. 1959. The viability of near-normal irradiated chromosomes. *Internat. J. Rad. Biol.* 1: 170–180. [12]

Bateson, P. 1983. *Mate choice.* Cambridge Univ. Press, Cambridge, UK. [9]

Beadle, G. W. 1939. Teosinte and the origin of maize. *J. Heredity* 30: 245–247. [5,15]

Beadle, G. W. 1980. The ancestry of corn. *Sci. Am.* 242: 112–119, 162. [15]

Beardmore, J. A. 1960. Developmental stability in constant and fluctuating environments. *Heredity* 14: 411–422. [6]

Beardmore, J. A. 1970. Viral components in the genetic background? *Nature* 226: 766–767. [12]

Beardmore, J. A., and S. A. Shami. 1976. Parental age, genetic variation and selection. *In* S. Karlin and E. Nevo (eds.), *Population genetics and ecology,* pp. 3–22. Academic Press, NY. [6]

Beardmore, J. A., F. Lints, and A. L. F. Al-Baldawi. 1975. Parental age and heritability of sternopleural chaeta number in *Drosophila melanogaster. Heredity* 34: 71–82. [6]

Beavis, W. D. 1994. The power and deceit of QTL experiments: lessons from comparative QTL studies. *In 49th Annual Corn and Sorghum Research Conference.* pp. 252–268. American Seed Trade Association, Washington, D.C. [15]

Beavis, W. D., E. Pollak, and K. J. Frey. 1987. A theoretical model for quantitatively inherited traits influenced by nuclear-cytoplasmic interactions. *Theor. Appl. Genet.* 74: 571–578. [23]

Beavis, W. D., D. Grant, M. Albertsen, and R. Fincher. 1991. Quantitative trait loci for plant height in four maize populations and their associations with qualitative genetic loci. *Theor. Appl. Genet.* 83: 141–145. [14]

Becker, H. C. 1981. Correlations among some statistical measures of phenotypic stability. *Euphytica* 30: 835–840. [22]

Becker, H. C., and J. Léon. 1988. Stability analysis in plant breeding. *Plant Breeding* 101: 1–23. [22]

Beckmann, J. S., and M. Soller. 1983. Restriction fragment length polymorphisms in genetic improvement: methodologies, mapping and costs. *Theor. Appl. Genet.* 67: 35–43. [14]

Beckmann, J. S., and M. Soller. 1986a. Restriction fragment length polymorphisms and genetic improvement in agricultural species. *Euphytica* 35: 111–124. [14]

Beckmann, J. S., and M. Soller. 1986b. Restriction fragment length polymorphisms in plant genetic improvement. *Oxford Surveys Plant Mol. Cell Biol.* 3: 196–250. [14]

Beckmann, J. S., and M. Soller. 1988. Detection of linkage between marker loci and loci affecting quantitative traits in crosses between segregating populations. *Theor. Appl. Genet.* 76: 228–236. [15]

Belknap, J. K., P. Metten, M. L. Helms, L. A. O'Toole, S. Angeli-Gade, J. C. Crabbe, and T. J. Phillips. 1993. Quantitative trait loci (QTL) applications to substances of abuse: physical dependence studies with nitrous oxide and ethanol in B × D mice. *Behav. Genet.* 23: 213–222. [14]

Bell, A. E. 1977. Heritability in retrospect. *J. Heredity* 68: 297–300. [7]

Bell, B. R., B. T. McDaniel, and O. W. Robison. 1985. Effects of cytoplasmic inheritance on production traits of dairy cattle. *J. Dairy Sci.* 68: 2038–2051. [23]

Bell, G. 1990. The ecology and genetics of fitness in *Chlamydomonas.* I. Genotype-by-environment interaction among pure strains. *Proc. Royal Soc. Lond.* B 240: 295–321. [22]

Bell, G. 1991. The ecology and genetics of fitness in *Chlamydomonas.* III. Genotype-by-environment interaction within strains. *Evolution* 45: 668–679. [22]

Bell, G., and V. Koufopanou. 1986. The cost of reproduction. *Oxford Surv. Evol. Biol.* 3: 83–131. [21]

Bell, J. I., and G. M. Lathrop. 1996. Multiple loci for multiple sclerosis. *Nature Genetics* 13: 377–378. [16]

Bengtsson, B. O., and G. Thomson. 1981. Measuring the strength of associations between HLA antigens and diseases. *Tissue Antigens* 18: 356–363. [14].

Bennett, J. H. 1954. Panmixia with tetrasomic and hexasomic inheritance. *Genetics* 39: 150–158. [5]

Bennetzen, J., and M. Freeling. 1993. Grasses as a single genetic system: genome composition, collinearity and compatibility. *Trends Genet.* 9: 259–261. [14]

Benson, D. L., and A. R. Hallauer. 1994. Inbreeding depression rates in maize populations before and after recurrent selection. *J. Heredity* 85: 122–128. [10]

Bentolila, S., C. Guitton, N. Bouvet, A. Sailland, S. Nykaz, and G. Freyssinet. 1991. Identification of an RFLP marker tightly linked to the *Ht1* gene in maize. *Theor. Appl. Genet.* 82: 393–398. [14]

Bentsen, H. B., and G. Klemetsdal. 1991. The use of fixed-effects models and mixed models to estimate single-gene associated effects on polygenic traits. *Genet. Sel. Evol.* 23: 407–419. [14]

Berenbaum, M. R., A. R. Zangerl, and J. K. Nitao. 1986. Constraints on chemical coevolution: wild parsnips and the parsnip webworm. *Evolution* 40: 1215–1228. [18]

Bereskin, B., C. E. Shelby, K. E. Rowe, W. E. Urban, Jr., C. T. Blunn, A. B. Chapman, V. A. Garwood, L. N. Hazel, J. F. Lasley, W. T. Magee, J. W. McCarty, and J. A. Whatley, Jr. 1968. Inbreeding and swine productivity traits. *J. Anim. Sci.* 27: 339–350. [10]

Berke, T. G., and T. R. Rocheford. 1995. Quantitative trait loci for flowering, plant height, and kernel traits in maize. *Crop Sci.* 35: 1542–1549. [14,15]

Beyer, W. H. 1968. *CRC handbook of tables for probability and statistics, 2nd Ed.* CRC Press, Boca Raton, FL. [11]

Biémont, C. 1983. Homeostasis, enzymatic heterozygosity and inbreeding depression in natural populations of *Drosophila melanogaster. Genetica* 61: 179–189. [6]

Billewicz, W. Z. 1972. A note on birth weight correlation in full-sibs. *J. Biosoc. Sci.* 4: 455–460. [7]

Bingham, P. M., R. Levis, and G. M. Rubin. 1981. Cloning of DNA sequences from the white locus of *Drosophila melanogaster* by a general and novel method. *Cell* 25: 693–704. [14]

Birnbaum, A. 1972. The random phenotype concept, with applications. *Genetics* 72: 739–758. [13]

Bishir, J., and G. Namkoong. 1987. Unsound seeds in conifers: estimation of numbers of lethal alleles and of magnitudes of effects associated with the maternal plant. *Silvae Genetica* 36: 180–185. [10]

Bishop, D. T., and J. A. Williamson. 1990. The power of identity-by-state methods for linkage analysis. *Am. J. Hum. Genet.* 46: 254–265. [16]

Bishop, D. T., C. Cannings, M. Skolnick, and J. A. Williamson. 1983. The number of polymorphic DNA clones required to map the human genome. *In* B. S. Weir (ed), *Statistical analysis of DNA sequence data,* pp. 181– 200. Marcel Dekker, NY. [14].

Bishop, G. R. 1992. Phenotypic variability of polygenic traits. *Acta Zool. Fennica* 191: 133–136. [6]

Bittles, A. H., and J. V. Neel. 1994. The costs of human inbreeding and their implications for variations at the DNA level. *Nature Genetics* 8: 117–121. [10]

Blackwelder, W. C., and R. C. Elston. 1974. Comment on Dr. Roberston's communication. *Behav. Genet.* 4: 97–99. [16]

Blackwelder, W. C., and R. C. Elston. 1982. Power and robustness of sib-pair linkage tests and extension to larger sibships. *Commun. Stat. Theor. Meth.* 11: 449–484. [16]

Blackwelder, W. C., and R. C. Elston. 1985. A comparison of sib-pair linkage tests for disease susceptibility loci. *Genet. Epidem.* 2: 85–97. [16]

Blanco, L. G., and J. A. Sanchez-Prado. 1986. Enzymatic heterozygosity and morphological variance in synthetic populations of *Drosophila melanogaster. Genet. Sel. Evol.* 18: 417–426. [6]

Blangero, J., and L. W. Konigsberg. 1991. Multivariate segregation analysis using the mixed model. *Genet. Epidem.* 8: 299–316. [13]

Blangero, J., J. W. MacCluer, C. M. Kammerer, G. E. Mott, T. D. Dyer, and H. C. McGill. 1990. Genetic analysis of apolipoprotein A1 in two dietary environments. *Am. J. Hum. Genet.* 47: 414–428. [13]

Blangero, J., S. Williams-Blangero, and J. E. Hixson. 1992. Assessing the effects of candidate genes on quantitative traits in primate populations. *Am. J. Primatol.* 27: 119–132. [14]

Boag, P. T. 1983. The heritability of external morphology in Darwin's ground finches *(Geospiza)* on Isla Daphne Major, Galápagos. *Evolution* 37: 877–894. [17]

Boag, P. T., and P. R. Grant. 1978. Heritability of external morphology in Darwin's finches. *Nature* 274: 793–794. [17]

Boag, P. T., and A. R. van Noordwijk. 1987. Quantitative genetics. *In* F. Cooke and P. A. Buckley (eds.), *Avian genetics,* pp. 45–77. Academic Press, London. [17]

Boake, C. R. B. (ed.) 1994. *Quantitative genetic studies of behavioral evolution.* Univ. Chicago Press, Chicago. [1]

Bodmer, W. F. 1986. Human genetics: the molecular challenge. *Cold Spring Harbor Symp. Quant. Biol.* 51: 1–13. [14]

Boehnke, M., and P. P. Moll. 1989. Identifying pedigrees segregating at a major locus for a quantitative trait: an efficient strategy for linkage analysis. *Am. J. Hum. Genet.* 44: 216–224. [16]

Boerwinkle, E., and C. F. Sing. 1986. Bias of the contribution of single-locus effects to the variance of a quantitative trait. *Am. J. Hum. Genet.* 39: 137–144. [14]

Boerwinkle, E., and C. F. Sing. 1987. The use of measured genotype information in the analysis of quantitative phenotypes in man. III. Simultaneous estimation of the frequencies and effects of apolipoprotein E polymorphism and residual polygenetic effects on cholesterol, betalipoprotein and triglyceride levels. *Ann. Hum. Genet.* 51: 211–226. [14]

Boerwinkle, E., R. Chakraborty, and C. F. Sing. 1986. The use of measured genotype information in the analysis of quantitative phenotypes in man. I. Models and analytical methods. *Ann. Hum. Genet.* 50: 181–194. [13,14]

Bohidar, N. R. 1964. Derivation and estimation of variance and covariance components associated with covariance between relatives under sex-linked transmission. *Biometrics* 20: 505–521. [24]

Bohren, B. B., H. E. McKean, and Y. Yamada. 1961. Relative efficiencies of heritability estimates based on regression of offspring on parent. *Biometrics* 17: 481–491. [17]

Bohrnstedt, G. W., and A. S. Goldberger. 1969. On the exact covariance of products of random variables. *J. Am. Stat. Assoc.* 64:1439–1442. [A1]

Boichard, D., L. R. Schaeffer, and A. J. Lee. 1992. Approximate restricted maximum likelihood and approximate predictor variance of the Mendelian sampling effect. *Genet. Sel. Evol.* 24: 331–343. [27]

Bondari, K., R. L. Willham, and A. E. Freeman. 1978. Estimates of direct and maternal genetic correlations for pupa weight and family size of *Tribolium*. *J. Anim. Sci.* 47: 358–365. [23]

Bonierbale, M. W., R. L. Plaisted, and S. D. Tanksley. 1988. RFLP maps based on a common set of clones reveal modes of chromosomal evolution in potato and tomato. *Genetics* 120: 1095–1103. [14,15]

Bonney, G. E. 1984. On the statistical determination of major gene mechanisms in continuous human traits: regressive models. *Am. J. Med. Genet.* 18: 731–749. [13]

Bonney, G. E. 1992. Compound regression models for family data. *In* J. Ott (ed.), *Models and methods for the genetic analysis of pedigree data,* pp. 28– 41. Karger, Basel. [13]

Bonney, G. E., G. M. Dunston, and J. Wilson. 1989. Regressive logistic models for ordered and unordered polychotomous traits: application to affective disorders. *Genet. Epidem.* 6: 211–215. [13]

Booth, C. L., D. S. Woodruff, and S. J. Gould. 1990. Lack of significant associations between allozyme heterozygosity and phenotypic traits in the land snail *Cerion*. *Evolution* 44: 210–213. [6]

Borecki, I. B., M. A. Province, and D. C. Rao. 1995. Inferring a major gene for quantitative traits by using segregation analysis with tests on transmission probabilities: how often do we miss? *Am. J. Hum. Genet.* 56: 319–326. [13]

Botstein, D., R. L. White, M. Skolnick, and R. W. Davis. 1980. Construction of a genetic linkage map in man using restriction fragment length polymorphisms. *Am. J. Hum. Genet.* 32: 314–331. [14,16]

Bouchard, T. J., Jr., D. T. Lykken, M. McGue, N. L. Segal, and A. Tellegen. 1990. Sources of human psychological differences: the Minnesota study of twins reared apart. *Science* 250: 223–228. [19]

Boucher, W. 1988. Calculation of the inbreeding coefficient. *J. Math. Biol.* 26: 57–64. [7]

Bovenhuis, H., and J. I. Weller. 1994. Mapping and analysis of dairy cattle quantitative trait loci by maximum likelihood methodology using milk protein genes as genetic markers. *Genetics* 137: 267–280. [14]

Bowman, J. C., and D. S. Falconer. 1960. Inbreeding depression and heterosis of litter size in mice. *Genet. Res.* 1: 262–274. [10]

Bowman, K. O., and L. R. Shenton. 1975. Omnibus test contours for departures from normality based on $\sqrt{b_1}$ and b_2. *Biometrika* 62: 243–250. [11]

Boyle, C. R., and R. C. Elston. 1979. Multifactorial genetic models for quantitative traits in humans. *Biometrics* 35: 55–68. [13]

Box, G. E. P., and D. R. Cox. 1964. An analysis of transformations. *J. Royal Stat. Soc.* B 26: 211–252. [11,13]

Bradford, G. 1972. The role of maternal effects in animal breeding. VI. Maternal effects in sheep. *J. Anim. Sci.* 35: 1315–1325. [23]

Bradley, B. P. 1978. Genetic and physiological adaptation of the copepod *Eurytemora affinia* to seasonal temperatures. *Genetics* 90: 193–205. [7]

Bradshaw, A. D. 1965. Evolutionary significance of phenotypic plasticity in plants. *Adv. Genetics* 13: 115–155. [22]

Bradshaw, H. D., Jr., S. M. Wilbert, K. G. Otto, and D. W. Schemske. 1995. Genetic mapping of floral traits associated with reproductive isolation in monkeyflowers (*Mimulus*). *Nature* 376: 762–765. [15]

Brady, R. H. 1979. Natural selection and the criteria by which a theory is judged. *Syst. Zool.* 28: 600–621. [1]

Breese, E. L. 1956. The genetical consequences of assortative mating. *Heredity* 10: 323–343. [7]

Breese, E. L., and K. Mather. 1957. The organization of polygenic activity within a chromosome in *Drosophila*. I. Hair characters. *Heredity* 11: 373–395. [14]

Breese, E. L., and K. Mather. 1960. The organization of polygenic activity within a chromosome in *Drosophila*. II. Viability. *Heredity* 14: 375–399. [14]

Bridges, W. C., Jr., and S. J. Knapp. 1987. Probabilities of negative estimates of genetic variances. *Theor. Appl. Genet.* 74: 269–274. [18,20]

Brinkman, M. A., and K. J. Frey. 1977. Yield component analysis of oat isolines that produce different grain yields. *Crop Sci.* 17: 165–168. [14]

Brodie, E. D., III. 1989. Genetic correlation between morphology and antipredator behavior in natural populations of the garter snake *Thamnophis ordinoides*. *Nature* 342: 542–543. [5,21]

Brodie, E. D., III. 1993. Homogeneity of the genetic variance-covariance matrix for antipredator traits in two natural populations of the garter snake *Thamnophis ordinoides*. *Evolution* 47: 844–854. [21]

Broemeling, L. D. 1969. Confidence intervals for measures of heritability. *Biometrics* 25: 424–427. [18]

Brooks, L. D. 1988. The evolution of recombination rates. *In* R. E. Michod and B. R. Levin (eds.), *The evolution of sex*, pp. 87–105. Sinauer Assoc., Sunderland, MA. [14]

Brown, A. F. 1991. Outbreeding depression as a cost of dispersal in the harpacticoid copepod, *Tigriopus californicus*. *Biol. Bull.* 181: 123– 126. [9]

Brown, A. H. D. 1975. Sample sizes required to detect linkage disequilibrium between two or three loci. *Theor. Pop. Biol.* 8: 184–201. [5]

Brown, D. L., M. B. Gorin, and D. E. Weeks. 1994. Efficient strategies for genomic searching using the affected-pedigree-member method of linkage analysis. *Am. J. Hum. Genet.* 54: 544–552. [16]

Brown, G. H. 1969. An empirical study of the distribution of the sample genetic correlation coefficient. *Biometrics* 22:63–72. [21]

Browne, R. A., S. E. Sallee, D. S. Grosch, W. O. Segreti, and S. M. Purser. 1984. Partitioning of genetic and environmental components of reproduction and lifespan in *Artemia*. *Ecology* 65: 949–960. [19]

Bruce, A. B. 1910. The Mendelian theory of heredity and the augmentation of vigor. *Science* 32: 627–628. [10]

Brückner, D. 1976. The influence of genetic variability on wing asymmetry in honeybees (*Apis mellifera*). *Evolution* 30: 100–108. [6]

Bryant, E. H., L. M. Meffert, and S. A. McCommas. 1990. Fitness rebound in serially bottlenecked populations of the house fly. *Am. Nat.* 136: 542–549. [10]

Bubeck, D. M., M. M. Goodman, W. D. Beavis, and D. Grant. 1993. Quantitative trait loci controlling resistance to gray leaf-spot in maize. *Crop Sci.* 33: 838–847. [15]

Bucher, K. D., H. G. Schrott, W. R. Clarke, and R. M. Lauer. 1982. The Muscatine cholesterol family study: distribution of cholesterol levels within families of probands with high, low and middle cholesterol levels. *J. Chron. Dis.* 35: 385–400. [13]

Buckler, A. J., D. D. Chang, S. L. Graw, J. D. Brook, D. A. Haber, P. A. Sharp, and D. E. Housman. 1991. Exon amplification: a strategy to isolate mammalian genes based on RNA splicing. *Proc. Natl. Acad. Sci. USA* 88: 4005– 4009. [14]

Bull, J. J., R. C. Vogt, and M. G. Bulmer. 1982. Heritability of sex ratio in turtles with environmental determination. *Evolution* 36: 333–341. [25]

Bulmer, M. G. 1957. Approximate confidence limits for components of variance. *Biometrika* 44: 159–167. [18]

Bulmer, M. G. 1970. *The biology of twinning in man.* Clarendon Press, Oxford, UK. [19]

Bulmer, M. G. 1971. The effects of selection on genetic variability. *Am. Nat.* 105: 210–211. [5,12]

Bulmer, M. G. 1972. The genetic variability of polygenic characters under optimizing selection, mutation, and drift. *Genet. Res.* 19: 17–25. [12]

Bulmer, M. G. 1974. Linkage disequilibrium and genetic variability. *Genet. Res.* 23: 281–289. [5]

Bulmer, M. G. 1976. Regressions between relatives. *Genet. Res.* 28: 199–203. [17]

Bulmer, M. G. 1980. *The mathematical theory of quantitative genetics.* Oxford Univ. Press, NY. [1,7,10,12,17, 18,20]

Bulmer, M. G., and J. J. Bull. 1982. Models of polygenic sex determination and sex ratio control. *Evolution* 36: 13–26. [25]

Bultman, S. J., E. J. Michaud, and R. P. Woychik. 1992. Molecular characterization of the mouse *agouti* locus. *Cell* 71: 1195–1204. [14]

Bürger, R., G. P. Wagner, and F. Stettinger. 1989. How much heritable variation can be maintained in finite populations by mutation-selection balance? *Evolution* 43: 1748–1766. [12]

Burns, T. L., P. P. Moll, and M. A. Schork. 1984. Comparisons of different sampling designs for the determination of genetic transmission mechanisms in quantitative traits. *Am. J. Hum. Genet.* 36: 1060–1074. [13]

Burr, B., and F. A. Burr. 1991. Recombinant inbred lines for molecular mapping in maize. *Trends Genet.* 7: 55–60. [14]

Burton, G. W. 1951. Quantitative inheritance in pearl millet *(Pennisetum glaucum). Agron. J.* 43: 409–417. [9]

Burton, G. W., and E. H. DeVane. 1953. Estimating heritability in tall fescue *(Festuca arundinacea)* from replicated clonal material. *Agron. J.* 45: 478–481. [19]

Burton, R. S. 1987. Differentiation and integration of the genome in populations of *Tigriopus californicus.. Evolution* 41: 504–513. [9]

Burton, R. S. 1990a. Hybrid breakdown in physiological response: a mechanistic approach. *Evolution* 44: 1806–1813. [9]

Burton, R. S. 1990b. Hybrid breakdown in developmental time in the copepod *Tigriopus californicus. Evolution* 44: 1814–1822. [9]

Busch, R. H., K. A. Lucken, and R. C. Frohberg. 1971. F_1 hybrids versus random F_6 line performance and estimates of genetic effects in spring wheat. *Crop Sci.* 11: 357–361. [10]

Bush, R. M., and P. E. Smouse. 1991. The impact of electrophoretic genotype on life history traits in *Pinus taeda. Evolution* 45: 481–498. [10]

Butlin, R. K., I. L. Read, and T. H. Day. 1982. The effects of a chromosomal inversion on adult size and male mating success in the seaweed fly, *Coelopa frigida. Heredity* 49: 51–62. [14]

Caballero, A., and P. D. Keightley. 1994. A pleiotropic nonadditive model of variation in quantitative traits. *Genetics* 138: 883–900. [12]

Caballero, A., P. D. Keightley, and W. G. Hill. 1995. Accumulation of mutations affecting body weight in inbred mouse lines. *Genet. Res.* 65: 145–149. [12]

Caballero, A. P., M. A. Toro, and C. López-Fanjul. 1991. The response to artificial selection from new mutations in *Drosophila melanogaster. Genetics* 127: 89–102. [12]

Cabot, E. L., A. W. Davis, N. A. Johnson, and C.-I. Wu. 1994. Genetics of reproductive isolation in the *Drosophila simulans* clade: complex epistasis underlying hybrid male sterility. *Genetics* 137: 175–189. [14]

Calder, W. A. 1984. *Size, function, and life history.* Harvard Univ. Press, Cambridge, MA. [11]

Caligari, P. D. S., and K. Mather. 1988. Competitive interactions in *Drosophila melanogaster*. IV. Chromosome assay. *Heredity* 60: 355–366. [14]

Cannings, C., and E. A. Thompson. 1977. Ascertainment in the sequential sampling of pedigrees. *Clin. Genet.* 12: 208–212. [13]

Cannings, C., E. A. Thompson, and M. H. Skolnick. 1976. The recursive derivation of likelihoods on complex pedigrees. *Adv. Appl. Prob.* 8: 622–625. [13]

Cannings, C., E. A. Thompson, and M. H. Skolnick. 1978. Probability functions on complex pedigrees. *Adv. Appl. Prob.* 10: 26–61. [13]

Carbonell, E. A., and T. M. Gerig. 1991. A program to detect linkage between genetic markers and nonadditive quantitative trait loci. *J. Heredity* 82: 435. [15]

Carbonell, E. A., W. E. Nyquist, and A. E. Bell. 1983. Sex-linked and maternal effects in the Eberhart-Gardner general genetics model. *Biometrics* 39: 607–619. [20]

Carbonell, E. A., T. M. Gerig, E. Balansard, and M. J. Asins. 1992. Interval mapping in the analysis of nonadditive quantitative trait loci. *Biometrics* 48: 305–315. [15]

Carbonell, E. A., M. J. Asins, M. Baselga, E. Balansard, and T. M. Gerig. 1993. Power studies in the estimation of genetic parameters and the localization of quantitative trait loci for backcross and doubled haploid populations. *Theor. Appl. Genet.* 86: 411–416. [15]
Cardellino, R. A., and T. Mukai. 1975. Mutator factors and genetic variance components of viability in *Drosophila melanogaster*. *Genetics* 80: 567–583. [12]
Cardon, L. R., and D. W. Fulker. 1994. The power of interval mapping of quantitative trait loci, using selected sib pairs. *Am. J. Hum. Genet.* 55: 825–833. [16]
Cardon, L. R., S. D. Smith, D. W. Fulker, W. J. Kimerling, B. F. Pennington, and J. C. DeFries. 1994. Quantitative trait locus for reading disability on chromosome 6. *Science* 266: 276–279. [16]
Carey, G., and J. Williamson. 1991. Linkage analysis of quantitative traits: increased power by using selected samples. *Am. J. Hum. Genet.* 49: 786–796. [14,16]
Carmelli, D., S. Karlin, and R. Williams. 1979. A class of indices to assess major gene versus polygenic inheritance of distributed variables. *In* C. F. Sing and M. Skolnick (eds.), *The genetic analysis of common diseases: applications to predict factors in coronary heart disease,* pp. 259–270. Alan R. Liss, NY. [13]
Carpenter, J. R., H. Grüneberg, and E. S. Russell. 1957. Genetical differentiation involving morphological characters in an inbred strain of mice. II. American branches of the C57BL and C57BR strains. *J. Morphology* 100: 377–388. [12]
Carson, H. L., and R. Lande. 1984. Inheritance of a secondary sexual character in *Drosophila silvestris*. *Proc. Natl. Acad. Sci. USA* 81: 6904–6907. [9]
Carson, H. L., and A. R. Templeton. 1984. Genetic revolutions in relation to speciation phenomena: the founding of new populations. *Ann. Rev. Ecol. Syst.* 15: 97–131. [14]
Carter, C. O. 1961. The inheritance of congenital pyloric stenosis. *Brit. Med. Bull.* 17: 251. [25]
Carter, C. O. 1965. The inheritance of common congenital malformations. *Prog. Med. Genet.* 4: 59–84. [25]
Carter, C. O. 1969. Genetics of common disorders. *Brit. Med. Bull.* 32: 21–26. [25]
Castle, W. E. 1906. The origin of a polydactylous race of guinea pigs. *Carnegie Inst. Wash.* Publ. No. 241: 3–55. [25]
Castle, W. E. 1921. An improved method of estimating the number of genetic factors concerned in cases of blending inheritance. *Proc. Natl. Acad. Sci. USA* 81: 6904–6907. [9]
Catchside, D. G. 1977. *Genetics of recombination.* Univ. Park Press, Baltimore, MD. [14]
Caten, C. E. 1979. Quantitative genetic variation in fungi. *In* J. N. Thompson, Jr. and J. M. Thoday (eds.), *Quantitative genetic variation,* pp. 35–60. Academic Press, NY. [5]
Caten, C. E., and J. L. Jinks. 1976. Quantitative genetics. *In* K. D. MacDonald (ed.), *Second international symposium on the genetics of industrial microorganisms,* pp. 93–111. Academic Press, NY. [5]
Cavalier-Smith, T. 1978. Nuclear volume control by nucleoskeletal DNA, selection for cell volume and cell growth rate, and the solution of the DNA C-value paradox. *J. Cell Sci.* 34: 247–268. [12]
Cavalier-Smith, T. (ed.) 1985. *The evolution of genome size.* John Wiley & Son, NY. [12]
Cavalli, L. L. 1952. An analysis of linkage in quantitative inheritance. *In* E. C. R. Reeve and C. H. Waddington (eds.), *Quantitative inheritance,* pp. 135–144. His Majesty's Stationary Office, London. [9]
Cavalli-Sforza, L. L., and W. F. Bodmer. 1971. *The genetics of human populations.* W. H. Freeman and Co., San Francisco, CA. [13]
Ceranka, B., A. Dobek, and H. Kielczewska. 1987. The analysis of partial diallel crosses. *Biom. J.* 29: 455–460. [20]
Chai, C. K. 1956. Analysis of quantitative inheritance of body size in mice. II. Gene action and segregation. *Genetics* 41: 165–178. [9]
Chai, C. K. 1957. Developmental homeostasis of body growth in mice. *Am. Nat.* 85: 49–55. [6]
Chakraborty, R. 1987. Biochemical heterozygosity and phenotypic stability of polygenic traits. *Heredity* 59: 19–28. [6]
Chakraborty, R., and M. Nei. 1982. Genetic differentiation of quantitative characters between populations or species. I. Mutation and random genetic drift. *Genet. Res.* 39: 303–314. [12]
Chakraborty, R., and N. Ryman. 1983. Relationship of mean and variance of genotypic values with heterozygosity per individual in a natural population. *Genetics* 103: 149–152. [6]
Champoux, M. C., G. Wang, S. Sarkarung, D. J. Mackill, J. C. O'Toole, N. Huang, and S. R. McCouch. 1995. Locating genes associated with root morphology and drought avoidance in rice via linkage to molecular markers. *Theor. Appl. Genet.* 90: 969–980. [15]
Changjian, J., P. Xuebiao, and G. Minghong. 1994. The use of mixture models to detect effects of major genes on quantitative characters in plant breeding experiments. *Genetics* 136: 383–394. [13]
Charlesworth, B. 1974. The Hardy-Weinberg law with overlapping generations. *Adv. Appl. Prob.* 6: 4–6. [4]
Charlesworth, B. 1990. Mutation-selection balance and the evolutionary advantage of sex and recombination. *Genet. Res.* 55: 199–221. [12]
Charlesworth, B. 1994. *Evolution in age-structured populations.* 2nd Ed. Cambridge Univ. Press, Cambridge, UK. [4]
Charlesworth, B., R. Lande, and M. Slatkin. 1982. A neo-Darwinian commentary on macroevolution. *Evolution* 36: 474–498. [9,14]
Charlesworth, B., J. A. Coyne, and N. H. Barton. 1987. The relative rates of evolution of sex chromosomes and autosomes. *Am. Nat.* 130: 113–146. [14]
Charlesworth, B., D. Charlesworth, and M. T. Morgan. 1990. Genetic loads and estimates of mutation rates in highly inbred plant populations. *Nature* 347: 380–382. [10,12]

Charlesworth, D., and B. Charlesworth. 1979. Selection on recombination in a multi-locus system. *Genetics* 91: 575–580. [5]

Charlesworth, D., and B. Charlesworth. 1987. Inbreeding depression and its evolutionary consequences. *Ann. Rev. Ecol. Syst.* 18: 237–268. [10]

Charlesworth, D., M. T. Morgan, and B. Charlesworth. 1992. The effect of linkage and population size on inbreeding depression due to mutational load. *Genet. Res.* 59: 49–61. [12]

Charlesworth, D., M. T. Morgan, and B. Charlesworth. 1993. Mutation accumulation in finite outbreeding and inbreeding populations. *Genet. Res.* 61: 39–56. [12]

Charlesworth, D., E. E. Lyons, and L. B. Litchfield. 1994. Inbreeding depression in two highly inbreeding populations of *Leavenworthia. Proc. Royal Soc. Lond.* B 258: 209–214. [10,12]

Chevalet, C. 1988. Control of genetic drift in selected populations. *In* B. S. Weir, E. J. Eisen, M. M. Goodman, and G. Namkoong (eds.), *Proceedings of the second international conference on quantitative genetics*, pp. 379–394. Sinauer Assoc., Sunderland, MA. [5]

Cheverud, J. M. 1982. Relationships among ontogenetic, static, and evolutionary allometry. *Am. J. Phys. Anthr.* 59: 139–149. [11,21]

Cheverud, J. M. 1984. Evolution by kin selection: a quantitative genetic model illustrated by maternal performance in mice. *Evolution* 38: 766–777. [23]

Cheverud, J. M. 1988. A comparison of genetic and phenotypic correlations. *Evolution* 42: 958–968. [21]

Cheverud, J. M. 1989. A comparative analysis of morphological variation patterns in the papionins. *Evolution* 43: 1737–1747. [21]

Cheverud, J. M. 1995. Morphological integration in the saddle-back tamarin *(Saguinus fuscicollis)* cranium. *Am. Nat.* 145: 63–89. [21]

Cheverud, J. M., and E. J. Routman. 1995. Epistasis and its contribution to genetic variance components. *Genetics* 139: 1455–1461. [5]

Cheverud, J. M., L. Leamy, W. R. Atchley, and J. J. Rutledge. 1983. Quantitative genetics and the evolution of ontogeny. I. Ontogenetic changes in quantitative genetic variance components in randombred mice. *Genet. Res.* 42: 65–75. [23]

Cheverud, J. M., G. P. Wagner, and M. M. Dow. 1989. Methods for the comparative analysis of variation patterns. *Syst. Zool.* 38: 201–213. [21]

Cheverud, J. M., E. J. Routman, F. A. M. Durante, B. van Swinderen, K. Cothran, and C. Perel. 1996. Quantitative trait loci for murine growth. *Genetics* 142: 1305–1319. [15]

Choo, T. M. 1981. Doubled haploids for studying the inheritance of quantitative characters. *Genetics* 99: 525–540. [9]

Choo, T. M. 1983. Doubled haploids for locating polygenes. *Can. J. Genet. Cytol.* 25: 425–429. [14]

Choo, T. M., and E. Reinbergs. 1982. Estimation of the number of genes in doubled haploid populations of barley *(Hordeum vulgare). Can. J. Genet. Cytol.* 24: 337–341. [9]

Chovnick, A., and A. S. Fox. 1953. The problem of estimating the number of loci determining quantitative variation in haploid organisms. *Am. Nat.* 87: 263–267. [9]

Christian, J. C., K. W. Kang, and J. A. Norton, Jr. 1974. Choice of an estimate of genetic variance from twin data. *Am. J. Hum. Genet.* 26: 154–161. [19]

Christian, J. C., S. W. Cheung, K. W. Kang, F. P. Harmuth, D. J. Huntzinger, and R. C. Powell. 1976. Variance of plasma free and esterified cholesterol in adult twins. *Am. J. Hum. Genet.* 28: 174–178. [19]

Christie, B. R., and V. I. Shattuck. 1992. The diallel cross: design, analysis, and use for plant breeders. *Plant Breed. Rev.* 9: 9–36. [20]

Churchill, G. A., and R. W. Doerge. 1994. Empirical threshold values for quantitative trait mapping. *Genetics* 138: 963–971. [13,15]

Clare, H. J., and L. S. Luckinbill. 1985. The effects of gene-environment interaction on the expression of longevity. *Heredity* 55: 19–29. [6]

Clark, A. G. 1987. Senescence and the genetic-correlation hang-up. *Am. Nat.* 129: 932–940. [21]

Clark, A. G., L. Wang, and T. Hulleberg. 1995a. *P*-element-induced variation in metabolic regulation in *Drosophila. Genetics* 139: 337– 348. [12]

Clark, A. G., L. Wang, and T. Hulleberg. 1995b. Spontaneous mutation rate of modifiers of metabolism in *Drosophila. Genetics* 139: 767–779. [12]

Clark, P. J. 1956. The heritability of certain anthropometric characters as ascertained from measurements of twins. *Am. J. Hum. Genet.* 7: 49–54. [19]

Clarke, G. M. 1992. Fluctuating asymmetry: a technique for measuring developmental stress of genetic and environmental origin. *Acta Zool. Fennica* 191: 31–35. [6]

Clarke, G. M., G. W. Brand, and M. J. Whitten. 1986. Fluctuating asymmetry: a technique for measuring development stress caused by inbreeding. *Aust. J. Biol. Sci.* 39: 145–153. [6]

Clarke, G. M., and J. A. McKenzie. 1987. Developmental stability of insecticide resistant phenotypes in blowfly: a result of canalizing natural selection. *Nature* 325: 345–346. [11]

Clayton, G. A., and A. Robertson. 1955. Mutation and quantitative variation. *Am. Nat.* 89: 151–158. [12]

Clayton, G. A., and A. Robertson. 1964. The effects of X-rays on quantitative characters. *Genet. Res.* 5: 410–422. [12]

Clayton, G. A., J. A. Morris, and A. Robertson. 1957. An experimental check on quantitative genetical theory. I. Short-term responses to selection. *J. Genetics* 55: 131–151. [7]

Cleghorn, T. E. 1960. MNSs gene frequencies in English blood donors. *Nature* 187: 701. [5]

Clerget-Darpoux, F., C. Bonaïti-Pellié, and J. Hochez. 1986. Effects of misspecifying genetic parameters in Lod score analysis. *Biometrics* 42: 393–399. [16]

Cloninger, C. R., J. Rice, and T. Reich. 1979a. Multifactorial inheritance with cultural transmission and assortative mating. II. A general model of combined polygenic and cultural inheritance. *Am. J. Hum. Genet.* 31: 176–189. [7]

Cloninger, C. R., J. Rice, and T. Reich. 1979b. Multifactorial inheritance with cultural transmission and assortative mating. III. Family structure and analysis of separation experiments. *Am. J. Hum. Genet.* 31: 366–388. [7]

Cock, A. G. 1964. Dosage compensation and sex-chromatin in non-mammals. *Genet. Res.* 5: 354–365. [24]

Cockerham, C. C. 1954. An extension of the concept of partitioning hereditary variance for analysis of covariances among relatives when epistasis is present. *Genetics* 39: 859–882. [5,7,21]

Cockerham, C. C. 1963. Estimation of genetic variances. *In* W. D. Hanson and H. F. Robinson (eds.), *Statistical genetics and plant breeding*, pp. 53–94. Natl. Acad. Sci., Natl. Res. Council Publ. No. 982, Washington, D.C. [20,22]

Cockerham, C. C. 1980. Random and fixed effects in plant genetics. *Theor. Appl. Genet.* 56: 119–131. [9]

Cockerham, C. C. 1986. Modifications in estimating the number of genes for a quantitative character. *Genetics* 114: 659–664. [9]

Cockerham, C. C., and B. S. Weir. 1968. Sib mating with two linked loci. *Genetics* 60: 629–640. [7]

Cockerham, C. C., and B. S. Weir. 1973. Descent measures for two loci with some applications. *Theor. Pop. Biol.* 4: 300–330. [7]

Cockerham, C. C., and B. S. Weir. 1977a. Digenic descent measures for finite populations. *Genet. Res.* 30: 121–147. [5,7]

Cockerham, C. C., and B. S. Weir. 1977b. Quadratic analyses of reciprocal crosses. *Biometrics* 33: 187–203. [20]

Cockerham, C. C., and Z.-B. Zeng. 1996. Design III with marker loci. *Genetics*. 143: 1437–1456. [10,15]

Coen, E. S., R. Carpenter, and C. Martin. 1986. Transposable elements generate novel spatial patterns of gene expression in *Antirrhinum majus*. *Cell* 47: 285–296. [12]

Coles, J. F., and D. P. Fowler. 1976. Inbreeding in neighboring trees in two white spruce plantations. *Silvae Genet.* 25: 29–34. [10]

Collins, F. S. 1992. Positional cloning: let's not call it reverse anymore. *Nature Genetics* 1: 3–6. [14]

Collins, F. S. 1995. Positional cloning moves from perditional to traditional. *Nature Genetics* 9: 347–350. [14]

Collins, J. P., and J. E. Cheek. 1983. Effect of food and density on development of typical and cannibalistic salamander larvae in *Ambystoma tigrinum*. *Am. Zool.* 23: 77–84. [6]

Collins, R. L. 1967. A general nonparametric theory of genetic analysis. I. Application to the classical cross. *Genetics* 56: 551. [13]

Collins, R. L. 1968. A general nonparametric theory of genetic analysis. II. Digenic models with linkage for the classical cross. *Genetics* 60: 169–170. [13]

Collins, R. L. 1973. Reply to Whitney and Klein. *Genetics* 74: 382–383. [13]

Commenges, D. 1994. Robust genetic linkage analysis based on a score test of homogeneity: The weighted rank pairwise correlation statistic. *Genet. Epidem.* 11: 198–200. [16]

Comstock, R. E., and R. H. Moll. 1963. Genotype-environment interactions. *In* W. D. Hanson and H. F. Robinson (eds.), *Statistical genetics and plant breeding*, pp. 164–196. Natl. Acad. Sci., Natl. Res. Council Publ. No. 982, Washington, D.C. [22]

Comstock, R. E., and H. F. Robinson. 1948. The components of genetic variance in populations of biparental progenies and their use in estimating the average degree of dominance. *Biometrics* 4: 254–266. [20]

Comstock, R. E., and H. F. Robinson. 1952. Estimation of average dominance of genes. *In* J. W. Gowen (ed.), *Heterosis*, pp. 494–516. Iowa State College Press, Ames. [5,15,20]

Conneally, P. M., J. H. Edwards, K. K. Kidd, J.-M. Lalouel, N. E. Morton, J. Ott, and R. White. 1985. Report of the committee on methods of linkage analysis and reporting. *Cytogen. Cell Genet.* 40: 356–359. [15]

Connor, J., and S. Via. 1993. Patterns of phenotypic and genetic correlations among morphological and life-history traits in wild radish, *Raphanus raphanistrum*. *Evolution* 47: 704–711. [21]

Connor, J. L., and M. J. Bellucci. 1979. Natural selection resisting inbreeding depression in captive wild housemice *(Mus musculus)*. *Evolution* 33: 929–940. [10]

Cooke, P., and K. Mather. 1962. Estimating the components of continuous variation. II. Genetics. *Heredity* 17: 211–236. [14]

Cooper, D. W. 1971. Directed genetic change model for X chromosome inactivation in eutherian mammals. *Nature* 230: 292–294. [24]

Copeland, N. G., N. A. Jenkins, D. J. Gilbert, J. T. Eppig, L. J. Maltais, J. C. Miller, W. F. Dietrich, A. Weaver, S. E. Lincoln, R. G. Steen, L. D. Stein, J. H. Nadeau, and E. S. Lander. 1993. A genetic linkage map of the mouse: current applications and future prospects. *Science* 262: 57–66. [14].

Copeman, J. B., and 15 others. 1995. Linkage disequilibrium mapping of a type 1 diabetes susceptibility gene (*IDDM7*) to chromosomes 2q31-q33. *Nature Genetics* 9: 80–85. [14]

Corder, E. H., A. M. Saunders, W. J. Strittmatter, D. E. Schmechel, P. C. Gaskell, G. W. Small, A. D. Roses, J. H. Haines, and M. A. Pericak-Vance. 1993. Gene dose of apolipoprotein E type 4 allele and the risk of Alzheimer's disease in late onset families. *Science* 261: 921–923. [14]

Cornelius, P. L., and J. W. Dudley. 1974. Effects of inbreeding by selfing and full-sib mating in a maize population. *Crop Sci.* 14: 815–819. [10]

Cothran, E. G., J. W. MacCluer, L. R. Weitkamp, and S. A. Guttormsen. 1986. Genetic variability, inbreeding, and reproductive performance in standardbred horses. *Zoo Biol.* 5: 191–201. [10]

Cotterman, C. W. 1940. A calculus for statistico-genetics. Ph. D. Thesis, Ohio State Univ., Columbus. [7]

Cotterman, C. W. 1954. Estimation of gene frequencies in nonexperimental populations. *In* O. Kempthorne, T. A. Bancroft, J. W. Gowen, and J. L. Lush (eds.), *Statistics and mathematics in biology*, pp. 449–465. Iowa State College Press, Ames. [7]

Cowan, C. M., M. R. Detine, R. L. Ax, and L. A. Schuler. 1990. Structural variation around prolactin gene linked to quantitative traits in an elite Holstein sire family. *Theor. Appl. Genet.* 79: 577–582. [14]

Cowen, N. M. 1988. The use of replicated progenies in marker-based mapping of QTLs. *Theor. Appl. Genet.* 75: 857–862. [14]

Cowley, D. E. 1991. Genetic prenatal maternal effects on organ size in mice and their potential contribution to evolution. *J. Evol. Biol.* 4: 363–382. [23]

Cowley, D. E., and W. R. Atchley. 1988. Quantitative genetics of *Drosophila melanogaster.* II. Heritabilities and genetic correlations between sexes for head and thorax traits. *Genetics* 119: 421–433. [24]

Cowley, D. E., and W. R. Atchley. 1990. Developmental and quantitative genetics of correlation structure among body parts of *Drosophila melanogaster. Am. Nat.* 135: 242–268. [21]

Cowley, D. E., and W. R. Atchley. 1992. Quantitative genetic models for development, epigenetic selection, and phenotypic evolution. *Evolution* 46:495–518. [21]

Cowley, D. E., W. R. Atchley, and J. J. Rutledge. 1986. Quantitative genetics of *Drosophila melanogaster.* I. Sexual dimorphism in genetic parameters for wing traits. *Genetics* 114: 549–566. [24]

Cowley, D. E., D. Pomp, W. R. Atchley, E. J. Eisen, and D. Hawkins-Brown. 1989. The impact of maternal uterine genotype on postnatal growth and adult body size in mice. *Genetics* 122: 193–203. [23]

Cox, T. S., D. J. Cox, and K. J. Frey. 1987. Mutations for polygenic traits in barley under nutrient stress. *Euphytica* 36: 823–829. [12]

Coyne, J. A. 1983. Genetic basis of differences in genital morphology among three sibling species of *Drosophila. Evolution* 37: 1101–1118. [14]

Coyne, J. A. 1984. Genetic basis of male sterility in hybrids between two closely related species of *Drosophila. Proc. Natl. Acad. Sci. USA* 81: 4444–4447. [14]

Coyne, J. A. 1985. Genetic studies of three sibling species of *Drosophila* with relationship to theories of speciation. *Genet. Res.* 46: 169–192. [14]

Coyne, J. A. 1992. Genetics and speciation. *Nature* 355: 511–515. [14]

Coyne, J. A., and E. Beecham. 1987. Heritability of two morphological characters within and among natural populations of *Drosophila melanogaster. Genetics* 117: 727–737. [7,17]

Coyne, J. A., B. Charlesworth, and H. A. Orr. 1991. Haldane's rule revisited. *Evolution* 45: 1710–1714. [14]

Coyne, J. A., and R. Lande. 1985. The genetic basis of species differences in plants. *Am. Nat.* 126: 141–145. [9]

Coyne, J. A., and H. A. Orr. 1989a. Patterns of speciation in *Drosophila. Evolution* 43: 362–381. [14]

Coyne, J. A., and H. A. Orr. 1989b. Two rules of speciation. *In* D. Otte and J. A. Endler (eds.), *Speciation and its consequences*, pp. 180–207. Sinauer Assoc., Sunderland, MA. [14]

Coyne, J. A., and H. A. Orr. 1993. Further evidence against the meiotic drive model of hybrid sterility. *Evolution* 47: 685–687. [14]

Coyne, J. A., and H. A. Orr. 1997. "Patterns of speciation in *Drosophila*" revisited. *Evolution* 51: 295–303. [14]

Crabbe, J. C., J. K. Belknap, and K. J. Buck. 1994. Genetic animal models of alcohol and drug abuse. *Science* 264: 1715–1723. [14]

Crittenden, L. B. 1961. An interpretation of familial aggregation based on multiple genetic and environmental factors. *Ann. New York Acad. Sci.* 91: 769–780. [25]

Crnokrak, P., and D. A. Roff. 1995. Dominance variance: associations with selection and fitness. *Heredity* 75: 530–540. [7]

Croft, J. H., and J. L. Jinks. 1977. Aspects of the population genetics of *Aspergillus nidulans. In* J. E. Smith and J. A. Pateman (eds.), *Genetics and physiology of* Aspergillus, pp. 339–360. Academic Press, NY. [5]

Croft, J. H., and G. Simchen. 1965. Natural variation among monokaryons of *Collybia velutipes. Am. Nat.* 99: 451–462. [9]

Crow, J. F. 1948. Alternative hypotheses of hybrid vigor. *Genetics* 33: 477–487. [10]

Crow, J. F. 1952. Dominance and overdominance. *In* J. E. Gowen (ed.), *Heterosis.* Iowa State College Press, Ames. [10]

Crow, J. F. 1954. Random mating with linkage in polysomics. *Am. Nat.* 88: 431–434. [4]

Crow, J. F. 1958. Some possibilities for measuring selection intensities in man. *Human Biol.* 30: 1–13. [10]

Crow, J. F. 1992. Mutation, mean fitness, and genetic load. *Oxford Surv. Evol. Biol.* 9: 3–42. [12]

Crow, J. F. 1993a. Francis Galton: count and measure, measure and count. *Genetics* 135: 1–4. [1]

Crow, J. F. 1993b. How much do we know about spontaneous human mutation rates? *Environ. Mol. Mutagenesis* 21: 122–129. [12]

Crow, J. F., and J. Felsenstein. 1968. The effect of assortative mating on the genetic composition of a population. *Eugenics Quart.* 15: 85–97. [7]

Crow, J. F., and M. Kimura. 1970. *An introduction to population genetics theory.* Harper & Row, NY. [1,4,7,10]

Crow, J. F., and M. Kimura. 1979. Efficiency of truncation selection. *Proc. Natl. Acad. Sci. USA* 76: 396–399. [12]

Crow, J. F., and M. J. Simmons. 1983. The mutation load in *Drosophila In* M. Ashburner, H. L. Carson, and J. N. Thompson, Jr. (eds.), *The genetics and biology of Drosophila. Volume 3c*, pp. 1–35. Academic Press, NY. [12]

Crowley, P. H. 1992. Resampling methods for computation-intensive data analysis in ecology and evolution. *Ann. Rev. Ecol. Syst.* 23: 405–448. [18]

Cullis, C. A. 1981. Environmental induction of heritable changes in flax: defined environments inducing changes in rDNA and peroxidase isozyme band pattern. *Heredity* 47: 87–94. [12]

Cullis, C. A. 1985. Sequence variation and stress. *In* B. Hohn and E. S. Dennis (eds.), *Genetic flux in plants*, pp. 157–168. Springer-Verlag, NY. [12]

Cunningham, E. P. 1982. The genetic basis of heterosis, pp. 190–205. *Proc. Second World Cong. on Genetics Applied to Livestock Production.* Madrid, Spain. [10]

Curnow, R. N. 1961. The estimation of repeatability and heritability from records subject to culling. *Biometry* 17: 553–566. [27]

Curnow, R. N. 1963. Sampling the diallel cross. *Biometrics* 19: 287–306. [20]

Curnow, R. N. 1972. The multifactorial model for the inheritance of liability to disease and its implications for relatives at risk. *Biometrics* 28: 931–946. [25]

Curnow, R. N., and C. Smith. 1975. Multifactorial models for familial diseases in man. *J. Royal Stat. Soc.* A 138: 131–169. [25]

Curtis, D. 1996. Genetic dissection of complex traits. *Nature Genetics* 12: 356–357. [16]

Curtis, D., and P. C. Stam. 1994. Using risk calculation to implement an extended relative pair analysis. *Am. J. Hum. Genet.* 58: 151–162. [16]

Curtis, D., and P. C. Stam. 1995. Model-free linkage analysis using likelihoods. *Am. J. Hum. Genet.* 57: 703–716. [13,16]

Curtsinger, J. W., P. M. Service, and T. Prout. 1994. Antagonistic pleiotropy, reversal of dominance, and genetic polymorphism. *Am. Nat.* 144: 210–228. [21]

D'Agostino, R. B. 1971. An omnibus test of normality for moderate and large size samples. *Biometrika* 58: 341–348. [11]

Daly, M. J., and E. S. Lander. 1996. The importance of being independent: sib pair analysis in diabetes. *Nature Genetics* 14: 131–132. [16]

Damerval, C., A. Maurice, J. M. Josse, and D. de Vienne. 1994. Quantitative trait loci underlying gene product variation: a novel perspective by analyzing regulation of genome expression. *Genetics* 137: 289–301. [15]

Daniels, S. B., M. McCarron, C. Love, and A. Chovnick. 1985. Dysgenesis-induced instability of rosy locus transformation in *Drosophila melanogaster*: analysis of excision events and the selective recovery of control element deletions. *Genetics* 109: 95–117. [12]

Dapkus, D., and D. J. Merrell. 1977. Chromosomal analysis of DDT-resistance in a long-term selected population of *Drosophila melanogaster*. *Genetics* 87: 685–697. [14]

Darvasi, A. and M. Soller. 1992. Selective genotyping for determination of linkage between a marker locus and a quantitative trait locus. *Theor. Appl. Genet.* 85: 353–359. [14,15]

Darvasi, A. and M. Soller. 1994a. Selective DNA pooling for determination of linkage between a marker locus and a quantitative trait locus. *Genetics* 138: 1365–1373. [14]

Darvasi, A. and M. Soller. 1994b. Optimum spacing of genetic markers for determining linkage between marker loci and quantitative trait loci. *Theor. Appl. Genet.* 89: 351–357. [15]

Darvasi, A., and M. Soller. 1995. Advanced intercross lines, an experimental population for fine genetic mapping. *Genetics* 141: 1199–1207. [15]

Darvasi, A., A. Weinreb, V. Minke, J. I. Weller, and M. Soller. 1993. Detecting marker-QTL linkage and estimating QTL gene effect and map location using a saturated genetic map. *Genetics* 134: 943–951. [15]

Darwin, C. 1859. *The origin of species by means of natural selection.* Murray, London. [1]

Darwin, C. 1876. *The effects of cross and self-fertilization in the vegetable kingdom.* Appleton, NY. [9,10]

Davenport, C. B. 1908. Degeneration, albinism, and inbreeding. *Science* 28: 454–455. [10]

David, P., B. Delay, P. Berthou, and P. Jarne. 1995. Alternative models for allozyme-associated heterosis in the marine bivalve *Spisula ovalis*. *Genetics* 139: 1719–1726. [10]

Davies, J. L., and 15 others. 1994. A genome-wide search for human type 1 diabetes susceptibility genes. *Nature* 371: 130–136. [16]

Davies, R. W. 1971. The genetic relationship of two quantitative characters in *Drosophila melanogaster*. II. Location of the effects. *Genetics* 69: 363–375. [14]

Davies, R. W., and P. L. Workman. 1971. The genetic relationship of two quantitative characters in *Drosophila melanogaster*. I. Response to selection and whole chromosome analysis. *Genetics* 69: 353–361. [14]

Davis, A. W., and C.-I. Wu. 1996. The broom of the sorcerer's apprentice: the fine structure of a chromosomal region causing reproductive isolation between two sibling species of *Drosophila*. *Genetics* 143: 1287–1298. [14]

Davis, S., M. Schroeder, L. R. Goldin, and D. E. Weeks. 1996. Nonparametric simulation-based statistics in detecting linkage in general pedigrees. *Am. J. Hum. Genet.* 58: 867–880. [16]

Dawson, P. S. 1965. Estimation of components of phenotypic variance for development rate in *Tribolium*. *Heredity* 20: 403–417. [18,20]

Day, N. E. 1969. Estimating the components of a mixture of normal distributions. *Biometrika* 56: 463–474. [13]

Day, N. E., and M. J. Simmons. 1976. Disease susceptibility genes — their identification by multiple case family studies. *Tissue Antigens* 8: 109–119. [16]

de Jong, G. 1990. Quantitative genetics of reaction norms. *J. Evol. Biol.* 3: 447–468. [22]

de Jong, G. 1995. Phenotypic plasticity as a product of selection in a variable environment. *Am. Nat.* 145: 493–512. [22]

de la Chapelle, A. 1993. Disease gene mapping in isolated human populations: the example of Finland. *J. Med. Genet.* 30: 857–965. [14]

Demenais, F., and G. E. Bonney. 1989. Equivalence of the mixed and regressive models for genetic analysis. Continuous traits. *Genet. Epidem.* 6: 597–617. [13]

Demenais, F., M. Lathrop, and J. M. Lalouel. 1986. Robustness and power of the unified model in the analysis of quantitative measurements. *Am. J. Hum. Genet.* 38: 228–234. [13]

Demenais, F., M. Lathrop, and J. M. Lalouel. 1988. Detection of linkage between a quantitative trait and a marker locus by the lod score method: sample size and sampling considerations. *Am. J. Hum. Genet.* 52: 237–246. [16]

DeMoivre, A. 1738. *The doctrine of chances.* 2nd Ed. (reprinted 1967, Frank Cass, London). [2]

Dempfle, L. 1990. Problems in the use of the relationship matrix in animal breeding. *In* D. Gianola and K. Hammond (eds.), *Statistical methods for genetic improvement of livestock,* pp. 454–473. Springer-Verlag, NY. [26]

Dempster, A. P., N. M. Laird, and D. B. Rubin. 1977. Maximum likelihood from incomplete data via the EM algorithm. *J. Royal Stat. Soc.* 39: 1–38. [27,A4]

Dempster, E. R., and I. M. Lerner. 1950. Heritability of threshold characters. *Genetics* 35: 212–236. [25]

Deng, H.-W. 1997. Decrease of developmental stability upon inbreeding in *Daphnia. Heredity* 78: 182–189. [6]

Deng, H.-W., and M. Lynch. 1996a. Change of genetic architecture in response to sex. *Genetics* 143: 203–212. [5,9]

Deng, H.-W., and M. Lynch. 1996b. Estimation of deleterious-mutation parameters in natural populations. *Genetics* 144: 349–360. [12]

Denniston, C. 1974. An extension of the probability approach to genetic relationships; one locus. *Theor. Pop. Biol.* 6: 58–75. [7]

Dentine, M. R., and C. M. Cowan. 1990. An analytical model for the estimation of chromosome substitution effects in the offspring of individuals heterozygous at a segregating marker locus. *Theor. Appl. Genet.* 79: 775–780. [16]

Deol, M. S., H. Grüneberg, A. G. Searle, and G. M. Truslove. 1957. Genetical differentiation involving morphological characters in an inbred strain of mice. I. A British branch of the C57BL strain. *J. Morphology* 100: 345–375. [12]

DeSalle, R., and A. R. Templeton. 1986. The molecular through ecological genetics of *abnormal abdomen* in *Drosophila mercatorum.* III. Tissue- specific differential replication of ribosomal genes modulates the *abnormal abdomen* phenotype in *Drosophila mercatorum. Genetics* 112: 877–886. [12]

DeSalle, R., J. Slightom, and E. Zimmer. 1986. The molecular through ecological genetics of *abnormal abdomen* in *Drosophila mercatorum.* II. Ribosomal DNA polymorphism is associated with the *abnormal abdomen* syndrome in *Drosophila mercatorum. Genetics* 112: 861–875. [12]

deVicente, M. C., and S. D. Tanksley. 1993. QTL analysis of transgressive segregation in an interspecific tomato cross. *Genetics* 134: 585–596. [15]

deVries, R. R. P., R. F. M. Fat, A. Lai, L. E. Nijenhuis, and J. J. Van Rood. 1976. HLA-linked genetic control of host response to *Mycobacterium leprae. Lancet* ii: 1328–1330. [16]

Dhondt, A. A. 1982. Heritability of blue tit tarsus length from normal and cross-fostered broods. *Evolution* 36: 418–419. [17]

Dhrymes, P. J. 1978. *Mathematics for econometrics.* Springer-Verlag, NY. [A3]

Dickerson, G. E. 1947. Composition of hog carcasses as influenced by heritable differences in rate and economy of gain. *Iowa Agric. Exp. Stn. Res. Bull.* 354: 492–524. [23]

Dickerson, G. E. 1962. Implications of genetic-environmental interaction in animal breeding. *Anim. Prod.* 4: 47–63. [22]

Dickerson, G. E. 1969. Techniques for research in quantitative animal genetics. *In* Am. Soc. Animal Sci., *Techniques and procedures in animal science research,* pp. 36–79. Albany, NY. [9,18,A1]

Dickerson, G. E., C. T. Blunn, A. B. Chapman, R. M. Kottman, J. L. Krider, E. J. Warwick, J. A. Whatley, Jr., M. L. Baker, J. L. Lush, and L. M. Winters. 1954. Evaluation of selection in developing inbred lines of swine. *Mo. Agric. Exp. Sta. Res. Bull.* 551. [10]

Dickinson, A. G., and J. L. Jinks. 1956. A generalized analysis of diallel crosses. *Genetics* 41: 65–78 [20].

Dinkel, C. A., D. A. Busch, J. A. Minyard, and W. R. Trevillyan. 1968. Effects of inbreeding on growth and conformation of beef cattle. *J. Anim. Sci.* 27: 313–322. [10]

Dirlewanger, E., P. G. Issac, S. Rande, M. Belajouza, R. Cousin, and D. de Vienne. 1994. Restriction fragment length polymorphism analysis of loci associated with disease resistance genes and developmental traits in *Pisum sativum* L. *Theor. Appl. Genet.* 88: 17–27. [15]

Dobzhansky, T. 1936. Studies on hybrid sterility. II. Localization of sterility factors in *Drosophila pseudoobscura* hybrids. *Genetics* 21: 113–135. [14]

Dobzhansky, T. 1948. Genetics of natural populations. XVIII. Experiments on chromosomes of *Drosophila pseudoobscura* from different geographical regions. *Genetics* 33: 588–602. [9]

Dobzhansky, T., and H. Levene. 1955. Developmental homeostasis in natural populations of *Drosophila pseudoobscura. Genetics* 40: 797–808. [6]

Dobzhansky, T., and B. Spassky. 1954. Genetics of natural populations. XXII. A comparison of the concealed variability in *Drosophila prosaltans* with that in other species. *Genetics* 39: 472–487. [6]

Dobzhansky, T., and B. Spassky. 1963. Genetics of natural populations. XXXIV. Adaptive norm, genetic load and genetic elite in *Drosophila pseudoobscura. Genetics* 48: 1467-1485. [10]

Dobzhansky, T., H. Levene, B. Spassky, and N. Spassky. 1959. Release of genetic variability through recombination. III. *Drosophila prosaltans. Genetics* 44: 75–92. [5]

Dobzhansky, T., B. Spassky, and T. Tidwell. 1963. Genetics of natural populations. XXXII. Inbreeding and the mutation and balanced genetic loads in natural populations of *Drosophila pseudoobscura*. *Genetics* 48: 361–373. [10]

Doebley, J. 1992. Mapping the genes that made maize. *Trends Genet.* 8: 302–307. [15]

Doebley, J., and A. Stec. 1991. Genetic analysis of the morphological differences between maize and teosinte. *Genetics* 129: 285–295. [15]

Doebley, J., and A. Stec. 1993. Inheritance of the morphological differences between maize and teosinte: comparison of results for two F_2 populations. *Genetics* 134: 559–570. [15]

Doebley, J., A. Stec, and C. Gustus. 1995a. *teosinte branched1* and the origin of maize: evidence for epistasis and the evolution of dominance. *Genetics* 141: 333–346. [5,15]

Doebley, J., A. Stec, and L. Hubbard. 1997. The evolution of apical dominance in maize. *Nature* 386: 485–488. [15]

Doebley, J., A. Stec, and B. Kent. 1995b. *Suppressor of sessile spikelets1 (sos1)*: a dominant mutant affecting inflorescence development in maize. *Am. J. Bot.* 82: 571–577. [5]

Doebley, J., A. Stec, J. Wendell, and M. Edwards. 1990. Genetic and morphological analysis of maize-teosinte F_2 population: implications for the origin of maize. *Proc. Natl. Acad. Sci. USA* 87: 9888–9892. [5,15]

Doebley, J., A. Bacigalupo, and A. Stec. 1994. Inheritance of kernel weight in two maize-teosinte hybrid populations: implications for crop evolution. *J. Heredity* 85:191–195. [15]

Doerge, R. W., and G. A. Churchill. 1996. Permutation tests for multiple loci affecting a quantitative character. *Genetics* 142: 285–294. [15]

Doerge, R. W., and A. Rebaï. 1996. Significance thresholds for QTL interval mapping tests. *Heredity* 76: 459–464. [15]

Dole, J., and K. Ritland. 1993. Inbreeding depression in two *Mimulus* taxa measured by multigenerational changes in the inbreeding coefficient. *Evolution* 47: 361–373. [10]

Doris-Keller, H., and 32 other authors. 1987. A genetic linkage map of the human genome. *Cell* 51: 319–337. [14]

Dorn, L. A., and T. Mitchell-Olds. 1991. Genetics of *Brassica campestris*. 1. Genetic constraints on evolution of life-history characters. *Evolution* 45: 371–379. [21]

Dorweiler, J., A. Stec, J. Kermicle, and J. Doebley. 1993. *Teosinte glume architecture1*: a genetic locus controlling a key step in maize evolution. *Science* 262: 233–235. [5,6,15]

Dover, G. A., and R. B. Flavell (eds.) 1982. *Genome evolution.* Academic Press, NY. [4]

Dragani, T. A., Z.-B. Zeng, F. Canzian, M. Gariboldi, M. T. Ghilarducci, G. Manenti, and M. A. Pierotti. 1995. Mapping of body weight loci on mouse chromosome X. *Mammalian Genome* 6: 778–781. [15]

Drake, J. W. 1991. A constant rate of spontaneous mutation in DNA-based microbes. *Proc. Natl. Acad. Sci. USA* 88: 7160–7164. [12]

Ducrocq, V. 1992. Solving animal model equations through an approximate incomplete Cholesky decomposition. *Genet. Sel. Evol.* 24: 193–209. [26]

Ducrocq, V., and B. Besbes. 1993. Solutions of multiple trait animal models with missing data on some traits. *J. Anim. Breed. Genet.* 110: 81–89. [26]

Dudash, M. R. 1990. Relative fitness of selfed and outcrossed progeny in a self-compatible, protandrous species, *Sabatia angularis* L. *(Gentianaceae)*: a comparison in three environments. *Evolution* 44: 1129–1139. [10]

Dudley, J. W. 1977. 76 generations of selection for oil and protein percentage in maize. *In* E. Pollak, O. Kempthorne, and T. B. Bailey, Jr. (eds.), *Proceedings of the international conference on quantitative genetics*, pp. 459–473. Iowa State Univ. Press, Ames. [15]

Dudley, J. W. 1992. Theory for identification of marker locus-QTL associations in population by line crosses. *Theor. Appl. Genet.* 85: 101–104. [15]

Dudley, J. W. 1993. Molecular markers in plant improvement: manipulation of genes affecting quantitative traits. *Crop Sci.* 33: 660–668. [14]

Dudley, J. W., and R. J. Lambert. 1992. Ninety generations of selection for oil and protein in maize. *Maydica* 37: 1–7. [15]

Dun, R. B., and A. S. Fraser. 1958. Selection for an invariant character, vibrissa number, in the house mouse. *Nature* 181: 1018–1019. [11]

Dun, R. B., and A. S. Fraser. 1959. Selection for an invariant character, vibrissa number, in the house mouse. *Aust. J. Biol. Sci.* 12: 506–523. [11]

Duyk, G. M., S. Kim, R. M. Myers, and D. R. Cox. 1990. Exon trapping: a genetic screen to identify candidate transcribed sequences in cloned mammalian genomic DNA. *Proc. Natl. Acad. Sci. USA* 87: 8995–8999. [14]

Eanes, W. F. 1978. Morphological variance and enzyme heterozygosity in the monarch butterfly. *Nature* 276: 263–264. [6]

East, E. M. 1908. Inbreeding in corn. *Report Conn. Agric. Exp. Sta.* 1907, pp. 419–428. [10]

East, E. M. 1910. A Mendelian interpretation of variation that is apparently continuous. *Am. Nat.* 44: 65–82. [1]

East, E. M. 1911. The genotype hypothesis and hybridization. *Am. Nat.* 45: 160–174. [1]

East, E. M. 1916. Studies on size inheritance in *Nicotiana*. *Genetics* 1: 164–176. [1]

Easteal, S., and C. Collet. 1994. Consistent variation in amino-acid substitution rate, despite uniformity of mutation rate: protein evolution in mammals is not neutral. *Mol. Biol. Evol.* 11: 643–647. [12]

Eaves, L. 1976. The effect of cultural transmission on continuous variation. *Heredity* 37: 41–57. [7]

Eaves, L. J. 1984. The resolution of genotype × environment interaction in segregation analysis of nuclear families. *Genet. Epidem.* 1: 215–228. [13]

Eaves, L. J., K. A. Last, P. J. Young, and N. G. Martin. 1978. Model-fitting approaches to the analysis of human behavior. *Heredity* 41: 249–320. [19]

Eaves, L. J., J. K. Hewitt, and A. C. Heath. 1988. The quantitative genetic study of human developmental change: a model and its limitations. *In* B. S. Weir, E. J. Eisen, M. M. Goodman, and G. Namkoong (eds.), *Proceedings of the second international conference on quantitative genetics*, pp. 297–311. Sinauer Assoc., Sunderland, MA. [7]

Eberhart, S. A., and C. O. Gardner. 1966. A general model for genetic effects. *Biometrics* 22: 864–881 [20].

Eberhart, S. A., and W. A. Russell. 1966. Stability parameters for comparing varieties. *Crop Sci.* 6: 36–40. [22]

Ebers, G. C., and 29 others. 1996. A full genome search for multiple sclerosis. *Nature Genetics* 13: 472–476. [16]

Ebert, R. H., V. A. Cherkasova, R. A. Dennis, J. H. Wu, S. Ruggles, T. E. Perring, and R. J. Shmookler-Reis. 1993. Longevity-determining genes in *Caenorhabditis elegans*: Chromosomal mapping of multiple noninteractive loci. *Genetics* 135: 1003–1010. [14]

Eckert, C. G., and S. C. H. Barrett. 1994. Inbreeding depression in partially self-fertilizing *Decodon verticillatus* (Lythrace-ae): population-genetic and experimental analyses. *Evolution* 48: 952–964. [10]

Edmunds, G. F., and D. N. Alstad. 1978. Coevolution in insect herbivores and conifers. *Science* 199: 941–945. [9]

Edwards, A. W. F. 1992. *Likelihood.* Expanded edition. Johns Hopkins Press, Baltimore, MD. [A4]

Edwards, J. H. 1969. Familial predisposition in man. *Brit. Med. Bull.* 25: 58–63. [25]

Edwards, M. D., C. W. Stuber, and J. F. Wendel. 1987. Molecular-marker-facilitated investigations of quantitative-trait loci in maize. I. Numbers, genomic distributions and types of gene action. *Genetics* 116: 113–125. [15]

Edwards, M. D., T. Helentjaris, S. Wright, and C. W. Stuber. 1992. Molecular-marker-facilitated investigations of quantitative-trait loci in maize. 4. analysis based on genome saturation with isozyme and restriction fragment length polymorphism markers. *Theor. Appl. Genet.* 83: 765–774. [14,15]

Efron, B. 1979. Bootstrap methods: another look at the jackknife. *Ann. Stat.* 7: 1–26. [15]

Efron, B. 1982. *The jackknife, the bootstrap and other resampling plans.* SIAM, Philadelphia. [15,18]

Ehiobu, N. G., M. E. Goddard, and J. F. Taylor. 1989. Effect of rate of inbreeding on inbreeding depression in *Drosophila melanogaster. Theor. Appl. Genet.* 77: 123–127. [10]

Eisen, E. J. 1967. Mating designs for estimating direct and maternal genetic variances and direct-maternal covariances. *Can. J. Genet. Cytol.* 9: 13–22. [23]

Eisen, E. J., and J. E. Legates. 1966. Genotype-sex interaction and the genetic correlation between the sexes for body weight in *Mus musculus. Genetics* 54: 611–623. [24]

Eisenhart, C. 1947. The assumptions underlying the analysis of variance. *Biometrics* 3: 1–21. [26]

Elsen, J. M., S. Knott, P. Le Roy, and C. S. Haley. 1997. Comparison between some approximate maximum-likelihood methods for quantitative trait locus detection in progeny test designs. *Theor. Appl. Genet/* 95: 236–2456. [16]

Elston, R. C. 1980. Segregation analysis. *In* J. H. Mielke and M. H. Crawford (eds.), *Current developments in anthropological genetics. Vol. 1: Theory and methods*, pp. 327–354. Plenum, NY. [13]

Elston, R. C. 1981a. Segregation analysis. *Adv. Hum. Genet.* 11: 63–120. [13]

Elston, R. C. 1981b. Testing one- and two-locus hypotheses for the genetic difference of a quantitative trait between two homozygous lines. *In* E. S. Gershon, S. Matthysse, X. O. Breakefield, and R. D. Ciaranello (eds.), *Genetic research strategies for psychobiology and psychiatry*, pp. 283–293. Plenum, NY. [13]

Elston, R. C. 1984. The genetic analysis of quantitative trait differences between two homozygous lines. *Genetics* 108: 733–744. [13]

Elston, R. C. 1990a. Models for discrimination between alternative modes of inheritance. *In* D. Gianola and K. Hammond (eds.), *Advances in statistical methods for genetic improvement of livestock*, pp. 41–55. Springer-Verlag, Berlin. [13]

Elston, R. C. 1990b. A general linkage method for the detection of major genes. *In* D. Gianola and K. Hammond (eds.), *Advances in statistical methods for genetic improvement of livestock*, pp. 495–506. Springer-Verlag, Berlin. [16]

Elston, R. C., and D. C. Rao. 1978. Statistical modeling and analysis in human genetics. *Ann. Rev. Biophys. Bioeng.* 7: 253–286. [13]

Elston, R. C., and E. Sobel. 1979. Sampling considerations in the gathering and analysis of pedigree data. *Am. J. Hum. Genet.* 31: 62–69. [13]

Elston, R. C., and J. Stewart. 1971. A general model for the genetic analysis of pedigree data. *Hum. Hered.* 21: 523–542. [13]

Elston, R. C., and J. Stewart. 1973. The analysis of quantitative traits for simple genetic models from parental, F_1 and backcross data. *Genetics* 73: 695–711. [13]

Elston, R. C., K. K. Nasmboodiri, H. V. Nino, and W. S. Pollitzer. 1974. Studies on blood and urine glucose in Seminole Indians: indications for segregation of a major gene. *Am. J. Hum. Genet.* 26: 13–34. [13]

Elston, R. C., K. K. Nasmboodiri, C. J. Glueck, R. Fallat, R. Tsang, and V. Leuba. 1975. Studies of the genetic transmission of hypercholesterolemia and hypertriglyceridemia in a 195 member kindred. *Ann. Hum. Genet.* 39: 67–87. [13]

Elston, R. C., J. E. Bailey-Wilson, G. E. Bonney, B. J. Keats, and A. F. Wilson. 1986. S. A. G. E. — a package of computer programs to perform statistical analysis for genetic epidemiology. Paper presented at the Seventh International Congress of Human Genetics, Berlin, September 22-26, 1986. [13]

Emerson, R. A. 1910. The inheritance of sizes and shapes in plants. *Am. Nat.* 44: 739–746. [1]

Emerson, R. A., and E. M. East. 1913. The inheritance of quantitative characters in maize. *Bull. Agric. Exp. Sta. Neb.* 2. [1,9]

Emik, L. O., and C. E. Terrill. 1949. Systematic procedures for calculating inbreeding coefficients. *J. Heredity* 40: 51–55. [26]

Emlen, J. M., D. C. Freeman, and J. H. Graham. 1993. Nonlinear growth dynamics and the origin of fluctuating asymmetry. *Genetica* 89: 77–96. [6]

Enfield, F. D., R. E. Comstock, and O. Braskerud. 1966. Selection for pupa weight in *Tribolium castaneum*. I. Parameters in base populations. *Genetics* 54: 523–533. [24]

Engelke, D. R., P. A. Hoener, and F. S. Collins. 1988. Direct sequencing of enzymatically amplified human genomic DNA. *Proc. Natl. Acad. Sci. USA* 85: 544–548. [14]

Eshed, Y., and D. Zamir. 1995. An introgression line population of *Lycopersicon pennellii* in the cultivated tomato enables the identification and fine mapping of yield-associated QTL. *Genetics* 141:1147–1162. [14]

Eshed, Y., and D. Zamir. 1996. Less-than-additive epistatic interactions of quantitative trait loci in tomato. *Genetics* 143: 1807–817. [15]

Everitt, B. S., and D. J. Hand. 1981. *Finite mixture distributions*. Chapman and Hall, London. [13]

Everitt, P. F. 1910. Tables of the tetrachoric functions for fourfold correlation tables. *Biometrika* 7: 437–451. [25]

Ewens, W. J., and N. C. E. Shute. 1986. A resolution of the ascertainment sampling problem. *Theor. Pop. Biol.* 30: 388–412. [13]

Ewens, W. J., and R. S. Spielman. 1995. The transmission/disequilibrium test: history, subdivision, and admixture. *Am. J. Hum. Genet.* 57: 455–464. [14]

Fain, P. R. 1978. Characteristics of simple sibship variance tests for the detection of major loci and application to height, weight and spatial performance. *Ann. Hum. Genet.* 42: 109–120. [13]

Falconer, D. S. 1952. The problem of environment and selection. *Am. Nat.* 86: 293–298. [22]

Falconer, D. S. 1965a. Maternal effects and selection response. *Proc. XIth Internat. Cong. Genetics* 3: 763–774. [23]

Falconer, D. S. 1965b. The inheritance of liability to certain diseases, estimated from the incidence among relatives. *Ann. Hum. Genet.* 29: 51–71. [25]

Falconer, D. S. 1985. A note on Fisher's 'average effect' and 'average excess'. *Genet. Res.* 46: 337–347. [4]

Falconer, D. S. 1989. *Introduction to quantitative genetics.* 3rd Ed. Longman Sci. and Tech., Harlow, UK. [7,21]

Falconer, D. S., and T. F. C. Mackay. 1996. *Introduction to quantitative genetics.* 4th Ed. Longman Sci. and Tech., Harlow, UK. [1,6]

Falconer, D. S., and R. C. Roberts. 1960. Effects of inbreeding on ovulation rate and foetal mortality in mice. *Genet. Res.* 1: 422–430. [10]

Falconer, D. S., I. K. Gauld, and R. C. Roberts. 1978. Cell numbers and cell sizes in organs of mice selected for large and small body size. *Genet. Res.* 31: 287–301. [21]

Falk, C. T., and P. Rubinstein. 1987. Haplotype relative risks: an easy reliable way to construct a proper control sample for risk calculations. *Ann. Hum. Genet.* 51: 227–233. [14]

Famula, T. R. 1986. Identifying single genes of large effect in quantitative traits using best linear unbiased prediction. *J. Animal Sci.* 63: 68–76. [13]

Faraway, J. J. 1993. Improved sib-pair linkage test for disease susceptibility loci. *Genet. Epidem.* 10: 225–233. [16]

Fatokun, C. A., D. I. Meanacio-Hautea, D. Danesh, and N. D. Young. 1992. Evidence for orthologous seed weight genes in cowpea and mungbean. *Genetics* 132: 841–846. [14]

Feingold, E., P. O. Brown, and D. Siegmund. 1993. Gaussian models for genetic linkage analysis using complete high-resolution maps of identity by descent. *Am. J. Hum. Genet.* 53: 234–251. [15]

Feldman, M. W., and R. C. Lewontin. 1975. The heritability hang-up. *Science* 190: 1163–1168. [7]

Feldman, M. W., F. B. Christiansen, and L. D. Brooks. 1980. Evolution of recombination in a constant environment. *Proc. Natl. Acad. Sci. USA* 77: 4838–4841. [5]

Feldmann, K. A., M. D. Marks, M. L. Christianson, and R. S. Quatrano. 1989. A dwarf mutant of *Arabidopsis* generated by T-DNA insertional mutagensis. *Science* 243: 1351–1354. [14]

Felley, J. 1980. Analysis of morphology and asymmetry in the bluegill sunfish *(Lepomis macrochirus)* in the southeastern United States. *Copeia* 1980: 18–29. [6]

Felsenstein, J. 1965. The effect of linkage on directional selection. *Genetics* 52: 349–363. [5]

Felsenstein, J. 1973. Estimation of number of loci controlling variation in a quantitative character. *Genetics* 74: s78–s79. [13]

Felsenstein, J. 1974. The evolutionary advantage of recombination. *Genetics* 78: 737–756. [5]

Felsenstein, J. 1979. A mathematically tractable family of genetic mapping functions with different amounts of interference. *Genetics* 91: 769–775. [14]

Fenster, C. B., and K. Ritland. 1994. Quantitative genetics of mating system divergence in the yellow monkeyflower species complex. *Heredity* 73: 422–435. [9]

Ferguson, M. M. 1986. Developmental stability of rainbow trout hybrids: genomic coadaptation or heterozygosity? *Evolution* 40: 323–330. [6]

Fernando, R. L., and D. Gianola. 1990. Statistical inferences in populations undergoing selection or nonrandom mating. *In* D. Gianola and K. Hammond (eds.) *Statistical Methods for Genetic Improvement of Livestock*, pp. 437–453. Springer-Verlag, NY. [26,27]

Fernando, R. L., C. Stricker, and R. C. Elston. 1993. An efficient algorithm to compute the posterior genotypic distribution for every member of a pedigree without loops. *Theor. Appl. Genet.* 87:89–93. [13]

Fernando, R. L., C. Stricker, and R. C. Elston. 1994. The finite polygenic mixed model: an alternative formulation for the mixed model of inheritance. *Theor. Appl. Genet.* 88: 573–580. [13]

Ferrari, J. A. 1987. Components of genetic variation associated with second and third chromosome gene arrangements in *Drosophila melanogaster. Genetics* 116: 87–97. [14]

Ferreira, M. E., J. Satagopan, B. S. Yandell, P. H. Williams, and T. C. Osborn. 1995. Mapping loci controlling vernalization requirement and flowering time in *Brassica napus. Theor. Appl. Genet.* 90: 727–732. [15]

Festing, M. F. W. 1973. A multivariate analysis of subline divergence in the shape of the mandible in C57BL/Gr mice. *Genet. Res.* 21: 121–132. [12]

Finch, C. E. 1990. *Longevity, senescence, and the genome.* Univ. Chicago Press, Chicago. [6]

Findlay, C. S., and F. Cooke. 1983. Genetic and environmental components of clutch size variance in a wild population of lesser snow geese *(Anser caerulescens caerulescens). Evolution* 37: 724–734. [17]

Finlay, K. W., and G. N. Wilkinson. 1963. The analysis of adaptation in a plant breeding programme. *Aust. J. Agri. Res.* 14: 742–754. [22]

Finnegan, D. J. 1992. Transposable elements. *In* D. L. Lindsley, and G. G. Zimm (eds.), *The genome of Drosophila melanogaster,* pp. 1096–1107. Academic Press, NY. [12]

Fisch, R. D., M. Ragot, and G. Gay. 1996. A generalization of the mixture model in the mapping of quantitative trait loci for progeny from a biparental cross of inbred lines. *Genetics* 143: 571–577. [15]

Fisher, R. A. 1918. The correlation between relatives on the supposition of Mendelian inheritance. *Trans. Royal Soc. Edinburgh* 52: 399–433. [1,2,4,7,18]

Fisher, R. A. 1921. On the mathematical foundations of statistics. *Phil. Trans. Royal Soc. Lond.* B 222: 309–368. [A4]

Fisher, R. A. 1925. *Statistical methods for research workers.* Hafner, NY. [1,18]

Fisher, R. A. 1928a. The possible modification of the response of the wild type to recurrent mutations. *Am. Nat.* 62: 115–126. [4]

Fisher, R. A. 1928b. Two further notes on the origin of dominance. *Am. Nat.* 62: 571–574. [4]

Fisher, R. A. 1929. The evolution of dominance: a reply to Professor Sewall Wright. *Am. Nat.* 63: 553–556. [4]

Fisher, R. A. 1935. *The design of experiments.* 8th Ed. Hafner, NY. [1]

Fisher, R. A. 1941. Average excess and average effect of a gene substitution. *Ann. Eugen.* 11: 53–63. [4]

Fisher, R. A. 1947. The theory of linkage in polysomic inheritance. *Phil. Trans. Royal Soc. Lond.* B 233: 55–87. [4]

Fisher, R. A. 1956. *Statistical methods and scientific inference.* 13th Ed. Hafner, NY. [1]

Fisher, R. A. 1958. *The genetical theory of natural selection.* Dover Publ., NY. [4,7,11]

Fisher, R. A., and E. B. Ford. 1947. The spread of a gene in natural conditions in a colony of the moth *Panaxia dominula* L. *Heredity* 1: 143–174. [4]

Fisher, R. A., and K. Mather. 1936. Verification in mice of the possibility of more than fifty per cent recombination. *Nature* 137: 362–363. [5]

Fishman, P., B. K. Suarez, S. E. Hodge, and T. Reich. 1978. A robust method for the detection of linkage in familial diseases. *Am. J. Hum. Genet.* 30: 308–321. [16]

Fleagle, J. G. 1985. Size and adaptation in primates. *In* W. L. Jungers (ed.), *Size and scaling in primate biology,* pp. 1–19. Plenum, NY. [11]

Fletcher, R. 1987. *Practical methods of optimization.* 2nd Ed. John Wiley, NY. [13]

Flux, J. E. C., and M. M. Flux. 1982. Artificial selection and gene flow in wild starlings, *Sturnus vulgaris. Naturwissenschaften* 69: 96–97. [17]

Foltz, D. W. 1986. Null alleles as possible causes of heterozygote deficiencies in the oyster *Crassostrea virginica* and other bivalves. *Evolution* 40: 869–870. [10]

Foolad, M. R., and R. A. Jones. 1992. Models to estimate maternally controlled genetic variation in quantitative seed characters. *Theor. Appl. Genet.* 83: 360–366. [23]

Forbes, S. H., and F. W. Allendorf. 1991. Mitochondrial genotypes have no detectable effects on meristic traits in cutthroat trout hybrid swarms. *Evolution* 45: 1350–1359. [23]

Ford, E. B. 1975. *Ecological genetics,* 4th Ed. John Wiley & Sons, NY. [12]

Ford, M. J., and C. F. Aquadro. 1996. Selection on X-linked genes during speciation in the *Drosophila athabasca* complex. *Genetics* 144: 689–703. [14]

Foulley, J. L. 1993. A simple argument showing how to derive restricted maximum likelihood. *J. Dairy Sci.* 76: 2320–2324. [27]

Fowler, D. P., and Y. S. Park. 1983. Population studies of white spruce. I. Effects of self-pollination. *Can. J. For. Res.* 13: 1133–1138. [10]

Fowler, K., and M. C. Whitlock. 1994. Fluctuating asymmetry does not increase with moderate inbreeding in *Drosophila melanogaster. Heredity* 73: 373–376. [6]

Fox, C. W. 1993. A quantitative genetic analysis of oviposition preference and larval performance on two hosts in the bruchid beetle, *Callosobruchus maculatus. Evolution* 47: 166–175. [22]

Frahm, R., and K.-I. Kojima. 1966. Comparison of selection response on body weight under divergent larval density conditions in *Drosophila pseudoobscura. Genetics* 54: 625–637. [14]

Frank, S. A. 1991. Haldane's rule: a defense of molecular drive theory. *Evolution* 45: 1714–1717. [14]

Frankel, W. N. 1995. Taking stock of complex trait genetics in mice. *Trends Genet.* 11: 471–477. [15]

Frankham, R. 1977. The nature of quantitative genetic variation in *Drosophila.* III. Mechanism of dosage compensation for sex-linked abdominal bristle polygenes. *Genetics* 85:185–191. [24]

Frankham, R. 1988. Exchanges in the rRNA multigene family as a source of genetic variation. *In* B. S. Weir, E. J. Eisen, M. M. Goodman, and G. Namkoong (eds.), *Proceedings of the second international conference on quantitative genetics*, pp. 236–242. Sinauer Assoc., Sunderland, MA. [12]

Frankham, R., and R. K. Nurthen. 1981. Forging links between population and quantitative genetics. *Theor. Appl. Genet.* 59: 251–263. [13]

Frankham, R., D. A. Briscoe, and R. K. Nurthen. 1980. Unequal crossing over at the rRNA tandon as a source of quantitative genetic variation in *Drosophila. Genetics* 95: 727–742. [24]

Franklin, E. C. 1972. Genetic load in loblolly pine. *Am. Nat.* 106:262–265. [10]

Franklin, I., and R. C. Lewontin. 1970. Is the gene the unit of selection? *Genetics* 65: 707–734. [5]

Fraser, A. 1963. Variation of scutellar bristles in *Drosophila*. I. Genetic linkage. *Genetics* 48: 497–514. [11]

Fraser, A. 1967. Variation of scutellar bristles in *Drosophila*. XV. Systems of modifiers. *Genetics* 57: 919–934. [11]

Fraser, A. S. 1968. Specificity of modifiers of *scute* and *extravert* expression. *Genetics* 60: 179. [12]

Fraser, A. 1970. Variation of scutellar bristles in *Drosophila*. XVI. major and minor genes. *Genetics* 65: 305–309. [11]

Fraser, A. S., and B. M. Kindred. 1960. Selection for an invariant character, vibrissa number, in the house mouse. II. Limits to variability *Aust. J. Biol. Sci.* 13: 48–58. [11]

Freeman, D. C., J. H. Graham, and J. M. Emlen. 1993. Developmental stability in plants: symmetries, stress, and epigenesis. *Genetica* 89: 97–119. [6]

Freeman, F., and J. W. Lundelius. 1982. The developmental genetics of dextrality and sinistrality in the gastropod *Lymnaea peregra. Roux's Arch. Devel. Biol.* 191: 69–83. [6]

Freeman, G. H. 1973. Statistical methods for the analysis of genotype-environment interactions. *Heredity* 31: 339–354. [22]

Fripp, Y. J., and C. E. Caten. 1973. Genotype-environmental interactions in *Schizophyllum commune.* III. The relationship between mean expression and sensitivity to change in environment. *Heredity* 30: 341–349. [22]

Fry, J. D. 1992. The mixed-model analysis of variance applied to quantitative genetics: biological meaning of the parameters. *Evolution* 46: 540–550. [22]

Fry, J. D. 1993. The "general vigor" problem: can antagonistic pleiotropy be detected when genetic covariances are positive? *Evolution* 47: 327–333. [22]

Fry, J. D., K. A. deRonde, and T. F. C. Mackay. 1995. Polygenic mutation in *Drosophila melanogaster:* genetic analysis of selection lines. *Genetics* 139: 1293–1307. [12]

Frydenberg, O. 1963. Population studies of a lethal mutant in *Drosophila melanogaster.* I. Behavior in populations with discrete generations. *Hereditas* 50: 89–116. [10]

Fu, Y.-B., and K. Ritland. 1994. Evidence for the partial dominance of viability genes contributing to inbreeding depression in *Mimulus guttatus. Genetics* 136: 323–331. [10]

Fulker, D. W., and L. R. Cardon. 1994. A sib-pair approach to interval mapping of quantitative trait loci. *Am. J. Hum. Genet.* 54: 1092–1103 (errata 55: 419). [16]

Fulker, D. W., S. S. Cherry, and L. R. Cardon. 1995. Multipoint interval mapping of quantitative trait loci, using sib pairs. *Am. J. Hum. Genet.* 56: 1224–1233. [16]

Futuyma, D. J., and T. E. Philippi. 1987. Genetic variation and covariation in responses to host plants by *Alsophila pometaria* (Lepidoptera: Geometridae). *Evolution* 41: 269–279. [22]

Fyfe, J. L., and N. Gilbert. 1963. Partial diallel crosses. *Biometrics* 19: 278–286. [20]

Gabriel, W., M. Lynch, and R. Bürger. 1993. Muller's ratchet and mutational meltdowns. *Evolution* 47: 1744–1757. [12]

Gaffney, P. M. 1990. Enzyme heterozygosity, growth rate, and viability in *Mytilus edulis:* another look. *Evolution* 44: 204–210. [10]

Gaffney, P. M., T. M. Scott, R. K. Koehn, and W. J. Diehl. 1990. Interrelationships of heterozygosity, growth rate and heterozygote deficiencies in the coot clam, *Mulina lateralis. Genetics* 124: 687–699. [10]

Gail, M., and R. Simon. 1985. Testing for qualitative interactions between treatment effects and patient subsets. *Biometrics* 41: 361–372. [22]

Galen, C., J. S. Shore, and H. Deyoe. 1991. Ecotypic divergence in alpine *Polemonium viscosum:* genetic structure, quantitative variation, and local adaptation. *Evolution* 45: 1218–1228. [9]

Gallais, A. 1974. Covariances between arbitrary relatives with linkage and epistasis in the case of linkage disequilibrium. *Biometrics* 30: 429–446 (Correction 33: 766). [5,7]

Gallais, A., and M. Rives. 1993. Detection, number and effects of QTLs for a complex character. *Argonomi* 13: 723–738. [15]

Galton, F. 1869. *Hereditary genius.* (reprinted 1962, Meridian Books, NY). [1]

Galton, F. 1875. The history of twins as a criterion of the relative powers of nature and nurture. *J. Royal Anthro. Inst.* 5: 391–406. [19]

Galton, F. 1879. The geometric mean in vital and social statistics. *Proc. Royal Soc. Lond.* 29: 365–367. [11]

Galton, F. 1889. *Natural inheritance.* Macmillan, London. [1,3]

Ganders, F. R. 1979. The biology of heterostyly. *New Zealand J. Bot.* 17: 607–635. [5]

Garbutt, K., and A. R. Zangerl. 1983. Application of genotype-environment interaction analysis to niche quantification. *Ecology* 64: 1292–1296. [22]

Garcia, N., C. López-Fanjul, and A. Garcia-Dorado. 1994. The genetics of viability in *Drosophila melanogaster:* effects of inbreeding and artificial selection. *Evolution* 48: 1277–1285. [10]

Gardner, C. O. 1963. Estimates of genetic parameters in cross-fertilizing plants and their implications in plant breeding. *In* W. D. Hanson and H. F. Robinson (eds.), *Statistical genetics and plant breeding,* pp. 225–252. Natl. Acad. Sci., Natl. Res. Council Publ. 982.

Washington, D. C. [5,20]

Gardner, C. O., and S. A. Eberhart. 1966. Analysis and interpretation of the variety cross diallel and related populations. *Biometrics* 22: 439–452. [20]

Garnett, I., and D. S. Falconer. 1975. Protein variation in strains of mice differing in body size. *Genet. Res.* 25: 45–57. [14]

Gärtner, K., and E. Baunack. 1981. Is the similarity of monozygotic twins due to genetic factors alone? *Nature* 292: 646–647. [19]

Gauch, H. G. 1988. Model selection and validation for yield trials with interaction. *Biometrics* 44: 705–715. [22]

Gavrilets, S., and A. Hastings. 1993. Maintenance of genetic variability under strong stabilizing selection: a two-locus model. *Genetics* 134: 377–386. [5]

Gavrilets, S., and A. Hastings. 1994a. Dynamics of genetic variability in two- locus models of stabilizing selection. *Genetics* 138: 519–532. [5]

Gavrilets, S., and A. Hastings. 1994b. A quantitative-genetic model for selection on developmental noise. *Evolution* 48: 1478–1486. [6]

Gavrilets, S., and A. Hastings. 1995. Dynamics of polygenic variability under stabilizing selection, recombination, and drift. *Genet. Res.* 65: 63–74. [5]

Gauss, C. F. 1809. *Theoria motus corporum coelestium.* (English translation, 1857, Little, Brown and Co., Boston, MA). [2]

Gebhardt, M. D., and S. C. Stearns. 1988. Reaction norms for developmental time and weight at eclosion in *Drosophila mercatorum*. *J. Evol. Biol.* 1: 335–354. [21]

Gebhardt-Henrich, S. G., and A. J. van Noordwijk. 1991. Nestling growth in the great tit. I. Heritability estimates under different environmental conditions. *J. Evol. Biol.* 3: 341–362. [7,23]

Geiringer, H. 1944. On the probability theory of linkage in Mendelian heredity. *Ann. Math. Stat.* 15: 25–57. [5]

Geldermann, H. 1975. Investigations on inheritance of quantitative characters in animals by gene markers. I. Methods. *Theor. Appl. Genet.* 46: 319–330. [12]

Geldermann, H. 1975. Investigations on inheritance of quantitative characters in animals by gene markers. I. Methods. *Theor. Appl. Genet.* 46: 319–330. [16]

Gelfand, A. E., and A. F. M. Smith. 1990. Sampling-based approaches to calculating marginal densities. *J. Am. Stat. Assoc.* 85: 398–409. [13]

Georges, M., D. Nielsen, M. Mackinnon, A. Mishra, R. Okimoto, A. T. Pasquino, L. S. Sargeant, A. Sorensen, M. R. Steele, Z. Zhao, J. E. Womack, and I. Hoeschele. 1995. Mapping quantitative trait loci controlling milk production in dairy cattle by exploiting progeny testing. *Genetics* 139: 907–920. [16]

Gerats, A. G. M., E. Farcy, M. Wallroth, S. P. C. Groot, and A. Schram. 1984. Control of anthocyanin synthesis in *Petunia hybrida* by multiple allelic series of the genes *AN1* and *AN2*. *Genetics* 106: 501–508. [12]

German, S., and D. German. 1984. Stochastic relaxation, Gibbs distributions and Bayesian restoration of images. *IEEE Transactions on Pattern Analysis and Machine Intelligence* 6: 721–741. [13]

Gershon, E. S., and C. R. Cloninger. 1994. *Genetic approaches to mental disorders.* American Psychiatric Press, Washington, D. C. [16]

Gershon, E. S., W. E. Bunney, Jr., J. F. Leckman, M. Van Eerdewegh, and B. A. DeBauche. 1976. The inheritance of affective disorders: a review of data and of hypotheses. *Behav. Genet.* 6: 227–261. [25]

Gessler, M., A. Poustka, W. Cavenee, R. L. Neve, S. H. Orkin, and G. A. P. Burns. 1990. Homozygous deletion in Wilms tumours of a zinc-finger gene identified by chromosome jumping. *Nature* 343: 774–778. [14]

Gev, D., N. Roguin, and E. Freundlich. 1986. Consanguinity and congenital heart disease in the rural Arab population in northern Israel. *Hum. Hered.* 36: 213–217. [10]

Ghosh, S., S. M. Palmer, N. R. Rodrigues, H. J. Cordell, C. M. Hearne, R. J. Cornall, J.-B. Prins, P. McShane, G. M. Lathrop, L. B. Peterson, L. S. Wicker, and J. A. Todd. 1993. Polygenic control of autoimmune diabetes in nonobese diabetic mice. *Nature Genetics* 4: 404–409. [15]

Gianola, D. 1982. Theory and analysis of threshold characters. *J. Anim. Sci.* 54: 1079–1096. [25]

Gianola, D., and R. L. Fernando. 1986. Bayesian methods in animal breeding theory. *J. Anim. Sci.* 63: 217–244. [27]

Gianola, D., and K. Hammond (eds.) 1990. *Advances in statistical methods for genetic improvement of livestock.* Springer-Verlag, Berlin. [1]

Gianola, D., S. Im, and R. L. Fernando. 1988. Prediction of breeding values under Henderson's selection model: a revisitation. *J. Dairy Sci.* 71: 2790–2798. [26,27]

Gibbs, H. L. 1988. Heritability and selection on clutch size in Darwin's medium ground finches *(Geospiza fortis)*. *Evolution* 42: 750–762. [17,23]

Gibson, G., and D. S. Hogness. 1996. Effect of polymorphism in the *Drosophila* regulatory gene *Ultrabithorax* on homeotic stability. *Science* 271: 200–203. [12]

Giesel, J. T., and E. E. Zettler. 1980. Genetic correlations of life historical parameters and certain fitness indices in *Drosophila melanogaster:* r_m, r_s, diet breadth. *Oecologia* 47: 299–302. [21]

Gilbert, N. E. G. 1958. Diallel cross in plant breeding. *Heredity* 12: 477–492. [20]

Gill, J. L., and E. L. Jensen. 1968. Probability of obtaining negative estimates of heritability. *Biometrics* 24: 517–526. [18]

Gill, P. E., Murray, W. and Wright, M. E. 1981. *Practical optimization.* Academic Press, NY. [13]

Gillespie, J. H. 1991. *The causes of molecular evolution.* Oxford Univ. Press, NY. [4]

Gillespie, J. H., and M. Turelli. 1989. Genotype-environment interactions and the maintenance of polygenic variation. *Genetics* 121: 129–138. [22]

Gillois, M. 1964. La relation d'identité génétique. Thesis, Faculté des Sciences, Paris. [7]

Gimelfarb, A. 1981. A general linear model for the genotypic covariance between relatives under assortative mating. *J. Math. Biol.* 13: 209–226. [7]

Gimelfarb, A. 1984. Quantitative characters under assortative mating: gametic model. *Theor. Pop. Biol.* 25: 312–330. [7]

Gimelfarb, A. 1985. Is offspring-midparent regression affected by assortative mating of parents? *Genet. Res.* 47: 71–75. [17]

Gimelfarb, A. 1986. Offspring-parent genotypic regression: how linear is it? *Biometrics* 42: 67–71. [17]

Gimelfarb, A. 1994. Additive-multiplicative approximation of genotype-environment interaction. *Genetics* 138: 1339–1349. [22]

Gimelfarb, A., and J. H. Willis. 1994. Linearity versus nonlinearity of offspring-parent regression: an experimental study of *Drosophila melanogaster. Genetics* 138: 343–352. [17]

Ginzburg, É. Kh. 1983. Possible localization of genes controlling a quantitative character in self-pollinators. *Genetica* 19: 577–583. [15]

Giovannoni, J. J., R. A. Wing, M. W. Ganal, and S. D. Tanksley. 1991. Isolation of molecular markers from specific chromosomal intervals using DNA pools from existing mapping populations. *Nucl. Acids Res.* 19: 6553–6558. [14]

Go, R. C. P., R. C. Elston, and E. B. Kaplan. 1978. Efficiency and robustness of pedigree segregation analysis. *Am. J. Hum. Genet.* 30: 28–37. [13]

Goffinet, B. 1983. Selection on selected records *Genet. Sel. Evol.* 15: 91. [26]

Goldberger, A. S. 1962. Best linear unbiased predictors in the generalized linear regression model. *J. Am. Stat. Assoc.* 57: 369–375. [26]

Goldgar, D. E. 1990. Multipoint analysis of human quantitative genetic variation. *Am. J. Hum. Genet.* 47: 957–967. [16]

Goldin, L. R., and E. S. Gershon. 1988. Power of the affected-sib-pair method for heterogeneous disorders. *Genet. Epidem.* 5: 35–42. [16]

Goldin, L. R., and D. E. Weeks. 1993. Two-locus models of disease: comparison of likelihood and nonparametric linkage methods *Am. J. Hum. Genet.* 53: 908–915. [16]

Goldin, L. R., K. K. Kidd, S. Matthysse, and E. S. Gershon. 1981. The power of pedigree segregation analysis for traits with incomplete penetrance. *In* E. S. Gershon, S. Matthysse, X. O. Breakefield, and R. D. Ciaranello (eds.), *Genetic research strategies for psychobiology and psychiatry*, pp. 305–317. Plenum, NY. [13]

Goldin, L. R., N. J. Cox, D. L. Pauls, E. S. Gershon, and K. K. Kidd. 1984. The detection of major loci by segregation and linkage analysis: a simulation study. *Genet. Epidem.* 1: 285–296. [16]

Golding, B. (ed.) 1994. *Non-neutral evolution.* Chapman and Hall, NY. [4].

Goldman, I. L., T. R. Rocheford, and J. W. Dudley. 1993. Quantitative trait loci influencing protein and starch concentration in the Illinois long term selection maize strains. *Theor. Appl. Genet.* 87: 217–224. [15]

Goldman, I. L., T. R. Rocheford, and J. W. Dudley. 1994. Molecular markers associated with maize kernel oil concentration in an Illinois high protein × Illinois low protein cross. *Crop Sci.* 34: 908–915. [15]

Goldman, I. L., I. Paran, and D. Zamir. 1995. Quantitative trait locus analysis of a recombinant inbred line population derived from a *Lycopersicon esculentum* × *Lycopersicon cheesmanii* cross. *Theor. Appl. Genet.* 90: 925–932. [14]

Gomulkiewicz, R., and M. Kirkpatrick. 1992. Quantitative genetics and the evolution of reaction norms. *Evolution* 46: 390–411. [22]

Gonyon, D. S., R. E. Mather, H. C. Hines, G. F. W. Haenlein, C. W. Arave, and S. N. Gaunt. 1987. Associations of bovine blood and milk polymorphisms with lactation traits: Holsteins. *Theor. Appl. Genet.* 70: 2585–2598. [16]

Good, R. L., and A. R. Hallauer. 1977. Inbreeding depression in maize by selfing and full-sibbing. *Crop Sci.* 17: 935–940. [10]

Goodwill, R., and F. D. Enfield. 1971. Heterozygosity in inbred lines of *Tribolium castaneum. Theor. Appl. Genet.* 41: 5–12. [12]

Goodwin, R. H. 1944. The inheritance of flowering time in a short-day species, *Solidago sempervirens* L. *Genetics* 29: 503–519. [9]

Goradia, T. M., K. Lange, P. L. Miller, and P. M. Nadkarni. 1992. Fast computations of genetic likelihoods on human pedigree data. *Hum. Hered.* 42: 42– 62. [13]

Gordon, I. L., D. E. Byth, and L. N. Balaam. 1972. Variance of heritability ratios estimated from phenotypic variance components. *Biometrics* 28: 401–415. [22]

Gottlieb, L. D. 1984. Genetics and morphological evolution in plants. *Am. Nat.* 123: 681–709. [9]

Götz, K. U., and L. Ollivier. 1992. Theoretical aspects of applying sib-pair linkage tests to livestock species. *Genet. Sel. Evol.* 24: 29–42. [16]

Gould, S. J. 1966. On the scaling of tooth size in mammals. *Am. Zool.* 15: 351–362. [11]

Gould, S. J. 1980. Is a new and general theory of evolution emerging? *Paleobiol.* 6: 119–130. [9]

Gould, S. J., and R. C. Lewontin. 1979. The spandrels of San Marco and the Panglossian paradigm: a critique of the adaptationist programme. *Proc. Royal Soc. Lond.* B 205: 581–598. [1]

Govind, C. K., and J. Pearce. 1986. Differential reflex activity determines claw and closer muscle asymmetry in developing lobsters. *Science* 233: 354–356. [6]

Grafen, A. 1985. A geometric view of relatedness. *Oxford Surveys Evol. Biol.* 2: 28–89. [7]

Graham, A. 1981. *Kronecker products and matrix calculus with applications.* Halsted Press, NY. [A3]

Graham, J. H., and J. D. Felley. 1985. Genomic coadaptation and developmental stability within introgressed populations of *Enneacanthus gloriosus* and *E. obesus* (Pisces, Centrarchidae). *Evolution* 39: 104–114. [6]

Graser, H.-U., S. P. Smith, and B. Tier. 1987. A derivative-free approach for estimating variance components in animal models by restricted maximum likelihood. *J. Anim. Sci.* 64: 1362–1370. [27]

Gray, A., and A. Tait. 1993. Identification of ornithine decarboxylase as a trait gene for growth in replicated mouse lines divergently selected for lean body mass. *Genet. Res.* 62: 31–37. [14]

Graybill, F. A. 1961. *An introduction to linear statistical models.* McGraw- Hill, NY. [18]

Graybill, F. A., and C. M. Wang. 1979. Confidence intervals for proportions of variability in two-factor nested variance component models. *J. Am. Stat. Assoc.* 74: 368–374. [18]

Graybill, F. A., F. Martin, and G. Godfrey. 1956. Confidence intervals for variance ratios specifying genetic heritability. *Biometrics* 12: 99–109. [18]

Green, C. V. 1931. Linkage in size inheritance. *Am. Nat.* 65: 502–511. [14]

Green, C. V. 1933. Further evidence of linkage in size inheritance. *Am. Nat.* 67: 377–380. [14]

Green, J. R., and S. Shah. 1993. Power comparison of various sibship tests of association. *Ann. Hum. Genet.* 57: 151–158. [16]

Green, J. R., and J. C. Woodrow. 1977. Sibling method for detecting HLA-linked genes in disease. *Tissue Antigens* 9: 31–35. [16]

Green, M. M. 1959. The discrimination of wild-type isoalleles at the white locus of *Drosophila melanogaster. Proc. Natl. Acad. Sci. USA* 45: 549–553. [12]

Greenberg, D. A. 1986. The effect of proband designation on segregation analysis. *Am. J. Hum. Genet.* 39: 329–339. [13]

Greenberg, R., and J. F. Crow. 1960. A comparison of the effect of lethal and detrimental chromosomes from *Drosophila* populations. *Genetics* 45: 1153–1168. [10]

Gregorius, H.-R. 1976. Convergence of genetic compositions assuming infinite population size and overlapping generations. *J. Math. Biol.* 3: 179–186. [4]

Gregory, W. C. 1965a. Mutation frequency, magnitude of change, and the probability of improvement in adaptation. *Rad. Botany* 5 (Suppl.): 429–441. [12]

Gregory, W. C. 1965b. Mutation breeding. *In* K. J. Frey (ed.), *Plant breeding,* pp. 189–218. Iowa State Univ. Press, Ames. [12]

Grewal, M. S. 1962. The rate of genetic divergence of sublines in the C57BL strain of mice. *Genet. Res.* 3: 226–237. [12]

Griffing, B. 1956. Concept of general and specific combining ability in relation to diallel crossing systems. *Aust. J. Biol. Sci.* 9: 463–493. [20]

Griffiths, A. J. F., J. H. Miller, D. T. Suzuki, R. C. Lewontin, and W. M. Gelbart. 1996. *An introduction to genetic analysis.* 6th Ed. W. H. Freeman, NY. [12]

Groeneveld, E., and M. Kovac. 1990a. A generalized computing procedure for setting up and solving mixed linear models. *J. Dairy Sci.* 73: 513–531. [26]

Groeneveld, E., and M. Kovac. 1990b. A note on multiple solutions in multivariate restricted maximum likelihood covariance component estimation. *J. Anim. Sci.* 73: 2321–2329. [27]

Grossman, M. 1970. Sampling variance of the correlation coefficients estimated from analyses of variance and covariance. *Theor. Appl. Genet.* 40: 357–359. [21]

Grossman, M., and E. J. Eisen. 1989. Inbreeding, coancestry, and covariance between relatives for X-chromosomal loci. *Heredity* 80: 137–142. [24]

Grossman, M., and G. A. E. Gall. 1968. Covariance analysis with unequal subclass numbers: component estimation in quantitative genetics. *Biometrics* 24:49–59. [21]

Grossman, M., and H. W. Norton. 1974. Simplification of the sampling variance of the correlation coefficients. *Theor. Appl. Genet.* 44: 332. [21]

Grossman, M., and H. W. Norton. 1981. An approximation of the minimum-variance estimator of heritability based on variance component analysis. *Genetics* 98: 417–426. [18]

Grüneberg, H. 1952. Genetical studies on the skeleton of the mouse. IV. Quasi-continuous variations. *J. Genet.* 51: 95–114. [25]

Grüneberg, H. 1970. Is there a viral component in the genetic background? *Nature* 225: 39–14. [12]

Guo, S. W. 1994a. Computation of identity-by-descent proportions shared by two siblings. *Am. J. Hum. Genet.* 54: 1104–1109. [16]

Guo, S. W. 1994b. Proportion of genes survived in offspring conditional on inheritance of flanking markers. *Genetics* 138: 953–962. [16]

Guo, S. W., and E. A. Thompson. 1992. A monte carlo method for combined segregation and linkage analysis. *Am. J. Hum. Genet.* 51: 1111–1126. [13]

Guo, S. W. ,and E. A. Thompson. 1994. Monte carlo estimation of mixed models for large complex pedigrees. *Biometrics.* 50: 417–432. [13]

Gustafsson, L. 1986. Lifetime reproductive success and heritability: empirical support for Fisher's fundamental theorem. *Am. Nat.* 128: 761–764. [17]

Gustafsson, L., and J. Merilä. 1994. Foster parent experiment reveals no genotype-environment correlation in the external morphology of *Ficedula albicollis,* the collared flycatcher. *Heredity* 73: 124–129. [23]

Hackett, C. A., and J. I. Weller. 1995. Genetic mapping of quantitative trait loci for traits with ordinal distributions. *Biometrics* 51: 1252–1263. [15]

Haenlein, G. F. W., D. S. Gonyon, R. E. Mather, and H. C. Hines. 1987. Associations of bovine blood and milk polymorphisms with lactation traits: Guernseys. *Theor. Appl. Genet.* 70: 2599–2609. [16]

Hagger, C., C. Stricker, R. C. Elston, and G. Stranzinger. 1995. Lack of evidence for segregation of a single dominant major gene as the cause of the difference in egg weight between two highly inbred lines of chickens. *Theor. Appl. Genet.* 90: 120–123. [13]

Haig, D. 1993. Genetic conflicts in human pregnancy. *Quart. Rev. Biol.* 68: 495–532. [23]

Hailman, J. P. 1986. The heritability concept applied to wild birds. *Current Ornith.* 4: 71–95. [17]

Halbach, U., and J. Jacobs. 1971. Seasonal selection as a factor in rotifer cyclomorphosis. *Naturwissenschaften* 57: 1–2. [11]

Haldane, J. B. S. 1919. The combination of linkage values, and the calculation of distance between the loci of linked factors. *J. Genetics* 8: 299–309. [14]

Haldane, J. B. S. 1922. Sex ratio and unisexual sterility in hybrid animals. *J. Genetics* 12: 101–109. [14]

Haldane, J. B. S. 1927. Mathematical theory of natural and artificial selection. V. Selection and mutation. *Proc. Cambridge Phil. Soc.* 23: 838–844. [10]

Haldane, J. B. S. 1935. The rate of spontaneous mutation of a human gene. *J. Genetics* 33: 317–326. [12]

Haldane, J. B. S. 1937. The effect of variation on fitness. *Am. Nat.* 71: 337–349. [12]

Haldane, J. B. S. 1947. The rate of mutation of the gene for hemophilia and its segregation ratios in males and females. *Ann. Eugenics* 13: 262–271. [12]

Haldane, J. B. S. 1955. The measurement of variation. *Evolution* 9:484–486. [2]

Haldane, J. B. S., and C. H. Waddington. 1931. Inbreeding and linkage. Genetics 16: 357–374. [15]

Haley, C. S. 1991. Use of DNA fingerprints for the detection of major genes for quantitative traits in domestic species. *Anim. Genet.* 22: 259–277. [16]

Haley, C. S., and S. A. Knott. 1992. A simple regression method for mapping quantitative trait loci in line crosses using flanking markers. *Heredity* 69: 315–324. [15,16]

Haley, C. S., and K. Last. 1981. The advantages of analysing human variation using twins and twin half-sibs and cousins. *Heredity* 47: 221–236. [19]

Haley, C. S., J. L. Jinks, and K. Last. 1981. The monozygotic twin half-sib method for analysing maternal effects and sex-linkage in humans. *Heredity* 46: 227–238. [19]

Haley, C. S., S. A. Knott, and J.-M. Elsen. 1994. Mapping quantitative trait loci in crosses between outbred lines using least squares. *Genetics* 136: 1195–1207. [15]

Hall, J. G. 1990. Genomic imprinting: review and relevance to human diseases. *Am. J. Hum. Genet.* 46: 857–873. [24]

Hallauer, A. R., and J. B. Miranda. 1981. *Quantitative genetics in maize breeding.* Iowa State Univ. Press, Ames. [1,10,20]

Hallauer, A. R., and J. H. Sears. 1973. Changes in quantitative traits associated with inbreeding in a synthetic variety of maize. *Crop Sci.* 13: 327–330. [10]

Hamilton, W. D. 1964. The genetical evolution of social behavior. *J. Theor. Biol.* 7: 1–16. [23]

Hammond, K., and J. W. James. 1970. Genes of large effect and the shape of the distribution of a quantitative character. *Aust. J. Biol. Sci.* 23: 867–876. [13]

Hammond, K., and F. W. Nicholas. 1972. The sampling variance of the correlation coefficients estimated from two-fold nested and offspring-parent regression analyses. *Theor. Appl. Genet.* 42: 97–100. [18,21]

Hanis, C., I., and 33 others. 1996. A genome-wide search for human non-insulin-dependent (type 2) diabetes genes reveals a major susceptibility locus on chromosome 2. *Nature Genetics* 13: 161–166. [16]

Hanrahan, J. P., and E. J. Eisen. 1973. Sexual dimorphism and direct and maternal genetic effects on body weight in mice. *Theor. Appl. Genet.* 43: 39–45. [23,24]

Hanson, L., R. C. Elston, D. J. Petitt, P. H. Bennett, and W. C. Knowler. 1995. Segregation analysis of non-insulin-dependent diabetes mellitus in Pima indians: evidence for a major-gene effect. *Am. J. Hum. Genet.* 57: 160–170. [14]

Hanson, W. D. 1959a. The theoretical distribution of lengths of parental gene blocks in the gametes of an F_1 individual. *Genetics* 44: 197–209. [14]

Hanson, W. D. 1959b. Early generation analysis of lengths of heterozygous chromosome segments around a locus held heterozygous with backcrossing or selfing. *Genetics* 44: 833–837. [14]

Hanson, W. D. 1959c. Theoretical distributions of the initial linkage block lengths intact in the gametes of a population intermated for n generations. *Genetics* 44: 839–846. [14]

Hanson, W. D. 1959d. The breakup of initial linkage blocks under selected mating systems. *Genetics* 44: 857–868. [14]

Hanson, W. D., and H. F. Robinson (eds.) 1963. *Statistical genetics and plant breeding.* Natl. Acad. Sci., Natl. Res. Council Publ. 982. Washington, D. C. [1]

Harada, K. 1995. A quantitative analysis of modifier mutations which occur in mutation accumulation lines in *Drosophila melanogaster.* Heredity 75: 589–598. [12]

Hard, J. J., W. E. Bradshaw, and C. M. Holzapfel. 1992. Epistasis and the genetic divergence of photoperiodism between populations of the pitcher-plant mosquito, *Wyeomyia smithii.* *Genetics* 131: 389–396. [9]

Hardwick, R. C., and J. T. Wood. 1972. Regression methods for studying genotype-environmental interactions. *Heredity* 28: 209–222. [22]

Hardy, G. H. 1908. Mendelian proportions in a mixed population. *Science* 28: 49–50. [4]

Harris, D. L. 1964. Genotypic covariances between inbred relatives. *Genetics* 50: 1319–1348. [26]

Harrison, G. A., and J. J. T. Owen. 1964. Studies on the inheritance of human skin colour. *Ann. Hum. Genet.* 28: 27–37. [9]

Hartl, D. L., and A. G. Clark. 1989. *Principles of population genetics.* 2nd Ed. Sinauer Assoc., Sunderland, MA. [1,10]

Hartl, L., H. Weiss, F. J. Zeller, and A. Jahoor. 1993. Use of RFLP markers for identification of alleles of the *Pm3* locus conferring powdery mildew resistance in wheat *Triticum aestivum* L.) *Theor. Appl. Genet.* 86: 959–963. [14]

Hartley, H. O., and J. N. K. Rao. 1967. Maximum-likelihood estimation for the mixed analysis of variance model. *Biometrika* 54: 93–108. [27]

Harville, D. A. 1977. Maximum likelihood approaches to variance component estimation and to related problems. *J. Am. Stat. Assoc.* 72: 320–338. [27]

Harville, D. A. and T. P. Callanan. 1990. Computational aspects of likelihood-based inference for variance components. *In* D. Gianola and K. Hammond (eds.) *Statistical Methods for Genetic Improvement of Livestock,* pp. 136–176. Springer-Verlag, NY. [27]

Harville, D. A., and A. P. Fenech. 1985. Confidence intervals for a variance ratio, or for heritability, in an unbalanced mixed linear model. *Biometrics* 41: 137–152. [18]

Haseman, J. K., and R. C. Elston. 1970. The estimation of genetic variance from twin data. *Behav. Genetics* 1: 11–19. [19]

Haseman, J. K., and R. C. Elston. 1972. The investigation of linkage between a quantitative trait and a marker locus. *Behav. Genet.* 2: 3–19. [16]

Hasselblad, V. 1966. Estimating the parameters for a mixture of normal distributions. *Technometrics* 8: 431–444. [13]

Hasson, E., J. J. Fanara, C. Rodriguez, J. C. Vilardi, O. A. Reig, and A. Fontdevila. 1992. The evolutionary history of *Drosophila buzzattii.* XXIV. Second chromosome inversions have different average effects on thorax length. *Heredity* 68: 557–563. [14]

Hasstedt, S. J., and P. E. Cartwright. 1979. PAP-pedigree analysis package. Tech. Report 13, Dept. of Medical Biophysics and Computing, Univ. of Utah, Salt Lake City, UT. [13]

Hästbacka, J., A. de la Chapelle, I. Kaitila, P. Sistonen, A. Weaver, and E. Lander. 1992. Linkage disequilibrium mapping in isolated founder populations: diastrophic dysplasia in Finland. *Nature Genetics* 2: 204–211. [14]

Hästbacka, J., A. de la Chapelle, M. M. Mahtani, G. C. Lines, M. P. Reeve-Daly, M. Daly, B. A. Hamilton, K. Kusumi, B. Trivedi, A. Weaver, A. Coloma, M. Lovett, A. Buckler, I. Kaitila, and E. S. Lander. 1994. The diastrophic dysplasia gene encodes a novel sulfate transporter: positional cloning by fine-structure linkage disequilibrium mapping. *Cell* 78: 1073–1087. [14]

Hastings, A. 1986. Multilocus population genetics with epistasis. II. Equilibrium properties of multilocus models: what is the unit of selection? *Genetics* 112: 157–171. [14]

Hauser, T. P., and V. Loeschcke. 1994. Inbreeding depression and mating-distance dependent offspring fitness in large and small populations of *Lychnis floscuculi* (Caryophyllaceae). *J. Evol. Biol.* 7: 609–622. [10]

Havel, J. E. 1987. Predator-induced defenses: a review. *In* W. C. Kerfoot and A. Sih (eds.), *Predation: direct and indirect impacts on aquatic communities,* pp. 263–278. Univ. Press New England, Hanover, NH. [6,25]

Hayman, B. I. 1954. The theory and analysis of diallel crosses. *Genetics* 39: 789–809. [20]

Hayman, B. I. 1960a. The separation of epistatic from additive and dominance variation in generation means. *Genetica* 31: 371–390. [9]

Hayman, B. I. 1960b. Maximum likelihood estimation of genetic components of variation. *Biometrics* 16: 369–381. [9]

Hazel, L. N. 1943. The genetic basis for constructing selection indices. *Genetics* 28: 476–490. [21]

Heath, A. C., N. G. Martin, L. J. Eaves, and D. Loesch. 1984. Evidence for polygenic epistatic interactions in man? *Genetics* 106: 719–727. [19]

Hébert, D., S. Fauré, and I. Olivieri. 1994. Genetic, phenotypic, and environmental correlations in black medic, *Medicago lupulina* l., grown in three different environments. *Theor. Appl. Genet.* 88: 604–613. [21]

Hedrick, P. W. 1987a. Gametic disequilibrium measures: proceed with caution. *Genetics* 117: 331–341. [5]

Hedrick, P. W. 1987b. Genetic load and the mating system in homosporous ferns. *Evolution* 41: 1282–1289. [10]

Hedrick, P. W. 1994. Purging inbreeding depression and the probability of extinction: full-sib mating. *Heredity* 73: 363–372. [10]

Hedrick, P. W., S. Jain, and L. Holden. 1978. Multilocus systems in evolution. *Evol. Biol.* 11: 101–182. [5]

Hegmann, J. P., and H. Dingle. 1982. Phenotypic and genetic covariance structure in milkweed bug life history traits. *In* H. Dingle and J. P. Hegmann (eds.), *Evolution and genetics of life histories,* pp. 177–185. North Holland, Amsterdam, Netherlands. [21]

Henderson, C. R. 1949. Estimates of changes in herd environment. *J. Dairy Sci.* 32: 706. [26,27]

Henderson, C. R. 1950. Estimation of genetic parameters. *Ann. Math. Stat.* 21: 309–310. [26]

Henderson, C. R. 1953. Estimation of variance and covariance components. *Biometrics* 9: 226–252. [20,27]

Henderson, C. R. 1963. Selection index and the expected genetic advance *In* W. D. Hanson and H. F. Robinson (eds.), *Statistical genetics and plant breeding,* pp. 141–163. Natl. Acad. Sci., Natl. Res. Council Publ. No. 982, Washington, D. C. [26].

Henderson, C. R. 1973. Sire evaluation and genetic trends. *In* Proceedings of the animal breeding and genetics symposium in honor of Dr. J. L. Lush, pp. 10–41. American Society of Animal Science, Champaign, IL. [26]

Henderson, C. R. 1975. Best linear unbiased estimation and prediction under a selection model. *Biometrics* 31: 423–447. [26]

Henderson, C. R. 1976. A simple method for the inverse of a numerator relationship matrix used in prediction of breeding values. *Biometrics* 32: 69–83. [26]

Henderson, C. R. 1977a. Prediction of future records. *In* E. Pollak, O. Kempthorne, and T. B. Bailey, Jr. (eds.), *Proceedings of the international conference on quantitative genetics,* pp. 615–638. Iowa State Univ. Press. [26].

Henderson, C. R. 1977b. Best linear unbiased prediction of breeding values not in the model for records. *J. Dairy Sci.* 60: 783–787. [A3]

Henderson, C. R. 1977c. Prediction of merit for single crosses. *Theor. Appl. Genet.* 49: 273–282 [26]

Henderson, C. R. 1984a. *Applications of linear models in animal breeding.* Univ. Guelph, Guelph, Ontario. [1,26,A3]

Henderson, C. R. 1984b. Estimation of variances and covariances under multiple trait models. *J. Dairy Sci.* 67: 1581–1589. [27]

Henderson, C. R. 1985a. Best linear unbiased prediction of nonadditive genetic merits in noninbred populations. *J. Anim. Sci.* 60: 111–117. [26]

Henderson, C. R. 1985b. MIVQUE and REML estimation of additive and nonadditive genetic variances. *J. Anim. Sci.* 61: 113–121. [26,27]

Henderson, C. R. 1985c. Equivalent linear models to reduce computations. *J. Dairy Sci.* 68: 2367–2377. [A3]

Henderson, C. R. 1986. Recent developments in variance and covariance estimation. *J. Anim. Sci.* 63: 208–216. [27]

Henderson, C. R. 1988a. Progress in statistical methods applied to quantitative genetics since 1976. *In* B. S. Weir, E. J. Eisen, M. M. Goodman, and G. Namkoong (eds.), *Proceedings of the second international conference on quantitative genetics*, pp. 85–90. Sinauer Assoc., Sunderland, MA. [26].

Henderson, C. R. 1988b. Use of an average numerator relationship matrix for multiple-sire joining. *J. Anim. Sci.* 66: 1614–1621. [26]

Henderson, C. R. 1990. Accounting for selection and mating biases in genetic evaluations. *In* D. Gianola and K. Hammond (eds.), *Statistical methods for genetic improvement of livestock*, pp. 413–436. Springer-Verlag, NY. [26].

Henderson, C. R., O. Kempthorne, S. R. Searle, and C. M. von Krosigk. 1959. The estimation of environmental and genetic trends from records subject to culling. *Biometrics* 15: 192–218. [27]

Henderson, C. R., and R. L. Quaas. 1976. Multiple trait evaluation using relatives' records. *J. Anim. Sci.* 43: 1188–1197. [26]

Herández, M., J. M. Larruga, A. M. González, and V. M. Cabrera. 1993. Association among quantitative, chromosomal and enzymatic traits in a natural population of *Drosophila melanogaster*. *Genet. Sel. Evol.* 25: 229–248. [14]

Heuch, I., F. H. F. Li. 1972. Pedig — a computer program for calculation of genotype probabilities using phenotype information. *Clin. Genet.* 3: 501–504. [13]

Heywood, J. S. 1993. Biparental inbreeding depression in the self- incompatible annual plant *Gaillardia pulchella* (Asteraceae). *Am. J. Bot.* 80: 545–550. [10]

Highton, R. 1960. Heritability of geographic variation in trunk segmentation in the red-backed salamander, *Plethodon cinereus*. *Evolution* 14: 351–360. [17]

Hill, A. 1975. Quantitative linkage: a statistical procedure for its detection and estimation. *Ann. Hum. Genet.* 38: 439–449. [16]

Hill, J. 1964. Effects of correlated gene distributions in the analysis of diallel crosses. *Heredity* 19: 27–46. [14]

Hill, J. 1966. Recurrent backcrossing in the study of quantitative inheritance. *Heredity* 21: 85–120. [9]

Hill, J. 1975. Genotype-environment interactions — a challenge for plant breeding. *J. Agri. Sci.* 85: 477–493. [22]

Hill, W. G. 1970. Design of experiments to estimate heritability by regression of offspring on selected parents. *Biometrics* 26: 566–571. [17]

Hill, W. G. 1974. Estimation of linkage disequilibrium in randomly mating populations. *Heredity* 33: 229–239. [5]

Hill, W. G. 1975. Linkage disequilibrium among multiple neutral alleles produced by mutation in finite populations. *Theor. Pop. Biol.* 8: 117–126. [5]

Hill, W. G. 1982a. Dominance and epistasis as components of heterosis. *Z. Tierzüchtg. Züchtgsbiol.* 99: 161–168. [9,10]

Hill, W. G. 1982b. Rates of change in quantitative traits from fixation of new mutations. *Proc. Natl. Acad. Sci. USA* 79: 142–145. [12]

Hill, W. G. 1982c. Predictions of response to artificial selection from new mutations. *Genet. Res.* 40: 255–278. [12]

Hill, W. G. 1990. Considerations in the design of animal breeding experiments. *In* D. Gianola and K. Hammond (eds.), *Advances in statistical methods for genetic improvement of livestock*, pp. 59–76. Springer-Verlag, Berlin. [17]

Hill, W. G., and P. J. Avery. 1978. On estimating number of genes by genotype assay. *Heredity* 40: 397–403. [9]

Hill, W. G., and P. D. Keightley. 1988. Interrelations of mutation, population size, artificial and natural selection. *In* B. S. Weir, E. J. Eisen, M. M. Goodman, and G. Namkoong (eds.), *Proceedings of the second international conference on quantitative genetics*, pp. 57–70. Sinauer Assoc., Sunderland, MA. [12]

Hill, W. G., and S. Knott. 1990. Identification of genes with large effects. *In* D. Gianola and K. Hammond (eds.), *Advances in statistical methods for genetic improvement of livestock*, pp. 477–494. Springer-Verlag, Berlin. [13]

Hill, W. G., and T. F. C. Mackay (eds.) 1989. *Evolution and animal breeding*. CAB International, Wallingford, UK. [1]

Hill, W. G., and A. Robertson. 1966. The effect of linkage on limits to artificial selection. *Genet. Res.* 8: 269–294. [5]

Hill, W. G., and A. Robertson. 1968. Linkage disequilibrium in finite populations. *Theor. Appl. Genet.* 38: 226–231. [5,14]

Hill, W. G., and R. Thompson. 1978. Probabilities of non-positive definite between-group or genetic covariance matrices. *Biometrics* 34: 429–439. [21]

Hill, W. G., and B. S. Weir. 1994. Maximum-likelihood estimation of gene location by linkage disequilibrium. *Am. J. Hum. Genet.* 54: 705–714. [14]

Hinkelmann, K. 1975. Design of genetical experiments. *In* J. N. Srivastava (ed.), *A survey of statistical design and linear models*, pp. 243–269. North Holland, Amsterdam, Netherlands. [20]

Hinkelmann, K. 1976. Diallel and multicross designs: what do they achieve? *In* E. Pollak, O. Kempthorne, and E. B. Bailey, Jr. (eds.), *Proceedings of the international conference on quantitative genetics*, pp. 659–676. Iowa State Univ. Press, Ames. [20]

Hinze, K., R. D. Thompson, E. Ritter, F. Salamini, and P. Schulz-Lefert. 1991. Restriction fragment length polymorphism-mediated targeting of the *ml-o* resistance locus in barley (*Hordeum vulgare*. *Proc. Natl. Acad. Sci. USA* 88: 3691–3695. [14]

Hocking, R. R. 1985. *The analysis of linear models.* Brooks/Cole, Monterey, CA. [22]

Hodge, S. E. 1984. The information contained in multiple sibling-pairs. *Genet. Epidem.* 1: 109–122. [16]

Hodge, S. E. 1988. Conditioning on subsets of the data: applications to ascertainment and other genetic problems. *Am. J. Hum. Genet.* 43: 364–373. [13]

Hoeschele, I. 1988. Statistical techniques for detection of major genes in animal breeding data. *Theor. Appl. Genet.* 76: 311–319. [13,26]

Hoeschele, I., and T. R. Meinert. 1990. Association of genetic defects with yield and type traits: the weaver locus effect on yield. *J. Dairy Sci.* 73: 2503–2515. [14]

Hoeschele, I., and P. M. VanRaden. 1993a. Bayesian analysis of linkage between genetic markers and quantitative trait loci. I. Prior knowledge. *Theor. Appl. Genet.* 85: 953–960. [16]

Hoeschele, I., and P. M. VanRaden. 1993b. Bayesian analysis of linkage between genetic markers and quantitative trait loci. II. Combining prior knowledge with experimental evidence. *Theor. Appl. Genet.* 85: 946–952. [16]

Hofer, A., and B. W. Kennedy. 1993. Genetic evaluation for a quantitative trait controlled by polygenes and a major locus with genotypes not or only partly known. *Genet. Sel. Evol.* 25: 537–555. [26]

Hoffmann, A. A., and P. A. Parsons. 1991. *Evolutionary genetics and environmental stress.* Oxford Univ. Press, NY. [6,7,10]

Hoffmann, A. A., and M. Turelli. 1988. Unidirectional incompatibility in *Drosophila simulans:* inheritance, geographic variation and fitness effects. *Genetics* 119: 435–444. [6]

Hohenboken, W. D. 1985. Genotype × environment interaction. *In* A. B. Chapman (ed.), *General and quantitative genetics,* pp. 151–165. Elsevier Science Publ., Amsterdam, Netherlands. [22]

Hoi-Sen, Y. 1972. Is subline differentiation a continuing process in inbred strains of mice? *Genet. Res.* 19: 53–59. [12]

Hollocher, H., A. R. Templeton, R. DeSalle, and J. S. Johnston. 1992. The molecular through ecological genetics of *abnormal abdomen* in *Drosophila mercatorum.* III. Components of genetic variation in a natural population of *Drosophila mercatorum. Genetics* 130: 355–366. [12]

Hollingsworth, M. J., and J. Maynard Smith. 1955. The effects of inbreeding on rate of development and on fertility in *Drosophila subobscura. J. Genetics* 53: 295–314. [10]

Holsinger, K. E. 1988. Inbreeding depression doesn't matter: the genetic basis of mating-system evolution. *Evolution* 42: 1235–1244. [10]

Holt, S. B. 1955. Genetics of dermal ridges: frequency distribution of total finger ridge-count. *Ann. Hum. Genet.* 20: 270–281. [19]

Holt, S. B. 1968. *The genetics of dermal ridges.* Charles Thomas Publ., Springfield, IL. [19]

Holtsford, T. P., and N. C. Ellstrand. 1990. Inbreeding effects in *Clarkia tembloriensis* (Ongraceae) populations with different natural outcrossing rates. *Evolution* 44: 2031–2046. [10]

Hopkins, C. G. 1899. Improvement in the chemical composition of the corn kernel. *Ill. Agric. Exp. Stn. Bull.* 55: 205–240. [15]

Hopper, J. L., M. C. Hannah, and J. D. Mathews. 1984. Genetic analysis workshop II: Pedigree analysis of a binary trait without assuming an underlying liability. *Genet. Epidem.* 1: 183–188. [16]

Houle, D. 1989a. Allozyme-associated heterosis in *Drosophila melanogaster. Genetics* 123: 789–801. [10]

Houle, D. 1989b. The maintenance of polygenic variation in finite populations. *Evolution* 43: 1767–1780. [12]

Houle, D. 1991. Genetic covariance of fitness correlates: what genetic correlations are made of and why it matters. *Evolution* 45: 630–648. [21]

Houle, D. 1992. Comparing evolvability and variability of quantitative traits. *Genetics* 130: 195–204. [7]

Houle, D. 1994. Adaptive distance and the genetic basis of heterosis. *Evolution* 48: 1410–1417. [10]

Houle, D., D. K. Hoffmaster, S. Assimacopoulus, and B. Charlesworth. 1992. The genomic mutation rate for fitness in *Drosophila. Nature* 359: 58–60. [12]

Houle, D., D. K. Hoffmaster, S. Assimacopoulus, and B. Charlesworth. 1994a. Correction: the genomic mutation rate for fitness in *Drosophila. Nature* 371: 358. [12]

Houle, D., K. A. Hughes, D. K. Hoffmaster, J. Ihara, S. Assimacopoulos, D. Canada, and B. Charlesworth. 1994b. The effects of spontaneous mutation on quantitative traits. I. Variance and covariance of life history traits. *Genetics* 138: 773–785. [12]

Houle, D., B. Morikawa, and M. Lynch. 1996. Comparing mutational heritabilities. *Genetics* 143: 1467–1483. [12]

Hu, S., A. M. L. Pattatucci, C. Patterson, L. Li, D. W. Fulker, S. S. Cherny, L. Kruglyak, and D. H. Hamer. 1995. Linkage between sexual orientation and chromosome Xq28 in males but not in females. *Nature Genetics* 11: 248–256. [16]

Hu, Z., X. Zhang, C. Xie, G. R. MacDaniel, and D. L. Kuhlers. 1995. A correlation method for detecting and estimating linkage between a marker locus and a quantitative trait locus using inbred lines. *Theor. Appl. Genet.* 90: 1074–1078. [15]

Huey, R. B., and A. E. Dunham. 1987. Repeatability of locomotor performance in natural populations of the lizard *Sceloporus merriami. Evolution* 41: 1116–1119. [6]

Hughes, K. A. 1995. The inbreeding decline and average dominance of genes affecting male life-history characters in *Drosophila melanogaster. Genet. Res.* 65: 41–52. [10]

Huidong, M. 1988. Genetic expression of endosperm traits. *In* B. S. Weir, E. J. Eisen, M. M. Goodman, and

G. Namkoong (eds.), *Proceedings of the second international conference on quantitative genetics*, pp. 478–487. Sinauer Assoc., Sunderland, MA. [23]

Hull, F. H. 1946. Overdominance and corn breeding where hybrid seed is not feasible. *J. Am. Soc. Agron.* 38: 1100–1103. [10]

Hunt, G. J., R. E. Page, Jr., M. K. Fondrk, and C. J. Dullum. 1995. Major quantitative trait loci affecting honey bee foraging behavior. *Genetics* 141: 1537–1545. [15]

Husband, B. C., and D. W. Schemske. 1996. Evolution of the magnitude and timing of inbreeding depression in plants. *Evolution* 50: 54–70. [10]

Hutchings, J. A., and M. M. Ferguson. 1992. The independence of enzyme heterozygosity and life-history traits in natural populations of *Salvelinus fontinalis* (brook trout). *Evolution* 69: 496–502. [6,10]

Hutchison, D. W., and J. M. Cheverud. 1995. Fluctuating asymmetry in tamarin *(Saguinus)* cranial morphology: intra- and interspecific comparisons between taxa with varying levels of genetic heterozygosity. *J. Heredity* 86: 280–288. [6]

Hutchinson, G. E. 1967. *A treatise on limnology. Vol. II. Introduction to lake biology and the limnoplankton.* John Wiley & Sons, NY. [6]

Huxley, J. S. 1932. *Problems of relative growth.* Methuen, London. [11]

Hyde, J. S. 1973. Genetic homeostasis and behavior: analysis, data, and theory. *Behav. Genet.* 3: 233–245. [6]

Hyer, R. N., C. Julier, J. D. Buckley, M. Trucco, J. Rotter, R. Spielman, A. Barnett, S. Bain, C. Boitard, I. Deschamps, J. A. Todd, J. I. Bell, and G. M. Lathrop. 1991. High resolution linkage mapping for susceptibility genes in human polgenic diseases: insulin-dependent diabetes mellitus and chromosome 11q. *Am. J. Hum. Genet.* 48: 243–257. [16]

Hyne, V., and M. J. Kearsey. 1995. QTL analysis: further uses of 'marker regression'. *Theor. Appl. Genet.* 91: 471–476. [15]

Itoh, Y. and H. Iwaisaki. 1990. Restricted best linear unbiased prediction using canonical transformation. *Genet. Sel. Evol.* 23: 339–347. [26].

Itoh, Y., and Y. Yamada. 1990. Relationships between genotype × environment interaction and genetic correlation of the same trait measured in different environments. *Theor. Appl. Genet.* 80: 11–16. [22]

Jackson, J. F. 1973. A search for the population asymmetry parameter. *Syst. Zool.* 22: 166–170. [6]

Jackson, N. 1983. Effect of ignoring full sib relationships when making half sib estimates of heritability. *Theor. Appl. Genet.* 65: 61–66. [18]

Jacob, H. J., K. Lindpainter, S. E. Lincoln, K. Kusumi, R. K. Bunker, Y.-P. Mao, D. Ganten, V. J. Dzau, and E. S. Lander. 1991. Genetic mapping of a gene causing hypertension in the stroke-prone spontaneously hypertensive rat. *Cell* 67: 213–224. [14]

Jacquard, A. 1974. *The genetic structure of populations.* Springer-Verlag, NY. [7]

Jacquard, A. 1983. Heritability: one word, three concepts. *Biometrics* 39: 465–477. [7]

Jaenike, J. 1987. Genetics of oviposition-site preference in *Drosophila tripunctata. Heredity* 59: 363–369. [9]

Jakubczak, J. L., W. D. Burke, and T. H. Eickbush. 1991. Retroposable elements R1 and R2 interrupt the rRNA genes of most insects. *Proc. Natl. Acad. Sci. USA* 88: 3295–3299. [12]

James, J. W. 1971. Frequency in relatives for an all-or-none trait. *Ann. Hum. Genet.* 35: 47–49. [16,25]

James, J. W. 1973. Covariances between relatives due to sex-linked genes. *Biometrics* 29: 584–588. [24]

Jansen, R. C. 1992. A general mixture model for mapping quantitative trait loci by using molecular markers. *Theor. Appl. Genet.* 85: 252–260. [15]

Jansen, R. C. 1993a. Maximum likelihood in a generalized linear finite mixture model by using the EM algorithm. *Biometrics* 49: 227–231. [15]

Jansen, R. C. 1993b. Interval mapping of multiple quantitative trait loci. *Genetics* 135: 205–211. [15]

Jansen, R. C. 1994a. Mapping of quantitative trait loci by using genetic markers: an overview of biometrical models. *In* J. W. van Ooijen and J. Jansen (eds.), *Biometrics in plant breeding: applications of molecular markers*, pp. 116–124. CPRO-DLO, Netherlands. [15]

Jansen, R. C. 1994b. Controlling the type I and type II errors in mapping quantitative trait loci. *Genetics* 138: 871–881. [15]

Jansen, R. C. 1996. A general Monte Carlo method for mapping multiple quantitative trait loci. *Genetics* 142: 305–311. [15]

Jansen, R. C. and P. Stam. 1994. High resolution of quantitative trait into multiple loci via interval mapping. *Genetics* 136: 1447–1455. [15]

Jansen, R. C., J. W. van Ooijen, P. Stam, C. Lister, and C. Dean. 1995. Genotype-by-environment interaction in genetic mapping of multiple quantitative trait loci. *Theor. Appl. Genet.* 91: 33–37. [15]

Janss, L. L. G., and J. H. J. Van Der Werf. 1992. Identification of a major gene in F_1 and F_2 data when alleles are assumed fixed in parental lines. *Genet. Sel. Evol.* 24: 511–526. [13]

Janss, L. L. G., R. Thompson, and J. A. M. Van Arendonk. 1995. Applications of Gibbs sampling for inference in a mixed model gene-polygenic inheritance model in animal populations. *Theor. Appl. Genet.* 91: 1137–1147. [13]

Janssen, G. M., G. de Jong, E. N. G. Joosse, and W. Scharloo. 1988. A negative maternal effect in springtails. *Evolution* 42: 828–833. [23]

Jayakar, S. D. 1970. On the detection and estimation of linkage between a locus influencing a quantitative character and a marker locus. *Biometrics* 26: 451–464. [16]

Jenkin, F. 1867. Origins of species. *North British Review* 46: 277–318. [1]

Jennrich, R. I., and P. F. Sampson. 1976. Newton-Raphson and related algorithms for maximum likelihood variance component estimation. *Technometrics* 18: 11–17. [27,A4]

Jensen, J. 1989. Estimation of recombination parameters between a quantitative trait locus (QTL) and two marker gene loci. *Theor. Appl. Genet.* 78: 613–618. [14,15]

Jensen, J., and I. L. Mao. 1988. Transformation algorithms in analysis of single trait and of multitrait models with equal design matrices and one random factor per trait: a review. *J. Anim. Sci.* 66: 2750–2761. [26,27]

Jiang, C., and Z.-B. Zeng. 1995. Multiple trait analysis of genetic mapping for quantitative trait loci. *Genetics* 140: 1111–1127. [15]

Jiménez, J. A., K. A. Hughes, G. Alaks, L. Graham, and R. C. Lacy. 1994. An experimental study of inbreeding depression in a natural habitat. *Science* 266: 271–273. [10]

Jinks, J. L. 1954. The analysis of continuous variation in a diallel of *Nicotiana rustica* varieties. *Genetics* 39: 767–788. [20]

Jinks, J. L., and V. Connolly. 1973. Selection for specific and general response to environmental differences. *Heredity* 30: 33–40. [22]

Jinks, J. L., and K. Mather. 1955. Stability in development of heterozygotes and homozygotes. *Proc. Royal Soc. Lond.* B 143: 561–578. [6]

Jinks, J. L., and J. M. Perkins. 1969. The detection of linked epistatic genes for a metrical trait. *Heredity* 24: 465–475. [9]

Jinks, J. L., and J. M. Perkins. 1970. A general method for the detection of additive, dominance and epistatic components of variation. III. F_2 and backcross populations. *Heredity* 25: 419–429. [20]

Jinks, J. L., and H. S. Pooni. 1982. Determination of the environmental sensitivity of selection lines of *Nicotiana rustica* by the selection environment. *Heredity* 49: 291–294. [22]

Jinks, J. L., and P. Towey. 1976. Estimating the number of genes in a polygenic system by genotype assay. *Heredity* 37: 69–81. [9]

Jinks, J. L., C. E. Caten, G. Simchen, and H. J. Croft. 1966. Heterokaryon incompatibility and variation in wild populations of *Aspergillus nidulans*. *Heredity* 21: 227–239. [5]

Johannsen, W. 1903. *Über Erblichkeit in Populationen und in Reinen Linien.* Gustav Fischer, Jena, Germany. [1]

Johannsen, W. 1909. *Elemente der exakten Erblichkeitslehre.* Gustav Fischer, Jena, Germany. [1]

John, J. A., and N. R. Draper. 1980. An alternative family of transformations. *Appl. Stat.* 29: 190–197. [11]

Johnson, M. S., and J. R. G. Turner. 1979. Absence of dosage compensation for a sex-linked enzyme in butterflies *(Heliconius). Heredity* 43: 71–77. [24]

Johnson, N. L., and S. Kotz. 1970a. *Continuous univariate distributions – 1.* John Wiley & Sons, NY. [2]

Johnson, N. L., and S. Kotz. 1970b. *Continuous univariate distributions – 2.* John Wiley & Sons, NY. [2,A5]

Johnson, N. L., and S. Kotz. 1972. *Continuous multivariate distributions.* John Wiley & Sons, NY. [2]

Johnson, R. A., and D. W. Wichern. 1988. *Applied multivariate statistical analysis.* 2nd Ed. Prentice-Hall, NJ [8]

Johnston, M. O. 1992. Effects of cross and self-fertilization on progeny fitness in *Lobelia cardinalis* and *L. siphilitica. Evolution* 46: 688–702. [10]

Johnston, M. O., and D. J. Schoen. 1994. On the measurement of inbreeding depression. *Evolution* 48: 1735–1741. [10]

Johnston, M. O., and D. J. Schoen. 1995. Mutation rates and dominance levels of genes affecting total fitness in two angiosperm species. *Science* 267: 226–229. [10,12]

Jones, D. F. 1917. Dominance of linked factors as a means of accounting for heterosis. *Genetics* 2: 466–479. [10]

Jones, D. F. 1918. The effects of inbreeding and cross-breeding upon development. *Conn. Agric. Exp. Sta. Bull.* 207: 5–100. [10]

Jorde, L. B. 1995. Linkage disequilibrium as a gene-mapping tool. *Am. J. Hum. Genet.* 56: 11–14. [14]

Jorde, L. B., W. S. Watkins, M. Carlson, J. Groden, H. Albertsen, A. Thliveris, and M. Leppert. 1994. Linkage disequilibrium predicts physical distance in the adenomatous polyposis coli region. *Am. J. Hum. Genet.* 54: 884–898. [14]

Jowett, D. 1972. Yield stability parameters for sorghum in East Africa. *Crop Sci.* 12: 314–317. [6]

Kackar, R. N., and D. A. Harville. 1981. Unbiasedness of two-stage estimation and prediction for mixed linear models. *Comm. Stat. Theor. Meth.* A10: 1249–1261. [26]

Kacser, H., and J. A. Burns. 1981. The molecular basis of dominance. *Genetics* 97: 639–666. [4]

Kang, K. W., J. P. Lindemann, J. C. Christian, W. E. Nance, and J. A. Norton, Jr. 1974. Sampling variances in twin and sibling studies of man. *Hum. Hered.* 24: 363–372. [19]

Kang, M. S., and H. G. Gauch, Jr. (eds.) 1995. *Genotype-by-environment interaction.* CRC Press, Boca Raton, FL. [22]

Kaplan, N. L., and B. S. Weir. 1995. Are moment bounds on the recombination fraction between a marker and a disease locus too good to be true? Allelic association mapping revisited for simple genetic diseases in the Finnish population. *Am. J. Hum. Genet.* 57: 1486–1498. [14]

Kaplan, N. L., W. G. Hill, and B. S. Weir. 1995. Likelihood methods for locating disease genes in nonequilibrium populations. *Am. J. Hum. Genet.* 56: 18–32. [14]

Kaprio, J., R. E. Ferrell, B. A. Kottke, M. I. Kamboh, and C. F. Sing. 1991. Effects of polymorphisms in apolipoproteins E, A-IV, and H on quantitative traits related to risk for cardiovascular disease. *Arterioscler. Thromb.* 11: 1330–1348. [14]

Karban, R. 1989. Fine-scale adaptation of herbivorous thrips to individual host plants. *Nature* 340: 60–61. [9]

Karigl, G. 1981. A recursive algorithm for the calculation of identity coefficients. *Ann. Hum. Genet.* 45: 299–305. [7]

Karlin, S. 1982. Theoretical aspects of genetic map functions in recombination processes. *In* A. Chakravarti (ed.), *Human population genetics: the Pittsburgh symposium*, pp. 209–228. Van Nostrand Reinhold, NY. [14]

Karlin, S., and P. T. Williams. 1981. Structured Exploratory Data Analysis (SEDA) for determining mode of inheritance of quantitative traits. II. Simulation studies on the effect of ascertaining families through high-valued probands. *Am. J. Hum. Genet.* 33: 282–292. [13]

Karlin, S., D. Carmelli, and R. Williams. 1979. Index measures for assessing the mode of inheritance of continuously distributed traits: I. Theory and justifications. *Theor. Pop. Biol.* 16: 81–106. [13]

Karlin, S., E. C. Cameron, and P. T. Williams. 1981. Sibling and parent-offspring correlation estimation with variable family size. *Proc. Natl. Acad. Sci. USA* 78: 2664–2668. [18]

Karowe, D. N. 1990. Predicting host range evolution: colonization of *Coronilla varia* by *Colias philodice* (Lepidoptera: Pierida). *Evolution* 44: 1637–1647. [22]

Kat, P. W. 1982. The relationship between heterozygosity for enzyme loci and developmental homeostasis in peripheral populations of aquatic bivalves (Unionidae). *Am. Nat.* 119: 824–832. [6]

Kearsey, M. J., and V. Hyne. 1994. QTL analysis: a simple 'marker regression' approach. *Theor. Appl. Genet.* 698–702. [15]

Kearsey, M. J., and J. L. Jinks. 1968. A general method of detecting additive, dominance and epistatic variation for metrical traits. *Heredity* 23: 403–409. [20]

Kearsey, M. J., and K.-I. Kojima. 1967. The genetic architecture of body weight and egg hatchability on *Drosophila melanogaster*. *Genetics* 56: 23–37. [14]

Kearsey, M. J., and H. S. Pooni. 1996. *The genetical analysis of quantitative traits.* Chapman and Hall, London. [1,6]

Keeble, F., and C. Pellew. 1910. The mode of inheritance of stature and of time of flowering in peas *Pisum sativum*. *J. Genetics* 1: 47–56. [10]

Keightley, P. D. 1989. Models of quantitative variation in flux in metabolic pathways. *Genetics* 121: 869–876. [5]

Keightley, P. D. 1994. The distribution of mutation effects on viability in *Drosophila melanogaster*. *Genetics* 138: 1315–1322. [9,12]

Keightley, P. D., and G. Bulfield. 1993. Detection of quantitative trait loci from frequency changes of marker alleles under selection *Genet. Res.* 62:195–203. [14]

Keightley, P. D., and W. G. Hill. 1990. Estimating new mutational variation in growth rate of mice. *In* W. G. Hill, R. Thompson, and J. A. Woolliams (eds.), *Proceedings of the 4th world congress on genetics applied to livestock production.* Edinburgh. [12]

Keightley, P. D., and W. G. Hill. 1992. Quantitative genetic variation in body size of mice from new mutations. *Genetics* 131: 693–700. [12]

Keightley, P. D., T. F. C. Mackay, and A. Caballero. 1993. Accounting for bias in estimates of the rate of polygenic mutation. *Proc. Royal Soc. Lond.* B 253: 291–296. [12]

Keightley, P. D., T. Hardge, L. May, and G. Bulfield. 1996. A genetic map of quantitative trait loci for body weight in mouse. *Genetics* 142: 227–235. [14]

Keller, E. C., Jr., and D. F. Mitchell. 1962. Interchromosomal genotypic interactions in *Drosophila*. I. An analysis of morphological characters. *Genetics* 47: 1557–1571. [14]

Keller, E. C., Jr., and D. F. Mitchell. 1964. Interchromosomal genotypic interactions in *Drosophila*. II. An analysis of viability characters. *Genetics* 49: 293–307. [14]

Keller, K. R., and S. T. Likens. 1955. Estimates of heritability in hops, *Humulus lupulus* L. *Agron. J.* 47: 518–521. [19]

Kelley, R. L., and M. I. Kuroda. 1995. Equality for X chromosomes. *Science* 270: 1607–1610. [24]

Kempthorne, O. 1953. The correlation between relatives in a simple autotetraploid population. *Genetics* 40: 168–174. [7]

Kempthorne, O. 1954. The correlation between relatives in a random mating population. *Proc. Royal Soc. Lond.* B 143: 103–113. [5,7,21,24]

Kempthorne, O. 1955. The correlation between relatives in random mating populations. *Cold Spring Harbor Symp. Quant. Biol.* 20: 60–78. [23]

Kempthorne, O. 1957. *An introduction to genetic statistics.* John Wiley & Sons, NY. [4,5,7]

Kempthorne, O., and R. N. Curnow. 1961. The partial diallel cross. *Biometrics* 17: 229–250. [20]

Kempthorne, O., and R. H. Osborne. 1961. The interpretation of twin data. *Am. J. Hum. Genet.* 13: 320–339. [19]

Kempthorne, O., and O. B. Tandon. 1953. The estimation of heritability by regression of offspring on parent. *Biometrics* 9: 90–100. [17]

Kendall, M., and A. Stuart. 1977. *The advanced theory of statistics. Vol. 1. Distribution theory.* 4th Ed. Macmillan, NY. [2,11,A1]

Kendall, M., and A. Stuart. 1979. *The advanced theory of statistics. Vol. 2. Inference and relationship.* 4th Ed. Macmillan, NY. [A4]

Kennard, W. C., and M. J. Harvey. 1995. Quantitative trait analysis of fruit quality in cucumber: QTL detection, confirmation, and comparison with mating-design variation. *Theor. Appl. Genet.* 91: 53–61. [15]

Kennedy, B. W. 1991. C. R. Henderson: the unfinished legacy. *J. Dairy Sci.* 74: 4067–4081. [26]

Kennedy, B. W., and D. A. Sorensen. 1988. Properties of mixed-model methods for prediction of genetic merit. *In* B. S. Weir, E. J. Eisen, M. M. Goodman, and G. Namkoong (eds.), *Proceedings of the second international conference on quantitative genetics*, pp. 91–103. Sinauer Assoc., Sunderland, MA. [26]

Kennedy, B. W., M. Quinton, and J. A. M. van Arendonk. 1992. Estimation of effects of single genes on quantitative traits. *J. Anim. Sci.* 70: 2000–2012. [14,26]

Kennedy, J. S. 1956. Phase transformation in locust biology. *Biol. Rev.* 31: 349–370. [6]

Kennedy, W. J., Jr., and J. E. Gentle. 1980. *Statistical computing.* Marcel Dekker, NY. [A1]

Kerem, B.-S., J. M. Rommens, J. A. Buchanan, D. Markiewicz, T. K. Cox, A. Chakravarti, M. Buchwald, and L.-C. Tsui. 1989. Identification of the cystic fibrosis gene: genetic analysis. *Science* 245: 1073–1080. [14]

Kerfoot, W. C. 1988. Defensive spines: inverse relationship between coefficients of variation and size. *Limnol. Oceanogr.* 33: 1412–1429. [11]

Kermicle, J. L. 1969. Androgenesis conditioned by a mutation in maize. *Science* 166: 1422–1424. [9]

Kestilä, M., M. Männikkö, C. Holmberg, G. Gyapay, J. Weissenbach, E.-R. Savolainen, L. Peltonen, and K. Tryggvason. 1994. Congenital nephrotic syndrome of the Finnish type maps to the long arm of chromosome 19. *Am. J. Hum. Genet.* 54: 757–764. [14]

Khambanonda, I. 1950. Quantitative inheritance of fruit size in red pepper *(Capsicum frutescens* L.) *Genetics* 35: 322–343. [9]

Khatib, H., A. Darvasi, Y. Plotski, and M. Soller. 1994. Determining relative microsatellite allele frequencies in pooled DNA samples. *PCR Methods Appl.* 4: 13–19. [14]

Kibota, T. T. 1996. Spontaneous mutations influencing fitness in *Escherichia coli.* Ph. D. Thesis, Univ. Oregon, Eugene, OR. [12]

Kibota, T. T., and M. Lynch. 1996. Estimate of the genomic mutation rate deleterious to overall fitness in *Escherichia coli. Nature* 381: 694–696. [12]

Kidwell, J. F., and M. M. Kidwell. 1966. The effects of inbreeding on body weight and abdominal chaeta number in *Drosophila melanogaster. Can. J. Genet. Cytol.* 8: 207–215. [10]

Kieser, J. A., and H. T. Groeneveld. 1991. Fluctuating odontometric asymmetry, morphological variability, and genetic monomorphism in the cheetah *Acinonyx jubatus. Evolution* 45: 1175–1183. [6]

Killick, R. J. 1971. The biometrical genetics of autotetraploids. 1. Generations derived from a cross between two pure lines. *Heredity* 27: 331–346. [5]

Kimura, M. 1983. *The neutral theory of molecular evolution.* Cambridge Univ. Press, UK. [4,12]

Kimura, M., and T. Maruyama. 1966. The mutational load with epistatic gene interactions in fitness. *Genetics* 54: 1337–1351. [12]

King, D. P. F. 1984. Enzyme heterozygosity associated with anatomical character variance and growth in the herring (*Clupea harengus* L.). *Heredity* 54: 289–296. [6]

Kinghorn, B. P., B. W. Kennedy, and C. Smith. 1993. A method for screening for genes of major effect. *Genetics* 134: 351–360. [26]

Kinzer, S. M., S. J. Schwager, and M. A. Mutschler. 1990. Mapping of ripening-related or -specific cDNA clones of tomato (*Lycopersicon esculentum*). *Theor. Appl. Genet.* 79: 489–496. [14]

Kirkpatrick, M., and R. Lande. 1989. The evolution of maternal characters. *Evolution* 43: 485–503. [23]

Kleczkowski, A. 1949. The transformation of local lesion counts for statistical analysis. *Ann. Appl. Biol.* 36: 139–152. [11]

Klein, T. W. 1974. Heritability and genetic correlation: statistical power, population comparisons, and sample size. *Behav. Genetics* 4: 171–189. [17,21]

Klein, T. W., J. C. DeFries, and C. T. Finkbeiner. 1973. Heritability and genetic correlation: standard errors of estimates and sample size. *Behav. Genetics* 3: 355–364. [17]

Kluge, A. G., and W. C. Kerfoot. 1973. The predictability and regularity of character divergence. *Am. Nat.* 107: 426–442. [11]

Kluge, R., and H. Geldermann. 1982. Effects of marked chromosome sections on quantitative traits in the mouse. *Theor. Appl. Genet.* 62: 1-4. [14]

Knapp, S. J. 1991. Using molecular markers to map multiple quantitative trait loci: models for backcross, recombinant inbred, and doubled-haploid progeny. *Theor. Appl. Genet.* 81: 333-338. [14,15]

Knapp, S. J., and W. C. Bridges, Jr. 1988. Parametric and jackknife confidence interval estimators for two-factor mating design genetic variance ratios. *Theor. Appl. Genet.* 76: 385–392. [18]

Knapp, S. J., and W. C. Bridges Jr. 1990. Using molecular markers to estimate quantitative trait locus parameters: power and genetic variances for unreplicated and replicated progeny. *Genetics* 126: 769–777. [14,15]

Knapp, S. J., W. C. Bridges Jr., and D. Birkes. 1990. Mapping quantitative trait loci using molecular marker linkage maps. *Theor. Appl. Genet.* 79: 583–592. [14,15]

Knott, S. A. 1994. Prediction of the power of detection of marker-quantitative trait locus linkages using analysis of variance. *Theor. Appl. Genet.* 89: 318–322. [16]

Knott, S. A., J. M. Elsen, and C. S. Haley. 1996. Methods for multiple-marker mapping of quantitative trait loci in half-sib populations. *Theor. Appl. Genet.* 93: 71–80. [16]

Knott, S. A., and C. S. Haley. 1992a. Aspects of maximum likelihood methods for the mapping quantitative trait loci in line crosses. *Genet. Res.* 60: 139–151. [13,14,15]

Knott, S. A., and C. S. Haley. 1992b. Maximum likelihood mapping of quantitative trait loci using full-sib families. *Genetics* 132: 1211–1222. [13,16]

Knott, S. A., C. S. Haley, and R. Thompson. 1990. Approximations to segregation analysis for the detection of major genes. *In* W. G. Hill, R. Thompson, and J. A. Woolliams (eds.), *Proc. 4th World Congr. Genet. Appl. Livestock Prod.*, Vol. 13 pp. 504–507. Edinburgh. [13]

Knott, S. A., C. S. Haley, and R. Thompson. 1991a. Methods of segregation analysis for animal breeding data: parameter estimates. *Heredity* 68: 313–320. [13]

Knott, S. A., C. S. Haley, and R. Thompson. 1991b. Methods of segregation analysis for animal breeding data: a comparison of power. *Heredity* 68: 299–311. [13]

Knowler, W. C., R. C. Williams, D. J. Pettitt, and A. G. Steinberg. 1988. $Gm^{3;5,13,14}$ and type 2 diabetes mellitus: an association in American indians with genetic admixture. *Am. J. Hum. Genet.* 43: 520–526. [14]

Knowles, P., and M. C. Grant. 1981. Genetic patterns associated with growth variability in ponderosa pine. *Am. J. Bot.* 68: 942–946. [6]

Knowles, P., and J. B. Mitton. 1980. Genetic heterozygosity and radial growth variability in *Pinus contorta. Silvae Genet.* 29: 114–118. [6]

Kobyliansky, E., and G. Livshits. 1983. Relationship between levels of biochemical heterozygosity and morphological variability in human populations. *Ann. Hum. Genet.* 47: 215–223. [6]

Koch, R. M. 1972. The role of maternal effects in animal breeding. VI. Maternal effects in beef cattle. *J. Anim. Sci.* 35: 1316–1323. [23]

Koch, R. M., and R. T. Clark. 1955. Genetic and environmental relationships among economic characters in beef cattle. I. Correlation among paternal and maternal half-sibs. *J. Anim. Sci.* 14: 775–785. [23]

Koehn, R. K., W. J. Diehl, and T. M. Scott. 1988. The differential contribution by individual enzymes of glycolysis and protein catabolism to the relationship between heterozygosity and growth rate in the coot clam, *Mulina lateralis. Genetics* 118: 121–130. [10]

Koester, R. P., P. H. Sisco, and C. W. Stuber. 1993. Identification of quantitative trait loci controlling days to flowering and plant height in two near isogenic lines of maize. *Crop Sci.* 33: 1209–1216. [14,15]

Kohn, J. R., and J. E. Biardi. 1995. Outcrossing rates and inferred levels of inbreeding depression in gynodioecious *Cucurbita foetidissima* (Cucurbitaceae). *Heredity* 75: 77–83. [10]

Kohn, L. A., and W. R. Atchley. 1988. How similar are genetic correlation structures? Data from mice and rats. *Evolution* 42:467–481. [21]

Kojima, K.-I., and T. M. Kelleher. 1963. A comparison of purebred and crossbred selection schemes with two populations of *Drosophila pseudoobscura. Genetics* 48: 57–72. [14]

Kondrashov, A. S. 1988. Deleterious mutations and the evolution of sexual reproduction. *Nature* 336: 435–440. [12]

Kondrashov, A. S. 1995. Contamination of the genome by very slightly deleterious mutations: why have we not died 100 times over? *J. Theor. Biol.* 175: 583–594. [12]

Kondrashov, A. S., and D. Houle. 1994. Genotype-environment interactions and the estimation of the genomic mutation rate in *Drosophila melanogaster. Proc. Royal Soc. Lond.* B 258: 221–227. [12]

Kondrashov, A. S., and M. Turelli. 1992. Deleterious mutations, apparent stabilizing selection and the maintenance of quantitative variation. *Genetics* 132: 603–618. [12]

Konigsberg, L. W., and J. M. Cheverud. 1992. Uncertain paternity in primate quantitative genetic studies. *Am. J. Primat.* 27: 133–143. [26]

Konigsberg, L. W., C. M. Kammerer, and J. W. MacCluer. 1989. Segregation analysis of quantitative traits in nuclear families: comparison of three program packages. *Genet. Epidem.* 6:713–726. [13]

Konigsberg, L. W., J. Blangero, C. M. Kammerer, and G. E. Mott. 1991. Mixed model analysis of LDL-C concentration with genotype-covariate interaction. *Genet. Epidem.* 8: 69–80. [13]

Koots, K. R., J. P. Gibson, and J. W. Wilson. 1994. Analyses of published genetic parameter estimates for beef production traits. 2. Phenotypic and genetic correlations. *Anim. Breed. Absts.* 62: 825–853. [21]

Korol, A. B., I. A. Preigel, and N. I. Bocharnikova. 1987. Linkage between loci of quantitative characters and marker loci. V. Combined analysis of several markers and quantitative characters. *Genetika* 23: 1421–1431. [15]

Korol, A. B., Y. I. Ronin, Y. Tadmor, A. Bar-Zur, V. M. Kirzhner, and E. Nevo. 1996. Estimating variance effects of QTL: an important prospect to increase the resolution power of interval mapping. *Genet. Res.* 67: 187–194. [15]

Korol, A. B., A. A. Zhuchenko, and A. P. Samovol. 1981. Linkage between loci of quantitative characters and marker loci. III. The bias of estimates during disturbance of the original hypothesis. *Genetika* 17: 1234–1247. [15]

Korol, A. B., A. A. Zhuchenko, and I. A. Preigel. 1983. Linkage between loci of quantitative characters and marker loci. IV. Evaluation of parameters by the least-squares method. *Genetika* 19: 594–601. [15]

Kosambi, D. D. 1944. The estimation of map distances from recombination values. *Ann. Eugen.* 12: 172–175. [14]

Koski, V. 1971. Embryonic lethals of *Picea abies* and *Pinus sylvestris. Comm. Inst. For. Fenn.* 75:1–30. [10]

Kosuda, K. 1993. A further study of interchomosomal epistatic interaction in male mating activity of *Drosophila melanogaster. Heredity* 70: 370–375. [14]

Kreitman, M., and M. Aguade. 1986. Genetic uniformity in two populations of *Drosophila melanogaster* revealed by filter hybridization of four-nucleotide-recognizing restriction enzyme digests. *Proc. Natl. Acad. Sci. USA* 83: 3562–3566. [14]

Krimbas, C. B. 1961. Release of genetic variability through recombination. VI. *Drosophila willistoni. Genetics* 46: 323–1334. [5]

Kruglyak, L. 1996. Thresholds and sample sizes. *Nature Genetics* 14: 132–133. [16]

Kruglyak, L., and E. S. Lander. 1995a. High-resolution genetic mapping of complex traits. *Am. J. Hum. Genet.* 56: 1212–1223. [16]

Kruglyak, L., and E. S. Lander. 1995b. Complete multipoint sib-pair analysis of qualitative and quantitative traits. *Am. J. Hum. Genet.* 57: 439–454. [16]

Kruglyak, L., and E. S. Lander. 1995c. A nonparametric approach for mapping quantitative trait loci. *Genetics* 139: 1421–1428. [15]

Kudo, A., K. Ito, and K. Tanaka. 1972. Genetic studies on inbreeding in some Japanese populations. X. The effects of parental consanguinity on psychometric measurements, school performances and school attendance in Shizuoka school-children. *Jap. J. Hum. Genetics* 17: 231–248. [10]

Kwon, J. M., M. Boehnke, T. L. Burns, and P. P. Moll. 1990. Commingling and segregation analyses: comparisons of results from a simulation study of a quantitative trait. *Genet. Epidem.* 7: 57–68. [13]

Lacy, R. C., A. Petric, and M. Warneke. 1993. Inbreeding and outbreeding in captive populations of wild animal species. *In* N. W. Thornhill (ed.), *The natural history of inbreeding and outbreeding: theoretical and empirical perspectives,* pp. 352–374. Univ. Chicago Press, Chicago. [10]

Lai, C. , R. F. Lyman, A. D. Long, C. H. Langley, and T. F. C. Mackay. 1994. Naturally occurring variation in bristle number associated with DNA sequence polymorphisms at the *scabrous* locus of *Drosophila melanogaster. Science* 266: 1697–1702. [12,14]

Lalouel, J. M. 1992. Linkage analysis in human genetics. *In* J. S. Beckmann and T. C. Osborn (eds.), *Plant genomes: methods for genetic and physical mapping.* pp. 167–180. Kluwer Academic, Boston. [14]

Lalouel, J. M., and N. E. Morton. 1981. Complex segregation analysis with pointers. *Hum. Hered.* 31: 312–321. [13]

Lalouel, J. M., D. C. Rao, N. E. Morton, and R. C. Elston. 1983. A unified model for complex segregation analysis. *Am. J. Hum. Genet.* 35: 816–826. [13]

Lamberson, W. R., and D. L. Thomas. 1984. Effects of inbreeding in sheep: a review. *Anim. Breeding Abst.* 52: 287–297. [10]

Lamkey, K. R., and O. S. Smith. 1987. Performance and inbreeding depression of populations representing several eras of maize breeding. *Crop Sci.* 27: 695–699. [10]

Lamy, M., J. Frézal, J. deGrouchy, and J. Kelley. 1957. Le nombre de dermatoglyphes dans un échantillon de jumeaux. *Ann. Hum. Genet.* 21: 374–396. [19]

Lande, R. 1975. The maintenance of genetic variation by mutation in a polygenic character with linked loci. *Genet. Res.* 26: 221–235. [12]

Lande, R. 1976. Natural selection and random genetic drift in phenotypic evolution. *Evolution* 30: 314–334. [12]

Lande, R. 1977. On comparing coefficients of variation. *Syst. Zool.* 26: 214–217. [11]

Lande, R. 1978. Evolutionary mechanism of limb loss in tetrapods. *Evolution* 32: 73–92. [25]

Lande, R. 1979. Quantitative genetic analysis of multivariate evolution, applied to brain:body allometry. *Evolution* 33: 402–416. [11,21]

Lande, R. 1981. The minimum number of genes contributing to quantitative variation between and within populations. *Genetics* 99: 541–553. [9]

Lande, R. 1985. Genetic and evolutionary aspects of allometry. *In* W. L. Jungers (ed.), *Size and scaling in primate biology,* pp. 21–32. Plenum Publ. Corp., NY. [21]

Lande, R. 1988. Quantitative genetics and evolutionary theory. *In* B. S. Weir, E. J. Eisen, M. M. Goodman, and G. Namkoong (eds.), *Proceedings of the second international conference on quantitative genetics,* pp. 71–84. Sinauer Assoc., Sunderland, MA. [1]

Lande, R. 1994. Risk of population extinction from new deleterious mutations. *Evolution* 48: 1460–1469. [12]

Lande, R., and S. J. Arnold. 1983. The measurement of selection on correlated characters. *Evolution* 37: 1210–1226. [8]

Lande, R., and T. Price. 1989. Genetic correlations and maternal effect coefficients obtained from offspring-parent regression. *Genetics* 122: 915–922. [21,23]

Lande, R., and D. W. Schemske. 1985. The evolution of self-fertilization and inbreeding depression in plants. I. Genetic models. *Evolution* 39: 24–40. [10]

Lander, E. S. 1993. Finding similarities and differences among genomes. *Nature Genetics* 4: 5–6. [14]

Lander, E. S., and D. Botstein. 1989. Mapping Mendelian factors underlying quantitative traits using RFLP linkage maps. *Genetics* 121: 185–199 (Correction 136: 705). [9,14,15]

Lander, E. S., and L. Kruglyak. 1995. Genetic dissection of complex traits: guidelines for interpreting and reporting linkage results. *Nature Genetics* 11: 241–247. [16]

Lander, E., and L. Kruglyak. 1996. Genetic dissection of complex traits. *Nature Genetics* 12: 357–358. [16]

Lander, E. S., and N. J. Schork. 1994. Genetic dissection of complex traits. *Science* 265: 2037–2048. [16]

Lange, K. 1986a. The affected sib-pair method using identity by state relations. *Am. J. Hum. Genet.* 39: 148–150. [16]

Lange, K. 1986b. A test statistic for the affected sib-set method. *Ann. Hum. Genet.* 50: 283–290. [16]

Lange, K., and M. Boehnke. 1982. How many polymorphic genes will it take to span the human genome? *Am. J. Hum. Genet.* 34: 842–845. [14]

Lange, K., and M. Boehnke. 1983. Extensions to pedigree analysis. V. Optimal calculation of Mendelian likelihoods. *Hum. Hered.* 33: 291–301. [13]

Lange, K, and R. C. Elston. 1975. Extensions to pedigree analysis. I. Likelihood calculations for simple and complex pedigrees. *Hum. Hered.* 25: 95–105. [13]

Lange, K., D. Weeks, and M. Boehnke. 1988. Programs for pedigree analysis: MENDEL, FISHER, and dGENE. *Genet. Epidem.* 5: 471–472. [13]

Lange, K., J. Westlake, and A. M. Spence. 1977. Extensions of pedigree analysis. III. Variance components by the scoring method. *Ann. Hum. Genet.* 39: 485–491. [27]

Langley, C. H. 1977. Nonrandom associations between allozymes in natural populations of *Drosophila melanogaster. In* F. B. Christiansen and T. M. Fenchel (eds.), *Measuring selection in natural populations,* pp. 265–273. Springer-Verlag, Berlin. [5]

Langley, C. H., D. B. Smith, and F. M. Johnson. 1978. Analysis of linkage disequilibria between allozyme loci in natural populations of *Drosophila melanogaster. Genet. Res.* 32: 215–230. [5]

Lannan, J. E. 1980. Broodstock management of *Crassostrea gigas.* I. Genetic and environmental variation in survival in the larval rearing system. *Aquaculture* 21: 323–336. [20]

Lansing, A. I. 1947. A transmissible, cumulative and reversible factor in aging. *J. Gerontol.* 2: 228–239. [6]

Lansing, A. I. 1948. Evidence for aging as a consequence of growth cessation. *Proc. Natl. Acad. Sci. USA* 34: 304–310. [6]

LaPlace, P. S. 1778. *Memoire sur les probabilites.* Histoire de l'Academie Royale de Sciences, Annee 1778. pp. 227–332. [2]

Lark, K. G., K. Chase, F. Alder, L. M. Mansur, and J. F. Orf. 1995. Interactions between quantitative trait loci in soybean in which trait variation at one locus is conditional upon a specific allele at another. *Proc. Natl. Acad. Sci. USA* 92: 4656–4660. [15]

Larsson, K. 1993. Inheritance of body size in the barnacle goose under different environmental conditions. *J. Evol. Biol.* 6: 195–208. [7]

Latta, R., and K. Ritland. 1994. The relationship between inbreeding depression and prior inbreeding among populations of four *Mimulus* taxa. *Evolution* 48: 806–817. [10]

Latter, B. D. H. 1965. The response to artificial selection due to autosomal genes of large effect. I. Changes in gene frequency at an additive locus. *Aust. J. Biol. Sci.* 18: 585–598. [13]

Latter, B. D. H., and A. Robertson. 1960. Experimental design in the estimation of heritability by regression methods. *Biometrics* 16: 348–353. [17]

Latter, B. D. H., and A. Robertson. 1962. The effects of inbreeding and artificial selection on reproductive fitness. *Genet. Res.* 3: 110–138. [10]

Latter, B. D. H., and J. A. Sved. 1994. A reevaluation of data from competitive tests shows high levels of heterosis in *Drosophila melanogaster. Genetics* 137: 509–511. [10]

Latter, B. D. H., J. C. Mulley, D. Reid, and L. Pascoe. 1995. Reduced genetic load revealed by slow inbreeding in *Drosophila melanogaster. Genetics* 139: 287–297. [10]

Law, C. N. 1966. The location of genetic factors affecting a quantitative character in wheat. *Genetics* 53: 487–498. [14]

Law, C. N., and M. D. Gale. 1979. Cytological markers and quantitative variation in wheat. *In* J. N. Thompson, Jr., and J. M. Thoday (eds.), *Quantitative genetic variation*, pp. 275–293. Academic Press, NY. [14]

Leamy, L. 1984. Morphometric studies in inbred and hybrid house mice. V. Directional and fluctuating asymmetry. *Am. Nat.* 123: 579–593. [6]

Leamy, L. 1992. Morphometric studies in inbred and hybrid house mice. VII. Heterosis in fluctuating asymmetry at different ages. *Acta Zool. Fennica* 191: 111–119. [6]

Leamy, L., and R. S. Thorpe. 1984. Morphometric studies in inbred and hybrid house mice. Heterosis, homeostasis and heritability of size and shape. *Biol. J. Linn. Soc.* 22: 233–241. [6]

Leary, R. F., F. W. Allendorf, and R. L. Knudson. 1983. Developmental stability and enzyme heterozygosity in rainbow trout. *Nature* 301: 71–72. [6]

Leary, R. F., F. W. Allendorf, and R. L. Knudson. 1984. Superior developmental stability of heterozygotes of enzyme loci in salmonid fishes. *Am. Nat.* 124: 540–551. [6]

Leary, R. F., F. W. Allendorf, and R. L. Knudson. 1985. Inheritance of meristic variation and the evolution of developmental stability in rainbow trout. *Evolution* 39: 308–314. [6]

Leary, R. F., F. W. Allendorf, and K. L. Knudson. 1987. Differences in inbreeding coefficients do not explain the association between heterozygosity at allozyme loci and developmental stability in rainbow trout. *Evolution* 41: 1413–1415. [6]

Leary, R. F., F. W. Allendorf, and K. L. Knudson. 1992. Genetic, environmental, and developmental causes of meristic variation in rainbow trout. *Acta Zool. Fennica* 191: 79–95. [6]

Lebowitz, R. J., M. Soller, and J. S. Beckmann. 1987. Trait-based analyses for the detection of linkage between marker loci and quantitative trait loci in crosses between inbred lines. *Theor. Appl. Genet.* 73: 556–562. [14]

Leberg, P. L. 1993. Strategies for population reintroduction: effects of genetic variability on population growth and size. *Cons. Biol.* 7: 194–199. [9]

Ledig, F. T., R. P. Guries, and B. A. Bonefeld. 1983. The relation of growth to heterozygosity in pitch pine. *Evolution* 37: 1227–1238. [6,10]

Lehesjoki, A.-E., M. Koskiniemi, R. Norio, S. Tirrito, P. Sistonen, E. Lander, and A. de la Chapelle. 1993. Localization of the *EMP1* gene for progressive myoclonus epilepsy on chromosome 21: linkage disequilibrium allows high resolution mapping. *Human Mol. Genet.* 2: 1229–1234. [14]

Legates, J. E. 1972. The role of maternal effects in animal breeding. IV. Maternal effects in laboratory species. *J. Anim. Sci.* 35: 1294–1302. [23]

Leone, F. C., and L. S. Nelson. 1966. Sampling distributions of variance components. I. Empirical studies of balanced nested designs. *Technometrics* 8: 457–468. [18]

Lerner, I. M. 1954. *Genetic homeostasis.* Oliver and Boyd, London. [6,9]

Le Roy, P., and J. M. Elsen. 1992. Simple test statistics for major gene detection: a numerical comparison. *Theor. Appl. Genet.* 83: 6325–644. [13]

Le Roy, P., and J. M. Elsen. 1995. Numerical comparison between powers of maximum likelihood analysis of variance methods for QTL detection in progeny test designs: the case of monogenic inheritance. *Theor. Appl. Genet.* 90: 65–72. [16]

Le Roy, P., J. Naveau, J. M. Elsen, and P. Sellier. 1990. Evidence for a new major gene influencing meat quality in pigs. *Genet. Res.* 55: 33-40. [13]

Lessells, C. M., and P. T. Boag. 1987. Unrepeatable repeatabilities: a common mistake. *Auk* 104: 116–121. [6]

Levin, D. A. 1989. Inbreeding depression in partially self-fertilizing *Phlox*. *Evolution* 43: 1417–1423. [10]

Lewis, D. 1954. A relationship between dominance, heterosis, phenotypic stability and variability. *Heredity* 8: 333–356. [6]

Lewontin, R. C. 1957. The adaptations of populations to varying environments. *Cold Spring Harbor Symp. Quant. Biol.* 22: 395–408. [6]

Lewontin, R. C. 1964. The interaction of selection and linkage. II. Optimal model. *Genetics* 50: 757–782. [5]

Lewontin, R. C. 1974. *The genetic basis of evolutionary change*. Columbia Univ. Press, NY. [10]

Lewontin, R. C. 1988. On measures of gametic disequilibrium. *Genetics* 120: 849–852. [5]

Lewontin, R. C., and L. C. Birch. 1966. Hybridization as a source of variation for adaptation to new environments. *Evolution* 20: 315–336. [15]

Lewontin, R. C., and C. C. Cockerham. 1959. The goodness of fit test for detecting natural selection in random mating populations. *Evolution* 13: 561–564. [4]

Lewontin, R. C., J. A. Moore, W. B. Provine, and B. Wallace. 1981. *Dobzhansky's genetics of natural populations, I-XLIII*. Columbia Univ. Press, NY. [14]

Lewontin, R. C., S. Rose, and L. J. Kamin. 1984. *Not in our genes: biology, ideology, and human nature*. Pantheon, NY. [19]

Li, C. C. 1975. *Path analysis – a primer*. Boxwood, Pacific Grove, CA. [A2]

Li, W.-H., and D. Graur. 1991. *Fundamentals of molecular evolution*. Sinauer Assoc., Sunderland, MA. [12]

Li, W.-H., and M. Nei. 1972. Total number of individuals affected by a single deleterious mutation in a finite population. *Am. J. Hum. Genet.* 24: 667–679. [12]

Li, Z., S. R. M. Pinson, J. W. Stansel, and W. D. Park. 1995. Identification of quantitative trait loci (QTLs) for heading date and plant height in cultivated rice (*Oryza sativa* L.). *Theor. Appl. Genet.* 91: 374–381. [15]

Liao, T. F. 1994. *Interpreting probability models : logit, probit, and other generalized linear models*. Sage University Papers Series on Quantitative Applications in the Social Sciences. No. 07-101. Sage, Thousand Oaks, CA. [13]

Lin, C. S., M. R. Binns, and L. P. Lefkovitch. 1986. Stability analysis: where do we stand? *Crop Sci.* 26: 894–900. [22]

Lin, Y.-R., K. F. Schertz, and A. H. Paterson. 1995. Comparative analysis of QTLs affecting plant height and maturity across the Poaceae, in reference to an interspecific sorghum population. *Genetics* 141: 391–411. [14,15]

Lindsley, D. L., and K. T. Tokuyasy. 1980. Spermatogenesis. *In* M. Ashburner, and T. Wright (eds.), *The genetics and biology of Drosophila*, pp. 226–294. Academic Press, NY. [12]

Lindstrom, E. W. 1924. A genetic linkage between size and color factors in the tomato. *Science* 60: 182–183. [14]

Lindstrom, E. W. 1931. Genetic tests for linkage between row number and certain qualitative genes in maize. *Res. Bull. Iowa State Coll. Agric.* 142: 250–288. [14]

Lints, F. A. 1978. *Genetics and ageing*. S. Karger, Basel, Switzerland. [6]

Lints, F. A., and S. Baeten. 1981. Studies on the descendency of four populations of Koekelaere pines: *Pinus nigra* Arnold, subsp. *laricio* Maire, cv. *Koekelaere*. *Gerontology* 27: 20–31. [6]

Lints, F. A., and P. Parisi. 1981. The variations of heritability as a function of parental age. *Twin Research* 3: 225–230. [6]

Lisitsyn, N. 1995. Representational difference analysis: find the difference between genomes. *Trends Genet.* 11: 303–307. [14]

Lisitsyn, N., N. Lisitsyn, and M. Wigler. 1993. Cloning the difference between two complex genomes. *Science* 259: 946–951. [14]

Little, R. J. A., and D. B. Rubin. 1987. *Statistical analysis with missing data*. John Wiley & Sons, NY. [18,27]

Liu, J., J. M. Mercer, L. F. Stam, G. C. Gibson, Z.-B. Zeng, and C. C. Laurie. 1996. Genetic analysis of a morphological shape difference in the male genitalia of *Drosophila simulans* and *D. mauritiana*. *Genetics* 142: 1129–1145. [14]

Liu, S.-C., S. P. Kowalski, T.-H. Lan, K. A. Feldmann, and A. H. Paterson. 1996. Genome-wide high-resolution mapping by recurrent intermating using *Arabidopsis thaliana* as a model. *Genetics* 142: 247–258. [15]

Livesay, E. A. 1930. An experimental study of hybrid vigor or heterosis in rats. *Genetics* 15: 17–54. [6]

Livshits, G., and E. Kobyliansky. 1984. Comparative analysis of morphological traits in biochemically homozygous and heterozygous individuals from a single population. *J. Hum. Evol.* 13: 161–171. [6]

Livshits, G., and P. E. Smouse. 1993. Relationship between fluctuating asymmetry, morphological modality and heterozygosity in an elderly Israeli population. *Genetica* 89: 155–166. [6]

Lofsvold, D. 1986. Quantitative genetics of morphological differentiation in *Peromyscus*. I. Tests of the homogeneity of genetic covariance structure among species and subspecies. *Evolution* 40:559–573. [21]

Loisel, P., B. Goffinet, H. Monod, and G. M. De Oca. 1994. Detecting a major gene in an F2 population. *Biometrics* 50: 512–516. [13]

Long, A. D., S. L. Mullaney, L. A. Reid, J. D. Fry, C. H. Langley, and T. F. C. Mackay. 1995. High resolution mapping of genetic factors affecting abdominal bristle number in *Drosophila melanogaster*. *Genetics* 139: 1273–1291. [14,15]

López, M. A., and C. López-Fanjul. 1993. Spontaneous mutation for a quantitative trait in *Drosophila melanogaster*. I. Response to artificial selection. *Genet. Res.* 61: 107–116. [12]

López-Fanjul, C., and B. Jódar. 1977. The genetic properties of egg laying of virgin females of *Tribolium castaneum*. *Heredity* 39: 251–258. [10]

Lowry, D. C., and F. Shultz. 1959. Testing associations of metric traits and marker genes. *Ann. Hum. Gen.* 23: 83–90. [16]

Lucchesi, J. C. 1978. Gene dosage compensation and the evolution of sex chromosomes. *Science* 202: 711–716. [24]

Luckinbill, L. S., J. L. Graves, A. H. Reed, and S. Koetsawang. 1988. Localizing genes that defer senescence in *Drosophila melanogaster*. *Heredity* 60: 367–374. [9,14]

Luo, Z. W. 1993. The power of two experimental designs for detecting linkage between a marker locus and a locus affecting a quantitative character in a segregating population. *Genet. Sel. Evol.* 25: 249–261. [16]

Luo, Z. W., and M. J. Kearsey. 1991. Maximum likelihood estimation of linkage between a marker gene and a quantitative locus. II. Application to backcross and doubled haploid populations. *Heredity* 66: 117–124. [14]

Luo, Z. W., and M. J. Kearsey. 1992. Interval mapping of quantitative trait loci in an F_2 population. *Heredity* 69: 236–242. [15]

Luo, Z. W., and J. A. Wolliams. 1993. Estimation of genetic parameters using linkage between a marker gene and a locus underlying a quantitative character in F_2 populations. *Heredity* 70: 245–253. [15]

Luria, S. E., and M. Delbrück. 1943. Mutations from bacteria from virus sensitivity to virus resistance. *Genetics* 28: 491–511. [14]

Lush, J. L. 1937. *Animal breeding plans.* Iowa State Univ. Press, Ames. [1]

Lush, J. L., W. F. Lamoreux, and L. N. Hazel. 1948. The heritability of resistance death in the fowl. *Poultry Sci.* 27: 375–388. [25]

Lyman, R. F., F. Lawrence, S. V. Nuzhdin, and T. F. C. Mackay. 1996. Effects of single *P* element insertions on bristle number and viability in *Drosophila melanogaster. Genetics* 143: 277–292. [12]

Lynch, C. B. 1977. Inbreeding effects upon animals derived from a wild population of *Mus musculus. Evolution* 31: 526–537. [10]

Lynch, M. 1984. The limits to life history evolution in *Daphnia. Evolution* 38: 465–482. [5]

Lynch, M. 1985. Spontaneous mutations for life-history characters in an obligate parthenogen. *Evolution* 39: 804–818. [12,19]

Lynch, M. 1987. Evolution of intrafamilial interactions. *Proc. Natl. Acad. Sci. USA* 84: 8507–8511. [23]

Lynch, M. 1988a. Design and analysis of experiments on random drift and inbreeding depression. *Genetics* 120: 791–807. [10]

Lynch, M. 1988b. The rate of polygenic mutation. *Genet. Res.* 51: 137–148. [12]

Lynch, M. 1988c. Estimation of relatedness by DNA fingerprinting. *Mol. Biol. Evol.* 5: 584–599. [27]

Lynch, M. 1988d. Path analysis of ontogenetic data. *In* L. Persson and B. Ebenman (eds.), *The dynamics of size-structured populations,* pp.29–46. Springer-Verlag, Berlin. [A2]

Lynch, M. 1991. The genetic interpretation of inbreeding depression and outbreeding depression. *Evolution* 45: 622–629. [9,10]

Lynch, M. 1994. The neutral theory of phenotypic evolution. *In* L. Real (ed.), *Ecological genetics,* pp. 86–108. Princeton Univ. Press, Princeton, NJ. [12]

Lynch, M. 1996. A quantitative-genetic perspective on conservation issues. *In* J. Avise and J. Hamrick (eds.), *Conservation genetics: case histories from nature,* pp. 471–501. Chapman and Hall, NY. [10]

Lynch, M., and H.-W. Deng. 1994. Genetic slippage in response to sex. *Am. Nat.* 144: 242–261. [5,9]

Lynch, M., and R. Ennis. 1983. Resource availability, maternal effects, and longevity. *Exp. Gerontol.* 18: 147–165. [6]

Lynch, M., and W. Gabriel. 1983. Phenotypic evolution and parthenogenesis. *Am. Nat.* 122: 745–764. [5]

Lynch, M., and W. Gabriel. 1990. Mutation load and the survival of small populations. *Evolution* 44:1725–1737. [12]

Lynch, M., and W. G. Hill. 1986. Phenotypic evolution and neutral mutation. *Evolution* 40: 915–935. [12]

Lynch, M., R. Bürger, D. Butcher, and W. Gabriel. 1993. The mutational meltdown in asexual populations. *Heredity* 84: 339–344. [12]

Lynch, M., J. Conery, and R. Bürger. 1995a. Mutation accumulation and the extinction of small populations. *Am. Nat.* 146: 489–518. [10,12]

Lynch, M., J. Conery, and R. Bürger. 1995b. Mutational meltdowns in sexual populations. *Evolution* 49: 1067–1088. [10,12]

MacCluer, J. W., and C. M. Kammerer. 1984. Power of sibship variance tests to detect major genes. *In* A. Chakravarti (ed.), *Human population genetics: the Pittsburgh symposium,* pp. 125–141. Van Nostrand Reinhold, NY. [13]

MacCluer, J. W., D. K. Wagner, and R. S. Spielman. 1983. Genetic analysis workshop: segregation analysis of simulated data. *Am. J. Hum. Genet.* 35: 784–792. [13]

Mackay, T. F. C. 1981. Genetic variation in varying environments. *Genet. Res.* 37: 79–93. [7]

Mackay, T. F. C. 1985a. A quantitative genetic analysis of fitness and its components in *Drosophila melanogaster. Genet. Res.* 47: 59–70. [10]

Mackay, T. F. C. 1985b. Transposable element-induced response to artificial selection in *Drosophila melanogaster*. *Genet. Res.* 48: 77–87. [12]

Mackay, T. F. C. 1987. Transposable element-induced polygenic mutations in *Drosophila melanogaster. Genet. Res.* 49: 225–233. [12]

Mackay, T. F. C. 1988. Transposable element induced quantitative genetic variation in *Drosophila. In* B. S. Weir, E. J. Eisen, M. M. Goodman, and G. Namkoong (eds.), *Proceedings of the second international conference on quantitative genetics*, pp. 219–235. Sinauer Assoc., Sunderland, MA. [12]

Mackay, T. F. C. 1989. Mutation and the origin of quantitative variation. *In* W. G. Hill, and T. F. C. Mackay (eds.), *Evolution and animal breeding,* pp. 113–119. CAB International, Wallingford, UK. [12]

Mackay, T. F. C. 1995. The genetic basis of quantitative variation: numbers of sensory bristles of *Drosophila melanogaster* as a model system. *Trends Genet.* 11: 464–470. [14]

Mackay, T. F. C. 1996. The nature of quantitative genetic variation revisited: lessons from *Drosophila* bristles. *BioEssays* 18: 113–121. [14]

Mackay, T. F. C., and C. H. Langley. 1990. Molecular and phenotypic variation in the *achaete-scute* region of *Drosophila melano-gaster*. *Nature* 348: 64–66. [12,14]

Mackay, T. F. C., R. F. Lyman, and M. S. Jackson. 1992a. Effects of P element insertion on quantitative traits in *Drosophila melanogaster*. *Genetics* 130: 315–332. [12]

Mackay, T. F. C., R. F. Lyman, M. S. Jackson, C. Terzian, and W. G. Hill. 1992b. Polygenic mutation in *Drosophila melanogaster*: estimates from divergence among inbred strains. *Evolution* 46: 300–316. [12,26]

Mackay, T. F. C., J. D. Fry, R. F. Lyman, and S. V. Nuzhdin. 1994. Polygenic mutation in *Drosophila melanogaster:* estimates from response to selection of inbred strains. *Genetics* 136: 937–951. [12]

Mackay, T. F. C., R. F. Lyman, and W. G. Hill. 1995. Polygenic mutation in *Drosophila melanogaster:* nonlinear divergence among unselected strains. *Genetics* 139: 849–859. [12,26]

Mackinnon, M. J., and M. A. Georges. 1992. The effects of selection on linkage analysis of quantitative traits. *Genetics* 132: 1177–1185. [16]

Mackinnon, M. J., and J. I. Weller. 1995. Methodology and accuracy of estimation of quantitative trait loci parameters in a half-sib design using maximum likelihood. *Genetics* 141: 755–770. [16]

MacLean, C. J., N. E. Morton, and R. Lew. 1975. Analysis of family resemblance. IV. Operational characteristics of segregation analysis. *Am. J. Hum. Genet.* 27: 365–384. [13]

MacLean, C. J., N. E. Morton, R. C. Elston, and S. Yee. 1976. Skewness in commingled distributions. *Biometrics* 32: 695–699. [13]

Macnair, M. R., and Q. J. Cumbes. 1989. The genetic architecture of interspecific variation in *Mimulus*. *Genetics* 122: 211–222. [9]

MacNeil, M. D., D. D. Dearborn, L. V. Cundiff, C. A. Dinkel, and K. E. Gregory. 1989. Effects of inbreeding and heterosis in Hereford females on fertility, calf survival and preweaning growth. *J. Anim. Sci.* 67: 895–901. [10]

Magnus, P. 1984. Causes of variation in birth weight: a study of offspring of twins. *Clinical Genet.* 25: 15–24. [7]

Malécot, G. 1948 *Les mathématiques de l'hérédité.* Masson, Paris. [7]

Malina, R. M., and P. H. Buschaung. 1984. Anthropometric asymmetry in normal and mentally retarded males. *Ann. Hum. Biol.* 11: 515–531. [6]

Malogolowkin-Cohen, C., H. Levene, N. P. Dobzhansky, and A. S. Simmons. 1964. Inbreeding and the mutational and balanced loads in natural populations of *Drosophila willistoni*. *Genetics* 50: 1299–1311. [10]

Mange, A. P. 1964. Growth and inbreeding of a human isolate. *Human Biol.* 36: 104–133. [10]

Mangin, B., B. Goffient, and A. Rebaï. 1994a. Constructing confidence intervals for QTL location. *Genetics* 138: 1301–1308. [15]

Mangin, B., B. Goffient, and A. Rebaï. 1994b. Constructing confidence intervals for QTL location. *In* J. W. van Ooijen and J. Jansen (eds.), *Biometrics in plant breeding: applications of molecular markers*, pp. 147–152. CPRO-DLO, Netherlands. [15]

Maniatis, T., E. F. Frisch, and J. Sambrook. 1982. *Molecular cloning*. Cold Spring Harbor Press, Cold Spring Harbor, NY. [14]

Manly, B. F. J. 1991. *Randomization and Monte Carlo methods in biology*. Chapman and Hall, London. [18]

Mansur, L. M., J. Orf, and K. G. Lark. 1993. Determining the linkage of quantitative trait loci to RFLP markers using extreme phenotypes of recombinant inbred lines of soybeans (*Glycine max* L. Merr.). *Theor. Appl. Genet.* 86: 914–918. [14]

Mansur, L. M., K. G. Lark, H. Kross, and H. Olliveira. 1993. Interval mapping of quantitative trait loci for reproductive, morphological, and seed traits of soybean (*Glycine max* L.). *Theor. Appl. Genet.* 86: 907–913. [15]

Marinkovic, D. 1967. Genetic loads affecting fertility in natural populations of *Drosophila pseudoobscura*. *Genetics* 57: 701–709. [10]

Marsden, J. E., S. J. Schwager, and B. May. 1987. Single-locus inheritance in the tetraploid treefrog *Hyla versicolor* with an analysis of expected progeny ratios in tetraploid organisms. *Genetics* 116: 299–311. [4]

Martin, G. B., J. G. K. Williams, and S. D. Tanksley. 1991. Rapid identification of markers linked to a *Pseudomonas* resistance gene in tomato by using random primers and near-isogenic lines. *Proc. Natl. Acad. Sci. USA* 88: 2336–2340. [14]

Martin, G. B., S. H. Brommonschenkel, J. Chunwongse, A. Frary, M. W. Ganal, R. Spivey, T. Wu, E. D. Earle, and S. D. Tanksley. 1993. Map-based cloning of a protein kinase gene conferring disease resistance in tomato. *Science* 262: 1432–1436. [14]

Martin, N. G., L. J. Eaves, M. J. Kearsey, and P. Davies. 1978. The power of the classical twin study. *Heredity* 40: 97–116. [19]

Martínez, O., and R. N. Curnow. 1992. Estimating the locations and the sizes of the effects of quantitative trait loci using flanking markers. *Theor. Appl. Genet.* 85: 480–488. [15]

Martínez, O., and R. N. Curnow. 1994a. Missing markers when estimating quantitative trait loci using regression mapping. *Heredity* 73: 198–206. [15]

Martínez, O., and R. N. Curnow. 1994b. Three marker scanning of chromosomes for QTL in neighboring intervals. *In* J. W. van Ooijen and J. Jansen (eds.), *Biometrics in plant breeding: applications of molecular markers*, pp. 153–162. CPRO-DLO, Netherlands. [15]

Martins, E. 1991. Individual and sex differences in the use of the push-up display by the sagebrush lizard, *Sceloporus graciosus*. *Anim. Behav.* 41: 403–416. [6]

Mather, K. 1941. Variation and selection of polygenic characters. *J. Genetics* 41: 159–193. [12]

Mather, K. 1942. The balance of polygenic characters. *J. Genetics* 43: 309–336. [5]

Mather, K. 1943. Polygenic inheritance and natural selection. *Biol. Rev.* 18: 32–64. [5]

Mather, K. 1944. The genetic activity of heterochromatin. *Proc. Royal Soc. Lond.* B 132: 308–332. [12]

Mather, K. 1953. Genetical control of stability in development. *Heredity* 7: 297–336. [6]

Mather, K., and J. L. Jinks. 1982. *Biometrical genetics.* 3rd Ed. Chapman and Hall, NY. [1,5,9,20]

Matsuda, E. 1973. Genetic studies on total finger ridge count among Japanese. *Jap. J. Hum. Genet.* 17: 293–318. [19]

Maynard Smith, J. 1978. Optimization theory in evolution. *Ann. Rev. Ecol. Syst.* 9: 31–56. [1]

Maynard Smith, J. 1982. *Evolution and the theory of games.* Cambridge Univ. Press, Cambridge, UK. [1]

Mayo, O. 1980. *The theory of plant breeding.* Clarendon Press, Oxford, UK. [1]

Mayo, O. 1989. Identification of genes which influence quantitative traits. *In* W. G. Hill and T. F. C. Mackay (eds.), *Evolution and animal breeding*, pp. 141–146. CAB International, Wallingford, UK. [13]

Mayo, O., T. W. Hancock, and P. A. Baghurst. 1980. Influence of major genes on variance within sibships for a quantitative trait. *Ann. Hum. Genet.* 43: 419–421. [13]

Mayr, E. 1983. How to carry out the adaptationist program? *Am. Nat.* 121: 324–334. [1]

McAndrew, B. J., R. D. Ward, and J. A. Beardmore. 1982. Lack of relationship between morphological variance and enzyme heterozygosity in the plaice, *Pleuranectus platessa. Heredity* 48: 117–125. [6]

McBride, G., and A. Robertson. 1963. Selection using assortative mating in *D. melanogaster*. *Genet. Res.* 4: 356–369. [7]

McCarthy, J. C. 1967. The effects of inbreeding on the components of litter size in mice. *Genet. Res.* 10: 73–80. [10]

McDonald, J. F., and F. J. Ayala. 1978. Genetic and biochemical basis of enzyme activity variation in natural populations. I. Alcohol dehydrogenase in *Drosophila melanogaster*. *Genetics* 89: 371–388. [14]

McGraw, J. B. 1987. Experimental ecology of *Dryas octapetala* ecotypes. IV. Fitness response to reciprocal transplanting in ecotypes with differing plasticity. *Oecologia* 73: 465–468. [9]

McGue, M., I. I. Gottesman, and D. C. Rao. 1983. The transmission of schizophrenia under a multifactorial threshold model. *Am. J. Hum. Genet.* 35: 1161–1178. [25]

McGuffin, P., and P. Huckle. 1990. Simulation of mendelism revisited: the recessive gene for attending medical school. *Am. J. Hum. Genet.* 46: 994- 999. [13]

McKenzie, J. A., and G. M. Clarke. 1988. Diazinon resistance, fluctuating asymmetry and fitness in the Australian sheep blowfly, *Lucilia cuprina. Genetics* 120: 213–220. [6]

McKenzie, J. A., and J. L. Yen. 1995. Genotype, environment and the asymmetry phenotype. Dieldrin-resistance in *Lucilia cuprina* (the Australian sheep blowfly). *Heredity* 75: 181–187. [6]

McLachlan, G. J., and K. E. Basford. 1988. *Mixture models*. Marcel Dekker, NY. [13]

McMillan, I., and A. Robertson. 1974. The power of methods for the detection of major genes affecting quantitative characters. *Heredity* 32: 349–356. [14,15]

Meagher, T. R. 1992. The quantitative genetics of sexual dimorphism in *Silene latifolia* (Caryophyllaceae). *Evolution* 46: 445–457. [24]

Mendell, N. R., and R. C. Elston. 1974. Multifactorial qualitative traits: genetic analysis and prediction of recurrence risks. *Biometrics* 30: 41–57. [25]

Mérat, P. 1968. Distributions de frequencies, interpretation du determinisme genetique des characters quantitatifs et recherche de 'genes majeours'. *Biometrics* 24: 277–293. [13]

Meredith, W. R., Jr., R. R. Bridge, and J. F. Chism. 1970. Relative performance of F_1 and F_2 hybrids from doubled haploids and their parent varieties in upland cotton, *Gossypium hirsutum* L. *Crop Sci.* 10: 295–298. [6]

Merilä, J., and L. Gustafsson. 1993. Inheritance of size and shape in a natural population of collared flycatchers, *Ficedula albicollis. J. Evol. Biol.* 6: 375–395. [17]

Merrick, M. J. 1975. The inheritance of penicillin titre in crosses between lines of *Aspergillus nidulans* selected for increased productivity. *J. Gen. Microbiol.* 91: 287–294. [5]

Meyer, H. H., and F. D. Enfield. 1975. Experimental evidence on limitations of the heritability parameter. *Theor. Appl. Genet.* 45: 268–273. [17]

Meyer, K. 1983. Maximum likelihood procedures for estimating genetic parameters for later lactations of dairy cattle. *J. Dairy Sci.* 66: 1988–1997. [27]

Meyer, K. 1985. Maximum likelihood estimation of variance components for a multivariate mixed model with equal design matrices. *Biometrics* 41: 153–165. [26,27]

Meyer, K. 1989a. Approximate accuracy of genetic evaluation under an animal model. *Livestock Prod. Sci.* 21: 87–100. [26]

Meyer, K. 1989b. Estimation of genetic parameters. *In* W. G. Hill and T. F. C. Mackay (eds.), *Evolution and animal breeding*, pp. 161–167. CAB International, Wallingford, UK. [27]

Meyer, K. 1991. Estimating variances and covariances for multivariate animal models by restricted maximum likelihood. *Genet. Sel. Evol.* 24: 67–83. [27]

Meyer, K., and W. G. Hill. 1992. Approximation of sampling variances and confidence intervals for maximum likelihood estimates of variance components. *J. Anim. Breed. Genet.* 109: 264–280. [13]

Meyer, K., and R. Thompson. 1984. Bias in variance and covariance component estimators due to selection on a correlated trait. *Z. Tierzüchtg. Züchtgsbiol.* 101:33–50. [21]

Mi, M. P., and M. N. Rashad. 1975. Genetic parameters of dermal patterns and ridge counts. *Hum. Hered.* 25: 249–257. [19]

Mi, M. P., M. Earle, and J. Kagawa. 1986. Phenotypic resemblance in birth weight between first cousins. *Ann. Hum. Genet.* 50: 49–62. [7]

Michelmore, R. W., L. Paran, and R. V. Kesseli. 1991. Identification of markers linked to disease-resistance genes by bulked segregant analysis: a rapid method to detect markers in specific genomic regions by using segregating populations. *Proc. Natl. Acad. Sci., USA* 88: 9828–9832. [14]

Michod, R. E., and W. D. Hamilton. 1980. Coefficients of relatedness in sociobiology. *Nature* 288: 694–697. [7]

Migeon, B. R. 1994. X-chromosome inactivation: molecular mechanisms and genetic consequences. *Trends Genet.* 10: 230–235. [24]

Milkman, R. D. 1970. The genetic basis of natural variation. X: Recurrence of *cve* polygenes. *Genetics* 65: 289–303. [12]

Milkman, R. D. 1978. Selection differentials and selection coefficients. *Genetics* 88: 391–403. [12]

Miller, P. S. 1994. Is inbreeding depression more severe in a stressful environment? *Zoo Biol.* 13: 195–208. [10]

Miller, R. G. 1968. Jackknifing variances. *Ann. Math. Stat.* 39: 567–582. [18]

Miller, R. G. 1974. The jackknife — a review. *Biometrika* 61: 1–17. [18]

Milliken, G. A., and D. E. Johnson. 1984. *Analysis of messy data. Vol. 1. Designed experiments.* Van Nostrand Reinhold, NY. [18,20]

Misztal, I., and D. Gianola. 1987. Indirect solutions of mixed model equations. *J. Dairy Sci.* 70: 716–724. [26]

Misztal, I., D. Gianola, and L. R. Schaeffer. 1987. Extrapolation and convergence criteria with Jacobi and Gauss-Seidel iteration in animal models. *J. Dairy Sci.* 70: 2577–2584. [26]

Mitchell-Olds, T. 1986. Quantitative genetics of survival and growth in *Impatiens capensis. Evolution* 40: 107–116. [18,21]

Mitchell-Olds, T., and J. Bergelson. 1990. Statistical genetics of an annual plant, *Impatiens capensis.* I. Genetic basis of quantitative variation. *Genetics* 124: 407–415. [13,18]

Mitchell-Olds, T., and J. J. Rutledge. 1986. Quantitative genetics in natural plant populations: a review of the theory. *Am. Nat.* 127: 379–402. [5,17]

Mitton, J. B. 1978. Relationship between heterozygosity for enzyme loci and variation of morphological characters in natural populations. *Nature* 273: 661–662. [6]

Mitton, J. B., and M. C. Grant. 1980. Observations on the ecology and evolution of quaking aspen, *Populus tremuloides,* in the Colorado front range. *Am. J. Bot.* 67: 202–209. [6]

Mitton, J. B., and M. C. Grant. 1984. Associations among protein heterozygosity, growth rate, and developmental homeostasis. *Ann. Rev. Ecol. Syst.* 15: 479–499. [10]

Mitton, J. B., P. Knowles, K. B. Sturgeon, Y. B. Linhart, and M. Davis. 1981. Associations between heterozygosity and growth rate variables in three western forest trees. *USDA Gen. Tech. Rep. PSW* 48:27–34. [10]

Miyashita, N., and C. C. Laurie-Ahlberg. 1984. Genetic analysis of chromosomal interaction effects of the activities of the glucose 6-phosphate and 6- phosphogluconate dehydrogenases in *Drosophila melanogaster. Genetics* 106: 655–668. [14]

Mode, C. G., and H. F. Robinson. 1959. Pleiotropism and the genetic variance and covariance. *Biometrics* 15: 518–537. [21]

Mode, C. J., and D. L. Gasser. 1972. A distribution free test for major gene differences in quantitative inheritance. *Math. Biosci.* 14: 143–150. [13]

Modi, W. S., R. K. Wayne, and S. J. O'Brien. 1987. Analysis of fluctuating asymmetry in cheetahs. *Evolution* 41: 227–228. [6]

Mohamed, A. H. 1959. Inheritance of quantitative characters in *Zea mays.* I. Estimation of the number of genes controlling the time of maturity. *Genetics* 44: 713–724. [9]

Mohamed, A. H., and A. S. Hanna. 1964. Inheritance of quantitative characters in rice. I. Estimation of the number of effective factor pairs controlling plant height. *Genetics* 49: 81–93. [9]

Moll, P. P., T. D. Berry, W. H. Weidman, R. Ellefson, H. Gordon, and B. A. Kottke. 1984. Detection of genetic heterogeneity among pedigrees through complex segregation analysis: an application to hypercholestererolemia. *Am. J. Hum. Genet.* 36: 197–211. [13]

Moll, R. H., J. H. Longquist, J. V. Fortuno, and E. C. Johnson. 1965. The relationship of heterosis and genetic divergence in maize. *Genetics* 52: 139–144. [9,20]

Møller, A. P. 1992. Parasites differentially increase the degree of fluctuating asymmetry in secondary sexual characters. *J. Evol. Biol.* 5: 691–699. [6]

Møller, A. P. 1993. Sexual selection in the barn swallow *Hirundo rustica.* III. Female tail ornaments. *Evolution* 47: 417–431. [24]

Møller, A. P., and M. Eriksson. 1994. Patterns of fluctuating asymmetry in flowers: implications for sexual selection in plants. *J. Evol. Biol.* 7: 97–113. [6]

Monteiro, L., and D. Falconer. 1966. Compensatory growth and sexual maturity in mice. *Anim. Prod.* 8: 179–192. [23]

Moran, N., and P. Baumann. 1994. Phylogenetics of cytoplasmically inherited microorganisms of arthropods. *Trends Ecol. Evol.* 9: 15–20. [6]

Moran, P. A. P., and C. A. B. Smith. 1966. Commentary on R. A. Fisher's paper on "The correlation between relatives on the supposition of Mendelian inheritance." *Eugen. Lab. Mem.* 41, Cambridge Univ. Press, Cambridge, UK. [4,7]

Moreno, G. 1994. Genetic architecture, genetic behavior, and character evolution. *Ann. Rev. Ecol. Syst.* 25: 31–44. [5]

Morley-Jones, R. 1965. Analysis of variance of the half-diallel table. *Heredity* 20: 117–121. [20]

Morrison, D. F. 1976. *Multivariate statistical methods.* McGraw-Hill, NY. [8, A3]

Morton, N. E. 1955a. The inheritance of human birth weight. *Ann. Hum. Genet.* 20: 125–134. [7]

Morton, N. E. 1955b. Sequential tests for the detection of linkage. *Am. J. Hum. Genet.* 7: 277–318. [15,16]

Morton, N. E. 1958. Empirical risks in consanguineous marriages: birth weight, gestation time, and measurements of infants. *Am. J. Hum. Genet.* 10: 344–349. [10]

Morton, N. E. 1959. Genetic tests under incomplete ascertainment. *Am. J. Hum. Genet.* 11: 1–16 [13]

Morton, N. E. 1978. Effect of inbreeding on IQ and mental retardation. *Proc. Natl. Acad. Sci. USA* 75:3906–3908. [10]

Morton, N. E. 1984. Trials of segregation analysis by deterministic and macro simulation. *In* A. Chakravarti (ed.), *Human population genetics: the Pittsburgh symposium*, pp. 83–107. Van Nostrand Reinhold, NY. [13]

Morton, N. E., and C. J. MacLean. 1974. Analysis of family resemblance. III. Complex segregation of quantitative traits. *Am. J. Hum. Genet.* 26: 489–503. [13]

Morton, N. E., J. F. Crow, and H. J. Muller. 1956. An estimate of the mutational damage in man from data on consanguineous matings. *Proc. Natl. Acad. Sci. USA* 42: 855–863. [10,12]

Morton, N. E., D. C. Rao, and J. M. Lalouel. 1983. *Methods in genetic epidemiology*. S. Karger, Basel, Switzerland. [13]

Mousseau, T. A., and D. A. Roff. 1987. Natural selection and the heritability of fitness components. *Heredity* 59: 181–197. [7,10,21]

Mrode, R. A. 1996 *Linear models for the prediction of animal breeding values.* CAB International, Wallingford, UK. [26]

Muehlbauer, G. J., J. E. Specht, M. A. Thomas-Compton, P. E. Staswick, and R. L. Bernard. 1988. Near-isogenic lines — a potential resource in the integration of conventional and molecular marker linkage maps. *Crop Sci.* 28: 729–735. [14]

Muir, W. M. 1986a. Estimation of response to selection and utilization of control populations for additional information and accuracy. *Biometrics* 42: 381–391. [10]

Muir, W. M. 1986b. Efficient design and analysis of selection experiments. *In* G. E. Dickerson and R. K. Johnson (eds.), *Proceedings of the 3rd world congress on genetics applied to livestock production.* Agric. Comms., Univ. Nebraska, Lincoln, Nebraska. [10]

Muir, W., W. E. Nyquist, and S. Xu. 1992. Alternative partitioning of the genotype-by-environment interaction. *Theor. Appl. Genet.* 84: 193–200. [22]

Mukai, T. 1964. The genetic structure of natural populations of *Drosophila melanogaster*. I. Spontaneous mutation rate of polygenes controlling viability. *Genetics* 50: 1–19. [12]

Mukai, T. 1969. The genetic structure of natural populations of *Drosophila melanogaster.* VII. Synergistic interaction of spontaneous mutant polygenes controlling viability. *Genetics* 61:749–761.[12]

Mukai, T. 1979. Polygenic mutation. *In* J. N. Thompson, Jr., and J. M. Thoday (eds.), *Quantitative genetic variation*, pp. 177–196. Academic Press, NY. [12]

Mukai, T., and C. C. Cockerham. 1977. Spontaneous mutation rates at enzyme loci in *Drosophila melanogaster*. *Proc. Natl. Acad. Sci. USA* 74: 2514–2517. [12]

Mukai, T., and O. Yamaguchi. 1974. The genetic structure of natural populations of *Drosophila.* XI. Genetic variability in a local population. *Genetics* 82: 63–82. [10,12]

Mukai, T., and T. Yamazaki. 1968. The genetic structure of natural populations of *Drosophila melanogaster.* V. Coupling-repulsion effect of spontaneous mutant polygenes controlling viability. *Genetics* 59: 513–535. [12]

Mukai, T., and T. Yamazaki. 1971. The genetic structure of natural populations of *Drosophila melanogaster.* X. Developmental time and viability. *Genetics* 69: 385–398. [12]

Mukai, T., S. I. Chigusa, L. E. Mettler, and J. F. Crow. 1972. Mutation rate and dominance of genes affecting viability in *Drosophila melanogaster.* *Genetics* 72: 335–355. [10,12]

Mukai, T., S. I. Chigusa, and I. Yoshikawa. 1965. The genetic structure of natural populations of *Drosophila melanogaster.* III. Dominance effect of spontaneous mutant polygenes controlling viability in heterozygous genetic backgrounds. *Genetics* 52: 493–501. [12]

Mukai, T., R. Cardellino, T. K. Watanabe, and J. F. Crow. 1974. The genetic variance for viability and its components in a local population of *Drosophila melanogaster.* *Genetics* 78: 1195–1208. [10]

Mulitze, D. K., and R. J. Baker. 1985a. Evaluation of biometrical methods for estimating the number of genes. 1. Effect of sample size. *Theor. Appl. Genet.* 69: 553–558. [9]

Mulitze, D. K., and R. J. Baker. 1985b. Evaluation of biometrical methods for estimating the number of genes. 2. Effect of type I and type II statistical errors. *Theor. Appl. Genet.* 69: 559–566. [9]

Muller, H. J. 1932. Further studies on the nature and causes of gene mutations. *Proc. 6th Internat. Cong. Genet.* 1: 213–255. [24]

Muller, H. J. 1935. On the incomplete dominance of the normal allelomorphs of white in *Drosophila. J. Genet.* 30: 407–414. [12]

Muller, H. J. 1939. Reversibility in evolution considered from the standpoint of genetics. *Biol. Rev.* 14: 261–280. [14]

Muranty, H. 1996. Power of tests for quantitative trait loci detection using full-sib families in different schemes. *Heredity* 76: 156–165. [16]

Murray, J., and B. Clarke. 1968. Inheritance of shell size in *Partula. Heredity* 23: 189–198. [17]

Nadeau, J. H. 1989. Maps of linkage and synteny homologies between mouse and man. *Trends Genet.* 5: 82–86. [14]

Nagai, J., A. J. Lee, and C. G. Hickman. 1971. Preweaning growth of inbred, F_1 hybrid, and random-bred mice as a measure of mother's lactation. *Can. J. Genet. Cytol.* 13: 20–28. [10]

Nagylaki, T. 1978. The correlation between relatives with assortative mating. *Ann. Hum. Genet.* 42: 131–137. [7]

Nagylaki, T. 1982. Assortative mating for a quantitative character. *J. Math. Biol.* 16: 57–74. [7]

Nakamura, R. R., and M. L. Stanton. 1989. Embryo growth and seed size in *Raphanus sativus:* maternal and paternal effects in vivo and in vitro. *Evolution* 43: 1435–1443. [23]

Namkoong, G., and J. H. Roberds. 1974. Choosing mating designs to efficiently estimate genetic variance components for trees. *Silvae Genet.* 23: 43–53. [20]

Nance, W. E. 1976. Note on the analysis of twin data. *Am. J. Hum. Genet.* 28: 297–299. [19]

Nance, W. E. 1979. The role of twin studies in human quantitative genetics. *Prog. Med. Genet.* 3: 73–107. [19]

Nance, W. E., and L. A. Corey. 1976. Genetic models for the analysis of data from the families of identical twins. *Genetics* 83: 811–826. [19]

Nance, W. E., A. A. Kramer, L. A. Corey, P. M. Winter, and L. J. Eaves. 1983. A causal analysis of birth weight in the offspring of monozygotic twins. *Am. J. Hum. Genet.* 35: 1211–1223. [7]

Nason, J. D., and N. C. Ellstrand. 1995. Lifetime estimates of biparental inbreeding depression in the self-incompatible annual plant *Raphanus sativus. Evolution* 49: 307–316. [10]

Nassar, R., and M. Hühn. 1987. Studies on estimation of phenotypic stability: test of significance for nonparametric measures of phenotypic stability. *Biometrics* 43: 45–53. [22]

Naveira, H., and A. Barbadilla. 1992. The theoretical distribution of lengths of intact chromosome segments around a locus held heterozygous with backcrossing in a diploid species. *Genetics* 130: 205–209. [14]

Neel, J. V., W. J. Schull, M. Yamamoto, S. Uchida, T. Yanase, and N. Fujiki. 1970. The effects of parental consanguinity and inbreeding in Hirado, Japan. II. Physical development, tapping rate, blood pressure, intelligence quotient, and school performance. *Am. J. Hum. Genet.* 22: 263–286. [10]

Nei, M. 1967. Modification of linkage intensity by natural selection. *Genetics* 57: 625–641. [5]

Neimann-Sørensen, A., and A. Robertson. 1961. The association between blood groups and several production characteristics in three Danish cattle breeds. *Acta Agric. Scand.* 11: 163–196. [16]

Nelson, S. F., J. H. McCusker, M. A. Sander, Y. Kee, P. Modrich, and P. O. Brown. 1993. Genomic mismatch scanning: a new approach to genetic linkage mapping. *Nature Genetics* 4: 11–18. [14]

Nilsson-Ehle, H. 1909. Kreuzungsuntersuchungen an Hafer und Weizen. *Lunds Univ. Årsskrift,* n. s., series 2, vol. 5, no. 2: 1–122. [1] 666

Nishida, A. 1972. Some characteristics of parent-offspring regression in body weight of *Mus musculus* at different ages. *Can. J. Genet. Cytol.* 14: 292–303. [17]

Nishida, A., and T. Abe. 1974. The distribution of genetic and environmental effects and the linearity of heritability. *Can. J. Genet. Cytol.* 16: 3–10. [17]

Nitzsche, W., and G. Wenzel. 1977. *Haploids in plant breeding.* Paul Parey, Hamburg. [9]

Nodari, R. O., S. M. Tsai, P. Guzmán, R. L. Gilbertson, and P. Gepts. 1993. Toward an integrated linkage map of common bean. III. Mapping genetic factors controlling host-bacteria interactions. *Genetics* 134: 341–350. [15]

Norman, J. K., A. K. Sakai, S. G. Weller, and T. E. Dawson. 1995. Inbreeding depression in morphological and physiological traits of *Schiedea lydgatei* (Caryophyllaceae) in two environments. *Evolution* 49: 297–306. [10]

Nuzhdin, S. V., J. D. Fry, and T. F. C. Mackay. 1995. Polygenic mutation in *Drosophila melanogaster:* the causal relationship of bristle number to fitness. *Genetics* 139: 861–872. [12]

Nuzhdin, S. V., P. D. Keightley, and E. G. Pasyukova. 1993. The use of retrotransposons as markers for mapping genes responsible for fitness differences between related *Drosophila melanogaster* strains. *Genet. Res.* 62: 125–131. [14]

Nuzhdin, S. V., and T. F. C. Mackay. 1995. The genomic rate of transposable element movement in *Drosophila melanogaster. Mol. Biol. Evol.* 12: 180–181. [12]

O'Brien, S. J. (ed.) 1990. *Genetic maps,* 5th Ed. Cold Spring Harbor Press, Cold Spring Harbor, NY. [9]

O'Donald, P. 1971. The distribution of genotypes produced by alleles segregating at a number of loci. *Heredity* 26: 233–241. [13]

O'Donald, P., and M. E. N. Majerus. 1985. Sexual selection and the evolution of preferential mating in ladybirds. I. Selection for high and low lines of female preference. *Heredity* 55: 401–412. [25]

Ohnishi, O. 1977. Spontaneous and ethyl methanesulfonate induced mutations controlling viability in *Drosophila melanogaster.* II. Homozygous effect of polygenic mutations. *Genetics* 87: 529–545. [12]

Ohta, T. 1995. Synonymous and nonsynonymous substitutions in mammalian genes and the nearly neutral theory. *J. Mol. Evol.* 40: 56–63. [12]

Oka, H. I., J. Hayashi, and I. Shiojiri. 1958. Induced mutations of polygenes for quantitative characters in rice. *J. Heredity* 49: 11–14. [12]

Ollivier, L., and L. L. G. Janns. 1993. A note on the estimation of the effective number of additive and dominant loci contributing to quantitative variation. *Genetics* 135: 907–909. [9]

Olson, J. M. 1994. Some empirical properties of an all-relative-pairs linkage test. *Genet. Epidem.* 10: 87–102. [16]

Olson, J. M. 1995. Multipoint linkage analysis using sib pairs: an interval mapping approach for dichotomous outcomes *Am. J. Hum. Genet.* 56: 788–798. [16]

Olson, J. M., and E. M. Wijsman. 1993. Linkage between quantitative trait and marker loci: methods using all relative pairs. *Genet. Epidem.* 10: 87–102. [16]

Orkin, S. H. 1986. Reverse genetics and human disease. *Cell* 47: 845–850. [14]

Orlove, M. J., and C. L. Wood. 1978. Coefficients of relationship and coefficients of relatedness in kin selection: a covariance form for the RHO formula. *J. Theor. Biol.* 73: 679–686. [7]

Orozco, F. 1976. Heterosis and genotype-environment interaction: theoretical and experimental aspects. *Bull. Tech. , Dept. Genet. Anim., Inst. Natl. Recherche Agron.* 24: 43–52. [10]

Orr, H. A. 1987. Genetics of male and female sterility in hybrids of *Drosophila pseudoobscura* and *D. persimilis*. *Genetics* 116: 555–563. [14]

Orr, H. A. 1991. A test of Fisher's theory of dominance. *Proc. Natl. Acad. Sci. USA* 88: 11413–11415. [4]

Orr, H. A. 1993a. A mathematical rule of Haldane's rule. *Evolution* 47: 1606–1611. [14]

Orr, H. A. 1993b. Haldane's rule has multiple genetic causes. *Nature* 361: 532–533. [14]

Orr, H. A. 1995. The population genetics of speciation: the evolution of hybrid incompatibilities. *Genetics* 139: 1805–1813. [14]

Orr, H. A. 1996. Dobzhansky, Bateson, and the genetics of speciation. *Genetics* 144: 133–135. [14]

Orr, H. A., and J. A. Coyne. 1989. The genetics of postzygotic isolation in the *Drosophila virilis* group. *Genetics* 121: 527–537. [14]

Orr, H. A., and J. A. Coyne. 1992. The genetics of adaptation: a reassessment. *Am. Nat.* 140: 725–742. [9]

Orr, H. A., and M. Turelli. 1996. Dominance and Haldane's rule. *Genetics* 143: 613–616. [14]

Osborn, T. C., D. C. Alexander, and J. F. Fobes. 1987. Identification of restriction fragment length polymorphisms linked to genes controlling soluble solids content in tomato fruit. *Theor. Appl. Genet.* 73: 350–356. [14]

Osborne, R., and W. S. B. Paterson. 1952. On the sampling variance of heritability estimates derived from variance analyses. *Proc. Royal Soc. Edinburgh* 64: 456–461. [18]

Ott, J. 1979. Maximum likelihood estimation by counting methods under polygenic and mixed models in human pedigrees. *Am. J. Hum Genet.* 31: 161–175. [13,27]

Ott, J. 1991. *Analysis of human genetic linkage*. Revised Edition. Johns Hopkins, Baltimore, MD. [14,16]

Ouborg, N. J., and R. Van Treuren. 1994. The significance of genetic erosion in the process of extinction. IV. Inbreeding load and heterosis in relation to population size in the mint *Salvia pratensis*. *Evolution* 48: 996–1008. [10]

Pacek, P., A. Sajantila, and A.-C. Syvänen. 1993. Determination of allele frequencies at loci with length polymorphism by quantitative analysis of DNA amplified from pooled samples. *PCR Methods Appl.* 2: 313–317. [14]

Packard, G. C., and T. J. Boardman. 1987. The misuse of ratios to scale physiological data that vary allometrically with body size. *In* M. E. Feder, A. F. Bennett, W. W. Burggren, and R. B. Huey (eds.), *New directions in ecological physiology*, pp. 216–239. Cambridge Univ. Press, Cambridge, UK. [11]

Palmer, A. R., and C. Strobeck. 1986. Fluctuating asymmetry: measurement, analysis, patterns. *Ann. Rev. Ecol. Syst.* 17: 391–421. [6]

Palmer, A. R., and C. Strobeck. 1992. Fluctuating asymmetry as a measure of developmental stability: implications of non-normal distributions and power of statistical tests. *Acta Zool. Fennica* 191: 57–72. [6]

Palmer, A. R., C. Strobeck, and A. K. Chippindale. 1993. Bilateral variation and the evolutionary origin of macroscopic asymmetries. *Genetica* 89: 201–218. [6]

Palopoli, M. F., and C.-I. Wu. 1994. Genetics of hybrid male sterility between *Drosophila* sibling species: a complex web of epistasis is revealed in interspecific studies. *Genetics* 138: 329–341. [14]

Pamilo, P., M. Nei, and W.-H. Li. 1987. Accumulation of mutations in sexual and asexual populations. *Genet. Res.* 49: 135–146. [12]

Pani, S. N., and J. F. Lasley. 1972. Genotype × environment interactions in animals. *Res. Bull., Agri. Exp. Sta., Univ. Missouri*, No. 992. [22]

Papa, K. E. 1970. Inheritance of growth rate in *Neurospora crassa:* crosses between previously selected lines. *Can. J. Genet. Cytol.* 12: 1–9. [5]

Paran, I., R. Kesseli, and R. Michelmore. 1991. Identification of restriction fragment length polymorphism and random amplified polymorphic DNA markers linked to downy mildew resistance genes in lettuce, using near-isogenic lines. *Genomes* 34: 1021–1027. [14]

Park, Y. S., and D. P. Fowler. 1982. Effects of inbreeding and genetic variances in a natural population of tamarack (*Larix laricina* (Du Roi) K. Koch) in eastern Canada. *Silvae Genetica* 31: 21–26. [10]

Park, Y. S., and D. P. Fowler. 1984. Inbreeding in black spruce (*Picea mariana* (Mill.) B. S. P.): self-fertility, genetic load and performance. *Can. J. For. Res.* 14: 17–21. [10]

Parker, M. A. 1992. Outbreeding depression in a selfing annual. *Evolution* 46: 837–841. [9]

Parkhurst, S. M., and P. M. Meneely. 1994. Sex determination and dosage compensation: lessons from flies and worms. *Science* 264: 924–932. [24]

Partridge, L., and P. Harvey. 1985. Costs of reproduction. *Nature* 316: 20–21. [21]

Partridge, L., T. F. C. Mackay, and S. Aitken. 1985. Male mating success and fertility in *Drosophila melanogaster. Genet. Res.* 46: 279–285. [10]

Pascoe, L., and N. E. Morton. 1987. The use of map functions in multipoint mapping. *Am. J. Hum. Genet.* 40: 174–183. [14]

Paterson, A. H., E. S. Lander, J. D. Hewitt, S. Peterson, S. E. Lincoln, and S. D. Tanksley. 1988. Resolution of quantitative traits into Mendelian factors by using a complete RFLP linkage map. *Nature* 335: 721–726. [14,15]

Paterson, A. H., J. W. DeVerna, B. Lanini, and S. D. Tanksley. 1990. Fine mapping of quantitative trait loci using selected overlapping recombinant chromosomes, in an interspecies cross of tomato. *Genetics* 124: 735–742. [14,15]

Paterson, A. H., S. D. Tanksley, and M. E. Sorrells. 1991. DNA markers in plant improvement. *Adv. Agron.* 46: 39–90. [14]

Paterson, A. H., S. Damon, J. D. Hewitt, D. Zamir, H. D. Rabinowitch, S. E. Lincoln, E. S. Lander, and S. D. Tanksley. 1991. Mendelian factors underlying quantitative traits in tomato: comparison across species, generations, and environments. *Genetics* 127: 181–197. [15]

Paterson, A. H., Y.-R. Lin, Z. Li, K. F. Schertz, J. F. Doebley, S. R. M. Pinson, S.-C. Liu, J. W. Stansel, and J. E. Irvine. 1995. Convergent domestication of cereal crops by independent mutations at corresponding genetic loci. *Science* 269: 1714–1718. [14]

Patnaik, P. B. 1949. The noncentral χ^2 and F distributions and their approximations. *Biometrika* 36: 202–232. [A5]

Patterson, H. D., and R. Thompson. 1971. Recovery of interblock information when block sizes are unequal. *Biometrika* 58: 545–554. [27]

Paulsen, S. M. 1994. Quantitative genetics of butterfly wing color patterns. *Devel. Gen.* 15: 79–81. [21]

Payne, F. 1918. The effect of artificial selection on bristle number in *Drosophila ampelophila* and its interpretation. *Proc. Natl. Acad. Sci. USA* 4: 55–58. [1,14]

Pearson, K. 1896. Contributions to the mathematical theory of evolution. III. Regression, heredity and panmixia. *Phil. Trans. Royal Soc. Lond.* A 187: 253–318. [8,A1]

Pearson, K. 1897. Mathematical contributions to the theory of evolution — on a form of spurious correlation which may arise when indices are used in measurement of organs. *Proc. Royal Soc. Lond.* 60: 489–498. [11]

Pearson, K. 1900. Mathematical contributions to the theory of evolution. VII. On the correlation of characters not quantitatively measurable. *Phil. Trans. Royal Soc. Lond.* A 190:1–47. [25]

Pearson, K. 1903. Mathematical contributions to the theory of evolution. XI. On the influence of natural selection on the variability and correlation of organs. *Phil. Trans. Royal Soc. Lond.* A 200: 1–66. [1,8,21]

Pearson, K. 1904. Mathematical contributions to the theory of evolution. XII. On a generalized theory of alternative inheritance, with special reference to Mendel's laws. *Phil. Trans. Royal Soc. Lond.* A 203: 53–86. [13]

Pearson, K. 1910. Darwinism, biometry, and some recent biology. *Biometrika* 7: 368–385. [2]

Pearson, K. 1920. Notes on the history of correlation. *Biometrika* 13: 25–45. [8]

Pearson, K., and A. G. Davin. 1924. On the biometric constraints of the human skull. *Biometrica* 16: 328–363. [11]

Pearson, K., and A. Lee. 1903. On the laws of inheritance in man. I. Inheritance of physical characters. *Biometrika* 2: 357–462. [7]

Pederson, D. G. 1968. Environmental stress, heterozygote advantage and genotype-environment interaction in *Arabidopsis*. *Heredity* 23: 127–138. [6]

Pederson, D. G. 1972. A comparison of four experimental designs for the estimation of heritability. *Theor. Appl. Genet.* 42: 371–377. [20]

Pedhazur, E. J. 1982. *Multiple regression in behavioral research.* 2nd Ed. Holt, Rinehart and Winston, Fort Worth, TX. [A2]

Penner, G. A., J. Chong, M. Lévesque-Lemay, S. J. Molnar, and G. Fedak. 1993. Identification of a RAPD marker linked to the oat stem rust gene *Pg3*. *Theor. Appl. Genet.* 85: 702–705. [14]

Penrose, L. S. 1935. The detection of autosomal linkage in data which consist of pairs of brothers and sisters of unspecified parentage. *Ann. Eugen.* 6: 133–138. [16]

Penrose, L. S. 1954a. Some recent trends in human genetics. *Carylogia* 6 (suppl.): 521–530. [7]

Penrose, L. S. 1954b. The general purpose sib-pair linkage test. *Ann. Eugen.* 18:120–124. [16]

Penrose, L. S. 1969. Effects of additive genes at many loci compared with those of a set of alleles at one locus in parent-child and sib correlations. *Ann. Hum. Genet.* 35: 15–21. [13]

Pereira, M. G., and M. Lee. 1995. Identification of genomic regions affecting plant height in sorghum and maize. *Theor. Appl. Genet.* 90: 380–388. [14]

Perez, D. E., and C.-I. Wu. 1995. Further characterization of the *Odysseus* locus of hybrid sterility in *Drosophila simulans*: one gene is not enough. *Genetics* 140: 201–206. [14]

Perez, D. E., C.-I. Wu, N. A. Johnson, and M.-L. Wu. 1993. Genetics of reproductive isolation in the *Drosophila simulans* clade: DNA marker-assisted mapping and characterization of a hybrid-male sterility gene, *Odysseus (Ods)*. *Genetics* 133: 261–275. [14]

Pericak-Vance, M. A., and J. L. Haines. 1995. Genetic susceptibility to Alzheimer disease. *Trends Genet.* 11: 504–508. [14]

Perkins, J. M., and J. L. Jinks. 1968. Environmental and genotype environmental components of variability. *Heredity* 23: 339–356. [22]

Perrins, C. M., and P. J. Jones. 1974. The inheritance of clutch size in the great tit *(Parus major* L.) *Condor* 76: 225–229. [17]

Peters, R. H. 1976. Tautology in evolution and ecology. *Am. Nat.* 110: 1–12. [1]

Peters, R. H. 1983. *The ecological implications of body size.* Cambridge University Press, Cambridge, UK. [11]

Pfahler, P. L. 1966. Heterosis and homeostasis in rye (*Secale cereale* L.). I. Individual plant production of varieties and intervarietal crosses. *Crop Sci.* 6: 397–401. [6]

Piepho, H.-P. 1994. Application of a generalized Grubbs' model in the analysis of genotype-environment interaction. *Heredity* 73: 113–116. [22]

Piepho, H.-P. 1995. Robustness of statistical tests for multiplicative terms in the additive main effects and multiplicative interaction model for cultivar trials. *Theor. Appl. Genet.* 90: 438–443. [22]

Piper, L. R., and B. M. Bindon. 1988. The genetics and endrocrinology of the Booroola sheep *F* gene. *In* B. S.

Weir, E. J. Eisen, M. M. Goodman, and G. Namkoong (eds.), *Proceedings of the second international conference on quantitative genetics*, pp. 270–280. Sinauer Assoc., Sunderland, MA. [4,13]

Pirchner, F. 1983. *Population genetics in animal breeding.* 2nd Ed. Plenum, NY. [1]

Pisani, J. F., and W. E. Kerr. 1961. Lethal equivalents in domestic animals. *Genetics* 46: 773–786. [10]

Platenkamp, G. A. J., and R. G. Shaw. 1992. Environmental and genetic constraints on adaptive population differentiation in *Anthoxanthum odoratum*. *Evolution* 46: 341–352. [21,22,27]

Plomin, R., G. E. McClearn, G. Gora-Maslak, and J. M. Neiderhiser. 1991. Use of recombinant inbred strains to detect quantitative trait loci associated with behavior. *Behav. Genet.* 21: 99–116. [14]

Plomion, C., N. Bahrman, C.-E. Durel, and D. M. O'Malley. 1995. Genomic mapping in *Pinus pinaster* (maritime pine) using RAPD and protein markers. *Heredity* 74: 661–668. [9]

Pogson, G. H., and E. Zouros. 1994. Allozyme and RFLP heterozygosities as correlates of growth rate in the scallop *Placopecten magellanicus:* a test of the associative overdominance hypothesis. *Genetics* 137: 221–231. [10]

Pollak, P. E. 1991. Cytoplasmic effects on components of fitness in tobacco hybrids. *Evolution* 45: 785–790. [23]

Pomp, D., D. E. Cowley, E. J. Eisen, W. R. Atchley, and D. Hawkins-Brown. 1989. Donor and recipient genotype and heterosis effects on survival and prenatal growth of transferred mouse embryos. *J. Repro. Fert.* 86: 493–500. [23]

Ponzoni, R. W., and J. W. James. 1978. Possible biases in heritability estimates from intraclass correlation. *Theor. Appl. Genet.* 53: 25–27. [18]

Pooni, H. S., J. L. Jinks, and J. F. F. de Toledo. 1985. Predicting and observing the properties of second cycle hybrids using basic generations and inbred line $\times F_1$ crosses. *Heredity* 54: 121–129. [9]

Pooni, H. S., P. S. Virk, D. T. Coombs, and M. K. U. Chowdhury. 1994. The genetical basis of hybrid vigour in a highly heterotic cross of *Nicotiana tabacum*. *Theor. Appl. Genet.* 89: 1027–1031. [10]

Popper, K. 1978. Natural selection and the emergence of mind. *Dialectica* 32: 339–355. [1]

Powell, W., P. D. S. Caligari, W. T. B. Thomas, and J. L. Jinks. 1985a. The effects of major genes on quantitatively varying characters in barley. 2. The *denso* and daylength response loci. *Heredity* 54: 349–352. [14]

Powell, W., W. T. B. Thomas, P. D. S. Caligari, and J. L. Jinks. 1985b. The effects of major genes on quantitatively varying characters in barley. 1. The GP *ert* locus. *Heredity* 54: 343–348. [14]

Powell, W., W. T. B. Thomas, D. M. Thompson, J. S. Swanston, and R. Waugh. 1992. Association between rDNA alleles and quantitative traits in doubled haploid populations of barley. *Genetics* 130: 187–194. [12]

Powers, L. 1942. The nature of the series of environmental variances and the estimation of the genetic variances and the geometric means in crosses involving species of *Lycopersicon*. *Genetics* 27: 561–575. [9]

Powers, L. 1951. Gene analysis by the partitioning method when interactions of genes are involved. *Bot. Gaz.* 113: 1–23. [9]

Prabhakaran, V. T., and J. P. Jain. 1987. Probability of inadmissible estimates of heritability from regression and half-sib analyses. *Biom. J.* 2: 219–230. [18]

Pray, L. A., J. M. Schwartz, C. J. Goodnight, and L. Stevens. 1994. Environmental dependency of inbreeding depression: implications for conservation biology. *Cons. Biol.* 8: 562–568. [10]

Price, B. 1950. Primary biases in twin studies: a review of prenatal and natal difference-producing factors in monozygotic pairs. *Amer. J. Hum. Genet.* 2: 293–352. [19]

Price, D. K., and N. T. Burley. 1993. Constraints on the evolution of attractive traits: genetic (co)variance of zebra finch bill colour. *Heredity* 71: 405–412. [24]

Price, G. R. 1970. Selection and covariance. *Nature* 227: 520–521. [3]

Price, G. R. 1972. Extension of covariance selection mathematics. *Ann. Hum. Genet.* 35: 485–490. [3]

Price, T. D., and P. R. Grant. 1985. The evolution of ontogeny in Darwin's Finches: a quantitative genetics approach. *Am. Nat.* 125: 169–188. [23]

Price, T., and D. Schluter. 1991. On the low heritability of life-history traits. *Evolution* 45: 853–861. [7]

Pringle, R. M., and A. A. Rayner. 1971. *Generalized inverse matrices with applications to statistics.* Griffin, London. [A3]

Pritchard, C., D. R. Cox, and R. M. Myers. 1991. The end in sight for Huntingtons disease? *Am. J. Hum. Genet.* 49: 1–6. [14]

Prout, T., and J. S. F. Barker. 1989. Ecological aspects of the heritability of body size in *Drosophila buzzatii*. *Genetics* 123: 803–813. [17]

Provine, W. B. 1971. *The origins of theoretical population genetics.* Univ. Chicago Press, Chicago. [1]

Quaas, R. L. 1976. Computing the diagonal elements and inverse of a large numerator relationship matrix. *Biometrics* 32: 949–953. [26]

Quaas, R. L., and E. J. Pollak. 1980. Mixed model methodology for farm and ranch beef cattle testing programs. *J. Anim. Sci.* 51: 1277–1287. [26,A3]

Quaas, R. L., and E. J. Pollak. 1981. Modified equations for sire models with groups. *J. Dairy Sci.* 54: 1868–1872. [26]

Queller, D. C., and K. F. Goodnight. 1989. Estimating relatedness using genetic markers. *Evolution* 43: 258–275. [27]

Rafalski, J. A., and S. V. Tingey. 1993. Genetic diagnostics in plant breeding: RAPDs, microsatellites and machines. *Trends Genet.* 9: 275–280. [14]

Raff, R. A., and T. C. Kaufman. 1983. *Embryos, genes, and evolution.* Macmillan, NY. [21]

Ragot, M., and D. A. Hoisington. 1993. Molecular markers for plant breeding: comparisons of RFLP and RAPD genotyping costs. *Theor. Appl. Genet.* 86: 975–984. [14]

Ragot, M., P. H. Sisco, D. A. Hoisington, and C. W. Stuber. 1995. Molecular-marker-mediated characterization of favorable exotic alleles at quantitative trait loci in maize. *Crop Sci.* 35: 1306–1315. [15]

Ralls, K., and J. Ballou. 1982a. Effect of inbreeding on juvenile mortality in some small mammal species. *Lab. Anim.* 16: 159–166. [10]

Ralls, K., and J. Ballou. 1982b. Effects of inbreeding on infant mortality in captive primates. *J. Primatol.* 3: 491–505. [10]

Ralls, K., K. Brugger, and J. Ballou. 1979. Inbreeding and juvenile mortality in small populations of ungulates. *Science* 206: 1101–1103. [10]

Rao, D. C., N. E. Morton, and S. Yee. 1974. Analysis of family resemblance. II. A linear model for familial correlation. *Am. J. Hum. Genet.* 26: 331–359. [7]

Rao, C. R., and S. K. Mitra. 1971. *Generalized inverse of matrices and its applications.* John Wiley & Sons, NY. [A3]

Rao, P. S. S., and S. G. Inbaraj. 1980. Inbreeding effects on fetal growth and development. *J. Med. Genet.* 17: 27–33. [10]

Rao, S. R. V., and S. Ali. 1982. Insect sex chromosomes. IV. A presumptive hyperactivation of the male X chromosome in *Acheta domesticus* (L.) *Chromosoma* 74: 241–252. [24]

Rao, S. R. V., and P. Arora. 1979. Insect sex chromosomes. IIII. Differential susceptibility of homologous X chromosomes of *Gryllotalpa fossor* to ^{3}H-Urd-induced aberrations. *Chromosoma* 74: 241–252. [24]

Rasmuson, M. 1952. Variation in bristle number of *Drosophila melanogaster. Acta Zoologica* 33: 1–31. [10]

Rasmusson, J. 1927. Genetically changed linkage values in *Pisum. Hereditas* 10: 1–152. [14]

Rausher, M. D. 1983. Variability for host preference in insect populations: mechanistic and evolutionary models. *J. Insect Physiol.* 31: 873–889. [5]

Rausher, M. D. 1984. Trade-offs in performance on different hosts: evidence from within and between site variation in the beetle *Deloyala guttata. Evolution* 38: 582–595. [22]

Rausher, M. D., and E. L. Simms. 1989. The evolution of resistance to herbivory in *Ipomoea purpurea.* I. Attempts to detect selection. *Evol.* 43: 563–572. [21]

Read, A., and S. Nee. 1991. Is Haldane's rule significant? *Evolution* 45: 1707–1709. [14]

Rebaï, A., and B. Goffinet. 1993. Power of tests for QTL detection using replicated progenies derived from a diallel cross. *Theor. Appl. Genet.* 86: 1014–1022. (Correction 92: 128–129). [15]

Rebaï, A., B. Goffinet, and B. Mangin. 1994b. Approximate thresholds of interval mapping tests for QTL detection. *Genetics* 138: 235–240. [15]

Rebaï, A., B. Goffinet, and B. Mangin. 1995. Comparing power of different methods of QTL detection. *Biometrics* 51: 87–99. [15]

Rebaï, A., B. Goffinet, B. Mangin, and D. Perret. 1994a. Detecting QTLs with diallel schemes. *In* J. W. van Ooijen and J. Jansen (eds.), *Biometrics in plant breeding: applications of molecular markers,* pp. 170–177. CPRO-DLO, Netherlands. [15]

Redner, R. A., and H. F. Walker. 1984. Mixture densities, maximum likelihood and the EM algorithm. *SIAM Review* 26: 195–239. [13]

Reed, E. S. 1981. The lawfulness of natural selection. *Am. Nat.* 118: 61–71. [1]

Reed, T., M. M. Evans, J. A. Norton, Jr., and J. C. Christian. 1979. Maternal effects on fingertip dermatoglyphics. *Am. J. Hum. Genet.* 31: 315–323. [19]

Reed, T., F. R. Sprague, K. W. Kang, W. E. Nance, and J. C. Christian. 1975. Genetic analysis of dermatoglyphic patterns in twins. *Hum. Hered.* 25: 263–275. [19]

Reed, T., I. A. Uchida, J. A. Norton, Jr., and J. C. Christian. 1978. Comparisons of dermatoglyphic patterns in monochorionic and dichorionic monozygotic twins. *Am. J. Hum. Genet.* 30: 383–391. [19]

Reeve, E. C. R. 1955. The variance of the genetic correlation coefficient. *Biometrics* 11: 357–374. [21]

Reeve, E. C. R. 1960. Some genetic tests on asymmetry of sternopleural chaeta number in *Drosophila. Genet. Res.* 1: 151–172. [6]

Reeve, E. C. R. 1961. A note on non-random mating in progeny tests. *Genet. Res.* 2: 195–203. [7,17]

Reeve, H. K., and P. W. Sherman. 1993. Adaptation and the goals of evolutionary research. *Quart. Rev. Biol.* 68: 1–32. [1]

Reich, T., J. W. James, and C. A. Morris. 1972. The use of multiple thresholds in determining the mode of transmission of semi-continuous traits. *Ann. Hum. Genet.* 36: 163–184. [25]

Reich, V. H., and R. E. Atkins. 1970. Yield stability of four population types of grain sorghum, *Sorghum bicolor* (L.) Moench, in different environments. *Crop Sci.* 10: 511–517. [6]

Rendel, J. M. 1965. Bristle pattern in *scute* stocks of *Drosophila melanogaster. Am. Nat.* 99: 25–32. [11]

Rendel, J. M. 1977. Canalisation in quantitative genetics. *In* E. Pollak, O. Kempthorne, and T. B. Bailey, Jr. (eds.), *Proceedings of the international conference on quantitative genetics,* pp. 23–28. Iowa State Univ. Press, Iowa. [11]

Rendel, J. M. 1979. Canalization and selection. *In* J. N. Thompson, Jr., and J. M. Thoday (eds.), *Quantitative genetic variation,* pp. 139–156. Academic Press, NY. [11]

Rendel, J. M., and B. L. Sheldon. 1960. Selection for canalization of the scute phenotype in *Drosophila melanogaster. Aust. J. Biol. Sci.* 13: 36–47. [11]

Rendel, J. M., B, L. Sheldon, and D. E. Finlay. 1966. Selection for canalization of the scute phenotype. II. *Am. Nat.* 100: 13–31. [11]

Reznick, D. 1981. "Grandfather effects": the genetics of interpopulation differences in offspring size in the mosquito fish. *Evolution* 35: 941–953. [6]

Reznick, D. 1982. Genetic determination of offspring size in the guppy *(Poecilia reticulata). Am. Nat.* 120: 181–188. [6]

Reznick, D. 1985. Costs of reproduction: an evaluation of the empirical evidence. *Oikos* 44: 257–267. [21]

Reyment, R. A. 1991. *Multidimensional palaeobiology.* Pergamon Press, Elmsford, NY. [21]

Rice, W. R. 1989. Analyzing tables of statistical tests. *Evolution* 43: 223–225. [21]

Rich, S. S., A. E. Bell, D. A. Miles, and S. P. Wilson. 1984. An experimental study of genetic drift for two quantitative characters in *Tribolium*. *J. Hered.* 75: 191–195. [10]

Rinchik, E. M., L. B. Russell, N. G. Copeland, and N. A. Jenkins. 1986. Molecular genetic analysis of the *dilute-short ear (D-SE)* region of the mouse. *Genetics* 112: 321–342. [14]

Risch, H. 1979. The correlation between relatives under assortative mating for an X-linked and autosomal trait. *Ann. Hum. Genet.* 43: 151–165. [24]

Risch, N. 1984. Segregation analysis incorporating linkage markers. I. Single-locus models with an application to type I diabetes. *Am. J. Hum. Genet.* 36: 363–386. [16]

Risch, N. 1987. Assessing the role of HLA-linked and unlinked determinants of disease. *Am. J. Hum. Genet.* 40: 1–14. [16]

Risch, N. 1990a. Linkage strategies for genetically complex traits. I. Multilocus models. *Am. J. Hum. Genet.* 46: 229–241. [16]

Risch, N. 1990b. Linkage strategies for genetically complex traits. II. The power of affected relative pairs. *Am. J. Hum. Genet.* 46: 229–241. [16]

Risch, N. 1990c. Linkage strategies for genetically complex traits. III. The effect of marker polymorphism on analysis of affected relative pairs. *Am. J. Hum. Genet.* 46: 242–253 (Correction 51: 673–675). [16]

Risch, N. 1993. Exclusion mapping of complex diseases. *Am. J. Hum. Genet.* 53: A185. [16]

Risch, N., and D. Botstein. 1996. A manic depressive history. *Nature Genetics* 12: 351–353. [16]

Risch, N., D. de Leon, L. Ozelius, P. Kramer, L. Almasy, B. Singer, S. Fahn, X. Breakefield, and S. Bressman. 1995. Genetic analysis of idiopathic torsion dystonia in Ashkenazi Jews and their recent descent from a small founder population. *Nature Genetics* 9: 152–159. [14]

Rise, M. L., W. N. Frankel, J. M. Coffin, and T. N. Seyfired. 1991. Genes for epilepsy mapped in the mouse. *Science* 253: 669–673. [14]

Riska, B., and W. R. Atchley. 1985. Genetics of growth predict patterns of brain-size evolution. *Science* 229: 668–671. [21]

Riska, B., W. R. Atchley, and J. J. Rutledge. 1984. A genetic analysis of targeted growth in mice. *Genetics* 107: 79–101. [11,23]

Riska, B., J. J. Rutledge, and W. R. Atchley. 1985. Covariance between direct and maternal genetic effects in mice, with a model of persistent environmental influences. *Genet. Res.* 45: 287–297. [23]

Riska, B., T. Prout, and M. Turelli. 1989. Laboratory estimates of heritabilities and genetic correlations in nature. *Genetics* 123: 865–871. [17]

Ritland, K. 1990a. Gene identity and the genetic demography of plant populations. *In* A. H. D. Brown, M. T. Clegg, A. L. Kahler, and B. S. Weir (eds.), *Plant population genetics, breeding, and genetic resources*, pp. 181–199. Sinauer Assoc., Sunderland, MA. [10]

Ritland, K. 1990b. Inferences about inbreeding depression based on changes of the inbreeding coefficient. *Evolution* 44: 1230–1241. [10]

Ritland, K. 1996a. Estimators for pairwise relatedness and inbreeding coefficients. *Genet. Res.* 67: 175–186. [27]

Ritland, K. 1996b. A marker-based method for inferences about quantitative inheritance in natural populations. *Evolution* 50: 1062–1073. [27]

Ritland, K., and C. Ritland. 1996. Inferences about quantitative inheritance based on natural population structure in the yellow monkeyflower, *Mimulus guttatus*. *Evolution* 50: 1074–1082. [27]

Roach, D. A., and R. D. Wulff. 1987. Maternal effects in plants. *Ann. Rev. Ecol. Syst.* 18: 209–236. [6]

Roberts, D. F., W. Z. Billewicz, and I. A. McGregor. 1978. Heritability of stature in a west African population. *Ann. Hum. Genet.* 42: 15–24. [7]

Robertson, A. 1955. Selection in animals: synthesis. *Cold Spring Harbor Symp. Quant. Biol.* 20: 225–229. [7]

Robertson, A. 1959a. Experimental design in the evaluation of genetic parameters. *Biometrics* 15: 219–226. [18]

Robertson, A. 1959b. The sampling variance of the genetic correlation coefficient. *Biometrics* 15:469–485. [21,22,24]

Robertson, A. 1966. A mathematical model of the culling process in dairy cattle. *Anim. Prod.* 8: 95–108. [3]

Robertson, A. 1967. The nature of quantitative genetic variation. *In* R.A. Brink and E. D. Styles (eds.), *Heritage from Mendel*, pp. 265–280. Univ. Wisconsin Press, Madison, WI. [12]

Robertson, A. 1973. Linkage between marker loci and those affecting a quantitative trait. *Behav. Genet.* 3: 389–391. [16]

Robertson, A. 1977a. The non-linearity of the offspring-parent regression. *In* E. Pollak, O. Kempthorne, and E. B. Bailey, Jr. (eds.), *Proceedings of the international conference on quantitative genetics*, pp. 297–304. Iowa State Univ. Press, Ames. [17]

Robertson, A. 1977b. The effect of selection on the estimation of genetic parameters. *Z. Tierzüchtg. Züchtgsbiol.* 94:131–135. [21]

Robertson, A., and I. M. Lerner. 1949. The heritability of all-or-none traits: viability of poultry. *Genetics* 34: 395–411. [25]

Robertson, D. 1989. Understanding the relationship between qualitative and quantitative genetics. *In* T. Helentjaris and B. Burr (eds.), *Development and application of molecular markers to problems in plant genetics*, pp. 81–87. Cold Spring Harbor Press, Cold Spring Harbor, NY. [14]

Robertson, F. W. 1954. Studies in quantitative inheritance V. Chromosome analyses of crosses between selected and unselected lines of different body sizes in *Drosophila melanogaster*. *J. Genetics* 52: 494–520. [14]

Robertson, F. W., and E. C. R. Reeve. 1952. Heterozygosity, environmental variation and heterosis. *Nature* 170: 286. [6]

Robertson, F. W., and E. C. Reeve. 1953. Studies in quantitative inheritance. IV. The effects of substituting chromosomes from selected lines into different genetic backgrounds. *J. Genetics* 51: 586–610. [14]

Robertson, H. M., C. R. Preston, R. W. Phillis, D. Johnson-Schlitz, W. K. Benz, and W. R. Engels. 1988. A stable genomic source of *P* element transposase in *Drosophila melanogaster. Genetics* 118: 461–470. [12]

Robinson, D. L. 1987. Estimation and use of variance components. *The Statistician* 36: 3–14. [27]

Robinson, G. K. 1991. That BLUP is a good thing: the estimation of random effects. *Stat. Sci.* 6: 15–51. [26]

Robinson, H. F., and R. E. Comstock. 1955. Analysis of genetic variability in corn with reference to probable effects of selection. *Cold Spring Harbor Symp. Quant. Biol.* 20: 127–136. [5]

Robison, O. W. 1972. The role of maternal effects in animal breeding. V. Maternal effects in swine. *J. Anim. Sci.* 35: 1303–1315. [23]

Robson, E. B. 1955. Birth weight in cousins. *Ann. Hum. Genet.* 19: 262–268. [7]

Rocheford, T. R., J. C. Osterman, and C. O. Gardner. 1990. Variation in the ribosomal DNA intergenic spacer of a maize population mass-selected for high grain yield. *Theor. Appl. Genet.* 79: 793–800. [12]

Rodolphe, F., and M. Lefort. 1993. A multiple-marker model for detecting chromosomal segments displaying QTL activity. *Genetics* 134: 1277–1288. [15]

Roff, D. A. 1986. The genetic basis of wing dimorphism in the sand cricket, *Gryllus firmus,* and its relevance to the evolution of wing dimorphism in insects. *Heredity* 57: 221–231. [25]

Roff, D. A. 1994. The evolution of dimorphic traits: predicting the genetic correlation between environments. *Genetics* 136: 395–401. [25]

Roff, D. A. 1995. The estimation of genetic correlations from phenotypic correlations: a test of Cheverud's conjecture. *Heredity* 74: 481–490. [21]

Roff, D. A. 1996. The evolution of genetic correlations: an analysis of patterns. *Evolution* 50: 1392–1403. [21]

Roff, D. A., and T. A. Mousseau. 1987. Quantitative genetics and fitness: lessons from *Drosophila. Heredity* 58: 103–118. [7]

Roff, D. A., and R. Preziosi. 1994. The estimation of the genetic correlation: the use of the jackknife. *Heredity* 73: 544–548. [21]

Rogers, A. R., and A. Mukherjee. 1992. Quantitative genetics of sexual dimorphism in human body size. *Evolution* 46: 226–234. [24]

Roginskii, Y. Y. 1959. Some results of using the quantitative method to study morphological variability. (In Russian). *Arkhiv. Anat. Gistol. Embriol.* 36: 83–89. [11]

Rohlf, F. J., A. J. Gilmartin, and G. Hart. 1983. The Kluge-Kerfoot phenomenon — a statistical artifact. *Evolution* 37: 180–202. [11]

Romeo, G., and V. A. McKusick. 1994. Phenotypic diversity, allelic series and modifier genes. *Nature Genetics* 7: 451–453. [12]

Rommens, J. M., M. C. Iannuzzi, B.-S. Kerem, M. L. Drumm, G. Melmer, M. Dean, R. Rozmahel, J. L. Cole, D. Kennedy, N. Hidaka, M. Zsiga, M. Buchwald, J. R. Riordan, L.-C. Tsui, and F. S. Collins. 1989. Identification of the cystic fibrosis gene: chromosome walking and jumping. *Science* 245: 1059–1065. [14]

Ronald, P. C., B. Albano, R. Tabien, L. Abenes, K.-S. Wu, S. McCouch, and S. D. Tanksley. 1992. Genetic and physical analysis of the rice bacterial blight disease resistance locus, *Xa21. Mol. Gen. Genet.* 236: 113–120. [14]

Ronin, Y. I., V. M. Kirzhner, and A. B. Korol. 1995. Linkage between loci for quantitative traits and marker loci: multi-trait analysis with a single marker. *Theor. Appl. Genet.* 90: 776–786. [15]

Rose, M. R. 1982. Antagonistic pleio-tropy, dominance, and genetic variation. *Heredity* 48: 63–78. [21]

Rose, M. R. 1984. Genetic covariation in *Drosophila* life history: untangling the data. *Am. Nat.* 123:565–569. [21]

Rosen, D. 1978. Darwin's demon. *Syst. Zool.* 27: 370–373. [1]

Rothschild, M. F., C. R. Henderson, and R. L. Quaas. 1979. Effects of selection on variances and covariances of simulated first and second lactations. *J. Dairy Sci.* 62: 996–1002. [27]

Rowe, D. C. 1994. *The limits of family influence.* Guilford Press, Elmsford, NY. [19]

Rowe, P. R., and R. H. Anderson. 1964. Phenotypic stability for a systematic series of corn genotypes. *Crop Sci.* 6: 563–566. [6]

Royer-Pokora, B., L. M. Kunkel, A. P. Monaco, S. C. Goff, P. E. Newburger, R. L. Baehner, F. S. Cole, J. T. Curnutte, and S. H. Orkin. 1986. Cloning the gene for an inherited human disorder — chronic granulomatous disease — on the basis of chromosomal location. *Nature* 322: 32–38. [14]

Rubinstein, P., M. Walker, C. Carpenter, C. Carrier, J. Krassner, C. Falk, and F. Ginsberg. 1981. Genetics of HLA disease associations: the use of the haplotype relative risk (HRR) and the "haplo-delta" (Dh) estimates in juvenile diabetes from three racial groups. *Hum. Immun.* 3: 384. [14]

Ruiz, A., M. Santos, A. Barbadilla, J. E. Quezada-Diaz, E. Hasson, and A. Fontdevila. 1991. Genetic variance for body size in a natural population of *Drosophila buzzatii. Genetics* 128: 739–750. [14,17]

Russell, W. A., G. F. Sprague, and L. H. Penny. 1963. Mutations affecting quantitative characters in long-term inbred lines of maize. *Crop Sci.* 3: 175–178. [10, 12]

Rutledge, J. J., O. W. Robison, E. J. Eisen, and J. E. Legates. 1972. Dynamics of genetic and maternal effects in mice. *J. Anim. Sci.* 35: 911–918. [23]

Ryder, E. J. 1958. The effects of complementary epistasis on the inheritance of a quantitative character, seed size in lima beans. *Agron. J.* 50: 298–301. [9]

Saghai-Maroof, M. A., K. M. Soliman, R. A. Jorgensen, and R. W. Allard. 1984. Ribosomal DNA spacer length polymorphism in barley: Mendelian inheritance chromosomal location and population dynamics. *Proc. Nat. Acad. Sci. USA* 81: 8014–8018. [12]

Saiki, R. K., S. Scharf, F. Faloona, K. B. Mullis, G. T. Horn, H. A. Erlich and N. Arnheim. 1985. Enzymatic amplification of β-globin genomic sequences and restriction site analysis for diagnosis of sickle cell anemia. *Science* 230: 1350–1354. [14]

Sakai, K.-I., and A. Suzuki. 1964. Induced mutation and pleiotropy of genes responsible for quantitative characters in rice. *Rad. Botany* 4: 141–151. [12]

Santiago, E., J. Albornoz, A. Dominguez, M. A. Toro, and C. López-Fanjul. 1992. The distribution of spontaneous mutations on quantitative traits and fitness in *Drosophila melanogaster*. *Genetics* 132: 771–781. [12]

Sarfatti, M., J. Katan, R. Fluhr, and D. Zamir. 1989. An RFLP marker linked to the *Fusarium oxysporum* resistance gene *I2*. *Theor. Appl. Genet.* 78: 755–759. [14]

Sarkar, S. 1991. Haldane's solution of the Luria-Delbrück distribution. *Genetics* 127: 257–261. [14]

Satagopan, J. M., B. S. Yandell, M. A. Newton, and T. C. Osborn. 1996. A Bayesian approach to detect quantitative trait loci using Markov chain monte carlo. *Genetics* 144: 805–816. [15]

Satterthwaite, F. E. 1946. An approximate distribution of estimates of variance components. *Biometrics Bull.* 2: 110–114. [18,19,20]

Savolainen, O., and P. Hedrick. 1995. Heterozygosity and fitness: no association in Scots pine. *Genetics* 140: 75–766. [10]

Savolainen, O., K. Kärkkäinen, and H. Kuittinen. 1992. Estimating numbers of embryonic lethals in conifers. *Heredity* 69: 308–314. [10]

Sax, K. 1923. The association of size differences with seed-coat pattern and pigmentation in *Phaseolus vulgaris*. *Genetics* 8: 552–560. [14]

Sawamura, K. 1996. Maternal effects as a cause of exceptions for Haldane's rule. *Genetics* 143: 609–611. [14]

Schaal, B. A. 1984. Life-history variation, natural selection, and maternal effects in plant populations. *In* R. Dirzo and J. Sarukhán (eds.), *Perspectives on plant population ecology*, pp. 188–206. Sinauer Assoc., Sunderland, MA. [6]

Schachermayr, G., H. Siedler, M. D. Gale, H. Winzeler, M. Winzeler, and B. Keller. 1994. Identification and localization of molecular markers linked to the *Lr9* leaf rust resistance gene of wheat. *Theor. Appl. Genet.* 88: 110–115. [14]

Schachermayr, G., M. M. Messmer, C. Feuillet, H. Winzeler, M. Winzeler, and B. Keller. 1995. Identification and localization of molecular markers linked to the *Agropyron elongatum*-derived leaf rust resistance gene *Lr24* in wheat. *Theor. Appl. Genet.* 90: 982–990. [14]

Schaeffer, L. R. 1986. Estimation of variances and covariances within the allowable parameter space. *J. Dairy Sci.* 69: 187–194. [26,27]

Schaeffer, L. R. 1991. C. R. Henderson: contribution to predicting genetic merit. *J. Dairy Sci.* 74: 4052–4066. [26]

Schaeffer, L. R., and B. W. Kennedy. 1986. Computing strategies for mixed model equations. *J. Dairy Sci.* 69: 575–579. [26]

Schaeffer, L. R. and H. Song. 1978. Selection bias and REML variance-covariance component estimation. *J. Dairy Sci.* 61: 91–92. [27]

Schaeffer, L. R., B. W. Kennedy, and J. P. Gibson. 1989. The inverse of the gametic relationship matrix. *J. Dairy Sci.* 72: 1266–1272. [26]

Schaffer, H. E., D. Yardley, and W. W. Anderson. 1977. Drift or selection: a statistical test of gene frequency variation over generations. *Genetics* 87: 371–379. [14]

Schaid, D. J., and T. G. Nick. 1990. Sib-pair linkage tests for disease susceptibility loci: common tests vs. the asymptotically most powerful test. *Genet. Epidem.* 7: 359–370. [16]

Schaid, D. J., and S. S. Sommer. 1994. Comparison of statistics for candidate- gene association studies using cases and parents. *Am. J. Hum. Genet.* 55: 402–409. [14]

Scharf, S. J., G. T. Horn, and H. A. Erlich. 1986. Direct cloning and sequence analysis of enzymatically amplified genomic sequences. *Science* 233: 1076–1078. [14]

Scharloo, W. 1988. Selection on morphological patterns. *In* G. de Jong (ed.), *Population genetics and evolution*, pp. 230–250. Springer-Verlag, NY. [11]

Scheffé, H. 1959. *The analysis of variance.* John Wiley & Sons, NY. [18,A5]

Scheinberg, E. 1966. The sampling variance of the correlation coefficients estimated in genetic experiments. *Biometrics* 22: 187–191. [21]

Scheiner, S. M. 1993. Plasticity as a selectable trait: reply to Via. *Am. Nat.* 142: 371–373. [22]

Scheiner, S. M., R. L. Caplan, and R. F. Lyman. 1989. A search for trade-offs among life history traits in *Drosophila melanogaster. Evol. Ecol.* 3: 51–63. [21]

Scheiner, S. M., R. L. Caplan, and R. F. Lyman. 1991. The genetics of phenotypic plasticity. III. Genetic correlations and fluctuating asymmetries. *J. Evol. Biol.* 4: 51–68. [6]

Scheiner, S. M., and R. F. Lyman. 1991. The genetics of phenotypic plasticity. II. Response to selection. *J. Evol. Biol.* 4: 23–50. [22]

Schemske, D. W. 1983. Breeding system and habitat effects in three neotropical *Costus* (Zingiberaceae). *Evolution* 37: 523–539. [10]

Schemske, D. W. 1984. Population structure and local selection in *Impatiens* (Balsaminaceae), a selfing annual. *Evolution* 37: 523–539. [9]

Schemske, D. W., and R. Lande. 1985. The evolution of self-fertilization and inbreeding depression in plants. II. Empirical observations. *Evolution* 39: 41–52. [10]

Schiefelbein, J. W., D. B. Furtek, H. K. Dooner, and O. E. Nelson, Jr. 1988. Two mutations in a maize *bronze-1* allele caused by transposable elements of the *Ac-Ds* family alter the quantity and quality of gene product. *Genetics* 120: 767–777. [12]

Schlichting, C. D. 1986. The evolution of phenotypic plasticity in plants. *Ann. Rev. Ecol. Syst.* 17: 667–693. [22]

Schlichting, C. D., and M. Pigliucci. 1993. Control of phenotypic plasticity via regulatory genes. *Am. Nat.* 142: 366–370. [22]

Schluter, D., and L. Gustafsson. 1993. Maternal inheritance of condition and clutch size in the collared flycatcher. *Evolution* 47: 658–667. [23]

Schmalhausen, I. I. 1949. *Factors of evolution: the theory of stabilizing selection.* Blakiston, Philadelphia. [22]

Schmidt, J. 1919. La valeur de l'individu átitre de générateur appréciée suivant la méthode du croisement diallèle. *Compte Rend. Lab. Carlsberg* 14. [20]

Schmidt-Nielsen, K. 1984. *Scaling: Why is animal size so important*? Cambridge Univ. Press, Cambridge, UK. [11]

Schmitt, J., and D. W. Ehrhardt. 1990. Enhancement of inbreeding depression by dominance and suppression in *Impatiens capensis. Evolution* 44: 269–278. [10]

Schmitt, J., and S. E. Gamble. 1990. The effect of distance from the parental site on offspring performance and inbreeding depression in *Impatiens capensis:* a test of the local adaptation hypothesis. *Evolution* 44: 2022– 2030. [9]

Schneeberger, R. G., and C. A. Cullis. 1991. Specific alterations associated with the environmental induction of heritable changes in flax. *Genetics* 128: 619–630. [12]

Schnell, F. W. 1961. Some general formulations of linkage effects in inbreeding. *Genetics* 46: 947–957. [7]

Schnell, F. W. 1963. The covariance between relatives in the presence of linkage. *In* W. D. Hanson and H. F. Robinson (eds.), *Statistical genetics and plant breeding,* pp. 468–483. Natl. Acad. Sci., Natl. Res. Council Publ. 982, Washington, D.C. [7]

Schnell, F. W., and C. C. Cockerham. 1992. Multiplicative vs. arbitrary gene action in heterosis. *Genetics* 131: 461–469. [9]

Schön, C. C., M. Lee, A. E. Melchinger, W. D. Guthrie, and W. L. Woodman. 1993. Mapping and characterization of quantitative trait loci affecting resistance against second-generation European corn borer in maize with the aid of RFLPs. *Heredity* 70: 648–659. [15]

Schork, N. J. 1991. Efficient computation of patterned covariance matrix mixed models in quantitative segregation analysis. *Genetic Epidem.* 8: 29–46. [13]

Schork, N. J. 1992. Extended pedigree patterned covariance matrix mixed models for quantitative phenotype analysis. *Genetic Epidem.* 9: 73–86. [13]

Schork, N. J. 1993. Extended multipoint identity-by-descent analysis of human quantitative traits: efficiency, power, and modeling considerations. *Am. J. Hum. Genet.* 53: 1306–1319. [16]

Schork, N. J., and M. A. Schork. 1988. Skewness and mixtures of normal distributions. *Comm. Stat. Theor. Meth.* 17: 3951–3969. [13]

Schork, N. J., and M. A. Schork. 1989. Testing separate families of segregation hypotheses: bootstrap methods. *Am. J. Hum. Genet.* 45: 803–813. [13]

Schroeder, M., D. L. Brown, and D. E. Weeks. 1994. Improved programs for the affected-pedigree-member meth-od of linkage analysis. *Genet. Epidem.* 11: 68–74. [16]

Schull, W. J. 1962. Inbreeding and maternal effects in the Japanese. *Eugen. Quart.* 9:14–22. [10]

Schull, W. J., and J. V. Neel. 1965. *The effects of inbreeding on Japanese children.* Harper & Row, NY. [10]

Schull, W. J., and J. V. Neel. 1972. The effects of parental consanguinity and inbreeding in Hirado, Japan. V. Summary and interpretation. *Am. J. Hum. Genet.* 24: 425–453. [10]

Schull, W. J., H. Nagano, M. Yamamoto, and I. Komatsu. 1970. The effects of parental consanguinity and inbreeding in Hirado, Japan. I. Stillbirths and prereproductive mortality. *Am. J. Hum. Genet.* 22: 239–262. [10]

Schüller, C., G. Backes, G. Fischbeck, and A. Jahoor. 1992. RFLP markers to identify the alleles on the *Mla* locus conferring powdery mildew resistance in barley. *Theor. Appl. Genet.* 84: 330–338. [14]

Schultz, S. T., and F. R. Ganders. 1996. Evolution of unisexuality in the Hawaiian Islands: a test of microevolutionary theory. *Evolution* 50: 842–855. [10]

Schultz, S. T., and J. H. Willis. 1995. Individual variation in inbreeding depression: the roles of inbreeding history and mutation. *Genetics* 141: 1209–1223. [10]

Schultz, S. T., J. H. Willis, and M. Lynch. (in prep.) Spontaneous deleterious mutation in *Arabidopsis.* [12]

Scott, J. P., and J. L. Fuller. 1965. *Genetics and social behavior of the dog.* Univ. Chicago Press, Chicago. [10]

Seager, R. D., and F. J. Ayala. 1982. Chromosome interactions in *Drosophila melanogaster*. I. Viability studies. *Genetics* 102: 467–483. [14]

Searle, S. R. 1961. Phenotypic, genetic and environmental correlations. *Biometrics* 17:474–480. [21]

Searle, S. R. 1971. *Linear models.* John Wiley & Sons, NY. [8,A3,A5]

Searle, S. R. 1982. *Matrix algebra useful for statistics.* John Wiley and Sons, NY. [27,A3]

Searle, S. R. 1987. *Linear models for unbalanced data.* John Wiley and Sons, NY. [27]

Searle, S. R., G. Casella, and C. E. McCulloch. 1992. *Variance components.* John Wiley and Sons, NY. [18,19,20,22,26,27]

Sears, E. R. 1953. Nullisomic analysis in common wheat. *Am. Nat.* 87: 245–252. [14]

Service, P. M., and M. R. Rose. 1985. Genetic covariation among life history components: the effect of novel environments. *Evolution* 39:943–945. [21]

Severo, N. C., and M. Zelen. 1960. Normal approximation to the chi-square and noncentral *F* probability distributions. *Biometrika* 47: 411–416. [A5]

Shah, S., and J. R. Green. 1994. The distribution of IQ: a powerful sibship test of association. *Am. J. Hum. Genet.* 58: 163–173. [16]

Shank, D. B., and M. W. Adams. 1960. Environmental variability within inbred lines and single crosses of maize. *J. Genetics* 57: 119–126. [6]

Shapiro, S. S., and M. B. Wilk. 1965. An analysis of variance tests for normality (complete samples). *Biometrika* 52: 591–611. [11]

Sharp, P. M. 1984. The effect of inbreeding on competitive male mating ability in *Drosophila melanogaster. Genetics* 106: 601–612. [10]

Sharp, P. M., and W.-H. Li. 1989. On the rate of DNA sequence evolution in *Drosophila. J. Mol. Evol.* 28: 398–402. [12]

Shaw, R. G. 1987. Maximum-likelihood approaches to quantitative genetics of natural populations. *Evolution* 41: 812–826. [27]

Shaw, R. G. 1991. The comparison of quantitative genetic parameters between populations. *Evolution* 45: 143–151. [21]

Shaw, R. G. 1992. Comparison of quantitative genetic parameters: reply to Cowley and Atchley. *Evolution* 46: 1967–1969. [21]

Shaw, R. G., and N. M. Waser. 1994. Quantitative genetic interpretations of postpollination reproductive traits in plants. *Am. Nat.* 143: 617–635. [23]

Sheldon, B. L., J. M. Rendel, and D. E. Finlay. 1964. The effect of homozygosity on developmental stability. *Genetics* 49: 471–484. [11]

Sheppard, W. E. 1898. On the calculation of the most probable values of frequency constants from data arranged according to equidistant divisions of a scale. *Proc. Lond. Math. Soc.* 29: 353–380. [2]

Sheridan, A. K. 1981. Crossbreeding and heterosis. *Anim. Breed. Abst.* 49: 131–144. [9,10]

Shi, M. J., D. Laloë, F. Ménissier, and G. Renand. 1993. Estimation of genetic parameters of preweaning performance in the French Limousin cattle breed. *Genet. Sel. Evol.* 25: 177–189. [23]

Shields, R. 1989. Moving in on plant genes. *Nature* 337: 308. [14]

Shields, W. M. 1982. *Philopatry, inbreeding, and the evolution of sex.* State Univ. of New York Press, Albany, NY. [9,10]

Shrimpton, A. E., and A. Robertson. 1988a. The isolation of polygenic factors controlling bristle score in *Drosophila melanogaster*. I: Allocation of third chromosome sterno-pleural bristle effects to chromosome sections. *Genetics* 118: 437–443. [14]

Shrimpton, A. E., and A. Robertson. 1988b. The isolation of polygenic factors controlling bristle score in *Drosophila melanogaster*. II: Distribution of third chromosome bristle effects within chromosome sections. *Genetics* 118: 445–459. [14]

Shukla, G. K. 1972. Some statistical aspects of partitioning genotype-environmental components of variability. *Heredity* 29: 237–224. [22]

Shukla, G. K. 1982. Testing the heterogeneity of variances in a two-way classification. *Biometrika* 69: 411–416. [22]

Shull, G. H. 1908. The composition of a field of maize. *Rpt. Am. Breed. Assoc.* 4: 296–301. [1,10]

Shull, G. H. 1914. Duplicate genes for capsule form in *Bursa bursapastoris. Z. Ind. Abstr. Ver.* 12: 97–149. [9]

Shute, N. C. E., and W. J. Ewens. 1988a. A resolution of the ascertainment sampling problem. II. Generalizations and numerical results. *Am. J. Hum. Genet.* 43: 374–386. [13]

Shute, N. C. E., and W. J. Ewens. 1988b. A resolution of the ascertainment sampling problem. III. Pedigrees. *Am. J. Hum. Genet.* 43: 387–395. [13]

Siegel, M. I., and W. J. Doyle. 1975a. Stress and fluctuating limb asymmetry in various species of rodents. *Growth* 39: 363–369. [6]

Siegel, M. I., and W. J. Doyle. 1975b. The differential effects of prenatal and postnatal audiogenic stress on fluctuating dental asymmetry. *J. Exp. Zool.* 191: 211–214. [6]

Siegel, M. I., and W. J. Doyle. 1975c. The effects of cold stress on fluctuating asymmetry in the dentition of the mouse. *J. Exp. Zool.* 191: 211–214. [6]

Siegel, M. I., and M. P. Mooney. 1987. Perinatal stress and increased fluctuating asymmetry of dental calcium in the laboratory rat. *Am. J. Phys. Anthropol.* 73: 267–270. [6]

Silander, J. A., Jr. 1985. The genetic basis of the ecological amplitude of *Spartina patens*. II. Variance and correlation analysis. *Evolution* 39: 1034–1052. [6]

Simmons, L. W., and P. I. Ward. 1991. The heritability of sexually dimorphic traits in the yellow dung fly *Scathophaga stercoraria* (L.) *J. Evol. Biol.* 4: 593–601. [24]

Simmons, M. J., and J. F. Crow. 1977. Mutations affecting fitness in *Drosophila* populations. *Ann. Rev. Genet.* 11: 49–78. [10,12]

Simmonds, N. W. 1981. Genotype (*G*), environment (*E*) and *GE* components of crop yields. *Exp. Agric.* 17: 355–362. [22]

Simons, A. M., and D. A. Roff. 1994. The effect of environmental variability on the heritabilities of traits of a field cricket. *Evolution* 48: 1637–1649. [7]

Simons, A. M., and D. A. Roff. 1996. The effect of a variable environment on the genetic correlation structure in a field cricket. *Evolution* 50: 267–275. [21]

Simpson, E., G. Bulfield, M. Brenan, W. Fitzpatrick, C. Hetherington, and A. Blann. 1982. *H-2* associated differences in replicated strains of mice divergently selected for body weight. *Immunogenetics* 15: 63–70. [14]

Simpson, S. P. 1989. Detection of linkage between quantitative trait loci and restriction fragment length polymorphism using inbred lines. *Theor. Appl. Genet.* 77: 815–819. [14,15]

Simpson, S. P. 1992. Correction: detection of linkage between quantitative trait loci and restriction fragment length polymorphisms using inbred lines. *Theor. Appl. Genet.* 85: 110-111. [15]

Sing, C. F., R. H. Moll, and W. D. Hanson. 1967. Inbreeding in two populations of *Zea mays* L. *Crop Sci.* 7: 631–636. [10]

Singh, S. M., and E. Zouros. 1978. Genetic variation associated with growth rate in the American oyster *(Crassostrea virginica). Evolution* 32: 342–353. [10]

Sittmann, K., H. Abplanalp, and R. A. Fraser. 1966. Inbreeding depression in Japanese quail. *Genetics* 54: 371–379. [10]

Skrøppa, T. 1984. A critical evaluation of methods available to estimate the genotype × environment interaction. *Studia Forestalia Suecica* 166: 3–14. [22]

Slatis, H. M. 1960. An analysis of inbreeding in the European bison. *Genetics* 45: 275–287. [10]

Slatis, H. M., and R. E. Hoene. 1961. The effects of consanguinity on the distribution of continuously variable characters. *Am. J. Hum. Genet.* 13: 28–31. [10]

Slatis, H. M., R. H. Reis, and R. E. Hoene. 1958. Consanguineous marriages in the Chicago region. *Am. J. Hum. Genet.* 10: 446–464. [10]

Slatkin, M. 1972. On treating the chromosome as the unit of selection. *Genetics* 72: 157–168. [5]

Smith, C. 1970. Heritability of liability and concordance in monozygotic twins. *Ann. Hum. Genet.* 34: 85–91. [25]

Smith, C. 1971. Discrimination between different modes of inheritance in genetic disease. *Clinical Genetics* 2: 303–314. [25]

Smith, C., D. S. Falconer, and L. J. P. Duncan. 1972. A statistical and genetical study of diabetes. II. Heritability of liability. *Ann. Hum. Genet.* 35: 281–299. [25]

Smith, C. A. B. 1953. The detection of linkage in human genetics. *J. Royal Stat. Soc. Ser. B* 15: 153–184. [16]

Smith, C. A. B. 1956. On the estimation of intraclass correlation. *Ann. Hum. Genet.* 21: 363–373. [18]

Smith, C. A. B. 1959. Some comments on the statistical methods used in linkage investigation. *Am. J. Hum. Genet.* 11: 289–304. [16]

Smith, H. H. 1937. The relation between genes affecting size and color in certain species of *Nicotiana. Genetics* 22: 361–375. [9,14]

Smith, H. H. 1952. Fixing transgressive vigour in *Nicotiana rustica. In* J. W. Gowen (ed.), *Heterosis,* pp. 161–164. Iowa State College Press, Ames. [10]

Smith, J. N. M., and A. A. Dhondt. 1980. Experimental confirmation of heritable morphological variation in a natural population of song sparrows. *Evolution* 34: 1155–1158. [17,23]

Smith, J. N. M., and R. Zach. 1979. Heritability of some morphological characters in a song sparrow population. *Evolution* 33: 460–467. [17]

Smith, S. P., and H. U. Graser. 1986. Estimating variance components in a class of mixed models by restricted maximum likelihood. *J. Dairy Sci.* 69: 1156–1165. [27]

Smith, S. P., and A. Mäki-Tanila. 1990. Genotypic covariance matrices and their inverses for models allowing dominance and inbreeding. *Genet. Sel. Evol.* 23: 65–91. [26]

Smouse, P. E. 1986. The fitness consequences of multiple-locus heterozygosity under the multiplicative overdominance and inbreeding depression models. *Evolution* 40: 946–958. [10]

Snape, J. W., C. N. Law, and A. J. Worland. 1977. Whole chromosome analysis of height in wheat. *Heredity* 38: 25–26. [14]

Snape, J. W., A. J. Wright, and E. Simpson. 1984. Methods for estimating gene numbers for quantitative characters using doubled haploid lines. *Theor. Appl. Genet.* 67: 143–148. [9]

Sober, E. 1984. *The nature of selection.* The M. I. T. Press, Cambridge, MA. [1]

Sokal, R. R. 1976. The Kluge-Kerfoot phenomenon reexamined. *Am. Nat.* 110: 1077–1091. [11]

Sokal, R. R., and F. J. Rohlf. 1995. *Biometry.* 2nd Ed. W. H. Freeman and Co., NY. [11,13,14,15,16,A4]

Sokolowski, M. B. 1980. Foraging strategies of *Drosophila melanogaster*:a chromosomal analysis. *Behav. Genet.* 10: 291–302. [14]

Soller, M., and J. S. Beckmann. 1987. Cloning quantitative trait loci by insertional mutagensis. *Theor. Appl. Genet.* 74: 369–378. [14]

Soller, M., and J. S. Beckmann. 1988. Genomic genetics and the utilization for breeding purposes of genetic variation between populations. *In* B. S. Weir, E. J. Eisen, M. M. Goodman, and G. Namkoong (eds.), *Proceedings of the second international conference on quantitative genetics,* pp. 161–188. Sinauer Assoc., Sunderland, MA. [14]

Soller, M., and J. S. Beckmann. 1990. Marker-based mapping of quantitative trait loci using replicated progenies. *Theor. Appl. Genet..* 80: 205–208. [14,15]

Soller, M., and A. Genizi. 1978. The efficiency of experimental designs for the detection of linkage between a marker locus and a locus affecting a quantitative trait in segregating populations. *Biometrics* 34: 47–55. [15,16]

Soller, M., T. Brody, and A. Genizi. 1976. On the power of experimental designs for the detection of linkage between marker loci and quantitative loci in crosses between inbred lines. *Theor. Appl. Genet.* 47: 35–39. [15]

Solter, D. 1988. Differential imprinting and expression of maternal and paternal genomes. *Ann. Rev. Genet.* 22: 127–146. [24]

Sondhi, K. C. 1961. Selection for a character with a bounded distribution of phenotypes in *Drosophila subobscura. J. Genetics* 57: 193–221. [11]

Sorensen , D. A., and B. W. Kennedy. 1984. Estimation of genetic variances from unselected and selected populations. *J. Anim. Sci.* 59: 1213–1233. [27]

Sorensen, F. 1969. Embryonic genetic load in coastal Douglas fir *Pseudotsuga menziesii* var. *menziesii. Am. Nat.* 103:389–398. [10]

Sorensen, F. 1970. Self-fertility of a central Oregon source of ponderosa pine. *USDA Forest Serv. Res. Pap. PNW-109.* Pac. Northwest Forest and Range Exp. Stn., Portland, OR. [10]

Sorensen, F. C., J. F. Franklin, and R. Woolard. 1976. Self-pollination effects on seed and seedling traits in noble fir. *Forest. Sci.* 22: 155–159. [10]

Soulé , M. E. 1979. Heterozygosity and developmental stability: another look. *Evolution* 33: 396–401. [6]

Soulé, M. E. 1982. Allomeric variation. 1. The theory and some consequences. *Am. Nat.* 120: 751–764. [6,11]

Soulé , M. E., and J. Cuzin-Roudy. 1982. Allomeric variation. 2. Developmental instability of extreme phenotypes. *Am. Nat.* 120: 765–786. [6]

Southwood, O. I., B. W. Kennedy, K. Meyer, and J. P. Gibson. 1989. Estimation of additive maternal and cytoplasmic genetic variances in animal models. *J. Dairy Sci.* 72: 3006–3012. [26,27]

Spassky, B., N. Spassky, H. Levene, and T. Dobzhansky. 1958. Release of genetic variability through recombination. I. *Drosophila pseudoobscura. Genetics* 43: 844–867. [5]

Spickett, S. G. 1963. Genetic and developmental studies of a quantitative character. *Nature* 199: 870–873. [14]

Spickett, S. G., and J. M. Thoday. 1966. Regular response to selection. 3. Interaction between located polygenes. *Genet. Res.* 7: 96–121. [14]

Spielman, R. S., R. E. McGinnis, and W. J. Ewens. 1993. Transmission test for linkage disequilibrium: the insulin gene region and insulin-dependent diabetes mellitus (IDDM). *Am. J. Hum. Genet.* 52: 506–516. [14]

Spiess, E. B. 1959. Release of genetic variability through recombination. II. *Drosophila persimilis. Genetics* 44: 43–58. [5]

Spiess, E. B., and A. C. Allen. 1961. Release of genetic variability through recombination. VII. Second and third chromosomes of *Drosophila melano-gaster. Genetics* 46: 1531–1553. [5]

Spitze, K., J. Burnson, and M. Lynch. 1991. The covariance structure of life-history characters in *Daphnia pulex. Evolution* 45: 1081–1090. [21]

Sprague, G. F. 1983. Heterosis in maize: theory and practice. *In* R. Frankel (ed.), *Heterosis,* pp. 47–70. *Monog. Theor. Appl. Genet.* Springer-Verlag, Berlin. [9,10]

Sprague, G. F., and B. Brimhall. 1949. Quantitative inheritance of oil in the corn kernel. *Agron. J.* 41: 30–33. [9]

Sprague, G. F., and W. T. Federer. 1951. A comparison of variance components in corn yield trials. II. Error, year × variety, and variety components. *Agron. J.* 43: 535–541. [22]

Sprague, G. F., and L. A. Tatum. 1942. General vs. specific combining ability in single crosses of corn. *J. Am. Soc. Agron.* 34: 923–932. [20]

Sprague, G. F., W. A. Russell, and L. H. Penny. 1960. Mutations affecting quantitative traits in the selfed progeny of double monoploid maize stocks. *Genetics* 45: 855–866. [12]

Spuhler, J. N. 1968. Assortative mating with respect to physical characteristics. *Eugen. Quart.* 15: 128–140. [7]

Sribney, W. M., and M. Swift. 1992. Power of sib-pair and sib-trio linkage analysis with assortative mating and multiple disease loci. *Am. J. Hum. Genet.* 51: 773–784. [16]

Stam, P. 1991. Some aspects of QTL analysis. *Proceedings of the Eighth Meeting of the Eucarpia Section Biometrics on Plant Breeding,* Brno, Czechoslovakia. Pp. 24–32. [15]

Stam, P., and A. C. Zeven. 1981. The theoretical proportion of the donor genome in near-isogenic lines of self-fertilizers based on backcrossing. *Euphytica* 30: 227–238. [14]

Stark, A. E. 1976. On the method of Penrose of estimating the number of effective factors contributing to a character. *Ann. Hum. Genet.* 39: 465–470. [13]

Stearns, S. C., and T. J. Kawecki. 1994. Fitness sensitivity and the canalization of life-history traits. *Evolution* 48: 1438–1450. [7]

Stern, C., and E. W. Schaeffer. 1943. On wild-type isoalleles in *Drosophila melanogaster*. *Proc. Natl. Acad. Sci. USA* 29: 361–367. [12]

Stigler, S. M. 1986. *The history of statistics.* Harvard Univ. Press, Cambridge, MA. [1,8]

Stolk, J. M., G. Vantini, R. B. Guchhait, J. H. Hurst, B. D. Perry, D. C. U'Prichard, and R. C. Elston. 1984. Inheritance of adrenal phenylethanolamine *N*-methyltransferase activity in the rat. *Genetics* 108: 633–649. [13]

Stouthamer, R., J. A. J. Breeuwer, R. F. Luck, and J. H. Werren. 1993. Molecular identification of microorganisms associated with parthenogenesis. *Nature* 361: 66–68. [6]

Stratton, D. A. 1994. Genotype-by-environ-ment interactions for fitness of *Erigeron annuus* show fine-scale selective heterogeneity. *Evolution* 48: 1607–1618. [22]

Strauss, R. E. 1985. Evolutionary allometry and variation in body forms in the South American catfish genus *Corydoras* (Callichthyidae). *Syst. Zool.* 34: 381–396. [11]

Strauss, R. E. 1991. Correlations between heterozygosity and phenotypic variability in *Cottus* (Teleostei: Cottidae): character components. *Evolution* 45: 1950–1956. [6]

Strauss, S. H. 1986. Heterosis at allozyme loci under inbreeding and crossbreeding in *Pinus attenuata. Genetics* 113:115–134. [10]

Strauss, S. H. 1987. Heterozygosity and developmental stability under inbreeding and crossbreeding in *Pinus attenuata. Evolution* 41: 331–339. [6]

Strauss, S. H., and W. J. Libby. 1987. Allozyme heterosis in radiata pine is poorly explained by overdominance. *Am. Nat.* 130: 879–890. [10]

Strauss, S. Y., and R. Karban. 1994. The significance of outcrossing in an intimate plant-herbivore relationship. I. Does outcrossing provide an escape from herbivores adapted to the parent plant? *Evolution* 48: 454–464. [6]

Streisinger, G., C. Walker, N. Dower, D. Knauber, and F. Singer. 1981. Production of clones of homozygous diploid zebra fish (*Brachydanio rerio*). *Nature* 291: 293-296. [14]

Strickberger, M. W. 1972. Viabilities of third chromosomes of *Drosophila pseudoobscura* differing in relative competitive fitness. *Genetics* 72: 679–689. [10]

Stricker, C., R. L. Fernando, and R. C. Elston. 1995a. An algorithm to approximate the likelihood for pedigree data with loops by cutting. *Theor. Appl. Genet.* 91: 1054–1063. [13]

Stricker, C., R. L. Fernando, and R. C. Elston. 1995b. Linkage analysis with an alternative formulation for the mixed model of inheritance: the finite polygenic mixed model. *Genetics* 141: 1651–1656. [13,16]

Struhl, K. 1987. Promoters, activator proteins, and the mechanism of transcriptional initiation in yeast. *Cell* 49: 295–297. [12]

Stuber, C. W., R. H. Moll, M. M. Goodman, H. E. Schaffer, and B. S. Weir. 1980. Allozyme frequency changes associated with selection for increased grain yield in maize. *Genetics* 95: 225–236. [14]

Stuber, C. W., M. M. Goodman, and R. H. Moll. 1982. Improvement of yield and ear number resulting from selection at allozyme loci in a maize population. *Crop Sci.* 22: 737–740. [14]

Stuber, C. W., M. D. Edwards, and J. F. Wendel. 1987. Molecular-marker-facilitated investigations of quantitative-trait loci in maize. II. Factors influencing yield and its component traits. *Crop Sci.* 27: 639–648. [15]

Stuber, C. W., S. E. Lincoln, D. W. Wolff, T. Helentjaris, and E. S. Lander. 1992. Identification of genetic factors contributing to heterosis in a hybrid from two elite inbred lines using molecular markers. *Genetics* 132: 823–839. [10,15]

Suarez, B. K. 1978. The affected sib pair IBD distribution for HLA disease susceptibility genes. *Tissue Antigens* 12: 87–93.[16]

Suarez, B. K., and P. V. Eerdewegh. 1984. A comparison of three affected-sib-pair scoring methods to detect HLA-linked disease susceptibility genes. *Am. J. Med. Genet.* 18: 135–146. [16]

Suarez, B. K., J. Rice, and T. Reich. 1978. The generalized sib pair IBD distribution: its use in the detection of linkage. *Ann. Hum. Genet.* 42: 87–94. [16]

Suarez, B. K., C. L. Hampe, and P. Van Eerdewegh. 1994. Problems of replicating linkage claims in psychiatry. *In* E. S. Gershon and C. R. Cloninger (eds.), *Genetic approaches to mental disorders*, pp. 23–46. American Psychiatric Press, Washington, D. C. [16]

Sulisalo, T., J. Klockars, O. Mäkitie, C. A. Francomano, A. de la Chapelle, I. Kaitila, and P. Sistonen. 1994. High-resolution linkage-disequilibrium mapping of the cartilage-hair hypoplasia gene. *Am. J. Hum. Genet.* 55: 937–945. [14]

Sultan, S. E., and F. A. Bazzaz. 1993. Phenotypic plasticity in *Polygonum persicaria*. III. The evolution of ecological breadth for nutrient environment. *Evolution* 47: 1050–1071. [22]

Swaddle, J. P., M. Witter, and I. C. Cuthill. 1994. The analysis of fluctuating asymmetry. *Anim. Behav.* 48: 986–989. [6]

Tachida, H., and C. C. Cockerham. 1990. Evolution of neutral quantitative characters with gene interaction and mutation. *In* N. Takahata and J. F. Crow (eds.), *Population biology of genes and molecules*, pp. 233–249. Baifukan, Tokyo. [12]

Takahashi, J. S., L. H. Pinto, and M. H. Vitaterna. 1994. Forward and reverse genetic approaches to behavior in the mouse. *Science* 264: 1724–1733. [14]

Takano, T., S. Kusakabe, and T. Mukai. 1987. The genetic structure of natural populations of *Drosophila melanogaster*. XX. Comparison of genotype-environment interaction in viability between a northern and a southern population. *Genetics* 117: 245–254. [22]

Tallis, G. M. 1959. Sampling errors of genetic correlation coefficients, calculated from the analyses of variance and covariance. *Aust. J. Stat.* 1: 35–43. [21]

Tan, W. Y., and W. C. Chang. 1972. Convolution approach to genetic analysis of quantitative characters of self-fertilized populations. *Biometrics* 28: 1073–1090. [13]

Tan, W. Y., and H. D'Angelo. 1979. Statistical analysis of joint effects of major genes and polygenes in quantitative genetics. *Biom. J.* 21: 179–192. [13]

Tanksley, S. D., and J. Hewitt. 1988. Use of molecular markers in breeding for soluble solids content in tomato — a re-examination. *Theor. Appl. Genet.* 75: 811–823. [14]

Tanksley, S. D., N. D. Young, A. H. Paterson, and M. W. Bonierbale. 1989. RFLP mapping in plant breeding: new tools for an old science. *Biotechnol.* 7: 257–264. [14]

Tanksley, S. D., M. W. Ganal, and G. B. Martin. 1995. Chromosome landing: a paradigm for map-based gene cloning in plants with large genomes. *Trends Genet.* 11: 63– 68. [14]

Tanner, J. M. 1949. Fallacy of per-weight and per-surface area standards and their relation to spurious correlation. *J. Appl. Physiology* 2: 1–15. [11]

Tantawy, A. O. 1957. Genetic variance of random-inbred lines of *Drosophila melano-gaster* in relation to coefficients of inbreeding. *Genetics* 42: 121–136. [10]

Tantawy, A. O., and E. C. R. Reeve. 1956. Studies in quantitative inheritance. IX. The effects of inbreeding at different rates in *Drosophila melanogaster*. *Z. indukt. Abst. Ver. bungslehre* 87: 648–667. [10]

Taylor, J. F., B. Bean, C. E. Marshall, and J. J. Sullivan. 1985. Genetic and environmental components of semen production traits of artificial insemination Holstein bulls. *J. Dairy Sci.* 68: 2703–2723]. [27]

Templeton, A. R. 1977. Analysis of head shape differences between two interfertile species of Hawaiian *Drosophila*. *Evolution* 31: 330–341. [9]

Templeton, A. R. 1980. The theory of speciation by the founder principle. *Genetics* 92: 1011–1038. [14]

Templeton, A. R. 1981. Mechanisms of speciation — a population genetic approach. *Ann. Rev. Ecol. Syst.* 12: 23–48. [9]

Templeton, A. R. 1986. Coadaptation and outbreeding depression. *In* M. E. Soulé (ed.), *Conservation biology: the science of scarcity and diversity*, pp. 105–116. Sinauer Assoc., Sunderland, MA. [9]

Templeton, A. R. 1987. The general relationship between average effect and average excess. *Genet. Res.* 49: 69–70. [4]

Templeton, A. R. 1995. A cladistic analysis of phenotypic associations with haplotypes inferred from restriction endonuclease mapping. V. Analysis of case/control sampling designs. Alzheimer's disease and the apoprotein E locus. *Genetics* 140: 403–409. [14]

Templeton, A. R., H. Hollocher, S. Lawler, and J. S. Johnston. 1989. Natural selection and ribosomal DNA in *Drosophila. Genetics* 31: 296–303. [12]

Templeton, A. R., and M. A. Rankin. 1978. Genetic revolutions and control of insect populations. *In* R. H. Richardson (ed.), *The screwworm problem,* pp. 83–112. Univ. Texas Press, Austin, TX. [12]

Templeton, A. R., and B. Read. 1983. The elimination of inbreeding depression in a captive herd of Speke's gazelle. *In* C. M. Schonewald-Cox, S. M. Chambers, F. MacBryde, and L. Thomas (eds.), *Genetics and conservation: a reference for managing wild animal and plant populations,* pp. 241–261. Benjamin/Cummings, Menlo Park, CA. [10]

Templeton, A. R., and B. Read. 1984. Factors eliminating inbreeding depression in a captive herd of Speke's gazelle *(Gazella spekei). Zoo Biol.* 3: 177–199. [10]

Templeton, A. R., C. F. Sing, and B. Brokaw. 1976. The unit of selection in *Drosophila mercatorum.* I. The interaction of selection and meiosis in parthenogenetic strains. *Genetics* 82: 349–376. [9]

Templeton, A. R., E. Boerwinkle, and C. F. Sing. 1987. A cladistic analysis of phenotypic associations with haplotypes inferred from restriction endonuclease mapping. I. Basic theory and an analysis of alcohol dehydrogenase activity in *Drosophila. Genetics* 117: 343–351. [14]

Templeton, A. R., C. F. Sing, A. Kessling, and S. Humphries. 1988. A cladistic analysis of phenotypic associations with haplotypes inferred from restriction endonuclease mapping. II. The analysis of natural populations. *Genetics* 120: 1145–1154. [14]

Templeton, A. R., K. A. Crandall, and C. F. Sing. 1992. A cladistic analysis of phenotypic associations with haplotypes inferred from restriction endonuclease mapping. III. Cladogram estimation. *Genetics* 132: 619–633. [14]

Terwilliger, J. D. 1995. A powerful likelihood method for the analysis of linkage disequilibrium between trait loci and one or more polymorphic marker loci. *Am. J. Hum. Genet.* 56: 777–787. [14]

Terwilliger, J. D., and J. Ott. 1992. A haplotype-based haplotype relative risk statistic. *Hum. Hered.* 42: 337–346. [14]

Terwilliger, J. D., and J. Ott. 1994. *Handbook of human genetic linkage.* Johns Hopkins Univ. Press, Baltimore, MD. [16]

Teutonico, R. A., and T. C. Osborn. 1994. Mapping of RFLP and qualitative trait loci in *Brassica rapa* and comparison to the linkage maps of *B. napus, B. oleracea,* and *Arabidopsis thaliana. Theor. Appl. Genet.* 89: 885–894. [14]

Thaller, G., L. Dempfle, and I. Hoeschele. 1996. Maximum likelihood analysis of rare binary traits under different modes of inheritance. *Genetics* 143: 1819–1829. [13]

Thiele, T. N. 1889. *Almindelig iagltagelseslaere.* (reprinted in *Ann. Math. Stat.* 2: 165–308). [2]

Thoday, J. M. 1953. Components of fitness. *Symp. Soc. Exp. Biol.* 7: 96–113. [6]

Thoday, J. M. 1961. Location of polygenes. *Nature* 191: 368–370. [14]

Thoday, J. M. 1979. Polygene mapping: uses and limitations. *In* J. N. Thompson, Jr., and J. M. Thoday (eds.), *Quantitative genetic variation,* pp. 219–233. Academic Press, NY. [14]

Thoday, J. M., and J. N. Thompson, Jr. 1976. The number of segregating genes implied by continuous variation. *Genetica* 46: 335–344. [13,14]

Thoday, J. M., J. B. Gibson, and S. G. Spickett. 1964. Regular responses to selection. II. Recombination and accelerated response. *Genet. Res.* 5: 1–19. [14]

Thomas-Orillard, M., and B. Jeune. 1985. Gene actions involved in determining the number of ovarioles and sternite chaetae in freshly collected strains of *Drosophila melanogaster. Genetics* 111: 819–829. [20]

Thompson, D'A. W. 1917. *On growth and form.* Cambridge Univ. Press, Cambridge, UK. [11]

Thompson, D'A. W. 1943. *On growth and form* (revised ed.) Cambridge Univ. Press, Cambridge, UK. [11]

Thompson, E. A. 1975. The estimation of pairwise relationships. *Ann. Hum. Genet.* 39: 173–188. [27]

Thompson, E. A., and R. G. Shaw. 1990. Pedigree analysis for quantitative traits: variance components without matrix inversion. Biometrics 46: 399–413. [27]

Thompson, E. A., and R. G. Shaw. 1992. Estimating polygenic models for multivariate data on large pedigrees. *Genetics* 131: 971–978. [27]

Thompson, E. A., and S. W. Guo. 1991. Evaluation of likelihood ratios for complex genetic models. *IMA J. Math. Appl. Med. Biol.* 8: 149–169. [13]

Thompson, E. A., S. Lin, A. B. Olshen, and E. M. Wijsman. 1993. Monte carlo analysis on a large pedigree. *Genet. Epidem.* 10: 677–682. [13]

Thompson, G. 1986. Determining the mode of inheritance of RFLP-associated diseases using the affected sib-pair method. *Am. J. Hum. Genet.* 39: 207–221. [16]

Thompson, H. R. 1951. Truncated lognormal distributions. *Biometrika* 38: 414–422. [11]

Thompson, J. N., Jr. 1973. General and specific effects of modifiers of mutant expression. *Genet. Res.* 22: 211–215. [12]

Thompson, J. N., Jr. 1974. Studies of the nature and function of polygenic loci in *Drosophila.* I. Comparison of genomes from selected lines. *Heredity* 33: 373–387. [12]

Thompson, J. N. 1988. Evolutionary ecology of the relationship between oviposition preference and performance of offspring in phytophagous insects. *Entomol. Exp. Appl.* 47: 3–14. [5]

Thompson, J. N., Jr., and W. E. Spivey. 1984. Organization of polygenic system: cell death modifiers from natural populations of *Drosophila melanogaster. Genet. Res.* 44: 261–269. [12]

Thompson, J. N., Jr., and J. M. Thoday. 1972. Modification of dominance by selection in the homozygote. *Heredity* 29: 285–292. [12]

Thompson, R. 1973. The estimation of variance and covariance components with an application when records are subject to culling. *Biometrics* 29: 527–550. [27]

Thompson, R. 1976. The estimation of maternal genetic variances. *Biometrics* 32: 903–917. [23]

Thompson, R. 1977. The estimation of heritability with unbalanced data. II. Data available on more than two generations. *Biometrics* 33: 496–504. [26]

Thompson, R., and W. G. Hill. 1990. Univariate REML analysis for multivariate data with the animal model. *In* W. G. Hill, R. Thompson, and J. A. Woolliams (eds.), *Proc. 4th World Congr. Genet. Appl. Livestock Prod.*, Vol. 13, pp. 472–475. Edinburgh. [27]

Thornhill, N. W. (ed.) 1993. *The natural history of inbreeding and outbreeding.* Univ. Chicago Press, Chicago. [10]

Tier, B. 1990. Computing inbreeding coefficients quickly. *Genet. Sel. Evol.* 23: 419–430. [26]

Tier, B., and J. Sölkner. 1993. Analysing gametic variation with an animal model. *Theor. Appl. Genet.* 85: 868–872. [26]

Tiku, M. L. 1965. Laguerre series forms for noncentral χ^2 and F distributions. *Biometrika* 52: 415–428. [A5]

Tiret, L., L. Abel, and R. Rakotovao. 1993. Effect of ignoring genotype- environment interaction on segregation analysis of quantitative traits. *Genet. Epidem.* 10: 581–586. [13]

Titterington, D. M., A. F. M. Smith, and U. E. Makov. 1985. *Statistical analysis of finite mixture distributions.* John Wiley & Sons, NY. [13]

Tourjee, K. R., J. Harding, and T. G. Byrne. 1995. Complex segregation analysis of *Gerbera* flower color. *Heredity* 74: 303–310. [13]

Touzet, P., R. G. Winkler, and T. Helentjaris. 1995. Combined genetic and physiological analysis of a locus contributing to quantitative variation. *Theor. Appl. Genet.* 91: 200-205. [14]

Towey, P., and J. L. Jinks. 1977. Alternative ways of estimating the number of genes in a polygenic system by genotype assay. *Heredity* 39: 399–410. [9]

Trivers, R. 1974. Parent-offspring conflict. *Am. Zool.* 14: 249–264. [23]

Trow, A. H. 1913. Forms of reduplication: primary and secondary. *J. Genetics* 2: 313–324. [14]

True, J. R., J. M. Mercer, and C. C. Laurie. 1996. Differences in crossover frequency and distribution among three sibling species of *Drosophila. Genetics* 142: 507–523. [14]

Trustrum, G. B. 1961. The correlations between relatives in a random mating diploid population. *Proc. Cambridge Phil. Soc.* 57: 315–320. [7]

Tsubota, S., and P. Schedl. 1986. Hybrid dysgenesis-induced revertants of insertions at the 5′ end of the *rudimentary* gene in *Drosophila melanogaster*: Transposon-induced control mutations. *Genetics* 114: 165–182. [12]

Tukey, J. W. 1956. Variance of variance components. I. Balanced designs. *Ann. Math. Stat.* 27: 722–736. [18]

Tukey, J. W. 1957. Variance of variance components. II. The unbalanced single classification. *Ann. Math. Stat.* 28: 43–56. [18]

Turelli, M. 1984. Heritable genetic variation via mutation-selection balance: Lerch's zeta meets the abdominal bristle. *Theor. Pop. Biol.* 25: 138–193. [12]

Turelli, M., and H. A. Orr. 1995. The dominance theory of Haldane's rule. *Genetics* 140: 389–402. [14]

Turner, J. R. G. 1977. Butterfly mimicry: the general evolution of adaptation. *Evol. Biol.* 10: 163–206. [12]

Turner, J. R. G. 1981. Adaptation and evolution in *Heliconius*: a defense of neo-Darwinism. *Ann. Rev. Ecol. Syst.* 12: 99–121. [12]

Turton, J. D. 1981. Crossbreeding of dairy cattle — a selective review. *Anim. Breed. Abst.* 49: 293–300. [9,10]

Uddin, M. N., F. W. Ellison, L. O'Brien, and B. D. H. Latter. 1994. The performance of pure lines derived from heterotic bread wheat hybrids. *Aust. J. Agric. Res.* 45: 591–600. [10]

Uimari, P., and B. W. Kennedy. 1990. Mixed model methodology to estimate additive and dominance genetic values under complete dominance. *In* W. G. Hill, R. Thompson, and J. A. Woolliams (eds.), *Proc. 4th World Cong. Genet. Appl. Livestock Prod.*, Vol. 13: 297–300. [26]

Uimari, P., G. Thaller, and I. Hoeschele. 1996. The use of multiple markers in a Bayesian method for mapping quantitative trait loci. *Genetics* 143: 1831–1842. [16]

Underhill, D. K. 1969. Heritability of some linear body measurements and their ratios in the leopard frog *Rana pipiens. Evolution* 23: 268–275. [17]

Utz, H. F., and A. E Melchinger. 1994. Comparison of different approaches to interval mapping of quantitative trait loci. *In* J. W. van Ooijen and J. Jansen (eds.), *Biometrics in plant breeding: applications of molecular markers*, pp. 195–204. CPRO-DLO, Netherlands. [15]

Uyenoyama, M. K. 1993. Genetic incompatibility as a eugenic mechanism. *In* N. W. Thornhill (ed.), *The natural history of inbreeding and outbreeding: theoretical and empirical perspectives*, pp. 60–73. Univ. Chicago Press, Chicago. [10]

Uyenoyama, M. K., K. E. Holsinger, and D. M. Waller. 1994. Ecological and genetic factors directing the evolution of self-fertilization. *Oxford Surv. Evol. Biol.* 9: 327–381. [10]

Valentine, D. W., and M. E. Soulé . 1973. Effects of $p{\cdot}p' - DDT$ on developmental stability of pectoral fin rays in the grunnion, *Leuresthes tenuis.* Fish. Bull. U. S. 71: 921–926. [6]

Vallejos, C. E., and S. D. Tanksley. 1983. Segregation of isozyme markers and cold tolerance of an interspecific backcross in tomato. *Theor. Appl. Genet.* 66: 241–247. [15]

Vallejos, C. E., N. S. Sakiyama, and C. D. Chase. 1992. A molecular marker-based linkage map of *Phaseolus vulgaris* L. *Genetics* 131: 733–740. [9]

Van Aarde, I. M. R. 1975. The covariance of relatives derived from a random mating population. *Theor. Pop. Biol.* 8: 166–183. [7]

van Arendonk, J. A. M., C. Smith, and B. W. Kennedy. 1989. Method to estimate genotype probabilities at individual loci in farm livestock. *Theor. Appl. Genet.* 78: 735–740. [13]

Vandenberg, S. G. 1972. Assortative mating, or who marries whom? *Behav. Genetics* 2: 127–157. [7]

van der Beck, S., J. A. M. van Arendonk, and A. F. Groen. 1995. Power of two- and three-generation QTL mapping experiments in an outbred populations containing full-sib or half-sib families. *Theor. Appl. Genet.* 91: 1115–1124. [16]

van der Werf, J. H. J. 1990. A note on the use of conditional models to estimate additive genetic variance in selected populations. *In* W. G. Hill, R. Thompson, and J. A. Woolliams (eds.), *Proc. 4th World Congr. Genet. Appl. Livestock Prod.*, Vol. 13, pp. 476–479. Edinburgh. [27]

van der Werf, J. H. J., and I. J. M. de Boer. 1990. Estimation of additive genetic variance when base populations are selected. *J. Anim. Sci.* 68: 3124–3132. [27]

van der Werf, J. H. J., and R. Thompson. 1992. Variance decomposition in the estimation of genetic variance with selected data. *J. Anim. Sci.* 70: 2975–2985. [27]

Van Eerdewegh, P., C. L. Hampe, B. K. Suarez, and T. Reich. 1993. Alzheimer's disease: a piscatorial trek. *Genet. Epidem.* 10: 395–400. [16]

van Noordwijk, A. J. 1984. Quantitative genetics in natural populations of birds illustrated with examples from the Great Tit, *Parus major.* si In K. Wöhrmann and V. Loeschcke (eds.), *Population biology and evolution*, pp. 67–79. Springer-Verlag, NY. [23]

van Noordwijk, A. J., and G. de Jong. 1986. Acquisition and allocation of resources: their influence on variation in life-history tactics. *Am. Nat.* 128: 137–142. [21]

van Noordwijk, A. J., and W. Scharloo. 1981. Inbreeding in an island population of the great tit. *Evolution* 35: 674–688. [10]

van Noordwijk, A. J., J. H. van Balen, and W. Scharloo. 1981. Genetic and environmental variation in the clutch size of the great tit *(Parus major). Neth. J. Zool.* 31: 342–372. [17]

van Ooijen, J. W. 1992. Accuracy of mapping quantitative trait loci in autogamous species. *Theor. Appl. Genet.* 84: 803–811. [15]

van Ooijen, J. W. 1994. Comparison of a single-QTL model with an approximate multiple-QTL model for QTL mapping. *In* J. W. van Ooijen and J. Jansen (eds.), *Biometrics in plant breeding: applications of molecular markers*, pp. 205–212. CPRO-DLO, Netherlands. [15]

Van Treuren, R., R. Bijlsma, N. J. Ouborg, and W. Van Delden. 1993. The effect of population size and plant density on outcrossing rates in *Salvia pratensis. Evolution* 47: 1094–1104. [10]

Van Valen, L. 1962. A study of fluctuating asymmetry. *Evolution* 16: 125–142. [6]

Van Vleck, L. D. 1968. Selection bias in estimation of the genetic correlation. *Biometrics* 24: 951–962. [21]

Van Vleck, L. D. 1971. Estimation of the heritability of threshold characters. *J. Dairy Sci.* 55: 218–255. [25]

Van Vleck, L. D. 1978. A genetic model involving fetal effects on traits of the dam. *Biometrics* 34: 123–127. [23]

Van Vleck, L. D., and C. R. Henderson. 1961. Empirical sampling estimates of genetic correlations. *Biometrics* 17:359–371. [21]

Veldboom, L. R., M. Lee, and W. L. Woodman. 1994. Molecular marker- facilitated studies in an elite maize population. I. Linkage analysis and determination of QTL for morphological traits. *Theor. Appl. Genet.* 88: 7–16. [14,15]

Venable, D. L., and A. Búrquez. 1990. Quantitative genetics of size, shape, life-history, and fruit characteristics of the seed heteromorphic composite *Heterosperma pinnatum.* II. Correlation structure. *Evolution* 44: 1748–1763. [21]

Vesely, J., and O. Robison. 1971. Genetic and maternal effects on preweaning growth and type score in beef calves. *J. Anim. Sci.* 32: 825–831. [24]

Vetta, A. 1976. Dominance variance in Fisher's model of assortative mating. *Ann. Hum. Genet.* 39: 447–453. [7]

Vetta, A., and C. A. B. Smith. 1974. Comments on Fisher's theory of assortative mating. *Ann. Hum. Genet.* 38: 243–248. [7]

Via, S. 1984. The quantitative genetics of polyphagy in an insect herbivore. I. Genotype-environment interaction in larval performance on different host plant species. *Evolution* 38: 881–895. [6,22]

Via, S. 1991. The genetic structure of host plant adaptation in a spatial patchwork: demographic variability among reciprocally transplanted pea aphid clones. *Evolution* 45: 827–852. [22]

Via, S. 1993. Adaptive phenotypic plasticity: target or by-product of selection in a variable environment? *Am. Nat.* 142: 352–365. [22]

Vieland, V. J., and S. E. Hodge. 1995. Inherent intractability of the ascertainment problem for pedigree data: a general likelihood framework. *Am. J. Hum. Genet.* 56: 33–43. [13]

Virdee, S. 1993. Unraveling Haldane's rule. *Trends Ecol. Evol.* 8: 185–187. [14]

Visscher, P. M., and C. S. Haley. 1996. Detection of putative quantitative trait loci in line crosses under infinitesimal genetic models. *Theor. Appl. Genet.* 93: 691–702. [15]

Visscher, P. M., C. S. Haley, and S. A. Knott. 1996a. Mapping QTLs for binary traits in backcross and F_2 populations. *Genet. Res.* 68: 55–63. [15]

Visscher, P. M., R. Thompson, and C. S. Haley. 1996b. Confidence intervals in QTL mapping by bootstrapping. *Genetics* 143: 1013–1020. [15]

Vogel, S. 1981. *Life in moving fluids: the physical biology of flow.* Princeton Univ. Press, Princeton, NJ. [11]

Vogel, S. 1989. *Life's devices: the physical world of animals and plants.* Princeton Univ. Press, Princeton, NJ. [11]

Vrijenhoek, R. C., and S. Lerman. 1982. Heterozygosity and developmental stability under sexual and asexual breeding systems. *Evolution* 36: 768–776. [6]

Waddington, C. H. 1949. Canalization of development and the inheritance of acquired characters. *Nature* 150: 563–565. [11]

Waddington, C. H. 1952. Selection of the genetic basis for an acquired character. *Nature* 169: 278. [11]

Waddington, C. H. 1953. Genetic assimilation of an acquired character. *Evolution* 7: 118–126. [11]

Waddington, C. H. 1957. *The strategy of the genes.* Macmillan, NY. [11,12]

Waddington, C. H. 1959. Canalization of development and genetic assimilation of acquired characters. *Nature* 183: 1654–1655. [11]

Wade, M. J., and N. W. Chang. 1995. Increased male fertility in *Tribolium confusum* beetles after infection with the intracellular parasite *Wolbachia. Nature* 373: 72–74. [6]

Wade, M. J., N. A. Johnson, and G. Wardle. 1994. Analysis of autosomal polygenic variation for the expression of Haldane's rule in flour beetles. *Genetics* 138: 791–799. [14]

Wahlsten, D. 1990. Insensitivity of the analysis of variance to heredity-environment interaction. *Behav. Brain Sci.* 13: 109–120. [22]

Wainwright, S. A., W. D. Biggs, J. D. Currey, and J. M. Gosline. 1982. *Mechanical design in organisms.* Princeton Univ. Press, Princeton, NJ. [11]

Wald, A. 1943. Tests of statistical hypotheses concerning several parameters when the number of observations is large. *Trans. Am. Math. Soc.* 54: 426–482. [13, A4]

Waller, D. M. 1993. The statics and dynamics of mating system evolution. *In* N. W. Thornhill (ed.), *The natural history of inbreeding and outbreeding: theoretical and empirical perspectives,* pp. 97–117. Univ. Chicago Press, Chicago. [10]

Walters, D. E., and J. R. Morton. 1978. On the analysis of variance of a half diallel table. *Biometrics* 34: 91–94. [20]

Wang, C. S., B. S. Yandell, and J. J. Rutledge. 1991. Bias of maximum likelihood estimator of intraclass correlation. *Theor. Appl. Genet.* 82: 421- -424. [18]

Wang, C. S., B. S. Yandell, and J. J. Rutledge. 1992. The dilemma of negative analysis of variance estimators of intraclass correlation. *Theor. Appl. Genet.* 85: 79–88. [18]

Ward, P. J. 1993. Some developments on the affected-pedigree-member meth-od of linkage analysis. *Am. J. Hum. Genet.* 52: 1200-1215. [16]

Ward, P. J. 1994. Parent-offspring regression and extreme environments. *Heredity* 72: 574–581. [22]

Waser, N. M. 1993a. The statics and dynamics of mating system evolution. *In* N. W. Thornhill (ed.), *The natural history of inbreeding and outbreeding: theoretical and empirical perspectives,* pp. 97–117. Univ. Chicago Press, Chicago. [10]

Waser, N. M. 1993b. Population structure, optimal outcrossing, and assortative mating in angiosperms. *In* N. W. Thornhill (ed.), *The natural history of inbreeding and outbreeding: theoretical and empirical perspectives,* pp. 1–13. Univ. Chicago Press, Chicago. [9]

Waser, N. M., and M. V. Price. 1983. Optimal and actual outcrossing in plants, and the nature of plant-pollinator interaction. *In* C. E. Jones, and R. J. Little (eds.), *Handbook of experimental pollination biology,* pp. 341–359. Van Nostrand Reinhold, NY. [9]

Waser, N. M., and M. V. Price. 1985. Reciprocal transplant experiments with *Delphinium nelsonii* (Ranunculaceae). Evidence for local adaptation. *Am. J. Bot.* 72: 1726–1732. [9]

Waser, N. M., and M. V. Price. 1989. Optimal outcrossing in *Ipomopsis aggregata:* seed set and offspring fitness. *Evolution* 43: 1097–1109. [9]

Waser, N. M., and M. V. Price. 1994. Crossing-distance effects in *Delphi-nium nelsonii:* outbreeding and inbreeding depression in progeny fitness. *Evolution* 48: 842–852. [9]

Wassermann, G. D. 1978. Testability of the role of natural selection within theories of population genetics and evolution. *Brit. J. Phil. Sci.* 29: 223–242. [1]

Watanabe, T. K., O. Yamaguchi, and T. Mukai. 1976. The genetic variability of third chromosomes in a local population of *Drosophila melanogaster. Genetics* 82: 63–82. [10]

Waters, N. F. 1931. Inheritance of body weight in domestic fowls. *Rhode Island Agric. Exp. Sta. Bull.* 228: 7–103. [9]

Wayne, R. K., W. S. Modi, and S. J. O'Brien. 1986. Morphological variability and asymmetry in the cheetah *(Acinonyx jubatus),* a genetically uniform species. *Evolution* 40: 78–85. [6]

Wearden, S. 1964. Alternative analyses of the diallel cross. *Heredity* 19: 669–680. [20]

Weber, K. E. 1992. How small are the smallest selectable domains of form? *Genetics* 130: 345–353. [12]

Weber, K. E., and L. T. Diggins. 1990. Increased selection response in larger populations. II. Selection for ethanol vapor resistance in *Drosophila melano-gaster* at two population sizes. *Genetics* 125: 585–597. [12]

Weeks, D. E., and K. Lange. 1988. The affected-pedigree-member method of linkage analysis. *Am. J. Hum. Genet.* 42: 315–326. [16]

Weeks, D. E., and K. Lange. 1992. A multilocus extension of the affected-pedigree-member method of linkage analysis. *Am. J. Hum. Genet.* 50: 859–868. [16]

Weeks, D. E., and G. M. Lathrop. 1995. Polygenic disease: methods for mapping complex disease traits. *Trends Genet.* 11: 513–519. [16]

Wehrhahn, C. and R. W. Allard. 1965 The detection and measurement of the effects of individual genes involved in the inheritance of a quantitative character. *Genetics* 51: 109–119. [9]

Weigensberg, I., and D. A. Roff. 1996. Natural heritabilities: can they be reliably estimated in the laboratory? *Evolution* 50: 2149–2157. [7,17]

Weinberg, W. 1908. Ueber den Nachweis der Vererbung beim Menschen. *Jh. Ver. Vaterl. Naturk. Wurttemb.* 64: 368–382. [4]

Weinberg, W. 1927. Mathematische grundlagen der probandenmethode. *Z. Induktive Abstammungs*

Veterbung-slehre 48: 179–228. [13]
Weir, B. S. 1979. Inferences about linkage disequilibrium. *Biometrics* 35: 235–254. [5]
Weir, B. S. 1996. *Genetic data analysis.* Sinauer Assoc., Sunderland, MA. [4,5,A4]
Weir, B. S., and C. C. Cockerham. 1968. Pedigree mating with two linked loci. *Genetics* 61: 923–940. [7]
Weir, B. S., and C. C. Cockerham. 1969. Group inbreeding with two linked loci. *Genetics* 63: 711–742. [7]
Weir, B. S., and C. C. Cockerham. 1973. Mixed self and random mating at two loci. *Genet. Res.* 21: 247–262. [7]
Weir, B. S., and C. C. Cockerham. 1974. Behavior of pairs of loci in finite monoecious populations. *Theor. Pop. Biol.* 6: 323–354. [7]
Weir, B. S., and C. C. Cockerham. 1977. Two-locus theory in quantitative genetics. *In* E. Pollak, O. Kempthorne, and T. B. Bailey, Jr. (eds.), *Proceedings of the international conference on quantitative genetics,* pp. 247–269. Iowa State Univ. Press, Ames. [5,7]
Weir, B. S., and C. C. Cockerham. 1979. Estimation of linkage disequilibrium in randomly mating populations. *Heredity* 42: 105–111. [5]
Weir, B. S., and C. C. Cockerham. 1989. Complete characterization of disequilibrium at two loci. *In* M. W. Feldman (ed.), *Mathematical evolutionary theory,* pp. 86–110. Princeton Univ. Press, Princeton, NJ. [7]
Weir, B. S., C. C. Cockerham, and J. Reynolds. 1980. The effects of linkage and linkage disequilibrium on the covariances of noninbred relatives. *Heredity* 45: 351–359. [5,7]
Weiss, K. M. 1993. *Genetic variation and human disease: principles and evolutionary approaches.* Cambridge Univ. Press, Cambridge, UK. [13,16]
Weller, J. I. 1986. Maximum likelihood techniques for the mapping and analysis of quantitative trait loci with the aid of genetic markers. *Biometrics* 42: 627–640. [15]
Weller, J. I. 1987. Mapping and analysis of quantitative trait loci in *Lycopersicon* (tomato) with the aid of genetic markers using approximate maximum likelihood methods. *Heredity* 59: 413–421. [15]
Weller, J. I. 1990. Experimental designs for mapping quantitative trait loci in segregating populations. *In* W. G. Hill, R. Thompson, and J. A. Woolliams (eds.), *Proc. 4th World Congr. Genet. Appl. Livestock Prod.,* Vol. 13, pp. 113–116. Edinburgh. [16]
Weller, J. I., and A. Wyler. 1992. Power of different sampling strategies to detect quantitative trait loci variance effects. *Theor. Appl. Genet.* 83: 582–588. [15]
Weller, J. I., M. Soller, and T. Bordy. 1988. Linkage analysis of quantitative traits in an interspecific cross of tomato (*Lycopersicon esculentum* × *Lycopersicon pimpinellifolium* by means of genetic markers. *Genetics* 118: 329–339. [15]
Weller, J. I., Y. Kashi, and M. Soller. 1990. Power of daughter and granddaughter designs for determining linkage between marker loci and quantitative trait loci in dairy cattle. *J. Dairy Sci.* 73: 2525–2537. [16]
Weller, S. G. 1976. The genetic control of tristyly in *Oxalis* section *Ionoxalis. Heredity* 37: 387–393. [5]
Werren, J. H., S. W. Skinner, and E. L. Charnov. 1981. Paternal inheritance of a daughterless sex ratio factor. *Nature* 293: 467–468. [6]
Werren, J. H., S. W. Skinner, and A. M. Huger. 1986. Male-killing bacteria in a parasitic wasp. *Science* 231: 990–992. [6]
Westcott, B. 1986. Some methods of analyzing genotype-environment interaction. *Heredity* 56: 243–253. [22]
Wetherill, G. B. 1986. *Regression analysis with applications.* Chapman and Hall, London. [11]
Wexelsen, H. 1933. Linkage between quantitative and qualitative characters in barley. *Hereditas* 17: 323–341. [14]
Wexelsen, H. 1934. Quantitative inheritance and linkage in barley. *Hereditas* 18: 307–348. [14]
White, J. M. 1972. Inbreeding effects upon growth and maternal ability in laboratory mice. *Genetics* 70: 307–317. [10]
White, M. J. D., and L. E. Andrew. 1960. Cytogenetics of the grasshopper *Moraba scurra.* V. Biometric effects of chromosomal inversions. *Evolution* 14: 284–291. [14]
Whitlock, M. 1993. Lack of correlation between heterozygosity and fitness in forked fungus beetles. *Heredity* 70: 574–581. [10]
Whittaker, J. C., R. Thompson, and P. M. Visscher. 1996. On the mapping of QTL by regression of phenotypes on marker type. Heredity 77: 23–32. [15]
Wiener, G., G. J. Lee, and J. A. Woolliams. 1992a. Effects of rapid inbreeding and of crossing of inbred lines on the body weight growth of sheep. *Anim. Prod.* 55: 89–99. [10]
Wiener, G., G. J. Lee, and J. A. Woolliams. 1992b. Effects of rapid inbreeding and of crossing inbred lines on the growth of linear body dimensions of sheep. *Anim. Prod.* 55: 101–114. [10]
Wiener, G., G. J. Lee, and J. A. Woolliams. 1992c. Effects of rapid inbreeding and of crossing of inbred lines on conception rate, prolificacy and ewe survival of sheep. *Anim. Prod.* 55: 115–121. [10]
Wienhues, F. 1968. Long-term yield analyses of heterosis in wheat and barley: variability of heterosis, fixation of heterosis. *Euphytica* 17: 49–62. [10]
Wilcockson, R. W., C. S. Crean, and T. H. Day. 1995. Heritability of a sexually selected character expressed in both sexes. *Nature* 374: 158-159. [14]
Wilkens, H. 1971. Genetic interpretation of regressive evolution processes: studies on hybrid eyes of two *Astyanax* cave populations (Characidae, Pisces). *Evolution* 25: 530–544. [9]
Wilkinson, G. S., K. Fowler, and L. Partridge. 1990. Resistance of genetic correlation structure to directional selection in *Drosophila melanogaster. Evolution* 44: 1990–2003. [21]
Willham, R. L. 1963. The covariance between relatives for characters composed of components contributed by related individuals. *Biometrics* 19: 18–27. [23,26]

Willham, R. L. 1972. The role of maternal effects in animal breeding. III. Biometrical aspects of maternal effects in animals. *J. Anim. Sci.* 35: 1288–1293. [23]

Williams, G. C. 1966. *Adaptation and natural selection.* Princeton Univ. Press, Princeton, NJ. [1]

Williams, J. G. K., A. R. Kubelik, K. J. Livak, J. A. Rafalski, and S. V. Tingey. 1990. DNA polymorphisms amplified by arbitrary primers are useful as genetic markers. *Nucl. Acids Res.* 18: 6531–6535. [14]

Williams, J. S. 1962. A confidence interval for variance components. *Biometrika* 49: 278–281. [18]

Williams, W. 1959. The isolation of pure lines from F_1 hybrids of tomato and the problem of heterosis in inbreeding crop species. *J. Agric. Sci.* 53: 347–353. [10]

Williams, W. 1960. Relative variability of inbred lines and F_1 hybrids in *Lycopersicon esculentum. Genetics* 45: 1457–1465. [6]

Willis, J. H. 1993. Effects of different levels of inbreeding on fitness components in *Mimulus guttatus. Evolution* 47: 864–876. [10]

Willis, J. H., J. A. Coyne, and M. Kirkpatrick. 1991. Can one predict the evolution of quantitative characters without genetics? *Evolution* 45: 441–444. [21]

Wills, C. 1966. The mutational load in two natural populations of *Drosophila pseudoobscura. Genetics* 53: 281–294. [10]

Willson, M. F. 1981. Ecology and science. *Bull. Ecol. Soc. Am.* 62: 4–12. [1]

Willson, M. F., and N. Burley. 1983. *Mate choice in plants: tactics, mechanisms, and consequences.* Monog. Pop. Biol. 19. Princeton Univ. Press, Princeton, NJ. [10]

Wilson, A. C. 1976. Gene regulation in evolution. *In* F. J. Ayala (ed.), *Molecular evolution*, pp. 225–236. Sinauer Assoc., Sunderland, MA. [12]

Wilson, A. F., and R. C. Elston. 1993. Statistical validity of the Haseman-Elston sib-pair test in small samples. *Genet. Epidem.* 10: 593–598. [16]

Wilson, S. R. 1973. The correlation between relatives under the multifactorial model with assortative mating. *Ann. Hum. Genet.* 37: 189–215. [7]

Winkelman, D. C., and R. B. Hodgetts. 1992. RFLPs for somatotropic genes identify quantitative trait loci for growth in mice. *Genetics* 131: 929–937. [14]

Witte, J. S., R. C. Elston, and N. J. Schork. 1996. Genetic dissection of complex traits. *Nature Genetics* 12: 355–356. [16]

Wolfe, K. H., W.-H. Li, and P. M. Sharp. 1987. Rates of nucleotide substitution vary greatly among plant mitochondrial, chloroplast, and nuclear DNAs. *Proc. Natl. Acad. Sci. USA* 84: 9054–9058. [12]

Wolfe, L. M. 1993. Inbreeding depression in *Hydrophyllum appendiculatum:* role of maternal effects, crowding, and parental mating history. *Evolution* 47: 374–386. [10]

Woltereck, R. 1909. Weitere experimentelle Untersuchungen über Art-veränderung, speziell über das Wesen quantitativer Artunterschiede bei Daphniden. *Verh. Deutsch. Tsch. Zool. Ges.* 1909: 110–172. [22]

Woolf, B. 1955. On estimating the relation between blood group and disease. *Ann. Hum. Genet.* 19: 251–253. [14]

Woolf, C. M., and A. D. Gianas. 1976. Congenital cleft lip and fluctuating dermatoglyphic asymmetry. *Am. J. Hum. Genet.* 28: 400–403. [6]

Wray, N. R. 1990. Accounting for mutation effects in the additive genetic variance-covariance matrix and its inverse. *Biometrics* 46: 177–186. [26,27]

Wricke, G. 1962. Uber eine Methode sur Erfassung der ökologischen Streubreite in Feldversuchen. *Zeit. für Pflanzenzüchtung* 47: 92–96. [22]

Wricke, G., and W. E. Weber. 1986. *Quantitative genetics and selection in plant breeding.* Walter de Gruyter and Co., NY. [1,4,5]

Wright, A. J. 1971. The analysis and prediction of some two factor interactions in grass breeding. *J. Agric. Sci.* 76: 301–306. [22]

Wright, A. J. 1976a. The significance for breeding of linear regression analysis of genotype-environment interactions. *Heredity* 37: 83–93. [22]

Wright, A. J. 1976b. Bias in the estimation of regression coefficients in the analysis of genotype-environment interaction. *Heredity* 37: 299–303. [22]

Wright, A. J. 1985. Diallel designs, analyses, and reference populations. *Heredity* 54: 307–311. [20]

Wright, A. J., and R. P. Mowers. 1994. Multiple regression for molecular-marker, quantitative trait data from large F_2 populations. *Theor. Appl. Genet.* 89: 305–312. [15]

Wright, S. 1918. On the nature of size factors. *Genetics* 3: 367–374. [A2]

Wright, S. 1921a. Correlation and causation. *J. Agric. Res.* 20: 557–585. [1,A2]

Wright, S. 1921b. Systems of mating. I. The biometric relations between parents and offspring. *Genetics* 6: 111–123. [1,7,A2]

Wright, S. 1921c. Systems of mating. II. The effects of inbreeding on the genetic composition of a population. *Genetics* 6: 111–123. [1]

Wright, S. 1921d. Systems of mating. III. Assortative mating based on somatic resemblance. *Genetics* 6: 144–161. [1,7]

Wright, S. 1922. Coefficients of inbreeding and relationship. *Am. Nat.* 56: 330–339. [7]

Wright, S. 1926. Effects of age of parents on characteristics of the offspring. *Am. Nat.* 60: 552–559. [6]

Wright, S. 1929a. Fisher's theory of dominance. *Am. Nat.* 63: 274–279. [4]

Wright, S. 1929b. The evolution of dominance: comment on Dr. Fisher's reply. *Am. Nat.* 63: 556–561. [4]

Wright, S. 1932. General, group and special size factors. *Genetics* 17: 603–619. [A2]

Wright, S. 1934a. Physiological and evolutionary theories of dominance. *Am. Nat.* 68: 25–53. [4]

Wright, S. 1934b. Professor Fisher on the theory of dominance. *Am. Nat.* 68: 562–565. [4]

Wright, S. 1934c. An analysis of variability in number of digits in an inbred strain of guinea pigs. *Genetics* 19: 506–536. [25]

Wright, S. 1934d. The results of crosses between inbred strains of guinea pigs, differing in number of digits. *Genetics* 19: 537–551. [25]

Wright, S. 1934e. The method of path coefficients. *Ann. Math. Stat.* 5: 161–215. [A2]

Wright, S. 1952. The genetics of quantitative variability. *In* Agric. Res. Council, *Quantitative inheritance,* pp. 5–41. Her Majesty's Stationery Office, London. [13,17]

Wright, S. 1968. *Evolution and the genetics of populations. I. Genetic and biometric foundations.* Univ. Chicago Press, Chicago. [1,5,9,11,12,13,21,A2]

Wright, S. 1969. *Evolution and the genetics of populations. II. The theory of gene frequencies.* Univ. Chicago Press, Chicago. [10]

Wright, S. 1978. *Evolution and the genetics of populations. III. Experimental results and evolutionary deductions.* Univ. Chicago Press, Chicago. [10]

Wright, S. 1983. On "Path analysis in genetic epidemiology: a critique". *Am. J. Hum. Genet.* 35: 757–768. [A2]

Wright, S. 1984. Diverse uses of path analysis. *In* A. Chakravarti (ed.), *Human population genetics: the Pittsburgh symposium,* pp. 1–34. Van Nostrand Reinhold, NY. [A2]

Wu, C. F. J. 1986. Jackknife, bootstrap, and other resampling methods in regression analysis. *Ann. Stat.* 14: 1261–1295. [18]

Wu, C.-I., and A. T. Beckenbach. 1983. Evidence for extensive genetic differentiation between the sex-ratio and the standard arrangement of *Drosophila pseudoobscura* and *D. persimilis* and identification of hybrid sterility factors. *Genetics* 105: 71–86. [14]

Wu, C.-I., and A. W. Davis. 1993. Evolution of postmating reproductive isolation: the composite nature of Haldane's rule and its genetic bases. *Am. Nat.* 142: 187–212. [14]

Wu, C.-I., and M. F. Palopoli. 1994. Genetics of postmating reproductive isolation in animals. *Ann. Rev. Ecol. Syst.* 27: 283–308. [9]

Wu. W. R., and W. M. Li. 1994. A new approach for mapping quantitative trait loci using complete genetic marker linkage maps. *Theor. Appl. Genet.* 89: 535–539. [15]

Wu. W. R., and W. M. Li. 1996. Model fitting and model testing in the method of joint mapping of quantitative trait loci. *Theor. Appl. Genet.* 92: 477–482. [15]

Xiao, J., J. Li., L. Yuan, and S. D. Tanksley. 1995. Dominance is the major genetic basis of heterosis in rice as revealed by QTL analysis using molecular markers. *Genetics* 140: 745–754. [10]

Xu, S. 1995. A comment on the simple regression method for interval mapping. *Genetics* 141: 1657–1659. [15]

Xu, S., and W. R. Atchley. 1995. A random model approach to interval mapping of quantitative trait loci. *Genetics* 141: 1189–1197. [16]

Xu, S., and W. R. Atchley. 1996. Mapping quantitative trait loci for complex binary diseases using line crosses. *Genetics* 143: 1417–1424. [15]

Yablokov, A. V. 1974. *Variability of mammals.* Amerind, New Delhi, India. [11]

Yaghoobi, J., I. Kaloshian, Y. Wen, and V. M. Williamson. 1995. Mapping a new nematode resistance locus in *Lycopersicon peruvianum. Theor. Appl. Genet.* 91: 457–464. [14]

Yamada, Y. 1962. Genotype × environment interaction and genetic correlation of the same trait under different environments. *Jap. J. Genetics* 37: 498–509. [22]

Yamada, Y., Y. Itoh, and I. Sugimoto. 1988. Parametric relationships between genotype × environment interaction and genetic correlation when two environments are involved. *Theor. Appl. Genet.* 76: 850–854. [22]

Yamaguchi, M., T. Yanase, H. Nagano, and N. Nakamoto. 1970. Effects of inbreeding on mortality in Fukuoka population. *Am. J. Hum. Genet.* 22: 145–155. [10]

Yampolsky, L. Y., and S. M. Scheiner. 1994. Developmental noise, phenotypic plasticity, and allozyme heterozygosity in *Daphnia. Evolution* 48: 1715–1722. [6]

Yates, F. 1947. Analysis of data from all possible reciprocal crosses between a set of parental lines. *Heredity* 1: 287–301. [20]

Yates, F., and W. G. Cochran. 1938. The analysis of groups of experiments. *J. Agri. Sci.* 28: 556–580. [22]

Yezerinac, S. M., S. C. Lougheed, and P. Handford. 1992. Morphological variability and enzyme heterozygosity: individual and population level correlations. *Evolution* 46: 1959–1964. [6]

Yoon, C. H. 1955. Homeostasis associated with heterozygosity in the genetics of time of vaginal opening in the house mouse. *Genetics* 40: 297–309. [6]

Yoshimaru, H., and T. Mukai. 1985. Relationships between the polygenes affecting the rate of development and viability in *Drosophila melanogaster. Jap. J. Genetics* 60: 307–334. [12]

Young, E. C. 1965. General development in British Corixidae. *Proc. Royal Entomol. Soc. Lond.* A 40: 159–168. [6]

Young, N. D., and S. D. Tanksley. 1989a. RFLP analysis of the size of chromosomal segments retained around the *Tm-2* locus of tomato during backcross breeding. *Theor. Appl. Genet.* 77: 353–359. [14]

Young, N. D., and S. D. Tanksley. 1989b. Restriction fragment length polymorphism maps and the concept of graphical genotypes. *Theor. Appl. Genet.* 77: 95–101. [14]

Young, N. D., D. Zamir, M. W. Ganal, and S. D. Tanksley. 1988. Use of isogenic lines and simultaneous probing to identify DNA markers tightly linked to the *Tm-2a* gene in tomato. *Genetics* 120: 579–585. [14]

Yu, Z. H., D. J. Mackill, J. M. Bonman, and S. D. Tanksley. 1991. Tagging genes for blast resistance in rice via linkage to RFLP markers. *Theor. Appl. Genet.* 81: 471–476. [14]

Yule, G. U. 1902. Mendel's laws and their probable relation to intra-racial heredity. *New Phytol.* 1: 193–207, 222–238. [1]

Zakharov, V. M. 1981. Fluctuating asymmetry as an index of developmental homeostasis. *Genetica (Pol.)* 13: 241–256. [6]

Zakharov, V. M. 1992. Population phenogenetics: analysis of developmental stability in natural populations. *Acta Zool. Fennica* 191: 7–30. [6]

Zapata, C., and G. Alvarez. 1992. The detection of gametic disequilibrium between allozyme loci in natural populations of *Drosophila. Evolution* 46: 1900–1917. [5]

Zapata, C., and G. Alvarez. 1993. On the detection of nonrandom associations between DNA polymorphisms in natural populations of *Drosophila. Mol. Biol. Evol.* 10: 823–841. [5]

Zeng, L.-W. 1996. Resurrecting Muller's theory of Haldane's rule. *Genetics* 143: 603–607. [14]

Zeng, Z.-B. 1992. Correcting the bias of Wright's estimates of the number of genes affecting a quantitative character: a further improved method. *Genetics* 131: 987–1001. [9]

Zeng, Z.-B. 1993. Theoretical basis for separation of multiple linked gene effects in mapping of quantitative trait loci. *Proc. Natl. Acad. Sci. USA* 90: 10972–10976. [15]

Zeng, Z.-B. 1994. Precision mapping of quantitative trait loci. *Genetics* 136: 1457–1468. [15]

Zeng, Z.-B., D. Houle, and C. C. Cockerham. 1990. How informative is Wright's estimator of the number of genes affecting a quantitative character? *Genetics* 126: 235–247. [9]

Zerba, K., R. E. Ferrell, and C. F. Sing. 1996. Genotype-environment interaction: apolipoprotein E (*ApoE*) gene effects and age as an index of time and spatial context in the human. *Genetics* 143: 463–478. [14]

Zhang, X. F., J. A. Mosjidis, and Z. L. Hu. 1992. Methods for detection and estimation of linkage between a marker locus and quantitative trait loci. *Plant Breeding* 109: 35–39. [15]

Zhao, H., and T. P. Speed. 1996. On genetic map functions. *Genetics* 142: 1369–1377. [14]

Zhu, J., and B. S. Weir. 1994a. Analysis of cytoplasmic and maternal effects. I. A genetic model for diploid plant seeds and animals. *Theor. Appl. Genet.* 89: 153–159. [26]

Zhu, J., and B. S. Weir. 1994b. Analysis of cytoplasmic and maternal effects. II. Genetic model for triploid endosperms. *Theor. Appl. Genet.* 89: 160–166. [26]

Zhuchenko, A. A., A. P. Samovol, A. B. Korol, and V. K. Andryushchenko. 1978. Linkage between loci of quantitative characters and marker loci. I. Model. *Genetika* 14: 771–778. [15]

Zhuchenko, A. A., A. P. Samovol, A. B. Korol, and V. K. Andryushchenko. 1979. Linkage between loci of quantitative characters and marker loci. II. Influence of three tomato chromosomes on variability of five quantitative characters in backcross progeny. *Genetika* 15: 672–683. [15]

Zink, R. M., M. F. Smith, and J. L. Patten. 1985. Association between heterozygosity and morphological variance. *J. Heredity* 76: 415–420. [6]

Zouros, E., and D. W. Foltz. 1987. The use of allelic isozyme variation for the study of heterosis. *Isozymes* 13: 1–59. [10]

Zouros, E., S. M. Singh, and H. E. Miles. 1980. Growth rate in oysters: an overdominant phenotype and its possible explanations. *Evolution* 34: 856–867. [6,10]

Zouros, E., M. Romero-Dorey, and A. L. Mallet. 1988. Heterozygosity and growth in marine bivalves: further data and possible explanations. *Evolution* 42: 1332–1341. [10]

Zuberi, M. I., and J. S. Gale. 1976. Variation in wild populations of *Papaver dubium.* X. Genotype-environment interaction associated with differences in soil. *Heredity* 36: 359–368. [22]

Author Index

Organism and Trait Index

Organisms are listed by Latin names, with the exception of domesticated plants and animals, which are listed by common name.

Subject Index

Particular organisms and characters are listed in the Organism and Trait index